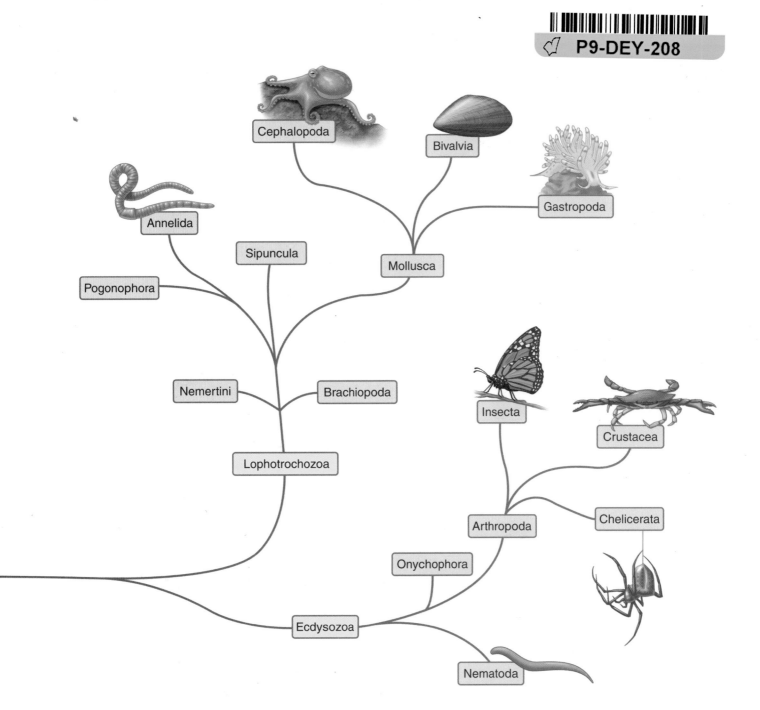

Principles of

Animal Physiology

SECOND EDITION

Christopher D. Moyes, Ph.D.
Queen's University

Patricia M. Schulte, Ph.D.
University of British Columbia

PEARSON

Benjamin
Cummings

San Francisco   Boston   New York
Capetown   Hong Kong   London   Madrid   Mexico City
Montreal   Munich   Paris   Singapore   Sydney   Tokyo   Toronto

Publisher: Frank Ruggirello
Development Manager: Claire Alexander
Senior Acquisitions Editor: Deirdre Espinoza
Senior Project Editor: Marie Beaugureau
Assistant Editor: Emily Portwood
Art Development Editor: Laura Southworth
Artists: Precision Graphics, Steve McEntee
Managing Editor, Production: Deborah Cogan
Production Supervisor: Caroline Ayres
Copy Editor: Jan McDearmon

Text Designer: Patrick Devine
Cover Designer: Rokusek Design
Cover Illustrator: Quade Paul, Echo Medical Media
Cover Image: Tim Flach/Getty Images/Stone Allstock,
Ron Sanford/CORBIS
Photo Research: Kristin Piljay
Senior Manufacturing Buyer: Stacey Weinberger
Project Coordination: Carlisle Publishing Services
Compositor: Carlisle Publishing Services
Marketing Manager: Gordon Lee

Library of Congress Cataloging-in-Publication Data
Moyes, Christopher D.
  Principles of animal physiology / Christopher D. Moyes, Patricia M. Schulte.—2nd ed.
      p. cm.
  Includes bibliographical references and index.
  ISBN 13: 978-0-321-50155-4
  1. Physiology. I. Schulte, Patricia M. II. Title.
QP31.2.M69 2007
571.1--dc22

                                                      2007017463

2 3 4 5 6 7 8—CRK—12 11 10 09 08

www.aw-bc.com

ISBN 10: 0-321-50155-1
ISBN 13: 978-0-321-50155-4

Brief Contents

About the Authors

Christopher D. Moyes, Ph.D.
Queen's University

Chris Moyes received his Ph.D. in Zoology from the University of British Columbia in the area of comparative muscle physiology. After postdoctoral fellowships in molecular physiology at the U.S. National Institutes of Health and Simon Fraser University, he took a position at Queen's University, where he is a Full Professor in the Departments of Biology and Physiology. He teaches a spectrum of courses in animal physiology, comparative biochemistry, and cell biology. Using a wide range of comparative and traditional models, his research addresses questions in molecular physiology and metabolic biochemistry. One major theme of his research is the study of the evolutionary and developmental origins of variation in muscle structure and function. Another major area of his research is the response of animals to environmental stress. In all of his research he emphasizes the integration of physiological processes, from molecular to organismal levels.

Dr. Moyes is a recipient of the Ontario Premier's Research Excellence Award. He is a member of the American Physiological Society and the Canadian Society of Zoologists and has served on research grant panels for the Natural Science and Engineering Research Council of Canada and the U.S. National Science Foundation. He is also a member of the Editorial Board of *Comparative Biochemistry and Physiology*.

He has published more than 70 peer-reviewed papers, including contributions to four books. Among his recent papers are Moyes, C. D., N. Fragoso, M. Musyl, and R. Brill, 2006, Predicting post-release survival in large pelagic fish, *Transactions of the American Fisheries Society* 135: 1389–1397; Dalziel, A. C., S. E. Moore, and C. D. Moyes, 2005, Mitochondrial enzyme content in the muscles of high performance fish: Evolution and variation among fiber-types, *American Journal of Physiology* 288: R163–R172; and Moyes, C. D., and D. Hood, 2003, Origins and consequences of mitochondrial variation, *Annual Reviews in Physiology* 65: 177–201.

More of his research is detailed on his homepage at http://biology.queensu.ca/~moyesc.

Patricia M. Schulte, Ph.D.
University of British Columbia

Trish Schulte received her Ph.D. in Biological Sciences from Stanford University in the area of evolutionary physiology. Her thesis work addressed the role of changes in gene expression in physiological evolution. After completing her postdoctoral studies she took a position as an assistant professor in the Department of Biology at the University of Waterloo. She is currently an Associate Professor in the Department of Zoology at the University of British Columbia in Vancouver. Her research interests center on the relationship between genetic variation, performance, and fitness in variable environments, using fish as experimental models to address these questions. Dr. Schulte's research group also performs applied research in fisheries, aquaculture, and aquatic toxicology.

Dr. Schulte is a recipient of the Ontario Premier's Research Excellence Award and several teaching awards, including the UBC Science Undergraduate Society Award for Excellence in Teaching, and the Faculty of Science Achievement Award for Teaching, for her teaching of animal physiology courses. She is the president of the Canadian Society of Zoologists (2007–2008), a member of the Society for Integrative and Comparative Biology, and an associate editor for *Physiological and Biochemical Zoology*.

She has published over 50 peer-reviewed papers, including contributions to several books. Among her recent papers are Schulte, P. M., 2007, Responses to environmental stressors in an estuarine fish: Interacting stressors and the impacts of local adaptation, *Journal of Thermal Biology* 32: 152–161; Fangue, N. A., M. Hofmeister, and P. M. Schulte, 2006, Intraspecific variation in thermal tolerance and heat shock protein gene expression in common killifish, *Fundulus heteroclitus, Journal of Experimental Biology* 209: 2859–2872; and Todgham, A. E., G. K. Iwama, and P. M. Schulte, 2006, Effects of the natural tidal cycle and artificial temperature cycling on Hsp levels in the tidepool sculpin, *Oligocottus maculosus, Physiological and Biochemical Zoology* 79: 1033–1045.

More of her research is detailed on her homepage at www.zoology.ubc.ca/person/schulte.

Contents

CHAPTER 3
CELL SIGNALING AND ENDOCRINE REGULATION 90

Overview 92

The Biochemical Basis of Cell Signaling 93

CHAPTER 4
**NEURON STRUCTURE
AND FUNCTION**　142

CHAPTER 5
CELLULAR MOVEMENT AND MUSCLES 196

PART TWO
INTEGRATING PHYSIOLOGICAL SYSTEMS 247

CHAPTER 6
SENSORY SYSTEMS 248

CHAPTER 7
FUNCTIONAL ORGANIZATION OF NERVOUS SYSTEMS   306

CHAPTER 8
CIRCULATORY SYSTEMS   348

CHAPTER 10
ION AND WATER BALANCE   470

CHAPTER 11
DIGESTION 526

CHAPTER 12
LOCOMOTION 572

CHAPTER 14
REPRODUCTION   662

Preface

The 21st century is an incredibly exciting time to be a biologist. Animal biologists now have access to data from a range of complete animal genomes, including those of many species of mammals, fish, tunicates, insects, and nematode worms—taxa covering a broad spectrum of the diversity of animals. But the fundamental questions about how the genes in these genomes work together to allow animals to perform their diverse physiological functions and go about their daily lives are still largely unanswered. Physiologists are at the forefront of integrating this new genome sequence information into a functional and evolutionary framework as part of their efforts to understand how animals work. Our goal in writing this textbook is to convey a sense of this excitement to students who are approaching the study of animal physiology for the first time.

One of the challenges that students face when they approach their first course in physiology is the great breadth and diversity of the subject matter. Physiology is among the most integrative of the life sciences, drawing on ideas from chemistry, physics, mathematics, molecular biology, and cell biology for its conceptual underpinnings. In addition, to fully appreciate the physiological diversity of animals, students must have a working knowledge of environmental biology, ecology, systematics, and evolutionary biology. We have written this book to give students a well-organized and engaging treatment of the fundamental principles of animal physiology. Throughout the book, we integrate concepts from all levels of biological organization to explore the nature of diversity in cells, physiological systems, and whole animals. We think that this approach will spark the interest of all students, whatever their background preparation.

Key Themes

Students are sometimes so focused on remembering the "facts" of physiology that they are unable to place these facts into a well-developed conceptual framework. In order to help students get past this difficult barrier we have organized this book around several key themes and fundamental principles that are highlighted in each chapter throughout this book, and have striven to present this material in an accessible fashion that engages student learning.

A Focus on Unifying Principles. In chapter 1, we introduce four unifying themes in animal physiology:

- Physiological processes have their basis in the laws of chemistry and physics.
- Physiological processes are homeostatically regulated.
- Physiological processes are the product of both the genotype and the environment.
- Physiological diversity among animals is the result of evolutionary processes.

These key themes are revisited in every chapter of the book, providing a unifying thread that ties together our concept of animal physiology.

An Emphasis on Animal Diversity and Evolution. We are strongly committed to the importance of teaching about the physiological diversity of animals, because we feel that this diversity is a fundamental property of the natural world. As a result, we have included extensive discussion of physiological processes in both vertebrates and invertebrates throughout the book and have attempted to fully interweave evolutionary thinking into these discussions.

We also believe that books focusing only on humans can cause students to form the erroneous impression that physiological processes in humans are typical of those in all animals, and thus we have striven to provide diverse examples in their evolutionary context.

Our commitment to diversity is emphasized by our inclusion of the phylogenies of the major animal taxa at the front and back of this book.

Attention to the Integrative Nature of Animal Physiology. Throughout the book we emphasize the integrative nature of physiology, and this theme is highlighted in a number of ways. Each chapter begins with an opening essay that places the system under discussion into its environmental or evolutionary framework. To reinforce this theme of integration, each chapter in Part Two concludes with a section entitled "Integrating Systems" in which we examine how the physiological system covered in the chapter interacts with one or more other systems in response to an environmental challenge. Together, these features help to build student understanding of how physiological systems interrelate and depend on each other.

Integration of Physiology with Cell and Molecular Biology. We have divided this book into two main sections. In Part One, we discuss the cellular basis of animal physiology. The goal of Part One is to provide students with a general context for understanding animal physiology and to show how, at a cellular level, animals are both similar to and different from other organisms. We hope that this treatment will help students begin to see how the somewhat abstract processes that they study in other courses have direct relevance to the understanding of animal physiology.

We feel that providing a strong foundation in cellular and molecular physiology is critical for students because our understanding of animal physiology has changed dramatically in the last 10 years due to advances in fields such as genomics, cell biology, and molecular biology, and a solid understanding of these disciplines is central to the modern concept of physiology.

In Part Two of the book we discuss how cells and tissues interact to form the integrative physiological systems of animals. We consider each of the major physiological systems in turn, building on the twin themes of conservation and diversity to address the question: how do different animals use fundamentally similar building blocks to construct unique physiological systems to meet the challenges imposed by the environment? Throughout the second part of this book we have attempted to integrate discussion of the cellular and molecular processes that underpin physiological processes, at a depth that is appropriate to help students understand the relevance of these disciplines to animal physiology.

Integrated Treatment of Endocrine Regulation. One unique element in the organization of this book is our treatment of endocrine systems. Rather than relegating these systems to a single isolated chapter, we discuss endocrinology in the first part of the book in the context of the various means of cellular signaling and communication, and then integrate the presentation of its various physiological roles throughout the chapters in the second part of the book. We find that students better understand how hormones control systems once they have been introduced to all the diverse ways in which cells send and receive signals. By establishing the foundation of cellular control early in the text, we are able to discuss the impact of specific hormones and glands in the context of each physiological system, increasing the integrative nature of the discussion. We feel that this approach places the endocrine system in its appropriate evolutionary framework—as one of several means of intercellular communication that are available to multicellular organisms—and clearly demonstrates how communication and coordination are critical for the functioning of essentially every organ system.

New for the Second Edition

Building upon the text's strengths, we made several significant revisions for the second edition of *Principles of Animal Physiology*. Our coverage of the endocrine system is further integrated and developed; the reviews of basic chemistry and cell physiology have been combined into one chapter (Chapter 2); our unique and beautiful art program has been strengthened; and we paid close attention to the revision of our important Integrating Systems essays, which can be found at the end of each chapter in Part 2.

In addition, we made it a goal to expand the pedagogic features throughout the text. New for the second edition, you will find the following in each chapter.

- Concept check questions at the end of major sections, which help students develop their understanding of important themes as they work through the material.

- An expanded selection of end-of-chapter questions, including new quantitative questions, to assess mastery of concepts.

- Bulleted chapter summaries to ease review of material.

- *For Further Reading* sections moved to the end of each chapter to better direct students to the animal physiology literature.

Each chapter contains new figures and revised content. In particular, the second edition features:

Chapter 1

- Expanded discussion of allometric scaling, including a more historical perspective.

Chapter 2

- Chapters 2 and 3 of the previous edition have been combined into a new chapter, titled "Chemistry, Biochemistry, and Cell Physiology." The new chapter integrates material on macromolecule structure and metabolism, with a reduced emphasis on chemical structures and greater reliance on schematics.
- New chapter opening essay on the origins of animal life.
- Revised section on membrane potentials.
- Revised Mathematical Underpinnings box on the Goldman equation.
- New feature on measurement of metabolic rate in living animals.

Chapter 3

- New chapter opening essay that focuses on the evolution of cell signaling.
- Improved discussion of the effects of caffeine and heroin on receptor numbers.
- New section on the termination of ligand receptor signaling.
- The introduction to "Signal Transduction via G-protein-coupled receptors" has been reorganized to better explain the role of the β subunit of G proteins in signaling.
- Expanded discussion of nitric oxide and other gaseous signaling molecules.
- Expanded section on endocrine regulation reflected in the new chapter title, "Cell Signaling and Endocrine Regulation."
- Additional examples of invertebrate hormones have been included.
- Content on environmental estrogens has been moved from an Applications box into the text.
- New section added on the vertebrate stress response.

Chapter 4

- New opening paragraph emphasizes that neurons use electrical signals to transmit information, but that they are not the only type of excitable cell.
- Clearer references to the discussion of resting membrane potential in Chapter 2 have been included.

- The section has been slightly expanded to provide a stronger review of electrical concepts, and a new subheading, "Gated ion channels allow neurons to alter their membrane potentials," has been added.

Chapter 5

- Expanded discussion of pacemaker cell function in cardiac muscle.
- Discussion of smooth muscle has been moved forward in the chapter to highlight its importance in all animals.
- Discussion of shortening and lengthening contractions has been clarified, with commentary on the use of the common alternative terminology of concentric/eccentric.
- Clarification of the nature of tonic muscles in humans and non-humans.
- More emphasis on diversity in muscle types of invertebrates, including new sections on obliquely-striated muscles and mollusc catch muscle.

Chapter 6

- Expanded information on logarithmic encoding in sensory systems, including a new figure.
- Reorganized and expanded section on arthropod proprioception and hearing.
- Additional coverage of signal transduction in hair cells.
- Revised figures of hair cells.
- Clarified signal transduction in photoreceptors.
- Completely revised section on signal processing in vertebrate retinas.
- Completely revised section on brain processing of visual signals in mammals.

Chapter 7

- Revised Integrating Systems feature.

Chapter 8

- New Overview section that emphasizes the limitations on diffusion and the importance of bulk flow and circulatory transport.
- Chapter reorganized to reduce repetition and emphasize important concepts.

- New section on transmural pressure and the Law of LaPlace.
- New section on the cardiac cycle in fish hearts.
- Section on cardiac action potentials clarified to better contrast the distinction between the mechanisms underlying pacemaker potentials and action potentials in contractile cardiomyocytes.

Chapter 9

- Clarified section on boundary layers.
- Clarified section on buccal-opercular pump of teleost fish.
- Clarified section on ventilation in amphibians.
- New figure on the ventilatory cycle in birds.
- New section on carbon dioxide and pH regulation, including discussion of the Hendersson-Hasselbalch equation and pH-bicarbonate plots (Davenport diagrams).

Chapter 10

- A new feature on life without water.
- Expanded discussion of the shark rectal gland function.
- Expanded coverage of cellular specializations in kidney tubules.
- More attention to the role of the kidney in control of pH.
- A greater focus on the role of hormones in regulating kidney function.
- Improved coverage of the kidney equivalents in invertebrates.

Chapter 11

- Revised section on the hypothalamic control of appetite.
- New Integrating Systems feature on obesity.

Chapter 12

- Expanded discussion of changes in locomotor physiology in relation to amphibious life.

Chapter 13

- Expanded discussion on the nature of thermotolerance in homeotherms and poikilotherms.
- New Integrating Systems essay on the cellular and systemic origins of fever.

Chapter 14

- Reorganized treatment of steroids and gonadotropins in vertebrates and invertebrates.
- Expanded discussion of temperature-dependent sex determination.
- More detailed discussion on the integration of ovulatory and uterine cycles in mammals.
- New Genetics and Genomics feature on prolactin.

For more information about this text, see the following pages for a visual presentation of the complete array of special topics boxes, the art program, text features, and the supplements package.

We hope that you enjoy using this textbook. Please feel free to contact us at the email addresses below if you have any comments or suggestions on how we could make this book an even better tool to help you learn or teach animal physiology.

Chris Moyes
Queen's University
chris.moyes@queensu.ca

Trish Schulte
University of British Columbia
pschulte@zoology.ubc.ca

An Emphasis on Animal Diversity and Evolution

The second edition of *Principles of Animal Physiology* continues a strong commitment to highlighting the diversity of physiological processes in animals, and placing this diversity in its evolutionary context.

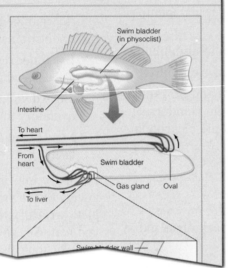

BOX 9.2 **EVOLUTION AND DIVERSITY**
Root-Effect Hemoglobins and Swim Bladders

Fish tissues are somewhat more dense than either freshwater or seawater, largely as a result of the high density of the skeleton, so without some form of buoyancy compensation, fish tend to sink. Many teleost fish use a gas-filled organ called a **swim bladder** to maintain their vertical position in the water. Swim bladders are located just above the gut, and below the vertebral column and kidneys. The walls of the swim bladder are largely impermeable to gas, since they are poorly vascularized and composed of a thick layer of connective tissues. In some species the wall of the swim bladder is coated with a layer of guanine crystals, which further decrease gas permeability. The gas content of swim bladders varies among species, but in most species O_2 is the principal gas.

The buoyancy provided by a swim bladder depends on the volume of this organ. Since swim bladders are soft-walled and filled with gas, they change volume as pressure changes. Atmospheric pressure increases with depth, as a result of the pressure of the overlying water. In fact, pressure increases by approximately 1 atmosphere (atm) for every 10 meters beneath the surface. If a fish descend...

Evolution and Diversity Boxes

Represented by Darwin's Galapagos finch, **Evolution and Diversity** boxes offer in depth exploration of unusual animals or systems in relation to specific chapters and provide an opportunity to explore the evolution of selected physiological processes.

Unique Animal Examples

Unique animal examples are used throughout the text and figures to show features and concepts found only in animals. The movement of pigment granules in the African clawtoed frog is shown using art and photomicrographs to convey many levels of information.

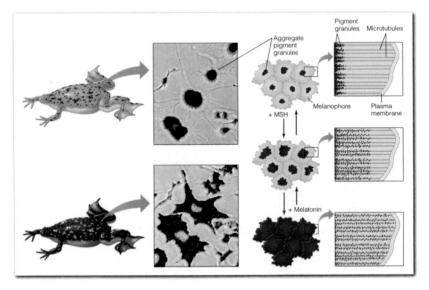

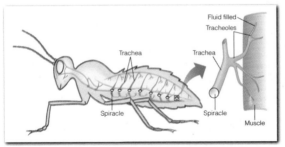

Non-mammalian Examples

Art is used extensively to improve understanding of anatomy and physiology in context, particularly in the case of the many non-mammalian examples presented throughout the book. Here we show the tracheal system of an insect.

Integrative Nature of Animal Physiology

Physiology is a fundamentally integrative science. It integrates levels of biological organization from molecules to cells to organisms, as well as across organ systems to study the coordinated responses of animals to their environment.

Integrating Systems | The Physiology of Diving

A variety of air-breathing vertebrates, including some mammals, birds, and reptiles, have adopted a fully or partially aquatic mode of life. However, all of these animals remain dependent on air as a respiratory medium, and must be able to actively hunt prey underwater while relying on the oxygen stores that they carry with them as they dive below the surface. The physiology of diving in these animals provides an ideal example of the ways in which the respiratory and circulatory systems are integrated to allow animals to function in their environment.

Sperm whales are the champion divers among the marine mammals, with recorded dives to a depth of more than 2000

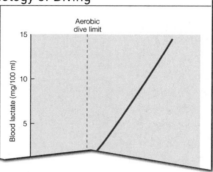

Integrating Systems Sections

To emphasize this integration, each chapter in the second half of the book ends with a section titled: **Integrating Systems**. These special features highlight the ways in which multiple physiological systems work together to allow animals to respond to environmental challenges.

Genetics and Genomics Boxes

Genetics and Genomics boxes are represented with the fruit fly that is so critical to genetic experimentation. These boxes help students to integrate their knowledge of cell and molecular biology into animal physiology concepts, by highlighting how genome sequencing projects and the high-throughput methods of genomics and proteomics have clarified our understanding of many physiological processes.

BOX 5.2 GENETICS AND GENOMICS
Muscle Differentiation and Development

It is difficult to discuss the origins of muscle diversity without also considering how muscle is made. Muscle synthesis is really two related processes: muscle differentiation, or myogenesis, and muscle development. Our understanding of muscle differentiation and development has benefited from research in animal model systems, including *Drosophila*, *C. elegans*, *Xenopus*, zebrafish, and mice, as well as cultured myoblasts. Each type of muscle follows its own path to the final phenotype. The control of muscle formation is best understood in skeletal muscle.

One reason why we understand muscle differentiation so well is that many of the processes can be studied in cell culture. Myoblasts from chickens and rodents are most useful because they can be grown for hundreds and thousands of "generations" without marked deterioration of their properties. Neonatal muscle, and to a lesser extent adult muscle, has a population of muscle precursor cells called **satellite cells**. These cells can be harvested and grown in culture. They rapidly proliferate but do not differentiate when grown under the right condition

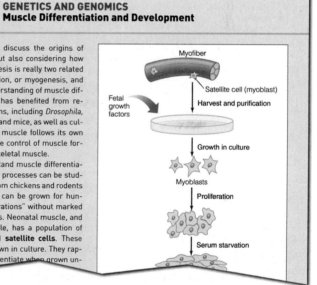

BOX 2.3 MATHEMATICAL UNDERPINNINGS
The Goldman Equation

The Goldman-Hodgkin-Katz Constant Field equation (often simply referred to as the Goldman equation or GHK), allows the estimation of the membrane potential based on the concentrations, valences, and relative permeabilities of a series of ions. Since most plasma membranes under resting conditions have appreciable permeability only to postassium, sodium, and chloride, the Goldman equation can be written as follows:

$$E_m = \frac{RT}{F} \ln \frac{P_K [K^+]_o + P_{Na} [Na^+]_o + P_{Cl} [Cl^-]_i}{P_K [K^+]_i + P_{Na} [Na^+]_i + P_{Cl} [Cl^-]_o}$$

where P_{ion} is the permeability of the membrane to that ion and $[ion]_o$ and $[ion]_i$ represent the extracellular and intracellular concentrations

The two terms for K^+ permeability also cancel out, leaving an equation that is equivalent to the Nernst equation for potassium (with the exception of the valence term, z, which is neglected because potassium has a valance of $+1$). Notice, however, that if we do the same exercise assuming that the membrane is permeable only to chloride, we end up with an equation that is similar to the Nernst equation, except that it neglects the valence and has the intracellular concentration in the numerator and the extracellular concentration in the denominator, rather than the other way around. Recall that one of the rules of logarithms is that $-\ln x = \ln (1/x)$. By inverting the ratio of the concentrations of chloride, the Goldman equation takes into account the fact that chloride has a valance of -1.

Mathematical Underpinnings Boxes

Because physiological processes have their basis in the laws of chemistry and physics, it is often impossible to understand these processes without an appreciation of the underlying physics, mathematics, or chemistry. **Mathematical Underpinnings** boxes provide additional background in these areas where needed. The nautilus, which embodies Fibonacci numbers, represents the underlying mathematical order present in nature.

The Dynamic Nature of Animal Physiology

Throughout the text, we also show students how physiological research is performed to give them a sense of physiology as a dynamic and growing research area. In addition, wherever possible, we discuss some of the medical, agricultural, or environmental applications of physiological research to demonstrate that physiology has relevance to daily life.

Methods and Model Systems Boxes

Methods and Model Systems boxes provide the opporturnity to highlight important experimental methods, and to explore how experiments using model systems can be used to test physiological principles, emphasizing the dynamic nature of physiology as a science. The roundworm *C. elegans*, which serves as the box icon, is a widely used model system in neurobiology and muscle biology.

BOX 4.1 METHODS AND MODEL SYSTEMS
Studying Ion Channels

One of the most widely used methods for studying ion channels in single cells is the **voltage clamp**. The basic idea of a voltage-clamp experiment is to hold the voltage across a membrane at a constant level by injecting current into the cell via a **microelectrode** any time the voltage across the membrane changes (as a result, the voltage is said to be clamped at a particular value). For example, suppose that the resting potential of the cell is -70 mV, and you set the voltage-clamp apparatus to hold the membrane potential at -70 mV. If the cell is at the resting membrane potential, then no current will be injected through the microelectrode. But suppose you introduce into the fluid bathing the cell a neurotransmitter that binds to a specific Na^+ channel. The neurotransmitter will bind and cause the Na^+ channel to open. Na^+ ions will enter the cell, causing a depolarization. The apparatus takes a measurement of the membrane potential and injects current to hold the membrane at the resting membrane potential, despite the influx of Na^+. In this way, a voltage-clamp apparatus is analogous to a thermostatically controlled heater operating by negative feedback.

The amount of injected current is a direct measure of natural ionic movements across the membrane. Neurophysiologists use voltage-clamp experiments to describe the electrical properties of intact membranes or whole cells.

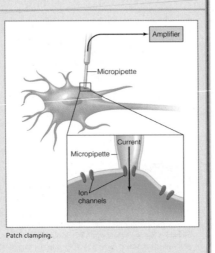

Patch clamping.

then voltage-clamp this small region of membrane and record the extremely small currents generated by a single ion channel (they are measured in picoamperes, $pico = 10^{-12}$)...allows neurobiologists

Applications Boxes

Applications boxes focus on studies of physiological diversity that provide insights into the ways in which systems function and, thus, lead to applications that can affect daily life in fields such as human health, the environment, and agriculture. A lab rat is one of the common vehicles for these studies.

BOX 7.1 APPLICATIONS
Split-Brain Syndrome

Mammalian brains are divided into two hemispheres, with the right hemisphere receiving sensory input from, and controlling, the left half of the body, and the left hemisphere receiving sensory input from, and controlling, the right half of the body. Ordinarily, the left and right sides of the brain coordinate their actions by communicating information via the corpus callosum, a region of white matter connecting the two hemispheres. But what would happen if the corpus callosum ceased to function? Roger Sperry was the first to investigate this question, when he performed experiments in which he cut the corpus callosum and optic chiasm in cats, and tested the effects of the surgery on the animals' behavior. Each cat was apparently normal following the surgery, but when Sperry covered its left eye and then taught it a simple conditioned behavior, the cat could not perform this task when its right eye was covered instead of the left. It was as if only one side of the brain learned to perform the task, and could not communicate this learning to the other side of the brain. Sperry termed this phenomenon the split-brain syndrome.

Similar observations have been made in human patients following brain surgery designed to reduce the severity of epi...

ered reported seeing the word *ring* that had been projected to the right visual field and processed by the left hemisphere. They were entirely unaware that the word *key* had been presented to the left visual field and processed by the right hemisphere, although some subjects occasionally reported that they saw a flash of light on the left side of the screen.

In most humans, the ability to communicate using language is localized in the left hemisphere of the brain, while the right hemisphere lacks language ability. Thus, the right hemisphere was unable to communicate that the light observed in the left visual field represented a word. Control subjects could verbalize both the words *key* and *ring* because the intact corpus callosum could transfer the information between the two hemispheres. This difference between normal subjects and "split-brain" patients is not obvious in everyday life because we seldom look at objects using only one eye. We can easily move our eyes or turn our heads so that both halves of the brain receive complete sensory information.

Although the right hemisphere does not have the ability to speak, it can still reason and communicate in other ways. For example, Sperry asked the split-brain ...choose the object

Promoting Effective Learning

The second edition of *Principles of Animal Physiology* incorporates many tools to improve student learning.

Overview Figures

Each chapter begins with an **Overview figure** that gives students an introduction to the chapter content. These figures help to encapsulate the key concepts of the chapter and provides a guide for review and self study.

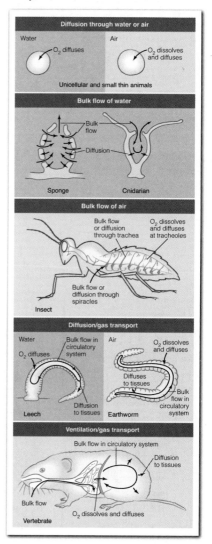

Conceptual Headings

The chapters are divided into manageable sections by **conceptual headings**, which succinctly explain what's coming next and can be used for efficient review.

> **Chordates have both open and closed circulatory systems**
>
> The vertebrates belong to the phylum Chordata, which also contains the invertebrate urochordates (the tunicates) and cephalochordates (the lancelets). Urochordates have a simple tubular heart that ~~~~~~~~~~~~ ~~~ through ~~~~~~~~ of well-

Concept Check Questions

Major sections in each chapter end with a series of **Concept Check Questions** that allow students to review material before they move on through the chapter. Answers to the Concept Checks can be found on the text's Companion Website, www.aw-bc.com/ animalphysiology.

> ⊙ | CONCEPT CHECK
>
> 1. Compare and contrast hydrophilic and hydrophobic messengers in terms of the three main steps of indirect signaling.
> 2. Are amines hydrophilic or hydrophobic messengers? How does this affect their release, transport, and signaling?
> 3. Why do some cells respond to a chemical messenger while other cells ignore it?
> 4. Compare and contrast receptor up-regulation and down-regulation. How do these phenomena help to maintain homeostasis?

Review, Synthesis, and Quantitative Questions

End of chapter review questions, synthesis questions, and quantitative questions provide increasingly sophisticated and challenging ways for students to test their comprehension of important concepts. **Quantitative Questions** are all-new for this edition. Answers to the end-of-chapter questions can be found on the text's Companion Website, www.aw-bc.com/ animalphysiology.

> ### Quantitative Questions
>
>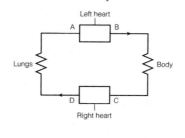
>
> **1.** Below is a schematic diagram of the mammalian cardiovascular system.
>
> If mean pressure at A = 2 mm Hg, B = 80 mm Hg, C = 5 mm Hg ~~~~~~~~~~~~~~
>
>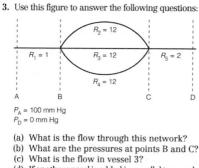
>
> **3.** Use this figure to answer the following questions:
>
> $P_A = 100$ mm Hg
> $P_D = 0$ mm Hg
>
> (a) What is the flow through this network?
> (b) What are the pressures at points B and C?
> (c) What is the flow in vessel 3?
> (d) If another vessel is added in parallel to vessels 2–4, with a resistance $R_6 = 18$ mm ~~~~~~~~~ through the

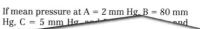

Innovative Art Program

Illustrations in this textbook have been carefully designed to highlight the fundamental principles of physiology and walk students through each step of complex processes.

"Stepped" Art

"Stepped" art figures are used extensively to break complex cellular and molecular processes into manageable pieces for greater comprehension.

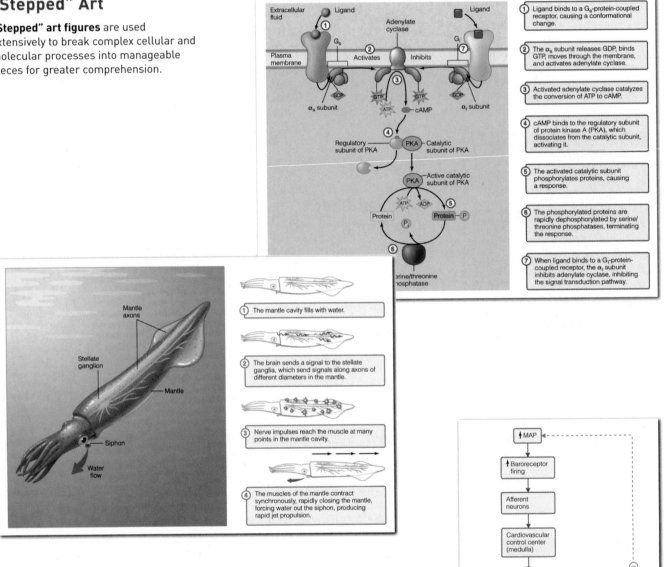

Flowcharts

We have included a large number of **flowcharts**, which help students understand the dynamic and multilayered nature of physiological processes. Flowcharts are used extensively to demonstrate the principles of negative feedback and homeostatic regulation and to emphasize how multiple control loops can regulate physiological processes.

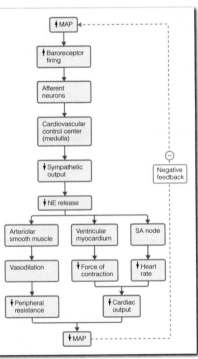

Superior Anatomy and Physiology Illustrations

The book contains many examples of art to help students make the link between **anatomical structures** and **physiological functions** in a diverse range of organisms. All anatomical art includes orientation clues to help students locate the organ within the organism. Every piece of art has been carefully checked, and many pieces have been updated for the second edition.

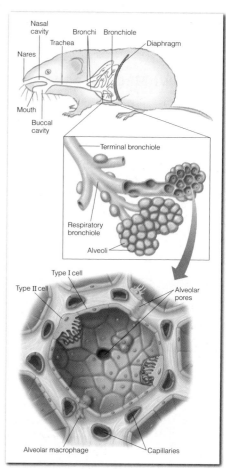

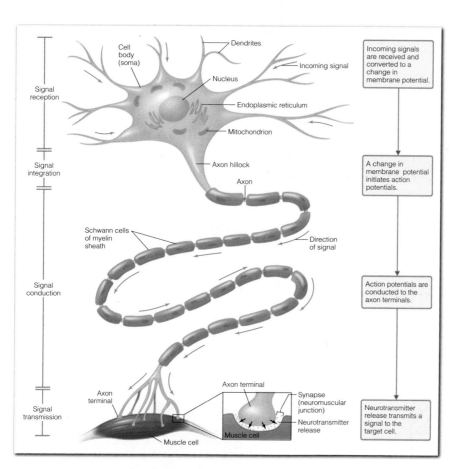

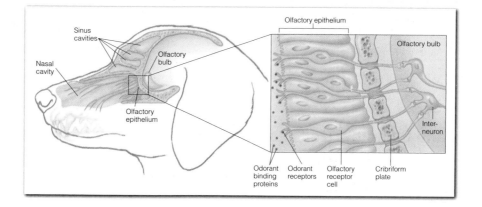

Supplements

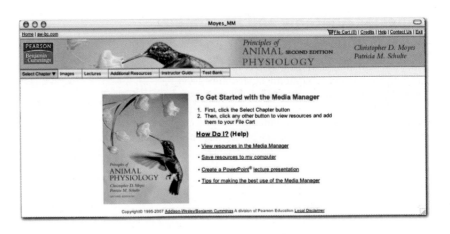

New! Media Manager

New, for the second edition is a complete **Media Manager** for instructors. A comprehensive teaching resource, this dual-platform CD-ROM contains JPEG and PowerPoint files for all of the art and tables and selected photos from the text. In addition, the Media Manager contains PowerPoint lecture slides for each chapter of the book, with appropriate figures and tables embedded.

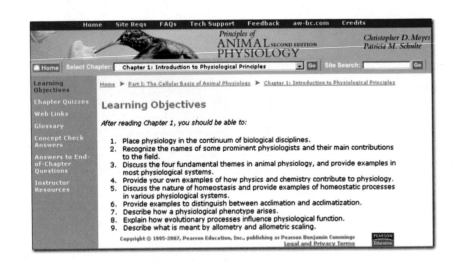

Companion Website

This student resource contains chapter objectives, answers to Review Questions and Concept Checks, chapter-specific quizzes, links to physiology labs and relevant online sites, and an interactive glossary. It is accessed through www.aw-bc.com/animalphysiology.

New! Printed and Computerized Test Bank

The brand new **Test Bank** provides 75–100 multiple choice, true/false, short answer, and essay questions for each chapter. Answers and page references are provided for all questions. The **Computerized Test Bank** is created in TestGen, our user-friendly cross-platform CD-ROM. Instructors can create and customize quizzes and tests using pre-written questions or inserting questions of their own.

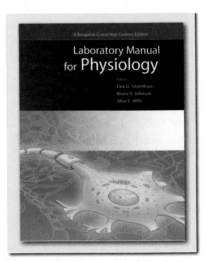

Laboratory Manual for Physiology

Create a custom lab manual from this collection of class-tested exercises edited by Dee Silverthorn, Bruce Johnson, and Alice Mills. Among the choices are 77 labs specifically designed for animal physiology. Go to www.aw-bc.com/integrate to view the lab exercises and build your customized lab manual.

Acknowledgments

Development of this textbook required a dedicated team of people who helped translate our manuscripts and sketches into this final textbook. We are exceedingly grateful to Catherine Murphy, the developmental editor for the first edition of this book, who had the onerous responsibility of making our text manuscript intelligible—a task that we hope got easier with time. Laura Southworth, our art development editor, was able to interpret our crude artistic efforts and convert them into something recognizable and useful. Her efforts on both the first and second editions of this book have been invaluable, and it has been nothing short of a treat to see our unsophisticated sketches morph into a truly beautiful and integrated art program for the book. Susan Malloy and Marie Beaugureau, our project editors at Benjamin Cummings, provided the encouragement and direction that enabled the whole group to work happily together (and more or less on schedule) through both the first and second editions of this book. The entire project team showed remarkable patience and a gift for coping with the personality quirks of the authors, which managed to make the process of writing and revising this book seem like fun. We would also like to thank the other people at Benjamin Cummings who helped bring the book to publication: Deirdre Espinoza, Claire Alexander, Amy Teeling, Caroline Ayres, and Leslie Berriman.

The brand-new Test Bank was written by Tracy Wagner (Washburn University), Kelly Shoemaker (Queen's University), and Alan F. Smith (Mercer University). It was accuracy checked by Linda Ogren (University of California Santa Cruz). The PowerPoint lecture outlines for the Instructor's Media Manager were written by Dr. Stephen Gehnrich (Salisbury University), and the clicker questions were written by Dr. Alan F. Smith (Mercer University). The student quizzes for the Companion Website were contributed by Jose de Ondarza (SUNY–Plattsburg) and Timothy D. Maze (Lander University). The answers to the end-of-chapter questions have been written by Patrice Boily (Western Connecticut State University) and the answers to the Concept Checks provided by Diara D. Spain (Dominican University of California). We thank them all.

We could not have written this textbook without support from our academic and research colleagues who fielded our questions and discussed points of confusion and contention. In particular, we are indebted to colleagues at Queen's University (Mike Adams, Bill Bendena, Adam Chippindale, Peter Davies, Steve Lougheed, Don Maurice, Bob Montgomerie, Gerry Morris, Mel Robertson, Stephen Scott, Bruce Tufts, Virginia Walker, and Kathy Wynne-Edwards) and the University of British Columbia (Vanessa Auld, Colin Brauner, Tony Farrell, John Gosline, Bob Harris, Wayne Maddison, Bill Milsom, and Jeff Richards). We also thank other colleagues for comments and contributions to specific features: Ted Garland, University of California–Riverside (artificial selection); Steve Hand, Louisiana State University (high-energy bonds, anhydrobiosis); Carlos Martinez Del Rio, University of Wyoming (gut reactor theory); Jim Staples, University of Western Ontario (thermoregulation); and Raul Suarez, University of California–Santa Barbara, for reminding us of the relationship between blind men, elephants, and physiologists. We also thank the many professors, listed below but anonymous at the time, who spent extraordinary amounts of time reviewing and class-testing chapters.

Finally, we thank our families, friends, and the trainees in our research labs for their patience during the past couple of years, when they couldn't quite get as much attention as they deserved. We thank our own research mentors for their support in our careers. In particular, Peter Hochachka, an important graduate mentor for both of us, was one of our most enthusiastic supporters when we began this textbook project. Sadly, he passed away before the first edition of this book was completed—we dedicate this book to his memory.

Reviewers

Rod Allrich, *Purdue University–West Lafayette*
Eli Asem, *Purdue University*
Todd Backes, *SUNY–Fredonia*
James S. Ballantyne, *University of Guelph*
John Berges, *University of Wisconsin–Milwaukee*
Jay Blundon, *Rhodes College*

Patrice Boily, *Western Connecticut State University*

Winnifred Bryant, *University of Wisconsin–Eau Claire*

Sara Hiebert Burch, *Swarthmore College*

Mark Burleson, *University of Texas–Arlington*

Dennis Claussen, *Miami University of Ohio*

Craig Clifford, *Northeastern State University*

Randy Cohen, *California State University–Northridge*

Thomas Cox, *Southern Illinois University*

Kenneth Crawford, *Western Kentucky University*

Joseph Crivello, *University of Connecticut*

Peter Daniel, *Hofstra University*

Dale Erskine, *Lebanon Valley College*

David Evans, *University of Florida**

Anthony Farrell, *University of British Columbia**

Michael S. Finkler, *Indiana University–Kokomo*

Kim Fredericks, *Viterbo University*

David Froman, *Oregon State University*

Stephen Gehnrich, *Salisbury State University*

Alice Gibb, *Northern Arizona University*

Stuart Goldstein, *University of Minnesota*

Neil Hadley, *University of North Carolina–Wilmington*

John D. Harder, *Ohio State University*

Jean Hardwick, *Ithaca College*

Michael S. Hedrick, *California State University–Hayward**

Steve Hempleman, *Northern Arizona University*

Raymond P. Henry, *Auburn University*

Kelly Johnson, *Ohio University*

Valerie Kalter, *Wilkes University*

Kevin S. Kinney, *DePauw University*

Heather Koopman, *University of North Carolina–Wilmington*

David T. Kurjiaka, *University of Arizona*

Dominic Lannutti, *El Paso Community College*

Gary Laverty, *University of Delaware*

John P. Leonard, *University of Illinois–Chicago*

John Lepri, *University of North Carolina–Greensboro*

James Long, *Boise State University*

David Mallory, *Marshall University*

Duane McPherson, *SUNY–Geneseo*

Scott Mills, *Purdue University–West Lafayette*

Sarah Milton, *Florida Atlantic University*

Michael O'Donnell, *McMaster University**

David O'Drobinak, *Valdosta State University*

Linda Ogren, *University of California–Santa Cruz*

Sanford Ostroy, *Purdue University*

Peggy Shadduck Palombi, *Transylvania University*

John Parrish, *Georgia Southern University*

Kathryn Ponnock, *Delaware Valley College*

Stephen Reid, *University of Toronto at Scarborough**

J. Larry Renfro, *University of Connecticut*

Gerald Robinson, *Towson State University*

Max G. Sanderford, *Tarleton State University*

Jason Schreer, *SUNY–Potsdam*

Donal Skinner, *University of Wyoming*

Alan Smith, *Mercer University–Macon*

James Staples, *University of Western Ontario*

Phil Stephens, *Villanova University*

Jonathon Stillman, *University of Hawaii–Manoa*

Ann E. Stuart, *University of North Carolina–Chapel Hill**

Sterling Sudweeks, *Brigham Young University*

Mark T. Sugalski, *Southern Polytechnic State University*

Steven Swoap, *Williams College*

Douglas A. Syme, *University of Calgary**

Malcolm H. Taylor, *University of Delaware*

Mark Varner, *University of Maryland–College Park*

Samuel Velez, *Dartmouth College*

Teresa Viancour, *University of Maryland–Baltimore County*

Susan Whittemore, *Keene State College*

Paul Wagner, *Washburn University*

Tracy Wagner, *Washburn University*

Jean-Michel Weber, *University of Ottawa*

Brian Witz, *Nazareth College*

William A. Woods, *University of Massachusetts–Boston*

Patricia Wright, *University of Guelph*

*Special topics reviewers.

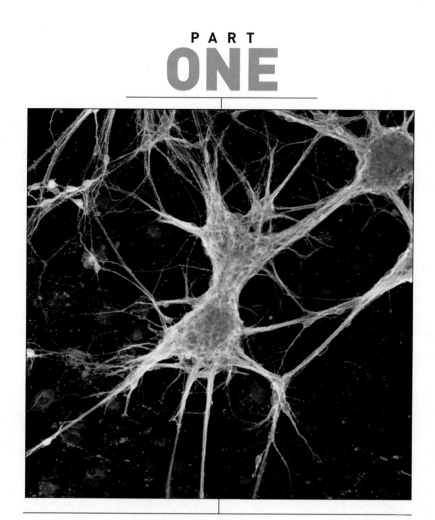

The Cellular Basis of Animal Physiology

In a classic poem, J. G. Saxe wrote of six blind men exploring the wonder of an elephant. The first man touches the side of the elephant and envisions a wall, while the next grasps the tusk and thinks it a spear. The third touches the trunk and is reminded of a snake, while the forth holds a leg and imagines a tree. The fifth grasps the ear and perceives a fan, and the last holds the tail and conjures a rope. Saxe's moral was that the only way to understand complex issues is to consider them from multiple perspectives. An understanding of physiology requires that it be investigated through multiple disciplines: physics, chemistry, mathematics, engineering, genetics, cell biology, ecology, and evolution. A second lesson of the poem, also relevant to physiologists, is that even the most unusual animal traits are in many ways familiar. In fact, the blind men were quite perceptive. The side of the elephant really is a wall, its trunk does move like a snake, and the tusk is used as a spear and the tail, like a rope. Physiologists often use analogies and comparisons, to mechanical structures and to other animals, to help them understand the structure and function of animals. One of the wonders of physiology is this paradox of unity within diversity: even the most unusual traits have features that share similarities with those of other organisms.

Part I of this text introduces how molecular and cellular processes constitute the underpinnings of physiology. Just as an animal is built from a single cell, physiology is constructed from molecular and cellular building blocks. In Chapter 1, we introduce the science of physiology and discuss the origins of physiological diversity. Chapter 2 reviews the essentials of cellular function: the influence of physical and chemical forces on physiological function and the basics of biochemistry and cell biology. Though many of these cellular features are similar among plants, fungi, and animals, there are three main ways in which animal cellular function is distinct from that of the other eukaryote taxa: cell communication, muscles, and neurons. In Chapter 3 we explore the complex ways that animal cells and tissues communicate with each other. Neurons (Chapter 4) and muscle cells (Chapter 5) are found only in animals and underlie all physiological systems. In these last two chapters of this section, we set the foundation for the role of these unique animal cells in complex physiological systems, focusing on how these cells work and how variations in their properties arise in evolution and development. ◉

Introduction to Physiological Principles

Physiological research exploded in the 1960s as a result of several related events. Advances in diverse technologies, from nuclear medicine to molecular genetics, paved the way for new approaches to studying animal diversity. Population demographics led to massive hiring of university-based scientists, creating a critical mass of researchers interested in understanding the physiological diversity of animals. It became commonplace to see international teams of researchers collaborate as multidisciplinary teams, exploring projects that would otherwise be prohibitative. The ease of travel and growth of the worldwide research community created opportunities for physiologists to study unusual animals in exotic places.

It was during this period that Dr. Per Scholander, a renowned animal physiologist and director of the Scripps Institute of Oceanography (University of California, San Diego),

Research vessel, *Alpha Helix.*

Elephant seal.

spearheaded an initative that would allow teams of international researchers to work together to study biological problems in remote locations. After many years of effort and negotiation with researchers, universities, and government agencies, the Alpha Helix program was launched.

The *Alpha Helix* was an oceanic research vessel named after the structural model of DNA proposed by Watson and Crick only 10 years earlier. It was purchased in 1964 by the Scripps Institute of Oceanography through a $1.5 million grant from the U.S. National Science Foundation. The ship was built to provide technically sophisticated laboratories for experimental biologists to use as the ship explored the unusual natural habitats of the world. Although launched and funded by the U.S. government, it supported the research of both American and international scientists. On her maiden voyage in 1966, the *Alpha Helix* carried 12 crew members and 10 scientists around the world on a "quest of biological and medical knowledge." The *Alpha Helix* program was a career-changing experience that inspired a whole generation of animal physiologists.

The ship was the floating laboratory for three or four expeditions each year, bringing together teams of scientists with complementary interests and expertise. The inaugural expedition of the *Alpha Helix* was a six-month expedition to the Great Barrier Reef, where researchers studied coral reefs, tropical mangrove forests, and the animals that lived in the sea and on land. Three months after returning from Australian waters, a new expedition was launched to South America. The *Alpha Helix* traveled up the Amazon River to study the behavioral and evolutionary properties of fish and terrestrial animals in the neotropics.

The trip home passed through the Galapagos Islands, where researchers studied the same animals that Darwin studied a century earlier. The cruises over the next 15 years took researchers back to these same locations, and to others such as the Bering Sea (cold-water fishes), New Guinea (tropical animals), Guadalupe Island (fish and elephant seals), Antarctica (polar animals), the eastern Pacific (reef animals, sharks, whales), Australia (sea snakes), Hawaii (deep-sea fishes), and the Philippines (nautilus). Many of these animals had never been studied, and their physiological properties were largely mysterious at that time.

The *Alpha Helix* exemplified the explosion of work in animal physiology that began in the 1960s. The Alpha Helix program continued until 1980, at which point government support ended and the ownership of the vessel was transferred to the National Science Foundation. The vessel itself remains in active duty, based at the University of Alaska, and is used for international oceanographic research. The Alpha Helix program gave hundreds of researchers the opportunity to learn firsthand about the diversity of the natural world and how organisms function in their various environments. ⊙

Overview

In the words of the renowned physiologist Knut Schmidt-Nielsen, animal physiology is *"the study of how animals work."* Animal physiologists study the structure and function of the various parts of an animal, and how these parts work together to allow animals to perform their normal behaviors and to respond to their environments. One hallmark of animal physiology is diversity. More than a million different species of animals live on Earth, each of which has acquired through evolution countless unique properties. Each physiological process is a product of the activities of complex tissues, organs, and systems that can arise through complex patterns of genetic regulation of countless cells.

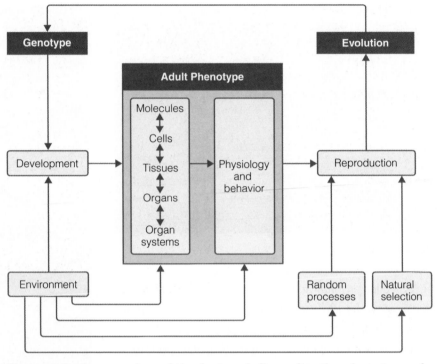

Figure 1.1   An overview of the factors influencing the phenotype of adult animals

Despite this great diversity, there are many commonalities within physiology—unifying themes that apply to all physiological processes. First of all, physiological processes obey physical and chemical laws. Second, physiological processes are regulated to maintain internal conditions within acceptable ranges. This internal constancy, known as *homeostasis,* is maintained though feedback loops that sense conditions and trigger an appropriate response. Third, the physiological state of an animal is part of its *phenotype,* which arises as the product of the genetic makeup, or *genotype,* and its interaction with the environment. Fourth, the genotype is a product of evolutionary change in a group of organisms—populations or species—over many generations.

Most physiological studies examine how various processes affect the physiological phenotype of an animal (Figure 1.1). Both the genotype of an organism and its environment interact through development to produce the phenotype of the adult organism. The phenotype is itself the product of processes at many levels of biological organization, including the biochemical, cellular, tissue, organ, and organ system levels. Together these processes interact to produce complex behaviors and physiological responses. The environment can, in turn, influence the adult phenotype. Organisms may change their behavior as a result of learning, or alter their physiological responses through modification of their phenotypes. Ultimately, the phenotype (morphology, physiology, and behavior) of an animal influences its reproductive success. Differential survival of organisms with distinct phenotypes may result in evolutionary change in the genotype of a population over many generations.

Physiology: Past and Present

Modern animal physiology is a discipline concerned with the whole range of processes that affect animal function. Although animal physiology is an experimental science that can trace its roots back more than two millennia to the ancient Greeks, it plays an important role in modern biology as the intellectual glue that holds disparate biological fields together.

A Brief History of Animal Physiology

Although Greek thinkers such as Hippocrates (460–*circa* 377 B.C., the father of medicine) and Aristotle (384–322 B.C., the father of natural history) were not primarily experimental physiologists, Hippocrates' emphasis on the importance of careful observation in the treatment of disease and Aristotle's emphasis on the relationship between structure and function make them important figures in the history of physiology. Claudius Galenus (A.D. 129–*circa* 199), known as Galen, was the first to use systematic and carefully designed experiments to probe the function of the body. Galen made extensive use of dissection and vivisection of nonhuman primates such as Barbary apes and other mammals to test his physiological ideas. For example, Galen performed experiments in which he tied off the ureters (the tubes leading from the kidney to the bladder), and observed that the kidneys swelled. From this observation he concluded that the kidneys play a role in the formation of urine. Similarly, he tied off the laryngeal nerve (which leads to the vocal cords) of a living pig, at which point the pig stopped squealing. From this experiment, he concluded that the brain and nerves regulate the voice. This experimental work, combined with his practice as a physician to the Roman gladiators, allowed him to formulate detailed descriptions of anatomy and elucidate the basis of many physiological processes. Although much of Galen's work was fundamentally incorrect when viewed from a modern perspective, his emphasis on careful observation and experimentation makes him the founder of physiology.

During the Middle Ages the medical traditions of the ancient Greeks were practiced and further developed by physicians in the Muslim world, most notably Ibn al-Nafis (1213–1288), who was the first to correctly describe the anatomy of the heart, the coronary circulation, the structure of the lungs, and the pulmonary circulation. He was also the first to describe the relationship between the lungs and the aeration of the blood.

The Renaissance brought a new flowering of physiological research in the Western world. Jean-Francois Fernal (1497–1558) outlined the current state of knowledge of human health and disease. Andreas Vesalius (1514–1564), author of the first modern anatomy textbook, demonstrated that Galen had made many errors in both anatomy and physiology. Because Galen was thought to have done everything that was necessary to understand the workings of the body, many medical practitioners of the time shunned physiological research. Thus, by showing that Galen was not entirely correct, Vesalius's work triggered the modern study of anatomy and physiology.

William Harvey (1578–1657) identified the path of blood through the body, and showed that contractions of the heart power this movement. Although Harvey could not see the fine capillaries that connect arteries and veins using the crude magnifying glasses that were available at the time, he postulated that they must exist to form a closed circulation for the blood around the body. Harvey showed how dissections, close observation of living organisms, and careful experiments could be combined to teach us about the functions of the body.

Prior to the 18th century, physiologists fell into one of two camps. The *iatrochemists* believed that body function involved only chemical reactions, whereas *iatrophysicists* believed that only physical processes were involved. In the late 17th and early 18th centuries a Dutch physician, Hermann Boerhaave, and his Swiss pupil, Albrecht von Haller, proposed that bodily functions were a combination of both chemical and physical processes. By uniting these two approaches, these researchers were among the earliest proponents of physiology as we understand it today.

In the 19th century physiological knowledge began to accumulate at a rapid rate. For example, in 1838 Matthias Schleiden and Theodor Schwann formulated the "cell theory," which states that organisms are made up of units called cells, a discovery that paved the way for modern physiology. Claude Bernard (1813–1878) discovered that hemoglobin carries oxygen, that the liver contains glycogen, that nerves can regulate blood flow, and that ductless glands produce internal secretions (hormones) that are carried in the blood and influence distant tissues. One of Bernard's most important contributions was his concept of the **milieu interieur** (internal environment); he postulated that living organisms preserve a distinct internal environment despite changes in the external environment. This concept—the ability to maintain a constant internal environment—was later developed more fully by the American physiologist Walter B. Cannon (1871–1945), who coined the term *homeostasis*.

Until the 20th century, physiologists made little distinction between animal physiology and medical physiology. Most physiological experiments on animals were performed with the goal of improving the understanding of the human body both in health and in illness. But in the 20th century biologists became interested in applying the newly emerging physiological knowledge to animals living in diverse environments, and in trying to understand the nature of physiological diversity.

Per Scholander (1905–1980) was one of the first and most influential of these comparative physiologists. Scholander studied a remarkable diversity of physiological responses, including the mechanisms involved in diving vertebrates, the responses of warm-blooded animals to cold environments, and how fish fill their swim bladders (air-filled organs that fish use for buoyancy). Scholander also organized the influential expeditions of the *Alpha Helix* in the research program described in the opening essay of this chapter.

The contributions of C. Ladd Prosser (1907–2002) include the discovery of so-called **central pattern generators**. These groups of neurons coordinate many rhythmic behaviors, including breathing and walking. Prosser also discovered the relationship between muscle diameter and conduction speed, and during World War II he worked on the effects of radiation on animal life as part of the Manhattan Project.

Knut Schmidt-Nielsen (1915–2007) devoted his career to understanding how animals live in harsh and unusual environments. In his classic early work on the adaptations of the camel to desert life, he showed that the camel's nose contains a countercurrent exchanger that allows it to recapture moisture from exhaled air, resulting in an almost 60% reduction in water loss compared to other mammals.

George Bartholomew (1923–2006) is the founder of the field of ecological physiology, or the study of how an organism interacts with its environment. Bartholomew combined the study of animal behavior, ecology, and physiology to assess the evolutionary significance of adjustments or adaptations in animals to their environments. He identified the individual as the principal unit of natural selection, and emphasized the importance of variation in physiology.

Peter Hochachka (1937–2002) and George Somero (1941–) founded the field of adaptational biochemistry. By applying the concepts and techniques of biochemistry to the questions of comparative physiology, they have extended to the subcellular level our understanding of how animals adapt to hostile environments, providing insights into the biochemical mechanisms that allow animals to live in habitats as diverse as the deep sea, the Antarctic oceans, high mountain peaks, and tropical rain forests.

Any attempt to survey the major figures in the history of animal physiology excludes countless other researchers who have made important contributions to the field. As we proceed through this book, we introduce many of the other important figures in animal physiology and discuss their specific contributions in detail.

Subdisciplines in Physiological Research

Modern physiological knowledge is the product of the efforts of multitudes of scientists with diverse interests and expertise. Typically, an animal physiologist specializes in one or two subdisciplines of physiology, with an awareness of the central issues in other, related subdisciplines. There are three main ways to categorize physiological subdisciplines: by the biological level of organization, by the nature of the process that causes physiological variation, and by the ultimate goals of the research.

Physiological subdisciplines can be distinguished by the biological level of organization

Since physiology is concerned with biological function at many levels of organization (Figure 1.2), one of the most common ways to distinguish branches of physiology is by reference to these levels.

- *Cell and molecular physiologists* study phenomena that occur at the cellular level, although these effects have important consequences for higher levels of organization. Cell and molecular physiologists might include researchers studying molecular genetics, signal transduction, metabolic biochemistry, or membrane biophysics.

- Many physiologists focus their efforts on specific physiological systems. A *systems physiologist* is concerned with how cells and tissues interact to carry out specific responsibilities within the whole animal. In fact, each of the chapters in Part Two of this text is focused on a physiological system. Thus, there are respiratory physiologists, sensory physiologists, and so on.

- An *organismal physiologist* is most often concerned with the way an intact animal un-

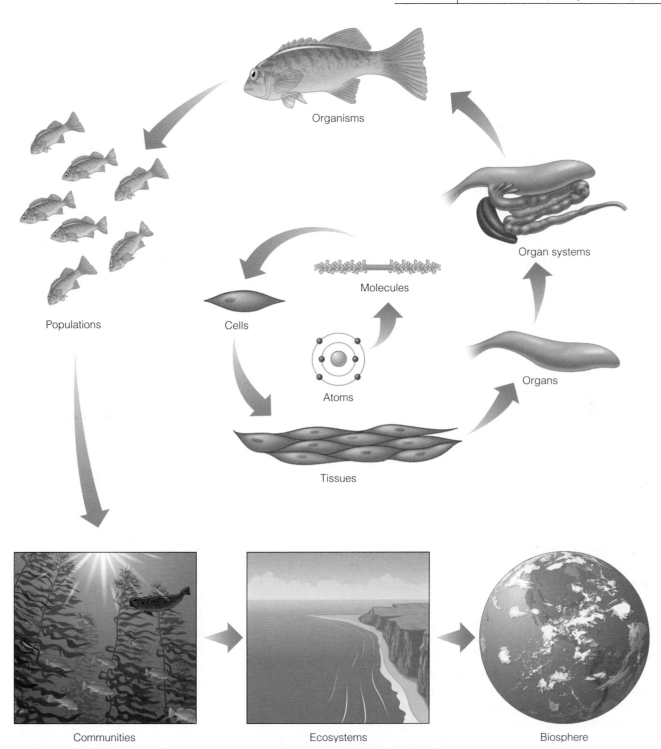

Figure 1.2 Levels of biological organization Chemists and biochemists study the properties of atoms and molecules. Molecular biologists study the properties of molecules and cells. Physiologists study the interactions between molecules, cells, tissues, organs, and organ systems to understand the structure and function of an organism. Ecologists study the interactions of organisms, populations and communities to understand the properties of ecosystems, and ultimately the biosphere.

dertakes a specific process or behavior. For example, an organismal physiologist might study changes in animal metabolic rate in response to a stressor, such as temperature.

An organismal characteristic such as metabolic rate is the product of multiple physiological systems interacting in complex ways. Some organismal physiologists specialize in particular groups of animals: thus, there are marine mammal physiologists, avian physiologists, fish physiologists, and so on.

- An *ecological physiologist* studies how the physiological properties of an animal influence the distribution and abundance of a species or population. For example, an ecological physiologist may study how the nutrient distribution in the environment influences the growth rate of an animal. While organismal physiologists may focus their research on an interesting group of animals, ecological physiologists are more concerned with how an interesting environment affects diverse animals within that environment.

- An *integrative physiologist* attempts to understand physiological processes at a variety of levels of biological organization and across multiple physiological systems. For example, an integrative physiologist might study how variation in hemoglobin genes contributes to differences in oxygen delivery and how those differences in the ability to extract oxygen from the environment contribute to the geographical distribution of the species.

Of course, there is a great deal of overlap among these subdisciplines, and making distinctions among them is often difficult. In fact, few physiological researchers confine themselves exclusively to investigating a single level of biological organization. Often a physiologist interested in a process at one level of organization also studies its function at the next lower level. This approach, known as **reductionism**, assumes that we can learn about a system by studying the function of its parts. Although a reductionist approach can be extremely illuminating, and has been the basis of many important biological discoveries, ultimately many processes have characteristics that are not apparent simply by examining the component parts. This feature of complex systems is called **emergence**, which is just another way of saying that the whole is often more than the sum of its parts. The emergent properties of a system are due to the interactions of the component parts of the system, and can be difficult to predict by studying each part in isolation. Physiologists are usually interested in these emergent properties, and thus physiologists study how molecules, cells, and tissues interact to produce the complex system that is an organism.

Physiological subdisciplines can be distinguished by the process that generates variation

Many physiologists are interested in how biological functions change over time or in response to changes in the environment. Thus, physiology can also be divided on the basis of the mechanism by which changes or differences in physiological processes arise.

- A *developmental physiologist* studies how structures and functions change as animals grow through the various life stages from embryo to reproductive maturity, to senescence and death. These developmental pathways transform omnipotent stem cells into specialized cells, forming multicellular tissues and physiological systems. To understand the diversity in animal morphology and function, it is important to appreciate how these structures arise in development.

- An *environmental physiologist* assesses how animals mount physiological responses to environmental challenges. For example, changes in temperature have the potential to affect many physiological systems in complex ways. An environmental physiologist is concerned with the way an individual animal organizes or reorganizes its physiology to survive the environmental challenge.

- An *evolutionary physiologist* is primarily concerned with explaining how specific physiological traits arise within lineages over the course of multiple generations. Thus, evolutionary physiologists may be interested in the origins of variation within populations of a single species, or the basis of differences between closely related groups of animals.

Animal physiology can be a pure or an applied science

Physiological research can be distinguished on the basis of the ultimate goal. The research of an *applied physiologist* is intended to achieve a specific, practical goal. For example, physiologists study some animals because of their economic importance. Thus, veterinary medicine relies on physiological research to improve the health of agricultural animals and household pets. Simi-

BOX 1.1   **METHODS AND MODEL SYSTEMS**
August Krogh Models in Animal Physiology

A model species is an organism that is studied by a wide community of researchers because (1) it has features that are conducive to experimentation and (2) understanding a process in the model provides insight into how the process works in other species of interest. Each model species has been chosen because it demonstrates a combination of features that make it well suited to some studies, though not all studies. This approach of using an animal model with features that are favorable for scientific study is known as the **August Krogh principle**: *For every biological problem there is an organism on which it can be most conveniently studied.*

The importance of specific model systems changes over time, as technologies advance and genomic databases expand. An animal that was inconvenient to study in the past may be much easier to study now. For example, mice became more useful models in developmental physiology when transgenic mouse technologies became readily available.

Knowledge gained from model systems is useful only if that information is relevant to other species. Most commonly, a model was originally chosen because of parallels with human biology. Although the major model animals are quite different in appearance, much of the genetic and structural machinery that underlies development is similar among animals. The early patterns of embryonic development are similar in most vertebrate models, such as zebra fish, chickens, mice, and humans. However, there are always concerns about the phylogenetic distance between model systems. For this reason, each taxon has one or more species that have been trumpeted as a model.

Some animals are useful models because they have unusual anatomical features. Perhaps the most famous example of such a model system is the squid giant axon. Squid are relatively simple animals that have certain axons large enough to be easily seen and readily manipulated. The oocytes of the African clawed frog (*Xenopus laevis*) are useful as models for the expression of foreign proteins. *Xenopus* oocytes are large, so scientists can easily introduce foreign RNA by microinjection. The RNA is then translated and the protein sent to the appropriate location. For example, microinjection of RNA coding for membrane proteins causes the oocyte to translate the protein and insert it into the membrane where its functional properties can be assessed.

Many animals are useful models because of their developmental biology. Nematodes are small animals composed of only a few thousand cells. The development from fertilized egg to adult has been studied to the point that the fate of each cell has been mapped. Researchers can microinject substances into a designated cell at a specific developmental stage, knowing that specific cell will divide and differentiate into a specific tissue or organ. Zebra fish are useful models because the embryo grows quickly and remains transparent for much of its early development. This allows researchers to follow complex cellular changes in living animals. Such studies are aided by transfection of genes encoding fluorescent proteins that can be monitored more easily.

One important factor that determines the utility of a model species is the ease with which genes can be modified. The ability to generate mutations that result in the gain or loss of a function allows physiologists to explore the importance of structural features. For many years, random mutagenesis was the only way to generate mutants. During this period, invertebrates and small fish were used because it was possible to conduct large-scale screening projects to identify interesting mutants. More recently, genetic approaches to physiology have been facilitated by two trends. First, the proliferation of techniques for targeted mutagenesis makes it easier to work on animals with long generation times, because large-scale screening is not needed. Second, there is a rapid growth in the number of species for which we have genomic information. Models become a lot more convenient to use in genetic studies when we know their entire genome.

larly, much physiological research is aimed at understanding the human body. Although the ultimate goal of medical physiology is to understand human disease, medical physiology relies on other species as model systems (see Box 1.1, Methods and Model Systems: August Krogh Models in Animal Physiology).

In contrast to a medical physiologist, who uses animals to understand the human condition, a *comparative physiologist* studies animals to explore the origins and nature of physiological diversity. Comparative animal physiology thrives on the breadth of physiological diversity, all the while searching for unifying themes.

⊙ | CONCEPT CHECK

1. How would you define physiology?
2. Who were some of the major figures in physiology prior to the 20th century?
3. How do 20th-century comparative physiologists differ from earlier physiologists?
4. What are the subdisciplines of physiology?
5. How do some of the other sciences help us to understand physiological processes?

Unifying Themes in Physiology

In spite of the vast and diverse nature of animal physiology, several unifying themes and principles apply to all of its subdisciplines (Table 1.1). Throughout this book, we return to these themes as we examine animals at the cellular and system levels.

Physics and Chemistry: The Basis of Physiology

To understand physiology you need a basic understanding of chemistry and physics. Animals are constructed from natural materials and thus obey the same physical and chemical laws that apply to everything that we see around us. Temperature, for example, exerts its effects on physiology by altering the nature of chemical bonds in biomolecules, or solubility of gases in solution. Physiologists often borrow concepts and techniques from the physical and chemical sciences, including engineering, to help them understand how animals work.

Mechanical theory helps us understand how organisms work

Each material has physical properties that are useful in some contexts but not others. It would be a mistake for an engineer to design a skyscraper from Styrofoam, or a kite of concrete. Likewise, biological materials, or biomaterials—proteins, carbohydrates, and lipids—also have characteristic physical properties that make them useful for some processes but not others. For example, some proteins are rigid and inflexible whereas others readily deform. The physicochemical characteristics of these biomaterials are determined by their molecular properties. For example, a network of proteins can be made more rigid by additional bonds that cross-link proteins together. Cells use enzymatic reactions to fine-tune the physical properties of macromolecules. The macromolecules combine to form cells, which are collected together to form tissues. Thus, the mechanical properties of a tissue, such as bone, are conferred

Table 1.1 Unifying themes in animal physiology.	
Unifying theme	**Examples**
Physiological processes obey the laws of physics and chemistry.	Mechanical engineering rules apply to physical properties of animals. Chemical laws, including the effects of temperature, govern interactions between biological molecules. Electrical laws describe membrane function of all cells, including excitable cells. Body size affects many physiological processes.
Physiological processes are usually regulated.	Homeostasis is the maintenance of internal constancy. Negative feedback loops help maintain homeostasis. Positive feedback loops generate an explosive response.
The physiological phenotype is a product of the genotype and the environment.	Even identical genotypes can result in different phenotypes. Phenotype changes with normal development. Phenotype changes with environmental and physiological challenges. Phenotypic plasticity is the ability of a phenotype to change in response to environmental conditions.
A genotype is the product of evolution, acting through natural selection and other evolutionary processes.	The definition of adaptation is context dependent. In the strictest evolutionary sense, adaptation refers to a trait that confers an increase in reproductive success. Adaptation can also refer to phenotypic changes that improve the performance of a physiological system, without underlying evolutionary change. Not all physiological differences are adaptations.

by the molecular properties of the components of the bone-forming cells, the nature of the connections between cells, and the interactions between tissues.

In addition to mechanical properties, other engineering concepts such as flow, pressure, resistance, stress, and strain play important roles in physiology. An engineer designing a system to pump water from a deep well takes into consideration factors such as the pressure gradients, fluid dynamics, the power of the pump, and resistance in the plumbing. A cardiovascular physiologist has the same concerns in trying to understand how the heart delivers blood through the blood vessels.

Electrical potentials are a fundamental physiological currency

Just as we use electricity to power many of the machines we use in our daily lives, animals use electricity to power cellular activities. Cells establish a charge difference across biological membranes by moving ions and molecules to create ion and electrical gradients. All cells and many organelles within cells rely on this potential difference, or membrane potential, to drive processes that are needed for survival. Animals also use changes in electrical potentials to send signals within and between cells, helping to coordinate the complex processes of the body. Muscles and neurons, two cell types that are found only in animals, use changes in membrane potential to send signals. Thus, electrical theory has played an important role in helping physiologists to understand the way that neurons and muscles work.

Biochemical and physiological patterns are influenced by body size

From tiny zooplankton weighing less than a milligram to blue whales weighing over 100,000 kg, animals vary greatly in body size, and these differences have profound effects on physiological processes. One reason is that the ratio of the surface area to volume changes with body size. Consider an animal shaped like a sphere. With a radius r, its mass, or rather its volume $(V) = (4/3)\pi r^3$ and its surface area $(A) = 4\pi r^2$. Since surface area increases by the power of two, and volume increases by the power of three, the surface area is proportional to the volume to the 2/3 power, or $V^{0.67}$. This relationship between surface area and volume has an important influence on thermal biology. Heat is

produced by tissue metabolism, and thus the metabolic rate of the animal as a whole depends on the mass of tissues. Metabolic heat is lost across the surface of the body. Since heat production varies with body mass and heat loss varies with body surface area, a larger animal has more difficulty shedding metabolic heat than does a smaller animal.

It has long been known that metabolic rate of animals does not increase proportionately with body mass. That is, animals differing 10-fold in body size differ less than 10-fold in metabolic rate (Figure 1.3). In the late 1800s, Max Rubner reported that the metabolic rate of dogs of various sizes was constant when body surface area was taken into account. For many years, it was thought that the relationship between body mass and metabolic rate was related to the ratio of surface area to volume. In the 1930s, Max Kleiber examined the influence of body size on metabolic rate of birds and mammals. Based on these data, he formulated the **allometric scaling** equation, relating body mass (M) and metabolic rate (y)

$$y = aM^b$$

where a is the normalization coefficient and b the **scaling coefficient**. Kleiber's work suggested the value for b was closer to 0.75 (3/4) rather than the value of 0.67 (2/3) expected from Rubner's studies. These data, and many studies since, suggest that allometric scaling of metabolism is not easily explained by simple differences in ratio of surface area to volume. Despite the complexity of size-dependent physiological properties, or perhaps

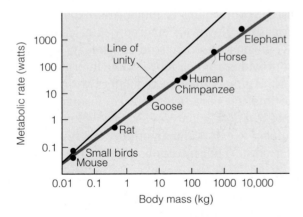

Figure 1.3 Metabolic rate of various birds and mammals plotted against body weight on a double logarithmic scale The line of unity shows the relationship that would be expected in the absence of allometric scaling. In this figure, an animal that is 10 times larger than another has a metabolic rate that is only about 7 times greater.
(Source: Adapted from Schmidt-Nielsen, 1984)

because of the complexity, allometric scaling remains one of the dominant themes in comparative animal physiology. Normally reticent physiologists have been inspired to engage in animated, and sometimes vitriolic arguments about both the exact value of b and the underlying mechanisms.

Physiological Regulation

Most organisms are faced with environmental variation. Temperature, food availability, and the physiochemical environment around an animal may change with the time of day, the season, or the movement of an animal across the landscape. Multicellular animals can be classified according to the strategies they use to cope with changing conditions.

Conformers allow internal conditions to change when faced with variation in external conditions. For example, the body temperature of a fish will be low in cold water and high in warm water. Thus, each of the cells in a fish's body must cope with the effects of changes in external temperature.

Regulators maintain relatively constant internal conditions regardless of the conditions in the external environment. Your body temperature is likely to be approximately 37°C whether you are in a warm room or standing outside on a very cold day. Your body has mechanisms to maintain its internal temperature, and thus the vast majority of the cells in your body do not have to cope with the effects of changes in ambient temperature.

Each strategy has its benefits and costs. Because physiological responses demand metabolic energy, conforming is much less expensive than regulating. However, environmental changes can have deleterious effects on physiology, so regulating provides a much more stable internal environment. Animals may be regulators with respect to one internal parameter, but conformers with respect to another parameter. For example, lizards conform to external temperature but regulate their internal salt concentrations within a narrow range.

Homeostasis is the maintenance of internal constancy

The maintenance of internal conditions in the face of environmental perturbations is referred to as **homeostasis**. The word *homeostasis* does not imply that there is no change in the organism, only that the animal initiates specific responses to control or

regulate a particular variable. For example, your body temperature remains relatively constant only because numerous physiological processes actively change, adjusting the rates of heat production and heat loss. For example, when you stand in the cold air, your muscles may shiver to produce heat that replaces the heat lost to the environment. Thus, muscle activity changes in order to maintain constant body temperature.

The nature of the physiological response to an environmental change depends on many factors. Short-term challenges can often be dealt with using existing physiological systems. When a dog is too hot, it can move to a cooler location or pant to shed heat in its breath. These are effective short-term behavioral and physiological approaches to reducing thermal stress. However, they are not effective long-term strategies. A dog hiding in the shade cannot hunt for food, and panting conflicts with oxygen delivery during running. Instead, dogs cope with long-term changes in temperature, such as seasonal cycles, by growing fur in the autumn and shedding fur in the spring.

This example illustrates several principles that govern physiological changes. First, some physiological strategies are effective in the short term but less useful for the long term. Holding your breath may be fine for diving to the bottom of a lake, but it will not help you cope with low oxygen levels while you climb Mount Everest. Second, some strategies require a significant investment in resources and need longer to take effect. Hair growth, for instance, is a relatively slow process that requires metabolic energy. Third, some stressors are sufficiently predictable that animals remodel physiology in anticipation of the stress, and often in predictable cycles. Many physiological processes change daily, showing a **circadian rhythm**. Some changes are seasonal, such as the growth and shedding of fur. Other patterns, such as human reproductive cycles, are linked to the lunar cycle. In some cases, cyclical physiological changes proceed without any environmental input, but generally they arise in response to specific environmental cues, such as temperature or photoperiod.

Feedback loops control physiological pathways

To maintain homeostasis, animals must (1) detect external conditions and (2) if necessary initiate compensatory responses that (3) keep vital areas buffered against unfavorable change. Animals most often maintain homeostasis using a **reflex control pathway**. A change in the internal or external environment provides a stimulus. The stimulus then causes a response. For instance, when you depress the gas pedal of your car (the stimulus), the car accelerates (the response). If you take your foot off the gas (remove the stimulus), the car will slow down.

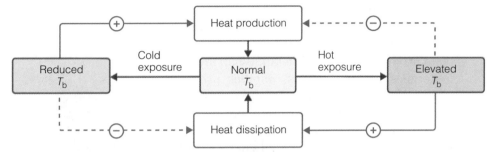

Figure 1.4 Antagonistic controls Your body temperature is held relatively constant by antagonistic loops. If cold conditions cause a decrease in your body temperature, this triggers an increase in heat production and a reduction in heat dissipation. When body temperature increases, heat production pathways are inhibited and heat dissipation pathways are stimulated, correcting body temperature.

Animals fine-tune physiological responses by using **antagonistic controls**: independent regulators that exert opposite effects on a step or pathway. In the car analogy above, the gas pedal and the brake are examples of antagonistic controls. You can cause the car to decelerate by taking your foot off the gas or depressing the brake, but the car's response will be greater if you use both in combination. Animals control body temperature by regulating both heat production and heat dissipation (Figure 1.4). Hormones mediate many antagonistic controls. As discussed in Chapter 3, insulin and glucagon are antagonistic controllers of glucose levels.

Negative feedback loops maintain homeostasis

In a **negative feedback loop**, the response sends a signal back to the stimulus, reducing the intensity of the stimulus. For example, when you eat, the incoming food causes the stomach to swell. The change in stomach volume and early digestion products trigger a negative feedback loop, acting through your brain, to reduce your appetite.

Many physiological systems have a **set-point**, a preferred physiological state defended through feedback loops. Your body temperature has a set-point of approximately 37°C. When temperature rises, your body may sweat to cool you down, whereas a decrease in body temperature may trigger shivering to warm you back to your set-point. Although the set-point for human body temperature is near 37°C, the exact body temperature set-point varies between individuals and changes throughout the day.

Positive feedback loops cause explosive responses

Some physiological systems are controlled by **positive feedback loops**. Unlike negative feedback, which minimizes changes in the regulated variable, positive feedback loops maximize changes in the regulated variable. For example, the muscles in the stomach are normally regulated to contract and relax in a regular pattern to gently mix food. However, when a toxin is detected, a positive feedback loop is triggered to induce forceful contractions that propel the food back up the esophagus to induce vomiting. Pathways involving positive feedback loops begin slowly but rapidly increase in intensity. In a positive feedback loop there must also be a signal that allows the animal to stop the process at the proper time, so that the action does not spiral out of control.

Phenotype, Genotype, and the Environment

The physiological properties of an animal are aspects of the animal's **phenotype**. Physiological traits, like other characteristics of animals, are determined in large part by the genes of the genome—the **genotype**—but are also influenced by the way the genes are regulated, particularly in response to external conditions.

An individual genotype has the capacity to produce considerable variation in cellular properties. Although the same genes are found in each cell, they are regulated in combinations to allow animals to develop distinct tissues. During this process of tissue formation, called *morphogenesis,* networks of genes are turned on and off in precise patterns to create the appropriate phenotype. For example, when the fertilized egg of a frog develops into a tadpole, a developmental program is turned on to produce the gills and a tail. When the tadpole undergoes metamorphosis, another program is triggered that results in the formation of the lungs and legs, and death of the cells in the gills and tail. In addition to orchestrating the normal developmental program, the genotype controls the way animals can alter their phenotype in response to physiological and environmental conditions. For example, changes in the expression of genes allow your muscles to change in size and strength in relation to exercise training. The differences in genotype among animals are central to the phenotypic variation upon which natural selection acts. Every individual genotype has a capacity to differ in complex, often unpredictable ways because of the way the genes respond to external conditions.

A single genotype results in more than one phenotype

A single genotype can result in multiple phenotypes, depending on the environmental conditions that the animal experiences. For example, if identical twins were raised in different places, it is possible that one twin might grow taller than the other due to differences in diet. This ability of a single genotype to generate more than one phenotype, depending on environmental conditions, is called **phenotypic plasticity**. We observe this phenomenon most commonly at the population level, where individuals with similar genotypes can have different phenotypes depending on environmental conditions. The term *phenotypic plasticity* encompasses a wide range of changes in phenotype, some reversible and some irreversible. Developmental plasticity, or **polyphenism**, is a form of phenotypic plasticity in which development under different conditions results in alternative phenotypes in the adult organism that cannot be reversed by subsequent changes in the environment. The similar concept of a **reaction norm**, or the range of phenotypes produced by a particular genotype in different environments, applies to phenotypes that exist as a continuum. For example, when water fleas (*Daphnia pulex*) are reared in the presence of predators (or even chemical extracts of predators) they develop large, armored, helmet-shaped heads and an elongated spiny tail. When they are reared in the absence of predators,

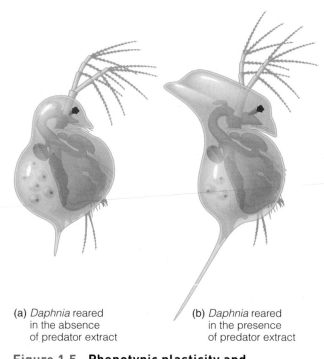

(a) *Daphnia* reared in the absence of predator extract

(b) *Daphnia* reared in the presence of predator extract

Figure 1.5 Phenotypic plasticity and polyphenism Alternative morphs of the water flea, *Daphnia pulex*. When genetically identical individuals are reared in the absence of predator extracts, these features are absent. When reared in the presence of chemical extracts of predators, *Daphnia pulex* have a large helmet-shaped head and a long spiky tail.

they develop with much smaller heads and a shorter, less spiky tail (Figure 1.5). Adult water fleas retain these morphologies even if the predator extracts are removed from the water.

Acclimation and acclimatization result in reversible phenotypic changes

Most animals are able to remodel their physiological machinery in response to external conditions. Physiologists use the related terms **acclimation** and **acclimatization** when referring to processes that cause reversible changes in the phenotype of an organism in response to an environmental change. The word *acclimation* refers to the process of change in response to a controlled environmental variable (usually in a laboratory setting), while the word *acclimatization* refers to the process of change in response to natural environmental variation. For example, if you take a fish from water at 15°C and leave it in water at 5°C, you will observe a variety of changes in muscle biochemistry, metabolic rate, and other physiological parameters. This process would be referred to as acclimation. In contrast, if you compare a fish that you capture in the summer from a lake with a mean temperature of 15°C with a fish that you capture in winter from a lake at 5°C, you will observe many of the same changes, but in this case the process would be termed acclimatization. Acclimatization may be the result not just of the temperature change, but also of changes in day length, food availability, and any other environmental parameters that vary between summer and winter. In general, both acclimation and acclimatization are reversible physiological changes.

Physiology and Evolution

One of the fundamental challenges of animal physiology is to understand and account for the great diversity of animal body forms and the strategies that animals use to cope with their environments. Consider the neck of a giraffe, which, in relation to its body size, is far longer than the neck of its closest living relative, the okapi. When a physiologist thinks about the neck of a giraffe, what question first springs to mind? A respiratory physiologist might wonder: how can a giraffe breathe through such a long neck? A cardiovascular physiologist might wonder: how can a giraffe's heart pump blood all the way up to its head? These mechanistic questions are amenable to the experimental methods of physiology and can be addressed using many of the techniques and conceptual approaches we discuss in this book. In contrast, an evolutionary physiologist might wonder: why does a giraffe have a long neck? This question actually encompasses two different kinds of thinking. If we wish to address the **proximate cause** of the giraffe's long neck, we might examine the genes that specify the size or number of vertebrae in the skeleton. Alternatively, we might wish to understand the **ultimate cause** of the giraffe's long neck: whether long necks provided an evolutionary advantage to the ancestors of the giraffe. To address these ultimate questions we need to consider the impact of evolutionary change and the adaptive significance of the physiological traits that we study.

What is adaptation?

Adaptation has two distinct meanings within the context of physiology. The most common usage refers to the product or process of evolution by nat-

ural selection, that is, a change in a population or group of organisms over evolutionary time. Many evolutionary biologists argue that the word *adaptation* should *only* be used in this context. However, physiologists often use the word *adaptation* as a synonym for the word *acclimation*. One usage is in the context of phenotypic plasticity: a beneficial change in an individual's physiology that occurs over the course of its lifetime. For example, a medical physiologist might discuss exercise adaptations: the changes in the muscles and heart that occur during exercise training. In this book, *adaptation* is used in the context of evolutionary adaptation, but it is important that you learn to make the distinction between this definition and the way the term is used by other scientists and the general community.

To an evolutionary physiologist, an adaptation is a trait that arose via a process such as natural selection and conveys an increase in reproductive success. Thus, an evolutionary adaptation is the result of processes that occur over the course of many generations, rather than within the lifetime of a single individual. The evolution of insecticide resistance in insects provides an excellent example of the principles of adaptive evolution. Over the last 50 years, chemical insecticides have been used to kill insects that harm crops or carry disease. For instance, organophosphates have been used for decades to control populations of insects, such as the common house mosquito *Culex pipiens*. Organophosphates kill mosquitoes by inhibiting acetylcholinesterase, an enzyme that is vital for neuronal transmission. The insecticides kill off all or most of the susceptible mosquitoes, but those few rare individuals with beneficial mutations survive and reproduce. This differential survival changes the structure of the population.

Resistant populations of *Culex pipiens* have evolved in two ways. Some mosquitoes have mutations in the acetylcholinesterase gene, which makes the enzyme more tolerant of the insecticide. Other mosquitoes have extra copies of the *esterase* gene, which encodes an enzyme that converts the organophosphate into a less toxic form. These mutations are vital for survival in the presence of the insecticide, but in the absence of insecticide the individuals carrying these mutations are at a disadvantage. Those that overproduce esterase use energy that could serve other physiological functions; those with the mutated acetylcholinesterase have an enzyme that does not function quite as well as the nonmutant (or *wild type*). Thus, these genotypes are superior to wild-type genotypes only when the insecticides are present.

We can distill several general principles about the process of evolutionary adaptation from the evolution of insecticide tolerance in mosquitoes:

1. For evolution to occur, there must be variation among individuals in the trait under consideration.

2. The trait must be heritable—genetically determined and passed on to offspring.

3. The trait must increase fitness—the reproductive success of the individuals that have the trait.

4. The relative fitness of the different genotypes depends on the environment. If the environment changes, the trait may no longer be beneficial.

Not all differences are evolutionary adaptations

Not all evolution is adaptive. For example, **genetic drift**, or random changes in the frequency of particular genotypes in a population over time, can result in substantial differences in the phenotype of two populations, independent of any adaptive evolution. Genetic drift is most likely to occur in small populations and is a result of happenstance, not of differences in fitness. If a forest fire kills most of the individuals of a population, the few survivors may happen to display a different genotype frequency than the ancestral population. After a number of generations, the derived population may differ from the ancestral population, but not for any reason related to natural selection and fitness. This example of genetic drift is known as the **founder effect**.

Evolutionary relationships influence morphology and physiology

Although it is easy to be overwhelmed by the diversity in animal form and function, animal biologists strive to understand the nature of this diversity. One of the best ways to understand how an animal works is to establish in which ways the animal is similar to other organisms. Some animal traits are shared among all organisms, some among all animals, and some among related animals (lineages). Other traits are truly unique to the species under study.

When a new species of insect is discovered deep in the heart of the Amazon jungle, we already know many of its features. Like all eukaryotic organisms, it will possess a genome of DNA, proteins of the same 20 amino acids, and phospholipid membranes. Like other animals, its cells will be connected to each other with proteins such as collagen and elastin, and it will have nerves and muscles that allow it to sense the world around it and move from place to place. Like other invertebrates, it will lack a spinal cord. Like other arthropods, it will have an exoskeleton of chitin. Like other insects, it will have six legs and paired wings. We can be reasonably certain of these features because the new species of insect has an evolutionary history that included, at some point in the last billion years, ancestors that it shared with other insects, invertebrates, metazoans, and ultimately all eukaryotic organisms. Thus, species that are closely related to each other are likely to share more common features than do species that are distantly related. Throughout this book we will refer to the phylogenetic relationships illustrated on the inside cover to help make sense of the physiological diversity among animals.

⊚ | CONCEPT CHECK

6. What are the major unifying themes in physiology?

7. What is homeostasis?

8. Compare and contrast negative feedback with positive feedback.

9. What is a phenotype?

10. What are some of the ways in which an individual's phenotype can change?

Summary

Physiology: Past and Present

→ Throughout history, physiological advances have been made because of detailed observations of living and dead animals, united with carefully planned experiments to elucidate how animals work.

→ Advances in physiology have followed in step with advances in physics, chemistry, and molecular biology, which have allowed physiologists to gain an ever-increasing understanding of animal structure and function.

→ Physiology can be divided into several subdisciplines. It can be divided based on the level of biological organization that the researcher studies, what kind of physiological variation the researcher studies, or the purpose of the research.

→ Physiological processes can be reduced to their component parts at a lower level of biological organization and each part studied in isolation. Emergent properties of the system result from the interactions of these parts and are not always evident when the parts are studied in isolation.

Unifying Themes in Physiology

→ A number of important unifying themes apply across all of the subdisciplines of physiology. (1) Physiological processes obey physical and chemical laws. (2) Physiological processes are often regulated. (3) Genotype and phenotype are linked. (4) Phenotypes are the product of evolution.

→ Physiological processes obey physical and chemical laws, and physiologists often use theories and ideas from physics, chemistry, biochemistry, and molecular biology to help them understand how organisms work.

→ Body size can have a profound influence on animal physiology.

→ Conformers allow their internal environment to change, whereas regulators maintain their internal environment relatively constant in the face of external change.

→ Homeostasis is the maintainance of an internal state that is constant or within tolerable limits.

→ Homeostasis is maintained by reflex control pathways that include antagonistic controls, negative feedback, and positive feedback.

→ Negative feedback loops tend to minimize the change in the regulated variable, while positive feedback loops tend to amplify the change.

→ An animal's phenotype is the result of a complex interaction between its genotype and its environment. Because of phenotypic plasticity, one genotype may produce many phenotypes, depending on the effects of the environment.

→ Polyphenism is a type of phenotypic plasticity in which the environment that an organism encounters as it develops influences the phenotype of the adult. These changes are usually irreversible.

→ Acclimation and acclimatization are types of phenotypic plasticity in which the environment causes reversible changes in an organism's phenotype.

→ The genotype of an animal is the result of evolutionary processes, including adaptation and genetic drift.

Synthesis Questions

1. Home heating systems such as a furnace are regulated via negative feedback. Describe how such a system might work.

2. A herpetologist (a biologist who studies reptiles and amphibians) brings two frogs into your lab. One frog is blue, and the other is green. For each of the main subdisciplines that were described in this chapter, outline the kinds of investigations you might perform to help the herpetologist understand the nature of this variation.

3. What physical, chemical, or physiological constraints may lead to allometric scaling?

4. Why do physiologists need to understand evolution?

5. Compare and contrast adaptive evolution and genetic drift.

For Further Reading

See the Additional References section at the back of the book for more references related to the topics in this chapter.

Physiology: Past and Present

The following four papers are essays on the history of important subdisciplines in physiology. Each provides insight into the major questions and suggestions for the future of the field.

Costa, D. P., and B. Sinervo. 2004. Field physiology: Physiological insights from animals in nature. *Annual Review of Physiology* 66: 209–238.

Feder, M. E., A. F. Bennett, and R. B. Huey. 2000. Evolutionary physiology. *Annual Review of Ecology and Systematics* 31: 315–341.

Somero, G. N. 2000. Unity in diversity: A perspective on the methods, contributions, and future of comparative physiology. *Annual Review of Physiology* 62: 927–937.

Tracy, C. R., and J. S. Turner. 1982. What is physiological ecology: A collection of commentaries by noted physiological ecologists. *Bulletin of the Ecological Society of America* 63: 340–346.

These two short books summarize the history of medical physiology and some of the most important contributors to its development.

Franklin, K. J. 1949. *A short history of physiology*. London: Staples Press.

Leake, C. D. 1956. *Some founders of physiology*. Washington, DC: American Physiological Society.

This entertaining autobiography provides a personal glimpse into the life and work of one of the 20th century's greatest animal physiologists.

Schmidt-Nielsen, K. 1998. *The camel's nose: Memoirs of a curious scientist*. Washington, DC: Island Press/ Shearwater Books.

Unifying Themes in Physiology

The following works summarize some of the important unifying themes in animal physiology.

Feder, M. E., A. F. Bennett, W. W. Burggren, and R. B. Huey. 1987. *New directions in ecological physiology*. Cambridge: Cambridge University Press.

Mangum, C. P., and P. W. Hochachka. 1998. New directions in comparative physiology and biochemistry: Mechanisms, adaptations, and evolution. *Physiological Zoology* 71: 471–484.

This fascinating book attempts to unify the disparate fields of developmental biology and evolution. It contains many important themes that are relevant to animal physiology.

Gerhart, J., and M. Kirschners. 1997. *Cells, embryos and evolution: Toward a cellular and developmental understanding of phenotypic variation and evolutionary adaptability*. Malden, MA: Blackwell Science.

This book of the collected essays of J. B. S. Haldane includes his famous essay on the problems of biological scaling.

Haldane, J. B. S. 1985. *On being the right size and other essays*, J. M. Smith, ed. Oxford: Oxford University Press.

These works address the concept of phenotypic plasticity from a variety of viewpoints.

Piersma, T., and J. Drent. 2003. Phenotypic flexibility and the evolution of organismal design. *Trends in Ecology and Evolution* 18: 228–233.

Pigliucci, M. 2001. *Phenotypic plasticity— Beyond nature and nurture*. Baltimore: Johns Hopkins University Press.

West-Eberhard, M. J. 2003. *Developmental plasticity and evolution*. Oxford: Oxford University Press.

This brief review summarizes the mechanisms of insecticide resistance in mosquitoes.

Raymond, M., C. Berticat, M. Weill, N. Pasteur, and C. Chevillon. 2001. Insecticide resistance in the mosquito *Culex pipiens:* What have we learned about adaptation? *Genetica* 112–113: 287–296.

This readable book discusses the physiological implications of animal size.

Schmidt-Nielsen, K. 1984. *Scaling: Why is animal size so important?* Cambridge: Cambridge University Press.

This book provides a comprehensive introduction to physics and engineering principles as applied to physiology.

Vogel, S. 1988. *Life's devices*. Princeton, NJ: Princeton University Press.

Chemistry, Biochemistry, and Cell Physiology

About 4.5 billion years ago the planet Earth coalesced from clumps of debris floating through space after the Big Bang. For another billion years Earth's surface was a harsh place; asteroid bombardment and volcanic eruptions were constantly remodeling the face of the planet. Yet it was during this tumultuous period that life on Earth began. Some researchers believe that organic molecules arose from a primordial soup of methane, ammonia, and water, energized by atmospheric electrical discharges. Others believe that the first organic molecules arose from chemical reactions of products of deep-sea volcanoes. Regardless of the origins of the first small organic molecules, the pathway to living organisms required the formation of larger macromolecules

with the capacity for catalysis and self-replication. At some point around 4 billion years ago, these purely chemical processes produced the earliest life-form, the progenote. The progenote was likely a chemoautolithotroph, capable of surviving without oxygen and living on inorganic sources of energy and carbon. The closest living relatives to the progenote are likely the archaea. These modern prokaryotes can survive in the harshest environments that now exist on Earth, such as sulfuric hot springs and deep-sea vents.

The progenote was the ancestor to all organisms on the planet and, as a result, it is likely that many of the ubiquitous biological features arose in the progenote. The dependence on water, the role of nucleic acids, the use of only 20 amino

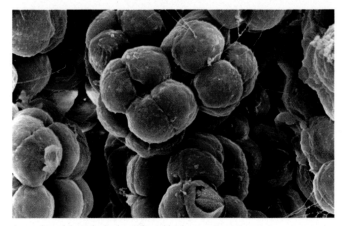

A species of domain Archaea found in deep-sea vents.

acids in proteins, and the basic pathways of intermediary metabolism are shared attributes of all living organisms.

Within the first billion years, the progenote gave rise to three distinct types of organisms: eubacteria, archaea, and eukaryotes. Each lineage diversified independently over the next 3 billion years. The two prokaryote lineages, eubacteria and archaea, remained single-cell organisms with little intracellular organization. In contrast, the ancestral eukaryotes experienced evolutionary changes that resulted in the production of membranous, subcellular compartments, thereby increasing intracellular organization. This began when the earliest eukaryotes found a way to package their DNA into a membrane-bound compartment: the nucleus. Later, around 3 billion years ago, a eukaryote engulfed a bacterium that resembled a modern purple bacterium. Although the purple bacterium was probably ingested as food, it developed a symbiotic relationship with its host, replicating with the host cell. Over time, the bacterial endosymbiont lost its capacity to exist outside the cell, and the host cell became reliant on the metabolic contributions of the endosymbiont, the ancestor of mitochondria. By 2 billion years ago the diverse groups of protists were established. The protists include organisms like the euglena (with features of both animals and plants), the trypanosomes (the single-cell flagellate parasites of blood that cause malaria), and amoebas (ciliated cells that are the namesake of amoeboid movement). These protists were once called **protozoans** because they were considered to be primitive animals, but we now recognize the protists to be a group of over 50 different phyla that emerged prior to the origins of the three main eukaryote kingdoms: plants, fungi, and animals. The term **metazoan**, which arose to distinguish multicellular animals from the single-cell protozoans, is now used synonymously with "animal," although some taxonomists separate sponges, the most primitive animals, from true metazoans (*eumetazoans*).

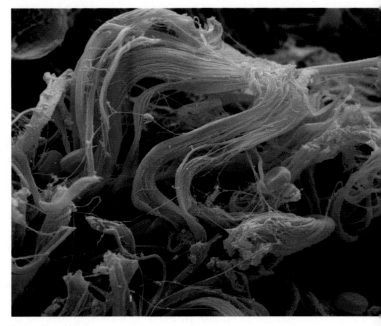

Connective tissue.

The transition from single-cell organisms to multicellular organisms occurred independently in the ancestors of plants, fungi, and animals. Each lineage found different solutions to the challenge of building multicellular tissues. The strategy used by fungi and plants relies on a cell wall for resistance to osmotic swelling and intercellular connections. Animal cells, in contrast, found other solutions to these physical challenges. Na^+/K^+ ATPase appeared early in animal evolution, enabling animal cells to regulate cell volume, ionic balance, and osmotic balance. Collagen, one of the vital proteins used to construct tissues, also arose very early in metazoan evolution. Once these physical associations were established, more elaborate pathways for intercellular communication became possible and necessary. Even plants and fungi use chemical messengers to communicate, but animals possess much more complicated mechanisms for cell-to-cell signaling.

We cannot understand the basis of animal diversity without an awareness of the evolutionary origins of animals. On the one hand, many cellular processes are similar across broad taxa, so what we learn from studies on model species of fungi and plants tells us a lot about how these features work in animals. On the other hand, each lineage evolved novel ways of using similar machinery to face the chemical and physical stresses imposed by the environment. By understanding how different taxa solved similar problems, we can better understand the constraints on animal cell function and evolution. Modern animal physiology builds upon studies of organisms in diverse taxa to understand the cellular origins of diversity in animals.◉

Overview

Physiology is the study of how animals work and how they solve the challenges of surviving in the natural environment. Though we often think of animal physiology as a study of organs, systems, and whole animals, it is important to recognize that reasons for many of these features can be traced back to underlying rules of chemistry, biochemistry, and cell biology. Many of the properties of organs and systems emerge from regulation of cellular processes, such as energy production, membrane transport, cellular anatomy, and gene expression (Figure 2.1). While the physiology of an animal is much more than the sum of these molecular and cellular processes, an awareness of how cells work is vital to understanding complex physiological processes.

Chemistry

In the purely chemical world, chemical reactions proceed according to the rules of thermodynamics. The first law of thermodynamics, also called the law of conservation of energy, states that energy can be converted from one form to another but the total amount of energy in the universe is constant. The second law, also called the law of

entropy, states that the universe is becoming more chaotic. Both laws describe the constraints that exist when energy is transferred between systems. With any spontaneous transfer of energy, some energy is diverted in a way that increases the entropy of a system, another form of energy. Although each chemical reaction conforms to these principles, living organisms are able to delay the inevitable increase in chaos, or entropy. The survival of living organisms depends upon an ability to obstruct the natural processes that lead to chemical breakdown.

Energy

Energy is the ability to do work. In our world, gasoline is an important form of chemical energy. We know that if the fuel tank of a car is full of gasoline, we have the potential to use this fuel to move the car from place to place. Burning the gasoline causes the pistons in the engine to move, turning the drive-shaft and ultimately the wheels. This familiar analogy illustrates many important principles that govern energy transfers or **energetics**. The gasoline in the tank has **potential energy** trapped within its chemical structure. When gasoline is ignited, the resulting explosion releases heat and carbon dioxide, moving the piston in its

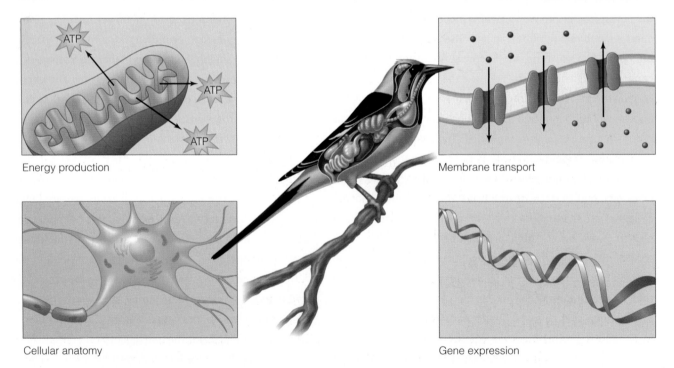

Energy production

Membrane transport

Cellular anatomy

Gene expression

Figure 2.1 Cells and tissues Many cellular processes underlie physiological systems.

cylinder. This type of energy is **kinetic energy**, the energy of movement.

The standard SI (Système International) unit of energy is the joule, although the imperial unit, the calorie, persists in the scientific literature. A joule is defined many ways, depending on the circumstances. In electrical terms, a joule (J) is the amount of energy used when 1 watt of power (W) is expended for the period of 1 second (1 J = 1 W · sec). Conversely, a watt is defined as a joule per second. You are probably most familiar with the units of energy from your household electrical bill, with energy consumption expressed in kilowatt hours (1 kW · h = 3.6×10^6 J). In more biological terms, a piece of toast with butter has about 300 kJ of energy, which is enough energy to allow you to run for about 6 minutes or light a 100-watt bulb for about 1 hour.

All energy is kinetic energy, potential energy, or a combination of both. However, in the context of biological systems it is more useful to classify types of energy by other categories.

- **Radiant energy** is energy that is released from an object and transmitted to another object by waves or particles. The sun is the most obvious source of radiant energy, emitting light that serves as an energy source for photosynthetic organisms. Other forms of radiant energy occur in animals, such as the infrared radiation given off from warm-bodied objects. Radiant energy is important in the thermal biology of animals, and is discussed in more detail in Chapter 13: Thermal Physiology.

- **Mechanical energy** is a combination of potential and kinetic energy that can be used to move objects from place to place. A flying bird uses its wings to produce the mechanical energy necessary for flight. A kangaroo uses its legs to store mechanical energy in the form of **elastic storage energy**. Recoil of the springs helps the kangaroo hop. Many forms of mechanical energy have important roles in animal locomotion, which is discussed in more detail in Chapter 12: Locomotion.

- **Electrical energy** is a combination of potential and kinetic energy that results from the movement of charged particles down a charge gradient.

- **Thermal energy** is a form of kinetic energy that is reflected in the movement of particles, and serves to increase temperature.

- **Chemical energy** is a form of potential energy that is held within the bonds of chemical structures.

Food webs transfer energy

Most biological processes are essentially transfers of energy from one form to another. When you see and smell a rose, the perception is essentially a cascade of chemical and electrical energy transfers between the sensory system and the brain. We are more familiar with the concept of energy transfers in the context of food webs. Plants capture the energy of photons and use it to create sugars. Herbivorous animals eat the plants, and carnivores eat the herbivores. At each level, some potential energy in the diet is assimilated to form animal tissues. Some potential energy is converted to heat, which is either lost to the environment or retained within the animal. Dietary potential energy is also transferred to kinetic energy, when animals use nutrients to fuel locomotion. A portion of the potential energy in the diet is locked in chemical structures that can't be liberated by the animal, and is excreted in waste products. Light is the ultimate source of dietary energy for most animals; it also provides the energy that allows animals to use vision and perceive color.

The chemical energy transferred between trophic levels is stored within molecules in the bonds between atoms. Chemical reactions liberate energy from one bond in order to produce other bonds. We discuss the nature of chemical bonds, and the role of energy in bond formation, a bit later in this chapter. However, other forms of energy are also critical components of biological function.

Energy is stored in electrochemical gradients

Molecules within a system tend to disperse or *diffuse* randomly within the available space. Imagine starting a dozen spinning tops in the center of a box. The tops collide frequently at first, eventually dispersing randomly throughout the box.

Two aspects of **diffusion** govern the properties of many biological processes. First, diffusion is certain to lead to a random distribution of molecules, but the rate of diffusion can be slow. Many physiological systems function to reduce the reliance on slow rates of diffusion. Second, the tendency of molecules to diffuse is a source of energy that cells can use to drive other processes. Living organisms can invest energy to delay the inevitable tendency toward randomness. In the previous example, you could prevent the spinning tops from randomly distributing by moving each top back to the center position. Your efforts to reverse the random distribution reflect an energetic investment on your part. Similarly, biological systems can invest energy to move molecules out of a random distribution. The resulting diffusion gradient is a form of energy storage that the cell can use for other purposes. Of particular importance are the gradients established across biological membranes. Transmembrane gradients created by cells differ in terms of the nature of the molecules (Figure 2.2). A **chemical gradient** arises when one type of molecule occurs at a higher concentration on one side of a membrane. The magnitude of the chemical gradient is expressed as a ratio of the concentrations of the specific molecule on either side of the membrane. For example, you might say that a given molecule is 10-fold more concentrated outside the cell. The second type of gradient, an **electrical gradient**, arises if the distribution of charged molecules is unequal on either side of an electrical barrier in a circuit. The electrical gradient across the barrier is dependent on the distributions of all the charged molecules combined. The strength of the electrical gradient is expressed in the electrical unit of volts. In cells, membranes are the electrical barrier and the electrical gradient is called the *membrane potential*.

The nature of the molecule determines whether the potential energy of the gradient is primarily electrical or chemical. If a molecule is uncharged, then it can only form a chemical gradient. A charged molecule can form a chemical gradient and influence the electrical gradient. For instance, if the concentration of Na^+ is greater outside the cell than inside, there is both an electrical gradient (more positive charges outside the cell) and a chemical gradient (more Na^+ ions outside the cell). Consequently, these gradients are often discussed as **electrochemical gradients**.

Thermal energy is the movement of molecules

It is impossible to ignore the importance of thermal energy, or heat energy, in discussing chemical or biological processes. When a system gains thermal energy, there is an increase in the movement of molecules within that system. This type of movement has a profound effect on molecular reactivity and the rate of chemical reactions.

Most chemical reactions involve changes in thermal energy. **Exergonic reactions** release energy and **endergonic reactions** absorb energy. To understand the reasons for these changes in energy, let's consider a simple reaction in which a single substrate, S, becomes a single product, P:

$$S \rightarrow P$$

At any given time, each molecule of S is vibrating in solution, experiencing subtle changes in its structure. A single molecule of S at times moves quickly (lots of kinetic energy) and at other times moves slowly (less kinetic energy). Occasionally, a molecule of S has so much kinetic energy that it is able to assume a specific structure that is vulnerable to a more significant change. This structure, intermediate between P and S, is called the **transition state**. The energy required for a molecule to reach the transition state is the **activation energy**, or E_A. Once a molecule reaches the transition state, it is equally likely to revert to

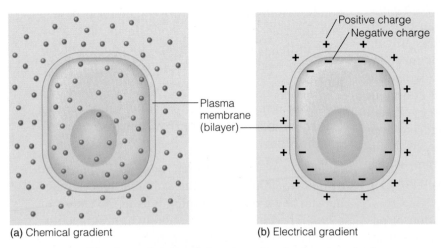

(a) Chemical gradient

(b) Electrical gradient

Positive charge
Negative charge

Plasma membrane (bilayer)

Figure 2.2   Storage of potential energy in electrochemical gradients   Animals can use the energy stored as **(a)** chemical gradients or **(b)** electrical gradients, or membrane potential.

the substrate, S, or convert to the product, P. The progression of the reaction from S to P, expressed in terms of energy content, is shown in Figure 2.3. Since the free energy content, (G), of S is greater than the free energy of P, the chemical reaction leads to a change in free energy (ΔG), calculated as

$$\Delta G = G_{\text{products}} - G_{\text{substrates}}$$

In this reaction (Figure 2.3), P has a lower free energy, making ΔG negative. S is converted to P, and the difference in free energy, or ΔG, is released to the environment, primarily as heat. Thus, an exergonic reaction is defined in thermodynamic terms as a reaction with a negative ΔG. An endergonic reaction has a positive ΔG.

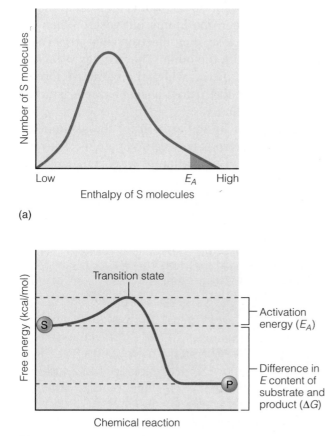

(a)

(b)

Figure 2.3 Chemical reactions, substrates, products, and thermal energy (a) A collection of substrate molecules, S, possesses an average energy level and an average arrangement of electrons around its nucleus. But at any given time, some S molecules are energy-rich and others energy-poor. (b) Occasionally a molecule of S might absorb enough energy from the surroundings (perhaps from a molecular collision) that it achieves the transition state (S*). At this point it could revert to S, or change into a novel product P.

All chemical reactions are reversible under the right conditions. The reaction of S to P is favored only because the activation energy barrier is lower for S than it is for P. Because free energy was released when S was converted to P, free energy must be absorbed if P is to be converted to S. The reverse reaction, with its positive ΔG, is an endergonic reaction. If both S and P are present, at any point in time both forward and reverse reactions occur simultaneously. The net reaction is the difference between the forward rate and the reverse rate. Because energy is released in one direction and absorbed in the other, the balance between forward and reverse directions depends on temperature. At high temperatures, endergonic reactions become more feasible. Thus, temperature influences chemical reactions in two ways. Increasing temperature allows more molecules to reach activation energy, and increases the likelihood of endergonic reactions.

Chemical Bonds

Most biologically available energy is stored in the form of chemical bonds. **Covalent bonds** hold individual atoms together to form a molecule. These strong bonds involve the sharing of electrons between two atoms. Noncovalent bonds organize molecules into three-dimensional structures. In general, noncovalent bonds are called **weak bonds** or sometimes weak interactions to further distinguish them from strong bonds.

Covalent bonds involve shared electrons

Each element has a characteristic arrangement of electrons that influences the types of bonds it can form. Specifically, for the six common biological elements, each atom has at least one unpaired electron in its outer electron shell. Atoms with unpaired electrons can readily form covalent bonds with other atoms with unpaired electrons. These atoms share electrons so readily that they are rarely present in elemental form. Atoms with more than one unpaired electron can form multiple covalent bonds. For instance, molecular oxygen has two oxygen atoms joined by a double covalent bond. Many atoms are covalently bonded to more than one other atom. Methane, for example, is composed of four hydrogen atoms covalently bound to a single carbon atom. Each type of covalent bond has a characteristic bond energy, the energy required to either form or break the bond. The greater the bond energy, the stronger the bond. Multiple bonds possess more bond energy than single bonds. Large molecules are built from a collection of individual atoms attached by covalent bonds. Functional groups are combinations of atoms and bonds that recur in biological molecules (Figure 2.4).

Weak bonds control macromolecular structure

Weak bonds arise between atoms with asymmetrical distributions of electrons either within the atom or between atoms. Four types of weak bonds can be distinguished based on how they form molecular interactions: van der Waals forces, hydrogen bonds, ionic bonds, and hydrophobic bonds (Figure 2.5).

The electrons in a bond between two atoms can be shared unequally. This asymmetry in electron distribution creates a polarity, or transient dipole, within the molecular structure. One region is slightly negative (δ^-), and the other is slightly positive (δ^+).

When an atom with a transient dipole encounters another atom, the distribution of electrons in the second atom is altered. The weak interaction between the two dipoles is the **van der Waals interaction**. Van der Waals interactions are effective only over a very narrow range of atomic distances. When two atoms are far away, the dipole of one atom has no effect on the electron cloud of the other. As the atoms approach, the attraction between the atoms increases. When the atoms get too close, their electron shells repel each atom away from the other. The van der Waals radius is the distance at which the attractive force is at its greatest. Each atom has a characteristic van der Waals radius.

Hydrogen bonds arise from the asymmetric sharing of electrons between two atoms. They are critical to the organization of water molecules. In a single water molecule, each hydrogen atom is covalently linked to the oxygen atom. However, the oxygen atom is just a bit better at attracting the

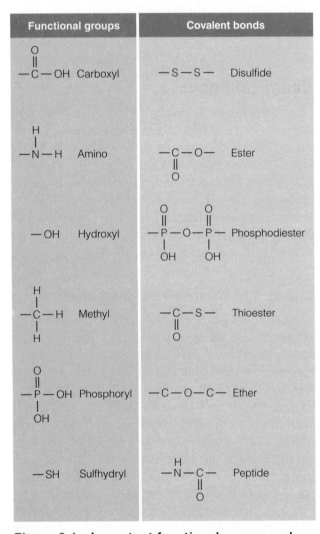

Figure 2.4 Important functional groups and bonds Although there are many types of bonds and functional groups, those illustrated here are particularly common in macromolecular structure.

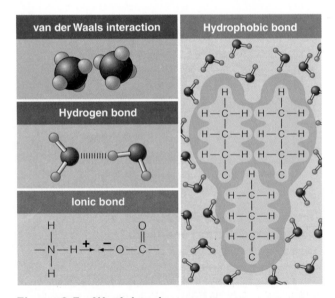

Figure 2.5 Weak bonds Four types of weak bonds are involved in building macromolecules: hydrogen bonds, ionic bonds, van der Waals forces, and hydrophobic interactions.

electron of hydrogen. More precisely, hydrogen's electron spends a bit more time closer to the oxygen atom than the hydrogen atom. Consequently, the hydrogen is slightly positive (δ^+), and the oxygen atom slightly negative (δ^-). The attraction between the δ^+ of hydrogen in one water molecule and the δ^- of oxygen in another water molecule constitutes a hydrogen bond (Figure 2.6).

In some cases, a nucleus is so good at attracting electrons that when a bond breaks, an electron from one atom remains with the other to create ions. Electronegative ions, or **anions**, possess extra electrons, whereas electropositive ions, or **cations**, have lost electrons. Anions and cations can interact to form an **ionic bond**. Most of the molecules we think of as salts, acids, and bases rely on ionic bonds to join anions and cations.

Van der Waals forces, hydrogen bonds, and ionic interactions form on the basis of mutual attraction between two charged or slightly charged atoms. However, **hydrophobic bonds** form between atoms because of a mutual aversion to water. Whole molecules or specific regions of large molecules can be hydrophobic ("water-fearing"). The bonds within hydrophobic molecules share electrons equally and therefore do not possess significant dipoles. With little internal charge, they cannot interact effectively with the more polar molecules such as water.

Weak bonds are sensitive to temperature

Bond energy reflects the amount of thermal energy required to break (or form) a bond. Weak bonds are more vulnerable to the effects of temperature because their bond energies are much lower than the bond energies of covalent bonds. Whereas covalent bonds have energies of formation of 200–900 kcal/mol, weak bonds have energies of formation less than 5 kcal/mol. The three-dimensional macromolecular structures of proteins, membranes, and DNA, which primarily depend upon weak bonds, are also sensitive to temperature. As a result, rising temperature can cause macromolecules to unfold, or *denature,* when these weak bonds break. However, not all weak bonds are affected by temperature the same way. Hydrogen bonds, ionic bonds, and van der Waals forces each have positive energy of formation and tend to break when temperature increases. In contrast, hydrophobic bonds have negative energy of formation and are strengthened by thermal energy.

◉ | CONCEPT CHECK

1. What are the five main forms of energy used by animals? Provide biological and nonbiological examples of processes that represent conversion of energy from one form to another.
2. What are the four types of weak bonds, and how do they differ from each other and from covalent bonds?
3. What is the difference between (a) thermal energy and temperature and (b) exergonic and exothermic?

Properties of Water

Most cells are composed primarily of water. Aquatic organisms also live in water, and even the cells of terrestrial organisms are bathed in the aquatic environment of their extracellular fluids. Many physiological processes arose to meet the challenges of the physical and chemical properties of water.

The properties of water are unique

A **solvent** is the most abundant molecule in a liquid, whereas the other molecules within the liquid are **solutes**. Collectively, the solutes and solvents constitute the **solution**. In biological systems, the solvent is usually water. Water's unusual combination

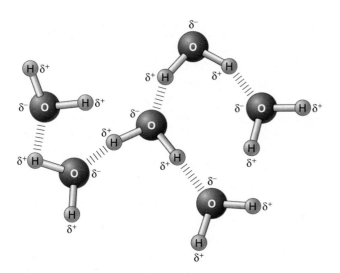

Figure 2.6 Water dipole and hydrogen bonds
Oxygen atoms in water strongly attract the electron of the hydrogen atom. The result is a small charge difference (δ). The hydrogen atom is slightly positive (δ^+) and the oxygen atom is slightly negative (δ^-). These charges influence the way that water molecules interact.

of physicochemical properties, which can be attributed to its ability to form hydrogen bonds, have special significance in biological processes and constrain the direction of biological evolution. Liquid water is actually a network of interconnected water molecules. Each water molecule interacts strongly with other water molecules, creating internal cohesiveness. At the interface between air and water, the attraction between water molecules creates a force called **surface tension**. This prevents most water molecules from spontaneously escaping to the air. Many animals take advantage of surface tension to move over water (Figure 2.7). Their mass exerts a force on the water, but it is not great enough to disrupt the molecular interactions between water molecules.

The organization of water molecules changes in relation to temperature. At high temperatures, the water molecules possess enough thermal energy to escape the restraining force of surface tension. At this point, the water "boils," and water molecules can escape as gaseous water (steam). In contrast, low temperatures stabilize water structure as a result of the formation of additional hydrogen bonds. Water solidifies, or freezes, when each water molecule forms four hydrogen bonds to create a stable lattice of water molecules.

Changes in temperature also influence the density of water. Although frozen water molecules incorporate more hydrogen bonds, the geometry is such that the water molecules are held further apart than in liquid water. Consequently, ice is less dense than liquid water and tends to float. These physical properties of water have important effects on aquatic ecosystems. In temperate regions of Earth, a layer of ice forms on the surface of lakes in early winter. The ice layer insulates the lake water from the air conditions, creating a more stable environment for aquatic organisms. Temperature also alters the density of liquid water. Because the density of water is greatest at 4°C, the deepest parts of large water bodies tend to be a constant 4°C, whereas surface waters can be colder or warmer, depending on the latitude and season.

Other physical properties of water have an important impact on biological processes. Water has a higher melting point (0°C) and a higher boiling point (100°C) than other solvents. In most habitable locations on Earth, then, water is a very stable liquid. Water's high heat of vaporization, the amount of energy required to cause liquid water to boil or evaporate, makes sweating an effective cooling strategy for mammals. A great deal of energy is absorbed when liquid water vaporizes. Water on the skin absorbs a lot of thermal energy from the body in the process of evaporation.

Solutes influence the physical properties of water

Many solutes can dissolve in water because they can form hydrogen bonds with water molecules. Water-soluble molecules in solution are often surrounded by a coat of water molecules called the **hydration shell**. The hydration shell increases the functional size of the molecule, and influences how the solute interacts with other molecules in complex biological systems.

In the tissues of most animals, the most common solutes are inorganic ions. K^+ is the most abundant cation inside cells, and Na^+ is the most abundant cation in the extracellular fluid. However, in some species, particularly marine animals, the most abundant solutes are organic ones such as urea, amino acids, and sugars. Each type of solute can exert specific, distinct effects on the chemical properties of other molecules within the solution. However, all solutes, regardless of their chemical nature, exhibit four basic properties, known as **colligative properties**. Solutes reduce the freezing point of the solution, and increase the boiling point, the vapor pressure, and the osmotic pressure of the solution. The colligative properties depend only on the concentration of solutes, not their size or charge.

Figure 2.7 Surface tension The basilisk lizard is able to run across the surface of water. The surface tension of water can support the lizard because the force is distributed over the large surface area of the feet.

In a solution with high concentrations of solutes, cooling to 0°C will not induce freezing. The thermal energy of the system is low enough to form the extra hydrogen bonds, but the solutes block the formation of hydrogen bonds necessary to form the ice crystal. When solutes are present, the solution must be cooled below 0°C before the extra hydrogen bonds can form. The freezing point of biological fluids, such as cytoplasm or blood, is always lower than freshwater, and sometimes even as low as seawater. The difference in the freezing point of body fluids and the aquatic environment has important ramifications for aquatic animals.

Similar mechanisms are responsible for the effects of solutes on the vapor pressure and boiling point of water. A water molecule can escape liquid water only at the water-gas interface. When solute molecules are also present at the surface, they reduce the likelihood that a water molecule will escape. We discuss the fourth colligative property, osmotic pressure, after we discuss a related concept: diffusion.

Solutes move through water by diffusion

The direction of diffusion of molecules in a solution depends on the concentration gradient, but the rate of diffusion depends on many additional factors. Molecules move more rapidly when the gradients are steeper. The properties of the solute itself also influence the rate of diffusion. If solute molecules are relatively large, they have a more difficult time moving through the restrictive structure of water. Large molecules like proteins diffuse much more slowly than small molecules like K^+. The hydration shell that forms around many molecules enlarges the functional size of the molecule, restricting its mobility. Other factors that influence how the solute interacts with the solvent, such as charge and solubility, also affect the rate of diffusion. Each solute has an experimentally determined **diffusion coefficient** (D_s), which is influenced by the structural properties of the solute. The rate of diffusion of a solute (dQ_s/dt) depends on the diffusion coefficient of the solute (D_s), the diffusion area (A), and the concentration gradient (dC/dX). The relationship between these parameters is defined by the Fick equation:

$$\frac{dQ_s}{dt} = D_s \times A \times \frac{dC}{dX}$$

Small solutes, such as inorganic ions, are able to traverse the width of the cell, typically about 10 μm, in a fraction of a second. The time required for a molecule to diffuse a given distance increases with the square of the distance. If a molecule takes 1 sec to diffuse 0.1 mm, it would take about 3 h to diffuse 1 cm. Many biological processes depend on diffusion, such that physiological and anatomical strategies have evolved to prevent these processes from becoming "diffusion limited."

Solutes in biological systems impose osmotic pressure

The *semipermeable membranes* of cells allow some molecules to cross while restricting the movement of others. Imagine a situation where two identical solutions of pure water exist on either side of a membrane that allows free movement of water molecules only (Figure 2.8). On each

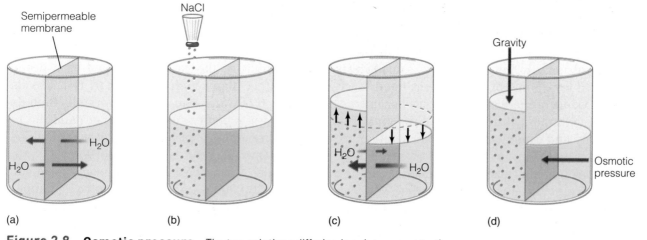

Figure 2.8 Osmotic pressure   The two solutions differing in solute concentration are separated by a semipermeable membrane. The movement of water creates osmotic pressure. Movement will continue until the force of gravity is equal to the osmotic pressure.

side of the membrane is approximately 55.5 mol of water per liter. Water molecules freely cross the membrane in both directions. If you added NaCl to one side of the membrane, a concentration gradient would be created for Na^+ and Cl^-. Since the membrane is permeable to water alone, only water molecules could move across to equalize the concentration gradients. There would be a net movement of water molecules from the side with pure water to the side with solutes. This would increase the volume on the side with solutes. Eventually, the net movement of water would stop when the force generated by the movement of water equaled the force of gravity, which prevents the water column from getting any higher. In cells, the movement of water is restricted not by gravity but by the flexibility of the cell membrane. In either case, the force associated with the movement of water is the **osmotic pressure**, the fourth colligative property of solutes.

The ability of solutions to induce water to cross a membrane is expressed as the **osmolarity**, expressed in units of osmoles per liter (OsM). Osmolarity is analogous in many respects to molarity (M). Whereas molarity is a reflection of the concentration of specific molecules in a solution, osmolarity depends on the total concentration of particles in solution. The osmolarity of a solution of known molarity can be calculated on the basis of the number of particles derived from each molecule. If a solution has only one solute, and that solute does not dissociate, then molarity and osmolarity are equivalent. For instance, 1 mol/l (or 1 M) glucose solution has an osmolarity of 1 osmol/l (or 1 OsM). Some solutes dissociate into multiple particles. Each mole of NaCl produces 1 mol of Na^+ and 1 mol of Cl^-. Thus, a 1 M NaCl solution has an osmolarity of 2 OsM. Knowledge of the concentration and valency of the solutes would allow you to estimate osmolarity, but in reality the osmolarity is somewhat less. Some of the salt does not dissociate, and some of the water molecules become associated with the hydration shell of the ions. The osmolarity and osmotic pressure of a solution are physical properties of a solution. However, in a biological setting the absolute osmolarity is often less important than the osmolarity of an extracellular fluid relative to the osmolarity of the intracellular fluid (Figure 2.9a). If a cell is placed in a solution with greater osmolarity, then the solution is considered **hyperosmotic** (relative to the cell). Similarly, if a cell is placed in pure water, the solu-

tion is **hyposmotic**. When the osmolarity is the same on both sides of the cell membrane, the solution is **isosmotic**.

Differences in osmolarity can alter cell volume

Biologists usually make distinctions between osmolarity, which is related to the osmotic pressure, and **tonicity**, which is the effect of a solution on cell volume. Tonicity depends on differences in osmolarity, but also on the types of solutes and the permeability of the membrane to those solutes.

To understand the distinction between osmolarity and tonicity, consider the following example (Figure 2.9b). A cell that is placed in an isosmotic salt solution neither shrinks nor swells (an **isotonic** solution). If more salt is added, the cell loses water and shrinks. Thus, this solution is both hyperosmotic and **hypertonic**. Imagine now that small amounts of urea, a permeant solute, are added to the isotonic salt solution. The urea would equilibrate across the cell membrane, and thus prevent the net movement of water in or out of the cell; this is an isotonic solution. Of course, if the cell were placed in a solution containing only urea, the movement of urea into the cell, combined with the high internal salt concentration, would draw water into the cell, causing it to swell or even burst; this is a hypotonic solution.

pH and the Ionization of Water

A small proportion of the H_2O molecules in any solution dissociates into ions by breaking one of the covalent bonds between oxygen and hydrogen. In reality, water is in equilibrium with itself.

$$H_2O + H_2O \rightleftharpoons H_3O^+ + OH^-$$

For simplicity, the cation is treated as a proton (H^+) rather than a hydronium ion (H_3O^+). The dissociation of water into ions is reversible. Both the forward reaction (water dissociation) and the reverse direction (water formation) occur simultaneously. Only a very small proportion of water molecules are dissociated at any given time, about 1 in 55,500,000 water molecules at room temperature (25°C). Under these conditions (pure water at 25°C), the concentration of protons arising from water dissociation is 10^{-7} M. For the sake of convenience, the concentration of protons is usually converted to the **pH scale**. The pH of a solution is

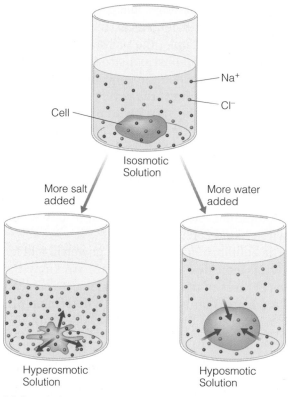

(a) Osmolarity

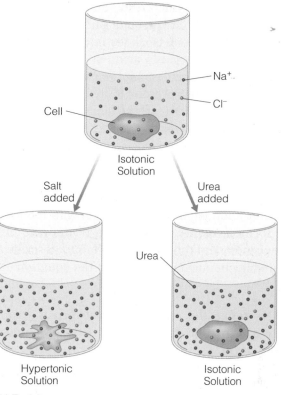

(b) Tonicity

Figure 2.9   Osmolarity versus tonicity **(a)** A cell is in an isosmotic solution when the solution has an osmotic pressure equal to that of the cell cytoplasm. If salt is added to the solution it becomes a hyperosmotic solution. Water leaves the cell, causing the cell volume to shrink, until the osmotic pressures are again equal. If the salt concentration is reduced, as would be the case if more water was added, the solution becomes hyposmotic. Water flows into the cell, causing it to swell. **(b)** The effects of solutes on cell volume depend on the ability of the solute to enter the cell. If NaCl is added to an isosmotic solution, the cell shrinks and the solution is considered hypertonic. If urea is added to the solution, there is little change in cell volume because urea can cross the cell membrane. Thus, adding urea to this solution makes it hyperosmotic, but it is also isotonic.

calculated as the negative logarithm of proton concentration (denoted as $[H^+]$). Thus, the pH of pure water at 25°C is pH 7 ($-\log 10^{-7}$). As we see later in this chapter, the negative logarithmic scale, designated by the prefix p, is also a convenient way to express low concentrations of other ions, such as pOH for $[OH^-]$ and pCa for $[Ca^{2+}]$.

Neutrality is not always at pH 7

A solution is considered neutral when $[H^+] = [OH^-]$, or pH = pOH. Pure water at 25°C possesses 10^{-7} M concentrations of both H^+ and OH^-: pH = 7 and pOH = 7. The temperature of a solution of pure water alters the proportion of water molecules with enough thermal energy to break the covalent O–H bond. For instance, at 45°C almost twice as many H_2O molecules dissociate, lowering the pH to 6.72. At 5°C, about half as many H_2O molecules dissociate, raising the pH to 7.28. In each of these situations, water remains neutral, but the pH at neutrality, or pN, varies inversely with temperature. In practice, pure water changes its pH at a rate of -0.014 units per degree Celsius.

Acids and bases alter the pH of water

Pure water is never anything but neutral. However, ionizable solutes can influence the pH of a solution. An **acid** releases one or more protons. Hydrochloric acid (HCl) is an acid because it dissociates into H^+ and Cl^-. A **base** causes a reduction in the $[H^+]$ of the solution. When the base sodium hydroxide (NaOH) is dissolved into water, it rapidly dissociates into Na^+ and OH^-. The extra OH^- arising from NaOH dissociation rapidly interacts with H^+ to form H_2O, reducing the $[H^+]$ and increasing pH.

The degree to which acids and bases change the pH of a solution depends on the ease with which the molecule dissociates under physiological conditions. Inorganic acids such as HCl and H_2SO_4 are considered *strong acids* because they readily release their protons to the solution. Similarly, NaOH and KOH are *strong bases* because they readily dissociate to release OH^-. Many biological molecules are weak acids or weak bases, which are only partially ionized under physiological conditions.

If an acid is defined as something that releases a proton, then we can discuss acids with the general formula of HA. Dissociation of the acid HA produces H^+ and the anion, A^-. We can describe a reversible chemical reaction with the equation

$$HA \rightleftharpoons H^+ + A^-$$

We define the relationship between the substrate (HA) and products (H^+ and A^-) as the **mass action ratio**, using the equation

$$\text{Mass action ratio} = \frac{[H^+] \times [A^-]}{[HA]}$$

To understand how these parameters change, consider an experiment where the acid HA is added to pure water. When first added to water, HA remains intact and $[A^-]$ is equal to zero; the mass action ratio is also close to zero. However, very quickly at least some of the acid dissociates. There is an increase in both $[H^+]$ and $[A^-]$, and as a result an increase in the mass action ratio. At some point the reaction slows, with [HA] reaching a minimum and $[H^+]$ and $[A^-]$ reaching a maximum. When this occurs, the reaction is at equilibrium. It is important to recognize that although there is no net change in the concentrations of reactants, both forward and reverse reactions continue, but at equal rates. When the reaction is at equilibrium, the mass action ratio attains a specific value, K_{eq}, the **equilibrium constant**. Under most circumstances, the equilibrium constant is converted to its negative log ($-\log_{10} K_{eq}$), analogous to the way we converted $[H^+]$ to pH. Thus, the equilibrium equation can be rewritten after log transformation as

$$pK = pH - \log \frac{[A^-]}{[HA]}$$

Put another way, the pK is the pH at which half the acid is dissociated, $[A^-] = [HA]$, $[A^-]/[HA] = 1$ and $\log [A^-]/[HA] = 0$.

This simple equation is useful for understanding many different biochemical and physiological principles. For example, the pK value reflects the strength of acids and bases. A strong acid will give up its proton even when the concentration of protons in the surrounding area is very high (low pH). Thus, the pH must be very low to prevent a strong acid from dissociating. The pK value is low for a strong acid, less than 3 for hydrochloric acid and sulfuric acid. Similarly, strong bases, such as sodium hydroxide and ammonium hydroxide, have pK values greater than 11. The pK values for some common biological acids and bases are shown in Table 2.1.

The equilibrium equation is a powerful tool for analyzing biological solutions. Once we know the values of three of the four parameters, we can calculate the one that is unknown. To determine pH, we can rearrange the equation into the form

$$pH = pK + \log \frac{[A^-]}{[HA]}$$

This rearrangement is known as the **Henderson-Hasselbalch equation**, named after the researchers who used the relationship to explain the behavior of CO_2 (HA) and HCO_3^- (A^-), which is important in respiratory physiology.

Table 2.1 Acids and bases.

Acid	Reaction	pK
Carbonic acid	$H_2CO_3 \rightarrow HCO_3^- + H^+$	3.8
	$HCO_3^- \rightarrow CO_3^- + H^+$	10.2
Phosphoric acid	$H_3PO_4 \rightarrow H_2PO_4^- + H^+$	3.1
	$H_2PO_4^- \rightarrow HPO_4^{2-} + H^+$	6.9
	$HPO_4^{2-} \rightarrow PO_4^{3-} + H^+$	12.4
Ammonium	$NH_4^+ \rightarrow NH_3 + H^+$	9.3
Acetic acid	$CH_3COOH \rightarrow CH_3COO^- + H^+$	4.8
Glycine (amino)	$R-NH_3^+ \rightarrow R-NH_2 + H^+$	2.3
(carboxy)	$R-COOH \rightarrow R-COO^- + H^+$	9.6
Histidine		6.0

Both pH and temperature affect the ionization of biological molecules

Changes in pH can alter the dissociation of other molecules with ionizable groups. Let's look at the amino acid glycine to explore how pH affects its structure (Figure 2.10). Glycine has a carboxyl group that can be protonated (–COOH) or deprotonated (–COO⁻). It has an amino group that can be deprotonated (–NH₂) or protonated (–NH₃⁺). The protonation state of the carboxyl and amino groups in a molecule of glycine depends on the pH of the solution.

We can observe the effects of pH on glycine structure by performing a titration, where an acid or base is added to a solution. We start our titration by dissolving glycine in an acidic solution. At very low pH, where [H⁺] is high, both amino and carboxyl groups are protonated. The carboxyl group is uncharged (–COOH) and the amino group has a positive charge (–NH₃⁺), giving glycine a net positive charge. When we add base to the solution to increase pH, the protonation state of both of these groups begins to change. First, the carboxyl groups become deprotonated (–COOH → –COO₂⁻ + H⁺). At pH 2.3, exactly half of the carboxyl groups in the glycine molecule are ionized. This pH value is the equilibrium constant for the carboxyl group

of the glycine molecule, or its pK_{COOH}. Adding more base causes deprotonation of the amino group. At pH 9.6, exactly half the amino groups are protonated. This pH value is the equilibrium constant for the amino group of glycine, or its pK_{NH2}. At still higher pH values, the carboxyl groups remain charged and the amino groups are fully deprotonated, giving the glycine molecule a net negative charge. Midway between the two pK values, the glycine molecule has no net charge, as the charges on the carboxyl group (COO⁻) balance the charges on the amino group (NH₃⁺). Glycine and other molecules that have both negative and positive charges are called **zwitterions**.

The ionization state of molecules is very sensitive to temperature. Let's return to the previous example where we titrated the ionizable groups of glycine. As pH rose to equal pK_{COOH}, protons became so scarce that half of the carboxyl groups lost their proton. If we repeated the titration at lower temperature, the pK value would change because the lower temperature increases the strength of the bond holding the proton to the carboxyl group. At any given pH the colder glycine would be more protonated. Put another way, a higher pH would be required to lure half the protons off the carboxyl group. Thus, the pK value increases as temperature decreases. Each ionizable group has a characteristic sensitivity to temperature, expressed as ΔpK/°C. For example, the ionization of phosphoric acid is relatively insensitive to temperature (ΔpK/°C = −0.005), whereas the ionization of the imidazole group of histidine is more sensitive to temperature (ΔpK/°C = −0.017).

The protonation state of many molecules can have important effects on molecular processes. Many of the effects of temperature and pH on cells can be traced to the effects on the protonation state of critical molecules. For example, many proteins form structures that depend on particular ionization states of amino acids. Changes in pH or temperature can affect how these proteins fold and function. By actively regulating

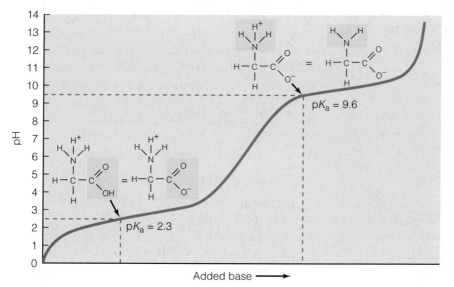

Figure 2.10 Changing pH and the ionization state of acids and bases The amino acid glycine occurs in several different ionization states that change with pH. At low pH (high [H⁺]) both the amino group and the carboxyl group are protonated. As pH increases, the carboxyl loses its proton first, becoming half ionized at pH 2.3, its pK_a value. The amino group does not lose its proton until much higher pH values are reached. Near neutral pH, glycine is primarily neutral.

temperature and pH, animals diminish the debilitating effects of changes in protonation state.

Buffers limit changes in pH

A variety of mechanisms help cells regulate pH. The first level of defense is a **buffer**. A buffer is a chemical found in a solution that dampens the effect of added acid or base on the pH of the solution. Buffers are often described as if they were single molecules. In reality we should think of them as buffer systems, because they are mixtures of at least two forms of a molecule, typically protonated and deprotonated.

If we add a buffer to the solution, the protons liberated from the acid can associate with the buffer. As a result, the addition of acid has less effect on pH than it does in the absence of buffer. Most buffer systems rely on weak acids, present in both the acid form (HA) and the anion form (A$^-$). Furthermore, a buffer works only over a particular range of pH values. Acetic acid is a weak acid that can be used as a buffer. The effects of an acetic acid/acetate buffer are illustrated by the titration curve shown in Figure 2.11. If you started your titration at low pH, most of the acetic acid would be in the protonated form (HA). If you added small volumes of NaOH, the pH would increase proportionately. Below pH 3.75, acetic acid would remain mostly protonated (HA). If you add more base, the increase in pH would induce some acetic acid (HA) to become deprotonated (HA → H$^+$ + A$^-$). Since some of the protons are liberated from acetic acid, the added NaOH has a reduced effect on the pH of the solution. This buffering effect is evident over the pH range of about 3.75 to 5.75, where the titration curve is quite shallow. Once pH reaches 5.75, most of the acetic acid is in the deprotonated form (A$^-$), which cannot act as a buffer. This pH range corresponds to the greatest buffering capacity of the solution, and is centered on the pK value for the buffer, about 4.75, where about half of the buffer is protonated (HA) and half deprotonated (A$^-$).

Animals use a variety of different molecules as buffers. The best buffers in animal cells have pK values that approach the pH of the compartment in which they are used. Phosphate ($H_2PO_4^-$/HPO_4^{2-}) is an important buffer in the cytoplasm of most cells, with a pK_A of 6.9. The amino acid histidine contributes to buffering in many animal cells because the pK value of its imidazole side chain is very close to intracellular pH. Histidine residues within large proteins help buffer the cytoplasm against changes in pH. Many species use amino acids with imidazole groups to produce dipeptides that serve as important intracellular buffers. The dipeptides carnosine (histidine and β-alanine), anserine (1-methylhistidine and β-alanine), and ophidine (3-methylhistidine and β-alanine) are important buffers in the muscle of many species.

In air-breathing animals, the most important extracellular buffer is bicarbonate/CO_2, but it works by a different mechanism than a simple A$^-$/HA buffer pair. In a closed test tube bicarbonate/CO_2 would have little buffering capacity at physiological pH because its pK is much too low (3.8). It works as a biological buffer because animals can expire CO_2. As [H$^+$] increases, bicarbonate is consumed and carbonic acid is produced (H_2CO_3), which in turn forms H_2O and CO_2.

$$H^+ + HCO_3^- \rightarrow H_2CO_3 \rightarrow H_2O + CO_2$$

When an animal expires CO_2 as a gas, it is essentially eliminating a weak acid from the body, buffering against a change in pH. You will learn more about the interaction between CO_2 and acid-base balance in Chapter 9: Respiratory Systems.

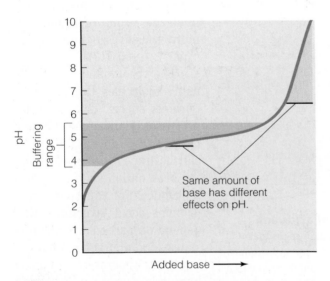

Figure 2.11 Effects of buffers on changes in pH
Buffers blunt the effects of added bases (or acid) on the pH of a solution.

CONCEPT CHECK

4. What is the relationship between pK and pH?
5. How does temperature influence the pK of water? What might this mean for animals that experience changes in body temperature?

6. What change in pH has a greater effect on proton concentration: pH 6 to 7 or pH 7 to 8?
7. What properties of a particle influence its rate of diffusion across a membrane? What membrane properties influence this rate?

Biochemistry

Animals control the inner workings of cells through the use of enzymes, which interconvert macromolecules to create building blocks and control the flow of chemical energy. A metabolic pathway is a series of consecutive enzymatic reactions that catalyze the conversion of substrates to products, with multiple stable intermediates. Flow through the pathway is called **metabolic flux**. Metabolic pathways can be either synthetic (**anabolic**), degradative (**catabolic**), or a combination of both (**amphibolic**). **Energy metabolism** revolves around production of ATP and other energy-rich molecules. **Metabolism** is the sum of all these metabolic pathways within the cell, tissue, or organism. Many metabolic pathways span multiple cellular compartments, allowing cells to create distinct microenvironments. For example, the mitochondria are specialized compartments with a major role in energy metabolism.

In the following sections, we discuss the nature of enzymes and metabolic energy, and the metabolism of three of the four major classes of biological macromolecules: proteins, carbohydrates, and lipids. The fourth class of macromolecules, nucleic acids, is discussed later in this chapter when we consider genetics.

Enzymes

Enzymes are biological catalysts that convert a substrate to a product. Enzymes, like other types of catalysts, have three properties: (1) they are active at very low concentrations within the cell; (2) they increase the rate of reactions but they themselves are not altered in the process; (3) they do not change the nature of the products.

Although some enzymes, called ribozymes, are made of RNA, most enzymes are composed of protein. Many enzymes possess nonprotein components, called **cofactors**. A cofactor that is covalently bonded into the enzyme is called a **prosthetic group**. Some enzymes use cofactors that are metals, such as copper, iron, magnesium, zinc, and selenium. Organic cofactors, or **coenzymes**, are usually derived from vitamins; coenzyme A is derived from panthothenic acid, FAD from riboflavin, and NAD from niacin. Many of the life-threatening diseases we associate with vitamin deficiencies can be traced back to perturbations of metabolism due to loss of function of specific enzymes.

Enzymes accelerate reactions by reducing the reaction activation energy

The laws of thermodynamics that govern chemical reactions in test tubes also apply to chemical reactions in living cells (see Box 2.1, Mathematical Underpinnings: Thermodynamics). Enzymes do not determine whether or not a chemical reaction is thermodynamically possible. However, enzymes do have the ability to accelerate thermodynamically feasible reactions by factors of 10^8 to 10^{12}.

Previously we discussed how substrate molecules in an uncatalyzed reaction must obtain sufficient energy to meet the activation energy barrier (E_A). Once the E_A is met, the substrate can adopt the transition state and then spontaneously change into the product. Although enzymatic reaction uses the same substrate and yields the same product as an uncatalyzed reaction, it produces a different intermediate at the transition state. First, the enzyme (E) and substrate (S) bind to form the ES complex. After conversion to transition states (ES*, EP*), the final product (P) is formed and then is released by the enzyme. This is represented as shown:

$$S + E \rightleftharpoons ES \rightleftharpoons ES^* \rightleftharpoons EP^* \rightleftharpoons EP \rightleftharpoons E + P$$

The energy required to reach this intermediate state is lower than in the uncatalyzed reaction (Figure 2.12). With a lower energy barrier, more of the substrate molecules possess enough energy to reach the transition state, and the reaction is accelerated. Like other chemical reactions, enzyme reactions are reversible, proceeding through the same set of reaction intermediates.

An enzymatic reaction begins with the substrate binding at a specific location called the **active site**. Think of the active site as a pocket into which the substrate fits. The enzyme can bind the substrate only if it possesses the proper conformation. The three-dimensional folding of the enzyme, maintained by weak bonds, forms the active site.

BOX 2.1 MATHEMATICAL UNDERPINNINGS
Thermodynamics

All chemical reactions, whether they occur in test tubes or biological systems, are governed by the laws of thermodynamics. The first law of thermodynamics deals with conservation of energy. The energy within a substrate is either transferred to the product or released. The first law doesn't tell us if the reaction will go forward or backward, only that the energy transformations must be balanced. The second law of thermodynamics provides a way of predicting if a reaction is likely to occur. It says that spontaneous processes occur in the direction that will increase randomness, or entropy (ΔS). Throughout this chapter we discuss many examples of increases in entropy. When table salt dissolves in water or ice melts, the molecules that were once in a well-ordered crystal begin to disperse. Diffusion also illustrates the principle of spontaneous increases in entropy. Solutes at high concentration tend to disperse to regions of lower concentration. Collectively, these laws tell us that the total energy of the universe is constant but that it tends toward randomness.

What does this mean for chemical reactions? As we discovered earlier, spontaneous chemical reactions liberate thermal energy. Some of this thermal energy is used within the system to increase randomness or entropy. The remainder of the thermal energy is called **free energy** (ΔG) because it is available for other purposes. The equation relating enthalpy (ΔH), entropy (ΔS), free energy (ΔG), and temperature (T) was first proposed by J. Willard Gibbs in 1878.

$$\Delta H = \Delta G + T\Delta S$$

From this equation, we see two factors that influence biological systems. First, the change in energy associated with randomness is dependent upon temperature. This is because the potency of a fixed amount of thermal energy, or its ability to induce randomness, depends on temperature. Thermal energy is more effective at inducing entropy at low temperature. The second principle is more apparent if we rearrange the equation to isolate ΔG.

$$\Delta G = \Delta H - T\Delta S$$

The amount of free energy available in a reaction is the difference between the total energy change and the amount of energy associated with the change in randomness. This equation allows us to predict if a reaction will occur spontaneously. If a reaction is to occur spontaneously, the amount of energy potentially released by a reaction (ΔH) must be greater than the energy used to increase entropy ($T\Delta S$). Recall that exothermic reactions, those that release heat, have a negative ΔH. Similarly, reactions that release free energy have a negative

ΔG. All reactions that occur spontaneously possess a negative ΔG; free energy was released.

Chemists evaluate these parameters under **standard conditions**. The standard free energy, or $\Delta G°$ is assessed at 25°C, with each reactant, including H^+, present at a concentration of 1 M. The proton concentration used by chemists equates to pH 0, which is not relevant to biological systems. When we use the laws of thermodynamics to discuss biological systems, the parameters must be altered to reflect normal cellular conditions. When biochemists adjust ΔG for standard conditions, including a pH of 7.0, the symbol $\Delta G°'$ is used.

It is important to distinguish between ΔG and $\Delta G°'$ when discussing chemical reactions. The value of $\Delta G°'$ is a constant. It tells how much free energy is available when a reaction begins under standard conditions. The value of ΔG, the actual free energy of a reaction in a cell, depends upon the concentrations of reactants. If a reaction is close to equilibrium, then ΔG equals zero. For the reaction

$$A + B \rightleftharpoons Y + Z$$

the relationship between ΔG, $\Delta G°'$, and concentration is defined by the following equation where R is the gas constant:

$$\Delta G = \Delta G°' + RT \ln \frac{[Y][Z]}{[A][B]}$$

When the reaction is at equilibrium $\Delta G = 0$ and the mass action ratio is equal to K_{eq}, the equation is reduced to

$$0 = \Delta G°' + RT \ln K_{eq}$$

or

$$\Delta G°' = -RT \ln K_{eq}$$

We can measure K_{eq} directly by letting the reaction reach equilibrium. The value of $\Delta G°'$ can be calculated from the equation above. Knowing K_{eq} and $\Delta G°'$, we can calculate the amount of actual free energy ΔG available for a reaction at any concentration of reactants.

Remember that ΔG represents the maximal amount of free energy theoretically available from a reaction, under a constant temperature and pressure that approximates conditions found in the cell. Cells use enzymes to mediate chemical reactions and transfer as much energy as possible to other useful forms. Some enzymes mediate reactions that store energy as chemical energy, such as ATP or NADH. Free energy can also be used to create electrochemical gradients. The ability to divert free energy into useful forms is central to the success of living organisms.

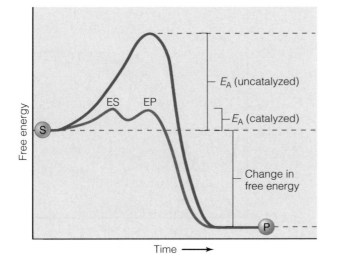

Figure 2.12 Enzymes and E_A Enzymes are biological catalysts that accelerate reactions without changing the nature of the product. When the substrate (S) binds the enzyme (E), the enzyme-substrate complex (ES) is formed. The enzyme alters the substrate through a series of transition states, ultimately releasing the product (P). The rate is faster than the noncatalyzed rate because of the lower activation energy (E_A).

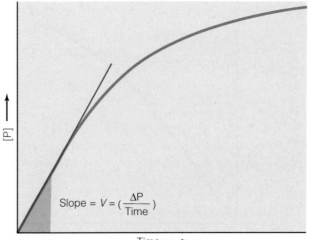

Figure 2.13 Time course of enzyme reaction Enzyme assays begin with the addition of substrate to the reaction. The enzyme rapidly converts the substrate (S) to product (P). The buildup of [P] eventually slows the reaction, as P competes with S for the active site. The initial velocity (V) is the fastest because P has not yet accumulated.

Once it binds the substrate, the enzyme induces a change in the molecular structure of the substrate, perhaps as subtle as a shift in the distribution of electrons across a particular bond or a twist in the substrate molecule. By inducing these subtle structural changes in the substrate, the enzyme makes the substrate more likely to spontaneously undergo more significant changes. Many enzymes require two or more substrates. These enzymes accelerate reactions by bringing destabilized reactants in close proximity. All together, these changes increase the probability that the substrate will undergo a major change in structure toward the formation of EP*.

Enzyme kinetics describe enzymatic properties

Enzymes accelerate reaction rates, and make possible reactions that would not normally occur at a useful rate. However, cells must ensure that enzymatic reactions occur not at the fastest possible rate, but at the appropriate rate. Thus, enzyme activity is regulated within complex metabolic pathways. The conditions that influence the rate of enzymatic reactions are referred to as *enzyme kinetics*.

The simplest way to influence an enzymatic reaction is to change the concentration of substrates

(S) or products (P). We use the reaction S → P to illustrate the importance of substrate concentration ([S]) in two experimental scenarios.

The first scenario illustrates how the buildup of [P] influences the rate of the forward reaction (Figure 2.13). When the reaction begins, there is no product ([P] = 0). As it proceeds, molecules of P accumulate and eventually compete with molecules of S for the same active site. Finally, the reaction approaches **equilibrium**, where the forward and reverse reaction rates are equal and the mass action ratio equals K_{eq}. We can determine the initial velocity of the forward reaction (V) from the slope of the curve before P accumulates.

The second scenario illustrates how the initial [S] influences the enzymatic rate (Figure 2.14). The experiment previously described is repeated many times using a wide range of starting [S]. Increasing [S] from a low concentration to a higher concentration causes a proportional increase in V. Under these conditions, a higher [S] increases the frequency with which molecules of S find the active site. However, after a point, increases in [S] no longer cause a proportional increase in V. The higher abundance of S molecules still increases the probability of a collision with E. However, if S encounters E in the midst of a reaction cycle, the enzyme is unable to bind S. Eventually, E is *saturated* with S molecules and further increases in [S] do not increase V beyond a maximal rate

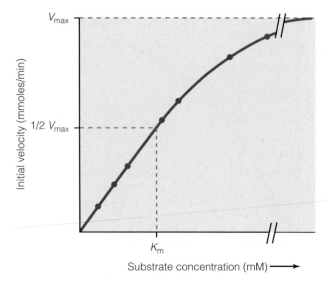

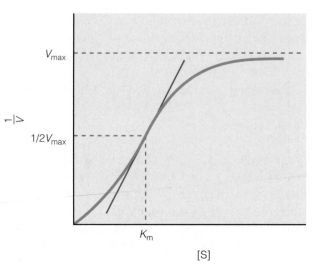

Figure 2.14 The Michaelis-Menten rectangular hyperbola Each point on the curve represents the initial velocity (V) calculated as shown in Figure 2.16. The maximal velocity (V_{max}) is the velocity at which the curve reaches an asymptote. The K_m is the [S] required to reach a velocity that is one-half of the maximal velocity.

Figure 2.15 Homotropic enzymes and sigmoidal kinetics Not all enzymes obey Michaelis-Menten kinetics. Homotropic enzymes show sigmoidal kinetics. The enzymes usually possess multiple active sites. When the enzyme binds one molecule of S, the changes in conformation increase the ability to bind a second molecule of S. The slope of the linear range of the curve indicates the degree of cooperativity. The slope of this region provides the Hill coefficient.

(V_{max}). When enzymes are at V_{max}, each molecule of enzyme has a characteristic number of catalytic cycles per unit time, known as the turnover number or k_{cat}.

A high rate of enzymatic activity could be achieved by a cell, in principle, in either of two ways. Some enzymes work very fast, and show a high k_{cat}. The cell does not need many molecules of the enzyme because each molecule works quickly. The fastest enzymes can undergo more than 40,000,000 catalytic cycles each second. Alternatively, cells can make many copies of an enzyme with a low k_{cat}. The relative importance of each strategy—faster enzymes versus more enzymes—depends on the nature of the reaction and the design of the enzyme.

The relationship between [S] and V was first described mathematically by the biochemists Leonar Michaelis and Maud Menten as a rectangular hyperbola. The **Michaelis-Menten equation** is

$$V = V_{MAX} \times \frac{[S]}{[S] + K_m}$$

The value for the Michaelis-Menten constant (K_m) is the concentration of substrate [S] required to obtain an initial velocity (V) that is half the maximal velocity (V_{max}). K_m is an indicator of the affinity of an enzyme for a substrate. A low K_m means that

the enzyme has high affinity for the substrate, and little substrate is needed to drive the reaction at a high rate.

Not all enzymes demonstrate hyperbolic Michaelis-Menten kinetics. For instance, homotropic enzymes show a sigmoidal relationship between V and [S] (Figure 2.15). Homotropic enzymes typically have multiple subunits that can each bind a substrate molecule. At low [S], each active site has a low affinity for S. The enzyme does not bind S very well and the reaction velocity is slow. Once one subunit binds one molecule of S, it undergoes a change in conformation that in turn alters the ability of other subunits to bind a substrate molecule. As a result, doubling of [S] more than doubles V, a phenomenon called *cooperativity*. The degree of cooperativity is described by the *Hill coefficient*, which is the slope of relationship at the point of inflection.

Enzyme kinetics are assessed under carefully controlled experimental conditions that do not approximate normal cellular conditions. Interpreting the impact of enzyme kinetics in living cells is often difficult. The conditions necessary to evaluate V_{max} require [P] to be zero, which never occurs in living cells. Thus, enzymes in cells almost never could proceed at V_{max}. As with other chemical reactions, the rate and direction of the enzymatic reaction depend on the difference between the mass action ratio, which is calculated from the actual [S] and [P], and the K_{eq} value, which is the expected [S] and [P] when the reaction reaches equilibrium. In a **near-equilibrium reaction**, the mass action ratio is close to K_{eq}; the forward and reverse directions continue at equal rates, with little net change in [S] or [P]. Most enzyme reactions are far from equilibrium in cells. If the mass action ratio is lower than K_{eq}, then the reaction will proceed in the forward direction. When the mass action ratio is higher than K_{eq}, the reaction will tend to favor the reverse direction. By altering the concentrations of substrates and products, cells can regulate enzyme activities and metabolic pathways.

The physicochemical environment alters enzyme kinetics

Every enzyme has a characteristic optimal activity under a specific set of environmental conditions (Figure 2.16). Enzyme kinetics are influenced by environmental conditions, such as temperature, pH, salt concentration, and hydrostatic pressure. While these factors generally have little impact on your metabolism, such environmental factors can influence the metabolic biochemistry of other species.

Some enzymes function optimally under conditions that resemble normal cellular conditions. For instance, mammalian enzymes often function optimally at normal body temperatures of 37–40°C. However, the optimal conditions for many enzymes bear little similarity to normal cellular conditions; the optimal temperature for some mammalian enzymes is well above normal body temperatures.

Environmental conditions typically influence enzyme kinetics through effects on weak bonds. First, changes in weak bonds can alter the three-dimensional structure of the enzyme. For instance, warm temperatures could break bonds that are necessary to form the active site. Second, environmental conditions can alter the ionization state of critical amino acids within the active site. For instance, the amino acid histidine is important in many active sites, and changes in pH can alter its protonation state and consequently substrate affinity (K_m). Any environmentally induced change in K_m, either an increase or a decrease, can be disruptive to a cell. Third, environmental conditions can alter the ability of the enzyme to undergo structural changes necessary for catalysis. Enzymes must be rigid enough to maintain the

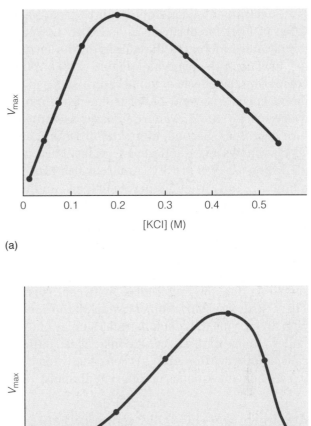

(a)

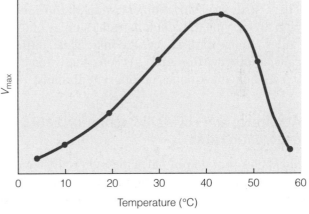

(b)

Figure 2.16   Effects of salt and temperature on enzyme kinetics   Most enzymes function optimally under physiologically-realistic conditions. **(a)** The activity of mammalian enzymes changes in response to the concentration of the salt KCl. Maximal activity occurs at concentrations that approximate those found within the cell (100–150 mM K^+). **(b)** Increasing temperature accelerates enzymes. Beyond an optimal temperature, the enzyme denatures and loses catalytic activity.

proper conformation, but flexible enough to incur conformational changes during catalysis.

Many of the studies assessing the effects of environmental conditions have focused on the effects of temperature on the enzyme lactate dehydrogenase (LDH). This enzyme has an important role in glucose metabolism, which we discuss in more detail later in this chapter. It catalyzes the following reversible reaction:

Pyruvate + $NADH^+$ + H^+ $\rightleftharpoons$ lactate + NAD^+

Environmental conditions can change the K_m value of LDH for pyruvate and NADH. Lowering temperature increases the affinity of the enzyme for its substrate pyruvate (Figure 2.17). When comparing the effects of temperature on K_m in different species, several patterns emerge. First, in every species, the K_m value decreases as temperature decreases. Second, at any temperature, each species shows a very different K_m value. For example, when assayed at 15°C, Antarctic fish LDH has a high K_m, LDH from temperate fish has an intermediate K_m, and desert lizard LDH has a low K_m. Third, when the LDH from each species is assayed at its normal body temperature, the resulting K_m values fall within a narrow range, from 0.1 to 0.3 mM. Evolutionary variation in LDH structure is responsible for the differences between species. These structural variations provide all the species with an enzyme that demonstrates similar kinetics under their natural conditions. This pattern, called **conservation of K_m**, is common when comparing enzyme kinetics of different animals.

Allosteric and covalent regulation control enzymatic rates

Molecules that do not participate directly in catalysis can also alter enzyme kinetics. **Competitive inhibitors** are molecules that can bind to the active site, preventing substrate molecules from binding (Figure 2.18a). The effectiveness of a competitive inhibitor depends on [S]. When [S] is low, the inhibitor outcompetes S for the active site, reducing the reaction rate. At a very high [S], the inhibition by the competitor is greatly reduced. Thus, a competitive inhibitor increases K_m but doesn't affect V_{max}.

Allosteric regulators are molecules that alter enzyme kinetics by binding to the protein at locations far away from the active site. The allosteric regulator alters the three-dimensional structure of the enzyme, inducing complex changes in enzyme kinetics. For example, an allosteric activator could increase the affinity of the enzyme for the substrate, as depicted in Figure 2.18b. Allosteric effectors can activate or inhibit enzyme activity, changing either K_m or V_{max}. Enzymes often possess multiple sites for different allosteric regulators. Enzymes controlled by allosteric regulators are often larger and more complex than other enzymes. Typically, each metabolic pathway is regulated by one or more key allosteric enzymes.

Enzymes can also be regulated by the covalent modification of critical amino acid residues within the protein. The most common type of covalent modification is protein phosphorylation, where a specific **protein kinase** transfers the phosphate group from ATP to an amino acid of the target enzyme. For instance, tyrosine kinase is a regulatory enzyme that phosphorylates target proteins at specific tyrosine residues. Another common class of protein kinases is specific for threonine and serine residues. Protein phosphorylation is reversible. Cells possess suites of **protein phosphatases** that cleave phosphate groups from phosphorylated amino acid residues. Phosphorylation might stimulate an enzyme, as depicted in Figure 2.18c, or inhibit it.

Enzymes convert nutrients to reducing energy

Enzymes transfer energy from nutrients to molecules that function as energy stores. These energy-rich molecules are a type of energy currency, acting as substrates and products for hundreds of different enzymes. Cells store chemical energy in two main forms: reducing energy and high-energy molecules.

Many enzymatic reactions capture energy in the form of **reducing equivalents**: NAD and NADP. The enzymes that use reducing equivalents are called oxidoreductases and include enzymes

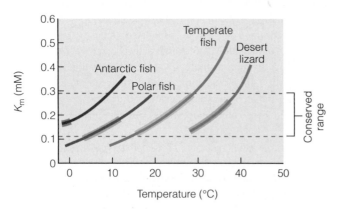

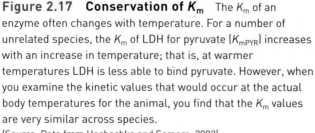

Figure 2.17   Conservation of K_m   The K_m of an enzyme often changes with temperature. For a number of unrelated species, the K_m of LDH for pyruvate (K_{mPYR}) increases with an increase in temperature; that is, at warmer temperatures LDH is less able to bind pyruvate. However, when you examine the kinetic values that would occur at the actual body temperatures for the animal, you find that the K_m values are very similar across species.
(Source: Data from Hochachka and Somero, 2002)

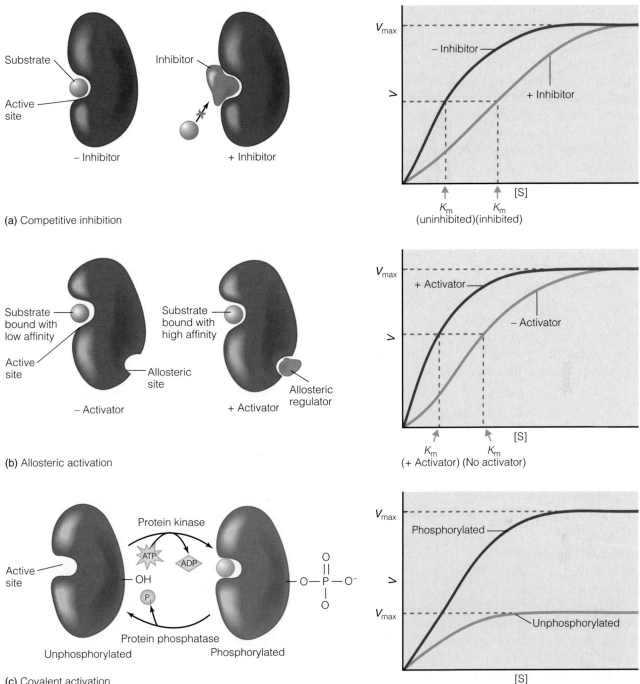

(a) Competitive inhibition

(b) Allosteric activation

(c) Covalent activation

Figure 2.18 Enzyme regulation   (a) Competitive inhibitors are able to bind to the active site of enzymes, thereby preventing the real substrate from binding. At low [S], the inhibitor outcompetes the substrate. However, if [S] is increased to very high levels, the true substrate outcompetes the inhibitor, and thus these regulators have no effect on V_{max}. **(b)** Allosteric enzymes are regulated by molecules that bind at sites distant from the active site. The resulting structural change in the enzyme alters its kinetic properties.

In this figure, the allosteric regulator activates the enzyme by increasing the affinity for the substrate, shown in the graph as a decrease in the K_m. **(c)** Many enzymes are controlled by phosphorylation-dephosphorylation. Protein kinases phosphorylate the target enzyme, transferring a phosphate group from ATP to specific hydroxy groups. Protein phosphatases remove the phosphate group. In this figure, the enzyme is activated by phosphorylation, greatly increasing the V_{max}.

with the common names dehydrogenase, reductase, and oxidase. When an enzymatic reaction transfers an electron to NAD⁺ (or NADP⁺), the reduced NADH (or NADPH) that is formed can be used to drive other reactions. In other words, energy can be stored by reducing a molecule and this energy can be recovered, in part, by oxidizing the reduced compound.

Consider the nonenzymatic and enzymatic reactions for lactate oxidation. Without an enzyme, lactate is oxidized to form pyruvate with the following reaction:

$$\text{Lactate}^- \rightarrow \text{pyruvate}^- + 2H^+ + 2e^-$$
$$\Delta G^{\circ\prime} = -36 \text{ kJ/mol}$$

The negative standard free energy ($\Delta G^{\circ\prime}$) means that energy is liberated in this reaction, and without an enzyme the energy released would be lost as heat. Cells possess the enzyme LDH, introduced earlier in this chapter, which couples lactate oxidation to NADH reduction. The NAD reduction reaction has a positive $\Delta G^{\circ\prime}$.

$$NAD^+ + 2e^- + 2H^+ \rightarrow NADH + H$$
$$\Delta G^{\circ\prime} = +62 \text{ kJ/mol}$$

By coupling lactate oxidation to NAD reduction, the enzymatic reaction captures free energy from lactate oxidation in the form of NADH.

$$\text{Lactate} + NAD^+ \rightarrow NADH + H^+ + \text{pyruvate}^-$$
$$\Delta G^{\circ\prime} = +26 \text{ kJ/mol}$$

Note that the enzymatic reaction for lactate oxidation has a positive $\Delta G^{\circ\prime}$, which means the reverse direction of this reaction (lactate formation) is normally favored.

The most important reducing equivalent in energy metabolism is NADH. The reducing energy within the cell, or **redox status**, is best expressed as [NADH]/[NAD⁺]. This ratio is high when a cell is rich in reducing energy, and low when cells are energy poor. NAD is a reactant in many enzymes of energy metabolism, but other enzymes are allosterically regulated by NAD. Whether acting through mass action effects or allosteric regulation, enzymes sensitive to [NADH]/[NAD⁺] allow metabolic pathways to respond to the energy state.

ATP is a carrier of free energy

Cells use many types of molecules to store energy (Figure 2.19), but ATP is the most versatile of these high-energy molecules and participates in count-

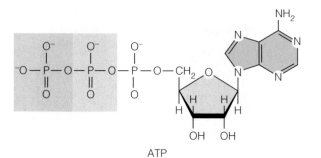

ATP

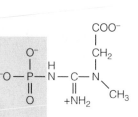

Phosphocreatine

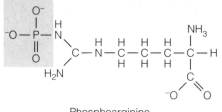

Phosphoarginine

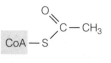

Acetyl CoA

Figure 2.19 High-energy molecules Cells use several energy-rich molecules, such as ATP, phosphocreatine, phosphoarginine, and acetyl CoA, as energy currency.

less reactions. ATP synthesis requires energy, and ATP breakdown liberates energy.

$$ADP^{3-} + HPO_4^{2-} + H^+ \rightarrow ATP^{4-} + H_2O$$
$$\Delta G^{\circ\prime} = +30.5 \text{ kJ/mol}$$

ATP possesses two phosphodiester bonds (–P–O–P–). Some enzymes break the bond between the second and third phosphate groups, forming ADP. In some cases the inorganic phosphate (P_i) is released as a product, but often the P_i is transferred to another molecule. Other enzymes target the bond between the first and second phosphate groups, forming AMP and pyrophosphate (PP_i). Because these energy exchange reactions involve a breakdown of a phosphodiester bond, they are often called high-energy bonds. It is important to realize that the energy is not stored in the bond per se, but is released when ATP hydrolysis occurs—a reaction with large, negative free energy.

The importance of utilizing a metabolite like ATP is, first, to avoid high concentrations of other metabolites; participation of ATP permits reactions that otherwise would be thermodynamically unfavorable. Second, ATP links major metabolic pathways that require cellular energy, such as endergonic pathways of biosynthesis, with those that generate energy, such as the exergonic process of carbohydrate catabolism.

The relative abundance of ATP reflects the energy status of a cell. The absolute concentration of ATP is unimportant; what counts is the relative proportion of the adenylate pool (ATP + ADP + AMP) that exists in the energy-rich forms ATP and ADP. The ATP status of the cell is best expressed by the **phosphorylation potential** (ΔG_p), the free energy associated with ATP hydrolysis (ATP → ADP + P_i):

$$\Delta G_p = \Delta G^{\circ\prime} + RT \ln \frac{[\text{ADP}]\,[P_i]}{[\text{ATP}]}$$

ATP is the most common form of energy currency, but the other nucleotides—GTP, TTP, and CTP—have the same energetic value, although only GTP is commonly used in energy metabolism.

Phosphorylated guanidine derivatives are important energy stores in many animals. Vertebrates use phosphocreatine and invertebrates use phosphoarginine, phosphoglycocyamine, phosphotaurocyamine, or phospholombricine. Phosphoguanidine compounds, each with a –P–N– bond, are useful energy stores because they do not participate in many reactions within the cell. Consequently, cells can accumulate very high concentrations of phosphoguanidines without affecting other pathways. The concentration of ATP, in contrast, is kept low and relatively constant. Major changes in ATP concentration would have kinetic consequences for countless enzymes that use ATP as a substrate or product. For instance, the ATP concentration in vertebrate muscle is typically about 5 mM, whereas phosphocreatine concentrations might be 10–50 mM. Animal tissues use these high-energy compounds when the need for ATP temporarily outstrips the capacity to produce ATP. When ATP levels decline, the energy within phosphoguani-

dine is transferred to ADP to form ATP. In vertebrates, creatine phosphokinase (CPK) catalyzes this reaction.

Phosphocreatine + ADP ⇌ ATP + creatine

Acetyl coenzyme A, or acetyl CoA, is another important high-energy store. Energy is released in reactions that hydrolyze its thioester bond (–O–S–). As we see later in this chapter, many pathways of biosynthesis and energy metabolism intersect at acetyl CoA. Collectively, reducing energy and high-energy compounds provide the energetic support for many cellular processes.

Proteins

Proteins play many important roles in cell structure and function. Almost all enzymes are proteins (though many have nonprotein components). Proteins form the internal skeleton of a cell (cytoskeleton) as well the extracellular matrix needed to organize cells into complex tissues. The diversity in protein structure is afforded by the use of 20 amino acids that can be strung together in countless combinations. The blueprint for all proteins in a cell is in the form of DNA, which is transcribed into RNA and translated to form the appropriate proteins at the right time.

Proteins are polymers of amino acids

Animals build proteins from combinations of 20 amino acids. As the name implies, amino acids share the general structure of an amino group (–NH_2) and a carboxylic acid group (–COOH). They are called α-amino acids because both the amino and carboxyl groups are located on the first, or α, carbon.

Amino acids are distinguished from one another by their side groups (R). The R groups of polar amino acids form hydrogen bonds with water. Some polar amino acids are uncharged at physiological pH values (serine, threonine, cysteine, tyrosine, asparagines, glutamine), while others possess R groups with side chains that can become charged. Acidic amino acids (aspartate, glutamate) are negatively charged at physiological pH when carboxyl groups become deprotonated (–COOH → –COO^- + H^+). Basic amino acids (arginine, lysine) take on a positive charge when amino groups become protonated (–NH_2 + H^+ → –NH_3^+).

Many amino acids are nonpolar because their R groups are aliphatic chains (alanine, valine, leucine, isoleucine, methionine) or aromatic rings (phenylalanine, tryptophan) that do not readily interact with water. The collection of amino acids, with their unique properties of side chain length, shape, charge, and polarity, provides cells with the building blocks necessary to construct thousands of different proteins.

Proteins are folded into three-dimensional shapes

Amino acids are polymerized into linear chains by covalent peptide bonds that link the amino group ($-NH_3$) of one amino acid to the carboxyl group ($-COOH$) of another amino acid.

$$R_1\text{-N-H} + \text{H-O-C-}R_2 \rightarrow R_1\text{-N-C} - R_2 + H_2O$$

Two amino acids in a chain is a dipeptide. Polypeptides are longer chains of amino acids. At one end of the polymer, called the C terminus, the amino acid has an unbonded carboxyl group. At the other end, the N terminus, the amino acid has an unbonded amino group. The linear sequence of amino acids in a protein is called the **primary structure**.

Once the primary structure is established, proteins are organized into more complex three-dimensional conformations (Figure 2.20). First, the protein folds onto itself to assume its **secondary structure**. The information for proper folding is contained directly in the primary structure. The size, charge, and polarity of the side groups influence the interactions between amino acids in the chain. Secondary structures arise when side groups of amino acids interact to form a structure that is more stable than the simple linear conformation. The two most common protein secondary structural motifs are the α-helix and the β-sheet (Figure 2.21). In the α-helix, the protein is twisted into a spiral with 3.6 amino acids per turn and side chains extending outward. The structure is stabilized in two ways. First, hydrogen bonds form between the C=O of one amino acid and the N–H of the amino acid four positions along the chain. Second, the α-helix structure is stabilized when opposing side chains can interact. With the period of 3.6 amino acids, a side chain is exposed to the side chain of the amino acid three or four positions away. For example, if two aromatic amino acids are three positions apart, when the protein twists into an α-helix the structure will be stabilized by the hydrophobic interactions between the side chains. Similarly, negatively charged amino acids are often found three residues away from positively charged amino acids. Their electrostatic interactions stabilize the protein. The other common type of secondary structure, the β-sheet, forms when linear regions of a protein align side by side and form hydrogen bonds. In this conformation, the side chains extend above and below the face of the sheet.

Once a protein forms its secondary structure, the different regions fold together to create its

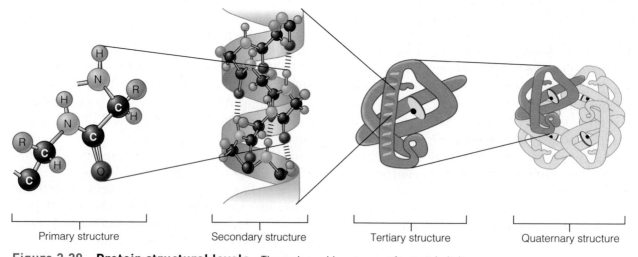

Primary structure Secondary structure Tertiary structure Quaternary structure

Figure 2.20 Protein structural levels The amino acid sequence of a protein is its primary structure. This polypeptide can then be folded and organized into three-dimensional conformations.

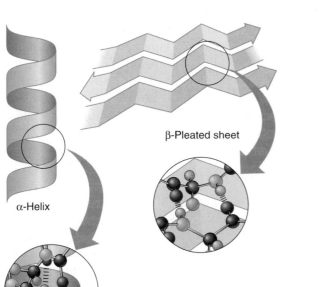

Figure 2.21 Protein secondary structure: The α-helix and β-sheet The most common secondary structures in proteins are the α-helix and α-sheet. Weak bonds stabilize both types of secondary structures. The information that is used to fold the protein is contained within the primary sequence.

tertiary structure (Figure 2.22). If the protein folds in a way that allows two adjacent cysteine residues to come into close proximity, their sulfhydryl groups (–SH) can form a covalent bond (–S–S–) called a disulfide bond or bridge. Multiple weak bonds link various amino acids and side chains to stabilize three-dimensional structure. Many proteins assume a globular structure when hydrophobic interactions form between regions scattered throughout the protein. By pulling together hydrophobic regions, a hydrophobic core is formed that stabilizes the structure of the protein.

A protein achieves its quaternary structure when multiple subunits, or polypeptide chains, are brought together. Proteins with two subunits are called dimers—a **homodimer** if the monomers are identical, otherwise a **heterodimer**. Proteins can be composed of even larger numbers of subunits, such as trimers (three subunits) and tetramers (four subunits).

Molecular chaperones help proteins fold

Proteins can function properly only when they are folded into the correct conformation. Many proteins can use the information within the primary

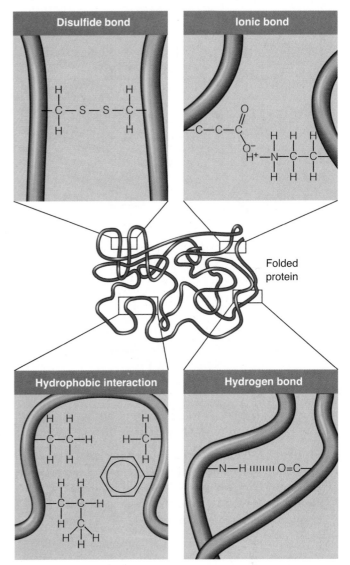

Figure 2.22 Weak bonds and protein tertiary structure Both covalent bonds and weak bonds contribute to protein three-dimensional structures.

sequence to fold spontaneously, but others require the help of **molecular chaperones**. Each cell contains different types of chaperones to ensure that proteins are properly folded. They work by forcing the protein into a conformation that allows the appropriate weak bonds to form.

Environmental conditions, such as temperature, can alter weak bonds and disrupt three-dimensional protein structure. Increasing temperature weakens the hydrogen bonds that stabilize α-helices and β-sheets. High temperature can cause the protein to unfold, or **denature**. Once denatured, a protein can no longer perform its proper function and may even damage cells. Therefore, a partially denatured protein must be

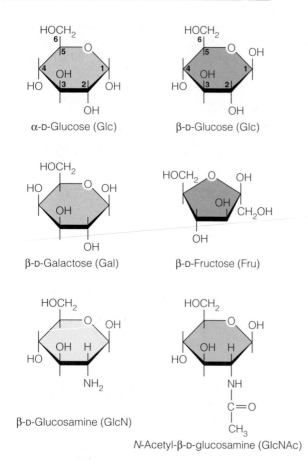

Figure 2.23 Common monosaccharides These structural models of monosaccharides show how side groups extend above and below the plane of the ring structures. The α and β forms of glucose differ in the orientation of the hydroxy group on C-1.

refolded or destroyed before it can damage the cell. Molecular chaperones bind to denatured proteins, folding them into the proper configuration. During heat stress, cells increase the levels of molecular chaperones called **heat shock proteins** to cope with the increased number of denatured proteins.

Carbohydrates

Carbohydrates share a preponderance of hydroxyl (–OH), or alcohol, groups, and for this reason they are often called *polyols*. For any animal, the diet is a vital source of the carbohydrates used to build and fuel cells. Glucose, the most common carbohydrate in animal diets, is central to cellular energy metabolism and biosynthesis because of its metabolic versatility. Cells can break glucose down for energy, or store it for later consumption, or use it to build other carbohydrates needed by the cell.

Animals use monosaccharides for energy and biosynthesis

Monosaccharides are small carbohydrates that have from three to seven carbons. The most common monosaccharides are the six-carbon sugars (hexoses) including glucose, fructose, and galactose (Figure 2.23). Glucose and galactose, as well as mannose, can be modified by the addition of acidic groups, amino groups, and modified amino groups. These sugar derivatives serve many purposes in the cell, primarily as modifications of other macromolecules, including proteins, lipids, and nucleic acids.

Many of the sugars that animals obtain in the diet are **disaccharides**, two monosaccharides connected by a covalent bond (Figure 2.24). In order to use disaccharides, animals first break them down into monosaccharides. Animals can also produce disaccharides such as lactose, an important component of milk in mammalian mammary secretions, and trehalose, an energy store and solute.

The addition of carbohydrates to other macromolecules is called **glycosylation**. Glycosylated lipids (**glycolipids**) and proteins (**glyco-**

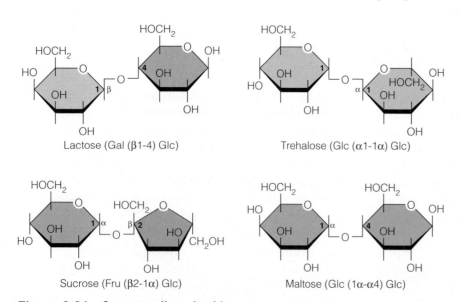

Figure 2.24 Common disaccharides Both trehalose and maltose are made from two glucose molecules but with bonds forming between different pairs of carbons. Sucrose and maltose are synthesized in plants; animals obtain them by eating the plants.

proteins) are common in the plasma membrane of cells. A glycosylated macromolecule displays an altered molecular profile, changing how it interacts with other macromolecules and reducing its susceptibility to degradation.

Complex carbohydrates perform many functional and structural roles

Complex carbohydrates, or **polysaccharides**, are larger polymers of carbohydrates that serve in energy storage and structure. Polysaccharides can be composed of long chains of a single type of monosaccharide or combinations of two alternating monosaccharides. Common polysaccharides important in metabolism and structure are shown in Figure 2.25. Starch is a general term for the glucose polysaccharides used by plants and animals for energy storage. Plant starch, a mixture of amylose and amylopectin, is an important dietary source of energy for many animals. Animal starch, or **glycogen**, is central to animal energy metabolism, acting as an internal energy store for most animals and a nutrient for animals that eat other animals. Amylose, amylopectin, and glycogen differ in the linkage between glucose molecules and the nature of the branching pattern. Cellulose, another plant-derived glucose polymer, is essentially indigestible in animals because of the nature of the bonds between glucose units. Cellulose, in most animals, provides dietary fiber. However, some animals, such as ruminants and termites, possess gastrointestinal symbionts that can degrade cellulose for energy.

Polysaccharides are also critical structural components of animal cells. Arthropods build their exoskeletons with **chitin**, a polysaccharide of N-acetyl-glucosamine. Vertebrates secrete hyaluronate, a polymer of N-acetyl-glucosamine and glucuronic acid, into the extracellular space, where its gel-like properties act as a spacer between cells and tissues. Hyaluronate is a member of a class of compounds called **glycosaminoglycans** that include chondroitin sulfate and keratan sulfate. These compounds are important components of animal tissues, such as cartilage.

In order to use glycogen as an energy store, animals control the balance between glycogen synthesis (**glycogenesis**) and glycogen breakdown (**glycogenolysis**). Glycogen phosphorylase initiates glycogenolysis, releasing glucose in the form of glucose 1-phosphate. When glucose is

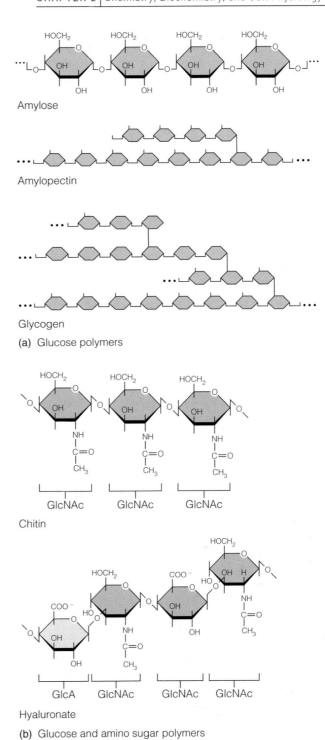

Amylose

Amylopectin

Glycogen

(a) Glucose polymers

Chitin

Hyaluronate

(b) Glucose and amino sugar polymers

Figure 2.25 Polysaccharides **(a)** Plants and animals use polymers of glucose as energy stores. Amylose and amylopectin are the two polysaccharides that compose starch, an important dietary source of energy for animals. Animals produce glycogen, which resembles the plant polysaccharides but with much greater branching.
(b) Animals build many polysaccharides from combinations of monosaccharides and amino sugars, such as N-acetyl-glucosamine (GlcNAc). Chitin is a polymer of N-acetyl-glucosamine, whereas hyaluronate is a polymer of N-acetyl-glucosamine and glucuronic acid (GlcA).

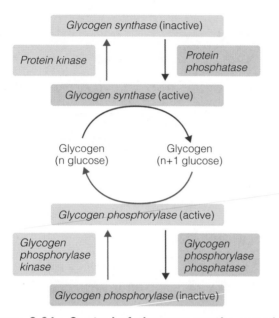

Figure 2.26 Control of glycogen synthase and glycogen phosphorylase Under conditions in which glycogen breakdown is desirable, both glycogen synthase and glycogen phosphorylase are phosphorylated by protein kinases. Phosphorylation inhibits glycogen synthase but stimulates glycogen phosphorylase. Similarly, dephosphorylation of these two enzymes by protein phosphatases favors glycogen synthesis.

abundant, glycogen synthase is activated and glucose 1-phosphate is used to increase the size of the glycogen particle. Protein kinases and protein phosphatases regulate both glycogen synthase and glycogen phosphorylase (Figure 2.26).

Gluconeogenesis builds glucose from noncarbohydrate precursors

Glucose is essential for energy metabolism and biosynthesis. When dietary glucose is inadequate or when glycogen stores are compromised, animals can produce glucose from noncarbohydrate precursors via **gluconeogenesis**. The gluconeogenic pathway (Figure 2.27) using mitochondrial pyruvate as a starting point has the following overall reaction:

$$2\text{pyruvate} + 4\text{ATP} + 2\text{GTP} + 2\text{NADH} + 4\text{H}_2\text{O} \rightarrow$$
$$\text{glucose} + 4\text{ADP} + 2\text{GDP} + 6\text{Pi} + 2\text{NAD}^+ + 2\text{H}^+$$

Figure 2.27 Gluconeogenesis Cells convert pyruvate to glucose and glycogen using the enzymes of gluconeogenesis. The exact route of phosphoenolpyruvate synthesis depends upon tissue and species. Some species use a mitochondrial PEPCK to produce phosphoenolpyruvate.

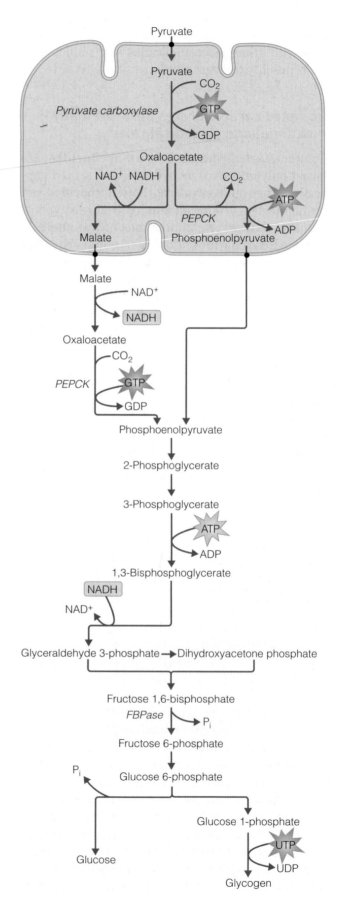

Gluconeogenesis begins in the mitochondria, where pyruvate carboxylase converts pyruvate to oxaloacetate, the substrate for PEP carboxykinase (PEPCK). In species with a mitochondrial PEPCK, PEP is transported to the cytoplasm; if PEPCK is cytoplasmic, the mitochondria convert oxaloacetate to malate, export it, and then resynthesize oxaloacetate within the cytoplasm. A series of reactions produces glucose 6-phosphate, which can be used to produce glycogen, or in some tissues converted to glucose by glucose 6-phosphatase. Because gluconeogenesis requires a great deal of energy, cells stimulate gluconeogenesis only when they have excess energy available. The metabolic indicators of energy status, such as acetyl CoA and adenylates (AMP, ADP, and ATP), regulate the gluconeogenic rate. The pathway is controlled mainly by availability of gluconeogenic substrates and allosteric regulation of pyruvate carboxylase and fructose 1,6-bisphosphatase (FBPase).

Glycolysis is a low-efficiency, high-velocity pathway

Glycolysis is the pathway that breaks down glucose obtained from the blood and glucose 6-phosphate derived from processing of the glucose 1-phosphate liberated from stored glycogen. This pathway is a vital source of ATP because it can proceed in the absence of oxygen (anoxia) and can produce ATP very rapidly (albeit for brief periods).

Although glycolysis is usually discussed from the perspective of glucose or glycogen breakdown, other carbohydrates derived from the diet are also processed into hexoses that can enter glycolysis. Disaccharides are first broken down into monosaccharides: trehalose into two glucose, lactose into glucose and galactose, sucrose into glucose and fructose. The glycolytic pathway (Figure 2.28) using glucose as initial substrate has the following overall reaction:

$$\text{Glucose} + 2\text{ADP} + 2\text{NAD}^+ \rightarrow$$
$$2\text{ATP} + 2\text{pyruvate} + 2\text{NADH} + 2\text{H}^+$$

When glucose is carried into the cell, the enzyme hexokinase rapidly phosphorylates it, using a molecule of ATP. Since glucose 6-phosphate is not readily transported across the cell membrane, phosphorylation of glucose traps glucose within the cell. The next steps in glycolysis are a series of enzymatic reactions that convert the glucose backbone to fructose, which is then hydrolyzed to form two trioses that are ultimately converted to pyruvate.

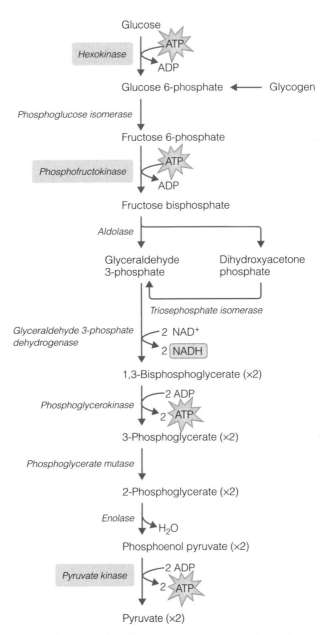

Figure 2.28 Glycolysis Glycolysis is a series of cytoplasmic enzymes that breaks down glucose or glycogen to produce ATP. Because ATP is required by hexokinase, glycolysis from glycogen produces more ATP (three ATP per glucosyl) than it does from glucose (two ATP per glucose). The other important products of glycolysis are pyruvate and NADH. The three irreversible reactions are highlighted.

Seven of the ten glycolytic reactions are freely reversible, and catalyzed by the enzymes shared with the gluconeogenic pathway. The three irreversible glycolytic reactions—hexokinase, phosphofructokinase (PFK), and pyruvate kinase (PK)—are important sites of regulation for the pathway, acting via mass action effects, allosteric regulation, and covalent modification. During periods of high energy demand, much of the ATP is broken down to ADP and

AMP, affecting the mass action ratios for all three regulatory enzymes. Both ADP and AMP are powerful activators of PFK enzymatic activity, whereas ATP inhibits PFK, as well as PK. When cells do not need energy, glycolysis is inhibited at PFK and PK. With PFK inhibited, glucose 6-phosphate is diverted into glycogen synthesis. Thus, the fate of the glucose 6-phosphate—glycolysis or glycogen synthesis—is linked to energy status through regulation of PFK. This is an example of *negative feedback regulation*, where an increase in the concentration of products inhibits the pathway.

In addition to the 2 mol of ATP per glucose, glycolysis produces 2 mol of pyruvate and NADH. Glycolysis can continue only if the cell can remove the pyruvate and NADH produced. The fate of these products depends on two factors: the metabolic demands of the cell and the availability of oxygen.

Mitochondria oxidize glycolytic pyruvate and NADH under aerobic conditions

When energy is required and oxygen abundant, pyrvuate produced in glycolysis enters the mitochondria for further oxidation. First, the enzyme pyruvate dehydrogenase (PDH) produces acetyl CoA, which is further oxidized to produce CO_2. The reducing energy (4 NADH and 1 $FADH_2$) and nucleotides (1 GTP) allow mitochondria to produce the equivalent of 15 ATP from pyruvate. Since the cytoplasmic production of pyruvate produces only 1 ATP per pyruvate, considerably more energy is produced by glucose oxidation (glucose $\rightarrow CO_2$) than by glycolysis (glucose $\rightarrow$ pyruvate).

Mitochondria also dispose of the cytoplasmic NADH produced in glycolysis. Although they cannot oxidize NADH directly, mitochondria use two **redox shuttles** to obtain the reducing energy of cytoplasmic NADH: the α-glycerophosphate shuttle and the malate-aspartate shuttle (Figure 2.29). In the α-glycerophosphate shuttle, cytoplasmic NADH is first oxidized by the enzyme α-glycerophosphate dehydrogenase (α-GPDH), embedded within the mitochondrial inner membrane. Oxidation of glycolytic NADH in the α-glycerophosphate shuttle generates two ATP. The malate-aspartate shuttle uses pairs of enzymes that are located in both the cytoplasm and the mitochondria. First, cytoplasmic malate dehydrogenase oxidizes NADH. This transfers the reducing energy of NADH to malate, which

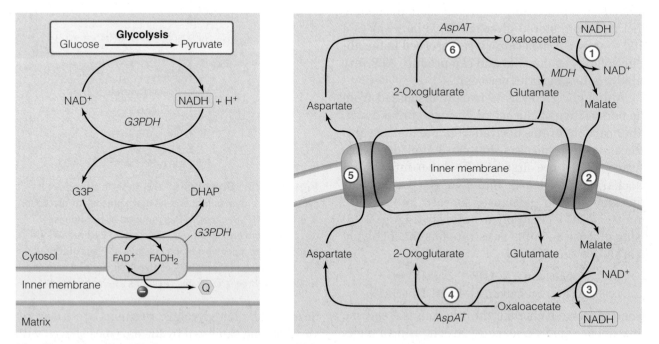

(a) α Glycerophosphate shuttle

(b) Malate-aspartate shuttle

Figure 2.29   Redox shuttles   (a) Glycolytic NADH can be oxidized by the combined actions of cytoplasmic and mitochondrial forms of glycerol 3-phosphate dehydrogenase. The complete α-glycerophosphate shuttle leads to transfer of the reducing power of NADH to mitochondrial $FADH_2$. **(b)** The malate-aspartate shuttle, shown as a series of reactions from 1 to 6, results in net transfer of NADH into the mitochondria with complete cycling of the reactants. The enzymes malate dehydrogenase (MDH) and aspartate aminotransferase (AspAT) are located in both the cytoplasm and mitochondria. This cycle also requires specific transporters capable of carrying metabolites across the mitochondrial membrane.

enters the mitochondria and is oxidized by malate dehydrogenase. The remainder of the cycle ensures that the cytoplasm is provided with adequate oxaloacetate and the mitochondrial oxaloacetate is removed. The enzyme aspartate aminotransferase converts oxaloacetate to aspartate in the mitochondria. Aspartate is exported to the cytoplasm where another aspartate aminotransferase regenerates oxaloacetate. The other substrates and products in the aspartate aminotransferase reaction, glutamate and 2-oxoglutarate, are required in the transport of malate and aspartate.

Terminal dehydrogenases oxidize NADH under anaerobic conditions

Since mitochondria require oxygen to process pyruvate and NADH, the nature of the end products of glycolysis depends on the availability of oxygen. *Environmental hypoxia* arises when external oxygen levels fall below critical levels for prolonged periods. Intertidal bivalves may close their shells during tidal cycles, inducing hours of hypoxia, whereas parasites of the gastrointestinal tract live perpetually under hypoxia. *Functional anoxia* can arise when tissue oxygen demands outstrip oxygen delivery from the blood. For example, muscle can become hypoxic during intense exercise. Diving animals gradually deplete their onboard oxygen stores, causing short-term hypoxia in some tissues. In each of these situations, animals depend on glycolysis for energy and must be able to oxidize NADH to allow glycolysis to continue. One of the most common pathways for NAD^+ regeneration is through the activity of LDH, an enzyme we introduced earlier in this chapter.

$$\text{Pyruvate} + \text{NADH} + \text{H}^+ \rightleftharpoons \text{lactate} + \text{NAD}^+$$

This reaction regenerates NAD^+ and disposes of pyruvate, permitting glycolysis to continue. For many species, lactate production is a good indication of glycolytic flux. Once produced in the LDH reaction, lactate can either be retained in the tissue or exported from the cell into extracellular fluid. Although lactate is slightly toxic, it can be tolerated for short periods. When the anoxia bout ends, lactate is metabolized, and is often used as a substrate to regenerate glucose and glycogen.

The most hypoxia-tolerant and anoxia-tolerant animals use three general mechanisms to extend survival. One is to reduce their metabolic demands to extend the life of their energy stores by entering some form of dormancy or reducing the metabolic demands of specific tissues. For example, a turtle can depress its metabolic rate at the onset of a dive. Second, animals can extend hypoxic survival by storing high levels of glycogen. For some bivalve molluscs, for example, almost half their dry weight is glycogen, providing many days of anoxia tolerance. Third, some anoxia-tolerant organisms alter the nature of glycolysis to produce an alternative end product that is less toxic than lactate. Some molluscs produce strombine, alanopine, or octopine. Some species of fish can convert lactate to ethanol, which is then excreted into the water. This additional reaction spares them the toxicity of lactate, but in the process they lose a valuable source of energy. Bivalve molluscs and some endoparasites also produce alternate end products, but gain extra energy along the way. Phosphoenolpyruvate (PEP) can be diverted from glycolysis to produce succinate (4 ATP/glucose) or propionate (6 ATP/glucose). The various alternative pathways and anaerobic end products are summarized in Figure 2.30.

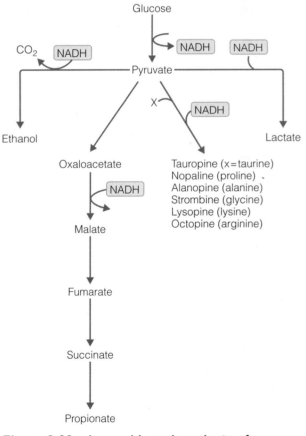

Figure 2.30 Anaerobic end products of glycolysis Animals collectively have many different ways to oxidize NADH when oxygen is limiting. Many animals and tissues rely on lactate dehydrogenase, but other pathways occur in hypoxia-tolerant animals. Some anaerobic end products can lead to production of extra ATP.

Collectively, these variations of glycolysis allow animals to succeed in anoxic environments that are toxic to other species.

Lipids

Lipids are a class of hydrophobic organic molecules including fatty acids, triglycerides, phospholipids, steroids, and steroid derivatives. They have many roles in animal cells, acting as substrates for energy production, building blocks for membranes, and signaling molecules.

Fatty acids are long aliphatic chains produced from acetyl CoA

Fatty acids are long chains of carbon atoms (aliphatic) ending with a carboxyl group (Figure 2.31). They can vary in chain length from two carbons, as with acetate, to more than 30 carbons. The shortest fatty acids are often called volatile fatty acids, or VFAs, because they readily evaporate from solution. VFAs are produced by ruminants with the bacterial fermentation of cellulose.

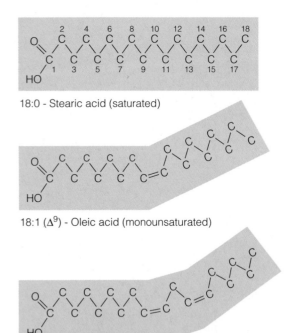

18:0 - Stearic acid (saturated)

18:1 (Δ^9) - Oleic acid (monounsaturated)

18:2 ($\Delta^{9,12}$) - Linoleic acid (polyunsaturated)

Figure 2.31 Fatty acids Saturated fats such as stearic acid (18:0) are linear in structure. Addition of a double bond, as shown with the monounsaturated fatty acid oleic acid (18:1), introduces a bend in the structure. The second double bond shown in the polyunsaturated acid linoleic acid (18:2) causes further disruption of the linear structure.

Medium chain fatty acids (MCFAs) and long chain fatty acids (LCFAs) are common in energy stores and as part of phospholipids that make up cell membranes. Fatty acids also differ in the number and position of double bonds between carbon atoms. *Saturated* fatty acids have no double bonds and are linear in structure. The introduction of a double bond into a linear fatty acid causes a bend in the chain, which has important consequences for membrane structure. Monounsaturated fatty acids have one double bond. Polyunsaturated fatty acids, or PUFAs, possess multiple double bonds.

Fatty acid nomenclature considers both the chain length and the number of bonds. Palmitic acid is denoted as 16:0, meaning it is 16 carbons long and has no double bonds. There are two naming systems to denote fatty acids, which differ in how they identify the location of the first double bond. In the delta (Δ) system, a number corresponds to location of the double bond relative to the carboxyl carbon; in the omega (ω) system, the number refers to the distance from the methyl end of the fatty acid. Thus, the 18-carbon fatty acid oleic acid can be either 18:1 Δ^9 or 18:1 ω9. Linoleic acid is denoted as either 18:2 $\Delta^{9,12}$ or 18:2 ω6.

Animals can produce many fatty acids using the enzyme *fatty acid synthase*, which cyclically adds two-carbon units to the fatty acid. Though fatty acids grow by adding acetyl groups, malonyl CoA, a three-carbon activated fatty acid is the actual substrate for the enzyme fatty acid synthase. Malonyl CoA is produced by acetyl CoA carboxylase. Using the reducing energy of NADPH, fatty acid synthase repeatedly adds acetyl CoA groups to the fatty acid. After seven cycles, when the fatty acid has grown to 16 carbons, palmitate has been produced and is released by the enzyme. The overall reaction for palmitate synthesis is

$$\text{Acetyl CoA} + 7\text{malonyl CoA} + 7\text{NADPH} + 7\text{H}^+ \rightarrow \text{palmitate} + 7\text{NADP}^+$$

Though palmitate is the product of fatty acid synthase, accessory enzymes process much of it to produce other fatty acids. These enzymes elongate the carbon chain and introduce double bonds to produce the other important fatty acids, such as oleic acid. Many animals can produce all of the fatty acids needed for growth, but some animals are incapable of producing specific fatty acids and must obtain these in the diet. For example, humans have a dietary requirement for linoleic acid (18:2 ω6) and linolenic acid (18:3 ω3).

Fatty acids are oxidized in mitochondrial β-oxidation

Fatty acids are an important fuel for many tissues, such as the mammalian heart, which typically derives more than 70% of its energy from fatty acid oxidation. The fatty acid oxidation pathway occurs primarily in the mitochondria and results in the production of acetyl CoA. Depending on the conditions, this acetyl CoA can be oxidized by mitochondria or be diverted to other pathways. Fatty acids can have many structures, differing in chain length, branching patterns, and desaturation. These variations require side reactions to convert the fatty acids to forms that can enter β-oxidation. We will focus on the pathway for oxidation of palmitate, but along the way we will identify some of the alternate pathways used to process other fatty acids.

Because the actual substrate for β-oxidation is fatty acyl CoA, cells must first convert fatty acids to their CoA esters using a fatty acyl CoA synthase. Short and medium chain fatty acids are able to enter the mitochondria directly, where they are activated by a mitochondrial fatty acyl CoA synthase. Palmitate, which cannot cross into mitochondria, is oxidized by the mitochondria by a multistep process involving activation and transport. The fatty acid is converted to fatty acyl CoA. Next, the enzyme carnitine palmitoyl transferase-1, or CPT-1, replaces the CoA with carnitine, forming fatty acyl carnitine, which is carried into the mitochondria, where another enzyme, CPT-2, converts it back to fatty acyl CoA. This elaborate transport scheme provides an extra level of control over long chain fatty acid oxidation. By regulating the activity of CPT-1, cells control how much fatty acid can enter the mitochondria for catabolism.

Once inside the mitochondria, fatty acids enter the β-oxidation pathway (Figure 2.32). This is a cyclical pathway that sequentially cuts pairs of carbons off the end of the fatty acid in the form of acetyl CoA. The shortened fatty acid returns to the pathway, and the cycle is repeated until the entire fatty acid is broken down to acetyl CoA. With each trip through the pathway, reducing equivalents are produced at two enzymatic steps: fatty acyl CoA dehydrogenase produces $FADH_2$, and β-hydroxyacyl CoA dehydrogenase produces NADH. About 30% of the energy liberated from fatty acids is derived from the reducing equivalents produced in β-oxidation. The remaining 70% derives from oxidation of acetyl CoA in the TCA cycle.

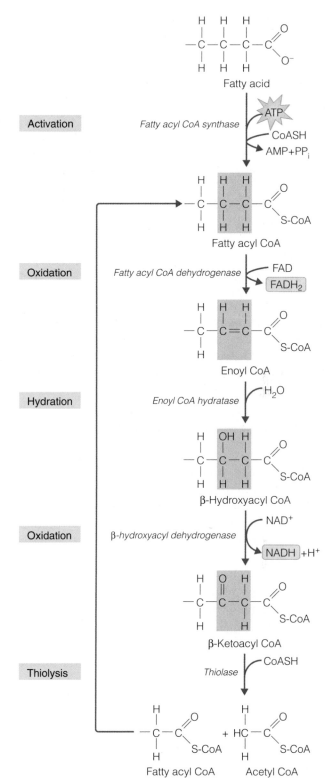

Figure 2.32 Fatty acid oxidation Fatty acids are activated in the cytoplasm to form fatty acyl CoA. Once transported into mitochondria, fatty acyl CoA enters the β-oxidation pathway. Oxidation, hydration, oxidation, and thiolysis produce acetyl CoA, reducing equivalents, and a fatty acyl CoA shortened by two carbons. The shortened fatty acyl CoA reenters the β-oxidation pathway and the cycle repeats until the fatty acid is reduced to acetyl CoA units.

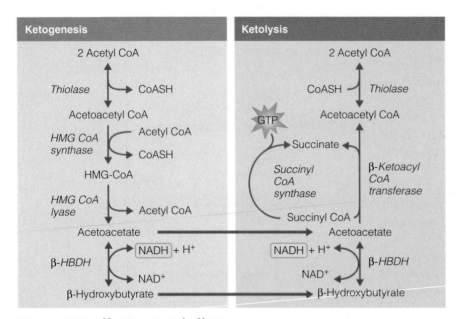

Figure 2.33   Ketone metabolism   Acetyl CoA can be converted to the ketone bodies acetoacetate and β-hydroxybutyrate. This reaction normally occurs in specific ketogenic tissues, such as liver. Ketone bodies are released through the blood for uptake by ketolytic tissues, such as brain. Acetyl CoA is resynthesized at the cost of one GTP to regenerate the substrate succinyl CoA.

Fatty acids can be converted to ketone bodies

Fatty acids are valuable sources of energy, but under some conditions they must first be processed into ketone bodies: acetone, acetoacetate, and β-hydroxybutyrate (Figure 2.33). Ketone bodies provide a fuel for tissues that cannot use fatty acids directly. The mammalian brain usually relies on glucose oxidation for energy, but after an extended period of food deprivation, glucose levels may decline, forcing tissues to rely more on lipid stores. Since the brain cannot use fatty acids directly, the liver converts the fatty acids to ketone bodies, which can be transported into the brain and oxidized.

The first step in ketone body synthesis, or **ketogenesis**, is the production of acetoacetyl CoA from two molecules of acetyl CoA, catalyzed by thiolase. This is the same enzyme used in the final step of β-oxidation, but in ketogenesis it operates in the reverse direction. After condensation with another acetyl CoA and subsequent hydrolysis, acetoacetate is formed. Acetoacetate can then be converted to β-hydroxybutyrate by the enzyme β-hydroxybutyrate dehydrogenase (β-HBDH), or it can break down spontaneously to form acetone. In the target tissues, **ketolysis** reconverts β-hydroxybutyrate and acetoacetate to acetyl CoA. Acetoacetate is ac-

tivated into the CoA ester, then hydrolyzed by thiolase to form two acetyl CoA molecules.

Ketone bodies are a useful alternative to fatty acids for many animals. Although some energy is lost in the complete cycle of ketogenesis and ketolysis, for some tissues, particularly under starvation conditions, ketone bodies are the only metabolic energy source available. Chondrichthians (sharks and their relatives) in fact appear biochemically predisposed to using ketone bodies as their "lipid" fuel. Unlike in other vertebrates, their muscles are unable to use fatty acids directly, instead relying on ketone bodies as a fuel for energy.

Triglyceride is the major form of lipid storage

Most fatty acids in animal cells are esterified to a glycerol backbone. A monoacylglyceride has a single fatty acid esterified to glycerol, typically at the first position of the glycerol molecule. Diacylglyceride has fatty acids esterified to the first and second position of glycerol. Triglyceride has three fatty acids esterified to the glycerol backbone (Figure 2.34). Each of these terms—monoacylglycerides, diacylglycerides, and triglycerides—refers to a class of molecules. For example, hundreds of chemically distinct triglyceride molecules can be constructed by using different fatty acids in each of the three positions on the glycerol backbone.

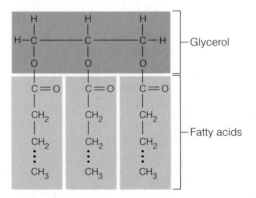

Figure 2.34   Triglycerides   Triglycerides are composed of three fatty acids esterified to a glycerol backbone. Fatty acids can vary in chain length and number of double bonds.

Triglycerides are vital energy stores for animals, and can be found in high concentrations in lipid storage tissues in the form of lipid droplets. In insects, a tissue called the fat body is the main site of lipid storage. Many other invertebrates, such as molluscs and crustaceans, store lipid in a large hepatopancreas. Vertebrates store triglyceride in liver, muscle, and adipose tissue, such as blubber. Adipocytes, the cells within adipose tissue, store triglyceride when an animal is well fed, then release lipids when the animal needs extra fuel.

Triglyceride synthesis, or **lipogenesis**, is a multistep process overlapping with phospholipid synthesis (Figure 2.35). Each fatty acid is activated into its CoA ester by fatty acyl CoA synthase. Starting with glycerol 3-phosphate, the fatty acids are added sequentially to carbon 1, then carbon 2, forming a phosphatide. After removal of the phosphate group, diacylglycerol is formed. Addition of a third fatty acid completes the triglyceride molecule.

Triglyceride breakdown, or **lipolysis**, requires enzymes called **lipases** that attack the triglyceride molecule, breaking the bond between the fatty acid and the glycerol backbone. Hormone-sensitive lipase liberates fatty acids from triglycerides and diacylglycerides. Another lipase, monocyglyceride lipase, completes the breakdown of the triglyceride by releasing the last fatty acid from the glycerol backbone. The liberated fatty acids are either used directly within the cell or transferred to the circulation for uptake by other tissues that use them for energy metabolism.

The balance between triglyceride synthesis and degradation is carefully controlled within animals. Lipolysis does not directly generate energy, but lipogenesis requires energy. As with other pathways we have discussed, if both synthesis and degradation occurred simultaneously, cells would waste energy.

Phospholipids predominate in biological membranes

In addition to their role in energy metabolism, fatty acids are vital components of the phospholipids used to produce biological membranes. Animal cells produce two classes of phospholipids: phosphoglycerides and sphingolipids (Figure 2.36).

Phosphoglycerides are constructed from phosphatides, an intermediate in triglyceride synthesis. In phospholgylceride synthesis, the phosphate group links the phosphatide to a polar head group, such as serine, choline, ethanolamine, and inositol. The physical properties of a phosphoglyceride are determined by the properties of both the fatty acids (chain length, saturation) and the polar head group.

Although **sphingolipids** are chemically very different from phosphoglycerides, they have similar three-dimensional shapes. The backbone of a sphingolipid is sphingosine. With its long aliphatic

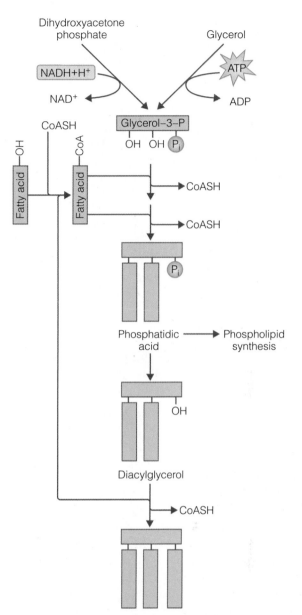

Figure 2.35   Triglyceride synthesis   Glycerol 3-phosphate, produced from glycolysis (dihydroxyacetone phosphate) or glycerol, is the acceptor for two sequential additions of activated fatty acids (fatty acyl CoA). The formation of triglyceride requires dephosphorylation and addition of another fatty acid group to the third and last position on the glycerol backbone.

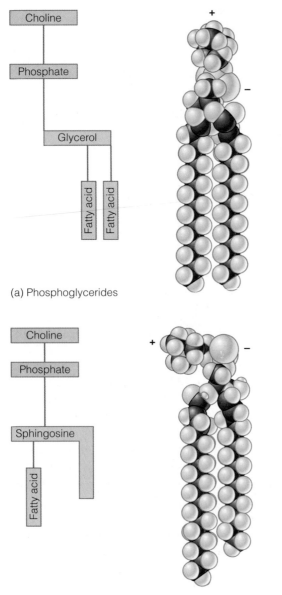

(a) Phosphoglycerides

(b) Sphingolipids

Figure 2.36 Phospholipids Phospholipids, including **(a)** phosphoglycerides and **(b)** sphingolipids, share a similar three-dimensional structure. They are built on different backbones: glycerol for phosphoglycerides and sphingosine for sphingolipids.

chain, its structure is similar to monoacylglycerol. Ceramide is formed when a fatty acid is esterified to sphingosine, creating a structure that resembles diacylglycerol. Many different types of sphingolipid can be constructed by attaching different polar head groups to ceramide. When phosphocholine is attached to ceramide, sphingomyelin is formed. Carbohydrates can be attached to ceramide to form neutral glycolipids and gangliosides. Sphingolipids are most often found on the extracellular side of the cell membrane. Each cell makes a specific combination of sphingolipids, providing a kind of cellular signature that other cells can recognize. During development, when cells move throughout the body to begin the process of tissue formation, sphingolipid signatures provide migrating cells with landmarks. When the migrating cells find the correct sphingolipid signature, they cease migration and differentiate to form tissues.

Phospholipids are broken down by **phospholipases**, many of which are important in cell signaling cascades. Each type of phospholipase attacks a specific region of a phospholipid molecule. Phospholipase A (PLA) breaks the ester bonds that connect the fatty acids to the glycerol backbone. PLA_1 releases the fatty acid from the first carbon of glycerol, whereas PLA_2 releases the fatty acid from the second position. Phospholipases B and C break different phosphodiester bonds between the polar head group and the glycerol backbone. When PLB attacks phosphatidyl inositol, inositol and phosphatide are produced. When PLC attacks the same phospholipid, inositol phosphate and diacylglyceride are released. Regulation of PLC, an important enzyme in signal transduction pathways of many cells, will be discussed in more detail in Chapter 3: Hormones and Cell Signaling.

Steroids share a multiple ring structure

Steroids are a collection of lipid molecules that share a basic aromatic structure of four hydrocarbon rings. The steroid cholesterol is found in many cellular membranes and is part of the lipoprotein complexes that transport lipids through the blood. It is also a precursor for synthesis of the vertebrate steroid hormones. Although invertebrates don't possess steroid hormones, some use a steroid-like hormone, ecdysone, to control maturation and development.

The pathways of steroid synthesis involve nonsteroid intermediates (Figure 2.37). Steroid synthesis begins when acetate is used to produce mevalonate, the precursor for activated isoprene. Activated isoprene is the precursor for many familiar molecules, such as carotenoids and vitamins A, E, and K. Ubiquinone, a type of quinone, is an important component in mitochondrial energy production. Activated isoprene is also used to produce isoprenoids that act as hormones, including insect juvenile hormone and pheromones. Further along the pathway of cholesterol synthesis is squalene, a steroid used by sharks to aid in buoyancy. Cholesterol is the precursor for the many steroid hormones that we discuss in later chapters.

Mitochondrial Metabolism

Mitochondria process metabolites generated in the cytoplasm, breaking them down to capture their chemical energy in the form of ATP. The main point of entry for mitochondrial energy-producing pathways is acetyl CoA, which as you've learned earlier in this chapter is produced in many pathways (Figure 2.38). Acetyl CoA enters the cyclical **tricarboxylic acid cycle (TCA cycle)** and is oxidized to form reducing equivalents (NADH, FADH$_2$), which provide the fuel for mitochondrial ATP production.

The TCA cycle uses acetyl CoA to generate reducing equivalents

Once acetyl CoA is produced within mitochondria, its fate depends on intracellular conditions. When cells need energy, acetyl CoA enters the TCA cycle, where its oxidation ultimately leads to ATP production. The TCA cycle (Figure 2.39) consists of eight enzymes that collectively catalyze the following reaction:

$$\text{Acetyl CoA} + 3\text{NAD}^+ + \text{GDP} + \text{P}_i + \text{FAD} \rightarrow$$
$$2\text{CO}_2 + 3\text{NADH} + \text{FADH}_2 + \text{GTP}$$

The four dehydrogenases in the TCA cycle produce reducing equivalents: NADH is produced by isocitrate dehydrogenase, 2-oxoglutarate dehydrogenase, and malate dehydrogenase; FADH$_2$ is produced by succinate dehydrogenase. Most of the ATP produced through acetyl CoA oxidation comes from the subsequent oxidation of NADH and FADH$_2$. The TCA cycle also produces one molecule of GTP, which is energetically equivalent to ATP. This reaction, catalyzed by succinyl CoA synthase, is an example of **substrate-level phosphorylation**. The TCA cycle is not really an isolated pathway so much as a collection of enzymes acting on a pool of metabolites exchanged with other pathways. When intermediates are removed for other reactions, cells use **anaplerotic pathways** to regenerate the intermediates.

Cells control the rate of the TCA cycle in three ways: by regulating the concentrations of reactants (substrates and products), the levels of the

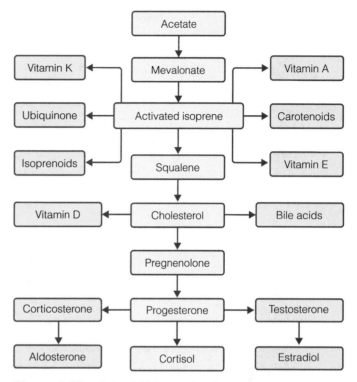

Figure 2.37 Steroid biosynthesis This simplified pathway of steroid synthesis illustrates the many intermediates that are used by cells.

enzymes, and the catalytic activity of enzymes. Tissues that use a lot of energy, such as heart and brain, have high levels of the TCA enzymes. In many tissues, the flux through the TCA cycle is affected by the levels of acetyl CoA and oxaloacetate, as well as other intermediates in the cycle. When tissues have abundant energy, they typically use acetyl CoA and intermediates as biosynthetic substrates. Acetyl CoA is an important substrate in fatty acid synthesis, and oxaloacetate is a substrate for glucose synthesis. When biosynthetic reactions deplete these substrates, the rate of the TCA cycle declines. Allosteric effectors also regulate the TCA cycle. Calcium, frequently elevated

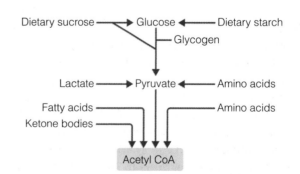

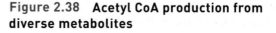

Figure 2.38 Acetyl CoA production from diverse metabolites

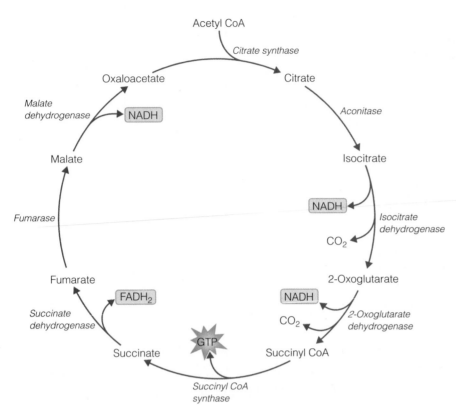

Figure 2.39 Tricarboxylic acid cycle The enzymes of the TCA cycle oxidize acetyl CoA to release its energy in the form of reducing equivalents (3 NADH, 1 FADH$_2$) and 1 GTP.

during periods of high metabolic demand, stimulates isocitrate dehydrogenase and 2-oxoglutarate dehydrogenase, increasing the rate of NADH production to help the cell meet its energy demands.

The ETS generates a proton gradient, heat, and reactive oxygen species

Mitochondria use reducing equivalents as the substrate for **oxidative phosphorylation**, a complex pathway that combines oxidation by the **electron transport system (ETS)** with phosphorylation (Figure 2.40). The ETS builds an electrochemical gradient that can be used to drive ATP synthesis and energy-dependent transport processes. Found within the inner mitochondrial membrane, the ETS consists of four multisubunit proteins (complexes I, II, III, and IV) and two electron carriers (ubiquinone, cytochrome *c*).

Although electrons can enter the ETS in several ways, each pathway converges on the first mobile carrier, ubiquinone. The NADH produced in the TCA cycle passes electrons to complex I, which in turn reduces ubiquinone. Several FADH$_2$-linked enzymes found in the inner mitochondrial membrane

pass electrons directly to ubiquinone. For example, the TCA cycle enzyme succinate dehydrogenase is actually complex II of the ETS. An FAD group within its structure becomes reduced during oxidation of succinate, forming FADH$_2$. The enzyme in turn passes the electrons from FADH$_2$ to ubiquinone. Once reduced, ubiquinone transfers its electrons to complex III, which in turn transfers electrons to cytochrome *c*. Complex IV, or cytochrome *c* oxidase, accepts electrons from cytochrome *c* and passes them on to molecular oxygen. Complete reduction of oxygen (O$_2$) requires four electrons from cytochrome *c* and consumes four protons to produce two water molecules.

Complexes I, III, and IV are also proton pumps. When these complexes transfer energy to the next carrier, enough free energy remains to pump protons out of the mitochondrial matrix. Fewer protons are pumped in the oxidation of FADH$_2$ because these enzymes pass their electrons directly to ubiquinone, bypassing the proton-pumping complex I. The proton gradient formed by the ETS has both electrical and chemical components: a pH gradient (ΔpH) and a membrane potential ($\Delta\Psi$). This **proton motive force** is potential energy that can be used to drive other processes, such as ATP synthesis.

The ETS converts much of the energy liberated from NADH oxidation to the proton motive force. Some energy is "lost" in the formation of two by-products: heat and **reactive oxygen species (ROS)**. Conditions that increase electron flow and oxygen consumption also increase heat production. ROS production is an inevitable consequence of electron movement throught the ETS. Usually more than 99% of the electrons that enter the ETS make the journey all the way to the end of the chain, forming water. However, a few are stolen from the ETS by molecular oxygen to form superoxide (O$_2^-$), a potent ROS that can attack chemical bonds, damaging macromolecules such as lipids, proteins, and DNA. Cells possess vigorous antioxidant defense mechanisms to inactivate superoxide before it can cause damage. The enzyme

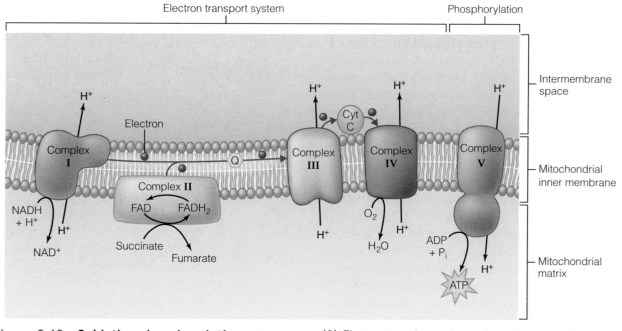

Figure 2.40 **Oxidative phosphorylation** Complex I collects electrons from NADH produced by various mitochondrial dehydrogenases. Complex II, or succinate dehydrogenase, transfers electrons from succinate to FAD. Both complex I and II, as well as other FAD-link dehydrogenases not shown, transfer electrons to ubiquinone (Q). Electron transfer continues through complex III, cytochrome *c*, and finally complex IV, cytochrome oxidase. During electron transport, complexes I, III, and IV also pump protons out of the mitochondrial matrix, creating the proton motive force (Δp). The mitochondrial F_1F_O ATPase (complex V) uses Δp to fuel ATP synthesis.

superoxide dismutase, or SOD, consumes superoxide to produce hydrogen peroxide (H_2O_2), which is less toxic than superoxide. Other antioxidant enzymes, such as catalase and glutathione peroxidase, consume H_2O_2, preventing it from causing cellular damage.

The F_1F_O ATPase uses the proton motive force to generate ATP

To this point we have discussed how mitochondria build Δp but not how it is used to produce ATP. Mitochondria possess an ATP synthase, usually called the F_1F_O ATPase, that uses the energy contained in Δp to drive the phosphorylation of ADP. (Although it normally functions in the direction of ATP synthesis, the F_1F_O ATPase is reversible and able to break down ATP under some conditions.) The F_1F_O ATPase possesses a proton-pumping region and a catalytic region. When protons pass through the enzyme, which spans the mitochondrial inner membrane, the energy is used to catalyze the synthesis of ATP. Oxidative phosphorylation is the combination of oxidation by the ETS and phosphorylation by F_1F_O ATPase. Note that there is no physical linkage between oxidation and phosphorylation; the two processes are functionally coupled through a mutual dependence on Δp.

The rate of ATP synthesis by the F_1F_O ATPase depends on the magnitude of Δp and the availability of the substrates ADP and inorganic phosphate (P_i). When cells are rapidly hydrolyzing ATP, [ADP] and [P_i] increase, accelerating the rate of the F_1F_O ATPase reaction. To understand how this process is regulated, consider what happens in a muscle that goes from rest to exercise. At rest, the rate of ATP breakdown is slow and [ATP] builds up while [ADP] and [P_i] decline. The ETS builds Δp to its maximum because the ATPase, the major drain on the gradient, is inhibited. With little flux through the ETS, the rate of respiration is low. This is the biochemical reason why you breathe less when resting. When muscle activity begins, ATP is hydrolyzed and the concentrations of ADP and P_i increase. With the stimulation of the ATP synthase, Δp is depleted and ETS accelerates to replenish the gradient. We increase our oxygen consumption during exercise because of this linkage between ATP synthesis and oxidation.

The functional linkage between oxidation and phosphorylation depends on the integrity of the inner mitochondrial membrane. All membranes are somewhat permeable to protons, but the inner mitochondrial membrane is relatively resistant to proton leak. If the protons pumped out of the mitochondria by the ETS were to leak back into the mitochondria, Δp would be dissipated, causing two effects on oxidative phosphorylation. First, the ETS would continue at a high rate, pumping protons and consuming oxygen in a futile effort to rebuild Δp. Second, the reduction in Δp would prevent the mitochondria from producing ATP. Mitochondria that show high rates of respiration with no ATP production are considered uncoupled. While this state is disastrous for energy production, it is an important mechanism to produce heat. Some mammals have specific proteins that facilitate the movement of protons across the inner membrane. As you will learn in Chapter 13: Thermal Physiology, these *uncoupling proteins* are important in mammals that experience cold stress, such as newborns and hibernators.

Creatine phosphokinase enhances energy stores and transfer

Some of the energy stored first as ATP is used to produce other high-energy phosphate compounds of equivalent energy, such as phosphocreatine. A vertebrate muscle, for example, may have 5–10 times more phosphocreatine than ATP, serving as an energy store. When the muscle begins to work at high intensity, ATP is consumed to support muscle activity. The resulting decrease in [ATP] and increase in [ADP] drive the creatine phosphokinase (CPK) reaction in the direction of ATP production, at the expense of phosphocreatine. As muscle activity continues, the phosphocreatine pool is gradually depleted, allowing muscles to preserve ATP at normal levels for longer periods. Whereas the existing pool of ATP is sufficient for only a few seconds of activity, the large phosphocreatine pool allows muscle to maintain ATP levels and sustain contractions for a much longer duration.

In addition to bolstering energy stores, phosphocreatine is a component of the phosphocreatine shuttle, a pathway that improves the efficiency of energy transfer within the cell (Figure 2.41). The cycle begins with ATP produced by the mitochondria. CPK on the outer mitochondrial membrane uses this ATP to phosphorylate creatine. The phosphocreatine diffuses from the mitochondria to the myofibrils where another CPK uses the phosphocreatine to regenerate ATP. This CPK

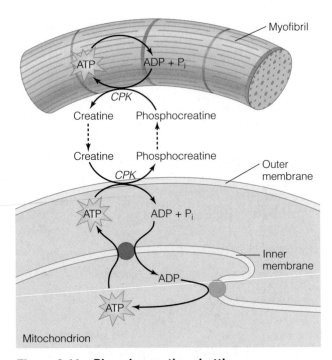

Figure 2.41 Phosphocreatine shuttle
Phosphocreatine is produced by the mitochondrial enzyme creatine phosphokinase (CPK) and diffuses to the myofibrils, where another CPK enzyme regenerates the ATP.

shuttle improves the efficiency of energy transfer two ways. First, creatine and phosphocreatine are smaller molecules than the adenylates and have higher diffusion coefficients. Second, the absolute concentrations of creatine and phosphocreatine are much greater than those of the adenylates, allowing much steeper intracellular gradients to form, which accelerates the rates of diffusion.

Integration of Pathways of Energy Metabolism

The metabolic traits exhibited by whole animals can be traced back to the cellular pathways of energy metabolism. Put simply, in order to remain in energetic balance, cells must produce ATP at rates that match the ATP demand. Consider the situation in a mammalian heart cell. We know from rates of oxygen consumption that the heart consumes ATP at a rate of about 30 µmol/min/g. Since the concentration of ATP doesn't change during normal activity, the heart cell must also produce ATP at the same rate (30 µmol/min/g). Since the concentration of ATP is only about 5 µmol/g, at a metabolic rate of 30 µmol/min/g, the heart turns over the entire ATP pool several times per minute.

At this **turnover rate**, a cell that produces ATP at a rate only 10% less than the rate of demand would be depleted of ATP within two minutes. At the cellular level, the balance between energy-consuming pathways and energy-producing pathways is orchestrated by the diverse regulators of metabolism, such as adenylates, redox balance, Ca^{2+}, and carbon supply. These regulators "inform" the cell of energy needs, and the metabolic pathways respond accordingly. What is interesting about multicellular animals is that the needs of the cell are often superceded by the needs of the whole organism. A liver cell, for example, not only responds to its own metabolic needs but produces energy substrates for the entire animal. When glucose levels are low, the liver increases the rates of gluconeogenesis and glycogenolysis, releasing glucose to the blood for use in other tissues. This altruistic response is imposed on the liver cell by hormones that affect the catalytic properties of the enzymes of intermediary metabolism, largely through covalent modification, such as phosphorylation. We will discuss the nature of hormones in the next chapter, Hormones and Cell Signaling.

The sum of the cellular metabolic properties yields the whole animal metabolic patterns. Variations in animal metabolic rate are central to many problems in animal physiology. You learned in Chapter 1 about *allometric scaling* of metabolic rate: large animals have lower mass-specific metabolic rates than small animals. In later chapters you will learn how animals control metabolic rate to survive challenging conditions, such as environmental hypoxia, hypothermia, and dehydration. In the accompanying feature, we describe the ways physiologists measure whole animal metabolic rate in the lab and in the field (see Box 2.2, Methods and Model Systems: Metabolic Rate).

Metabolic strategies in animals address the constant fluctuations in nutrient availability, energy demand, and environmental conditions. Ensuring the correct flow of energy requires exquisite control of the pathways of intermediary metabolism. Opposing pathways must be reciprocally regulated to avoid a **futile cycle**: simultaneous synthesis and degradation of a metabolite. Similarly, the various alternatives must be utilized in a way that takes into consideration the advantages and disadvantages of each class of fuel. These choices are also influenced by long-term and short-term metabolic priorities of the cells and organisms.

Oxygen and ATP control the balance between glycolysis and OXPHOS

Glycolysis produces 2 mol of ATP for every mol of glucose, but glucose oxidation yields 36 mol of ATP per mol of glucose. Despite the differences in energy yield, glycolysis and oxidative metabolism both play important roles in energy metabolism. Glycolysis, in addition to being able to operate without oxygen, can produce ATP at much greater rates than can oxidative metabolism. The conditions that allow such high rates also require the pathway to be somewhat inefficient. In contrast, oxidative metabolism is very efficient in conserving chemical energy, but to do so, it is necessarily slow. Think of glycolysis as a high-performance sports car—useful to get somewhere fast but not the best car for gas mileage. Oxidative metabolism, by contrast, is the fuel-efficient compact car. Like some suburban families, the cell maintains both types of engines in working order, to be called upon for different needs.

Physical properties of fuels influence fuel selection

Each of the major metabolic fuels displays physical properties that influence how the fuel is stored and used. Carbohydrate is stored as large granules of glycogen, coated with water molecules that make up its hydration shell. Glycogen particles can be so large that they interfere with cellular processes. Some tissues, such as the mantle of bivalve molluscs, can safely accumulate high levels of glycogen, but if glycogen reached this high level in a muscle it would prevent the muscle from contracting normally. Although glycogen is readily mobilized, its physical properties prevent most cells from storing high levels. In contrast, lipid is stored at much higher levels in the form of anhydrous, amorphous droplets of triglyceride. These physical differences, coupled with the energy content of a given mass of stored fuel, affect fuel selection. Cells can obtain about 10 times more ATP from lipid than from the same mass of hydrated glycogen particles. Given the advantage of lipid as an energy store, you might wonder why animals use glycogen at all. Recall that glycogen can be mobilized much faster than lipid, and plays a vital role under conditions in which energy is required very quickly, as in the "fight-or-flight" response. Most cells use a combination of lipid and carbohydrate fuels to balance the advantages and disadvantages of each fuel.

BOX 2.2 METHODS AND MODEL SYSTEMS
Metabolic Rate

If *metabolism* is literally the sum of all chemical reactions, then what exactly is meant by metabolic rate and how is it measured? Metabolic rate is the overall flux through the pathways of energy production, which matches the rate of energy consumption. In the presence of oxygen, the pathways that lead to production of ATP consume carbon fuels and oxygen (O_2), and produce heat, carbon dioxide (CO_2), and water. Many physiological questions revolve around changes in metabolic rate, and many approaches have been developed to measure the substrates and products of metabolism. **Direct calorimetry** assesses metabolic rate in terms of heat production, measured in energy units (joules). For purely pragmatic reasons (direct calorimetry requires expensive, specialized equipment), this is the least common way to measure metabolic rate. A more common approach to measuring metabolic rate is **indirect calorimetry**, in which the researcher measures the rate of O_2 consumption or CO_2 production. To infer a metabolic rate from these measurements, it is important to recognize where O_2 is consumed (largely in the electron transport system) and where CO_2 is produced (primarily in the TCA cycle). The quantitative relationship between these three estimates of metabolic rate—heat production, O_2 consumption, and CO_2 production—depends on many factors. You learned earlier in this chapter that the ratio of CO_2 produced/O_2 consumed reflects the metabolic fuel. Likewise, the oxycaloric relationships (O_2 consumed/joules released) depend on the nature of the fuel.

Each of these approaches for measuring metabolic rate requires that the researcher hold an animal under controlled conditions and ensure that it remains in a constant, stable condition. A researcher interested in **resting metabolic rate (RMR)** must ensure that the animal is unstressed and inactive at a neutral ambient temperature and has digested its most recent meal. While this approach yields important information about the physiological hardwiring of an animal, it may not be the best estimate of the animal's metabolic rate under normal conditions. Ecological physiologists are often more interested in long-term metabolic rate of free-ranging animals in the natural setting. One of the most common approaches to measuring field metabolic rate is the **doubly labeled water** method. Most water in the body is composed of the most common isotopes of hydrogen (^1H) and oxygen (^{16}O). To initiate a doubly labeled water experiment, the animal of interest is captured and injected small volumes of water composed of less common isotopes of hydrogen (e.g., ^2H) and oxygen (^{18}O). Over time, the labeled hydrogen is lost from the body primarily as water, through evaporation, respiration, and excretion. Likewise, labeled oxygen is also lost in water, but some is exchanged with the O in CO_2. Thus, the difference between the loss of labeled oxygen and labeled hydrogen reflects the rate of CO_2 production. This method works very well in air-breathing animals, but in water breathers, the rates of water flux are much too great to detect the impact of CO_2 production on isotope ratios.

References

○ Hulbert, A. J., and P. L. Else. 2004. Basal metabolic rate: History, composition, regulation, and usefulness. *Physiological and Biochemical Zoology* 77: 869–876.

○ Nagy, K. A. 2005. Field metabolic rate and body size. *Journal of Experimental Biology* 208: 1621–1625.

The main way the cells regulate the balance between fatty acids and carbohydrates is through the mitochondrial enzyme pyruvate dehydrogenase (PDH). This enzyme is regulated allosterically by ATP, acetyl CoA, and NADH. When cells have fatty acids available, their oxidation increases concentrations of ATP, NADH, and acetyl CoA. These metabolites inhibit PDH, sparing pyruvate for gluconeogenesis. When energy stores are depleted, the concentrations of NADH, ATP, and acetyl CoA tend to decrease, which lessens the inhibition of PDH. These same metabolites also influence the phosphorylation state of PDH by regulating the activities of PDH kinase (PDHK) and PDH phosphatase (PDHP). ATP, NADH, and acetyl CoA each activate PDHK, causing PDH to be converted to its inactive, phosphorylated form. The activity of PDHP, in contrast, is governed primarily by Ca^{2+}. High $[Ca^{2+}]$ stimulates PDHP, converting PDH to its active dephosphorylated form.

Fuel selection can be calculated from the respiratory quotient

Each pathway for oxidation of fuels demonstrates characteristic relationships between the amount of (1) ATP produced, (2) oxygen consumed, and (3) CO_2 generated. The reason these parameters differ among fuels can be traced back to the pathways of degradation. Ratios of different combinations of these three parameters provide important information about fuel selection. The differences in the ratio of ATP produced to oxygen consumed (the ATP/O ratio) can be traced to the reliance on FAD-linked enzymes. Each time an NADH molecule is produced in the mitochondria, oxidative phosphorylation can produce 3 molecules of ATP and consume 1 atom of oxygen (ATP/O = 3). When a molecule of $FADH_2$ is produced, only 2 molecules of ATP can be generated while consuming the same 1 atom of oxygen (ATP/O = 2). Carbohydrate oxidation uses predominantly NADH-linked enzymes, whereas lipid oxidation relies more heavily on FAD-linked enzymes. Because of this difference, carbohydrate yields more ATP for a given volume of oxygen. This difference has an effect on the fuel preference of at least some animals that live at low oxygen levels. For example, the heart of most humans uses lipid as a major fuel. In contrast, humans that have adapted to high altitude, such as the natives of the high Andes, rely more heavily on glucose oxidation. Of course, in more extreme hypoxia and anoxia, animals have little choice but to rely on glycolysis.

Differences in the ratio of CO_2 produced to O_2 consumed, known as the **respiratory quotient (RQ)**, arise from the pathways of oxidation. Glucose has six carbons, and oxidizing it completely to 6 CO_2 yields 2 NADH in the cytoplasm, 2 NADH via PDH, 6 NADH via the TCA cycle, and 2 $FADH_2$ via succinate dehydrogenase. The 12 reducing equivalents consume 12 atoms of oxygen, or 6 molecules of O_2. Thus, carbohydrate oxidation yields an RQ of 1 (6 mol of CO_2 to 6 mol of O_2). In contrast, oxidation of fatty acids generates an RQ of about 0.7, although the exact number depends on the specific fatty acid. Consider the pathway of palmitate oxidation. As a 16-carbon fatty acid, it generates 16 mol of CO_2 per mol of palmitate. Seven cycles of β-oxidation are required to break palmitate into 8 molecules of acetyl CoA, yielding 7 $FADH_2$ and 7 NADH. Oxidation of the 8 acetyl CoA in the TCA cycle yields 24 NADH and 8 $FADH_2$. Oxidation of the 46 reducing equivalents consumes 23 mol of O_2, giving an RQ of 0.7 (16 mol of CO_2 to 23 mol of O_2). Because of these characteristic relationships between RQ and fuel oxidation, measurement of CO_2 production and O_2 consumption of whole animals can provide important insight into the pathways that are being used to support energy metabolism.

Energetic intermediates regulate the balance between anabolism and catabolism

A cell activates pathways of energy production when it needs energy, but when energy is abundant it stimulates anabolic pathways, storing nutrients or producing building blocks for cell growth or cell division. How do cells actually sense the need for energy and regulate the transition between catabolism and anabolism? Many of the pathways we have discussed are sensitive to the cellular indices of energetic status, primarily acetyl CoA, adenylates, and NADH. Changes in the concentration of these products reflect energy status and cause compensatory changes in metabolic pathways. When cells are "energy-rich," the concentrations of acetyl CoA, NADH, and ATP are relatively high and the concentrations of CoA, NAD^+, ADP, and AMP are low. Consider how these metabolites stimulate gluconeogenesis while inhibiting glycolysis (Figure 2.42). By matching the

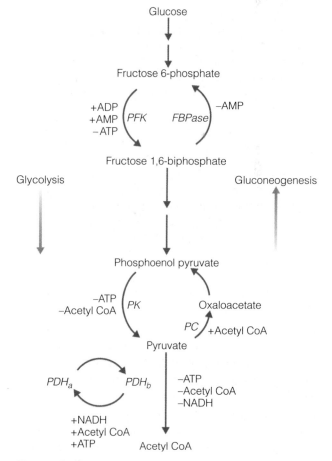

Figure 2.42 Reciprocal regulation of glucose metabolism High-energy compounds, such as ATP, acetyl CoA, and NADH, inhibit glycolysis and stimulate gluconeogenesis at key regulatory enzymes.

rates of ATP synthesis with rates of ATP utilization, cells are able to defend ATP concentrations within a narrow range.

Animals also use other metabolites to reciprocally regulate opposing pathways. When metabolic conditions induce cells to commit to fatty acid synthesis, the increases in the levels of malonyl CoA block fatty acid oxidation by inhibiting CPT-1. Cells further separate anabolism and catabolism using tissue specializations. The liver has the enzymes for ketogenesis but cannot break down ketone bodies. Muscles have the enzymes for ketolysis but cannot synthesize ketone bodies. In fact, the control of energy metabolism in complex animals reflects a division of labor such that some tissues become servile to others. Liver and adipose tissue perform important functions for whole animal metabolic balance, giving lower priority to their own cellular and metabolic needs.

⊙ | CONCEPT CHECK

8. Compare the four types of macromolecules. Discuss how variation in structure arises in each class.

9. Distinguish between the following types of reactions: anabolic, catabolic, amphibolic, anaplerotic.

10. How is oxidation coupled to phosphorylation in mitochondrial oxidative phosphorylation?

11. Compare the pathways of glucose metabolism (synthesis and degradation) under (a) high versus low energy conditions and (b) normal versus low oxygen conditions.

12. Under what conditions is it more advantageous to use carbohydrate rather than lipid as a metabolic fuel?

Cell Physiology

In the opening essay, we discussed how the evolutionary origins of animals required a new relationship between cells to foster multicellular relationships. Cellular membranes[1] perform two main roles that are central to multicellularity and cellular function. First, they allow cells to isolate

---

[1] Cellular membranes refer to all of the membranes within a cell, including the plasma membrane (or cell membrane) that surrounds the cell and the membranes that form the organelles.

themselves from the environment, giving them control over intracellular conditions. Second, membranes help cells organize intracellular pathways into discrete subcellular compartments, including organelles. Separation of processes also requires specific mechanisms to transfer molecules across the membranes. Many complex physiological properties and responses depend on cellular transport and transfers between subcellular compartments. Ultimately, the cellular properties that determine physiological function are controlled through regulation of gene expression.

Membrane Structure

Cellular membranes are composed of phospholipids, other lipids, and diverse proteins. The fluid mosaic model, shown in Figure 2.43, illustrates the structural features of biological membranes. Each of the two layers is a sheet of lipid molecules arranged side by side. The surfaces of the lipid bilayer are composed of the polar head groups of phospholipids, and the internal hydrophobic core is composed of long fatty acid chains of the phospholipids attached through van der Waals forces.

The lipid profile affects membrane properties

Animals produce specialized membranes with unique molecular signatures by varying the structures and proportions of the different types of lipid. Much of the variation can be attributed to the profile of *phosphoglycerides*, the most abundant of which are phosphatidylcholine (PC), phosphatidylserine (PS), and phosphatidylethanolamine (PE). Recall from earlier in this chapter that each of these phospholipids is really a class of molecules with constituent fatty acids that differ in chain length and saturation. Although phosphoglycerides are the most abundant molecules in the bilayer, membranes also possess other lipids, including sphingolipids, glycolipids, and cholesterol, as well as many proteins. Glycolipids resemble phospholipids, but with complex carbohydrate modifications that impart a negative charge to the polar head group. Nerve cells possess high concentrations of sphingolipids and glycolipids because they alter the electrical properties of the membrane. Glycolipids are also important in communication between cells.

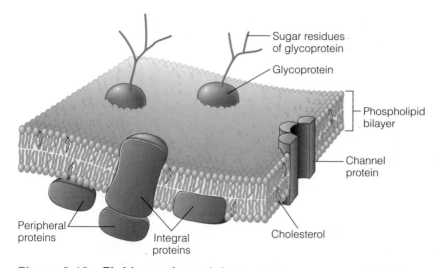

Figure 2.43 Fluid mosaic model Membranes are composed of lipids such as phospholipids, cholesterol, and glycolipids. Proteins can be embedded in the lipid bilayer. Each of the elements moves laterally within the membrane, giving it a functional fluidity.

Cholesterol has an unusual role in membranes. Although it is absent from some membranes, such as mitochondrial membranes, cholesterol can compose almost half the lipid component of other membranes. Cholesterol influences membrane properties in complex ways because of the way it integrates into the lipid bilayer (Figure 2.44). One end of the molecule interacts with phospholipids near the polar head groups, filling gaps between phospholipids to reduce the permeability to low-molecular-weight solutes. Cholesterol also disrupts the interactions between fatty acids, enhancing membrane fluidity.

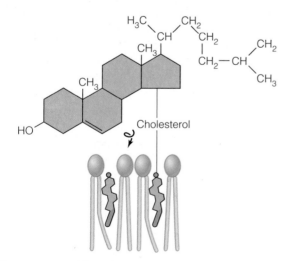

Figure 2.44 Unusual membrane properties of cholesterol Cholesterol strengthens the interactions between phospholipid polar head groups while disrupting the interactions between fatty acid tails.

The unusual ability of cholesterol to increase fluidity while decreasing permeability provides animals with an important mechanism for controlling membrane properties.

Lipid membranes are heterogeneous

While the fluid mosaic model illustrates the general relationships between lipid and protein variation in membrane proteins, it underemphasizes the spatial variation seen in membrane lipids (Figure 2.45). The inner and outer layers of the phospholipid bilayer typically possess different types of lipids. PE and PS are found almost exclusively in the inner leaflet, whereas PC is concentrated in the outer leaflet. Glycolipids are found only in the outer leaflet of the membrane. Membranes also possess discrete regions that are enriched in cholesterol and glycolipids. These **lipid rafts** serve two important functions. Their molecular composition causes a slight thickening of the lipid bilayer, which recruits phospholipids with longer chain fatty acids and proteins with relatively long transmembrane domains. Because of the distinct molecular composition of lipid rafts, they can act as microcompartments within the cell, providing an additional way to spatially organize pathways.

Environmental stress can alter membrane fluidity

An essential component of the fluid mosaic model is the ability of the constituents to move throughout the membrane. Phospholipids and proteins can rotate in position as well as move laterally through the membrane. Membrane fluidity depends on the properties of the membrane lipids, which are influenced by the physical environment. Cells regulate the fluidity of the membrane by controlling the nature of lipids to achieve the appropriate degree of molecular movement. Low temperature, for example, can strengthen the van der Waals forces between membrane lipids, restricting molecular movement within the membrane (Figure 2.46). Since this can adversely affect membrane function, many animals actively

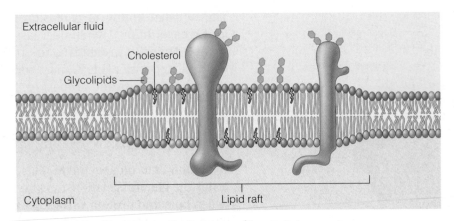

Figure 2.45 **Membrane heterogeneity** Cellular membranes are heterogeneous in composition. Most cells maintain distinct profiles in inner and outer monolayers, sometimes exchanging phospholipids between layers. Lipid rafts are regions of the plasma membrane that accumulate cholesterol and glycolipids, thickening the membrane. These thicker regions preferentially recruit proteins with longer transmembrane domains.

Membrane proteins have important structural and regulatory roles within cells. They contribute to structural support by linking the intracellular cytoskeleton to the extracellular matrix. Many of the intrinsic membrane proteins are receptors that are part of complex signaling pathways. Because membranes are physical barriers to the free movement of many vital organic and inorganic solutes, cells use integral proteins to transport molecules across membranes.

remodel their membranes to compensate for the effects of the physical environment. By altering the membrane lipid profile, they can keep membrane fluidity constant. We will discuss this pattern of membrane regulation, called *homeoviscous adaptation*, in Chapter 13: Thermal Physiology.

Membranes possess integral and peripheral proteins

Protein is an important constituent of most cellular membranes, in some cases making up more than half the mass of the membrane. **Integral membrane proteins** are tightly bound to the membrane, either embedded in the bilayer or spanning the entire membrane. **Peripheral membrane proteins** have a weaker association with the lipid bilayer, typically binding to integral membrane proteins or glycolipids. The different relationships between the bilayer and membrane proteins are shown in Figure 2.47.

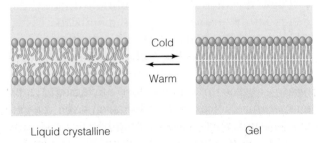

Figure 2.46 **Temperature and membrane fluidity** Temperature alters the fluidity of membranes by changing the interactions between phospholipids.

Transport Across Cellular Membranes

Many cellular processes depend on the ability to move molecules across membranes. A cell must be able to bring nutrients across its plasma membrane and expel end products. Hormones synthesized within the cell must be processed and packaged for secretion. All animal cells must be able to move specific ions across the plasma membrane to control their ionic and osmotic properties. Specific integral membrane proteins mediate these transport processes. The kinetic properties of a transporter are similar to those of enzymes, with an affinity constant (K_m) and a maximal velocity (J_{max}). The three main classes of membrane transport are passive diffusion, facilitated diffusion, and active transport. These classes of transport are distinguished by the direction of transport, the nature of the carriers, and the role of energy in the process (Figure 2.48).

Lipid-soluble molecules cross membranes by passive diffusion

Although membranes are barriers to the movement of many molecules, some molecules cross membranes without a transporter. The ability of a molecule to freely cross a biological membrane depends on its hydrophobicity. Many molecules, such as steroid hormones, are freely soluble in lipid. When these molecules encounter a cell membrane, they dissolve into the lipid bilayer and escape to the other side. Both influx and efflux occur simultaneously, but the net movement (influx minus efflux) depends on the concentration gradient. The net movement of molecules is from high concentration to low concentration. The steeper the concentration gradient, the greater the rate of movement across the membrane. This type of transport is called **passive diffusion**. No specific transporters are required, and no energy, beyond the concentration gradient itself, is required.

Membrane proteins can facilitate the diffusion of impermeant molecules

Hydrophilic molecules cross membranes by other pathways that involve specific transport proteins. If the concentration gradient is favourable, the molecule may cross the membrane by **facilitated diffusion**. As with passive diffusion, no energy beyond that of the concentration gradient is required to drive transport, but with facilitated diffusion a protein is required to carry the molecule across the membrane. Three main types of proteins carry out facilitated diffusion: ion channels, porins, and permeases (see Figure 2.49).

Ion channels are membrane proteins that form pores through which only specific ions may pass, and only when the channel is open. The channels are specific to one or sometimes two ions. Ca^{2+} channels, for instance, possess a structure that allows the free movement of Ca^{2+}, but does not allow other cations such as Mg^{2+}, K^+, or Na^+ to cross at appreciable rates. The specificity of transport is due to a structural component of the channel known as the *selectivity filter*. The channel can be opened in response to cellular conditions. **Ligand-gated channels** are opened when specific regulatory molecules are present. One important ligand-gated channel is the Ca^{2+} channel sensitive to inositol triphosphate (IP_3); this channel induces the release of Ca^{2+} stores when its ligand,

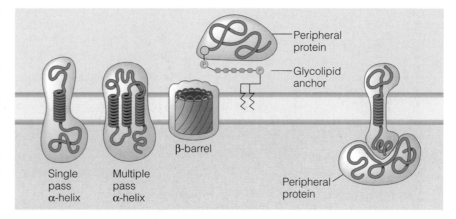

Figure 2.47 Integral membrane proteins Membrane proteins can demonstrate many different types of relationships with membranes. Each membrane protein has within its structure hydrophobic regions that interact favorably with the bilayer. Depending on the protein, these regions can be α-helices or β-barrels. Transmembrane proteins span the entire bilayer, exposing regions to both sides of the membrane. Often these exposed regions possess modifications, such as the carbohydrate chains of glycoproteins. Peripheral membrane proteins are not embedded within the membrane but associate with exposed regions of integral membrane proteins.

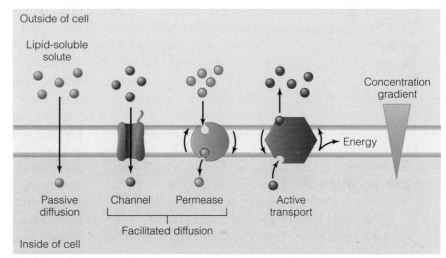

Figure 2.48 Modes of membrane transport The mechanisms of transport across cellular membranes depend on the lipid solubility of the solute as well as the direction and magnitude of the concentration gradient. Passive diffusion needs no carrier, as lipid-soluble solutes move freely across the membrane. Facilitated diffusion carries impermeant solutes across the membrane on protein carriers, including channels (either ion channels or porins) and permeases. Solutes can also be transported by active transport, which can move molecules against a concentration gradient.

IP_3, is present. **Voltage-gated channels** are opened or closed in response to membrane potential. For example, K^+ channels in muscle and neurons open when the membrane depolarizes. **Mechanogated channels** are regulated through interactions with the subcellular proteins that make up the cytoskeleton. Changes in cell shape, such as cell swelling, alter the arrangement of the

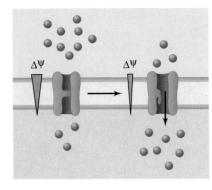
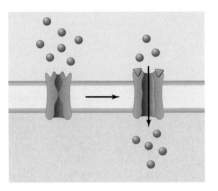
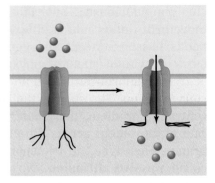

(a) Voltage-gated   (b) Ligand-gated   (c) Mechanogated

Figure 2.49 Carriers involved in facilitated diffusion Channels are proteins
that mediate facilitated diffusion of ions and other metabolites. Channels typically exist in either
a closed or an open conformation. They are opened by specific triggers, including specific
ligands, voltage conditions, or physical associations with structural elements.

cytoskeleton. Upon sensing the changes in the
cytoskeleton, mechanogated channels may open
or close.

Porins are large channels that function in
similar ways to ion channels but permit the passage of much larger molecules. Mitochondria have
a porin in the outer membrane that facilitates the
transfer of low-molecular-weight molecules from
the cytoplasm to the mitochondria. **Aquaporins**
are water channels in the plasma membranes;
each aquaporin molecule can transport 3 billion
water molecules per second. Some aquaporins,
called *aquaglyceroporins*, are also capable of
transporting nonwater molecules, such as glycerol
and possibly urea. Aquaporins may also be involved in transport of gases across membranes.

The third type of protein that facilitates diffusion is a **permease**. Rather than creating a pore for
a molecule, a permease functions more like an enzyme. It binds the substrate and then undergoes a
conformation change that causes the carrier to release the substrate to the other side. Several tissues
possess glucose permease, a transporter that allows
glucose to enter cells, passing from high concentration to low concentration. Unlike porins and ion
channels, permeases can become saturated with
substrate at high concentration, such that the transport process depends on how quickly the permease
can carry its substrate across the membrane.

Active transporters use energy to pump molecules against gradients

In passive and facilitated diffusion, molecules can
move only from high concentration to low concen-

tration. In contrast, cells use **active transport** to
move molecules across membranes against concentration gradients. Two main forms of active
transport are distinguished by the source of the energy that drives the process. In **primary active
transport**, the carrier protein uses an exergonic reaction to provide the energy to transport a molecule.
The other form of active transport, called **secondary
active transport**, couples the movement of one
molecule to the movement of a second molecule.

The most common primary active transporters
use the hydrolysis of ATP to provide the necessary
energy. Three general classes of ATP-dependent
transporters, or ATPases, mediate primary active
transport: P-type ATPases, F-type (or V-type) AT-
Pases, and ABC transporters. P-type ATPases use
ATP hydrolysis to pump specific ions across membranes. For example, animal cells have a Na^+/K^+
ATPase in the cell membrane that extrudes Na^+
from the cell in exchange for K^+. Many tissues have
Ca^{2+} ATPases to transport Ca^{2+} across membranes.

F-type and V-type ATPases are structurally related ATPases that pump H^+ across membranes
using the energy of ATP hydrolysis. The mitochondrial F-type ATPase operates in reverse, using H^+
movements down electrochemical gradients to
provide the energy for ATP synthesis. The V-type
ATPases allow cells and organelles to extrude protons to acidify a compartment, such as the lumen
of the lysosome or the inside of the stomach.

The ABC transporters carry large organic mol-
ecules across the cell membrane. Cells often use
ABC transporters to export toxins from the cell.
The multidrug resistance protein, an important
ABC transporter, is often linked to types of cancers

that become resistant to chemotherapy. Some cancerous cells survive chemotherapy by transporting the toxic drug out of the cell before the chemotherapeutic agent can kill it.

Secondary active transport uses the energy held in the electrochemical gradient of one molecule to provide the energy to drive another molecule against its gradient. If the molecules move in opposite directions, the carrier is called an **antiport**, or **exchanger**. For example, red blood cells use a Cl^-/HCO_3^- exchanger (also called *band 3*) to drive the transport of these ions across the membrane. The direction of ion movement by this carrier depends on the relative gradients of the two ions. Alternatively, a **symport**, or **cotransporter**, is used to move molecules in the same direction. For example, intestinal cells use a Na^+-glucose cotransporter to import glucose against its concentration gradient, driven by a greater inward Na^+ electrochemical gradient.

All of these transport processes influence chemical gradients across membranes, but only a subset of transporters affect the electrical gradient. Carriers that transport uncharged molecules, such as the glucose permease, are termed **electroneutral** carriers. Similarly, carriers such as the Cl^-/HCO_3^- exchanger that exchange two ions of the same charge are also electroneutral. In contrast, the carriers that transfer a charge across the membrane are called **electrogenic** carriers. When we discuss Na^+/K^+ ATPase throughout this text, keep in mind that it is an electrogenic carrier because it exchanges 3 Na^+ for 2 K^+ ions.

Membrane Potential

All animal cells maintain a voltage difference across their cell membranes, as well as some organelle membranes, such as the mitochondrial membrane. This voltage difference represents a source of potential energy that cells can harness to move molecules across membranes. The voltage difference is termed the *resting membrane potential difference*, or the **membrane potential**, for short. In addition to using the membrane potential as a source of energy, **excitable cells** use changes in membrane potential as communication signals. As we discuss in Chapters 4 and 5, this property is particularly important for nerve and muscle cells, and thus the membrane potential is critical for allowing the coordinated movements of cells and or-

ganisms. Electrical signaling is not, however, a unique property of nerve and muscle cells. Several other types of cells use electrical signals, including fertilized eggs and hormone-secreting cells.

The interior of the membrane is electronegative at rest

Membrane potential can be measured using a microelectrode. Microelectrodes consist of a thin recording electrode encased in a very fine-tipped glass pipette that can be inserted through the cell membrane into the cell. The microelectrode is connected via a voltmeter to a reference electrode that is immersed in the solution outside the cell. The voltmeter measures the voltage drop across the circuit caused by the membrane potential (V_m). In most animal cells, the membrane potential is between -5 and -100 mV. By convention, the membrane potential is expressed relative to the voltage outside the cell. Thus, the negative value for V_m means that the interior of the cell membrane is more electronegative than the exterior of the cell membrane.

Ionic concentration gradients and permeability establish membrane potential

Only two factors are required to establish a potential difference across a membrane: a concentration gradient for an ion and a membrane that is permeable to that ion. Consider a situation where two solutions are separated by a membrane that is impermeable to ions (Figure 2.50). Assume that the interior of the cell contains 100 mM KCl and 10 mM NaCl, and the extracellular fluid contains 100 mM NaCl and 10 mM KCl. The concentration gradient for K^+ (100 mM inside the cell and 10 mM outside the cell) favors outward movement, whereas the concentration gradient for Na^+ (100 mM outside and 10 mM inside the cell) favors inward movement. There is no gradient for the movement of Cl^- (because the concentration of Cl^- is 110 mM both inside and outside the cell). The solutions on either side of the membrane are also electroneutral, with equal numbers of anions and cations.

To create a membrane potential, we insert channels that allow the passage of K^+, but no other ion. The concentration gradient will cause K^+ to move out of the cell along the concentration gradient, creating a local region of electronegativity on

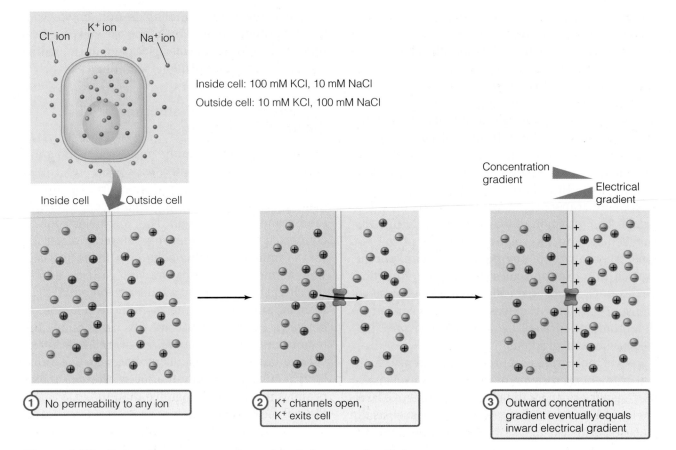

Inside cell: 100 mM KCl, 10 mM NaCl
Outside cell: 10 mM KCl, 100 mM NaCl

Concentration gradient
Electrical gradient

① No permeability to any ion

② K⁺ channels open, K⁺ exits cell

③ Outward concentration gradient eventually equals inward electrical gradient

Figure 2.50 Potassium movements and membrane potential

the inner face of the membrane (where K⁺ left) and a local region of electropositivity on the outer face of the membrane (where K⁺ appeared). This excess negative charge at the inside face of the membrane generates an electrical force that tends to draw positive charges back into the cell. As more K⁺ leaves the cell, the electrical force gradually increases to a level that exactly balances the driving force from the K⁺ concentration gradient. Potassium ions continue to move across the membrane, but their inward and outward fluxes exactly balance each other. The potential difference across the membrane under these equilibrium conditions is termed the **equilibrium potential** for that ion (E_{ion}). Because only a single ion can move across the membrane in this hypothetical example, the equilibrium potential is equivalent to the resting membrane potential ($E_{ion} = V_m$).

For a given concentration gradient, it is possible to calculate the equilibrium potential for an ion using the Nernst equation,

$$E_{ion} = \frac{R\,T}{z\,F} \ln \frac{[X]_{outside}}{[X]_{inside}}$$

where R is the gas constant, T is the temperature (Kelvin), z is the valence of the ion, F is the Faraday constant (23,062 cal/V-mol), and [X] is the molar concentration of the ion. In our example, $[K^+]_{outside} = 10$ mM and $[K^+]_{inside} = 100$ mM, resulting in $E_K = -60$ mV. In other words, the force driving the outward movement of K⁺ resulting from its 10-fold concentration gradient can be exactly balanced by a 60 mV excess of negative charge inside the membrane (a membrane potential of -60 mV).

The equilibrium potential for a particular ion is often termed its **reversal potential**, because the direction of ion movement across the membrane changes when the voltage difference across the membrane exceeds this level. In the case of our hypothetical example, net K⁺ flux was down its concentration gradient from the inside to the outside of the cell until the membrane potential difference reached -60 mV. If the membrane potential were to become even more negative, the net movement of K⁺ would be from the outside of the cell back to the inside—against its concentration gradient. Note that the actual number of ions that need to move across the membrane before reaching the equilibrium potential is actually very small (less than 1/100,000 of the total K^+ ions within a typical cell), and does not result in a measurable change in the overall K^+ concentration either inside or outside the cell.

It is important to emphasize that the membrane potential is the result of extremely small differences in the number of charged molecules immediately adjacent to the membrane—a difference that is too small to detectably affect the overall ion concentration of the cytoplasm or extracellular fluid. The localization of the charge difference immediately adjacent to the membrane arises because the cell membrane acts as a *capacitor*. A capacitor is a device containing two electrically conductive materials separated by an *insulator*, a very thin layer of a nonconducting material. Electrical charges can interact with each other across the insulator if the layer is sufficiently thin. In a cell, the cytoplasm and the extracellular fluid are conducting materials, whereas the lipid bilayer of the cell membrane is the insulator. The excess positive charge along the outside of the membrane attracts the excess negative charge along the intracellular face of the membrane. These electrical interactions can only occur across very small distances, and do not affect ions in the bulk phase of the cytoplasm or extracellular fluid. Thus, the membrane potential occurs only in the area immediately adjacent to the membrane.

Potassium plays the major role in establishing membrane potential

In our hypothetical example neither Na^+ nor Cl^- affected the membrane potential because there was no concentration gradient for Cl^- and because the membrane was not permeable to either of these ions. Of course, the situation in real cells is not so simple, since there are several ions that differ in concentration between the inside and the outside of the cell, and real membranes have varying degrees of permeability to multiple ions. For most cells, the primary ions that affect the membrane potential are K^+, Na^+, and Cl^- because they can move across membranes and there are differences in their intracellular and extracellular concentrations. A modification of the Nernst equation, the Goldman-Hodgkin-Katz Constant Field equation (usually referred to as the **Goldman equation**) can be used to calculate the resting membrane potential based on the concentrations

and permeabilities of all of the relevant ions (see Box 2.3, Mathematical Underpinnings: The Goldman Equation). The Goldman equation essentially represents the sum of the equilibrium potentials for all of the relevant ions, with a weighting factor that takes into account the relative permeabilities of the ions. The influence of each ion over the overall membrane potential is thus proportional to its permeability. For example, resting neurons are more permeable to K^+ than to the other ions, and as a result, K^+ plays the major role in setting the resting membrane potential of neurons.

The Na^+/K^+ ATPase establishes concentration gradients

Active pumping of Na^+ and K^+ ions by the electrogenic Na^+/K^+ATPase is responsible for establishing the concentration gradients for these ions across the cell membrane. Ultimately it is these concentration gradients (along with the selective permeability of the membrane) that establish the membrane potential. The Na^+/K^+ATPase is also responsible for maintaining the resting membrane potential. Although most membranes are only sparingly permeable to Na^+ at rest, a small amount of Na^+ does leak into the cell down its electrochemical gradient, while K^+ leaks out. Without appropriate compensation, these ion movements would result in the dissipation of the Na^+ and K^+ concentration gradients that are needed to establish the membrane potential. Cells use the Na^+/K^+ ATPase to compensate for the leakage of Na^+ and K^+ ions. If you poison the Na^+/K^+ ATPase with a drug called ouabain, the membrane potential difference of the cell slowly decays over the course of a few hours, eventually reaching a value of 0 mV.

Changes in membrane permeability alter membrane potential

Because ion permeability is a major factor involved in establishing the resting membrane potential, changes in ion permeability cause changes in membrane potential. Excitable cells such as neurons and muscle cells alter the permeability of their membranes to generate changes in membrane potentials. We can use the Nernst equation to predict the nature of the ion movements following changes in membrane permeabilities. For example, in mammalian neurons the concentration

BOX 2.3 MATHEMATICAL UNDERPINNINGS
The Goldman Equation

The Goldman-Hodgkin-Katz Constant Field equation (often simply referred to as the Goldman equation or GHK), allows the estimation of the membrane potential based on the concentrations, valences, and relative permeabilities of a series of ions. Since most plasma membranes under resting conditions have appreciable permeability only to postassium, sodium, and chloride, the Goldman equation can be written as follows:

$$E_m = \frac{RT}{F} \ln \frac{P_K [K^+]_o + P_{Na} [Na^+]_o + P_{Cl} [Cl^-]_i}{P_K [K^+]_i + P_{Na} [Na^+]_i + P_{Cl} [Cl^-]_o}$$

where P_{ion} is the permeability of the membrane to that ion and $[ion]_o$ and $[ion]_i$ represent the extracellular and intracellular concentrations, repectively, of a given ion.

From the Goldman equation, the impact of ion permeability on the membrane potential is clear. Any ion with a low permeability has little effect on the membrane potential, even if there is a large concentration gradient across the membrane for that ion. Notice that if the permeability of the membrane for an ion is zero, then the term for that ion drops out of the equation. For example, consider the case of a membrane that is impermeable to Na^+ and Cl^-. In this case, the Goldman equation simplifies to

$$E_m = \frac{RT}{F} \ln \frac{P_K [K^+]_o}{P_K [K^+]_i}$$

The two terms for K^+ permeability also cancel out, leaving an equation that is equivalent to the Nernst equation for potassium (with the exception of the valence term, z, which is neglected because potassium has a valance of $+1$). Notice, however, that if we do the same exercise assuming that the membrane is permeable only to chloride, we end up with an equation that is similar to the Nernst equation, except that it neglects the valence and has the intracellular concentration in the numerator and the extracellular concentration in the denominator, rather than the other way around. Recall that one of the rules of logarithms is that $-\ln x = \ln (1/x)$. By inverting the ratio of the concentrations of chloride, the Goldman equation takes into account the fact that chloride has a valence of -1.

In addition to providing an estimate of the resting membrane potential, the Goldman equation allows the estimation of the membrane potential during electrical signaling. For example, when a large number of Na^+ channels open within the membrane (as is the case during signaling in nerve cells), the permeability of the membrane to Na^+ increases greatly. In the case of neural signaling, this increase in Na^+ permeability is so large that P_{Na} becomes much greater than P_K and P_{Cl}. Under these conditions, the Goldman equation is dominated by the term for Na^+, and the membrane potential approaches the equilibrium potential for Na^+ as calculated by the Nernst equation.

of Na^+ is typically about 10-fold greater outside the cell, so $E_{Na} = +58$ mV. In contrast, K^+ concentration outside the cell is only about 1/40 of that inside the cell, so $E_K = -90$ mV. The resting membrane potential of neurons is typically about -70 mV. If the Na^+ permeability of the membrane increases (as a result of the opening of Na^+ channels), Na^+ will enter the cell, because both the electrical and concentration gradients favor inward Na^+ movement until the membrane potential reaches the equilibrium potential for Na^+ of $+58$ mV. The resulting inward Na^+ movement causes a reduction in the membrane potential termed a **depolarization** (Figure 2.51). In contrast, if the K^+ permeability of the membrane increases (as a result of the opening of K^+ channels) K^+ will move out of the cell, because both its concentration and electrical gradients favor outward

K^+ movement until the membrane potential reaches the equilibrium potential for K^+ of -90 mV. The loss of positive charges from the interior of the cell results in a **hyperpolarization**. As we discuss in later chapters, many cells use cycles of depolarization, hyperpolarization, and repolarization as communication signals.

Subcellular Organization

Eukaryotes rely on complex intracellular organization to orchestrate the many processes required for life. Central to the diversity in physiological function is the ability of individual cells to perform specific roles for the tissue and animal. We gain a clearer understanding of complex physiological systems by studying the role of the various cellular compartments that contribute to the process.

Experimentally, it is easier to measure the relative permeability of ions, rather than the absolute permeability. Hence, the Goldman equation is often rewritten after dividing each term by P_K.

$$E_m = \frac{RT}{F} \ln \frac{[K^+]_o + P_{Na}/P_K\,[Na^+]_o + P_{Cl}/P_K\,[Cl^-]_i}{[K^+]_i + P_{Na}/P_K\,[Na^+]_i + P_{Cl}/P_K\,[Cl^-]_o}$$

For a cell such as a squid giant axon, the following values can be used to calculate the membrane potential:

$$[K^+]_i = 400 \text{ mM and } [K^+]_o = 20 \text{ mM}$$
$$[Na^+]_i = 50 \text{ mM and } [Na^+]_o = 440 \text{ mM}$$
$$[Cl^-]_i = 51 \text{ mM and } [Cl^-]_o = 560 \text{ mM}$$
$$P_{Na}/P_K = 0.04$$
$$P_{Cl}/P_K = 0.45$$

Substituting these values into the Goldman equation predicts the membrane potential of this squid giant axon to be −60 mV at rest, which is a good approximation of the measured resting membrane potential.

Returning to the Nernst equation, we can also calculate the equilibrium potentials for each of these ions. Using the concentrations relevant to the squid giant axon, the equilibrium potential is −75 mV for K^+, +55 mV for Na^+, and −60 mV for Cl^-. These equilibrium potentials establish the "boundary conditions" for the membrane potential. That is, the membrane potential cannot be more negative than −75 mV or more positive than +55 mV because there are no chemical gradients large enough to produce larger membrane potential differences. At rest, the membrane does not quite reach the equilibrium potential for K^+ because of the competing effects of Na^+, but because Na^+ permeability is relatively low its influence is small, and the membrane potential is close to the K^+ equilibrium potential. Note that the squid giant axon also has appreciable permeability to Cl^- (about half that of K^+). In fact, some cell membranes (e.g., in muscle cells) are more permeable to Cl^- than they are to K^+. However, even in this case, K^+ plays the major role in establishing the membrane potential. The Na^+/K^+ ATPase actively pumps Na^+ and K^+ ions to establish their concentration gradients. The K^+ concentration gradient sets the resting membrane potential difference, and Cl^- ions passively distribute themselves across the membrane in response. Thus, in the case of Cl^- ions, the intracellular and extracellular Cl^- levels are a consequence rather than a cause of the resting membrane potential.

Reference

○ Hodgkin, A. L., and A. F. Huxley. 1952. A quantitative description of membrane current and its application to conduction and excitation in the nerve. *Journal of Physiology* 117: 500–544.

Mitochondria are the powerhouse of the cell

Mitochondria are complex organelles, possessing intricate networks of membranes (Figure 2.52). The innermost compartment is the mitochondrial matrix, delimited by the inner mitochondrial membrane. The outer mitochondrial membrane surrounds the organelle and creates another compartment called the intermembrane space. Each of these compartments has its own complement of enzymes and performs different functions for the mitochondria and the cell. The matrix houses the enzymes and metabolites of the TCA cycle. The inner mitochondrial membrane, which is often highly convoluted, holds the enzymes of oxidative phosphorylation and all the transporters necessary to move metabolites in and out of the mitochondria. About 80% of the mass of the inner membrane is protein, the highest protein content of any biological membrane in animals. Mitochondria organize the inner membrane into layers, or lamellae, that are tightly folded. In some tissues, as much as 70 m^2 of mitochondrial inner membrane can be folded into a 1-cm^3 volume of mitochondria.

Mitochondrial structure varies greatly among cell types. Many cells, such as liver, contain hundreds of individual oblong mitochondria scattered throughout the cell. These individual mitochondria are rapidly transported throughout the cell. Some cells organize their mitochondria into networks of interconnected organelles called the *mitochondrial reticulum*, which is constantly remodeled by enzymes that mediate its fission and fusion.

Earlier in this chapter you learned that mitochondria possess the enzymes of oxidative phosphorylation, and make most of the ATP a cell

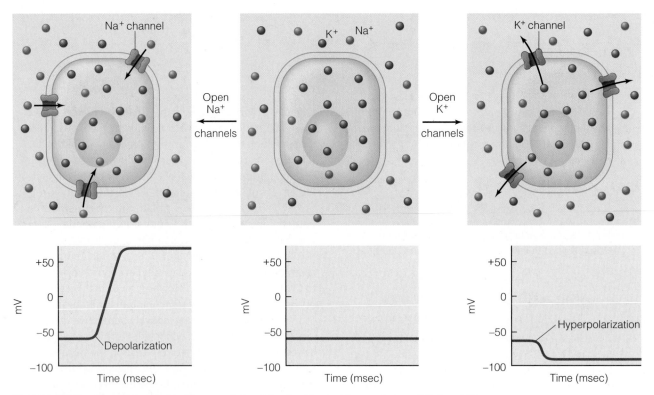

Figure 2.51 Hyperpolarization and depolarization The gradients of Na^+ and K^+ across the cell membrane largely determine the resting membrane potential. When specific ion channels open, the movement of ions changes the membrane potential. If K^+ moves out of the cell, the magnitude of the membrane potential increases (hyperpolarization). If Na^+ moves into the cell, the magnitude of the membrane potential decreases (depolarization).

requires. Cells frequently respond to changes in energy demand by altering their levels of mitochondria, using both biosynthetic and degradative pathways. Most of the genes required for synthesis of mitochondrial proteins are located in the nucleus. Mitochondrial biogenesis requires that each of these genes be expressed in unison to produce the hundreds of proteins needed for new mitochondria or an extension of the mitochondrial reticulum. Mitochondrial biogenesis also requires replication of mitochondrial DNA (mtDNA) and synthesis of additional mitochondrial membranes. Degradative pathways control the levels of mitochondria and mitochondrial proteins. Damaged mitochondrial fragments are engulfed by autophagosomes and degraded in lysosomes. Cells that fail to destroy defective mitochondria suffer energy shortfalls and eventually cell death.

The cytoskeleton controls cell shape and directs intracellular movement

The **cytoskeleton** is a network of protein-based fibers that extends throughout the cell (Figure 2.53).

It has an important role in maintaining cell structure, acting as a frame upon which the cell membrane is mounted. It gives the cell its characteristic external shape and also supports and organizes intracellular membranes. Organelle networks such as the endoplasmic reticulum and Golgi apparatus are mounted on the cytoskeleton. The cytoskeleton is dynamic in structure, under constant reorganization. Apart from its structural roles, the cytoskeleton is an important participant in many cellular processes, including signal transduction.

The cytoskeleton is constructed from three types of fibers: **microfilaments**, **microtubules**, and **intermediate filaments**. These proteins are long strings of monomers connected end-to-end to form a polymer. Microtubules are large, stiff tubes composed of the protein **tubulin**. Microfilaments are small, flexible chains of **actin**. Intermediate filaments, so named because they are intermediate in size, are composed of many types of monomers. Most cells possess each of these cytoskeletal elements, but many cells are richer in one particular type. For example, the tails of sperm are largely microtubules, muscles are largely actin

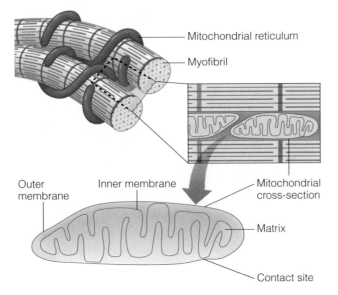

Figure 2.52 Mitochondrial structure Mitochondria are found in almost every cell type, but with many different appearances. Muscle mitochondria exist as a network extending throughout the muscle myofibrils. In cross-section they appear as individual organelles, but three-dimensional reconstructions show the reticulum structure. Inside the mitochondria the highly folded inner membrane can be seen.

polymers, and skin is rich in the intermediate filament keratin.

Other proteins work in conjunction with the cytoskeleton to conduct many types of movement. These proteins, called **motor proteins**, are mechanoenzymes that use the energy of ATP hydrolysis to walk along the cytoskeleton. **Myosin** is the motor protein that walks along actin polymers; **kinesin** and **dynein** move on microtubules. In Chapter 5: Cellular Movement and Muscles, we discuss the structure and function of the cytoskeleton and motor proteins in the context of cellular and intracellular movement.

The endoplasmic reticulum and Golgi apparatus mediate vesicular traffic

Cells have layers of membranous organelles extending around the nucleus to the periphery of the cell (Figure 2.54). The first layer, the **endoplasmic reticulum (ER)**, is the gateway to the other compartments. Proteins are made in the ER, folded, and then sent to their final destinations in the plasma membrane, the Golgi apparatus, lysosomes, and endosomes. The vehicle that carries proteins between compartments is a **vesicle**, a small membrane-bound organelle. Some vesicles are surrounded by

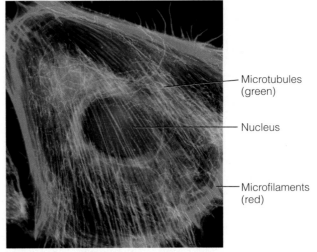

(a)

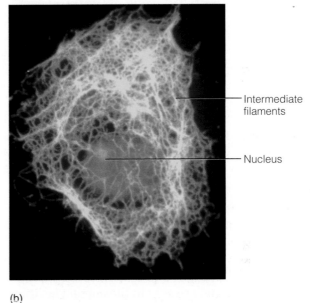

(b)

Figure 2.53 Three protein fibers of the cytoskeleton Panel **(a)** shows microtubules (green) and microfilaments (red). Panel **(b)** shows intermediate filaments.

a shell of coat proteins, such as clathrin, coat protein complex I (COP-I), and COP-II. These proteins help form the vesicle, but they also have an important influence on where the vesicle is sent.

Cells are often illustrated in ways that suggest that vesicles drift freely throughout the cytoplasm. In reality, vesicles are carried throughout the cell by motor proteins moving on cytoskeletal tracks. For example, vesicles coated with COP-I may be carried toward the Golgi apparatus, whereas vesicles coated with COP-II may be sent to the ER. Coat proteins and other vesicle membrane proteins influence which motor protein is

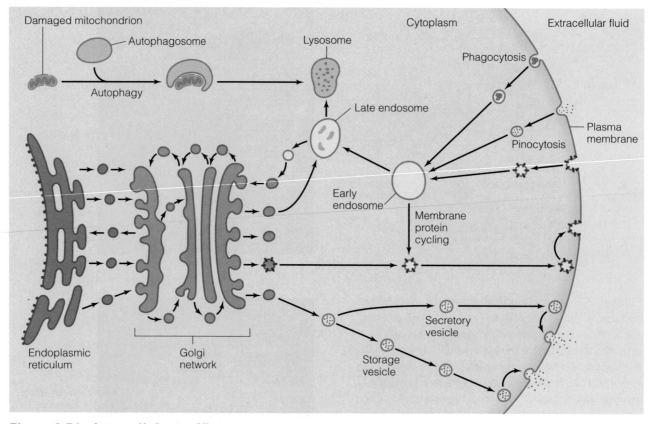

Figure 2.54 Intracellular traffic Vesicles move throughout the cell, transferring membranes and vesicle contents between compartments.

bound. If a vesicle binds myosin it will be carried on microfilaments, but if it binds dynein it will be carried on microtubules. Protein kinases and protein phosphatases regulate vesicular traffic by altering the cytoskeleton, motor proteins, or vesicle proteins. These processes ensure that vesicles and their contents are sent to the correct location at the correct time.

Many types of intracellular sorting pathways use the ER-Golgi network. Most cells produce proteins, and sometimes other molecules, for release from the cell. This process, called **exocytosis**, begins in the ER. Proteins are made here and packaged into vesicles that move through the Golgi apparatus, ultimately fusing to the plasma membrane to release the vesicle contents to the extracellular space. In the reverse pathway, **endocytosis**, vesicles form at the plasma membrane, engulfing liquid droplets (**pinocytosis**) or large particles (**phagocytosis**). The same pathways of endocytosis and exocytosis regulate the proteins found in the plasma membrane, such as membrane transporters and channels. When transporters are no longer needed, they can be removed from the membrane and stored in vesicles until needed again.

Conversely, when a secretory vesicle fuses to the plasma membrane, its internal contents are expelled but the vesicle membrane, both lipid and integral proteins, disperses into the plasma membrane. Cells control the numbers and types of proteins in the plasma membrane through endocytosis and exocytosis. Vesicles rich in transporters fuse to the plasma membrane to increase transport capacity. Conversely, regions of the plasma membrane are extracted during vesicle formation to remove transporters for storage or degradation. Vesicles in transit can be directed to other compartments to assist in processing their contents. Endosomes act as clearinghouses for vesicles, collecting them and then redistributing their contents and membrane proteins into new vesicles that are sent to their correct locations. They send damaged proteins and foreign materials to lysosomes for proteolytic degradation. Once vesicles reach their destination, another series of proteins mediate the fusion of vesicles with target membranes.

The pathways of intracellular sorting allow animal cells to control many of the processes we have considered throughout this chapter, including secretion, ingestion, and membrane transport.

Another function of these pathways, specifically the secretory pathway, is to build and maintain a fibrous network outside the cells: the extracellular matrix.

The extracellular matrix mediates interactions between cells

Cells are organized into a three-dimensional tissue by a network of fibers called the **extracellular matrix**. The proteins used to build the matrix are synthesized by the ER, packaged into vesicles, and sent out of the cell using the secretory pathway. During transit through the Golgi apparatus, suites of enzymes modify the proteins, adding branched chains of sugars. As you learned earlier in this chapter, glycosylation alters the properties of the proteins in many ways. In the extracellular matrix, water binds to the hydrophilic sugars to create a gel-like coating that fills the space between cells.

Extracellular matrix macromolecules can be proteins, simple glycoproteins, glycosaminoglycans, or combinations of both, known as proteoglycans (Figure 2.55). **Collagen** is a long, stiff fiber formed as a triple helix of three separate collagen glycoprotein monomers. Elastin is a small protein that is linked together into an intricate web. When the network is stretched it acts like a rubber band, providing the tissue with elasticity. Many extracellular matrix components are linked together by the glycoprotein fibronectin. Each fibronectin molecule binds other fibronectins as well as different matrix components to form a fibrous network.

Hyaluronan is a glycosaminoglycan composed of thousands of repeats of the disaccharide glucuronic acid-N-acetylglucosamine. With its hydration shell, it forms a noncompressible gel that acts as a cushion between cells. Hyaluronan fills the spaces between joints of land animals, easing movement. Other glycosaminoglycans, such as chondroitin sulfate and keratan sulfate,[2] are cova-

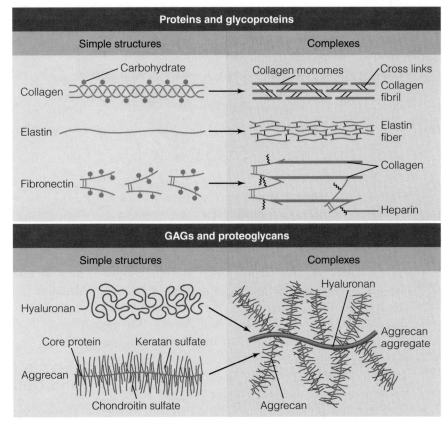

Figure 2.55 Extracellular matrix components The extracellular matrix is composed of combinations of proteins and glycoproteins, glycosaminoglycans (GAGs) and proteoglycans. Many of the individual molecules, shown in the left column, can be combined into more complex macromolecules, shown on the right. The protein components are shown in green and the GAG components in blue.

lently attached to proteins to form **proteoglycans**. Cartilage is composed primarily of aggrecan, a proteoglycan that incorporates more than 100 glycosaminoglycans into its structure. Many proteoglycans link the different extracellular matrix proteins together to form a network.

The extracellular matrix can be simple in structure and composed of only a few proteins, or it can be organized into an extensive network. The extracellular matrix is more than just the cement that connects cells together. Many specialized structures such as the insect exoskeleton, vertebrate skeleton, and molluscan shells are modified extracellular matrices secreted by specific cells. For example, bone and cartilage are tissues formed from the extracellular matrix of osteoblasts and chondroblasts, respectively. The **basal lamina** (Figure 2.56), or basement membrane, is a type of extracellular matrix found in many tissues, where it acts as a solid support that helps anchor cells. It is designed and maintained primarily by specialized cells called fibroblasts.

---

[2] *Keratan* is a GAG of the extracellular matrix, whereas *keratin* is an intermediate filament protein of the cytoskeleton.

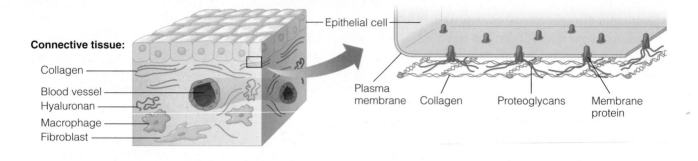

Figure 2.56 Basal lamina In many tissues, fibroblasts produce a thick layer of extracellular matrix called the basal lamina. Some cells use the basal lamina as a foundation, but other cells and blood vessels use it as a porous frame.

Cells use various strategies to modulate both the matrix properties and their relationship with the matrix. First, most types of extracellular matrix components can be made many ways. For instance, mammals have 20 different collagen genes, so in principle a collagen trimer can be constructed 8000 (20^3) ways. Even though most of these possible variants are never constructed, it illustrates the potential for variation in one of the many components of the extracellular matrix. Second, variations occur in the type and position of carbohydrate groups of simple glycoproteins and proteoglycans. Each variation influences the physical properties of the extracellular matrix protein. By controlling which proteins are made and how they are modified by glycosylation, cells determine which building blocks are available to build the extracellular matrix. Cells control which proteins are released to the extracellular space using the secretory pathway discussed in the previous section.

Secreting the extracellular matrix components from the cell is really only one step in building a tissue. The cells also produce integral membrane proteins called matrix receptors to connect them to the extracellular matrix. **Integrins** are an important class of plasma membrane receptors that bind the cytoskeleton on the inside of cells and bind the extracellular matrix on the outside of cells. A cell changes its association with the extracellular matrix by changing the types of integrins in its membrane, mediated by endocytosis and exocytosis.

Cells can also break down the extracellular matrix by secreting proteases called matrix metalloproteinases. By controlling both the production of the matrix and its degradation, cells can regulate their ability to move throughout a tissue. For example, when blood vessels grow, they use ma-

trix metalloproteinases to break down the extracellular matrix of the local cells to allow the blood vessels to penetrate into new regions of the tissue.

Physiological Genetics and Genomics

The nature of physiological diversity, whether in the response of an individual or in the variations arising over evolutionary time, resides in the genes: how they differ between species and how they are regulated in individual cells. Homeostatic regulation depends on the ability of the cell to put the right protein in the proper place at the proper time with the appropriate activity. Cells have many mechanisms to control the rates of synthesis of specific proteins. RNA polymerases read the genes, producing mRNA in the process of **transcription**. Once RNA is made, it is used as a template to produce protein in the process of **translation**. Cells can control the levels of both RNA and protein using mechanisms that target rates of synthesis and degradation.

Nucleic acids are polymers of nucleotides

The two types of nucleic acids, deoxyribonucleic acid (DNA) and ribonucleic acid (RNA), are structurally similar but perform different functions within the cell. DNA is the genetic blueprint for building cells. RNA reads the information encoded by the DNA and interprets it to make proteins. Cells produce three main forms of RNA: transfer RNA (tRNA), ribosomal RNA (rRNA), and messenger RNA (mRNA). Certain molecules of RNA complex with proteins to form riboproteins.

Both RNA and DNA are polymers of nucleotides. All nucleotides are composed of a nitrogenous base attached to a sugar linked to a phosphate. RNA and DNA differ in the type of sugar in the nucleotide: ribonucloetides contain ribose whereas deoxyribonucleotides possess deoxyribose. Both RNA and DNA are synthesized from combinations of four types of nucleotides that differ in the nature of their nitrogenous bases. Three of the four nitrogenous bases, the pyrimidine cytosine and the purines adenine and guanine, are found in nucleotides of both RNA and DNA. The fourth nitrogenous base is another pyrimidine: uracil in RNA and thymine in DNA. The ribonucleotides are ATP, UTP, CTP, and GTP. The deoxyribonucleotides are dATP, dTTP, dCTP, and dGTP. In many cases, the nucleotide sequence in DNA and RNA is represented using one-letter codes. Thus, A refers to the residue derived from the nucleotide ATP (in RNA) or dATP (in DNA), C is CTP/dCTP, G is GTP/dGTP, T is dTTP, and U is UTP.

Nucleic acids form from long polymers of nucleotides linked by phosphodiester bonds that form between the phosphate of one nucleotide and the sugar of the adjacent nucleotide. The end of the polymer that terminates with a phosphate group is deemed the 5-prime end (5′); the other end terminates with a sugar and is the 3′ end. The nucleic acid has a polarity, conferred by its 5′ and 3′ ends, that is an important consideration when discussing the biochemical processes involved in nucleic acid function.

Double-stranded DNA twists into an α-helix with two topological features: a minor groove and a major groove. The two strands of DNA appear as ridges, separated by a trough. These contours between two strands compose the minor groove. The major groove results from the twisting pattern of the α-helix. Every 10 base pairs, a distance of about 3.6 nm, the helix completes a full turn, forming the major groove that resembles a saddle. Variations in nucleotide sequence cause subtle regional alterations in the shape of DNA and the topology of the major and minor grooves. This structural variation is information that is used by the DNA-binding proteins to attach to the correct location to regulate expression of specific genes.

The DNA in animal cells is highly compressed into tight structures with the aid of DNA-binding proteins called **histones**. If you were to unwind the DNA in a single mammalian cell, the strands would stretch several meters. The long strands of DNA wrap twice around the barrel-shaped histones until a structure resembling a strand of pearls is formed. These strands are then twisted and folded into highly compressed arrangements, which has two main advantages to cells. First, it allows the cell to fit large amounts of DNA into the small volume. Second, coating DNA with histones helps reduce the damage caused by radiation and chemicals. However, in this compressed configuration DNA is biochemically inert; it cannot function as a template for RNA synthesis (transcription)

DNA is a double-stranded α-helix packaged into chromosomes

DNA usually exists within cells as a double-stranded polymer (Figure 2.57) in which hydrogen bonds connect the two strands. Each specific nucleotide can form hydrogen bonds with only one other nucleotide. Three hydrogen bonds form between G and C, whereas two hydrogen bonds form between A and T. When one strand of DNA encounters another complementary strand, hydrogen bonds form between the strands, creating a double-stranded molecule. The two strands *anneal* in an antiparallel arrangement, with the 5′ end of one strand associated with the 3′ end of the other strand.

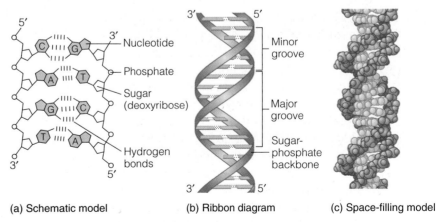

(a) Schematic model   (b) Ribbon diagram   (c) Space-filling model

Figure 2.57 DNA structure Each strand of DNA binds to another, complementary strand. Hydrogen bonds form between specific base pairs. Two bonds form between A and T. Three bonds form between C and G. The double-stranded DNA is twisted into an α-helix, forming a minor groove between strands. The major groove reflects the period of the twisting of the helix.

or DNA synthesis (replication). Cells must use histone-modifying enzymes to release histones from DNA, thereby regulating gene expression.

DNA is organized into genomes

The entire collection of DNA within a cell is called the **genome**. Within the nucleus, the genome is divided into separate segments of DNA called **chromosomes**. Within chromosomes are the genes, which possess the DNA sequences that are used to produce all the different types of RNA, including the mRNA that encodes proteins. Each gene also possesses regions of DNA called promoters that determine when the gene is expressed. Many genes are divided into multiple sections on the same chromosome. The sections that encode RNA are known as **exons**, and the interspersed DNA sections are called **introns** (Figure 2.58).

In most animals, genes account for less than half of the genome. The majority of the genome is a mixture of different types of random and repetitious DNA, much of which serves no known function and is often called junk DNA.

Across the animal kingdom, genome size ranges more than 6000-fold (Figure 2.59). The smallest genome is found in one of the simplest animals; placozoans, a relative of sponges, have only about 0.02 pg of DNA per cell. The largest genome in animals, about 133 pg/cell, belongs to the African marbled lungfish. Surprisingly, there is lit-

tle relationship between the size of the genome and the complexity of the animal. For example, both the largest and the smallest vertebrate genomes are found in fish. The pufferfish genome is only about 0.3% the size of the lungfish genome. There is also no relationship between the number of chromosomes and the complexity of the animal. Humans possess 46 chromosomes. Some deer have only 6, whereas carp may have more than 100.

Transcriptional control acts at gene regulatory regions

The rate of synthesis for many proteins is proportional to the levels of mRNA. Historically, mRNA

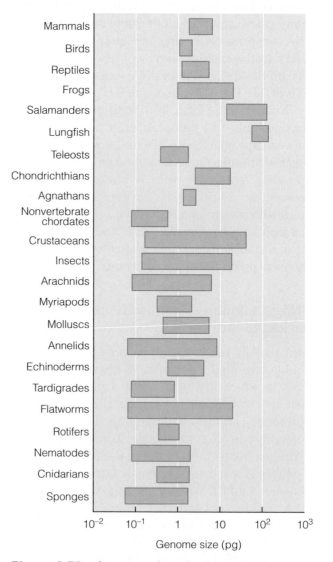

Figure 2.59 Genome sizes in the animal kingdom The genome size in animals can vary widely, and there is no relationship between genome size and complexity. Bar lengths reflect the range in the sizes of genomes measured in picograms (pg).

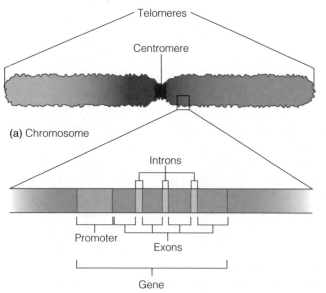

(a) Chromosome

(b) Gene

Figure 2.58 Chromosomes and genes Chromosomes possess structural regions, such as centromeres and telomeres, in addition to noncoding regions and genes.

levels were measured using northern blots, but recent advances in genomics and engineering have led to the development of techniques for assessing complex changes in the levels of mRNA for thousands of genes simultaneously.

At any point in time, most of the genome of a cell is wrapped around histones and rolled into nucleosomes (Figure 2.60). Under these conditions the genes are quiescent, unable to bind the transcriptional machinery. When the gene product is required, the chromatin must be remodeled to allow transcriptional activators access to the regulatory regions of the gene. Transcriptional regulators, both DNA-binding proteins and coactivators, associate with each other to form regulatory complexes on the promoter. The transcription initiation complex assembles near a specific region of the promoter designated as the transcription start site, typically a sequence of TATA (the TATA-box). Once the complex assembles, the process of mRNA synthesis can begin.

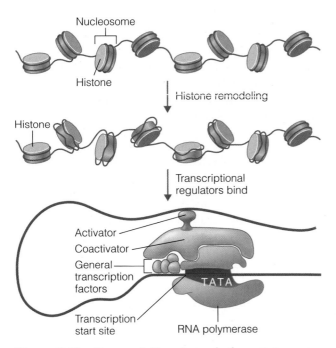

Figure 2.60 Transcriptional regulation Quiescent DNA is tightly wrapped around histones. Remodeling of chromatin gives DNA-binding proteins access to gene control regions. The general transcription factors allow RNA polymerase II to bind to initiate transcription. Other DNA regulatory proteins, such as the activators and coactivators shown here, increase the likelihood that the transcriptional machinery will assemble.

Cells can regulate the rate of mRNA synthesis by altering the conformation of the gene and changing the ability of the transcriptional machinery to assemble. Sometimes gene expression is induced by stimulation of the enzymes that remodel chromatin. These enzymes work by altering the structure of the histones that organize DNA into nucleosomes. Histones can be modified by acetylation, methylation, and phosphorylation. For example, when a histone acetyl transferase (HAT) adds an acetyl group to a critical lysine in a histone, this induces a change in structure that permits remodeling of chromatin to favor gene expression. The gene can be silenced by a histone deacetylase (HDAC) that removes the acetyl group.

Once the regulatory regions within the gene are exposed, the transcriptional machinery is able to assemble. Transcription factors may bind to sites close to, or distant from, the transcriptional start site. Some transcription factors introduce bends into the DNA that bring critical regions of the gene in close proximity. Other transcription factors bind coactivators, which serve as docking sites for other proteins. Eventually, the general transcription factors are assembled, the RNA polymerase is recruited, and the process of transcription can begin. The entire process depends critically on the interactions between dozens of proteins. Consequently, cells can fine-tune the process by regulating the ability of different proteins to interact, typically by changes in protein phosphorylation. The phosphorylation state can affect the transfer of a transcription factor between the cytoplasm and the nucleus. It can also alter the ability of transcriptional regulators to interact with DNA or other proteins, both stimulatory and inhibitory proteins. Since each gene is regulated by dozens of transcription factors, the combinations of regulatory conditions are endless.

The primary mRNA transcript possesses sequences that will eventually code for the protein (exons) as well as other sequences that are interspersed between exons (introns). It must first be processed in a way that removes introns and splices together exons. Next, the spliced RNA must be polyadenylated; long strings of 200 or more ATP residues are added to the 3′ end of the transcript to produce the poly A$^+$ tail that is characteristic of mRNA. Once these post-transcriptional modifications are completed, the mature mRNA is exported to the cytoplasm.

RNA degradation influences RNA levels

Controlling transcription is one important mechanism for cells to alter RNA levels; another is to vary the rate of RNA degradation. RNA is degraded by nucleases called **RNases**. An RNase can attack the end of the RNA (exonucleases) or internal sites (endonucleases), preventing the mRNA from acting as a template for protein synthesis.

Cells have ways to preferentially degrade or protect individual mRNAs. A long poly A^+ tail protects an mRNA from degradation. Soon after release into the cytoplasm, exonucleases nibble off the ends of the poly A^+ tail. The mRNA can still be translated into protein at this point. Once the exonucleases shorten the tail to about 30 bases, the RNA is attacked by an endonuclease, causing enough damage to prevent the protein from being translated.

Other processes accelerate the rate of mRNA degradation. Some mRNAs are unstable, existing in the cytoplasm for only a few minutes before becoming degraded. These unstable mRNAs have long stretches of A and U bases within their 3′ untranslated regions (3′ UTR). These AU-rich regions recruit proteins that accelerate mRNA degradation. The ability to accelerate RNA degradation is essential in many cells, particularly those that produce regulatory proteins. Once a signaling protein is no longer needed, the RNase machinery can rapidly degrade the mRNA to prevent it from being translated.

Cells can also reduce the rate of RNA degradation. Stabilizing proteins can bind to specific regions in the poly A^+ tail or other regions of the mRNA to prevent RNase attack. This allows the cell to maintain a pool of preformed mRNA available for immediate use if cellular conditions demand the gene product.

Global changes in translation control many pathways

Once an mRNA arrives in the cytoplasm, the process of translation can begin with the assistance of ribosomes and amino acyl tRNAs. Ribosomes, complexes of rRNA and proteins, catalyze the formation of peptide bonds between amino acids in the growing protein. The amino acids are provided in the form of amino acyl tRNA. Each amino acid uses a specific tRNA that can bind to a specific set of three nucleotides on the mRNA called a codon. The 5′ end of the mRNA recruits proteins called initiation factors, in combination with a methionine tRNA ($tRNA_{MET}$) and a ribosome. The complex moves down the mRNA chain until it reaches the sequence AUG, which is the start codon. Another amino acyl tRNA is recruited, and the ribosome catalyzes the formation of a peptide bond between the amino acids to begin the process of elongation. In most circumstances, proteins called elongation factors enter the ribosome and accelerate the catalytic cycle. In a typical animal cell, each individual ribosome can add an amino acid to the chain at a rate of one to two per second. The process continues until the ribosomal complex reaches a stop codon, a nucleotide sequence that is incapable of binding any amino acyl tRNA. At any point in time, a single mRNA may be translated by many ribosomes bound all along the mRNA.

Cells can control the rate of translation using nonspecific mechanisms that affect all translation within the cell, as well as specific mechanisms that influence only a subset of mRNAs. Many of the initiation factors and elongation factors are regulated through protein phosphorylation. In addition, each of these factors can bind inhibitory proteins. Such mechanisms allow cells to mount global changes in translation rates. Many types of mRNA possess sequences that act to regulate their translation. For example, sequences in the 3′ UTR and 5′ UTR bind proteins that alter the ability of the mRNA to be translated.

Cells rapidly reduce protein levels through protein degradation

Once proteins are synthesized, they remain in the cell until they are degraded. Just as cells use degradation to control mRNA levels, they use protein degradation to control protein levels. Some proteins are removed only when they sustain enough damage to become dysfunctional. The structural changes in damaged proteins recruit enzymes that mark the protein for degradation. These enzymes transfer a small protein called **ubiquitin** to the damaged protein. Once the ubiquitination machinery has attached a ubiquitin chain to the damaged protein, the protein is bound by a multiprotein complex called the **proteasome**. Proteolytic enzymes within the proteasome degrade the ubiquitin-tagged proteins to amino acids.

Earlier we discussed how some types of mRNA are preferentially degraded. Many of these unstable mRNAs encode proteins that are also subject to accelerated degradation. Proteins such as cell cycle regulators and transcription factors can be ubiquitinated even in the absence of structural damage. Characteristic amino acid sequences within the proteins recruit the ubiquitination machinery. Often the recognition sequences can be phosphorylated, altering their ability to be subjected to rapid degradation.

Collectively, cells use these regulatory processes to control the levels of mRNA and protein. They enable cells to modify cellular properties in response to changing environmental and physiological conditions. Cells are also able to modulate their physiological response by altering the types of proteins they express. Animals, particularly vertebrates, can draw upon isoforms of proteins with subtly different properties that provide cells with alternative strategies to meet environmental and physiological challenges.

Protein variants arise through gene duplications and rearrangements

Protein isoforms provide a cell with flexibility in structure and function. A suite of proteins can be created with distinct properties. Isoforms can be produced through multiple mechanisms involving single genes, different alleles, or different genes (Figure 2.61).

Variations in protein structure can arise when the primary mRNA from a gene is connected together using different combinations of exons, a process known as **alternative splicing**. For example, more than 40 different isoforms of fibronectin can result from a single gene. Each isoform of fibronectin binds different combinations of extracellular matrix molecules.

Within any population of animals, there is some variation in the exact sequence of specific genes. As a consequence, a diploid individual may possess two different versions of the same gene, one arising from the mother and one from the father. These different forms of the same gene are **alleles**. If the gene encodes an enzyme, the isoforms are also called **allozymes**. Often the differences in allozyme structure have little effect on function. Because they are functionally neutral, natural selection does not remove them from the population. However, in some cases the regulatory

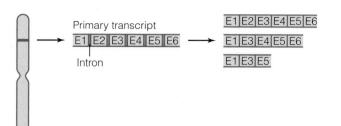

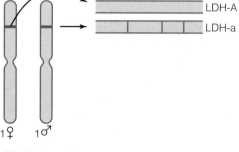

(a) Alternate splicing

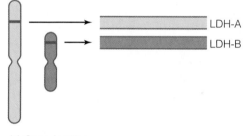

(b) Allelic variation

(c) Gene families

Figure 2.61   Origins of protein variants   Cells are able to produce protein isoforms in many different ways. Cells can splice exons in different combinations to create distinct proteins. Often the same gene can occur in different sequences within a population. Some individuals can have two different versions of the same gene (*A* or *a*) on chromosomes inherited from each parent. Gene duplications can lead to extra gene copies in different loci. These genes can diverge to encode different enzymes (A and B).

or catalytic properties of allozymes may be subtly different. Often different allozymes predominate in two populations of animals. For example, if a specific allozyme functions better in the cold, that gene might occur at a higher frequency in populations of animals exposed to the cold.

Other types of isoforms are encoded by separate genes that arose from ancestral **gene duplications**. Figure 2.62 shows some of the ways that genes can become duplicated. During the process of meiosis, long stretches of DNA may be

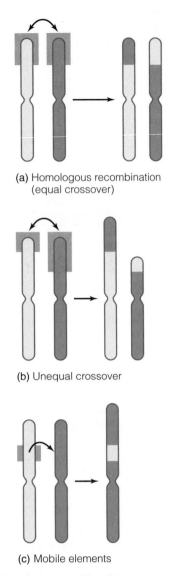

(a) Homologous recombination
(equal crossover)

(b) Unequal crossover

(c) Mobile elements

Figure 2.62 Gene duplications Gene recombination can provide cells with extra copies of genes. In contrast to equal crossover, **(a)** where homologous regions of chromosomes are exchanged, unequal crossover **(b)** provides one chromosome with extra genetic material. **(c)** Cells also possess many different kinds of mobile elements that can move or duplicate genes between chromosomes.

transferred from one chromosome to another. In most cases, two chromosomes exchange homologous regions and no gain or loss of genes occurs. This process of shuffling gene combinations is one of the advantages of sexual reproduction. Occasionally, the machinery of homologous recombination misidentifies homologous regions. Unequal crossover results, and one chromosome donates an end to another chromosome. The progeny derived from the gamete that lost the chromosomal region would not likely survive. However, the progeny from the recipient gamete

will be endowed with extra copies of the duplicated genes. These extra copies could kill the cell or, if neutral or beneficial, get transmitted to the next generation.

Another way that genes can become duplicated is through **mobile elements**. Many organisms possess genes that are capable of jumping from one chromosome to another. In most cases, the mobile element encodes a transposase, the enzyme required to cut the DNA from one strand and insert it into another. Occasionally, other genes become trapped in the mobile elements. When the mobile elements move, the other genes are carried along, endowing the recipient chromosome with the extra copy.

Genetic recombination does not always lead to production of extra copies of entire genes. In some cases, fragments of genes are moved from one gene and inserted into a completely different gene. A protein may possess domains within its structure that resemble regions of otherwise unrelated proteins. For instance, hundreds of different proteins can bind ATP using a protein structure called an **ATP-binding cassette**. This structure, which appears in all living organisms, probably arose only once, or a few times, billions of years ago. Its appearance in so many different genes and in all taxa is likely due to genetic recombination events that moved this region from one gene to another.

Ancient genome duplications contribute to physiological diversity

Gene duplications provide organisms with extra copies of redundant DNA that can accumulate mutations and diverge to endow the organisms with novel capacities. The key to achieving the opportunity for specialization is obtaining the raw material: a nonlethal extra copy of a gene. At several points in the evolution of animals, whole genomes were duplicated. Many of the duplicated genes were eventually lost, but many were retained and diverged to form gene families. Many of the anatomical and functional specializations of vertebrates are a result of these genomic duplications.

Often, if a particular gene is found in a single copy in an invertebrate, there are four isoforms in vertebrates. This "rule-of-four" reflects ancestral genome duplications; each single gene locus was duplicated, giving two copies of all genes, then reduplicated, giving four copies of all genes. The individual genes within the duplicated genomes underwent mutation, selection, and drift to diverge into distantly related genes. After a period of divergence, some individual genes duplicated again. The newly duplicated genes were more closely related to each other than to their distant ancestors, creating gene clusters. When did these genome duplications occur? A possible answer comes from phylogenetic analyses of a family of genes involved in development, the *Hox* family. The first genome duplication probably occurred just before the jawless vertebrates, or agnathans, diverged from the vertebrate lineage. The second duplication coincided with the development of jaws. The primitive chordates such as amphioxus have a single cluster of *Hox* genes, the agnathan lamprey has two or sometimes three clusters, and the more recent jawed vertebrates, from sharks to humans, possess at least four clusters of *Hox* genes. In each case, genome duplications coincided with important revolutions in morphological and physiological complexity.

These original genome duplications in the vertebrate lineage probably occurred more than 300 million years ago. Many modern animals have experienced relatively recent genome duplications, including many examples of frogs and fish that gained an extra set of chromosomes to become tetraploids. In some cases, tetraploid populations exist within diploid species; not nearly enough time has passed within the tetraploid lineage for the duplicated genes to diverge. The common carp, however, became tetraploid about 15 million years ago. Its closest relative, the grass carp, has half the number of chromosomes. Many genes that are in single copy in other vertebrates are found in pairs in common carp. While the pairs have diverged in structure, they have not yet become different in function.

Over many generations, the duplicated genes can follow many fates. The duplicated gene might incur mutations in the promoter or coding region that prevent it from being transcribed, rendering it a pseudogene. In some cases, one copy of the gene mutates and diverges, resulting in a protein with distinct properties. In other cases, both copies mutate and diverge, resulting in a pair of proteins with overlapping functions.

These genetic processes, originating early in animal evolution and operating at the level of individual cells, provide animals with physiological flexibility. The integration of different cell types into complex physiological systems is an important reason why animals have radiated into so many diverse species over the course of evolution.

⊙ | CONCEPT CHECK

13. Compare the categories of membrane transport in terms of energy requirements and direction of transport in relation to chemical gradients.
14. Discuss the composition of biological membranes. What are the unique properties of each type of lipid?
15. How can cells alter the fluidity of membranes, and why is this capacity important to cellular function?
16. Summarize the roles of the different subcellular compartments within a cell, and discuss how they influence physiological function.
17. Discuss the origins of genetic variation. How does genetic variation provide physiological flexibility?

Summary

Chemistry

→ All biological systems depend on kinetic and potential energy.

→ Food webs are essentially transfers of chemical energy between organisms.

→ Molecules possess thermal energy, which is reflected in molecular movement, and many metabolic processes in cells are mechanisms for capturing and transferring this energy.

→ Cells can also store energy in the form of electrochemical gradients. Gravitational energy and elastic storage energy are used in locomotion.

→ Covalent bonds, which arise when two atoms share electrons, are strong in comparison to weak bonds, including hydrogen bonds, van der Waals forces, and hydrophobic interactions.

→ Weak bonds control the three-dimensional structure of macromolecules. They form and break in response to modest changes in temperature.

→ Solute concentration imposes osmotic challenges. Organisms must modulate their biological solutions to regulate the ionization of water into H^+ and OH^-.

→ Changes in proton concentration, or pH, alter many molecular properties. As a result, animals have many physiological mechanisms to regulate pH, including pH buffers.

Biochemistry

→ Enzymes are organic catalysts, usually proteins, that speed reactions by reducing the activation energy barrier.

→ Enzyme reaction velocity (V) and substrate affinity (K_m) depend on the physicochemical environment, such as the temperature, ion composition, and pH of the solution.

→ Cells control reaction rates by changing the concentration of reactants, the levels or activities of enzymes, or the concentration of substrates and products.

→ Competitive inhibitors compete for the enzyme active site. Allosteric regulators bind at locations distant from the active site, altering enzyme kinetics. Many enzyme and nonenzyme proteins are regulated by covalent modification. For example, protein kinases use ATP to attach phosphate groups to specific amino acid residues, and protein phosphatases remove phosphate groups.

→ Cells use combinations of enzymes and enzymatic regulation to construct and maintain complex metabolic pathways.

→ Proteins, carbohydrates, and lipids have important roles in structure and metabolism. Animals store excess carbohydrate as glycogen. Glucose can be produced from noncarbohydrate precursors using gluconeogenesis. Glucose can be broken down to pyruvate (glycolysis) or further oxidized to CO_2.

→ Most animals use lactate dehydrogenase to balance redox and dispose of pyruvate. Anoxia-tolerant animals can use other pathways for oxidizing NADH in the absence of oxygen, some of which provide additional ATP.

→ Phospholipids, including phosphoglycerides and sphingolipids, are used to make cell membranes. Steroids and their precursors fulfill many roles within cells, and steroid hormones are particularly important in cell signaling.

→ Cells oxidize fatty acids for energy using the mitochondrial β-oxidation pathway, which generates reducing equivalents and acetyl CoA. The rate of β-oxidation is governed by the availability of fatty acids and the rate of transport into the mitochondria using the carnitine shuttle.

→ Fatty acids can be synthesized by the enzyme fatty acid synthase, for use in biosynthesis or energy storage. When energy is needed, lipases can break down triglycerides to release the fatty acids.

→ Under some conditions, such as starvation, fatty acids can be converted to ketone bodies for use in tissues that cannot use fatty acids directly.

→ Most oxidative fuels can be converted to acetyl CoA within mitochondria. When acetyl CoA enters the tricarboxylic acid cycle, acetyl CoA is oxidized to produce reducing equivalents, NADH and $FADH_2$.

→ Oxidation of reducing equivalents by the electron transport system generates a proton gradient, heat, and reactive oxygen species.

→ The mitochondria F_1F_O ATPase, or ATP synthase, uses the proton motive force to generate ATP. Phosphorylation is coupled to oxidation through a shared dependence on the proton motive force.

→ Under some circumstances, mitochondria can become uncoupled, leading to the production of heat instead of ATP.

→ The balance between biosynthesis and catabolism is regulated by energetic intermediates such as ATP, NADH, and acetyl CoA. Without this regulation, the two processes could occur simultaneously, leading to loss of energy in futile cycles.

→ Metabolic regulation also determines which fuels are oxidized under which conditions.

Cell Physiology

→ Membranes allow cells to create permeability barriers that help them to define environments. Membranes are heterogeneous combinations of phospholipids, cholesterol, and numerous integral and peripheral proteins.

→ The nature of the lipid membrane influences fluidity, an important determinant of protein function.

→ While some hydrophobic molecules can cross membranes by passive diffusion, membrane proteins are required for transport of most molecules.

→ Some transporters, such as ion channels, facilitate the diffusion of impermeant molecules down concentration gradients by creating pores.

→ Active transporters use energy to pump molecules against gradients.

→ The electrochemical gradients that exist across cellular membranes are produced by active transporters and used to drive diverse physiological processes.

→ The interior of the plasma membrane is electronegative, with a membrane potential between −5 and −100 mV. Potassium gradients are the most important component of the resting membrane potential. Changes in membrane permeability alter the membrane potential in ways that cells use to communicate.

→ Many aspects of animal physiology can be traced back to cellular processes.

→ The basic structure of cells—including the mitochondria, cytoskeleton, extracellular matrix, and secretory networks—can be regulated and remodeled to serve many purposes.

→ The ability to follow developmental programs, or respond to physiological and environmental challenges, resides in the genes. Physiological change begins in many cases with the ways cells control genes.

→ Cells and tissues are remodeled using processes from transcriptional control to post-translational regulation.

→ Evolutionary processes, including gene and genome duplications, provide the raw material for achieving physiological diversity.

Review Questions

1. How does the density of water change in relation to temperature? How do these properties affect animals that live in marine and freshwater environments?

2. If the enzymatic reaction $A + B \rightleftharpoons C + D$ is near equilibrium, then the mass action ratio is close to the equilibrium constant. What happens to the mass action ratio if you add more enzyme? What happens when you add more of A? What do you need to know to predict what would happen if temperature changed?

3. What metabolic conditions can affect the values of the respiratory quotient?

4. What metabolic conditions affect the relationship between ATP produced and oxygen consumed?

5. Trace the path of a protein hormone, such as insulin, from its gene in the nucleus to secretion out of the cell.

6. Discuss the mechanism by which cells can use transporters to change their osmotic and ionic properties.

Synthesis Questions

1. A type of protein comes in six different forms. Each form can dimerize with the other. How many unique homodimers and heterodimers can be formed from these six proteins?

2. Many animals maintain metabolites at concentrations near the K_m value for metabolic enzymes. For example, the concentration of pyruvate is often close to the K_m value for LDH. Why might this be advantageous, in terms of kinetic regulation?

3. Describe, in chemical terms, how antacids work.

4. Why do your hands get wrinkled if you spend too much time in the bathtub? Would the same

thing happen when you swim in the ocean? Describe these environments using the terminology of osmolarity and tonicity.

5. Many physiological processes require a change in the levels of proteins, such as membrane transporters. Discuss the processes that cells can use to change the protein levels. Discuss how the subcellular compartment influences this pathway.

6. Other physiological processes require changes in the *activities* of proteins. While this can arise through changes in the *levels* of proteins, it can also change through regulation of protein function. Discuss the various ways that cells can alter the activity of enzymes or transporters.

7. Discuss the ways in which a cell is able to alter its interactions with other cells.

Quantitative Questions

1. What is the proton concentration of a solution at pH 7.4? At what temperature would this solution be neutral?

2. Calculate the basis for an RQ = 1 for carbohydrate oxidation. Why does palmitate oxidation give an RQ = 0.7?

3. What rate of oxygen consumption would you expect in a tissue with a metabolic rate of 30 μmol ATP/ min?

For Further Reading

See the Additional References section at the back of the book for more references related to the topics in this chapter.

Chemistry

These texts provide good overviews of the chemical and physical underpinnings of cell biology and biochemistry.

Becker, W. M., L. J. Kleinsmith, and J. Hardin. 2003. *The world of the cell*, 5th ed. San Francisco: Benjamin Cummings.

Lehninger, A. L., D. L. Nelson, and M. M. Cox. 1999. *Principles of biochemistry*, 3rd ed. New York: Worth.

This text is a good primer for understanding the factors that affect protein structure.

Branden, C., and J. Tooze. 1991. *Introduction to protein structure*. New York: Garland Science.

These publications provide good background on the interactions between energy, chemical bonds, and water.

Bryant, R. G. 1996. The dynamics of water-protein interactions. *Annual Review of Biophysics and Biomolecular Structure* 25: 29–53.

Thornton, R. M. 1998. *The chemistry of life*. Menlo Park, CA: Benjamin Cummings.

Westof, E. 1993. *Water and biological macromolecules*. Boca Raton, FL: CRC Press.

This book looks at how animals and other organisms alter macromolecules in relation to environmental stress.

Hochachka, P. W., and G. N. Somero. 2002. *Biochemical adaptation*. Oxford: Oxford University Press.

These two books present differing views of the history of the discovery of the structure of DNA.

Sayre, A. 1975. *Rosalind Franklin & DNA*. New York: Norton, 1975.

Watson, J. 2001. *The double helix: A personal account of the discovery of the structure of DNA*. New York: Touchstone Books.

Biochemistry

This book, written by two pioneers in comparative biochemistry, explores the metabolic basis of biological diversity. Although the focus is on animals, they also consider other organisms that exemplify biochemical strategies for survival in adverse environments.

Hochachka, P. W., and G. N. Somero. 2002. *Biochemical adaptation*. Oxford: Oxford University Press.

Arthur Kornberg's autobiography gives his perspective on the history of the study of metabolic biochemistry.

Kornberg, A. 1991. *For the love of enzymes: The odyssey of a biochemist*. Cambridge, MA: Harvard University Press.

Lehninger is one of the standard undergraduate textbooks in biochemistry, with particularly good sections on metabolism and metabolic regulation.

Nelson, D. L., and M. M. Cox. 2000. *Lehninger principles of biochemistry*. New York: Worth.

Cell Physiology

These two textbooks cover the breadth of cell and molecular biology, with excellent illustrations. The strength of Alberts is its comprehensive nature, while Becker is very readable.

Alberts, B., A. Johnson, J. Lewis, M. Raff, K. Roberts, and P. Walter. 2002. *Molecular biology of the cell*. New York: Garland Science.

Becker, W. M., L. J. Kleinsmith, and J. Hardin. 2002. *The world of the cell*. San Francisco: Benjamin Cummings.

This comprehensive review of the ATP synthase does an excellent job of explaining how the enzyme works in the context of structural models of its function.

Boyer, P. D. 1997. The ATP synthase—A splendid molecular machine. *Annual Review of Biochemistry* 66: 717–749.

This book discusses the nature of evolutionary and physiological variation from the perspective of cell and developmental biology.

Gerhart, J., and M. Kirschner. 1997. *Cells, embryos and evolution*. New York: Blackwell Science.

Ohno's early book outlines his perspective on the importance of gene duplication in the evolution of biological diversity. More recently, in a series of papers, a number of authors bring the field up-to-date, incorporating recent evidence of the role of genome duplications in origins of gene families and cellular diversity.

Ohno, S. 1970. *Evolution by gene duplication*. Heidelberg: Springer Verlag.

Various authors. 1999. Gene duplication in development and evolution. *Seminars in Cell and Developmental Biology* 10: 515–563.

An excellent overview of transport and transporters.

Stein, W. D. 1990. *Channels, carriers and pumps: An introduction to membrane transport*. San Diego: Academic Press.

The original book by Sir D'Arcy Wentworth Thompson, written in 1917, was one of the first to examine how physiology was influenced by mathematics and physics.

Thompson, D. W. 1961. *On growth and form*. Abridged edition edited by J. T. Bonner. Cambridge: Cambridge University Press.

Cell Signaling and Endocrine Regulation

At every level of organization, life depends on communication. Animals send signals to each other in the form of sounds, scents, and visual cues. Within an animal, organs, tissues, and cells communicate with each other using chemical and electrical signals. Even within a single cell there is constant communication of information among organelles. Two of the most familiar types of cellular communication in animals involve the nervous system and the endocrine system. Although the nervous and endocrine systems may appear to be quite different, they are part of a continuum of cellular communication systems that share many important similarities.

In all organisms, cellular communication systems involve sending and receiving a signal, often in the form of a chemical. We can see the fundamentals of these mechanisms even in prokaryotes. For example, the marine bacterium *Vibrio fischeri* is capable of producing light, but does so only when the bacteria are present at high density. When the bacteria are at low densities, they produce a chemical called an autoinducer that diffuses across the membrane

into the environment, but its concentration remains low. When many bacteria are present within a small area, however, the environmental concentration of autoinducer rises. At high concentrations, the autoinducer binds to a specific receptor with the bacterial cell, causing the receptor to change shape and act as a transcription factor that induces the transcription of the genes involved in light production. Thus, when the bacteria are present at high densities, the light-producing genes are induced and the bacteria glow in the dark.

V. fischeri seldom reach high enough densities to glow when they are free-living, but these bacteria are also found in a mutualistic relationship with a species of squid—*Euprymna scolopes* shown in the photograph above. The bacteria colonize specialized light organs on the underside of the squid. The squid's light organs provide an ideal home for the bacteria, allowing them to grow to very high density and produce light. This light, which glows from the underside of the squid, allows the predatory squid to blend in with the light descending through the water from the surface, mak-

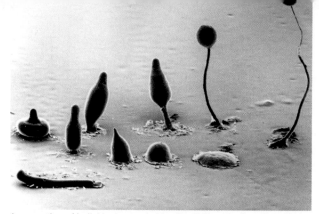

Aggregation of individual amoeboid cells of *Dictyostelium discoideum* into a colony is activated by chemical signals.

The brightly colored rumps of female Hamadryas baboons are the result of chemical signaling.

ing them invisible from below. Thus, the glowing bacteria act as camouflage that helps the squid to catch their prey. The complex mutualistic relationship between the bacteria and the squid depends on cellular signaling among the bacteria via the autoinducer, and between the bacteria and the squid because squid reared in the laboratory in the absence of the bacteria do not develop a complete light organ.

Another example of cellular signaling can be found in unicellular eukaryotes such as *Dictyostelium discoideum*, a species of cellular slime mold. Much of the time slime molds function as individual, independent, amoeboid cells. These cells move through their environment phagocytosing other cells for food. But when conditions are poor, slime mold cells begin to secrete a signaling molecule called cyclic adenosine monophosphate (cAMP). When a slime mold cell encounters environmental cAMP, the cAMP binds to a receptor on the surface of the slime mold cell, causing the receptor to undergo a conformational change. The conformational change of the receptor activates two different intracellular signaling pathways. The first signaling pathway activates an enzyme called adenylate cyclase, which catalyzes the production of cAMP in the recipient cell, causing cAMP secretion. The second signaling pathway causes the recipient cell to release intracellular calcium, which acts on the proteins of the cytoskeleton to induce amoeboid movements. Together, these two intracellular responses cause an amoeboid slime mold cell that encounters environmental cAMP to move up the cAMP gradient toward the signaling individual, and to add to the secreted cAMP in the environment. As more and more cells respond to the signal and begin to aggregate into a small area, the cAMP signal intensifies, attracting even more amoeboid cells, and increasing the size of the aggregation of cells. Eventually, the group of cells forms a migratory blob termed a *pseudoplasmodium*. The pseudoplasmodium moves through its environment until it finds a suitable spot, and then differ-

entiates to form a complex structure consisting of a stalk and a fruiting body. The fruiting body produces spores that are capable of surviving extremely harsh conditions. The spores can also break away from the fruiting body to be carried by the wind to other locations. Once conditions improve, the spores germinate into individual amoeboid cells, starting the cycle over again.

These two examples of cellular signaling, in a prokaryote and a unicellular eukaryote, illustrate the fundamental features of cellular communication in all living things: the production of a signal in one cell, the transport of that signal to a target cell, and the transduction of that signal into a response in the target cell.

The complexity of animal physiology and behavior requires an enormous diversity in signaling mechanisms. Nowhere is this diversity more obvious than in the endocrine system. In most animals, the endocrine system is involved in controlling and regulating almost every physiological process including growth, development, metabolism, and ion and water balance. The endocrine system's role in reproduction and development is one of the most obvious manifestations of cellular signaling. For example, when female Hamadryas baboons are ready to mate, a variety of endocrine signals cause the development of a characteristic patch of red, swollen skin around their genitals. These swellings act as a visual signal to attract males, helping to promote reproduction. Thus, the endocrine system is responsible for inducing a very large and obvious change in the phenotype of these females. In insects, the endocrine system controls the metamorphosis from larva to butterfly. Despite their diversity and complexity, however, the mechanisms of endocrine signaling in animals share many features in common with the signaling systems of unicellular organisms. In this chapter, and throughout the remainder of this book, you will see the critical role of these cellular communication mechanisms in allowing animals to perform their complex functions.◉

Overview

Everything that an animal does involves communication among cells. Moving, digesting food, and even reading this book all require the coordinated action of thousands of individual cells engaging in constant communication. Communication between cells occurs when a *signaling cell* sends a signal to a *target cell*, usually in the form of a *chemical messenger*. Figure 3.1 summarizes the principal types of cell signaling in animals. Adjacent cells can communicate directly through aqueous pores in the membrane called gap junctions, but the majority of cells have no direct contact with each other. Thus, most cell signaling is indirect, and begins when one cell releases a chemical messenger into its environment. The chemical messenger then travels through the extracellular fluids until it reaches the target cell. At the target cell, the chemical messenger binds to a **receptor**, changing the shape of the receptor and activating **signal transduction pathways** that cause a response within the target cell. Interactions between chemical messengers, receptors, and signal transduction mechanisms allow cells to communicate with each other.

Chemical messengers can travel from a signaling cell to nearby target cells by diffusion in a process called **paracrine** communication. These messengers can even affect the signaling cell, in a process called **autocrine** communication. But the rate of diffusion is limited by distance, and thus

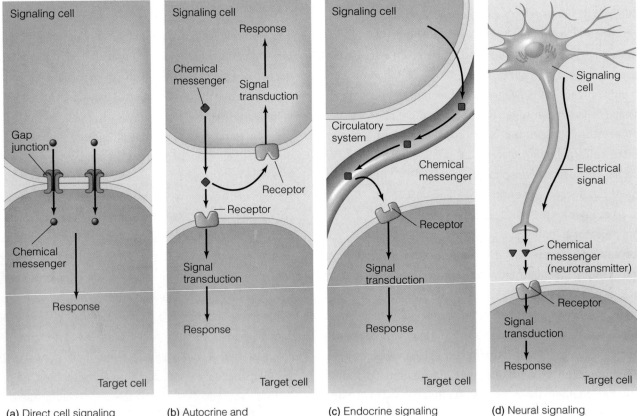

(a) Direct cell signaling

(b) Autocrine and paracrine signaling

(c) Endocrine signaling

(d) Neural signaling

Figure 3.1 An overview of cell signaling Cells communicate either directly, via aqueous pores that connect adjacent cells, or indirectly when the signaling cell releases a chemical messenger into the extracellular environment. **(a)** Direct cell signaling can occur through pores called gap junctions. **(b)** Paracrine signaling occurs when chemical messengers diffuse from the signaling cell to nearby target cells where they bind to receptors and initiate signal transduction pathways that cause a response. Autocrine signaling is similar except that the chemical messenger causes a response in the signaling cell. **(c)** Endocrine signaling occurs when chemical messengers called hormones travel long distances via the circulatory system. When the hormone reaches the target cell it binds to a receptor, initiates signal transduction pathways, and causes a response. **(d)** In neural signaling, electrical signals travel across long distances within a single cell. The electrical signal then either passes directly to the target cell via gap junctions, or triggers the release of a chemical messenger called a neurotransmitter. The neurotransmitter carries the signal to the target cell by diffusing across a short distance, where it binds to receptors on the target cell, initiates signal transduction pathways, and causes a response.

diffusion is insufficient to carry signals to distant target cells. For long-distance cell-to-cell communication, animals use the **endocrine system** and **nervous system**. In the endocrine system, the chemical messenger travels from the signaling cell to the target cell carried by the circulatory system. These endocrine messengers are called **hormones**. In the nervous system, an electrical signal travels across a long distance within a single cell (the neuron), and is transferred to the target cell over a very short distance, often in the form of a chemical messenger called a **neurotransmitter**. Animals can even send chemical messengers between individuals, a system termed *exocrine* communication.

Although these systems appear to be rather distinct, they actually share many features in common at the biochemical level. In this chapter, we first examine the biochemical basis of cell signaling, outlining the shared features of different signaling systems. We look at how cells release chemical messengers, how these messengers travel to the target cell, how they bind to receptors, and how they exert their effects through signal transduction pathways. We devote much of this chapter to a discussion of the fundamental properties of receptors and signal transduction mechanisms, not only because these processes are involved in the regulation of every physiological system, but also because you will encounter receptors and signal transduction mechanisms many times throughout this book. We then step back from the cellular details of communications mechanisms to take a closer look at one of the important cellular communication systems in animals: the endocrine system.

The Biochemical Basis of Cell Signaling

Cells are separated from their environment by a phospholipid membrane. Thus, any chemical messenger traveling between two cells must first pass from the aqueous cytoplasm of the signaling cell, through its lipid membrane, and into the aqueous extracellular fluid. At the target cell the messenger must then signal across the lipid membrane of the target cell into its aqueous cytoplasm. Since most chemicals are either soluble in aqueous solutions (hydrophilic) or soluble in lipids (hydrophobic), sending a chemical messenger from one cell to another presents a substantial challenge. For example, hydrophobic chemical messengers can pass through cell membranes, but do not dissolve well in aqueous fluids such as cytoplasm or blood. Hydrophilic chemical messengers are soluble in the cytoplasm and extracellular fluids, but do not pass through cell membranes. These fundamental chemical properties pose a problem that cells must solve in order to communicate with each other.

General Features of Cell Signaling

Cells can circumvent the problem of moving a hydrophilic chemical messenger through the lipid environment of the membrane by communicating via **gap junctions**. Gap junctions are specialized protein complexes that create an aqueous pore between the cytoplasms of two adjacent cells (Figure 3.2). Gap junctions are composed of interlocking cylindrical proteins (called *connexins* in vertebrates, or *innexins* in invertebrates) assembled in groups of four or six to form doughnut-like pores (*hemichannels* or *connexons*) in the cell membrane. The hemichannels of two adjacent cells come together to form a hollow tube, connecting the two cells via an aqueous bridge. Thus, chemical messengers can travel from the signaling

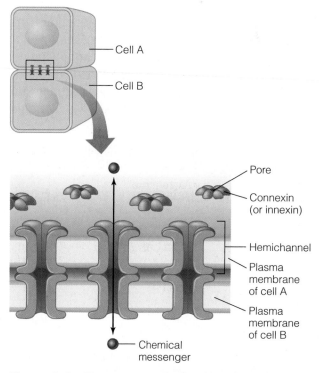

Cell A

Cell B

Pore

Connexin
(or innexin)

Hemichannel

Plasma
membrane
of cell A

Plasma
membrane
of cell B

Chemical
messenger

Figure 3.2 The structure of gap junctions Gap junctions are protein complexes that form aqueous pores between adjacent cells. Proteins called connexins (in vertebrates) or innexins (in invertebrates) form the structure of the gap junction.

cell to the target cell via gap junctions without ever leaving an aqueous environment.

We can demonstrate that two cells are connected via gap junctions by injecting a fluorescent dye that cannot cross the cell membrane into one of the cells. If gap junctions connect two cells, dye that is injected into one cell will diffuse through the gap junctions into the adjacent cell (if the dye is small enough to pass through the pore), and both cells will start to fluoresce. If no gap junctions are present, the dye will remain in the first cell because it is unable to cross the membrane, and the second cell will not fluoresce.

In most physiological situations, direct communication via gap junctions involves the movement of ions between cells. The movement of ions into or out of a cell can act as a signal by causing a change in the membrane potential (see Chapter 2: Chemistry, Biochemistry, and Cell Physiology) that triggers a response in the target cell. This rapid communication of signals between adjacent cells is a simple way to coordinate cellular responses. As we see in later chapters, the movement of ions through gap junctions helps to coordinate the contraction of smooth and cardiac muscle, and is involved in the transmission of electrical signals between some nerve cells. Other small molecules can also move between cells via gap junctions, including a variety of intracellular signaling molecules such as cyclic adenosine monophosphate (cAMP). Thus, gap junctions play a critical role in coordinating physiological responses at the tissue level. Gap junctions are not just passive channels between adjacent cells. They can be opened and closed to regulate communication of substances between cells. Increased intracellular calcium and decreased intracellular pH both cause gap junctions to close. The number of gap junctions connecting two cells can also be regulated on a physiological time scale.

Direct communication via gap junctions is a very efficient way to send signals, but gap junctions can only form between adjacent cells. Animals need other strategies for sending signals to more distant cells, or to neighboring cells that are not connected by gap junctions. This kind of signaling is called *indirect cell signaling*, and involves three steps:

1. Release of a chemical messenger from the signaling cell into the extracellular environment
2. Transport of the chemical messenger through the extracellular environment to the target cell
3. Communication of the signal to the target cell via receptor binding

Indirect signaling systems form a continuum

Although the systems that animals use for indirect signaling are often discussed as if they were quite different from each other, they are actually just specialized ways of achieving the same result. In fact, at the biochemical level they share a great deal in common. Table 3.1 shows some of the similarities and differences between the various types of cellular communication. In general, autocrine,

	Table 3.1 Comparison of systems for cell-to-cell communication.			
Feature	**Autocrine/ Paracrine**	**Nervous**	**Endocrine**	**Exocrine**
Secretory cell	Various	Neural	Endocrine	Various
Target cell	Most cells in body	Neuron, muscle, endocrine	Most cells in body	Sensory and neural
Signal type	Chemical	Electrical and chemical	Chemical	Chemical
Maximum signaling distance	Short	Long intracellularly, short across synapse	Long	Very long
Transport	Extracellular fluid	Synapse	Circulatory system	External environment
Speed	Rapid	Rapid	Slower	Various
Duration of response	Short	Short	Longer	Various

paracrine, neural, endocrine, and exocrine communication systems differ largely in the type of cell involved in messenger secretion and in the way that the messenger is transported to the target cell. In contrast, the mechanisms governing the release of the chemical messenger from the signaling cell, the types of chemical messengers utilized, and the mechanisms for communicating the signal to the target cell are actually very similar among systems.

The most important distinction among the different systems for cellular communication is the distance across which the chemical messenger must travel. In autocrine and paracrine communication, the chemical messenger simply diffuses through the extracellular fluid from the signaling cell to the target cell. Because the rate of diffusion is limited by distance, autocrine and paracrine signals are localized, affecting only those target cells that are within a short distance of the signaling cell. Intercellular signaling also occurs across short distances in the nervous system, at a structure called the **synapse**, a region where the signaling cell and the target cell are very close together. Signals can move from cell to cell across the synapse via gap junctions, if they are present, in a form of direct cell-to-cell communication. Alternatively, neurotransmitters can carry a signal across the synapse by diffusing from the signaling cell to the target cell, where they bind to receptors. Because neurotransmitters diffuse from the signaling cell to the target cell across the synapse, this mechanism of synaptic communication is similar to paracrine communication. Although cell-to-cell communication in the nervous system can only occur across short distances, nervous signals can be communicated across very long distances. Unlike other forms of cellular communication, however, long-distance nervous communication occurs within a single cell. The unique structure of neurons allows electrical signals to be propagated across a long distance within a single cell without degrading. We do not discuss the physiology of neurons in detail in this chapter, but instead devote Chapter 4: Neuron Structure and Function to an examination of their special properties. The endocrine system can regulate the activities of distant cells, tissues, and organs by sending chemical signals through the blood in the form of hormones. Because they are carried by the circulatory system, rather than moving only by diffusion, hormones can quickly travel across long distances through the body. In exocrine communication, a chemical termed a **pheromone**

is released by one individual and travels through the external environment (e.g., air or water) to exert its effects on a different individual.

The differences in the mechanisms that the various types of communication systems use to transport chemical messengers from the signaling cell to the target cell result in differences in the speed of communication of these systems. Autocrine and paracrine communication are very rapid, because chemical signals need only diffuse across very small distances. Diffusion is a rapid process at these scales, so autocrine and paracrine communication occurs on a time scale of milliseconds to seconds. Nervous communication is similarly rapid. Propagation of electrical signals within a neuron occurs on a millisecond scale, and diffusion of a neurotransmitter across the synapse is also rapid. In contrast, endocrine communication is usually slower, because it relies on transport of hormones in the circulatory system. Depending on the organism, blood may require several seconds to minutes to make a complete circuit around the body. In addition, endocrine hormones are often longer-lived in extracellular fluids than are paracrines or neurotransmitters, increasing the length of time over which they can have an effect.

Only neurons act as the secretory cells in nervous communication, but a variety of cell types can be involved in exocrine and endocrine communication. The distinction between nervous and endocrine signaling is, however, somewhat blurry. Some neurons can secrete neurotransmitters directly into the circulatory system, in which case the messenger is termed a **neurohormone**, because it is secreted by a neuron but acts like a hormone. The secretory cells of the exocrine and endocrine tissues are often grouped into structures called **glands** (Figure 3.3). Endocrine glands release their secretions (hormones) directly into the circulatory system. The endocrine cells within these glands are typically very specialized for their secretory function. However, there are many hormones that are not secreted by endocrine glands. For example, cells within the atria of the heart release a hormone called *atrial natriuretic peptide* that is involved in the regulation of blood pressure. Thus, the distinction between endocrine communication and other types of intercellular communication can be difficult to elucidate when viewed from the perspective of the signaling cell.

Exocrine glands release their secretions into ducts that lead to the surfaces of the body (including

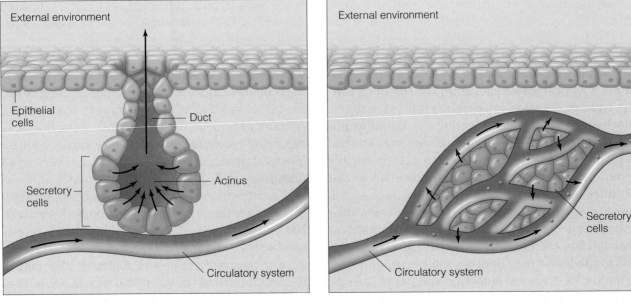

(a) Exocrine gland

(b) Endocrine gland

Figure 3.3 The structure of exocrine and endocrine glands Exocrine glands secrete chemicals into ducts that lead to the surface of the body, whereas endocrine glands secrete hormones directly into the circulatory system.

the skin, respiratory surfaces, and the surface of the gut). Exocrine secretions that contain pheromones are involved in animal-to-animal communication, but exocrine secretions can also participate in many processes in addition to communication, including locomotion, digestion, and prey capture. For example, exocrine mucus secretions form a protective layer over many epithelia, including the gills of fish and the lungs of terrestrial animals. Mucus secretions can also help in locomotion, as in the slime trails of slugs and snails. Saliva produced in the mouth of mammals begins digestion, and helps food to slide down the esophagus. Spiders make silk, a very specialized exocrine secretion, to trap prey.

Because all of the different forms of cell signaling share so many features in common, in the next sections we begin our consideration of the biochemical basis of cell signaling without separating the different types of signaling used by animals. In this way, we can clearly see how cells have solved the general problem of sending chemical signals across the cell membrane when direct communication is not possible.

The structure of the messenger determines the type of signaling mechanism

The chemical structure of the messenger is the critical property that affects the way in which indirect signaling is accomplished. Hydrophobic messengers use different mechanisms for signaling than do hydrophilic messengers, because hydrophobic messengers can diffuse freely across cell membranes whereas hydrophilic messengers cannot. Table 3.2 summarizes the similarities and differences between hydrophilic and hydrophobic chemical messengers in each step of indirect cell signaling.

There are six main classes of chemicals that are known to participate in cellular signaling in animals: peptides, steroids, amines, lipids, purines, and gases. All known hormones are peptides, steroids, or amines, whereas there are examples of all six classes of messengers acting as autocrines, paracrines, or neurotransmitters. In the next sections we look at each of these main classes of chemical messengers to see how their biochemical properties affect their release from the signaling cell, transport through the extracellular fluid, and actions on the target cell.

Peptide Messengers

Amino acids, peptides, and proteins can all act as signaling molecules. Amino acids typically act as neurotransmitters, whereas peptides and proteins may be autocrines, paracrines, neurotransmitters, neurohormones, hormones, or pheromones. Peptide and protein messengers consist of two or

Table 3.2   A comparison of hydrophilic and hydrophobic chemical messengers.

Feature	Hydrophilic messengers	Hydrophobic messengers
Storage	Intracellular vesicles	Synthesized on demand
Secretion	Exocytosis	Diffusion across membrane
Transport	Dissolved in extracellular fluids	Short distances: dissolved in extracellular fluid Long distances: bound to carrier proteins
Receptor	Transmembrane	Intracellular or transmembrane
Effects	Rapid	Slower or rapid

more amino acids linked in series, and range in size from 2 to 200 amino acids in length. Chains of fewer than 50 amino acids are usually called peptides, while the word *protein* is used for longer chains. Peptide and protein messengers are hydrophilic chemicals that cannot diffuse across the membranes, but are soluble in aqueous solutions.

Peptide messengers are released by exocytosis

Peptide and protein messengers are synthesized on the rough endoplasmic reticulum along with most of the other proteins destined for secretion from the cell. The peptides are then packaged into vesicles for either immediate release, or storage for later use. Most of the peptide hormones and neurotransmitters are synthesized in advance and stored for later release, whereas paracrine peptides such as the **cytokines** are synthesized only on demand. We can see the importance of regulated exocytosis of stored messenger by examining the effects of botulinum toxin, a protein produced by the bacterium *Clostridium botulinum*. This protein blocks the regulated exocytosis of neurotransmitters traveling between nerves and muscles, preventing muscle contraction and causing paralysis. Exposure to a large dose of this toxin causes the disease botulism, which is characterized by weakness and paralysis, generally starting in the area of the head and progressing to paralysis of the muscles of the rest of the body, including those involved in swallowing and breathing. If untreated, an individual with a severe case of botulism is likely to die of respiratory failure. Although the botulinum toxin (also called botox) is one of the most potent poisons known, it can be used as a medical therapy. Injecting small amounts of botox directly into a muscle leads to local paralysis, and can be used to treat muscle spasms. It is also used in cosmetic medicine to reduce facial wrinkles such as frown lines.

Peptide hormones are often synthesized as large, inactive polypeptides called **preprohormones** (Figure 3.4). Preprohormones contain not only one or more copies of a peptide hormone or hormones, but also a signal sequence that targets the polypeptide for secretion. The signal sequence is cleaved from the preprohormone prior to being packaged into secretory vesicles, forming the **prohormone**, which like the preprohormone is usually inactive. The secretory vesicle contains proteolytic enzymes that cut the prohormone into the active hormone or hormones. The signaling cell then releases the active peptide hormone by exocytosis.

Figure 3.5 shows an example of a preprohormone, the one containing arginine vasopressin (AVP), also known as antidiuretic hormone (ADH). Ribosomes on the exterior of the rough endoplasmic reticulum translate the preprovasopressin mRNA into protein. The signal peptide directs the newly synthesized polypeptide to the interior of the rough endoplasmic reticulum. The signal peptide is then cleaved off, forming provasopressin, which is packaged into secretory vesicles. In the secretory vesicles it is cleaved into three different peptides: vasopressin, neurophysin, and a glycoprotein. Arginine vasopressin is a hormone that acts on the kidney to regulate the reabsorption of water (see Chapter 10: Ion and Water Balance). The functions of neurophysin and the glycoprotein are not yet well understood, but they may be involved in the proper sorting and secretion of arginine vasopressin.

Peptide messengers dissolve in extracellular fluids

Once released from the signaling cell, a chemical messenger must move through the extracellular fluid to the target cell. Hydrophilic chemical

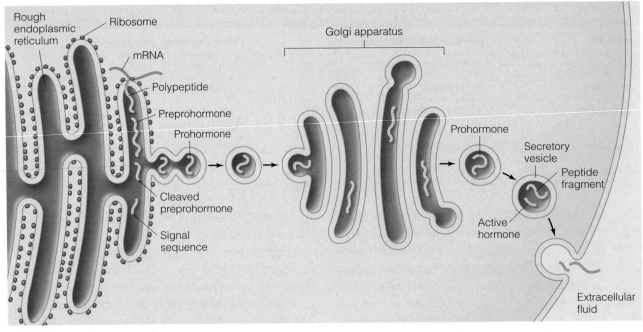

Figure 3.4 Synthesis of peptide hormones

Peptide hormones are synthesized by ribosomes on the rough endoplasmic reticulum, often as large preprohormones. The preprohormone enters the rough endoplasmic reticulum, where the signal sequence is cleaved off. The resulting prohormone is packaged into vesicles that move to the Golgi apparatus for further processing and sorting. In the Golgi apparatus, the prohormone is packaged into secretory vesicles, where it is cleaved into active hormone and one or more peptide fragments. The secretory vesicle fuses with the plasma membrane, releasing its contents by exocytosis.

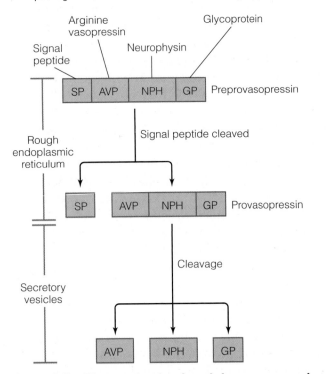

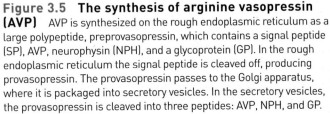

Figure 3.5 The synthesis of arginine vasopressin (AVP)

AVP is synthesized on the rough endoplasmic reticulum as a large polypeptide, preprovasopressin, which contains a signal peptide (SP), AVP, neurophysin (NPH), and a glycoprotein (GP). In the rough endoplasmic reticulum the signal peptide is cleaved off, producing provasopressin. The provasopressin passes to the Golgi apparatus, where it is packaged into secretory vesicles. In the secretory vesicles, the provasopressin is cleaved into three peptides: AVP, NPH, and GP.

messengers such as peptides and proteins dissolve well in aqueous solutions and can easily move from the signaling cell to the target cell, either by diffusion or carried by the circulatory system. Peptides messengers are usually broken down and removed from extracellular fluids by proteolytic enzymes. The rate of this breakdown can be measured as the messenger's **half-life**—the time taken to reduce the concentration of the messenger by half. Peptide messengers generally have half-lives ranging from a few seconds to a few hours. As a result of these short half-lives, the signaling cell must continually produce messenger in order to cause a sustained response in a target cell.

Peptides bind to transmembrane receptors

Hydrophilic signaling molecules such as peptides and proteins cannot pass through the membrane of the target cell, but instead bind to **transmembrane receptors** (Figure 3.6). The extracellular portion of a transmembrane receptor contains the *ligand-binding domain*. **Ligand** is the general term for any small molecule that binds specifically to a protein. Thus, a peptide chemical messenger acts as a

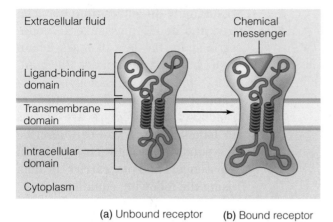

(a) Unbound receptor **(b)** Bound receptor

Figure 3.6 The structure of a transmembrane receptor **(a)** Transmembrane receptors have an extracellular ligand-binding domain, a membrane-spanning domain, and an intracellular domain. **(b)** When the messenger (ligand) binds to the receptor, the conformation of the receptor changes.

ligand for a transmembrane receptor protein. Transmembrane receptors also have a membrane-spanning (*transmembrane*) domain and an *intracellular domain.* When a ligand binds to the ligand-binding domain of a transmembrane receptor, the receptor changes shape, communicating the signal carried by the ligand across the cell membrane, without the ligand itself needing to cross the lipid-rich membrane. Transmembrane receptors activate cytoplasmic signal transduction pathways that cause rapid changes in the activity of the target cell, usually by altering membrane potential, or phosphorylating and modifying the activity of existing proteins.

Steroid Messengers

Steroids are derived from the molecule cholesterol, and are important hormones in both vertebrates and invertebrates. Steroids can also act as paracrine and autocrine signals in some tissues, and are important pheromones involved in communication among animals. Figure 3.7 shows a generalized synthetic pathway for some of the important steroid hormones in vertebrates. The enzymes for steroid biosynthesis are located in the smooth endoplasmic reticulum or mitochondria. There are three major classes of steroid hormones in the vertebrates. **Mineralocorticoids** are involved in regulating sodium uptake by the kidney, and are important for fluid and electrolyte balance

in the body. Aldosterone is the primary mineralocorticoid in mammals. We discuss the mineralocorticoids in more detail in Chapter 10: Ion and Water Balance. The **glucocorticoids** (cortisol, cortisone, and corticosterone), also called the stress hormones, have widespread actions including increasing glucose production, increasing the breakdown of proteins into amino acids, increasing the release of fatty acids from adipose tissue, and regulating the immune system and inflammatory responses. The *reproductive hormones* (estrogens, progesterone, testosterone), which we discuss in Chapter 14: Reproduction, regulate sex-specific characteristics and reproduction.

The principal steroids in invertebrates are the *ecdysteroids,* which play an important role in the regulation of molting in the arthropods (see Box 3.1, Evolution and Diversity: Ecdysone: An Arthropod Steroid Hormone). Much less is known about the role of these steroids in other invertebrate phyla, but they are thought to play a role in development and reproduction.

Because all steroids contain several carbon rings, some synthetic chemicals with similar structures bind to steroid receptors and mimic or block the action of the natural hormone. Environmental exposures to chemicals such as the insecticide DDT have been associated with low sperm counts and increased incidence of breast and prostate cancer in humans, developmental abnormalities such as reduced penis size and feminization in animals including fish and alligators, and interference with molting in crustaceans. Environmental chemicals such as DDT bind to and activate the receptor for estrogen, and other chemicals such as some pesticides interfere with other aspects of steroid metabolism—a phenomenon called *endocrine disruption.*

Steroids bind to carrier proteins

Because steroids can easily pass through biological membranes, they cannot be stored within the cell, and thus must be synthesized on demand. They then diffuse across the membrane of the signaling cell and into the extracellular fluid. Steroids can diffuse across short distances dissolved in extracellular fluids, but for long-distance transport they are usually bound to **carrier proteins**. Some steroids have specific carrier proteins (termed binding globulins), while others bind nonspecifically to generalized carrier proteins, such as **albumin**, the

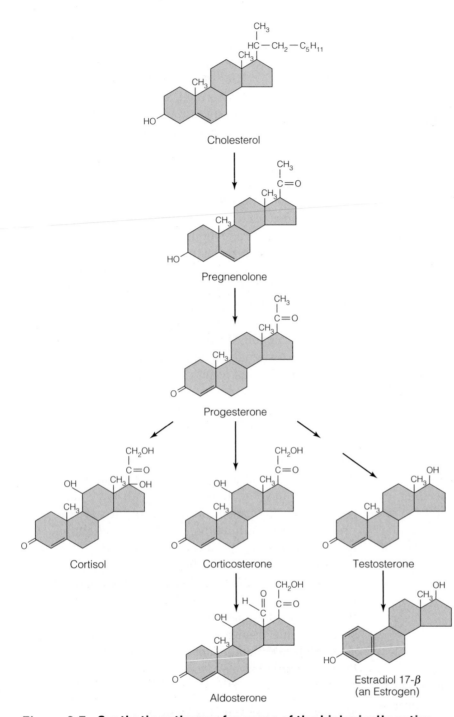

Figure 3.7 Synthetic pathways for some of the biologically active steroids in vertebrates Cholesterol is the precursor for the three main classes of vertebrate steroids: the glucocorticoids (including cortisol and corticosterone), the mineralocorticoids (including aldosterone), and the sex steroids (including testosterone and estradiol).

resulting in an equilibrium between free and bound messengers. As described by the law of mass action (see Chapter 2: Chemistry, Biochemistry, and Cell Physiology), in an equilibrium system the amounts of reactants and products are always in balance. Thus, you can describe the equilibrium between a chemical messenger and its carrier protein using the following equation:

$$M + C \rightleftharpoons M\text{-}C$$

where M is the concentration of unbound messenger, C is the concentration of carrier protein, and M-C is the concentration of messenger bound to carrier protein. If the amount of free messenger increases, the equilibrium will shift to the right, increasing the amount of messenger bound to carrier protein. If the amount of free messenger decreases, the equilibrium will shift to the left, decreasing the amount of messenger bound to carrier protein.

The binding of a hydrophobic messenger to its carrier proteins is outlined in Figure 3.8. When a signaling cell releases a chemical messenger into the extracellular fluid, the free concentration of the messenger is high in the local environment, and the messenger will tend to bind to its carrier protein. For most hydrophobic chemical messengers, greater than 99% of the messenger binds to its carrier protein, but a small fraction of the messenger is always free in solution. Both free and bound messenger travel through the circulatory system to the target cell. At the target cell, the free messenger diffuses into the cell and binds to its receptor. The binding of the messenger to its receptor reduces the concentration of free messenger in the extracellular fluid adjacent to the target cell. The resulting low concentration of free messenger causes the bound messenger to dissociate from the carrier protein (because of the law of mass action), delivering the messenger to the target cell.

principal carrier protein in vertebrate blood. Carrier proteins help hydrophobic chemical messengers dissolve in aqueous solutions by surrounding the hydrophobic messenger and isolating it from the aqueous solution. Hydrophobic chemical messengers bind reversibly to their carrier proteins,

BOX 3.1 **EVOLUTION AND DIVERSITY**
Ecdysone: An Arthropod Steroid Hormone

All arthropods have a rigid exoskeleton, a hard outer covering that provides both protection and support. In order to grow, an arthropod must shed its exoskeleton in a process called molting. The hormones that regulate molting have been intensively studied in the insects, and one of the most important is a steroid hormone called ecdysone. Most insects molt several times during larval development. In the hemimetabolous insects, the larval stages and adult all resemble each other, with each stage simply being larger than the preceding one. The younger stages may differ slightly in shape or color from the adults, and lack sexual organs, but there is no major change in body form. In contrast, the adults of the holometabolous insects differ radically in shape from their larvae. Caterpillars and butterflies, for example, are the larval and adult stages of the holometabolous Lepidopteran insects. Holometabolous insects have an additional developmental stage, called a pupa, between the larva and the adult, during which they undergo the process of metamorphosis—a complete remodeling of their body structures.

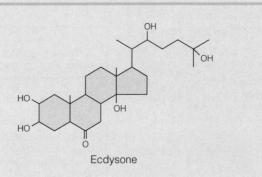

Ecdysone

As discussed in Chapter 14, the steroid hormone ecdysone can stimulate an insect larva to molt to form a larger larva, a pupa, or an adult, depending on the level of an additional hormone, juvenile hormone. When juvenile hormone levels are high, ecdysone stimulates molting from one larval stage to another. When juvenile hormone levels are low, ecdysone triggers the formation of the pupa in holometabolous insects. When juvenile hormone is absent, ecdysone triggers the emergence of the adult insect.

The structure of ecdysone is similar to that of the vertebrate steroid hormones, but it contains more hydroxyl groups. Ecdysone secretion is regulated by a neurohormone called prothoracicotropic hormone (PTTH) produced by the insect brain. This neurohormone stimulates the prothoracic gland to secrete ecdysone. Ecdysone is actually a prohormone, and is rapidly converted to the active hormone 20-hydroxyecdysone (also called ecdysterone) by enzymes found in the hemolymph and various peripheral tissues. 20-Hydroxyecdysone binds to an intracellular receptor that regulates gene expression by binding to a hormone responsive element, as do the vertebrate steroid hormones.

Although 20-hydroxyecdysone is structurally similar to the vertebrate steroids, it does not appear to be biologically active in vertebrates, and does not have detectable effects on the reproductive system. However, a few studies have reported that ecdysterone has anabolic effects in vertebrates, increasing muscle growth and lean muscle mass.

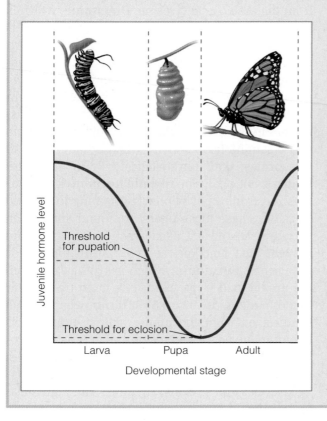

References

○ Gilbert, L. I., R. Rybczynski, and J. T. Warren. 2002. Control and biochemical nature of the ecdysteroidogenic pathway. *Annual Review of Entomology* 47: 883–916.

○ Slama, K., K. Koudela, J. Tenora, and A. Mathova. 1996. Insect hormones in vertebrates: Anabolic effects of 20-hydroxyecdysone in Japanese quail. *Experientia* 52: 702–706.

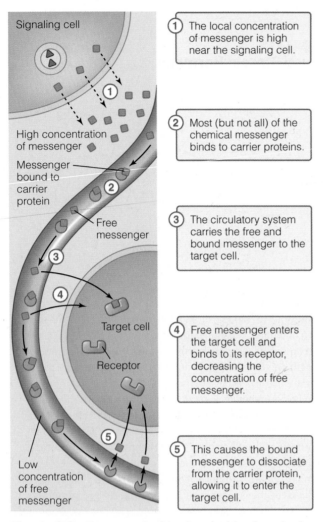

Signaling cell

High concentration of messenger

Messenger bound to carrier protein

Free messenger

Target cell

Receptor

Low concentration of free messenger

① The local concentration of messenger is high near the signaling cell.

② Most (but not all) of the chemical messenger binds to carrier proteins.

③ The circulatory system carries the free and bound messenger to the target cell.

④ Free messenger enters the target cell and binds to its receptor, decreasing the concentration of free messenger.

⑤ This causes the bound messenger to dissociate from the carrier protein, allowing it to enter the target cell.

Figure 3.8 Transport of hydrophobic chemical messengers

Because free and bound chemical messengers are in equilibrium, changes in the concentration of any of the reactants or products influence the concentrations of the others. Thus, increases in the amount of messenger that is released from the signaling cell will increase the amount of messenger delivered to the target cell. Conversely, increases in the concentration of the carrier protein will tend to decrease the concentration of free chemical messenger, whereas decreases in the concentration of carrier protein will increase the concentration of free messenger. As we discuss in the next section, the amount of free messenger influences the response of the target cell. Thus, changes in both the amount of messenger and the amount of carrier protein can affect cell signaling.

Steroids bind to intracellular receptors

The lipophilic steroids can easily cross the membrane of the target cell, and thus they can bind either to transmembrane receptors or to receptors inside the cell. The intracellular receptors are the best-studied class of steroid receptor. Intracellular steroid receptors act as transcription factors, controlling the expression of target genes. Because this pathway relies on changes in transcription and translation, there is a detectable lag time between binding of the messenger and observation of the initial effects. In contrast, when a steroid messenger binds to a transmembrane receptor, it activates a cytoplasmic signal transduction pathway, which causes rapid *non-genomic* effects that do not require changes in transcription or translation.

Biogenic Amines

Amines are chemicals that possess an amine (–NH$_2$) group attached to a carbon atom. Amines that function in cellular signaling are termed **biogenic amines**. Many amines are synthesized from amino acids. The **catecholamines** (dopamine, norepinephrine, and epinephrine) are synthesized from the amino acid tyrosine. **Dopamine**, which is found in all animal taxa, acts as a neurotransmitter. **Norepinephrine** and **epinephrine** are known only from vertebrates, and can act as neurotransmitters, paracrines, and hormones. *Octopamine* and *tyramine*, which are also synthesized from the amino acid tyrosine, are important neurotransmitters in the invertebrates. Although octopamine and tyramine have activity when administered to vertebrates, their physiological role in the vertebrates is not clear. The **thyroid hormones** are synthesized from a polypeptide containing the amino tyrosine. These messengers are found only in the vertebrates, and act as hormones. They are not thought to function as neurotransmitters or paracrines. **Serotonin**, which is synthesized from the amino acid tryptophan, is a neurotransmitter found in all animal taxa. **Melatonin**, which is also synthesized from the amino acid tryptophan, is found in almost all organisms and acts as a neurotransmitter and a hormone. In the vertebrates, melatonin plays a critical role in regulating sleep-wake cycles and seasonal rhythms. Although melatonin is found in most invertebrate taxa, its role in these organisms is not well understood. As is the case in vertebrates, most evidence suggests that it is involved in the regulation of activity patterns. **Histamine** is synthesized from the amino acid histidine. This biogenic amine functions as a neurotransmitter and a paracrine signaling molecule in both vertebrates and invertebrates. Histamine plays an important role in immune responses and allergic reactions. **Acetylcholine**, a neurotransmitter found in all animals, is synthesized from choline, an amine that is not an amino acid, and acetyl-coenzyme A. It is the primary neurotransmitter at the neuromuscular junction of vertebrates, and because of its importance, and the fact that it is not synthesized from an amino acid, it is sometimes classified separately from the other biogenic amines.

Most biogenic amines are hydrophilic molecules that are packaged into vesicles and released into the extracellular fluid by exocytosis. They can either be synthesized on demand or be stored for later release. Because they are important neurotransmitters, we discuss the mechanisms for the synthesis and release of the catecholamines, acetylcholine, and serotonin in more detail in Chapter 4. In this chapter we focus on the thyroid hormones, which are an interesting exception to the general rules governing the synthesis and release of biogenic amines.

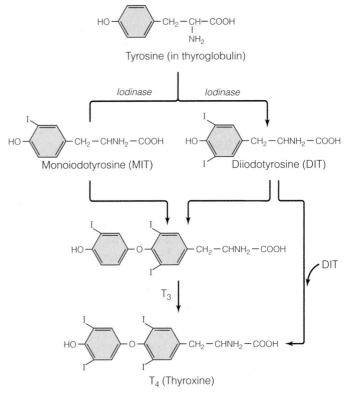

Figure 3.9 Synthetic pathways for the thyroid hormones Thyroid hormone synthesis begins when the enzyme iodinase adds one or more iodine molecules to the amino acid tyrosine within the protein thyroglobulin. Monoiodotyrosine (MIT) has a single iodine molecule added per tyrosine residue; diiodotyrosine (DIT) has two iodine molecules per tyrosine residue. If one molecule of MIT and one molecule of DIT combine, they form triiodothyronine (T_3). Adding an additional molecule of DIT forms tetraiodothyronine (T_4), also known as thyroxine. Collectively, T_3 and T_4 are termed the thyroid hormones.

Thyroid hormones diffuse across the membrane

Thyroid hormone synthesis begins when the enzyme iodinase adds one or more iodine molecules to tyrosine residues in the protein thyroglobulin (Figure 3.9). If a particular tyrosine residue is iodinated once, the resulting compound is called monoiodotyrosine (MIT). If a particular tyrosine residue is iodinated twice, the resulting compound is called diiodotyrosine (DIT). The iodinated tyrosine residues in the thyroglobulin molecule are then coupled via a covalent bond. If two DIT groups combine, the result is 3,5,3′,5′-tetraiodothyronine, called T_4 (or thyroxine). Alternatively, if one DIT group and one MIT group combine, the result is 3,5,3′-triiodothyronine, called T_3. Collectively, T_3 and T_4 are called the thyroid hormones. At this point, the T_3 and T_4 are still part of the thyroglobulin protein, which is packaged into vesicles. The vesicles then fuse with the lysosome, an organelle that contains proteinases (or proteases). The pro-

teinases digest the thyroglobulin, releasing the T_3 and T_4.

Although thyroid hormones are derived from a hydrophilic precursor (a protein), the thyroid hormones are hydrophobic and thus easily diffuse out of the lysosome and cross the plasma membrane of the signaling cell.

Thyroid hormones are hydrophobic messengers

The hydrophobic thyroid hormones are carried in the blood bound to a carrier protein, and bind to an intracellular receptor in the target cell. Like all intracellular receptors for chemical messengers, the thyroid-hormone receptor acts as a transcription factor when bound to thyroid hormone, altering the transcription of target genes. Thus, although thyroid hormones are derived from a protein, they behave more like steroid hormones than like peptide hormones. Thyroid hormones play an important role in setting metabolic rate and regulating body temperature in mammals (Chapter 13: Thermal Physiology).

Other Classes of Messenger

All hormones are peptides, steroids, or amines, but a number of other classes of molecules can act as neurotransmitters or paracrine chemical messengers, including certain lipids, purines, and even gases. Many of these molecules have only recently been identified as important chemical signaling molecules, but research in these areas is extremely active, because these molecules are involved in many important disease-related processes in humans, including inflammation, pain, and vascular disease.

Eicosanoids are lipid messengers

A class of lipids known as the **eicosanoids** can act as neurotransmitters and paracrine chemical messengers. The hydrophobic eicosanoids diffuse out of the membrane of the signaling cell and diffuse to the target cell, where they bind to transmembrane receptors. Most eicosanoids have an extremely short half-life in extracellular fluids, and degrade within a few seconds. As a result, they cannot be transported across long distances, and thus cannot act as hormones. Most eicosanoids are derivatives of arachidonic acid, a 20-carbon fatty acid common in plasma membrane phospholipids. The pathway for eicosanoid synthesis is shown in Figure 3.10. Eicosanoid synthesis proceeds through either the lipoxygenase pathway, which produces the *leukotrienes* and *lipoxins*, or the cyclooxygenase pathway, which produces *prostaglandins*, *prostacyclins*, and *thromboxanes*. Prostaglandins are one of the most studied groups of eicosanoids because they are involved in pain perception. Many common pain-

killers (including aspirin and ibuprofen) work by blocking prostaglandin synthesis.

Eicosanoids can also function as neurotransmitters. For example, one of the eicosanoids is thought to bind to the cannabinoid receptor in the brain. These receptors were so-named because they also bind to the drug tetrahydrocannabinoid (THC), a lipid that is the bioactive component of the marijuana plant, *Cannabis sativa*.

Nitric oxide is a gaseous chemical messenger

Only three gases are known to act as chemical messengers in animals: nitric oxide, carbon monoxide, and hydrogen sulfide. **Nitric oxide** (NO) was the first gas identified as a chemical messenger, and a great deal is now known about its mechanisms of action. Nitric oxide is produced by the enzyme *nitric oxide synthase* (NOS), which catalyzes the reaction of the amino acid arginine with oxygen to produce nitric oxide and citrulline (another amino acid). Animals have several isoforms of NOS, some of which are inducible (synthesized in response to specific signals), and some of which are consititutive (present all the time). Like the eicosanoids, nitric oxide has an extremely short half-life (2–30 seconds) in extracellular fluids and thus can act as a paracrine

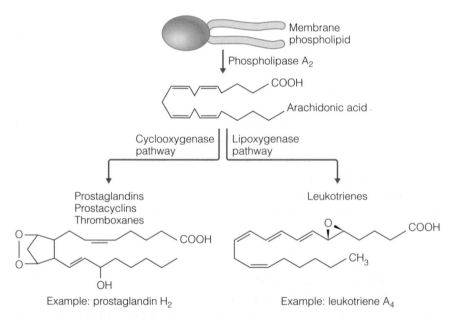

Figure 3.10 Synthetic pathway for eicosanoids Phospholipase A_2 catalyzes the conversion of membrane phospholipids into arachidonic acid, the substrate for eicosanoid synthesis. The cyclooxygenase pathway produces prostaglandins, prostacyclins, and thromboxanes. The lipoxygenase pathway produces leukotrienes.

messenger or neurotransmitter but cannot act as a hormone. Nitric oxide plays a critical role in regulating many physiological functions because it is a *vasodilator*. It causes the smooth muscle around blood vessels to relax, increasing the diameter of the blood vessel and causing more blood to flow into the local area. Nitric oxide is also important for paracrine communication in the immune system.

Because it is a gas, nitric oxide can freely diffuse across the cell membrane from the signaling cell to the target cell. Nitric oxide can act within the cell in several ways. One important action of nitric oxide is to activate the intracellular enzyme guanylate cyclase. Guanylate cyclase catalyzes the formation of cyclic cGMP, which then activates a specific protein kinase, which goes on to phosphorylate a variety of target proteins. The cGMP produced by guanylate cyclase is quickly removed from the cell by a series of enzymes termed phosphodiesterases (PDE), thus terminating the nitric oxide signal. Drugs such as Viagra block the isoform of PDE that is found in the smooth muscle cells surrounding blood vessels of the penis. Blocking PDE results in prolonged elevation of cGMP within the cell, causing the cells to relax and the blood vessels to vasodilate. The net result of this vasodilation is increased blood flow to the penis, which (as we discuss in Chapter 14) is necessary to sustain erection.

Purines can act as neurotransmitters and paracrines

A variety of purines including adenosine, adenosine monophosphate (AMP), adenosine triphosphate (ATP), and the guanine nucleotides are known to act as neurotransmitters, *neuromodulators,* or paracrines. A neuromodulator is a cellular signaling molecule that alters the activity of other signaling molecules, such as neurotransmitters. Purines have a very wide range of functions. For example, adenosine acts on the immune system to promote wound healing, can change the rhythm of the heartbeat in vertebrates, and is a potent calming neurotransmitter in the brain. Purines are released from signaling cells via a variety of mechanisms. Adenosine can be moved across the membrane by specific proteins termed nucleoside transporters. Other purines are packaged into secretory vesicles, often along with other classes of neurotransmitters, and released by exocytosis.

When involved in cellular signaling, purines bind to transmembrane receptors known as *purinergic* receptors.

Communication of the Signal to the Target Cell

Each of the classes of chemical messengers described above exerts its effects by binding to a receptor protein. When a ligand binds to its receptor, the receptor undergoes a conformational change. This change in the shape of the receptor sends a signal to the target cell. Hydrophilic ligands bind to transmembrane receptors, and the conformational change of that receptor communicates the signal to the inside of the cell without the need for the ligand to cross the membrane. Hydrophobic ligands can either bind to transmembrane receptors, or pass through the membrane of the target cell and bind to intracellular receptors. Because intracellular receptors are located within the cell (either in the cytoplasm or nucleus), changes in the shape of intracellular receptors can easily be communicated to other biochemical pathways inside the cell.

Ligand-receptor interactions are specific

Ligand-receptor interactions are extremely specific, because the ligand-binding site of a receptor has a particular shape, allowing only molecules sharing related structures to bind efficiently to the receptor. Just as only the correctly shaped key will open the lock on your door, only the correctly shaped ligand can bind to a given receptor (Figure 3.11a). Some chemicals with structures similar to the natural ligand can mimic the action of a ligand on its receptor. Chemicals that bind to and activate receptors are termed receptor **agonists** (Figure 3.11b). Chemicals that bind to but do not activate receptors are termed receptor **antagonists** (Figure 3.11c). Many drugs are receptor agonists or antagonists. For example, tubocurarine is a plant compound that is the active ingredient in poison darts used by South American indigenous hunters to paralyze their prey. Tubocurarine binds to a receptor at the neuromuscular junction. Because tubocurarine is a receptor antagonist, binding of tubocurarine blocks the receptor, which prevents communication from nerves to muscles and causes paralysis.

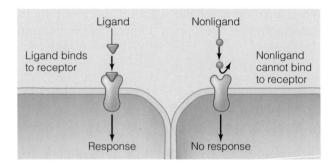

(a) Ligand binding causes a response

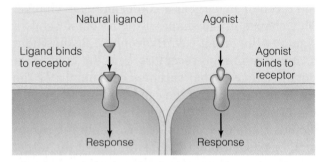

(b) Agonist binding causes a response

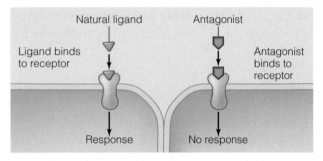

(c) Antagonist binding does not cause a response

Figure 3.11 Ligand-receptor interactions
A ligand is a small molecule that binds specifically to a larger macromolecule such as a receptor, causing a response in the target cell. Both agonists and antagonists can bind to a receptor, but only agonists cause a response.

Receptor type determines the cellular response

A target cell can respond to a ligand only if the appropriate receptor is expressed on or in the target cell. Two cells side by side in the body may be bathed in a chemical signal, but only the cell that possesses the appropriate receptor will respond. Thus, chemical signaling is a bit like a radio signal. Two people jogging side by side along a city street are both exposed to radio waves, but only the person who has a portable radio (the appropriate receptor for radio waves) will receive the information signal carried by the broadcast.

The hundreds of chemical messengers found in animals can be used in millions of combinations. But any given cell responds to only a fraction of these signals, depending on the types of receptors that are present in the cell. Although no cells in the body are capable of responding to all possible ligands, most cells express receptors for many types of ligands. It is the particular combination of receptors expressed by a cell that generates the specificity of cellular responses to different combinations of chemical signals.

Receptors have several domains

Receptors are large proteins that are composed of several domains. The ligand-binding domain contains the binding site for the chemical messenger. The remaining domains of the protein convey its functional activity by interacting with signal transduction molecules within the cell. The structure of the ligand-binding domain determines the nature of the ligands that can interact with the receptor. The remaining functional domains determine the nature of the effects of that receptor on the target cell. For many receptors it is possible to construct recombinant proteins with the ligand-binding domain of one receptor and the functional domains of another. The types of functional domains present in the recombinant protein determine the nature of the response in the target cell, not the type of ligand-binding domain.

A ligand may bind to more than one receptor

Many receptors are part of large gene families. These genes are transcribed into similar proteins, termed isoforms, with distinct properties (see Chapter 2). Receptor isoforms often share similar ligand-binding domains, but differ in their functional domains. The presence of these isoforms allows the same signaling molecule to have very different effects on different target cells. For example, epinephrine causes the smooth muscle cells surrounding the bronchioles of the lung to relax, but causes the smooth muscle cells surrounding the blood vessels leading to the intestine to contract. The smooth muscle cells in these different locations express different adrenergic receptor isoforms (different versions of the receptor for epinephrine).

Ligand-receptor binding obeys the law of mass action

Like the binding of chemical messengers to their carrier proteins, ligand-receptor interactions are governed by the law of mass action (see Chapter 2). Natural ligands usually bind reversibly to their receptors; thus, the following equation represents the binding of a ligand to its receptor:

$$L + R \rightleftharpoons L\text{-}R \rightarrow \text{response}$$

where L is the free ligand, R is the receptor, and L-R is the bound ligand-receptor complex. Binding of the ligand to its receptor causes a response in the target cell. As ligand concentration increases, the balance shifts to the right, and the proportion of receptor bound to ligand increases. The more receptor that is bound to ligand, the greater the response in the target cell. However, the amount of ligand bound to the receptors on a cell cannot increase indefinitely. Instead, the receptors eventually become saturated with ligand, once all the available receptors are bound to ligand (Figure 3.12). Once the saturation point is reached, adding more ligand will not increase the response in the target cell.

Receptor number can vary

Target cells vary in the number of receptors they possess. The more receptors on a cell, the more likely it is that a ligand will bind to the receptor at any given concentration of ligand (Figure 3.13a), and the greater the response in the target cell. Target cells with high concentrations of receptors will be more sensitive to the presence of the ligand than target cells with lower concentrations of the receptor.

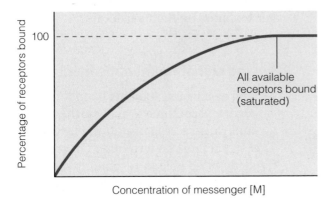

Concentration of messenger [M]

Figure 3.12 **Effects of messenger concentration** As messenger concentration increases, the percentage of receptors bound to messenger increases up to the saturation point, at which all available receptors are bound to messenger.

The number of receptors on a target cell can change over time. These effects can easily be observed following the administration of certain drugs. For example, opiate drugs (opium, morphine, codeine, and heroin) bind to and activate opiate receptors that are found on cells throughout the body, and particularly in the brain. The normal function of these receptors is to induce pleasure and block pain. When a person regularly consumes a drug such as heroin, the number of opiate receptors on the target cells decreases in an attempt to reduce the intensity of the pleasure signal and maintain homeostasis—a phenomenon termed **down-regulation**. As a result, heroin users must consume more and more of the drug in order to achieve the same effects. When a habitual heroin user stops taking the drug, the low levels of opiate receptors in the brain reduce the brain's sensitivity to endorphins, the natural ligands of the opiate receptors. The reduced signal from the endorphins causes withdrawal symptoms, including nausea, vomiting, muscle pain, and bone pain, when an addicted individual stops taking the drug. After a period of time without heroin, receptor number returns to normal, and the withdrawal symptoms gradually abate.

Receptors can also be **up-regulated**. For example, caffeine (the active ingredient in coffee) binds to receptors for the neurotransmitter adenosine. Adenosine is an inhibitory neurotransmitter, so when it binds to its receptor it tends to reduce brain activity, producing a calming effect. Caffeine is an antagonist for these receptors, binding but not activating them. The net result is that caffeine acts as a stimulant by removing the calming effects of adenosine. The brain responds to the removal of this calming signal by increasing the number of adenosine receptors on these brain cells. Up-regulation results in increased sensitivity to the naturally occurring adenosine, and thus homeostatically regulates brain activity, restoring brain activity to normal by balancing out the effects of the ingested caffeine. As a result of this up-regulation, coffee drinkers must drink more and more coffee over time to obtain the same stimulatory effect. A habitual coffee drinker may need several cups of coffee just to fully wake up in the morning because they need higher levels of caffeine to cancel out the effects of adenosine on their highly sensitive up-regulated target cells. If a habitual coffee drinker attempts to suddenly stop drinking

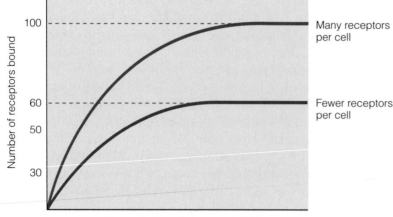

(a) Effect of receptor concentration

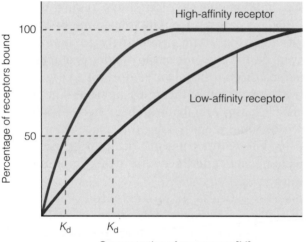

(b) Effect of receptor affinity

Figure 3.13 Effects of receptor concentration and affinity on the percentage of bound receptors **(a)** Cells that have a higher concentration of receptors have a larger number of bound receptors at any given concentration of messenger, and these cells respond to the messenger more strongly than cells with fewer receptors. **(b)** At a given concentration of messenger, cells with high-affinity receptors have a higher percentage of bound receptors, and a greater response than cells with low-affinity receptors, as long as the messenger concentration is low. At messenger concentrations where all receptors are saturated, there is no difference in response between the cells if the total number of receptors is similar. The dissociation constant (K_d), or the concentration of messenger at which the receptor is 50% saturated, is an indication of the affinity of the receptor for the messenger. Receptors with high K_d have low affinity for the messenger, whereas receptors with low K_d have high affinity for the messenger. Alternatively, affinity can be expressed using the affinity constant (K_a), which is the inverse of K_d.

coffee, the high levels of adenosine receptor in the brain cause these individuals to be unusually sensitive to naturally occurring adenosine in their system, and thus they will tend to feel sleepy without their morning coffee.

Receptor affinity for ligand can vary

Receptors can also vary in the strength with which they bind a ligand. The strength of binding between a ligand and a receptor can be expressed using the **dissociation constant (K_d)** for that receptor. The dissociation constant is defined as the concentration of messenger at which half of the receptors on the cell surface are bound to ligand (Figure 3.13b). Thus receptors with high affinity have a low dissociation constant, and receptors with low affinity have a high dissociation constant. Alternatively, we can express the strength of receptor-ligand interactions with the **affinity constant (K_a)**, which is defined as the inverse of the dissociation constant (or the inverse of the concentration of messenger at which half of the receptors are bound). The larger the K_a value, the higher the affinity, and the more tightly the ligand binds to the receptor. Figure 3.13b illustrates the effects of differences in the affinity constant. A high-affinity receptor causes greater activity in the target cell at low-ligand concentration than does a low affinity receptor. High-affinity receptors also saturate at lower ligand concentrations.

The affinity constants of some hormone receptors are very large (>10^8 l/mol). Thus, a receptor can bind to these messengers even when they are present at very low concentrations. In contrast, the affinity constants for some neurotransmitter and paracrine receptors are lower (~10^4 l/mol), requiring higher concentrations of messenger to stimulate their receptors.

Ligand signaling must be inactivated

As long as a ligand remains bound to its receptor, it will continue to activate that receptor and cause a response in the target cell. This signal must be terminated in order for the body to be able to respond to changing conditions. The activity of ligand-receptor complexes can be regulated in a variety of ways (Figure 3.14). The simplest way to terminate signaling is to remove the ligand from the extracellular fluid. For example, enzymes in the liver and kidney degrade many circulating hormones. When hor-

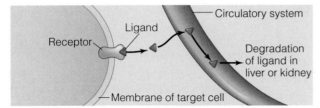

(a) Ligand removed by distant tissues

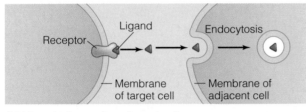

(b) Ligand taken up by adjacent cells

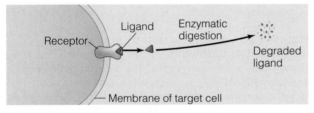

(c) Ligand degraded by extracellular enzymes

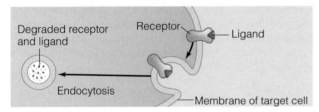

(d) Ligand-receptor complex removed by endocytosis

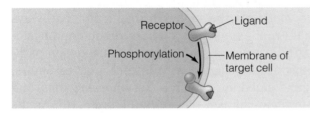

(e) Receptor inactivation

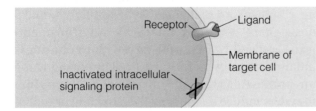

(f) Inactivation of signal transduction pathway

Figure 3.14 Termination of ligand-receptor signaling

mone levels fall in the blood, they will also fall in the fluid surrounding a cell, causing bound hormone to dissociate from its receptor (according to the law of mass action). When the receptor is no longer bound to the hormone, signaling terminates.

Removal of a hormone from the blood is a relatively slow process, requiring several minutes to hours. Most signaling molecules must be regulated over much shorter time periods. These molecules can be inactivated or removed in several ways. Adjacent cells can take up signaling molecules from the extracellular fluid, thus reducing the concentration of the signaling molecule and causing them to dissociate from the receptor. This is a common mechanism for the removal of neurotransmitters from the synapse. This process is the target of a number of drug therapies. For example, drugs called selective serotonin reuptake inhibitors (SSRIs) inhibit the reuptake of serotonin from the synapse, increasing the concentration of serotonin in the synapse, which causes increased binding of serotonin to its receptors. SSRIs are commonly used to treat depression. An alternative means of removing a signaling molecule from its receptor is to use enzymes that degrade the signaling molecule. The ligand-receptor complex can also be removed from the membrane by endocytosis. Internalized receptors can then either be degraded (resulting in receptor down-regulation) or recycled to the cell membrane, once the ligand has been removed. Intracellular enzymes can degrade hydrophobic chemical messengers that diffuse into cells. Receptors can also be inactivated by phosphorylation or other similar mechanisms. Inactivation of components of the signaling pathways within the cell that are stimulated by ligand-receptor complexes can also be used to terminate signaling.

⊙ | CONCEPT CHECK

1. Compare and contrast hydrophilic and hydrophobic messengers in terms of the three main steps of indirect signaling.

2. Are amines hydrophilic or hydrophobic messengers? How does this affect their release, transport, and signaling?

3. Why do some cells respond to a chemical messenger while other cells ignore it?

4. Compare and contrast receptor up-regulation and down-regulation. How do these phenomena help to maintain homeostasis?

Signal Transduction Pathways

So far, we have seen that the type and concentration of both the ligand and receptor can affect the response of the target cell, but we have not yet discussed the details of how the binding of the ligand to the receptor causes a response in the target cell. When a ligand binds to a receptor, the receptor undergoes a conformational change. But how does a simple signal like the change in the shape of a protein get converted into a complex response in the target cell? The cell uses signal transduction pathways to convert the change in the shape of a receptor to a complex response. *Transducers* are devices that convert signals from one form to another. Signal transduction in the cell is analogous to signal transduction in familiar transducing devices like a radio. All transducers have four important components: a receiver, a transducer, an amplifier, and a responder. In the cell, the ligand-binding domain of the receptor acts as a receiver, receiving the signal by binding to the incoming chemical messenger. The ligand-binding domain, together with other domains within the receptor, acts as a transducer by undergoing a conformational change that activates a signal transduction pathway. The signal transduction pathway acts as an amplifier, increasing the number of molecules affected by the signal.

All signal transduction pathways have the same general structure (Figure 3.15). When a ligand binds to its receptor, the receptor undergoes a conformational change. The conformational change in the receptor acts as a signal that converts an inactive substance (A) to its active form. The activated substance A in turn activates substance B, which activates substance C, and so on, until the end of the cascade. The change in conformation of a single receptor caused by the binding of a single molecule of chemical messenger can result in the conversion of many molecules of substance A to their active forms. Each one of these many molecules of substance A can then go on to activate many molecules of substance B, and so on down the chain, potentially producing millions of molecules of the final product. As a result, signal transduction cascades greatly amplify the original signal caused by binding of a molecule of chemical messenger. The longer the signal transduction cascade, the greater the degree of signal amplification.

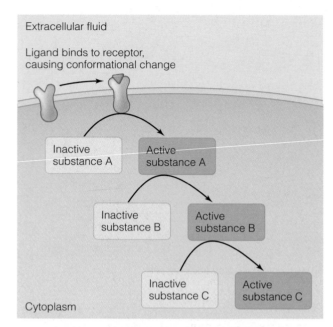

Figure 3.15 Amplification by signal transduction pathways When a single molecule of ligand binds to a single receptor, the receptor undergoes a conformational change. The change in shape of the receptor converts inactive substance A to active substance A. As long as the ligand remains bound to the receptor, it will continue to activate substance A. Thus, a single molecule of ligand can activate many molecules of substance A. Substance A then goes on to activate substance B, and so on down the chain. At each step, one molecule of a substance can activate many molecules of the next substance in the chain. Thus, signal transduction cascades can greatly amplify the signal.

Cells have many signal transduction pathways, some of them very complex. In this book we focus on the signal transduction pathways that are the most important in regulating physiological processes. These signal transduction pathways are associated with intracellular receptors, ligand-gated ion channels, receptor-enzymes, and G-protein-coupled receptors (Figure 3.16). As the name suggests, **intracellular receptors** are located inside the cell, and interact with hydrophobic chemical messengers. Hydrophilic chemical messengers generally interact with transmembrane receptors. **Ligand-gated ion channels** initiate a response in the target cell by changing the ion permeability of the membrane. **Receptor-enzymes** induce a response by activating or inactivating intracellular enzymes. **G-protein-coupled receptors** send signals to an associated **G protein**, which then initiates a signal transduction pathway that causes a response in the target cell.

Intracellular Receptors

When a ligand binds to an intracellular receptor, the receptor changes shape and becomes activated (Figure 3.17). Activated intracellular receptors act as transcription factors that regulate the transcription of target genes by binding to specific DNA sequences, and increasing or decreasing mRNA production from the target gene. Intracellular receptors have three domains: a *ligand-binding domain*, a *DNA-binding domain*, and a *transactivation domain*, each of which performs specific steps in signal transduction. Once a hydrophobic ligand has diffused across the cell membrane, the ligand binds to the ligand-binding site. Ligand binding causes a conformational change in the receptor that activates it. Some intracellular receptors are located in the cytoplasm, and only move to the nucleus once they bind to the ligand. Other intracellular receptors are found in the nucleus, already bound to DNA and ready to be activated.

The DNA-binding domain of an intracellular receptor binds to specific sequences, termed *responsive elements*, adjacent to their target genes. Because the DNA-binding domain of each intracellular receptor recognizes a specific responsive sequence, and only the intended target genes contain appropriate responsive element sequences, intracellular receptors bind only to their target genes and not to other genes in the genome. Once the receptor is bound to the responsive element, the transactivation domain of the receptor interacts with other transcription factors to regulate the transcription of the target genes, increasing or decreasing the production of mRNA. Together, the DNA-binding domain and the transactivation domain act as the transducer in this signal transduction pathway. Many important endocrine hormones bind to intracellular receptors, including estrogen, testosterone, and the glucocorticoid stress hormones.

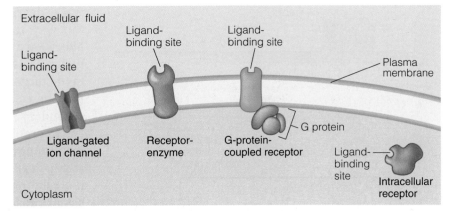

Figure 3.16   Types of receptors in animals   Some of the physiologically important receptors in animals are intracellular receptors, ligand-gated ion channels, receptor-enzymes, and G-protein-coupled receptors.

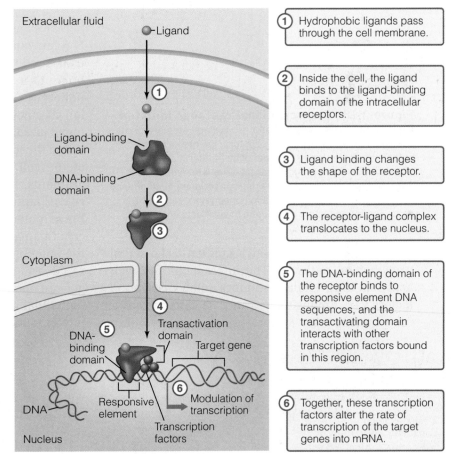

1. Hydrophobic ligands pass through the cell membrane.

2. Inside the cell, the ligand binds to the ligand-binding domain of the intracellular receptors.

3. Ligand binding changes the shape of the receptor.

4. The receptor-ligand complex translocates to the nucleus.

5. The DNA-binding domain of the receptor binds to responsive element DNA sequences, and the transactivating domain interacts with other transcription factors bound in this region.

6. Together, these transcription factors alter the rate of transcription of the target genes into mRNA.

Figure 3.17   Signal transduction by intracellular receptors

The changes in transcription initiated by the binding of a ligand to its receptor set off a cascade of events within the target cell (Figure 3.18). The first step of the response is often activation of a small number of specific genes, usually coding for other transcription factors. The gene products then go on to activate other genes. This cascade of

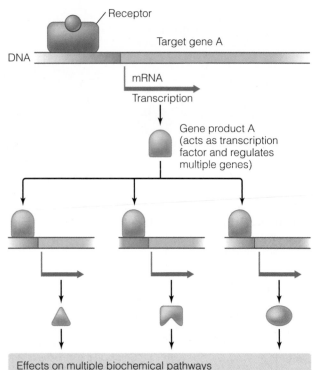

Figure 3.18 Transcriptional cascades initiated by intracellular receptors In the first step of signal transduction by intracellular receptors, the messenger-receptor complex binds to target gene A, altering its transcription. The product of target gene A then goes on to interact with DNA and regulate the transcription of additional genes. The products of these additional genes may also act as transcription factors or go on to have effects on many biochemical pathways.

gene regulation acts as the amplifier in the signal transduction pathway. The interactions between activated intracellular receptors and transcription factors vary among genes, and the same receptor may increase the transcription of some genes

while decreasing the transcription of others. In this way, a hydrophobic ligand can have complex effects on a target cell. Because these ligands exert their effects by altering transcription, the response of the target cell is generally slow, with the first effects detectable within about 30 minutes and the secondary effects occurring over hours or days.

Hydrophobic ligands can also bind to transmembrane receptors, in which case the responses within the target cell are very rapid, because they do not rely upon transcription. However, the specific signal transduction pathways involved in these rapid *non-genomic* responses are not well understood.

Ligand-Gated Ion Channels

Signal transduction by ligand-gated ion channels is relatively simple and direct compared to signal transduction by other receptors. When a ligand binds to a ligand-gated ion channel, the protein undergoes a conformational change, opening an ion channel within the protein—a route for ions to move across the cell membrane (Figure 3.19). When the ion channel opens, ions move into or out of the cell, as dictated by their electrochemical gradients, altering the membrane potential of the cell (Chapter 2). The resulting change in membrane potential acts as a signal within the target cell. Changes in membrane potential as a result of the opening of ligand-gated ion channels are very rapid, and a single molecule of chemical messenger can open an ion channel that could allow many individual ions to cross the cell membrane, allowing for some signal amplification.

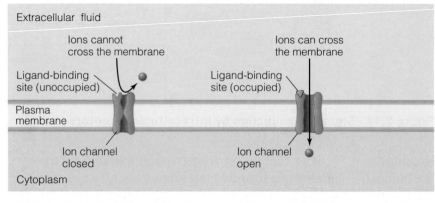

(a) Unbound ligand-gated ion channel   (b) Bound ligand-gated ion channel

Figure 3.19 The structure and function of ligand-gated ion channels **(a)** When no ligand is bound to the receptor, the ion channel is closed and ions cannot cross the membrane. **(b)** When a ligand binds to the ion channel, the channel changes conformation and the ion channel opens, allowing ions to cross the membrane.

Signal Transduction via Receptor-Enzymes

Receptor-enzymes contain an extracellular ligand-binding domain, a transmembrane domain, and an intracellular catalytic domain (Figure 3.20a). The ligand-binding domain contains a region that binds specifically to a chemical messenger. When ligand binds to the ligand-binding domain, the receptor changes shape, and the transmembrane domain transmits this shape change across the membrane, activating the cat-

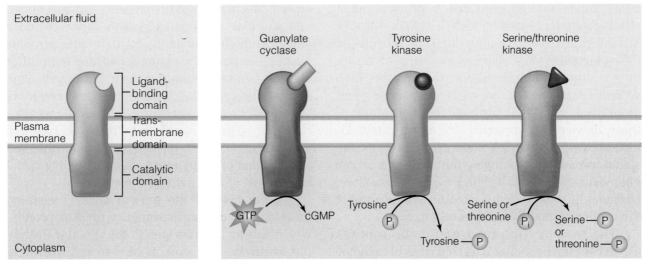

(a) Structure of a receptor-enzyme

(b) Types of receptor-enzymes

Figure 3.20 Receptor-enzymes **(a)** Structure of a receptor-enzyme. A receptor-enzyme has an extracellular ligand-binding domain, a transmembrane domain, and an intracellular catalytic (enzyme) domain. **(b)** Three types of receptor-enzymes in animals. Receptor guanylate cyclases convert GTP to cGMP. Receptor tyrosine kinases phosphorylate tyrosine residues in proteins. Receptor serine/threonine kinases phosphorylate serine or threonine residues in proteins.

alytic domain of the enzyme. The catalytic domains of receptor-enzymes act as enzymatic catalysts that initiate the next link in the signal transduction cascade. The signal transduction pathways of receptor-enzymes involve *phosphorylation cascades* in which proteins at each step phosphorylate or dephosphorylate other proteins within the target cell. Phosphorylation cascades amplify the original signal, causing a response in the target cell.

Receptor-enzymes are named based on the reaction catalyzed by the intracellular catalytic domain. In this book, we discuss three types of receptor-enzymes: (1) receptor guanylate cyclases, (2) receptor tyrosine kinases, and (3) receptor serine/threonine kinases (Figure 3.20b). In animals, the majority of known receptor-enzymes are tyrosine kinases. Animals also have many forms of receptor serine/threonine kinase, some of which play important roles in growth and development and in the response to environmental stressors. Only a few receptor guanylate cyclases are known.

Receptor guanylate cyclases generate cyclic GMP

When a ligand binds to a receptor guanylate cyclase, the receptor undergoes a conformational change, activating the guanylate cyclase domain of the receptor (Figure 3.21). The activated guanylate cyclase produces cyclic guanosine monophos-

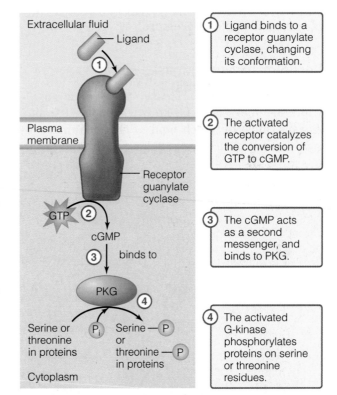

1. Ligand binds to a receptor guanylate cyclase, changing its conformation.

2. The activated receptor catalyzes the conversion of GTP to cGMP.

3. The cGMP acts as a second messenger, and binds to PKG.

4. The activated G-kinase phosphorylates proteins on serine or threonine residues.

Figure 3.21 Signal transduction via guanylate cyclase receptor-enzymes

phate (cGMP). The cGMP acts as a **second messenger** within the cell. Second messengers are low-molecular-weight diffusible molecules that act as part of signal transduction pathways to communicate signals within the cell. The second messenger cGMP binds to and activates a protein called cGMP-dependent protein kinase (PKG). Kinases are enzymes that phosphorylate other proteins. PKG phosphorylates proteins at serine or threonine residues. The phosphorylated proteins then go on to activate other proteins, propagating and amplifying the signal through the cell. Many of these downstream proteins are also protein kinases that phosphorylate other proteins. Thus, the signal transduction pathway initiated by receptor guanylate cyclases is termed a *phosphorylation cascade*. Each step in this cascade acts to amplify the original signal from the receptor.

The receptors for atrial natriuretic peptides (ANPs) are the best-characterized class of receptor guanylate cyclases. ANPs are a group of closely related peptides that are produced by muscle cells in the heart in response to increases in blood pressure. As we see in Chapter 10, ANPs trigger vasodilation and induce the kidney to reduce blood volume. Both of these responses lower blood pressure, returning it to normal. Thus, ANPs are part of the negative feedback system that homeostatically regulates blood pressure.

Receptor tyrosine kinases signal through Ras proteins

There are more than 50 known receptor tyrosine kinases, most of which bind to chemical messengers that are critical for cellular growth and prolif-

eration, such as insulin, epidermal growth factor, and vascular endothelial growth factor. When a chemical messenger binds to a receptor tyrosine kinase, the bound receptor associates with other tyrosine kinase receptors in the membrane to form dimers (Figure 3.22). The dimerized receptors then phosphorylate each other on multiple tyrosine residues, a process called *autophosphorylation*. The phosphorylated receptors interact with and activate one of many intracellular signaling molecules, most of which are protein kinases.

In the case of the growth factor receptors, these activated kinases signal to the Ras protein, which acts as the next step in the signal transduction pathway. Ras proteins bind to and hydrolyze GTP and function as switches by cycling between the active state, when GTP is bound, and the inactive state, when GDP is bound. GTPase-activating proteins (GAPs) and guanine nucleotide–releasing proteins (GNRPs) catalyze the transition between active and inactive Ras. Receptor tyrosine kinases signal through GAPs and GNRPs to regulate Ras.

Ras activates a serine/threonine phosphorylation cascade that sends a signal through the cell. There are many serine/threonine phosphorylation cascades in animal cells, but one particularly important one involves the MAP kinases (Figure 3.23). Activated Ras signals to a MAP-kinase-kinase-kinase (MAPKKK), which phosphorylates a MAP-kinase-kinase (MAPKK). In turn, the MAP-kinase-kinase phosphorylates a MAP kinase (MAPK). The MAP

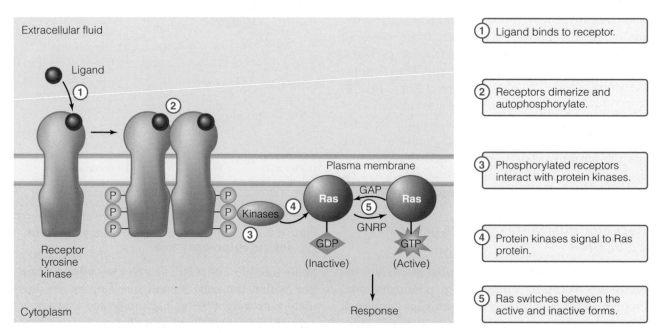

Figure 3.22 Signal transduction via receptor tyrosine kinases

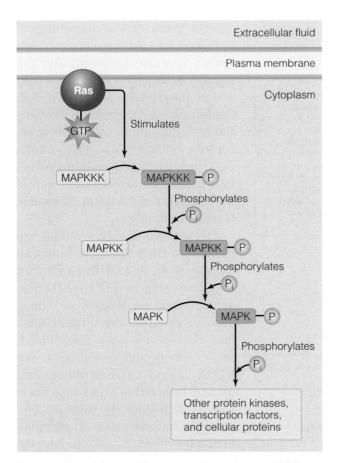

Figure 3.23 **Signal transduction via the MAP-kinase phosphorylation cascade** The Ras proteins that are activated by receptor tyrosine kinases phosphorylate MAP-kinase-kinase-kinase, and the phosphorylated MAPKKK then phosphorylates a MAP-kinase-kinase, which in turn phosphorylates a MAP kinase, which then phosphorylates other protein kinases, transcription factors, and diverse cellular proteins.

kinase then phosphorylates other protein kinases, cellular proteins, and the transcription factors Elk-1 and Jun. These transcription factors regulate the transcription of other transcription factors, which regulate the transcription of various genes. Thus, the phosphorylation cascades triggered by receptor tyrosine kinases greatly amplify the original chemical signal. Because they activate extensive phosphorylation cascades within the cell, the Ras proteins have wide-ranging effects on cellular growth and metabolism. Approximately 30% of human cancers involve mutations in the genes encoding Ras. These mutations turn the Ras protein "on" constitutively so that it is active even in the absence of a ligand. The activated Ras sends a strong signal to the cell, stimulating it to grow and divide uncontrollably, causing cancer.

The insulin receptor is another example of a critically important tyrosine kinase receptor. Like other tyrosine kinase receptors, the insulin receptor functions as a dimer. Binding of insulin to the extracellular ligand-binding domain (also called the alpha subunit) causes the receptor to dimerize. Dimerization causes the intracellular domains (also called the beta subunits) to autophosphorylate each other on their tyrosine residues and become active. The intracellular domains of the receptor dimer also act as a tyrosine kinase that phosphorylates other target proteins on their tyrosine residues. The best-known target of the insulin receptor is called the insulin receptor substrate (IRS). The insulin receptor substrate acts as a docking protein that binds to other intracellular signaling proteins that participate in insulin signal transduction.

Receptor serine/threonine kinases directly activate phosphorylation cascades

Receptor serine/threonine kinases directly activate phosphorylation cascades, without working through Ras proteins. When a ligand binds to a receptor serine/threonine kinase, the conformational change in the receptor directly activates a serine/threonine kinase (Figure 3.24a). The activated serine/threonine kinase then phosphorylates other proteins, activating a phosphorylation cascade. The signaling pathways activated by receptor serine/threonine kinases are not yet fully understood, but are similar to the pathways used by receptor tyrosine kinases in that they involve phosphorylation cascades that greatly amplify the signal in the target cell.

The *transforming growth factor* β (TGF-β) receptors are among the most intensively studied receptor serine/threonine kinases, because mutations in TGF-β receptors and the associated signal transduction pathways have been implicated in the development of human cancers. TGF-β receptors are present as a complex consisting of two distinct proteins, called TGF-β Type I and Type II receptors (Figure 3.24b). When TGF-β binds, the Type I and Type II receptors associate with each other. The Type II receptor then phosphorylates the Type I receptor, activating the intracellular catalytic domain of the Type I receptor. The catalytic domain of the activated Type I receptor then phosphorylates a series of target proteins called SMADs on specific serine and threonine residues. The phosphorylated SMADs move to the nucleus where they interact with other proteins to regulate the transcription of target genes.

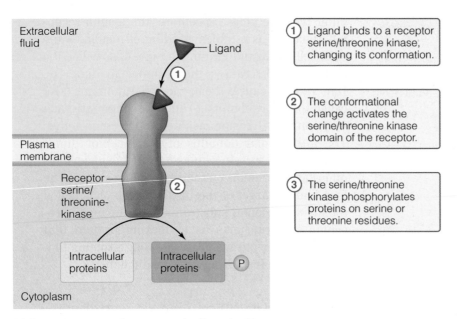

(1) Ligand binds to a receptor serine/threonine kinase, changing its conformation.

(2) The conformational change activates the serine/threonine kinase domain of the receptor.

(3) The serine/threonine kinase phosphorylates proteins on serine or threonine residues.

(a) General structure of receptor serine/threonine kinase

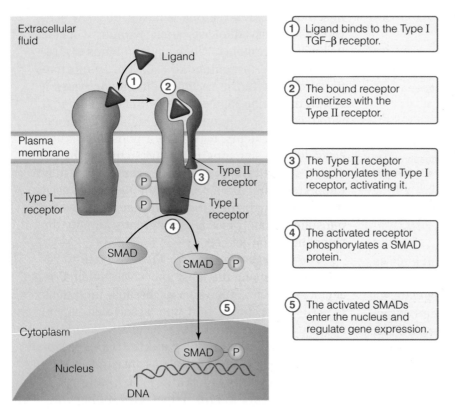

(1) Ligand binds to the Type I TGF–β receptor.

(2) The bound receptor dimerizes with the Type II receptor.

(3) The Type II receptor phosphorylates the Type I receptor, activating it.

(4) The activated receptor phosphorylates a SMAD protein.

(5) The activated SMADs enter the nucleus and regulate gene expression.

(b) Signal transduction by TGF-β receptors

Figure 3.24 Signal transduction via receptor serine/threonine kinases

Signal Transduction via G-Protein-Coupled Receptors

G-protein-coupled receptors are a large family of membrane-spanning proteins with seven trans-membrane domains. G-protein-coupled receptors control many critical physiological functions, and there is enormous diversity in these receptors and the signal transduction pathways with which they interact (see Box 3.2, Evolution and Diversity: G-Protein-Coupled Receptors). All of these receptors, however, share a common first step in their signal transduction pathways: activation of one of the members of the **heterotrimeric G protein** family.

Heterotrimeric G proteins are named for their ability to bind and hydrolyze GTP, and the fact that they are composed of three different subunits (α, β, and γ). The α subunit contains the binding sites for the guanosine nucleotides, while the β and γ subunits are tightly bound to each other, and usually referred to as a single functional group, the βγ subunit. The general features of the signaling pathways from G-protein coupled receptors via G proteins to amplifier enzymes are outlined in Figure 3.25. When a ligand binds to a G-protein-coupled receptor, the receptor changes shape, sending a signal to the α subunit of the G protein, inducing a conformational change in the G protein. The conformational change causes the α subunit of the G protein to release GDP, bind a molecule of GTP, and become active. The activated α subunit then dissociates from the βγ subunit. Both the βγ and α subunits can then go on to interact with downstream targets.

The best-characterized targets of the βγ subunit are ion channels. Interaction with the βγ subunit causes these ion channels to open, allowing ions to move into or out of the cell, depending on their electrical and concentration gradients. Ion movements cause changes in membrane potential, which act as signals within the cell. Thus, G protein signaling via ion channels is a relatively direct pathway to generate a response in the cell.

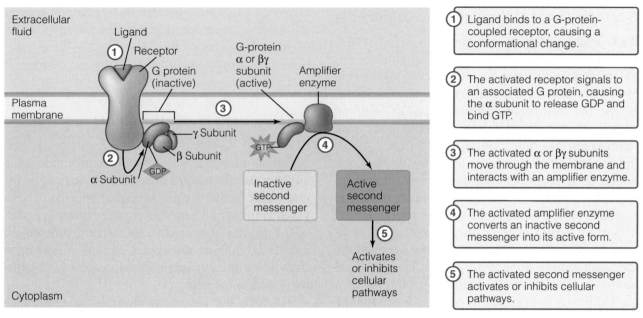

1. Ligand binds to a G-protein-coupled receptor, causing a conformational change.

2. The activated receptor signals to an associated G protein, causing the α subunit to release GDP and bind GTP.

3. The activated α or βγ subunits move through the membrane and interacts with an amplifier enzyme.

4. The activated amplifier enzyme converts an inactive second messenger into its active form.

5. The activated second messenger activates or inhibits cellular pathways.

Figure 3.25
Signal transduction via G-protein-coupled receptors

BOX 3.2 EVOLUTION AND DIVERSITY
G-Protein-Coupled Receptors

G protein signaling is involved in cell-to-cell communication in a wide variety of organisms, including fungi, plants, and animals, but the number and diversity of G-protein-coupled receptors have greatly increased during the evolution of the metazoans. The single-cell budding yeast *Saccharomyces cerevisiae* has only three G-protein-coupled receptors, but even relatively simple metazoans have hundreds of different G-protein-coupled receptors. For example, the genome of the nematode *Caenorhabditis elegans* contains almost 1100 different genes with sequences similar to G-protein-coupled receptors. Although we do not yet know whether all of these sequences encode functional receptors, it is likely that at least several hundred of these genes function in cell-to-cell communication in nematodes. This high diversity of G-protein-coupled receptors is not unique to nematodes. In fact, in all of the animal genomes that have been fully sequenced to date, somewhere between 1% and 5% of the entire protein-coding part of the genome consists of sequences similar to G-protein-coupled receptors.

G-protein-coupled receptors recognize many different ligands and stimuli including light, odors, and chemical messengers, and thus play an important part in both environmental sensing and cell-to-cell commu-

nication in multicellular organisms. The human genome, for example, contains approximately 1000 sequences related to the G-protein-coupled receptors. Approximately 700 of these are involved in the senses of smell and taste, or other chemosensory functions. The remaining 300 likely interact with chemical signaling molecules, and are thus involved in cell-to-cell communication. Of the G-protein-coupled receptors that are involved in cell signaling, approximately 140 have no known ligand or function, and are termed **orphan receptors**. Evolutionary analyses suggest that all of the G-protein-coupled receptor genes in animals have a common ancestor, and arose by duplication and descent with modification over evolutionary time to perform different roles in complex multicellular animals.

References

○ Bockaert, J., S. Claeysen, C. Becamel, S. Pinloche, and A. Dumuis. 2002. G-protein-coupled receptors: Dominant players in cell-cell communication. *International Review of Cytology* 212: 63–132.

○ Jones, A. M. 2002. G-protein-coupled signaling in *Arabidopsis*. *Current Opinion in Plant Biology* 5: 402–407.

○ Pierce, K. L., R. T. Premont, and R. J. Lefkowitz. 2002. Seven-transmembrane receptors. *Nature Reviews. Molecular Cell Biology* 3: 639–650.

G proteins can act through Ca²⁺-calmodulin

If a G protein interacts with and opens a Ca^{2+} channel, the increase in cytoplasmic $[Ca^{2+}]$ initiates signal transduction cascades within the target cell. Most Ca^{2+}-mediated signal transduction cascades act through the protein **calmodulin**, a Ca^{2+}-binding protein that is present in every eukaryotic cell. Calmodulin has four binding sites for Ca^{2+}. Binding of Ca^{2+} to all four sites activates the protein, which then interacts with numerous other proteins. Calmodulin is known to interact with and regulate over 100 different cellular proteins. One important group of these target proteins is a diverse family of serine/threonine kinases called the Ca^{2+}-calmodulin-dependent protein kinases (CaM kinases). One of the best-studied examples of a CaM kinase is CaM kinase II, which is found in high concentration in neurons that secrete neurotransmitters called catecholamines. When cytoplasmic Ca^{2+} increases in these neurons, the change in Ca^{2+} concentration activates CaM kinase II. CaM kinase II phosphorylates tyrosine hydroxylase (one of the key enzymes in catecholamine biosynthesis). CaM kinases play many other important roles in animals. For example, one of the CaM kinase genes is implicated in the process of learning and memory. Transgenic mice that have this CaM kinase gene knocked out have altered brain activity and are unable to learn how to swim through a water maze.

G proteins can interact with amplifier enzymes

In addition to acting via ion channels, the βγ and α subunits of G proteins can also interact with a variety of other kinds of target molecules here given the generic name "amplifier enzyme." The activated G protein subunits alter the activity of the amplifier enzyme, either increasing or decreasing its activity (depending on the particular G protein involved in the signaling). These amplifier enzymes then go on to initiate signal transduction pathways that result in diverse indirect effects within the target cell.

Amplifier enzymes alter the concentration of second messengers

Amplifier enzymes catalyze the conversion of a small molecule **second messenger** between its inactive and active forms. A single molecule of activated amplifier enzyme can catalyze the conversion of thousands of molecules of second messenger, greatly amplifying the signal. Second messengers then go on to activate or inhibit a variety of pathways within the cell.

Despite the enormous diversity of G-protein-coupled receptors, all G proteins act through one of only four second messengers: Ca^{2+}, cyclic GMP, phosphatidylinositol, and **cyclic adenosine monophosphate (cAMP)**. Table 3.3 summarizes the similarities and differences between these second messenger cascades. All of these cascades amplify the signal within the target cell, inducing responses that may occur in milliseconds or hours.

Guanylate cyclase generates cGMP

Most of the G proteins that use cGMP as a second messenger activate the amplifier enzyme guanylate cyclase, which catalyzes the conversion of GTP to cGMP. The cGMP then goes on to activate PKG, which goes on to phosphorylate many other proteins. In addition, some G-protein-coupled receptors use a different signal transduction pathway. When a ligand binds to these G-protein-coupled receptors, the α subunit of the associated G protein moves laterally through the membrane and binds

Table 3.3 Second messengers.

Second messenger	Synthesized by the enzyme	Action	Effects
Ca²⁺	None	Binds to calmodulin	Alters enzyme activity
cGMP	Guanylate cyclase	Activates protein kinases (usually protein kinase G)	Phosphorylates proteins Opens and closes ion channels
cAMP	Adenylate cyclase	Activates protein kinases (usually protein kinase A)	Phosphorylates proteins Opens and closes ion channels
Phosphatidyl inositol	Phospholipase C	Activates protein kinase C Stimulates Ca^{2+} release from intracellular stores	Alters enzyme activity Phosphorylates proteins

to and activates the amplifier enzyme phosphodiesterase. The activated phosphodiesterase catalyzes the conversion of cGMP to GMP, causing cGMP levels in the cytoplasm to drop. The decrease in cytoplasmic cGMP causes cGMP to dissociate from Na^+ channels in the membrane, closing them. The closing of the Na^+ channels prevents Na^+ from entering the cell, which changes the membrane potential and thus transduces the chemical signal into an electrical signal. This signal transduction pathway plays a part in vertebrate vision. We discuss signal transduction in the cells of both invertebrate and vertebrate eyes in more detail in Chapter 6: Sensory Systems.

Phospholipase C generates phosphatidylinositol

The inositol-phospholipid signaling pathway (Figure 3.26) was first discovered as the signal

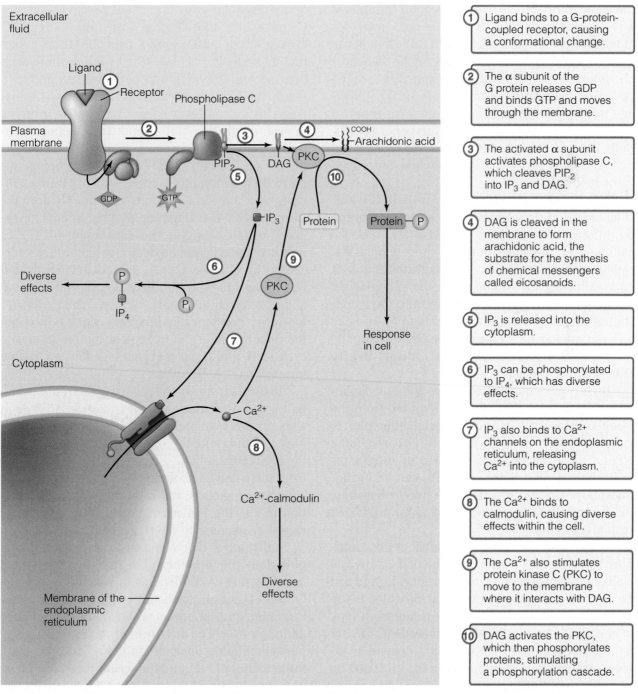

Figure 3.26 The inositol-phospholipid signaling pathway

transduction pathway responsible for regulating secretion from the salivary glands of insects, but a huge variety of G-protein-coupled receptors that signal through the inositol-phospholipid pathway are now known from most animal taxa. These pathways regulate a diversity of physiological functions including smooth muscle contraction, glycogen degradation in the liver, water reabsorption by the vertebrate kidney, and many aspects of immune function.

When a chemical messenger binds to one of these receptors, the activated receptor stimulates a G protein called G_q, which in turn activates inositide-specific phospholipase C (phospholipase C-β). In less than a second, this enzyme cleaves a phosphorylated membrane phospholipid, called phosphatidylinositol bisphosphate (PIP_2). Cleavage of PIP_2 produces two products: **inositol trisphosphate (IP$_3$)** and **diacylglycerol (DAG)**. Both IP$_3$ and DAG act as second messengers in two branches of the phosphatidylinositol signal transduction cascade.

The IP$_3$ produced by PIP_2 hydrolysis is water soluble and rapidly leaves the plasma membrane by diffusion. IP$_3$ binds to IP$_3$-gated Ca^{2+} release channels in the membrane of the endoplasmic reticulum, activating them. The activated channels open, allowing Ca^{2+} efflux from the endoplasmic reticulum. The increased cytoplasmic Ca^{2+} concentration further activates the channel, causing an even greater Ca^{2+} efflux. Increases in cytoplasmic Ca^{2+} act as a third messenger, causing diverse effects within the cell.

IP$_3$ is rapidly inactivated by specific dephosphorylases, and the Ca^{2+} is quickly removed from the cytoplasm by active transport, terminating the response. The actions of IP$_3$ generally last less than a second after the chemical messenger dissociates from the receptor. Some of the IP$_3$ can be further phosphorylated to form 1,3,4,5-tetrakisphosphate (IP$_4$), which mediates slower and more prolonged responses in the cell.

DAG, the other cleavage product of PIP_2, initiates two different signal transduction pathways. Unlike IP$_3$, DAG remains in the membrane and can be cleaved to form arachidonic acid, which is the substrate for the synthesis of eicosanoids—a type of chemical messenger. Alternatively, DAG can activate protein kinase C (PKC), a Ca^{2+}-dependent kinase. An increase in cytoplasmic Ca^{2+} (caused by signals from IP$_3$) triggers PKC to move to the membrane, where it interacts with DAG. At the membrane, DAG activates PKC. Activated PKC phosphorylates serine and threonine residues on a variety of proteins including MAP kinase, which we have already discussed. Through these pathways, activated PKC can alter the activities of existing proteins and influence the transcription of genes and thus the production of new proteins.

Cyclic AMP was the first second messenger to be discovered

Many physiologically important processes involve G proteins that signal via the adenylate cyclase–cyclic AMP system, using cAMP as a second messenger. Cyclic AMP was the first intracellular second messenger to be identified, and as a result we know a great deal about these signal transduction pathways. Two types of G proteins interact with the cAMP signal transduction pathway: stimulatory G proteins (G_s) and inhibitory G proteins (G_i) (Figure 3.27). G_s and G_i proteins differ in their α subunits, although their β and γ subunits can be similar. Both G_i and G_s proteins interact with the amplifier enzyme adenylate cyclase, which catalyzes the conversion of ATP to cAMP. When a ligand binds to a receptor that interacts with a G_s protein, the α_s subunit of the activated G_s protein binds to and activates the membrane-bound enzyme adenylate cyclase. When a ligand binds to a receptor that interacts with a G_i protein, the α_i subunits of the G_i protein inhibit adenylate cyclase. G_i and G_s proteins act together to regulate intracellular cAMP levels.

In the next step of the cAMP signal transduction pathway, cAMP binds to protein kinase A (PKA) at sites on the regulatory subunit of the inactive kinase. Binding of cAMP alters the conformation of the regulatory subunits, causing them to dissociate from the catalytic subunits. The unbound catalytic subunits are active, and catalyze the phosphorylation of specific proteins. Protein phosphorylation causes a response in the target cell.

Cells have mechanisms to rapidly dephosphorylate the proteins phosphorylated by PKA, ensuring that cAMP-dependent signals persist only for short periods (seconds to minutes). Serine/threonine phosphatases remove the phosphates added by PKA. The activity of proteins regulated by phosphorylation depends on the balance between the activities of PKA and the serine/threonine phosphatases. When cAMP stimulates PKA activity, the balance of the reaction tends to swing toward

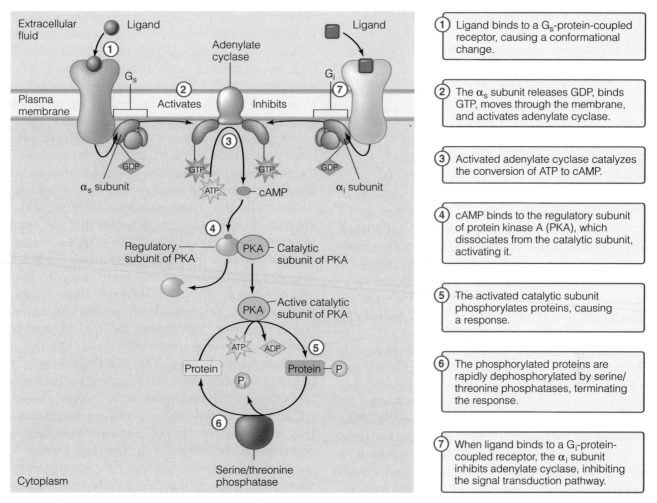

Figure 3.27 G-protein signal transduction via adenylate cyclase G-protein-coupled signal transduction through adenylate cyclase can be either stimulatory or inhibitory.

phosphorylation of the target proteins. In contrast, when cAMP levels are low, the balance of the reaction tends to swing toward the dephosphorylation of the target proteins.

Signal transduction pathways can interact

Ca^{2+} and cAMP signal transduction pathways interact with each other at several levels. For example, Ca^{2+}-calmodulin interacts with adenylate cyclase. Adenylate cyclase is the first amplifier enzyme of the cAMP-mediated signal transduction pathway, and catalyzes the production of cAMP. Similarly, Ca^{2+}-calmodulin also interacts with cAMP phosphodiesterase, the enzyme that breaks down cAMP. Therefore, Ca^{2+} plays a role in regulating the cAMP signaling pathway. Conversely, PKA, one of the steps in the cAMP signaling pathway, can phosphorylate Ca^{2+} channels and pumps, altering their activity. Thus, the cAMP signaling pathway can regulate the Ca^{2+}-calmodulin

pathway. Both protein kinase A and CaM kinase often phosphorylate different sites on the same target proteins. From this example, it is clear that signal transduction cascades in the cell are not simple linear connections from the binding of a chemical messenger, through several amplification steps, culminating in a cellular response. Instead, signal transduction in the cell acts more like a network of intertwined threads that combine to generate complex responses. In vivo, the network is even more complex, because cells may receive multiple signals, many of which may have interacting effects.

⊙ | CONCEPT CHECK

5. Compare and contrast intracellular and transmembrane receptors.

6. Compare and contrast the different types of membrane receptors and their signal transduction pathways.

7. What are second messengers and what is their functional importance?

8. Explain how signal transduction pathways cause signal amplification. Select one signal transduction pathway, and outline specific examples of amplification.

Introduction to Endocrine Systems

As we discussed in Chapter 1: Introduction to Physiological Principles, one of the unifying themes in physiology is that physiological processes are often highly regulated. The cellular signaling pathways that we have discussed in this chapter are a critical component of the complex control systems that monitor and adjust the activity of essentially all physiological systems.

Like mechanical control systems, biological control systems are composed of a *sensor* that detects the state or level of a *regulated variable,* a controller that acts as an *integrating center* by evaluating the incoming information and sending out a signal that provokes an appropriate response in an *effector*—a target tissue that causes (effects) a change in the regulated variable. In a negative feedback loop the effector brings the variable back toward a predetermined **set point**. Thus, negative feeback loops help to maintain homeostasis by maintaining a regulated variable within a small range around the set point. In a positive feedback loop the system responds to a change in the regulated variable by causing further deviation from the set point. Positive feedback is less common in physiological systems than is negative feedback, but it is sometimes used to reinforce and amplify signals. Feedback loops require close communication between the cells and tissues that act as sensors, integrating centers, and effectors, and thus the correct functioning of feedback loops depends upon cellular signaling pathways.

Feedback Regulation

Feedback regulation occurs both at a local level and across long distances in animals. Paracrine and autocrine signals are responsible for local physiological control. The regulation of local blood flow provides an example of the importance of paracrine signals in local feedback regulation.

When a cell or tissue becomes very active metabolically, its oxygen consumption increases. For example, when oxidative muscles (see Chapter 5: Cellular Movement and Muscles) contract, they use aerobic metabolism to generate energy, which causes the mitochondria to become more active and consume more oxygen, leading to a drop in local oxygen concentration. When local oxygen concentration drops in the contracting muscle cells, the endothelial cells that line the blood vessels that supply oxygen to the tissues secrete paracrine signaling molecules. These paracrine signaling molecules act on the smooth muscle cells that surround the blood vessels, causing them to relax. Relaxation of the smooth muscle cells causes the blood vessels to dilate, and allows more blood to flow into the tissue. The flow of blood delivers additional oxygen, restoring local oxygen concentrations, removing the signal for the release of paracrine signaling molecules, thus forming a negative feedback loop. The exact nature of the paracrine signals that cause local vasodilation is hotly debated, because multiple factors change within a tissue when oxygen concentrations drop. It is likely that several of these variable factors contribute to the vasodilation. The purine messenger adenosine is one candidate for an important paracrine signaling molecule involved in vasodilation. As we have already discussed, adenosine is a purine that binds to a G-protein-coupled receptor. The adenosine receptor on smooth muscle cells in most vascular beds is of the A2 subtype, which signals via a cAMP-mediated signal transduction pathway. Activation of the receptor causes the smooth muscle to relax, causing vasodilation of the blood vessel. Vasodilation brings more blood (and thus oxygen) to the local area, turning off the signals for adenosine release.

Reflex control mediates long-distance regulation

In animals, the nervous and endocrine systems are responsible for regulating physiological systems across long distances. The simplest types of these long-distance regulatory systems involve only the endocrine system and are termed direct feedback loops (Figure 3.28a). In a direct feedback loop the endocrine cell itself senses a change in the extracellular environment and releases a chemical messenger that acts on target cells elsewhere in the body. Thus, the endocrine cell acts as the inte-

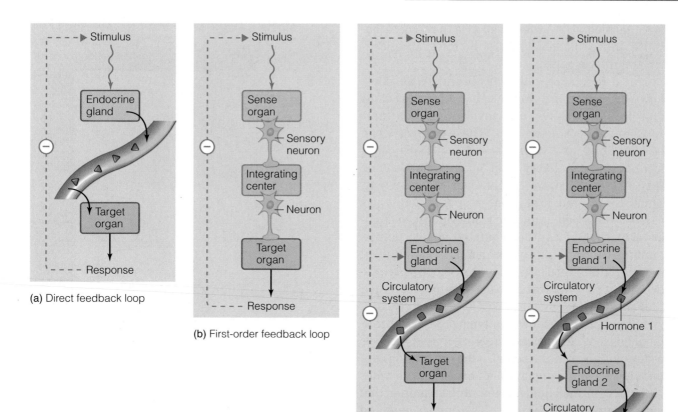

(a) Direct feedback loop

(b) First-order feedback loop

(c) Second-order feedback loop

(d) Third-order feedback loop

Figure 3.28 Feedback regulatory systems in animals **(a)** Direct feedback loops involve only the endocrine system, and the endocrine gland acts both as the integrating center and as the tissue that communicates with the target organ. **(b)** First-order feedback loops have one step (a neuron that releases a neurotransmitter or a neurohormone) between the integrating center and the target organ. **(c)** Second-order feedback loops have two steps (a neuron and an endocrine gland) between the integrating center and the target organ. **(d)** Third-order feedback loops contain an additional endocrine gland in the pathway, providing a third point of feedback regulation.

grating center that interprets the change in the stimulus variable. The response of the target cell to the secreted hormone then brings the stimulus variable back into the normal range. Atrial natriuretic peptide (ANP), a hormone that we discussed earlier in the chapter, is an example of a hormone involved in a direct feedback loop. Stretch-sensitive cells in the atrium of the mammalian heart can sense the increased tension in the cell membranes of atrial cells caused by increased blood pressure within the atrium. These cells then secrete ANP, which travels to target cells in the blood vessels and kidneys and causes responses that lower

blood pressure. The lowered blood pressure feeds back by reducing the tension on the atrial cells, reducing the release of ANP.

First-order feedback loops provide a slightly more sophisticated level of regulation, involving the nervous system (Figure 3.28b). In these types of pathways, a sensory organ perceives a stimulus and sends a signal via the nervous system to an integrating center (such as the brain) that interprets the signal. Neurons then transmit the signal (in the form of either a neurotransmitter or a neurohormone) to a specific target organ, causing a response. Neural and neurohormonal pathways are both termed first-order response pathways because only a single step links the integrating center and the response.

Most regulatory pathways in the vertebrates, however, are more complicated than direct or first-order pathways, and involve both the nervous and endocrine systems. These pathways can be classified as either second- or third-order feedback loops. Every step in a response loop may act as a control point over the pathway. Thus, direct and first-order response pathways can be regulated at only one control point, second-order pathways can be regulated at two points, and third-order pathways can be regulated at three points. Third-order feedback loops provide the most sophisticated and tightly regulated feedback.

Figure 3.28c shows a typical second-order feedback loop. In this case, a sense organ perceives a stimulus and sends a signal to the integrating center, which sends a signal via a neuron that secretes either a neurohormone or a neurotransmitter that acts on an endocrine gland. The endocrine gland then secretes a hormone into the blood. The hormone travels to the target cell, causing a response.

In third-order feedback loops (Figure 3.28d), a sense organ perceives a stimulus and sends a signal to the integrating center. The integrating center then sends a signal via a neuron that secretes either a neurohormone or a neurotransmitter that acts on an endocrine gland. The endocrine gland then secretes a hormone that binds to a receptor on a second endocrine gland and triggers the secretion of a second hormone, which then induces a response in the target cells.

Pituitary hormones provide examples of several types of feedback loops

The vertebrate pituitary gland secretes many important hormones that regulate growth, reproduction, and metabolism. Because these hormones regulate many physiological functions, we encounter them again and again throughout this book. The pituitary gland is closely associated with a part of the brain called the hypothalamus (Figure 3.29) and is connected to the hypothalamus by a narrow stalk called the infundibulum. The pituitary gland is divided into two distinct sections called the **anterior pituitary** (or adenohypophysis) and the **posterior pituitary** (or neurohypophysis). The pituitary also has a third

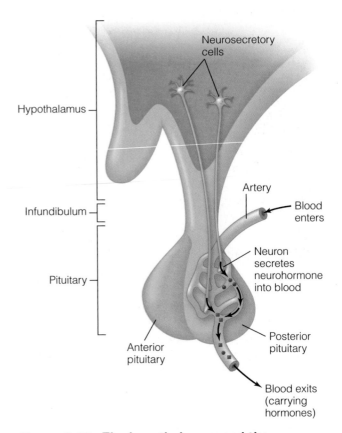

Figure 3.29 The hypothalamus and the posterior pituitary gland The pituitary gland is located at the base of the brain, and is divided into the anterior pituitary and the posterior pituitary. The infundibulum connects the hypothalamus—a part of the brain—and the posterior pituitary, which is made up of the endings of neurons that originate in the hypothalamus. The nerve endings of the posterior pituitary secrete neurohormones into the blood. The anterior pituitary secretes hormones into the blood, under the control of neurohormones released by the hypothalamus.

division, called the intermediate lobe, located between the anterior and posterior pituitary, which secretes melanocyte-stimulating hormone (MSH). In adult mammals, this region is simply a thin sheet of cells that cannot be easily distinguished from the anterior lobe of the pituitary, but it can be quite large in other vertebrates.

The posterior pituitary secretes neurohormones

The posterior pituitary is not really an independent organ but is instead an extension of the hypothalamus. Neurons that originate in the hypothalamus terminate in the posterior pituitary (Figure 3.29). In the hypothalamus, the cell bodies of these neurons synthesize the hormones oxytocin and vasopressin and package them into secretory vesicles. The vesicles are transported along the neuron via a process called axonal transport, which we discuss in more detail in Chapter 5. The neural endings, in the posterior pituitary, secrete these hormones into the blood. Because only a single step (a hypothalamic neuron that secretes a neurohormone) connects the integrating center and the effector organs, oxytocin and vasopressin are examples of neurohormones involved in first-order feedback loops.

Oxytocin is involved in a positive feedback loop

Most hormonal regulation involves negative feedback loops, but oxytocin is an example of a hormone that is involved in a positive feedback pathway. Oxytocin has a wide range of functions, and is both a neurotransmitter and a hormone. In mammals, one of its important endocrine functions involves regulation of uterine contraction (see Chapter 14). At the onset of *parturition,* the process of expelling a fetus from the uterus at birth, the fetus changes position, putting pressure on the cervix (the opening of the uterus). Stretch-sensitive cells in the cervix send a signal to the brain that causes the release of oxytocin from the posterior pituitary. Oxytocin binds to receptors on the smooth muscle cells of the uterus, causing them to contract. Uterine contractions push the fetus against the cervix, increasing the stimulus on the stretch-sensitive cells. This increases the signal and causes even more oxytocin to be released. This positive feedback loop continues until the fetus is delivered, releasing the pressure on the cervix and terminating the signal.

Hypothalamic neurohormones regulate anterior pituitary hormones

The hypothalamus controls the secretion of hormones from the anterior pituitary by secreting neurohormones into a specialized microcirculation, called the **hypothalamic-pituitary portal system** (Figure 3. 30).[1] The portal system carries the neurohormones secreted by the hypothalamus to the anterior pituitary, where they stimulate or inhibit the release of pituitary hormones. The hypothalamic-pituitary portal system allows neurohormones to be carried from the hypothalamus

[1] A portal system is a specialized arrangement of blood vessels with two capillary beds separated by a portal vein.

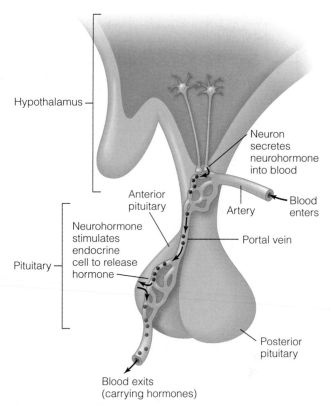

Figure 3.30 The anterior pituitary and the hypothalamic-pituitary portal system Neurons from the hypothalamus secrete neurohormones into the hypothalamic-pituitary portal circulatory system. The portal vein carries the neurohormones to the anterior pituitary where they stimulate endocrine cells to release hormone into the blood. The blood exits from the pituitary, carrying the hormones throughout the body via the circulatory system.

to the pituitary without being diluted in the general circulation.

Figure 3.31 shows the relationship between the hypothalamic neurohormones and the hormones of the anterior pituitary. Prolactin is best known for regulating the secretion of milk from the mammary glands in mammals, but it also has diverse effects on sexual behavior and growth. It is also involved in the regulation of larval development and ion and water balance in some nonmammalian vertebrates. Prolactin is the one anterior pituitary hormone that only functions as part of a second order feedback loop. The brain acts as the integrating center that regulates the secretion of prolactin, stimulating the hypothalamus to release the neurohormones prolactin-releasing hormone or prolactin-inhibiting hormone into the hypothalamic-pituitary portal system. These neurohormones regulate the release of prolactin, which has direct effects on its target tissues such as the breast.

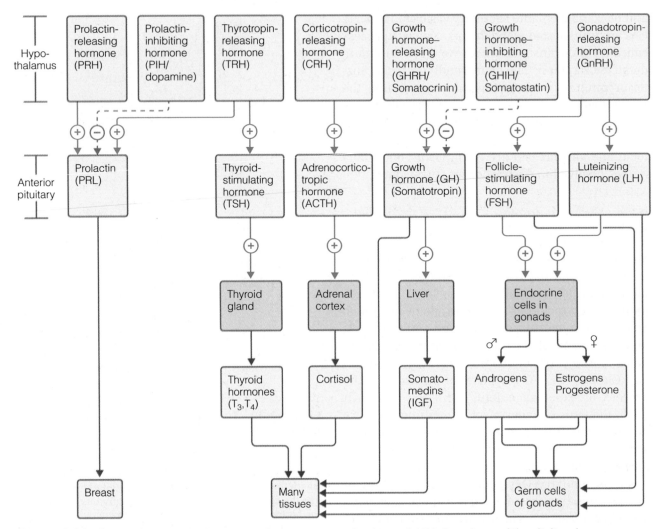

Figure 3.31 The relationship between the hypothalamic hormones and the hormones of the anterior pituitary The hypothalamus secretes releasing or inhibiting neurohormones into the hypothalamic-pituitary portal system. These neurohormones act on the endocrine cells of the anterior pituitary to stimulate or inhibit the release of the pituitary hormones. The circulatory system carries these hormones to their target tissues, causing a response. Some of these target tissues are endocrine glands, which secrete hormones into the blood. The circulatory system carries these hormones to their target tissues, causing a response.

Many anterior pituitary hormones participate in third-order pathways

In contrast to prolactin, the majority of anterior pituitary hormones can go on to regulate the release of yet more hormones, and thus they participate in third order feedback loops. Hormones that cause the release of other hormones are called **tropic (or trophic) hormones**, from the Greek root *tropos*, "to turn toward." (The alternate term, which is often heard, is from the Greek root *trophikos*, "nourishment.") For example, the neurohormone corticotropin releasing hormone from the hypothalamus regulates the secretion of adrenocorticotropic hormone (ACTH) from the pituitary, which in turn causes the release of gluco-corticoid hormones from the adrenal cortex, which goes on to affect the activity of many target tissues. Third-order feedback loops are subject to very complex regulation because change in the concentration of any of the hormones in the hypothalamic-pituitary axis can regulate the concentrations of other hormones in the pathway, generally via negative feedback.

Regulation of Glucose Metabolism

Hormones are involved in regulating almost all physiological processes, and within the scope of this chapter it would be impossible to examine all of the hormones in all animal groups in any detail.

Instead, we defer treatment of most hormones to later chapters, where they will be discussed in the context of their role in the homeostatic regulation of specific physiological systems. Thus, endocrine regulation is a recurring theme throughout this book. In this chapter we discuss two processes that involve endocrine regulation as "case-studies" to illustrate some of the important principles of endocrine regulation.

In this section, we focus on the endocrine regulation of glucose metabolism as a case study in hormonal regulation. As we discussed in Chapter 2, the metabolism of a cell can be divided into a series of catabolic and anabolic processes involving the breakdown or buildup of biological macromolecules. Hormones regulate the balance between anabolism and catabolism in the body and thus help cells to maintain homeostasis between energy supply and energy demand. By mediating these processes, the endocrine system regulates the metabolic activity of essentially every cell in the body.

The actions of insulin illustrate the principle of negative feedback

Most animals maintain some level of homeostatic regulation over the concentration of sugars in their extracellular fluids. Mammals have particularly precise control over the glucose levels in their blood, because the mammalian brain is entirely reliant on glucose as a fuel. If glucose levels fall too low, the brain cannot function. In contrast, if glucose levels rise too high, the osmotic balance of the blood will be disturbed. This precise homeostatic regulation is governed by negative feedback control.

Insulin is one of several hormones involved in the homeostatic regulation of blood glucose in mammals. In mammals, a gland called the **pancreas** secretes the peptide hormone insulin when blood glucose rises. The pancreas is a complex gland with both exocrine and endocrine functions (Figure 3.32). The exocrine pancreas secretes digestive enzymes into the gut (see Chapter 11: Digestion). Dispersed among the exocrine tissue are small clumps of cells, termed the **islets of Langerhans**, which perform the endocrine functions of the pancreas. **Pancreatic β cells** within these islets secrete insulin when blood glucose rises.

Increases in blood glucose cause the metabolic rate of the β cell to increase, resulting in an increase in ATP levels within the cell. The increased

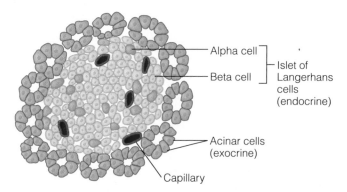

Figure 3.32 The mammalian pancreas The pancreas consists of both exocrine and endocrine tissues. The islets of Langerhans contain cells called beta cells that secrete the hormone insulin and cells called alpha cells that secrete the hormone glucagon.

[ATP] sends a signal to an ATP-dependent potassium (K_{ATP}) channel, causing it to close. Closing of a K^+ channel will cause the cell to depolarize (see Chapter 2). This change in membrane potential causes a voltage-gated Ca^{2+} channel to open, causing Ca^{2+} to enter the cell. The increase in intracellular Ca^{2+} acts as a signal to cause the exocytosis of vesicles containing insulin. The insulin released from the β cell travels through blood to target cells such as liver, adipose tissue, and muscle. At the target cell, insulin binds to and activates its receptor, which, as we have already discussed, is a receptor tyrosine kinase. The activated receptor is then autophosphorylated, initiating a complex network of signal transduction pathways. The ultimate effect of these signal transduction pathways is to promote the uptake and storage of glucose, resulting in a decrease in blood glucose levels. The decrease in blood glucose removes the signal for the pancreatic β cell to release insulin, and insulin levels decline, in an example of negative feedback regulation. In humans, defects in insulin signal transduction cause the disease diabetes mellitus (see Box 3.3, Applications: Cell-to-Cell Communication and Diabetes).

Multiple types of feedback control can regulate blood glucose

The regulation of blood glucose by insulin is an example of a direct feedback loop because the pancreas secretes insulin when it senses increases in blood glucose, without involving integrating centers like the brain. But insulin secretion can be regulated in multiple ways (Figure 3.33). Stretch receptors in the gut can detect the presence of food

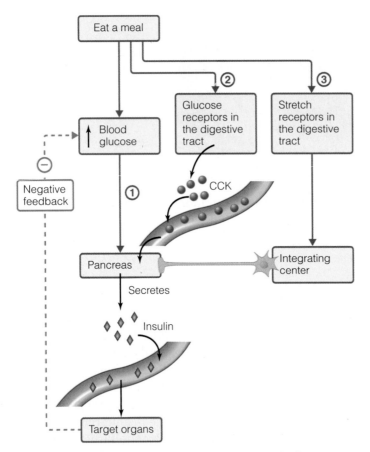

Figure 3.33 Interaction of pathways regulating insulin secretion Insulin is an example of a hormone that is regulated by several feedback pathways. A direct stimulus-response pathway (pathway 1) regulates insulin synthesis. The pancreas senses increases in blood glucose and secretes insulin into the bloodstream. Insulin binds to receptors on target organs, causing responses that reduce blood glucose, reducing the stimulus for insulin secretion in a direct feedback loop. Insulin is also part of a second-order control pathway 3, in which stretch receptors in the digestive tract sense the change in gut volume caused by eating a meal. The stretch receptors send a signal to an integrating center in the neurons surrounding the digestive tract. This integrating center sends a signal via the nervous system to the pancreas to release insulin. At the same time, in pathway 2, glucose receptors in the digestive tract cause the digestive tract to release the hormone cholecystokinin (CCK). The circulatory system carries CCK to the pancreas, stimulating it to secrete insulin.

in the digestive tract, and send a signal to an integrating center in the enteric nervous system (the neurons surrounding the digestive system; see Chapter 7: Functional Organization of Nervous Systems). The enteric nervous system then sends a neural signal directly to the pancreas, causing an increase in the secretion of insulin even before blood glucose starts to rise. This kind of direct control of hormone secretion by the nervous system is an example of a second-order feedback loop. In addition, the pancreas secretes insulin in response to the hormone cholecystokinin (CCK), which is se-

creted by the gut. The gut releases CCK when glucose-sensitive cells in the gut detect the presence of glucose in a meal. Note that the CCK-mediated pathway does not fit neatly into our classification of pathway types, since it involves two hormones but does not utilize the nervous system. This example emphasizes the concept that pathways of feedback regulation represent a continuum of design, rather than discrete organizational systems, and that many of these pathways can interact to form even more complex regulatory networks. In the case of insulin, many pathways interact to regulate insulin secretion and provide for homeostatic regulation of blood glucose.

Insulin and glucagon illustrate the principle of antagonistic control

The second major hormone involved in glucose homeostasis in mammals is the peptide hormone **glucagon**. Glucagon is secreted from α cells in the pancreatic islets of Langerhans. When blood glucose falls, α cells release glucagon into the circulation, where it binds to receptors on target cells, initiating pathways that cause them to release glucose, thus causing blood glucose to rise (another example of negative feedback). Glucagon binds to a G-protein-coupled receptor that stimulates an adenylate cyclase–mediated signal transduction pathway and activates protein kinase A (PKA). PKA phosphorylates a variety of target proteins, causing biochemical changes that ultimately promote the release of glucose into the blood. Thus insulin and glucagon have opposite, or *antagonistic*, effects on blood glucose (Figure 3.34).

Both insulin and glucagon act as important feedback controllers of blood glucose concentration. When blood glucose concentration rises above the set point, the pancreas secretes insulin, causing target cells to take up and store glucose, lowering blood glucose levels. When blood glucose falls below the set point, the pancreas secretes glucagon, causing target cells to release stored glucose, increasing blood glucose levels. The relationship between insulin and glucagon is termed an antagonistic pairing, in which one hormone increases the rate of glucose production and the other decreases it. Antagonistic pairings also control many familiar mechanical devices. For example, the gas pedal and the brake in a car are an antagonistic pairing of control devices. When you depress the gas pedal, the car speeds up, and

BOX 3.3 APPLICATIONS
Cell-to-Cell Communication and Diabetes

Diabetes, one of the most common diseases in the Western world, results when the body fails to either secrete or respond to insulin. There are two major types of diabetes: type 1 (or juvenile onset) and type 2 (or adult onset). In type 1 diabetes the body does not produce sufficient insulin in response to increases in blood glucose. In type 2 diabetes, the target cells do not fully respond to insulin, even if it is present. So both types of diabetes result from failures in cell-to-cell communication, although in type 1 diabetes the primary defect is in the signaling cell, whereas in type 2 diabetes the problem is in the target cells. Type 2 diabetes is by far the more common of the two types. Currently, over 90% of North Americans with diabetes have type 2, and the incidence of type 2 diabetes in Western populations is growing as millions of additional people with type 2 diabetes are diagnosed every year. Particularly alarming is the rapid rate of increase in type 2 diabetes in teenagers.

Type 2 diabetes is a progressive disease that begins with defects in the signal transduction pathway for insulin. The initial symptoms of the disease are usually mild, and may involve frequent urination, thirst, and fatigue. In the early stages of the disease, diabetes can be controlled with a careful diet and a limited intake of glucose, but as the disease progresses, the pancreas secretes more and more insulin in an attempt to signal to the target tissues. Eventually, the pancreas loses its ability to secrete high amounts of insulin, and insulin levels fall. At this point, the disease must be treated with injections of insulin to regulate blood glucose. Untreated diabetes has many serious complications including blindness, vascular disease, kidney failure, heart attack, and stroke.

The signal transduction pathways for insulin are rather complex and have only recently been identified.

When insulin binds to its receptor (a tyrosine kinase), the receptor is autophosphorylated and the tyrosine kinase domain then phosphorylates a protein called the insulin receptor substrate (IRS). Phosphorylated IRS activates the phosphatidylinositol and MAP-kinase signal transduction pathways. The phosphatidylinositol pathway stimulates glucose uptake from the blood, while the MAP-kinase pathway stimulates cell growth. Because so many different proteins are involved in insulin signal transduction, the precise defect associated with type 2 diabetes is not yet known, and may vary from person to person or among tissues.

Obesity is a major risk factor for type 2 diabetes, and most patients with type 2 diabetes are obese when diagnosed. Lack of exercise and a diet high in simple carbohydrates such as sugars also predispose a person to type 2 diabetes. Genetic factors also contribute to type 2 diabetes, so having a close relative with type 2 diabetes indicates an increased risk that a person will develop the disease. Scientists do not yet understand why obesity is related to increased risk of type 2 diabetes, but studies in mice have shown that adipocytes (fat cells) release a hormone called resistin, and levels of resistin are elevated in obese mice. Resistin is thought to downregulate the insulin signal transduction pathway, suggesting the possibility of a link between obesity and type 2 diabetes.

References

○ Bevan, P. 2001. Insulin signalling. *Journal of Cell Science* 114: 1429–1430.

○ Steppan, C. M., S. T. Bailey, S. Bhat, E. J. Brown, R. R. Banerjee, C. M. Wright, H. R. Patel, R. S. Ahima, and M. A. Lazar. 2001. The hormone resistin links obesity to diabetes. *Nature* 409: 307–312.

○ White, M. F. 2002. IRS proteins and the common path to diabetes. *American Journal of Physiology: Endocrinology and Metabolism* 283: E413–422.

---

when you depress the brake the car slows down. It would be possible to design a car with only a gas pedal, but this kind of car would only be able to come to a gradual stop, which would make it much more difficult to control. Similarly, insulin and glucagon allow rapid and precise regulation of blood glucose by speeding up or slowing down glucose release into the blood. By acting together, insulin and glucagon maintain blood glucose levels within a very narrow range. Many hormones are grouped into antagonistic pairs, allowing the endocrine system to exert extremely precise control over physiological functions.

Hormones can demonstrate additivity and synergism

Like glucagon, the hormones **epinephrine** (also called adrenalin) and cortisol can increase blood glucose. Figure 3.35 illustrates the results of an experiment in which glucagon, epinephrine, cortisol, or combinations of these hormones were

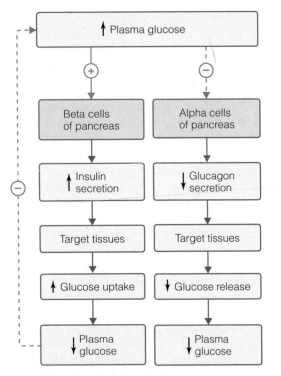

Figure 3.34 Antagonistic regulation of blood glucose by insulin and glucagon Increases in plasma glucose stimulate the beta cells of the pancreas to increase insulin secretion. At the same time, this causes the alpha cells of the pancreas to decrease glucagon secretion. Increased insulin stimulates its target tissues to increase glucose uptake. Decreased glucagon causes its target tissues to decrease glucose release. Together these actions decrease plasma glucose in a negative feedback loop. Similarly, if plasma glucose declines, insulin secretion decreases and glucagon secretion increases, stimulating glucose release into the plasma.

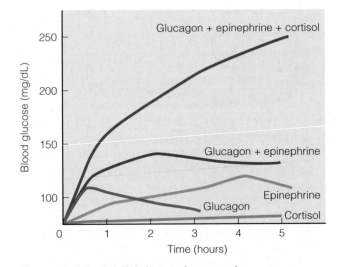

Figure 3.35 Additivity and synergism Infusion of cortisol, glucagon, or epinephrine into dogs results in an increase in blood glucose. These effects are larger when the hormones are injected in combination. Infusion of epinephrine and glucagon results in additive effects on blood glucose. Infusion of all three hormones in combination has a synergistic effect.
(Data from Eigler et al., 1979)

injected into dogs. Alone, injection of glucagon, epinephrine, or cortisol causes an increase in blood glucose. When both glucagon and epinephrine are injected together, the increase in blood glucose is larger, and is equivalent to the sum of the increase in blood glucose in response to epinephrine plus the increase in blood glucose in response to glucagon. This phenomenon is termed *additivity*.

Epinephrine binds to a G-protein-coupled receptor on the liver. This receptor signals via an adenylate cyclase–mediated signal transduction pathway that activates PKA. This is similar to glucagon signaling, which also occurs via activation of PKA. Thus, although these two hormones bind to different G-protein-coupled receptors, they both activate PKA, and the effect of the hormones in combination is equal to the sum of the actions of each hormone alone.

Cortisol is a steroid hormone that is involved in the stress response. As can be seen from Figure 3.35, cortisol causes an increase in blood glucose, but this effect is smaller than the responses to glucagon or epinephrine. As a steroid hormone, cortisol interacts with an intracellular receptor, and thus exerts its effects through a different signal transduction pathway than does either epinephrine or glucagon, and typically acts more slowly. When cortisol, glucagon, and epinephrine are injected in combination, however, the net effect is much greater than the sum of the effects observed when any one hormone is injected alone. This is an example of a phenomenon called *synergism*.

Hyperglycemic hormones control extracellular glucose in arthropods

As is the case in the vertebrates, many invertebrates have mechanisms to regulate extracellular glucose. For example, in crustaceans (crabs, prawns, and shrimp) a neurohormone termed *crustacean hyperglycemic hormone* (CHH) plays a principal role in glucose regulation. CHH was first discovered when researchers injected crabs with extracts of tissues from the eyestalks of other crabs and found that these extracts caused hyperglycemia—an increase in circulating glucose. CHH is synthesized in the cell bodies of secretory neurons that are clustered into an area termed the X-organ within the crustacean eyestalk. Projections from these cell bodies extend into a region called the sinus gland, which acts as a storage and release site for the neurohormone. Because CHH is released by neural tissue, it is considered to be a neurohormone or neuropeptide. The sinus gland releases CHH into the circulatory system, which carries the neurohormone to target cells throughout the body. At the target cell, CHH binds to a transmembrane receptor that activates guanylate cyclase and increases the concentration of cGMP within the target cell. The cGMP acts as a second messenger, activating a signaling pathway that results in release of glucose from the target cell into the circulatory system, causing hyperglycemia.

CHH regulates blood glucose via a negative feedback mechanism (Figure 3.36). When blood glucose levels are high, a K^+ channel on the membrane of the neurosecretory cells within the sinus gland is in the open conformation, allowing K^+ to leave the cell. This hyperpolarizes the membrane (makes the inside of the cell more negative; see Chapter 2). When blood glucose levels drop, this K^+ channel closes, and the cell depolarizes. Depolarization causes the cells to release CHH. The CHH then travels through the circulatory system and causes target cells to release glucose into the circulation, causing glucose levels to return to normal. As is the case for many hormones, other factors can also modulate the release of CHH. For example, inputs from the nervous system alter the activity of the sinus gland cells in response to external cues including season, time of day, temperature, and changes in environmental salinity. CHH also has other functions in addition to the regulation of circulating glucose, including the regulation of lipid metabolism.

Although CHH is primarily regulated via negative feedback from circulating glucose levels, crustacean hyperglycemic hormone can also be regulated by positive feedback. When CHH binds to its receptor on target cells, the activated receptor increases flux through glycolysis. One of the end products of glycolysis is a three-carbon unit called lactate (see Chapter 2). When stimulated by CHH, target cells produce lactate, which is released into the circulation. The neurosecretory cells of the X-organ–sinus gland complex are sensitive to circulating lactate, which causes them to release more CHH in a positive feedback loop. The signals from lactate and glucose work together to regulate CHH secretion.

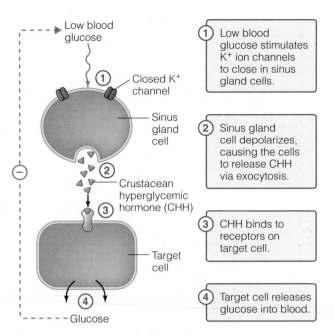

① Low blood glucose stimulates K^+ ion channels to close in sinus gland cells.

② Sinus gland cell depolarizes, causing the cells to release CHH via exocytosis.

③ CHH binds to receptors on target cell.

④ Target cell releases glucose into blood.

Figure 3.36 Regulation of circulating glucose by crustacean hyperglycemic hormone

The Vertebrate Stress Response

Our second case study in endocrine regulation is the vertebrate stress response, because it provides an example of the ways in which the nervous and endocrine systems work together to regulate physiological responses. Because of its importance, we return to aspects of the stress response in various chapters throughout this book.

When the sense organs of a vertebrate perceive an alarming stimulus (such as the presence of a predator), the organism initiates a complex set of behavioral and physiological responses that are often called the "fight-or-flight" response. The fight-or-flight response involves both the endocrine system and the nervous system acting together to coordinate this complex but critically important behavioral and physiological response (Figure 3.37).

Stressful stimuli activate the sympathetic nervous system

When an animal detects the presence of an alarming stimulus (such as a predator), sensory nerves send a signal to the brain. The brain acts as an integrating center that takes information from the various senses and makes a decision regarding the "threat level" of the stimulus. If the brain decides that the stimulus represents a threat, it sends out a signal via motor neurons, which causes muscles to contract, causing the animal to

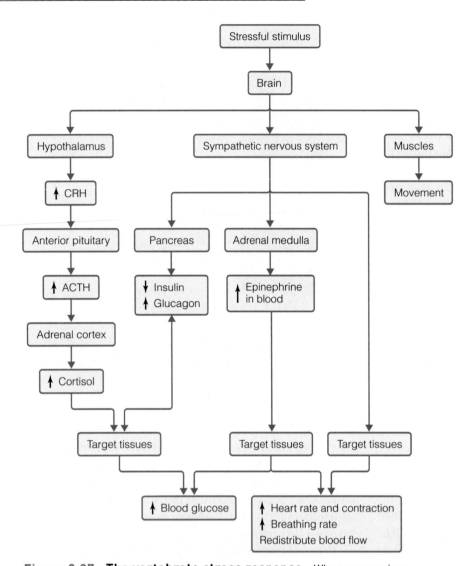

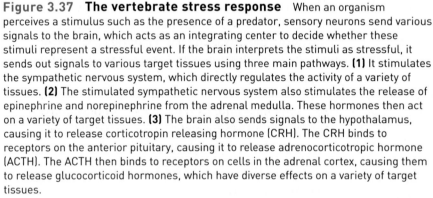

Figure 3.37 The vertebrate stress response When an organism perceives a stimulus such as the presence of a predator, sensory neurons send various signals to the brain, which acts as an integrating center to decide whether these stimuli represent a stressful event. If the brain interprets the stimuli as stressful, it sends out signals to various target tissues using three main pathways. **(1)** It stimulates the sympathetic nervous system, which directly regulates the activity of a variety of tissues. **(2)** The stimulated sympathetic nervous system also stimulates the release of epinephrine and norepinephrine from the adrenal medulla. These hormones then act on a variety of target tissues. **(3)** The brain also sends signals to the hypothalamus, causing it to release corticotropin releasing hormone (CRH). The CRH binds to receptors on the anterior pituitary, causing it to release adrenocorticotropic hormone (ACTH). The ACTH then binds to receptors on cells in the adrenal cortex, causing them to release glucocorticoid hormones, which have diverse effects on a variety of target tissues.

and redirect it toward the working muscles and away from tissues such as the gut. The sympathetic nervous system also increases the rate and depth of breathing. Together these responses help to provide the skeletal muscles with the oxygen they need to contract and thus engage in the fight-or-flight response.

The sympathetic nervous system stimulates the adrenal medulla

In addition to the target tissues discussed above, the sympathetic nervous system also affects the activity of several endocrine glands. For example, stimulation of the sympathetic nervous system reduces the release of insulin from the pancreas and increases the release of glucagon. Target tissues respond to the change in insulin and glucagon levels by increasing blood glucose, which can be used as an energy source during the fight-or-flight response. The sympathetic nervous system also stimulates the **adrenal glands**. In mammals, the adrenal glands are compact organs located adjacent to each kidney, which consist of two types of tissue. The **adrenal cortex**, on the outside of the gland, is composed of interrenal tissue, and secretes mineralocorticoid and glucocorticoid hormones such as aldosterone and cortisol. The inside of the adrenal gland is called the **adrenal medulla** and is composed of **chromaffin cells** that secrete the catecholamines, epinephrine and norepinephrine.

The sympathetic nervous system releases the neurotransmitter acetylcholine onto chromaffin cells of the adrenal medulla. These cells then release either norepinephrine or epinephrine into the circulatory system. The ratio of norepinephrine to epinephrine that is released varies among species. In dogfish sharks, norepinephrine is the only catecholamine released by chromaffin cells,

run away or fight, as necessary. At the same time, the hypothalamus activates a portion of the nervous system termed the *sympathetic nervous system* (see Chapter 7: Functional Organization of Nervous Systems). The sympathetic nervous system sends out signals to target organs including the heart, vascular smooth muscle, and other tissues. These responses help to increase blood flow

whereas in frogs norepinephrine makes up about 55–70% of the released catecholamines. In contrast, mammals release mostly epinephrine.

As we have already discussed, epinephrine and norepinephrine bind to members of a family of G-protein-coupled receptors, termed the adrenergic receptors, that activate signal transduction pathways that alter the activity of existing proteins. Thus, epinephrine and norepinephrine have very rapid effects within their target cells. Epinephrine and norepinephrine interact with many target organs including the heart, lungs, and muscles to galvanize the body into action.

The hypothalamo-pituitary axis stimulates the adrenal cortex

The fight-or-flight response also involves the activation of the hypothalamo-pituitary endocrine response. When the hypothalamus is activated by a stressful stimulus, it increases the secretion of corticotropin-releasing hormone (CRH) into the hypothalamic-pituitary portal system. CRH binds to its receptors on target cells in the anterior pituitary and causes them to release adrenocorticotropic hormone (ACTH) into the bloodstream. ACTH binds to G-protein-coupled receptors in the membranes of cells in the adrenal cortex. Activation of this receptor stimulates adenylate cyclase, which catalyzes the formation of cAMP. The cAMP activates protein kinase A, which phosphorylates and activates an enzyme that causes cholesterol to be released from intracellular stores. This cholesterol is transported to the mitochondria where it is used as a substrate for the synthesis of glucocorticoid hormones. In humans and fish, cortisol is the primary glucocorticoid hormone, whereas the structurally similar corticosterone is the primary glucocorticoid hormone in rats and mice. In all these species, however, the effects of glucocorticoids in the stress response are similar.

As hydrophobic hormones, glucocorticoids bind to an intracellular receptor located in the cytoplasm of target cells. Glucocorticoid binding induces a conformational change that causes the hormone-receptor complex to move to the nucleus and regulate transcription. Glucocorticoids have diverse functions, including the breakdown of lipids and proteins, and increasing blood glucose. Because these effects are mediated through changes in transcription and translation, in contrast to the rapid effects of epinephrine, which acts through cytoplasmic signal transduction pathways, the effects of glucocorticoids are much slower, and are involved in recovery from the effects of the immediate fight-or-flight response. The glucocorticoids' metabolic functions help the body to restore energy balance following the energetically costly fight-or-flight response.

The structure of adrenal tissue varies among vertebrates

The catecholamines and the glucocorticoids are involved in the stress response in all vertebrates, but there is substantial diversity among taxa in the structure of the tissue that secretes these hormones (Figure 3.38). Mammals have a compact and highly organized adrenal gland. The adrenal glands of reptiles and birds are also quite compact, as they are in mammals, but the interrenal (glucocorticoid-secreting) and chromaffin (epinephrine-secreting) tissues are intermingled, rather than being separated into a distinct cortex and medulla. The interrenal and chromaffin cells of amphibians are intermingled in a diffuse stripe along the kidney. In elasmobranch fishes, the interrenal cells form a fairly compact organ that is located on the kidney, but the chromaffin cells are found in the body cavity anterior to the kidney, grouped into loose clusters. Bony fish entirely lack a discrete adrenal gland; their interrenal cells are generally located in a single layer around the blood vessels of the anterior kidney, while the chromaffin cells vary in location, often being associated with interrenal cells. However, despite these differences in the structure of the target organs among vertebrates, the overall organization and functions of the stress response are similar. This transition from a dispersed group of hormone-secreting cells toward a compact and organized gland is a general trend in the evolution of endocrine systems in both the vertebrates and the invertebrates.

Evolution of Endocrine Systems

Cellular signaling plays an important role in the maintenance of homeostasis and the coordination of reproduction, growth, and development in all animals. As we discuss in later chapters, there are substantial similarities in the structure and function of nervous systems across all taxa (see Chapter 5 and Chapter 7). In contrast, the organization of

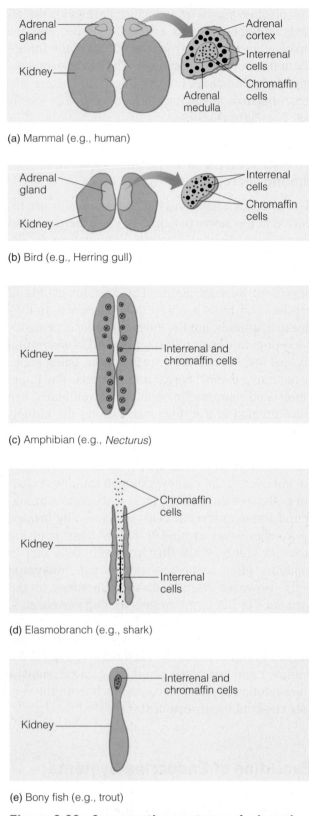

(a) Mammal (e.g., human)

(b) Bird (e.g., Herring gull)

(c) Amphibian (e.g., *Necturus*)

(d) Elasmobranch (e.g., shark)

(e) Bony fish (e.g., trout)

Figure 3.38 Comparative anatomy of adrenal tissues in the vertebrates Chromaffin cells (shown in gray) and interrenal cells (shown in black) are associated with the kidneys of vertebrates. In mammals, birds, and reptiles they form discrete adrenal glands, while in fishes and amphibians the cells are in isolated clusters.

endocrine systems is quite diverse. Unlike nervous systems, which were present very early in the evolution of animals, endocrine systems could only arise following the evolution of a circulatory system that could carry hormones from one part of the body to another. Because circulatory systems are thought to have arisen independently several times in different animal groups, we can conclude that endocrine systems have arisen multiple times and the endocrine systems of, for example, vertebrates and arthropods are not closely related.

Although there are substantial differences in the organization of animal endocrine systems, there are also substantial similarities. These similarities likely stem from the evolution of endocrine systems from a shared set of basic signal transduction mechanisms involved in paracrine communication in the ancestral metazoans. Over time, however, animal cell-to-cell communication mechanisms have diverged and diversified into the complex endocrine systems we see in various taxa. In all animals, however, endocrine systems rely upon a similar set of chemical messengers, receptors, and signal transduction pathways. For example, the insulin receptor and its associated signal transduction pathways are present in both vertebrates and invertebrates. Both invertebrates and vertebrates have steroid receptors, phospholipase C, and adenylate cyclase.

All vertebrates, including the jawless lampreys and hagfish, use a series of related steroid hormones as chemical messengers, including estrogens, androgens, and glucocorticoids. Of these hormones, only estrogen has been found in invertebrates. Instead, insects and crustaceans use a different series of steroid hormones related to ecdysone. Like the vertebrate steroid hormones, ecdysone binds to an intracellular receptor that interacts with DNA and regulates gene transcription.

Despite the similarities at the molecular level, the organization of endocrine systems varies between invertebrates and vertebrates. Invertebrates have relatively few endocrine glands, and most endocrine signaling utilizes neurohormones rather than hormones. In general, in both vertebrates and invertebrates, there is a correlation between the complexity of the endocrine system and the complexity of body form or organization. For example, the so-called lower invertebrates (such as cnidarians and platyhelminths) have a limited number of neurohormones that are mostly involved with regulating growth and development.

They appear to have few physiologically active hormones. In contrast, the "higher" invertebrate phyla (such as the annelids, molluscs, and arthropods), the cephalochordates, and the vertebrates have many complex endocrine pathways that regulate most physiological processes. This increase in complexity of the endocrine system is related to the increase in complexity of the circulatory system that allows hormones to be transported across long distances in these groups.

Because of the great variation in endocrine systems among organisms, it would be impossible to discuss all of the hormones and their regulatory pathways within the scope of a single chapter, but you will find references to endocrine systems and their hormones throughout this book. Table 3.4 summarizes some of the major hormones of vertebrates, and provides a guide to the chapters in which these hormones are discussed.

⊙ CONCEPT CHECK

9. What are the primary functions of endocrine systems?
10. What are antagonistic pairings? What are the advantages of this organization of control systems?
11. Compare and contrast negative feedback and positive feedback. Which type of control allows maintenance of homeostasis?
12. Compare and contrast additivity and synergism.
13. Provide an example of a hormone controlled by a third-order endocrine pathway, and outline each step in the regulatory cascade.

Table 3.4 Major hormones of the vertebrates.

Secretory tissue	Hormone	Chemical class	Effects	For more details see
Pineal gland	Melatonin	Amine	Circadian and seasonal rhythms	Chapter 6
Hypothalamus (clusters of secretory neurons)	Trophic hormones (see Figure 3.31)	Peptides	Regulation of anterior pituitary	Chapter 3 Chapter 7 Chapter 10 Chapter 13
Posterior pituitary (extensions of hypothalamic neurons)	Oxytocin	Peptides	Breast and uterus in mammals; also involved in social bonding and behavior	Chapter 6 Chapter 14
	Vasopressin		Water reabsorption in excretory system	Chapter 10
	Vasotocin (fish, amphibians, birds)		Activities similar to both oxytocin and vasopressin	
Anterior pituitary gland	Prolactin (PRL)	Peptides	Milk production in mammals, osmoregulation, growth, metabolism	Chapter 14 Chapter 11
	Growth Hormone (GH)		Growth, metabolism	Chapter 11
	Adrenocorticotropic hormone (ACTH)		Release of corticosteroids	Chapter 3 Chapter 7 Chapter 13 Chapter 14
	Thyroid stimulating hormone (TSH)		Synthesis and release of thyroid hormones	Chapter 13
	Follicle stimulating hormone (FSH)		Egg or sperm production; sex hormone production	Chapter 14
	Luteinizing hormone (LH)		Sex hormone production; egg or sperm production	Chapter 14

(continued)

Table 3.4 Major hormones of the vertebrates *(continued).*

Secretory tissue	Hormone	Chemical class	Effects	For more details see
Thyroid gland	Triiodothyronine (T$_3$) and thyroxine (T$_4$) Calcitonin	Iodinated amines Peptide	Metabolism, growth, and development Regulation of plasma Ca^{2+} (in non-human vertebrates)	Chapter 13
Parathyroid gland	Parathyroid hormone	Peptide	Regulates plasma Ca^{2+} and phosphate	
Thymus gland	Thymosin, thymopoitin	Peptides	Immune system	
Heart (individual cells in atrium)	Atrial natriuretic peptide (ANP)	Peptide	Regulation of sodium levels and blood pressure	Chapter 8 Chapter 10
Liver (various cells)	Angiotensinogen Insulin-like growth factors (IGF)	Peptides	Regulation of aldosterone; regulation of blood pressure Growth and metabolism	Chapter 8
Stomach and small intestine (various cells)	Gastrin, cholecystokinin (CCK), secretin, ghrelin, and many others	Peptides	Digestion and absorption of nutrients; regulation of food intake	Chapter 11
Pancreas	Insulin, glucagon, somatostatin, pancreatic polypeptide	Peptides	Regulation of blood glucose and other nutrients; regulation of metabolism	Chapter 3 Chapter 11
Adrenal gland (cortex) in mammals; dispersed cells in other vertebrates	Steroids	Aldosterone (mammals only) Corticosteroids (e.g., cortisol) Androgens	Ion regulation Stress response; metabolism Sex drive in females; bone growth at puberty in males	Chapter 10 Chapter 3 Chapter 7 Chapter 13 Chapter 14
Adrenal gland (medulla) in mammals; chromaffin cells in other vertebrates	Amines	Epinephrine, norepinephrine	Stress response; regulation of cardiovascular system	Chapter 3 Chapter 7 Chapter 8 Chapter 13 Chapter 14
Kidney (various cells)	Peptide	Erythropoietin (EPO)	Red blood cell production	Chapter 8
Adipose tissue (various cells)	Peptides	Leptin and others	Food intake, metabolism, reproduction	Chapter 11
Testes (male)	Steroids	Androgens	Sperm production; secondary sexual characteristics	Chapter 12
Ovaries (female)	Steroids	Estrogens and progesterone	Egg production; secondary sexual characteristics	Chapter 14
Placenta (pregnant female mammals only)	Steroids	Estrogens and progesterone, chorionic somatomammotropin (CS), chorionic gonadotropin (CG)	Fetal and maternal development	Chapter 14

Summary

The Biochemical Basis of Cell Signaling

→ There are many types of cell-to-cell communication in animals, including direct, autocrine, paracrine, neural, endocrine, and exocrine. These types of communication vary in the distance that the chemical messengers travel from one cell to another.

→ Chemical messengers involved in indirect cell signaling must travel through both aqueous and lipid environments. Thus, hydrophobic and hydrophilic chemical messengers face different challenges during cell signaling.

→ Hydrophilic messengers can travel between adjacent cells via gap junctions, but more complex mechanisms are required for indirect cell-to-cell communication between cells that are not adjacent.

→ Indirect cell signaling involves three steps: (1) release of the messenger from the signaling cell, (2) transport through the extracellular environment, and (3) communication with the target cell.

→ The mechanisms involved in these steps of indirect cell signaling differ depending on whether the signaling molecule is hydrophobic or hydrophilic.

→ Hydrophilic messengers are often peptides. They are released from the signaling cell by exocytosis, and bind to transmembrane receptors on the target cell.

→ Peptide hormones are often synthesized as large preprohormones that are processed within the signaling cell prior to the release of the active hormone.

→ Hydrophobic messengers are synthesized on demand and diffuse out of the signaling cell. Carrier proteins transport them to target cells where they bind to intracellular receptors. Some hydrophobic messengers also bind to transmembrane receptors. Hydrophobic messengers are often steroids.

→ Steroids are derived from cholesterol. The primary vertebrate steroid hormones are the mineralocorticoids, the glucocorticoids, and the reproductive hormones. The primary invertebrate steroid hormones are the ecdysteroids.

→ Amines can be paracrines, hormones, or neurotransmitters. These messengers are derived from hydrophilic amino acids or peptides, and are often hydrophilic messengers, but the amine thyroid hormones are hydrophobic.

→ Chemical messengers involved in indirect cell signaling bind specifically to specific receptor proteins on or in the target cell. Thus, chemical messengers act as specific ligands for those receptors.

→ Hydrophobic chemical messengers can interact with intracellular receptors or transmembrane receptors. Hydrophilic chemical messengers can only interact with transmembrane receptors.

→ Ligand-receptor binding obeys the law of mass action, and exhibits saturation. The affinity constant describes the tightness of binding between a ligand and a receptor.

→ Ligand-receptor signaling must be terminated for signaling to be effective. Signal termination can be accomplished in a variety of ways, including removal of the ligand, removal of the receptor from the membrane, inhibition of the receptor, or inhibition of downstream signaling pathways.

Signal Transduction Pathways

→ Each step in a signal transduction pathway can amplify the signal.

→ Signal transduction via ligand-gated ion channels is relatively simple and direct, but the other signal transduction pathways have many steps.

→ Intracellular receptors regulate gene transcription.

→ Receptor-enzymes activate intracellular phosphorylation cascades.

→ G-protein-coupled receptors interact with heterotrimeric G proteins. G proteins can signal to ion channels or to amplifier enzymes that activate small molecules called second messengers.

→ G-protein-coupled receptors use four different second messengers: Ca^{2+}, cGMP, inositol phosphates, and cAMP. Each of these messengers links to a different signal transduction cascade.

→ Cells have numerous types of transmembrane and intracellular receptors, and thus several signal transduction cascades can be activated at any given time. Thus, signal transduction cascades in living cells operate as complex networks that integrate the various signals and convert them into appropriate physiological responses.

Introduction to Endocrine Systems

→ The organization of endocrine systems varies among animals. Endocrine communication in the invertebrates generally involves neurohormones, whereas hormones are more common in the vertebrates.

→ Endocrine systems are responsible for maintaining homeostatis and regulating growth, development, and reproduction.

→ Negative feedback systems allow the maintenance of homeostasis (e.g., control of blood glucose by insulin).

→ Positive feedback allows explosive responses.

→ Hormones are often grouped into antagonistic pairs that allow extremely precise homeostatic regulation (e.g., insulin and glucagon).

→ Hormones can also work additively or synergistically (e.g., glucagon, cortisol, and epinephrine).

→ Hormones can regulate other hormones by negative feedback, in regulatory loops of varying complexity (e.g., the hormones of the vertebrate anterior pituitary).

→ Hormones can also be involved in positive feedback regulation (e.g., oxytocin).

→ The vertebrate stress response is an example of coordination of physiological functions by multiple signaling systems.

Review Questions

1. What are the three major steps involved in indirect chemical signaling?

2. Would you expect the chemical signaling molecules involved in signaling via gap junctions to be hydrophilic or hydrophobic? Justify your answer.

3. Compare and contrast autocrine, paracrine, endocrine, and neural communication.

4. You read an article in the newspaper about the discovery of a new steroid hormone. What can you predict about how it is synthesized and/or stored by the signaling cell, how it is transported through the blood, and how it acts on the target cell?

5. If the newspaper article were about a peptide hormone, how would your predictions change?

6. What is the difference between a neurohormone and a neurotransmitter?

7. From the perspective of the target cell, is there a fundamental difference between a paracrine signal and an endocrine signal? Why or why not?

8. List the main classes of chemicals involved in indirect cell signaling in animals. Which of these classes are utilized for endocrine communication?

9. Why are peptide messengers released by exocytosis?

10. What are the three main domains of a transmembrane receptor, and what are their functions?

11. Describe the phenomenon of "endocrine disruption."

12. Compare and contrast the functions of intracellular and transmembrane steroid receptors.

13. How do the thyroid hormones differ from all of the other biogenic amines?

14. What would be the effect of increasing receptor number on the response of a target cell to the ligand for that receptor?

15. How do selective serotonin reuptake inhibitors (SSRIs) affect the response of a target cell to serotonin?

16. Compare and contrast the signal transduction cascades initiated by intracellular receptors and G-protein coupled receptors.

17. Compare and contrast the function of heterotrimeric G proteins and a small soluble G protein such as Ras.

18. What is the difference between signaling through G_s and G_i?

19. What are the major parts of any control system (mechanical or biological)? Choose an example of a biological control system and show how it fits the general description of control systems that you provided.

20. Compare and contrast positive and negative feedback. Provide a biological example for each type of feedback.

21. Compare and contrast the anterior and posterior pituitary.

22. Compare and contrast the insulin/glucagon system for blood glucose regulation in the vertebrates with the function and regulation of crustacean hypoglycemic hormone (CHH).

23. Outline the major steps of the vertebrate stress response.

Synthesis Questions

1. Epinephrine and glucagon both act to increase blood glucose, but they act on a different subset of tissues. What characteristics are likely to determine whether a particular tissue responds to epinephrine, glucagon, or to both hormones?

2. People who do not regularly drink coffee often feel much greater effects when they ingest modest doses of caffeine than do heavy coffee drinkers. Explain at a molecular level why this might be so.

3. The anticancer drug tamoxifen binds to the estrogen receptor. Tamoxifen inhibits the growth of breast tissue but promotes growth of uterine tissues, thus reducing the risk of breast cancer but potentially increasing the risk of uterine cancer. Explain how the same chemical messenger could have opposite effects in two different tissues.

4. What are the advantages of a multistep signal transduction pathway in cell-to-cell communication?

5. Epinephrine binds to a G-protein-coupled receptor that signals via G_s. Acetylcholine binds to a G-protein-coupled receptor that signals via G_i. You construct a recombinant receptor with the extracellular domain of the acetylcholine receptor and the intracellular domain of the epinephrine receptor, and transfect it into cultured cells. Your preliminary experiments indicate that the receptor is processed correctly, and inserted into the plasma membrane. If you applied acetylcholine to your transfected cells, what would you expect to happen to intracellular cAMP levels? What would happen if you applied epinephrine? Explain your answers

Quantitative Questions

1. The graph below outlines the results of an experiment to determine the binding characteristics of a ligand to its receptor on the surface of adipocytes (fat cells).

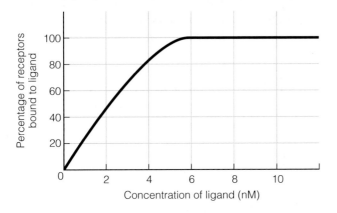

(a) What is the minimum concentration of ligand at which the receptor is saturated?

(b) What is the affinity constant of the receptor?

(c) If the receptor number on the adipocytes were doubled, what would be the predicted maximum binding of the ligand?

(d) If the receptor number on the adipocytes were doubled, would the affinity constant of the receptor change?

2. In insects, the Malpighian tubules are involved in the maintenance of ion and water balance. When the peptide hormone diuretic hormone is applied to Malpighian tubules isolated from the blood sucking insect *Rhodnius prolixus*, the tubule epithelium begins to secrete fluid at a rate of approximately 5 nL/min. The biogenic amine serotonin has similar effects, causing secretion at a rate of approximately 4 nL/min. When both chemical messengers are applied together, however, fluid secretion occurs at a rate of approximately 45 nL/min. Is this an example of additivity, synergism, or antagonism? Justify your answer.

For Further Reading

See the Additional References section at the back of the book for more readings related to the topics in this chapter.

The Biochemical Basis of Cell Signaling

The review paper below highlights some of the important features of exocytosis in the release of peptides and other cell signaling molecules

Burgoyne, R. D., and A. Morgan. 2003. Secretory granule exocytosis. *Physiological Reviews* 83: 581–632.

The review below outlines the role of carrier proteins in the regulation of steroid hormone activity.

Bruener, C. W., and M. Orchinik. 2002. Beyond carrier proteins: Plasma binding proteins as mediators of corticosteroid action in vertebrates. *Journal of Endocrinology* 175: 99–112.

Written by the scientist who discovered the physiological actions of nitric oxide, the paper below provides a fascinating look at the process of scientific discovery.

Moncada, S. 2006. Adventures in vascular biology: A tale of two mediators. *Philosophical Transactions of the Royal Society of London: Biological Sciences* 361: 735–759.

This book provides a good general introduction to signal transduction pathways.

Sitaramayya, A. 1999. Introduction to cellular signal transduction. Boston: Birkhauser.

Signal Transduction Pathways

These reviews and others from a special issue of *Science* magazine entitled "Mapping Cellular Signaling" summarize recent advances in signal transduction research.

Attisano, L., and J. Wrana. 2002. Signal transduction by the TGF-β superfamily. *Science* 296: 1646–1647.

Neeves, S. R., P. T. Ram, and R. Iyengar. 2002. G protein pathways. *Science* 296: 1636–1639.

These reviews highlight some of the recent findings regarding signal transduction via steroid hormones, through both intracellular and transmembrane receptors.

Cheskis, B. J. 2004. Regulation of cell signalling cascades by steroid hormones. *Journal of Cellular Biochemistry* 93: 20–27.

Losel, R., and M. Wehling. 2003. Nongenomic actions of steroid hormones. *Nature Reviews: Molecular Cell Biology* 4: 46–56.

Losel, R. M., E. Falkenstein, M. Feuring, A. Schultz, H-C. Tillmann, K. Rossol-Haseroth, and M. Wehling. 2003. Nongenomic steroid action: Controversies, questions, and answers. *Physiological Reviews* 83: 965–1016.

The paper below discusses the evolution of signal transduction pathways and their role in the evolution of the earliest metazoans.

Suga, H., M. Koyanagi, D. Hoshiyama, K. Ono, N. Iwabe, K-I. Kuma, and T. Miyata. 1999. Extensive gene duplication in the early evolution of animals before the parazoan-eumetazoan split demonstrated by G proteins and protein tyrosine kinases from sponge and hydra. *Journal of Molecular Evolution* 48: 646–653.

Introduction to Endocrine Systems

This book contains a collection of essays outlining the development of concepts in endocrinology from Aristotle to the present, telling the stories of some of the scientists involved in discovering the fundamental principles of endocrinology.

McCann, S. M., ed. 1997. *Endocrinology: People and ideas: An American Physiological Society book*. New York: Oxford University Press.

This highly accessible book provides a review of the function of the endocrine system in humans.

Neal, M. J. 2001. *How the endocrine system works*. New York: Blackwell Science.

This book provides an introduction to the extensive literature on invertebrate endocrinology.

Nijhout, H. F. 1998. *Insect Hormones*. New York: Princeton University Press.

This paper summarizes the current state of knowledge on CHH.

Fanjul-Moles, M. L. 2006. Biochemical and functional aspects of crustacean hyperglycemic hormone in decapod crustaceans: Review and update. *Comparative Biochemistry and Physiology C: Toxicology and Pharmacology* 142: 390–400.

These reviews provide some additional insights into the stress response and its regulation in the vertebrates.

Flik, G., P. H. Klaren, E. H. Van den Burg, J. R. Metz, and M. O. Huising. 2006. CRF and stress in fish. *General and Comparative Endocrinology* 146: 36–44.

DeRijk, R., de and E. R. Kloet. 2005. Corticosteroid receptor genetic polymorphisms and stress responsivity. *Endocrine* 28: 263–270.

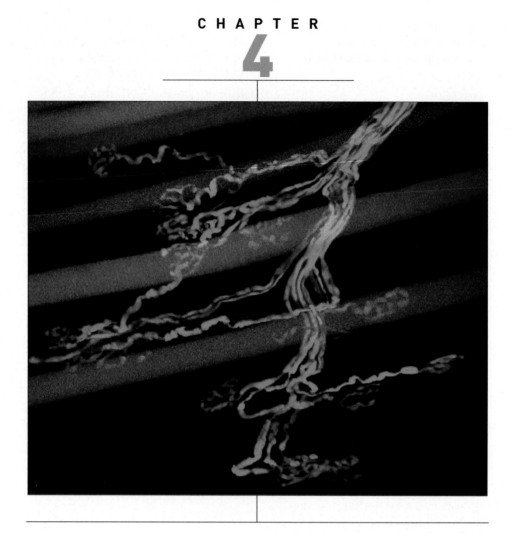

Neuron Structure and Function

With an anxiety that almost amounted to agony, I collected the instruments of life around me, that I might infuse a spark of being into the lifeless thing that lay at my feet.

Frankenstein (or, the Modern Prometheus)

Mary Shelley, 1818

It was no coincidence that Mary Shelley chose to use electricity as the animating force that brought Frankenstein's monster to life. In writing her novel, Mary Shelley was influenced by the work of the scientist Luigi Galvani, who had demonstrated about 30 years earlier what we now understand as electrical transmission in the nervous system. Galvani showed that the muscles of a dead frog twitch when ʸ ⁱ apply an electrical current to the frog's nerves. ʸ careful experiments, Galvani concluded that ᶦᵗ transmit "an animal electricity" that was somehow involved in controlling the activities of the body. This was a revolutionary idea, since at the time nerves were thought to be similar to pipes or canals that carried fluid. Although Galvani's interpretation of animal electricity ultimately proved to be incorrect (because he thought it was a unique property of living things and distinct from other electrical phenomena), his pioneering discoveries led the way to the modern study of neurophysiology.

We now know that nerves are composed of groups of cells termed neurons that are specialized for processing and conveying information in the form of electrical signals rapidly and precisely across long distances. Neurons perform this function by coding incoming information into changes in the electrical potential across the cell membrane. In particular, neurons use a specialized form of

Galvani's experiments on frog's legs.

Venus flytrap.

electrical impulse, called the action potential, to transmit electrical signals across long distances. Along with muscle cells, another class of electrically excitable cells, which we discuss in Chapter 5: Cellular Movement and Muscles, neurons allow animals to sense and respond to their environments in ways that no other organisms can.

Only metazoans have neurons, but they are not the only organisms capable of generating rapid and coordinated responses to their environments. Charles Darwin was fascinated with plants such as the Venus flytrap, which he called the most wonderful plants in the world. The leaves (or lobes) of the Venus flytrap resemble a set of open jaws. When an insect or other small animal lands on the lobe of the Venus flytrap and disturbs one of the trigger hairs on the surface of the lobes, the jaws snap shut, trapping the insect. The Venus flytrap then digests the trapped insect, and this digested material provides the plant with a supplementary source of the nitrogen and minerals that are lacking in their boggy habitats. Because of the plant's ability to actively catch animal prey, Darwin hypothesized that the Venus flytrap must possess neurons similar to those in animals. He contacted John Burdon-Sanderson, an eminent medical physiologist at the University College London, to test his hypothesis about the basis of movement in the Venus flytrap. Burdon-Sanderson placed electrodes on the lobes of the Venus flytrap, and recorded what happened when he touched one of the trigger hairs. He found that the plant responded by generating an electrical signal that was very similar to an action potential.

In fact, the Venus flytrap does not have neurons. Instead, action potentials travel through the structural tissues of the plant, transferred from one plant cell to another via plasmodesmata—intercellular connections analogous to gap junctions in animals. This kind of conduction is very slow compared to conduction in neurons. In plants, action potentials typically travel at speeds between 1 and 3 centimeters per second, whereas in animals action potentials can travel along neurons at speeds up to 100 *meters* per second (or 10,000 cm/sec). The high-speed conduction of action potentials is a unique property of animals. In this chapter, we will see how the special properties of neurons allow the rapid and precise conduction of electrical signals that is the hallmark of animal life.⊙

Overview

As we discussed in Chapter 2: Chemistry, Biochemistry, and Cell Physiology, animal cells have a voltage difference across their cell membranes, termed the *resting membrane potential*. This voltage difference, together with concentration gradients across the membrane, results in an electrochemical gradient that acts as a form of potential energy that cells can harness to move substances across the membrane. In addition to using this electrochemical potential as a form of energy, cells can also use changes in the membrane potential as communication signals. In fact, certain classes of cells, termed **excitable cells**, can rapidly alter their membrane potential in response to an incoming signal.

Animals have a variety of types of excitable cells, including some endocrine cells (see Chapter 3: Cell Signaling and Endocrine Regulation), muscle cells (see Chapter 5: Cellular Movement and Muscles), and cells such as fertilized eggs. However, the best-known excitable cells are **neurons**—cells that are specialized to carry electrical signals, often across long distances.

Neurons vary in their structure and properties, but all neurons use the same basic mechanisms to send signals. Figure 4.1 illustrates the structure and function of some representative neurons. Each part of the neuron plays a different role in neural signaling. At one end of the cell is a zone specialized to receive incoming signals. Farther along the cell is a zone that integrates these signals. The next zone of the neuron is specialized to conduct these integrated signals along the neuron, potentially across long distances. Finally, the fourth zone of the neuron is specialized to transmit signals to other cells. As a result of this organization, neurons typically have a specific polarity: signals are transmitted from one end of the neuron to the other, but not in the opposite direction.

In the first half of this chapter we examine how these four functional zones of neurons participate in cell-to-cell communication. Using a vertebrate motor neuron as an example, we follow a signal as it travels from one end of the motor neuron to the other, discussing the features of the electrical signals in each part of the cell, and how the neuron transmits signals to its target cells, vertebrate skeletal muscles. We conclude the first half of the chapter with a brief discussion of how the muscle responds to these signals.

In the second half of the chapter, we look at how each of these steps has been modified and specialized in different neurons and in neurons from different kinds of organisms. We first discuss variation in the structure of neurons, and then address variation in the functional properties of neurons. We end the chapter with a discussion of the evolution of neurons.

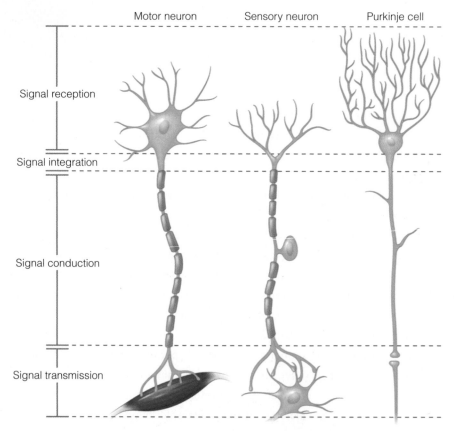

Figure 4.1 An overview of neuron structure and function Neurons vary in size and shape, but most neurons are divided into four functional regions, each specialized for a particular task: signal reception, signal integration, signal conduction, or signal transmission to other cells.

Signaling in a Vertebrate Motor Neuron

Figure 4.2 illustrates the structure and function of a vertebrate **motor neuron**—a type of neuron that sends signals from the central nervous system to skeletal muscles and is thus involved in controlling animal movement. In the first half of this chapter, we use this vertebrate motor neuron to illustrate some of the fundamental characteristics of neu-

rons and neural signaling, because these neurons provide a good example of the mechanisms underlying signal conduction.

A motor neuron, like most neurons, can be divided into four distinct zones, and each of these zones plays a somewhat different role in neural signaling. In motor neurons, the first zone, which is specialized for signal reception, consists of the **dendrites** and **cell body** (or *soma*) of the neuron. Dendrites are fine, branching extensions of the

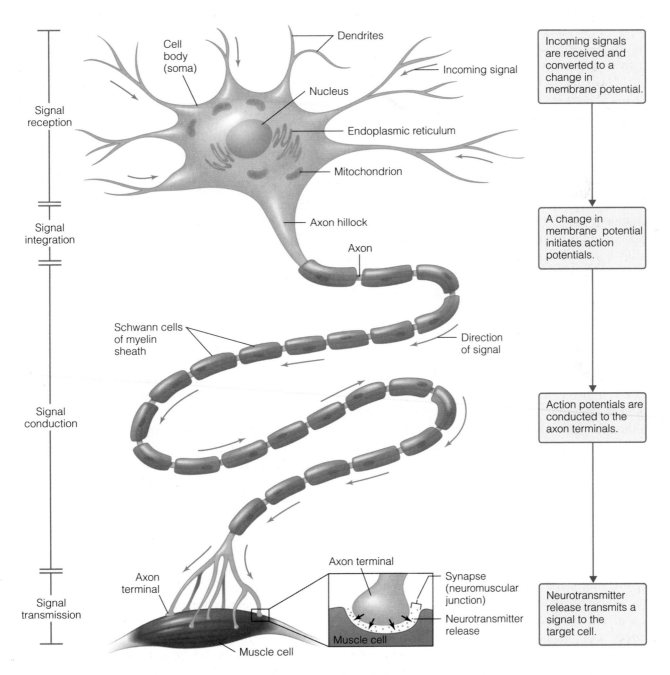

Figure 4.2 Structure and function of a typical vertebrate motor neuron
Like other neurons, motor neurons can be divided into four functional zones.

neuron, originating at the cell body. The word *dendrite* is derived from the Greek word for tree (*dendron*) because of the highly branched appearance of the dendrites of many neurons. The dendrites are responsible for sensing incoming signals, converting these signals to an electrical signal in the form of a change in the membrane potential, and transmitting the signal to the cell body. The cell body contains the nucleus and the protein synthetic machinery of the cell, as well as most of the organelles, such as mitochondria and the endoplasmic reticulum. The cell body performs all of the routine metabolic functions of the neuron—synthesizing and degrading proteins, providing energy, and helping to maintain the structure and function of the neuron. Like the dendrites, the plasma membrane of the cell body often also contains receptors, and thus can participate in detecting incoming signals.

The second zone of the motor neuron, which is specialized for signal integration, consists of the **axon hillock**. The axon hillock is located at the junction of the cell body and the **axon**. Incoming signals from dendrites and the cell body are conducted to the axon hillock. If the signal at the axon hillock is sufficiently large, an electrical signal, termed the **action potential**, is initiated. Action potentials occur in the axon, a long slender extension leading off the cell body at the axon hillock.

The axon forms the third functional zone of the neuron, and is specialized for signal conduction. Axons are often quite short (just a few millimeters), but the axons of some neurons, such as motor neurons in large mammals, can be several meters long. Each neuron has only a single axon, although the axon may branch into several *collaterals*. Vertebrate motor neurons are wrapped in a **myelin sheath** that aids in the conduction of nerve impulses to the **axon terminal**.

The axon terminals make up the fourth functional zone of the neuron, which is specialized for signal transmission to target cells. In a motor neuron, the end of the axon branches to form several axon terminals. Each axon terminal is a swelling of the end of the axon that forms a **synapse** with the target skeletal muscle cell. At the axon terminal of a motor neuron the electrical signal is transduced into a chemical signal in the form of a chemical neurotransmitter. The neurotransmitter diffuses across the synapse and binds to specific receptors on the muscle cell membrane, initiating

a signal in the muscle cell and causing the muscle to contract.

Thus, the overall process of signaling in a motor neuron involves receiving an incoming signal, converting that signal to a change in the membrane potential, triggering action potentials that conduct the signal across long distances, and then transmitting the signal to target cells in the form of a neurotransmitter. In the following sections we examine each of these processes in detail, first considering the general properties of electrical signals in neurons, and then looking at the types of signals that occur in each of the functional zones of a motor neuron.

Electrical Signals in Neurons

As excitable cells, neurons can rapidly alter their membrane potential in response to an incoming signal, and these changes in membrane potential can act as electrical signals. As we discussed in Chapter 2, neurons are not the only excitable cells. Muscle cells, fertilized eggs, some types of plant cells, and many unicellular organisms also have the capacity to rapidly alter their membrane potentials. But it is this property of excitability that gives neurons the ability to store, recall, and distribute information, and which is the main subject of this chapter.

Most neurons have a resting membrane potential of approximately −70 mV, meaning that when the neuron is at rest and not involved in sending an electrical signal, the inside of the cell membrane is about 70 mV more negatively charged than the outside of the membrane (Figure 4.3). During **depolarization**, the charge difference between the inside and outside of the cell membrane decreases, and the membrane potential becomes less negative. Either positively charged ions entering the cell or negatively charged ions moving out of the cell can make the inside of the cell membrane less negatively charged, causing depolarization. During **hyperpolarization**, the membrane potential becomes more negative. Either negatively charged ions entering the cell or positively charged ions moving out of the cell can make the inside of the cell membrane more negative, causing hyperpolarization. During repolarization, the cell membrane returns to the resting membrane potential, following a depolarization or hyperpolarization.

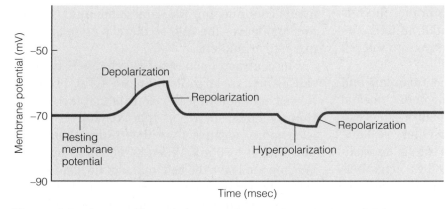

Figure 4.3 A recording of changes in membrane potential in a neuron Resting membrane potential of a neuron is usually about −70 mV. During depolarization, the membrane potential becomes less negative. During hyperpolarization, membrane potential becomes more negative. During repolarization, the membrane returns to the resting membrane potential.

The Goldman equation describes the resting membrane potential

As we discussed in Chapter 2, three factors contribute to establishing the membrane potential of a cell: the distribution of ions across the plasma membrane, the relative permeability of the membrane to these ions, and the charges on these ions. These factors are included in the Goldman equation, which describes the effects of each of these factors on the membrane potential (see Box 2.3). To review, the Goldman equation takes the form

$$E_m = \frac{RT}{F} \ln \frac{P_K[K^+]_o + P_{Na}[Na^+]_o + P_{Cl}[Cl^-]_i}{P_K[K^+]_i + P_{Na}[Na^+]_i + P_{Cl}[Cl^-]_o}$$

where E_m represents the membrane potential, R is the gas constant, T is the temperature (Kelvin), z is the valence of the ion, F is the Faraday constant, $[ion]_o$ and $[ion]_i$ represent the concentration of that ion outside and inside the cell, respectively, and P_K, P_{Na}, and P_{Cl} are the permeabilities of the membrane to the respective ions. This form of the Goldman equation considers only Na^+, K^+, and Cl^-, because most neurons under resting conditions are only permeable to these ions to any measurable degree.

Gated ion channels allow neurons to alter their membrane potentials

From the Goldman equation it is easy to see that if the membrane is not permeable to an ion, that ion does not contribute to the membrane potential. Alternatively, if the membrane is highly permeable to an ion, that ion makes a large contribution to the membrane potential. Thus, like other electrically excitable cells such as muscle cells, neurons depolarize or hyperpolarize by selectively altering the permeability of their membranes to ions, which they do by opening and closing gated ion channels in the membrane. This change in permeability alters the membrane potential and generates electrical signals.

As we discussed in Chapter 3, gated ion channels open and close in response to a stimulus, such as the binding of a neurotransmitter. It is possible to record the changes in membrane potential as ion channels open and close (see Box 4.1, Methods and Model Systems: Studying Ion Channels), and these techniques have been crucial in developing an understanding of the functions of neurons. When a gated ion channel opens, the membrane becomes much more permeable to that ion than it is to the other ions. Under these conditions, the Goldman equation can be simplified, and becomes essentially identical to the Nernst equation, which can be written as

$$E_{ion} = \frac{RT}{zF} \ln \frac{[X]_{outside}}{[X]_{inside}}$$

where $[X]$ is the molar concentration of the ion. As we discussed in Chapter 2, the Nernst equation can be used to calculate the equilibrium potential (also called the reversal potential) for a particular ion.

The equilibrium potential is the membrane potential at which the electrical and chemical gradients favoring the movement of a particular ion exactly balance each other, and there is no net movement of that ion across the membrane. If the membrane potential is far from the equilibrium potential of an ion, and gated ion channels for that ion open, ions will tend to move across the membrane because under these circumstances the electrochemical driving force for movement of that ion is large. When the membrane potential reaches the equilibrium potential, net ion movement stops because there is no electrochemical driving force for ion movement. Although ions continue to move across the membrane, there is no overall change in the distribution of ions, because the same number of ions move into the cell as move out.

For example, we can use the Goldman and Nernst equations to calculate that under normal conditions in a vertebrate motor neuron, the resting membrane potential is approximately −70 mV, and the equilibrium potential for Na$^+$ is approximately +60 mV. When ligand-gated Na$^+$ channels open, Na$^+$ ions tend to enter the cell, because of the large difference between the membrane potential and the equilibrium potential for Na$^+$, which provides a large electrochemical driving force for Na$^+$ entry (Figure 4.4a). As Na$^+$ enters the cell, the inside of the cell becomes more and more positively charged until the membrane has depolarized from the resting membrane potential of −70 mV to approximately +60 mV. At this point there is no electrochemical gradient driving Na$^+$ entry, and net Na$^+$ movement stops (ion move-

ments continue, but the same amount of Na$^+$ enters and leaves the cell, so there is no net change in ion distribution).

In contrast, opening of K$^+$ channels typically causes hyperpolarization (Figure 4.4b). From the Nernst equation we can calculate that the equilibrium potential for K$^+$ is approximately −90 mV, even more negative than the resting membrane potential of −70 mV. When K$^+$ channels open, K$^+$ ions tend to leave the cell, making the inside of the cell more negative, until the membrane has hyperpolarized from the resting membrane potential of −70 mV to approximately −90 mV, at which point net K$^+$ movement ceases.

As we discussed in Chapter 2, it is important to once again emphasize that these changes in membrane potential occur as a result of the movement of relatively small numbers of ions across the membrane, and thus a single depolarization or hyperpolarization does not measurably alter the overall concentrations of ions inside or outside of the cell (because of the very large number of ions inside and outside of the cell, and the relatively small number of ions that move across the membrane during a typical depolarization or hyperpolarization). Thus, it is changes in membrane permeability rather than measurable changes in ion concentration that cause the membrane potential to deviate from the resting membrane potential during electrical signals.

In the following sections we see how depolarization, repolarization, and hyperpolarization of the membrane as a result of changes in membrane permeability are involved in sending a signal along a vertebrate motor neuron from the dendrites to the axon terminal.

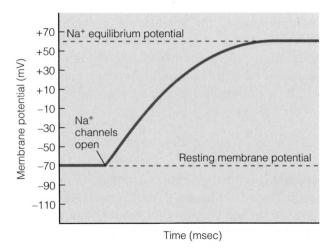

(a) Opening of Na$^+$ channels depolarizes the membrane

Signals in the Dendrites and Cell Body

Vertebrate motor neurons receive incoming signals in the form of a chemical neurotransmitter. Membrane-bound receptors in the dendrites or cell body transduce (convert) this incoming chemical signal into an electrical signal in the form of a change in the membrane potential. In Chapter 3 we discussed how receptors in many cells, including neurons, transduce incoming chemical signals into electrical signals. Recall that binding of neurotransmitter to a specific ligand-gated receptor causes ion channels in the membrane to open or close, changing the permeability of the membrane

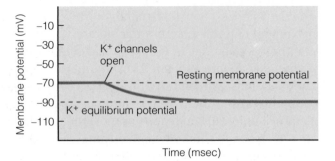

(b) Opening of K$^+$ channels hyperpolarizes the membrane

Figure 4.4 Depolarization or hyperpolarization due to opened ion channels

BOX 4.1 **METHODS AND MODEL SYSTEMS**
Studying Ion Channels

One of the most widely used methods for studying ion channels in single cells is the **voltage clamp**. The basic idea of a voltage-clamp experiment is to hold the voltage across a membrane at a constant level by injecting current into the cell via a **microelectrode** any time the voltage across the membrane changes (as a result, the voltage is said to be clamped at a particular value). For example, suppose that the resting potential of the cell is −70 mV, and you set the voltage-clamp apparatus to hold the membrane potential at −70 mV. If the cell is at the resting membrane potential, then no current will be injected through the microelectrode. But suppose you introduce into the fluid bathing the cell a neurotransmitter that binds to a specific Na$^+$ channel. The neurotransmitter will bind and cause the Na$^+$ channel to open. Na$^+$ ions will enter the cell, causing a depolarization. The apparatus takes a measurement of the membrane potential and injects current to hold the membrane at the resting membrane potential, despite the influx of Na$^+$. In this way, a voltage-clamp apparatus is analogous to a thermostatically controlled heater operating by negative feedback.

The amount of injected current is a direct measure of natural ionic movements across the membrane. Neurophysiologists use voltage-clamp experiments to describe the electrical properties of intact membranes or whole cells.

Neurobiologists also use **patch clamping** to study ion channel function. The patch clamp is particularly useful for studying the properties of single channels. In patch clamping, the experimenter fuses the tip of a glass micropipette to the plasma membrane to act as a recording electrode. The region of membrane within the patch is extremely small (often less than 1 micron), and usually contains a relatively low number of ion channels. In fact, some of the patches will contain only a single ion channel, as shown in the figure. The experimenter can

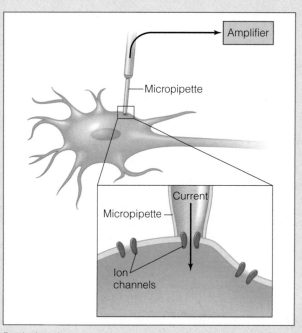

Patch clamping.

then voltage-clamp this small region of membrane and record the extremely small currents generated by a single ion channel (they are measured in picoamperes, pico = 10^{-12}). Patch clamping allows neurobiologists to study the properties of a single ion-channel molecule, while voltage clamping a whole cell, or a large region of a membrane, provides information about the behavior of populations of ion channels.

References

○ Neher, E., and B. Sakmann. 1976. Single-channel currents recorded from membrane of denervated frog muscle fibres. *Nature* 260: 799–802.

○ Neher, E., and B. Sakmann. 1992. The patch clamp technique. *Scientific American* 266: 44–51.

and altering the movement of ions. This change in permeability alters the membrane potential and causes an electrical signal. In the dendrites and cell bodies of neurons, these electrical signals are called **graded potentials**.

Graded potentials vary in magnitude

Graded potentials vary in magnitude (are graded) depending on the strength of the stimulus. A strong stimulus, such as a high concentration of neurotransmitter, increases the probability that a given ion channel will open, thus causing more ion channels to open, and keeping them open for a longer time. If more ion channels open (or stay open longer), more ions will move across the plasma membrane, causing a larger change in membrane potential. Figure 4.5 illustrates what happens when different concentrations of neurotransmitter are present near the dendrite of a

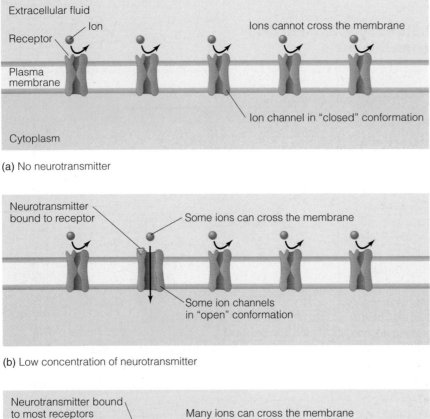

(a) No neurotransmitter

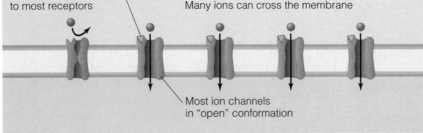

(b) Low concentration of neurotransmitter

(c) High concentration of neurotransmitter

Figure 4.5 Stimulus strength and graded potentials

neuron. When no neurotransmitter is present, the ligand-gated ion channels on the surface of the dendrite remain closed, no ions can move across the membrane through those channels, and the membrane potential stays the same. When the neurotransmitter is present at low concentrations, a few ion channels open, allowing a small number of ions to cross the membrane, causing a small change in membrane potential. When a high concentration of the neurotransmitter is present, many ion channels open, and stay open longer, allowing more ions to cross the membrane, causing a large change in membrane potential. Thus, the amplitude of the graded potential directly reflects the strength of the incoming stimulus.

As we discussed above, graded potentials can either hyperpolarize or depolarize the cell, depending on the type of ion channel that is opened or closed. The most important ion channels in the dendrites and cell body of a neuron are Na^+, K^+, Cl^-, and Ca^{2+} channels. From the Nernst equation we can calculate that opening Na^+ or Ca^{2+} channels will depolarize a neuron, while opening K^+ or Cl^- channels will hyperpolarize a neuron.

Graded potentials are short-distance signals

Graded potentials can travel through the cell, but they decrease in strength as they get farther away from the opened ion channel, a phenomenon called *conduction with decrement*. Figure 4.6 shows a neuron with a ligand-gated Na^+ channel on the membrane. When neurotransmitter (the ligand) binds to a ligand-gated Na^+ channel, the channel opens and Na^+ ions move into the cell. Na^+ entry causes a local depolarization in a small area of the membrane surrounding the opened channel. This positive charge then spreads along the inside of the membrane, causing depolarization, a phenomenon termed *electrotonic current spread*. The extent of this depolarization decreases as it moves farther and farther from the opened channels, just as ripples in a pond decrease in strength as they move farther away from their source. The signal is conducted, but it gets fainter and fainter as it travels. With ripples in a pond, the ripples decrease in size with distance due to the frictional resistance of the water. As we discuss in more detail later in the chapter, several features of the neuron influence why a graded potential decreases as it travels through the cell, including leakage of charged ions across the cell membrane, the electrical resistance of the cytoplasm, and the electrical properties of the membrane. As a result of these features, although graded potentials can travel the short distance from the dendrites to the axon hillock, they cannot travel longer distances without dying away.

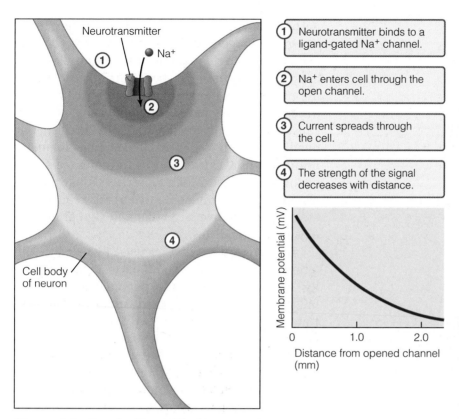

Neurotransmitter

Na+

① Neurotransmitter binds to a ligand-gated Na+ channel.

② Na+ enters cell through the open channel.

③ Current spreads through the cell.

④ The strength of the signal decreases with distance.

Cell body of neuron

Membrane potential (mV)

Distance from opened channel (mm)

Figure 4.6 Conduction with decrement

Because graded potentials cannot be transmitted across long distances without degrading, neurons use another type of electrical signal, the **action potential**, to transmit information across distances of more than a few millimeters. Action potentials are triggered by the net graded potential at the membrane of the axon hillock. The axon hillock is sometimes called the trigger zone of the neuron because it acts in a way similar to the trigger on a gun. If you pull the trigger on a gun hard enough, the gun will fire. If you don't pull the trigger hard enough the gun will not fire. Similarly, if a graded potential causes the membrane potential at the axon hillock to depolarize beyond the **threshold potential**, the axon will "fire" an action potential. If the membrane potential at the axon hillock does not reach the threshold potential, the axon will not initiate an action potential (Figure 4.7). In many neurons, the threshold potential is approximately −55 mV. Thus, the axon hillock must depolarize by more than 15 mV from the resting membrane potential of −70 mV in order to initiate an action potential. A graded potential that is not large enough to trigger an action potential is called a *subthreshold potential*.

Graded potentials that are even larger than needed to trigger an action potential are called *suprathreshold potentials*.

Because the axon hillock must reach the threshold potential in order to generate an action potential, graded potentials can either increase or decrease the likelihood of an action potential firing in the axon. A depolarizing graded potential moves the membrane potential at the axon hillock closer to the threshold potential. A hyperpolarizing graded potential moves the membrane potential at the axon hillock farther from the threshold potential. A depolarizing graded potential is called an **excitatory potential** because it makes an action potential more likely to occur by bringing the membrane potential closer to the threshold potential. A hyperpolarizing graded potential makes an action potential less likely to occur (by taking the membrane potential farther from the threshold potential), and so is called an **inhibitory potential**.

Graded potentials are integrated to trigger action potentials

The dendrites and cell body of a neuron have receptors at many sites on the membrane, and each neuron may have multiple kinds of receptors and ion channels. Thus, neurons can generate many graded potentials simultaneously. Graded potentials from different sites can interact with each other to influence the net change in membrane potential at the axon hillock; this phenomenon is called **spatial summation**. In the example of spatial summation shown in Figure 4.8, a neurotransmitter opens ligand-gated Na+ channels in one dendrite, causing Na+ to enter the dendrite, and depolarizing that area of the membrane, but alone this depolarization is not sufficient to trigger an action potential. Similarly, in the other dendrite a neurotransmitter also opens a ligand-gated Na+ channel, but again this depolarization is not

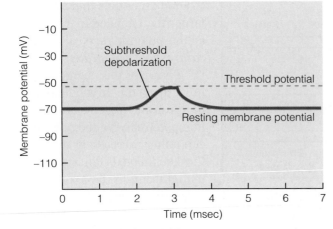

(a) Subthreshold graded potential

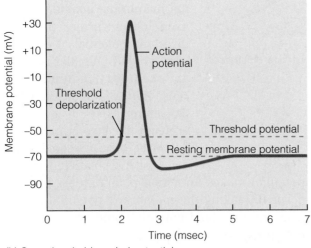

(b) Suprathreshold graded potential

Figure 4.7 Subthreshold and suprathreshold potentials The resting membrane potential of most neurons is around −70 mV and the threshold potential is −55 mV. **(a)** Subthreshold graded potentials (less than +15 mV) do not trigger an action potential. **(b)** Graded potentials that are at or above the threshold potential (greater than +15 mV) trigger an action potential.

sufficient to trigger an action potential. Both of these depolarizations travel to the axon hillock, and when they meet, they sum together to result in a net depolarization that exceeds the threshold potential and triggers an action potential. It is important to note that the phenomenon of spatial summation can also prevent action potential generation. Imagine a situation in which a suprathreshold depolarization as the result of the opening of a ligand-gated Na^+ channel occurs at the same time that, in the other dendrite, a neurotransmitter opens ligand-gated K^+ channels. Opening of K^+ channels causes K^+ to leave the dendrite, and hy-

perpolarizes that area of the membrane. These two graded potentials travel through the cell to the axon hillock. In this example there is no change in membrane potential at the axon hillock despite the changes in membrane potential in the dendrites, because change in membrane potential caused by the movement of Na^+ into the cell in one dendrite exactly balances the change in membrane potential caused by the movement of K^+ out of the cell in the other dendrite. Thus, the net change in membrane potential at the axon hillock reflects the relative strengths and sign of the signals in the dendrites.

Depolarizations that occur at two slightly different times can also combine to determine the net change in membrane potential at the axon hillock, a phenomenon called **temporal summation** (Figure 4.9). Consider two depolarizations, E_1 and E_2, each of 10 mV. If depolarization E_2 occurs after depolarization E_1 has died out, then the maximum depolarization is 10 mV, which is not large enough to trigger an action potential. In contrast, if depolarization E_2 occurs before E_1 has died out, the two depolarizations build on each other and result in an increased net depolarization to a maximum of 20 mV, bringing the cell from the resting membrane potential of −70 mV beyond the threshold potential of −55 mV, triggering an action potential.

The axon hillock acts as a decision point for the neuron. The neuron will fire an action potential in the axon only if the combination of all the graded potentials in the dendrites and cell body causes the axon hillock to depolarize beyond threshold. Spatial and temporal summation of graded potentials allow a neuron to integrate inputs from many different stimuli, and determine whether the axon hillock is depolarized beyond threshold and if an action potential will occur in the axon.

⊙ | CONCEPT CHECK

1. List the structures of a typical neuron, and summarize their functions.
2. Describe how membrane permeability and ion concentrations affect the membrane potential.
3. What is a gated ion channel? Why are gated ion channels important in neural signaling?
4. How do graded potentials code information about the intensity of the incoming signal?

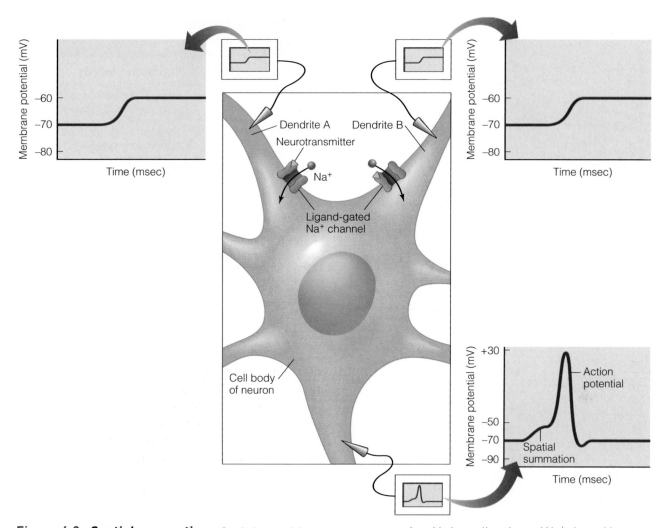

Figure 4.8 Spatial summation Graded potentials from different locations can interact to influence the net change in membrane potential at the axon hillock. In the neuron shown above, neurotransmitter binds to a ligand-gated Na^+ channel in dendrite A, opening the channel, and causing a subthreshold depolarizing graded potential. At the same time, neurotransmitter binds to a ligand-gated Na^+ channel in dendrite B, also causing a subthreshold depolarizing graded potential. Both graded potentials travel electrotonically through the cell. At the axon hillock, these subthreshold depolarizations add together, causing a suprathreshold depolarization that triggers an action potential in the axon.

Signals in the Axon

Action potentials can be transmitted across long distances without degrading, and differ from graded potentials in many respects (Table 4.1). Action potentials typically have three phases (Figure 4.10a). The **depolarization phase** of the action potential is triggered when the membrane potential at the axon hillock reaches threshold (as a result of the net graded potential at the axon hillock). Once the axon hillock reaches threshold, the adjacent axonal membrane quickly depolarizes, reaching a positive membrane potential of about +30 mV. The depolarization phase is followed by a **repolarization phase**, during which the membrane potential rapidly returns to the resting membrane potential. Following repolariza-tion, the membrane potential becomes even more negative than the resting membrane potential, and may approach the K^+ equilibrium potential. The duration and size of this **after-hyperpolarization phase** varies greatly among neurons, typically lasting between 2 and 15 msec, at which point the membrane returns to the resting membrane potential.

The ability of an axon to generate new action potentials varies during the phases of the action potential. During the **absolute refractory period**, which coincides with the depolarization and repolarization phases, the axon is incapable of generating a new action potential, no matter how strong the stimulus. During the **relative refractory period**, which coincides with the after-hyperpolarization phase, a new action potential can be generated, but only by very large stimuli.

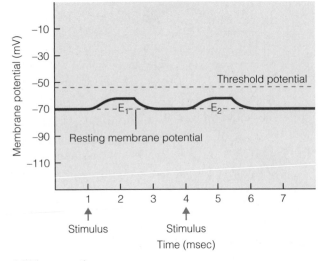

(a) No summation

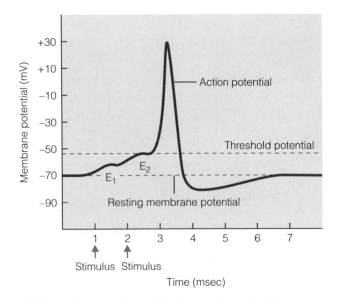

(b) Temporal summation resulting in an action potential

Figure 4.9 Temporal summation Graded potentials occurring at slightly different times can interact to influence the net graded potential. **(a)** Subthreshold depolarizations (E_1 and E_2) of 10 mV that do not overlap in time do not trigger an action potential. **(b)** Subthreshold depolarizations that occur at slightly different times may sum, if they overlap in time. If the net change in membrane potential exceeds the threshold, they will trigger an action potential.

Voltage-gated channels shape the action potential

Opening and closing of **voltage-gated ion channels** cause the characteristic phases of the action potential. Just as the binding of a neurotransmitter changes the shape of a ligand-gated ion channel, changes in membrane potential change the shape

Table 4.1	Differences between graded potentials and action potentials.
Graded potentials	**Action potentials**
Vary in magnitude	Always the same magnitude (in a given cell type)
Vary in duration	Always the same duration (in a given cell type)
Decay with distance	Can be transmitted across long distances
Occur in dendrites and cell body	Occur in axons
Caused by opening and closing of many kinds of ion channels	Caused by opening and closing of voltage-gated ion channels

of voltage-gated ion channels, allowing ions to move across the membrane.

Because there is some variation in the ion channels involved in the action potential in axons from different species, here we concentrate on the model of the action potential developed for the giant axon of the squid. The squid giant axon, which we discuss in more detail later in this chapter, sends signals from the central nervous system to the muscle of the squid's mantle cavity, and thus is part of an invertebrate motor neuron. Opening of voltage-gated Na^+ channels initiates the depolarization phase of the action potential, and opening of voltage-gated K^+ channels initiates the repolarization phase in the squid giant axon. When the membrane potential at the axon hillock approaches the threshold potential (typically around -55 mV), voltage-gated Na^+ channels in the axon hillock begin to open, changing the permeability of the membrane to Na^+ ions (Figure 4.10b), allowing Na^+ ions to move across the membrane. The probability that a given voltage-gated Na^+ channel will be open (termed the open probability of the channel) depends on the size of the graded potential. An excitatory graded potential that depolarizes the membrane toward the threshold potential increases the probability that a voltage-gated Na^+ channel will be open. Thus, at the threshold potential, more voltage-gated Na^+ channels will be open than when the axon hillock is at the resting membrane potential, increasing the permeability of the membrane to Na^+.

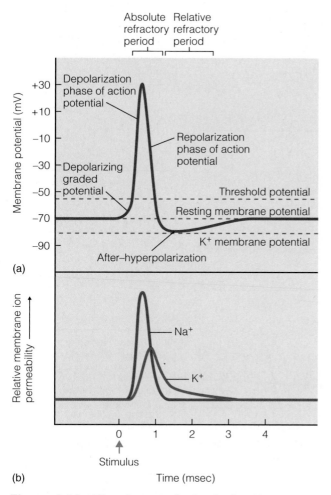

Absolute Relative
refractory refractory
period period

Figure 4.10 The phases of a typical action potential (a) Changes in membrane potential during an action potential. (b) Changes in membrane permeability during an action potential.

axon, action potentials generally occur in the axon, not in the cell body or dendrites of a neuron.

If voltage-gated Na$^+$ channels remained open indefinitely, Na$^+$ ions would enter the cell until the membrane potential reached approximately $+60$ mV (the equilibrium potential for Na$^+$). However, shortly before the membrane reaches this point, the voltage-gated Na$^+$ channels close, terminating the depolarization phase of the action potential.

In addition to increasing the open probability of voltage-gated Na$^+$ channels, threshold depolarization of the membrane at the axon hillock increases the probability that voltage-gated K$^+$ channels will open. But voltage-gated K$^+$ channels open more slowly than voltage-gated Na$^+$ channels. In fact, voltage-gated K$^+$ channels only begin to open in substantial numbers shortly before the voltage-gated Na$^+$ channels close. When voltage-gated K$^+$ channels open, the permeability of the membrane to K$^+$ ions increases (Figure 4.10b), and K$^+$ ions leave the cell in response to their electrochemical driving force, making the intracellular side of the membrane more negative, and causing the repolarization phase of the action potential. The difference in the time it takes for voltage-gated Na$^+$ channels and voltage-gated K$^+$ channels to open in response to a threshold depolarization explains why repolarization occurs after depolarization.

Following the repolarization phase, the voltage-gated K$^+$ channels close slowly, and may stay open even after the membrane has reached the resting membrane potential of approximately -70 mV. Because the electrochemical potential for K$^+$ is -90 mV, K$^+$ ions continue to move out of the cell until the membrane is slightly hyperpolarized, as long as the channels remain open, accounting for the after-hyperpolarization phase of action potentials such as those in the squid giant axon.

Voltage-gated Na$^+$ channels have two gates

Figure 4.11 summarizes a model of the changes in the conformation of voltage-gated Na$^+$ channels during the action potential. When the membrane of the neuron is at the resting membrane potential (step 1), there is a high probability that a given voltage-gated Na$^+$ channel will be closed, preventing movement of Na$^+$ ions across the membrane. When the membrane potential at the axon hillock reaches the threshold potential, the probability that

The Na$^+$ influx from the first voltage-gated channels to open in response to the graded potential further depolarizes the local region of the membrane, further increasing the probability that voltage-gated Na$^+$ channels will open, causing even more voltage-gated Na$^+$ channels to open, further increasing the permeability of the membrane, allowing even more Na$^+$ ions to enter the cell. This positive feedback loop of Na$^+$ entry reinforces itself, resulting in the extremely rapid change in membrane Na$^+$ permeability shown in Figure 4.10b, and accounting for the rapid depolarization phase of the action potential. The density of voltage-gated Na$^+$ channels in the membrane must be high in order for the positive feedback mechanism of the action potential to function. Since voltage-gated Na$^+$ channels are usually present at high concentration only in the

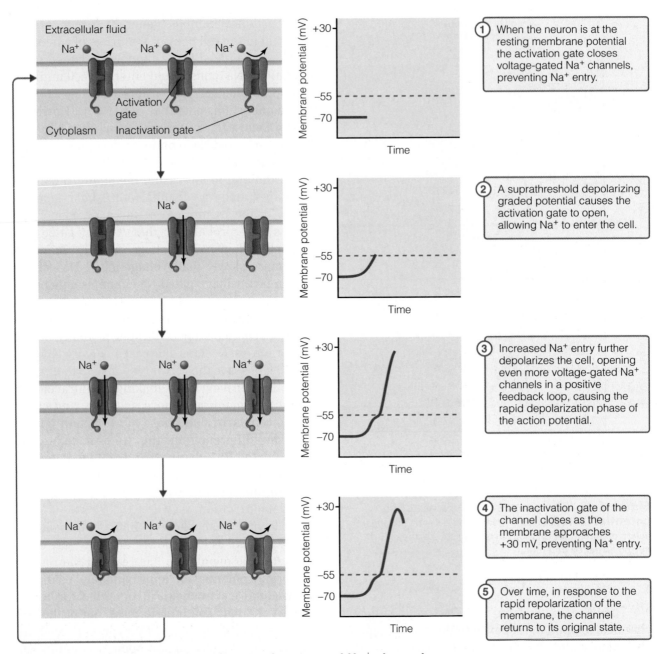

Figure 4.11 A model for the action of voltage-gated Na⁺ channels

the channel will open increases greatly. To open, the Na⁺ channel undergoes a conformational change that opens an **activation gate**, allowing Na⁺ ions to move across the membrane (step 2). The opening of the activation gate increases the permeability of the membrane to Na⁺. As Na⁺ enters the cell, more and more voltage-gated Na⁺ channels open, and the axonal membrane potential rapidly becomes less negative (step 3), depolarizing the cell toward the equilibrium potential for Na⁺ ions (approximately +60 mV). As the membrane potential approaches

the equilibrium potential for Na⁺, the electrochemical gradient that acts as a driving force for Na⁺ movement decreases and Na⁺ entry slows.

Meanwhile, a time-dependent conformational change occurs in the channel, closing an **inactivation gate** (step 4). With the inactivation gate closed, no more Na⁺ can enter the cell, terminating the depolarization phase of the action potential. Over several milliseconds, in response to changes in the membrane potential caused by the actions of the voltage-gated K⁺ channels, the inactivation gate resets, and the channel returns to its initial conformation (activation gate closed, inactivation gate open), ready to initiate another action potential.

Figure 4.12 summarizes the relationship between the voltage-gated Na^+ and K^+ channels and how they produce the action potential. When the axon hillock depolarizes beyond the threshold potential, both the Na^+ and K^+ channels receive a signal to open. The voltage-gated Na^+ channels open very quickly, allowing Na^+ to enter the cell, causing further depolarization. This greater depolarization opens even more Na^+ channels, causing even greater depolarization in a positive feedback cycle. As the axon hillock approaches the equilibrium potential for Na^+, ion entry slows, and the voltage-gated Na^+ channels close, preventing further Na^+ entry, and terminating the positive feedback loop of the depolarization phase. At about the same time, the voltage-gated K^+ channels begin to open, K^+ leaves the cell, and the intracellular side of the membrane becomes more negative, initiating the repolarization phase of the action potential.

At the end of an action potential, some Na^+ ions have entered the cell and some K^+ ions have moved out, leaving the cell in a slightly different state from the starting point. From the preceding discussion, you might think that large numbers of ions must move across the cell membrane during an action potential. In fact, the number of ions moving across the membrane is extremely small compared to the total number of ions in the intracellular and extracellular fluids. As a result, the changes in membrane potential during the action potential are not associated with any measurable changes in ion concentrations inside or outside the cell. However, even though only relatively small numbers of ions actually move across the membrane during a single action potential, thou-

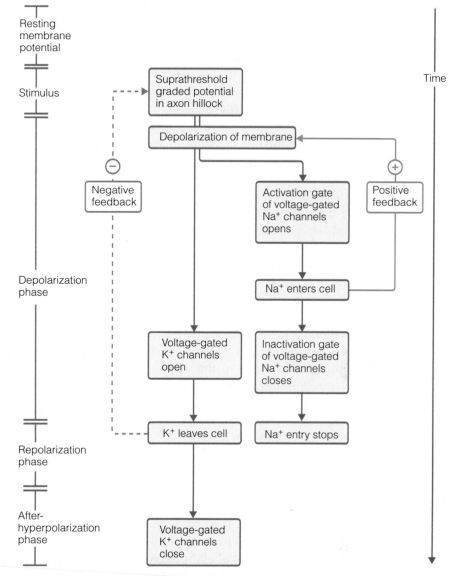

Figure 4.12 **Relationship of voltage-gated Na^+ and K^+ channels during an action potential** A suprathreshold graded potential stimulates both Na^+ and K^+ channels to open. Na^+ channels open immediately, and the resulting influx of Na^+ causes even more Na^+ channels to open, in a positive feedback loop. K^+ channels open more slowly, becoming fully opened around the time that the Na^+ channels close and causing an efflux of K^+ ions that repolarizes the membrane. K^+ ions may continue to leave the cell and cause the membrane to hyperpolarize. Repolarization and hyperpolarization remove the stimulus to open K^+ channels, causing them to close.

sands of repeated action potentials would eventually cause the Na^+ and K^+ gradients of the resting cell membrane to dissipate, changing the resting membrane potential of the cell, unless ion gradients were restored. As you might expect from its role in establishing the resting membrane potential, the Na^+/K^+ ATPase, which we discussed in Chapter 2, plays a primary role in restoring ion gradients following repeated action potentials.

Action potentials transmit signals across long distances

Up to this point we have discussed how an action potential occurs at the axon hillock, but we have not considered how action potentials travel along the axon. One property of the action potential, which is sometimes termed its *"all-or-none"* nature, is crucial in allowing neurons to transmit electrical signals across long distances. Action potentials are often described as all-or-none phenomena because once an action potential has been initiated (by the opening of a sufficient number of voltage-gated Na$^+$ channels), it always proceeds to its conclusion; it never stops halfway through, or fails to reach its peak depolarization. But how does this property allow action potentials to travel along the axon, potentially across long distances?

In fact, individual action potentials do not actually travel along the axon. Instead, an action potential in one part of the axon triggers other action potentials in adjacent areas of the axonal membrane. The transmission of an action potential is similar to what happens when you knock over the first in a long line of dominoes. The first domino that is knocked over starts the next domino falling, which starts the next domino, and so on down to the end of the line. In neurons, the first action potential at the axon hillock causes another action potential farther down the axon, and so on down to the axon terminal. Just as the last domino in a series of falling dominoes is identical to the first domino, the last action potential at the axon terminal is identical to the first action potential at the axon hillock. Thus, action potentials can be conducted across long distances without decaying.

Figure 4.13 summarizes the mechanism of action potential conduction along the axon. During an action potential, the Na$^+$ ions entering via the voltage-gated Na$^+$ channels depolarize the section of the membrane immediately surrounding the channel. This depolarization can then spread along the axon by electrotonic current spread, just as the depolarizations associated with graded potentials can spread through the dendrites and cell body. When the membrane in the adjacent region of the axon reaches the threshold potential, voltage-gated Na$^+$ channels in this region open and trigger another action potential. The cycle of ion entry, current spread, and triggering of an action potential continues along the axon from the axon hillock to the axon terminal, causing a wave of depolarization to spread along the axon. Thus, conduction of

an action potential along the axon represents a combination of action potentials occurring at specific points along the axon, and local flow of ions and electrical current along the axon, which triggers action potentials further downstream. Because of the all-or-none nature of the action potential, each action potential that is generated along the axon is essentially identical to all the other action potentials along the axon. In this way, electrical signals can be transmitted across long distances along the axon without degrading.

Vertebrate motor neurons are myelinated

The axons of vertebrate motor neurons are wrapped in an insulating layer of **myelin** (Figure 4.14). Specialized lipid-rich cells called **Schwann cells** form the myelin sheath by wrapping in a spiral pattern around the axon of the neuron. Schwann cells are one of a large class of cells that are collectively known as **glial cells**, which we discuss later in this chapter. Several Schwann cells may wrap long axons, separated by areas of exposed axonal membrane called **nodes of Ranvier** that contain high densities of voltage-gated channels. In contrast, the myelinated regions of the axons are termed the **internodes**. In myelinated axons, current spreads electrotonically through internodes, while action potentials occur only in the nodes of Ranvier. This mode of action potential propagation is termed **saltatory conduction** from the Latin word *saltare* (to leap or dance) because the action potential appears to jump from node to node along the axon. As we discuss in more detail later in the chapter, all else being equal, saltatory conduction along a myelinated axon is more rapid than conduction along an unmyelinated axon. This is because electrotonic currents can travel farther with less degradation through the internodes than through an equivalent region of unmyelinated axon, and electrotonic current spread is much faster than generating an action potential.

Axons conduct action potentials unidirectionally

If you electrically stimulate an axon halfway along its length, action potentials will be generated in both directions (toward the axon hillock and toward the axon terminal). In a natural action potential, however, the stimulus always starts at the axon hillock and travels toward the axon terminal, with little or no conduction in the reverse direction. Since the de-

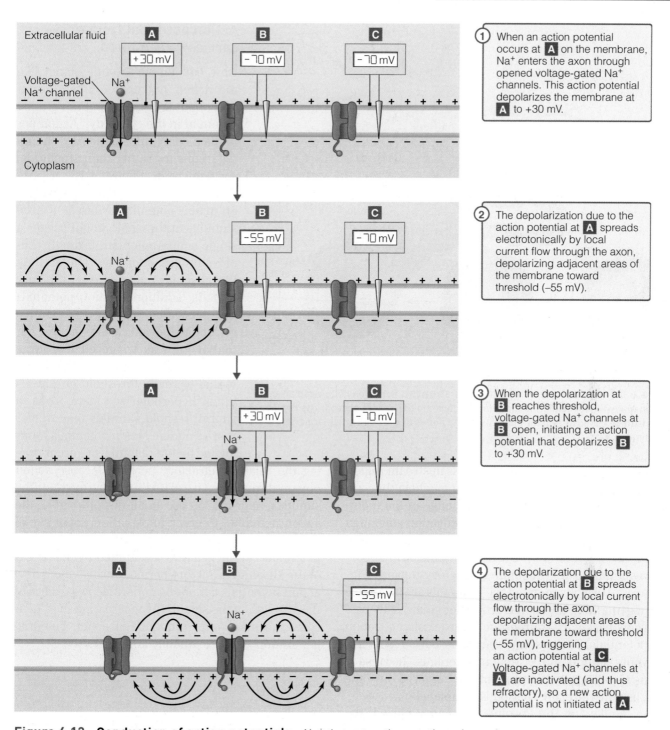

① When an action potential occurs at **A** on the membrane, Na⁺ enters the axon through opened voltage-gated Na⁺ channels. This action potential depolarizes the membrane at **A** to +30 mV.

② The depolarization due to the action potential at **A** spreads electrotonically by local current flow through the axon, depolarizing adjacent areas of the membrane toward threshold (–55 mV).

③ When the depolarization at **B** reaches threshold, voltage-gated Na⁺ channels at **B** open, initiating an action potential that depolarizes **B** to +30 mV.

④ The depolarization due to the action potential at **B** spreads electrotonically by local current flow through the axon, depolarizing adjacent areas of the membrane toward threshold (–55 mV), triggering an action potential at **C**. Voltage-gated Na⁺ channels at **A** are inactivated (and thus refractory), so a new action potential is not initiated at **A**.

Figure 4.13 Conduction of action potentials   Na⁺ that enters the axon through voltage-gated Na⁺ channels induces a local depolarization. This local depolarization spreads along the axon via electrotonic conduction, triggering additional action potentials further down the axon. This process of electrotonic current spread and new action potential initiation continues down to the end of the axon. Each action potential is essentially the same as the preceding ones, resulting in conduction without decrement.

polarization caused by the Na⁺ entering through voltage-gated Na⁺ channels spreads in all directions along the axon, why do action potentials occur only in the downstream direction (toward the axon terminal) rather than also spreading backward toward the axon hillock? If you examine a natural action po-

tential (that started at the axon hillock and is being transmitted toward the axon terminal) at any point along the membrane, the region just upstream of the point you are observing must have recently produced an action potential (since action potentials are initiated at the axon hillock). As a result, the

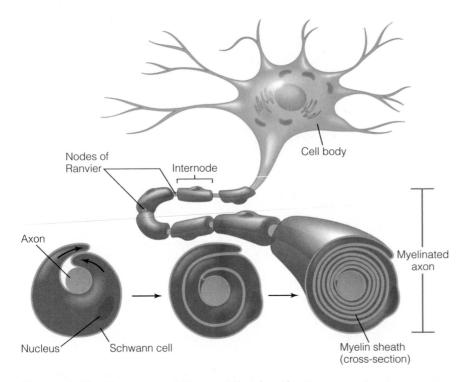

Figure 4.14 Structure of the myelin sheath Schwann cells wrap around the axon many times, insulating the axon and forming the myelin sheath. The myelin sheath is interrupted at regular intervals by the nodes of Ranvier, which are areas of unmyelinated axon.

Labels: Nodes of Ranvier, Internode, Cell body, Axon, Myelinated axon, Nucleus, Schwann cell, Myelin sheath (cross-section)

voltage-gated Na^+ channels in this upstream region of the axon are in a conformation in which they are unable to open in response to change in the membrane potential (i.e., with their activation gate open and their inactivation gate closed, as illustrated in Figure 4.11, step 4). During this time, which corresponds to the absolute refractory period (Figure 4.10), voltage-gated Na^+ channels are incapable of generating additional action potentials. This prevents backward (*retrograde*) transmission of action potentials. The absolute refractory period also prevents summation of action potentials, because a new action potential can only be triggered once the absolute refractory period is completed.

Following the absolute refractory period, the membrane enters the relative refractory period (Figure 4.10). During the relative refractory period, the voltage-gated Na^+ channels have reset and are capable of initiating another action potential, but new action potentials are more difficult to generate because the membrane is hyperpolarized. As a result, a larger depolarization is required to reach threshold. Only a very strong stimulus can cause an action potential during the relative refractory period. Together, the absolute and relative refractory periods prevent retrograde transmission of action potentials.

Action potential frequency carries information

How can an all-or-none signal like the action potential carry information about the strength of the graded potential in the cell body? Action potentials carry information by changing frequency rather than amplitude. As shown in Figure 4.15, a subthreshold stimulus does not trigger an action potential, whereas a brief stimulus at threshold might trigger a single action potential. If the threshold stimulus continues longer than the absolute and relative refractory periods, additional action potentials are generated. During the relative refractory period, a new action potential may be triggered if a large stimulus causes the membrane potential to reach threshold despite its initial hyperpolarized state. Thus, a suprathreshold stimulus may trigger more frequent action potentials by allowing action potentials to occur during the relative refractory period. Because action potential frequency is related to the strength of the stimulus, neurons can use an all-or-none signal, the action potential, to carry information about signal strength. The maximum frequency at which action potentials can be generated is limited by the length of the absolute refractory period, during which new action potentials cannot be generated regardless of the strength of the signal. In most mammalian neurons, the maximum frequency of action potential generation is approximately 500–1000/sec.

⊙ | CONCEPT CHECK

5. Compare and contrast action potentials and graded potentials.

6. How do action potentials code information about the intensity of the incoming signal?

7. Why does the membrane potential become positive during the depolarization phase of the action potential?

8. Why can action potentials be conducted across long distances along the axon without degrading, when graded potentials die out within a few millimeters?

9. What limits the frequency of action potentials?

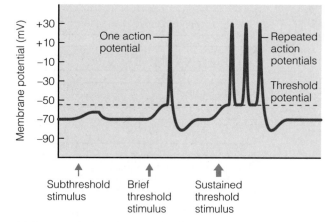

(a) A weak stimulus triggers a low frequency of action potentials

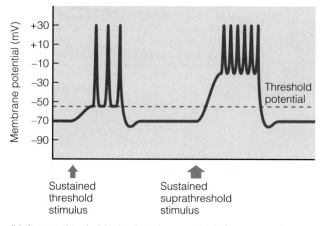

(b) A suprathreshold stimulus triggers a high frequency of action potentials

Figure 4.15   Frequency of action potentials
Action potential frequency relates to stimulus frequency. **(a)** A weak stimulus triggers a low frequency of action potentials. **(b)** A sustained suprathreshold stimulus triggers more frequent action potentials. A sufficiently large suprathreshold stimulus can trigger a new action potential during the relative refractory period of the previous action potential. The maximum frequency of action potentials is limited by the absolute refractory period of the voltage-gated Na^+ channels.

Signals Across the Synapse

Once the action potential reaches the axon terminal, the fourth important functional zone of a neuron, the neuron must transmit the signal carried by the action potential across the synapse to the target cell. The cell that transmits the signal is referred to as the **presynaptic cell**, and the cell receiving the signal is called the **postsynaptic cell**. The space between the presynaptic and postsynaptic cell is referred to as the **synaptic cleft**. Together, these three components make up the synapse. Neurons can form synapses with themselves, with other neurons, and with many other kinds of postsynaptic cells, including muscle and endocrine cells. The synapse between a motor neuron and a skeletal muscle cell, which we discuss in detail in this part of the chapter, is termed the **neuromuscular junction**.

Intracellular Ca^{2+} regulates neurotransmitter release

Much of what we know about the biochemical events at the synapse has been learned from studying the neuromuscular junction. The mechanism of synaptic transmission at the neuromuscular junction is outlined in Figure 4.16. When an action potential reaches the membrane of the presynaptic axon terminal of the neuromuscular junction, the resulting depolarization triggers the opening of voltage-gated Ca^{2+} channels on the cell membrane of the axon terminal. The concentration of Ca^{2+} inside the neuron is much lower than the concentration of Ca^{2+} outside the neuron, the equilibrium potential for Ca^{2+} is $+130$ mV (as calculated using the Nernst equation), and the resting membrane potential is -70 mV. Thus, both concentration and electrical gradients favor the movement of Ca^{2+} into the cell. The resulting increased Ca^{2+} concentration inside the axon terminal acts as a signal to neurotransmitter-containing **synaptic vesicles**. These vesicles are not randomly distributed within the synapse. Instead, they are grouped into at least two distinct pools: a readily releasable pool, and a storage pool. The readily releasable pool of vesicles is located at the active zone of the synapse, bound to docking proteins at the synaptic membrane, ready to release their contents by exocytosis. The storage pool, in contrast, consists of vesicles bound to the cytoskeleton, and not docked to the membrane. The Ca^{2+} signal causes vesicles from the readily releasable pool to fuse with the plasma membrane and release their contents by regulated exocytosis, in a process similar to the release of other intercellular signaling molecules, as described in Chapter 3. The Ca^{2+} signal also causes vesicles from the storage pool to move to the active zone of the plasma membrane and bind to docking proteins, ready for release following subsequent action potentials.

Each vesicle contains many molecules of neurotransmitter, and the number of molecules of neurotransmitter within a vesicle is similar for all vesicles within a neuron. With increasing action

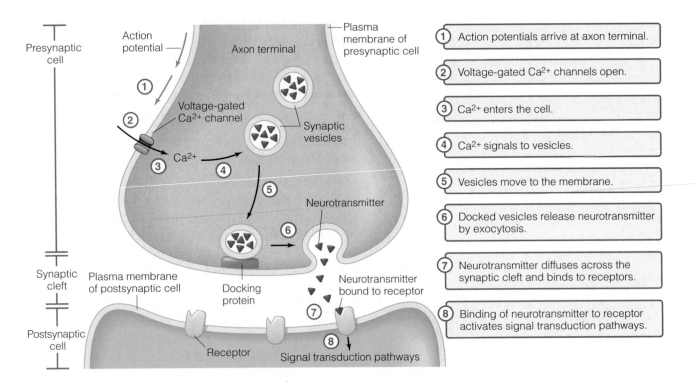

Figure 4.16 **Events of signal transmission at a chemical synapse**

potential frequency, more and more vesicles move to the membrane and release their contents by exocytosis. Because each vesicle contains many molecules of neurotransmitter, the amount of neurotransmitter a neuron releases increases in a steplike fashion, with each step corresponding to the contents of a vesicle, rather than increasing in a smoothly graded fashion as would happen if neurotransmitter were released one molecule at a time. This pattern of release is termed the *quantal* release of neurotransmitter. However, under normal physiological conditions most neurons release many synaptic vesicles when stimulated, so the quantal release of transmitter is not generally apparent.

Action potential frequency influences neurotransmitter release

The amount of neurotransmitter released at a synapse is related to the frequency of action potentials at the axon terminal. Weak signals, resulting from low-frequency action potentials, cause fewer synaptic vesicles to release their contents, whereas strong signals, resulting from high-frequency action potentials, cause more synaptic vesicles to release their contents. But how is action potential frequency coupled to the extent of neurotransmit-

ter release? After the arrival of a single action potential at the axon terminal, Ca^{2+} enters the cell through activated voltage-gated Ca^{2+} channels. This Ca^{2+} is, however, quickly bound up by intracellular buffers or removed from the cytoplasm by Ca^{2+} ATPases, keeping intracellular Ca^{2+} concentration low and limiting the release of neurotransmitter. In contrast, when action potentials arrive at the axon terminal at high frequency, the processes removing Ca^{2+} from the cell cannot keep up with the influx of Ca^{2+} through the activated channels, and the intracellular Ca^{2+} concentration increases. This increased intracellular Ca^{2+} provides a stronger signal for exocytosis. Thus, the signal intensity that was coded by action potential frequency is translated into differences in the amount of neurotransmitter released by the neuron.

Acetylcholine is the primary neurotransmitter at the vertebrate neuromuscular junction

Although the motor neurons of invertebrates release other neurotransmitters, vertebrate motor neurons release the neurotransmitter **acetylcholine** (ACh) into the synapse. ACh is a biogenic amine (see Chapter 3) that is synthesized from the amino acid choline. ACh synthesis occurs in the axon terminal in a reaction catalyzed by the enzyme *choline acetyl transferase*:

$$\text{Acetyl CoA} + \text{choline} \rightarrow \text{ACh} + \text{CoA}$$

Acetyl CoA from the mitochondria is combined with the amino acid choline to form ACh and coenzyme A. The ACh is packaged into synaptic vesicles and stored until an action potential arriving at the axon terminal triggers its release. ACh then diffuses into the synapse and binds to receptors on the postsynaptic cell membrane.

Signaling is terminated by acetylcholinesterase

As we discussed in Chapter 3, the signaling between a ligand such as a neurotransmitter and its receptor must be terminated in order to be effective. A specific enzyme in the synapse, called **acetylcholinesterase**, removes the ACh from its receptor, breaking the ACh down into choline and acetate (Figure 4.17). The choline is taken up by the presynaptic neuron and reused to form ACh, while the acetate diffuses out of the synaptic cleft.

Acetylcholinesterase plays an important role in regulating the strength of the signal to the postsynaptic cell by regulating the concentration of neurotransmitter at the synapse.

Postsynaptic cells express specific receptors

The responses of postsynaptic cells to neurotransmitters are similar to the responses of target cells to the hormones and other chemical messengers discussed in Chapter 3. Postsynaptic cells detect neurotransmitters using specific cell-surface receptors. When a neurotransmitter binds to its receptor, the receptor changes shape. This change in shape of the receptor acts as a signal in the target cell. Skeletal muscle cells express a class of receptor called **nicotinic ACh receptors**, which were named because of their ability to bind to the drug nicotine (the active ingredient in tobacco). Nicotinic ACh receptors are ligand-gated ion channels. When ACh binds to a nicotinic receptor, the receptor changes shape, opening a pore in the middle of the receptor that allows ions to cross the membrane. Nicotinic ACh receptors contain a

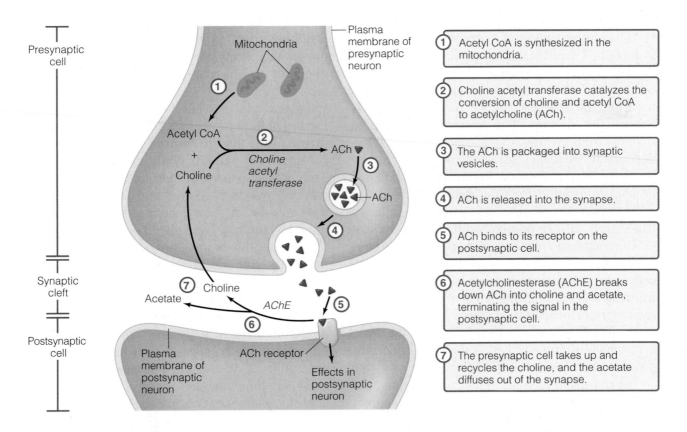

Figure 4.17   Synthesis and recycling of acetylcholine (ACh) at the synapse

relatively nonselective channel that is permeable to Na^+, K^+, and to a lesser extent Ca^{2+}; however, graded potentials in the postsynaptic cell caused by these channels are dominated by Na^+ ions because of the high driving force for Na^+ influx relative to K^+ efflux (as predicted by the Nernst equation). ACh binding to nicotinic receptors on skeletal muscle cells always causes a rapid excitatory postsynaptic potential because the resulting influx of Na^+ depolarizes the postsynaptic muscle cell. As we discuss in more detail in Chapter 5, these excitatory potentials initiate muscle contraction.

Neurotransmitter amount and receptor activity influence signal strength

As in the ligand-receptor interactions that we discussed in Chapter 3, both the amount of neurotransmitter present in the synapse and the number of receptors on the postsynaptic cell influence the strength of signal in the target cell. Small amounts of neurotransmitter provoke relatively small responses in the postsynaptic cell. As neurotransmitter concentration increases, the response of the postsynaptic cell increases up to the point that all of the available receptors are saturated.

The concentration of neurotransmitter in the synapse is a result of the balance between the rate of neurotransmitter release from the presynaptic cell and the rate of removal of the neurotransmitter from the synapse. As we have already discussed, the amount of neurotransmitter that is released from the presynaptic cell is largely a function of the frequency of action potentials at the presynaptic axon terminal. In contrast, the removal of neurotransmitter from the synapse depends on three main processes: (1) Neurotransmitters can simply diffuse passively out of the synapse. (2) Surrounding cells, including presynaptic neurons, can also take up neurotransmitter. These cells act as important regulators of many neurotransmitters. (3) Enzymes present in the synapse can degrade neurotransmitters. As we have already discussed, at the neuromuscular junction, acetylcholinesterase activity is the most important determinant of ACh concentration.

At any given amount of neurotransmitter, the response of the postsynaptic cell is also dependent on the number of receptors present on the target cell. As you would expect, a postsynaptic cell can only respond if it has the appropriate receptors in the cell membrane. If there is a very low density of receptors on the postsynaptic membrane, neurotransmitter will cause a weak response. If the density of receptors on the postsynaptic membrane is very high, the response will be larger. The density of receptors on the postsynaptic cell can be regulated by a variety of factors, including genetic variation among individuals, the metabolic state of the postsynaptic cell, and specific drugs and disease states.

The human disorder *myasthenia gravis* is an example of a disease state caused by alterations in receptor number on muscle cells. People with myasthenia gravis experience muscle weakness and increased susceptibility to muscle fatigue, particularly in muscles that are used repeatedly. These symptoms are the result of an autoimmune condition in which antibodies from a person's immune system destroy ACh receptors at the neuromuscular junction. The decrease in receptor number reduces the intensity of the signal in the postsynaptic muscle cell at any given level of acetylcholine release, which reduces the strength of muscle contractions and causes muscle weakness.

The symptoms of myasthenia gravis can be treated with a class of drugs called *acetylcholinesterase inhibitors*. By partially inhibiting the enzyme acetylcholinesterase, these drugs reduce the rate of removal of ACh from its receptors, increasing the concentration of ACh in the synapse. This increase in ACh prolongs the effects of this neurotransmitter, partially compensating for the decreased number of ACh receptors in patients with myasthenia gravis. Thus, these drugs can help to reduce the symptoms of muscle weakness and fatigue. However, the dosage of acetylcholinesterase inhibitors must be carefully controlled because at high levels they can be deadly. Indeed, organophosphate pesticides and chemical weapons such as the nerve gas sarin are acetylcholinesterase inhibitors. Like the drugs used to treat myasthenia gravis, these chemicals work by inhibiting the degradation of ACh by acetylcholinesterase. At high doses, these agents greatly increase the concentration of ACh in the synapse. At the neuromuscular junction, these large increases in ACh lead to overexcitation of the muscle, causing twitching and other forms of uncoordinated muscle contraction, potentially leading to muscle fatigue, paralysis, severe difficulty in breathing, and ultimately death because of increasing fatigue and paralysis of the respiratory muscles. Lower doses of these agents also cause a range of other symptoms because, as we discuss in the second half of the chapter, ACh acts as a neurotransmitter not just at the neuromuscular junction but also at many other synapses, causing a wide variety of effects.

◉ | CONCEPT CHECK

10. Describe the relationship between action potential frequency and neurotransmitter release.
11. What determines whether a neurotransmitter will depolarize or hyperpolarize a postsynaptic cell?
12. Why does increasing the amount of neurotransmitter increase the response of the postsynaptic cell? Why does the response reach a maximum, and not increase even when additional neurotransmitter is added?

Diversity of Neural Signaling

Now that we have examined how signals travel from one end of a motor neuron to the other, we can begin to address some of the enormous diversity in these processes among neurons from a single organism, and among neurons from different kinds of organisms. The diversity of neuron structure and function allows neurons to play many roles. Some neurons (including the motor neurons that we have already discussed) are specialized to transmit signals very rapidly across long distances, while other neurons are specialized to integrate many incoming signals and process them to produce a response. We begin this section by examining the structural diversity of neurons, looking at how neuron structure relates to neuron function. We then look at some of the important processes performed by neurons to see how they vary among neurons that perform different physiological roles in a variety of animal species.

Neurons perform three distinct functions. They receive and integrate incoming signals, they conduct these signals through the cell, and they transmit these signals to other cells. In the first part of the chapter we discussed how vertebrate motor neurons detect incoming signals in the

form of neurotransmitters. Many chemical substances can act as neurotransmitters, and we discuss some of this diversity later in the chapter in our consideration of the diversity of synaptic transmission. But neurons are also capable of detecting many kinds of incoming signals in addition to chemical signals in the form of neurotransmitters. Some neurons are specialized to detect incoming signals such as temperature, pressure, light, or environmental chemicals. The mechanisms that neurons use to detect these signals are extremely diverse, but they share one fundamental characteristic. Whatever the incoming signal, membrane-bound receptors in the dendrites of the sensory neuron receive the signal and transduce it into an electrical signal in the form of a change in the membrane potential. Because of the diversity and complexity of these processes, we do not consider them in detail here. Instead, we devote Chapter 6: Sensory Systems to these fascinating issues. In this chapter we focus on the diversity of signal conduction and transmission, looking first at the diversity of the action potential and the conduction velocity of action potentials along the axon. Then we examine some of the enormous diversity of synaptic transmission. We conclude the chapter with a discussion of the evolution of neurons.

Structural Diversity of Neurons

Although most neurons have dendrites, a cell body, and an axon, the details of neuron structure vary greatly at the cellular level. Some neurons have relatively simple structures, while others have complex, highly branched structures (Figure 4.18a). There is no clear correlation between the complexity of an organism and the complexity of its neurons. Instead, the structure of a neuron relates to the function of that particular neuron. For example, neurons within the mammalian brain typically have large numbers of dendrites, but may lack an obvious axon. The many dendrites of these neurons allow them to integrate an enormous number of incoming signals from other neurons. In contrast, the dendrites and axon of a motor neuron can easily be distinguished, as the axon is typically much longer than the dendrites. These neurons are specialized for rapid, long-distance, electrical signaling.

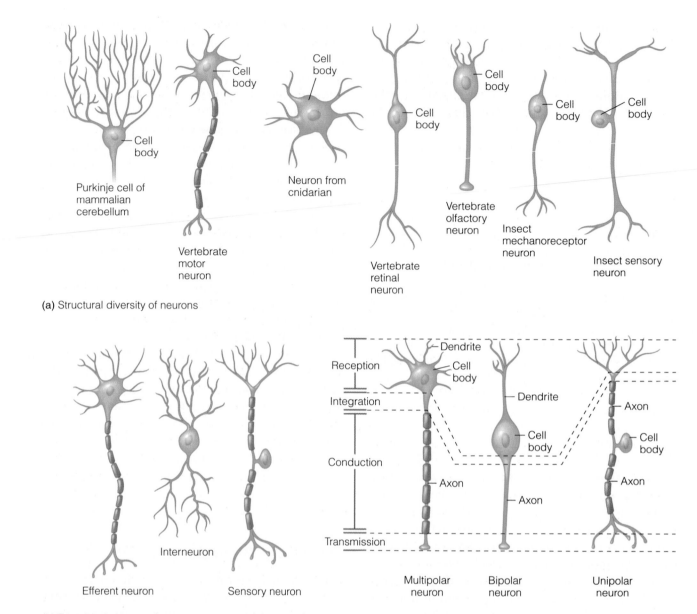

(a) Structural diversity of neurons

(b) Functional classes of neurons

(c) Structural classes of neurons

Figure 4.18 Variation in neuron structure and function (a) Structural diversity of neurons. Neurons always have a cell body, one axon, and at least one dendrite, but the number of dendrites, the position of the cell body, and the length of the axon can vary. (b) Functional classes of neurons. Sensory neurons detect incoming signals. Interneurons form connections among neurons. Efferent neurons convey signals from the nervous system to effector organs. (c) Structural classes of neurons. Multipolar neurons have one obvious axon and multiple dendrites. Bipolar neurons have a single branched dendrite and an obvious axon. Unipolar neurons have a single large axon that branches into two main processes. Note the variation in the location of the integrating center among these neurons.

Neurons can be classified based on their function

As we discuss in more detail in Chapter 7: Functional Organization of Nervous Systems, neurons can be divided into one of three classes, depending on their functions (Figure 4.18b). **Sensory** (or **afferent**) **neurons** convey sensory information from the body to the central nervous system (which consists of the brain and spinal cord in the vertebrates). **Interneurons** are located within the central nervous system, and convey signals from one neuron to another. **Efferent neurons** convey signals from the central nervous system to effector organs. The motor neurons that we have already discussed are one class of efferent neuron. In the case of motor neurons the effector is always a skeletal muscle, but other types of efferent neurons communicate with a variety of effector organs including smooth muscles and endocrine glands.

Neurons can be classified based on their structure

Although there is substantial diversity in the structure of neurons, most of this diversity falls within one of three major structural types (Figure 4.18c). The vertebrate motor neurons that we discussed in the first part of this chapter are examples of **multipolar neurons**. These neurons have many cellular extensions (or processes) leading from the cell body. Only one of these processes is an axon, whereas the remaining processes are dendrites. Multipolar neurons are the most common type of neuron in the vertebrates. **Bipolar neurons** have two main processes extending from the cell body, one of which is highly branched and conveys signals to the cell body, and thus is functionally similar to a dendrite, and the other of which conveys signals away from the cell body, and thus acts as an axon. As discussed in Chapter 6, some sensory neurons, such as retinal cells and olfactory cells, are bipolar neurons. However, few other vertebrate neurons have this form, and bipolar neurons are thus the least common type of neuron in the vertebrate nervous system. A **unipolar neuron** has a single process from the cell body. In most unipolar neurons, however, this process splits into two main branches. As a result, these cells are sometimes termed *pseudo-unipolar*. One of these two branches conveys signals toward the cell body, and the other conveys signals away from the cell body. Unipolar neurons are generally sensory neurons that are involved in detecting environmental signals and conveying this information to the rest of the nervous system.

From Figure 4.18c you can see that the cell body, dendrites, and axon are arranged differently in each of these types of neuron. This change in arrangement has important implications for the functions of each of the zones of the neuron. In a multipolar neuron, such as the vertebrate motor neurons that we have already discussed, receptors in the dendrites and cell body detect incoming signals and transduce them into an electrical signal in the form of a graded potential. Incoming graded potentials are conducted electrotonically to the axon hillock, which acts as the integrating center for the neuron. If the graded potential at the axon hillock exceeds the threshold potential, it triggers action potentials, which are conducted along the axon to the axon terminal. This general scheme fits well for most multipolar neurons, although not all multipolar neurons generate action potentials. In some multipolar neurons with very short axons, electrotonic current spread is sufficient to convey information along the axon.

In a bipolar neuron, just as in a multipolar neuron, receptors in the membrane at the end of one of the processes detect incoming signals and transduce them into a graded potential. This graded potential spreads electrotonically to the cell body, where it triggers action potentials in the second process, which acts as an axon. The exact location of the trigger zone varies among bipolar neurons, and (like multipolar neurons) some kinds of bipolar neurons do not use action potentials to convey signals along the axon.

In a unipolar neuron, the dendrites detect incoming signals and transduce them into graded potentials, as in the other types of neurons. These graded potentials do not, however, travel directly to the cell body. Instead, they travel only as far as the beginning (or initial segment) of the process that leads to the cell body. If the graded potential in this initial segment exceeds threshold, it will trigger an action potential. These action potentials then travel toward the cell body, and onward to the axon terminal. As a result of this arrangement, there has been some disagreement as to whether to call the first of these long extensions of a unipolar neuron an axon or a dendrite, because it is functionally similar to an axon in that it can generate action potentials, but it conducts impulses toward the cell body rather than away from the cell body and thus is functionally similar to a dendrite. For the purposes of this book, we will refer to both of the processes of a unipolar neuron as axons. The important point to keep in mind, however, is that the integrating center is located in a very different position in a unipolar neuron compared to a multipolar neuron.

Neurons from invertebrates can also be grouped into these main structural classifications, and are organized in ways similar to the vertebrate neurons that we have discussed so far. In the invertebrates, however, unipolar neurons are more common than they are in the vertebrates. Indeed, invertebrate motor neurons are often unipolar, rather than multipolar. Whether in an invertebrate or a vertebrate, however, most neurons share the common property of polarity. One end of the neuron receives incoming signals, and the other end of the neuron transmits signals to other cells. Cnidarians, including sea anemones and jellyfish, provide an exception to this rule. Some cnidarian neurons lack polarity. That is, they are capable of sending and receiving signals at either end and can conduct signals in either direction along the neuron. As we see in Chapter 7, this difference has important implications for the unique organization of the nervous system in cnidarians.

Neurons are associated with glial cells

As we mentioned in the first half of this chapter, vertebrate motor neurons are associated with a type of glial cell called a Schwann cell. But Schwann cells are not the only type of glial cell in the vertebrates. In fact, glial cells far outnumber neurons in most organisms. For example, 90% of

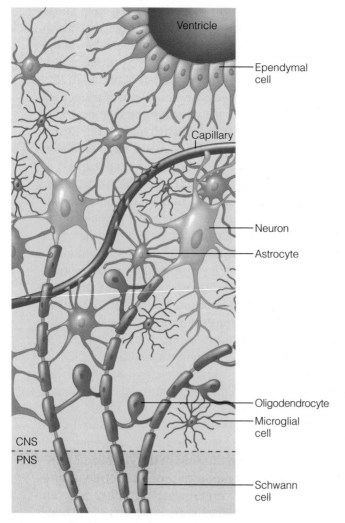

Figure 4.19 The primary glial cells of vertebrates

the cells in the human brain are glia. Until recently, these glial cells were believed to play a rather passive role in the nervous system, and their name (which is derived from the Greek word *glia* = glue) reflects this view. However, we now know that glial cells play a wide variety of critically important roles in the nervous system.

In the vertebrates, there are five main types of glial cells (Figure 4.19). Schwann cells, which form the myelin sheath, are associated with motor neurons and many sensory neurons. Schwann cells play an important role in neural signaling by increasing the conduction speed of action potentials along the axon. They are also essential for the regeneration and regrowth of damaged sensory and motor neurons. When a neuron is damaged, Schwann cells digest the damaged axon and provide a pathway for neuronal regrowth. **Oligodendrocytes** form a myelin sheath for neurons in the central nervous system (CNS). A single oligodendrocyte may wrap around the axons of several neurons, and thus differs from a Schwann cell, which always enwraps a single neuron. **Astrocytes** have large stellate (star-shaped) cell bodies and many processes. They are located in the central nervous system and play a variety of roles including transporting nutrients to neurons, removing debris, guiding neuronal development, and regulating the contents of the extracellular space around neurons (including regulating synaptic neurotransmitter levels). In fact, astrocytes in the brain often enwrap synapses and may play an important role in regulating synaptic communication by regulating neurotransmitter levels. **Microglia** are involved in neuronal maintenance. Microglia are the smallest glial cells. They are similar to the macrophages of the immune system, and they function to remove debris and dead cells from the central nervous system. Microglia are most active following trauma or during disease. **Ependymal cells** line the fluid-filled cavities of the central nervous system. They often have cilia, which they use to circulate the *cerebrospinal fluid* that bathes the central nervous system of vertebrates.

Although glial cells maintain a resting membrane potential, they do not generate action potentials, nor do they form obvious chemical synapses. However, despite their lack of obvious chemical synapses, glial cells can take up and release neurotransmitters, and thus may have important effects on neurons. In addition, some glial cells in the central nervous system, such as astrocytes, form connections with each other and with neurons via gap junctions. Astrocytes actively communicate with each other through these gap junctions using intracellular Ca^{2+} and other signaling molecules. The presence of gap junctions suggests a complex interchange of signals between neurons and glia, which may be important in regulating the function of the nervous system.

Glial cells in invertebrates have a wide range of morphologies, depending on their location in the organism and the species being examined. Invertebrate glial cells are often termed **gliocytes**, and appear to be functionally similar to astrocytes, as they intimately ensheathe synapses. Invertebrates lack a true myelin sheath, but axons of peripheral neurons may still be wrapped in several layers of glial cell membrane. Box 4.2, Evolution and Diversity—The Evolution of Myelin Sheaths, provides a comparison of the structure and function of the myelin sheaths in the vertebrates and the invertebrates. Overall, the functions of glial cells are thought to be similar in both vertebrates and invertebrates.

From the preceding discussion, it is clear that neurons are structurally diverse, and can form complex associations with each other and with the surrounding glial cells. Neurons are also diverse in their functions. In the next sections of the chapter we revisit the primary functions of the neuron that we discussed in the context of a typical vertebrate motor neuron, examining the diversity of signal conduction and signal transmission in neurons.

CONCEPT CHECK

13. Is a typical vertebrate efferent (motor) neuron (as shown in Figure 4.18b) multipolar, bipolar, or unipolar?

14. Answer the same question for vertebrate interneurons and sensory neurons.

15. Describe the primary types of glial cells in the vertebrates. What are their functions?

Diversity of Signal Conduction

We have already seen that axons can conduct signals either electrotonically or using a combination of electrotonic current spread and regenerating action potentials, but there is additional diversity in signal conduction among neurons that we have yet

to consider. Both the shape of the action potential and the speed of action potential conduction along the axon vary among neurons. In the first half of the chapter we considered the shape of an action potential in a squid giant axon, and most action potentials conform to this general form. However, the exact shape of the action potential can vary among neurons from different organisms, between types of neurons from the same organism, and even among action potentials within the same neuron under different physiological conditions. The variations in the shapes of these action potentials are the result of the diversity of the molecular properties of the voltage-gated Na^+ and K^+ channels among these neurons. In fact, some neurons entirely lack voltage-gated K^+ channels. In these neurons the repolarization phase of the action potential is carried out by K^+ movements through K^+ leak channels that are open at all times. As you might expect, neurons of this type do not exhibit an after-hyperpolarization phase following the action potential.

Voltage-gated ion channels are encoded by multiple genes

Many ion channels exist as multiple isoforms: slightly different molecular variants of the same protein, encoded by different genes (Chapter 2). Sequence variation among the isoforms of voltage-gated ion channels can lead to functional differences that change the way neurons work. In mammals at least 18 separate genes encode voltage-gated K^+ channels, and over 50 distinct types of voltage-gated K^+ channels have been characterized among all animal species. Voltage-gated K^+ channels cause the repolarizing phase of the action potential in most neurons. Thus, their diverse isoforms result in a diversity of shapes during the repolarizing phase of the action potential in different cells, tissues, and organisms. Voltage-gated K^+ channels also have a strong influence on the excitability of the cell, action potential duration, and action potential rate. For example, voltage-gated K^+ channels that open extremely quickly in response to depolarization tend to make action potentials more difficult to generate, because K^+ ions leave the cell at the same time that Na^+ ions are entering, countering the depolarization due to voltage-gated Na^+ channels. In contrast, some voltage-gated K^+ channels are referred to as *delayed rectifiers*, because they respond relatively slowly to changes in membrane potential, increasing the length of the action

BOX 4.2 EVOLUTION AND DIVERSITY
The Evolution of Myelin Sheaths

Certain invertebrate neurons, including giant nerve fibers in the ventral nerve cord of earthworms, crabs, and shrimp, are wrapped in multiple layers of cell membranes, in a pattern that looks superficially like the myelin sheaths that are wrapped around the axons of neurons in the vertebrates. Protein complexes termed *septate junctions* hold together the cells that wrap around axons with these myelin-like structures in the invertebrates. Septate junctions form a tight seal between the cells that enwrap the neuron, and these septate junctions function to isolate the nerve from the extracellular fluid. This observation suggests that invertebrate wrappings may play a role that is similar to the insulating function of myelin. There are, however, a number of differences between the myelin sheaths of vertebrates and the wrappings of invertebrates. For example, note in the figure that the location of the nucleus of the ensheathing cell is generally next to the axon, rather than in the outer layer of the sheath as in the vertebrates.

Although there is substantial diversity among invertebrates in the morphology of these neuronal wrappings, in general the layers of membrane in invertebrate wrapping are not as closely stacked as they are in a vertebrate myelin sheath. Also, the proteins involved in the structure of the myelin sheath of vertebrates and the wrappings of invertebrates differ. For example, certain proteins are known to be critically important for the function of the vertebrate myelin sheath. If these proteins are defective or present in reduced levels, the rate of action potential conduction decreases. None of these proteins are found in invertebrates. This observation suggests that the molecular machinery involved in invertebrate wrappings is fundamentally different from that of the myelin sheath of vertebrates, and likely evolved independently.

Recently, however, it has been shown that some of the proteins involved in the interaction between the myelin sheath and the axon are found in both vertebrates and invertebrates. For example, in the fruit fly *Drosophila* (an invertebrate), proteins called neurexins are found at high concentrations in the septate junctions of the cells that ensheathe axons. A similar protein in mammals is found in the myelin sheath near the nodes of Ranvier. This part of the myelin sheath forms junctions similar to the septate junctions found in invertebrates. Taken together, these observations suggest that there are underappreciated similarities between the myelin sheath of the vertebrates and the wrappings surrounding invertebrate neurons.

Other similarities between the neurons of invertebrates and vertebrates may provide some insight into the evolution of the vertebrate myelin sheath. For example, in the invertebrate sea slug *Aplysia*, voltage-gated Na^+ channels are clustered at distinct locations along the axon. Similarly, voltage-gated Na^+ channels in vertebrates are clustered in the nodes of Ranvier in a myelinated neuron. Clustering of voltage-gated channels is likely important to optimize action potential conduction, even in unmyelinated axons. At present, however, it is not clear whether these similarities between invertebrate wrappings and vertebrate myelination are independent evolutionary events that represent convergence to an optimal design, or were instead present in the common ancestors of all animals.

References

○ Waehneldt, T. V. 1990. Phylogeny of myelin proteins. *Annals of the New York Academy of Sciences* 605: 15–28.

○ Weatherby, T. M., A. D. Davis, D. K. Hartline, and P. H. Lenz. 2000. The need for speed. II. Myelin in calanoid copepods. *Journal of Comparative Physiology*. A. 186: 347–357.

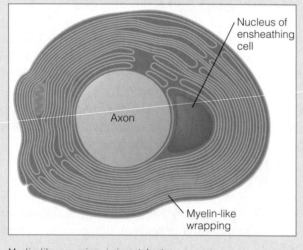

Myelin-like wrappings in invertebrates.

potential. Table 4.2 lists some examples of the diversity of K$^+$ channels. The significance of this diversity for the functioning of the whole organism is not yet fully understood, but it clearly influences the functional diversity of neurons.

Compared to voltage-gated K$^+$ channels, voltage-gated Na$^+$ channels are much less diverse. Mammals express at least 11 isoforms of the voltage-gated Na$^+$ channel, but the functional differences among these Na$^+$ channel isoforms are rather minor. There are measurable differences in the exact time required for them to open, the length of time they stay open, and their inactivation characteristics, but the importance of these differences is not yet understood. Only two voltage-gated Na$^+$ channel genes have been identified in *Drosophila* and squid, compared to the 11 isoforms in mammals. The significance of the increase in isoform number as the complexity of the nervous system increases is also poorly understood, but it may be important in the functioning of complex mammalian nervous systems.

The density of voltage-gated Na$^+$ channels also has a profound effect on the function of a neuron. All else being equal, neurons that have a higher density of voltage-gated Na$^+$ channels will have a lower threshold than neurons with a lower density of voltage-gated Na$^+$ channels. A higher density means more Na$^+$ channels are available to open at a given stimulus intensity, and more Na$^+$ will enter the cell. As a result, the balance point between the dissipation and influx of Na$^+$ ions is more easily reached at a lower level of depolarization. Thus, a smaller graded potential can excite a neuron with high densities of voltage-gated Na$^+$ channels. Similarly, the density of voltage-gated Na$^+$ channels can also influence the length of the relative refractory period. Neurons with higher densities of voltage-gated Na$^+$ channels tend to have shorter relative refractory periods because of the decrease in the threshold potential.

The many isoforms of voltage-gated channels have only recently been identified, and neurobiologists still do not entirely understand the role that these isoforms play in generating functional diversity in the nervous system. In general, there is a correlation between the complexity of the nervous system and the total number of isoforms of voltage-gated ion channels, which suggests (but does not prove) that more diverse voltage-gated channels are required to build a highly complex nervous system. Variants in ion channels can be mixed and matched to generate even larger numbers of combinations. There are millions of possible combinations of isoforms of voltage-gated channels, neurotransmitters, and receptors, and thus millions of possible types of neuron. The human nervous system, one of the most complex nervous systems of any animal, contains billions of individual neurons, many with unique properties and functions. Biologists are only just beginning to probe the complexities of these interactions, and many important questions have yet to be addressed. The role of isoforms in generating diversity in neural signaling is thus an area of intensive current research.

Voltage-gated Ca^{2+} channels can also be involved in action potentials

In some neurons, voltage-gated Ca^{2+} channels are involved in the action potential. In neurons that have voltage-gated Ca^{2+} channels in the axon,

---

Table 4.2 Diversity of K$^+$ channels.

Channel type	Function
Delayed rectifier	Opens slowly in response to changes in membrane potential, closes slowly, responsible for repolarizing axonal membrane following an action potential
A channel (K$_A$ channel)	Opens when membrane is depolarized, closes rapidly, influences neuron excitability
Inward rectifier (K$_{IR}$ channel)	Opens when membrane is hyperpolarized, influences duration of action potential
Ca^{2+} activated (K$_{Ca}$ channel)	Opens in the presence of Ca^{2+}, influences excitability of neuron
M channel (K$_M$ channel)	Opens when membrane is depolarized, closes slowly, regulated by neurotransmitters
ACh channel (K$_{ACh}$ channel)	Opens when membrane is exposed to ACh, involved in regulating heartbeat

these channels open at the same time as (or instead of) voltage-gated Na$^+$ channels. This results in Ca^{2+} entry into the cell, causing a depolarization. Generally, the depolarization caused by Ca^{2+} influx is slower and more sustained than the depolarization from Na$^+$ influx. A sustained depolarization phase slows down the rate at which action potentials can be generated by prolonging the refractory period. For example, the action potentials that control rhythmic swimming in jellyfish have a sustained depolarization phase due to Ca^{2+} influx, and last about 10 times longer than a typical vertebrate action potential. As we discuss in Chapter 5, voltage-gated Ca^{2+} channels are also important in establishing the shape of action potentials in excitable tissues other than neurons, including cardiac muscle.

Conduction speed varies among axons

In addition to differences in the shape of the action potential, the speed of action potential conduction along the axon varies greatly among neurons (Table 4.3). Some neurons conduct action potentials very quickly, while action potentials in other neurons are conducted rather slowly. Animals use two main strategies for increasing the speed of action potential conduction: myelination and increasing the diameter of the axon. The axons of some neurons, including the vertebrate motor neurons that we have already mentioned, are myelinated. Other neurons with high conduction velocity have unusually large-diameter axons termed **giant axons**. The fastest nerve conduction is always observed in either large-diameter or myelinated neurons. In the next sections we examine how the properties of the axon influence conduction speed, and see how these properties are modified in giant axons and myelinated axons.

The cable properties of the axon influence current flow

To understand how the properties of the axon influence the speed of action potential conduction, we need to review some basic physics and take a closer look at electrical currents in the axon. The physical principles that govern the extent of current flow along an axon are similar to the physical principles governing the transmission of electrical current through transatlantic telephone cables. Thus, the properties of the axon that dictate current flow along the axon are often called the **cable properties** of the axon. Current, whether in an electrical wire or in an axon, is simply a measure of the amount of charge moving past a point in a given amount of time, and is a function of the drop in voltage across the circuit and the resistance of the circuit. Ohm's law (a principle that you should be familiar with from introductory physics courses) describes this relationship between current and voltage. Ohm's law is often written in the form

$$V = IR$$

where I is the current, V is the voltage drop across the circuit, and R is the resistance of the circuit. Voltage is a measure of the energy carried by a unit of charge. Thus, the difference in voltage between two points is a measure of the energy avail-

Table 4.3 **Conduction velocities in axons from various species.**			
Organism/nerve	**Diameter (µm)**	**Myelination**	**Speed of propagation (m/sec)**
Squid/giant axon	50–1000	No	30 (at 15°C)
Crayfish/leg	36	No	8 (at 20°C)
Lumbricus (worm) lateral	60	No	11.3 (at 20°C)
Frog/sciatic nerve, A fibers	18	Yes	42 (at 20°C)
Frog/sciatic nerve, B fibers	2	Yes	4 (at 20°C)
Frog/sciatic nerve, C fibers	2.5	No	0.3 (at 20°C)
Cat/saphenous nerve, A fibers	22	Yes	120 (at 37°C)
Cat/saphenous nerve, B fibers	3	Yes	15 (at 37°C)
Cat/saphenous nerve, C fibers	1	No	2 (at 37°C)

able to move charge from one point to the other, just as potential energy is a measure of the energy available to move an object from one point to another. In contrast, resistance is a measure of the force opposing the flow of electrical current. Thus, by rearranging the equation, you can see that current is proportional to the voltage drop across a circuit, and inversely proportional to the resistance.

Current flows through an electrical circuit only when the circuit is complete. You can think of an axon as behaving like a simple electrical circuit in which current flows as shown in Figure 4.20a. Ions moving through voltage-gated channels cause a current across the membrane. This introduced current spreads electrotonically along the axon. Some of this current leaks out of the axon, and a current flows "backward" along the outside of the axon, completing the circuit. Each compartment of the axon has an associated resistance, which impedes the flow of the current. Thus, we can think of each small area of the axon as consisting of an electrical circuit with three resistors (the extracellular fluid, the membrane, and the cytoplasm) as shown in Figure 4.20b.

Notice that in addition to the membrane resistance (designated R_m), intracellular resistance (designated R_i), and extracellular resistance (designated R_e), there is an additional element in this circuit diagram, designated C_m. The parallel bar symbol in the circuit diagram indicates the presence of a capacitor. Thus, the part of the circuit that crosses the membrane is actually represented by a resistor and a capacitor arranged in parallel. Capacitors are devices for storing electrical charge that consist of two conducting materials separated by an insulating layer. In the case of the cell membrane, the intracellular fluid and extracellular fluid are the conducting layers of the capacitor, while the phospholipids of the cell membrane are the insulating layer.

The circuit shown in Figure 4.20b describes what is happening in a small patch of the axon, but recall that axons can be very long. In order to fully model the axon, we need to think about the axon as a series of these small circuits connected together to form a much larger electrical circuit along the axon (Figure 4.20c). With this simplified model of the axon as an electrical circuit in mind, we can begin to see how these circuit elements affect the speed of action potential conduction along the axon.

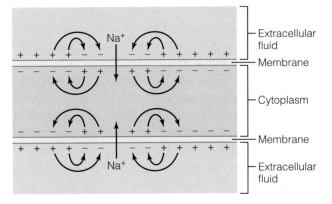

(a) Current flow in axon

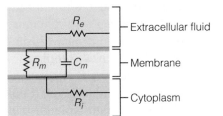

(b) An electrical circuit model for a patch of membrane

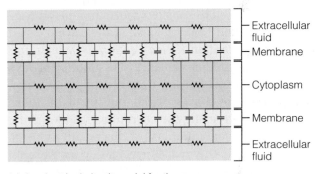

(c) An electrical circuit model for the axon

Figure 4.20 Model of the current flow in an axon **(a)** Electrotonic current spread. Introduced current (for example, due to Na^+ influx) spreads electrotonically through the axon, but some of this current leaks out through the membrane and flows "backward" along the outside of the axon. **(b)** An electrical circuit model for a patch of membrane. The axon consists of three compartments: the extracellular fluid, the membrane, and the cytoplasm with an associated electrical resistance. The cell membrane also acts as a capacitor, and can be modeled as a resistor and capacitor arranged in parallel. **(c)** An electrical circuit model for a segment of an axon. An actual axon can be modeled as a series of smaller circuits connected together.

Intracellular and membrane resistance influence conduction speed

When you depolarize a region of the membrane, the inside of the membrane becomes more positively charged than adjacent regions of membrane, while the outside of the membrane becomes more negatively charged than adjacent regions. As a result, current spreads along the axon (on both the inner and outer surfaces) by electrotonic conduction. As this electrotonic current spreads along the axon, it depolarizes these adjacent regions of the membrane. However, as we mentioned in the first part of the chapter, the change in membrane potential (measured as the voltage drop across the membrane) decreases with distance, a phenomenon called *conduction with decrement*. But why does voltage decrease with distance? Recall that resistance is a force that impedes current flow. Thus, in a simple electrical conductor the decrease in voltage is a direct result of the resistance of the material. Since resistance is cumulative with distance, we would expect to see the voltage drop with distance, according to Ohm's law. For electrotonic current spread we need to consider the resistance of both the extracellular and intracellular fluids. If the resistance of these materials is high, voltage will drop quickly with distance.

However, an axon is not just a simple conductor, and we need to consider more than just the intracellular and extracellular resistances. Most membranes contain K^+ *leak channels*, which, unlike voltage-gated channels, are essentially always open. Thus, as current travels along the axon, some positive charge leaks out through these channels, decreasing the current as it flows along the axon. The extent of loss of this positive charge depends upon the resistance of the membrane. When membrane resistance is high, current flow across the membrane will be low, and less charge will be lost. When membrane resistance is low, current flow across the membrane will be large, and more charge will be lost, resulting in greater dissipation of the axonal current with distance.

The effects of membrane resistance, extracellular resistance, and intracellular resistance on the distance an electrical signal can travel are summarized by a parameter termed the *length constant* (λ) of the membrane. The length constant is defined as the distance over which a change in membrane potential will decrease to 37% of its original value. This may seem to be an arbitrary value, but it is a consequence of the fact that a change in membrane potential decreases exponentially with distance (37% is equivalent to $1/e$). As shown in Figure 4.21, when the length constant is large, the change in membrane potential degrades less with distance, whereas if the length constant is small, change in membrane potential

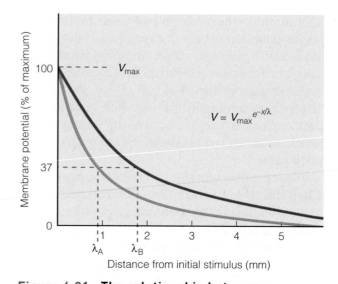

Figure 4.21 The relationship between membrane potential, distance along the axon, and the length constant When the length constant is large, the change in membrane potential as a result of an introduced current decays slowly with the distance traveled by electrotonic current spread. When the length constant is small, the membrane potential decays rapidly with distance. V_{max} = maximum change in membrane potential at the stimulus point. x = distance from the initial stimulus. V = change in membrane potential at a distance (x) from the initial stimulus. λ = length constant of the membrane.

degrades quickly with distance. The length constant can be calculated as follows:

$$\lambda = \sqrt{r_m/(r_i + r_o)}$$

where r_m = membrane resistance, r_i = intracellular resistance, and r_o = extracellular resistance. The extracellular resistance is usually assumed to be low and constant, and is often neglected in these calculations, so the equation can be rewritten as

$$\lambda = \sqrt{r_m/r_i}$$

From this equation, it is easy to see that the length constant of the membrane will be largest when membrane resistance is high and intracellular resistance is low.

So why does the length constant of a membrane influence the speed of conduction along the axon? Recall that conduction along an axon represents a combination of electrotonic conduction along the axon and action potential generation at specific points on the axon. Electrotonic conduction is very rapid compared to the speed of opening and closing voltage-gated channels during an action potential; in fact, for our purposes electrotonic currents can be considered to spread essentially instantaneously along the axon. A neuron that used only electrotonic current flow would transmit signals very rapidly. In fact, neurons with very short axons often use only electrotonic conduction to carry electrical signals, but electrotonic current spread is effective only up to a distance of 2 or 3 mm in most organisms. In longer axons, action potentials must be generated to "boost" the signal before it dies out because of the decrease in voltage with distance. However, the ability to signal over long distances using the action potential comes with a cost—a reduced speed of signal transmission. Since electrotonic currents spread along the axon extremely rapidly, the farther a threshold depolarization can spread along the axon, the shorter the length of time it will take for an impulse to reach the end of the axon. Thus, increasing the length constant of the axon increases the velocity of action potential conduction.

Membrane capacitance influences the speed of conduction

As we have already mentioned, biological membranes act as electrical capacitors. You can observe the presence of the membrane capacitor by examining what happens when you inject current into a neuron (Figure 4.22). A rectangular pulse of current does not result in an immediate change in the membrane potential of the cell. Instead, there is a lag caused by the presence of the membrane capacitor. When a capacitor is present in an electrical circuit (whether in a manufactured electrical circuit or a biological membrane), it will accumulate a charge difference across its insulating surface. For example, consider a simple electrical circuit that consists of a switch, a battery, a capacitor, and a resistor arranged in series. When we close the switch on the circuit, the voltage difference between the poles of the battery causes electrons to try to flow from the negative pole (cathode) of the battery to the positive pole (anode) of the battery. But the capacitor acts as an insulator, so negative charges cannot flow across the capacitor and instead "pile up" on one side. Recall from basic physics that like charges repel and opposite charges attract. As a result of this attraction (which occurs across the thin insulating layer of the capacitor), the negative charges on one side of the capacitor "pull" positive charges toward the capacitor and repel negative charges, causing current to flow through the circuit. Note that current does not actually flow across the insulating layer of the capacitor. Instead, electrostatic forces acting across the insulating layer of the capacitor induce a current in the circuit.

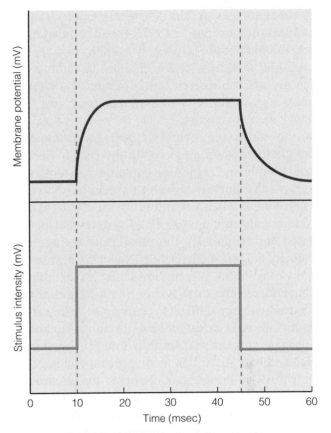

Figure 4.22 Response of a membrane to a rectangular pulse of introduced current When a neuronal membrane is exposed to a rectangular pulse of current, the membrane potential does not change instantaneously. Instead, due to the capacitance of the membrane, membrane potential increases gradually with injected current, and then decreases gradually when the stimulus is removed.

As more and more charges build up on the capacitor, they increasingly repel each other, and it becomes more and more difficult for additional charges to be deposited on the capacitor. Eventually, the charge on the capacitor will equal the driving force coming from the voltage drop across the battery, and no more current will flow. The point at which current stops flowing across a particular capacitor is determined by a parameter called *capacitance*. You can think of capacitance as the quantity of charge needed to create a potential difference between the two surfaces of the capacitor. Thus, a capacitor with high capacitance is able to store large amounts of charge, and a capacitor with low capacitance is only able to store relatively small amounts of charge. Capacitance depends on three features of the capacitor: the material properties of the capacitor, the area of the two conducting surfaces, and the thickness of the insulating layer. The electrical properties of biological membranes don't change that much from one cell to another, so we only need to consider the area and thickness of the membrane. The larger the area of the capacitor, the greater the capacitance, while the thicker the insulating layer, the lower the capacitance.

So why is the membrane capacitor important for the function of an axon? In the case of the axonal membrane, which we can model as a resistor and a capacitor arranged in parallel as shown in Figure 4.20, when you introduce an electrical current into an axon (for example, by opening voltage-gated Na^+ channels), the membrane voltage will change, but more slowly than expected because initially most of the current flows into the membrane capacitor. As the capacitor becomes fully charged, it becomes more difficult for current to flow into the capacitor, and once the membrane capacitor is charged, current will not flow into this portion of the circuit at all. At this point, current will begin to flow through the resistor, changing the membrane potential. Thus, there is a balance between current flowing through the membrane resistors and current flowing into the membrane capacitor.

The time needed for the membrane capacitor to charge can be described by the *time constant* (τ) of the membrane. The larger the time constant, the longer it will take for the membrane to reach a given membrane potential (Figure 4.23). The time constant is defined as the time taken for the membrane potential to decay to 37% of its original value (or to reach 63% of its maximal value). As was the case with the length constant of the membrane, these numbers are not arbitrary, but instead reflect the observation that there is an exponential increase in membrane potential. The relationship between electrical properties of the membrane and the time constant of the membrane is described as follows:

$$\tau = r_m \, c_m$$

where r_m = membrane resistance and c_m = membrane capacitance. Increases in either membrane resistance or membrane capacitance will increase the time constant of the membrane, delaying current flow across the membrane.

The time constant of the membrane has important consequences for temporal summation in neuronal cell bodies. Imagine two graded potentials occurring at the same time in a presynaptic cell that sum to provide a suprathreshold potential. What will happen if these two graded potentials occur at slightly different times? The time constant of the membrane helps us to determine the answer to this question. If the time constant is small, these potentials will decay rapidly, and they are less likely to be able to sum to provide a suprathreshold potential. In contrast, if the time constant is large, these potentials will decay slowly, making them more likely to overlap in time, and thus to sum to a suprathreshold potential.

It is clear that the time constant of the membrane is important in temporal summation, but how does changing the time constant of the membrane affect the speed of conduction along the axon? As current spreads electrotonically along the axon, some of the voltage must first be used in order to charge the membrane capacitor. Only once the capacitor is fully charged does current begin to flow across the membrane and alter membrane potential. As a result, electrotonic current spread is delayed. The smaller the time constant of the membrane, the faster the membrane can depolarize by a given amount, and the greater the rate of electrotonic current spread and action potential propagation.

To summarize our discussion so far, three main factors influence the speed of action potential propagation. The first factor is the kinetics of the voltage-gated channels. For example, all things being equal, action potentials typically propagate faster at higher temperatures than at lower temperatures

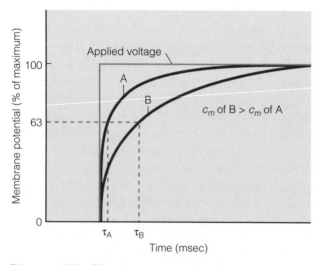

Figure 4.23 The time constant of the membrane When the time constant (τ) of the membrane is large, it takes longer for the membrane to reach the maximum potential difference. The time constant (τ) is a reflection of the capacitance (c_m). When capacitance is large, the time constant will be large.

(within physiological limits) because the channels open faster at warmer temperatures. This observation suggests that the speed of opening of the voltage-gated channels sets limits on the speed of action potential propagation. In fact, voltage-gated channels open and close very slowly compared to the speed of electrotonic current spread, so any factors that can increase the speed or distance of electrotonic current spread will increase the speed of conduction. Electrotonic current spread is, in turn, dependent on the length constant and the time constant of the axon. In the next sections we address how myelination and increasing the diameter of the axon, as in giant axons, alter these properties of the axon and thus conduction velocity.

Giant axons have high conduction speed

Giant axons have evolved independently many times, and are found in both vertebrates and invertebrates, although they are absent in mammals. Giant axons are easily visible to the naked eye and can be up to a millimeter in diameter, much larger than most mammalian axons, which are typically less than 5 µm in diameter. In squid, for example, giant axons are found in the neurons that stimulate muscle contraction around the mantle cavity (Figure 4.24). A squid can expand and contract its mantle, drawing water into the mantle cavity and rapidly expelling it through the siphon, providing a kind of jet propulsion. Jet propulsion allows the squid to move very fast, but for the jet propulsion to work properly, muscle fibers throughout the entire mantle must contract at almost the same time. Some parts of the mantle are much farther away from the central nervous system of the squid than others. In order to reach all parts of the mantle at the same time, action potentials must be conducted faster in the neurons that innervate the distant parts of the mantle than in neurons with

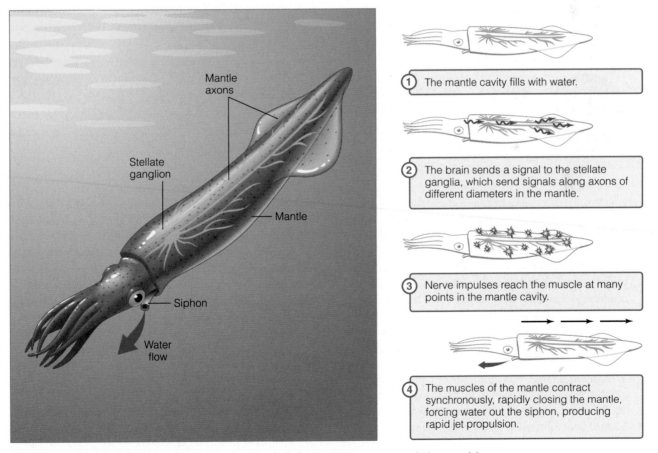

Figure 4.24 Schematic diagram of part of the nervous system of the squid *Loligo pealei* When squid want to move rapidly, they expel water out of their siphon by rapidly contracting the mantle muscles. To ensure that the entire mantle contracts rapidly in a coordinated way, axons of neurons that innervate distant parts of the mantle have much larger-diameter axons than neurons that innervate parts of the mantle close to the stellate ganglion. These giant axons conduct action potentials much more rapidly than smaller-diameter axons.

short axons. Axons that activate muscles at the far end of the mantle cavity have very large diameters, while axons that activate muscles in the region of the mantle cavity closest to the central nervous system have smaller diameters. Combining axons of varying diameters allows the near-simultaneous contraction of the entire mantle by speeding up transmission to the most distant part of the body. Because of its large size, the squid giant axon has been an important model system for neurobiology (Box 4.3, Methods and Model Systems: The Squid Giant Axon).

The effects of membrane resistance and intracellular resistance on the length constant of the membrane explain why large-diameter axons, such as giant axons, conduct signals more rapidly than small axons. Recall that the length constant of the membrane increases as membrane resistance increases, but decreases as intracellular resistance increases. So what happens to membrane resistance and intracellular resistance as axon diameter increases? Membrane resistance is inversely proportional to the surface area of the membrane. As surface area increases so does the number of leak

channels, allowing greater ion flow across the membrane so that membrane resistance decreases. Assuming that the axon is roughly cylindrical in shape, the surface area of the membrane is related to the diameter of the axon via the following formula:

$$\text{Surface area} = 2\pi rh$$

where r is the radius of the axon, and h is the length. Thus, the membrane resistance is proportional to the radius of the axon. As axon diameter increases, membrane resistance decreases.

Intracellular resistance, however, is related to the volume of the axon. As volume increases, intracellular resistance decreases. The volume of the axon can be approximated with the formula for the volume of a cylinder:

$$\text{Volume} = \pi r^2 h$$

Thus, intracellular resistance decreases in proportion to the radius of the axon *squared*. So what are the effects of membrane resistance and intracellular resistance on the length constant of the membrane? As axon radius increases, both membrane

---

BOX 4.3 **METHODS AND MODEL SYSTEMS**
The Squid Giant Axon

More than half a century ago, Alan Hodgkin and Andrew Huxley first showed that neurons send electrical signals by selectively allowing ions to cross the cell membrane in a voltage-dependent fashion. The Hodgkin-Huxley theory of the action potential is the basis for our current understanding of neurophysiology. When Hodgkin and Huxley were performing their groundbreaking experiments in 1939, while Huxley was still an undergraduate student, the only available recording electrodes were far too large to fit into a typical mammalian axon. Instead, Hodgkin and Huxley used the giant axon of the squid *Loligo pealei* as a model system to make electrical recordings from the inside of a single axon. The squid giant axon is the largest known axon in any animal. It is hundreds of times larger in diameter than a typical mammalian axon, and as much as 50 times larger than giant axons from other invertebrates. Only through detailed experimental work using the squid giant axon were Hodgkin and Huxley able to obtain the data that allowed them to formulate the mathematical models to describe the action potential. In 1952 they published a paper containing a mathematical model that could explain the action potential in

terms of known electrical theory. In 1963, they received the Nobel Prize for Medicine for this work.

Their research highlights the importance of selecting the right model system to address an experimental question. Hodgkin and Huxley chose the giant axon because its size allowed them to make recordings of the action potential that would not have been possible using any other animal. In addition, the action potential in the squid giant axon is a relatively simple one, shaped by only two ion channels: a voltage-gated K^+ channel and a voltage-gated Na^+ channel, with no complexities resulting from the presence of multiple isoforms. This property of the squid giant axon allowed Hodgkin and Huxley to develop their elegant mathematical model of the action potential, which would have been more difficult had the ion dynamics of the system been more complex. They could never have anticipated this property of the squid giant axon, so serendipity as well as planning played an important part in the success of their experiments.

References

Hodgkin, A. L., and A. F. Huxley. 1952. A quantitative description of membrane current and its application to conduction and excitation in nerve. *Journal of Physiology (London)* 117: 500–544.

resistance and intracellular resistance decrease. From the definition of the length constant ($\lambda = \sqrt{r_m/r_i}$), we can see that decreasing the intracellular resistance will increase the length constant of the membrane, increasing conduction speed. However, decreasing membrane resistance will tend to decrease the length constant, slowing conduction speed. So why do these two effects not simply cancel each other out? Remember that the intracellular resistance decreases in proportion to the radius of the axon squared, while membrane resistance decreases in direct proportion to the radius of the axon. Thus, increasing the radius of an axon has a much greater effect on the intracellular resistance than on the membrane resistance. Therefore, the net effect of increasing the radius of an axon is to increase the speed of conduction (Figure 4.25).

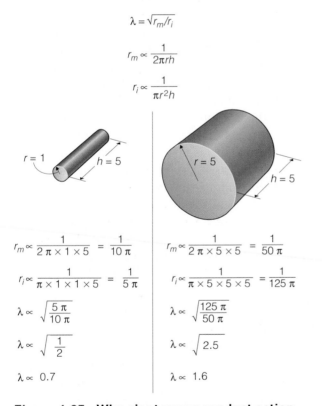

Figure 4.25 Why giant axons conduct action potentials rapidly   The geometry of the axon influences the length constant (λ) of the membrane and explains why larger-diameter axons conduct signals more rapidly than small-diameter axons. The length constant of the membrane is directly proportional to the membrane resistance (r_m), and inversely proportional to the intracellular resistance (r_i). The membrane resistance is inversely proportional to axon radius, whereas intracellular resistance is inversely proportional to axon radius *squared*. An axon with radius 1 will have a length constant proportional to 0.7, while an axon with radius 5 will have a length constant proportional to 1.6. A longer length constant means that local currents can flow farther without degrading, so signal conduction will be faster.

The capacitance of the membrane also changes as axon diameter increases, but this has only a marginal effect on the time constant of the membrane. We have already seen that membrane resistance decreases as membrane area increases. In contrast, membrane capacitance increases with membrane area. Thus, the effects of membrane resistance and membrane capacitance on the time constant of the membrane have a tendency to cancel each other out. Therefore, changes in the time constant of the membrane have a relatively small effect on local current flow as axon diameter increases.

Myelinated neurons evolved in the vertebrates

Although increasing axon diameter provides substantial increases in conduction velocity, there are two main disadvantages to using large axons to increase conduction velocity. Large axons take up more space, and this may limit the number of neurons that can be packed into the nervous system. Organisms such as mammals, with very complex nervous systems, do not have giant axons. Instead, they use myelination to increase the speed of action potential conduction. Large-diameter axons also have a much larger volume of cytoplasm per unit length, making them energetically expensive to produce and maintain. As a result, you would expect that giant axons would be used only when extremely high-speed conduction is a necessity for survival. In squid, giant axons are present only in the neurons controlling escape and prey-capture behaviors. Similarly, giant axons are associated with startle and escape responses in other organisms (including both vertebrates and invertebrates).

Myelinated neurons are found in most vertebrates. Only lampreys and hagfish (jawless vertebrates) lack multilayered myelin sheaths. As we have already discussed, certain invertebrate neurons also have axons that are wrapped in multiple layers of cell membrane, although these wrappings differ in structure from the true myelin sheath found in vertebrates, and may not be as effective in increasing the rate of signal conduction. The myelin sheath is an important evolutionary innovation, allowing rapid signal conduction in a compact space, which may have provided the conditions necessary for the evolution of the complex nervous systems of the vertebrates.

Myelination increases conduction speed

All else being equal, myelinated neurons conduct signals more rapidly than unmyelinated neurons because the myelin sheath acts as insulation for the axon, reducing current loss through leak channels and thus increasing membrane resistance. Reducing ion leak increases the length constant of the membrane, increasing the distance that local current can travel before degrading. Thus, reducing ion leak increases conduction velocity. The presence of the myelin sheath also decreases the capacitance of the membrane, because capacitance is inversely proportional to the thickness of the insulating layer in a capacitor. The many layers of cell membrane of the myelin sheath act together as a single insulator. Thus, although each membrane alone has the same thickness, the effective thickness of the many-layered myelin sheath is much greater. The increase in the thickness of the membrane decreases the capacitance, reducing the time constant of the membrane and thus increasing the speed of electrotonic conduction in the internodes.

Note that the placement of the nodes of Ranvier is critical for the function of a myelinated axon. The nodes cannot be placed too far apart or the signal will not be sufficient to depolarize the neuron beyond threshold at the next node, because current inevitably decreases with distance, although less so in a myelinated axon than in an unmyelinated axon. Typically, the length of the internodes is about 100 times the diameter of the axon, ranging from about 200 μm to 2 mm. Indeed, in some neurons, electrotonic spread can carry a suprathreshold depolarization past several nodes of Ranvier, which then appear to fire "simultaneously."

◉ | CONCEPT CHECK

16. What causes the shape of action potentials to vary among neurons?

17. What sets the speed of action potential conduction, and why?

18. Compare and contrast giant axons and myelinated axons as strategies for increasing the speed of signal conduction.

19. What factors would you expect to be important in determining the maximum spacing between nodes of Ranvier in a myelinated neuron, and why?

Diversity of Synaptic Transmission

Once the wave of depolarization reaches the axon terminal, this electrical signal must be transferred to the postsynaptic cell. In the first half of the chapter, we saw how vertebrate motor neurons release the neurotransmitter acetylcholine to send signals across the synapse. But synaptic transmission is incredibly diverse, and can be accomplished via a variety of mechanisms. For example, unlike the vertebrate motor neurons that we discussed in the first half of the chapter, some neurons do not release chemical neurotransmitters onto their target cells. Instead, these neurons have gap junctions that directly connect them to their target cells (Figure 4.26). As we discussed in Chapter 3, gap junctions are composed of a series of proteins that form small pores in the membranes of two adjacent cells, allowing ions and other small molecules to travel directly from cell to cell. Synapses in which the presynaptic and postsynaptic cells are connected via gap junctions are termed **electrical synapses**, because the electrical signal in the presynaptic cell is directly transferred to the postsynaptic cell through the gap junctions. Most neurons, however, do not form gap junctions with their target cells. Instead, these neurons form **chemical synapses**. As we saw in the case of a vertebrate motor neuron, at a chemical synapse the presynaptic neuron converts its electrical signal to a chemical signal in the form of a neurotransmitter, which diffuses across the synapse to the postsynaptic cell and binds to receptors on the postsynaptic membrane.

Electrical and chemical synapses play different roles

Electrical and chemical synapses differ in a number of respects. In a chemical synapse, the primary flow of information is from the presynaptic cell to the postsynaptic cell, and not in the reverse direction. Transmission across a chemical synapse is also relatively slow compared to the speed of propagation of an action potential because of the need for docking and fusion of synaptic vesicles, diffusion across the synapse, and signal transduction in the postsynaptic cell. Thus, transmission across a chemical synapse is associated with a synaptic delay of several milliseconds. In contrast, transmission across an electrical synapse is essentially instantaneous, since it occurs via electrotonic cur-

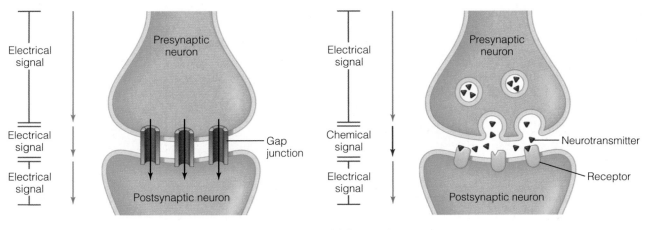

(a) Electrical synapse

(b) Chemical synapse

Figure 4.26   Electrical and chemical synapses   **(a)** In an electrical synapse, the electrical signal is directly transmitted from the presynaptic cell to the postsynaptic cell via gap junctions. **(b)** In a chemical synapse, the electrical signal in the presynaptic cell is converted to a chemical signal, in the form of a neurotransmitter, which crosses the synaptic cleft and binds to a receptor on the postsynaptic cell membrane. The receptor converts the chemical signal to an electrical signal in the postsynaptic cell.

rent spread, and thus is not associated with any significant synaptic delay. In addition, electrical synapses can easily convey information in either direction, because electrical currents or ions can move freely in either direction through the gap junctions connecting the cells (although some gap junctions have specialized structures that ensure unidirectional signal transmission).

Although signal transmission across an electrical synapse is much more rapid than across a chemical synapse, chemical synapses have one substantial advantage over electrical synapses. In an electrical synapse, the signal in the postsynaptic cell is always similar to the signal sent by the presynaptic cell, because direct transfer of ions or current causes the postsynaptic signal. In a chemical synapse, the signal in the postsynaptic cell is not necessarily the same as in the presynaptic cell. For example, a series of action potentials in a presynaptic cell could result in the release of a neurotransmitter that causes the postsynaptic cell to hyperpolarize, inhibiting it from firing action potentials. Chemical synapses provide an additional level of regulation for the nervous system; in comparison, direct electrical coupling across an electrical synapse limits the diversity of the signal in the postsynaptic cell.

Electrical synapses are present in neural pathways involved in escape behaviors in some organisms, presumably because they increase the speed of the escape response. For example, the neurons involved in the escape response of crayfish are connected via electrical synapses.

The proportion of electrical to chemical synapses in the nervous system also varies among organisms. For example, organisms with relatively simple nervous systems, such as cnidarians (jellyfish, sea anemones, and related animals), often have electrical synapses between their neurons, whereas organisms with more complex neural pathways generally make more use of chemical synapses. As we discuss in Chapter 7, from these more complex neural pathways and networks emerge more sophisticated and plastic animal behaviors. However, electrical synapses also play an important role in organisms with more complex nervous systems. In the mammalian brain, for example, electrical synapses among neurons may be important in synchronizing brain function.

Chemical synapses have diverse structures

There is substantial diversity in the morphology of chemical synapses (Figure 4.27a). We have already examined the morphology of the neuromuscular junction, the chemical synapse between a motor neuron and a muscle. The axon of a motor neuron splits into several terminal branches, and each branch terminates in a swelling called the axon terminal (or sometimes the *terminal bouton*

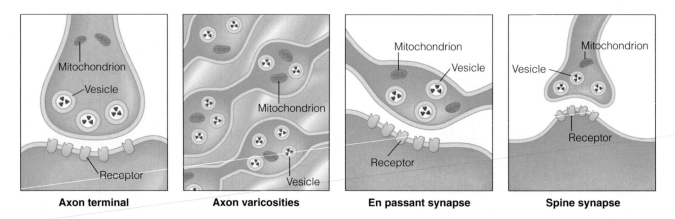

Axon terminal   **Axon varicosities**   **En passant synapse**   **Spine synapse**

(a) Types of synapses

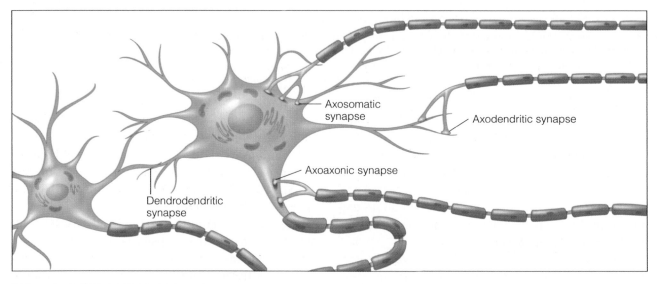

(b) Locations of neuron-to-neuron synapses

Figure 4.27 Variation in the structure and location of synapses
(a) Structural diversity of chemical synapses. There are four main types of chemical synapses.
(b) Diversity in the location of neuron-to-neuron synapses. Synapses can be axodendritic, axosomatic, dendrodendritic, or axoaxonic.

or *synaptic knob*). The synapses formed at axon terminals are highly structured, and the postsynaptic cell membrane contains increased densities of neurotransmitter receptors in close proximity to the axon terminal. Axon terminals are found at the ends of many types of neurons, in addition to the motor neurons that we have already encountered. Alternatively, some neurons form synapses at **axon varicosities**, or swellings along the axon that can be arranged like beads on a string. Each of these swellings contains vesicles filled with neurotransmitter, which are released onto the target cell. As we will see in Chapter 7, certain types of neurons in the peripheral nervous system, called *autonomic neurons*, form synapses with their effector organs at axon varicosities. These *neuroeffector junctions* differ from true

synapses in that the postsynaptic cell membrane at the junction is not specialized, and does not contain a high concentration of receptors. Instead, neurotransmitter diffuses broadly and contacts receptors located across large areas of the target organ. Neurons in the central nervous system can form a similar type of synapse, called an *en passant synapse,* that consists of a swelling along the axon of the presynaptic neuron. These synapses differ from neuroeffector junctions in that the postsynaptic membrane may be specialized and contain high densities of receptors. Another common type of synapse in the central nervous system is termed a *spine synapse*. In these synapses, the presynaptic cell connects with a specialized structure, termed a *dendritic spine*, on the dendrite of the postsynaptic cell.

Neuron-to-neuron synapses can form at a variety of locations (Figure 4.27b). *Axodendritic synapses* form between the axon terminal of one neuron and the dendrite of another, while *axosomatic synapses* form between the axon terminal of one neuron and the cell body of another. Axodendritic and axosomatic synapses are the most common types of neuron-to-neuron synapses. *Dendrodendritic synapses* form between the dendrites of two neurons, and are often electrical synapses that allow communication of information in both directions between neurons. *Axoaxonic synapses* form between an axon terminal of a presynaptic neuron and the axon of a postsynaptic neuron. Axoaxonic synapses most commonly occur near the axon terminal of the postsynaptic neuron, and play a role in regulating neurotransmitter release from the postsynaptic neuron, often by altering Ca^{2+} influx. We discuss some examples of axoaxonic synapses in Chapter 7. By modulating the release of neurotransmitter from neurons within the nervous system, these axoaxonic synapses play a role in regulating learned behaviors.

There are many types of neurotransmitters

Neurons that form chemical synapses with their target cells can communicate in diverse ways in part because of the large number of different chemical substances that act as neurotransmitters. Neurobiologists have discovered more than 50 substances that act as neurotransmitters (Table 4.4), and these neurotransmitters have diverse effects on postsynaptic cells.

To be classified as a neurotransmitter, a substance must meet several criteria. It must be synthesized in neurons. It must be released at the presynaptic cell membrane following depolarization, and it must bind to a postsynaptic receptor and cause a detectable effect. Neurobiologists often group neurotransmitters into five major classes: amino acids, neuropeptides, biogenic amines, acetylcholine, and a grab-bag class consisting of neurotransmitters that do not fit into any of the other groups. As we discussed in Chapter 3, many of these classes of molecules can also act as hormones or paracrine signals, and thus neurotransmission is part of a continuum of chemical communication systems in animals.

Four amino acids have been shown to act as neurotransmitters: glutamate, aspartate, glycine, and gamma-aminobutyric acid (GABA). Glutamate, aspartate, and glycine are also used for protein synthesis; GABA is a derivative of glutamate. Animals can synthesize all four of the amino acids that act as chemical messengers, although they may also obtain these amino acids from food. Once synthesized, amino acid neurotransmitters are packaged into vesicles, and stored until they are released by exocytosis.

The neuropeptides, also called neuroactive peptides or peptide neurotransmitters, are composed of short chains of amino acids. Neuropeptides are synthesized in the rough endoplasmic reticulum, which synthesizes all secreted peptides. In neurons, the rough endoplasmic reticulum is generally found in the cell body. Vesicles containing peptide neurotransmitters are then transported from the cell body to the axon terminal along a complex network of microtubules, via a process called **fast axonal transport**, which we discuss in Chapter 5. However, neurobiologists have recently discovered that some neurons in the brains of invertebrates such as snails can synthesize peptide neurotransmitters in both the axon and axon terminal, suggesting an additional layer of functional complexity.

Acetylcholine and the biogenic amines play particularly important roles in integrating physiological functions because they are important neurotransmitters that communicate with many kinds of tissues. You will encounter these neurotransmitters again and again as you read this book, since they are involved in the homeostatic regulation of many physiological systems. We have already discussed the role of acetylcholine at the neuromuscular junction, but this neurotransmitter plays many other roles in the nervous system. Because of their physiological importance, we discuss acetylcholine and the biogenic amines in more detail in later sections.

Some neurotransmitters do not fit into any simple chemical class. These neurotransmitters include purines such as ATP, which is important in energy metabolism, and the gas nitric oxide. The gaseous neurotransmitters, such as nitric oxide (NO), are not packaged into vesicles. Instead, after they are synthesized at the axon terminal, they diffuse freely out of the presynaptic neuron in all directions into every nearby cell. Because NO diffuses freely across membranes, it cannot be stored, and must be synthesized as needed.

Table 4.4 A summary of neurotransmitters.

Neurotransmitter	Receptor	Receptor type	Receptor location	Effect
Acetylcholine	Nicotinic	Ionotropic	Skeletal muscles, autonomic neurons, CNS (central nervous system)	Excitatory
	Muscarinic	Metabotropic	Smooth and cardiac muscle, endocrine and exocrine glands, CNS	Excitatory or inhibitory
Amino acids				
Glycine	Glycine	Ionotropic	CNS	Inhibitory
Aspartate	Aspartate	Ionotropic	CNS	Excitatory
Glutamate	AMPA	Ionotropic	CNS	Excitatory
	NMDA	Ionotropic	CNS	Excitatory
	mGlu1-8	Metabotropic	CNS	Excitatory or inhibitory
GABA	GABA-A	Ionotropic	CNS	Inhibitory
	GABA-B	Metabotropic	CNS	Generally inhibitory
Biogenic amines				
Dopamine	Dopamine	Metabotropic	CNS	Excitatory or inhibitory
Norepinephrine	α and β adrenergic	Metabotropic	CNS and peripheral nervous system (PNS), cardiac muscle, smooth muscle	Excitatory or inhibitory
Epinephrine	α and β adrenergic	Metabotropic	Cardiac muscle, smooth muscle, CNS	Excitatory or inhibitory
Peptides				
Endorphins	Opiate	Metabotropic	CNS	Generally inhibitory
Neuropeptide Y	NPY	Metabotropic	CNS	Excitatory or inhibitory
Other				
Adenosine	Purine	Metabotropic	CNS	Generally inhibitory
Nitric oxide	None	N/A	N/A	N/A

Neurotransmitters can be excitatory or inhibitory

As we discussed in Chapter 3, the response of a target cell depends on the type of receptors it expresses. Thus, depending on the nature of its receptor, a neurotransmitter can cause the postsynaptic cell to either depolarize or hyperpolarize. *Inhibitory neurotransmitters* generally cause hyperpolarization, making the postsynaptic cell less likely to generate an action potential. The resulting changes in membrane potential are often referred to as **inhibitory postsynaptic potentials (IPSPs)**. Excitatory neurotransmitters generally cause depolarization, making the postsynaptic cell more likely to generate an action potential. These depolarizations are termed **excitatory postsynaptic potentials (EPSPs)**.

Neurotransmitter receptors can be ionotropic or metabotropic

The binding of a neurotransmitter to its receptor can cause either a fast or a slow response within the postsynaptic cell, depending on the signal transduction cascade associated with the receptor. Neurotransmitter receptors are often classified as either ionotropic or metabotropic. **Ionotropic receptors** are ligand-gated ion channels. As we discussed in Chapter 3, when a neurotransmitter or other chemical signaling molecule binds to an ionotropic receptor, the conformation of the protein changes, opening a pore within the receptor protein that allows ions to move across the cell membrane (Figure 4.28a). Because binding of the neurotransmitter directly causes changes in the shape of the protein to result in ion movement, ionotropic receptors initiate rapid changes in the membrane potential of the postsynaptic cell. The nicotinic ACh receptors that we have already encountered are an example of an ionotropic receptor.

When a neurotransmitter binds to a **metabotropic receptor**, there is a change in the conformation of the receptor (Figure 4.28b) that sends a signal via a second messenger, initiating a signaling cascade within the postsynaptic cell. We have already discussed the organization and function of various signal transduction pathways in Chapter 3, and metabotropic receptors work through similar pathways. A signaling cascade activated by a metabotropic receptor ultimately sends a message to ion channel proteins, modulating the activity of ion channels on the postsynaptic cell membrane and thus altering membrane potential. Metabotropic receptors tend to cause slower-acting changes in the postsynaptic cell than ionotropic receptors because of the complex signaling pathways between binding of the neurotransmitter to the receptor and the opening of ion channels. Metabotropic receptors often also cause long-term changes in the postsynaptic cell by affecting the transcription or translation of receptors and ion channels.

Acetylcholine receptors can be ionotropic or metabotropic

We have already discussed the role of acetylcholine (ACh) in carrying signals across the neuromuscular junction, but acetylcholine is also a neurotransmitter at many other synapses, includ-

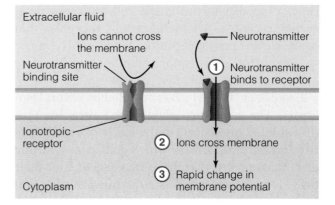

(a) Ionotropic receptors

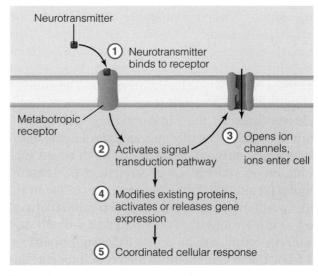

(b) Metabotropic receptors

Figure 4.28 Ionotropic and metabotropic receptors **(a)** Structure and function of an ionotropic receptor. When there is no neurotransmitter bound to an ionotropic receptor, the ion channel within the protein is closed, and ions cannot cross the cell membrane. When neurotransmitter binds to an ionotropic receptor, the gated ion channel opens, which allows ions to cross the membrane and cause a response in the postsynaptic cell. **(b)** Structure and function of a metabotropic receptor. When neurotransmitter binds to a metabotropic receptor, the receptor changes shape, sending a signal that activates a signal transduction pathway. The signal transduction pathway can open or close ion channels, modify existing ion channel proteins, or activate or repress gene expression, causing a coordinated cellular response.

ing synapses in the autonomic nervous system and the brain in vertebrates (see Chapter 7 for a discussion of the physiology of these systems). Receptors for acetylcholine are termed the **cholinergic receptors**. There are two major classes of cholinergic receptors: the *nicotinic* and the *muscarinic* receptors. As we have already discussed, nicotinic receptors are ionotropic receptors that cause a rapid response in the target cell, whereas muscarinic receptors are metabotropic receptors that cause slower responses in the target cell.

The nicotinic receptor is made up of a variety of combinations of the five possible subunits: α, β, γ, ε, and δ, each of which is encoded by several isoforms. The nicotinic acetylcholine receptor was first studied intensively in the electric organ of the ray *Torpedo californica*, which generates a strong electrical current that these rays use to stun their prey. The electric organ is a modified muscle that has high levels of the nicotinic acetylcholine receptor. Figure 4.29 shows the combination of subunits of the ACh receptor expressed in the *Torpedo* electric organ. These subunits are arranged like the staves of a barrel around a central pore. This muscle-type nicotinic ACh receptor has two binding sites for ACh that are located on the α subunit at the α-δ and α-γ subunit interfaces. The subunit composition of nicotinic receptors differs between skeletal muscle, the autonomic nervous system, and the brain. The nicotinic receptors in the autonomic nervous system are made up of an α3 subunit, an α5 subunit, an α7 subunit, a β2 subunit, and a β4 subunit, while the receptors in the brain are predominantly composed of combinations of α4 and β2 subunits. These different subunit and isoform combinations confer differing properties, adding to the complexity of the vertebrate nervous system.

Muscarinic ACh receptors are metabotropic receptors that are indirectly coupled to ion channels through G proteins. Muscarinic receptors are

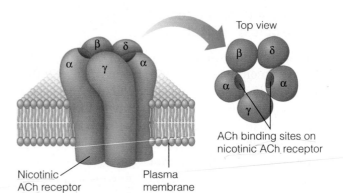

Figure 4.29 **A schematic diagram of a nicotinic ACh receptor from the electric organ of *Torpedo*** The nicotinic ACh receptor is an ionotropic receptor made up of five subunits arranged around a central pore that forms a Na$^+$ channel. Each receptor has two binding sites for ACh, formed by the α subunit at the junction of the γ or δ subunits.

named because the drug muscarine binds to them and not to nicotinic receptors. They are found on a variety of tissues, including the brain, the heart, the gut, and the bronchial passages. Stimulation of muscarinic receptors causes a slower response in the postsynaptic cell than do nicotinic receptors, and the response can be either excitatory or inhibitory, depending on the cell type. Thus, although metabotropic receptors (such as the muscarinic receptors) cause slower responses than ionotropic receptors (such as the nicotinic receptors), they are capable of generating more diverse responses. Table 4.5 summarizes some of the similarities and differences between types of cholinergic receptors.

Table 4.5	Cholinergic receptor subtypes.				
Receptor subtype	**Location**	**Effect of binding**	**Second messenger pathway**	**Agonists**	**Antagonists**
Nicotinic	Neuromuscular junctions, ganglionic neurons, adrenal medulla	Excitation	Ion influx	ACh, nicotine, carbachol	Curare
Muscarinic	Gut	Excitation	G-protein coupled	ACh, muscarine, carbachol	Atropine, scopolamine
	Heart	Inhibition			
	Bronchioles (lung)	Excitation			
	Sweat glands	Activation			
	Blood vessels of skeletal muscle	Inhibition			

The biogenic amines play diverse physiological roles

As we discussed in Chapter 3, amines are chemicals that possess an amino (–NH$_2$) group; those that can act as chemical messengers are referred to as the biogenic amines. Several biogenic amines act as neurotransmitters, including the **catecholamines** (dopamine, norepinephrine, and epinephrine), and serotonin. All of these biogenic amines are synthesized in the axon terminal using an amino acid as a precursor. Acetylcholine also contains an NH$_2$ group, and thus potentially could be considered a biogenic amine. But because ACh is not synthesized from an amino acid precursor, and because the NH$_2$ group is in the center of the molecule rather than at one end, ACh is usually classified separately from the biogenic amines.

The catecholamines are synthesized via a common pathway from the amino acid tyrosine (Figure 4.30). Serotonin is synthesized from the amino acid tryptophan via a common pathway with the hormone melatonin. Dopamine and serotonin are primarily involved in signaling within the central nervous system and are discussed in more detail in Chapter 7. Epinephrine and norepinephrine (also called adrenaline and noradrenaline) play an important role in the peripheral nervous system and are involved in regulating many important physiological processes, including heart rate and breathing, which we discuss in more detail in later chapters.

Receptors for norepinephrine and epinephrine are termed the **adrenergic receptors** (derived from the word *adrenaline*). There are two major classes of adrenergic receptors: alpha (α) and beta (β). Both norepinephrine and epinephrine bind to α receptors, although epinephrine binding to α receptors is weak. In contrast, β receptors bind strongly to both neurotransmitters. In mammals, several variants of each receptor type are present (α1, α2; β1, β2, etc.). The great diversity of receptor types allows norepinephrine and epinephrine to have opposing effects on different tissues, depending on the particular receptor that is present. For example, when norepinephrine binds to β2 receptors on the smooth muscles surrounding the bronchioles (passages leading to the lungs), the muscle relaxes. Muscle relaxation increases the diameter of the bronchiole, making it easier to breathe. In contrast, when norepinephrine binds to α2 receptors on the smooth muscles surrounding blood vessels, the muscles contract. Muscle contraction decreases

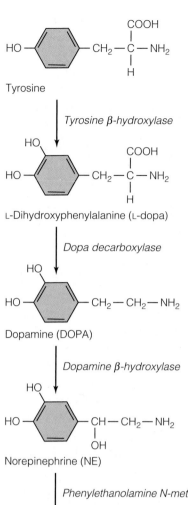

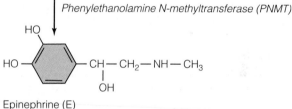

Figure 4.30 The synthetic pathway for the catecholamines The catecholamines norepinephrine (NE) and epinephrine (E) are synthesized via a common pathway with dopamine from the amino acid tyrosine. L-Dopa, DOPA, NE, and E are biogenic amines: chemical messengers containing an amine group (NH$_2$).

the diameter of the blood vessel, increasing blood pressure. The diversity of adrenergic receptors accounts for the opposing effects of norepinephrine and epinephrine on different tissues.

As we discussed in Chapter 3, isoforms of the same class of receptor may activate very different signal transduction cascades within a target cell. Figure 4.31 summarizes the major postsynaptic effects of norepinephrine binding to several classes of adrenergic receptor.

The binding of the neurotransmitter to α1 adrenergic receptors activates a signal transduction

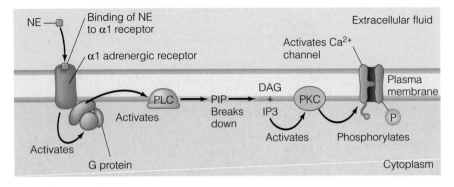

(a) Binding of NE to α1 adrenergic receptors

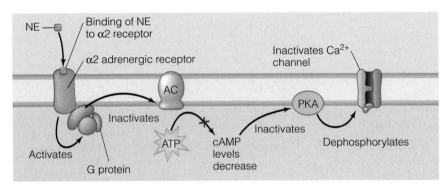

(b) Binding of NE to α2 adrenergic receptors

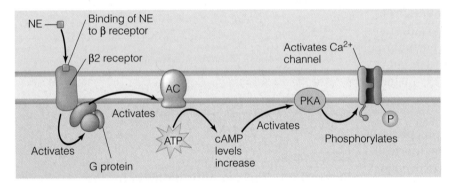

(c) Binding of NE to β receptors

Figure 4.31 Binding of norepinephrine to different receptor subtypes Norepinephrine can bind to several types of receptor, causing opposing responses in the target cell. **(a)** When norepinephrine binds to an α1 adrenergic receptor, the receptor changes shape and activates a G protein, signaling to the enzyme phospholipase C (PLC), which catalyzes the breakdown of phosphatidylinositol-phosphate (PIP) into diacylglycerol (DAG) and inositol triphosphate (IP3). The IP3 activates the enzyme protein kinase C, which then phosphorylates and activates Ca^{2+} channels. **(b)** When norepinephrine binds to an α2 adrenergic receptor, the receptor activates a G protein that inactivates the enzyme adenylate cyclase (AC). This reduces the production of cAMP from ATP, reducing intracellular cAMP levels. The reduced cAMP inactivates protein kinase A (PKA), dephosphorylating Ca^{2+} channels and inactivating them. **(c)** When norepinephrine binds to a β receptor, the change in shape of the receptor activates a G protein, which activates adenylate cyclase (AC), which increases the conversion of ATP to cAMP, increasing intracellular cAMP. The cAMP signals to protein kinase A, which then phosphorylates and activates Ca^{2+} channels. Thus, the same neurotransmitter can have opposing effects in different postsynaptic cells, depending on the type of receptor that is present.

cascade involving a G protein that activates phospholipase C, which in turn breaks down the molecule phosphoinositol-phosphate (PIP) into a molecule of diacylglycerol (DAG) and inositol triphosphate (IP3). The IP3 causes release of Ca^{2+} from intracellular stores and, with DAG, activates the enzyme protein kinase C, which phosphorylates voltage-gated Ca^{2+} channels, placing them in an activated conformation. So ultimately, the signal transduction cascade activates the target tissue by making Ca^{2+} channels easier to open. In contrast, the binding of the neurotransmitter to α2 adrenergic receptors activates a different G protein, which inactivates the enzyme adenylate cyclase, which causes cyclic AMP (cAMP) levels to decrease. This decrease in cAMP inactivates the enyzme protein kinase A, inactivating voltage-gated Ca^{2+} channels. Ultimately, this signal transduction cascade tends to inactivate the target tissue by making Ca^{2+} channels more difficult to open. The binding of the neurotransmitter to β adrenergic receptors activates a different G protein, which activates adenylate cyclase, causing cAMP to increase. The increased cAMP activates protein kinase A, which activates voltage-gated Ca^{2+} channels. This signal transduction pathway tends to activate target tissues by making voltage-gated Ca^{2+} channels easier to open. Thus, the effects of a single neurotransmitter can vary depending on the particular receptor that is present on the target tissue. Table 4.6 summarizes some of the characteristics of the major adrenergic receptors in humans.

Neurons can synthesize more than one kind of neurotransmitter

For many years it was believed that a neuron could secrete only a single kind of neurotransmitter, but now it

Table 4.6 Summary of some major adrenergic receptor subtypes.

Receptor subtype	Location (in humans)	Effect (in humans)	Second messenger system	Sensitivity
α1	Blood vessels of skin, gut, kidneys, salivary glands	Vasoconstriction	G protein activates phospholipase C	NE > E
α2	Membrane of adrenergic axon terminals	Inhibits release of NE	G protein inactivates adenylate cyclase, inhibits cAMP production	NE > E
β1	Heart	Increases heart rate and strength	G protein activates adenylate cyclase, activates cAMP production	NE = E
β2	Lungs	Dilates bronchial passages	G protein activates adenylate cyclase, activates cAMP production	E > NE

is known that a single neuron can secrete several different neurotransmitters. For example, many neurons synthesize both a small molecule neurotransmitter (like ACh or norepinephrine) and one or more neuropeptides. It is still not entirely clear how a neuron controls which neurotransmitter it releases, but different neurotransmitters appear to be released from a single axon terminal at different stimulus frequencies. For example, low-frequency stimulation might release ACh, whereas high-frequency stimulation might release a neuropeptide. It is likely that separate groups of synaptic vesicles reside in a single neuron, each containing a different neurotransmitter, and each releasing its contents in response to different stimulus frequencies.

Neurotransmitter release varies depending on physiological state

In addition to its substantial diversity among neurons and across species, synaptic transmission also varies within a single neuron, depending on the physiological state of that neuron. We have already discussed how action potential frequency relates to neurotransmitter release, but most neurons have another layer of functional complexity because neurotransmitter release can vary depending on the past history of action potentials at that axon terminal. As we will see in Chapter 7, this synaptic plasticity, or the ability of a synapse to change its function in response to patterns of use, underlies many important brain functions including learning and memory. The vast majority of neurons exhibit at least some degree of synaptic plasticity.

Figure 4.32 illustrates some features of synaptic plasticity at the neuromuscular junction. An increase in neurotransmitter release in response to repeated action potentials is termed **synaptic facilitation**. Synaptic facilitation occurs because the accumulation of intracellular Ca^{2+} following each action potential allows more neurotransmitter to be released by subsequent action potentials. In contrast, **synaptic depression**, which is a decrease in neurotransmitter release with repeated

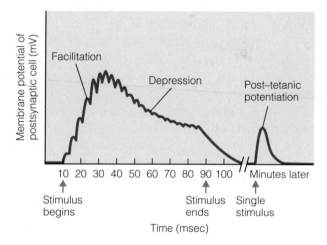

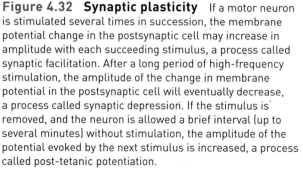

Figure 4.32 Synaptic plasticity If a motor neuron is stimulated several times in succession, the membrane potential change in the postsynaptic cell may increase in amplitude with each succeeding stimulus, a process called synaptic facilitation. After a long period of high-frequency stimulation, the amplitude of the change in membrane potential in the postsynaptic cell will eventually decrease, a process called synaptic depression. If the stimulus is removed, and the neuron is allowed a brief interval (up to several minutes) without stimulation, the amplitude of the potential evoked by the next stimulus is increased, a process called post-tetanic potentiation.

action potentials, occurs because of the progressive depletion of the readily accessible pool of synaptic vesicles that is available for fusion and exocytosis of neurotransmitter.

Post-tetanic potentiation (PTP) occurs after a train of high-frequency action potentials in the presynaptic neuron. For several seconds or minutes following a burst of action potentials, a subsequent action potential will result in increased release of neurotransmitter. The mechanisms underlying PTP differ from those involved in synaptic facilitation, and are thought to involve a Ca^{2+}-dependent increase in the available pool of neurotransmitter-containing vesicles. Synaptic facilitation and post-tetanic potentiation result in only brief changes in the activity of the synapse, but as we shall see in Chapter 7, neurons have other mechanisms that allow them to undergo long-term changes in synaptic activity.

Evolution of neurons

Only the metazoans have neurons, but electrical signaling in other organisms can provide clues as to the evolution of the metazoan neuron. Plants, which do not have neurons, do express voltage-gated channels, and use them for electrical communication. For example, algae from the family Characeae have giant cells that are capable of generating action potentials. Single cells in this species can be up to a millimeter in diameter and several centimeters in length. Early neurobiologists sometimes used this species as an experimental model when squid were not available, since these algae produce an action potential that has a shape similar to those observed in the squid giant axon. However, at a molecular level the action potential in *Chara* is very different from the action potential in animals. It results from ion movements through Cl^- channels that are activated in a Ca^{2+}-dependent manner. An increase in Ca^{2+} influx through a voltage-gated ion channel takes place at the beginning of the action potential, which initiates a signal transduction pathway that opens Cl^- channels, causing Cl^- ions to leave the cell. The efflux of Cl^- depolarizes the cell, resulting in an action potential. Therefore, the action potential in *Chara* is not due to a voltage-gated channel, although a voltage-gated channel triggers it. The action potential in *Chara* shares some features with metazoan action potentials, although it differs in many respects. It acts in an all-or-none

fashion, but is conducted about 1000 times more slowly than a typical vertebrate action potential.

Green plants, from tomatoes to trees, can also produce action potentials, although they do not have a specialized tissue for conducting these signals to specific locations over long distances. The action potential in plants appears to travel through the xylem or phloem vessels, and is a means of rapidly transmitting a signal to the entire plant. Most of the work on action potentials in plants has concentrated on the response to wounding, but even a stimulus as simple as turning on a light can provoke an action potential in plants such as a tomato. The nature and ionic basis of the plant action potential is not yet well understood because plant cells are more difficult to work with than animal cells, since they have a rigid cell wall and multiple intracellular compartments with varying ionic composition. However, it is known that action potentials are conducted without decrement in plants, and that the action potential may involve Ca^{2+} ions.

Similarly, voltage-gated channels are present in many unicellular organisms, including protists and prokaryotes. *Paramecium*, a ciliate protist, swims via the coordinated beating of the cilia that cover its exterior. If a *Paramecium* makes contact with a solid object while swimming, it will back up by reversing the direction in which the cilia beat. This reversal is the result of opening of voltage-gated Ca^{2+} channels, which causes an action potential. Mutant *Paramecium* that do not contain a functional copy of this voltage-gated Ca^{2+} channel can only swim forward. In general, action potentials in protists appear to be Ca^{2+} dependent; a single species, *Actinocoryne contractilis*, has been demonstrated to have both Ca^{2+}- and Na^+-dependent action potentials.

Only animals have voltage-gated Na⁺ channels

Only metazoans have voltage-gated Na^+ channels, and essentially all metazoans have at least one gene that codes for a voltage-gated Na^+ channel.[1] In fact, as we have already discussed, many metazoan genomes contain multiple genes that code for

---

[1] *C. elegans* provides one known exception to this rule. The *C. elegans* genome lacks voltage-gated Na^+ channels, and these animals do not produce action potentials. *C. elegans* is thought to have lost the ancestral voltage-gated Na^+ channels present in other animals. Graded potentials are sufficient to transmit information along the neurons of these small animals.

slightly different isoforms of voltage-gated Na^+ channels. The DNA sequences of voltage-gated Na^+ channel genes from all metazoans share many features, suggesting that the voltage-gated Na^+ channel arose only once, in a common ancestor of the metazoans. Current evidence suggests that the most likely ancestor of the voltage-gated Na^+ channel was a voltage-gated channel that generated both Na^+- and Ca^{2+}-dependent signals (perhaps a channel similar to the one discovered in *Actinocoryne contractilis*, previously discussed). This observation suggests that the separation of electrical conduction as a Na^+-based process may have been a key innovation in the evolution of multicellular animals. Ca^{2+} plays an important role in intracellular signaling in many cell types, and it is possible that this limits its utility as an ion that can be used to carry long-distance electrical signals.

Most organisms use chemicals for cell-to-cell communication

Although synaptic transmission shares many features with other modes of cell-to-cell communication, such as endocrine and paracrine communication (Chapter 3), current evidence suggests that synaptic transmission arose only once, since all living animals have similar mechanisms for converting electrical signals to chemical signals at the synapse. For example, jellyfish, which are very distantly related to vertebrates, have mechanisms of Ca^{2+}-induced neurotransmitter release from presynaptic neurons very similar to the mechanisms used by mammalian neurons.

Many neurotransmitters are simple molecules, such as amino acids, that are found in all living things. Even acetylcholine has been detected in bacteria, algae, protozoans, and plants (organisms that do not have nervous systems). So it is apparent that most neurotransmitters did not originally evolve to perform their neural signaling role. Metazoans appear to have taken ancient molecules and used them for a new function: cell-to-cell signaling in the nervous system.

As nervous systems have become more elaborate, the number and complexity of neurotransmitter-receptor interactions has increased. For example, *Bracheostoma lanceolatum*—the amphioxus—a cephalochordate (the sister group to the vertebrates), has only one catecholamine receptor gene, and uses dopamine but not norepinephrine as a neurotransmitter. Lampreys and hagfish have two catecholamine receptor genes, and both dopamine and norepinephrine are used as neurotransmitters. In contrast, in mammals there are five different dopamine receptors, nine α adrenergic receptors, and three β adrenergic receptors. As we shall see in Chapter 7, the increased complexity of neurotransmitter-receptor interactions may be involved in the evolution of increasing complexity in vertebrate nervous systems.

⊙ | CONCEPT CHECK

20. Compare and contrast electrical and chemical synapses.
21. Compare and contrast ionotropic and metabotropic receptors.
22. Compare and contrast the effect of norepinephrine binding to the different types of adrenergic receptors.
23. What are the fundamental evolutionary innovations that allow neural signaling in animals?

Summary

Signaling in a Vertebrate Motor Neuron

→ Neurons are excitable cells that use a combination of electrical and chemical signals to rapidly transmit information throughout the body.

→ Vertebrate motor neurons consist of dendrites, a cell body, an axon, and several axon terminals.

→ Membrane-bound receptors in the dendrites and cell body receive incoming signals that stimulate gated ion channels, which open or close, causing graded potentials.

→ Graded potentials travel across short distances through the cell body to the axon hillock where they are integrated, either by spatial summation or temporal summation, to alter the membrane potential of the axon hillock.

→ When the axon hillock depolarizes beyond the threshold potential, the graded potentials can trigger action potentials in the axon.

→ Action potentials occur only in axons, and differ from graded potentials in that they occur in an "all-or-none" fashion, and can be conducted across long distances without degrading.

→ Action potentials have three main phases: a depolarization phase, a repolarization phase, and an after-hyperpolarization phase.

→ The depolarization phase of the action potential is the result of the opening of voltage-gated Na^+ channels, which open in response to threshold or suprathreshold depolarization of the membrane.

→ Opening of voltage-gated K^+ channels causes the repolarization phase.

→ An action potential in one part of the axon triggers other action potentials in adjacent areas of the axonal membrane, allowing conduction without decrement.

→ When an action potential reaches the axon terminal, the signal must be transmitted to other cells across the synapse.

→ At electrical synapses, signals are transmitted directly from cell to cell via gap junctions.

→ At chemical synapses, the electrical signal encoded by the action potential is converted to a chemical signal, in the form of a neurotransmitter.

→ The binding of a neurotransmitter to its receptor generates a signal in the postsynaptic cell.

Diversity of Neural Signaling

→ The structure and function of neurons is diverse, and can vary among animals and among different neurons within a single animal.

→ Action potentials can vary in length and shape among different neurons, as a result of differ-

ences in the properties or density of voltage-gated Na^+ and K^+ channels.

→ The cable properties of the axon influence the speed at which action potentials are conducted along the axon.

→ Large-diameter axons and myelinated axons conduct signals more rapidly than small-diameter or unmyelinated axons.

→ Over 50 different substances are known to act as neurotransmitters.

→ Some neurotransmitters bind to several different receptors, and are thus able to have opposite effects on different postsynaptic cells.

→ Only metazoans have nervous systems, but many other organisms use changes in membrane potential as signals to convey information.

→ Voltage-gated channels are present in plants, algae, and protozoans, and are used to generate signals that can modify behavior or convey information between distant tissues. Voltage-gated Na^+ channels, however, are found only in animals.

→ Over the course of evolution, the number and diversity of ion channels in metazoans appears to have increased with increasing complexity of the nervous system.

→ The number and complexity of neurotransmitter-receptor interactions have also increased greatly in the metazoans.

Review Questions

1. What are the four main functional zones of a neuron?

2. Why does the opening of a Na^+ channel cause a neuron to depolarize?

3. Why do only the ions Na^+, K^+ and Cl^- appear in the Goldman equation as formulated for a neuron at rest?

4. Why can't graded potentials be propagated across long distances in neurons?

5. What is the difference between temporal and spatial summation? Can spatial summation occur without temporal summation?

6. Draw a diagram to illustrate the relationship between the states of the various voltage-gated ion channels, membrane permeability, and the phases and refractory periods of the action potential.

7. Explain in your own words why increasing the density of voltage-gated Na^+ channels decreases the threshold potential of a neuron.

8. What molecular properties of the ion channels involved in action potentials cause unidirectional propagation of action potentials along the axon, and why?

9. Why are acetylcholinesterase inhibitors effective in the treatment of myasthenia gravis?

10. Which type of neuron would you expect to have more dendrites, an afferent (sensory) neuron or an interneuron? Justify your answer.

11. Draw a diagram of the shape of an action potential in a neuron that expresses voltage-gated K^+ channels compared to the action potential in a neuron that does not express voltage-gated K^+ channels, assuming that all other factors are similar between the neurons. Explain the reasoning behind any differences that you indicate in shape between the two action potentials.

12. Explain why a myelinated neuron conducts signals more rapidly than an equivalent unmyelinated neuron.

13. Compare and contrast the signal transduction pathways initiated by binding of norepinephrine to the various types of adrenergic receptors.

Synthesis Questions

1. Ouabain is a poison that selectively binds to the Na^+/K^+ ATPase and inhibits it. What would happen over the course of a few hours to the resting membrane potential of a neuron that was poisoned with ouabain?

2. Immediately after the application of ouabain, would the neuron in question 1 still be able to generate an action potential? Why or why not?

3. A student is eating at the lab bench (in clear violation of laboratory policy), and mistakenly sprinkles tetrodotoxin on his fries. Given that this substance inhibits voltage-gated Na^+ channels, indicate whether the following statements concerning this student are true or false. Explain your answers, and consider the time course of the response.
   - It will be more difficult for the student's neurons to generate action potentials.
   - The student's neurons will fire more frequently, since membrane potential will be brought closer to threshold.
   - The effect on the membrane potential of the student's neurons could be predicted by the Nernst equation, which factors in the effects of both ion concentration and ion permeability.

4. Describe the relationship between the afterhyperpolarization phase of the action potential and the relative refractory period. Why is the relative refractory period important for neural signaling?

5. What would happen if you experimentally stimulated an axon close to both the axon hillock and the axon terminal at the same time?

6. What would happen to action potential generation in an axon if you applied a drug that caused voltage-gated K^+ channels to remain open constantly?

7. Imagine a postsynaptic neuron that is contacted by two different excitatory presynaptic neurons. One of these presynaptic neurons (neuron A) contacts the cell body of the postsynaptic cell next to the axon hillock, whereas the other presynaptic neuron (B) contacts a dendrite of the postsynaptic cell on the side of the cell body farthest away from the axon hillock. Explain why repeated firing of neuron A at slightly below the threshold potential could cause the postsynaptic neuron to initiate an action potential, while firing of neuron B at exactly the same intensity and frequency might not.

8. You have discovered a drug that blocks voltage-gated Ca^{2+} channels. What are the likely effects of this drug at the synapse?

9. Drugs called selective serotonin reuptake inhibitors (SSRIs), which affect the reuptake of neurotransmitter by presynaptic cells, are used for the treatment of depression. Serotonin normally causes an excitatory postsynaptic potential. What effect would the administration of an SSRI have on the response of these postsynaptic cells, and why?

Quantitative Questions

1. Use the table to the right and the Goldman equation to calculate the resting membrane potential of a neuron at 37°C. (Temperature in Kelvin = Temperature in °C + 273.15). Please report your answer in millivolts.

Ion	Intracellular concentration (mM)	Extracellular concentration (mM)	Membrane permeability at rest
K^+	140	4	1
Na^+	15	145	0.05
Cl^-	4	110	0.1
Ca^{2+}	0.0001	5	0

2. The neuron described in question 1 contains ligand-gated Ca^{2+} channels. What will happen to the membrane potential of this neuron if neurotransmitter binds to these channels? Be quantitative in your answer; what is the maximum possible change in membrane potential?

3. (a) During extreme dehydration, plasma K^+ can increase to as high as 10 mM. What would the membrane potential of this neuron be under these conditions? (Assume there are no other changes in ion concentrations.) (b) What would happen to the ability of this neuron to generate action potentials during extreme dehydration? Why might this be problematic?

4. Twelve neurons synapse on one postsynaptic neuron. At the axon hillock of the postsynaptic neuron, 10 of the presynaptic neurons produce EPSPs of 2 mV each and the other two produce IPSPs of 4 mV each. The threshold potential of the postsynaptic cell is −60 mV (resting membrane potential is −70 mV). Will an action potential be produced? Justify your answer.

5. Calculate the relative conduction velocities in two different axons, one with a diameter of 2 μm and another with a diameter of 50 μm, assuming that all other factors are the same between the two axons.

For Further Reading

See the Additional References section at the back of the book for more readings related to the topics in this chapter.

Signaling in a Vertebrate Motor Neuron

These excellent neurobiology textbooks provide a detailed look at the structure and function of neurons.

Kandel, E. R. 2000. *Principles of neural science.* New York: McGraw-Hill.

Levitan, I. B., and L. K. Kaczmarek. 2001. *The Neuron: Cell and molecular biology.* New York: Oxford University Press.

Matthews, G. 2001. *Neurobiology: Molecules, cells and systems.* New York: Blackwell Science.

This extremely informative book is an excellent reference for learning more about the molecular biology and physiology of ion channels.

Hille, B. 2001. *Ion channels of excitable membranes,* 3rd ed. Sunderland, MA: Sinauer Associates.

In these classic papers, Hodgkin and Huxley demonstrated the features of the action potential, and presented their theory of the underlying mechanism.

Hodgkin, A. L., and A. F. Huxley. 1939. Action potentials from inside a nerve fibre. *Nature* 144: 710–712.

Hodgkin, A. L., and A. F. Huxley. 1952. A quantitative description of membrane current and its application to conduction and excitation in nerve. *Journal of Physiology, London* 117: 500–544.

This autobiography provides a personal glimpse into the research that initiated the modern era of neurobiology.

Hodgkin, A. L. 1992. *Chance and design: Reminiscences of science in peace and war.* New York: Cambridge University Press.

This review outlines some of the shortcomings of the Hodgkin and Huxley model for the action potential, particularly for neurons with very complex sets of voltage-gated ion channels.

Meunier, C., and I. Segev. 2002. Playing the devil's advocate: Is the Hodgkin-Huxley model useful? *Trends in Neuroscience* 25: 558–563.

Neher and Sakmann shared the Nobel Prize in part for their development of the patch clamp technique. Using the patch clamp technique, Neher and Sakmann were able to demonstrate that ion channels actually exist, and to determine how they function. This discovery is one of the fundamental underpinnings of modern neurophysiology.

Neher, E., and B. Sakmann. 1976. Single-channel currents recorded from membrane of denervated frog muscle fibres. *Nature* 260: 799–802.

Neher, E., and B. Sakmann. 1992. The patch clamp technique. *Scientific American* 266: 44–51.

This review highlights the properties of the nodes of Ranvier, and summarizes some of the current literature demonstrating that different ion channels are localized at specific sites along the axon.

Salzer, J. L. 2002. Nodes of Ranvier come of age. *Trends in Neuroscience* 25: 2–5.

This paper demonstrates for the first time that action potentials may not actually be initiated at the axon hillock, but rather a little bit further down the axon, at least in pyramidal neurons. The authors are also able to demonstrate that differences in the properties of voltage-gated Na$^+$ channels, not just differences in their concentration, may be important for the occurrence of action potentials in axons and not the cell body.

Colbert, C. M., and E. Pan. 2002. Ion channel properties underlying axonal action potential initiation in pyramidal neurons. *Nature Neuroscience* 5: 533–538.

The paper below is an excellent overview of synaptic transmission.

Jessell, T. M., and E. R. Kandel. 1993. Synaptic transmission: A bidirectional and self-modifiable form of cell-cell communication. *Cell* 72 Suppl: 1–30.

In this review of his Nobel Prize–winning work, Bert Sakmann describes some of his experiments on the neuromuscular junction.

Sakmann, B. 1992. Elementary steps in synaptic transmission revealed by currents through single ion channels. *Science* 256: 503–512.

Diversity of Neural Signaling

This paper summarizes the diversity and evolution of the voltage-gated sodium channels that are the fundamental basis of neuronal action potentials.

Goldin, A. L. 2002. The evolution of voltage-gated Na$^+$ channels. *Journal of Experimental Biology* 205: 575–584.

These papers provide a discussion of the evolution of myelin and its role in increasing the speed of action potential conduction.

Waehneldt, T. V. 1990. Phylogeny of myelin proteins. *Annals of the New York Academy of Sciences* 605: 15–28.

Weatherby, T. M., A. D. Davis, D. K. Hartline, and P. H. Lenz. 2000. The need for speed. II. Myelin in calanoid copepods. *Journal of Comparative Physiology, Part A: Sensory, Neural, and Behavioral Physiology* 186: 347–357.

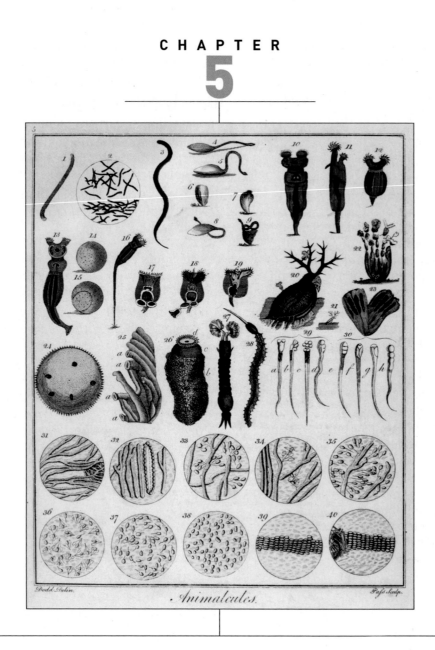

Animalcules.

Cellular Movement and Muscles

More than 300 years ago a Dutch dry-goods merchant named Anton van Leeuwenhoek became one of the earliest cell biologists. Utilizing his flair for glasswork, van Leeuwenhoek created a homemade lens that allowed him to discover the microscopic organisms inhabiting pond water. He was struck by how these small creatures swam forward and backward through the water. Even then, *movement was synonymous with life*, and he recognized that these microscopic "animalcules," as he called them, were alive. Over the next 200 years, the quality of microscopes improved. By the late 1800s, microscopists were able to look inside living cells, al-

lowing them to see organelles move rapidly throughout large algal cells. Even the cytoplasm itself seemed to flow beneath the margins of the plasma membrane.

We now realize that all eukaryotic organisms show some form of movement, either within cells, by cells, or by organisms. However, animals are the only group of multicellular organisms that are able to actively move from place to place, courtesy of a distinctive cell type found only in animals: the muscle cell. A study of the evolutionary and developmental origins of muscles reveals a paradox of unity and diversity. At the molecular level, most muscle proteins have

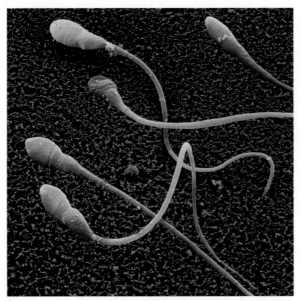

Sperm.

Antelope.

homologues in fungi, plants, and other eukaryotes. Although muscles are constructed from the same cytoskeletal elements shared by all organisms, the distinct features of the homologues in animals enable them to construct muscle.

By looking at the anatomical, physiological, and genetic properties of living animals, we can gain insight into the evolutionary origins of muscle. The simplest animals lack true muscles, although they do have specialized cells that contract. For example, sponges (phylum Porifera), the earliest multicellular animals, have pores that allow seawater to penetrate their bodies. Specialized contractile cells surround pores, controlling their diameter.

Musclelike cells first arose in cnidarians, such as the familiar *Hydra*. Myoepithelial cells combined to form fibers that worked in conjunction with their internal hydrostatic skeleton to extend the body stalk. True muscle first appeared in a related group of animals called ctenophores. These animals, which include sea walnuts and sea gooseberries, have true smooth muscle cells in the body wall.

The animals within the various worm phyla, including flatworms, nematodes, and annelids, have more elaborate muscle systems. Worms use complex longitudinal and circular smooth muscles for locomotion, nematodes have pharyngeal muscles used for feeding, and annelids have thickened regions of blood vessels that act as pumping hearts. Although these ancient animals have several discrete types of muscle, more complex recent animals display much greater diversity in muscle anatomy and physiology.

One of the most important factors driving the diversity of muscle types in more complex animals was the trend toward larger bodies. While small animals can survive using simple diffusion of respiratory gases, large animals have low surface area to volume ratios, and simple diffusion cannot meet their metabolic demands. Thus, the genes for muscle proteins evolved in combination with primitive respiratory and circulatory systems. For example, molluscs possess well-developed muscular hearts, and their multiple types of muscle are used in locomotion and feeding. Likewise, arthropods have complex muscles that control ventilation and movement.

The greatest diversity in muscle types, however, occurs in the vertebrates. More than 300 million years ago, the early vertebrate ancestors experienced two rounds of genome duplications. The extra copies of genes for critical muscle proteins allowed for the evolution of highly specialized muscle types. Instead of only having single genes for important muscle proteins, as found in the invertebrates and protochordates, genome duplication and later gene duplications in ancestors of more complex animals created extra copies of these genes, providing fertile ground for the evolution of specialized muscle protein isoforms. Whereas simple invertebrates must employ only one or two muscle myosin genes to build all muscles, vertebrates possess at least 15 different myosin genes. With the transition to land and the challenges of movement under the full weight of gravity, muscle genes rapidly evolved, allowing muscle specialization and diversification.

While this remarkable diversity in locomotor muscles is impressive, remember that muscles are built from the same components that enable intracellular movements in other eukaryotes. When you marvel at the athleticism of a cheetah sprinting, a tuna swimming, or a hummingbird hovering, remember that these impressive capacities depend upon cellular machinery not unlike that found in the fungus growing on your shower curtain. ⊙

Overview

Every physiological process, be it intracellular transport, changes in cell shape, cell motility, or muscle-dependent animal locomotion, depends in some way on movement. Regardless of the type of movement, the same intracellular machinery underlies each one: the *cytoskeleton* and its *motor proteins*. Recall from Chapter 2: Chemistry, Biochemistry, and Cell Physiology that eukaryotic cells possess a cytoskeleton composed of microtubules, microfilaments, and intermediate filaments. Of these, only microtubules and microfilaments have important roles in cellular movement. Microtubules work in conjunction with the motor proteins kinesin and dynein. Myosin, in contrast, is the actin-dependent motor protein. The diversity in cellular movement is possible because these basic elements can be arranged and used in countless combinations.

There are three general ways that cells use these elements to move (Figure 5.1). Most commonly, cells use the cytoskeleton as a roadway, where motor proteins act as trucks carrying cargo over the complex cytoskeletal networks. Just as the highway route controls traffic, cells mediate intracellular traffic by controlling where the roads go, which vehicles ride the road, and the nature of the cargo. For example, the precision of the cell signaling pathways we discussed in Chapter 3: Cell Signaling and Endocrine Regulation depends on motor proteins being able to carry secretory vesicles from sites of synthesis to the plasma membrane for exocytosis. If a vesicle is carried to the wrong place or released at the wrong time, dangerous miscommunications can result.

A second class of movement is driven by active reorganization of the cytoskeletal network. Rather than acting as a road, in this case the cytoskeletal fibers act as bulldozers that push the cellular contents forward. This type of movement, often called amoeboid movement, is most common in protists. Many metazoan cell types, such as leukocytes and macrophages, also use amoeboid movement. Motor proteins may or may not be involved in the process. Cells regulate this type of movement by controlling the rate and direction of growth of cytoskeletal fibers.

The third type of movement is analogous to a group of people pulling a rope. In this case, the motor protein pulls on the cytoskeletal rope. Cells then organize the cytoskeleton in a way that translates this tugging action into movement. As you will see

(a)

(b)

(c)

Figure 5.1 Three ways to use the cytoskeleton for movement (a) Cells can use their cytoskeleton as a road on which motor proteins move, often carrying intracellular cargo. (b) Some cells move by pushing the cytoskeleton forward, much like a bulldozer pushes earth ahead. (c) Movement sometimes resembles a tug-of-war, where motor proteins, depicted as people, can pull the cytoskeleton, symbolized by the rope.

later in this chapter, these cytoskeletal superstructures are the foundation of cilia, flagella, and muscle. Cells primarily regulate this type of movement by controlling the activity of the motor protein.

Cytoskeleton and Motor Proteins

The cytoskeleton and motor proteins work in conjunction to enable animals to mediate intracellular trafficking, changes in cell shape, and cellular movement. Three general explanations exist for the variations seen in the cellular movement in animal cells. First, most animals possess multiple isoforms of critical cytoskeletal and motor proteins. This arsenal of genetic variation allows metazoans to build specialized types of cells. Second, animal cells can use a single set of building blocks to organize the cytoskeleton in different ways. Third, animals can regulate an existing suite of proteins in real time; hormones bind to receptors, triggering regulatory cascades that alter enzyme activity that modifies the properties of the cytoskeleton and motor proteins. These three aspects of diversity account for the distinct ways animal cells build and use the cytoskeleton and motor proteins for movement. The capacity to be different at a cellular level is central to the animals' ability to generate specific types of cells, as well as to adapt to evolutionary challenges. As we proceed through this textbook, you will see that these cellular processes underlie many important physiological systems.

Microtubules

Cells can organize microtubules in many arrangements. Most cells gather the ends of microtubules near the nucleus of the cell at the **microtubule-organizing center (MTOC)** (Figure 5.2). The microtubules radiate from the MTOC like spokes of a wheel that extend to all margins of the cell. The outward ends of microtubules are anchored to integral proteins embedded within the plasma membrane. This microtubule network is vital to intracellular

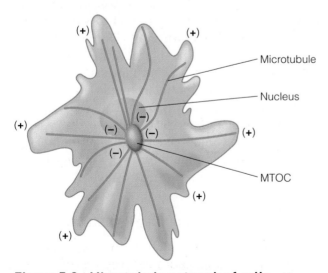

Figure 5.2 Microtubule network of cells Many cells organize microtubules into a network, with the minus ends gathered near the center of the cell at the microtubule-organizing center (MTOC).

traffic, as motor proteins can move either toward the central MTOC or to the periphery of the cell.

Cells use their microtubule network to control the movement of subcellular components, such as vesicles and organelles. Microtubule systems also mediate the rapid changes in skin color seen in some animals that use cryptic coloration, such as the African claw-toed frog, *Xenopus laevis* (Figure 5.3). Skin color is determined by the distribution of dark pigment granules within cells called *melanophores*. When the pigment granules are concentrated near the MTOC, the skin is pale in color. When the granules are dispersed throughout the cell, the skin darkens. Changes in the directional movement of pigment granules along microtubule tracks within the melanophore, controlled and triggered by hormones, create adaptive coloration in animals. A closer look at how microtubules are built will lay the foundation for understanding the role they play in vesicle traffic, pigment dispersal, and other types of intracellular and cellular movements that are central to physiological function.

Microtubules are composed of α-tubulin and β-tubulin

Microtubules, so named because of their tubelike appearance, are composed of long strings of the protein tubulin, itself a dimer of two closely related proteins: α-tubulin and β-tubulin. The evolutionary history of tubulin is intriguing and rich in paradoxes. For example, tubulin genes have changed very little since the earliest eukaryotes. The α-tubulin of yeast is very similar to your own; even α-tubulin and β-tubulin are nearly 40% identical in most species. Many animals have multiple tubulin isoforms that are expressed in different tissues. Because of the similarity in the structures of different isoforms, they were believed to be interchangeable: for example, one α-tubulin isoform could be replaced with another α-tubulin isoform without obvious consequences. The importance of the subtle differences in tubulin structure between species, as well as within a species, has only recently been appreciated. In one instance, when nematodes (*C. elegans*) were genetically modified to express a different isoform of β-tubulin in their touch neurons, the mutant worms had sensory dysfunction. These studies showed that even subtle differences in the structure of tubulin isoforms have important consequences for cellular function.

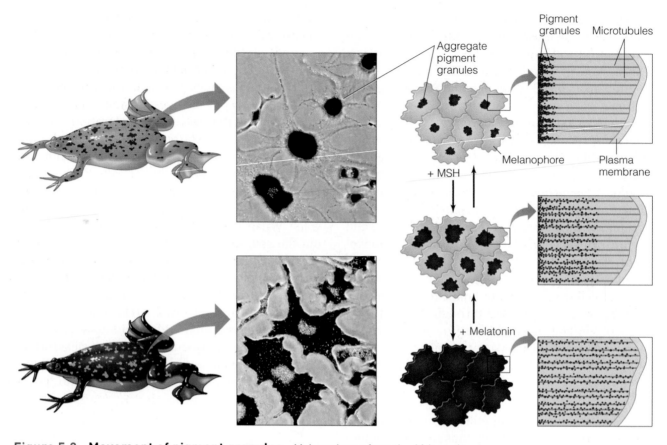

Figure 5.3 Movement of pigment granules Melanophores from the African claw-toed frog *Xenopus* allow rapid changes in color. Arrays of microtubules radiating from the central MTOC carry pigment granules throughout the cell. Actin filaments, not shown here, also play a role in controlling local pigment distribution. Pigment granules aggregate in response to the hormone melatonin, and disperse in response to melanophore stimulating hormone, MSH. (Micrographs courtesy of V. Gelfand, University of Illinois)

Unlike many large, complex proteins, microtubules form spontaneously within cells, a feature that is central to microtubule function. The first step of assembly (Figure 5.4) occurs when α-tubulin and β-tubulin combine to form tubulin. Prior to dimerization, both subunits bind to a single molecule of GTP. When tubulin forms, the GTP bound by β-tubulin may be hydrolyzed into GDP and phosphate. In contrast, the GTP bound by α-tubulin remains intact and bound within the tubulin structure. The α-tubulin, with its GTP intact, is on one end of the dimer; the β-tubulin, with its hydrolyzed GTP, is on the other end of the dimer. The difference between the two monomers creates structural asymmetries within tubulin, known as *polarity*. The α-tubulin subunit is at the so-called minus end (−) of the tubulin dimer, whereas β-tubulin is at the plus end (+). The polarity of tubulin has important ramifications in the subsequent steps of microtubule assembly.

The next step in microtubule assembly occurs when multiple tubulins assemble end-to-end. Like a line of magnets, the plus end of the growing chain attracts the minus end of a free dimer. The chain, or **protofilament**, grows until it reaches a critical length. The protofilaments then line up side by side to form a sheet that eventually rolls into a tube to form the microtubule. Because the angle between adjacent protofilaments is about 28°, 13 protofilaments are required to form a complete circular tube. Once the microtubule is formed, it can continue to grow by incorporating more dimers, or it may shrink by shedding them.

Microtubules show dynamic instability

Microtubule dynamics, such as the rates of growth and shrinkage, regulate many cellular functions. Any chemical that disrupts microtubule dynamics can become a potent poison. Some plants use mi-

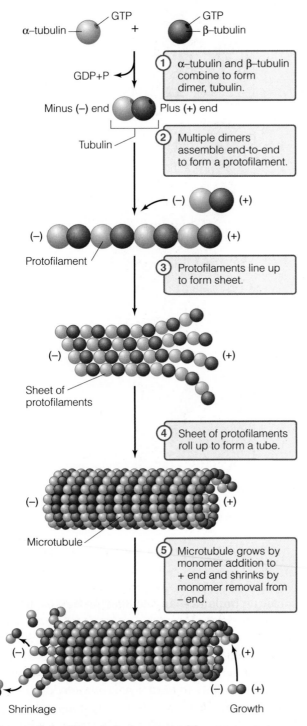

Figure 5.4 Microtubule assembly Microtubules are composed of repeating units of the protein tubulin, a dimer of two GTP-binding proteins, α-tubulin and β-tubulin. Tubulin dimers connect end-to-end to begin the construction of a protofilament. The protofilaments join side by side to start the formation of a sheet. Once the sheet reaches a critical width, it rolls into a tube to form the microtubules. Microtubules grow by adding tubulin and shrink by losing tubulin.

crotubule poisons as part of their defense against animal grazing; for example, the Pacific yew tree (*Taxus*) produces taxol, the periwinkle plant (*Vinca*) produces vinblastine, and the autumn crocus (*Crocus*) produces colchicine. Animals that graze on these plants are sickened as a result of the effects of these alkaloids on their own microtubule dynamics. Many of these plant defense agents have been developed as anticancer drugs because of their ability to kill rapidly dividing cells. These compounds are also very useful tools in the laboratory, as they allow researchers to dissect the processes that control microtubule dynamics.

The balance between growth and shrinkage determines the length of the microtubule (Figure 5.5). Many factors influence microtubule dynamics, but the most important is the local concentration of tubulin. If the end of the microtubule is exposed to a high concentration of tubulin, it will tend to grow. At low tubulin concentrations, however, microtubules tend to lose tubulin dimers and shrink. At a specific critical concentration (C_c), growth and shrinkage are in balance and there is no net change in length. However, several factors complicate this simple pattern of concentration-dependent regulation. First, the C_c value at the plus end is lower than at the minus end. This

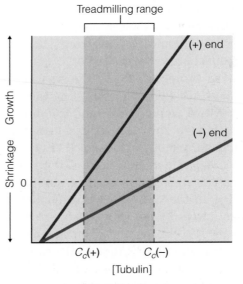

Figure 5.5 Microtubule dynamics Whether a microtubule grows or shrinks depends on tubulin concentration. Below a critical concentration (C_c) the microtubule is more likely to shrink. Above C_c it will likely grow. While both ends can add or lose tubulin, the plus end has a lower C_c. This means at any particular tubulin concentration, the plus end is more likely to grow than is the minus end.

means that if *both ends* are exposed to the *same tubulin concentration*, the plus end is more likely to grow and the minus end is more likely to shrink. Colchicine and vinblastine are toxic because they prevent microtubule growth. Colchicine binds to free tubulin and prevents it from incorporating into microtubules, while vinblastine prevents microtubule formation by causing free tubulin dimers to aggregate. This reduces the concentration of free tubulin, curtailing microtubule assembly.

The second feature that distinguishes microtubule growth is known as *dynamic instability*. Even when the tubulin concentration exceeds C_c, the microtubule will grow for a few seconds, then spontaneously shrink for a few seconds. This concentration-independent transition is due to a change in the GTP bound by β-tubulin. Once incorporated into a microtubule, the β-tubulin subunit may or may not hydrolyze GTP. As long as the GTP in β-tubulin remains intact, the microtubule tends to grow. Alternatively, if the GTP is hydrolyzed, the microtubule will tend to shrink. Microtubules maintain their constant length by balancing growth and shrinkage, while hydrolyzing a lot of GTP in the process. This may at first seem to be a waste of the cell's energy, but it is a necessary cost. Dynamic instability, despite its energetic costs, enhances the ability of the cell to regulate microtubule growth in space and time. Systems in motion are much easier to alter than static systems.

Microtubule dynamics are also regulated by **microtubule-associated proteins**, or MAPs (Figure 5.6). These proteins bind to the surface of microtubules, stabilizing or destabilizing the microtubule structure. Some MAPs bind to the plus end of microtubules and prevent the transition from growth to shrinkage. A group of MAPs called *stable-tubule only polypeptides,* or STOPs, are used by many cell types that need long, stable microtubules. For instance, STOPs are abundant in nerves where microtubules are important for the development of long axons and dendrites. Other MAPs act as protein cross-linkers. MAPs can join microtubules together into bundles, or link the microtubules to other cellular structures, such as membrane receptors. Taxol is a potent toxin because it stabilizes microtubules. However, not all MAPs stabilize microtubules. For example, *katanin* (Japanese for "sword") is a MAP that severs microtubules. Normal cell function depends on the regulation of both assembly and disassembly of microtubules. Preventing microtubules from dis-

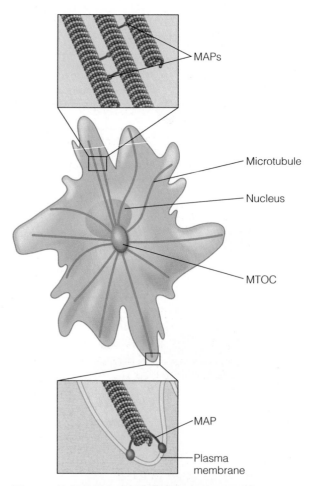

Figure 5.6 Microtubule-associated proteins
Microtubules are connected to each other and to membrane proteins by microtubule-associated proteins, or MAPs.

sociating impairs many cellular processes, including cell division.

The activities of MAPs are regulated by protein kinases and protein phosphatases. Changes in MAP phosphorylation can alter its subcellular location, change its ability to bind a microtubule, or alter its functional properties. Many signaling pathways target MAPs to alter microtubule structure. For example, the hormones that regulate cell division, known as cytokines, induce changes in microtubule structure by regulating the MAP structure and activity. The subsequent changes in the microtubule network ensure that cellular constituents are equally divided between daughter cells.

Temperature is another parameter that affects microtubule dynamics. Early experiments showed that isolated microtubules could assemble and disassemble spontaneously in test tubes. When micro-

BOX 5.1 **EVOLUTION AND DIVERSITY**
Thermal Adaptation in Microtubules

The thermal instability of microtubules presents a conundrum. If mammalian microtubules spontaneously disassemble at 25°C, what is different about the microtubules of animals that live at even colder temperatures? Many mammalian tissues can stabilize microtubules using a number of microtubule-binding proteins, such as STOPs (stable-tubule only proteins), MAPs, and capping proteins. Do cold-dwelling organisms use these same proteins to prevent thermal instability, or is there something different about tubulin itself? Insight into this question comes from studies using models in which differences arise from both natural selection and genetic engineering approaches.

For many cold-dwelling organisms, microtubule stability can be traced to the structure of tubulin itself. When first discovered, this was a bit surprising because the sequence of tubulin is extraordinarily conserved across animals. Isolated microtubule proteins from cold-water fish spontaneously assemble at lower temperatures than do those proteins from mammals. Antarctic fish have been isolated in Polar Seas for more than 10 million years. Over this time, the sequences of α-tubulins and β-tubulins have accumulated only a few amino acid variations, yet the microtubules from these fish are much more stable than microtubules from warm-water fish. When the genes for β-tubulin from a cold-tolerant cod were transfected into cultured human cells, the microtubules from the transgenic cells were stable in the cold. These studies show that very subtle differences in tubulin structure, even one or two amino acids, can result in profound differences in cold stability. Researchers studied microtubules produced by yeast in which the β-tubulin gene was subtly mutated; a single cysteine was mutated to alanine. This simple mutation made the microtubules cold-stable. Unfortunately for the yeast, the structural changes that increased cold-stability also dramatically impaired processes that depend on microtubule dynamic instability, such as growth and cell replication. These studies illustrate two important aspects of microtubules. First, microtubule function is critically dependent upon maintaining a dynamic balance between assembly and disassembly, or stability and instability. Second, even modest changes in microtubule structure, arising through evolution or genetic engineering, can produce a microtubule with very different properties. Whether these subtle mutations are adaptive or lethal depends on how the specific mutation affects the proteins, and how the structural change influences function in the context of environmental conditions.

References

○ Detrich, H. W., III, S. K. Parker, R. C. Williams Jr., E. Nogales, and K. H. Downing. 2000. Cold adaptation of microtubule assembly and dynamics. Structural interpretation of primary sequence changes present in the alpha- and beta-tubulins of Antarctic fishes. *Journal of Biological Chemistry* 275: 37038–37047.

○ Modig, C., M. Wallin, and P. E. Olsson. 2000. Expression of cold-adapted beta-tubulins confer cold-tolerance to human cellular microtubules. *Biochemical and Biophysical Research Communications* 269: 787–791.

○ Sidell, B. D. 2000. Life at body temperatures below 0 degrees C: The physiology and biochemistry of Antarctic fishes. *Gravity and Space Biological Bulletin* 13: 25–34.

---

tubules were cooled to 25°C, for example, they disassembled. Although this is a useful laboratory technique to study microtubule dynamics, what does it mean for the animals? Temperature-induced disassembly is not physiologically relevant for most endothermic animals, such as mammals and birds, because they maintain body temperatures well above the threshold temperature. However, many ectothermic animals must endure temperatures low enough to disrupt the microtubules of a mammal. In that case, how do animals that live in the cold avoid spontaneous disassembly of their microtubules? See Box 5.1, Evolution and Diversity: Thermal Adaptation in Microtubules for an explanation.

Microtubule polarity determines the direction of movement

The extensive microtubule networks within cells provide a complex roadway for the motor proteins. But how do motor proteins identify which road to ride? Once on the road, how do they decide which way to go? Recall that the orientation of the dimers endows a microtubule with a structural polarity, where microtubules have a plus end and a minus end. Since cells organize microtubules by collecting the minus ends at the MTOC, the plus ends are found at the periphery. Motor proteins recognize microtubule polarity, and each motor protein moves in a characteristic direction; kinesin

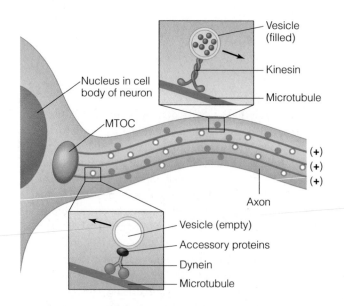

Figure 5.7 Vesicle traffic in a neuron Vesicle traffic depends on the polarity of the microtubules. Kinesin carries vesicles of neurotransmitters to the synapse, whereas dynein carries empty vesicles back to the MTOC.

moves along the microtubule in the plus direction, whereas dynein moves in the minus direction.

The polarity of the microtubules and the directional movement of the motor proteins allow cells to transport cargo to the right place. Consider how a neuron uses this network to transport neurotransmitter vesicles (Figure 5.7). Kinesin can pick up vesicles filled with neurotransmitters in the cell body, and walk along microtubules toward the plus ends at the synapse. Once the vesicles release their neurotransmitters, endocytosis returns empty vesicles to the cell. Dynein then carries the endocytic vesicle to the cell body, moving along the microtubule toward the minus end. This simple example illustrates why directional movements of neurosecretory vesicles are necessary for nerve function. Most cells possess countless types of vesicles that need to be transported to many locations. How do cells ensure that each of these diverse vesicles goes to the correct location? At least part of the answer lies in the structural diversity of motor proteins themselves. Large gene families encode multiple isoforms of kinesin, dynein, and their respective regulatory proteins. Each combination of isoforms imparts different transport characteristics.

Kinesin and dynein move along microtubules

Although kinesin and dynein are unrelated proteins, they work in similar ways. Both undergo conformational changes, where they stretch out to grab a tubulin dimer, then bend to pull themselves along the microtubule. Likewise, in both, the structural changes in the motor protein are fueled by ATP hydrolysis, the rate of movement of kinesin and dynein along the microtubule is determined primarily by the ATPase domain of the proteins, and regulatory proteins that associate with the motor protein control the rate of movement. Despite these similarities, kinesin and dynein have important differences that affect how cells use them to move along microtubules.

Let's first consider the structure and function of kinesin. Each kinesin molecule has a long neck, a fanlike tail, and a globular head that possesses ATPase activity. The tail is responsible for attaching to cargo, whereas the head attaches to the microtubule. Phylogenetic analyses have revealed a very large and diverse family of kinesins. Some members of the kinesin superfamily are active as monomers. Other kinesins assemble into dimers, either homodimers or heterodimers. These kinesin dimers may in turn interact with regulatory proteins called *kinesin-associated proteins*. Some kinesin-associated proteins can alter the kinetics of movement, such as the rate of ATP hydrolysis, while some influence the type of cargo kinesin binds. Many of these kinesin-associated proteins are themselves members of multigene families, which enable cells to fine-tune microtubule-based movements.

Like kinesin, dynein has a globular head, a neck, and a tail. Dynein is larger than kinesin, and can move along microtubules about five times faster. The many isoforms of dyneins fall into two classes: cytoplasmic and axonemal. *Cytoplasmic dyneins* are dimers of two identical subunits (heavy chains) with a number of associated smaller proteins. The dynein heavy chains possess the ATPase activity, and mediate binding to the microtubule. Unlike kinesin, dynein does not attach directly to vesicles. Instead, large multiprotein complexes of accessory proteins link dynein to its cargo, providing another layer of regulation of microtubule movement. *Axonemal dyneins* are the driving force behind movements generated by cilia and flagella.

Cilia and flagella are composed of microtubules

Cilia and flagella are similar structures with diverse roles in animal physiology. For example, flagella propel sperm toward the egg, while cilia allow epithelial cells to push mucus over the cell surface. Cilia differ from flagella in their arrangement and the way they move. Flagella normally occur singly or in pairs, whereas cilia are more abundant. In addition, flagella move in a whiplike manner, whereas cilia move with a wavelike motion. Microtubules in cilia and flagella are arranged into a structure called an *axoneme*, which is wrapped in an extension of the plasma membrane in the form of a membranous sheath.

A cross-section through a flagellum reveals a structure that resembles a wagon wheel (Figure 5.8). At the hub of the wheel are two single microtubules interconnected by a protein bridge. Around the edge are nine pairs of microtubules or doublets, connected to each other by the protein nexin. Protein spokes then radiate from the two singlets toward the nine doublets. Almost 10 years before microtubules were first identified, this "nine + two" arrangement in axonemes was seen to underlie the structure of flagella and cilia.

How does dynein power microtubule movement in cilia and flagella? Each doublet has a series of dynein motors that extend toward the neighboring doublet. At rest, the dynein sits inactive in this structure. When the cell receives a signal, protein kinases phosphorylate critical proteins associated with dynein to activate the ATPase. Once activated, dynein walks along the neighboring microtubule toward the minus end of the microtubule located at the base of the axoneme. The waving of cilia and whipping of flagella result from asymmetric activation of dyneins on opposing sides of the axoneme. When dyneins on one side of the axoneme are activated, the tip of the flagellum bends in that direction. These cycles of activation and inactivation of dynein along the entire length of the axoneme generate movement. If all of the dyneins were activated simultaneously, no movement would occur.

Table 5.1 summarizes some of the important roles microtubules play in diverse physiological functions.

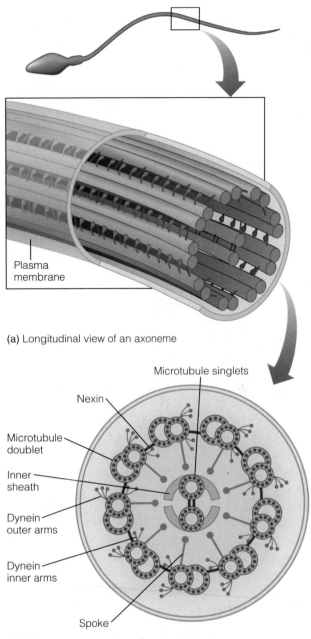

(a) Longitudinal view of an axoneme

(b) Enlarged cross-section of an axoneme

Figure 5.8 Structure of the flagellum The tail of a sperm is constructed from microtubules arranged into a complex network called an axoneme. The core structure is composed of nine doublets of microtubules, connected by the linker protein nexin. Radial spokes extend from this outer ring toward a central pair of single microtubules. Dynein arms extend from one doublet to the adjacent doublet.

Microfilaments

Microfilaments are the other type of cytoskeletal fiber used in movement. Like microtubules, microfilaments play important roles in the transport of vesicles throughout cells. In addition, microfilament-based movement also allows cells to change shape and move from place to place. The elements of microfilament-based movement, actin and its mo-tor protein myosin, are found in all eukaryotic cells; the organization of these elements enables diverse types of cellular movement. In some cases, cellular movement arises simply from the polymerization of actin. More often, however, actin-based movement involves myosin. Let's look at the many ways in which microfilaments drive movement.

Table 5.1 Microtubules and animal physiology.

Cellular process	Physiological function
Cytokinesis	Development and growth: All cells need to divide, and microtubules ensure that chromosomes are equally divided after mitosis.
Axon structure	Nervous system: Microtubules support the long axons.
Vesicle transport	Hormones and cell signaling: Microtubules carry hormones from sites of synthesis to sites of release.
Pigment dispersion	Adaptive coloration: Microtubules control the distribution of pigment granules throughout the cell to affect animal color.
Flagellar movement	Reproduction: Flagella allow sperm to swim toward the egg.
Ciliary movement	Respiration, digestion: Cilia propel mucus and other fluids over the epithelial surface.

Microfilaments are polymers of actin

Microfilaments are composed of long strings of the protein actin. These actin monomers are called *G-actin*, because of the globular structure of the protein. When G-actin assembles into filaments, however, it is referred to as *F-actin* (Figure 5.9).

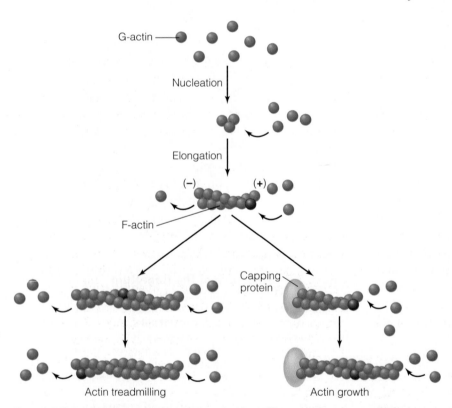

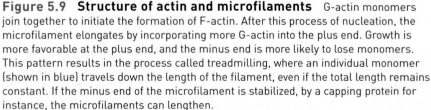

Figure 5.9 Structure of actin and microfilaments G-actin monomers join together to initiate the formation of F-actin. After this process of nucleation, the microfilament elongates by incorporating more G-actin into the plus end. Growth is more favorable at the plus end, and the minus end is more likely to lose monomers. This pattern results in the process called treadmilling, where an individual monomer (shown in blue) travels down the length of the filament, even if the total length remains constant. If the minus end of the microfilament is stabilized, by a capping protein for instance, the microfilaments can lengthen.

Actin can spontaneously assemble and disassemble without an energy investment. It polymerizes spontaneously when its concentration is above a threshold C_c. Each actin filament can grow from both plus and minus ends, but growth is six to ten times faster at the plus end. If the growth at the plus end exactly balances the shrinkage at the minus end, the total length of the microfilament is constant. However, if you were to follow the position of an individual actin monomer, you would see it move progressively from the plus end toward the minus end. This process is called *treadmilling*. As with microtubules, accessory proteins can modulate the rate of microfilament growth. One way that a cell increases the length of a microfilament is by stabilizing the minus end, preventing it from disassembling. To do that, cells use *capping proteins* that bind on the end of microfilaments to stabilize the structure.

Cells can arrange microfilaments in many ways, often with the help of actin-binding proteins that cross-link microfilaments (Figure 5.10). Microfilaments can be arranged in tangled networks, linked together by long, flexible actin-binding proteins such as *filamin*, or aligned in parallel into stiff bundles, cross-linked by short actin-binding proteins such as *fascin*. Actin bundles run throughout the cell, providing support. In some instances, these stiff actin fibers push the margins of the cell outward. For

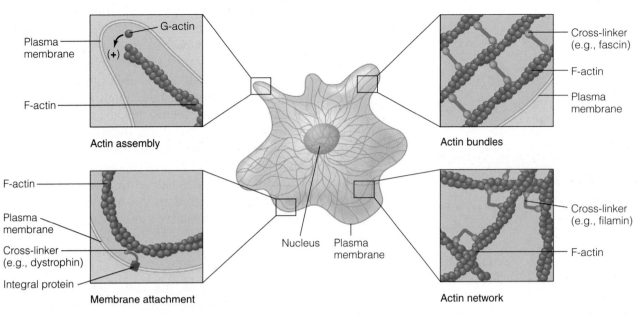

Figure 5.10 Actin networks Actin microfilaments can be arranged in many different conformations, often using cross-linking proteins for stabilization. Microfilaments can grow from their plus ends, causing cellular extensions. Actin bundles form when parallel arrays are cross-linked together. The microfilaments can be attached to integral membrane proteins by cross-linking proteins such as dystrophin. Actin can also be arranged into complex networks stabilized by cross-linking proteins such as filamin and fascin.

example, they provide the foundation for microvilli, the fingerlike extensions of digestive epithelia. The bundles and networks of microfilaments comprising the actin cytoskeleton are connected to the plasma membrane by specific anchoring proteins such as *dystrophin*.

Actin polymerization can generate movement

Although most types of microfilament-based movement rely on myosin, the actual polymerization of actin can mediate some forms of movement. While biologists do not yet fully understand how it works, actin polymerization is important in two kinds of amoeboid movement in animals. **Filapodia** are thin rodlike extensions of cells formed by actin fibers. Cells build filapodia for many purposes. For example, nerve cells use filapodia to make physical contact with neighboring cells, which is an important step in the embryonic development of the nervous system. Digestive epithelia use filapodia to build microvilli, protrusions that increase the surface area of the plasma membrane. In contrast, some metazoan cells move using actin-based extensions called lamellipodia. **Lamellipodia** resemble the *pseudopodia* found in

protists, but they are thinner and more sheetlike. The nature of the amoeboid protrusions in animals depends upon how the newly synthesized microfilaments are integrated into fibers. Filapodia result when the microfilaments are limited to simple fibers. Lamellipodia arise from sheetlike networks of microfilaments.

In a stationary cell, the actin network extends around the cell's periphery, attached at many points to plasma membrane receptors. When this cell is induced to move, it protrudes a region of the membrane forward. Underneath the plasma membrane, the plus ends of the microfilaments rapidly incorporate G-actin, pushing the membrane forward. At the trailing edge of the cell, the minus ends lose G-actin monomers. Actin-binding proteins regulate actin polymerization, and consequently amoeboid movement. At the leading edge, the protein *profilin* binds to free G-actin monomers, helping them integrate into the plus end of the microfilament. Another protein, *cofilin*, however, breaks microfilaments at the trailing edge to trigger disassembly.

Sperm also use actin polymerization during fertilization (Figure 5.11). The process of fertilization depends on the sperm's ability to control the growth of its actin cytoskeleton toward the egg.

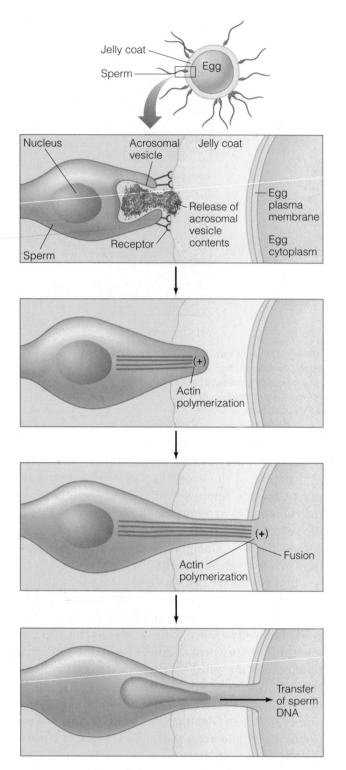

Figure 5.11 Acrosome of sperm Once the sperm find the egg, activation of membrane receptors in the sperm triggers the exocytosis of the acrosomal vesicle and the polymerization of microfilaments. The acrosomal enzymes help dissolve the physical barriers around the egg. The growing microfilaments push the sperm membrane through the jelly coat into contact with the egg plasma membrane. After membrane fusion, the sperm DNA moves into the egg to complete fertilization.

When a sperm encounters an egg, it uses surface receptors to form a tight bond with the egg's outer surface. Activation of these receptors triggers the formation of a structure called an **acrosome.** Within the acrosome, a vesicle full of hydrolytic enzymes is pushed to the cell surface. When it binds to the sperm plasma membrane, exocytosis of the acrosomal vesicle helps break down the egg's jelly coat. The sperm then uses actin polymerization to push an extension of the sperm plasma membrane through the softened jelly coat. Once the sperm plasma membrane fuses with the egg plasma membrane, the nuclear DNA of the sperm can be transferred into the egg.

Actin uses myosin as a motor protein

Although some cells use actin polymerization to generate movement, in most situations microfilaments are used in combination with myosin. Different arrangements of actin and myosin enable cells to transport vesicles and organelles, change shape, and even move from place to place. As was the case with microtubule-based movements, diversity in both the motor protein and associated regulatory proteins provides cells with the regulatory precision needed to control intracellular traffic. Many aspects of actin- and myosin-based movement are similar throughout eukaryotes. For example, muscle uses a unique arrangement of actin and myosin, in combination with novel isoforms of myosin and its regulatory proteins. Let's begin by examining myosin structure and consider how it controls movement.

The myosin gene family of eukaryotes is very large, with at least 17 different classes of myosins (I–XVII) distinguished by differences in their structural properties. The most common myosins studied in animals are in classes I, II, and IV. Myosin II is sometimes called *muscle myosin*, although it also occurs in nonmuscle tissues. Myosins I and V are most important in intracellular traffic. Most animals possess multiple isoforms of myosins within each class, adding to the repertoire of myosin functions available in animal cells.

Despite their structural differences, each myosin isoform shares a general organization, with a head, a tail, and a neck (Figure 5.12). The head possesses ATPase activity, which provides the energy for movement. The tail allows myosin to bind cargo, such as vesicles, organelles, or even the plasma membrane. In addition, the tail struc-

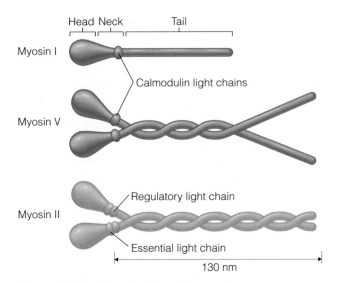

Head Neck   Tail

Myosin I

Calmodulin light chains

Myosin V

Regulatory light chain

Myosin II

Essential light chain

130 nm

Figure 5.12 **Myosin structures** Each myosin isoform possesses a catalytic head, a regulatory neck, and a tail region that interacts with other proteins. Regulatory proteins, such as light chains and calmodulin, can bind the neck region. Differences in structures of myosin and its regulatory proteins account for the specific properties of each isoform. Myosins I and V are used primarily in intracellular traffic. Myosin II is involved in cytokinesis and muscle contraction.

ture of some myosin isoforms can cause the individual myosin proteins to assemble into dimers. Whereas myosin I remains as a monomer, both myosin II and myosin V normally dimerize. The neck regulates the activity of the myosin head directly, and also mediates the effects of proteins that associate with the neck, known as **myosin light chains**. Myosin II, for example, has two different myosin light chains: *essential light chain* and *regulatory light chain*. Myosin light chains are regulated by reversible phosphorylation. Phosphorylation by **myosin light chain kinase (MLCK)** may alter the catalytic activity of the myosin head or induce a structural change that permits myosin to interact with actin. Many of the hormones that regulate myosin function either target MLCK or **myosin light chain phosphatase (MLCP)**, which dephosphorylates the myosin light chain.

The sliding filament model describes actino-myosin activity

Despite the great diversity in myosin, the basic mechanism that defines its interaction with microfilaments is shared by all isoforms. Myosin, like all the motor proteins we have discussed, is an ATPase that converts the energy released from ATP hydrolysis into mechanical energy. To understand this process we must consider both the chemical events associated with the enzymatic head of the myosin, as well as the structural changes throughout the myosin that culminate in movement. The two processes are integrated in the **sliding filament model**. This general model, first proposed almost 60 years ago by Hugh Huxley, shows how a myosin head walks along an actin polymer. This model can be used to explain all the different types of movement mediated by myosin. For example, a model involving a single myosin can be used to describe vesicular transport. The sliding filament model can also be used to describe how myosin and actin interact during muscle contraction, discussed later in this chapter.

Many of the principles explained by the sliding filament model can be illustrated through the following analogy. Imagine a rope stretching across the floor of a room. Now think about how you would pull yourself across the room using your arm. You start by extending your arm forward to grasp the rope, then bend your extended arm, pulling yourself forward. Next, you release the rope, extend your arm, grasp the rope again, and bend your arm. As you make your way across the room, your arm undergoes cycles of extension, grasping, and bending. Although each part of the cycle costs energy, the most demanding step in the cycle is when you bend your arm to pull yourself forward. In the sliding filament model, myosin acts very much like your arm, and actin is the equivalent of the rope. The myosin molecule extends by straightening its neck, pushing the head forward. The myosin head then forms a bond with actin, just as your hand grasps the rope. This strong interaction between myosin and actin is called a **cross-bridge**. Myosin bends, pulling the actin toward its tail. This step is called the **power stroke**. The *cross-bridge cycle* includes the formation of the cross-bridge, the power stroke, and the return to the resting, unattached position.

The mechanical changes in the cross-bridge cycle are driven by chemical and structural changes occurring within the myosin catalytic head (Figure 5.13). As previously discussed, myosin is an ATPase; the breakdown of ATP provides the energy for the mechanical changes. At the beginning of the cycle, myosin is tightly bound to actin and the ATP binding site is empty. If no ATP is available, the myosin remains firmly attached. However, once ATP binds, myosin loses its affinity for actin, and the cross-bridge is broken.

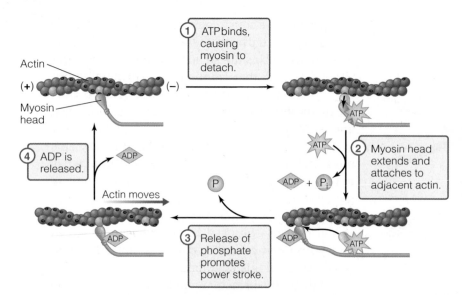

Figure 5.13 Sliding filament model In this figure, we follow a single myosin head as it progresses through a cross-bridge cycle. In the absence of ATP, the myosin head remains attached to the microfilament. Once ATP binds (step 1), myosin releases the microfilament. ATP hydrolysis induces myosin to extend toward the plus end of the microfilament (step 2), although the energy remains trapped in the myosin head. Upon release of the phosphate, the stored energy is used to bend myosin, pulling the filament back in the power stroke (step 3). Once the movement is complete, ADP is released (step 4) and the ATP binding site remains vacant until ATP binds to initiate another cross-bridge cycle.

Release of actin activates the myosin ATPase to break ATP down to ADP and phosphate. The hydrolysis of ATP causes myosin to extend forward to grasp further up the actin microfilament. (Although the ATP molecule within the myosin head has been chemically changed to ADP and phosphate, the energy that had been stored within the ATP remains stored within the myosin head as an energy-rich conformation.) Once myosin binds again, it first releases phosphate and then ADP. Upon phosphate release, myosin uses the stored energy to pull the actin microfilament in the power stroke. The myosin head remains attached to the actin until another ATP molecule finds its empty nucleotide-binding site and the cycle repeats. If no ATP is available, myosin remains firmly attached to actin, creating a condition known as **rigor**. When an animal dies, the ATP levels decline and muscles become locked in *rigor mortis*.

The actual movement that happens within the cell during a cross-bridge cycle depends upon the structural arrangements of actin and myosin, specifically which of the two is free to move. Returning to our earlier analogy, if the rope is tied to the wall, your arms pull you across the room. However, if the rope is not attached to the wall, your arm actions move the rope. Within the cell, actino-myosin movement depends on which of the elements, actin or myosin, is immobilized. If the actin microfilament is immobile, then myosin walks along the microfilament. This is analogous to myosin carrying a vesicle throughout the cell. Conversely, if myosin is immobile, the actin filament moves. In some cases, myosin is attached to the plasma membrane; in this situation, cross-bridge cycling pulls the actin microfilament over the surface of the plasma membrane. This arrangement allows cells to change shape. We will consider a third scenario later in this chapter when we discuss how the sliding filament model applies in muscle, where both the actin and myosin are organized into a three-dimensional superstructure.

Myosin activity is influenced by unitary displacement and duty cycle

The sliding filament model provides the context for understanding two features of actino-myosin–based movement: duty cycle and unitary displacement. These properties are most easily understood using the myosins involved in intracellular trafficking as an example.

Unitary displacement corresponds to the distance myosin steps during each cross-bridge cycle. Returning to our rope-climbing analogy, the unitary displacement is the distance you are able to move with each cycle of release, extend, grasp, and pull. In this analogy, the unitary displacement depends on the length of one's arm. With myosin, the step size depends on the length of the neck. Optical studies show that the actual distance moved with each step is not fixed; for example, the unitary displacement of a myosin V monomer may range anywhere from 5 nm to a maximum distance of about 20 nm. The myosin V dimer uses both of its monomers in tandem, walking along actin with an average unitary displacement of about 36 nm. This distance is related to an important structural characteristic of the actin microfilament.

To understand the relationship between unitary displacement and actin structure, consider the following analogy. Think of the actin filament as a spiral staircase, with each step representing an actin monomer. You, acting as myosin, have the challenge of climbing the stairs from the *outside of the staircase*. You can only use your two arms to climb. If you climbed the staircase one step at a time, your travels would carry you up the staircase in a spiral. How would your strategy change if you needed to stay *on the same side of the staircase* as you ascended? You would have to reach straight up as high as the stair directly above. This distance reflects the *period* of the spiral. Like the spiral staircase, microfilaments are spirals, twisted into a helix with a period of 36 nm (Figure 5.14). Because myosin walks with an average unitary displacement of 36 nm, it remains on the same side of the spiral as it travels along the microfilament. If it had a shorter or longer unitary displacement, it would spiral around the microfilament, creating a problem for myosin carrying a large vesicle or organelle, as its spiral trajectory would complicate movement through the dense cytoskeletal network. As you will see later in this chapter, muscle myosins do not have this 36-nm unitary displacement; nonetheless, they avoid these problems in other ways.

The second parameter that describes myosin activity is **duty cycle**, the proportion of time in each cross-bridge cycle that myosin is attached to actin. Most nonmuscle myosins have duty cycles of about 0.5. This means that myosin is tightly bound to actin for only half of each cross-bridge cycle. Why is duty cycle significant? Imagine climbing that spiral staircase using only one arm. If you released your grasp to reach the next step, you'd fall. Likewise, if vesicles were carried along microfilaments using only a single myosin head, they would float away from the actin track when the myosin reached the point in the cross-bridge cycle where it released actin.

Vesicles and organelles avoid falling off the microfilament in two ways. First, vesicles use dimers of myosin. When one myosin head attaches, the other can detach and extend forward, functioning much like you did when you climbed the staircase with two arms. The duty cycle of 0.5 would mean that each arm could only hold the stair half the time. Clearly, climbing the stairs or walking along a microfilament this way requires exquisite coordination, with the two myosins working in perfect synchrony. If at any point neither of the heads was attached, the vesicle would fall off the microfilament. In reality, the two heads are not perfectly coordinated and a second mechanism is required to ensure that the vesicle remains attached. Vesicles further reduce the risk of falling off the microfilament by engaging multiple myosin dimers. Imagine how much easier it would be to climb that spiral staircase if you could use two arms as well as two legs.

The sliding filament model was an important advancement in our understanding of how myosin moves along actin. Its general features apply to most types of actino-myosin activity in all eukaryotes. However, the exact values of duty cycle, unitary displacement, and other kinetic features of actino-myosin change in different situations. For example, the kinetics differ depending on whether myosin and actin are immobilized or free to move. The mechanical properties of actino-myosin influence the enzymatic features, and vice versa.

Actin and myosin perform diverse and important functions in animal cells (Table 5.2). Many of their responsibilities in animal cells are little different from their roles in other eukaryotes. Over hundreds of millions of years, animals evolved novel isoforms of myosin, and arranged actin and myosin in different ways, providing the foundation for a specialized contractile tissue: muscle.

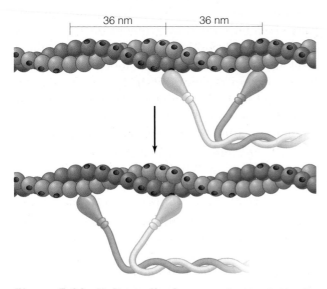

Figure 5.14 Unitary displacement Myosin V walks along the actin filament in steps of about 36 nm, which corresponds to the period of the actin filament.

Table 5.2 Actin and myosin function in animal physiology.

Cellular process	Physiological function
Vesicle transport	Hormones and cell signaling: Microfilaments carry hormones from sites of synthesis to sites of release.
Microvilli	Digestion: Actin supports the fingerlike extensions of the cells of the intestinal epithelium.
Amoeboid movement	Cardiovascular physiology: Blood cells use amoeboid movement to invade damaged tissue.
Skeletal muscle contraction	Locomotion: Muscles provide the contractile force for movement. Respiratory physiology: Trunk muscles help move air over the respiratory surface.
Cardiac muscle contraction	Circulatory physiology: Cardiac muscles pump blood.
Smooth muscle contraction	Circulatory physiology: Vascular smooth muscle controls the diameter of blood vessels. Digestion: Visceral smooth muscle forces food down the intestinal lumen.

⊙ | CONCEPT CHECK

1. Compare and contrast microtubules and microfilaments in terms of primary, secondary, tertiary, and quaternary structural levels.

2. What factors influence the assembly and disassembly of the cytoskeleton?

3. What is meant by polarity with respect to microfilaments and microtubules? Why is it important to structure and function?

4. Which plant alkaloids disrupt animal cytoskeleton function? What do they do for the plants? Why might they have no effect on the cytoskeleton in the plants?

Muscle Structure and Regulation of Contraction

Earlier in this chapter we discussed how the cytoskeleton and motor proteins mediate diverse types of intracellular and cellular movement. Animals use these same elements to build muscle cells, or **myocytes**. A "muscle," such as skeletal muscle or heart muscle, is composed of many types of cells, each of which contributes to muscle structure and function. In addition to the myocytes, which confer the contractile properties of muscle, there are also endothelial cells that make up capillaries, immune cells for defense, pluripotent stem cells to rebuild damaged myocytes, and fibroblasts to produce the extracellular matrix and connective tissue that holds the muscle together. In a heart, for example, there are more nonmuscle cells than muscle cells, though the larger myocytes make the greatest contribution to mass.

Muscles provide the contractile force needed in many multicellular tissues and physiological systems. We are most familiar with their role in animal locomotion, where skeletal muscles move the body trunk and appendages. However, muscles play many roles in animal physiology beyond locomotion. In the circulatory system, for example, muscles provide the pumping power of the heart and give blood vessels control over their diameter. In subsequent chapters, we will also discuss how muscles are used by the respiratory system to pump gases; by the digestive system to move food along the gut; and by the reproductive system to expel gametes and embryos.

The remainder of this chapter focuses on the *cellular aspects of muscle function*: how muscle cells are built, how they are controlled, and how the elements have been fine-tuned at the cellular level to achieve diversity in function. Although there is extraordinary diversity in the way muscles are constructed and used, some features are shared among all muscle types and species. First and foremost, the contractile elements of all muscles are polymers of myosin and actin. The myosin polymer forms the backbone of a multiprotein complex known as the **thick filament**. Analogous to the microfilaments of the cytoskeleton, muscle cells possess a **thin filament**, composed primarily of polymerized actin (Figure 5.15).

In most areas of cell biology, "myosin" refers to the motor protein itself. When physiologists discuss muscle, however, "myosin" refers to a hexamer consisting of two myosin II motor proteins, or **myosin heavy chains**, and four myosin light chains. About 150 myosins are collected together by the tail to create an assembly that resembles a bouquet of flowers; the thick filament is composed of two bouquets arranged end-to-end. The two ends of the thick filament appear bushy from the myosin heads extending outward, while the tails of the two bouquets are located in the center of the thick filament, in a region devoid of myosin heads. A thick filament is composed of about 300 myosin hexamers, providing 300 myosin heads on each end.

Thin filaments are similar in structure to cytoskeletal microfilaments, but they are constructed with different actin isoforms. Microfilaments are polymers of β-actin; thin filaments are made from α-actin. As we learned earlier in this chapter, microfilaments constantly assemble and disassemble. In contrast, thin filaments are stabilized in a way that prevents spontaneous growth or shrinkage. Each thin filament is capped by *tropomodulin* at the minus end and *CapZ* at the plus end, preventing changes in length. In some muscles, thin filaments are decorated at regular intervals with the proteins **troponin** and **tropomyosin**, which mediate the interaction between actin and myosin, thereby regulating contraction.

Another feature that is shared among all muscles is the basis of interaction between the thick and thin filaments; the sliding filament model, discussed earlier in this chapter, applies equally well to all actin-myosin interactions. However, the application of the sliding filament model to muscle is more complicated because of the unique properties of muscle myosin, its arrangement into a thick filament, and the integration of thick and thin filaments into a three-dimensional lattice.

As you will see later in this chapter, animals use these basic elements to produce many types of muscles with unique structural and functional features. One important dichotomy in muscle biology is the distinction between smooth and striated muscle (Figure 5.16). Muscles such as cardiac and skeletal muscle have a striped appearance, giving

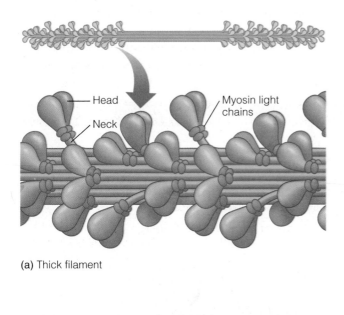

(a) Thick filament

(b) Thin filament

Figure 5.15 Thick and thin filaments Muscle is composed of thick filaments and thin filaments. **(a)** Thick filaments consist mainly of myosin molecules connected by the tail with heads extending radially. **(b)** Thin filaments are mainly actin, though numerous actin-binding proteins (not shown) influence thin filament function.

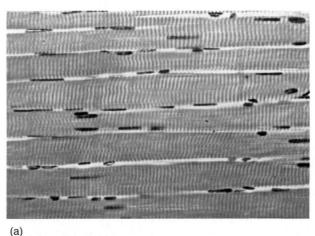

(a)

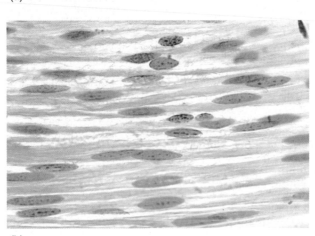

(b)

Figure 5.16 Smooth and striated muscle
(a) Cardiac and skeletal muscles are called striated muscle because of their striped microscopic appearance. **(b)** Blood vessels, respiratory tracts, and visceral linings possess muscle termed smooth muscle because it lacks striations.

rise to the name **striated muscle**. In contrast, the muscles that line blood vessels and viscera do not appear striped, and are called **smooth muscle**. The difference in microscopic appearance in these muscle types can be traced to the way thick and thin filaments are organized inside the cell. In the next section, we begin by discussing how striated muscle is constructed and regulated, returning to structure and function of smooth muscle later in this section. Although we focus on vertebrate muscles, most of the basic features apply equally well to invertebrates.

Structure of the Vertebrate Striated Muscle Contractile Apparatus

In the following discussion, we focus on the cellular basis of contraction in vertebrate striated muscle, which includes cardiac and skeletal muscles. Striated muscle can be used to describe the general features of muscle structure and regulation, but it also provides examples of functional diversity.

Striated muscle has been studied for nearly one hundred years by many researchers working on diverse models. This rich history has led to confusion in the way muscle types are described. A set of terms that is useful for distinguishing between muscle types of one species may be useless in distinguishing between muscle types in other species. Some of the other ways that animal physiologists categorize muscle into subtypes are summarized in Table 5.3.

Striated muscle thick and thin filaments are arranged into sarcomeres

Striated muscles arrange their thick and thin filaments in highly organized arrays. The end of each thick filament is surrounded by an array of thin filaments, typically six. This unit, called a **sarcomere,** is repeated in parallel and in series throughout the muscle cell. While the structure of the sarcomere is relevant to striated muscles, the principles of contraction apply broadly to all muscles.

The microscopic appearance of striated muscle is rooted in sarcomere structure (Figure 5.17). A protein plate called the **Z-disk** forms the end of each sarcomere. Thin filaments then extend from the Z-disk, with the minus end of the actin chain directed toward the center of the sarcomere. The double-headed thick filaments are arranged between Z-disks, spanning two opposing thin filament arrays. The region of a sarcomere where

Basis of category (muscle types)	**Distinction between muscle types**
Innervation (phasic and tonic)	Phasic (twitch) muscles have single innervation, whereas tonic muscles have multiple innervations. "Tonic" is sometimes (erroneously) used to describe the slowest twitch muscles that are continuously stimulated.
Rate of shortening (fast and slow)	Vertebrate skeletal muscles contract at different velocities, usually due to myosin isoform pattern.
Myosin isoforms (I, IIa, IIb, vs IIx/d)	Most vertebrate myosin heavy chains are formed from these four genes. Some lower vertebrates have more isoforms. Invertebrates have fewer myosin isoforms.
Metabolism (oxidative and glycolytic)	Fast-twitch skeletal muscles usually have few mitochondria and derive energy from glycolysis.
Myoglobin (red, white, and pink)	Slow-twitch oxidative fibers usually possess high levels of myoglobin, giving them a red appearance.
Morphology (fusiform and pinnate)	Myofibrils usually run perpendicular to the plane of contraction, but in pennate muscles the myofibrils run at an oblique angle.
EC coupling (synchronous and asynchronous)	Most striated muscles respond to a neural stimulus with a single contraction. Asynchronous muscles found in some invertebrates contract and relax repeatedly after a single stimulus.
Excitation (myogenic and neurogenic)	Myogenic muscles contract spontaneously, whereas neurogenic muscles contract in response to a nerve stimulus.

Table 5.3 Terminology used to classify striated muscle cell types.

thick filaments occur forms a dark region called the **A-band** (or **Anisotropic band**). The narrower **I-band** (or **Isotropic band**) region spans a Z-disk, and includes the portion of the thin filaments without overlap with thick filaments. The **M-line** is the central region of the sarcomere between the two minus ends of the thin filament. In this region, the thick filaments do not overlap with thin filaments.

Specific proteins maintain these structural relationships within the sarcomere. For example, *nebulin* runs along the length of the thin filament; the length of nebulin determines the length of the thin filament. The thick filament is held into position by the protein *titin*, which connects the end of the thick filaments to the Z-disks. Since the distance between the end of the thick filament and Z-disk changes with contraction, titin must be compressible. Although we discuss sarcomere features based on its two-dimensional microscopic

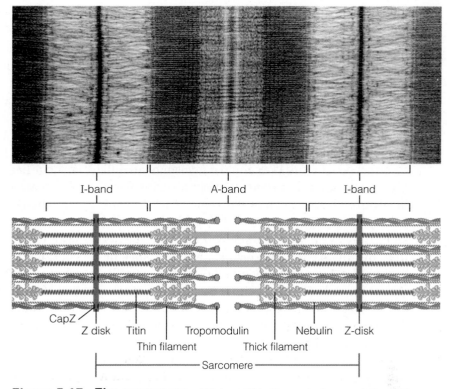

Figure 5.17 The sarcomere Thick and thin filaments, in association with structural proteins, comprise the sarcomere. Each thin filament is anchored into the Z-disk by the protein CapZ, and capped at the minus end by tropomodulin. Nebulin parallels the thin filament to establish the appropriate length of each filament. The thick filaments are held in position by titin, which anchors the thick filament to the Z-disk.

appearance, you should remember the three-dimensional arrangement of thick filament and thin filaments (Figure 5.18). The thin filaments are arrayed in a cylinder around the thick filament, while the thick filament is held at a constant location near the center of the thin filament array. In vertebrate striated muscle, six thin filaments surround each thick filament, each thin filament interacts with three separate thick filaments, and the resulting ratio of thick filaments to thin filaments is 1:2.

Myosin II has a unique duty cycle and unitary displacement

The sarcomeric structure, maintained by suites of proteins, ensures that bouquets of myosin heads are kept in a location where they are able to bind actin. The interaction between actin and myosin in muscle is very similar to the sliding filament model we discussed early in this chapter. However, the structural organization, coupled with unique properties of muscle myosin, complicates the simple model described earlier involving a single myosin head.

The distinct features of muscle actino-myosin activity are linked to the sarcomeric organization.

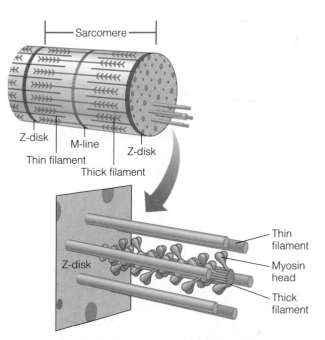

Figure 5.18 Arrangement of thick and thin filaments Within the sarcomere, thick filaments are surrounded by thin filaments. This arrangement ensures that myosin heads are able to find a microfilament at all times.

First, unlike the situation in vesicle traffic, when myosin detaches from actin it cannot drift away. Myosin heads on the thick filament are held in position opposite actin. Second, hundreds of myosin molecules are attached together in the thick filament. Consider these structural relationships in the context of duty cycle. If muscle myosin had the same duty cycle as vesicle myosins, roughly 0.5, then at any given time half of the myosins would be attached to actin. How could a myosin head pull the thin filament if dozens of other myosins were firmly attached to actin at the same time? In contrast to other myosins, muscle myosin II has a very short duty cycle, approximately 0.05. That means that during each cross-bridge cycle, a specific myosin head is physically attached to the actin filament for only 5% of the time. For the remainder of the cycle this myosin is unattached and therefore does not impede other myosins from pulling the thin filament.

Muscle myosin II activity is also unusual in its unitary displacement. Earlier in this chapter we discussed how a unitary displacement of 36 nm was critical for a vesicular myosin to walk along the one plane of an actin filament, much like an acrobat crosses a tightrope. In reality, muscle myosins behave much less like a tightrope walker than like an octopus pulling itself through a tube. Wherever the octopus reaches, it finds a wall to grasp. Since the thin filaments surround each thick filament, there is little risk of a myosin head failing to find a binding site on actin. As a result of the structural relationships between thick and thin filaments, myosin II is able to function with a much shorter unitary displacement, typically 5 to 15 nm. You can think of the molecular interactions in muscle actino-myosin activity as a series of myosin heads taking turns pulling along the thin filaments with small, quick tugs.

Sarcomeric organization determines contractile properties of the muscle cell

Thick filament movement is the sum of many individual cross-bridge events. Cross-bridges can only form where myosin heads are in a position that can contact the thin filament. Consequently, the degree of overlap between thick and thin filaments can influence contractile properties. For any muscle, the degree of overlap is reflected in the **sarcomere length**, measured as the distance between the Z-disks.

Most vertebrate striated muscles show a resting sarcomere length of about 2.0 μm. The amount of force generated by a sarcomere is maximal over this range because there is an optimal overlap between thick and thin filaments (Figure 5.19). Muscle cells can be stretched, however, changing sarcomere length enough to influence the degree of overlap. If a muscle cell is stretched beyond a sarcomere length of about 2.5 μm, some of the myosin heads near the midpoint of the thick filament cannot connect with the thin filament. If it is stretched to beyond about 3.5 μm, there is little overlap between thick and thin filaments; no cross-bridges can form and no shortening can occur. The contraction is also weakened if the sarcomere length is much shorter than about 2 μm. At this point, the thin filaments from adjacent Z-disks start to overlap, impeding efficient cross-bridge formation. Below a sarcomere length of about 1.65 m, thick filaments collide with the Z-disk and no further contraction is possible. Sarcomeric proteins such as titin and nebulin help maintain sarcomere lengths within a useful range.

When muscle cells are integrated into complex tissues, sarcomere length can be modulated to alter contractile properties. For example, cardiac myocytes are stretched when the heart fills with blood, but typically sarcomere lengths are shorter than the optimum length. When the volume of blood returning to the heart increases, as it may during exercise, the additional stretching increases sarcomere length, and allows the cardiomyocyte to

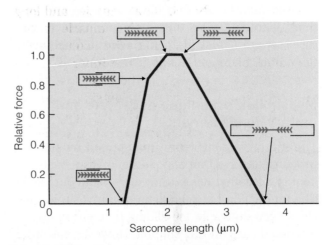

Figure 5.19 Sarcomere length-force relationship The ability of a sarcomere to contract depends upon the degree of overlap of thick and thin filaments. Maximal force can be generated within a narrow range of sarcomere lengths, characteristic of the muscle type. (Modified from Bers, 1991)

generate a stronger contraction. This phenomenon, known as Starling's Law, will be discussed in Chapter 8: Circulatory Systems.

Many of the processes we have discussed to this point can be understood in terms of the events happening in a single sarcomere. However, muscle cells incorporate hundreds or thousands of repeating sarcomeres into larger structures. A single continuous stretch of interconnected sarcomeres, called a **myofibril**, is usually 1 to 2 μm in diameter and stretches the length of the muscle cell. Although the myofibril is the contractile element within muscle cells, each type of muscle cell organizes its myofibrils in a particular three-dimensional pattern (Figure 5.20).

A cardiac muscle cell, or **cardiomyocyte**, possesses myofibrils that are typically about 100 sarcomeres in length. Thus, a typical mammalian ventricular cardiomyocyte is about 0.2 mm in length. Most vertebrate cardiomyocytes are individual cells, though some have undergone an additional round of the cell cycle (without cell division) and possess two nuclei.

A skeletal muscle cell, or **myofiber**, possesses much longer myofibrils, although they are much more variable in size and arrangement. Because myofibrils usually run the length of the muscle, they are short in small muscles and long in larger muscles. The smallest muscle in humans, the 1.3-mm-long stapedius, controls the movement of small bones in the inner ear. The longest muscle in humans is the sartorius, which stretches about 60 cm from the outside of the hip to the inside of the knee, winding around the thigh. The greater size of the skeletal myofiber is possible because it is produced by the fusion of many individual cells.

The differences in cellular organization of muscle cells can be traced back to the earliest stages of embryonic development, where muscle precursor cells are induced to differentiate into myocytes (see Box 5.2, Genetics and Genomics: Muscle Differentiation and Development).

The three-dimensional organization of sarcomeres and myofibrils influences the contractile properties of the striated muscle, such as force

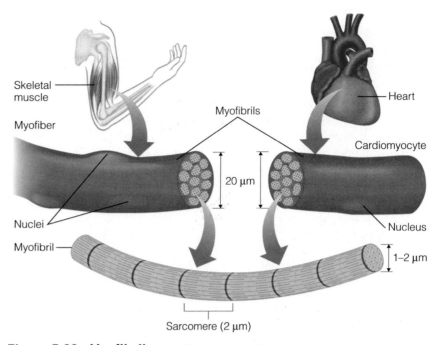

Figure 5.20 Myofibrils Myofibers and cardiomyocytes are used to construct the multicellular tissues we know as skeletal muscle and cardiac muscle. Myofibers, which are formed from the fusion of many different muscle cells, can use very long myofibrils and arrange many myofibrils in parallel. Cardiomyocytes, which are single cells, possess short myofibrils composed of about 100 sarcomeres. The myofibril is composed of sarcomeres connected end-to-end.

generation and contraction velocity. Each sarcomere is able to generate about 5 pN of force and can shorten about 0.5 μm; these properties are fairly constant among species and muscle types. Different types of muscle rely on specific arrangements of sarcomeres to carry out specific functions. A myofibril composed of 1000 sarcomeres in series would be about 2.5 mm long. It could shorten by about 0.5 mm, but only generate about 5 pN of force. What would happen if we arranged 1000 sarcomeres in parallel? The fiber would be only about 2.5 μm long and could shorten only 0.5 μm. However, it could generate 1000 times more force than the same number of sarcomeres arranged in series. These simple examples illustrate how anatomic variations can allow muscles to be optimized for different types of contraction: maximal shortening versus maximal force.

The contractile properties of muscle cells are determined by the molecular properties of the thick and thin filaments comprising the contractile element, the organization of the sarcomere, and the arrangement of the myofibrils. Variations in contraction properties help define distinct muscle types. We next turn our attention to the processes that control interaction between the sarcomeric proteins in muscle cells.

BOX 5.2 GENETICS AND GENOMICS
Muscle Differentiation and Development

It is difficult to discuss the origins of muscle diversity without also considering how muscle is made. Muscle synthesis is really two related processes: muscle differentiation, or myogenesis, and muscle development. Our understanding of muscle differentiation and development has benefited from research in animal model systems, including *Drosophila, C. elegans, Xenopus*, zebrafish, and mice, as well as cultured myoblasts. Each type of muscle follows its own path to the final phenotype. The control of muscle formation is best understood in skeletal muscle.

One reason why we understand muscle differentiation so well is that many of the processes can be studied in cell culture. Myoblasts from chickens and rodents are most useful because they can be grown for hundreds and thousands of "generations" without marked deterioration of their properties. Neonatal muscle, and to a lesser extent adult muscle, has a population of muscle precursor cells called **satellite cells**. These cells can be harvested and grown in culture. They rapidly proliferate but do not differentiate when grown under the right conditions. Most commonly, they are given a nutrient medium that is rich in fetal growth factors, such as TGF-β (transforming growth factor beta) and bFGF (basic fibroblast growth factor). If deprived of the growth factors, the cells begin to express their own signaling factors, such as IGF-II. These hormones induce the myoblasts to enter the myogenic program. Within the first day, the myoblasts begin to express a suite of proteins that act as transcriptional activators. These transcription factors, including proteins of the *myoD* and MEF families, in turn induce the expression of genes that encode the muscle-specific proteins, such as

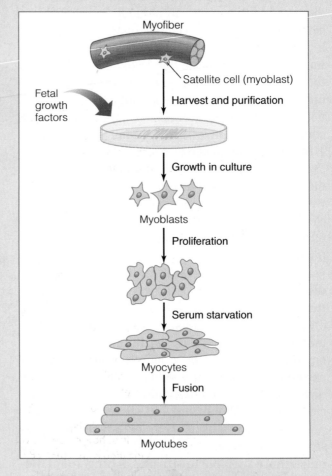

myosin II, α-actin, and troponins. Simultaneously, the hormone-induced pathways trigger individual myoblasts to line up in parallel and fuse together to form a multinucleated myotube. The myotubes have many of

Contraction and Relaxation in Vertebrate Striated Muscle

Muscle activity is a symphony of cellular processes, integrating membrane events with intracellular changes in ions and protein-protein interactions. Muscle activity is initiated by *excitation:* a depolarization of muscle plasma membrane, or **sarcolemma**. The translation of an excitatory signal at the sarcolemma into a stimulation of contraction is called **excitation-contraction coupling**, or EC coupling: a combination of physical and chemical changes within the myocyte that elevate calcium concentration ([Ca^{2+}]). This increase in intracellular

[Ca^{2+}] activates the actino-myosin machinery to induce *contraction*. *Relaxation* ensues when the [Ca^{2+}] falls to resting levels, which is only possible when the sarcolemma repolarizes.

In the previous section we discussed the structure of the contractile elements within muscle. Thus, we begin our discussion of the control of muscle activity by examining the processes that control how thick and thin filaments interact during contraction and relaxation. We will return to excitation and EC coupling later in this section. But before we begin, recognize that any general discussion of how muscle contraction occurs is challenged by diversity in muscle structure and the structural and functional features of myofibers, including the ability to contract. While in vitro myogenic models are useful for many purposes, the process of myogenesis is much more complicated in real animals.

For a brief time after fertilization, each cell within the embryo has the potential to become any type of cell. A subset of these embryonic stem cells takes the first step toward muscle formation to become myoblasts, which proliferate but do not yet express the genes required to make them into muscle. During this stage of embryogenesis, myoblasts use lamellipodia to crawl over other cells, following hormonal trails through the embryo. At their final destination, the community of cells produces a number of regulatory factors, including the protein called *sonic hedgehog*. The regulatory factors induce the migrating myoblasts to stop traveling and enter myogenesis. (The same regulatory proteins coordinate the differentiation and development of other cell types as well.) As occurs in cell culture models, the myoblasts initiate the same cascade of transcription factor activation and muscle gene expression. Once the cells commit to myogenesis, they continue to sense the surrounding hormones and neurotransmitters in the diverse multicellular neighborhood. The complex combination of stimulatory and inhibitory signals enables each muscle to develop the appropriate contractile phenotype. These pathways of differentiation and development control how much muscle is made, its fiber type, and its location within the body.

These pathways of muscle differentiation and development begin early in embryogenesis but continue to play important roles in adults. In an adult muscle, about 5% of the muscle nuclei are found in the satellite cells that are attached to the myofiber surface. These cells act as reserve cells, helping to repair and remodel muscle. When muscle is damaged, satellite cells sense chemical signals released from damaged muscle, migrate to the lesion, and enter myogenesis. Within hours, they activate the myogenic transcription factors to trigger expression of muscle-specific genes. The differentiating myoblasts can fuse with other myoblasts to form new myofibers, or become incorporated into adult muscle. Muscles can also activate myoblasts to remodel muscle. Muscles can grow bigger either by inducing each muscle cell to increase in size (**hypertrophy**) or by incorporating more myoblasts into the mature muscle (**hyperplasia**). The relative importance of these two mechanisms of muscle growth depends on the situation. Cardiac muscle grows under many conditions where cardiac output is elevated for long periods. For example, cardiac mass can grow by as much as 30% in response to exercise or hypertension. In most situations, cardiac mass increases by hypertrophy, with cardiomyocytes growing in size. Much of the early growth of fish is hyperplasic, as additional muscle fibers form within the trunk.

Although much of our understanding of muscle formation comes from work on model organisms, by studying these pathways we better understand the relationships between evolution and development, providing explanations for diversity in muscle phenotype.

References

○ Parker, M. H., P. Seale, and M. A. Rudnicki. 2003. Looking back to the embryo: Defining transcriptional networks in adult myogenesis. *Nature Reviews in Genetics* 4: 497–507.

○ Snider, L., and S. J. Tapscott. 2003. Emerging parallels in the generation and regeneration of skeletal muscle. *Cell* 113: 811–812.

function. The variability among muscle types and species is due in large part to the diversity in the genes that encode muscle proteins. Each animal can draw on a suite of protein isoforms to change the contractile properties of muscle. This genetic repertoire provides an animal with the ability to make different muscle types, such as fast-twitch and slow-twitch skeletal muscles. Isoform switching allows individual animals to remodel muscles in response to physiological changes, such as exercise, and environmental conditions, such as temperature. Evidence of evolutionary divergence in muscle genes is also apparent when you compare the muscle contractile properties of different species. The athleticism of some animals is due, in part, to the molecular specializations of contractile proteins.

Thin filament proteins confer Ca^{2+} sensitivity in striated muscle

Striated muscles contract when Ca^{2+} levels increase within the myofiber. The Ca^{2+} signal is transmitted to the contractile apparatus by the thin filament proteins troponin and tropomyosin (Figure 5.21). When $[Ca^{2+}]$ is low, the troponin-tropomyosin complex sits on the thin filament in a position that blocks actin's binding site for myosin.

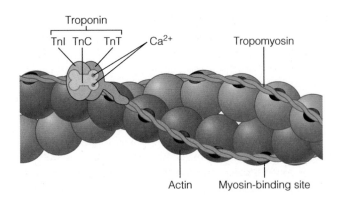

Figure 5.21 Troponin and tropomyosin
Troponin, a trimer of TnC, TnI, and TnT, binds to every seventh actin on the thin filament. Tropomyosin extends from troponin over seven actins. Its position on the thin filament in relation to the myosin binding site either permits or inhibits actino-myosin activity.

When [Ca^{2+}] rises, they roll out of the way, allowing myosin to bind to actin to initiate the cross-bridge cycle. To understand how these processes are regulated, we must consider in more detail the structures of troponin and tropomyosin, focusing on how they respond to [Ca^{2+}].

The troponin component is composed of three subunits: TnC, TnI, and TnT. Each subunit con-

tributes to Ca^{2+}-dependent regulation of contraction. The first subunit, TnC, is the Ca^{2+} sensor (the *C* in TnC stands for *calcium*). It is a member of a large family of Ca^{2+}-binding proteins. TnC is a dumbbell-shaped protein with four Ca^{2+}-binding sites, two in the N-terminal domain and two in the C-terminal domain. The two C-terminal sites have a very high Ca^{2+} affinity and are probably always occupied. They are often termed structural sites because they help physically anchor TnC into the troponin complex. The N-terminal Ca^{2+} binding sites trigger contraction, and are therefore referred to as the *regulatory sites*. TnI is the subunit that links troponin to actin, thereby inhibiting actino-myosin ATPase (*I* is for *inhibitory*). The third troponin subunit is TnT, an elongated protein that binds tropomyosin (*T* is for *tropomyosin*). Tropomyosin is a double-stranded protein that extends over approximately seven actin monomers and blocks the myosin-binding sites on actin. The entire troponin-tropomyosin complex acts as a unit, shifting its position on the thin filament in response to Ca^{2+} (Figure 5.22).

In a typical resting muscle, Ca^{2+} is maintained at a very low concentration, typically below 200 nM. At this concentration, the TnC regulatory sites are unable to bind Ca^{2+}. With the regulatory sites vacant, TnC assumes a particular structure that restricts its interactions with TnI. As a result, TnI binds actin, and the entire troponin-tropomyosin complex remains in an inhibitory position. When the muscle is activated, cytoplasmic Ca^{2+} levels can rise 100-fold. This allows the regulatory sites to bind Ca^{2+}, causing a structural change within TnC that exposes a hydrophobic region in the protein. Once uncovered, the hydrophobic patch on TnC can bind a corresponding hydrophobic region in TnI. Strengthening the TnC-TnI interaction causes a weakening in the TnI-actin interaction, allowing troponin to slide into the groove in actin. The strong TnT-tropomyosin interaction ensures that troponin and tropomyosin move as a complex. In this position, myosin is now free to bind actin and induce actino-myosin ATPase activity. Cross-bridge cycling can continue as long

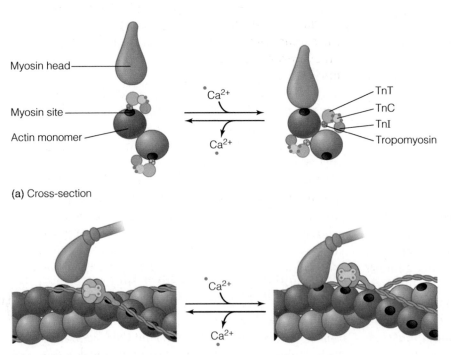

(a) Cross-section

(b) Longitudinal view

Figure 5.22 Regulation of actino-myosin contraction by thin filament proteins Calcium binding to the low affinity sites of TnC triggers a structural reorganization of troponin-tropomyosin, sliding it off the myosin-binding site of actin, into the major groove of the thin filament.

as the troponin-tropomyosin complex remains locked in this permissive position, and there is sufficient ATP to supply the actino-myosin ATPase.

The actino-myosin activity stops when [Ca^{2+}] falls to resting levels and the structural changes are reversed. The regulatory sites on troponin lose their Ca^{2+}. The TnC bends to hide its hydrophobic TnI binding site. TnI reestablishes its connection with actin, and the troponin-tropomyosin complex returns to its inhibitory position. The molecular processes involved in contraction are summarized in Figure 5.23.

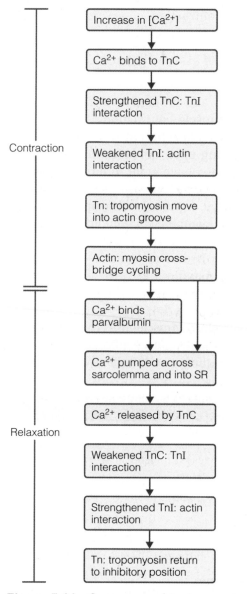

Figure 5.23 **Summary of ionic events in contraction** Contraction begins when the Ca^{2+} levels within the muscle cell cytoplasm rise in response to excitation. Relaxation begins when the cytosolic Ca^{2+} levels decline, through the actions of ion pumps.

This general model of Ca^{2+}-induced contraction applies to all striated muscles. However, there is a great deal of diversity in contraction kinetics. We attribute much of this diversity to the control of cytoplasmic [Ca^{2+}]. The strength of contraction depends on [Ca^{2+}] because it influences how many troponin-tropomyosin complexes are affected; the duration of contraction is influenced by how long [Ca^{2+}] remains elevated. Before discussing muscle excitation and the control of the Ca^{2+} transient, we will consider how muscle contraction kinetics are influenced by the properties of the contractile apparatus itself. Animals can build muscles with diverse functional properties by altering the composition of thick and thin filaments, and by using regulatory proteins to modify their kinetic properties.

The troponin-tropomyosin complex influences contraction kinetics

The troponin-tropomyosin complex plays a central role in the control of contraction. Animals possess multiple isoforms of these proteins, each with subtly different properties that influence contraction kinetics. By regulating the expression of these isoforms, an animal can fine-tune the regulatory properties of the muscle. Recent studies have shown how variation in the proteins of the troponin-tropomyosin complex impart unique kinetic properties that are well suited for specific muscle types or physiological circumstances. Animals draw upon these suites of thin filament proteins to create distinct muscle fiber types, or subtly alter the sensitivities to Ca^{2+}, pH, or temperature.

Muscle myofibrils differ widely in their sensitivity to Ca^{2+}. For example, mammalian cardiac muscle myofibrils are less sensitive to Ca^{2+} than are mammalian fast-twitch muscles. This difference in Ca^{2+} sensitivity is due in large part to the properties of TnC. Vertebrates possess two separate genes for TnC; one isoform is expressed in both slow-twitch skeletal muscle and cardiac muscle (s/cTnC) and the other in fast-twitch skeletal muscle (fTnC). The main difference between the isoforms is in the N-terminal domain. In the s/cTnC, an amino acid insertion inactivates the first of the two regulatory Ca^{2+} binding sites. The two isoforms demonstrate important differences in Ca^{2+} binding that influence muscle properties. The fTnC has a higher Ca^{2+} affinity than s/cTnC, and consequently fast-twitch muscle is more responsive to Ca^{2+}.

These differences in TnC also influence how vertebrate heart and muscle are affected by temperature. At low temperature, heart myofibrils become less sensitive to Ca^{2+}, whereas the Ca^{2+} sensitivity of skeletal muscle is unaffected. Much of the difference in the response to temperature can be traced back to molecular properties of TnC isoforms. Cold temperatures impair the ability of s/cTnC to bind Ca^{2+}, whereas Ca^{2+} affinity of fTnC is less affected. Transgenic mice that express fTnC in the heart exhibit the thermal properties of fast-twitch skeletal muscle. The Ca^{2+} sensitivity of myofibrils from these hearts was much less affected by cold temperature.

Troponin isoforms are also responsible for the pH sensitivity of hearts. Cardiac muscle of adult mammals is very sensitive to low pH, whereas fetal cardiac muscle is much less affected. Low pH reduces the Ca^{2+} affinity of the contractile apparatus, such that it needs more Ca^{2+} to reach maximal contraction. The benefits to the fetus are clear. Since its contractile apparatus is less sensitive to pH, it is better able to tolerate the frequent bouts of hypoxia experienced *in utero*. We can trace the differences in pH sensitivity back to the TnI isoforms. Mammals express three different isoforms of TnI: cardiac muscle, fast-twitch skeletal muscle, and slow-twitch skeletal muscle. Fetal hearts express the less pH-sensitive skeletal isoforms of TnI, switching to the cardiac isoform shortly after birth. When researchers created transgenic mice that expressed the fast-muscle TnI in the adult heart, they found that the hearts were less affected by pH. Changes in TnI isoform expression also occur during exercise. With sustained increases in

muscle activity, the muscle cell slowly replaces fast-muscle TnI with slow-muscle TnI. It is difficult to show the benefits of this type of TnI isoform switch because many other features of EC coupling also change during exercise training.

Much less is known about the functional properties of the different isoforms of TnT and tropomyosin. There are many isoforms of TnT arising from three genes (cardiac, fast skeletal, slow skeletal) with multiple forms produced by alternate mRNA splicing. Four tropomyosin genes generate a multitude of isoforms through alternate splicing. Transgenic studies show that tropomyosin isoforms can influence the rates of relaxation and contraction.

Thick filaments also influence contractile properties

The composition and properties of thick filaments also influence muscle contraction. Animals have the potential to build different types of thick filaments by drawing upon the large myosin II gene family. Vertebrates have eight different myosin II genes, each producing a myosin heavy chain with distinct structural or functional properties. Since muscle myosins combine as homodimers or heterodimers, vertebrates can potentially make 36 different myosin II dimers from the eight genes. Although 36 combinations are possible, each muscle cell normally expresses only a subset of the myosin II genes (Table 5.4).

Vertebrate heart muscle uses two myosin II genes (α and β) to make three different dimers (αα, αβ, ββ). Each of these combinations has a distinc-

Table 5.4	Myosin isoform properties in mammals.
Isoform	**Properties**
α	This fast cardiac isoform is expressed in cardiac muscle, in species with faster heart rates, or in response to activity.
β (= I)	This slow cardiac/slow oxidative isoform is expressed in cardiac muscle, in species with slower heart rates.
IIa	Found in fast oxidative-glycolytic fibers. ATPase rates intermediate between I and IIx/d.
IIx/d	Found in fast glycolytic fibers. ATPase rates intermediate between IIa and IIb.
IIb	Found in fast glycolytic fibers, this creates the fastest ATPase rates.
Embryonic	Expressed in skeletal muscles in early embryonic development, as well as some adult muscle.
Perinatal	Expressed in skeletal muscles in late embryonic development, as well as some adult muscle.
Extraoccular	Expressed in eye muscles.

tive actino-myosin ATPase rate: the αα combination has the fastest ATPase, whereas ββ has the slowest, and αβ has an intermediate rate. Animals alter the myosin heavy chain profile in response to changes in activity level. Exercise training may cause cardiac muscle to shift from ββ to αα myosin isoforms. The relationship with activity level is also reflected in interspecies comparisons. Some species, such as rabbits, typically express their β-myosin II genes, whereas species with higher heart rates, such as rats, express their α-myosin II genes.

Myosin isoform shifts also occur in skeletal muscle, which can express seven different myosins (I, IIa, IIb, IIx/d, perinatal, embryonic, and extraoccular). Many of these skeletal isoforms vary in their ATPase rates, whereas others differ in noncatalytic aspects of myosin function, such as the ability to interact with regulatory or structural proteins. As the names suggest, some isoforms are expressed at discrete points in development. As embryos develop, their skeletal muscle progresses from embryonic, through perinatal, and then finally to muscle-specific adult isoforms. It is not yet known how each myosin II isoform influences muscle function during development. In fact, some muscles in the jaw and neck continue to use embryonic or perinatal myosin II isoforms into adulthood. Adult skeletal muscles are categorized on the basis of the myosin II isoform as type I (or β), IIa, IIb, or IIx/d. The catalytic properties of myosins are matched to the contractile demands of the muscle. Slow-twitch skeletal muscle uses predominately β-myosin II, the "cardiac" isoform with low velocity and high efficiency. Fast-twitch skeletal muscle, in contrast, uses IIb-myosin II, which has faster velocity but lower efficiency. While each myofiber expresses a single myosin isoform, a muscle can be made up of myofibers expressing different myosin II isoforms.

Muscle contraction can generate force

Activation of actino-myosin ATPase in muscle can be considered in terms of molecular interactions, but in terms of animal physiology the important factor is how these molecular events translate into changes at the whole tissue level. The response of muscle upon activation is described in terms of degree of change in length, the rate of change in length, and the amount of force generated during contraction.

In reality, "contraction" is not the best choice of a term to describe an activated muscle because it implies that a contracting muscle gets smaller. When a muscle is activated, it may shorten, or remain the same length, or even lengthen. The changes in length depend a lot on how the muscle is connected to the rest of the body. In the most familiar situation, a contracting muscle shortens in length. A simple example of a shortening contraction is when your bicep contracts and your elbow bends. Alternatively, an activated muscle may remain at a fixed length in what is known as an **isometric contraction**. For example, many muscles in your back contract without much of a change in length, helping you to maintain posture. A third possibility is when an activated muscle actually lengthens. When you walk down stairs some leg muscles undergo lengthening contractions, slowing the rate of descent by acting like a brake. We will discuss specific examples of such muscles in Chapter 12: Locomotion.

In many fields of muscle biology, the terms *eccentric* and *concentric* are used to describe the nature of changes in length in relation to contraction. Concentric literally means "having the same center," whereas eccentric means "not having the same center." The problem with these terms is that, in some fields, they have been used in ways that can be misleading, given the strict definition of the term. Cardiovascular physiologists use concentric and eccentric in an appropriate way when describing the orientation of contraction with respect to the center of the chamber. For example, a normal heart produces a concentric contraction because it contracts symmetrically around the center of the chamber. If one wall of the heart hypertrophies and gets stronger, the contraction may be eccentric, or "off center." However, exercise physiologists use concentric to describe a shortening contraction, with the term chosen because the ends of the muscle move toward the center. Likewise, eccentric contractions are used synonymously with lengthening contractions, as if the term meant "away from center." While the use of these terms is more common in the exercise literature, they are more accurately used in describing cardiac physiology. To avoid confusion, we use the more descriptive terms: shortening, isometric, and lengthening contractions.

In the accompanying feature, we discuss how variation in shortening and force generation arises at the level of the sarcomeres and cross-bridge kinetics (Box 5.3, Mathematical Underpinnings: Sarcomeric Changes in Force Generation and Shortening).

BOX 5.3 **MATHEMATICAL UNDERPINNINGS**
Sarcomeric Changes in Force Generation and Shortening

Let's consider the following scenarios to illustrate how differences in contractile properties arise. Since we have already learned that sarcomere length influences contraction, in the following scenarios we will assume an optimal sarcomere length, with all 600 myosin heads able to interact with the thin filaments in each sarcomere. The contractile properties of a muscle cell depend on the number of myosin heads involved in the contraction. Each individual myosin head generates about 5 pN of force during a cross-bridge cycle. We might assume then that a thick filament with 600 myosin heads, each producing 5pN of force, would produce 3000 pN of force. However, when we measure the force of a single thick filament, we find that it generates only about 150 pN of force. A single thick filament *could* generate 3000 pN of force if every single myosin head pulled at the same time. In a cross-bridge cycle, each myosin head is attached only about 5% of the time (its duty cycle) and can only generate force during that part of the cycle when it is attached to the thin filament.

$$600 \times 5\% = 30 \text{ myosin heads generating force at any time}$$

$$30 \times 5 \text{ pN} = 150 \text{ pN force}$$

How can a sarcomere generate different amounts of force? Each activated myosin head generates the same amount of force (−5 pN), so it stands to reason that force must be dependent on the number of cross-bridges that can form. Many types of muscle cells can use changes in the magnitude of the Ca^{2+} signal to alter the number of cross-bridges. If the SR released few Ca^{2+} ions, few troponin-tropomyosin complexes would be induced to move, and few cross-bridges would form.

Muscle contractile elements show a sigmoidal relationship between $[Ca^{2+}]$ and muscle force.

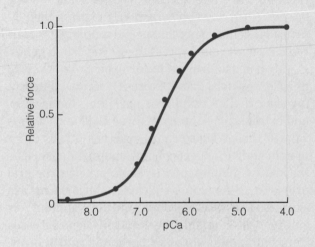

Recall from Chapter 2 that the prefix p, as in pCa or pH, signifies that the concentration is expressed as the negative logarithm (pCa 6 = 10^{-6} M Ca^{2+}, or 1 μM Ca^{2+}). This strategy of altering Ca^{2+} levels to regulate force is important in cardiac muscle, but most skeletal muscles release enough Ca^{2+} during each contraction to induce near-maximal force.

What is the relationship between force and the velocity of shortening? Or put another way, why can you lift a feather faster than a brick? About 70 years ago, A. V. Hill tried to explain how force influences the rate of muscle shortening. He determined experimentally how contraction velocity was affected by force, which he altered by changing the amount of weight a muscle lifted.

⊙ | CONCEPT CHECK

5. Describe duty cycle and unitary displacement in relation to nonmuscle and muscle myosin activity.

6. How does the organization of the sarcomere influence contractile force?

7. Compare the constraints on myosin function in vesicle traffic versus the contractile apparatus.

8. Is muscle activity more accurately described as cellular movement or a change in cell shape? What types of cells need to move within the vertebrate body?

Excitation and EC Coupling in Vertebrate Striated Muscle

So far, we have discussed the machinery involved in muscle contraction, but we have not yet discussed how muscle contraction is triggered. Excitation in most striated muscles occurs when depolarization of the sarcolemma induces an increase in cytosolic $[Ca^{2+}]$ to trigger contraction. Beyond that simple summary, it is difficult to make any generalizations about excitation and EC coupling in striated muscle. As you will see in the following sections, muscles differ in the cause of depolarization, the

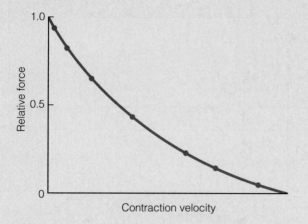

He developed the following equation relating contractile force (P) to velocity of shortening (V):

$$V = \frac{b\,(P_o - P)}{P + c}$$

where P_o is the maximal isometric tension of the muscle, b is a velocity constant, and c is a force constant. When you lift the feather (P approaching zero), the numerator is at its maximum ($b\,P_o$), the denominator approaches zero, and velocity is at its maximum. But what is the mechanistic basis of this relationship in terms of molecular events in the sarcomere? The same number of cross-bridges will be involved whether the situation generates maximal force or maximal velocity of shortening.

In 1957, Andrew Huxley explained the force-velocity relationship in terms of cross-bridge kinetics. The difference between force generation and shortening lies in the structural changes in the myosin head. When we think of the cross-bridge cycle in terms of a single myosin molecule, we see how it reaches forward and pulls the thin filament. When you factor into this model the hundreds of other myosin heads, things get a bit more complicated. Each individual myosin can only bind the thin filament when it reaches forward looking for a binding site on actin. Once it binds, several chemical steps must occur before the head can generate force in its power stroke. If shortening is fast, other myosin heads can pull the thin filament back before the myosin head has a chance to undergo its power stroke. The sliding filament bends myosin into the position that it would have assumed had it been given the time to undertake its power stroke. Although the chemical events in the power stroke (ADP and P_i release) still happen, the structural changes in the myosin head have already occurred. Consequently, this cross-bridge cycle generates no force. Put simply, high contraction velocity prevents many cross-bridges from generating force. Now consider what happens when a muscle generates its maximal force, such as when it lifts the heaviest object possible. During a cross-bridge cycle, the tension on the muscle prevents the thin filament from moving appreciably and each myosin head in a cross-bridge remains in a form that allows it to generate force.

These sarcomeric changes in force, length, and contractile velocity have important ramifications for muscle function. As we revisit muscles in later chapters, recall how these sarcomeric events contribute to muscle function.

References

○ Huxley, A. F. 2000. Cross-bridge action: Present views, prospects, and unknowns. *Journal of Biomechanics* 33: 1189–1195.

pattern of change in membrane potential over time, the propagation of depolarization along the sarcolemma, and the cellular origins of Ca^{2+}. Let's consider each of these factors in turn, focusing on the implications for muscle function and the basis of differences among muscle types and species.

Muscles are excited by an action potential

The action potential, first described in Chapter 2, is also the signal for contraction of most muscle cells. The resting membrane potential of the sarcolemma is about -70 mV. Upon activation, muscles experience a rapid depolarization, followed by repolarization and hyperpolarization. The properties of the action potential, such as rates of depolarization and repolarization and action potential duration, are determined by the density and activities of various channels in the sarcolemma.

As with other cell types, depolarization is induced when Na^+ channels are opened. The inward rush of Na^+ causes a rapid reduction in membrane potential. At this point, voltage-sensitive Ca^{2+} channels open, allowing the influx of Ca^{2+} into the cell from the extracellular space. After a period, Na^+ channels and Ca^{2+} channels begin to close and voltage-sensitive K$^+$ channels open, causing the cell to repolarize. The density and kinetic properties of these various ion channels determine the features of the action potential: the rate of depolarization, the rate of repolarization, and consequently, the duration of the action potential. Since the depolarization and repolarization leads to movement of ions, active transporters are responsible for reestablishing the ion gradients. As in most cells, the Na$^+$/K$^+$ ATPase is important in reestblishing Na$^+$ and K$^+$ gradients. In muscle, where the action potential also induces Ca^{2+} movement into the cell, a suite of Ca^{2+} transporters is also essential. This general pattern of an action potential, depolarization, and repolarization is similar among vertebrate striated muscles. However, muscles show very important differences in the time course of the change in membrane potential.

Cardiac and skeletal muscles have dramatic differences in the shape and duration of the action potential (Figure 5.24). Striated muscle cells cannot be depolarized again until the repolarization phase is near complete. This window of insensitivity is called the **effective refractory period** because the muscle cell cannot be induced to contract again by normal physiological regulators. Skeletal myofibers depolarize and repolarize very quickly, typically within about 5% of the time required to complete a contraction-relaxation cycle. Once a skeletal muscle membrane repolarizes, a second action potential can induce another contraction even if the muscle has not yet relaxed from the previous contraction. Cardiomyocytes also depolarize rapidly but take much longer to repolarize. The main reason for this is that the voltage-sensitive Ca^{2+} channels in cardiac muscle stay open for a much longer period. As a result, the duration of the action potential in cardiomyocytes is approximately half the duration of a contraction cycle. The prolonged effective refractory period of cardiomyocytes is critical to the function of the heart. Cardiomyocytes are connected into an electrical network that transmits an action potential between cells to create a wave of contraction. The long effective refractory period prevents the action potential from stimulating contraction in cardiomyocytes that are in the midst of a contractile cycle. Without the effective refractory period, contraction of individual cardiomyocytes or regions of cardiomyocytes could occur chaotically, a condition known as *arrhythmia*. In a typical vertebrate heart, there are many types of cardiomyocytes, each of which undergoes an action potential

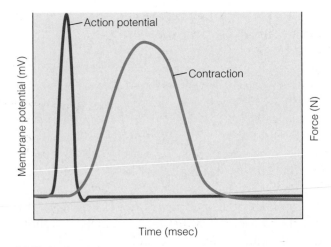

(a) Skeletal muscle

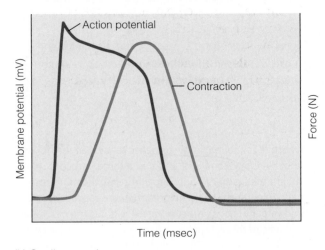

(b) Cardiac muscle

Figure 5.24 Action potentials in striated muscle   The time course of change in action potential and force are shown for a skeletal muscle **(a)** and cardiac muscle **(b)**. While the contraction profiles are similar, the action potential in cardiac muscle is prolonged. This is attributed to Ca^{2+} channels remaining open.

with characteristic amplitude, duration, and shape. In Chapter 8 you will see that the different types of cardiomyocytes, as well as their organization into heart muscle, are central to cardiac function.

The maximal contraction rate of striated muscle depends upon the rate at which the muscle cell can complete the action potential. Rapidly contracting striated muscles, such as fast-twitch skeletal muscle, complete an action potential within a few milliseconds. This prepares the muscle for another excitation and contraction soon after. Many types of skeletal muscle are distinguished by the maximal frequency of contraction, due in large part to differences in the rate of repolarization.

Much of the difference in repolarization rate can be attributed to the properties of the K^+ channels. Not surprisingly, K^+ channels are the targets of hormones and drugs that regulate contraction rate. Regulatory factors, such as acetylcholine and adenosine, act by modulating the properties of K^+ channels. Each muscle cell type has a characteristic profile of K^+ channels that confers different patterns of repolarization. As with skeletal muscle, many of the hormonal pathways that influence cardiac contractility exert their effects via the K^+ channels that determine the rate of repolarization, but the properties of the Ca^{2+} channels also influence the rate of repolarization.

Depolarization is the first step in vertebrate striated muscle excitation, but it can be induced in different ways. In the next section, we will distinguish two general classes of muscle based on the trigger for sarcolemmal depolarization. **Myogenic muscle cells** contract spontaneously, whereas **neurogenic muscle cells** are stimulated by the action of neurons.

Myogenic muscle cells spontaneously depolarize

The most common examples of myogenic myocytes are from the vertebrate heart. Because the entire heart contracts without neuronal input, each of the myocytes of the heart is considered a myogenic muscle. In the intact heart, some specialized myocytes depolarize spontaneously. These **pacemaker cells** transmit their electrical signal throughout the heart and cause other cardiomyocytes to depolarize and contract.

Pacemaker cells are unusual in that they show an unstable resting membrane potential. These cells possess an unusual ion channel, the funny channel or *f-channel,* that is permeable to both Na^+ and K^+. When the channel is open, an imbalance in Na^+ influx and K^+ efflux leads to a slow depolarization. Once the pacemaker cell membrane depolarizes to a critical voltage, the *threshold voltage,* voltage-sensitive Ca^+ channels open to initiate the action potential. Though the f-channels close during the action potential, hyperpolarization of the pacemaker cells at the end of the action potential reactivates the f-channels, causing the cells to slowly depolarize again. Many of the factors that regulate heart rate, such as adenosine, acetylcholine, and catecholamines, alter the kinetic properties of the f-channels.

The action potential of the pacemaker cells induces an action potential in the myocytes to which they are connected through gap junctions. In contrast to pacemaker cells, depolarization of non-pacemaker cardiomyocytes is due to the opening of voltage-dependent Na^+ channels, much like the situation seen in other excitable cells. Interestingly, a normal cardiomyocyte has the ability to contract spontaneously, much like a pacemaker cell. In an intact heart, these cardiomyocytes would receive an excitatory signal from a pacemaker before they would experience their own spontaneous contraction. However, if the pacemaker cells become damaged, other cardiomyocytes can become the pacemaker to determine the rate of cardiac contraction.

Neurogenic muscle is excited by neurotransmitters

Most vertebrate skeletal muscles are neurogenic muscles, and receive signals from a *motor neuron.* The motor neuron axon termini are located in a region of the sarcolemma called the **motor end plate** (Figure 5.25). The sarcolemma at the motor end

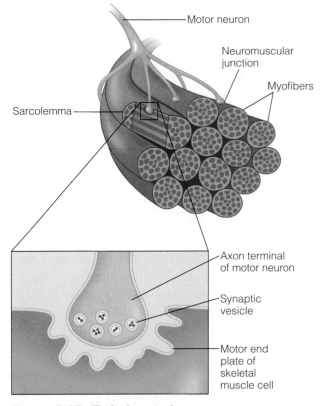

Figure 5.25 Twitch muscles Motor neurons innervate individual myofibers, making contacts at regions on the myofiber called motor end plates. Twitch muscles possess myofibers that are innervated by single motor neurons.

plate is rich in receptors for the neurotransmitter released by the motor neuron: acetylcholine. Upon stimulation of a motor neuron, acetylcholine is released from synaptic vesicles into the neuromuscular synapse. It crosses the synapse and binds nicotinic acetylcholine receptors within the sarcolemma. As we discussed in Chapter 4: Neuron Structure and Function, these ligand-gated ion channels are Na^+ channels. If enough nicotinic acetylcholine receptors are activated, the depolarization at the motor end plate initiates a wave of depolarization along the sarcolemma: the action potential. The passage of the action potential along the sarcolemma induces an all-or-none contraction.

Twitch muscles are neurogenic skeletal muscles that are innervated by one or, occasionally, a few motor neurons. In these muscles the action potential spreads rapidly along the sarcolemma, causing a uniform contraction along the length of the myofiber. Because of their electrical nature, action potentials move rapidly, but in many muscles passive conductance from the motor end plate is inadequate to ensure that the signal reaches the entire muscle essentially simultaneously. There are two main ways that muscles are able to ensure the entire sarcolemma is depolarized uniformly in space and time: through multiple innervations (tonic muscle) and through invaginations of the sarcolemma (T-tubules).

Tonic muscles have multiple innervations

One way in which the challenge of uniform contraction is met in some neurogeneic muscles is through multiple innervations. Vertebrate striated muscles with multiple innervations are called **tonic muscle**. When motor neurons are stimulated, neurotransmitter release occurs at many sites along the tonic muscle fiber. The fiber is then induced to contract in response to depolarization at multiple points along the fiber, reducing the dependency on action potential conductance. Tonic muscles contract slowly, but maintain tension for long periods. In contrast to twitch muscles, tonic muscles are not all-or-none. The level of depolarization of the sarcolemma depends on the number and frequency of stimulatory signals from the motor neuron.

Many researchers studying mammals use the term *tonic muscle* to describe muscles that exhibit a long duration of contraction. For example, human physiologists refer to the postural muscles of the back as tonic muscle. These muscles have only single innervation, and thus are more accurately described as slow twitch muscle. Mammals do have a few true tonic muscles, located around the eye (extraoccular), in the ear, and in the esophagus.

T-tubules enhance action potential penetration into the myocyte

Many twitch fibers depend on simple action potential conductance along the sarcolemmal surface. Myofibers can facilitate action potential conductance throughout the muscle with the help of extensive sarcolemmal invaginations called transverse tubules, or **T-tubules** (Figure 5.26). When the sarcolemma depolarizes, the action potential follows the T-tubules into the muscle fiber. The relative importance of the T-tubule system depends upon the nature of the muscle and the work it performs. Many muscles do not need to contract quickly. For instance, postural muscles, found in the body trunk of vertebrates, remain contracted for long periods without relaxing or fatiguing. However, the T-tubule system is extensive in large or quick-contracting muscles, such as vertebrate fast-twitch skeletal muscles. T-tubules also exist in the cardiac muscle of mammals and some birds, although it is generally less developed than in skeletal muscle of the same species.

Ca^{2+} for contraction comes from intracellular or extracellular stores

The regulation of the Ca^{2+} transient during muscle contraction involves many transporters and cellular compartments. In fact, many aspects of muscle con-

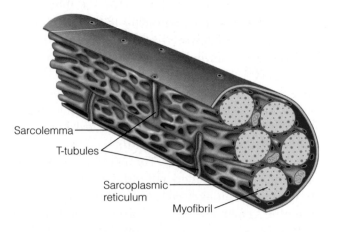

Sarcolemma
T-tubules
Sarcoplasmic reticulum
Myofibril

Figure 5.26 T-tubules Many types of muscle have T-tubules, invaginations of the sarcolemma that penetrate deep into the myofiber to speed the spread of the action potential.

tractile properties can be traced to the way these proteins are made and utilized to mediate Ca^{2+} transients. Most muscles respond to sarcolemmal depolarization by opening voltage-dependent Ca^{2+} channels. Because of the electrochemical gradient for Ca^{2+}, these channels open to allow Ca^{2+} to rush into the cell from the extracellular space. There are several types of Ca^{2+} channels in striated muscle sarcolemma, named for how long they remain open after activation. L-type Ca^{2+} channels are open for a **L**ong period with **L**arge conductance, whereas T-type Ca^{2+} channels are open **T**ransiently with **T**iny conductance. Animals also possess N-type Ca^{2+} channels that are **N**either L-type nor T-type. The predominant channel in the heart of most animals is the L-type Ca^{2+} channel, a protein that has been shown to bind a class of drugs called dihydropyridines. Consequently, the L-type Ca^{2+} channels are also known as **dihydropyridine receptors (DHPR)**, to distinguish them from other types of Ca^{2+} channels.

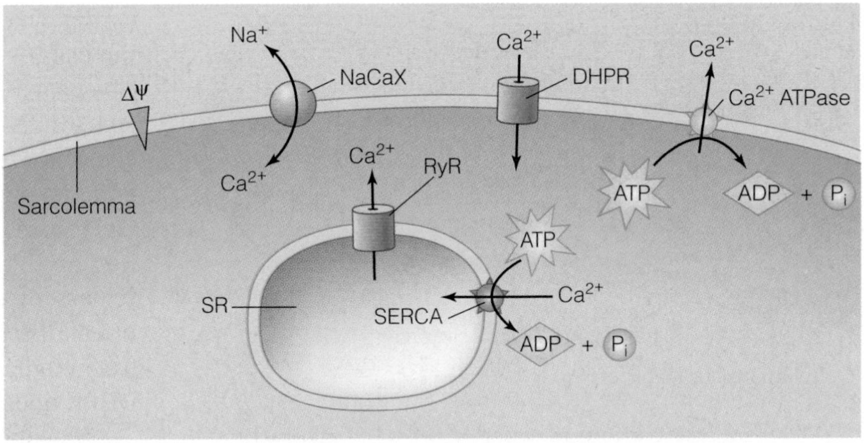

Figure 5.27 **Transporters and channels involved in EC coupling**
Ion movements during a contraction cycle are mediated by ion channels and pumps within the sarcolemma and the sarcoplasmic reticulum (SR). Ca^{2+} channels in the sarcolemma (the dihydropyridine receptor or DHPR) and the SR open to allow Ca^{2+} to flow into the cytoplasm. Ca^{2+} pumps in the SR, known as SERCA, as well as the sarcolemma use the energy of ATP to reverse the Ca^{2+} movements allowed by the channels. The Na^+/Ca^{2+} exchanger (NaCaX) facilitates the reversible exchange of Na^+ and Ca^{2+}. The membrane potential ($\Delta\psi$) influences many of these transport processes. The image in this figure shows the events in a cardiac muscle cell, where DHPR and RyR are physically separated.

Though important, the DHPR is only one of many types of Ca^{2+} transporter in muscle. The sarcolemma possesses Ca^{2+} ATPases and a Na^+/Ca^{2+} exchanger (NaCaX). The muscle endoplasmic reticulum, known as the **sarcoplasmic reticulum**, or SR, has its own Ca^{2+} channel (RyR) and Ca^{2+} ATPase (SERCA). The location of these Ca^{2+} transporters is summarized in Figure 5.27.

The rate of Ca^{2+} movement into the muscle cell upon depolarization depends upon many factors related to the structure and activity of DHPR. Since individual channels can respond to different voltages, the degree of depolarization can influence the number of Ca^{2+} channels that open. As with other membrane proteins, the number of channels in the sarcolemma can be changed using pathways of endocytosis and exocytosis. When animals experience prolonged periods of elevated activity, signaling pathways can induce synthesis of more DHPR. Intracellular signaling pathways can also influence how long each channel remains open. The structure of the DHPR influences electrical properties, such as voltage sensitivity and open time, or alter the sensitivity of the channel to regulatory proteins and ligands.

In some situations, Ca^{2+} delivery through DHPR is sufficient to induce contraction. For example, the hearts of most fish are able to deliver enough Ca^{2+} through the DHPR to initiate contraction. However, in most striated muscles, the Ca^{2+} delivery through DHPR is either too slow or too minor to achieve the contraction threshold. As a result, most striated muscles require more effective means of delivering more Ca^{2+} at much faster rates.

DHPR activation induces Ca^{2+} release from the SR

Most skeletal muscles, as well as the cardiac muscles of birds and mammals, use the sarcolemmal Ca^{2+} channels to signal the release of even greater amounts of Ca^{2+} from vast intracellular stores in the SR. Cardiac and skeletal muscles accumulate Ca^{2+} within their SR, ensuring the cell maintains a low intracellular $[Ca^{2+}]$. In striated muscle, the SR frequently has enlargements, called **terminal cisternae** (Figure 5.28) that increase the capacity for Ca^{2+} storage and localize it to discrete regions within the muscle cell. Because terminal cisternae ensure rapid Ca^{2+} delivery, they are well developed in muscles that contract quickly, such as fast-twitch skeletal muscle. Muscles are able to accumulate Ca^{2+} to very high levels within the SR. While some

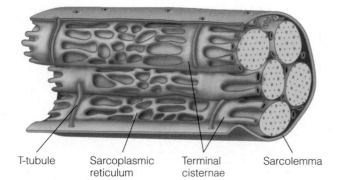

T-tubule   Sarcoplasmic   Terminal   Sarcolemma
            reticulum   cisternae

(a) Skeletal myofiber

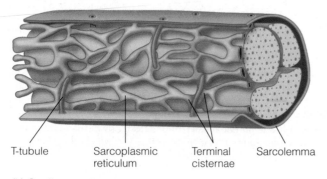

T-tubule   Sarcoplasmic   Terminal   Sarcolemma
            reticulum   cisternae

(b) Cardiomyocyte

Figure 5.28   Terminal cisternae   Many striated muscles possess enlargements of the sarcoplasmic reticulum (SR) near the region of the T-tubules. In mammals, the terminal cisternae are extensive in fast-twitch skeletal muscle **(a)** and less well developed in cardiac muscle **(b)**.

of the Ca^{2+} is free in solution, most is bound to **calsequestrin**, another member of the large Ca^{2+}-binding protein family that includes TnC.

During excitation, the SR releases its Ca^{2+} stores through a Ca^{2+} channel, frequently called the **ryanodine receptor** (or RyR) because it can bind the drug ryanodine, a plant alkaloid. Once the RyR is activated, free Ca^{2+} escapes the SR and flows into the cytoplasm. The loss of free Ca^{2+} favors the release of Ca^{2+} bound to calsequestrin.

The general features of this pathway apply equally to the many muscles that rely upon SR Ca^{2+} to trigger contraction. Activation of the sarcolemmal DHPR induces Ca^{2+} release through the RyR. This linkage is enhanced by the physical arrangement of the different Ca^{2+} channels. Within the sarcolemma, DHPRs are clustered in regions directly adjacent to the terminal cisternae. However, what differs between muscle types is the way DHPR activation is coupled to RyR activation.

Cardiac muscle uses a process called **Ca^{2+}-induced Ca^{2+} release** to link DHPR and RyR acti-

vation. Once DHPR open, extracellular Ca^{2+} enters the cell. Because DHPR are localized near terminal cisternae, local $[Ca^{2+}]$ can increase in the small space between the sarcolemma and the terminal cisternae (Figure 5.29). The high local $[Ca^{2+}]$ triggers the opening of cardiac muscle RyR, and the SR Ca^{2+} stores are released into the muscle cytoplasm. Researchers can demonstrate Ca^{2+}-induced Ca^{2+} release by manipulating the composition of the extracellular fluid. If cardiac muscle is bathed in Ca^{2+}-free media, depolarization and activation of the DHPR does not induce a contraction.

Skeletal muscle differs from cardiac muscle in how DHPR activation is coupled to RyR activation (Figure 5.30). As with cardiac muscle, sarcolemmal depolarization opens the DHPR and allows Ca^{2+} into the cell from the extracellular space. However, in skeletal muscle it is the voltage-dependent changes in the DHPR structure that trigger the opening of RyR. These two channels physically interact with each other to couple sarcolemmal depolarization with SR Ca^{2+} release. In this case, activation of RyR is not influenced by local accumulation of $[Ca^{2+}]$. Upon activation of DHPR, the RyR opens even if no Ca^{2+} ions move through the DHPR. This pattern of EC coupling is called **depolarization-induced Ca^{2+} release**.

Relaxation follows removal of Ca^{2+} from the cytoplasm

To this point, we have discussed the mechanisms that lead to depolarization and the subsequent increase in cytoplasmic $[Ca^{2+}]$ that induce contraction. These ion movements across membranes must be reversed to allow relaxation to occur. As mentioned previously, the duration of the action potential determines how quickly a muscle can relax. Once the membrane repolarizes, the muscle can start to reestablish Ca^{2+} gradients. In vertebrate striated muscle, relaxation requires a suite of transporters to pump Ca^{2+} out of the cytoplasm, back across the sarcolemma, or into the SR. Both the sarcolemma and the SR possess active Ca^{2+} ATPases that pump Ca^{2+} out of the cell using the energy of ATP hydrolysis. The sarcolemma also possesses a transporter that exchanges Na^+ for Ca^{2+}, called the Na/Ca exchanger or *NaCaX*. During excitation, this reversible exchanger can allow extracellular Ca^{2+} to enter the cell in exchange for intracellular Na^+. However, it is most important during relaxation, where Ca^{2+} efflux is coupled to Na^+ influx. As in other Na^+-driven transport processes, muscles ultimately use the Na^+/K^+ ATPase to reestablish Na^+ gradients.

The role of each specific Ca^{2+} transporter depends upon the way Ca^{2+} is used to induce contraction. Those muscles that primarily rely on sarcolemmal Ca^{2+} influx to initiate contraction, such as hearts of lower vertebrates, use the sarcolemmal NaCaX and Ca^{2+} ATPase to pump Ca^{2+} out of the cell. However, muscles that elevate cytoplasmic $[Ca^{2+}]$ using intracellular stores, such as most types of mammalian striated muscle, use the sarcoplasmic (endoplasmic) reticulum Ca^{2+} ATPase, or SERCA, to resequester Ca^{2+} in the SR.

In addition to the proteins involved in transporting Ca^{2+} across membranes, relaxation in many muscles also relies upon a cytosolic Ca^{2+} buffer called **parvalbumin**. By binding cytoplasmic Ca^{2+}, parvalbumin accelerates muscle relaxation. Not surprisingly, parvalbumin is found in muscle types that contract and relax very quickly. Its role in relaxation has been elegantly demonstrated using transgenic mice. One group of mice was engineered to prevent the expression of parvalbumin; the muscles of these *parvalbumin-null mutants* relaxed much more slowly than wild-type mice. Researchers have also engineered transgenic mice that express parvalbumin in muscles that normally lack parvalbumin. These mice had muscles that relaxed faster than wild-type mice.

In the natural world, parvalbumin levels differ between muscle types and species. The fastest muscles possess very high levels of parvalbumin to accommodate the high frequencies of contraction. The highest levels of parvalbumin are found in fish white muscle. Fish use white muscle to burst away from danger or to attack prey, strategies that require very rapid rates of muscle contraction. Although most vertebrates possess at least some parvalbumin in their skeletal muscles, humans do not appear to express parvalbumin. The genetic reasons for their loss of parvalbumin, and the physiological consequences, are not yet known.

Table 5.5 summarizes the general features of cardiac and skeletal muscle.

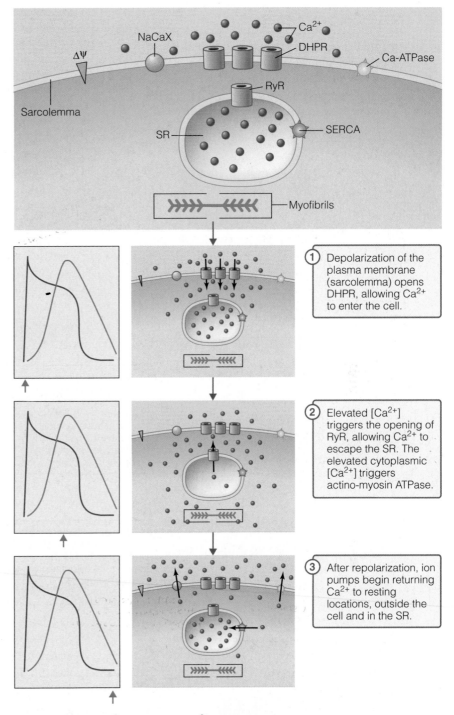

Figure 5.29 Ca^{2+}-induced Ca^{2+} release   At rest, the high membrane potential ($\Delta\psi$) keeps closed the cardiac sarcolemmal Ca^{2+} channel (DHPR) and intracellular Ca^{2+} levels are low. The graphs to the left reflect the patterns of action potential (purple line) and contraction (blue line), as shown in Figure 5.24.

1. Depolarization of the plasma membrane (sarcolemma) opens DHPR, allowing Ca^{2+} to enter the cell.

2. Elevated $[Ca^{2+}]$ triggers the opening of RyR, allowing Ca^{2+} to escape the SR. The elevated cytoplasmic $[Ca^{2+}]$ triggers actino-myosin ATPase.

3. After repolarization, ion pumps begin returning Ca^{2+} to resting locations, outside the cell and in the SR.

9. How does the action potential differ between cardiac and skeletal muscles? Why is this important to the function of the muscle?

10. Other than myocytes, what other cell types contribute to the makeup of a muscle?

11. What arrangement of thick and thin filaments allows each myosin to interact with six actins, and each actin to interact with three myosins?

12. What factors determine the rate of shortening of a muscle? What factors affect the rate of relaxation?

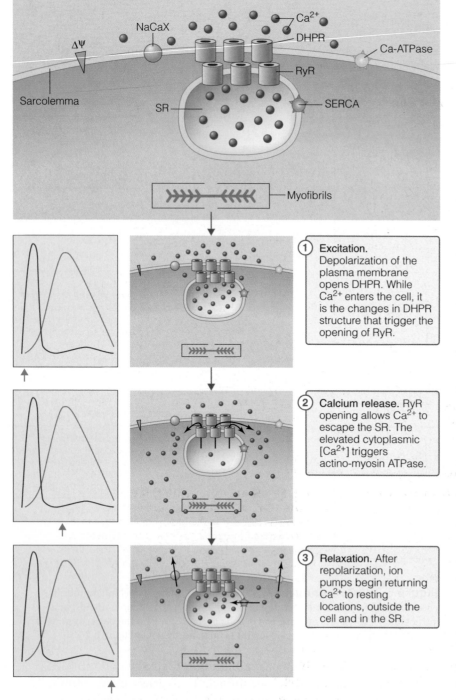

① **Excitation.** Depolarization of the plasma membrane opens DHPR. While Ca²⁺ enters the cell, it is the changes in DHPR structure that trigger the opening of RyR.

② **Calcium release.** RyR opening allows Ca²⁺ to escape the SR. The elevated cytoplasmic [Ca²⁺] triggers actino-myosin ATPase.

③ **Relaxation.** After repolarization, ion pumps begin returning Ca²⁺ to resting locations, outside the cell and in the SR.

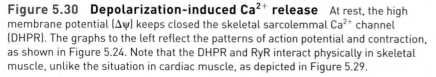

Figure 5.30 Depolarization-induced Ca²⁺ release At rest, the high membrane potential (Δψ) keeps closed the skeletal sarcolemmal Ca²⁺ channel (DHPR). The graphs to the left reflect the patterns of action potential and contraction, as shown in Figure 5.24. Note that the DHPR and RyR interact physically in skeletal muscle, unlike the situation in cardiac muscle, as depicted in Figure 5.29.

Smooth Muscle

Much of the discussion of muscle cell properties has, to this point, focused on striated muscle. Vertebrates, as well as many invertebrates, possess another form of muscle: smooth muscle. Many tissues use layers of smooth muscle to induce slow regular contractions, or maintain a degree of contraction for long periods. For example, smooth muscle lines the walls of blood vessels, controlling blood flow by regulating the diameter of the blood vessels. Smooth muscle works in a similar fashion in the respiratory system of terrestrial vertebrates to control the diameter of airways. Circular and longitudinal layers of smooth muscle in the digestive tract propel food down the gut and control the length of the gastrointestinal tract. Reproductive function also depends on smooth muscle to propel gametes or offspring along the reproductive tract. Although it shares many features with striated muscles, such as the basic interaction between actin and myosin, it has important differences that provide the smooth muscle cell with remarkable flexibility in contraction dynamics and distinct pathways of regulation of EC coupling.

Smooth muscle lacks organized sarcomeres

Although smooth muscle cells are composed of the same contractile elements as striated muscle, animals can organize and regulate smooth muscle in various ways. Striated muscles arrange their thick and thin

Table 5.5 **Comparing vertebrate cardiac and skeletal striated muscles.**

	Cardiac	Skeletal
Cell morphology	Single cells (cardiomyocytes) about 10 to 20 μm in diameter and 100 μm in length.	Multiple cells fused into large myofibers that are 10 to 100 μm in diameter and 1 to 100 mm in length.
Excitation	Myogenic and involuntary.	Neurogenic and usually voluntary.
Action potential	Slow repolarization, with long refractory period.	Fast repolarization, with short refractory period.
EC coupling	Ca^{2+}-induced Ca^{2+} release.	Depolarization-induced Ca^{2+} release.
Sarcoplasmic reticulum	Well-developed terminal cisternae in birds and mammals. Poorly developed SR in lower vertebrates.	Amount of terminal cisternae depends on fiber type.

filaments into sarcomeres, producing their characteristic striped appearance. Smooth muscle also has thin filaments and thick filaments, but they are not organized into sarcomeres. At the cellular level, smooth muscle is a collection of individual cells that are organized into a functional network. Gap junctions between smooth muscle cells allow them to communicate and exert a common response to local regulators, creating a functional group that acts as a unit. This cellular organization is reminiscent of the organization of cardiac muscle. One or more functional groups may be physically linked together by connective tissue, but regulated independently within that tissue. In the circulatory system, for example, a layer of smooth muscle surrounds the blood vessels. The smooth muscle cells may be induced to contract in unison in one region, while a neighboring region remains relaxed. Many organs have layers of smooth muscle arranged in a way that allows contraction in different planes. For example, the gastrointestinal tract has an inner layer of circular muscle that regulates circumference, and a layer of longitudinal muscle that regulates length.

The main difference between smooth and striated muscle is in the organization of the thick and thin filaments. Instead of parallel arrays of sarcomeres, smooth muscle scatters clusters of thick and thin filaments throughout the cytoplasm (Figure 5.31). The aggregated filaments interconnect with each other to form a network within the cytoplasm, and attach to the plasma membrane at specific regions called **adhesion plaques**. This three-dimensional arrangement of thick and thin filaments allows smooth muscle cells to contract in all dimensions. In contrast to striated muscle, with

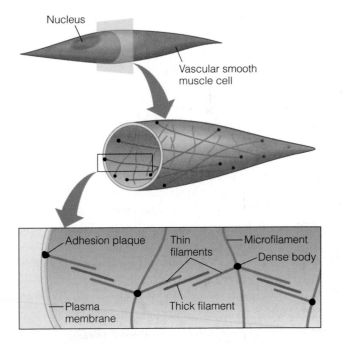

Figure 5.31 Smooth muscle thick and thin filaments Smooth muscle cells lack organized sarcomeres. Thick and thin filaments are arranged in complex networks throughout the cell: Thin filaments are fixed to the plasma membrane by adhesion plaques, while thick filaments overlap separate thin filaments. The thin filaments are integrated into the cytoskeletal network via dense bodies, which are points of attachment with microfilaments.

twice as many thin filaments as thick filaments, smooth muscle has about 15 thin filaments for each thick filament.

Smooth muscle also differs in structure from striated muscle in membrane organization. It lacks the elaborate sarcolemmal invaginations called T-tubules, and does not have an extensive sarcoplasmic reticulum. Since these structures aid in excitation and Ca^{2+} delivery, it should not be surprising that smooth muscle also differs from striated muscle in EC coupling.

Smooth muscle contraction is regulated by both thick and thin filament proteins

Regulation of contraction is much more complex in smooth muscle than in striated muscle. Smooth muscle contractility is regulated by nerves, hormones, and physical conditions, such as stretch. As in striated muscle activation, many regulators of smooth muscle contractility exert their effects by changing $[Ca^{2+}]$. In smooth muscle, however, $[Ca^{2+}]$ exerts its effect on both thick filaments and thin filaments. Furthermore, many types of smooth muscle alter contractility by changing the *sensitivity* to Ca^{2+}, rather than $[Ca^{2+}]$. In many of the subsequent chapters, we consider the specific mechanisms by which regulators influence smooth muscle contractility. In the next section, we consider in general terms some of the more common regulatory cascades that affect smooth muscle contraction through Ca^{2+}-dependent and Ca^{2+}-independent mechanisms.

In contrast to striated muscle, smooth muscle lacks troponin; the effects of Ca^{2+} are mediated via other regulatory proteins. **Caldesmon** is an actin-binding protein that binds to the thin filament and prevents myosin from binding to actin. In this sense, caldesmon in smooth muscle functionally replaces TnC. Caldesmon moves out of this inhibitory position in response to Ca^{2+}, but it does not directly bind Ca^{2+}. When the $[Ca^{2+}]$ increases, the soluble protein calmodulin binds to Ca^{2+}, then binds to caldesmon. The calmodulin-caldesmon complex dissociates from actin and allows the formation of a cross-bridge between myosin and actin. When Ca^{2+} levels fall, the Ca^{2+}-calmodulin-caldesmon complex dissociates and caldesmon returns to its inhibitory site on actin. Many hormones that act on smooth muscle mediate their effects by regulating the Ca^{2+}-dependent effects of caldesmon. These hormones alter signaling cascades that stimulate protein kinases and protein phosphatases. For instance, when caldesmon is phosphorylated by a *MAP kinase*, it is unable to bind to actin, even though Ca^{2+} levels may fall. Thus, caldesmon phosphorylation sustains contractions in a manner that is independent of Ca^{2+}.

Much of the regulation of vertebrate smooth muscle is mediated via the thick filament proteins. Recall that muscle myosin is a hexamer of two myosin heavy chains with four myosin light chains. In smooth muscle, the myosin light chains regulate the ability of the myosin heavy chain heads to form a cross-bridge. Many agents that alter smooth muscle contractility act by changing the phosphorylation state of myosin light chain. When phosphorylated by myosin light chain kinase (MLCK), the myosin light chain enhances the ability of myosin to bind to actin. When dephosphorylated by myosin light chain phosphatase (MLCP), myosin light chain prevents the myosin heavy chain from forming the cross-bridge, thereby allowing the smooth muscle to relax.

Many of the effectors that regulate smooth muscle contractility induce their effects via regulation of the activity of MLCK or MLCP. For example, Ca^{2+} can stimulate MLCK and thereby favor contraction. The effects of Ca^{2+} on MLCK are mediated indirectly by calmodulin. Thus, Ca^{2+} exerts effects on both the thin filament (Ca^{2+}-calmodulin-caldesmon) and the thick filament (MLCK-myosin light chain). The two main pathways of Ca^{2+}-dependent regulation of smooth muscle are summarized in Figure 5.32. Many of these factors alter Ca^{2+} levels in a very complex manner. One hormone may cause a small but rapid increase in Ca^{2+} throughout the cell, whereas another hormone might cause a greater Ca^{2+} increase that is localized near the plasma membrane. These complex spatial and temporal patterns of Ca^{2+}, known as Ca^{2+} signatures, affect different signaling cascades. Once a hormone binds to its receptor on the smooth muscle membrane, it may exert effects directly on one or more components of the smooth muscle signaling pathway.

Many of the hormones act in ways that do not cause changes in $[Ca^{2+}]$ by activating or inhibiting MLCK and MLCP. For example, nitric oxide stimulates smooth muscle relaxation by stimulating guanylate cyclase. The increase in cGMP levels activates cGMP-dependent protein kinase (PKG), which phosphorylates and activates MLCP.

In later chapters, we will discuss the pathways by which diverse neural and hormonal factors regulate smooth muscle function in specific physiological systems.

Latch cross-bridges maintain smooth muscle contraction for long periods

The contractile properties of smooth muscle differ widely in terms of force generation, as well as con-

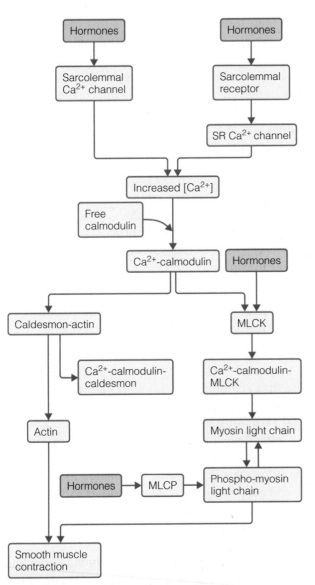

Figure 5.32 Control of smooth muscle contraction Smooth muscle contraction is regulated by pathways that target both thick and thin filament proteins.

Many smooth muscles can exhibit both tonic and phasic behavior, depending on the regulatory conditions. However, some smooth muscles have cellular specializations that favor one type of contraction. Tonic muscles are able to maintain contraction for long periods by forming a different type of cross-bridge. These latch cross-bridges alter the maximal velocity of shortening and expend less energy during isometric contraction. However, the mechanism for this difference in the latch state is not yet clear. As previously discussed, most smooth muscle contracts in response to myosin light chain phosphorylation. MLCK activates myosin light chain, triggering an increase in actino-myosin ATPase and force. However, in the latch state, force is maintained although myosin light chains are dephosphorylated and actino-myosin ATPase activity is low. This suggests that tonic muscles in the latch state are using the existing contractile machinery in a different way. At this point, we do not know for certain what factors are responsible for this different type of cross-bridge activity. Some researchers believe that the entire process of cross-bridge cycling slows. Others believe that the cytoskeleton itself interacts with actin and myosin to strengthen the physical interactions in this tonic state.

⊙ | CONCEPT CHECK

13. Does smooth muscle have actin and myosin? Does it have thick and thin filaments? Does it have sarcomeres?

14. Discuss the regulation of smooth muscle contractile properties through Ca^{2+} and Ca^{2+}-independent mechanisms.

Muscle Diversity in Vertebrates and Invertebrates

In this section, we discuss the origins of variation in muscle design. First, we will consider how individuals orchestrate changes in striated muscle fiber by changing the elements of contraction and EC coupling. Next, we will survey some of the diverse types of muscle seen in animals. Together, these sections address how animals achieve diversity in muscle structure and function within a single muscle over time (through remodeling), within an individual (through development), and between homologous muscles of different species (through evolution).

traction and relaxation rates. Smooth muscles are often broadly divided into tonic and phasic smooth muscles. Tonic muscles are those that remain contracted for a long period, whereas phasic muscles contract and relax frequently. Within the digestive system, for example, phasic muscles contract rhythmically to push the bolus of food down the gut, whereas tonic muscles in sphincters are usually contracted to prevent movement between compartments. Since these same terms are often used to distinguish types of skeletal muscle, it is important to keep in mind that these are simply descriptive terms. Tonic skeletal muscle has very different properties from tonic smooth muscle, although both exhibit long-term contraction.

Physiological and Developmental Diversity in Vertebrate Striated Muscle

Muscle performs many different functions in animals. This flexibility arises from the ability to make distinct muscle types, as well as the capacity to remodel muscle properties as required. Much of the diversity in muscle structure and contractile properties begins in embryogenesis. As we know from examining the regulation of contraction, the individual elements of EC coupling exist in many isoforms, including ion channels, pumps, Ca^{2+}-binding proteins, and contractile machinery. In principle, these genes could be expressed in countless combinations resulting in muscles with a myriad of contractile phenotypes. In actuality, most animals make only a few different types of muscle. Diversity in muscle types requires both genetic variation and an ability to express individual genes in specific combinations.

Animals make muscles of different fiber types

The genetic controls that determine isoform expression are used to produce muscles with distinct contractile properties, known as *muscle fiber types*. Some vertebrate skeletal muscles are specialized for burst activity (short duration and high intensity), whereas others are suited to endurance activity (long duration, low intensity). Various descriptive terms are used to distinguish between these fiber types. They may be called white and red muscle (based upon myoglobin content), fast twitch and slow twitch (based on the speed of contraction), glycolytic and oxidative (based on metabolic specialization), or type II and type I (based on myosin heavy chain isoforms).

Consider what is necessary to produce a specialized muscle that is used for low-frequency contractions. It contracts slowly, but it can continue contraction-relaxation cycles for long periods. Slow muscle cells express specific *types* of proteins: "slow" isoforms of thick filament proteins (myosin, myosin light chains), thin filaments (troponin, tropomyosin), and ion transport machinery. Slow muscle cells must also regulate the *amounts* of proteins involved in EC coupling, such as parvalbumin, ion channels, and ion pumps. Fiber-type specialization also demands the appropriate levels of metabolic proteins. Slow muscle fibers produce very high levels of myoglobin and mitochondrial enzymes to ensure that the ATP demands can be met by oxidative phosphorylation. In addition, the slow muscle cell must be integrated into a complex, multicellular muscle. The appropriate motor neurons make connections with the motor end plates. The blood vessels grow throughout the tissue to ensure an adequate blood supply. Finally, the slow muscle must also be connected into the necessary biomechanical framework of the skeleton. The contractile machinery is an important component of the muscle phenotype, but as you can see, many other cellular processes, both in the muscle cell itself and in surrounding cells, are necessary to construct a functional muscle.

Individuals alter fiber type in response to changing conditions

The contractile properties of muscle can be altered in response to changing physiological conditions. The first remodeling process occurs during early development, as embryonic skeletal muscles possess slow muscle isoforms of many proteins. As the fetus develops, fast muscle proteins gradually replace slow muscle isoforms in some muscles. Adult muscles can also be remodeled in response to changes in activity levels and environmental temperature. For example, exercise training can cause profound changes in both cardiac and skeletal muscle.

Both hormonal and nonhormonal mechanisms control muscle remodeling. Thyroid hormones have long been known to influence the pattern of myosin isoform expression. Thyroid hormone exerts its effects on gene expression using a specific nuclear receptor protein. The thyroid hormone receptor binds to the promoter regions that possess a thyroid hormone responsive element, or TRE. Once it binds to a hormone, the activated receptor recruits other proteins to form a multiprotein complex that can increase or decrease the rate of transcription. Thyroid hormone treatment has reciprocal effects on myosin gene expression in cardiac myocytes; it represses the expression of the β-myosin II gene, while inducing the α-myosin II gene. If the average levels of thyroid hormones remain high over a few weeks, the contractile machinery is gradually remodeled with α-myosin II replacing β-myosin II. As mentioned previously, αα-myosin dimers exhibit the fastest actino-myosin ATPase rates. Thyroid hormones regulate many of the genes involved in muscle synthesis, as well as many other genes in other tissues. By using a circulating endocrine hormone like thyroid hormone to respond to physiological challenges, animals are able to coordinate the remodeling of many tissues and physiological functions.

In contrast to endocrine control, many aspects of muscle remodeling occur in response to local signals induced by the muscle itself (Figure 5.33). Mechanoreceptors in muscle cells can detect physical changes in muscle shape and trigger changes in signaling pathways. When a muscle cell is stretched, it synthesizes regulatory proteins that influence muscle remodeling. One such protein is the protein *insulin-like growth factor II*, which is synthesized then secreted into the extracellular space. The IGF II binds to receptors on muscle plasma membranes to trigger signaling pathways that alter the expression of genes encoding muscle proteins. This type of autocrine stimulation, in combination with endocrine pathways, allows muscle to be remodeled in response to physiological challenges.

Sonic muscles produce rapid contractions but generate less force

Many animals use sound-producing organs in combination with muscles that are more specialized for high-frequency contraction. The muscles of the shaker organ in a rattlesnake tail contract 100 times per second (100 Hz). The cicada is an insect that buzzes by bending a region of its exoskeleton, called a tymbal, about 200 times a second. The toadfish produces a shrill, whistlelike sound using a sonic muscle that vibrates its swim bladder at more than 200 Hz. What is striking about each of these muscles is the way in which the animal modifies the muscle machinery to operate at such frequencies, often 10 times faster than the fastest locomotive muscles in the same animal. Surprisingly, the contractile machinery of sonic muscles is not very different from that of locomotive muscle. Typically, sonic muscles are built using fast skeletal isoforms of thick and thin filament proteins, resulting in cross-bridge cycling rates and ATPase rates that are similar to fast-twitch fibers. So what makes a sonic muscle able to contract and relax so quickly?

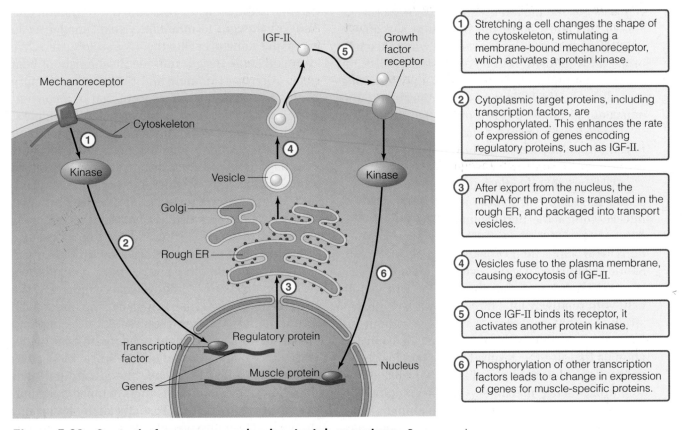

1. Stretching a cell changes the shape of the cytoskeleton, stimulating a membrane-bound mechanoreceptor, which activates a protein kinase.

2. Cytoplasmic target proteins, including transcription factors, are phosphorylated. This enhances the rate of expression of genes encoding regulatory proteins, such as IGF-II.

3. After export from the nucleus, the mRNA for the protein is translated in the rough ER, and packaged into transport vesicles.

4. Vesicles fuse to the plasma membrane, causing exocytosis of IGF-II.

5. Once IGF-II binds its receptor, it activates another protein kinase.

6. Phosphorylation of other transcription factors leads to a change in expression of genes for muscle-specific proteins.

Figure 5.33 Control of gene expression by stretch receptors   Some muscle cells sense the degree of stretch and respond by a cascade initiated by stretch receptors and culminating in changes in muscle gene expression.

First, the muscles have a very fast Ca^{2+} transient. Sonic muscles have very abundant SR. Upon excitation, the flood of Ca^{2+} from the SR rapidly saturates the regulatory sites of TnC to activate contraction. Flooding the cytoplasm with Ca^{2+} is a great way to speed contraction, but it presents a bit of a problem for relaxation. Sonic muscles speed relaxation by removing Ca^{2+} from the myofibril and the sarcoplasm very quickly. Though the structural basis remains unclear, sonic muscle troponin releases Ca^{2+} faster than skeletal muscle troponin. The SR has very active Ca^{2+} uptake machinery. These muscles also have very high levels of the Ca^{2+} buffer parvalbumin. Collectively, these processes allow for a very fast Ca^{2+} transient.

The second property necessary for rapid contraction rates is fast cross-bridge cycling. The myosin head must form a cross-bridge, undergo the power stroke, then detach. The slowest step in this cycle is the detachment of myosin from actin. Sonic muscle myosin detachment rates are about six times faster than toadfish fast-twitch fibers. The molecular basis of this difference in cross-bridge kinetics is not yet established.

Third, some muscles are able to shorten sarcomeres beyond the limit seen in most muscles. As shown in Figure 5.19, the minimum sarcomere length for most muscles is achieved when the ends of the thick filament butt up against the Z-disk. In some sonic muscles, the Z-disk has perforations that allow the thick filaments to penetrate into the adjacent sarcomeres. It is thought that this ability to change length to such a dramatic degree is important in achieving the high frequency–low force contraction in sonic muscles.

We know that the mechanical properties of the sound-producing structures also impinge on the muscle contractile performance. The muscle designs that enable these high-frequency contractions also limit their ability to generate force. Sound-producing organs use elements that are made in such a way that they can be vibrated or bent with relatively little force. They are dedicated structures that can change radically without affecting other physiological systems. In contrast, animals that use the respiratory system for vocalization face constraints on just how radically the sound-producing machinery can be modified in evolution. Any adaptations in these animals must adequately serve the dual purposes of the structures, namely respiration and sound production. It is possible that the specialized muscle properties seen in toadfish, rattlesnakes, and cicadas were made possible because they evolved in combination with dedicated sound-producing organs.

Heater organs and electric organs are modified muscles

Genetic diversity in contractile proteins affords animals the opportunity to produce muscles with unique contractile properties. These capacities arise through relatively modest changes in the profile or arrangement of muscle proteins. Although the diverse muscle fiber types may have differences in contractile properties, each muscle remains recognizable as a muscle. In some cases, a muscle may undergo *trans-differentiation*, in which it is diverted from a typical developmental program to create a tissue endowed with novel properties. Let's examine two situations that occur in fish, where embryonic muscle undergoes trans-differentiation to create a tissue with a non-contractile function.

This first example of a trans-differentiated muscle is found in billfish, a group that includes marlin and swordfish. These fish possess a trans-differentiated eye muscle that functions as a heater organ. By warming the optical sensory system, billfish are thought to maintain visual function even when pursuing prey into the deep, cold waters. We can gain some insight into the mechanism of heat generation by examining how the cellular structure of this heater organ differs from that of a conventional muscle. Heater organs have few myofibrils, but abundant SR and mitochondria. To understand how heater organs function, let's consider how normal muscles produce heat. All muscles produce some heat as a by-product of muscle metabolism, and all tissues produce heat in the reactions that lead to ATP production, as well as the reactions that lead to ATP hydrolysis. As in most tissues, considerable heat is produced by mitochondria during oxidative phosphorylation. In muscles, ATP is hydrolyzed by the ATPase reactions at the myofibrils during cross-bridge cycling, and at the ion-pumping ATPases required in EC coupling. Heater organs are thought to generate heat by cycling Ca^{2+} in and out of the SR (Figure 5.34). Activation allows Ca^{2+} to escape the SR through RyR into the cytoplasm. Ca^{2+} is then pumped back into the SR using the Ca^{2+} ATPase, fueled by mitochondrial ATP. The entire process of Ca^{2+} cycling and mitochondria energy metabolism generates enough heat to warm the eye and optical nerves.

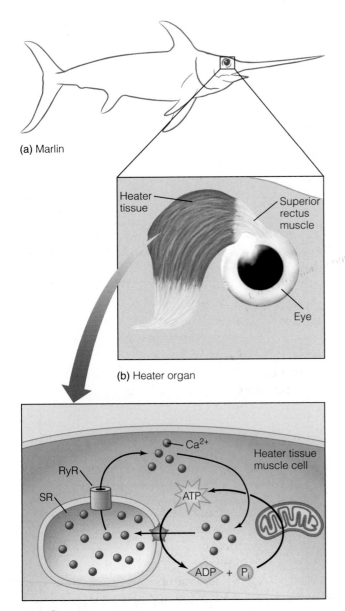

(a) Marlin

Heater tissue

Superior rectus muscle

Eye

(b) Heater organ

Ca²⁺

RyR

SR

ATP

Heater tissue muscle cell

ADP + Pᵢ

(c) Ca²⁺ cycling

Figure 5.34 Billfish heater organ Billfish, such as marlin and the swordfish **(a)**, possess heater organs. They are modified muscles found near the eye **(b)**, where they are thought to warm the optical system to maintain optical function in cold water. **(c)** Heat is generated by futile cycling of Ca²⁺ in and out of the SR, fueled by mitochondrial oxidative phosphorylation.

A second type of trans-differentiated muscle is the **electric organ**, a tissue with modified muscle cells called electrocytes. These cells produce an electrical discharge in response to neuronal stimulation. Large fish like the electric eel can produce enough electricity to shock a predator or stun its prey. Smaller species that live in dark, murky waters may use weak electrical signals to communicate. Electric organs have a polyphyletic origin, meaning they have arisen independently in many

distant groups of fish. Researchers have been able to follow the developmental processes that led to the production of electric organs. Muscle precursor cells called **myoblasts** cluster together to form a *blastema.* This ball of cells then begins to differentiate into muscle, expressing muscle-specific proteins and organizing sarcomeres. While the cells at the periphery of the blastema continue to differentiate into mature muscle, the central cells grow in size and then lose their sarcomeres. This transition probably occurs when the muscle becomes innervated by specialized electromotor neurons. These cells eventually become the electrocytes. We will revisit the electric organs in Chapter 6: Sensory Systems when we discuss their role in sensory pathways.

Invertebrate Muscles

All muscles share the features of myosin-based thick filaments and actin-based thin filaments, but the variation in the arrangement of filaments and regulation of contraction is much more pronounced in the invertebrates than the vertebrates. Researchers have studied the structural diversity in muscle of invertebrates for many years, identifying many variations in myofibrillar organization and muscle design. More recently, studies of common invertebrate model species (*Drosophila, C. elegans*) have furthered the understanding of the molecular basis of muscle development and regulation through functional genomics.

Invertebrates possess smooth, cross-striated, or obliquely striate muscle

As in vertebrates, some invertebrate muscles are smooth (lacking sarcomeres) or striated, with numerous sarcomeres attached end-to-end to form long myofibrils. Unlike vertebrates, invertebrates show many muscle forms that are intermediate between smooth and striated. There is also a great deal of variation in the arrangement of thick and thin filaments, with ratios ranging from 1:3 to 1:10 in different muscles and species.

Recall that cross-striated muscle is composed of sarcomeres attached end to end to form a myofibril that is attached to the sarcolemma at each end. The cross-striated pattern arises because the sarcomeres are attached side by side, perpendicular to the sarcolemma. **Obliquely striated muscle**, found in many invertebrates, differs from cross-striated muscle in two respects. First, the sarcomeres are not connected side by side, disrupting the pattern of cross striation. Second, instead of long myofibrils of sarcomeres, each individual sarcomere is attached to a pinnacle extending from the sarcolemma perpendicularly through the thin muscle cell. The protrusions are called dense bodies, and are similar in many respects to the dense bodies of smooth muscle. The dense bodies are attached to the inside of the sarcolemma, which is in turn connected via extracellular matrix proteins to the basal lamina, which in turn is connected to the cuticle. When obliquely striated muscle contracts, it pulls on the dense bodies, causing a local shortening of the body. In the case of *C. elegans*, the obliquely striated muscle runs under the cuticle, such that contraction causes the body to bend at that point.

Invertebrate muscles contract in response to graded excitatory postsynaptic potentials

The vertebrate striated muscles we have discussed to this point all contract when the sarcolemmal membrane potential briefly depolarizes. In the case of a neurogenic skeletal muscle, activation of the motor neurons controlling that myofiber induces depolarization of that cell and a subsequent contraction. Contraction of each fiber is "all-or-none" in response to the neuronal signal; a suprathreshold stimulus triggers massive depolarization and contraction. Vertebrate twitch muscles, which are composed of multiple myofibers, can produce graded contractions by recruiting different numbers of motor units. Strong contractions result when many motor neurons are stimulated to activate many myofibers within the muscle.

Some invertebrate muscles have a different way of translating excitatory information from the nervous system into a graded muscle contraction. Unlike vertebrate twitch muscle, these invertebrate myofibers do not contract in an all-or-none manner. In the simplest system, a single muscle fiber is innervated by a single motor neuron that controls the myofiber at multiple motor end plates, much like a vertebrate tonic muscle. When the neuron fires a single impulse, the muscle experiences a minor depolarization. The muscle responds with a small elevation of Ca^{2+} and a weak contraction. Because this depolarization induces an excitation of the muscle, it is called an **excitatory postsynaptic potential**, or EPSP (Figure 5.35). This system is able

to achieve a graded contraction because EPSPs can summate. When the nerve sends two rapid impulses, the neurotransmitters affect a broader area

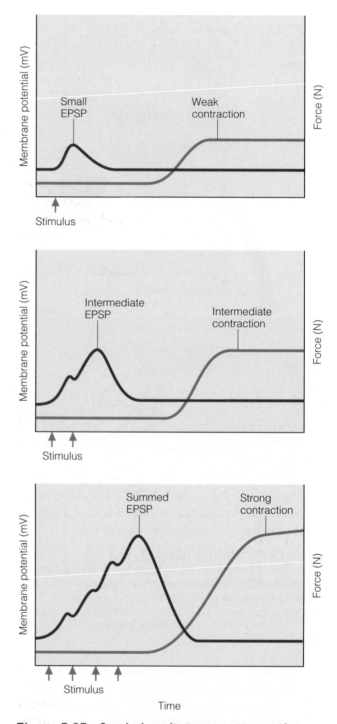

Figure 5.35 Graded excitatory postsynaptic potentials in invertebrate muscles Invertebrate muscles receive impulses from motor neurons. The degree of depolarization depends on the number of stimuli from the neurons. A single stimulus causes a small depolarization, or excitatory postsynaptic potential (EPSP), which is capable of triggering a small contraction. Multiple stimuli trigger a greater depolarization and stronger contraction.

of the sarcolemma and induce a greater depolarization, which in turn causes a greater release of Ca^{2+}. The strongest contractions result when multiple impulses trigger a very large depolarization and maximal Ca^{2+} release.

In many cases, these muscles are innervated by multiple neurons, each with a different effect on the muscle membrane potential. One excitatory neuron may induce a strong depolarization with a single impulse, acting in many ways like a motor neuron in a twitch fiber. Other excitatory neurons may innervate the same muscle cell but exert smaller effects on membrane potential, acting primarily through the summation of EPSPs. The muscle may also be innervated by inhibitory neurons. When these neurons fire, they hyperpolarize the membrane to make it more difficult to induce a contraction. In general, the invertebrates use complex innervation to control simple muscles, whereas vertebrates use a multiplicity of fibers with more straightforward innervation.

Asynchronous insect flight muscles do not use Ca^{2+} transients

As we have seen, many muscles rely on the Ca^{2+} transient to trigger cycles of contraction and relaxation. In the fastest of vertebrate skeletal muscles, the toadfish sonic muscle, Ca^{2+} transients occur as fast as one hundred times a second (100 Hz). However, vertebrate muscles cannot be induced to contract faster than this due to the limits of the vertebrate EC coupling machinery. The sonic muscles of the cicada are unusual in that their mode of EC coupling is fundamentally similar to that of vertebrate skeletal muscles, yet they are able to contract and relax much faster. The flight muscles of many insects are even faster. The high-frequency buzz of flying insects arises when the wings beat in the range of 250 to 1000 Hz. They are able to contract at these remarkable frequencies by using a different mode of EC coupling.

Recall that vertebrate muscles contract in response to a single spike of Ca^{2+} arising from a single action potential. To relax, these muscles must reduce Ca^{2+} to low levels to inactivate the actinomyosin ATPase. Insect flight muscles differ from this model in the linkage between neuronal stimulation and contraction (Figure 5.36). As with other neurogenic muscles, the insect first activates the flight muscle by a single neuronal stimulation. However, unlike other muscles, a single action po-

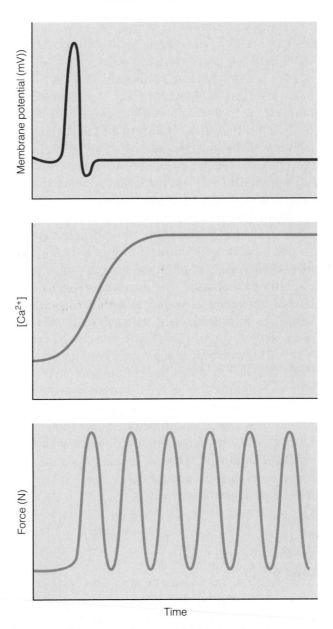

Figure 5.36 Stretch-activated asynchronous muscles Asynchronous muscles generate multiple cycles of contraction and relaxation in response to a single neuronal stimulation. During the period following excitation, Ca^{2+} levels likely remain high. Relaxation occurs in response to contraction-induced inaction. Contraction is in response to stretch activation.

tential is followed by a long series of contraction and relaxation cycles. During flight, multiple action potentials occur but the frequency is much lower than the wing beat frequency. This type of muscle is called **asynchronous flight muscle** because nervous stimulation is not synchronized with contraction. Most flying insects use asynchronous flight muscle to fly, although many also incorporate synchronous flight muscles, particularly to control the fine movements required for navigation.

The asynchronous flight muscle is able to contract and relax at high frequency because the transition between contraction and relaxation does not require a Ca^{2+} transient. Although it has not been measured, it is likely that the Ca^{2+} levels probably remain high in asynchronous muscles for the entire contraction-relaxation cycle and for the duration of flight. Once the muscle contracts, it becomes insensitive to Ca^{2+}, which is then released from TnC, allowing the muscle to relax. Once relaxed, the muscle is stretched by elastic elements in the flight apparatus. Once stretched, the myofibril regains its affinity for Ca^{2+}. Although the phenomenon of stretch activation–contraction inactivation has been recognized for decades, the molecular basis remains a bit obscure. Recent studies suggest that stretch activation is linked to peculiar structural variations in thin filament regulatory proteins. Insects with asynchronous flight muscle express a type of TnC with only a single Ca^{2+} binding site. The flight muscles in these insects possess myofibrils with combinations of both the normal two-site TnC and the unusual one-site TnC. The two-site TnC may be responsible for initiating contraction in response to the Ca^{2+} trigger induced by the action potential. The second form of TnC may be responsible for the pattern of stretch activation.

Mollusc catch muscles maintain contraction for long periods

Bivalve molluscs (clams, oysters, and mussels) possess a most remarkable muscle that is capable of generating long duration contractions while expending remarkably little energy. The muscles, often adductor muscles, are responsible for rapidly closing the shells and maintaining this state for very long periods, protecting the animal from predators or harsh external conditions.

These muscles possess a thick and thin filament structure similar in many respects to that of vertebrate smooth muscle, but with important differences. A large dimeric protein, *paramyosin,* forms the core of

the thick filament, around which a monolayer of myosin molecules is attached. Myosin itself is distinct from vertebrate myosins, and can be regulated directly by Ca^{2+}. (Recall that the thick filament of vertebrate smooth muscle is also regulated by Ca^{2+}, though indirectly.)

When the catch muscle is stimulated by cholinergic nerves, the acetylcholine triggers an increase in sarcoplasmic $[Ca^{2+}]$ (Figure 5.37). When Ca^{2+} binds myosin, cross-bridge cycling occurs and the muscle contracts. Sustained cholinergic activity for a time ensures that $[Ca^{2+}]$ remains elevated and force is generated. However, after a time, Ca^{2+} levels decline, yet the catch muscle remains contracted. It is not until serotonergic nerves release serotonin that the muscle relaxes, without changes in $[Ca^{2+}]$. Remarkably, during this period of sustained contraction, the muscle consumes very little energy, suggesting that cross-bridge cycling has ceased.

The mechanisms by which the catch muscle sustains contraction remain unclear, but it is thought that the changes are related to phosphorylation of another unusual protein, *twitchin*. This protein is related to titin, the enormous protein that controls the length of a sarcomere. When twitchin

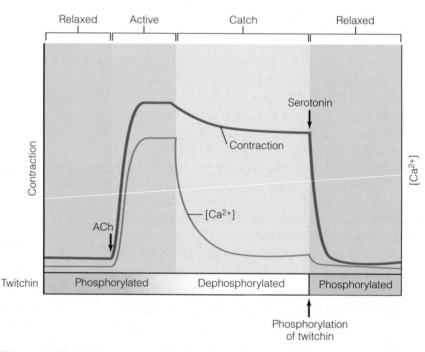

Figure 5.37 Molluscan catch muscle contraction and relaxation
Upon stimulation by cholinergic nerves, the increase in acetylcholine induces contraction of the mollusk adductor muscle. Even though Ca^{2+} levels decline, the muscle remains contracted in the catch state, where little energy is consumed. Relaxation ensues after serotonergic nerves fire. The changes in catch state coincide with changes in phosphorylation of the protein twitchin.
(Adapted from Funabara et al., 2005)

is phsophorylated, the muscle is capable of twitch activity: contracting and relaxing. However, when the catch muscle is engaged, twitchin becomes progressively dephosphorylated, likely via the action of a calmodulin-sensitive protein phosphatase calcineurin. The dephosphorylation of twitchin coincides with the entry into the catch state. Upon exit from the catch state, serotonin activates protein kinase A (PKA), which phosphorylates twitchin. It remains unclear how dephosphorylated twitchin works to attain the catch state. It is possible that the protein strengthens the actin-myosin cross-bridges or alternately creates other types of interactions between thick and thin filaments.

⊙ | CONCEPT CHECK

15. What are muscle fiber types and how are they produced?
16. Why are heater organs and electric organs considered modified muscles?
17. Compare and contrast EC coupling in synchronous and asynchronous insect flight muscles.

Summary

Cytoskeleton and Motor Proteins

→ The cytoskeleton (microtubules, microfilaments) in combination with motor proteins (dynein, kinesin, myosin) conducts many types of intracellular and cellular movement. Suites of accessory proteins control cytoskeletal assembly and disassembly.

→ Motor proteins, enzymes that use the energy of ATP to move in specific directions along cytoskeletal tracks, are encoded by large gene families, providing the cell with functional flexibility.

→ Microtubules, polymers of tubulin, assemble and disassemble spontaneously, subject to tubulin concentration, temperature, and microtubule-associated proteins or MAPs.

→ Two types of motor proteins use microtubules. Kinesin moves along microtubules in the positive direction, whereas dynein moves in the negative direction. Cilia and flagella are composed of microtubules and dynein.

→ Microfilaments are polymers of actin that work with myosin as a motor protein. Each myosin shares a general structure of a head, a neck, and a tail.

→ The sliding filament model describes how myosin forms cross-bridges with actin, then undergoes conformational changes that cause myosin to walk along the microfilament.

→ The distance myosin reaches with each cross-bridge cycle is called the unitary displacement. The duty cycle is the proportion of time in the cross-bridge cycle that each myosin is attached to actin.

Muscle Structure and Regulation of Contraction

→ Myocytes are contractile cells unique to animals that generate force in muscles, which have vital roles in many physiological systems.

→ Myocytes have thick filaments, mostly myosin, and thin filaments, mostly actin. Striated muscle filaments are arranged into sarcomeres. The sarcomere length reflects the degree of overlap, and consequently the ability to form cross-bridges.

→ Sarcomeres can be arranged in series or in parallel to achieve a favorable balance between the degree of shortening and force generation.

→ Contraction in striated muscle occurs in response to Ca^{2+}-dependent activation of thin filament regulatory proteins. At rest, the troponin-tropomyosin complex is located on the actin filament in a position that prevents myosin from binding. The increase in $[Ca^{2+}]$ upon excitation causes a structural change in TnC that initiates a chain reaction of structural changes in other troponin subunits and tropomyosin.

→ These thin filament regulatory proteins exist in many isoforms, each with subtle differences in sensitivity to physiological regulators such as Ca^{2+}, pH, and temperature.

→ Muscle myosin has an unusual unitary displacement and duty cycle. Animals have different myosin II isoforms that allow them to fine-tune actino-myosin ATPase activity.

→ EC coupling describes how extracellular processes trigger elevation of muscle $[Ca^{2+}]$.

→ Striated muscles contract in response to membrane depolarization, which can occur spontaneously (myogenic) or in response to nerve stimuli (neurogenic).

→ Tonic muscles are innervated at many locations along the muscle fiber, while twitch muscles are usually innervated at a single motor end plate.

→ Depolarization of muscle membranes, particularly in larger fibers, is facilitated by T-tubules, inward extensions of the sarcolemma deep into the fiber.

→ Depolarization opens voltage-dependent Ca^{2+} channels (DHPR) that trigger Ca^{2+} release from the SR. Muscles differ in how they link DHPR activation to SR Ca^{2+} release; heart muscle uses Ca^{2+}-induced Ca^{2+} release to trigger contraction, while skeletal muscle uses depolarization-induced Ca^{2+} release.

→ Relaxation requires removal of Ca^{2+} from the cytoplasm using pumps and exchangers in the sarcolemma and the SR.

Muscle Diversity in Vertebrates and Invertebrates

→ Animals possess large gene families for many thick and thin filament proteins, allowing them to make countless distinct muscle fiber types.

→ Individuals can alter muscle properties during development, as well as in response to physiological and environmental changes.

→ Heater organs and electric organs are examples of extreme modifications in striated muscle fiber design. They trans-differentiate, during development, losing muscle properties and gaining unique features.

→ Sonic muscles contract and relax rapidly, due in part to faster cross-bridge kinetics, faster Ca^{2+} transients, and Z-disks that allow shorter sarcomere length.

→ Smooth muscle lacks organized sarcomeres, scattering thick and thin filament arrays throughout the cell with a complex geometry.

→ Contraction of smooth muscle can be controlled by Ca^{2+}-dependent and Ca^{2+}-independent ways, acting at both the thick and thin filament.

→ Ca^{2+} levels influence both caldesmon regulation and myosin light chain phosphorylation, acting through the soluble Ca^{2+}-binding protein calmodulin.

→ A number of hormones affect myosin light chain phosphorylation by acting on MLCK or MLCP.

→ Tonic smooth muscle can alter the fundamental properties of cross-bridge reactions. They can reduce the velocity of shortening and the actinomyosin ATPase rate without sacrificing force. The mechanisms by which these long-lasting "latch" cross-bridges form is not yet clear.

→ Some insect muscles can exhibit graded contractions, with the strength of contraction dependent on summation of postsynaptic potentials, either excitatory (EPSP) or inhibitory (IPSP).

→ Some insect muscles contract so quickly that Ca^{2+} transients are not possible. These asynchronous flight muscles rely on stretch activation of opposing muscles to elevate and depress opposing wings.

→ Catch muscles strengthen cross-bridges to maintain tension while consuming minimal energy.

Review Questions

1. What is the role of energy in construction and use of the cytoskeleton?

2. How do animals use muscle in physiological systems?

3. Compare the contractile properties of sonic and locomotor muscles of fish.

4. Contrast the properties exhibited by myosins that walk on microfilaments versus thin filaments.

5. What is the difference between the latch state of vertebrate smooth muscle and the catch state of mollusc adductor muscle?

6. What are muscle fiber types? How do animals alter muscle fiber types in response to physiological challenges?

7. Discuss the role of Ca^{2+}-binding proteins in muscle contraction.

Synthesis Questions

1. What genomic and genetic events might have contributed to the expansion of the myosin II family in vertebrates?

2. What would happen if cells could only add or remove tubulin (or actin) from one end of the microtubule (or microfilament)?

3. Describe the molecular processes of neuromuscular excitation, from the sites of neurotransmitter synthesis to Ca^{2+} release within the muscle.

4. Hummingbird hearts beat at about 30 Hz. Predict what you would find if you examined the structure of a hummingbird cardiomyocyte.

5. In Chapter 2 we compared the main pathways of energy production: glycolysis and mitochondria. Discuss how these metabolic pathways integrate into the EC coupling patterns of different muscles.

6. Striated muscle cells are postmitotic and can live for the lifetime of the organism. Discuss how this property affects muscle biology, both normally and in disease.

Quantitative Questions

1. Many cellular structures require metabolic energy to build and maintain. Calculate the cost of building the microtubule support for the axon of a motor neuron. Assume that the axon is 1 m long, 1 μm in diameter with 50 microtubules aligned in parallel. If a tubulin monomer is 8 nm long, how many tubulins are needed to produce the microtubules of the axon? How many moles of GTP and GDP are tied up in the structure of this microtubule?

2. Most skeletal muscles generate about 20 N of force per cm^2 of cross-sectional area. If a myosin head generates 5 pN of force, and a thick filament has about 600 myosin heads, how many thick filaments appear per cm^2 of cross-sectional area?

For Further Reading

See the Additional References section at the back of the book for more readings related to the topics in this chapter.

Cytoskeleton and Motor Proteins

These two textbooks are good general references for cellular and molecular aspects of the cytoskeleton and motor proteins.

Alberts, B., A. Johnson, J. Lewis, M. Raff, K. Roberts, and P. Walter. 2002. *Molecular biology of the cell,* 4th ed. New York: Garland Science.

Becker, W. M., L. J. Kleinsmith, and J. Hardin. 2002. *The world of the cell,* 5th ed. San Francisco: Benjamin Cummings.

Muscle myosin evolution is responsible for much of the diversity in muscle function seen in animals. These reviews examine myosin evolution in the context of the broader roles of myosins in animals, including muscle function.

Berg, J. S., B. C. Powell, and R. E. Cheney. 2001. A millennial myosin consensus. *Molecular biology of the cell* 12: 780–794.

Sellers, J. R. 2000. Myosin: A diverse superfamily. *Biochimica et Biophysica Acta* 1496: 3–22.

These two articles examine the relationships between the structure and function of the three motor proteins that work in conjunction with the cytoskeleton.

Burgess, S. A., M. L. Walker, H. Sakakilbara, P. J. Knight, and K. Oiwa. 2003. Dynein structure and power stroke. *Nature* 421: 715–718.

Kull, F. J., R. D. Vale, and R. J. Fletterick. 1998. The case for a common ancestor: Kinesin and myosin motor proteins and G proteins. *Journal of Muscle Research and Cell Motility* 19: 877–886.

Muscle Structure and Regulation of Contraction

These excellent reviews address the molecular and genetic mechanisms that control changes in muscle contractile properties.

Baldwin, K. M., and F. Haddad. 2001. Plasticity in skeletal, cardiac and smooth muscle. *Journal of Applied Physiology* 90: 345–357.

Berchtold, M. W., H. Brinkmeier, and M. Muntener. 2000. Calcium ion in skeletal muscle: Its crucial role for muscle function, plasticity and disease. *Physiological Reviews* 80: 1216–1265.

Bers, D. M. 1991. *Excitation-contraction coupling and cardiac contractile force.* Dordrecht, the Netherlands: Kluwers Academic.

This review discusses the impact of structural variation in muscle myosin. It focuses on how structural variations in vertebrates affect function, particularly in relation to muscle diseases.

Reggiano, C., R. Bottinelli, and G. J. M. Stienen. 2000. Sarcomeric myosin isoforms: Fine-tuning of a molecular motor. *News in Physiological Sciences* 15: 26–33.

Muscle Diversity in Vertebrates and Invertebrates

These reviews discuss the ways specialized muscles are produced, and the importance of structural variation on muscle function.

Nahirney, P. C., J. G. Forbes, H. D. Morris, S. C. Chock, and K. Wang. 2006. What the buzz was all about: Superfast song muscles rattle the tymbals of male periodical cicadas. *FASEB Journal* 20: 2017–2026.

Rome, L. C., R. P. Funke, R. M. Alexander, G. Lutz, H. Aldridge, F. Scott, and M. Freadman. 1988. Why animals have different muscle fibre types. *Nature* 355: 824–827.

This interesting review discusses how animals control their modified muscles in relation to neuroethology.

Bass, A. H., and H. H. Zakon. 2005. Sonic and electric fish: At the crossroads of neuroethology and behavioral neuroendocrinology. *Hormones and Behavior* 48: 360–372.

PART
TWO

Integrating Physiological Systems

With an understanding of the cellular underpinnings of animal biology, we can explore the way groups of cells are integrated into systems. In each of the following chapters, we discuss the ways that these systems allow animals to solve physiological problems and environmental challenges. Animals detect environmental cues with the sensory systems (Chapter 6) and coordinate complex physiological and behavioral responses using nervous systems (Chapter 7). The circulatory system (Chapter 8) is responsible for transporting oxygen, carbon dioxide, and signaling molecules throughout the body to meet the demands of the tissues, and the respiratory system (Chapter 9) is responsible for exchanging gases with the environment. Animals meet osmotic and ionic challenges by controlling the levels of solutes, water, and nitrogenous wastes (Chapter 10). Nutrients must be extracted from a complex world, and processed into molecules that can be used as building blocks or energy (Chapter 11), and movement, central to animal function, is mediated by muscles and skeletal systems (Chapter 12). An example of the complex interactions needed for movement can be seen in the scales of a butterfly wing, shown here in a microscopic view. Environ-

mental temperature is a challenge that influences many aspects of animal physiology (Chapter 13). If an animal overcomes all of these challenges, it may have an opportunity to reproduce (Chapter 14).

As you will see in Part Two, each physiological system relies to some extent on neurons and muscle cells, and the signaling pathways that allow cells to communicate. Although Part Two is organized around physiological systems, you will see that this is somewhat arbitrary—each system interacts with others. For example, the locomotor system is regulated by the nervous system and fueled by the digestive system. The delivery of vital nutrients and gases to muscles is controlled by the cardiovascular and respiratory systems. Locomotor activity impinges on thermal biology (generating heat) and reproduction (finding and competing for mates). In reading the following chapters, keep in mind that the division of integrative physiology into these nine systems is an organizational tool that allows these complex systems to be discussed as if they were isolated units. To remind you of their integration, each chapter concludes with an essay that emphasizes how these systems interact. ◉

Sensory Systems

Animals are equipped with a diverse array of sensory systems that they use to monitor their internal and external environments. When we think of these sensory systems, we often imagine the complex ears of vertebrates, or the multifaceted eyes of insects. Complex sensory organs such as eyes and ears contain a large number of sensory cells and accessory tissues, but animal sensory systems may be as simple as an isolated sensory cell that sends information to the brain for processing. Indeed, at the level of the sensory cell, the sensory systems of multicellular animals have much in common with the sensory mechanisms used by unicellular organisms.

Consider a unicellular eukaryote, the paramecium. A paramecium swims by coordinated beating of its cilia. As discussed in Chapter 4: Neuron Structure and Function, if you gently touch a paramecium, it will back away from the touch stimulus by reversing the direction of beating of its cilia, turn slightly, and then proceed forward. Touching the surface of

the paramecium opens a mechanosensitive ion channel on the surface of the membrane, allowing ions to move across the membrane and depolarize the cell. This depolarization opens voltage-gated Ca^{2+} channels, causing an action potential that sends a signal to the cilia to reverse the direction of their beating. Paramecia can also detect environmental chemicals. They move toward some chemicals, but are repelled by others. Exposure to an attractant chemical hyperpolarizes the membrane, whereas exposure to a repellent chemical depolarizes the membrane, changing the beating of the cilia. A paramecium can also detect environmental temperature. If you acclimate a paramecium to a particular temperature, it will swim away from water that is either warmer or colder than the acclimation temperature. The cell membrane of the paramecium contains heat-sensitive and cold-sensitive Ca^{2+} channels. When these channels are activated, the resulting Ca^{2+} current changes the beating of the cilia, changing swimming behavior.

Deerfly eye.

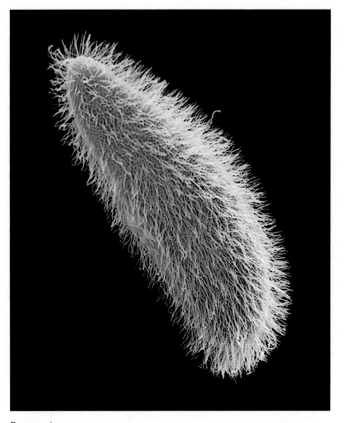

Paramecium.

Paramecia can also sense a variety of other environmental parameters using sensory mechanisms that we do not yet understand. For example, paramecia tend to swim up toward the surface of a container, but they lose their ability to orient toward the surface in low-gravity environments, which suggests that they can detect gravitational cues. Paramecia are also sensitive to electrical currents. When placed in an electrical current, a paramecium will swim toward the cathode and away from the anode. Electrical shocks are a noxious stimulus for a paramecium; when given an electric shock, it rapidly swims away. Together, these observations suggest that paramecia are sensitive to electric fields and discharges. Interestingly, paramecia do not ordinarily change their behavior in response to vibrations, but if an individual is conditioned by repeatedly exposing it to an electrical shock and a vibration, it will soon begin to respond to the vibration alone, which suggests that the organism can sense vibrations, even though they do not ordinarily respond to them.

The sensory mechanisms used by single-celled organisms like the paramecium have a lot to teach us about the sensory mechanisms used by animals. In fact, as we will see in this chapter, the role of a sensory system (whether it is in a paramecium or a complex multicellular animal) is to detect an environmental stimulus and transduce this signal into a change in the membrane potential of the sensory cell. This change in membrane potential then acts as a signal to the nervous system that can be interpreted and used by the organism to regulate physiological systems or behavior.◉

Overview

The sensory receptors of animals can be as simple as a single sensory neuron, or can involve complex sense organs, such as the eye, that contain multiple sensory receptor cells and accessory structures. Sensory receptor cells are typically specialized to detect a single type of stimulus, but no matter what kind of stimulus they detect, all sensory receptor cells work via mechanisms that are broadly similar to those used by cells to detect incoming chemical signals that we discussed in Chapter 3: Cell Signaling and Endocrine Regulation.

Sensory receptor cells take incoming stimuli of various types and *transduce* (convert) them into changes in membrane potential (Figure 6.1). In most sensory receptor cells, specialized receptor proteins in the membrane absorb the energy of the incoming stimulus and undergo a conformational change. The conformational change in the receptor protein then activates a signal transduction pathway that, directly or indirectly, opens or closes ion channels in the cell membrane, changing the membrane potential.

The change in membrane potential caused by the detection of the incoming stimulus ultimately sends a signal onward to integrating centers such as the brain. The integrating centers must then interpret this incoming sensory information and elicit appropriate responses. Thus, sensory reception is a process with many steps, including (1) reception of the signal, (2) transduction of the signal, (3) transmission of the signal to the integrating center, and (4) perception of the stimulus at the integrating center. In this chapter we begin by discussing some of the general features of sensory reception, and then we examine how specific sensory systems perform the steps of sensory reception.

General Properties of Sensory Reception

The terminology used in the field of sensory physiology can be confusing, because similar terms can be used for very different structures. In this chapter, we use the term *sense organ* to describe a

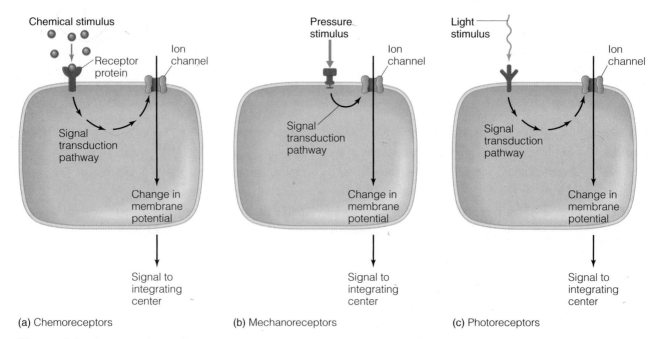

(a) Chemoreceptors   (b) Mechanoreceptors   (c) Photoreceptors

Figure 6.1 An overview of sensory receptors
Sensory receptors detect incoming stimuli of many kinds. **(a)** Chemoreceptors detect chemical stimuli. For most chemoreceptors, chemicals bind to the receptor, causing a conformational change and activating a signal transduction pathway that opens or closes ion channels, which alters the membrane potential of the sensory cell. **(b)** Mechanoreceptors detect stretch or tension on the cell membrane. When a pressure stimulus distorts the cell membrane, it changes the conformation of the mechanoreceptor protein, opening ion channels and changing the membrane potential of the sensory cell. **(c)** Photoreceptors detect light by absorbing the energy carried by the incoming light stimulus, and changing shape, activating a transduction pathway that opens or closes ion channels, resulting in a change in the membrane potential of the sensory cell.

complex structure consisting of multiple tissues that work together to allow an organism to detect an incoming stimulus. The eyes of vertebrates are an example of a sense organ. We use the term **sensory receptor** to refer to a cell that is specialized to detect incoming sensory stimuli. Sensory receptor cells can be found within complex sensory organs, as is the case for the light-sensitive cells in the retina of vertebrate eyes. Other sensory receptors are isolated cells embedded within a nonsensory tissue, as is the case for many of the touch-sensitive cells in the skin of vertebrates. The membranes of sensory receptor cells contain specific *receptor proteins* that are specialized to detect incoming sensory signals. A change in the conformation of these proteins activates signal transduction pathways within the sensory receptor, causing a change in membrane potential that can act as a signal in the nervous system.

Recall from Chapter 4 that **afferent neurons** send signals in the form of action potentials from the periphery to integrating centers such as the brain. Some sensory receptors are themselves afferent neurons. These afferent neurons detect incoming signals and transduce them into action potentials that can be sent to the integrating center. This type of sensory receptor is termed a *sensory neuron* (Figure 6.2a). Other sensory receptors are epithelial cells that send a signal to a separate afferent neuron that then sends signals in the form of action potentials to the integrating center (Figure 6.2b). In the case of a sensory neuron, a receptor protein in the dendrite of the neuron detects the incoming sensory signal, and changes conformation. The change in the conformation of the receptor protein alters the activity of a signal transduction pathway that ultimately results in a change in the membrane potential of the receptor. This change in membrane potential is a type of graded potential (see Chapter 4) that is termed a **generator potential**. The generator potential spreads along the membrane to the spike-initiating (trigger) zone of the neuron, where it will generate action potentials in the axon, if

the generator potential exceeds the threshold potential for the neuron. Recall from Chapter 4 that the spike-initiating zone of a sensory neuron need not be located in the axon hillock of the neuron. Sensory neurons are often bipolar or unipolar neurons, with their spike-initiating zones located at the distal end of the neuron between the dendrites and the axon. The action potentials are then conducted along the axon to the axon terminals of the neuron where they cause the release of a neurotransmitter. This neurotransmitter conveys the signal to other neurons and onward to integrating centers such as the brain, where they are interpreted.

When the sensory receptor cell is separate from the afferent sensory neuron, the initial graded potential in the sensory receptor cell is called a **receptor potential**. The receptor potential spreads across the sensory receptor cell to the site of the synapse with the afferent neuron, where it triggers the release of neurotransmitter. The neurotransmitter then binds to receptors on the

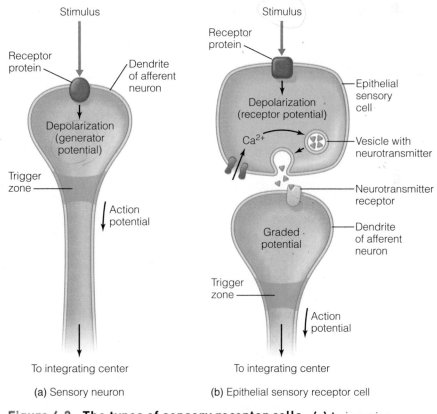

(a) Sensory neuron

(b) Epithelial sensory receptor cell

Figure 6.2 **The types of sensory receptor cells** **(a)** An incoming stimulus activates a receptor protein in the sensory neuron, causing a depolarization called a generator potential. The generator potential triggers action potentials in the axon of the neuron. **(b)** An incoming stimulus activates a receptor protein on the surface of the receptor cell, causing a receptor potential. The receptor potential opens voltage-gated Ca^{2+} channels, causing the release of neurotransmitter onto the primary afferent neuron. The stimulated afferent neuron generates action potentials that are conducted to integrating centers.

primary afferent neuron and causes a postsynaptic graded potential. This potential then travels to the trigger zone of the afferent neuron, where it initiates action potentials if it exceeds threshold. The action potentials are conducted along the axon to the axon terminals of the afferent neuron, causing the release of neurotransmitter and communicating the signal to the nervous system.

Whether a sensory receptor is a neuron or an epithelial cell, however, its function is to detect an incoming stimulus and transduce it into changes in membrane potential that convey information about the signal to integrating centers.

Classification of Sensory Receptors

Sensory receptors and sense organs can be classified in a number of different ways. In elementary school, you probably learned about the five senses (touch, taste, smell, hearing, and vision). This classification, first developed by Aristotle over 2000 years ago, is an explicitly human-centered system that focuses only on the senses that we consciously employ, ignores some obvious senses such as our ability to detect temperature changes, and entirely neglects sensory information that we are not consciously aware of, such as internal environmental parameters like blood pressure and blood oxygenation. This classification scheme also neglects the wide range of sensory systems in other animals. Many animals have senses that humans do not appear to possess, such as the ability to detect electric or magnetic fields. Similarly, some animals lack one or more of the five senses defined for humans.

Receptors can be classified based on stimulus location or modality

An alternative way of classifying sensory receptors is by the location of the stimulus. In this classification, *telereceptors* (or *teleceptors*) detect stimuli coming from locations at some distance from the body. Vision and hearing are good examples of telereceptive senses. *Exteroceptors* detect stimuli occurring on the outside of the body, such as pressure and temperature, and *interoceptors* detect stimuli occurring inside the body, such as blood pressure and blood oxygen. This classification is of limited utility to physiologists, however, because it tells us little or nothing about how the receptors work.

The most physiologically meaningful classification of sensory receptors is based on the type of stimulus that the receptor can detect, which is sometimes called the *stimulus modality*. **Chemoreceptors** detect chemical signals. They form the basis of the senses of smell and taste and are important in sensing components of the internal environment such as blood oxygen and pH. Pressure and movement stimulate **mechanoreceptors**, which are involved in the senses of touch, hearing, and balance, as well as in *proprioception*, or the sense of body position. Mechanoreceptors are also involved in detecting many important internal body parameters, such as blood pressure. **Photoreceptors** detect light, and are the basis for the sense of vision. **Thermoreceptors** sense temperature. **Electroreceptors** and **magnetoreceptors** sense electric and magnetic fields, respectively.

Receptors may detect more than one stimulus modality

Although most receptors have a preferred (or most sensitive) stimulus modality, called the **adequate stimulus**, some receptors can also be excited by other stimuli, if the incoming signal is sufficiently large. For example, if you press on your eyelid when your eye is closed, you may perceive a bright spot of light. Although light is the adequate stimulus for the photoreceptors of your eyes, sufficient pressure can also stimulate these photoreceptors, causing them to send a signal to your brain. Your brain interprets this signal as a light, because it has been programmed to interpret any signal coming from the photoreceptors of your eyes as a light stimulus.

A few types of receptors are naturally sensitive to more than one stimulus modality. For example, in the noses of sharks, sense organs called the **ampullae of Lorenzini** are sensitive to electricity, touch, and temperature. Receptors that can detect more than one class of stimulus are sometimes called **polymodal receptors**. The most common polymodal receptors in humans are the **nociceptors**, which detect extremely strong stimuli of various kinds, including temperature, pressure, and chemicals. Nociceptors are responsible for the sensation of pain in humans and many other animals. Although not all nociceptors are polymodal, many appear to be sensitive to a variety of tissue-damaging stimuli.

Stimulus Encoding in Sensory Systems

Whatever the type of stimulus, sensory receptors ultimately convert the signal to a series of action potentials in an afferent neuron. Since all action potentials are essentially the same, how can an organism differentiate among stimuli, or detect the strength of a signal? In order for an organism to interpret an incoming signal in a coherent way, a sensory receptor must be able to encode four important pieces of information about the stimulus into action potentials: stimulus modality, stimulus location, stimulus intensity, and stimulus duration.

Sensory pathways encode stimulus modality

One way in which sensory systems can encode stimulus modality is described by the *theory of labeled lines,* derived from the "law of specific nerve energies" proposed more than 150 years ago by Johannes Müller. Müller hypothesized that different kinds of nerves lead from sensory organs such as the ear or eye to the brain, and that each of these nerves has its own "specific nerve energy" that transmits information about a particular kind of stimulus. Thus, the optic nerve transmits the signal "light" whenever the eye is stimulated, even if the stimulus is actually pressure on the eyeball. While Müller was not quite correct in his theory (because all neurons use the same signal—the action potential), his hypothesis did outline some of the essential features of the labeled-line theory. Since most sensory receptors are maximally sensitive to only one type of stimulus, and a sensory receptor is part of or synapses with a particular afferent neuron, signals in that afferent neuron must represent a specific stimulus modality. Sensory systems are often organized into *sensory units* consisting of multiple sensory receptors that form synapses with a single afferent neuron. In general, all of the sensory receptors associated with a single afferent neuron are of the same type, and thus the theory of labeled-line perception can, in most cases, account for our ability to distinguish among different stimulus modalities.

The fundamental assumption of the labeled-line theory is that there is a discrete pathway from a sensory cell to the integrating center. However, it is clear that not all information about stimulus modality can be encoded in this way. For example,

recall the ampullae of Lorenzini, receptors in sharks that are sensitive to electricity, pressure, and temperature. How could such a receptor encode information regarding stimulus modality? A receptor sensitive to more than one sensory modality likely encodes information in the temporal pattern of its action potentials. For example, bursts of action potentials could convey a different message than a continuous series. In addition, the relative firing patterns of several adjacent sensory cells may carry information regarding stimulus modality. For example, imagine a situation in which each sensory cell is sensitive to more than one type of stimulus, but their relative sensitivities vary (for example, the first receptor might be very sensitive to stimulus A, but less sensitive to stimulus B, while a second receptor has the opposite pattern). By comparing the relative intensity of the signal coming from the two receptors, an afferent neuron could code information regarding the stimulus modality. The mechanisms underlying this "cross-fiber" coding of information are not yet entirely understood, but may be important for the coding of information from senses such as taste in the vertebrates.

Receptive fields provide information about stimulus location

Sensory systems must also encode the location of the stimulus. The task of encoding stimulus location varies among receptors. We discuss how sensory systems such as vision and hearing encode the location of a stimulus later in the chapter. But for many sensory systems, the main factor coding stimulus location is the location of the stimulated receptor on the body. Thus, the labeled-line theory, which in part accounts for coding of stimulus modality, can also explain how these sensory systems code for stimulus location.

In this section, we use the sense of touch in the vertebrates as an example of how a sensory system can encode the location of a stimulus. Afferent neurons involved in the sense of touch have a **receptive field**, which corresponds to the region of the skin that causes a response in that particular afferent neuron. The size of the receptive field varies among neurons. Neurons with large receptive fields detect stimuli across a larger area than neurons with small receptive fields, and thus neurons with small receptive fields provide more precise localization of the stimulus, or greater **acuity**, than neurons with large receptive fields. However, the information from a single afferent neuron can only signal whether a stimulus has occurred within the receptive field, and cannot provide more precise localization. Animals improve their ability to localize stimuli by having afferent neurons with overlapping receptive fields. A stimulus that causes both neuron A and neuron B to respond must be located within the area of overlap between the receptive fields of the two afferent neurons. This is an example of a phenomenon termed *population coding*, in which information about the stimulus is encoded in the pattern of firing of multiple neurons.

Many sensory systems take advantage of a phenomenon termed **lateral inhibition** to further improve acuity. In the simplified example shown in Figure 6.3, a weak stimulus such as a gentle touch across the receptive fields of neurons A, B, and C would cause each neuron to release a small amount of neurotransmitter onto its second-order neuron, stimulating all three of the second-order neurons

(A_2, B_2, and C_2). In contrast, a strong stimulus such as a pin pushing into the skin in the center of the receptive field of neuron B causes it to release a large amount of neurotransmitter onto its second-order neuron (B_2). This pin prick also causes the skin to bend slightly in the area of the receptive fields for the adjacent neurons A and C, weakly stimulating them. A weak stimulus to neurons A and C would ordinarily cause them to release a small amount of neurotransmitter onto their second-order neurons (A_2 and C_2). But in the example shown here, there are lateral interneurons that form synapses between the axon terminals of neuron B and neurons A and C. The strong response of neuron B causes it to release neurotransmitter onto these lateral interneurons. These interneurons release an inhibitory neurotransmitter that prevents the release of neurotransmitter from neurons A and C onto their second-order neurons. Thus, rather than exhibiting a weak response,

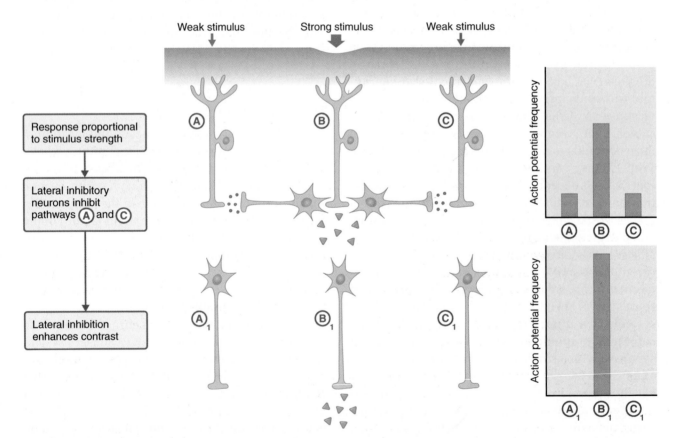

Figure 6.3 Determining the location of a stimulus with multiple receptors A stimulus at the center of the receptive field of neuron B strongly stimulates this neuron, and weakly stimulates the adjacent neurons A and C. Neuron B forms synapses with lateral interneurons that make connections with the axon terminals of neurons A and C. These lateral interneurons release an inhibitory

neurotransmitter onto neurons A and C, reducing the amount of neurotransmitter that they release. As a result, neuron B_2 receives a strong stimulus that triggers action potentials, while neurons A_2 and C_2 receive a weak stimulus that does not trigger action potentials. Lateral inhibition increases the contrast between the signals in neurons A_2, B_2, C_2, improving the ability to discriminate between stimuli.

neurons A_2 and C_2 do not fire. Lateral inhibition increases the contrast between the signals from neurons at the center of the stimulus and neurons on the edge, allowing finer discrimination.

Sensory receptors have a dynamic range

As discussed in Chapter 4, action potentials are all-or-none electrical events that do not usually code intensity through changes in magnitude. Instead, action potentials code stimulus intensity through changes in frequency. Strong stimuli typically trigger high-frequency series (or *trains*) of action potentials, whereas weaker stimuli trigger lower-frequency trains of action potentials.

Most sensory receptor cells are able to encode stimuli over a relatively limited range of intensities, called the **dynamic range** of the receptor (Figure 6.4a). The weakest stimulus that produces a response in a receptor 50% of the time is termed the *threshold of detection*. Many sensory receptors are extremely sensitive and can detect signals that are close to the theoretical detection limits for the stimulus. For example, some of the photoreceptors in the human eye can detect a single photon of light, and some mechanoreceptors on human fingertips can detect depression of the skin of less than 0.1 micron. Below the threshold stimulus intensity, the receptor cell fails to initiate action potentials. At the top of the dynamic range, the receptor cell is saturated and cannot increase its response even if the signal strength increases. In principle, any of the steps in sensory transduction can set the top of the dynamic range of a receptor. A receptor reaches the top of its dynamic range if all of the available receptor proteins become saturated. The receptor could also reach the top of its dynamic range if all available ion channels have

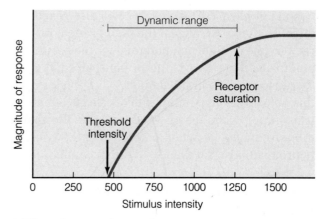

(a) Dynamic range of a receptor

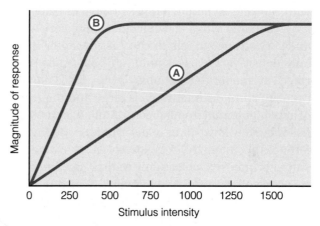

(b) Varying sensitivity

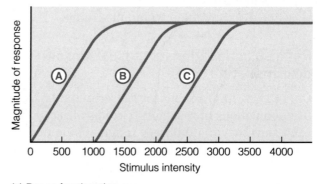

(c) Range fractionation

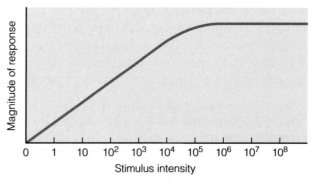

(d) Logarithmic encoding

Figure 6.4 Stimulus-response relationships in sensory receptors **(a)** Sensory receptors have a dynamic range over which the response of the receptor increases with increasing stimulus intensity. **(b)** Receptors with varying dynamic range. Receptor A is saturated at high intensity, but has a relatively small change in response for each change in stimulus intensity. Receptor B is saturated at low stimulus intensity, but has a large change in response for each change in stimulus intensity. **(c)** Using the strategy of range fractionation, several receptors can work together to provide fine discrimination across a wide range of stimulus intensities. **(d)** Some receptors encode signals logarithmically, allowing fine discrimination at low stimulus intensities and coarser discrimination at high stimulus intensities.

opened or closed. The receptor will also reach the top of its dynamic range if the membrane potential reaches the equilibrium potential for the particular ion involved in the receptor or generator potential (because no net ion movement will occur beyond this point). The maximum rate of release of neurotransmitter from the receptor cell, or the maximum frequency of action potentials in the afferent neuron can also set the top of the dynamic range.

There is a trade-off between dynamic range and discrimination

Figure 6.4b illustrates two hypothetical receptors with varying dynamic ranges. Receptor A has a large dynamic range and can detect both very weak and very strong stimuli. In contrast, receptor B can only detect very weak stimuli, and becomes saturated at moderate stimulus levels. Because the range of stimulus intensities is large and the range of action potential frequencies is limited, receptor A has relatively low power to discriminate among differences in intensity. A relatively large change in stimulus intensity causes only a small change in the response of receptor A, whereas a relatively small change in stimulus intensity causes a large change in the response of receptor B. Receptor B is sensitive to only a small portion of the possible range of stimulus intensities, but it has the ability to provide very fine discrimination within that range.

Range fractionation increases sensory discrimination

One way to improve sensory discrimination is to use populations of receptors. Groups of receptors, each sensitive to a different range of stimulus intensities, can work together to provide fine discrimination across a wider range of intensities. With this strategy, called **range fractionation**, individual receptor cells are sensitive to only a small portion of the possible range of intensities, but multiple receptors cover different parts of the range (Figure 6.4c). In a system designed in this way, stimulus intensity is actually coded through the behavior of populations of sensory receptors.

Sense organs can have a very large dynamic range

The upper limit of the frequency of action potentials is set by the refractory periods of the voltage-gated channels involved in the action potential

(see Chapter 4). The lowest action potential frequency that is likely to be physiologically meaningful is on the order of one per second, and the maximum frequency of action potentials in most neurons is around 1000 per second, yielding a dynamic range of approximately 1000-fold. In contrast, the intensity of many environmental stimuli varies across a much larger range. For example, a jet engine is about 1.4 million times as loud as the faintest sound that a human being can hear. So how can a sensory receptor code for such a wide range of stimulus intensities with such a small range of action potential frequencies? Range fractionation can extend the dynamic range of a receptor, but many receptors use another strategy.

Many receptors encode signals logarithmically

It is possible to encode a wide range of stimulus intensities using a single sensory receptor cell, without resorting to range fractionation. Figure 6.4d shows a hypothetical example of a receptor that encodes stimuli logarithmically so that the response increases linearly with the logarithm of the stimulus intensity. In this relationship there is a large increase in the response to changes in stimulus intensity when stimulus intensity is low, providing fine discrimination, but when stimulus intensity is high there is only a limited change in the response even when there is a very large change in the stimulus. Thus, there is only coarse discrimination at high stimulus intensities. This type of curve represents a compromise between a broad dynamic range and fine discrimination between similar stimulus intensities. Logarithmic coding allows a receptor to have a constant response to a given percentage change in stimulus intensity.

Many of our sensory systems employ this kind of strategy. For example, if you stand in a darkened room and light a candle, it is easy to notice the change in light intensity, but if you do the same thing in a bright room, you are unlikely to notice the difference. You have the ability to make fine discriminations between intensities at low light levels, but cannot make fine discriminations at high light levels. Similarly, if you help a friend to move furniture, you're unlikely to notice the change in weight if someone puts a book on top of the sofa, but you could easily detect the weight of the book if that was the only object you were holding. This logarithmic relationship between actual and perceived stimulus intensity is known as the Weber-Fechner relation-

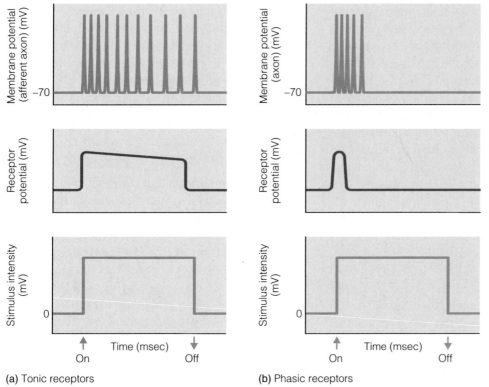

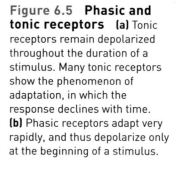

Figure 6.5 Phasic and tonic receptors **(a)** Tonic receptors remain depolarized throughout the duration of a stimulus. Many tonic receptors show the phenomenon of adaptation, in which the response declines with time. **(b)** Phasic receptors adapt very rapidly, and thus depolarize only at the beginning of a stimulus.

(a) Tonic receptors   **(b)** Phasic receptors

ship. Sensations such as brightness, loudness, and weight all obey the Weber-Fechner relationship.

Tonic and phasic receptors encode stimulus duration

Two functional classes of sensory receptors code stimulus duration (Figure 6.5). **Tonic receptors** fire action potentials as long as the stimulus continues, and thus can convey information about how long the stimulus lasts. However, most tonic receptors do not fire action potentials at the same frequency throughout the duration of a prolonged stimulus. Instead, action potential frequency often declines if the stimulus intensity is maintained at a constant level. This process is known as **receptor adaptation**. In fact, some receptors adapt so quickly that they produce action potentials only when the stimulus begins. These receptors, termed **phasic receptors**, code changes in the stimulus but do not explicitly encode stimulus duration.

We have all experienced the phenomenon of receptor adaptation. When you first step into a hot bath, the water may feel uncomfortably warm, but very soon you will no longer feel that the water is too hot. Similarly, if you walk into a house where someone has been cooking strong-smelling food, at first you may find the scent very noticeable, but after a while you may not detect the smell at all. Receptor adaptation is a physiologically critical mechanism because it allows animals to tune out unimportant information about factors in their environment that aren't changing, and to focus primarily on novel sensations.

⊙ | CONCEPT CHECK

1. For the following stimuli, determine whether the receptor involved is a mechanoreceptor, a chemoreceptor, or a photoreceptor: blood oxygen, acceleration, a light, sound waves, blood glucose.
2. Compare and contrast receptor potentials, generator potentials, and graded potentials.
3. How does lateral inhibition enhance contrast?
4. Explain the advantages of encoding sensory signals logarithmically.
5. Do you think the sensory receptor cells in the eye are tonic or phasic receptors? Justify your answer.

Chemoreception

Most cells can sense incoming chemical signals, and animals have many types of chemoreceptors that they use to sense their external and internal chemical environments. Here we focus on the senses of smell and taste, which multicellular organisms use to sense chemicals in their external environment. For terrestrial animals, **olfaction**, or the sense of smell, is generally defined as the detection of chemicals carried in air. Thus, olfaction provides the ability to sense chemicals whose source is located at some distance from the body. This is in contrast to the sense of taste, or **gustation**, which allows the detection of dissolved chemicals emitted from ingested food. Although it is easy to distinguish between gustation and olfaction for terrestrial organisms, it is more difficult to make this distinction in aquatic organisms. In aquatic vertebrates, gustation always involves detecting sensations involving food, whereas olfaction involves detecting a wide variety of environmental chemicals including those associated with food, predators, potential mates, and particular locations. In the vertebrates (whether aquatic or terrestrial), olfaction and gustation are also distinct from one another based on structural criteria; they are performed by different sense organs, use different signal transduction mechanisms, and separate integrating centers process the incoming information from the senses of taste and smell.

The Olfactory System

The ancestors of all animals undoubtedly possessed chemoreceptors, and vertebrates and insects share many similarities in the mechanisms of olfaction. However, current evidence suggests that the olfactory systems of the vertebrates and insects evolved independently. We first discuss the mechanisms underlying olfaction in the vertebrates. We then briefly compare and contrast the analogous mechanisms in the insects.

The vertebrate olfactory system can distinguish thousands of odorants

Vertebrate olfactory systems have an enormous capacity to distinguish among **odorants**, the chemicals detected by the olfactory system. Studies on humans indicate that most people can distinguish among tens of thousands of different odorants, and even a very small change in the structure of an odorant can cause a huge difference in the subjective perception of an odor. For example, humans perceive the compound octanol as smelling like oranges or roses, and describe the compound octanoic acid as smelling rancid or sweaty. The only difference between octanoic acid and octanol is that octanoic acid ends with a carboxylic acid group, whereas octanol ends in a hydroxyl group.

The vertebrate olfactory system is located in the roof of the nasal cavity (Figure 6.6). Olfaction begins when an odorant molecule comes in contact with the mucus layer that lines and moistens the olfactory epithelium of the nose. The mucus contains **odorant binding proteins**, which are thought to

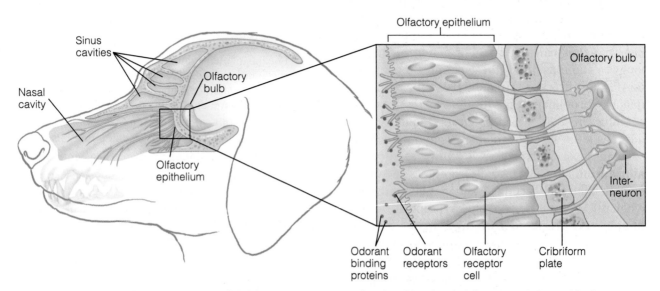

Figure 6.6 The olfactory organ of a dog The olfactory epithelium of mammals, located in the nasal cavity, contains supporting cells and olfactory receptor neurons. These bipolar sensory neurons have one end that forms synapses within the olfactory bulb of the brain. These neurons then pass through holes in the bony cribiform plate so that the ciliated end of the neuron is located in the olfactory epithelium. The cilia of the bipolar neurons contain the odorant receptor proteins that detect incoming chemical stimuli. These cilia project into a mucus layer containing odorant binding proteins that coats the olfactory epithelium.

be involved in allowing lipophilic odorant molecules to dissolve in the aqueous mucus layer. Vertebrate olfactory receptor cells are bipolar neurons with one end in the olfactory epithelium and another end that passes through holes in the bony cribiform plate and forms synapses with neurons in the olfactory bulb of the brain. On the outer surface of the olfactory epithelium the membrane of the olfactory receptor cell is highly modified and covered in cilia, which project into the mucus layer lining the inside of the nose. The cilia on the olfactory receptor neurons are nonmotile, and thus they do not beat, but they contain the **odorant receptor proteins**, which are the receptor proteins involved in detecting incoming chemical signals.

Odorant receptors are G protein coupled

Odorant receptor proteins are G-protein-coupled receptors, similar in many respects to those involved in hormonal communication (see Chapter 3). Odorant receptor proteins are members of a large multigene family, and all of the vertebrate genomes that have been sequenced so far contain many genes coding for odorant receptors (for example, the mouse genome contains at least 1000 potential odorant receptor genes). Each odorant receptor cell expresses only a single kind of odorant receptor protein out of this wide range of possible proteins.

When an odorant molecule binds to an odorant receptor, the receptor undergoes a conformational change that sends a signal to an associated G protein, G_{olf}. Activated G_{olf} signals via adenylate cyclase, activating a signal transduction pathway (shown in Figure 6.7) that ultimately causes a depolarizing generator potential. If the depolarization is sufficiently large, action potentials will be triggered in the dendrite of the olfactory receptor neuron. Note that these action potentials travel _toward_ the cell body of this bipolar neuron, in contrast to the arrangement found in a motor neuron, in which the action potential always travels <u>away</u> from the cell body. These action potentials are ultimately transmitted to the other end of the neuron, where the axon terminals form synapses with the neurons of the olfactory bulb in the brain.

Recent evidence suggests that additional signal transduction pathways may also play a role in

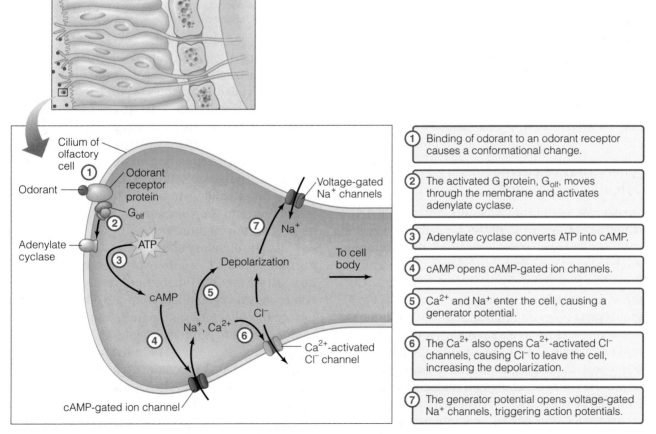

Figure 6.7 **Signal transduction in an olfactory receptor cell**

odorant detection in mammals. For example, some odorant receptors are coupled to G proteins that activate a phospholipase C (PLC)–mediated signal transduction cascade, in which PLC hydrolyzes phosphatidylinositol-4,5-bisphosphate (PIP_2) in the plasma membrane, producing inositol trisphosphate (IP_3) and diacylglycerol (DAG), which results in an increase in intracellular Ca^{2+}, causing plasma membrane Cl^- channels to open. However, just as with the cAMP-mediated signal transduction cascade, the ultimate result of the PLC-mediated signal transduction cascade is to depolarize the cell, triggering action potentials.

Although vertebrate genomes contain up to a thousand genes coding for odorant receptor proteins, the total number of odors that an animal can distinguish is even larger, possibly numbering in the tens of thousands. Experiments in mammals such as rats and humans indicate that each olfactory neuron expresses only one odorant receptor gene, but that each odorant receptor can recognize more than one odorant. Thus, a given odorant excites multiple olfactory neurons, but to different degrees. As a result, each odorant excites a unique combination of olfactory neurons. The number of distinct odorants that can be discriminated using such a combinatorial code is extremely large. Even if each odorant were coded by a combination of only three different receptors, there would be approximately 1 billion potential combinations. The code for each odor actually involves more than three receptors, and thus the potential for odor discrimination by the vertebrate olfactory system may be much larger than a billion.

An alternative chemosensory system detects pheromones

Terrestrial vertebrates use an organ called the vomeronasal organ to detect a particular class of environmental chemicals, termed **pheromones**. Pheromones are chemical signals that are released by an animal that affect the behavior of another animal of the same species (see Chapter 3). Pheromones play an important role in maintaining social hierarchies and stimulating reproduction in many animals (see Chapter 14: Reproduction). The vomeronasal organ is an accessory olfactory organ that is structurally and molecularly distinct from the primary olfactory epithelium (Figure 6.8). In mammals, the paired vomeronasal organs are found on each side of the base of the nasal cavity

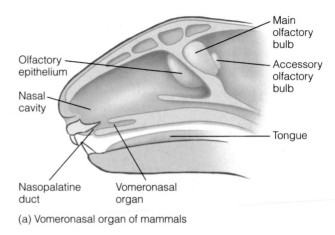

(a) Vomeronasal organ of mammals

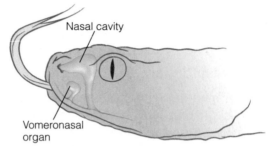

(b) Vomeronasal (Jacobson's) organ of reptiles

Figure 6.8 Vomeronasal organs **(a)** In mammals, the vomeronasal organ, which detects pheromones, is located at the base of the nasal cavity and is connected to the mouth via the nasopalatine duct. **(b)** In reptiles, the vomeronasal organ (called Jacobson's organ) is located in the palate.

near the nasal septum (the tissue that separates the two nostrils). In reptiles, the vomeronasal organ (often called Jacobson's organ, after the scientist who discovered it) is found on an analogous location on the palate. A narrow tube leads from the vomeronasal organ to either the oral cavity or the nasal cavity, depending on the species. For example, in snakes this tube is located in the oral cavity, and a snake can use its tongue to transfer pheromones to the vomeronasal organ by flicking its tongue into its mouth.

Like the olfactory epithelium, the epithelium of the vomeronasal organ expresses chemoreceptors. However, the pheromone receptors of the vomeronasal organ differ from the odorant receptors of the olfactory epithelium. Vomeronasal receptors activate a phospholipase C-based signal transduction system, while most olfactory receptors activate an adenylate cyclase–cAMP signal transduction pathway. The vomeronasal receptors have some similarity to the vertebrate receptors for bitter tastes, which we discuss in later sections of this chapter covering the gustatory system.

Invertebrate olfactory mechanisms differ from those in the vertebrates

Invertebrate olfactory organs are evolutionarily distinct from those in the vertebrates and can be located in many parts of the body, although they are most often concentrated at the anterior end, on or near the head. In arthropods (such as insects and crustaceans), the invertebrates for which olfaction has been most intensively studied, the primary olfactory organs are generally located on the antennae or antennules. The antennae are covered with hundreds of hairlike projections of the cuticle called **sensilla** (Figure 6.9). Sensilla are complex sensory organs that have a variety of morphologies and functions, including both mechanosensory and chemosensory transduction. Olfactory sensilla have a small pore at their tip to allow odorants to cross the exoskeleton. Olfactory sensilla also contain odorant receptor neurons. As in the vertebrates, these neurons express odorant receptor proteins.

The signal transduction mechanisms activated by odorant receptor proteins have been studied in only a few species of invertebrates, but they gener-ally involve cAMP as a second messenger, just as in the vertebrates. Similarly, odorant binding proteins and G-protein-coupled odorant receptors have been detected in every species of invertebrate examined so far. However, the odorant receptors of invertebrates share little sequence similarity with mammalian odorant receptors, and are likely independently derived from G-protein-coupled receptors found in the common ancestor of all animals. Even within the invertebrates, odorant receptors share little similarity among groups. For example, the odorant receptors in *Drosophila* (a fruit fly) are unlike those found in *Caenorhabditis elegans* (a nematode).

Although the odorant code has not yet been deciphered for any invertebrate, the mechanisms of signal processing likely differ among invertebrate groups. In *Drosophila*, as in the vertebrates, each olfactory neuron expresses a single odorant receptor, and olfactory neurons likely code odorant information combinatorially. In contrast, in *C. elegans,* each olfactory neuron expresses several different odorant receptors, and thus the "odorant code" cannot be a simple combinatorial system like that found in mammals.

Most invertebrate groups also produce and detect pheromones. Aquatic invertebrates are thought to use essentially the same system for detecting both odorants and pheromones, but in terrestrial invertebrates such as insects these two systems are separated. Insects have specialized pheromone-sensitive sensilla on their antennae that are similar in structure to those that detect odors, but their numbers and distributions differ between males and females. The sensory neurons of these sensilla are exceptionally sensitive and highly selective. In fact, the pheromone-sensitive sensilla of the silk moth *Bombyx mori* can detect as little as a single molecule of the pheromone bombykol.

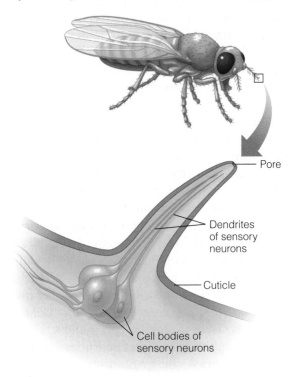

Figure 6.9 The structure of a chemosensitive sensillum   Insect sensilla are complex sensory organs that can contain both chemoreceptive and mechanoreceptive sensory neurons. Sensilla are involved in olfaction, detection of pheromones, gustation, and the senses of touch and hearing in insects.

The Gustatory System

Unlike the olfactory system, the gustatory system (or sense of taste) is not able to discriminate among thousands of different molecules. Instead, at least in humans, tastes can be grouped into one of five classes: salty, sweet, bitter, sour, and umami. Umami is a word coined by a Japanese scientist from the words *umai* (delicious) and *mi* (essence), and corresponds to a savory or meaty sensation. Sweet, umami, and salty tastes indicate nutritionally important carbohydrates, proteins, and ions, whereas bitter and sour tastes generally reflect potentially toxic substances.

Taste buds are vertebrate gustatory receptors

In terrestrial vertebrates, taste receptor cells are found on the tongue, soft palate, larynx, and esophagus and are clustered into groups known as **taste buds** (Figure 6.10). In aquatic vertebrates, taste buds can also be located on the external surface of the body. For example, many fish have taste buds on the barbells (whiskerlike projections from the lower jaw). The sea robin even has taste buds on the tips of its fins, which are useful because these fish use their fins to probe in the mud for food. Although the shapes, sizes, and distributions of taste buds vary among vertebrate species, all taste buds share certain common features. Taste buds are onion-shaped structures that contain multiple taste receptor cells (in humans each bud contains between 50 and 100 taste receptor cells), with a pore that opens out to the surface of the body. Dissolved chemicals from food, termed **tastants**, enter through this pore

and contact the taste receptor cell. The apical surface of the taste cell is folded into numerous microvilli, which contain the receptors and ion channels that mediate the transduction of the taste signal.

Vertebrate taste receptors use diverse signal transduction mechanisms

Figure 6.11 summarizes the signal transduction mechanisms used by taste receptor proteins for salty, sour, sweet/umami, and bitter tastes, respectively. Salty tastes are conveyed by Na^+ ions in food, while sour tastes are conveyed by H^+ ions. Sugars and related organic molecules convey sweet tastes, while amino acids and related molecules convey the sensation umami. In contrast, a wide range of organic molecules can convey a bitter taste, including compounds like caffeine, nicotine, and quinine.

The receptor protein for salty substances is not actually a receptor at all, but instead a Na^+ ion channel (Figure 6.11a). These Na^+ channels are also permeable to H^+ ions, and thus may play a role in the perception of sour tastes. Because Na^+ and H^+ compete for access to the channel, however, these channels are probably important for the perception of "sourness" only in species with relatively low Na^+ levels in their saliva. Thus, hamsters, which have low saliva Na^+, use these channels to detect sourness, while humans and rats, which have relatively high saliva Na^+, taste sourness through other mechanisms.

A number of sour-taste transduction mechanisms have been proposed, depending on the species being investigated. Figure 6.11b summarizes one of these potential mechanisms, which was first described in the taste receptor cells of salamanders. These taste receptor cells sense sourness via an apically localized K^+ channel that is blocked directly by protons. Blocking these K^+ channels leads to depolarization of the taste cells, by decreasing K^+ permeability and altering the resting membrane potential, as described by the Goldman equation as discussed in Chapters 3 and 4. This depolarization ultimately causes neurotransmitter release. In contrast, in frogs, taste cells contain H^+-gated Ca^{2+} channels and H^+ transporters that are believed to be involved in detecting sourness, although the specific proteins involved have not yet been sequenced. Recent molecular studies in mammals have suggested that acid-sensing ion channels (ASICs) may be important for detection of sourness.

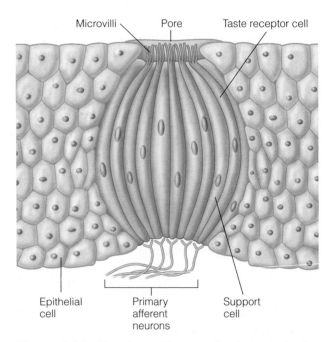

Microvilli   Pore   Taste receptor cell

Epithelial cell   Primary afferent neurons   Support cell

Figure 6.10   Structure of a vertebrate taste bud
A taste bud consists of a pore containing sensory receptor cells and support cells. The apical surface of the receptor cells is covered with microvilli that project into a pore open to the surface of the body. Receptor proteins on these microvilli detect tastants dissolved in saliva or other fluids.

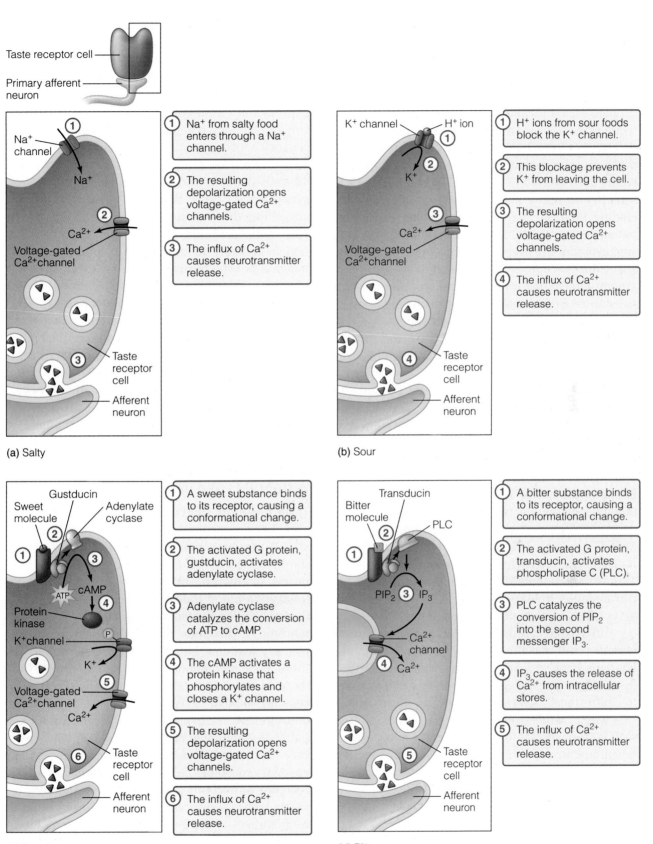

(a) Salty

① Na⁺ from salty food enters through a Na⁺ channel.

② The resulting depolarization opens voltage-gated Ca²⁺ channels.

③ The influx of Ca²⁺ causes neurotransmitter release.

(b) Sour

① H⁺ ions from sour foods block the K⁺ channel.

② This blockage prevents K⁺ from leaving the cell.

③ The resulting depolarization opens voltage-gated Ca²⁺ channels.

④ The influx of Ca²⁺ causes neurotransmitter release.

(c) Sweet

① A sweet substance binds to its receptor, causing a conformational change.

② The activated G protein, gustducin, activates adenylate cyclase.

③ Adenylate cyclase catalyzes the conversion of ATP to cAMP.

④ The cAMP activates a protein kinase that phosphorylates and closes a K⁺ channel.

⑤ The resulting depolarization opens voltage-gated Ca²⁺ channels.

⑥ The influx of Ca²⁺ causes neurotransmitter release.

(d) Bitter

① A bitter substance binds to its receptor, causing a conformational change.

② The activated G protein, transducin, activates phospholipase C (PLC).

③ PLC catalyzes the conversion of PIP₂ into the second messenger IP₃.

④ IP₃ causes the release of Ca²⁺ from intracellular stores.

⑤ The influx of Ca²⁺ causes neurotransmitter release.

Figure 6.11 Signal transduction in taste receptor cells **(a)** Signal transduction for salty substances. **(b)** Signal transduction for sour substances. **(c)** Signal transduction for sweet or umami substances. **(d)** Signal transduction for bitter substances.

These channels appear to be Na$^+$ channels that open in response to changes in pH.

The signal transduction pathway for sweet-taste receptors is summarized in Figure 6.11c. Sweet substances such as sugars bind to G-protein-coupled receptors at the apical cell surface, and activate the G protein **gustducin**, which signals through an adenylate cyclase signal transduction pathway. The receptors for "sweetness" have recently been identified in mice. These receptors are sensitive to many kinds of sweet substances, including monosaccharides, polysaccharides, high-potency sweeteners, and some amino acids. This suggests that the sweet-taste receptors are broad-spectrum receptors that do not discriminate among alternative sweet substances. Some sweet substances (in particular, strong artificial sweeteners such as saccharine) may also activate an IP$_3$-mediated signal transduction cascade, which leads to the closing of K$^+$ channels and depolarization of the receptor cell.

The taste umami, which is caused by L-glutamate and other amino acids present in foods, as well as the food additive MSG, can be detected by two different kinds of receptors, one that is similar to the receptors responsible for detecting sweetness and another that is similar to the glutamate receptors found in the brain. When glutamate binds to this modified glutamate receptor, the receptor undergoes a conformational change, activating an associated G protein. The G protein then activates a phosphodiesterase that degrades cAMP into AMP. The decreases in cAMP are thought to trigger neurotransmitter release, although the precise pathways involved have not yet been identified.

Bitter-taste receptors appear to be much more complex and specific than sweet-taste receptors. Humans have at least 25 genes coding for bitter-taste receptors, and each taste cell that is sensitive to "bitterness" expresses many of these genes. The way in which this complex pattern of expression is translated into the perception of bitterness is still unknown, although the signal transduction mechanisms within the bitter-taste receptor cells have been worked out (Figure 6.11d).

Coding differs between the olfactory and gustatory systems

There is considerable debate among sensory neurobiologists as to how the perception of a taste is coded in the brain. Taste receptor proteins act through a variety of signal transduction mechanisms, unlike odor receptor proteins, which are always coupled to G proteins. Each taste receptor cell expresses more than one kind of taste receptor protein, unlike olfactory neurons, which each express only a single olfactory receptor protein. Unlike olfactory receptor cells, which are bipolar sensory neurons, taste receptor cells are epithelial cells that release neurotransmitter onto a primary afferent neuron, and a single taste neuron may synapse with more than one taste receptor cell, suggesting that coding of taste information may be very complex. Thus, coding in the gustatory system is unlikely to operate via a mechanism in which a neuron is responsible for a single particular taste sensation. Instead, it is probable that each taste is coded by the complex pattern of activity across many neurons, and the code for perception of tastants must be quite different from the code for perception of odorants. However, despite the fact that olfaction and gustation are very different from a physiological perspective, they work together closely, and our perception of the taste of a substance is dependent on our sense of smell.

Taste reception differs between vertebrates and invertebrates

Taste receptors in arthropods are located in sensilla that are structurally similar to olfactory sensilla. Gustatory sensilla are found on many parts of the insect body including the outside of the proboscis or mouth, in the internal mouth parts (pharynx), along the wing margin, at the ends of the legs, and in the female vaginal plates. Like vertebrates, arthropods can distinguish among the primary tastants, but the mechanisms underlying these taste perceptions are quite different from those in the vertebrates. Arthropod taste receptor cells are bipolar sensory neurons, similar to the neurons involved in olfaction in the vertebrates, and unlike the epithelial cells that synapse with a sensory neuron in vertebrate gustation. In insects, the gustatory receptors belong to the G-protein-coupled receptor superfamily, similar to the olfactory receptors of vertebrates. There are approximately 60 members of the gustatory receptor gene family in the *Drosophila* genome, suggesting substantial functional complexity. In *Drosophila*, each gustatory neuron appears to express only a single receptor protein, quite unlike the situation in mammals in which each gustatory receptor cell expresses several different receptor proteins. These data suggest that, at least in *Drosophila*, the gustatory code may be combinatorial, similar to the olfactory code of mammals.

The mechanisms of gustation clearly differ between insects and vertebrates, and they differ among invertebrates as well. For example, in nematodes (the only other invertebrate for which the molecular basis of gustation has been worked out in detail), many receptor proteins are expressed in each neuron, similar to the situation in mammals, and different from the mechanisms in insects. The differences between the mechanisms of gustation in vertebrates and among invertebrates suggest that gustatory organs must have evolved independently several times.

ance, and it plays a critical role in regulating blood pressure in the vertebrates. Most mechanoreceptor cells are small and widely dispersed, making it challenging to use traditional biochemical approaches to isolate the proteins responsible for mechanosensory transduction. Thus, despite decades of investigation, the mechanisms by which a mechanoreceptor converts a mechanical stimulus to an electrical stimulus are only now being elucidated.

Genetic studies in *Drosophila* and *C. elegans* have demonstrated that there are two main types of mechanoreceptor proteins in animals: ENaC (epithelial sodium channels) and TRP (transient receptor potential) channels (Figure 6.12). Although these channels were first identified in invertebrates, they have recently been isolated from

6. Compare and contrast olfaction and gustation in the vertebrates.

7. What would happen to the ability to smell if a drug that inhibited adenylate cyclase were applied to the olfactory epithelium of a vertebrate? Would this drug affect the sensing of pheromones if applied to the vomeronasal epithelium? Justify your answer.

8. How would the response of a taste receptor cell differ between a food that is slightly salty and a food that is very salty? How would this affect action potential generation in the afferent neuron?

Mechanoreception

Mechanoreceptors are specialized cells or organs that can transform mechanical stimuli, such as pressure changes, into electrical signals that can then be interpreted by the rest of the nervous system. All organisms, and probably all cells, have the ability to sense and respond to mechanical stimuli. Mechanoreception is important for cell volume control, and the senses of touch, hearing, and bal-

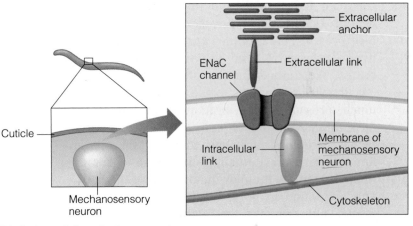

(a) ENaC channels in a *C. elegans* touch receptor

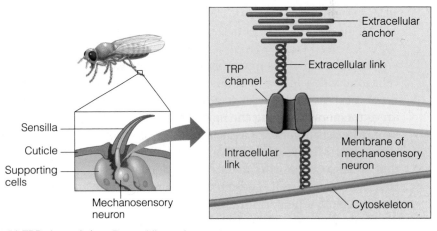

(b) TRP channels in a *Drosophila* touch receptor

Figure 6.12 Mechanosensory protein complexes **(a)** *C. elegans* touch receptors contain mechanosensory neurons with ENaC-type channels in their membranes. **(b)** *Drosophila* touch receptors contain mechanosensory neurons with TRP-type channels in their membranes. In both cases, mechanical stimuli cause the extracellular anchors to move relative to the cytoskeleton, pulling on the channel and causing a conformational change that opens or closes the channel, changing the membrane potential of the cell.

the mechanoreceptors in the ears and skin of vertebrates, suggesting that they play an important role in all forms of mechanoreception. Both ENaC and TRP-like mechanoreceptor proteins are attached to the cytoskeleton and to extracellular matrix proteins. Mechanical stimuli such as touch and pressure move the extracellular anchoring proteins, pulling on the ion channel and causing a conformational change that alters the movement of ions across the membrane, changing the membrane potential of the cell, and allowing the cell to transduce mechanical signals into electrical signals.

Touch and Pressure Receptors

The mechanoreceptors that detect touch and pressure can be grouped into three classes. **Baroreceptors** detect pressure changes in the walls of blood vessels, parts of the heart, and in the digestive, reproductive, and urinary tracts of vertebrates. We discuss baroreceptors in Chapter 8: Circulatory Systems. Tactile receptors detect touch, pressure, and vibration on the body surface. Both vertebrates and invertebrates have tactile receptors, although their structure and function vary substantially between these groups. **Proprioceptors** monitor the position of the body, and are found in both vertebrates and invertebrates, although like the tactile receptors, their structures vary greatly between these groups.

Vertebrate tactile receptors are widely dispersed

Vertebrate tactile receptors are isolated sensory cells embedded in the skin (Figure 6.13). Some of these receptors are simply free nerve endings that are interspersed among the epidermal cells of the skin, whereas others are associated with accessory structures. *Merkel's disks* are free nerve endings that are associated with an enlarged epidermal cell called the Merkel cell. These receptors have a very small receptive field, and are used for fine tactile discrimination. Both the free nerve endings and Merkel's disks are slowly adapting tonic receptors that are most sensitive to indentation of the skin, and are thus important for sensing light touch and pressure on the surface of the skin. We use the Merkel's disks in our skin when we perform tasks, such as reading Braille letters, that require very fine discrimination.

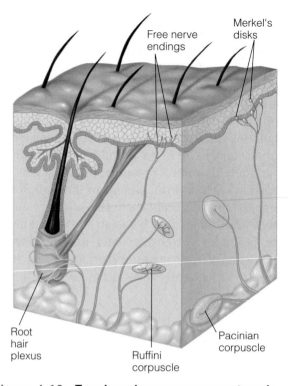

Figure 6.13 Touch and pressure receptors in vertebrate skin Skin mechanoreceptors may be free nerve endings or sensory neurons associated with complex accessory structures.

The nerve endings of the *root hair plexus*, which are wrapped around the base of hair follicles, monitor movements across the body surface. When a hair is displaced, the movement of the hair follicle causes the sensory nerve endings to stretch, stimulating mechanoreceptor proteins on the dendritic membrane. These receptors are rapidly adapting phasic receptors, so they are most sensitive to changes in movement. For example, you can often sense when an insect crawls across your skin, but you may not detect an insect that is not moving.

Pacinian corpuscles are located deep within the skin and in the muscles, joints, and internal organs. At almost a millimeter in length, they are actually visible to the naked eye in sections of skin, and a typical human hand contains as many as 400 of these receptors. Pacinian corpuscles contain a sensory dendrite surrounded by up to 70 layers of tissue with a viscous gel between them. These layers, called lamellae, are actually modified Schwann cells (see Chapter 4) and layers of connective tissue. When something presses on a Pacinian corpuscle, the lamellae change shape, changing the shape of the sensory dendrite and initiating a change in the membrane potential. The viscous gel quickly returns to its original position, even in the presence of continuous pressure, returning the membrane potential to its resting level. As a result, the sensation of pressure disappears even though the pressure is still present at the surface of the skin. When the pressure is removed, the connective tissue layers return to their normal shape, pulling on the nerve ending, which causes another change in membrane potential, and another stimulus. Thus, Pacinian corpuscles are rapidly adapting sensory receptors that are sensitive to both the beginning and end of a stimulus. This property makes Pacinian corpuscles especially sensitive to vibrations. So when you feel your cell phone vibrating, it is your Pacinian corpuscles that detect the incoming call. Pacinian corpuscles have relatively large receptive fields, and thus do not allow for fine-scale discrimination of touch sensations.

Ruffini corpuscles are located in the connective tissue of the skin and with the connective tissue of the limbs and joints. They are sensitive to stretching of the skin and movement of the joints as we move around. Ruffini corpuscles work together with other proprioceptors to help an animal determine the location of its body in space. When you hit the snooze button on your alarm clock without even opening your eyes, it is your Ruffini corpuscles that helped you do so!

Vertebrate proprioceptors monitor body position

In addition to touch and pressure receptors such as Ruffini corpuscles, there are three major groups of vertebrate proprioceptors associated with the joints and limbs:

1. *Muscle spindles* on the surface of skeletal muscles monitor the length of the muscle. Each muscle spindle consists of modified muscle fibers called intrafusal fibers enclosed in a connective tissue capsule.

2. *Golgi tendon organs* are located at the junction between a skeletal muscle and a tendon. These receptors are stimulated by changes in the tension in the tendon.

3. *Joint capsule receptors* are located in the capsules that enclose the joints. Several types of receptors are in this category, including receptors similar to free nerve endings, Pacinian corpuscles, and Golgi tendon organs. These receptors detect pressure, tension, and movement in the joint.

Proprioceptors typically do not adapt to stimuli, and thus constantly send information to the central nervous system regarding body position. Another class of more rapidly adapting receptors is responsible for detecting movement, and provides the sense of *kinesthesia*.

Insects have several types of tactile and proprioceptors

Insects and other arthropods are encased in a hard exoskeleton, so their sense of touch cannot function via free nerve endings in the body surface, as is the case for the touch receptors in the vertebrates. Instead, most insect touch receptors are grouped into complex organs called *trichoid sensilla* that consist of a hairlike projection of the cuticle associated with a bipolar sensory neuron (Figure 6.14a). When the hair bends in the socket of a trichoid sensillum (as a result of a touch or vibration), accessory structures transfer the movement

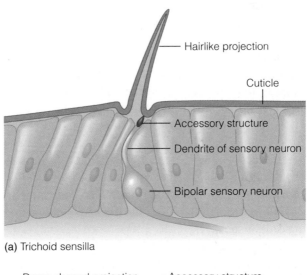

(a) Trichoid sensilla

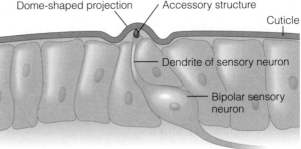

(b) Campaniform sensilla

Figure 6.14 Variation in the structure of insect sensilla **(a)** A trichoid sensillum is associated with a hairlike projection of the cuticle. **(b)** A campaniform sensillum is associated with a dome-shaped projection of the cuticle.

to the tip of the bipolar sensory neuron located beneath the hairlike projection. The movement opens stretch-sensitive TRP ion channels in the membrane of the mechanoreceptor neuron, changing the membrane potential, and sending a signal in the form of action potentials to the insect's nervous system. Trichoid sensilla can be extremely sensitive, detecting even small changes in air movements. Insects use their trichoid sensilla to detect the air movements caused by the motion of a predator, and can use this information to take evasive actions (explaining why it is so difficult to swat a fly!).

Insects use another type of sensillum on the external surface of the cuticle, called a *campaniform* sensillum, for proprioception (Figure 6.14b). Campaniform sensilla resemble trichoid sensilla except that they lack the hair shaft and instead are covered with a dome-shaped section of thin cuticle. They are usually found in clusters, particularly on or near the joints of the limbs, and detect the deformation of the cuticle as an insect moves. Thus, campaniform sensilla are critical in allowing an insect to make coordinated movements.

Insects also have a proprioceptor that can detect bending of the cuticle. These proprioceptors are organized into functional units called *scolopidia* (Figure 6.15), which consist of a specialized bipolar sensory neuron and a complex accessory cell (the *scolopale*) that surrounds the ciliated sensory dendrite at one end. This structure is attached to the cuticle via a ligament or attachment cell. These mechanoreceptors can exist as isolated cells or may be grouped to form complex organs called *chordotonal organs*, which (as we discuss later in the chapter) form the basis for the sense of hearing in some insects.

Insects also have a variety of internal mechanoreceptors that function as stretch receptors and proprioceptors. Unlike the mechanoreceptors associated with the cuticle, these receptors are not organized into complex organs, and do not contain ciliated bipolar neurons. Instead, these mechanoreceptors are usually isolated multipolar neurons associated with muscle and connective tissue. These mechanoreceptors use ENaC channels for signal transduction.

Equilibrium and Hearing

In addition to detecting touch, pressure, and the location of the limbs, mechanoreceptors are in-

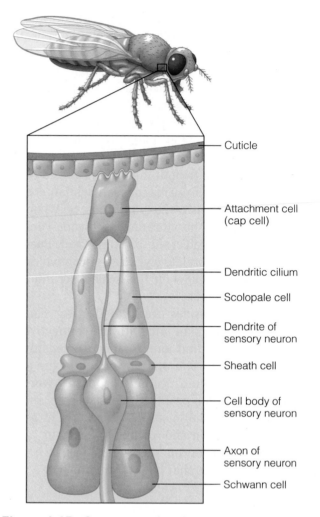

Figure 6.15 Structure of an insect scolopidium
Scolopidia are associated with the internal surface of the cuticle. The bipolar sensory neuron of the scolopidium is surrounded by sheath cells and scolopale cells. The attachment (or cap) cell links the complex to the cuticle.

volved in the senses of equilibrium and hearing. The sense of equilibrium, sometimes called the sense of balance in humans, involves detecting the position of the body relative to the force of gravity. The sense of hearing involves detecting and interpreting sound waves. In the vertebrates the ear is the organ responsible for both equilibrium and hearing. In invertebrates, however, the organs of equilibrium are entirely separate from the organs of hearing.

Statocysts are the organ of equilibrium for invertebrates

Many invertebrates have organs called **statocysts** that they use to detect the orientation of their bodies with respect to gravity (Figure 6.16). Statocysts

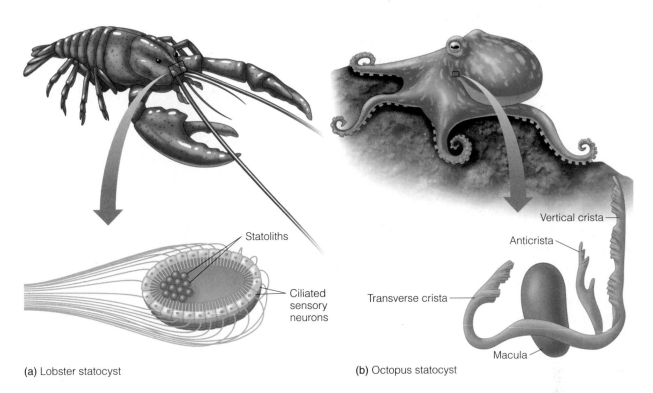

(a) Lobster statocyst

(b) Octopus statocyst

Figure 6.16 Invertebrate organs of equilibrium
Statocysts contain ciliated sensory neurons and calcified statoliths. When a mechanical stimulus such as a change in body orientation disturbs the statoliths, their motion stimulates receptor proteins on the cilia of the sensory neurons, depolarizing the cell. **(a)** Most invertebrates have simple statocysts. **(b)** Cephalopod molluscs have complex statocysts that consist of three cristae, oriented in different planes, with a sac called the macula at the base. The cristae detect angular acceleration, while the macula detects forward acceleration, providing the cephalopod with detailed information about body position and movement.

are hollow, fluid-filled cavities that are lined with mechanosensory neurons, and contain dense particles of calcium carbonate called **statoliths**. When the orientation of the animal changes, the statolith moves across the sheet of mechanoreceptors. This movement stimulates the mechanoreceptive cells, sending a signal to the nervous system. This signal provides a cue about the position of the body relative to gravity. Most marine invertebrates have relatively simple statocysts (as shown in Figure 6.16a), but cephalopod molluscs, such as the octopus, have a particularly complicated statocyst system (Figure 6.16b). An octopus has two statocysts, one on each side of the head. Each statocyst is composed of a globelike structure called the *macula*, and three *cristae*, each oriented in a different plane. The cristae and macula contain statoliths that move in response to mechanical stimuli. The crista detects angular acceleration, or the turning of the body, while the macula detects linear acceleration, or the degree of forward motion. This system is analogous to the organs of equilibrium in the vertebrates.

Insects use a variety of organs for hearing

There is a great deal of variation in the ability to hear among insect species; some species lack specialized organs for detecting sound, while others have specialized "ears" in several locations. The simplest type of insect ear is composed of groups of modified trichoid sensilla. Sound waves (which represent vibrations carried in air) cause these thin sensilla to bend, and send a signal to a bipolar sensory neuron. However, this type of ear is not particularly sensitive, and most insect ears are derived from the chordotonal organs that insects use for proprioception.

Many insects, including cockroaches, honeybees, and water striders use a modified chordotonal organ called the subgenual organ to detect vibrations carried through the ground (or the surface of the water in the case of a water strider), and in at least some species, these subgenual organs may also be able to detect sound waves (which represent vibrations carried in air). Subgenual organs are located inside the insect leg. Vibrations of the leg cause the subgenual organ to vibrate, opening a mechanosensitive ion channel on the sensory neuron within the chordotonal organ, initiating action potentials that send a signal to the integrating centers of the nervous systems.

An alternative type of insect ear is a modified chordotonal organ called the Johnston's organ, which is located at the base of the antennae of many insects, including moths, fruitflies, honeybees, and mosquitoes. Sound waves bend fine hairs on the antennae, stretching the membrane of the cells within the underlying chordotonal organ, opening mechanosensitive ion channels, and initiating action potentials in the mechanosensory neuron. These insects use Johnston's organ to detect sounds such as mating calls.

The most sensitive insect ears are called **tympanal organs**. A tympanal organ consists of a very thin region of the cuticle, called the *tympanum*, located over an air space similar to the air space in a drum. Sound waves cause the thin tympanum to vibrate, causing the air within the air space to vibrate. A chordotonal organ in this air space detects these vibrations, and sends signals in the form of action potentials to the nervous system. Tympanal organs are found on many locations on the insect body, including the legs, abdomen, thorax, and wing base.

Vertebrate organs of hearing and equilibrium contain hair cells

The vertebrate organs that are involved in the senses of hearing and equilibrium contain multiple mechanosensory cells and accessory structures. Unlike the mechanoreceptor cells that we have discussed so far, in these organs the mechanoreceptor cells are not themselves sensory neurons, but instead contain modified epithelial cells that synapse with a sensory neuron. These highly specialized sensory receptor cells have extensive extracellular structures associated with them and are termed **hair cells** because of the prominent cilia that extend from the apical end of each cell (Figure 6.17).

Most vertebrate hair cells have a single long cilium, the **kinocilium**, and many shorter projections, called **stereocilia**. Invertebrates also have mechanoreceptors that are similar to hair cells, but these cells can contain between 1 and 700 kinocilia. The kinocilium of a vertebrate hair cell is a true cilium with a 9 + 2 arrangement of microtubules (see Chapter 5: Cellular Movement and

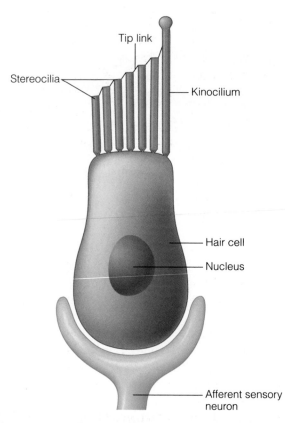

Figure 6.17 The structure of a vertebrate hair cell Vertebrate hair cells (except those in the ears of adult mammals) have a long kinocilium and several short stereocilia. The kinocilia and stereocilia are connected to each other via tip links and a variety of other structures that cause the stereocilia to work together as a bundle.

Muscles), although it is nonmotile, but the stereocilia are actually microvilli that contain polymerized actin molecules. There are hundreds of actin filaments along most of the length of a stereocilium, but there are far fewer (only a few dozen) at the base of the stereocilium. As a result, stereocilia taper at their bases, having the appearance of pencils balanced on their points.

The hair cells in the ears of adult mammals lack the kinocilium, suggesting that the kinocilium is not necessary for mechanoreception. Instead, the stereocilia play a critical role in mechanosensory transduction. The stereocilia and kinocilium (when present) are arranged in a tight bundle, with the shortest stereocilia placed farthest away from the kinocilium in the bundle, and with the stereocilia gradually becoming taller the closer they are to the kinocilium. The stereocilia are connected to each other and the kinocilium by a series of small fibers that cause the bundle of hair cells to act as a single unit. One particular type of these fibers, called a *tip link*, connects the top of each shorter stereocilium to the side of the adjacent taller one. These tip links are thought to play a critical role in sound transduction.

Mechanosensitive ion channels localized near the tips of the stereocilia are involved in sound transduction (Figure 6.18). These channels are thought to be members of the TRP family of channels, although the precise identity of the mechanosensitive channel in the vertebrate hair cell is currently somewhat debated. At rest, about 15% of these mechanosensitive ion channels are open, yielding a resting membrane potential of about -60 mV. Under these conditions, a modest number of voltage-gated Ca^{2+} channels are open on the hair cell, causing some release of neurotransmitter onto the primary afferent neuron, and a modest frequency of action potentials in the afferent sensory neuron.

When a hair cell is exposed to a mechanical stimulus such as a vibration, the stereocilia pivot about their bases, acting as rigid rods that do not bend. If the movement is toward the kinocilium (or longest stereocilium in the hair cells of the mammalian ear), mechanosensitive ion channels on the tips of the stereocilia open. These mechanosensitive channels are relatively nonselective, and allow the passage of a variety of ions, including K^+ and Ca^{2+}. However, at least in the hair cells of the vertebrate ear, the extracellular fluid around the hair cell is very high in K^+. As a result, K^+ enters the hair cell down its concentration gradient, causing the hair cell to depolarize by about 20 mV. This depolarization opens voltage-gated Ca^{2+} channels on the membrane of the hair cell, allowing additional Ca^{2+} to enter the cell (compared to the resting state), increasing the exocytosis of neurotransmitter from the hair cell onto the afferent neuron, and increasing the frequency of action potentials in the afferent neuron.

If the movement of the stereocilia is in the other direction, the mechanosensitive channels close. The closed channels prevent K^+ from entering the cell and cause the hair cell to hyperpolarize by about 5 mV (relative to the resting state), decreasing the release of neurotransmitter and the frequency of action potentials in the sensory neuron. Note that these sensory neurons associated with a hair cell fire action potentials all the time; neurotransmitter release from the hair cell simply increases or decreases the frequency of these action potentials depending on the direction that the stereocilia move. Thus, hair cells can detect not just movement, but the direction of that movement. The change in the membrane potential of the hair cell is also asymmetric—the change is larger in one direction than the other.

Tip links are critical for mechanosensory transduction

So far, we have not discussed how the mechanosensitive channels on the stereocilia are opened and closed by the pivoting movement of the stereocilia. Experiments using chemicals that destroy the tip links that connect adjacent stereocilia indicate that removing the tip links abolishes mechanosensory transduction, and that transduction is restored once the tip links regenerate. These results suggest that the tip links play a critical role in detecting mechanical stimuli. The tip links are proposed to function as part of a "gating spring" mechanism that physically pulls the channel open. When the stereocilia pivot in response to a mechanical stimulus, the vertical distance between the top of adjacent stereocilia changes; pivoting in one direction increases the distance, while pivoting in the other direction decreases the distance. The tip links are ideally placed to detect these changes. Increasing the vertical distance pulls on the tip links, whereas decreasing the vertical distance pushes on the tip links. The tip links are connected to the mechanically gated ion channels on the stereocilia via a series of elastic connector proteins that act as springs that either pull open the channel or push it closed, depending on the direction of movement of the stereocilia.

Hair cells are found in the lateral line and ears of fish

Hair cells are found in a variety of mechanosensitive organs. For example, fish, larval amphibians, and adult aquatic amphibians have structures called **neuromasts** that can detect water movements, such as those caused by potential predators or prey as they move through the water. Neuromasts consist of hair cells (from a few to over a hundred, depending on the species) and accessory supporting cells encased in a gelatinous cap (Figure 6.19). Neuromasts are found in the skin, either scattered over the body surface or grouped in particular areas (often at the anterior end of the animal). Most fish species (and some aquatic amphibians)

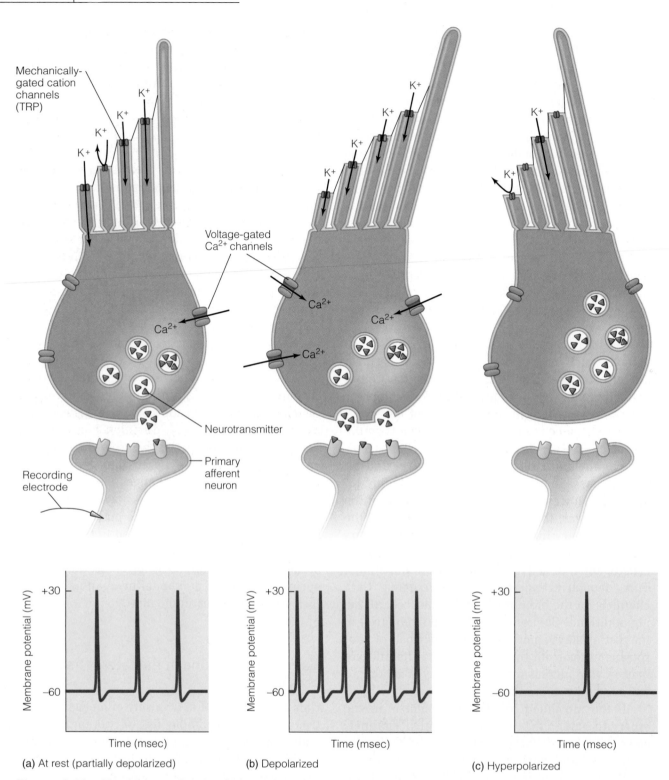

(a) At rest (partially depolarized) (b) Depolarized (c) Hyperpolarized

Figure 6.18 Signal transduction in a vertebrate hair cell **(a)** At rest the hair cell is slightly depolarized and releases moderate amounts of neurotransmitter onto the primary afferent neuron, causing an intermediate frequency of action potentials. **(b)** When a pressure signal causes the stereocilia to pivot toward the kinocilium, mechanically gated channels on the stereocilia open, allowing additional K$^+$ to enter the cell from the extracellular fluid, which has a high concentration of K$^+$. The resulting depolarization opens voltage-gated Ca^{2+} channels, allowing Ca^{2+} to enter the cell. The influx of Ca^{2+} causes increased release of neurotransmitter onto the primary afferent neuron, increasing the frequency of action potentials. **(c)** When a pressure signal causes the stereocilia to pivot away from the kinocilium, the mechanically gated channels on the stereocilia close, hyperpolarizing the cell and closing voltage-gated Ca^{2+} channels. The resulting reduction in intracellular Ca^{2+} decreases the release of neurotransmitter onto the primary afferent neuron, reducing the frequency of action potentials.

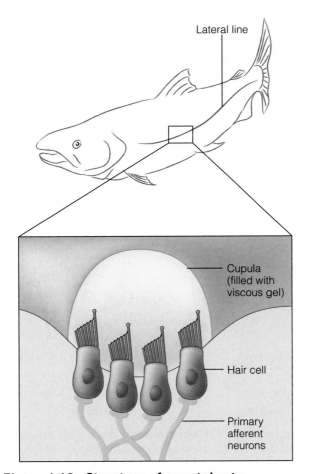

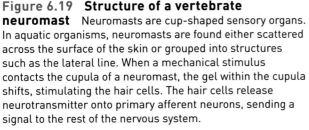

Figure 6.19 Structure of a vertebrate neuromast Neuromasts are cup-shaped sensory organs. In aquatic organisms, neuromasts are found either scattered across the surface of the skin or grouped into structures such as the lateral line. When a mechanical stimulus contacts the cupula of a neuromast, the gel within the cupula shifts, stimulating the hair cells. The hair cells release neurotransmitter onto primary afferent neurons, sending a signal to the rest of the nervous system.

have a conspicuous array of neuromasts arranged in a line along both sides of the body. This **lateral line system** consists of either pits (*ampullae*) or tubes running along the side of the animal's body and head. The lateral line system allows fish to detect changes in water pressure, such as those caused by the movements of other fish. In some species, the lateral line system has been modified to allow electroreception (see Box 6.1, Evolution and Diversity: Electroreception on page 279, for more details on the functions of electroreception).

Vertebrate ears function in hearing and equilibrium

Hair cells are also found within the ears of vertebrates, where they participate in the senses of hearing and equilibrium. Figure 6.20 shows the structure of a representative mammalian ear. The external structures are called the **outer ear** and in mammals consist of the *pinna*, which forms the distinctive shapes of mammalian ears, and the *auditory canal*. The auditory canal leads to the **middle ear**, which contains a series of small bones that transfer sound waves to the **inner ear**. The inner ear is embedded within the skull and consists of a series of fluid-filled membranous sacs and canals. Most nonmammalian vertebrates lack obvious outer ears, and fish lack both outer and middle ears, but all vertebrates have an inner ear. It is the inner ear that contains the mechanosensitive hair cells that play a role in hearing and the sense of equilibrium.

The vestibular apparatus is the organ of equilibrium in vertebrates

The **vestibular apparatus** of the inner ear detects movements or changes in body position with respect to gravity and is thus responsible for the sense of equilibrium or balance. In all craniates, except the lampreys and hagfish, the vestibular apparatus consists of three **semicircular canals** with an enlarged region at one end (called the *ampulla*), and

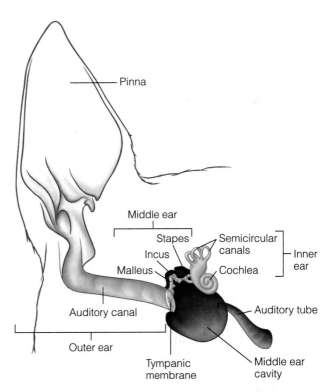

Figure 6.20 The structure of the mammalian ear Mammalian ears consist of an outer ear, a middle ear, and an inner ear.

two saclike swellings called the *utricle* and the *saccule* (Figure 6.21). In most vertebrates, the saccule also contains a small extension called the *lagena*. In birds and mammals, the lagena is greatly extended and is called the *cochlear duct* (in birds), or the **cochlea** (in mammals). The utricle, saccule, and the ampullae of the semicircular canals contain mechanoreceptive hair cells that are involved in the sense of equilibrium. The cochlea also contains hair cells, but it is involved in hearing and is not a part of the vestibular apparatus.

The mechanoreceptors of the ampullae and the vestibular sacs differ. The utricle and saccule contain a series of mineralized **otoliths** suspended in a gelatinous matrix above a membrane called the *macula* that is densely covered with more than 100,000 hair cells (Figure 6.22). The ampullae of the semicircular canals lack otoliths, and instead contain cristae that consist of hair cells located within a cup-shaped gelatinous mass called the cupula. The cristae of semicircular canals detect angular acceleration, and motion in circular patterns, such as when you shake your head. In contrast, the maculae of the vestibular sacs detect linear acceleration, or motion along a line, and are stimulated when the body is in a tilted position.

When you move your head to one side, the otoliths and the gelatinous masses of the maculae in the utricle and saccule induce a drag on the hair cells, stimulating them. The macula of the utricle is oriented horizontally in the ear, and can detect motion in the horizontal plane (Figure 6.23a–d). The macula of the saccule is oriented vertically, so it can detect motion in the vertical plane. Within the utricle and saccule, the hair cells are oriented in two different directions so that a single sheet of hair cells can detect motion forward and back or side-to-side, covering two dimensions of movement. The utricle can also detect tilting of the head (Figure 6.23e). When you tilt your head, gravity pulls on the gelatinous mass of the sacs, which stimulates particular subsets of the hair cells, depending on the direction of the tilt. Since different hair cells are stimulated by a forward and a backward tilt, the brain can determine the direction of the tilt. The intensity of the hair cell response is related to the angle of tilt, so the brain can also determine the degree of tilt. The vestibular sacs play an important role in maintaining the orientation of the body with respect to gravity. If your head and body start to tilt, the vestibular sacs send a signal to the brain, which automatically compen-

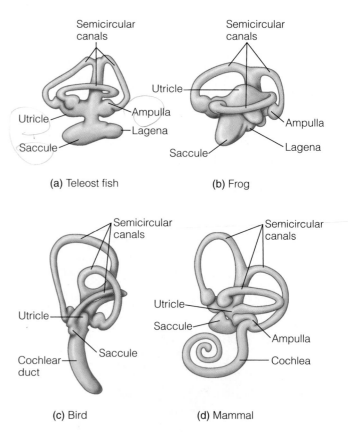

(a) Teleost fish   (b) Frog

(c) Bird   (d) Mammal

Figure 6.21 Vertebrate inner ears The inner ear in most vertebrates consists of three semicircular canals arranged in planes at right angles joined at their base by a swelling called the ampulla, and a series of sacs including the utricle and the saccule. In many vertebrates, the floor of the saccule contains a small pocket called the lagena. In birds and mammals, the lagena is greatly extended to form the cochlear duct or cochlea.

sates by altering posture in order to maintain your position.

In contrast to the vestibular sacs, which detect whether the body is tilted, the semicircular canals detect angular acceleration (Figure 6.24). Most vertebrates have three semicircular canals that are arranged perpendicular to each other, so that each canal detects acceleration in a single plane. When you turn your head in the plane of a particular canal, the fluid in that canal is set in motion. Because of the inertia of the fluid there is a difference between the movement of the fluid and the movement of the wall of the canal, causing the fluid to slosh against the ampulla, stimulating the hair cells. Because each canal is oriented in a different plane, acceleration of the fluid in a particular canal depends on the plane of the movement, allowing the vestibular system to sense the direction of movement by comparing the degree to which the hair cells in each canal are stimulated.

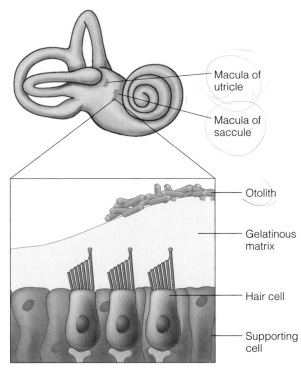

— Macula of utricle

— Macula of saccule

— Otolith

— Gelatinous matrix

— Hair cell

— Supporting cell

(a) Macula of an utricle or saccule

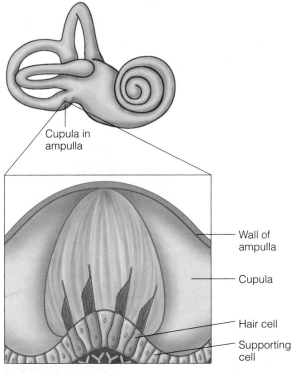

Cupula in ampulla

— Wall of ampulla

— Cupula

— Hair cell

— Supporting cell

(b) Crista of an ampulla

Figure 6.22 The mechanoreceptors of the inner ear (a) The mechanoreceptors of the utricle and saccule are found in structures called maculae. The hair cells of each macula are embedded in a gelatinous matrix that is overlain with a series of otoliths. **(b)** The mechanoreceptors of the semicircular canals are located in the ampullae in structures called cristae. Cristae are similar in structure to the neuromasts shown in Figure 6.19, consisting of hair cells embedded in a cup-shaped gelatinous mass called the cupula.

Balance and body orientation depend on inputs from the visual system, proprioceptors, and the inner ear. You can observe this effect if you ask someone to try to stand still with his or her eyes closed. It is almost impossible to do—you will notice that your subject makes small movements and rocks back and forth. The semicircular canals also play an important role in keeping your eyes oriented on a single point even when your head is moving. For example, if you try to read this text while nodding or shaking your head, you should have little difficulty reading the words. In contrast, if you quickly move the book in front of your face while holding your head still, you will likely find it difficult to read the words.

The inner ear detects sounds

In addition to detecting body position, the inner ear detects sounds. In fish, incoming sound waves cause the otoliths in the vestibular sacs to move, causing the stereocilia of the hair cells to pivot, and stimulating the auditory neurons. Some fish use their swim bladder to help amplify the sounds coming to the inner ear. The clupeids (fish in the herring family) have a gas duct that connects the swim bladder to the hearing system. Sounds cause the swim bladder to vibrate, and this vibration is passed through the gas duct to the ear. Clupeid fish such as shad use their excellent hearing to detect the echolocation sounds produced by whales and dolphins (their main predators).

In carp, the swim bladder is connected to the inner ear via a system of bones called the *Weberian ossicles* (Figure 6.25). Carp have excellent hearing because the Weberian ossicles transmit sounds to the inner ear.

In terrestrial vertebrates, hearing involves the inner, middle, and outer ears

Sound does not travel as well in air as in water, and much of the sound that travels through air is simply reflected when it contacts an object with much higher density, such as the body of an animal. As a result, sound transfers poorly between air and the fluid-filled inner ear. To compensate, the ears of terrestrial animals have a number of specializations to increase sound detection. In mammals, the pinna of the outer ear acts as a funnel that collects sound waves in the air from a large area, concentrating them onto the auditory

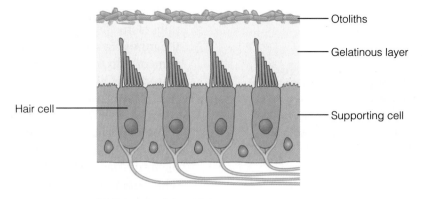

(a) Hair cells of the utricle

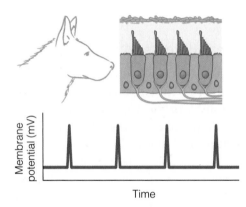

(b) Rest or constant motion

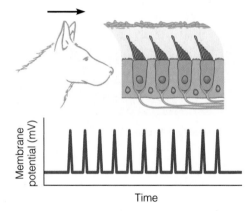

(c) Forward acceleration

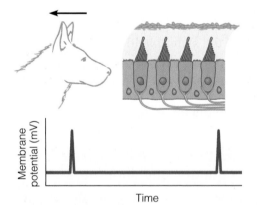

(d) Backward acceleration

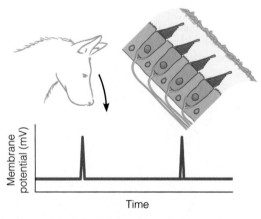

(e) Head tilted forward

Figure 6.23 Functions of the utricles in mammals **(a)** The hair cells of the utricles are overlain with a gelatinous layer topped with bony otoliths. **(b)** At rest or during constant motion, the hair cells are partially depolarized. **(c)** During forward acceleration, the hair cells pivot toward the longest stereocilium (recall that mammalian hair cells lack a kinocilium). This bending activates mechanogated channels on the stereocilia, which depolarizes the cell, increasing its release of neurotransmitter and thus increasing the frequency of action potentials in the primary afferent neurons. **(d)** During backward acceleration or **(e)** forward tilt of the head, the stereocilia pivot away from the longest stereocilium, reducing the frequency of action potentials.

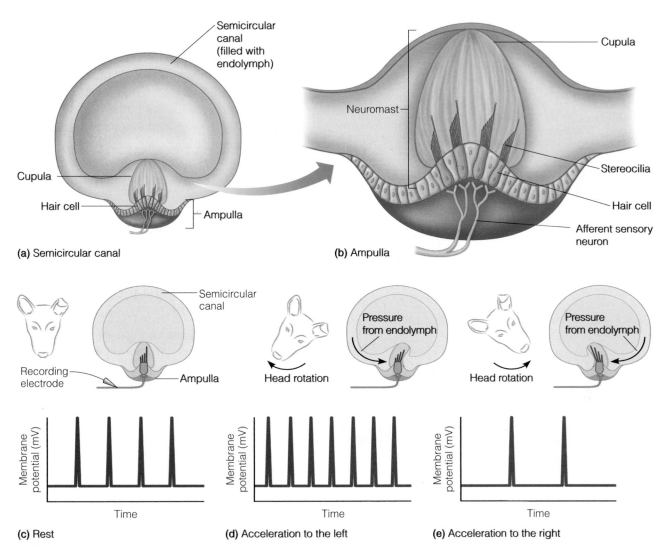

(a) Semicircular canal

(b) Ampulla

(c) Rest

(d) Acceleration to the left

(e) Acceleration to the right

Figure 6.24 Functions of the semicircular canals **(a)** A semicircular canal consists of a fluid-filled tube with a swelling, termed the ampulla, at the bottom. **(b)** The ampulla contains a neuromast that senses pressure. **(c)** At rest, the hair cells of the neuromast are partially depolarized. When the head is rotated in one direction, the fluid in the semicircular canal exerts pressure in the opposite direction, causing the stereocilia of the hair cells to pivot. Depending on the orientation of the hair cells, this will either **(d)** hyperpolarize the hair cell, decreasing the frequency of action potentials, or **(e)** depolarize the hair cell, increasing the frequency of action potentials.

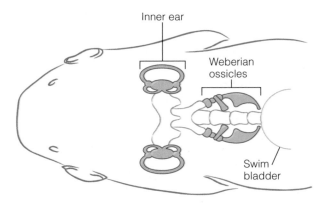

Figure 6.25 Structure of a carp ear The inner ear is connected to the swim bladder via a series of bones called the Weberian ossicles.

canal. Ears with a larger pinna capture more of the sound wave for a given sound intensity and hence receive more sound energy, so animals with large external ears typically have excellent hearing. While passing the pinna, sound also goes through a filtering process. For example, in humans sounds are enhanced in the frequency range where human speech is normally found. The filtering process also adds directional information.

The middle ear plays the most important role in improving detection of sounds in air. Although the details of middle ear structure vary substantially between groups of organisms, the fundamental design principles are similar. The air-filled

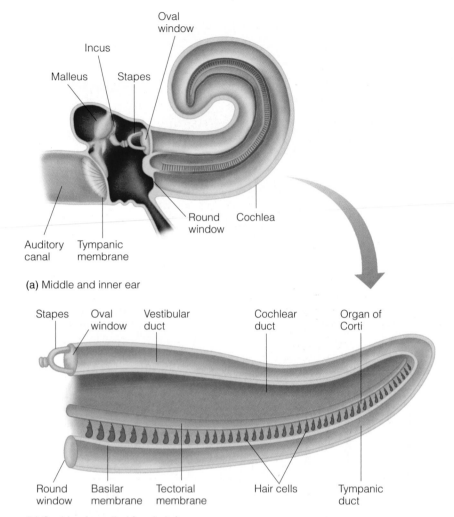

(a) Middle and inner ear

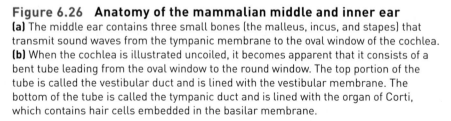

(b) Cochlea (uncoiled for clarity)

Figure 6.26 Anatomy of the mammalian middle and inner ear
(a) The middle ear contains three small bones (the malleus, incus, and stapes) that transmit sound waves from the tympanic membrane to the oval window of the cochlea. **(b)** When the cochlea is illustrated uncoiled, it becomes apparent that it consists of a bent tube leading from the oval window to the round window. The top portion of the tube is called the vestibular duct and is lined with the vestibular membrane. The bottom of the tube is called the tympanic duct and is lined with the organ of Corti, which contains hair cells embedded in the basilar membrane.

the oval window. Vibrations of the oval window transfer the sound stimulus to the fluid-filled inner ear. In mammals, the malleus, incus, and stapes are connected to each other with the biological equivalent of hinges, which tend to amplify the vibrations such that a vibration of the tympanic membrane as small as 0.1 angstrom (less than the size of a hydrogen atom) can cause a response large enough to stimulate the hair cells of the inner ear.

The inner ear of mammals has specializations for sound detection

The coiled cochlea of mammals is specialized for sound detection. Figure 6.26b shows the cochlea uncoiled, and from this diagram you can see that the two outer compartments (the *vestibular* and *tympanic ducts*) are actually one continuous tube, although early anatomists gave them two different names because they appear to be distinct structures in the tightly coiled cochlea. The vestibular and tympanic ducts are filled with a fluid called *perilymph*, which is similar in composition to other extracellular fluids. The *cochlear duct* is filled with a fluid called **endolymph** that is quite different from other extracellular fluids, being high in K^+ and low in Na^+. The **organ of Corti** contains the hair cells and sits on the **basilar membrane** that lines one side of the cochlear duct. Vertebrate inner ears contain several types of hair cells that perform slightly different auditory functions. In mammals, these types are called the **inner hair cells** and the **outer hair cells**. Inner hair cells detect sounds, and outer hair cells help to amplify sounds.

Incoming sounds cause the oval window of the inner ear to vibrate, causing waves in the perilymph of the vestibular duct. These waves in the perilymph push on the basilar membrane, causing it to vibrate. The stereocilia on the inner hair cells of the organ of Corti pivot in response to the vibra-

middle ear is separated from the outer ear by the **tympanic membrane** and from the fluid-filled inner ear by the **oval window** (Figure 6.26a). Within the middle ear are one or more small bones that together span the space from the tympanic membrane to the oval window. Mammals have three of these bones, called the **malleus** (hammer), the **incus** (anvil), and the **stapes** (stirrup). Sound waves traveling through the auditory canal cause the thin tympanic membrane to vibrate. Vibration of the tympanic membrane causes the first of the bones (the malleus in mammals) to vibrate. The vibration is transferred through the bones (from the malleus to the incus to the stapes in mammals) to

BOX 6.1   **EVOLUTION AND DIVERSITY**
Electroreception

Electroreception, the ability to sense electric fields or weak electrical discharges, is common in aquatic organisms. Aquatic environments are full of electric fields. For example, the flow of water over objects causes a static electrical discharge. Similarly, seawater moving over the magnetic field lines of the Earth causes a weak electric field. In addition, all animals produce weak electric fields as a result of the actions of their muscles and nerves. Thus, aquatic organisms can use electroreception for detecting both the abiotic features of their environment and the presence of other animals. Sharks have a particularly keen electrical sense. For example, the electrosensitive great hammerhead shark (*Sphyrna mokarran*) can detect buried stingrays by sweeping its wide head over the bottom of the ocean like a metal detector.

Some fish (the so-called weakly electric fish) have a specialized electric organ that can produce electrical discharges, and electroreceptors to detect these discharges, which they use to communicate with each other. They can also "electrolocate" in their murky environments. The discharges of the electric organ produce an electric field around the fish. Objects or other animals in the environment alter the shape of this electric field. Electric fish can detect these perturbations of the electric field and use this information to locate the object, in a process analogous to echolocation in bats.

In fishes, electroreceptors are modified from the lateral line; however, the hair cells of the electroreceptor are highly modified and lack cilia. These modified hair cells detect changes in electric fields rather than changes in pressure. Sharks have elaborate electrosen-

sory organs that are located in a series of pores distributed across the head. These pores, termed the ampullae of Lorenzini after the Italian anatomist who first described them in 1678, are filled with an electrically conductive jelly and lined with modified hair cells. A net negative charge inside the ampulla causes an electrical change in each hair cell, triggering the release of neurotransmitters to adjacent clusters of afferent sensory neurons. Some sharks, such as the scalloped hammerhead shark (*Sphyrna lewini*), can detect electric fields of less than 0.1 nV/cm, equivalent to the electric field of a flashlight battery connected to electrodes over 16,000 km apart in the ocean.

Although many fish have electroreceptors, and some species of amphibians are thought to have similar ones, none have yet been identified in any species of bird, reptile, or placental mammal. However, the monotremes (the egg-laying mammals, including the echidna and platypus) do have electroreceptors. In the platypus, the electroreceptors are located in the bill. They are bipolar sensory neurons, rather than modified epithelial cells as in fish, which suggests that the ability to sense electric fields has evolved independently several times.

References

⊙ Gibbs, M. A., and R. G. Northcutt. 2004. Development of the lateral line in the shovelnose sturgeon. *Brain, Behavior, and Evolution* 64: 70–84.

⊙ Kajiura, S. M., and K. N. Holland. 2002. Electroreception in juvenile scalloped hammerhead and sandbar sharks. *Journal of Experimental Biology* 205: 3609–3621.

⊙ Pettigrew, J. D. 1999. Electroreception in monotremes. *Journal of Experimental Biology* 202: 1447–1454.

tions of the basilar membrane. As with the hair cells in the lateral line of a fish, the tip links connecting the stereocilia pull open the mechanosensitive ion channels in the membrane of the inner hair cells, causing them to depolarize. The inner hair cells then release a neurotransmitter, glutamate, that excites sensory neurons to send nerve impulses down the auditory nerve. In this way, the cochlea transduces the pressure waves in the perilymph into electrical signals. The round window of the cochlea serves as a pressure valve, bulging outward as fluid pressure rises in the inner ear, which prevents the waves from doubling back through the fluid thus improving sound clarity.

The basilar membrane is stiff and narrow near its attachment point close to the round and oval windows (the proximal end), but wider and more flexible at the other (distal) end. This differential stiffness helps the cochlea to encode information about the frequency of a sound. Stiff objects vibrate at higher frequencies than flexible objects. The stiff proximal end of the basilar membrane vibrates most in response to high-frequency sounds, while the flexible distal end of the basilar membrane vibrates most in response to low-frequency sounds. Thus, different areas of the basilar membrane vibrate in response to sounds of different frequency, transforming a frequency signal carried by the sound waves into a spatial signal coded by location on the basilar membrane. Neurons from each part of the basilar membrane form synaptic connections with neurons in particular areas in the auditory cortex of the brain; therefore, specific areas of the auditory cortex respond to particular frequencies. This phenomenon is called *place coding*.

Outer hair cells amplify sounds

Inner hair cells code for sound loudness in much the same way as do other mechanosensory cells. Loud noises cause greater movement of the basilar membrane, and greater depolarization of the hair cell, which in turn generates a higher frequency of action potentials in the afferent sensory neurons. The outer hair cells also play an important role in the loudness of sounds. Current theories of sound transduction in the inner ear suggest that the outer hair cells amplify sounds by increasing the movement of the basilar membrane for a sound of a given loudness, thus causing a larger stimulus to the inner hair cells.

Outer hair cells perform this amplification function because, unlike inner hair cells, outer hair cells change *shape* in response to sound waves, rather than releasing neurotransmitter. When the stereocilia of an outer hair cell pivot in response to a sound wave, the mechanosensory channels on the stereocilia open, allowing K^+ to enter the cell. The resulting depolarization acts as a signal to a voltage-sensitive motor protein, which causes the cell to change shape and pull on the basilar membrane, increasing the amount the basilar membrane moves in response to a particular sound. The protein responsible for this change in shape of the outer hair cells has been identified, and if the gene that codes for this protein (called prestin) is knocked out in mice, the animals are born profoundly deaf. Certain types of deafness in humans are also caused by mutations in the prestin gene.

Outer hair cells make contact with very few afferent neurons that carry signals to the brain. Instead, they form synapses with efferent neurons that carry signals from the brain to the ear. These efferent neurons are part of a feedback loop; they release the neurotransmitter acetylcholine onto the outer hair cells in response to loud noises, reducing the response of the outer hair cells. Since outer hair cells normally amplify sounds, this feedback loop acts as a protective mechanism for the inner hair cells, which can be damaged by loud noises.

The ears can detect sound location

The brain uses information from both ears to estimate the location of the stimulus, including the time lag and differences in sound intensity. If a sound comes from one side, the sound waves will not reach both ears at the same time because the distance from the sound source is slightly different between the two ears. The brain registers the time lag, helping to localize the sound. Sounds coming from one side must also pass through the head to reach the other ear, altering the intensity of the sound in that ear. The discrepancy between the sound in the two ears helps to pinpoint the sound location. If a sound does not come from the sides, but rather from above, below, or immediately in front of the face, there is no time lag or discrepancy in intensity between the ears, and it is more difficult to determine the location of a sound. In mammals, the outer ears also help in localizing sounds. However, this mechanism is not particularly efficient, so most animals move their head or rotate their outer ears in order to better localize the source of a sound.

◎ | CONCEPT CHECK

9. What are possible advantages of having both tonic and phasic touch receptors in the skin of vertebrates?

10. Why do insects have complex touch organs, rather than isolated sensory neurons associated with the body surface as in mammals?

11. What is the functional significance of having a hairlike projection in the trichoid sensilla of insects? Why not simply have a touch receptor similar to the campaniform sensilla?

12. What would happen to sound transduction if the endolymph of the vertebrate inner ear had high $[Na^+]$ and low $[K^+]$?

13. How does the structure of the basilar membrane of the mammalian ear allow fine discrimination of different sound frequencies?

Photoreception

Photoreception is the ability to detect a small portion of the electromagnetic spectrum from the near ultraviolet to the near infrared, that is, wavelengths of approximately 300 nm to just greater than 1000 nm, although most species detect only a portion of this range (humans can only detect wavelengths from approximately 350 to 750 nm; Figure 6.27a). Animals lack the ability to detect other wavelengths of electromagnetic radiation such as radio waves. This concentration on a very narrow band of the electromagnetic spectrum supports the idea that animals evolved in water. The wavelengths that represent visible light travel relatively well through water, whereas water blocks most other wavelengths. Figure 6.27b shows the degree of attenuation, or the amount of signal lost, for an electromagnetic signal that passes through a meter of water. From this figure, you can see that water is relatively transparent to violet, blue, and green light, but that it quickly becomes rather opaque to yellow, orange, and particularly to red light. A meter of water almost completely blocks far red and near infrared light. Only at the other end of the electromagnetic spectrum, at very long wavelengths, are signals able to pass through water effectively. Thus, animals living in water can use only a narrow range of the electromagnetic spectrum. The degree of attenuation of light also varies depending on the presence of light-absorbing compounds in the water. Some aquatic animals, particularly those living in light-poor habitats have poor vision, and have instead developed the ability to sense electric fields (Box 6.1, Evolution and Diversity: Electroreception).

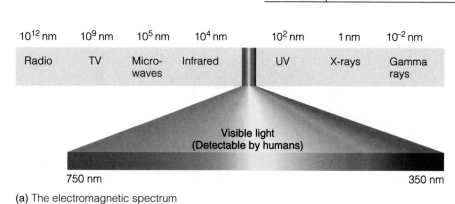

(a) The electromagnetic spectrum

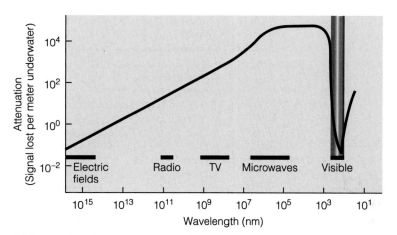

(b) Attenuation of electromagnetic radiation in water

Figure 6.27 Electromagnetic radiation and the electromagnetic spectrum **(a)** The types of electromagnetic radiation. **(b)** Most wavelengths of electromagnetic radiation do not travel well through water. Only visible light and very long wavelength electromagnetic radiation penetrate into deeper water. Animals detect a narrow band of the electromagnetic spectrum in the visible light range, which suggests the possibility that photoreceptors evolved in aquatic organisms.

Photoreceptors

Photoreceptive organs range in complexity from single light-sensitive cells to complex eyes that can form sharp, focused images. In this section, we first consider the structure of individual photoreceptive cells, and look at the signal transduction mechanisms they use to convert an incoming photon of light to a change in the membrane potential of the cell. Then we look at how these cells are put together into complex photoreceptive organs such as eyes. Finally, we examine how the interaction of multiple photoreceptive cells in complex eyes allows the formation of images and the detection of complex image properties such as color.

The structure of photoreceptor cells differs among animals

Two major types of photoreceptor cells are found in animals (Figure 6.28). **Ciliary photoreceptors** have a single cilium protruding from the cell, often with a highly folded ciliary membrane that forms lamellae or disks that contain **photopigments**, the molecules specialized for absorbing the energy coming from incoming photons. In contrast, in **rhabdomeric photoreceptors** (also called microvillus photoreceptors) the apical surface that contains the photopigments is elaborated into multiple outfoldings called microvillar projections.

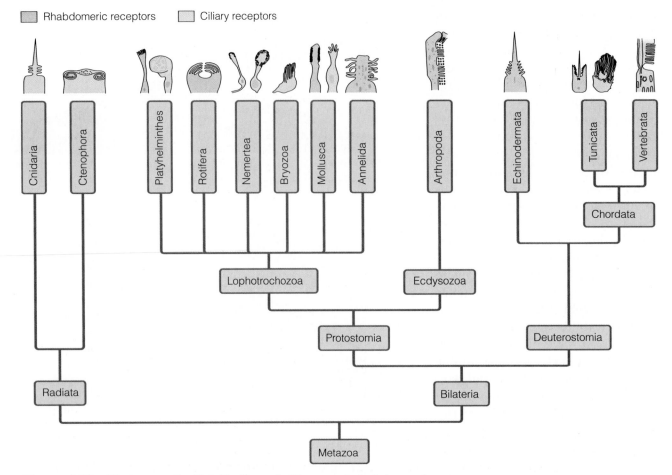

Figure 6.28 Phylogenetic distribution of ciliary and rhabdomeric photoreceptors There is no clear pattern in the phylogenetic distribution of ciliary photoreceptors (shown in orange) and rhabdomeric photoreceptors (shown in blue). Many groups have both kinds of photoreceptors. Vertebrates have only ciliary photoreceptors, and arthropods have only rhabdomeric photoreceptors.

In addition to these structural differences, ciliary and rhabdomeric photoreceptor cells also differ in that they use distinct signal transduction mechanisms for converting the energy carried by incoming photons to a change in the membrane potential of the receptor cell.

Both rhabdomeric and ciliary photoreceptors are widely distributed in most animal groups, but the pattern of the distributions of these types of photoreceptors among organisms presents a rather confusing picture (Figure 6.28). The majority of invertebrate groups have rhabdomeric photoreceptors in their eyes. Some invertebrate groups (such as the molluscs and the platyhelminths) also have some ciliary photoreceptors, but these are generally present only as small, isolated photoreceptors, or in very simple photoreceptive organs that are located outside the main eyes, or they are present only in larval forms and are absent from adult animals. The only known exceptions to the predominance of rhabdomeric eyes among the protostome invertebrates (worms, molluscs, and arthropods) are a few species of mollusc, such as the bay scallop *Pecten irradians* and the file clam *Lima scabra,* in which the adults have eyes that contain both rhabdomeric and ciliary photoreceptors. The picture in the deuterostomes (echinoderms, such as sea urchins, and chordates, such as the vertebrates) is also unclear. Most deuterostomes have rhabdomeric eyes, similar to those of the protostome invertebrates. The major exception to this rule is the vertebrates, which have only ciliary photoreceptors in their eyes. This pattern is also seen in the cnidarians (jellyfish and related organisms), which also have only ciliary photoreceptors. This phylogenetic pattern is difficult to interpret based on what we know about the relationships among living organisms. A recent discovery that rhabdomeric photoreceptors in some invertebrates pass through a developmental stage in which they have cilia suggests the possibility that all photoreceptor cells are derived from an ancestral ciliated cell. Alternatively, the bilateral ancestor of the protostomes and deuterostomes may have already possessed two types of photoreceptors, one of which may have been lost in some evolutionary lineages (such as the one leading to the vertebrates). Until the mechanisms of photoreception are studied in more animal taxa, particularly among the invertebrates, the evolution of animal photoreceptor cells is likely to remain an open question.

Mammals have two types of photoreceptor cells

Although all vertebrate photoreceptor cells are ciliary photoreceptors, in mammals they can be divided into two subclasses, **rods** and **cones** (Figure 6.29). Although rods and cones have different shapes, they share similar features. Both have an outer segment composed of a series of membranous disks that contain the photopigments. A connecting cilium joins the outer segment to the inner segment that contains the nucleus. The other end of this cell forms synaptic connections with other cells of the vertebrate eye.

In addition to their morphological differences, mammalian rods and cones differ functionally in a number of respects (Table 6.1). In comparison to cones, rods typically have more photopigment than do cones, have a much slower response time, and integrate signals over a longer period. As a result, rods have a very high sensitivity compared to cones, but saturate at relatively low light levels. Because of these differences between rods and cones, rods function best in dim light, while cones function best in bright light. In fact, in mammals, rods are so sensitive that they can respond even to a single photon. Many nocturnal mammals have

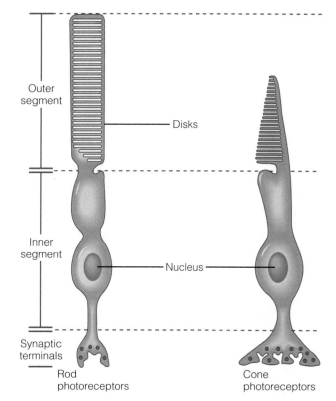

Figure 6.29 Structure of mammalian ciliary photoreceptors—the rods and cones Although they differ in shape, rods and cones have the same structural components: an outer segment consisting of a series of disks containing the photopigments, an inner segment containing the cell body, and synaptic terminals that make connections with neurons in the retina.

relatively higher numbers of rod cells in their eyes for better vision in dim light.

Many vertebrates have more than one type of cone photoreceptor, each having a slightly different photopigment that is maximally sensitive to a particular wavelength of light. As we discuss in detail later in the chapter, integrating centers compare the relative signals from these receptors to allow detection of colors. You have probably noticed that in dim light (such as at twilight), the world appears in shades of gray. You use your cones for color vision in bright light, and your rods for noncolor vision in dim light.

Table 6.1	Mammalian rods and cones.	
Feature	**Rods**	**Cones**
Class of photoreceptor	Ciliary	Ciliary
Shape	Outer segment rod shaped	Outer segment cone shaped
Sensitivity	Sensitive to very dim light	Sensitive to brighter light
Type of photopigment	One type	Up to three types in mammals

There is substantial diversity among vertebrates in the shape of the rods and cones (Figure 6.30). In fact, in many species it can be difficult to distinguish between rods and cones based on cell shape alone. For example, frogs have several types of rod-shaped photoreceptors in their eyes that they use to see colors. Thus, the shape of the photoreceptor cell is not the important characteristic that determines whether it is involved in color vision or dim-light vision. Instead, the properties of a photoreceptor cell depend on the properties of the photopigment that it contains.

Chromophores allow photoreceptors to absorb light

Photopigments consist of a pigment called a **chromophore** associated with a specific photoreceptor protein. In the vast majority of photoreceptors, the

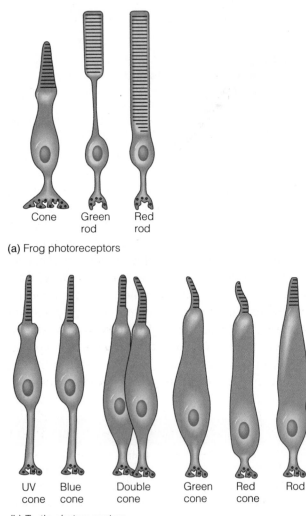

Cone   Green   Red
        rod   rod

(a) Frog photoreceptors

UV   Blue   Double   Green   Red   Rod
cone   cone   cone   cone   cone

(b) Turtle photoreceptors

Figure 6.30 Structural diversity of vertebrate photoreceptors

chromophore is a derivative of vitamin A, such as **retinal**, and the associated protein is a member of the **opsin** gene family. Opsins are G-protein-coupled receptors that are covalently linked to the chromophore. Depending on the particular photoreceptive cell, the photopigment complex is called by different names, including rhodopsin, iodopsin, porphyropsin, melanopsin, pinopsin, and VA opsin, among others. All of these photopigments, however, consist of a vitamin A-derived chromophore bound to a G protein in the opsin gene family. The sensitivity of the chromophore-opsin complex to particular parts of the light spectrum differs among these photopigments, as a result of differences in the amino acid sequence of the opsin protein. Differences in the spectral sensitivity of the chromophore-opsin combination underlie color vision.

Although the specific structures of the photopigments vary among photoreceptors, the general pattern of their chemical activation is similar. In the unactivated state, the chromophore is present in the *cis* conformation. When the chromophore absorbs the energy of incoming light, it undergoes a conformational change, rotating the molecule to an all-*trans* conformation. For example, absorbing light converts the chromophore 11-*cis* retinal to all-*trans* retinal (Figure 6.31). In the *cis* conformation, the chromophore binds to opsin, but when it is converted to the *trans* conformation, it no longer binds to opsin, and is released in a process known as **bleaching**. The chromophore is then reconverted back to the *cis* isomer by isomerase enzymes in an ATP-requiring process that takes several minutes. In the photoreceptors of vertebrates, the all-*trans* retinal is exported from the photoreceptor cell to nearby epithelial cells where it is converted to 11-*cis* retinal and then reimported into the photoreceptor, whereas in invertebrates this process typically takes place within the photoreceptor cell.

The mechanisms of phototransduction differ among organisms

When the chromophore dissociates from the opsin, the opsin undergoes a conformational change and becomes activated. Like other G-protein-coupled receptors, the activated opsin signals to an associated G protein that activates a downstream signal transduction cascade. Animal photoreceptors generally utilize one of two signal transduction cascades: either phospholipase C (PLC) or cGMP. The

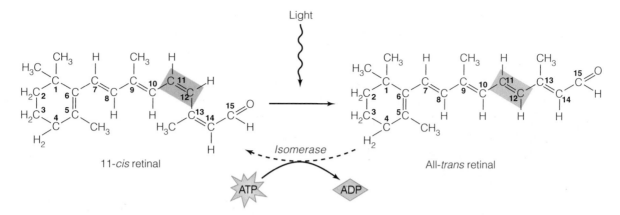

Figure 6.31 Isomerization of retinal The molecule 11-*cis* retinal absorbs a photon of light and rotates to form all-*trans* retinal.

opsins found in rhabdomeric photoreceptors, such as those present in most invertebrates, signal through a G_q protein that activates a phospholipase C (PLC)-mediated signal transduction cascade (Figure 6.32a). PLC catalyzes the breakdown of phosphatidyl-4,5-bisphosphate (PIP_2) into two intracellular messengers, inositol triphosphate (IP_3) and diacylglycerol (DAG). These signaling molecules initiate signal transduction pathways that open nonselective cation channels, and Ca^{2+} and Na^+ enter the cell, resulting in a depolarizing receptor potential. This depolarizing receptor potential causes an increase in neurotransmitter release from the photoreceptor, sending a signal to the nervous system that is ultimately interpreted as light.

In contrast, the opsins found in ciliary photoreceptors, such as those in the vertebrates, signal through an inhibitory G_i protein, called **transducin**, initiating a cyclic GMP-mediated signal transduction cascade (Figure 6.32b). Transducin activates a phosphodiesterase (PDE) enzyme that hydrolyzes cGMP to GMP. This decrease in cGMP concentration closes a cGMP-gated Na^+ channel in the photoreceptor membrane, and Na^+ influx slows or stops. Reduced Na^+ influx coupled with continuing K^+ efflux hyperpolarizes the cell, causing a receptor potential. The hyperpolarization decreases the release of neurotransmitter from the photoreceptor cell onto the associated afferent neuron, sending a signal to the nervous system that the brain ultimately interprets as light. In the dark, cGMP levels in the cell are high, cGMP binds to the channels, and most of the channels will be open, keeping the cell depolarized, and sending a constant signal to the afferent sensory neuron. Dim light causes a slight decrease in cGMP, causing a few channels to close, whereas bright light causes a larger decrease in cGMP, causing all or most of the Na^+ channels to close. Thus, the response of the cell is graded, depending on the light intensity.

The Structure and Function of Eyes

Although an individual photoreceptor cell can detect the relative brightness of a light source, an eye can obtain a great deal of additional information from an incoming light stimulus. The minimum criterion for calling a structure an eye, rather than simply a photoreceptor, is the ability to detect the direction from which light has entered the organ. Eyespots are single cells (or regions of a cell) that contain a photosensitive pigment and a shading pigment that helps provide directional information by shading light coming from some directions. For example, the eyespot of the protist *Euglena* is located at its anterior end, and consists of a light-sensitive swelling of the cell membrane that is associated with a red pigment. *Euglena*, which is a photosynthesizer, uses this eyespot to orient itself toward the light.

Eyes, however, are much more complex organs consisting of groups of cells specialized for different functions, and often include both multiple photoreceptor cells and separate pigment cells. Eyes can provide information such as light direction and contrasts between light and dark, and some eyes can form focused images. Among multicellular animals, there are four main types of eyes (Figure 6.33).

Flat-sheet eyes contain a layer of photoreceptor cells that form a primitive **retina** lined with a

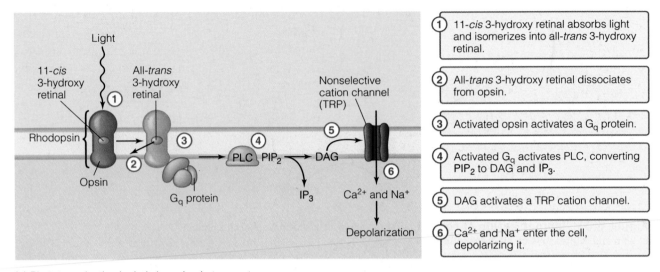

(1) 11-*cis* 3-hydroxy retinal absorbs light and isomerizes into all-*trans* 3-hydroxy retinal.

(2) All-*trans* 3-hydroxy retinal dissociates from opsin.

(3) Activated opsin activates a G_q protein.

(4) Activated G_q activates PLC, converting PIP_2 to DAG and IP_3.

(5) DAG activates a TRP cation channel.

(6) Ca^{2+} and Na^+ enter the cell, depolarizing it.

(a) Phototransduction in rhabdomeric photoreceptors

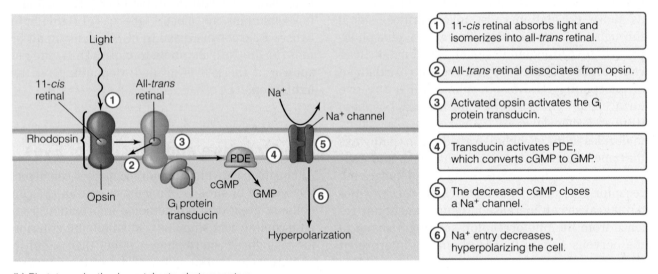

(1) 11-*cis* retinal absorbs light and isomerizes into all-*trans* retinal.

(2) All-*trans* retinal dissociates from opsin.

(3) Activated opsin activates the G_i protein transducin.

(4) Transducin activates PDE, which converts cGMP to GMP.

(5) The decreased cGMP closes a Na^+ channel.

(6) Na^+ entry decreases, hyperpolarizing the cell.

(b) Phototransduction in vertebrate photoreceptors

Figure 6.32 Phototransduction in the invertebrates and vertebrates

pigmented epithelium. These eyes provide some sense of light direction, and may allow the detection of contrasts between light and dark. Many animal groups have eyes of this type, although they are most often seen in larval forms or as accessory eyes in adults. However, the limpet *Patella* has a simple patch of pigmented cells that serve as its primary eyes.

Cup-shaped eyes (Figure 6.33b) are similar to flat-sheet eyes, except that the retinal sheet is folded to form a narrow aperture. These eyes provide much better discrimination of light direction and intensity, and allow improved detection of contrasts between light and dark. The most advanced cup-shaped eyes, such as those of the *Nautilus*, a cephalopod, have extremely small, pinhole-sized openings. The pinhole blocks most

of the light from entering the eye so that an incoming point light source illuminates a single point on the retina, forming an image. This design is similar to a primitive type of camera called a pinhole camera. Pinhole camera eyes can form images, although the resolution is poor and the image is dim. In order to form a crisp image, the aperture (pinhole) must be small, but a small aperture lets in only a small amount of light, resulting in a dim image. Thus, there is a compromise between image clarity and image intensity.

Vesicular eyes (Figure 6.33c) and modern cameras solve this conflict by inserting a lens into the pinhole aperture. A lens takes multiple sources of light and refracts them, focusing the light from a single source onto a single point on the retina. The challenge in developing a good vesicular eye

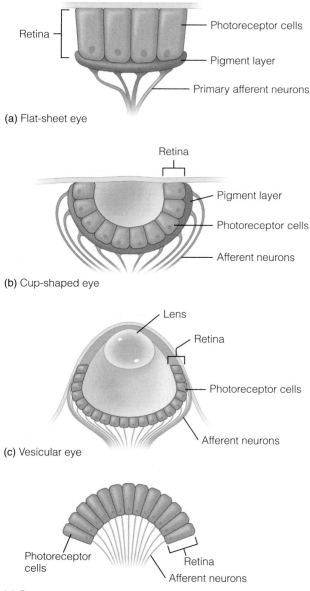

(a) Flat-sheet eye

Retina — Photoreceptor cells
— Pigment layer
— Primary afferent neurons

(b) Cup-shaped eye

Retina
— Pigment layer
— Photoreceptor cells
— Afferent neurons

(c) Vesicular eye

Lens
— Retina
— Photoreceptor cells
— Afferent neurons

(d) Convex eye

Photoreceptor cells
— Retina
— Afferent neurons

Figure 6.33   Structure of the major types of animal eyes

is that the lens must fit precise specifications in order to provide a clear image. However, even a bad lens is better than no lens at all, and provides an improvement over a pinhole camera-type eye.

Convex eyes (Figure 6.33d) are present in many annelids, molluscs, and arthropods. In these eyes, the individual photoreceptors radiate outward from the base, forming a convex, rather than a concave, light-gathering surface. The most complex convex eyes are the compound eyes of the arthropods (Figure 6.34). Compound eyes are composed of many **ommatidia** arranged radially to form the convex light-gathering surface. The

number of ommatidia in a compound eye varies greatly among species. For example, worker ants of the genus *Pomera* have only a single ommatidium per eye, while the eye of the dragonfly contains over 25,000 ommatidia arranged in a hexagonal pattern. The structure of an ommatidium also varies among species, although it generally consists of a modified region of the cuticle called the **cornea** overlying a crystalline cone that forms a lens. Immediately below this lens is a group of photoreceptive cells, called retinular cells, in a tubular arrangement. The retinular cells are rhabdomeric photoreceptor cells, as is typical for invertebrates. The microvilli of these photoreceptors project toward a central area called the rhabdom. Thus, in cross-section, the ommatidium resembles a slice through an orange.

Compound eyes form images in two rather different ways. *Apposition compound eyes*, which are found in many diurnal insects, consist of ommatidia that are each surrounded by a pigment cell. In an apposition compound eye each ommatidium operates essentially independently, and detects only a small part of the world directly in front of the ommatidium. However, the afferent neurons leading from the eye make many interconnections, so animals with apposition compound eyes are able to generate an integrated image. In contrast, *superposition compound eyes* have ommatidia that work together to produce a bright, superimposed image on the retina. Eyes of this type, found in nocturnal insects and crustaceans, function well in dim light. Compound eyes do not provide the resolving power of the camera eyes of vertebrates, but can still provide quite good visual discrimination.

There are two ways to increase the resolving power of a compound eye: reducing the size of each ommatidium or increasing the number of ommatidia. However, diffraction due to the wave properties of light limits the minimum size of an ommatidium. Once this size is reached, the only way to increase visual acuity is to increase the number of ommatidia, and thus the size of the compound eye. In fact, in order to have the average resolving power of the human eye, an insect eye would have to be nearly a meter in diameter.

Although insect eyes have limited resolving power, they are very good at capturing images from many directions. For example, a dragonfly can see almost 360° around itself, except for a small blind spot caused by its body. In addition, insects generally have very good close-up vision,

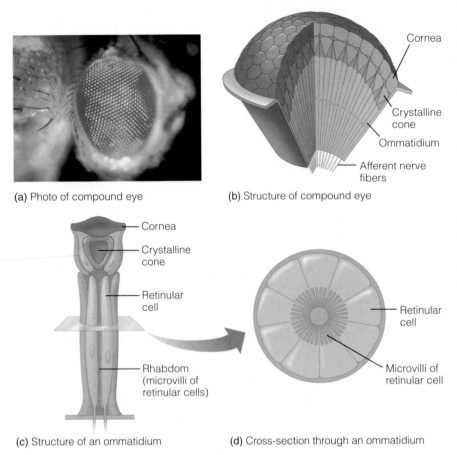

(a) Photo of compound eye

(b) Structure of compound eye

(c) Structure of an ommatidium

(d) Cross-section through an ommatidium

Figure 6.34 Structure of an insect compound eye and ommatidium
(a) The compound eye of *Drosophila melanogaster*. **(b)** A compound eye is composed of a cornea and many ommatidia. **(c)** Each ommatidium consists of a cornea, a crystalline cone, and several rhabdomeric photoreceptors called retinular cells. **(d)** The retinular cells are arranged radially, with their microvilli pointing inward to form a structure called the rhabdom.

and they can see objects for which we would need a microscope. However, most insects can see only a few millimeters away from their body. Dragonflies have the best distance vision among insects, and can see objects up to a meter away.

Because of the apparent complexity and enormous diversity of eyes, the evolution of eyes has been a topic of great interest to biologists. Recent discoveries in molecular developmental biology are providing some new insights into this classic problem (see Box 6.2, Genetics and Genomics: Molecular Similarity of Diverse Eyes).

The structure of the vertebrate eye relates to its function

The structure of the vertebrate eye allows the formation of a bright, focused image (Figure 6.35).

The outer surface of the mammalian eye consists of the **sclera**, a tough layer of connective tissue that makes up the "white" of the eye in humans, and the **cornea**, a transparent layer that allows light to enter the eye. At the front of the eye, just inside the cornea, are the **iris**, the **ciliary body**, and the **lens**. The iris consists of two layers of pigmented smooth muscle surrounding an opening called the pupil. The iris can constrict or dilate, controlling the amount of light that enters the eye. The iris dilates in dim light, increasing the size of the pupil, and allowing more light to enter the eye. In bright light, the iris constricts, reducing the size of the pupil, and limiting the amount of light that enters the eye. The lens is held in place behind the pupil by suspensory ligaments that are attached to the ciliary body, which contains the ciliary muscles. The iris and ciliary body divide the eye into two compartments. The anterior chamber contains a fluid called the *aqueous humor*. Aqueous humor is secreted by the ciliary body and circulates into the anterior chamber via the pupil. The lens is suspended in the posterior chamber, which contains a gelatinous mass called the *vitreous humor*. The vitreous humor assists in stabilizing the eye and provides support for the retina. Lining the inside surface of the eye is the retina, which contains the photoreceptor cells and several layers of interneurons that help to process the incoming visual signals. Immediately under the retina is the retinal pigment epithelium, which contains the cells that regenerate all-*trans* retinal back into the 11-*cis* conformation following light absorption. Just under the retinal pigment epithelium is a highly pigmented layer of tissue called the *choroid*. The choroid contains blood vessels, providing nourishment to the eye. In most diurnal animals, such as humans, the choroid also absorbs light that reaches the back of the eye so that it is not reflected, which might cause distortion of the visual image. The choroids of nocturnal animals such as

BOX 6.2 GENETICS AND GENOMICS
Molecular Similarity of Diverse Eyes

Although the structure of eyes appears to differ greatly among animals, ranging from a simple flat sheet to complex camera or compound eyes, at a molecular level the genes that control eye formation are surprisingly similar. For example, the gene *pax-6*, which codes for a transcription factor, has been isolated from humans, mice, chickens, zebrafish, sea urchins, and *Drosophila*. Loss-of-function mutations in this gene cause reduced or absent eye structures in both vertebrates and invertebrates. In humans, mutation of the *pax-6* gene causes the inherited disease aniridia, in which the iris of the eye is missing or misformed. In *Drosophila*, mutation of the *pax-6* gene causes the mutant phenotype called eyeless. Thus, the *pax-6* gene is responsible for the development of the eye in a wide variety of animals. In *Drosophila*, ectopic expression of *pax-6* (turning the gene on in tissues where it is not normally present) results in the formation of compound eyes in various parts of the body, including the legs, the antennae, and the wings. These ectopic eyes have been shown to respond to light, although they are not functional eyes because they are not correctly wired into the brain. Nevertheless, these experiments demonstrate that *pax-6* functions like an on switch, initiating a developmental cascade that results in eye formation, and acting as the master control gene for eye development. Since homologues of *pax-6* are found not just in *Drosophila*, which has compound eyes, but also in vertebrates, which have vesicular eyes, it is likely that all eyes share a common ancestor. This ancestral eye may have been just a single or a few photoreceptive cells whose development was controlled by *pax-6*. In fact, a homologue of *pax-6* is expressed in flatworms, which have a primitive cup-shaped eye consisting of a group of rhabdomeric photoreceptor cells surrounded by pigment cells.

References

○ Callaerts, P., A. M. Munoz-Marmol, S. Glardon, E. Castillo, H. Sun, W. H. Li, W. J. Gehring, and E. Salo. 1999. Isolation and expression of a Pax-6 gene in the regenerating and intact Planarian *Dugesia(G) tigrina. Proceedings of the National Academy of Sciences USA* 96: 558–563.

○ Gehring, W. J. 2002. The genetic control of eye development and its implications for the evolution of the various eye-types. *International Journal of Developmental Biology* 46: 65–73.

○ Salo, E., D. Pineda, M. Marsal, J. Gonzalez, V. Gremigni, and R. Batistoni. 2002. Genetic network of the eye in Platyhelminthes: Expression and functional analysis of some players during planarian regeneration. *Gene* 287: 67–74.

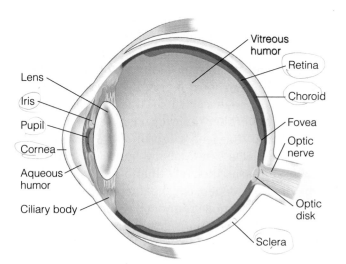

Figure 6.35 Structure of a mammalian eye
Light entering the eye passes through the cornea, the aqueous humor, the pupil, the lens, and the vitreous humor before striking the retina.

cats are slightly different from those of humans. They contain a layer called the *tapetum* that reflects light instead of absorbing it, amplifying the light and allowing noctural animals to see better than diurnal animals in dim light. Light reflected off the tapetum can make a cat's eyes appear to glow in the dark.

The lens focuses light on the retina

Both the cornea and lens have a convex shape, and thus act as converging lenses that focus the light on the retina (Figure 6.36). Converging lenses work by bending light rays toward each other, a process called refraction. Light refracts as it passes through objects of differing densities. In terrestrial vertebrates, the degree of refraction is much greater between the air and the cornea than between the cornea and the lens because of the large difference in density between the air and

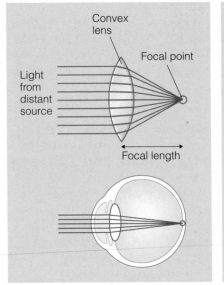

(a) Light rays from a distant object are parallel when they strike the eye, and focal length is short.

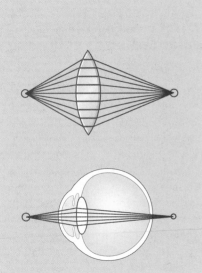

(b) Light rays from a nearby object are not parallel. Focal length increases and image is not focused on the retina.

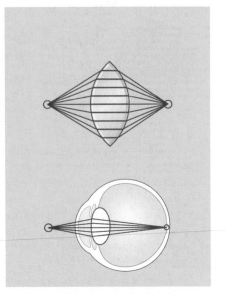

(c) Lens changes shape, altering focal length and bringing image of nearby object into focus on the retina in the process of accommodation.

Figure 6.36   Image formation and accommodation by the mammalian eye

corneal tissue. Thus, the cornea of terrestrial vertebrates plays the greatest role in focusing the image, whereas the lens only fine-tunes the focus. You can observe this effect for yourself; when you open your eyes underwater, you will find that it is difficult to bring objects into focus, because the cornea has a similar density to water and no longer refracts light in the same way as it does in the air. The cornea is less important than the lens for focusing images in the eyes of aquatic vertebrates because of this effect. The importance of the cornea in humans can be demonstrated by the success of laser eye surgery for correcting some vision problems.

The point at which the light waves converge after passing through a lens is called the *focal point*. The distance from the center of a lens to its focal point is called the *focal length*. A sharp image can be formed only at the focal point of a lens. Thus, incoming light rays must converge at the retina, not behind it or in front of it, in order to produce a clear image. The focal length of an image changes, depending on the distance between the object and the eye. As shown in Figure 6.36a, light rays reflected off a distant object are nearly parallel when they pass through the lens, but light rays reflected off a nearby object are not parallel when they pass through the lens (Figure 6.36b). As a result of this difference in angle, the focal

lengths for nearby and distant images differ. In order to produce focused images of objects at various distances, the eye must ensure that the focal point falls on the retina, a process termed **accommodation**. Because the location and shape of the cornea are fixed, the cornea does not participate in accommodation. Instead, the lens must either change position relative to the retina, or change shape.

Some polychaete worms change focal length by changing the volume of fluid in the eye, altering the size of the eye and thus the distance between the lens and the retina. Many invertebrates and vertebrates alter focal length by moving the lens forward or backward. In contrast, lizards, birds, and mammals alter their focal length by changing the shape of the lens (Figure 6.36). To focus on nearby objects, the ciliary muscles contract, which increases their width and loosens the tension on the suspensory ligaments, causing the lens to become more rounded. To focus on distant objects, the ciliary muscles relax. This reduces the width of the ciliary muscles, increasing tension on the suspensory ligaments, which pulls on the lens and flattens it. A more spherical lens aids in focusing on nearby objects, whereas a flatter lens brings distant objects into focus on the retina.

Vertebrate retinas have multiple layers

In addition to containing the photoreceptor cells that transduce incoming light energy into an electrical signal, vertebrate retinas contain many interneurons that play an important role in the processing of visual signals (Figure 6.37a). The rods and cones are actually located at the back of the retina, oriented with their tips embedded in the pigment epithelium at the back of the eye. The rods and cones form synapses with a layer of *bipolar cells*, and these bipolar cells in turn form synapses with a layer of *retinal ganglion cells*. In the same layers as the bipolar and ganglion cells are two additional classes of interneuron: the *horizontal cells* and the *amacrine cells*. The axons of the ganglion cells run along the surface of the retina, joining together to form the optic nerve, which exits the retina at a point slightly off the center of the retina. This area, called the optic disk, contains no photoreceptor cells, causing a "blind spot."

Because the photoreceptors of the vertebrate retina are located in its deepest layer, light entering the eye must travel through the ganglion and bipolar cells before reaching the photoreceptor cells. The only exception to this rule is an area called the *fovea* or the *visual streak*. The fovea is a circular region located roughly in the middle of the eye. Most nonmammalian vertebrates, as well as some mammals (including humans and other primates), have a fovea in each eye. In contrast, the majority of mammals, and some nonmammalian vertebrates, have a visual streak, which is a narrow strip along the retina arranged in the plane of the horizon. In both the fovea and the visual streak, the overlying bipolar and ganglion cells are pushed to one side, allowing light to strike the photoreceptors without passing through several layers of neurons. As a result, vision is sharpest in these regions.

The retina of cephalopods is arranged rather differently than the retina of vertebrates. In the cephalopods, the photoreceptors are located on

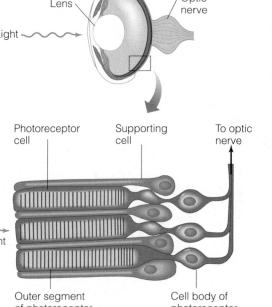

(a) Vertebrate eye and retina

(b) Cephalopod eye and retina

Figure 6.37 Organization of the retina in vertebrates and cephalopods (a) In the vertebrate retina, the photoreceptors are located toward the back. Light must pass through several layers of cells before striking the photoreceptors. The middle layers of the retina also contain interneurons that are important for signal processing within the vertebrate retina. (b) The cephalopod retina consists of a single layer of photoreceptor and supporting cells. Light entering the eye strikes the photoreceptors directly without passing through multiple retinal layers. There are no interneurons, and little or no signal processing occurs within the retina.

the surface of the retina, rather than at the back (Figure 6.37b). Supporting cells are located between the photoreceptor cells, but there are no additional layers of cells. The axons of the photoreceptors come together to form the optic nerve, rather than forming synapses with interneurons within the retina. Thus, the cephalopod retina has far fewer parts than a vertebrate retina, and little signal processing occurs in the retina itself.

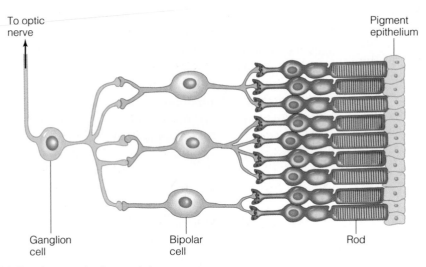

(a) Signal processing from rod photoreceptors

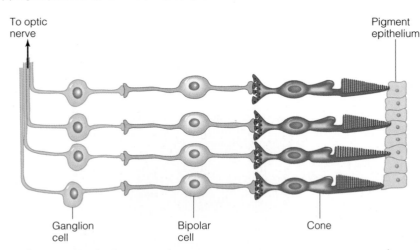

(b) Signal processing from cone photoreceptors

Figure 6.38 Convergence in the vertebrate retina **(a)** The signaling pathways of rods show convergence. Many rods can form synapses with one bipolar cell, and several bipolar cells may form synapses with a single ganglion cell. Thus, the receptive fields of these retinal ganglion cells include input from many photoreceptor cells. **(b)** The signaling pathways of cones in the fovea do not converge. A single cone forms a synapse with a single bipolar cell, which forms synapses with a single ganglion cell. Thus, the receptive fields of these ganglion cells include input from a single photoreceptor cell.

Information from rods and cones is processed differently

The vertebrate retina processes information coming from rods and cones differently (Figure 6.38). Rod signaling pathways are organized using the principle of *convergence*. Many rods synapse with a single bipolar cell, and many of these bipolar cells can synapse with a ganglion cell. As a result, as many as 100 rods may connect with a single retinal ganglion cell. In contrast, a cone located within the fovea connects to a single bipolar cell, and that bipolar cell connects to a single ganglion cell. Thus, a single pathway carries a signal from a cone cell to the visual centers of the brain. Toward the edge of the retina, cones participate in somewhat more convergent pathways, but never to the extent seen with rods. These differences in wiring result in differences in the size of the receptive fields of retinal ganglion cells. A retinal ganglion cell that is associated with only one or a few photoreceptors has a small receptive field, processing information from only a small area of the retina. In contrast, a retinal ganglion cell that is associated with many photoreceptors has a large receptive field, and processes information from a larger area of the retina. Thus, retinal ganglion cells that are associated with cones located in the fovea have very small receptive fields and can provide a detailed, high-resolution image. In contrast, the receptive field of a retinal ganglion cell that receives inputs from rod photoreceptors is much larger, and thus rods provide less detailed images.

Signal processing in the retina enhances contrast

Vertebrate retinas are organized such that they enhance the perception of borders and contrast, using the process of lateral inhibition that we discussed at the beginning of this chapter. In fact, a point light source causes a greater response in a retinal ganglion cell than does evenly distributed diffuse illumination of the same intensity. This phenomenon occurs because the receptive fields of retinal ganglion cells have a *center-surround* organization, consisting of a central region surrounded by a concentric ring that each have different responses to light (Figure 6.39). For example, an "on-center" retinal ganglion cell increases action potential frequency in response to illumination of the center of the receptive field, and decreases action potential frequency in response to illumination of the surround region of the receptive field. An "off-center" retinal ganglion cell shows the opposite response. The horizonatal and amacrine cells of the retina play the major role in establishing the center-surround organization of a retinal ganglion cell.

To see how this works, let's trace the events in the retina when light strikes the receptive field of a retinal ganglion cell with an on-center organization (Figure 6.39, left side). When a bright light is shone onto photoreceptors in the center region of the receptive field, the energy from the incoming light converts 11-*cis* retinal to all-*trans* retinal, activating the G protein transducin, which decreases cGMP within the photoreceptor cell. The decrease in cGMP closes Na^+ channels, hyperpolarizing the cell. This hyperpolarizing graded potential reduces the release of the neurotransmitter glutamate from the photoreceptor cell. Glutamate is an inhibitory neurotransmitter for the bipolar cell, so a decrease in the inhibitory neurotransmitter glutamate stimulates the bipolar cell, causing it to depolarize. The depolarization increases the release of neurotransmitter from the bipolar cell, stimulating the ganglion cell to depolarize.

Now let's look at what happens when a more diffuse light is shone onto the receptive field such that it illuminates photoreceptors in both the center and surround regions. In addition to forming synapses with bipolar cells, photoreceptors in the surround region of the receptive field form synapses with horizontal cells (Figure 6.40). When stimulated, these horizontal cells inhibit the activity of the bipolar cells that are connected to the photoreceptors at the center of the receptive field. Thus, bipolar cells that form synapses with photoreceptors in

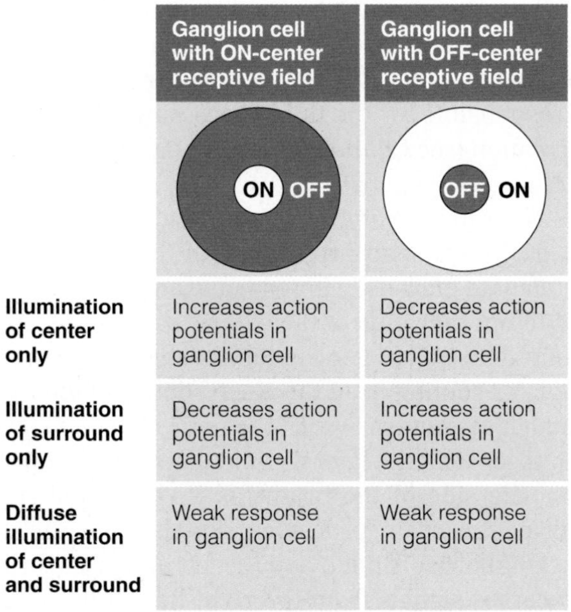

Ganglion cell with ON-center receptive field	Ganglion cell with OFF-center receptive field
ON OFF	OFF **ON**

	Ganglion cell with ON-center receptive field	Ganglion cell with OFF-center receptive field
Illumination of center only	Increases action potentials in ganglion cell	Decreases action potentials in ganglion cell
Illumination of surround only	Decreases action potentials in ganglion cell	Increases action potentials in ganglion cell
Diffuse illumination of center and surround	Weak response in ganglion cell	Weak response in ganglion cell

Figure 6.39 Receptive fields of retinal ganglion cells Retinal ganglion cells have complex receptive fields that are divided into regions with different responses to light. Ganglion cells with an on-center receptive field fire action potentials at higher frequency in response to light focused on the center of the receptive field and fire action potentials at a decreased frequency in response to light focused on the surrounding region of the receptive field. When light strikes both the center (on) region and the surround (off) region at the same time, the two effects partially cancel out and the frequency of action potentials increases only slightly. The opposite pattern holds for retinal ganglion cells with an off-center organization.

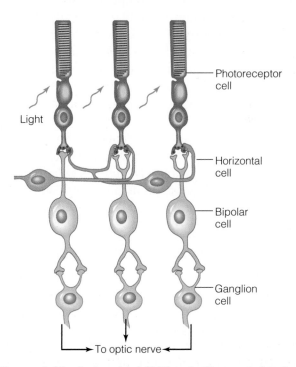

Figure 6.40 Lateral inhibition in the vertebrate retina Photoreceptors communicate with both bipolar cells and horizontal cells. Excited horizontal cells inhibit neighboring bipolar cells—the process of lateral inhibition.

the center of the receptive field receive two conflicting inputs: a stimulatory input from the center photoreceptors and an inhibitory input from the surround photoreceptors (via the horizontal cells). These two conflicting inputs cause the bipolar cell to send a much weaker signal to the retinal ganglion cell, reducing its response to diffuse light compared with a point of light in the center of the receptive field.

Similar processes occur for retinal ganglion cells with an off-center organization, but in this case, glutamate released from the photoreceptor cells acts as an excitatory neurotransmitter for bipolar cells connected to photoreceptors in the center of the receptive field. When light strikes these photoreceptors, it causes the photoreceptor to hyperpolarize and decrease the release of glutamate, just as in the case of a photoreceptor in an on-center receptive field. In the case of an off-center receptive field, however, this decrease in glutamate hyperpolarizes the bipolar cell and reduces the release of neurotransmitter onto the retinal ganglion cell, causing the frequency of action potentials in the retinal ganglion cell to decline. The difference in the response of the bipolar cell is caused by the presence of a different isoform of the glutamate receptor in bipolar cells associated with on-center and off-center receptive fields.

To add another layer of complexity, bipolar cells do not always form synapses directly with ganglion cells. Instead, bipolar cells form electrical synapses with amacrine cells. Depolarization of the bipolar cell is communicated directly to the amacrine cell via gap junctions. The amacrine cell integrates and modifies the inputs from several bipolar cells, ultimately altering the release of neurotransmitter from the amacrine cell onto the ganglion cell. These extremely complex relationships are particularly prevalent in the highly convergent pathways involved with rod photoreceptors.

The brain processes the visual signal

We can define a region called the **visual field**, which consists of the entire area that can be seen without moving the eyes. Depending on the position of the eyes on the head, each eye sees a somewhat different part of the visual field. In animals with their eyes on the sides of their heads, there is little overlap between the visual fields of the right and left eyes, whereas in animals with eyes placed toward the front of their heads there is a great deal of overlap between the visual fields of the right

and left eyes, in an area called the *binocular zone*. Figure 6.41 illustrates the visual field of a human. Human eyes are on the front of the head, and the binocular zone is large.

Each part of the retina detects a different portion of the visual field. Light from the left part of the visual field strikes the right part of the retina of each eye, whereas light from the right part of the visual field strikes the left part of the retina of each eye. In fact, we can divide the human retina down the middle (roughly at the fovea) and define two regions of each retina: the temporal half (toward the outside of the face) and the nasal half (toward the center of the face). The temporal retina of the right eye detects the left visual field, and the nasal retina detects the right visual field. In contrast, the temporal retina of the left eye detects the right visual field, and the nasal retina of the left eye detects the left visual field.

The two optic nerves carrying information from the right and left eyes converge in a region called the **optic chiasm** (Figure 6.41). Most of the neurons then form synapses in a part of the brain called the *lateral geniculate nucleus*, which in turn sends processes to the **visual cortex**, which is responsible for the final processing of visual information (see Chapter 7: Functional Organization of Nervous Systems).

Neurons coming from the temporal retina of the right eye send projections to the right lateral geniculate nucleus, whereas neurons coming from the temporal retina of the left eye send projections to the left lateral geniculate nucleus. In contrast, neurons coming from the nasal retinas of the right and left eyes cross over at the optic chiasm to form synapses with the lateral geniculate nucleus on the opposite side of the brain. As a result, the right half of the brain processes signals from the left part of the visual field, and the left half of the brain processes signals from the right half of the visual field. The right and left sides of the visual field overlap in the binocular zone, and thus signals from the binocular zone are processed on both sides of the brain. Animals can compare the properties of the images in the binocular zone coming from each eye to provide information such as the distance of an object from the body. This is one of the processes that underlies depth perception.

In general, the degree of crossing of neurons in the optic chiasm is related to the degree of overlap between the left and right visual fields. In fishes and amphibians with eyes located at the extreme sides of the head, the left and right visual fields do not overlap. These animals lack a binocular zone, and most of the neurons in the optic nerve from the right eye send projections to the left side of the brain, whereas the optic nerve from the left eye sends projections to the right side of the brain. Similarly, in mice, which also have a limited overlap between their right and left visual fields, about 97% of the fibers cross over to the other side of the brain, while only 3% of the fibers are uncrossed. Although animals (such as fish and rodents) with eyes on each side of the head tend to have poor depth perception, these animals have excellent panoramic vision, often having an almost 360° view of the world. Humans have a large binocular zone, and about 60% of the fibers in the optic nerve cross over to the other side of the brain at the optic chiasm, while 40% of the fibers are uncrossed. This cross-fiber organization plays a part in generating *stereopsis*, in which comparison of the information by the two eyes assists in depth perception. In general, animals with superior stereopsis tend to have roughly equal amounts of crossed and uncrossed fibers, allowing easy comparison of signals from each eye on both sides of the brain. Owls, which have eyes at the front of their heads, and excellent depth perception, are an exception to this rule because all of their optic neurons cross at the optic chiasm. The two sides of an owl's brain communicate with each other in other parts of the visual pathway, allowing both sides of the brain to process images from both eyes, and providing the necessary conditions for good depth perception.

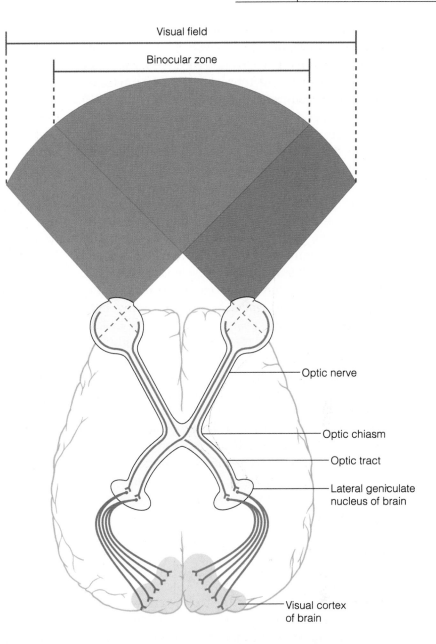

Figure 6.41 Visual processing In humans, about half of the neurons coming from each eye cross over each other in the optic chiasm. Neurons sending signals from the right side of the field of view from both the left and right eyes send processes to the left half of the brain, whereas neurons sending signals from the left side of the field of view from both the right and left eyes send processes to the right side of the brain. Thus, each side of the brain receives information from both eyes. Comparing these two views provides stereopsis, which enhances depth perception.

Color vision requires multiple types of photoreceptors

In addition to detecting shapes and movements, many animals are capable of detecting the wavelength of incoming light, a phenomenon we experience as color. In order to detect colors, an animal must be able to distinguish among different wavelengths. Animals accomplish this by having more than one type of photoreceptor cell, each containing a photopigment that is sensitive to light of specific wavelengths. Humans can distinguish about 1500 wavelengths between 400 nm (blue) and 700 nm (red), differences in wavelength of about 0.2 nm. This might suggest that humans would need several thousand different photopigments and photoreceptor cells; however, humans have only three different cone photoreceptors, with maximum sensitivities of approximately 440 nm (blue), 530 nm (green), and 565 nm (red) (Figure 6.42). Light of a given wavelength stimulates more than one type of cone, but to different degrees. The retina and brain then compare the output from each type of cone and infer the color of the stimulus.

Each cone photoreceptor is maximally sensitive to a particular wavelength of light, but can also be stimulated by light of other wavelengths. So how can the brain distinguish between a low-intensity stimulus at the peak wavelength and a strong stimulus at another wavelength? Clearly, a single cone photoreceptor cannot provide information about the wavelength of incoming light. The outputs of all types of cones must be used to estimate the wavelength of the incoming light. The first stage of this processing occurs in the horizontal and ganglion cells of the retina, where lateral inhibition by horizontal cells plays an important role in the initial processing of color information. This system, called *trichromatic color vision*, allows humans to see a wide range of colors using only three types of cone photoreceptors.

Birds, reptiles, and shallow-water fishes can be trichromatic, tetrachromatic, or even pentachromatic (depending on the species). It is difficult for us to understand the visual world of a pentachromatic animal. The additional photoreceptors likely allow these species to discriminate among colors that appear the same to humans, and some species can detect light in the ultraviolet (UV) or infrared ranges that humans cannot detect. Most mammals are *dichromats*, having only middle (green) and short (blue) wavelength cones (in addition to rods) in their retinas. Since dichromats lack the "red" cone, these animals cannot distinguish between red colors and green colors, similar to a human that is red/green color-blind. Many marine mammals and a few nocturnal rodents and carnivores have secondarily lost one of these pigments and become monochromats that cannot distinguish colors at all.

Because ancient reptile-like creatures with at least trichromatic color vision are the probable ancestors of the mammals, we can infer that mammals must have lost one or more of the ancestral photopigment genes. Mammals are thought to have evolved primarily as nocturnal creatures (first appearing during the time of the dinosaurs), and at that time some of the genes needed for color vision may have been lost because they were not needed for vision in dim light. Trichromacy was subsequently restored only in the primates. Interestingly, however, trichromatic color vision appears to have evolved independently in the Old World primates and the New World primates (see Box 6.3, Evolution and Diversity: The Evolution of Trichomatic Color Vision in Primates).

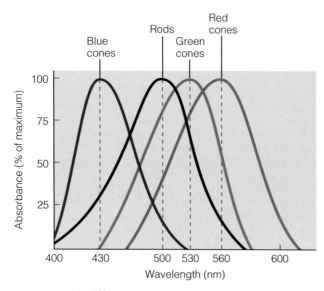

Figure 6.42   The absorbance spectra of human rods and cones   Humans typically have one type of rod photopigment and three types of cone photopigment. Although the absorbance spectra of the photopigments overlap, each has a unique absorbance maximum. By comparing the signals coming from each type of photoreceptor, the brain can distinguish over 1000 different wavelengths of light.

⊙ | CONCEPT CHECK

14. Compare and contrast phototransduction in rhabdomeric and ciliary photoreceptors.

15. What are the advantages of a vesicular eye compared to a pinhole-type eye?

16. Would you expect laser eye surgery (which affects the shape of the cornea) to be effective in an aquatic vertebrate? Why or why not?

17. Explain how lateral inhibition enhances contrast at the retina.

Thermoreception

Animals have central thermoreceptors, located in the hypothalamus of the brain, that monitor their internal temperature, and peripheral thermore-

BOX 6.3 **EVOLUTION AND DIVERSITY**
The Evolution of Trichromatic Color Vision in Primates

Mammals generally have much worse color vision than other vertebrates, and many species are entirely color-blind. Primates are one of the few exceptions to this rule. All of the Old World primates (humans, apes, and Old World monkeys) have trichromatic color vision similar to that found in humans. In contrast, the New World monkeys vary greatly in their ability to see colors. Most species are dichromatic, a few species have trichromatic females but dichromatic males, and only the howler monkeys are true trichromats. The genetics of these different visual systems have now been worked out, and their evolution has been studied in detail.

Humans and the other Old World primates have three opsin genes in the genome, one coding for a blue-sensitive photopigment, one coding for a green-sensitive photopigment, and one coding for a red-sensitive photopigment. The "green" and "red" opsins are coded by very similar DNA sequences, and differ by only 11 amino acids. This degree of differentiation suggests, based on the approximate mutation rate of genes in the vertebrates, that these genes began to diverge from each other about 40 million years ago. It appears that an ancestral "green" opsin gene was duplicated at that time, during the early evolution of the Old World primates, and the two genes began to diverge. In humans, these two genes are located very close together on the X chromosome, further suggesting that they arose through an ancestral duplication in this part of the genome.

Some species of New World primate, such as the owl monkey, a nocturnal animal, are monochromats and are thus color-blind. But most other species of New World monkeys have a form of trichromatic color vision. These monkeys have only two opsin genes in their genome—a "blue" opsin and a "green" opsin. As in the Old World primates, the "green" opsin gene is found on the X chromosome, but in this case the gene has not been duplicated. Instead, in some species of New World monkeys, two different alleles of this one gene are present in the population. One of the alleles is sensitive to green light, and the other is more sensitive to red light. An individual that is heterozygous for these alleles (that has one copy of the "green" allele and one copy of its "red" variant) is functionally trichromatic, expressing a "blue"

opsin, a "green" opsin, and a "red" opsin. Recall that the "green" opsin gene is found on the X chromosome. Male primates have only one copy of the X chromosone, and one copy of the Y chromosome. Thus, males of this species are always homozygous for the "green" opsin gene, and are functionally dichromatic. In these species, males are red/green color-blind, while females can be either color-blind or trichromats, depending on whether they are heterozygous for this gene.

Of all the New World primates, only the howler monkeys deviate from this system. In howler monkeys, the "green" opsin gene has been duplicated, similar to the situation in the Old World primates. Thus, both male and female howler monkeys are true trichromats, and have color vision similar to that in humans. Because the New World monkeys diverged from the Old World monkeys prior to the evolution of the primates, the gene duplication in the howler monkeys is independent from that shared by all of the Old World primates. It also appears to be somewhat more recent, since the "green" and "red" opsins of the howler monkeys differ from each other by only eight amino acids. Thus, true trichromacy has evolved at least twice in the primates, once in the lineage leading to the Old World primates, and once in the ancestors of the howler monkeys. Multiple independent evolution of a phenotypic trait strongly suggests that this trait has been selected over evolutionary time for some important function. For example, being able to distinguish many shades of red and green might allow primates to easily find ripe fruit in a background of leaves.

References

○ Dominy, N. J., and P. W. Lucas. 2001. Ecological importance of trichromatic vision to primates. *Nature* 410: 363–366.

○ Dulai, K. S., M. von Dornum, J. D. Mollon, and D. M. Hunt. 1999. The evolution of trichromatic color vision by opsin gene duplication in New World and Old World primates. *Genome Research* 9: 629–638.

○ Jacobs, G. H. 1996. Primate photopigments and primate color vision. *Proceedings of the National Academy of Sciences USA* 93: 577–581.

○ Orsorio, B., and M. Vorobyev. 1996. Colour vision as an adaptation to frugivory in primates. *Proceedings of the Royal Society of London (Series B: Biological Sciences)* 263: 593–599.

ceptors that monitor environmental temperature. There are three types of peripheral thermoreceptors: warm-sensitive thermoreceptors, cold-sensitive thermoreceptors, and thermoreceptors that are specialized for detecting painfully hot stimuli. In mammals, warm-sensitive neurons start to fire action potentials when the skin temperature is raised above 30°C, and firing frequency increases with increasing temperature up to a saturating value. In contrast, cold receptors are extremely sensitive to small (0.5°C) decreases in temperature, but they respond mostly to temperature change, rather than the absolute value of the temperature. The thermal nociceptors detect painful heat and burns, and start to fire only at higher, painful temperatures (starting at around 45°C in mammals). These neurons increase their firing frequency in parallel with increasing pain sensation.

Thermoreception begins when specific thermoreceptor proteins in the free nerve endings of thermoreceptor neurons are activated. These receptors, which are found in both vertebrates and invertebrates, are called thermoTRPs and, like some mechanoreceptors, are members of the TRP family of ion channels. Individual thermoTRPs are specialized to detect distinct temperature ranges; some thermoTRPs are activated by heat, others by cold. Capsaicin, the "hot" ingredient in peppers, stimulates warm-sensitive neurons, while menthol, the ingredient that makes mints taste "cool," stimulates cold-sensitive neurons. Further study of the responses of thermoreceptors to these chemicals should lead to clues about the gating properties of the thermoTRPs.

Some animals have highly specialized sensory organs that allow them to detect heat radiating from objects at a distance. For example, pit vipers (a group that includes rattlesnakes) have specialized **pit organs** that are found between the eye and nostril on either side of the head (Figure 6.43).

Other snakes such as the boa constrictor have labial pits along the upper and lower jaws. Pit organs and labial pits are extremely sensitive thermoreceptors that allow snakes to detect mammalian prey and to select thermally appropriate habitats. The thermoreceptive neurons in the pit organs can detect temperature changes as small as 0.003°C (compare this to the 0.5°C discrimination of human thermoreceptors). Little is known about the transduction mechanisms of pit organs. However, like the thermoTRP neurons in humans, pit organ thermoreceptors are sensitive to capsaicin, suggesting the possibility of a similar mechanism.

Magnetoreception

Magnetoreception, or the ability to detect magnetic fields, is widely distributed throughout the animal kingdom. Migratory birds, homing salmon, and many other organisms use the Earth's magnetic field to help them navigate, although humans apparently lack this sense. Magnetoreception has been extensively studied, but the mechanisms of magnetoreception are not understood for any animal, and it remains the most elusive of sensory modalities.

In one intriguing study, scientists identified specific neurons in the olfactory epithelium of rainbow trout that respond to magnetic fields. These neurons contain particles that resemble magnetite when examined under a microscope. Magnetite is a natural mineral that responds to magnetic fields, and thus could be the basis for magnetoreception in animals. The magnetite particles in trout olfactory neurons are arranged in a chain within the cell, similar to a compass needle, strongly suggesting that trout use a magnetite-based mechanism for detecting magnetic fields. A similar mechanism is used by some species of bacteria that can orient themselves in a magnetic field. However, the mechanism by which magnetoreceptive sensory neurons in trout respond to changes in the position of the magnetite is still unknown. Not all animals that can respond to magnetic fields have detectable magnetite crystals, so it is unlikely that this mechanism is found in all magnetoreceptors.

Figure 6.43 Pit organs of snakes The pit organs of this fer de lance pit viper are clearly visible between the nostril and the eye.

Integrating Systems | Sensory Systems and Circadian Rhythms

Circadian rhythms are predictable daily variations in physiological parameters that are linked with the daily cycle of light and dark. Almost every aspect of behavior and physiology undergoes a circadian rhythm, including processes such as metabolic rate, activity, and digestion. Circadian rhythms persist even when an organism is kept in constant darkness; however, without environmental cues these rhythms tend to be somewhat longer or shorter than 24 hours—giving rise to the name circadian (circa = about; dies = day). External environmental cues, such as the pattern of light and dark, help to keep this intrinsic circadian clock in sync with the natural environment.

In mammals, the circadian clock is located within a part of the brain called the hypothalamus, or more specifically within the suprachiasmatic nucleus, a grouping of about 10,000 neurons within the hypothalamus (Figure 6.44a). Very little light penetrates so deeply within the brain, so for many years scientists assumed that rod and cone photoreceptors must somehow communicate the incoming light information to the suprachiasmatic nucleus. This assumption was supported by the observation that mammals that lack eyes cannot reset their circadian clocks in response to light cues. However, genetically defective mammals that lack rods and cones but have otherwise intact eyes display normal circadian rhythms that respond to light cues, suggesting that the rods and cones cannot be the source of the light input to the circadian clock.

We now know that retinal ganglion cells play the critical role in sending light signals to the suprachiasmatic nucleus. Retinal ganglion cells make synaptic connections with neurons in the suprachiasmatic nucleus, providing a direct neural pathway between the sensory receptor cell (the retinal ganglion cell) and the integrating centers of the central nervous system. The nature of photoreceptor protein in these cells is still somewhat disputed, but most evidence points to an opsin-related protein called *melanopsin*.

The suprachiasmatic nucleus then communicates its rhythmic signal to other parts of the brain and to many physiological systems (Figure 6.44b). At present, most evidence suggests that the suprachiasmatic nucleus communicates with the rest of the body by secreting neuropeptides. For example, if you destroy the neurons in the suprachiasmatic nucleus, circadian rhythms disappear, but if you transplant suprachiasmatic nucleus neurons from another animal, the circadian rhythms return, even though the neurons do not form synaptic connections with other parts of the brain.

The suprachiasmatic nucleus communicates its circadian signal to other parts of the hypothalamus including the paraventricular nucleus, the ventromedial nucleus, and the periventricular nucleus. These nuclei are involved in regulating a large number of important physiological processes. The ventromedial nucleus regulates appetite and feeding behavior. The paraventricular nucleus synthesizes the hormones vasopressin (also called antidiuretic hormone or ADH) and oxytocin. Vasopressin regulates kidney function, whereas oxytocin influences milk ejection from the breast and the contraction of the uterus in mammals, and sexual behavior and pair-bonding in other animals. The periventricular nucleus secretes a large number of releasing hormones that regulate the hormones of the anterior pituitary. The pituitary hormone prolactin is best known for its effects on reproduction in mammals. It stimulates the growth of the mammary glands, causes milk production, and in some species it is important for maintaining pregnancy and stimulating reproductive behaviors such as nest building (see Chapter 14: Reproduction). Prolactin is also involved in a host of other physiological processes, including (1) water and electrolyte balance, (2) growth and development, and (3) immune function. Thyroid-stimulating hormone (TSH) released by the pituitary causes the release of thyroid hormones from the thyroid gland. Thyroid hormones play an important role in the regulation of metabolic rate. Adrenocorticotropic hormone (ACTH) causes the release of corticosteroids from the adrenal cortex. Corticosteroids, or the stress hormones, regulate many biological processes, particularly those involved in carbohydrate metabolism. Growth hormone (GH) released from the pituitary stimulates a variety of anabolic processes, and regulates the release of insulin-like growth factor from the liver, which in turn plays a role in regulating growth. Finally, follicle-stimulating hormone (FSH) and luteinizing hormone (LH) regulate the production of the sex hormones (androgens and estrogens). By modulating the activity of the hypothalamus, which influences pituitary function, the circadian clock can influence almost every function of the body.

The suprachiasmatic nucleus also sends signals to the pineal gland, in a neighboring part of the brain. The pineal gland secretes the hormone melatonin into the cerebrospinal fluid and the blood in a circadian rhythm. In humans, melatonin secretion is high at night and low during the day. Most tissues of the body have receptors for melatonin, so although the effects of this hormone

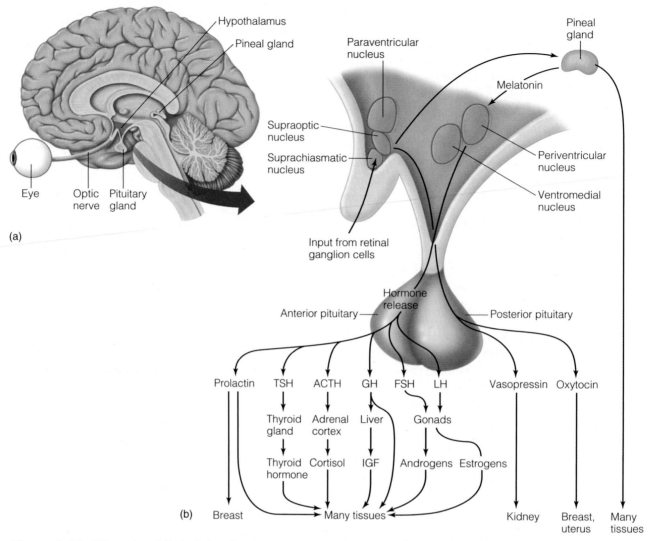

(a)

(b)

Figure 6.44 The role of light input in circadian rhythms in mammals **(a)** The organs involved in circadian rhythms in mammals. **(b)** The endocrine system and circadian rhythms. A light signal from the retinal ganglion cells entrains the circadian clock in the suprachiasmatic nucleus (SCN) of the hypothalamus. The SCN sends a signal to the pineal gland, altering the release of melatonin on a circadian cycle. Melatonin and secreted proteins from the SCN affect the other hypothalamic nuclei, causing circadian changes in the release of vasopressin and oxytocin from the posterior pituitary, and affecting the secretion of releasing hormones into the pituitary portal system. The releasing hormones in turn affect the secretion of the pituitary hormones, which go on to have direct effects on a variety of tissues, as well as influencing the release of hormones from other endocrine glands. Melatonin from the pineal gland also enters the bloodstream and has effects on many tissues. (TSH = thyroid-stimulating hormone; ACTH = adrenocorticotropic hormone; GH = growth hormone; FSH = follicle-stimulating hormone; LH = luteinizing hormone; IGF = insulin-like growth factor).

are not yet fully understood they are likely to be widespread. The suprachiasmatic nucleus and parts of the anterior pituitary have particularly high levels of melatonin receptors, so melatonin likely plays a role in feedback regulation of the circadian clock. In fact, administration of melatonin can shift the circadian clock, or improve entrainment to environmental cues. Because of these effects, melatonin is increasingly used as a nutritional supplement to reduce the severity of jet lag, although its effectiveness is controversial.

In most nonmammalian vertebrates, the pineal gland is directly sensitive to light and contains its own biological clock. In these animals, the pineal organ rests on top of the brain, and in some species the skull over the pineal gland is very thin, allowing substantial light to penetrate to the pineal organ. In fact, in some extinct vertebrates, the pineal organ apparently formed a third eye with a lens to focus light. In living organisms, only the lamprey and some lizards retain the remnants of this third eye.⊙

Summary

General Properties of Sensory Reception

→ Sensory receptors transduce the energy from incoming signals into changes in membrane potential that can be communicated to other parts of the nervous system.

→ Sensory receptors can be classified based on the type of stimulus that the receptor detects (the stimulus modality).

→ Chemoreceptors sense environmental chemicals in both the internal and external environments. Mechanoreceptors sense pressure changes. Photoreceptors detect light. Magnetoreceptors detect magnetic fields. Electroreceptors detect electrical currents, and thermoreceptors detect temperature.

→ Most receptors have an adequate stimulus, a specific stimulus that maximally excites the receptor, although other stimuli can excite these receptors, if they are sufficiently large.

→ Some polymodal receptors, including many pain receptors (nociceptors), have fairly broad specificity and can detect more than one type of stimulus.

→ Some sensory receptors are epithelia-derived cells. Incoming stimuli cause a receptor potential in these cells that causes the release of neurotransmitter onto an afferent neuron.

→ Some sensory receptors are neurons. Incoming stimuli cause a generator potential in these cells that triggers action potentials in the axon.

→ For many receptors, receptor location can encode stimulus modality and location.

→ Action potential frequency encodes stimulus intensity.

→ The beginning or ending of groups of action potentials can encode stimulus duration.

→ Range fractionation and logarithmic encoding can extend the dynamic range of a sense organ.

Chemoreception

→ Chemoreception is the process of detecting chemicals in the internal and external environments.

→ External chemoreception can be divided into olfaction, pheromone sensing, and gustation,

which are separate senses in the vertebrates. These distinctions are not as clear in the invertebrates.

→ Vertebrate olfactory receptors are bipolar neurons that express odorant receptor proteins that signal through an associated G protein.

→ The olfactory G protein (G_{olf}) signals via an adenylate cyclase signal transduction cascade.

→ Each olfactory neuron expresses one of several thousand odorant receptor genes, each of which is sensitive to a unique combination of odorants.

→ The combinatorial odorant code allows vertebrates to discriminate among hundreds of thousands of different odors using fewer than a thousand different odorant receptor proteins.

→ Vertebrates also use G-protein-coupled receptors to detect pheromones, but pheromone detection occurs in the vomeronasal organ, not the olfactory epithelium, and the G protein signals via a phospholipase C signal transduction cascade.

→ Invertebrate olfactory receptors also use odorant receptor proteins coupled to a G protein for signal transduction, but the genes for these receptors are not homologous to the vertebrate odorant receptors.

→ The vertebrate gustatory system detects five broad classes of chemical (sweet, umami, salty, sour, and bitter).

→ Vertebrate taste receptors are epithelial cells that form synapses with bipolar sensory neurons that use diverse signal transduction mechanisms, including ion channels and G-protein-coupled receptors.

→ Insect taste receptors are bipolar sensory neurons that signal through G-protein-coupled receptors.

Mechanoreception

→ Mechanoreceptors detect physical stimuli using stretch-sensitive ion channels.

→ Baroreceptors detect changes in blood pressure. Tactile (touch) receptors detect mechanical stimuli on the body surface. Proprioceptors monitor the position of the body.

→ Vertebrate tactile receptors are isolated sensory cells within the skin.

→ Insect tactile receptors are grouped into complex mechanosensory organs called sensilla.

→ Arthropods use statocysts as the organ of equilibrium. A variety of different structures involving modified chordotonal organs can be used as organs of hearing.

→ The ears are the organs of both equilibrium and hearing in the vertebrates.

→ Vertebrate inner ears contain specialized hair cells with ciliary projections that pivot in response to changes in pressure, opening a mechanosensitive ion channel that transduces the mechanical stimulus into a change in membrane potential.

→ The hair cells associated with the semicircular canals are involved in the sense of equilibrium.

→ The hair cells of the lagena, cochlear duct, or cochlea (depending on the species) are involved in hearing.

→ The middle ear of terrestrial vertebrates amplifies sounds.

Photoreception

→ Photoreception involves the transduction of the energy carried by light into a depolarization of a photoreceptor cell.

→ Animals have two classes of photoreceptor cells (ciliary and rhabdomeric photoreceptors).

→ Both types contain similar photoreceptor chromophores that are made up of the protein opsin and a pigment derived from vitamin A (such as retinal).

→ When light strikes the pigment in a chromophore, the pigment isomerizes and dissociates from opsin, changing opsin's conformation and signaling to an associated G protein.

→ In a rhabdomeric photoreceptor, the G protein signals via a phospholipase C signal transduction cascade that causes the cell to depolarize in response to light.

→ In a ciliary photoreceptor, the G protein signals via phosphodiesterase, causing the cell to hyperpolarize in response to light.

→ In most animals, photoreceptors are grouped in complex photoreceptors such as eyes that range in complexity from a simple flat sheet to complex vesicular eyes.

→ Despite the diversity of eye structure, the same genes are involved in the development of eyes in all animals.

→ A vesicular eye can focus images on the retina of objects at varying distances, a process called accommodation.

→ In the vertebrates, accommodation occurs via changes in the shape of the lens. The vertebrate retina performs substantial processing of the visual signal.

→ Comparison of signals from multiple photoreceptors with different photopigments allows color vision.

Thermoreception

→ Thermoreceptors may be warm sensitive, cold sensitive, or hot/pain sensitive.

→ Thermoreceptor proteins are in the transient receptor potential (TRP) family of ion channels.

→ Some chemicals (including capsaicin, from hot peppers) can stimulate temperature-sensitive TRP channels.

→ Some animals have specialized organs for detecting temperature, such as the pit organs of vipers.

Magnetoreception

→ Magnetoreception is present in many organisms, but is poorly understood.

→ Some magnetoreceptors contain particles of magnetite, a mineral that responds to magnetic fields.

Review Questions

1. What is the difference between a sense organ and a sensory receptor?

2. What are the primary stimulus modalities detected by animal sensory receptors?

3. What is a receptor potential? How does it differ from a generator potential?

4. Explain labeled-line coding and give an example of the kinds of sensory information that can be encoded by this method.

5. What is the relationship between the intensity of a stimulus and the response of the primary afferent neuron? How do neurons encode changes in stimulus intensity?

6. Many sensory systems encode stimuli logarithmically. Compare and contrast this approach with range fractionation.

7. What is sensory adaptation?

8. Compare and contrast the signal transduction mechanisms used by gustatory receptors to detect the primary types of tastants.

9. Using the vertebrate ear as an example, outline some of the ways in which sensory systems amplify environmental stimuli.

10. Outer hair cells respond to sounds, but they do not make synaptic connections with afferent neurons that carry sound information to the brain. What is their role in hearing?

11. Compare and contrast the rods and cones of mammals. Does this distinction apply to all vertebrates?

12. Explain the role of the following types of cells in the mammalian retina, using one or two sentences for each answer: rods, cones, horizontal cells, bipolar cells, amacrine cells, retinal ganglion cells.

Synthesis Questions

1. Mechanoreceptors do not depolarize in response to light, no matter how intense the stimulus, but the eye responds to a mechanical stimulus (such as pressing on the eyeball) if the stimulus is sufficiently large. Why might this be?

2. Do taste receptors use labeled-line coding? Why or why not?

3. Receptors for fine touch are typically located in the shallow layers of the skin, while receptors for stronger touch stimuli are typically located in deeper layers. Why might this be so?

4. Hair cells have prominent cilia on their apical surface. Why do these cilia increase the sensitivity of a hair cell to mechanical stimuli?

5. Why do the inner ears of most vertebrates have three semicircular canals and not just one?

6. Peripheral vision is the ability to detect objects outside the center of the visual field. Vertebrates vary in the extent of their peripheral vision. What differences would you expect in the retina of an animal with excellent peripheral vision, compared to one with poor peripheral vision?

7. Humans have only three types of cone photoreceptors, but can distinguish thousands of colors. How is this possible?

8. What predictions could you make about what would happen to vision in an individual with a degenerative disease that destroyed the horizontal cells of the retina?

Quantitative Questions

1. You are studying a sensory receptor and find that the amplitude of the receptor (generator) potential increases linearly with the log of the stimulus intensity. The generator potential results in a train of action potentials whose frequency increases linearly with increasing generator potential (above the threshold value). You also observe that above a certain level, additional increases in stimulus intensity do not result in increases in action potential frequency.

(a) Graph the results for generator potential amplitude and action potential frequency.

(b) What do these results tell you about how this receptor encodes stimulus intensity?

2. One way in which the vertebrate auditory system detects the location of a sound is to compare the time at which a sound reaches one ear to the time at which that sound reaches the other ear.

(a) How long would it take for a sound reaching the left side of the head to reach the right ear, assuming that the distance between the ears is approximately 12 cm and that the speed at which a sound travels through the head is approximately 1000 m/s?

(b) Neurotransmission takes approximately 10–20 milliseconds. Using this information and the value you calculated in part a, what are the implications for the localization of a sound?

3. The vertebrate olfactory system uses a combinatorial coding scheme in which each odorant receptor cell expresses only a single allele of a single gene of a G-protein-coupled odorant receptor, but in which each receptor can detect several odorants. You are a scientist working on a little-known vertebrate, the schmoo, and have discovered 100 functional olfactory G-protein coupled receptors in the schmoo genome. Assuming a simple combinatorial code, how many potential odorants could a schmoo distinguish if each receptor could detect 3 different odorants. What if each receptor could detect 5 different odorants? What is the minimum number of genes required to discriminate among 10,000 different odorants if each receptor can detect 2 different odorants?

For Further Reading

See the Additional References section at the back of the book for more readings related to the topics in this chapter.

General Properties of Sensory Reception

This excellent book summarizes the mechanisms by which sensory receptors transduce incoming sensory stimuli. The book is comprehensive but accessible, and includes examples from both vertebrates and invertebrates. This book is an ideal "next step" for students who want more information about sensory systems than can be presented in a single chapter in a physiology textbook.

Fain, G. L. 2003. *Sensory transduction.* Sunderland, MA: Sinauer Associates.

Chemoreception

These reviews discuss the mechanisms of olfaction and the evolution of olfactory systems in animals.

Breer, H. 2003. Sense of smell: Recognition and transduction of olfactory signals. *Biochemical Society Transactions* 31: 113–116.

Eisthen, H. 2002. Why are olfactory systems of different animals so similar? *Brain, Behavior, and Evolution* 59: 273–293.

Malnic, B., J. Hirono, T. Sato, and L. B. Buck. 1999. Combinatorial receptor codes for odors. *Cell* 96: 713–723.

Menashe, I., and D. Lancet. 2006. Variations in the human olfactory receptor pathway. *Cellular and Molecular Life Sciences* 63: 1485–1493.

The following reviews report recent findings regarding the sense of taste.

Gilbertson, T. A., and J. D. Boughter, Jr. 2003. Taste transduction: Appetizing times in gustation. *Neuroreport* 14: 905–911.

Reed, D. R., T. Tanaka, and A. H. McDaniel. 2006. Diverse tastes: Genetics of sweet and bitter perception. *Physiology and Behavior* 88: 215–226.

This interesting review discusses the evolution of olfaction, pheromone detection, taste, and the sense of vision in mammals, with a focus on the primates.

Liman, E. R. 2006. Use it or lose it: Molecular evolution of sensory signaling in primates. Pflugers Archiv: *European Journal of Physiology* 453: 125–131.

Mechanoreception

These reviews discuss the mechanisms of mechanoreception and the discovery of the molecules involved in transducing mechanical stimuli into changes in membrane potential.

Ernstrom, G. G., and M. Chalfie. 2002. Genetics of sensory mechanotransduction. *Annual Review of Genetics* 36: 411–453.

Gillespie, P. G., and R. G. Walker. 2001. Molecular basis of mechanosensory transduction. *Nature* 413: 194–202.

Goodman, M. B., and E. M. Schwarz. 2003. Transducing touch in *C. elegans. Annual Review of Physiology* 65: 429–452.

Nicolson, T. 2005. Fishing for key players in mechanotransduction. *Trends in Neuroscience* 28: 140–144.

This comprehensive review by LeMasurier and Gillespie provides an excellent introduction to the structure and function of the mammalian inner ear and its role in sound transduction.

LeMasurier, M., and P. G. Gillespie. 2005. Hair-cell mechanotransduction and cochlear amplification. *Neuron* 48: 403–415.

This review discusses the role of the outer hair cells in the mammalian ear, and the role of the prestin gene as part of their molecular motors.

Geleoc, G. S. G., and J. R. Holt. 2003. Auditory amplification: Outer hair cells *pres* the issue. *Trends in Neuroscience* 26: 115–117.

This paper is the first report of the identification of a mechanosensory channel in vertebrate hair cells.

Sidi, S., R. W. Friedrich, and T. Nicolson. 2003. NompC TRP channel required for vertebrate sensory hair cell mechanotransduction. *Science* 301: 96–99.

Photoreception

These reviews discuss the evolution of eyes and visual pigments in a variety of organisms.

Arendt, D. 2003. The evolution of eyes and photoreceptor cell types. *International Journal of Developmental Biology* 47: 563–571.

Briscoe, A. D., and L. Chittka. 2001. The evolution of color vision in insects. *Annual Review of Entomology* 46: 471–510.

Fernald, R. D. 2006. Casting a genetic light on the evolution of eyes. *Science* 313: 1914–1918.

Fernald, R. D. 2000. Evolution of eyes. *Current Opinion in Neurobiology* 10: 444–450.

Hisatomi, O., and F. Tokunaga. 2002. Molecular evolution of proteins involved in vertebrate phototransduction. *Comparative Biochemistry and Physiology, Part B: Biochemistry and Molecular Biology* 133: 509–522.

Yokoyama, S., and R. Yokoyama. 1996. Adaptive evolution of photoreceptors and visual pigments in vertebrates. *Annual Review of Ecology and Systematics* 27: 543–567.

Thermoreception

These reviews discuss the recently discovered TRP channels that are involved in temperature sensing.

Dhaka, A., V. Viswanath, and A. Patapoutian. 2006. TRP ion channels and temperature sensation. *Annual Review of Neuroscience* 29: 135–161.

Jordt, S. E., D. D. McKemy, and D. Julius. 2003. Lessons from peppers and peppermint: The molecular logic of thermosensation. *Current Opinion in Neurobiology* 13: 487–492.

Patapoutian, A., A. M. Peier, G. M. Story, and V. Viswanath. 2003. ThermoTRP channels and beyond: Mechanisms of temperature sensation. *Nature Reviews: Neuroscience* 4: 529–539.

Magnetoreception

This paper reports a possible mechanism for magnetoreception in vertebrates.

Diebel, C. E., R. Proksch, C. R. Green, P. Neilson, and M. M. Walker. 2000. Magnetite defines a vertebrate magnetoreceptor. *Nature* 406: 299–302.

This comprehensive review outlines the two mechanisms currently proposed to underlie magnetoreception in the vertebrates.

Wiltschko, R., and W. Wiltschko. 2006. Magnetoreception. *BioEssays* 28:157–168.

Circadian Rhythms

These reviews highlight some of the recent findings regarding circadian rhythms in mammals and other animals.

Macchi, M. M., and J. N. Bruce. 2004. Human pineal physiology and functional significance of melatonin. *Frontiers in Neuroscience* 25: 177–195.

Panda, S., and J. B. Hogenesch. 2004. It's all in the timing: Many clocks, many outputs. *Journal of Biological Rhythms* 19: 374–387.

Peirson, S., and R. G. Foster. 2006. Melanopsin: Another way of signaling light. *Neuron* 49: 331–339.

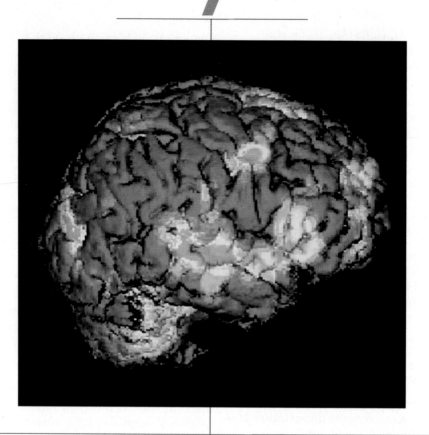

Functional Organization of Nervous Systems

The ancient Egyptians and Greeks considered the brain to be of little importance; when preparing a mummy for burial, the Egyptians would carefully preserve the heart but discard the brain, because they believed the heart to be the seat of consciousness. The Greek philosopher Aristotle, working in the 4th century B.C., thought that the brain acted as a sort of cooling system for the spirit (or soul), but that the soul was located in the heart. Five hundred years later, the Greek physician Galen disputed this finding, and concluded that mental activity occurred within the brain because of his observations of the effects of head injuries in Roman gladiators. However, in subsequent centuries, physiologists and anatomists made little progress beyond the observations of Galen in understanding the workings of the brain. The human brain appears to be rather uniform, composed of a soft amorphous tissue. Indeed, because of its gelatinous appearance, up until the middle of the 17th century most anatomists and physiologists felt that the brain was not subdivided into functional regions, but instead worked together as a whole, without any regional specialization.

By the 19th century, however, detailed observations of patients suffering from brain injuries, tumors, or strokes led physiologists to the conclusion that parts of the brain were specialized for particular functions. One of the most famous cases of this time was that of Phineas Gage. Gage, a railway worker, was injured on September 13, 1848, when a blasting charge that he was preparing accidentally exploded and drove a tamping iron into his skull. (A tamping iron is a tool similar to a crowbar that is used to compact an explosive charge into a borehole.) The resulting blast blew the tamping iron out of the borehole and into Mr. Gage's left cheek. It passed all the way through his head, exiting from the top of his skull and landing over 25 yards away. Mr. Gage survived the accident, but his brain injuries left his person-

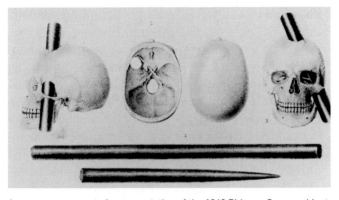
A computer-generated representation of the 1848 Phineas Gage accident.

A London taxi driver.

ality changed. Before the accident he had been efficient, well balanced, and highly intelligent. After the accident, he was reported to be irritable, profane, and unable to make decisions. He was still functional in many respects, but his personality was profoundly altered. Cases like that of Phineas Gage, and of patients with strokes or other brain damage, helped anatomists assign functions to various parts of the brain, and provided increasing insights into the way the brain works.

In this century, brain-imaging technology is revolutionizing the way in which physiologists study the functions of the brain and has revealed an astonishing level of plasticity. For example, scientists have been able to determine that the brains of taxi drivers working in London, England, differ from those of other people. In order to get a license to drive a taxi in London, drivers must pass a difficult test that assesses their ability to find their way. The streets of London are not laid out in a grid pattern, which makes navigating in London without a map difficult. London taxi drivers have an enlarged hippocampus, a part of the brain known to be involved in spatial relationships and memory.

But are these differences the result of training, or are people with these unusual brain structures simply attracted to professions in which they can excel? A new technique called functional magnetic resonance imaging (fMRI) is providing a way to address this question. Functional MRI is a modification of the MRI technique that was developed in the late 1970s. An MRI machine emits a powerful magnetic field that can be directed at the brain (or at other parts of the body). This magnetic field causes the hydrogen atoms in water molecules to realign with the mag-

netic field, just as a compass aligns with the Earth's magnetic field. The MRI machine then sends out a pulse of radio energy. This pulse briefly knocks the hydrogen atoms out of alignment. As the hydrogen atoms return to their aligned position they emit energy, which the MRI machine can detect and interpret. Because the amount of water (and hence hydrogen atoms) varies in different structures of the brain, an MRI machine can provide detailed brain images. Functional MRI is a simple modification of this technique. Parts of the brain that are working harder require more oxygen than parts of the brain that are resting, and thus tend to deplete the oxygen from the blood. It turns out that the MRI signal is a little different between oxygenated and deoxygenated blood, so the MRI signal changes as a subject uses different parts of the brain. If you make a series of MRI images while asking a subject to perform a mental task, you generate a so-called fMRI image, in which you can observe changes in blood flow (and thus changes in activity) in different parts of the brain. For example, listening to music activates a part of the brain involved in processing incoming auditory information, whereas speaking activates different parts of the brain. Studies using fMRI are revealing the truly dynamic nature of the brain. For example, there are observable changes in the brains of adults when they are taught a new alphabet.

As for the London taxi drivers, recent studies have shown that the differences in their brain structure and activity are a result of practice, not an accident of birth. The brain can alter its structure and function in response to training, and thus there is a physiological basis for the adage, "Practice makes perfect." ⊙

Overview

The nervous system is one of the body's homeostatic control systems, helping to regulate physiological processes and coordinate behavior. But how do the many individual neurons that make up the nervous system work together to perform these complex tasks? Like other homeostatic control systems, the nervous system contains sensors, integrating centers, and output pathways (Figure 7.1). In Chapter 6: Sensory Systems, we discussed how sensory receptors detect incoming stimuli and convert the signal to a change in membrane potential. Afferent sensory neurons carry these signals to one or more *integrating centers*, such as a brain or ganglion. Integrating centers typically contain many **interneurons**, which (as the name suggests) form synaptic connections among neurons. The more interneurons that are added to a neural pathway, the greater the possibilities for interconnections, and the greater the ability to integrate information. The complex behavioral and physiological control systems of animals result from these multistep neural pathways, which find their most elaborate form in large integrating centers such as the mammalian brain. For example, an average human brain contains more than 100 billion neurons connected via trillions of synapses. Integrating centers ultimately send an output signal via **efferent neurons** to effector organs, including skeletal muscles, glands, and internal organs. Thus, the nervous system acts to sense environmental information, integrate this information, and coordinate the response.

In this chapter we first examine the evolution of nervous systems and their organization. We then take a closer look at the functions of the principal integrating centers of vertebrates—the brain and spinal cord—using mammals as an example. Next, we focus on the peripheral nervous system, looking at the organization of the efferent pathways that carry signals to effector organs. Finally, we end the chapter with a consideration of the integrated functions of the nervous system, addressing how sensory receptors, afferent neurons, integrating centers, and efferent pathways work together to allow organisms to perform complex behaviors and maintain physiological homeostasis.

Organization of Nervous Systems

Most nervous systems are organized into three functional divisions: the afferent sensory division; integrating centers; and the efferent division. Only the cnidarians (a phylum that includes jellyfish and sea anemones, among others) have nervous

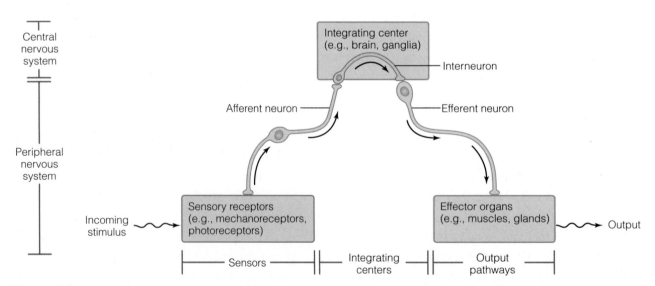

Figure 7.1 An overview of the nervous system
The nervous system contains sensors, integrating centers, and output pathways. Sensory receptors convert the energy from incoming stimuli of various kinds to changes in the membrane potential. Afferent neurons conduct these signals in the form of action potentials to integrating centers such as the brain or ganglia. Interneurons within the integrating centers process the information and send out signals via efferent neurons to effectors such as the muscles and internal organs, resulting in changes in behavior or physiological processes.

systems that depart from this general plan. Cnidarians are radially symmetrical animals with nervous systems that are interconnected into a large web (or nerve net) with neurons distributed throughout the body (Figure 7.2a). In general,

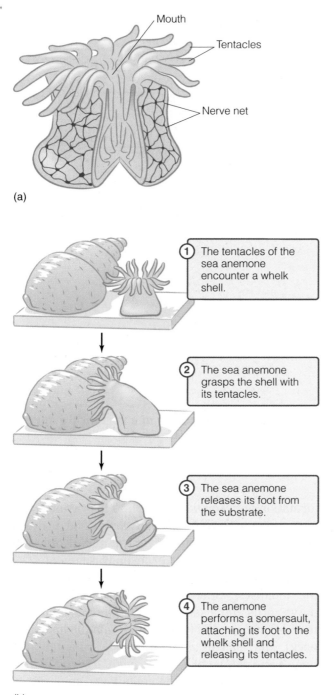

(a)

(b)

Figure 7.2 The nervous system of cnidarians
(a) Cutaway view of a sea anemone, showing the nervous system. The cnidarian nervous system is diffuse, composed of a loosely organized nerve net. **(b)** Shell-climbing behavior in a sea anemone, *Calliactis parasitica*. Despite their seemingly simple nervous systems, cnidarians can perform complex behaviors.

cnidarian neurons are not specialized but can function as sensory neurons, interneurons, or efferent neurons, and can communicate synaptically at several points along their length. Cnidarian neurons often form *en passant* synapses (see Chapter 4: Neuron Structure and Function), allowing information to be passed in either direction across the synapse. In fact, many cnidarian neurons are functionally bipolar in that a stimulus at any point on the organism triggers an impulse that radiates out from the stimulus site in every direction.

Despite having a seemingly simple nervous organization and no obvious single integrating center, cnidarians can perform some rather complex behaviors. For example, the sea anemone *Calliactis parasitica* attaches its tentacles onto a mollusc shell and somersaults onto the shell (Figure 7.2b), a behavior that involves detecting a shell, using its tentacles to grab onto the shell, detaching its foot from the substrate, making coordinated movements of the whole body to somersault up onto the shell, and reattaching its foot onto the shell. Thus, the apparent simplicity of the cnidarian nerve net must hide substantial complexities.

In some species the nerve net is broken down into several pathways with characteristic conduction speeds that control different behavioral responses. In addition, in some species neurons are concentrated around the oral opening, or into clusters in other locations. These groupings of neurons may act as integrating centers, providing additional layers of functional complexity to the nervous system. In fact, in many species of cnidarians epithelial cells can also generate action potentials, and are connected via gap junctions, adding yet another layer of complexity.

Evolution of Nervous Systems

Unlike the cnidarians, most animals are **bilaterally symmetrical**; they have an anterior and a posterior end and a right and left side. In bilaterally symmetrical organisms, sense organs tend to be concentrated at the anterior end of the body, close to the mouth, and the relatively unstructured netlike organization of the cnidarian nervous system is replaced by more complex groupings of neurons. For example, bilaterally symmetrical animals typically have one or more **ganglia**, which are groupings of neuronal cell bodies interconnected by synapses. Ganglia function as integrating centers for the nervous system. In many species, the ganglia in the anterior region of the body are grouped together into larger clusters forming a **brain**, a complex integrating center. Within the brain, groupings of neuronal cell bodies are termed *nuclei*, which are the functional equivalent of ganglia, and groupings of neuronal axons are called *tracts*. Outside of the integrating centers, the axons of afferent and efferent neurons are usually organized into structures called **nerves**, which are the functional equivalent of the tracts in the integrating centers.

Figure 7.3 illustrates the structure of a vertebrate nerve, which consists of parallel bundles of myelinated and unmyelinated axons enclosed in several layers of connective tissue. Within a nerve, individual axons and their myelin sheaths (if present) are surrounded by the *endoneurium*. Many axons are bundled together into structures called *fascicles* by another layer of connective tissue, the *perineurium*. Several fascicles and blood vessels are grouped together, enclosed by a fibrous layer of connective tissue called the *epineurium*, forming the nerve. Most nerves contain axons of both afferent and efferent neurons, and are thus termed *mixed nerves*, although there are some purely afferent or purely efferent nerves.

Bilaterally symmetrical animals exhibit cephalization

The pattern of locating sense organs and nervous integrating centers at the anterior end of the body, known as **cephalization**, becomes increasingly apparent in more complex nervous systems. The degree of cephalization varies greatly among the bilaterally symmetrical invertebrates, although most species have a well-developed brain, several ganglia, and one or more nerve cords (Figure 7.4). In the invertebrates, bundles of axons that connect ganglia or run between a ganglion and the brain are called *connectives* or *commissures*. Flatworms are the simplest of the bilaterally symmetrical animals. Some species of flatworms lack an obvious brain, while others have a well-developed brain that allows them to perform complex behaviors and even learn tasks such as navigating a maze. Nemertine, nematode, and annelid worms have a more structured nervous system than flatworms, with a well-developed brain, ganglia in each body segment, and one or more nerve cords that communicate in-

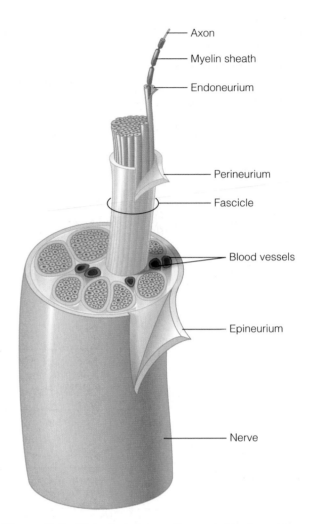

Figure 7.3 The structure of a vertebrate nerve
A nerve is composed of groups of axons from many neurons surrounded by successive layers of connective tissue (the endoneurium, perineurium, and epineurium).

formation between the tissues and the various integrating centers. Similarly, the arthropod nervous system contains a brain, a ventral nerve cord, and a large ganglion within each body segment. The nervous system varies greatly in complexity among the molluscs, although most species have dual nerve cords and a series of large ganglia, including the *cerebral ganglia* (which innervate the head and neck), the *buccal ganglia* (which innervate the mouth and stomach), and the *pedal ganglia* (which innervate the foot). In the cephalopod molluscs (a group that includes octopus and squid), the anterior pairs of ganglia are greatly expanded and placed close together to create a tightly packed mass that lies between the eyes and encircles the esophagus—in other words, a large and complex brain.

An octopus has a brain that is much larger relative to its body size than the brain of a fish or a

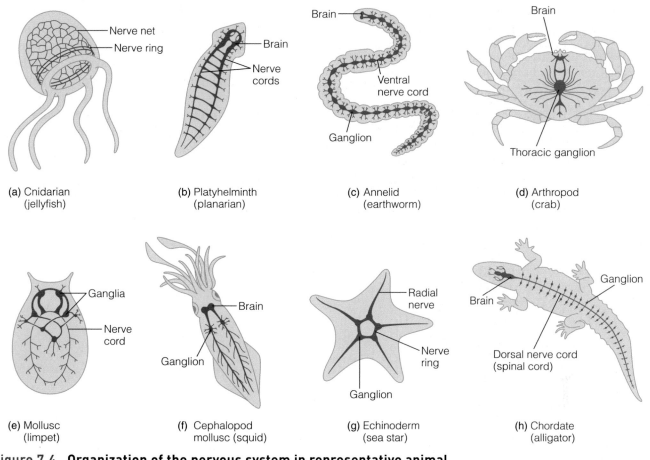

Figure 7.4 Organization of the nervous system in representative animal groups The cnidarians have a nerve net, while all other groups (with the exception of the radially symmetrical echinoderms) display some degree of cephalization.

reptile, suggesting the possibility of substantial intelligence. An octopus can learn to navigate a maze and distinguish between objects with different shapes, sizes, and degrees of brightness. Some studies indicate that an octopus can even learn by simply watching another octopus perform a task. Although an octopus has a very large brain, it has another important integrating center; each arm has a large ganglion that controls arm movements and that can function essentially independently of the brain. When researchers severed the connections between the brain and the arm of an octopus and then stimulated the skin on the arm, the arm behaved exactly as it would have in an intact octopus. Thus, the integrating center of an octopus is actually highly distributed and involves both the brain and the ganglia.

The echinoderms (sea stars and their relatives) are one of the few exceptions to the general trend of increasing cephalization in animals. These radially symmetrical animals lack an obvious brain, and instead have a series of ganglia and several nerve rings. Echinoderms are descended from a bilaterally symmetrical ancestor that likely had some cephalization. Presumably, present-day echinoderms lost this ancestral cephalization during the transition to a radially symmetrical body plan. In fact, many modern echinoderm groups have bilaterally symmetrical larvae that develop radial symmetry during metamorphosis to the adult form.

In general, organisms with more complex nervous systems have more neurons than organisms with less complex nervous systems. However, the total number of neurons is not necessarily larger in species with more complex integrating centers. For example, some species of flatworms have several thousand neurons, despite lacking an obvious brain. In contrast, the entire nervous system of the nematode *Caenorhabditis elegans* contains only 302 neurons and about 6000 synapses, despite having a clearly recognizable brain. Thus, the relationship between the number of neurons and the organization of the nervous system is not always clear-cut.

The vertebrate central nervous system is enclosed in a protective covering

Vertebrates are among the most highly cephalized organisms, and are unique in possessing a hollow dorsal nerve cord, rather than the solid ventral nerve cord seen in invertebrates (Figure 7.5). How the vertebrate nervous system evolved from an invertebrate ancestor is still a matter of considerable debate, although it has been suggested that protostome (e.g., worm, mollusc, and arthropod) and deuterostome (e.g., vertebrate and echinoderm) nervous systems evolved independently from a common ancestor with a nervous system similar to that of flatworms. One of the unique characteristics of the vertebrate nervous system is that a portion of the nervous system is encased within a cartilaginous or bony covering. This portion of the nervous system is termed the **central nervous system**, and is composed of the brain (located within the skull) and the **spinal cord** (located within the spine). The remainder of the nervous system, which is found throughout the rest of the body, is termed the **peripheral nervous system**.

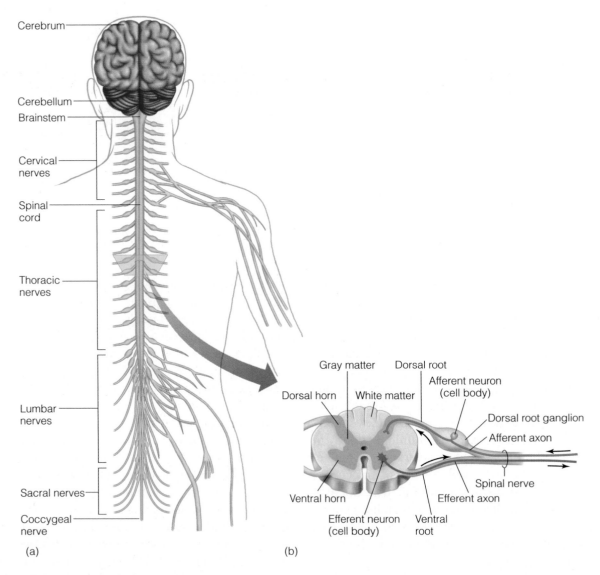

(a)   (b)

Figure 7.5   Structure of the vertebrate central nervous system   (a) The brain and spinal cord. The central nervous system is composed of the brain and spinal cord, enclosed in a cartilaginous or bony covering (the skull and spine). The cranial nerves emerge from the braincase, whereas the spinal nerves emanate from the spinal cord at regular intervals. These nerves are part of the peripheral nervous system. **(b)** Cross-section of a mammalian spinal cord. The spinal cord contains both gray and white matter. Afferent sensory neurons enter the spinal cord on the dorsal side, and efferent neurons exit the spinal cord on the ventral side.

The cranial and spinal nerves form synapses in the central nervous system

In the vertebrates, a series of nerves, called the **cranial nerves**, exit directly from the braincase, whereas the **spinal nerves** emerge from the spinal cord at regular intervals. For historical reasons, we often refer to the 12 pairs of cranial nerves, which are labeled with roman numerals, but in fact many vertebrates have 13 pairs of cranial nerves (Table 7.1). Some of the cranial nerves bring in afferent information from the sense organs, whereas other nerves send efferent signals out to effector organs, such as muscles, glands, and organs. The spinal nerves are named based on the region of the spine where they originate. The *cervical* spinal nerves emerge from the spinal cord in the region of the neck and innervate the head, neck, arms, hands, and diaphragm. The *thoracic* spinal nerves emerge from the spinal cord in the chest region, and innervate the intercostal muscles (involved in breathing) and the heart. The *lumbar*, *sacral*, and *coccygeal* spinal nerves emerge in the lower back and pelvis and innervate the legs, pelvis, bladder, and bowel. Although the spinal nerves emerge from the vertebral column along its entire length, the spinal cord itself does not reach all the way down into the lumbar region. Instead, the lumbar, sacral, and coccygeal nerves branch out from the spinal cord and travel down the vertebral column to the point where they exit. Thus, the bottom third of the vertebral column contains spinal nerves but no spinal cord.

Table 7.1 The cranial nerves.

Number	Name	Principal functions
0	Terminal	Function unclear. Probably neuromodulatory, regulating olfactory sensitivity and reproductive behavior. Absent in cyclostomes, birds, and humans.
I	Olfactory	Olfaction; chemosensory afferents.
II	Optic	Vision; integrates information from the retina. An extension of the central nervous system rather than a true nerve.
III	Oculomotor	Controls eye movements, constriction of pupil, and focusing of lens. Innervates most of the muscles of the eye and eyelid (where present).
IV	Trochlear	Eye movement. Innervates the superior oblique eye muscle.
V	Trigeminal	Motor nerve controlling muscles of chewing and mouth. Contains sensory afferents from most of head and parts of lateral line in fishes.
VI	Abducens	Eye movement. Innervates the lateral rectus muscle of the eye and the nictitating membrane (additional eyelid) of reptiles, birds, and some mammals.
VII	Facial	Taste and somatosensory afferents. Motor nerve controlling muscles of face. Controls lachrymal and some salivary glands. Serves parts of lateral line in fishes.
VIII	Vestibulocochlear (also called auditory)	Sensory afferents for hearing and equilibrium.
IX	Glossopharyngeal	Motor control of one of the muscles of swallowing. General and taste sensation from pharynx and posterior part of tongue. Innervates parts of lateral line in fishes.
X	Vagus	Motor control of larynx, pharynx, upper end of esophagus. Parasympathetic neurons controlling internal organs, including heart, breathing, and much of the alimentary canal. Also includes a sensory component. In fishes, contains sensory afferents from lateral line system of body and tail.
XI	Accessory	Motor control of some muscles that move the head. Not a true cranial nerve: results from fusion of parts of vagus and first two spinal nerves. Absent in fishes except the Crossopterygii.
XII	Hyopoglossal	Motor control of the muscles of the tongue. Absent in fishes except the Crossopterygii.

Both the brain and spinal cord contain two types of tissue, the so-called **gray matter** and **white matter**. White matter consists of bundles of axons and their associated myelin sheaths, while gray matter is composed of neuronal cell bodies and dendrites. In the spinal cords of all vertebrates, the white matter is located at the surface, and the gray matter is located inside. In cross-section, the gray matter of the spinal cord often has a butterfly-shaped appearance, a pattern that is particularly evident in humans (see Figure 7.5). The "wings" of this butterfly are termed the *dorsal* and *ventral horns*. Afferent sensory neurons from the periphery terminate in the dorsal horn where they synapse on interneurons or efferent neurons. The cell bodies of these bipolar sensory neurons are located outside the spinal cord in the *dorsal root ganglia*. Efferent neurons originate in the ventral horn of the spinal cord and exit through the *ventral root*.

The central nervous system is separated from the rest of the body

One or more protective layers of connective tissue called the **meninges** (singular meninx) surround the brain and spinal cord (Figure 7.6). Fish have only a single thin meninx, whereas amphibians, reptiles, and birds have two: a thick outer layer called the *dura mater* and a thin secondary meninx. Mammals have three meninges. Like the other tetrapods they have the dura mater, but the secondary meninx is divided into a weblike middle layer called the *arachnoid* and a thin inner layer called the *pia mater*. Within the meninges, the brain and spinal cord float in a plasma-like fluid called **cerebrospinal fluid (CSF)**, which acts as a shock absorber and cushions the delicate tissues of the central nervous system.

The vertebrate central nervous system is also physiologically separated from the rest of the body. The **blood-brain barrier**, which is formed by tight junctions between the endothelial cells lining the brain capillaries, prevents materials from leaking out of the bloodstream and into the central nervous system via paracellular pathways (between the cells). In addition, these cells do not perform pinocytosis, so the only ways that substances can move into the brain are by directly dissolving in the membrane or by catalyzed transport via a protein exchanger, channel, or pump. Small, lipid-soluble molecules such as ethanol and some barbiturate drugs can cross directly into the central nervous system, but most substances are excluded. However, a number of specialized carrier transport systems allow the brain to take up circulating nutrients such as glucose and amino acids. Thus, the blood-brain barrier protects the brain from harmful substances while allowing useful molecules to enter. There are several areas of the brain where the blood-brain barrier is more permeable. In particular, the regions around the pineal gland, the pituitary gland, and parts of the hypothalamus are quite permeable, allowing secreted molecules such as hormones to leave the brain and enter the circulatory system.

The vertebrate brain has three main regions

During embryonic development, both the brain and the spinal cord of vertebrates are formed from a simple hollow tube of ectoderm-derived cells called the *neural tube*. The posterior portion of the neural tube forms the spinal cord, while the anterior end of the neural tube develops three swellings that

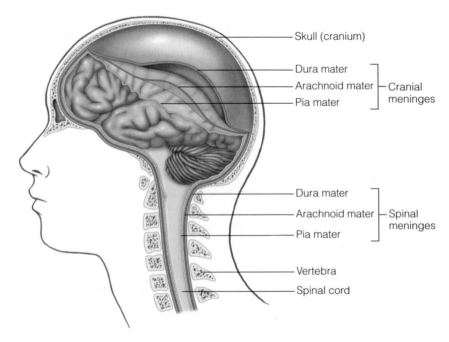

- Skull (cranium)
- Dura mater
- Arachnoid mater — Cranial meninges
- Pia mater
- Dura mater
- Arachnoid mater — Spinal meninges
- Pia mater
- Vertebra
- Spinal cord

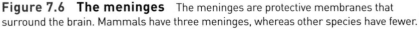

Figure 7.6 The meninges The meninges are protective membranes that surround the brain. Mammals have three meninges, whereas other species have fewer.

ultimately form the brain (Figure 7.7). These three regions, which are found in all vertebrate brains, are called the *rhombencephalon* (**hindbrain**), the *mesencephalon* (**midbrain**), and the *prosencephalon* (**forebrain**). Because the vertebrate brain is simply an extension of the spinal cord, it is also hollow on the inside. These central cavities are called the **ventricles**, and they are filled with cerebrospinal fluid. Ciliated ependymal cells circulate the cerebrospinal fluid through the ventricles and the spinal cord.

The hindbrain controls most reflex responses and regulates involuntary behaviors such as breathing and the maintenance of body position. The midbrain is predominantly involved in coordinating visual, auditory, and sensory information from touch and pressure receptors (although in mammals, as we shall see later in the chapter, it acts largely as a routing center rather than an integrating center per se). The forebrain is involved in processing olfactory information, integrating it with other sensory information, and regulating functions such as body temperature, reproduction, eating, sleeping, and emotion. The forebrain is also involved in learning and memory, and performs other complex processing tasks, particularly in mammals.

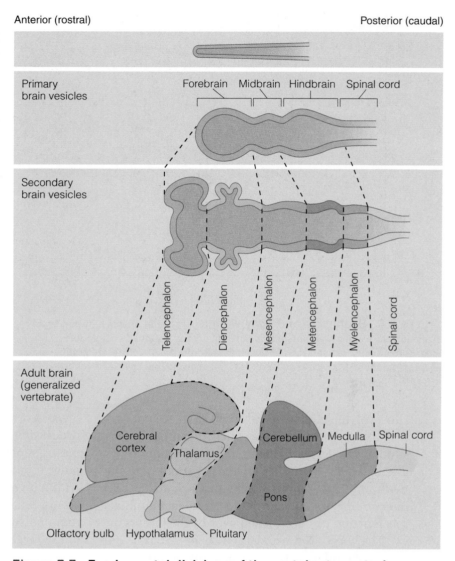

Figure 7.7 Fundamental divisions of the vertebrate central nervous system   During embryonic development, the neural tube quickly subdivides into the primary brain vesicles, which subsequently form the secondary brain vesicles and then the structures of the adult brain.

Brain size and structure vary among vertebrates

Brain size varies greatly among vertebrates (Figure 7.8), but much of this variation can be accounted for by differences in body size, because, within each group, larger animals tend to have larger brains. But at any given body size, brain size can differ substantially among taxa. In particular, birds and mammals have unusually large brains for their body size—six to ten times larger than those of similarly sized reptiles. Presumably, organisms with large brains compared to their body size have more complex integrating centers and an expanded repertoire of behaviors.

Variation in brain size among taxa is largely a result of changes in the relative sizes of different parts of the brain, rather than in the development of entirely new structures (Figure 7.9). For example, bony fishes and birds have a relatively large midbrain and cerebellum—the parts of the brain involved in the interpretation of sensory signals and coordinating motion. Fishes and birds live in a complex world that they move through in three dimensions, in contrast to terrestrial organisms that can move only along the ground. It has been suggested that fishes and birds use their enlarged midbrain and cerebellum in order to interpret complex sensory information and coordinate their body movements in this three-dimensional environment.

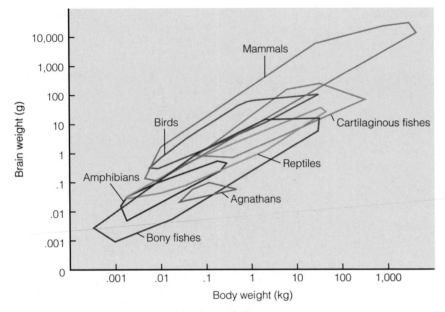

Figure 7.8 Brain size and body weight The relationship between brain size and body weight for representative animal groups is plotted on a double logarithmic scale. Each polygon encloses data from a major vertebrate group. For each group, the polygon rises toward the right, showing that brain size tends to increase with body size.

However, the midbrain and cerebellum are not particularly large in sharks, which presumably face similar challenges. In mammals, the midbrain is greatly reduced in size. In most vertebrates, the midbrain contains the regions that are involved in interpreting visual information, but in mammals this function has been taken over by the forebrain.

The forebrain is enlarged in both birds and mammals relative to the other major groups of vertebrates. In mammals, the outer layer of the forebrain is enlarged and reorganized, forming the **isocortex** (also called the *neocortex*). The isocortex is made up of gray matter, whereas the majority of the internal parts of the mammalian brain are made up of white matter, with the exception of structures called the **basal nuclei**—clusters of gray matter deep within the brain. Thus, the mammalian brain is fundamentally reorganized compared to the brains of other vertebrates, which have an outer layer of white matter surrounding an inner core of gray matter. Like mammals, birds have large forebrains; however, in birds the cortex is relatively thin and undeveloped. In contrast, other parts of the forebrain are enlarged, particularly in a structure called the *dorsoventricular ridge (DVR)*. The enlarged forebrains of birds and mammals presumably evolved independently, since the last common ancestor of birds and mammals would have had a small fore-

brain, as is typical for reptiles. The isocortex of mammals and the DVR of birds perform similar functions, and are thought to have evolved independently from similar structures in the reptilian brain. This subject is of more than just academic interest, because both birds and mammals are capable of performing complex, learned behaviors, and thus the evolution of brain structures may shed light on the evolution of intelligence, a process governed by the isocortex in mammals.

In addition to this variation among major groups of vertebrates, there is also substantial variation in brain size within groups. For example, the mormyrid fishes have unusually large midbrains relative to other fishes. Mormyrids are weakly electric fish that use electric fields for navigation and communication. The midbrain is involved in sensory processing, and processing electrosensory information likely requires sophisticated neural circuitry. Probably the most familiar example of variation in brain size within a group of organisms is the relatively large size of the human brain compared to that of other mammals—the result of a vast increase in the size of the forebrain in humans.

CONCEPT CHECK

1. Do cnidarians have clearly defined afferent neurons, interneurons, and efferent neurons?
2. What is cephalization?
3. What is the difference between a brain and a ganglion?
4. Compare and contrast gray matter and white matter.
5. What is the purpose of the blood-brain barrier?

Structure and Function of the Mammalian Brain

Table 7.2 lists the names and structures of the principal parts of the vertebrate brain.

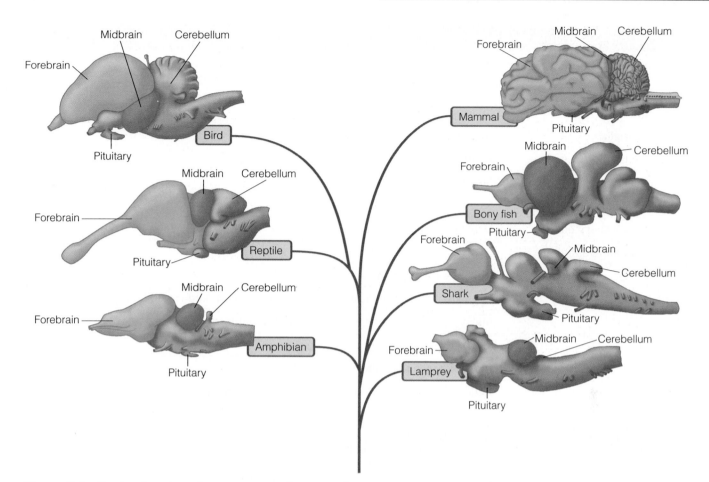

Figure 7.9 Brain structure in representative vertebrate groups Most groups of vertebrates have the same major brain structures, although these structures vary greatly in relative size.

The hindbrain supports basic functions

The hindbrain is located between the spinal cord and the remainder of the brain, and contains three structurally and functionally distinct regions, the *pons*, the *cerebellum*, and the *medulla oblongata*, which function collectively to support vital bodily processes such as breathing, circulation, and movement.

The **medulla oblongata** (often referred to simply as the medulla) is located at the top of the spinal cord, and contains reflex centers regulating breathing, heart rate, and the diameter of blood vessels, thus regulating blood pressure, as we discuss in detail in Chapter 8: Circulatory Systems. The medulla oblongata also contains neural pathways that communicate between the spinal cord and the brain. Many of these pathways cross over each other in the medulla such that the left side of the brain controls the right side of the body and the right side of the brain controls the left side of the body. Because it regulates such important sur-

vival systems, damage to the medulla is almost always fatal.

The **pons** (which means "bridge" in Latin) is located immediately above the medulla, and is an important pathway that communicates information between the medulla, the cerebellum, and the forebrain. The pons also contains centers that control alertness and initiate states such as sleep and dreaming, and it regulates reflex activities such as breathing by influencing the activity of the medulla oblongata.

The **cerebellum** is located at the back of the brain, and consists of two highly folded hemispheres. The cerebellum integrates sensory input from the eyes, ears, and muscle with motor commands from the forebrain, and thus is responsible for motor coordination. In humans, damage to this area during birth can cause cerebral palsy, a disorder characterized by uncontrollable tremors. The cerebellum may also play a role in speech, learning, emotions, and attention. Although the

Table 7.2 The parts of the brain.

Structure	Function
Forebrain: telencephalon	
Cerebrum	Information processing
Basal ganglia	Movements
Amygdala	Emotions
Hippocampus	Memory
Olfactory bulb	Sense of smell
Accessory olfactory bulb	Detection of pheromones
Forebrain: diencephalon	
Thalamus	Integrates sensory information
Hypothalamus, pituitary	Regulate body temperature, feeding, reproduction, and circadian rhythms
Epithalamus	Melatonin secretion, regulation of hunger and thirst
Midbrain	
Tectum (optic lobes)	Processes visual, auditory, and touch information
Tegmentum	Reflex responses to visual, auditory, and touch information
Hindbrain	
Medulla oblongata	Generates rhythmic breathing Regulates heart rate and blood pressure
Pons	Regulates breath-holding Integrates among areas
Cerebellum	Maintains body posture Coordinates locomotion Integrates information from proprioceptors

cerebellum makes up only 10% of the weight of the human brain, it contains as many neurons as the rest of the brain combined.

The midbrain is greatly reduced in mammals

In fish and amphibians, the midbrain coordinates reflex responses to auditory and visual stimuli and is the primary center for coordinating and initiating behavioral responses. In contrast, in mammals it is much smaller relative to the rest of the brain and primarily serves as a relay center. In non-mammalian vertebrates, the roof of the midbrain, called the **tectum**, contains a pair of brain centers called **optic lobes** that coordinate sensory input from the eyes. In mammals these regions are called the *superior colliculi,* and are much smaller than in other vertebrates, functioning only in re-

flex optical responses such as orienting the eyes toward visual stimuli or adjusting focus, while the forebrain takes over the majority of visual processing. The tectum also contains the paired *inferior colliculi,* nuclei that are involved in hearing. Neurons conducting signals from the inner ear form synapses in this region. The posterior part of the midbrain is called the *tegmentum,* and contains regions that help with fine control of muscles. Lesions in this area of the brain can lead to Parkinson's disease, a condition associated with muscle tremors. In mammals the midbrain is sometimes grouped together with the pons and medulla oblongata and termed the **brainstem**.

The forebrain controls complex processes

In mammals, the forebrain is involved in processing and integrating sensory information, and in coordinating behavior. The forebrain consists of the *cerebrum*, the *thalamus*, the *epithalamus*, and the *hypothalamus*. The cerebrum, whose outer layer is the cortex, is divided into two **cerebral hemispheres** (Figure 7.10). The left hemisphere controls the right half of the body, and the right hemisphere controls the left half of the body. Although the right and left hemispheres seem to be mirror images, they are not functionally identical. For example, in most humans the areas that control speech are located in the left hemisphere, and areas that govern perception of spatial relationships are found in the right hemisphere. Even though the two hemispheres have somewhat different functions, they do not work entirely independently. They are connected by a mass of white matter known as the **corpus callosum**, which allows the two hemispheres to communicate with each other. Damage to the corpus callosum prevents communication between the hemispheres and can lead to a variety of unusual symptoms (see Box 7.1, Applications: Split-Brain Syndrome).

The hypothalamus maintains homeostasis

The **hypothalamus** is located at the base of the forebrain and, as the name suggests, just below the thalamus. The hypothalamus controls the internal organs and interacts with the autonomic nervous system, which we discuss later in this chapter. In addition, it regulates the secretion of pituitary hormones (Chapter 3: Cell Signaling and Endocrine Regulation) and thus plays a role in regulating the endocrine system, serving as a crucial link between the nervous and endocrine systems. Indeed, the primary function of the hypothalamus is to maintain the body's homeostatic balance. The hypothalamus regulates body temperature, fluid balance, blood pressure, body weight, and many bodily sensations such as hunger, thirst, pleasure, and sex drive.

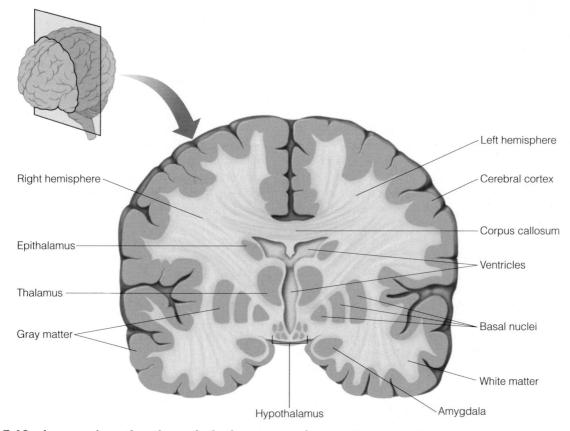

Figure 7.10 A coronal section through the human cerebrum   The cerebrum is divided into two hemispheres connected via the corpus callosum. A thin layer of gray matter (the cerebral cortex) surrounds a large mass of white matter. Embedded within this white matter are more areas of gray matter (the epithalamus, thalamus, hypothalamus, basal nuclei, and amygdala).

BOX 7.1 APPLICATIONS
Split-Brain Syndrome

Mammalian brains are divided into two hemispheres, with the right hemisphere receiving sensory input from, and controlling, the left half of the body, and the left hemisphere receiving sensory input from, and controlling, the right half of the body. Ordinarily, the left and right sides of the brain coordinate their actions by communicating information via the corpus callosum, a region of white matter connecting the two hemispheres. But what would happen if the corpus callosum ceased to function? Roger Sperry was the first to investigate this question, when he performed experiments in which he cut the corpus callosum and optic chiasm in cats, and tested the effects of the surgery on the animals' behavior. Each cat was apparently normal following the surgery, but when Sperry covered its left eye and then taught it a simple conditioned behavior, the cat could not perform this task when its right eye was covered instead of the left. It was as if only one side of the brain learned to perform the task, and could not communicate this learning to the other side of the brain. Sperry termed this phenomenon the split-brain syndrome.

Similar observations have been made in human patients following brain surgery designed to reduce the severity of epileptic seizures. In this surgery, a patient's corpus callosum is cut so that an epileptic seizure in one part of the brain cannot spread to the other hemisphere. Although it seems like rather radical surgery, cutting the corpus callosum is actually quite effective, and greatly reduces the severity of seizures with few apparent side effects. However, Sperry was able to demonstrate that these patients had a subtle form of split-brain syndrome. Sperry presented images or words to either the right or left visual field of these patients, and then asked the subjects a series of simple questions or had them perform basic tasks. For example, in one experiment, the word *key* was presented to the left visual field (which is processed by the right hemisphere of the brain), while the word *ring* was simultaneously presented to the right visual field (which is processed by the left hemisphere of the brain). Normal subjects report seeing the word *keyring*. Patients whose corpus callosum had been sev-

ered reported seeing the word *ring* that had been projected to the right visual field and processed by the left hemisphere. They were entirely unaware that the word *key* had been presented to the left visual field and processed by the right hemisphere, although some subjects occasionally reported that they saw a flash of light on the left side of the screen.

In most humans, the ability to communicate using language is localized in the left hemisphere of the brain, while the right hemisphere lacks language ability. Thus, the right hemisphere was unable to communicate that the light observed in the left visual field represented a word. Control subjects could verbalize both the words *key* and *ring* because the intact corpus callosum could transfer the information between the two hemispheres. This difference between normal subjects and "split-brain" patients is not obvious in everyday life because we seldom look at objects using only one eye. We can easily move our eyes or turn our heads so that both halves of the brain receive complete sensory information.

Although the right hemisphere does not have the ability to speak, it can still reason and communicate in other ways. For example, Sperry asked the split-brain subjects to reach behind a curtain and choose the object whose name had just been projected on the screen. They could not see the objects, but had to distinguish them by touch. When split-brain patients were asked to use their left hand (which is under the control of the right hemisphere), they chose the key, even though they had denied seeing the word. Thus, the right hemisphere had seen the word *key* and recognized its meaning, but was simply unable to communicate this information verbally. Interestingly, when asked to name the object they had just touched with their left hand, split-brain subjects responded by saying "ring"—the word observed by the left hemisphere.

Together, these and many subsequent studies have demonstrated that mammalian brains, and particularly the brains of humans, are highly lateralized with differing functions performed in each hemisphere.

The limbic system influences emotions

The hypothalamus is part of the **limbic system**, a network of connected structures that lie along the border between the cortex and the rest of the brain (Figure 7.11). These regions work together to influence many processes including emotions, motivation, and memory. Thus, the limbic system is sometimes called the "emotional brain" because it controls emotions, decisions, and motivation. The limbic system includes several structures in addition to the hypothalamus, including the *amygdala*, *hippocampus*, and *olfactory bulbs*.

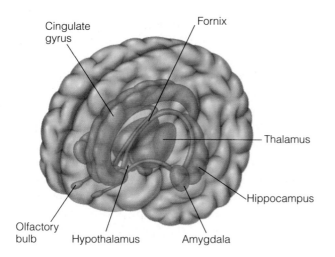

Cingulate gyrus

Fornix

Thalamus

Hippocampus

Amygdala

Hypothalamus

Olfactory bulb

Figure 7.11 Anatomy of the limbic system The limbic system consists of structures including the thalamus, hypothalamus, hippocampus, amygdala, and olfactory bulb.

The **amygdala** is involved in emotional responses, particularly those of aggression and fear. Electrical stimulation of the amygdala causes aggressive behavior, while removal of the amygdala results in decreased aggression and fear. For example, rats with damage to the amygdala will readily approach cats. Monkeys with damage to the amygdala are more eager to approach and interact with novel objects or unknown monkeys, suggesting that the amygdala controls fear reactions in primates. However, a different response is observed if the amygdala is damaged in infant monkeys. These monkeys are unable to develop normal social interactions, suggesting that the amygdala performs other roles in addition to simply regulating fear and aggression, at least in primates. For example, humans with damage to the amygdala are unable to accurately interpret facial expressions, particularly those associated with negative emotions such as fear or anger. The amygdala is also involved in maintaining memories of the emotional effects of an event.

The **hippocampus** converts short-term memories to long-term memories. For example, if you look up a telephone number, you can keep the number in your short-term memory by repeating it a few times, but the memory of this number usually fades quickly once you have placed the call. If you want to remember the number for a long time, the hippocampus must convert this short-term memory to a lasting one. A person with a damaged hippocampus cannot build lasting memories. He or she can remember new facts for a short time, but will forget them within a few minutes. In contrast, memories from before the time of damage are unaffected. We discuss how the hippocampus helps to form lasting memories at the end of this chapter.

The **olfactory bulb**, which also forms part of the limbic system, is important for the sense of smell. Sensory neurons from the olfactory epithelium connect directly to the olfactory bulb, rather than being routed through the midbrain, as is the case for most other incoming sensory information. The olfactory bulb then integrates the signals from the olfactory neurons and transmits them to the cortex for processing. As we discuss later in the chapter, all other sensory information is first processed by the thalamus before being sent to the cortex. In contrast, olfactory information bypasses the thalamus and instead takes a more direct route. The olfactory bulb is also connected to the amygdala and hippocampus, and thus odors tend to provoke strong emotions and memories in humans.

The thalamus acts as a relay station

The **thalamus** is a large grouping of gray matter located deep within the forebrain, immediately above the hypothalamus. The thalamic nuclei receive input from the limbic system and from every sensory modality except olfaction. In fact, some researchers consider it to be part of the limbic system itself. The thalamus integrates and relays this information to the cortex. The thalamus is part of a structure called the *reticular formation*. The reticular formation is a net of neurons extending from the thalamus down through the brainstem, including parts of the midbrain, pons, and medulla oblongata. The reticular formation acts as a filter for incoming sensory information. In fact, we do not consciously attend to the vast majority of incoming sensory information. Instead, it is filtered by the thalamus. We have all experienced this phenomenon. Imagine that you are at a party, surrounded by the buzz of many conversations. Suddenly, you hear your name spoken behind you and you become aware that someone is talking about you, despite not having noticed the conversation before. Although you were receiving sensory information about this conversation all along, your thalamus filtered out the unimportant information, and only triggered conscious attention by sending relays to the cortex when your name was mentioned.

The *epithalamus* is located above the thalamus, and contains the *habenular nuclei* and the **pineal complex**. The habenular nuclei communicate with the tegmentum of the midbrain, while the pineal is involved in establishing circadian rhythms and secretes the hormone melatonin (see Chapter 6).

The cortex integrates and interprets information

The outer layer of the mammalian cerebrum integrates and interprets sensory information and initiates voluntary movements, and thus has taken over many of the functions that are performed by the midbrain in other vertebrates. This region, called the cortex, is necessary for cognition and all other so-called higher functions, including the ability to concentrate, reason, and think in abstract form. In some mammals, the cortex is smooth, whereas in other species it is folded so that the surface of the brain has a walnut-like appearance (Figure 7.12b). The outer, visible regions of these folds are called **gyri** (singular: gyrus), and the grooves are called **sulci** (singular: sulcus). These folds greatly increase the surface area of the cortex, increasing the number of neurons and their interconnections, and thus increasing the functional complexity of the forebrain. The cortex varies in surface area by a factor of 125 between the least cortical mammals, such as hedgehogs, and the most cortical mammals, such as primates and cetaceans. The degree of folding of the cortex appears to be correlated with the functional complexity of the brain and the intelligence of the organism.

The cortex of mammals is rather distinct in structure compared to the cortex of other vertebrates. Because of its unusual organization, the mammalian cortex is often referred to as the neocortex or isocortex. The isocortex is organized into six functionally distinct layers with neuronal processes and cell bodies distributed within the layers in a specific fashion (Figure 7.13). The main visible difference between the layers is the shape and density of the neurons located in each layer. The outermost layer (I) contains few cell bodies and few connections among cells. Layers II and III are involved in integrating signals within the cortex, while the remaining layers contain neurons that communicate with other parts of the brain including the thalamus, brainstem, and spinal cord. The cortex is thought to be organized into functional units called *columns* that are oriented vertically within the cortex and extend through all six of the cortical layers, although the functional significance of this vertical organization is still a matter of debate. Indeed, the degree of columnar organization appears to vary among parts of the cortex and among species. Columns may be further broken down into *minicolumns* of less than a millimeter in diameter, containing only about 100 neurons. There are numerous interconnections between neurons within a column, and although there are fewer connections among columns, these connections can extend far across the cor-

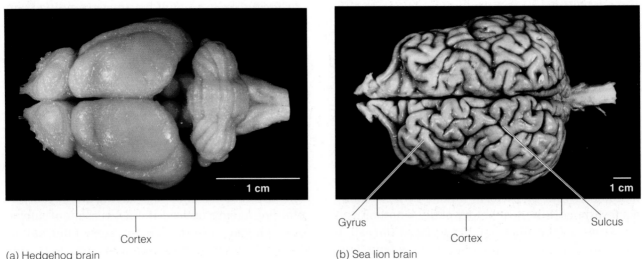

(a) Hedgehog brain

(b) Sea lion brain

Figure 7.12 Structural variation in mammalian brains In some species the cortex is folded into a series of elaborate gyri and sulci, whereas in other species it is relatively smooth.

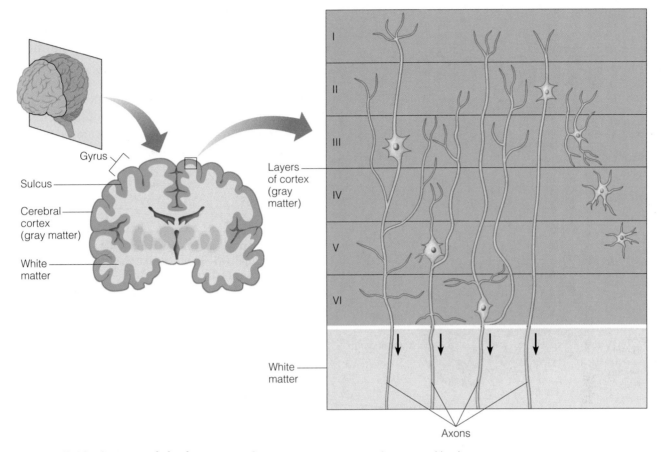

Figure 7.13 Layers of the human cortex The cerebral cortex is arranged in six distinct layers, although the cellular composition of these layers varies depending on the particular area of the cortex.

tex, or into subcortical areas such as the thalamus. Thus, the cortex may act as a massively parallel processor with each column acting as a semiautonomous unit.

Each of the cerebral hemispheres is divided into four regions, or *lobes*, that are defined based on the names of the overlying bones (Figure 7.14). The *frontal lobe* is involved in reasoning, planning, and some aspects of speech in humans. The *parietal lobe* is associated with movement, orientation, recognition, and perception of stimuli. The *occipital lobe* is involved with visual processing, and the *temporal lobe* is involved with perception and recognition of auditory stimuli, memory, and speech. Alternatively, the brain can be divided into areas that are specialized for different functions that roughly fall within the divisions defined by the lobes of the brain (Figure 7.14b).

Many of the functional regions of the cortex are organized *topographically*, such that specific areas of the cortex correspond to particular functions. This arrangement echoes the concept of la-

beled lines, which we encountered when discussing sensory systems, and applies to the visual cortex, the auditory cortex, the somatosensory cortex, and the motor cortex. The somatosensory cortex and primary motor cortex are particularly good examples of this topographic arrangement: each part of the cortex corresponds to the specific part of the body that it governs (Figure 7.15). Notice that the areas of the somatosensory cortex devoted to various parts of the body are disproportionate. For example, the face and hands take up more than half the map of both the sensory cortex and motor cortex in humans. The size of the cortical region typically reflects the number of sensory or motor neurons present in a particular body part, rather than the size of the body part itself. Thus, the amount of cortex devoted to inputs from a particular part of the body differs among species, reflecting the relative importance of various parts of the body for sensation and movement. For example, the nose takes up a disproportionate amount of the somatosensory cortex in the star-nosed

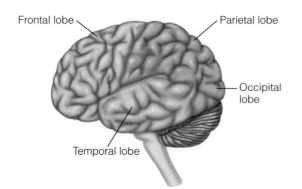

(a) Lobes of the brain

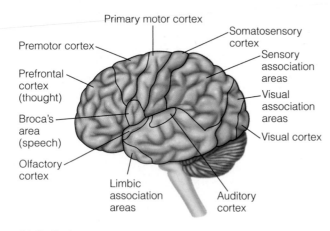

(b) Cortical areas

Figure 7.14 Lobes, cortices, and association areas of the human brain (a) The cerebrum can be divided into several lobes, each named after the overlying bones. (b) The cerebrum can also be divided into functional regions called cortical areas, each involved in coordinating a different function.

mole. These animals live in burrows and use their sensitive noses to probe their environments. This topographical organization of the cortex is maintained at even finer levels. For example, the neurons in the region of the somatosensory cortex of the star-nosed mole that is devoted to the nose are organized in a star shape, reflecting the shape of the nose.

In addition to the various sensory cortices, the brain also contains several *association areas* involved in higher-level cortical processing. Association areas receive input from adjacent cortical areas and further process and integrate this information. The functions of these association areas are not very well understood, because of the complex nature of their processing tasks and the difficulty of studying them experimentally. This difficulty is particularly acute for the human pre-

frontal association cortex, which is responsible for various skills such as language, logical thinking, planning, and judgment. The human prefrontal association cortex is six times the size of that in a chimpanzee, and this brain region is even less developed in other mammals, so experiments on nonhuman animals are likely to be of limited use in understanding the mechanisms underlying the complex behavior of humans.

⊙ | CONCEPT CHECK

6. What type of symptoms would you expect in an individual who had a stroke that damaged part of the cerebellum?
7. Compare and contrast the function of the midbrain in mammals with its function in other vertebrates.
8. What is the function of the corpus callosum?
9. What do somatosensory maps tell us about the organization of the cortex?

The Peripheral Nervous System

Afferent neurons carry sensory information to integrating centers (such as the vertebrate central nervous system) where it is processed. The integrating centers then send out signals via efferent pathways that govern physiological responses and behavior. Together, the afferent sensory neurons and the efferent neurons that send signals to effector organs make up the peripheral nervous system. We have already discussed the afferent branch of the peripheral nervous system in Chapter 6. The efferent branch of the peripheral nervous system is separated into two main divisions: the **autonomic division** and the **somatic motor division** (Figure 7.16). In the remainder of this section we discuss each of these divisions in turn.

Autonomic Pathways

The autonomic nervous system is involved in the homeostatic regulation of most physiological functions, including heart rate, blood pressure, breathing, and many other processes that are critical for life. These functions are not usually under conscious control, and thus this nervous system is sometimes referred to as the *involuntary nervous system*. The autonomic division can be differentiated into three branches. The **sympathetic nervous system** is most active during periods of stress or physical activity, whereas the **parasympathetic nervous system** is most active during periods of rest. Thus, the parasympathetic branch is sometimes referred to as the "resting and digesting" system, because it is mainly concerned with redirecting energy toward quiet activities such as digestion. In contrast, the sympathetic branch is sometimes called the "fight-or-flight" system. Stimulating the sympathetic nervous system causes increases in heart rate, deeper breathing, and diversion of blood from the digestive system to the working muscles. Although the action of the sympathetic branch is most obvious during the fight-or-flight response, which we discussed in Chapter 3, it also plays an important role in daily activities, in particular in regulating blood pressure and blood flow to tissues. The **enteric branch** of the autonomic nervous system operates independently of the other two branches, although the parasympathetic and sympathetic branches can regulate its activity. The enteric branch is entirely concerned with digestion, and innervates the gastrointestinal tract, pancreas, and gallbladder. We

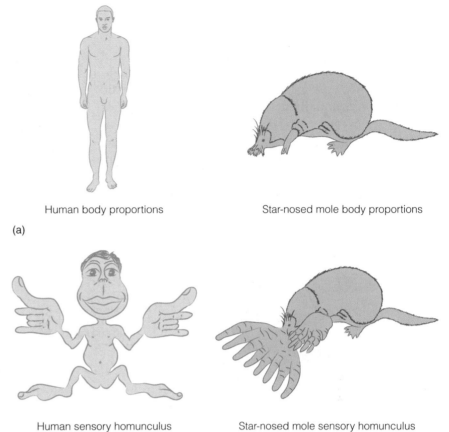

(a) Human body proportions   Star-nosed mole body proportions

(b) Human sensory homunculus   Star-nosed mole sensory homunculus

Figure 7.15 Somatosensory maps The area of the cortex devoted to a given body part depends on the importance of that body part to the organism. **(a)** Body proportions of a human and a star-nosed mole. **(b)** Proportion of the somatosensory cortex devoted to particular body parts. In humans, a disproportionate area of the cortex is devoted to sensory input from the hands and mouth. In star-nosed moles, a disproportionate amount of the cortex is devoted to the front paws and nose.

discuss the role of the enteric nervous system in Chapter 11: Digestion. For now, we concentrate on the sympathetic and parasympathetic branches of the autonomic nervous system.

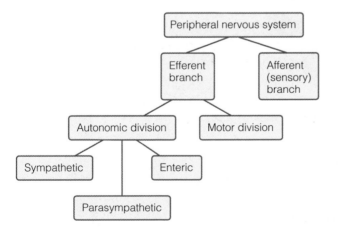

Figure 7.16 Major divisions of the vertebrate peripheral nervous system The vertebrate nervous system can be divided into the peripheral nervous system (consisting of afferent and efferent pathways), and the central nervous system (consisting of the brain and spinal cord). The efferent pathways of the peripheral nervous system can be further separated into two pathways: the motor division that initiates movement by stimulating skeletal muscles, and the autonomic division that regulates physiological functions. The autonomic division is divided into the sympathetic, parasympathetic, and enteric nervous systems.

The sympathetic and parasympathetic branches act together to maintain homeostasis

The autonomic nervous system maintains homeostasis by balancing the activity of the sympathetic and parasympathetic nervous systems and their effects on their target organs. Three important features of the autonomic nervous system underlie its ability to maintain homeostasis: dual innervation, antagonistic action, and basal tone. As you can see from Figure 7.17, most internal organs receive in-put from both the sympathetic and parasympathetic nervous systems. Through this process of dual innervation, the two branches can work together to regulate effector organs. The effects of the sympathetic branch and the parasympathetic branch are generally antagonistic—one stimulatory and the other inhibitory (Table 7.3). For example, stimulation of the parasympathetic nervous system causes the bronchioles of the lung to constrict by causing the associated smooth muscle to contract. In contrast, stimulation of the sympathetic nervous system causes bronchioles to dilate

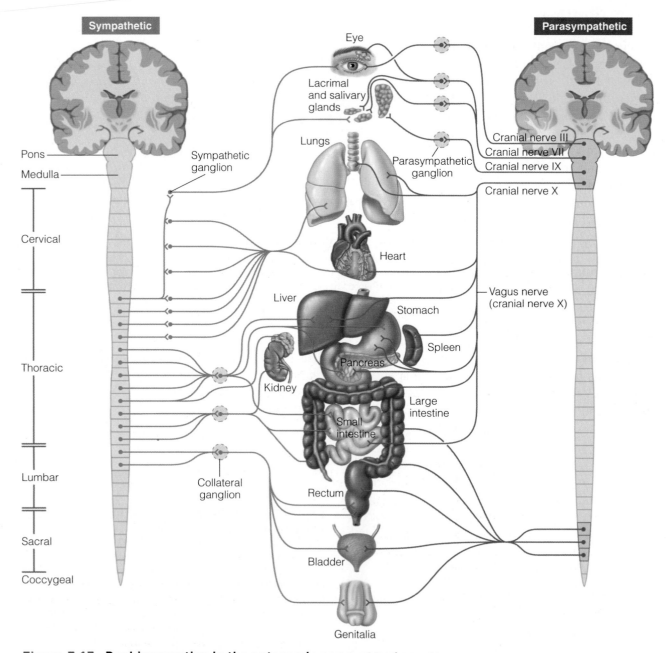

Figure 7.17 Dual innervation in the autonomic nervous system Most organs receive input from both the parasympathetic and sympathetic nervous systems.

Table 7.3 Actions of the sympathetic and parasympathetic nervous systems in humans.

Effector organ	Parasympathetic stimulation	Sympathetic stimulation	Adrenergic receptor
Pupil of eye	Constricts	Dilates	α
Lacrimal glands of eyes	Stimulates secretion	None	None
Salivary gland	Watery secretion	Thick secretion	α, β2
Heart	Slows heart rate	Increases rate and force of contraction	β1
Arterioles	None	Constricts	α
Nasal glands	Stimulates secretion	None	None
Bronchioles of lungs	Constricts	Dilates	β2
Digestive tract	Increased motility and secretion	Decreased motility and secretion	α, β2
Exocrine pancreas	Increases enzyme secretion	Decreases enzyme secretion	α
Endocrine pancreas	Stimulates insulin secretion	Inhibits insulin secretion	α
Adrenal medulla	None	Secretes epinephrine	None
Kidney	None	Increases renin secretion	β1
Bladder	Release of urine	Retention of urine	α, β2
Adipose tissue	None	Fat breakdown	β1
Sweat glands	General sweating	Localized sweating	α
Arrector pili muscles of skin	None	Contract, causing hair to stand on end	α
Male sex organs	Erection	Ejaculation	α
Uterus	Depends on stage of cycle	Depends on stage of cycle	α, β2

through relaxation of the associated smooth muscle. Finally, both the parasympathetic and sympathetic nervous systems have *basal tone* (or basal tonic activity), such that even under resting conditions autonomic neurons produce action potentials. Thus, both increases and decreases in action potential frequency can alter the response of the target organ, similar to a volume control on a radio. Together these three organizing principles allow the autonomic nervous system to exert precise control and to maintain homeostasis by balancing the input of the parasympathetic and sympathetic branches of the autonomic nervous system.

Autonomic pathways share some structural features

All autonomic pathways contain two neurons in series (Figure 7.18). The cell body of the first, or

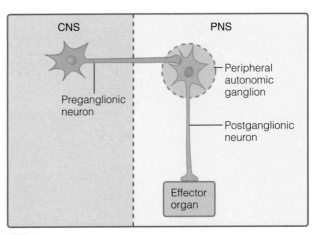

Figure 7.18   Structure of an autonomic pathway
Autonomic pathways consist of a two-neuron chain. The preganglionic neuron originates in the central nervous system, and forms a synapse at the peripheral ganglion. The postganglionic neuron originates at the peripheral ganglion and forms a synapse at the effector organ.

preganglionic, neuron is located within the central nervous system. This neuron synapses with a second, or **postganglionic**, efferent neuron in peripheral structures called **autonomic ganglia** that contain many such synapses. A single preganglionic neuron generally synapses with several postganglionic neurons, and may also make contact with *intrinsic neurons* that are located entirely within the ganglion, allowing for relatively complex integration of function within the ganglion itself. At the effector organ, the postganglionic neuron releases neurotransmitter from specialized structures called varicosities, as discussed in Chapter 4. The axons of postganglionic autonomic neurons have a series of swellings at their distal end arranged in series along the surface of the effector organ, like beads on a string. Each varicosity acts as a synapse with the effector organ, releasing neurotransmitter in response to action potentials. The underlying membrane of the effector organ is not specialized and does not contain high concentrations of receptors. Instead, the neuron simply releases neurotransmitter into the extracellular fluid. The neurotransmitter then diffuses to receptors distributed across the membrane of the effector organ.

The anatomy of the sympathetic and parasympathetic branches differ

There are three main anatomical differences between the sympathetic and parasympathetic branches of the autonomic nervous system. First, the cell bodies of preganglionic sympathetic and parasympathetic neurons are located in different regions of the central nervous system. Most sympathetic pathways originate in the thoracic and lumbar regions of the spinal cord, while most of the parasympathetic pathways originate either in the hindbrain or in the sacral region of the spinal cord (see Figure 7.17). Second, the locations of the ganglia differ between the sympathetic and parasympathetic branches of the autonomic nervous system. Sympathetic ganglia are found in a chain that runs close to the spinal cord, while parasympathetic ganglia are located close to the effector organ. Thus, most sympathetic pathways have short preganglionic neurons and long postganglionic neurons, while parasympathetic pathways have long preganglionic neurons and short postganglionic neurons (Figure 7.19). The final anatomical difference between the sympathetic and parasympathetic pathways lies in the relationship between the preganglionic and postgan-

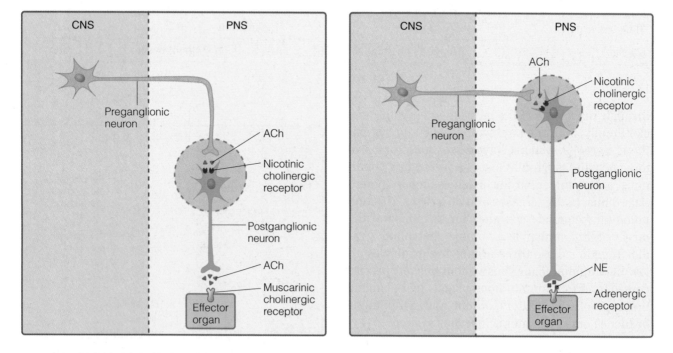

(a) Parasympathetic nervous system

(b) Sympathetic nervous system

Figure 7.19 Structure and neurotransmitters of the sympathetic and parasympathetic nervous systems The parasympathetic nervous system has a long preganglionic neuron and a short postganglionic neuron, while the sympathetic nervous system has a short preganglionic neuron and a long postganglionic neuron.

glionic neurons. In the sympathetic nervous system, on average, a preganglionic sympathetic neuron forms synapses with 10 or more postganglionic neurons. In contrast, in the parasympathetic system, an average preganglionic neuron forms synapses with three or fewer postganglionic neurons. Stimulation of a single sympathetic preganglionic neuron will thus have rather widespread effects, while stimulation of a preganglionic parasympathetic neuron typically causes a much more localized response.

The sympathetic and parasympathetic nervous systems can also be distinguished based on the neurotransmitters they release at the synapse with the effector organ. In both the sympathetic and parasympathetic divisions, the preganglionic neuron releases the neurotransmitter acetylcholine (ACh), and the postganglionic neuron has nicotinic receptors that bind the ACh (see Chapter 4). Nicotinic acetylcholine receptors are ligand-gated ion channels, and binding of ACh allows Na^+ to enter and rapidly depolarize the postganglionic cell. The effects of nicotinic receptors are always stimulatory.

In the parasympathetic nervous system, the postganglionic cell releases ACh, but the target organ has muscarinic rather than nicotinic ACh receptors. Muscarinic ACh receptors are coupled to G proteins, and thus typically cause somewhat slower responses than do nicotinic receptors. There are several types of muscarinic receptors, and binding of ACh can be either stimulatory or inhibitory, depending on the type of receptor present on the target cell.

In contrast, in the sympathetic nervous system, postganglionic cells typically release the neurotransmitter norepinephrine, which binds to α or β adrenergic receptors on the effector organ. The various types of adrenergic receptors work through different second messenger pathways and cause a variety of responses in the target cell (see Chapter 4). Differences in receptor subtypes among effector organs explain the diverse effects of sympathetic and parasympathetic stimulation of various tissues. As we discuss in Box 7.2, Applications: Receptor Subtype and Drug Design, these differences are important clinically in predicting

BOX 7.2 **APPLICATIONS**
Receptor Subtype and Drug Design

Asthma is a respiratory condition that affects up to 15 million people in the United States, many of them children. The symptoms of asthma include wheezing, coughing, and difficulty breathing. During an asthma attack, the smooth muscles surrounding the bronchioles contract, narrowing the passages that normally lead air into the lungs. The causes of asthma are not known, but a variety of fairly effective treatments are available to reduce the severity of an asthma attack.

Up until the mid-1980s, administering the drugs epinephrine or ephedrine was the main treatment for asthma. Both epinephrine and ephedrine bind to β adrenergic receptors, stimulating these receptors on the smooth muscles of the bronchioles, and causing them to relax. When the smooth muscle relaxes, the bronchioles dilate, opening the passages to the lungs, counteracting the effects of an asthma attack. Unfortunately, both ephedrine and epinephrine bind to many kinds of adrenergic receptors, including the α1, α2, β1, and β2 subtypes of the adrenergic receptors that are found on many tissues throughout the body. Recall that epinephrine is released as part of the fight-or-flight response of the sympathetic nervous system, increasing

heart rate and the force of cardiac contraction, redistributing blood flow to the active muscles, increasing alertness, and readying the body for action. Thus, ephedrine and epinephrine can have substantial side effects when used as drugs. For example, ephedrine can induce anxiety, tremors, irritability, sleeplessness, and a rapid or irregular heartbeat. Indeed, several countries have recently banned ephedrine as an ingredient in diet pills and food supplements because of these dangerous side effects.

Recently, antiasthma drugs have been developed that specifically target the β2 adrenergic receptors that are expressed on the smooth muscle of the bronchioles. For example, albuterol (which is marketed under such trade names as Ventolin) binds to β2 adrenergic receptors with approximately 500 times greater affinity than it does to β1 adrenergic receptors, and binds poorly if at all to α adrenergic receptors. Use of drugs like albuterol reduces the risks of serious cardiac side effects relative to epinephrine, because the heart expresses mostly β1 adrenergic receptors. But these drugs still provide good relief from asthma attacks because of their effects on the β2 receptors of the bronchiolar smooth muscle.

the effects of many drugs. In general, binding of norepinephrine to α receptors is stimulatory, while binding to β receptors is inhibitory. A few classes of postganglionic sympathetic neurons, including those innervating the sweat glands of the skin, release ACh rather than norepinephrine, but these neurons are much less numerous than the adrenergic neurons.

Table 7.4 summarizes some of the similarities and differences between the sympathetic and parasympathetic nervous systems.

Some effectors receive only sympathetic innervation

Although the principle of dual innervation applies to most of the target organs of the autonomic nervous system, some organs— including the sweat glands, the arrector pili muscles of the skin, the adrenal medulla, the kidneys, and most blood vessels—are only innervated by sympathetic neurons (see Table 7.3). The effects of sympathetic stimulation on the sweat glands and arrector pili muscles are obvious. Humans commonly sweat during stressful situations, and in many mammals fear causes the hair (or fur) to stand on end, because of the actions of the arrector pili muscles.

The adrenal medulla, the core of the adrenal gland, is also involved in the response to stressful situations. The adrenal glands are paired glands located immediately above the kidneys. The adrenal medulla is actually a highly modified sympathetic ganglion. Preganglionic sympathetic neurons terminate in the adrenal medulla, but the postganglionic neurons do not go on to innervate a target organ (Figure 7.20). Instead, they are modified into neurosecretory cells called chromaffin cells that release epinephrine and norepinephrine directly into the circulation, producing widespread excitatory effects. As we discussed in Chapter 3, we can easily see the origins of the adrenal glands as sympathetic ganglia by looking at fish, which lack a discrete adrenal gland. In the elasmobranchs (sharks and rays), these neurosecretory cells are directly associated with the autonomic ganglia. In bony fishes, these cells are dispersed throughout the anterior part of the kidney, similar to the location in mammals, although they are not grouped into a discrete gland. This progression from a clear ganglionic structure to dispersed cells to a non-ganglionic tissue (the adrenal medulla) suggests the likely evolutionary origin of this unusual structure.

The central nervous system regulates the autonomic nervous system

The central nervous system exerts control over the autonomic nervous system at several levels, including the spinal cord, brainstem, hypothalamus, and cortex. The relationship between these brain regions and the autonomic nervous system is outlined in Figure 7.21. Many of the inputs from the central nervous system reach the autonomic nervous system via the reticular formation, a set of neurons located throughout the brainstem, which we mention in relation to the thalamus. Although

Table 7.4 Similarities and differences between the sympathetic and parasympathetic nervous systems.		
Characteristic	**Sympathetic**	**Parasympathetic**
Number of neurons in chain	Two	Two
Location of cell bodies	Thoracic and lumbar regions of spinal cord	Hindbrain Sacral region of spinal cord
Location of ganglia	Close to spinal cord	Close to effector organ
Preganglionic neuron	Short	Long
Postganglionic neuron	Long	Short
Synapses per preganglionic neuron	Many	Few
Neurotransmitter released by preganglionic neuron	ACh	ACh
Neurotransmitter released by postganglionic neuron	NE	ACh

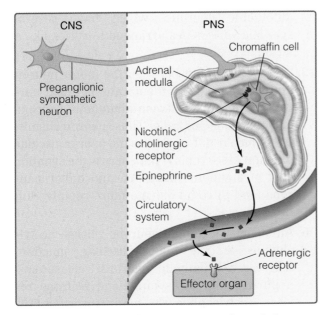

Figure 7.20 Sympathetic innervation of the adrenal medulla The adrenal medulla receives innervation from a preganglionic sympathetic neuron, and is thus equivalent to a sympathetic ganglion.

the reticular formation can itself act as an integrating center, its main role is to communicate signals coming from the cortex, the medulla oblongata, and the hypothalamus. The hypothalamus plays a dominant role in regulating the autonomic nervous system, and can communicate with the autonomic nervous system directly or via the reticular formation. The hypothalamus initiates the fight-or-flight response, which involves widespread activation of sympathetic neurons. The hypothalamus also contains regulatory centers for body temperature, food intake, and water balance, all of which are homeostatically regulated via the autonomic nervous system. The medulla oblongata contains centers that control heart rate, blood pressure, and breathing by influencing the activity of the autonomic nervous system.

Most of these changes in the activity of the autonomic nervous system occur at the unconscious level via **reflex arcs**, simple neural circuits that do not involve the conscious centers of the brain. Figure 7.22 shows an example of such a reflex arc, one involved in regulating blood pressure. When blood pressure falls, receptors located in various parts of the body detect the decrease. These receptors send

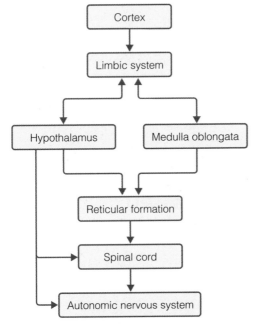

Figure 7.21 Regulation of the autonomic nervous system by the brain Many brain regions can modulate the activity of the autonomic nervous system. The reticular formation in the brainstem processes and communicates most of the descending information from higher brain centers to the autonomic nervous system. The hypothalamus is the most important of these brain regions and can communicate with the autonomic nervous system either directly or via the reticular formation.

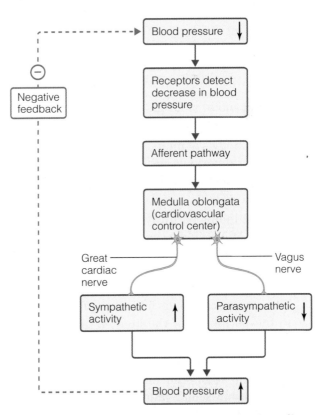

Figure 7.22 An example of an autonomic reflex arc: the reflex control of blood pressure

a signal to the cardiovascular control center in the medulla oblongata via afferent sensory neurons. The cardiovascular control center then influences the activity of the autonomic nervous system, increasing sympathetic activity and decreasing parasympathetic activity. These resulting changes in autonomic output cause adjustments in heart rate, stroke volume, and vasoconstriction, returning blood pressure back to normal in a negative feedback loop.

The limbic system, which governs emotions, also has a profound effect on the activity of the autonomic nervous system. Blushing, fainting at the sight of blood, and "butterflies" in the stomach are all examples of the response of the autonomic nervous system to emotions.

Somatic Motor Pathways

Somatic motor pathways control skeletal muscles, which are usually under conscious control. Thus, the motor pathways are sometimes called the "voluntary nervous system." However, some efferent motor pathways are not under conscious control, and instead represent reflex responses—rapid involuntary movements in response to a stimulus. For example, if you sit with your legs crossed and tap sharply just under your kneecap, your leg will kick out, in the patellar (knee-jerk) reflex. Efferent motor pathways can be distinguished from autonomic pathways in seven ways.

1. Efferent motor neurons control only one type of effector organ—skeletal muscle.

2. The cell bodies of motor neurons are located in the central nervous system in the vertebrates, and never within ganglia outside of the central nervous system.

3. Efferent motor pathways are monosynaptic—there is only a single synapse between the central nervous system and the effector organ. As a result, efferent motor neurons can be among the longest neurons in the vertebrate body. Their axons can reach from the spinal cord out to the periphery of the body, a distance that can span several meters in large mammals.

4. The morphology of the synapse differs between the autonomic and motor pathways. At the neuromuscular junction, a motor neuron splits into a cluster of axon terminals that branch out over the motor end plate, unlike

autonomic neurons, which have several synaptic varicosities arranged in series like a string of beads.

5. The synaptic cleft between the motor neuron and the muscle cell membrane is much narrower than that between autonomic neurons and their effector cells. Thus, neurotransmitters typically diffuse across the neuromuscular junction more rapidly than across the synaptic cleft of autonomic neurons, and motor neurons tend to communicate more rapidly with their effectors.

6. All vertebrate motor neurons release acetylcholine at the neuromuscular junction, whereas sympathetic neurons release epinephrine and parasympathetic neurons release acetylcholine. In many invertebrates, motor neurons release glutamate.

7. The effect of acetylcholine on skeletal muscle is always excitatory, whereas autonomic neurons may be excitatory or inhibitory. Stimulation of an efferent motor neuron leads to the contraction of skeletal muscle, and muscles relax only when the associated motor neurons are at rest.

◉ | CONCEPT CHECK

10. Compare and contrast the sympathetic and parasympathetic nervous systems.

11. What is the significance of having dual innervation of many organs by both the sympathetic and parasympathetic nervous systems?

12. What sort of receptors would you expect the neurosecretory chromaffin cells of the adrenal medulla to express?

13. What is a reflex arc?

Integrative Functions of Nervous Systems

Neurobiologists are only beginning to understand how integrating centers such as the brain take information from sensory systems and integrate this information to allow animals to respond to their environments in a dynamic way. In this section we discuss some of the important topics relating to how nervous systems function, beginning with simple behaviors, and then examining some of the more complex functions of the nervous system.

Coordination of Behavior

Multicellular animals are capable of diverse forms of behavior, which are made possible by the complexity of nervous system organization and function. Animal behaviors can be loosely grouped into three categories: reflex behaviors, rhythmic behaviors, and voluntary behaviors. *Reflex behaviors* are involuntary responses to stimuli, and are among the simplest types of animal behaviors. Many animals also have a series of *rhythmic behaviors*, and these rhythms underlie such important processes as locomotion, breathing, and the function of the heart. *Voluntary behaviors* range greatly in complexity, from apparently simple acts such as mating or fighting, to complex behaviors such as reading and writing. In this section we discuss each of these kinds of behaviors in turn, working from the simplest to the most complex.

Reflex arcs control many involuntary behaviors

The least complex integrated response of the nervous system is the reflex arc, which controls the simplest type of animal behavior—reflexes, or rapid involuntary responses to stimuli. In principle, a reflex arc could involve as few as two neurons (Figure 7.23): a sensory afferent neuron that detects the stimulus and an efferent neuron that carries the output to an effector cell (such as a muscle). This reflex arc is called a *monosynaptic reflex arc*, because it contains only a single neuron-to-neuron synapse in the chain from sensory neuron to effector neuron. A monosynaptic reflex arc may contain more than two neurons, as long as there is only one neuron-to-neuron synapse along any path from the stimulus to the response. In-

deed, most monosynaptic reflex arcs contain many neurons.

Neurons in a reflex arc can be arranged in two fundamentally different ways. Figure 7.24a illustrates the principle of **convergence**, in which multiple afferent neurons synapse with a single efferent neuron. A convergent arrangement of neurons allows spatial summation. For example, the activity of a single afferent neuron may be insufficient to excite the efferent neuron, but the simultaneous activity of many afferent neurons may be sufficient to cause a response. This effect occurs as a result of spatial summation. Convergence can also allow the comparison and integration of sensory signals from multiple parts of the body, increasing the complexity of information processing. For example, we have already discussed the significance of a convergent arrangement of neurons in the mammalian retina (see Chapter 6).

Figure 7.24b illustrates an alternative organization, called **divergence**. In this arrangement, a single afferent neuron forms synapses with more than one efferent neuron. Divergence allows a single signal to control multiple independent processes, and is a way to amplify the effect of a signal. Divergent functional arrangements allow the nervous system to engage in parallel processing, which allows very rapid integration of inputs and responses. The autonomic nervous system shows high levels

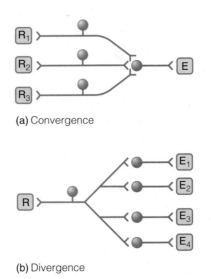

(a) Convergence

(b) Divergence

Figure 7.24 Convergence and divergence in a monosynaptic reflex arc **(a)** In a convergent arrangement, many presynaptic neurons interact with a single postsynaptic neuron. **(b)** In a divergent arrangement, a single presynaptic neuron forms synapses with many postsynaptic neurons. R = sensory receptor; E = effector organ.

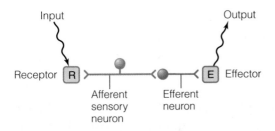

Figure 7.23 A two-neuron reflex arc

of divergence. A single neural pathway from the autonomic nervous system may make connections with many target organs, allowing a coordinated and amplified response.

Note that all of the reflex arcs illustrated in Figure 7.24 are monosynaptic reflex arcs, because they contain only a single synapse in the chain between stimulus and response. Most reflex arcs have a more complex structure, and are called *polysynaptic reflex arcs*, because they contain synapses between more than two types of neurons. A simple polysynaptic reflex arc is shown in Figure 7.25, and includes a sensory cell, an afferent sensory neuron, an interneuron, an efferent neuron, and an effector cell. This type of reflex arc is illustrated by the reflex response to touch in *C. elegans*, which is governed by six touch receptors, five pairs of interneurons, and 69 motor neurons. Adding interneurons to a reflex arc greatly increases the potential responses of the arc and the complexity of the processing.

Pattern generators initiate rhythmic behaviors

Pattern generators govern many important physiological processes and simple rhythmic behaviors such as chewing, walking, swimming, and breathing. Pattern generators are groups of neurons that produce self-sustaining patterns of depolarization, independent of sensory input. Pattern generators can be organized in two different ways. The simplest form of organization involves a **pacemaker cell**. A pacemaker cell generates a spontaneous rhythmic depolarization, and thus controls the firing of all the cells in the network. Pacemaker cells are common in biological systems. For example, as we discuss in Chapter 8, spontaneous pacemaker cells initiate the heartbeat in many kinds of animals. Pattern generators can also be made up of neurons that do not, as individuals, generate rhythmic depolarizations. Instead, the rhythm is an *emergent property* of the network that manifests itself because of the organization of the neurons in the network, rather than being an intrinsic property of the neurons themselves.

To get a sense of how pattern generators operate, consider a two-neuron pair. In this neuron pair, neither neuron generates a rhythm by itself, but when the first neuron (A) fires, it inhibits the other neuron (B) from firing until a defined period elapses, at which point neuron B fires. Neuron B then inhibits neuron A for a defined period of time, after which point it fires, and the loop continues. Imagine two robots programmed to hit if they are hit first. If robot A hits robot B, then robot B will respond by hitting back, which will cause robot A to hit back, and so on. The trick in this kind of network is getting it started in the first place. Once the chain of events is established, it will continue indefinitely, and it is no longer possible to determine where the behavior was initiated. Various mechanisms can start the rhythmic oscillations. Often, input from a sensory receptor is needed in order to start the rhythm. Thus, the distinction between reflex arcs and pattern generators is not precise. Instead, these two types of control pathways interact to produce the complex behavior and physiological responses of animals.

Pattern generators govern swimming behavior in the leech

One approach to understanding the neurobiology underlying complex behaviors is to study simple behaviors in organisms with less complex nervous systems than those found in mammals. One such organism is the medicinal leech, *Hirudo medicinalis*. Like other members of the phylum Annelida, leeches are segmented worms with a brain, a ventral nerve cord, and a series of ganglia located in each body segment. Each segmental ganglion contains approximately 400 neurons, and this simple nervous organization makes the leech an excellent experimental model system. Leeches are ectoparasitic—they attach themselves to vertebrate hosts and feed on their blood. When a leech bites into the skin it injects a local anesthetic and anticoagulant to keep the blood running freely and to avoid detection by the host. A leech can consume up to 15 ml of blood during a single blood

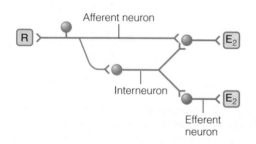

Afferent neuron

Interneuron

Efferent neuron

Figure 7.25 A polysynaptic reflex arc A polysynaptic reflex arc includes a sensory receptor (R), an afferent neuron, an interneuron, one or more efferent neurons, and one or more effector organs (E).

meal, or 10 times its unfed body size. Up to the middle of the 19th century, leeches were commonly used in a medical treatment called "bloodletting" in which physicians would apply leeches to the skin and allow them to suck the patient's blood. This therapy was thought to be helpful for a wide range of illnesses, including fever, headaches, and even obesity. Bloodletting is no longer a common therapy, but leeches are still occasionally used during surgical procedures, such as skin or tissue grafting. For example, leech therapy is particularly useful during finger or ear reattachment surgery to prevent pooling of blood, which can damage the newly grafted tissue.

In its natural habitat, a leech detects its prey by sensing the waves made by a prey animal as it moves about in the water. The leech then swims toward the potential prey, using a rhythmic undulatory motion. Over the last 30 years, neurobiologists have unraveled many components of the neural network that regulates this behavior (Figure 7.26). Swimming begins when mechanoreceptors in the skin sense a stimulus such as the waves made by a prey animal. These mechanoreceptors send an afferent sensory signal to the swim trigger interneuron, which makes a synaptic connection with the swim gating interneuron. When stimulated, the swim gating interneuron activates a network of neurons that form a central pattern generator called the swim oscillator. This central pattern generator sends out rhythmic signals to motor neurons that stimulate muscles in the body wall to initiate rhythmic swimming. The circuit diagram of the swim oscillator is not yet fully worked out, but it involves at least seven oscillator interneurons and four motor neurons. Leeches can also initiate swimming behavior in the absence of

a touch stimulus. An additional neuron in the circuit, sometimes called the swim excitor interneuron, can modulate the activity of the swim gating neuron or the central pattern generator itself in response to signals from the leech brain, but the pathways involved in this higher level of control are not yet understood.

Pattern generators and reflexes are involved in tetrapod locomotion

Four-limbed (tetrapod) vertebrates move by swinging their legs in stereotyped patterns that we call gaits (such as running, walking, or trotting). Gaits such as walking involve rhythmic back-and-forth movements of the legs. Even the seemingly simple movements involved in walking or running require the coordinated contraction of many muscles so that each joint moves just the right distance at just the right time. In some ways, the mechanisms underlying locomotion in four-limbed vertebrates bear a striking resemblance to the control of swimming behavior in leeches. The brainstem (particularly the pons and medulla) usually initiates the command to begin locomotion (Figure 7.27). The brainstem then sends a signal to a network of neurons in the spinal cord that acts as a central pattern generator, similar to the pattern generator in the leech ganglia. The pattern generator then sends coordinated motor output signals to the muscles that control the movement of the limbs, initiating rhythmic movements. Unlike the pattern generator that controls swimming in leeches, the structure and neural connections of this pattern generator are not yet known, and even their location within the spinal cord remains somewhat

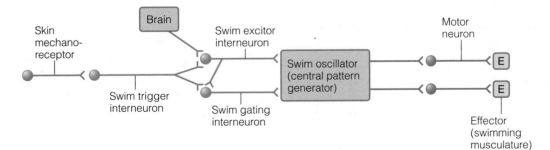

Figure 7.26 The neural circuit governing swimming behavior in the leech A sensory signal from skin mechanoreceptors stimulates a swim trigger interneuron that stimulates a swim gating interneuron and a swim excitor interneuron. These interneurons activate the group of neurons that make up the swim oscillator central pattern generator. The central pattern generator then sends out a rhythmic signal to the swimming musculature. The swim excitor interneurons also process descending information from the leech brain, allowing the leech to initiate swimming even in the absence of a touch stimulus.

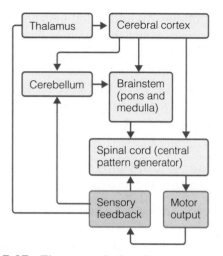

Figure 7.27 The neural circuit governing locomotion in mammals The brainstem sends a signal to the spinal cord central pattern generator. The central pattern generator then sends a rhythmic motor output signal to skeletal muscles. Sensory feedback from proprioceptors and vision travels to the pattern generator, the cerebellum, and the cerebral cortex (via the thalamus), modifying the output of the central pattern generator.

elusive. However, a variety of experiments have demonstrated that a pattern generator must exist within the spinal cord.

In addition to generating rhythmic limb movements, animals must be able to respond to obstacles as they walk or run by dynamically changing their movements in response to changes in the contours of the ground. Stretch receptors and proprioceptors in the limbs sense information about the position of the limbs and the impact of the feet on the ground during walking or running. These receptors send sensory feedback to the pattern generator, allowing the pattern generator to modify its output in response to changing environmental demands. Thus, reflex arcs play an important role in regulating locomotory movements. However, these afferent inputs are not necessary to initiate rhythmic locomotion. For example, if you apply a drug such as curare, which paralyzes the muscles without interfering with nervous system function, and then record electrical activity in the motor neurons leading to the limb musculature, you will observe a phenomenon called *fictive locomotion*. If you stimulate the central pattern generator, the motor neurons will produce rhythmic firing patterns much as they would during normal locomotion, even in the complete absence of any movement-related feedback from the paralyzed muscles. Thus, sensory feedback is not necessary to generate rhythmic lo-

comotory patterns, but simply modifies the output of the pattern generator.

The brain regulates and coordinates the activity of the spinal cord pattern generators, controlling the speed and smoothness of locomotion and adjusting locomotion in response to visual stimuli. Three parts of the brain (the brainstem, the cortex, and the cerebellum) have important roles to play in regulating locomotion. Centers in the brainstem regulate speed. By placing electrodes into the brains of experimental animals, neuroscientists have been able to demonstrate that weakly stimulating this part of the brain initiates walking. Increasing the stimulus intensity increases walking speed and eventually causes trotting and then galloping. The cortex plays an important role in guiding locomotion in complex environments, and in coordinating visual signals with locomotion. For example, a cat with damage to the premotor cortex can still walk on a smooth surface, or even on an inclined plane, but cannot step over objects. Sensory feedback from the working muscles and from other senses, such as vision, enters the cerebral cortex via the thalamus. The cerebral cortex then sends signals to the brainstem and spinal cord to modify locomotion.

The cerebellum fine-tunes locomotion by regulating the timing and intensity of signals to the spinal cord pattern generator. Humans or experimental animals with damage to the cerebellum walk in an uncoordinated way that resembles a drunken gait; their movements are jerky and uncoordinated, and they may stumble. In normal animals the cerebellum receives inputs from the stretch receptors and proprioceptors in the limbs, compares these signals to the intended movement, and then sends signals to the brainstem to correct the movement if necessary, thus coordinating locomotion.

The brain coordinates voluntary movements

Although reflex responses and central pattern generators play an important role in animal behavior, most vertebrates (and many invertebrates) can perform much more complex behavioral tasks. These voluntary behaviors are consciously planned and coordinated by the brain, and can be finely regulated in response to environmental circumstances. Figure 7.28 shows a schematic diagram of the parts of the vertebrate nervous system that are involved in regulating voluntary movements. First, an animal must decide to make a

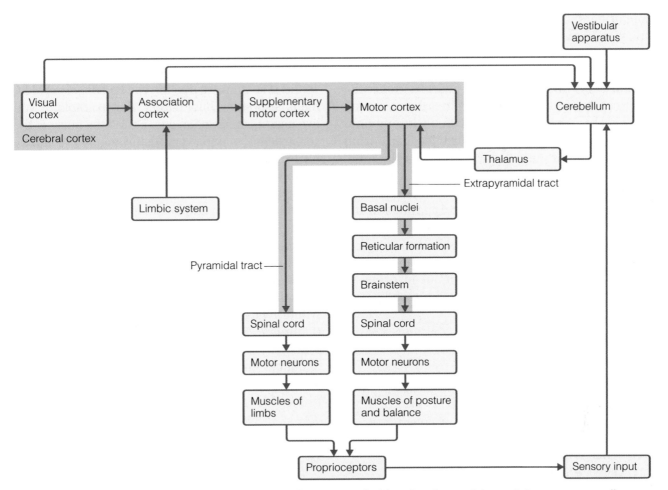

Figure 7.28 Control of voluntary movement in mammals The cerebral cortex initiates voluntary movements. The motor cortex then initiates a motor program by sending efferent signals via the direct pyramidal tract and the indirect extrapyramidal tract. The neurons of the pyramidal tract proceed directly from the cortex to the spinal cord without forming any intermediate synapses, sending a signal via motor neurons to the muscles of the limbs to control movement. The extrapyramidal tract is a multineuron chain that forms synapses in many brain areas before reaching the spinal cord and sending signals via motor neurons to the muscles of posture and balance.

motion. This decision is made in the cerebral cortex of the brain, and includes inputs from the supplementary motor cortex, the association cortex, the visual cortex, and the limbic system. The decision to move is then developed into a program for movement in the primary motor cortex. This motor program is independent of the actual muscles that execute the program. For example, a person who knows how to write his or her name can easily (although a little clumsily) write it by holding a pencil between the toes. Similar regions of the brain are activated in each case, demonstrating that the "program" for writing your name is independent of the specific controls of the muscles of your hands (or feet).

The primary motor cortex executes the motor program by sending signals along a series of tracts (groups of axons) to the spinal cord. Two main pathways are involved in voluntary movements. The pyramidal tracts are direct pathways from the primary motor cortex to the spinal cord and are so named because they pass through a portion of the medulla called the medullary pyramids. The pyramidal tracts play the major role in directing voluntary movements. These tracts cross over each other in the medulla, and thus the left side of the brain controls the right side of the body and vice versa. The extrapyramidal tracts are indirect pathways to motor neurons that, unlike the pyramidal tracts, make numerous synaptic connections within the brain prior to entering the spinal cord. They control the muscle groups that regulate posture and balance. For example, when you sign your name, the pyramidal tracts control the fine movements of your hands and arms, while the extrapyramidal tracts maintain your body position and orientation, although there is some overlap in function between the two systems.

The axons in the pyramidal and extrapyramidal tracts synapse with motor neurons within the spinal cord, and these motor neurons cause the appropriate muscles to contract in order to initiate movements. Just as with rhythmic locomotion, sensory afferent neurons return feedback from stretch receptors and proprioceptors in the muscles to the cerebellum. The cerebellum also receives sensory information from other sensory receptors such as the vestibular apparatus of the ear, which is involved in the sense of balance. The cerebellum integrates these inputs and sends a signal to the cortex (via the thalamus) to refine and adjust the descending motor output in order to complete the planned movement successfully.

⊙ | CONCEPT CHECK

14. What is the difference between a monosynaptic reflex arc and a polysynaptic reflex arc?
15. What kinds of behaviors involve pattern generators?
16. What is the location of the pattern generator governing walking in the vertebrates?

Learning and Memory

In addition to performing complex behaviors, most animals can remember experiences, and modify their behavior accordingly. Although learning and memory are related concepts, these words describe two distinct tasks. Learning refers to the process of acquiring new information, while memory refers to the retention and retrieval of that learned information. The vast majority of animals have the ability to form memories and to learn. Learning and memory are possible because of the **plasticity** of the nervous system—the ability to change both synaptic connections and functional properties of neurons in response to stimuli.

Invertebrates show simple learning and memory

The physiology of learning and memory is understood in the greatest detail in the sea slug *Aplysia californica*. Like other molluscs, *Aplysia* has a fairly

simple nervous system consisting of about 20,000 neurons organized into a series of ganglia. *Aplysia* demonstrates a simple kind of learning called **habituation**—a decline in the tendency to respond to a stimulus due to repeated exposure. Humans also show habituation. For example, if you live near a construction site, at first the noise of the construction may be very disturbing, and you may have difficulty concentrating or studying, but after a while you "get used to" the noise and easily ignore it—you have become habituated to the stimulus. Habituation is an important property of nervous systems, because it allows animals to ignore unimportant routine stimuli and pay more attention to novel, potentially dangerous ones. If you gently touch *Aplysia* on its siphon (a fleshy spout above the gill used to expel seawater), the animal will withdraw its gills and siphon into the mantle cavity (Figure 7.29). However, after repeated gentle touches, *Aplysia* will reduce gill withdrawal by about one-third. If you repeatedly touch the siphon

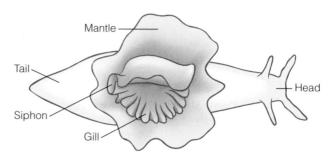

(a) *Aplysia californica*, dorsal view

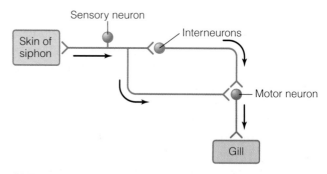

(b) The neural circuit governing the gill-withdrawal reflex

Figure 7.29 The gill-withdrawal reflex in *Aplysia californica* **(a)** Dorsal view of *Aplysia californica*. **(b)** The neural circuit governing the gill-withdrawal reflex. Sensory neurons in the skin of the siphon detect a mechanical stimulus. These sensory neurons form synapses with interneurons and motor neurons. These motor neurons send an efferent signal that causes the gill to withdraw. Habituation of the reflex occurs because of functional changes at the synapse between the sensory and motor neuron as a result of repeated stimulation.

10 or 15 times over the course of a few minutes, the habituation response lasts for about a day, a phenomenon called short-term habituation. If you repeat this stimulation protocol on several consecutive days, the habituation lasts for three or four weeks, a phenomenon called long-term habituation.

Habituation occurs because of functional changes at the synapse between the sensory neuron and the motor neuron. In short-term habituation, a Ca^{2+} channel in the membrane of the presynaptic axon terminal of the sensory neuron is inactivated. Touching the siphon still generates an action potential in the sensory neuron, but when the action potential reaches the axon terminal, less Ca^{2+} flows into the axon terminal, because of the partial inactivation of the voltage-gated Ca^{2+} channels. Neurotransmitter release depends on the influx of Ca^{2+} into the axon terminals, and therefore habituated animals release less neurotransmitter. In addition, there are some morphological changes in the presynaptic axon terminal, including changes in the number and location of neurotransmitter-containing vesicles. Long-term habituation results in similar changes in the presynaptic axon terminal, but to a greater degree. Although the molecular mechanisms involved in the inactivation of the voltage-gated Ca^{2+} channels and the changes in vesicle distribution are not yet known, it is clear that changes in the presynaptic axon terminal of sensory neurons that contact motor neurons cause habituation in *Aplysia*.

Aplysia also demonstrates a kind of learning called **sensitization** (Figure 7.30). In contrast to habituation, sensitization is an increase in the response to a gentle stimulus after exposure to a strong stimulus. For example, imagine being alone in your house in the middle of the night. You suddenly hear a loud noise coming from the basement. For the next little while you will probably be acutely aware of all the sounds around you—you will be sensitized to your environment. You can demonstrate the phenomenon of sensitization in *Aplysia* by delivering an electrical shock to the tail. If you gently touch *Aplysia* on its siphon after this electrical shock, the gill-withdrawal reflex will be much

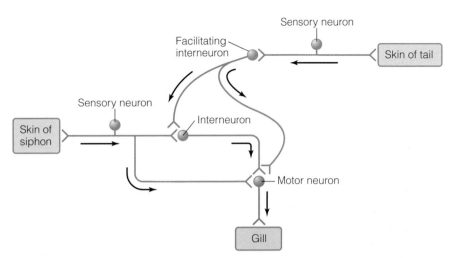

Figure 7.30 The neural network involved in sensitization in *Aplysia* Sensitization of the gill-withdrawal reflex involves a second neural circuit from the skin of the tail. An electrical shock to the tail sends an afferent signal along a sensory neuron that makes a synaptic connection with a facilitating interneuron. This facilitating interneuron makes synaptic connections with the neurons involved in the gill-withdrawal reflex, modifying their response to touch stimuli.

larger and last longer than in an unsensitized animal. The effects of a single shock die out after about an hour, but multiple strong shocks will affect the gill-withdrawal response for a week or more.

As with habituation, during sensitization physiological changes occur in the presynaptic axon terminal of the sensory neuron from the siphon. However, in the case of sensitization there is an increase in Ca^{2+} entry, and increased neurotransmitter release, rather than a reduction. The mechanism underlying this increase in neurotransmitter release involves a second neural circuit: a sensory neuron from the tail that makes a synaptic connection with several interneurons (only one interneuron is shown in Figure 7.30 for clarity). In turn, these interneurons make synaptic connections on the axon terminal of the sensory neuron involved in the gill-withdrawal response. An electrical shock to the tail sends an afferent signal to the interneurons, which then release the neurotransmitter serotonin onto the axon terminal of the sensory neuron involved in the gill-withdrawal response (Figure 7.31). Serotonin binds to a G-protein-coupled receptor that activates adenylate cyclase, which catalyzes the formation of the second messenger cAMP. The increase in cAMP activates protein kinase A (PKA), which phosphorylates voltage-gated K^+ channels in the membrane of the axon terminal, inactivating them. Voltage-gated K^+ channels are responsible for repolarizing the cell after the depolarization

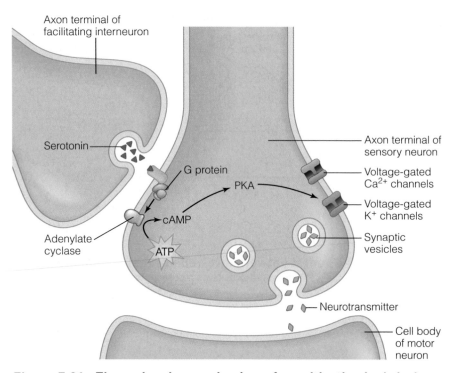

Figure 7.31 **The molecular mechanism of sensitization in *Aplysia***
The facilitating interneuron releases serotonin onto the axon terminal of the sensory neurons involved in the gill-withdrawal reflex. Serotonin binds to a G-protein-coupled receptor that increases intracellular cAMP, activating protein kinase A (PKA), which inactivates voltage-gated K^+ channels. When these K^+ channels are inactivated, action potentials last longer, leading to more Ca^{2+} influx through voltage-gated Ca^{2+} channels, and greater neurotransmitter release from the sensory neuron onto the cell body of the motor neuron.

phase of an action potential (see Chapter 4 for details of this process). When these K^+ channels are inactivated, action potentials last longer, leading to more Ca^{2+} influx through voltage-gated Ca^{2+} channels, and greater neurotransmitter release. The second messenger pathways activated when serotonin binds to its receptor also increase the number and location of neurotransmitter vesicles and activate another Ca^{2+} channel, allowing more Ca^{2+} to enter the cell, further increasing neurotransmitter release. These direct effects of serotonin are relatively short-lived and account for the short-term sensitization of the gill-withdrawal reflex.

Longer-term sensitization, such as occurs following repeated electrical shocks, involves more lasting changes to the neurons and neural circuitry. With repeated electrical shocks (and thus repeated release of serotonin onto the axon terminal of the sensory neuron in the withdrawal reflex), the levels of cAMP in the axon terminal become still higher, increasing the levels of activated PKA. Some of the activated PKA enters the nucleus and

phosphorylates a transcription factor called CREB-1 (<u>c</u>AMP <u>r</u>esponse <u>e</u>lement <u>b</u>inding protein <u>1</u>), which binds to cAMP-responsive sequences in the promoters of many genes, increasing their transcription. These activated genes code for protein products that fall into two classes: proteins involved in forming new synapses and proteins that increase PKA activity. Together these proteins increase the number of synaptic connections and their responsiveness, leading to long-term sensitization of the gill-withdrawal reflex.

The hippocampus is important for memory formation in mammals

Memory formation has also been extensively studied in mammals. For example, rats and mice can be trained to perform simple tasks such as locating a hidden object. If you place a mouse in an enclosure filled with murky water, but with a platform hidden below the surface, the mouse will swim around randomly until it encounters the platform, at which point it will climb onto the platform and remain there. With repetition, the mouse will learn to find the platform very quickly, by remembering its location relative to the walls of the enclosure. A mouse with a damaged hippocampus cannot learn to perform this task; however, if it learned the task prior to its brain damage, it performs as well as an undamaged mouse. Experiments such as these demonstrate that the hippocampus is involved in the formation of long-term memories, but the memories themselves appear to be stored elsewhere.

The cellular and molecular mechanisms underlying memory formation in the hippocampus have been examined in vitro using recording electrodes placed into thin slices of hippocampal tissue. In these preparations, repetitive stimulation of a particular presynaptic neuron eventually leads to an increase in the response of the postsynaptic neuron, a phenomenon called **long-term potentiation**. Over time, a particular level of presynaptic stimulation is converted to a larger postsynaptic output. Long-term potentiation is thought to be important in memory formation because it provides a mechanism in which repetitive activity of a particular neural pathway can leave a record of itself even after the activity has stopped. Although long-term potentiation can occur in several parts of the brain, it is easiest to demonstrate in the hippocampus, further suggesting that the hippocampus is important in memory formation.

Long-term potentiation likely occurs via several mechanisms, but the best-studied mechanism involves changes in certain specific postsynaptic neurons in the hippocampus, the so-called CA1 cells (Figure 7.32). Note that this is in contrast to habituation and sensitization in *Aplysia*, which involve changes in presynaptic neurons. These postsynaptic CA1 cells express two different types of receptors for the neurotransmitter glutamate:

AMPA receptors and NMDA receptors (which are named because they selectively bind the drugs AMPA and NMDA). NMDA receptors are ligand-gated Ca^{2+} channels, so when glutamate binds to NMDA receptors, Ca^{2+} enters the cell. AMPA receptors are ligand-gated Na^+ channels, so when glutamate binds to AMPA receptors, Na^+ enters the cell. Low-frequency stimulation of the presynaptic neuron causes moderate release of glutamate into the synapse, and only the AMPA receptors open, because Mg^{2+} blocks the NMDA ion channels (Figure 7.32a).

High-frequency stimulation of the presynaptic neuron causes greater release of glutamate, and the resulting greater depolarization of the postsynaptic membrane displaces the magnesium ions from the channel of the NMDA receptor (Figure 7.32b). With the Mg^{2+} gone and the ion channel open,

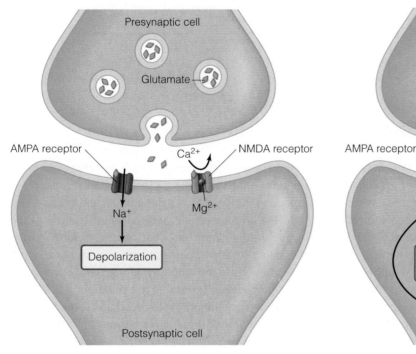

(a) Low-frequency stimulation

(b) High-frequency stimulation

Figure 7.32 Long-term potentiation in hippocampal neurons (a) Low-frequency stimulation of the presynaptic cell results in moderate release of glutamate. Glutamate released from the presynaptic cell binds to the AMPA and NMDA receptors on the postsynaptic cell. Na^+ enters through the AMPA receptor, causing depolarization, but the presence of Mg^{2+} in the NMDA receptor prevents Ca^{2+} from entering the cell. (b) High-frequency stimulation of the presynaptic cell results in greater release of glutamate. Glutamate binds to both receptor types on the postsynaptic cell. Increased glutamate causes increased Na^+ entry through AMPA receptors, causing a greater depolarization. This greater depolarization displaces Mg^{2+} from the NMDA receptor, allowing Ca^{2+} to enter the cell. The influx of Ca^{2+} activates protein kinases (CaMKII and PKC), phosphorylating the AMPA receptor, increasing its sensitivity to glutamate, and triggering the release of paracrine factors that cause the presynaptic cell to release more glutamate.

Ca^{2+} enters the postsynaptic cell via the NMDA receptor. The increase in intracellular calcium levels activates calcium-calmodulin-dependent protein kinase II (CaMKII) and protein kinase C (PKC), which phosphorylate a variety of proteins. For example, in CA1 cells CaMKII phosphorylates the AMPA receptor, making it more sensitive to glutamate, and also increases the number of AMPA receptors on the postsynaptic membrane by relocating receptors from intracellular stores. PKC activates a paracrine signaling pathway that causes the presynaptic cell to produce more glutamate. The net effect of these changes is more glutamate acting on more sensitive postsynaptic neurons, increasing the response to subsequent stimuli, and improving memory formation.

Transgenic mice have been used to test this mechanism of long-term potentiation and its relationship to memory formation. For example, transgenic mice that lack the CaMKII gene do not show long-term potentiation and have more trouble finding a hidden platform under murky water than do normal mice, while transgenic mice that produce too much CaMKII show greater long-term potentiation and perform better on hidden-platform tests and other tests of learning and memory. Similarly, transgenic mice that lack NMDA receptor expression in hippocampal neurons have more difficulty learning to find their way through a maze, or to find a hidden platform underwater. These results strongly indicate that long-term potentiation is involved in at least some kinds of memory formation in vertebrates.

◉ | CONCEPT CHECK

17. What is the difference between habituation and sensitization. Compare and contrast the mechanisms underlying these processes in the *Aplysia* tail withdrawal reflex.
18. What is long-term potentiation?
19. What kinds of evidence suggest that long-term potentiation is involved in learning and memory in mammals?

Integrating Systems | Stress and the Brain

The nervous system and the endocrine system work together to control and regulate the activity of essentially every physiological system. The stress response, which we first discussed in Chapter 3, provides an ideal example of this integration among multiple physiological systems.

Imagine that you are a baboon sitting beneath an acacia tree, calmly grooming one of the members of your troop. Suddenly, a hyena emerges from the long grass. Your nervous and endocrine systems leap into action, engaging the fight-or-flight response that activates the cardiovascular, respiratory, and musculoskeletal systems and causes you to flee. But how do these myriad physiological systems work together to perform this complex response? First, a baboon must perceive the stimulus, using its sense organs to see, hear, or smell the hyena. The stimulated sensory receptors then communicate this incoming sensory information to the brain in the form of action potentials conducted along primary afferent neurons. The brain integrates this information using two different pathways. In one pathway, the incoming sensory information travels from the thalamus to the sensory cortex where it is integrated. The brain concludes that this sensory information represents a hyena, and that hyenas are dangerous, and sends a signal to the limbic system, and more particularly to the amygdala—the seat of the fear response. At the same time, using a second pathway, the thalamus can send signals directly to the limbic system without any sophisticated processing, bypassing the sensory cortex. Thus, the baboon may initiate a fear response even before it concludes that the stimulus represents a hyena. For example, you may be startled and frightened by a sudden noise in the night, and a moment later realize that it was only the wind, but your heart will already be pounding, and you may have some trouble going back to sleep.

When activated, the amygdala sends a signal to the spinal cord, activating the autonomic nervous system. Within seconds, the postganglionic neurons of the sympathetic nervous system begin to release norepinephrine from their varicosities onto target tissues. Preganglionic sympathetic neurons that form synapses with the adrenal medulla release acetylcholine, causing the adrenal medulla to release the catecholamines (epinephrine and norepinephrine) into the circulatory system. The catecholamines from the adrenal medulla and the sympathetic nervous system bind to their receptors on many target tissues, increasing heart rate and the force of contraction, diverting blood flow from nonessential organs to the muscles, brain, heart, and lungs, and increasing blood pressure. Catecholamines dilate the bronchioles of the lungs and increase the rate and depth of breathing, readying the baboon for action. The sensory cortex also sends a signal to the motor cortex of the brain, making the decision that the best course of action when confronted with a hyena is to run away. The motor cortex then sends out a signal via motor neurons to the muscles, and the baboon runs up the tree, escaping from the hyena. Once the danger has passed, the sympathetic nervous system decreases its firing and the baboon begins to calm down.

At the same time that the amygdala is activating the sympathetic nervous system, another part of the limbic system—the hypothalamus—activates the endocrine system. The hypothalamus releases corticotropin-releasing hormone (CRH) into the hypothalamic-pituitary portal blood system, causing (perhaps five or ten seconds later) the pituitary to release adrenocorticotropic hormone (ACTH) into the circulation. The ACTH binds to receptors on the adrenal cortex (a part of the adrenal gland, surrounding the adrenal medulla). The adrenal cortex then releases glucocorticoid hormones, such as cortisol, into the blood. Cortisol plays a critical role in regulating carbohydrate and protein metabolism, causing the muscles to release amino acids and the liver to convert these amino acids to glucose and glycogen. Cortisol is a steroid hormone, and it mediates many of its actions by altering gene transcription in its target cells. Thus, cortisol typically acts fairly slowly, over the course of an hour or so—long after the baboon has escaped up the tree and returned to its normal activities. Because of the generally slow time-course of the cortisol response, the role of the glucocorticoid hormones in the immediate response to stress is not entirely understood. Cortisol may, however, be important in preparing an animal to respond to a subsequent stressor, or to recover from the previous one.

The fight-or-flight response is a vital survival tool that allows vertebrates to respond to stressful situations quickly and efficiently. But what happens if a stressful situation is prolonged? In the Serengeti, baboons encounter a wide variety of stressful situations, but one of their most important stressors may be social—as a result of interactions with other baboons. Baboons live in troops of between 20 and 200 individuals, with a complex social hierarchy. Female baboons inherit their position in this hierarchy—if a mother baboon is high-ranking, then her offspring will be of high rank as well. In contrast, male baboons fight for their position in the dominance hierarchy. Life can be rough for a low-ranking male baboon. High-ranking males pester low-ranking baboons and steal food from them, and the males can improve their rank only by fighting their way to the top. Thus, low-ranking baboons are chronically exposed to social stressors. In fact, low-ranking baboons have elevated levels of glucocorticoid hormones in their blood even under "resting" conditions, and their physiological response to a stressor differs from that of high-ranking baboons.

Chronic stress of this kind can have many deleterious effects, including a weakened immune system, elevated blood cholesterol levels, high blood pressure, and even impaired growth. In addition, chronic stress can affect the brain. In particular, chronic elevation of stress hormones interferes with long-term potentiation in the hippocampus—a process that is critically important in learning and memory. Indeed, long-term exposure to high levels of glucocorticoids can cause the hippocampus to atrophy, decreasing the total number of neurons in this area of the brain, eventually causing irreversible memory loss. So what is a low-ranking baboon to do? Researchers have shown that low-ranking baboons with strong social networks have reduced glucocorticoid levels compared with baboons with poor social networks. Perhaps making friends and having good relationships with siblings is an effective strategy to deal with the effects of chronic stress. ◉

Summary

Organization of Nervous Systems

→ Nervous systems consist of sensory pathways, integrating centers, and efferent pathways.

→ Cnidarians have relatively simple nerve nets with few obvious integrating centers, but over the course of evolution there has been a general trend toward cephalization, or the grouping of sensory organs and integrating centers at the anterior end of the body.

→ Integrating centers vary among the invertebrates but generally include one or more ganglia and a brain.

→ Vertebrate nervous systems have a clearly demarcated central nervous system, consisting of the brain and spinal cord, encased in either a cartilaginous or bony covering.

→ Additional integrating centers may be located in the peripheral nervous system in the form of ganglia.

→ The size of the brain varies greatly among major vertebrate groups, although the overall organization of the brain is similar among all vertebrates.

→ The three primary regions of the vertebrate brain are the forebrain, the midbrain, and the hindbrain.

→ The spinal cord acts as a reflex integrating center in the vertebrates.

→ The hindbrain controls essential functions, including breathing and heart rate.

→ The midbrain is greatly reduced in mammals relative to other vertebrates and acts largely as a relay station.

→ In mammals, the forebrain has taken over many of the sensory integration functions of the midbrain, and also controls more complex processes such as reasoning and the control of voluntary behavior.

→ The hypothalamus is a part of the forebrain that maintains homeostasis and helps to coordinate many aspects of the endocrine system.

→ The hypothalamus is part of the limbic system, a group of related structures that are involved in emotions and memory.

→ The cortex is the thin outer layer of the cerebral hemispheres, and is involved in integrating and interpreting sensory information.

→ The cerebral hemispheres are divided into lobes that are named based on the overlying bones.

→ The cerebral hemispheres can also be divided into a number of cortices and association areas that roughly correspond to functional regions of the brain.

→ Many of the cerebral cortices are organized topographically, with particular areas dealing with specific parts of the body.

The Peripheral Nervous System

→ The brain integrates and interprets sensory information and sends out signals via efferent pathways to effector organs.

→ Efferent pathways can be divided into the somatic motor division and the autonomic division.

→ Autonomic pathways are involved in maintaining homeostasis, and can be divided into the sympathetic, the parasympathetic, and the enteric nervous systems.

→ The sympathetic and parasympathetic systems work together, while the enteric nervous system is more autonomous.

→ The sympathetic nervous system is most active during periods of stress or physical activity, whereas the parasympathetic nervous system is most active during periods of rest.

→ Both the sympathetic and parasympathetic nervous systems exhibit basal tone, and are at least somewhat active under all conditions.

→ Sympathetic and parasympathetic pathways contain two neurons in series (a preganglionic and a postganglionic neuron).

→ In both the sympathetic and parasympathetic divisions, the preganglionic neuron releases acetylcholine.

→ Sympathetic postganglionic neurons release norepinephrine, whereas parasympathetic postganglionic neurons release acetylcholine.

→ Somatic (motor) pathways have only a single neuron between the spinal cord and the effector, and these neurons release only acetylcholine.

→ In addition, skeletal muscle is the only effector for motor pathways, whereas autonomic pathways innervate almost every organ in the body.

Integrative Functions of Nervous Systems

→ The afferent pathways, integrating centers, and efferent pathways of the nervous system work together to coordinate behavior and maintain physiological homeostasis.

→ Reflex arcs control many involuntary behaviors, and pattern generators initiate rhythmic behaviors, including apparently complex behaviors such as swimming in animals like leeches and locomotion in mammals.

→ Voluntary movements require coordination by more complex integrating centers such as the higher centers of the brain.

→ Animals are also able to modify their behavior based on experience.

→ The molecular basis of learning and memory has been worked out for a variety of model systems, and appears to involve changes in either presynaptic or postsynaptic neurons that are involved in reflex arcs, or in regions of the brain such as the hippocampus, which is important for memory formation in mammals.

Review Questions

1. What is a central pattern generator? Explain how a neural circuit can form a pattern generator.
2. What is the limbic system? How is it important in behavior?
3. Outline some of the ways in which the vertebrate brain can be subdivided.
4. Compare and contrast the somatic and autonomic nervous systems.
5. Would you expect the sympathetic or parasympathetic nervous system to be more active when you are (a) sitting quietly, (b) studying for an exam, (c) writing an exam?
6. Why is the autonomic nervous system sometimes termed the involuntary nervous system?
7. What is the importance of the phenomenon of basal tone in the autonomic nervous system?
8. Compare and contrast a neuron with a nerve. What is the difference between a nerve and a tract?

Synthesis Questions

1. You can surgically remove large parts of the forebrain from a mammal, and the animal will survive. However, destruction of even relatively small parts of the hindbrain usually causes death. Why might that be so?
2. What is the functional significance of the highly folded and grooved appearance of the surface of the brain in some mammals?
3. Nicotinic acetylcholine receptors (see Chapter 4) are found on muscle cells, and on postganglionic neurons in the sympathetic nervous system (among other places in the body). Use this information to explain why chewing nicotine-containing gum can cause a rapid heart rate and tremors in the hands of nonsmokers.
4. Would the autonomic nervous system function if the preganglionic neurotransmitters were different between the sympathetic and parasympathetic nervous systems, and the postganglionic neurotransmitters were the same?
5. Compare the role of presynaptic and postsynaptic mechanisms in habituation and sensitization.

For Further Reading

See the Additional References section at the back of the book for more readings related to the topics in this chapter.

Organization of Nervous Systems

The following is an engaging look at the scientists and the discoveries that shaped modern neuroscience and outlines the changing view of the brain from the time of the ancient Egyptians to the present.

Finger, S. 2000. *Minds behind the brain: A history of the pioneers and their discoveries.* New York: Oxford University Press.

This excellent and detailed book by Kandel covers a wide range of material in neuroscience, from the functions of single neurons to the functions of the nervous system as a whole.

Kandel, E. R. 2000. *Principles of neural science.* New York: McGraw-Hill.

This book provides an excellent review of how neurons work, and how these signals are integrated into the higher functions of the brain.

Nicholls, J. G., A. R. Martin, B. G. Wallace, and P. A. Fuchs. 2001. *From neuron to brain.* Sunderland, MA: Sinauer Associates.

Evolution of Nervous Systems

The following interesting book, although highly speculative, reviews the evolution of the brain and particularly addresses the question of the costs and benefits of having a large brain.

Allman, J. M. 1999. *Evolving brains.* New York: Scientific American Library/W. H. Freeman.

These reviews discuss the structure and evolution of the brain in vertebrates and invertebrates.

Ghysen, A. 2003. The origin and evolution of the nervous system. *International Journal of Developmental Biology* 47: 555–562.

Northcutt, R. G. 2002. Understanding vertebrate brain evolution. *Integrative and Comparative Biology* 42: 743–756.

These comprehensive reviews outline some of the complexities of the seemingly simple nervous systems of cnidarians.

Grimmelikhuijzen, C. J., and J. A. Westfall. 1995. The nervous systems of cnidarians. *EXS* 72: 7–24.

Mackie, G. O. 2004. Central neural circuitry in the jellyfish *Aglantha:* A model "simple nervous system." *Neurosignals* 13: 5–19.

Structure of the Mammalian Brain

This fascinating book discusses many important topics relating to the brain, including brain evolution, brain structure, and the nature of consciousness.

Bownds, M. D. 1999. *The biology of mind: Origins and structures of mind, brain, and consciousness.* Toronto: Wiley.

This article provides a clear and readable introduction to the structure of mammalian brains.

Nauta, W. J. H., and M. Feirtag. 1979. The organization of the brain. *Scientific American* 241: 88–111.

This comprehensive review discusses the evolution and development of the brain and tackles the complex issue of the nomenclature for the parts of the central nervous system.

Swanson, L. 2000. What is the brain? *Trends in Neuroscience* 23: 519–527.

The Peripheral Nervous System

This comprehensive review article and short book provide an excellent overview of the autonomic nervous system.

Donald, J. A. 1998. Autonomic nervous system. In *The physiology of fishes,* 2nd ed., D. H. Evans, ed., 407–493. Boca Raton, FL: CRC Press.

Robertson, D., P. A. Low, and R. J. Polinsky. 1996. *Primer on the autonomic nervous system.* San Diego: Academic Press.

Integrative Functions of Nervous Systems

This review provides an in-depth discussion of the neural circuitry involved in leech swimming behavior.

Brodfuehrer, P. D., and M. S. Thorogood. 2001. Identified neurons and leech swimming behavior. *Progress in Neurobiology* 63: 371–381.

This review by Dietz discusses the control of locomotion in the vertebrates.

Dietz, V. 2003. Spinal cord pattern generators for locomotion. *Clinical Neurophysiology* 114: 1379–1389.

Dr. Eric Kandel was awarded the Nobel Prize in Physiology or Medicine in 2000 for "discoveries concerning signal transduction in the nervous system." This paper is the published version of his lecture at the Nobel Prize ceremonies.

Kandel, E. R. 2001. The molecular biology of memory storage: A dialog between genes and synapses. *Bioscience Reports* 21: 565–611.

Stress and the Brain

These engaging and highly entertaining books outline the effects of stress hormones on the brain of mammals.

Sapolsky, R. M. 1992. *Stress, the aging brain, and the mechanisms of neuron death.* Cambridge, MA: MIT Press.

Sapolsky, R. M. 2004. *Why zebras don't get ulcers,* 3rd ed. New York: Owl Books.

8

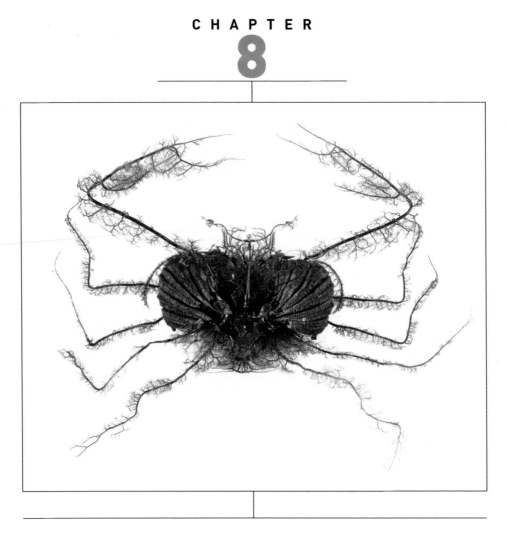

Circulatory Systems

Animal circulatory systems are structurally diverse, ranging in complexity from the relatively simple circulatory systems of insects to the highly branched circulatory systems of animals such as decapod crustaceans and vertebrates. Despite this structural diversity, animal circulatory systems share many features in common. Circulatory systems transport oxygen and nutrients to actively metabolizing tissues, and remove carbon dioxide and other waste products. They help to coordinate physiological processes by transporting signaling molecules from place to place within the body, and they assist in the defense of the body by transporting immune cells to the site of invasion by foreign organisms.

Although we now take this transport function for granted, the structure and function of circulatory systems remained obscure for many centuries. The correct pathway for the circulation of blood in mammals was first described by the 11th-century Muslim physician and theologian Ibn al Nafis. However, this insight was lost to the Western world, and it was not rediscovered until the early 17th century when the careful experimental work of William Harvey demonstrated that the heart circulates blood through the blood vessels to various parts of the body. In Harvey's words: "It is absolutely necessary to conclude that the blood in the animal body is impelled in a circle, and is in a state of ceaseless motion." We know now

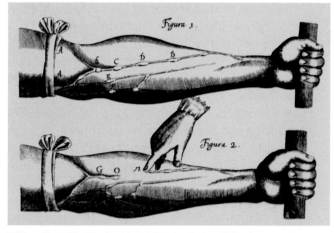

William Harvey's experiments on the circulation of blood.

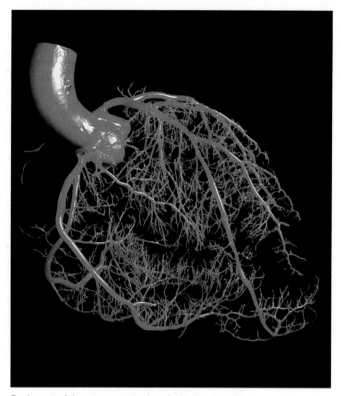

Resin cast of the coronary arteries of a human heart.

that the heart pumps blood through the arterial system and into the capillaries where exchange of materials with the tissues takes place. The blood then returns to the heart via the venous system. Because good compound microscopes were not available in the early 17th century, Harvey was unable to directly observe the capillaries that connect arteries and veins; however, he hypothesized their existence based on the results of his experiments. Three decades later, and four years after Harvey's death, the Italian physician and anatomist Marcello Malpighi used the newly available compound microscopes to identify tiny blood vessels in the lungs and kidneys, which he named capillaries. This discovery completed and confirmed the pioneering work of Harvey, and set the stage for our modern understanding of circulatory systems.

Consistent with the importance of the circulatory system, the heart is one of the first organs to form in a developing vertebrate embryo. For example, in zebrafish (*Danio rerio*, a common model system used by developmental biologists), the heart forms and begins to beat rhythmically during the first day following the fertilization of the egg. However, day-old zebrafish embryos are tiny, suggesting that diffusion should be more than adequate to deliver oxygen and nutrients to the tissues. Indeed, a variety of experiments have shown that zebrafish do not require a functioning circulatory system for oxygen transport during early development. But if the circulatory system is not needed for oxygen transport, why does the heart begin beating so early? Developmental physiologists have pro-

vided a partial answer to this puzzle. They have shown that the beating of the heart is critical for the appropriate development of the circulatory system.

By implanting a tiny bead at either the entrance or exit of the developing heart, researchers were able to reduce the flow of blood in the circulatory system. Blocking the flow of blood through the heart interfered with its development; the chambers of the heart were out of alignment or did not form properly, and the heart valves did not develop at all. Similarly, other researchers have shown that the force imposed by blood pulsating at the ends of developing blood vessels causes the vessels to grow and sprout, helping to form the circulatory system. Some of these processes continue to occur in adult animals, demonstrating that a circulatory system is not simply a passive plumbing system that transports substances around the body. Instead, circulatory systems are dynamic physiological systems whose structure and function are regulated in response to the ever-changing demands of the body's tissues.⊙

Overview

Unicellular organisms and some small metazoans lack circulatory systems and instead rely on diffusion to transport molecules from place to place. Although diffusion can be rapid over short distances (such as across a cell membrane or within a single cell), it is slow across long distances (Figure 8.1a). In fact, the time (t) needed for a molecule to diffuse between two points is proportional to the square of the distance (x) over which diffusion occurs ($t \propto x^2$). This relationship is a simplified form of Einstein's diffusion equation (which is also called the second law of diffusion). Using this equation, we can calculate that at 37°C a small molecule such as glucose in aqueous solution takes about 5 seconds to diffuse across 100 µm (the size of an average cell) but would take more than 60 years to diffuse across several meters (the distance from the heart to the feet and back again in an average-size human).

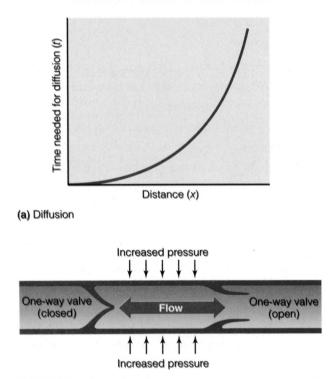

(a) Diffusion

(b) Bulk flow

Figure 8.1 **Diffusion and bulk flow** **(a)** Diffusion is rapid over short distances, but the time needed for diffusion increases exponentially with distance. To transport substances rapidly across long distances, animals use the bulk flow of fluids. **(b)** Increased local pressure in one area of the circulatory system drives flow from the area of high pressure to any adjacent areas of lower pressure, a phenomenon known as bulk flow. One-way valves ensure that this flow is unidirectional.

Because of this limitation on the rate of diffusion, larger animals move fluids through their bodies by a process called **bulk flow**, or *convective transport*. The bulk flow of fluids can transport substances across long distances far faster than would be possible by diffusion alone. For example, the human circulatory system can move a milliliter of blood from the heart to the feet and back again in about 60 seconds, rather than the 60 years needed for diffusion! The phenomenon of bulk flow is fundamental to many physiological processes, including respiration, digestion, and excretion.

As stated in Newton's second law of motion (force = mass × acceleration), if we exert sufficient force on an object, it will start moving (or accelerate, if it is already in motion). Thus, bulk flow of a fluid occurs when an external force is applied to the fluid, setting it in motion. In circulatory systems, the fluid is confined within a series of chambers and tubes (Figure 8.1b). By pressing down on this confined fluid, you increase the pressure in the immediate area. The fluid then flows from this area of high pressure to any adjacent areas of lower pressure. In many circulatory systems, one-way valves help to ensure that the fluid flows in one direction around the system. In this chapter, we examine the structure, function, and regulation of animal circulatory systems, in all their diversity, to see how they use bulk flow to perform their critical transport function.

Characteristics of Circulatory Systems

All animal circulatory systems play similar roles, carrying oxygen, carbon dioxide, nutrients, waste products, immune cells, and signaling molecules from one part of an animal to another. In some animals, the circulatory system even plays a role in temperature regulation, by conveying heat from the working muscles out to the surface of the body where it can be lost to the environment. Because of its important transport role, the circulatory system affects almost every physiological process that an animal performs.

Components of Circulatory Systems

Like all pumping systems, animal circulatory systems have three important components:

1. One or more pumps or other propulsive structures that apply a force to drive fluid flow, often in combination with one-way valves to ensure unidirectional flow

2. A system of tubes, channels, or other spaces through which the fluid can flow

3. A fluid that circulates through the system

There is, however, substantial diversity among animals in the structure and organization of each of these components.

Circulatory systems use diverse pumping structures

All circulatory systems have some type of pumping structure that propels fluids around the system. We are most familiar with the pumping action of contractile chambers such as the vertebrate **heart** (Figure 8.2). Chambered hearts are found in both vertebrates and invertebrates. Muscular contraction of the heart increases the pressure within the heart chambers. When the pressure in the heart exceeds that in the rest of the circulatory system, blood flows down this pressure gradient out into the circulatory system. One-way valves help to ensure unidirectional flow.

Chambered hearts often have more than one chamber. The chambers that the circulatory fluid first enters are typically called **atria**. Animal hearts may have one or more atria, and these chambers function both as reservoirs and as pumps. Fluid flows from the atria into an even more muscular chamber, called the **ventricle**, which acts as the primary pump.

Chambered hearts are not the only type of pumping structures found in animal circulatory systems. Organs that are not strictly associated with the circulatory system, such as skeletal muscles, can be used to develop pressure gradients (Figure 8.2b). For example, in terrestrial vertebrates the actions of the leg muscles help to push blood back to the heart. Similarly, in many arthropods, normal body movements propel blood around the body. In these systems, the blood vessels generally contain one-way valves to maintain the unidirectional flow of the circulatory fluid.

Pulsating or contractile blood vessels and tubelike hearts, which are found in some invertebrates and the early embryos of vertebrates, move blood by **peristalsis** (Figure 8.2c). Peristaltic contractions are rhythmic waves of muscle contrac-

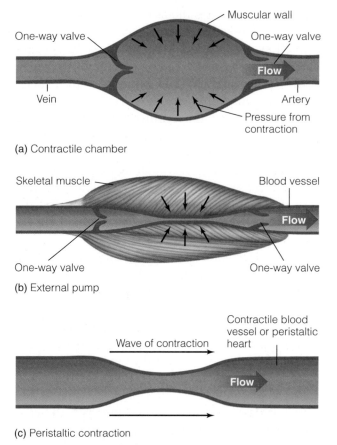

(a) Contractile chamber

(b) External pump

(c) Peristaltic contraction

Figure 8.2 Types of pumping structures in animal circulatory systems **(a)** Contractile chambers such as the vertebrate heart increase blood pressure in a closed chamber through contractions of their muscular walls. As pressure increases, valves open allowing fluid to flow down the resulting pressure gradient. One-way valves are required to ensure unidirectional flow. **(b)** Structures such as skeletal muscles can act as pumps. Contraction and relaxation of skeletal muscles alternatively compress and expand a blood vessel, forcing the fluid along the vessel. One-way valves ensure unidirectional flow. **(c)** Contractile blood vessels and peristaltic hearts push blood using waves of rhythmic contraction. These vessels may contain valves to ensure unidirectional flow, but the direction of contraction is often sufficient to cause flow to be largely unidirectional.

tion that proceed in a coordinated fashion from one end of a tube to the other. Similar to squeezing toothpaste from its tube, peristaltic contractions squeeze blood through the pumping structure and into the circulatory system. Because peristaltic contractions usually occur in a specific direction, these pumps can cause unidirectional flow even when no valves are present.

Circulatory systems can be open or closed

Circulatory fluids flow either through enclosed **blood vessels** that have walls with a specialized lining that separates the circulatory fluid from the tissues, or through open spaces called *sinuses* that allow the circulatory fluid to make direct contact with the tissues. In a **closed circulatory system**, the circulatory fluid remains within blood vessels at all points in the circulatory system. Thus, substances must diffuse across the walls of the blood vessels to enter the tissues in animals with closed circulatory systems. In an **open circulatory system**, the circulating fluid enters a sinus at least at one point in the circulatory system and thus comes into direct contact with the tissues, allowing the circulating fluid to mix with extracellular fluids.

Open circulatory systems usually contain both blood vessels and sinuses, and sinuses can have complex, highly branched structures. As a result, the difference between open and closed circulatory systems is not absolute. For example, the circulatory systems of decapod crustaceans such as crabs and lobsters are usually described as open, because they contain sinuses. However, these animals have some very small blood vessels across which diffusion to the tissue occurs, as is the case in a closed circulatory system, and in the sinuses blood flows through well-defined channels within the tissues as it returns to the heart. So on a functional basis, we might classify decapod circulatory systems as closed, although based on their structure, they are open systems.

Circulatory systems pump several types of fluids

There is some disagreement among comparative physiologists about the terminology that should be used for circulatory fluids, but for the purposes of this book we distinguish several major types of fluids. We use the term **interstitial fluid** for the extracellular fluid that directly bathes the tissues of either vertebrates or invertebrates. Even animals that lack a specialized circulatory system are usually able to propel interstitial fluid around their bodies by bulk flow. We define **blood** as the fluid that circulates within a closed circulatory system, such as that of a vertebrate. Blood is a complex tissue that has multiple components. It contains proteins and a variety of cells suspended in a fluid called **plasma**. We discuss the composition of blood in more detail at the end of this chapter.

Most vertebrates have a secondary circulatory system, in addition to the cardiovascular system,

that circulates a fluid called **lymph** around the body. Lymph is formed from blood by a process called **ultrafiltration** in the small blood vessels. The pressure difference across the walls of the small blood vessels forces fluid out of the blood and into the interstitial space where it mixes with the interstitial fluid. Blood cells and large molecules such as dissolved proteins cannot pass across the walls of most of the small blood vessels, but these walls are quite permeable to small molecules and water. The walls of the small blood vessels thus act as a filter, forming a lymphatic fluid that is similar in composition to plasma, but contains few proteins or cells. The **lymphatic system** pumps this ultrafiltrate through the body and returns it to the circulatory system.

Many fishes have an additional secondary circulation that is distinct from both the primary circulatory system and the lymphatic system. This secondary circulation is found in the gills and the skin of the fish, and may serve an osmoregulatory function. Like lymph, the fluid in the secondary circulatory system of fish is derived from blood, but it is not formed by ultrafiltration across the walls of the small blood vessels. Instead, fluid enters the secondary circulatory system through openings between the two circulatory systems. These openings allow plasma, proteins, and some cells to enter the secondary circulation. Thus, the fluid in this secondary circulation is very similar to blood, except that it has a lower concentration of blood cells.

Hemolymph is the circulating fluid of open circulatory systems. In an open circulatory system, hemolymph flows through blood vessels, but when it enters the sinuses it directly contacts the tissues, and thus is continuous with the interstitial fluid. As a result, it is difficult to distinguish between blood, lymph, and interstitial fluid in these organisms. Indeed, the word *hemolymph* was coined to imply this combination of blood and lymph (*hema* is the Greek root for blood). The sinuses of open circulatory systems are sometimes referred to collectively as the **hemocoel**.

Diversity of Circulatory Systems

There is substantial diversity in the structure of circulatory systems among animals. Animals such as sponges, cnidarians, and flatworms lack a circulatory system that transports an internal fluid, but all

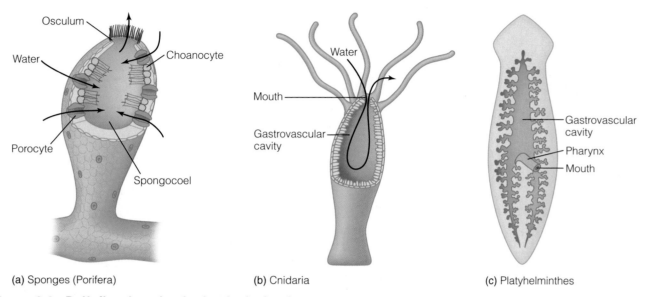

(a) Sponges (Porifera)   (b) Cnidaria   (c) Platyhelminthes

Figure 8.3 Bulk flow in animals that lack circulatory systems **(a)** The body wall of a sponge is full of pores that lead into an inner cavity called the spongocoel. The beating of flagellated choanocytes propels water through the pores into the spongocoel and out the osculum. **(b)** Cnidarians use muscular contractions to propel water into the mouth and through the gastrovascular cavity. **(c)** Platyhelminths and nematodes use contractions of a muscular pharynx to propel fluid through their gastrovascular cavity.

of these animals have mechanisms for propelling fluids around their bodies (Figure 8.3). For example, sponges propel water through their bodies using *choanocytes*, specialized cells with rhythmically beating flagellae. Cnidarians propel water from the external medium through their mouths into a **gastrovascular cavity** using muscular contractions, and pump the water down to their tentacles, carrying oxygen and digested food along with it. Flatworms also have a gastrovascular cavity, which in many species is lined with ciliated *flame cells* whose beating propels water containing food particles to all parts of the body. In all these species, the bulk flow of fluids is part of a combined respiratory, digestive, and circulatory system.

Nematodes (phylum Nematoda) and horsehair worms (phylum Nematomorpha) also lack specialized circulatory systems, but they can move interstitial fluid through their body cavity (called a *pseudocoelom*) by bulk flow powered by contractions of the muscles in their body walls. Nematodes and horsehair worms are seldom more than a millimeter thick (although some species can be up to 30 m long), and they obtain oxygen by diffusion across the entire body surface. As a result, these animals probably have little need for a circulatory system to transport oxygen. Instead, bulk flow of interstitial fluid is most important for transporting signaling molecules and immune cells.

Most annelids have closed circulatory systems

Phylum Annelida is divided into three main branches: class Polychaeta (e.g., tube worms), class Oligochaeta (e.g., earthworms), and class Hirudinea (leeches). The circulatory systems of leeches are different from those of the other annelids, and we do not discuss them further here. All polychaetes and oligochaetes are able to circulate interstitial fluid using either cilia or muscular contractions of the body wall. Some polychaetes rely solely on this system, but most polychaetes and oligochaetes have a system of blood vessels that circulates a specialized fluid containing oxygen carrier proteins. This system may have an open design, as in some polychaetes (Figure 8.4a), but the majority of annelids have closed circulatory systems that circulate blood through the body (Figure 8.4b).

Oligochaetes such as earthworms have a series of small blood vessels connecting the large dorsal and ventral blood vessels that run the length of the animal. The dorsal vessel is contractile, and moves blood toward the head using rhythmic waves of peristaltic contraction. The blood then flows through five pairs of muscular contractile tubes (or simple tubelike hearts) that pump blood from the dorsal to the ventral blood vessel. The blood travels back along the body

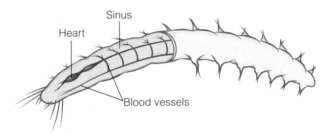

(a) Open circulatory system of annelid (polychaete)

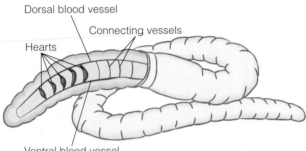

(b) Closed circulatory system of annelid (oligochaete)

Figure 8.4 Circulatory systems of annelid
(a) Some polychaetes have open circulatory systems.
(b) Oligochaetes have closed circulatory systems.

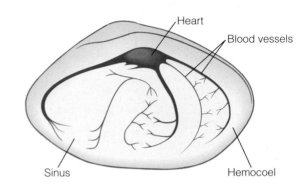

(a) Open circulatory system of a bivalve mollusc (clam)

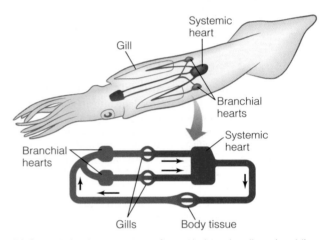

(b) Closed circulatory system of a cephalopod mollusc (squid)

Figure 8.5 Circulatory systems of molluscs
(a) The circulatory system of a bivalve such as a clam. Most molluscs have open circulatory systems. **(b)** The circulatory system of a cephalopod mollusc (squid). Most cephalopods have closed circulatory systems. The systemic heart pumps oxygenated blood to the body. The branchial hearts pump deoxygenated blood from the body through the gills.

through the ventral blood vessel. Small connecting blood vessels carry the blood from the tissues back to the dorsal vessel.

Most molluscs have open circulatory systems

The circulatory systems of molluscs are extremely diverse, consistent with the enormous diversity in body form within this phylum. All molluscs have hearts or contractile organs of some sort, and most groups have at least some blood vessels, with some species having extensive vascular networks. However, almost all molluscs have open circulatory systems (Figure 8.5). Only the cephalopods (squid, octopus, and cuttlefish) have completely closed circulatory systems.

The closed circulatory system of cephalopods evolved from an open circulatory system, likely one similar to those in ancient cephalopods such as the *Nautilus*. In nautiloids, blood returning from the gills enters the atria of the heart, and then is pumped by the ventricle through blood vessels that empty into a large sinus. Contractile blood vessels then pump blood across the gills and back to the heart. In contrast, squid and octopuses have a

closed circulatory system and three muscular chambered hearts (Figure 8.5b). The *systemic heart* pumps oxygenated blood to the body. After passing through the body tissues, the deoxygenated blood flows into the two *branchial hearts* that pump blood through the gills. From the gills, the oxygenated blood flows back into the systemic heart.

All arthropods have open circulatory systems

Almost all arthropods have one or more hearts and at least some blood vessels, but no arthropod lineages have evolved a completely closed circulatory system. Here we consider two of the major arthropod lineages: the crustaceans and the insects.

The circulatory systems of crustaceans vary from quite simple in smaller and less active species

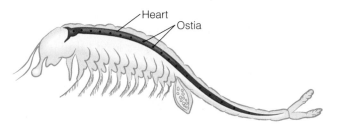

(a) Brachiopod crustacean (fairy shrimp)

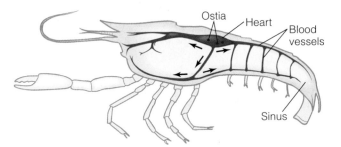

(b) Decapod crustacean (crayfish)

Figure 8.6   Circulatory systems in crustaceans
(a) Circulation in a brachiopod crustacean. Brachiopods such as fairy shrimp have simple circulatory systems with few blood vessels and a long tubular heart. **(b)** Circulation in a decapod crustacean. Decapod crustaceans have elaborate open circulatory systems with arteries and capillary beds and a muscular chamberlike heart. The heart pumps the circulatory fluid through the arteries into successively smaller blood vessels that drain into small channels within the head and body tissues. The fluid returns to the heart via a set of ostia.

to extremely complex in large, active species (Figure 8.6). The brachiopod crustaceans such as the fairy shrimp (also known as "sea monkeys" to generations of North American children) have a simple tubular heart that may extend almost the entire length of the body, and relatively few blood vessels. In contrast, decapod crustaceans such as lobsters, crabs, and crayfish have a very muscular heart that acts as a contractile chamber, and an extensive network of blood vessels (Figure 8.6b). These animals have a single heart encased in a sac called the *pericardial sinus*. Several branching arteries lead out of the heart to many parts of the body, ultimately emptying out into sinuses deep within the tissues. After passing through the tissues, the blood drains into a sinus located along the ventral side of the body. This sinus leads to the gills, where the blood is reoxygenated prior to its return to the heart. The blood passes into veins that empty into the pericardial sinus, entering the heart via small holes called *ostia* that can be opened and closed to regulate blood flow.

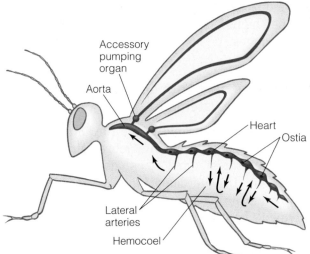

Figure 8.7   Circulatory system of insects
Insects have relatively simple open circulatory systems. The contractile dorsal blood vessel is elaborated into a series of hearts found along the body, often with one in each body segment. These hearts and the contractile dorsal blood vessel push blood using peristaltic contractions from the posterior end to the anterior end of the body. The circulatory fluid then discharges into the open hemocoel and percolates back through the sinuses of the body, assisted by normal body movements.

Decapods have among the most sophisticated open circulatory systems of any invertebrate, and many of their blood vessels have muscular valves that they can use to control the amount of blood flowing to particular tissues. The sinuses are very small in some species, and act functionally as blood vessels. Thus, although crustacean circulatory systems are structurally open, they may act as functionally closed systems.

In many insects the only obvious structure in the circulatory system is a large dorsal vessel that extends along most of the body (Figure 8.7). The posterior part of the dorsal vessel is contractile and is often divided into several discrete pumping organs that function as hearts, one per abdominal segment. The anterior part of the dorsal vessel is less muscular and is termed the **aorta**. The contractions of the hearts pump hemolymph toward the head. The hemolymph empties into a sinus in the region of the brain, and then percolates back to the abdomen, via another sinus. Normal body movements help to move the hemolymph through the sinuses, returning the blood to the heart via ostia, as in other arthropods. Many insects also have accessory pulsatile organs (simple hearts) in their antennae, wings, and limbs. In fact, some species have dozens of these small hearts, which help to propel hemolymph through their long, narrow appendages.

Chordates have both open and closed circulatory systems

The vertebrates belong to the phylum Chordata, which also contains the invertebrate urochordates (the tunicates) and cephalochordates (the lancelets). Urochordates have a simple tubular heart that propels fluid through a series of well-defined channels in the tissues. These channels lack walls, so the urochordate circulatory system is classified as open. The heart is located at the base of the digestive tract in the posterior part of the body and pumps fluid through the body using peristaltic contractions. In some tunicates such as *Ciona*, the direction of these contractions reverses periodically, causing the direction of blood flow to reverse. The physiological significance of this flow pattern is not yet understood, although some authors have suggested that it serves to disperse nutrient-gathering cells around the body.

Cephalochordates such as the lancelet (formerly called *Amphioxus*) lack an obvious chambered heart and instead have a long tubular heart or contractile blood vessel located at the base of the digestive tract and additional pulsatile blood vessels in other locations within the circulatory system that assist in pumping blood through the circulatory system. The circulatory system is largely closed, with blood vessels emptying into sinuses in only a few locations in the body.

Vertebrates have closed circulatory systems in which the blood remains within blood vessels at all points in its passage through the body.

Closed circulatory systems evolved multiple times in animals

From the examples outlined above, it is clear that there is substantial diversity in the structure and organization of animal circulatory systems, and that there are many alternate evolutionary solutions to the problem of moving fluids around the body by bulk flow. Figure 8.8 summarizes the properties of the circulatory systems of the major

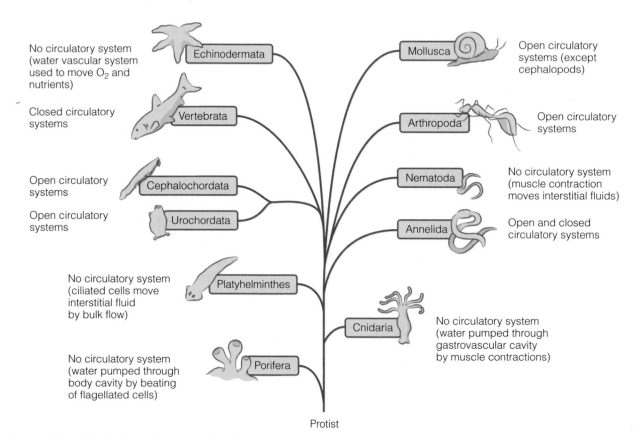

Figure 8.8 Evolution of animal circulatory systems

animal groups. Most systematists agree that animals evolved from flagellated protists resembling modern choanoflagellates. These small unicellular organisms lack circulatory systems, and rely on diffusion to transport substances through their bodies. Circulatory systems are thought to have first evolved to transport nutrients and other small molecules around the body, but very early in the evolution of animals the circulatory system began to serve a respiratory function, helping to transport oxygen to the actively metabolizing tissues. In most animal groups, this respiratory function has been a major force shaping the evolution of circulatory systems.

Although the earliest animal groups lack circulatory systems, most animals have them. Open systems are present in at least some representatives of most animal groups. Closed circulatory systems evolved independently from these ancestral open circulatory systems in several lineages of animals, including the vertebrates, cephalopod molluscs, and annelid worms. These closed circulatory systems differ in structure but are functionally similar, and are thus examples of convergent evolution. Closed circulatory systems provide several advantages over open circulatory systems, including the ability to generate high pressure and flow and the ability to better control and direct blood flow to specific tissues. These features are particularly important for oxygen delivery to actively metabolizing tissues. Consistent with this expectation, closed circulatory systems are usually found in highly active organisms with high demands for oxygen, or those living in oxygen-limited environments where oxygen supply is low.

Although insects can have extremely high metabolic rates, they have relatively simple open circulatory systems (see Figure 8.7). This pattern is in direct contrast to that observed in other groups in which closed or nearly closed circulatory systems are associated with highly active lifestyles. However, insects do not use the circulatory system as their primary means of gas transport. Instead, as we discuss in Chapter 9: Respiratory Systems, insects have a *tracheal system* that consists of a series of blind-ended air-filled tubes that conduct oxygen directly to the tissues in gaseous form, bypassing the circulatory system. Thus, in insects the circulatory system serves primarily to deliver nutrients, immune cells, and signaling molecules, rather than being critical for oxygen delivery, and high flow rates and pressure may not be required.

The Circulatory Plan of Vertebrates

Jawed vertebrates share a common circulatory plan (Figure 8.9), with a primary systemic heart that pumps blood to a large blood vessel termed an **artery**. The word *artery* is the general term for blood vessels that carry blood away from the heart. The large artery leading from the heart to the body is termed the *aorta*. The aorta branches into succeedingly smaller arteries, culminating in the *feed arteries* that lead to the tissues. Within the tissues, the arteries branch into **arterioles** that direct flow into the *capillary beds*. Capillary beds are made up of dense networks of thin-walled vessels called **capillaries**, which are the primary site of diffusion of materials into the tissues. At the end of the capillary beds, capillaries coalesce into small vessels called **venules**, which in turn coalesce into larger vessels called **veins** that return blood to the heart.

Although this general circulatory plan provides a good overview of the route of blood through the vertebrate circulatory system, actual circulatory systems are rather more complex. For example, arteries do not always simply branch to form progressively smaller vessels. Arteries can also form **anastomoses** (singular, anastomosis), which are connections from one blood vessel to another. Anastomoses provide an alternate pathway for blood to flow if one route is blocked. For example, the arteries in the joints contain numerous anastomoses, allowing blood to flow even if the bending of the joint closes off one of the arteries. Anastomoses become more frequent the farther you get

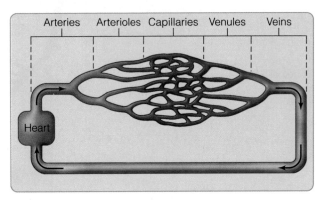

Figure 8.9 The vertebrate circulatory plan
Vertebrates share a common circulatory plan in which the heart pumps blood to a large artery, then through succeedingly smaller arteries to the arterioles that lead to the capillary beds, where substances diffuse to the tissues across the walls of the capillaries. Capillaries coalesce into venules and then veins, which return blood to the heart.

from the heart so that arterioles and capillaries tend to form dense interconnected networks. In addition, venous shunts and arterio-venous anastomoses allow blood to be redirected to avoid a particular capillary bed if necessary.

Vertebrate blood vessels have complex walls

Vertebrate blood vessels are hollow tubular structures consisting of a complex wall surrounding a central **lumen**. In the vertebrates, the walls of blood vessels are composed of up to three layers (Figure 8.10). The innermost layer of the blood vessel is the **tunica intima** or *tunica interna*. It consists of an inner lining called the **vascular endothelium**, made up of a smooth sheet of epithelial cells, and a basement membrane called the subendothelial layer, which supports the vascular endothelium. The *tunica media*, or middle layer, of a blood vessel is largely composed of smooth muscle and sheets of the extracellular matrix protein elastin that wrap around the tunica intima. Contraction and relaxation of the smooth muscle of the tunica media causes vasoconstriction and vasodilation. The outermost layer of the blood vessel wall is called the *tunica externa*, or *tunica adventitia*, and is composed largely of collagen fibers that support and reinforce the blood vessel.

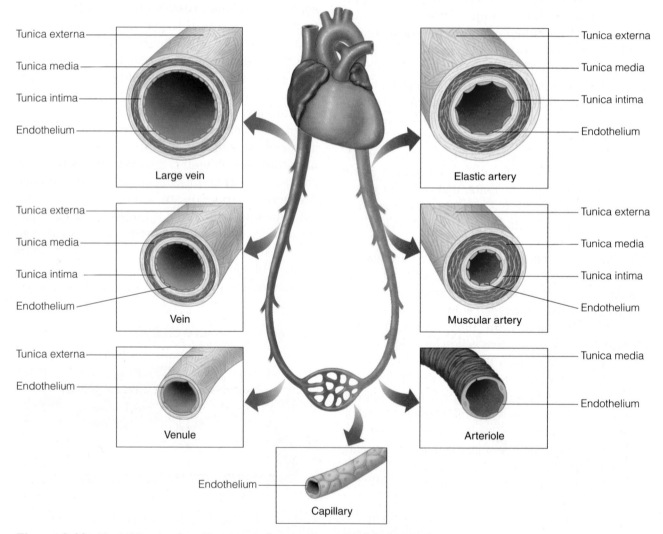

Figure 8.10 Variation in the structure of vertebrate blood vessels
Representative portions of blood vessels from the systemic circuit of a mammalian circulatory system are shown in cross-section. Arteries and veins are composed of three layers (the tunica externa, tunica media, and tunica intima) of varying thickness, lined with an endothelium. Smaller vessels such as arterioles, capillaries, and venules lack one or more of these layers.

Wall thickness varies among blood vessels

The thickness of the layers of the vessel walls varies greatly among types of blood vessels. Arteries are large-diameter, thick-walled blood vessels with a thick tunica externa and tunica media. The arteries closest to the heart have a particularly thick tunica externa, which makes them highly elastic. Arteries farther from the heart tend to have a thicker tunica media, and are sometimes called muscular arteries. Arterioles have thinner walls and lack an extensive tunica externa. Larger arterioles have a relatively extensive tunica media, composed of thick layers of smooth muscle, but in the smallest arterioles, the tunica media consists of a single layer of smooth muscle arranged in a spiral pattern around the endothelium. The smooth muscle cells allow the arterioles to vasoconstrict and vasodilate.

Capillaries lack the tunica media and tunica externa and have extremely thin walls composed of a single sheet of endothelial cells, wrapped in an occasional contractile *pericyte cell*. These thin walls allow substances to pass between the blood and the tissues. Substances can move across the capillary walls in several ways. Lipid-soluble substances can move across the cell membrane by simple diffusion. Vesicles transport large water-soluble substances such as proteins across the cell in a process called *transcytosis*. Small molecules such as water and ions can move across the capillary wall via a **paracellular pathway**, through pores between the cells of the capillary wall. Capillaries have very small diameters, and are often just large enough for blood cells to squeeze through.

The structure of the tunica intima varies among capillaries (Figure 8.11). The cells of the vascular endothelium of capillaries are held together with tight junctions. As we discussed in Chapter 7: Functional Organization of Nervous Systems, the capillaries of the central nervous system are particularly tightly joined, allowing few molecules to pass; this forms the blood-brain barrier. *Continuous capillaries* are found in the skin and muscle. The seal between the cells of a continuous capillary is not usually complete, leaving areas of unjoined membrane that allow fluids and small molecules to pass from the blood to the interstitial fluid. *Fenestrated capillaries* are similar to continuous capillaries except that the cells of the vascular endothelium contain numerous pores covered with a thin diaphragm. Small molecules

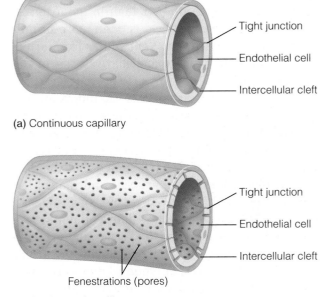

(a) Continuous capillary

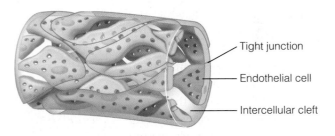

Fenestrations (pores)

(b) Fenestrated capillary

(c) Sinusoidal capillary

Figure 8.11   Variation in capillary structure
(a) In a continuous capillary, the endothelial cells are connected via tight junctions. **(b)** In a fenestrated capillary, the endothelial cells have many oval pores (fenestrations) that allow the regulated movement of solutes. **(c)** In a sinusoidal capillary, the endothelial cells are loosely linked and large molecules can move between the cells.

and fluids can pass easily through these pores, and thus fenestrated capillaries are found in areas of the body that are specialized for the exchange of substances, such as parts of the kidney, the endocrine organs, and the intestine. *Sinusoidal capillaries* are the most porous of all capillaries, and are found only in very specialized organs such as the liver and bone marrow. They have fewer tight junctions and more spaces between the cells. This structure allows large proteins to move across the capillary wall.

Capillaries empty into venules, which lead into the veins that return blood to the heart. A vein usually has a thinner wall and larger lumen than a similarly sized artery. As a result, veins can be easily stretched. In particular, the tunica media of the veins is much thinner than in the arteries. However, the tunica externa is often more prominent than in the arteries. Veins differ from arteries in that some veins (particularly those in the limbs) contain one-way valves to prevent backflow of blood. The valves are part of the tunica intima. For many years, physiologists assumed that the tunica intima was structurally similar in arteries and veins, differing only in thickness. But recent studies using zebrafish have shown that the vascular endothelium of the arteries and that of the veins express a different subset of genes, suggesting that they are functionally differentiated.

Note that we distinguish arteries and veins by whether they carry blood that is flowing toward or away from the heart, not whether they carry oxygenated or deoxygenated blood. For example, the *pulmonary artery* of mammals, which leads from the heart to the lungs, carries deoxygenated blood, while the *pulmonary vein*, which leads from the lungs to the heart, carries oxygenated blood. In contrast, the aorta carries oxygenated blood, while the *venae cavae* (the large veins leading from the body to the heart) carry deoxygenated blood.

Blood vessels undergo angiogenesis

During the embryonic development of vertebrates, the major vessels of the circulatory system grow into a network of arteries, arterioles, capillaries, venules, and veins, which remains fairly stable throughout adult life. Despite this overall stability, however, the minor vessels undergo constant remodeling throughout life, a process called **angiogenesis**. For example, as we will see in Chapter 14: Reproduction, in female mammals new blood vessels form each time the uterus develops during the estrous or menstrual cycle. In humans, this

occurs every month. Similarly, new vessels must form as wounds heal. We now know a great deal about the mechanisms involved in angiogenesis, and these findings are aiding in the search for a treatment for diseases including cancer and heart disease (see Box 8.1, Genetics and Genomics: Angiogenesis).

Vertebrate circulatory systems contain one or more pumps in series

Water-breathing fish have a single-circuit circulatory system in which blood flows from the heart through the gills to the body tissues and then back to the heart (Figure 8.12a). Because the heart must pump blood through the gills and tissues in series, some fish have a small accessory or caudal heart in the tail that assists blood flow back to the heart. In other species, normal movements of the body help venous return to the

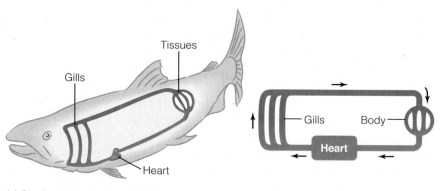

(a) Single-circuit circulatory system

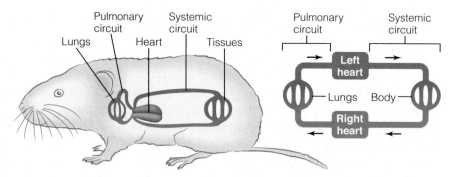

(b) Double-circuit circulatory system

Figure 8.12 Vertebrate circulatory systems The structure of vertebrate circulatory systems varies depending on the respiratory strategy of the animal. **(a)** In water-breathing fish, blood travels from the heart through the aorta to the gills and then to the body tissues, and returns to the heart. **(b)** Air-breathing tetrapods have a double circulatory system with two pumps arranged in series. Blood travels through the left heart to the aorta which leads to the systemic circuit through the body, returning to the right heart that pumps the blood via the pulmonary artery through the pulmonary circuit through the lungs.

BOX 8.1 GENETICS AND GENOMICS
Angiogenesis

Small blood vessels such as arterioles, capillaries, and venules undergo constant remodeling. This process, called angiogenesis, is controlled by both activator and inhibitor molecules that influence the rate of growth and division of vascular endothelial cells. Under normal circumstances, inhibitory factors are dominant, and vascular endothelial cells rarely divide, but when new blood vessels are needed (such as during wound healing), the body secretes angiogenic activator molecules that promote blood vessel growth.

Angiogenesis begins when cells in the region where the blood vessel will develop (the target site) secrete one or more angiogenic growth factors. These proteins are paracrine signaling molecules (see Chapter 3) that bind to receptors on the endothelial cells of existing blood vessels. Binding of the growth factor to its receptor activates a signal transduction cascade that helps to dissolve the basement membrane of the endothelium, and causes the endothelial cells to proliferate. The proliferating endothelial cells then migrate out through the holes dissolved in the wall of the existing vessel toward the target site. Specialized membrane proteins called integrins help to pull the sprouting blood vessel forward. Enzymes called matrix metalloproteases help to dissolve the tissues ahead of the advancing endothelial cells, making space for the new blood vessel. Once in place, the endothelial cells join together, forming the tube of the blood vessel, and the other cells of a blood vessel (smooth muscle, pericytes) are laid down, completing the development of the new vessel.

A number of factors, such as wounding and low oxygen levels (hypoxia) in tissue, can promote angiogenesis. When cells are hypoxic, levels of the protein hypoxia-inducible factor-1α (Hif-1α) increase. Hif-1α is part of a transcription factor complex. When the levels of Hif-1α increase, the transcription factor complex moves to the nucleus and binds to the promoters of a variety of hypoxia-inducible genes. One of these genes encodes an angiogenic activator protein called vascular endothelial growth factor (Veg-f). Veg-f binds to receptors on vascular endothelial cells and causes angiogenesis, increasing the density of the vasculature in the area. The increased vasculature can supply more oxygen to the tissues, reducing tissue hypoxia. Thus, the angiogenic response to tissue hypoxia acts as a negative feedback loop, maintaining tissue oxygen homeostasis.

Angiogenic activators and inhibitors are currently being studied as possible treatments for diseases such as cancer and coronary artery disease. Cancerous tumors secrete high levels of angiogenic activator molecules, causing new blood vessels to grow to supply the tumor with oxygen and nutrients. Tumor growth depends on this supply, so blocking angiogenesis can halt or slow tumor growth. Several dozen angiogenic-inhibiting drugs are currently being tested as possible cancer therapies. Some of these drugs are simply naturally occurring anti-angiogenic factors; others block the production of matrix metalloproteases, or aspects of the angiogenic signal transduction cascade.

Drugs that stimulate angiogenesis are also being tested for treatment of diverse diseases, including coronary artery disease and diabetes. In coronary artery disease, the arteries that supply oxygen to the working heart muscle become blocked by fatty deposits called plaques. These plaques inhibit blood flow to the heart muscle, depriving it of oxygen. Current treatments for coronary artery disease involve the surgical replacement of the blocked section of the blood vessel, but researchers are testing angiogenic growth factors as a way of making new blood vessels grow to supply the heart.

In late-stage diabetes, blood vessels begin to fail, and circulation to the feet can be very poor. As a result, the tissues can become oxygen deprived and die. Thus, one of the complications of untreated diabetes can be gangrene of the toes or foot, requiring amputation. Angiogenic growth factors may help slow the progress of this disease by promoting new blood vessel growth and helping to improve oxygen delivery. This treatment is not a cure, because it does not repair the underlying cause of blood vessel degeneration, but it may reduce the severity of symptoms. In addition, the role of angiogenesis in diabetic patients is complex, and varies from organ to organ. In fact, one complication of diabetes—diabetic retinopathy—is the result of excessive angiogenesis, which damages the tissues of the retina of the eye, potentially causing blindness.

---

heart. In contrast, tetrapods (amphibians, reptiles, birds, and mammals) have two circuits within their circulatory system. The right side of the heart pushes blood through the lungs in the **pulmonary circuit** of the circulatory system, whereas the left side of the heart pushes blood through body tissues in the **systemic circuit** of the circulatory system.

Mammals and birds have completely separated pulmonary and systemic circuits

Although the right and left sides of the heart are grouped together into a single organ, in mammals and birds these two sides of the heart are completely separated. As a result, a mammalian or bird circulatory system is conceptually similar to a single-circuit circulatory system with two pumps in series (Figure 8.12b). Oxygenated blood from the lungs flows to the left heart, which pumps the oxygenated blood to the body. The deoxygenated blood returning from the body flows into the right heart, which then pumps this deoxygenated blood to the lungs.

The completely separated systemic and pulmonary circuits of circulatory systems of mammals and birds are relatively inflexible, because blood cannot be diverted from one part of the system to the other. For example, when a mammal holds its breath, blood must still flow through the lungs, despite the fact that this tissue is not being utilized. However, because mammals and birds breathe more or less continuously, the ability to divert flow from the pulmonary circuit has not been an important force shaping the evolution of their circulatory systems.

Having completely separated pulmonary and systemic circuits has one important advantage: it allows pressures to be different in the pulmonary and systemic circuits. But why would having different pressures in the two circuits be an advantage? In the lungs, the capillaries must be very thin to allow good gas exchange, but if blood flows through these thin capillaries under high pressure, fluid will leak through the capillary walls. When this fluid accumulates it increases the diffusion distance and reduces the efficiency of gas exchange. Therefore, a low-pressure circulatory system through the lungs may be advantageous. In contrast, high pressures are needed to force blood through the long systemic circulatory system. Having separate pulmonary and systemic circuits allows these two differing demands to be met.

Many tetrapods have incompletely separated pulmonary and systemic circuits

Unlike mammals and birds, amphibians and most reptiles have an incompletely divided heart (Figure 8.13). Thus, it is possible for deoxygenated blood from the systemic circuit and oxygenated blood from the pulmonary circuit to mix. In many species the two streams of blood returning to the heart are kept fairly separate under most circumstances, although the mechanisms through which this separation is maintained are not fully understood. Because the ventricular chambers of the heart are interconnected, blood can be diverted from the systemic to the pulmonary circuit, or vice versa, if necessary. For example, these animals may divert blood from the pulmonary circuit to the systemic circuit during diving, allowing them to avoid perfusing the inactive lung.

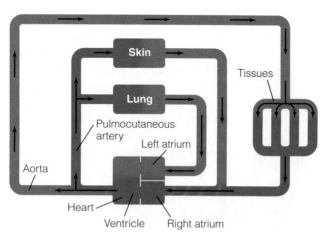

(a) Frog

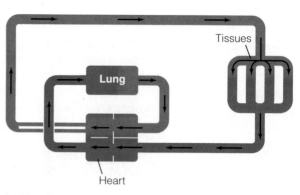

(b) Lizard

Figure 8.13 Circulatory patterns in amphibians and reptiles (a) Circulation in frogs. Deoxygenated blood flows to the pulmocutaneous artery that leads to the skin and lungs, and oxygenated blood flows via the aorta to the tissues, although some mixing may occur in the heart. (b) In reptiles, deoxygenated blood from the tissues enters the right atrium and is preferentially directed to the lungs. Oxygenated blood from the lungs enters the left atrium and is preferentially directed to the tissues. Oxygenated blood and deoxygenated blood are kept fairly separated under normal circumstances, although mixing is possible.

⊙ | CONCEPT CHECK

1. What is the difference between an open circulatory system and a closed circulatory system?
2. What is the major factor involved in the evolution of closed circulatory systems? Do all animals fit with this general rule?
3. What is the functional distinction between arteries and veins?
4. Which blood vessels have thicker walls, arteries or veins?
5. How can substances move across capillaries?

The Physics of Circulatory Systems

From the preceding sections it is clear that there is substantial variation in the organization and anatomy of animal circulatory systems. Despite this diversity, however, all circulatory systems use similar mechanisms to cause the bulk flow of fluids around the body. In order to understand these mechanisms, we must first review some of the fundamental physics of fluid flow. Recall from the beginning of the chapter that fluids flow down pressure gradients. Resistance due to friction opposes this movement. We can quantify the relationship between flow, pressure, and resistance in an equation called the *law of bulk flow*:

$$Q = \Delta P/R$$

where Q = flow, ΔP = the pressure gradient, and R = resistance. Note that flow is defined as the volume of fluid that moves past a given point per unit time, and has units such as liters per minute. You will see a variety of units of pressure used in the physiological literature. The SI unit for pressure is the pascal, or the force per unit area (in Newtons per meter squared). Physiologists and physicians often also use non-SI units to express pressure, including millimeters of mercury (mm Hg) and torr (where 1 torr = 1 mm Hg). These older units are the result of the use of mercury-filled manometers for the clinical measurement of blood pressure. Conversion factors among these units can be found in the appendix of this book. The units for resistance in a circulatory system are complex, and depend upon the units chosen for pressure and flow. For example, a unit for resistance could be kPa·min/L^{-1}. In medicine, the most common unit of resistance is the so-called peripheral resistance unit (PRU) in mm Hg·sec/ml^{-1}.

The radius of a tube affects its resistance

In circulatory systems, the circulating fluid is generally confined within a system of tubes or spaces, such as the blood vessels of vertebrates. We can begin to understand what sets the resistance of a blood vessel in the circulatory system by thinking about factors that affect flow through a drinking straw. Is it easier to drink liquids through a very long straw or a shorter straw? What is the difference between drinking through a narrow straw and a wider straw? What is the difference between drinking a milkshake and water (fluids with very different **viscosity**) through a straw? We can quantify these relationships mathematically as follows:

$$R = 8L\eta/\pi r^4$$

where R = the resistance of the tube, L = the length of the tube, η = the viscosity of the fluid, and r = the radius of the tube. Substituting this relationship into the law of bulk flow, we obtain **Poiseuille's equation**:

$$Q = \Delta P\pi r^4/8L\eta$$

Although real circulatory systems violate almost all of the assumptions of Poiseuille's equation (see Box 8.2, Mathematical Underpinnings: Poiseuille's Equation), it still provides a good conceptual summary of the factors that affect the flow of fluids through circulatory systems.

Because resistance is inversely proportional to radius to the fourth power, small changes in the radius of a tube result in large changes in its resistance. Many animals (both vertebrates and invertebrates) can control the flow through their organs by changing the radius of the blood vessels leading to those organs, a process called **vasoconstriction** or **vasodilation**. During vasoconstriction, the radius of the blood vessel decreases, increasing the resistance and reducing the flow through the vessel. During vasodilation the radius of the blood vessel increases, reducing the resistance and increasing the flow. Because small changes in radius cause large changes in resistance, even modest vasoconstriction and vasodilation can result in large changes in flow.

The total flow is constant across all parts of a circulatory system

The law of bulk flow is very similar to another basic physical principle—Ohm's law—that quantifies the behavior of charge in an electrical circuit.

BOX 8.2 **MATHEMATICAL UNDERPINNINGS**
Poiseuille's Equation

Although Poiseuille's equation provides a useful framework for thinking about the physics of circulatory systems, real circulatory systems violate almost all of its assumptions. For example, Poiseuille's equation assumes that the tubes in the system are unbranched and rigid, and that flow involves a simple fluid moving steadily through the tubes. In real circulatory systems, the vessels are branched and are distensible, changing their diameter as pressure changes; flow is often pulsatile, increasing and decreasing with the heartbeat; and the fluid is a complex mixture of plasma and cells.

The degree to which a blood vessel expands in response to increased pressure is called its **compliance**, C, and is equal to

$$C = \Delta V / \Delta P$$

where V = volume and P = pressure. Vessels with high compliance stretch easily when exposed to pressure, whereas vessels with low compliance stretch less. If we plot the change in volume against the change in pressure of a representative blood vessel, the slope of the line is the compliance of the vessel. The compliance of a blood vessel is not constant; compliance decreases at higher pressures and volumes—vessels become "stiffer" at high pressures. The compliance of a vessel is usually assessed under steady-state conditions, but blood vessels take some time to stretch, a phenomenon known as the *Windkessel effect*. In essence, blood vessels can store the potential energy imparted by pressure, and release it at a later time. As we see later in the chapter, this effect is important in the arteries.

Turbulent flow is relatively rare in the circulatory system, occurring in the heart and at some vessel branching points. In turbulent flow, the fluid moves in a complex pattern of eddies and whorls, oriented in various directions relative to the main axis of flow. In most blood vessels, flow is fairly laminar so that the fluid moves in a linear way along the blood vessel. But the velocity profile of the blood is not identical across the diameter of the vessel. Flow is slower near the walls because of the effects of friction. Poiseuille's equation ignores this effect. In larger vessels, flow is laminar but pulsatile, increasing when the heart contracts, and decreasing between contractions. The end result of this complex flow pattern is that the velocity profile is flatter, and the direction of flow changes as the heart beats.

The complex nature of blood has important effects on its viscosity. The viscosity of the aqueous component of the blood, called plasma, is low (about 1.8 times the viscosity of pure water), but whole blood has a viscosity about three to four times that of water because of the presence of blood cells. Because it is a mixture of components with different viscosities, blood acts as a non-Newtonian fluid; its viscosity varies depending on the size of the tube that it flows through, a phenomenon called the *Fahraeus-Lindqvist effect*. The Fahraeus-Lindqvist effect occurs because blood tends to separate in smaller blood vessels; in these smaller vessels, blood cells get swept into the higher-velocity flow at the center of the vessel, while the fluid close to the walls consists largely of plasma. The "high-viscosity" component at the center of the vessels has only minor interactions with the walls of the vessels, while the "low-viscosity" plasma interacts with the vessel walls, reducing the apparent viscosity of the fluid. In contrast, in very small vessels, blood cells fill almost the entire diameter of the vessel, and have to change shape to squeeze through the small space. Also, in these small vessels the blood cells tend to stick to each other and to the blood vessel walls, and together these three factors greatly increase the apparent viscosity of the fluid.

Despite these (and other) violations of its assumptions, Poiseuille's equation still provides a good conceptual model of flow through circulatory systems, and helps to explain the architecture of animal circulatory systems.

Ohm's law is usually written as $V = IR$ (where V = voltage, I = current, and R = resistance). If we rearrange this equation, we can write $I = V/R$. The electrical current (I) is simply the flow of electrons, and is thus equivalent to fluid flow (Q). The voltage drop across the circuit is the driving force for current movement, and is equivalent to the pressure gradient (ΔP). The electrical resistance is analogous to the frictional resistance of the blood vessels. Ohm's law and the law of bulk flow both quantify a fundamental physical phenomenon that is related to Newton's second law. Substances move because they are acted on by a force, and this movement is impeded by resistance. Because of this similarity, we can model circulatory systems as simple electrical circuits (Figure 8.14).

Like electrical resistors, blood vessels can be arranged in series or in parallel. The total resistance of a circuit with resistors arranged in series is the sum of the individual resistances, or

$$R_T = R_1 + R_2 \cdots$$

However, when resistors are arranged in parallel, the total resistance is determined as follows:

$$1/R_T = 1/R_1 + 1/R_2 + 1/R_3 \cdots$$

When you add resistors in series, the total resistance of the circuit increases, but when you add resistors in parallel, the total resistance of the circuit decreases. In circulatory systems, resistors are arranged both in series and in parallel.

Because of the law of conservation of mass, the flow through each segment of a circulatory system must be equal. For example, in Figure 8.14b, the total flow at point A and point B is the same. However, the amount of flow in each of the parallel blood vessels at point B need not be equal. The proportion of flow going through each of the parallel blood vessels depends upon the relative resistances of the blood vessels. As indicated by the law of bulk flow, blood tends to take the path of least resistance; more blood will flow through a low-resistance blood vessel than through one with high resistance. If we know the total flow and the resistance of each of the vessels in parallel, we can calculate the amount of flow going through each vessel, using the law of bulk flow.

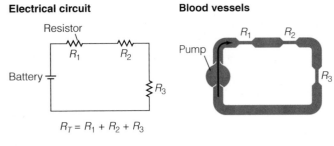

Electrical circuit   **Blood vessels**

$$R_T = R_1 + R_2 + R_3$$

(a) Resistors in series

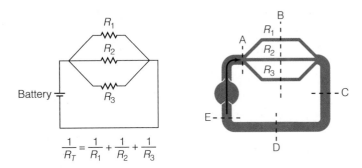

$$\frac{1}{R_T} = \frac{1}{R_1} + \frac{1}{R_2} + \frac{1}{R_3}$$

(b) Resistors in parallel

Figure 8.14 Resistors in series and parallel
Circulatory systems are analogous to electrical circuits with resistors arranged in both series and parallel. **(a)** The total resistance (R_T) of a circuit with resistors arranged in series is the sum of the individual resistances ($R_1 + R_2 + R_3$). **(b)** The total resistance of a group of resistors arranged in parallel decreases with increasing numbers of resistors. Total flow through each point of a circuit (A, B, C, D, E) is equal, but flow divides among the resistors arranged in parallel, depending on the resistance of each branch.

Velocity of flow is determined by pressure and cross-sectional area

Flow is, by definition, a rate—the volume of fluid transferred per unit time. But when fluid flows it also moves across a certain distance per unit time—that is, it has a velocity. The velocity of blood flow in a blood vessel is inversely related to the cross-sectional area of the blood vessel. You can visualize this by thinking about what happens to a volume of water as it flows through narrow and wide parts of the river. Because of the principle of the conservation of mass, the same amount of flow (volume per unit time) must pass through the narrow part of the river as the wide part of the river, but as a result its velocity (distance moved per unit time) must be greater in the narrow channel.

So what happens if a wide river splits into many small channels, such as you might encounter in a river delta? In this case, the velocity of flow in the small channels depends on the total cross-sectional area of the channels. Flow will split up among the channels, so mass will be conserved across the system as a whole, but all of the flow does not have to pass through any one smaller channel. The velocity of flow in the smaller channels will be inversely proportional to the total cross-sectional area of all the channels put together. If there are enough small channels, flow may be slower than in the wide part of the river. We can summarize these relationships as follows:

$$\text{Blood velocity} = Q/A$$

where A is equal to the summed cross-sectional area of the channels.

Exactly the same reasoning applies to circulatory systems. In areas where a single larger blood vessel splits into many small blood vessels arranged in parallel, the velocity of flow is likely to decrease as the blood enters the many small vessels (assuming that the total cross-sectional area of all the small vessels is greater than that of the single large vessel). This relationship between velocity and cross-sectional area is significant for a circulatory system, because it takes time for substances to diffuse between the blood and the tissues. Regions of the circulatory system that are involved in the exchange of materials have a very high total cross-sectional area, and so have very low flow velocities, which aids diffusion.

Pressure exerts a force on the walls of blood vessels

The blood pressure within a walled chamber such as a heart or blood vessel exerts a force on the walls of the chamber. This force can be quantified using the law of LaPlace (Figure 8.15), which states that the tension on the walls of a blood vessel is proportional to the blood pressure and the vessel radius according to the following equation:

$$T = aPr$$

where T is the tension on the walls (in N/cm), P is the **transmural pressure**, or the difference between the internal pressure and the external pressure (in Pa), r is the radius of the vessel, and a is a constant ($\frac{1}{2}$ for a cylindrical blood vessel or 1 for a spherical chamber). The law of LaPlace can also be rewritten to take into account the thickness of the wall of the vessel, as follows:

$$\sigma = Pr/w$$

where σ is the wall stress (in N/cm^2, or Pa), or the force per unit cross-sectional area of the wall, P is transmural pressure, r is the radius of the vessel, and w is the thickness of the wall.

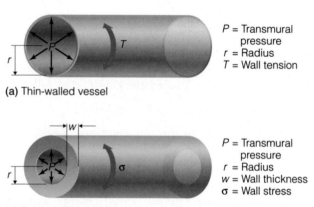

(a) Thin-walled vessel

P = Transmural pressure
r = Radius
T = Wall tension

(b) Thick-walled vessel

P = Transmural pressure
r = Radius
w = Wall thickness
σ = Wall stress

Figure 8.15 The Law of LaPlace (a) For a thin-walled vessel, the wall tension (T) is proportional to the transmural pressure (P) times the radius of the vessel. (b) For a thick-walled vessel, the wall stress σ is proportional to the transmural pressure (P) and the vessel radius (r), but inversely proportional to the wall thickness (w).

The law of LaPlace can be used to understand the structure and function of blood vessels and the heart. As can be seen from this relationship, as the thickness of a vessel increases, the stress in the wall of that vessel decreases. Thus, blood vessels such as the aorta, which are subjected to very high pressures, must be thicker, or made of stronger material, than vessels such as the arterioles, which are subjected to lower pressures. In addition, this relationship explains why when a vessel dilates (increases in radius), the stress or tension within the wall will increase, even if pressure remains the same. The law of LaPlace can also be used to understand the forces generated by the heart. A heart with a large radius must develop more tension within the heart wall to develop the same pressure within the heart (i.e., must undergo a stronger contraction) as would a heart with a smaller radius. Thus, we might expect a greater ratio of heart mass to heart volume in larger hearts.

In the next sections of the chapter, we use these physical principles to understand the functioning of animal circulatory systems. We begin by examining the pumping function of hearts, and then turn to an examination of the regulation of pressure and flow through the circulatory system.

◎ | CONCEPT CHECK

6. What physical force causes fluids to flow in circulatory systems?

7. What are the major factors that determine the resistance of a tube such as a blood vessel?

8. Imagine three identical blood vessels arranged either in series or in parallel. In which case will the total resistance be greatest?

9. What is transmural pressure?

Hearts

Because of the importance of hearts in animal circulatory systems, we devote a substantial part of this chapter to a discussion of their structure and function. The pumping action of a heart, which is called the **cardiac cycle**, is divided into two phases: contraction (**systole**) and relaxation (**diastole**). During systole, the heart contracts, increasing the pressure within the chambers of the heart and forcing blood out into the circulation. During diastole, the heart relaxes, reducing the pressure within the chambers of the heart and allowing blood to enter the heart from the circulatory system.

Because chambered hearts evolved from simple pulsatile blood vessels or tubular peristaltic hearts independently many times in different animal groups, it is not surprising that we find substantial differences in the structure and function of hearts among animals. However, recent evidence from developmental biology points to a surprising degree of similarity in the developmental program of hearts in very distantly related animal groups (see Box 8.3, Methods and Model Systems: Transcription Factors and Heart Development), suggesting that there is a fundamental unity of these diverse pumping structures.

BOX 8.3 METHODS AND MODEL SYSTEMS
Transcription Factors and Heart Development

All hearts perform similar functions, but they differ greatly in structure among taxa, ranging from simple contractile blood vessels to peristaltic tubular hearts and muscular contractile chambers. For many years biologists assumed that these diverse hearts had little in common other than their pumping function. However, recent work in model systems has revealed a surprising unity at the molecular level.

Using gene knockout technology, researchers are beginning to unravel the genetic program underlying the development of the heart in a range of organisms. In *Drosophila*, a gene called *Tinman* controls heart development. Researchers named this gene after the *Wizard of Oz* character of the tin woodsman, who lacks a heart. Flies that lack the gene *Tinman* never develop a heart. In mice, a gene in a family called *Nkx* is needed for heart development. Mice that have mutated versions of the *Nkx* genes have defects in cardiac development. *Nkx* and *Tinman* make very similar proteins, and these genes clearly share the same evolutionary origin. Similarly, an *Nkx/Tinman* homologue has been discovered in the lancelet (formerly called *Amphioxus*), an invertebrate chordate. This gene is expressed in the developing tube-like heart of these animals. This high degree of conservation suggests a common evolutionary origin of these very diverse hearts.

Nematodes such as *Caenorhabditis elegans* lack a heart and complex circulatory system, but they do have a digestive structure called the pharynx that contracts rhythmically and aids in feeding. The nematode pharynx shares a number of similarities with the hearts of other organisms. Like the hearts of insects and vertebrates, the pharyngeal muscles are myogenic—they contract without input from the nervous system. In addition, the development of the pharynx is controlled by the gene *ceh-22*, the molecular sequence of which is similar to *Tinman* and *Nkx*, suggesting that it is an evolutionary homologue of these genes.

Nematodes with a defective *ceh-22* gene do not develop a proper pharynx. However, if you introduce a copy of the mouse gene *Nkx2.5* into a nematode with a defective version of *ceh-22*, the pharynx develops normally. This experiment elegantly demonstrates that the genes controlling development of the pharynx in *C. elegans* and the heart of vertebrates are both structurally and functionally similar.

Presumably, the ancestral function of *ceh-22/Nkx/Tinman* was to specify the development of a rhythmically contracting structure. This developmental program was then co-opted during evolution to form the structurally diverse pumping organs of nematodes, fruitflies, and vertebrates. In fact, a gene related to *Nkx* has been detected in a cnidarian, the hydra (*Hydra magnipapillata*). The expression of this gene is localized around the base of the gastrovascular cavity, in a region that is involved in pumping fluids through the body. These observations suggest that the genes involved in heart development predate the evolution of circulatory systems, but have been involved in the development of pumping structures since the time of the earliest metazoans.

References

○ Bodmer, R. 1993. The gene tinman is required for specification of the heart and visceral muscles in *Drosophila*. *Development* 118: 719–729.

○ Haun, C., J. Alexander, D. Y. Stanier, and P. G. Okkema. 1998. Rescue of *Caenorhabditis elegans* pharyngeal development by a vertebrate heart specification gene. *Proceedings of the National Academy of Sciences USA* 95: 5072–5075.

○ Holland, N. D., T. V. Venkatesh, L. Z. Holland, D. K. Jacobs, and R. Bodmer. 2003. *AmphiNk2-tin*, an *amphioxus* homeobox gene expressed in myocardial progenitors: Insights into evolution of the vertebrate heart. *Developmental Biology* 255: 128–137.

○ Lints, T. J., L. M. Parsons, L. Hartley, I. Lyons, and R. P. Harvey. 1993. *Nkx-2.5*: A novel murine homeobox gene expressed in early heart progenitor cells and their myogenic descendants. *Development* 119: 419–431.

○ Shimizu, H., and T. Fujisawa. 2003. Peduncle of *Hydra* and the heart of higher organisms share a common ancestral origin. *Genesis* 36: 182–186.

Arthropod Hearts

Although the shape and size of the heart varies greatly among arthropods, their hearts share a number of features in common. Arthropod hearts generally pump hemolymph out into the circulation via arteries, and blood returns to the heart via a series of holes, or ostia. Valves within the ostia open and close, actively regulating the flow of hemolymph. The heart itself is suspended within the body cavity via a series of ligaments. Figure 8.16 illustrates the cardiac cycle in decapod crustaceans, which have particularly strong and muscular hearts. The hearts of most arthropods, including crustaceans, are **neurogenic**—they contract in response to signals from the nervous system (see Chapter 5: Cellular Movement and Muscles). The neurons of the cardiac ganglion, located on the surface of the heart and among the cardiomyocytes, are the primary rhythm generator. These neurons undergo spontaneous rhythmic depolarizations that initiate the rhythmic contraction of the heart (see Chapter 7 for a discussion of

neural rhythm generators). The neurons of the cardiac ganglion send a signal to close the ostia of the heart and initiate the heartbeat. As the cardiomyocytes contract, they decrease the volume of the heart chamber, exerting pressure on the circulatory fluid. This increase in pressure causes blood to squirt out of the heart and into the circulatory system via the arteries; the closed valves guarding the ostia prevent flow in the other direction.

The contraction of the heart also pulls on the ligaments that connect the heart to the body wall, stretching them. During diastole, when the heart relaxes, the ligaments spring back, pulling apart the walls of the heart. This elastic recoil increases the volume of the heart, reducing the pressure in the internal chambers. At this point, the valves of the ostia open, and the decrease in pressure sucks fluid into the heart via the ostia. Thus, arthropod hearts act as both suction and pressure pumps. They fill by suction, and they empty as a result of increasing pressure.

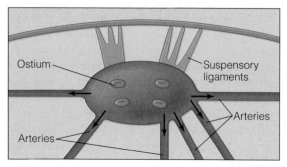

(a) Systole

(b) Diastole

Figure 8.16 The cardiac cycle in decapod crustaceans **(a)** Systole. When the heart contracts, the ostia close, and blood flows out via the arteries. The contraction pulls on the elastic suspensory ligaments, which store this potential energy. **(b)** Diastole. As the heart relaxes, the suspensory ligaments recoil, increasing the volume of the heart. The ostia open, and the low pressure sucks blood into the heart through the opened ostia.

Vertebrate Hearts

Vertebrate hearts have complex walls with four main parts (Figure 8.17). A sac called the **pericardium** surrounds the heart. In some species, such as elasmobranchs, the pericardium is relatively rigid, whereas in other species the pericardium is compliant, and stretches easily as the heart beats. The tough outer layer of the pericardium (the *parietal pericardium*) is made of connective tissue that protects the heart and anchors it to surrounding structures. The pericardium is filled with a small amount of fluid that acts as a lubricant, reducing friction as the heart beats.

The inner layer of the pericardium (the *visceral pericardium*) is continuous with the outer connective tissue of the heart, which is called the *epicardium*. If present, the nerves that regulate the heart and the **coronary arteries** that supply blood to the heart are located in the epicardium. These vessels extend into the next layer of the heart—the heart muscle, or **myocardium**. The myocardium is divided into several layers that can be distinguished based on the orientation of the **cardiomyocytes** (or cardiac muscle cells) in each layer. The innermost lining of the heart is called the **endocardium**, and is composed of a layer of connective tissue covered by a layer of epithelial cells, called the **endothelium**, that lines the chambers of the heart. This cardiac

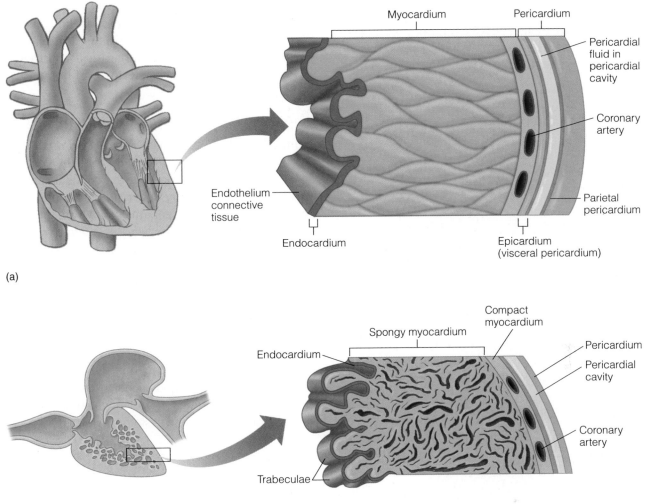

Figure 8.17 Structure of vertebrate hearts Vertebrate hearts have complex walls consisting of a pericardium, epicardium, myocardium, and endocardium. **(a)** Mammalian myocardium consists largely of compact myocardium. **(b)** In fishes and amphibians the myocardium is composed largely of spongy myocardium surrounded by a thin layer of compact myocardium. Spongy myocardium is poorly vascularized and receives oxygen from the blood flowing through the heart, whereas compact myocardium is supplied with oxygen by the coronary arteries.

endothelium is contiguous with the vascular endothelium that lines the blood vessels.

The myocardium can be spongy or compact

The ventricular muscle can be composed of two different types of myocardium: an outer layer of *compact myocardium*, made of tightly packed cells arranged in a regular pattern, and an inner layer of *spongy myocardium* consisting of a meshwork of loosely connected cells. However, the relative proportion of these two types of myocardium varies among species. In mammals the myocardium is al-

most entirely compact (Figure 8.17a), whereas in most fish and amphibians it is almost entirely spongy (Figure 8.17b). The compact myocardium on the outside of the heart is *vascularized* (contains blood vessels), but in many species the spongy myocardium does not contain blood vessels. In these species, the internal layer of the heart obtains its oxygen from the blood in the heart chambers. The spongy myocardium is often arranged into **trabeculae** that extend into the heart chambers. In fact, in some species the chambers of the heart are so filled with trabeculae that they resemble a sponge rather than the open chambers of the mammalian heart.

Fish heart chambers are arranged in series

The heart of a water-breathing fish consists of four chambers arranged in series (Figure 8.18a). Blood enters the heart via a thin-walled chamber called the **sinus venosus** and flows into the atrium and then into the muscular ventricle. The ventricle pumps the blood into either an elastic structure called the **bulbus arteriosus** (in most bony fishes) or a muscular **conus arteriosus** (in elasmobranchs). All of these chambers are contractile, except the elastic bulbus arteriosus of bony fish.

Amphibian hearts have three chambers

Amphibians have a three-chambered heart with two atria and one ventricle (Figure 8.18b). The ventricle of the heart pumps blood via the conus arteriosus into both the pulmonary and systemic circuits of the circulatory system. Oxygenated blood from the lungs returns to the left atrium via the pulmonary vein, while the deoxygenated blood from the systemic circuit returns via several veins that empty into the sinus venosus and then into the right atrium. The two atria then supply blood to the single ventricle. The trabeculae within the ventricle help to keep the oxygenated and deoxygenated blood separate, although the mechanisms by which they work are not yet fully understood. A *spiral fold* within the conus arteriosus directs deoxygenated blood to the pulmocutaneous artery leading to the lungs and skin and oxygenated blood to the systemic arteries.

Most reptiles have five heart chambers

The hearts of most non-crocodilian reptiles are composed of five chambers (Figure 8.19a). As in amphibians, there are two atria, but the ventricle is divided into three interconnected compartments (the *cavum venosum*, the *cavum pulmonale*, and the *cavum arteriosum*) by muscular ridges or septa. The conus arteriosus is divided to form the base of three large arteries: the pulmonary artery that leads to the lungs, and the right and left aortas that lead to the rest of the body. The pulmonary artery leads from the cavum pulmonale, whereas the aortas lead from the cavum venosum.

Despite their incompletely separated ventricles, reptiles generally maintain separation of oxygenated and deoxygenated blood. Deoxygenated blood enters the right atrium and flows into the cavum venosum and then across the muscular ridge into the cavum pulmonale and out the pulmonary artery. Oxygenated blood enters the left atrium and flows into the cavum venosum and then out the right and left aortas.

As mentioned earlier in this chapter, reptiles can also distribute blood selectively between the

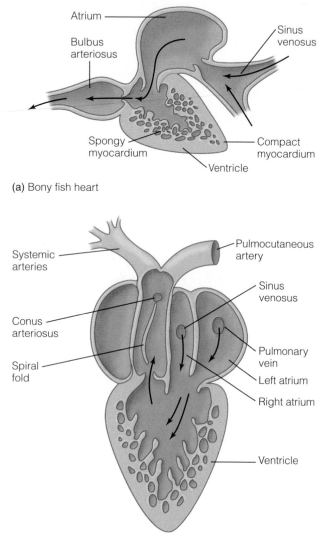

(a) Bony fish heart

(b) Amphibian heart

Figure 8.18 Cardiac anatomy of fish and frogs
(a) The heart of a fish is arranged in series. Blood enters the sinus venosus, which pumps blood into the atrium, and then into the muscular ventricle. The ventricle pumps blood via the bulbus arteriosus (in bony fish) or the conus arteriosus (in cartilaginous fish) to the body. **(b)** An amphibian heart has two atria and a single ventricle. Oxygenated blood from the lungs enters the left atrium via the pulmonary vein. Deoxygenated or partially oxygenated blood from the skin and tissues enters the right atrium via the sinus venosus. The atria pump blood into the single ventricle, but the oxygenated and deoxygenated blood are kept largely separate, by mechanisms that are not well understood. Oxygenated blood flows preferentially to the systemic arteries, whereas deoxygenated blood flows preferentially to the pulmocutaneous artery, directed by the spiral fold.

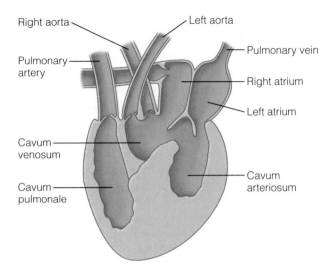

(a) Cardiac anatomy of non-crocodilian reptiles

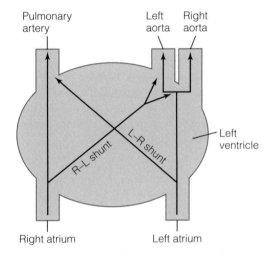

(b) Blood flow through the heart of non-crocodilian reptiles

Figure 8.19 Cardiac anatomy of non-crocodilian reptiles (a) Non-crocodilian reptiles have two atria and three incompletely separated ventricular chambers. **(b)** Diagrammatic view of blood flow through the heart of a non-crocodilian reptile. (Note that the shape of the heart has been "unfolded" so that the atria are shown at the bottom.) Under nonshunting conditions, blood flows from the right atrium to the pulmonary artery, and from the left atrium to the right and left aortas. During a right-to-left (R–L) shunt, some blood from the right atrium enters the aortas, bypassing the lungs. During a left-to-right (L–R) shunt, some blood from the left atrium enters the pulmonary artery, bypassing the tissues.

pulmonary and systemic circulation. This capacity to bypass either the pulmonary or systemic circuit is called a *shunt* (Figure 8.19b). In a right-to-left shunt (R–L), some fraction of the deoxygenated venous blood bypasses the pulmonary circulation and reenters the systemic circulation, thus causing oxygen-poor blood to circulate through the body. In a left-to-right shunt (L–R), some fraction of the pul-

monary blood reenters the pulmonary circuit rather than traveling to the body. Reptiles can regulate the degree and timing of these shunts, although the mechanisms involved are not yet understood and are likely to vary among species of reptile.

Reptiles are intermittent breathers, often holding their breath for long periods of time. During these periods, reptiles develop a pronounced R-L shunt, bypassing the pulmonary circulation and directing most of the blood to the body. R-L shunts are also associated with diving, particularly when a reptile dives to rest underwater. In contrast, L-R shunts have been proposed to aid oxygen delivery to the spongy myocardium of the right heart. The adaptive significance of shunts in reptiles is a matter of debate among comparative physiologists, and few of the proposed functions of shunting have been carefully evaluated experimentally.

Crocodilian reptiles (crocodiles, alligators, and caiman) have completely divided ventricles, and thus have four fully separated heart chambers. However, their pulmonary and systemic circuits are still connected, and these animals can shunt blood between them. (See Box 8.4, Evolution and Diversity: Shunting in Crocodiles.)

Birds and mammals have four heart chambers

The hearts of mammals and birds are composed of four unobstructed chambers with relatively smooth walls (Figure 8.20). The left side of the heart (shown on the right in this ventral view) consists of a thin-walled atrium and a thick-walled ventricle. The right side of the heart also consists of an atrium and ventricle, but the right ventricle has a much thinner wall than the left ventricle. The left ventricle, which pumps blood through the high-resistance systemic circulation, must pump more forcefully than the right ventricle, which pumps blood through the lower-resistance pulmonary circulation. A thick ridge called the intraventricular *septum* separates the two ventricles, while the interatrial septum separates the two atria. These septa are composed of muscle reinforced by connective tissue.

The *atrioventricular (AV) valves* are located between the atria and ventricles and allow blood to flow from the atrium to the ventricle, but not in the reverse direction. The right AV valve, also called the *tricuspid valve*, and the left AV valve, also called the *bicuspid valve*, are attached on the ventricular

BOX 8.4 **EVOLUTION AND DIVERSITY**
Shunting in Crocodiles

Crocodilian reptiles are unlike other reptiles in that they have a four-chambered heart with two atria and two ventricles that are completely divided by a muscular septum. However, like other reptiles, they have three major blood vessels leading away from the heart. The right aorta emerges from the left ventricle, whereas the pulmonary artery and the left aorta emerge from the right ventricle (see Figure A). The right aorta sends blood largely to the brain and anterior circulation, whereas the left aorta sends blood largely to the viscera and the posterior parts of the animal. The aortas are connected at two points in the circulatory system: the **foramen of Panizza**, a small opening located at the base of the aortas, near the heart, and an arterial anastomosis located in the abdomen.

Because of the complete separation of the ventricles, crocodilians cannot shunt blood from the systemic to the pulmonary circulation (a L-R shunt), but R-L shunts are possible (see Figure B). When blood pressure in the left and right ventricles is equal, such as might be expected in a resting crocodile breathing air, oxygenated blood from the left ventricle is directed via the right aorta to the brain, while deoxygenated blood flows via the left aorta to the visceral organs, where it may aid in digestion, because this acidic deoxygenated blood can counteract the alkalinization of the blood caused by secretion of digestive acids into the stomach. When the animal is active and breathing air, blood pressure is high in the left ventricle compared to the right ventricle. Oxygenated blood flows from the left ventricle both into

the right aorta and (via the foramen of Panizza and arterial anastomosis) into the left aorta because the pressure in the right aorta is high compared to the pressure in the left aorta. This prevents deoxygenated blood in the right ventricle from moving into the systemic circulation, and instead it flows almost entirely to the lungs.

The valve at the entrance of the pulmonary artery also helps to control the flow of blood between different parts of the circulatory system. Unlike the passive flap-like valves of other vertebrates, this valve has cog teeth made up of nodules of connective tissue. The cog teeth mesh together, forming a tight seal. The level of epinephrine in the bloodstream controls the position of the teeth in this valve, and thus the valve is controlled actively, rather than simply opening and closing passively in response to pressure changes in the heart. When the crocodile is at rest underwater, and levels of epinephrine are low, the cog teeth close, diverting blood away from the pulmonary artery. When the crocodile is active, the cog teeth open, allowing blood to flow into the lungs.

Crocodiles use the cog valve to shut off the pulmonary system when they dive below the water to rest, allowing them to remain submerged for several hours without perfusing their lungs.

References

○ Franklin, C. E., and M. Axelsson. 2000. An actively controlled heart valve. *Nature* 406: 847.

○ Hicks, J. W. 2002. The physiological and evolutionary significance of cardiovascular shunting patterns in reptiles. *News in Physiological Sciences* 17: 241–245.

○ Syme, D. A., K. Gamperl, and D. R. Jones. 2002. Delayed depolarization of the cog-wheel valve and pulmonary-to-systemic shunting in alligators. *Journal of Experimental Biology* 205: 1843–1851.

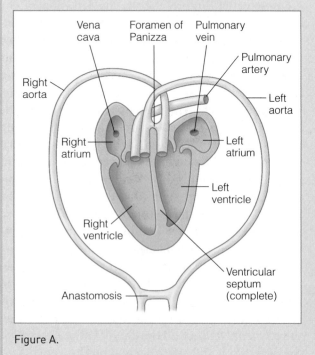

Figure A.

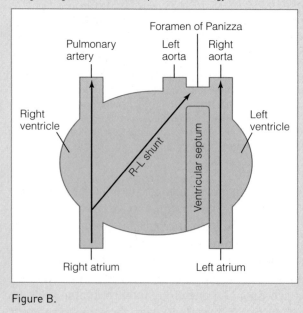

Figure B.

side to collagenous cords called the *chordae tendineae*. These cords anchor the valves to the *papillary muscles*, and prevent them from opening backward. The *semilunar valves*, located at the exit from the ventricles, prevent blood from flowing backward into the ventricles. The *pulmonary semilunar valve* is located between the right ventricle and the pulmonary artery leading to the lungs. The *aortic semilunar valve* is located between the left ventricle and the aorta, the artery leading to the systemic circulation.

Blood returning to the heart from the body first passes through the superior and inferior *venae cavae* (superior vena cava and inferior vena cava) into the right atrium. The blood then passes via the right AV, or tricuspid, valve into the right ventricle. The right ventricle pumps the blood through the pulmonary semilunar valve into the pulmonary artery leading to the lungs. The blood travels through the pulmonary capillary bed where it is oxygenated. It exits the lungs via the pulmonary veins that lead to the left atrium. The blood then travels from the left atrium past the left AV, or bicuspid, valve into the left ventricle. The left ventricle pumps the blood through the aortic semilunar valve into the aorta. The aorta branches into smaller arteries and then arterioles, finally leading to the capillary beds of the systemic circulation. From these capillary beds the blood travels through venules and veins, finally draining into the venae cavae, and returning to the right atrium.

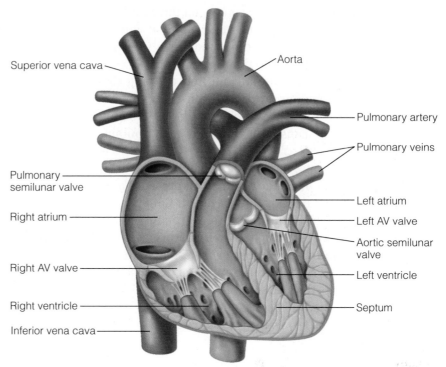

Figure 8.20 Internal anatomy of the mammalian heart Blood flows from the pulmonary veins into the left atrium and then the left ventricle. The left ventricle pumps blood to the aorta and the systemic circuit of the circulatory system. Blood from the tissues flows via the venae cavae to the right atrium and the right ventricle, which pumps blood to the pulmonary artery and the pulmonary circulation. One-way flow through the heart is ensured by two sets of valves.

The Cardiac Cycle

The vertebrate heart functions as an integrated organ, with each of the chambers contracting at appropriate points during the cardiac cycle. Only through this coordinated contraction can the heart pump blood effectively through the circulatory system. In this section we examine the cardiac cycle in fishes, birds, and mammals to explore how changes in pressure in the various heart chambers coupled with the opening and closing of one-way valves drive blood unidirectionally through the heart.

Fish hearts contract in series

During the cardiac cycle of a fish heart, each of the cardiac chambers contracts in series, starting with the sinus venosus. Contraction of the sinus venosus is unlikely to play an important role in propelling blood through the system because this thin-walled chamber is unable to develop substantial pressure and it lacks a one-way valve to prevent backflow into the circulation. Instead, the primary role of the sinus venosus is to initiate the heartbeat. Following the contraction of the sinus venosus, the atrium contracts, causing pressure to increase in this chamber. This increase in atrial pressure closes the valve to the sinus venosus, and opens the valve to the ventricle. It is important to note that the valves are passive structures that open and close in response to changes in pressure in the heart chambers, not as a result of active movements of the valves themselves.

The pressure difference between the contracting atrium and the relaxed ventricle then causes blood to flow through the opened valve into the ventricle. Next the muscular ventricle contracts, closing the valve to the atrium and opening the valve to the bulbus arteriosus. In bony fishes, blood flows from the ventricle into the elastic bulbus arteriosus, causing it to expand. The bulbus arteriosus acts as an elastic energy storage device that, as we discuss in more detail later in the chapter, acts to dampen changes in blood pressure and allow more continuous flow of blood. Contraction of the muscular ventricle plays the main role in propelling blood through the circulatory system. In elasmobranchs, blood flows from the ventricle into the conus arteriosus. Contraction of the conus arteriosus further assists in propelling blood through the body. The elasmobranch conus arteriosus contains several valves that help to ensure unidirectional flow.

The mammalian cardiac cycle is similar to that of fishes

Figure 8.21 illustrates the cardiac cycle of a mammalian heart. Because it is a cycle, we can arbitrarily begin our examination of the events at any point. Let's begin at the point labeled step 1. At this point, the atria and ventricles are relaxed and the AV valves are open, but the semilunar valves are closed. In mammals and birds, blood returning to the heart passes through the atria and enters the ventricles passively, without any pumping action of the heart. At step 2, the atria contract, but the ventricles are still relaxed. The pumping of the atria pushes some additional blood into the ventricles until they reach the **end-diastolic volume (EDV)**, the maximum volume of blood in the ventricle.

Next, at step 3, the ventricles begin to contract. The increased pressure caused by this contraction forces the AV valves shut. Since the semilunar valves are shut at this time, the ventricle is a completely sealed compartment and blood cannot flow out of it. Blood, like other liquids, is incompressible, so the volume of the ventricle does not change. Instead, the pressure inside the ventricle increases. Thus, the ventricles are said to undergo **isovolumetric contraction** (also known as isovolumic contraction) because the volume of the chamber does not change. Eventually, the pressure in the ventricles is sufficiently high that it forces

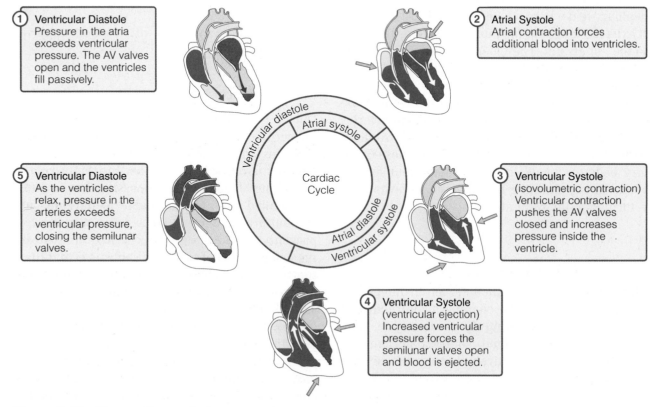

① Ventricular Diastole
Pressure in the atria exceeds ventricular pressure. The AV valves open and the ventricles fill passively.

② Atrial Systole
Atrial contraction forces additional blood into ventricles.

⑤ Ventricular Diastole
As the ventricles relax, pressure in the arteries exceeds ventricular pressure, closing the semilunar valves.

Cardiac Cycle

Ventricular diastole
Atrial systole
Atrial diastole
Ventricular systole

③ Ventricular Systole
(isovolumetric contraction) Ventricular contraction pushes the AV valves closed and increases pressure inside the ventricle.

④ Ventricular Systole
(ventricular ejection) Increased ventricular pressure forces the semilunar valves open and blood is ejected.

Figure 8.21 The cardiac cycle in mammals

open the semilunar valves, and blood flows out of the ventricles into the arteries in step 4 of the cardiac cycle: the *ventricular ejection* phase, in which the volume of blood in the ventricle declines. The chordae tendineae prevent the AV valves from being forced open, so blood cannot flow "backward" into the atrium. At this point, the ventricle has reached its minimum, or **end-systolic volume (ESV)**. The end-systolic volume is always greater than zero, as the heart does not completely empty itself with each beat. In a healthy human at rest, ESV can be as much as 50% of EDV.

At the end of the ventricular ejection phase, the ventricles begin to relax, causing the pressure in the ventricles to drop (step 5). Once the ventricular pressure drops below the pressure in the arteries, the backpressure forces the semilunar valves shut. Throughout ventricular systole, the atria have been in diastole; they have been relaxed and filling with blood. The pressure in the filled atria eventually exceeds the pressure in the relaxing ventricles, and the AV valves pop open, returning the heart to the configuration shown in step 1.

Some vertebrate hearts fill actively

The ventricles of birds and mammals fill passively during diastole, as a result of the relatively low pressure within the atria compared to the venous pressure, with only a small contribution from atrial contraction. But this is not the case for all vertebrates. For example, in fish and some amphibians, the ventricles are primarily filled by contraction of the atrium. In addition, some fishes, including the elasmobranchs, may utilize suction filling of the ventricle, analogous to that seen in the hearts of arthropods. Elasmobranchs have a relatively rigid pericardium. When the ventricle contracts, the volume of pericardial space occupied by the ventricle decreases. This increases pericardial volume and decreases the pressure inside the pericardial cavity. The sinus venosus and atrium are thin-walled chambers, and a very low pressure in the pericardium causes them to expand, reducing the pressure in the atrium and sucking blood into the heart. However, this mechanism will work only if pressure within the pericardium decreases below the pressure in the veins, and cardiovascular physiologists debate whether this mechanism actually operates in elasmobranchs under normal physiological

conditions, because it is difficult to measure the exact pressure within the pericardium of a swimming shark.

The right and left ventricles develop different pressures

During the cardiac cycle, the two ventricles of the mammalian heart contract simultaneously, but the left ventricle contracts much more forcefully than the right ventricle, and develops much higher pressure (Figure 8.22). Blood from the left ventricle travels via the aorta to the organs of the body, whereas blood from the right ventricle travels via the pulmonary artery to the lungs. The pulmonary circuit has relatively low total resistance because of the very large number of capillaries arranged in parallel and the relatively short distance traveled. Because the resistance of the circuit is low, the right side of the heart does not need to pump as forcefully to drive blood through the lungs, which protects the delicate blood vessels of the lungs.

⊙ CONCEPT CHECK

10. Which type of animal would you expect to have a higher proportion of spongy myocardium, a fish or a mammal?
11. Compare and contrast the heart of an amphibian and a mammal.
12. What is isovolumetric (or isovolumic) contraction?

Control of Contraction

From the preceding discussion it is clear that cardiac contraction must be precisely controlled in order to ensure coordinated unidirectional blood flow through the chambers of the heart. Unlike the neurogenic hearts of the invertebrates that we discussed previously, vertebrate hearts are **myogenic**; their cardiomyocytes can produce spontaneous rhythmic depolarizations that initiate contraction (see Chapter 5). But in order for the heart to contract in a coordinated way, cardiomyocytes must be electrically coupled via gap junctions so that the depolarization in one cell can spread to adjacent cells, triggering coordinated contractions. The rate of the spontaneous depolarizations varies among cardiomyocytes, with some having

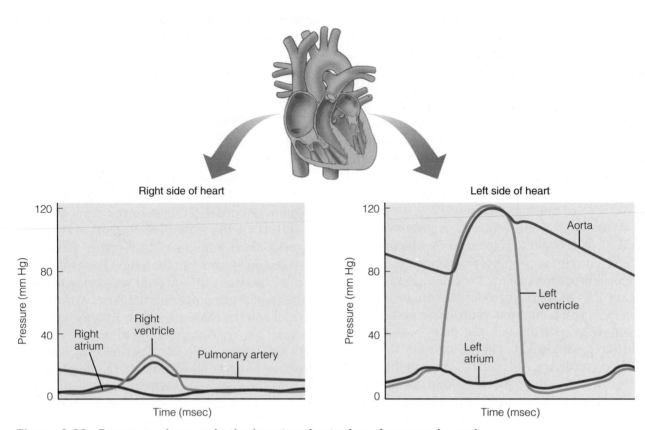

Right side of heart

Left side of heart

Figure 8.22 Pressure changes in the heart and arteries of mammals such as humans The left side of the heart, which supplies the systemic circuit, develops substantially greater pressures than the right side of the heart, which supplies the pulmonary circuit.

a relatively rapid rhythm and others depolarizing more slowly. The cells with the fastest intrinsic rhythm are termed the **pacemaker cells**, because they determine the contraction rate for the entire heart. In fish, the pacemaker cells are located in the sinus venosus, and in other vertebrates they are located in an area of the right atrium called the **sinoatrial (SA) node**, close to the point where the superior vena cava enters the right atrium. This structure is thought to be the remnant of the sinus venosus of fish.

Pacemaker cells initiate the heartbeat

Although derived from muscle cells, pacemaker cells are small with few myofibrils, mitochondria, or other organelles, and they do not contract. These cells have an unstable resting membrane potential (called the **pacemaker potential**) that slowly drifts upward from the starting potential of about −60 mV until it reaches threshold (about −40 mV) and initiates an action potential (Figure 8.23). This slow depolarization is, in part,

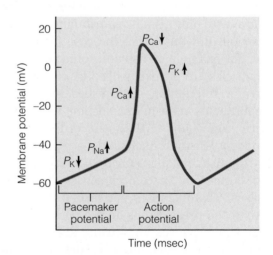

Figure 8.23 Pacemaker and action potentials In myogenic hearts, the pacemaker cells have an unstable resting membrane potential (the pacemaker potential). Nonselective cation ("funny") channels open, increasing the permeability (P) of the membrane to Na$^+$, which causes the membrane potential to increase gradually. As the membrane approaches threshold, T-type Ca^{2+} channels open, triggering an action potential. After about 200 msec these channels close and K$^+$ channels open, repolarizing the cell, and the cycle begins again.

the result of a slow inward movement of sodium, which is called the "funny" current (I_f) because of its unusual behavior. The funny current is the result of the opening of a nonselective cation channel (sometimes called the "funny channel," for consistency with the term *funny current*). This channel opens when the membrane is hyperpolarized, allowing Na^+ to enter the cell, and closes as the membrane gradually depolarizes. In addition, as in all other cells, there is a continuous leak of potassium ions at the resting membrane potential. In pacemaker cells, however, this potassium permeability decreases as the membrane depolarizes. The slow decrease in potassium movement contributes to the slow depolarization of the cell. The combination of reduced K^+ efflux and increased Na^+ influx causes pacemaker potential.

When the membrane potential of the pacemaker cell reaches threshold, T-type voltage-gated Ca^{2+} channels open and Ca^{2+} influx increases, causing a further, more rapid, depolarization. Opening of these T-type Ca^{2+} channels results in a depolarization phase that is much less steep than the depolarization of a neural action potential (caused by influx of Na^+ through voltage-gated Na^+ channels; see Chapter 4: Neuron Structure and Function), although it is faster than the depo-larization caused by the funny current. About 200 milliseconds after they open, these T-type Ca^{2+} channels begin to close, and K^+ channels open, initiating the repolarization phase of the action potential in the pacemaker cell.

The nervous and endocrine systems can modulate the rate of pacemaker potentials

The rate of action potentials in the pacemaker sets the heart rate. In most vertebrates, the nervous and endocrine systems can control heart rate by altering the rate of pacemaker potentials in the cells of the sinoatrial node or sinus venosus. Norepinephrine released from sympathetic neurons and epinephrine released from the adrenal medulla bind to β adrenergic receptors on the pacemaker cells (Figure 8.24). The receptors stimulate a cAMP-mediated signaling pathway that alters the transport properties of the ion channels in the cell membranes. Funny and Ca^{2+} channels open, increasing the influx of Na^+ and Ca^{2+} ions and increasing the rate of depolarization of the cell. The increased depolarization rate increases the frequency of action potentials in the pacemaker cells, which ultimately increases heart rate. These effects of epinephrine and norepinephrine

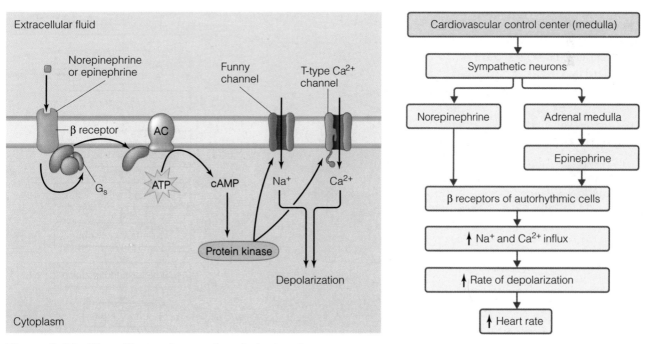

Figure 8.24 The effects of norepinephrine on heart rate Norepinephrine increases heart rate by binding to β adrenergic receptors, activating an adenylate cyclase (AC) signal transduction pathway that opens cation (funny) and T-type Ca^{2+} channels, increasing the rate of depolarization of the pacemaker potential.

on the pacemaker cells explain the dangerous side effects of drugs such as ephedrine and the herbal supplement ephedra, which can bind to adrenergic receptors and cause a rapid heart rate.

Acetylcholine, released from parasympathetic neurons, binds to muscarinic receptors on the pacemaker cells of the heart (Figure 8.25). These receptors stimulate a signal transduction pathway that ultimately leads to increased K^+ permeability. The increased K^+ efflux causes the pacemaker cell to hyperpolarize. The pacemaker potential starts at a more negative value, and thus takes longer to reach threshold potential. In addition, binding of acetylcholine to its receptor leads to decreased Ca^{2+} permeability, slowing the rate of the depolarization during a pacemaker potential. Together these effects decrease the number of depolarizations per unit time, and thus slow the heart rate.

Pacemaker depolarizations can spread via gap junctions

Cardiac cells are electrically connected to each other via gap junctions. Thus, the rhythmic depolarization initiated in the pacemaker cells of the sinus venosus or sinoatrial node can spread from cell to cell via electrotonic current spread (see Chapter 4). In the adjacent cells, this depolariza-tion triggers action potentials, which can then spread to adjacent cells, propagating the impulse throughout the heart.

Cardiac action potentials have an extended depolarization phase

The action potential of a contractile cardiomyocyte differs from the pacemaker potentials seen in the pacemaker cells of the sinus venosus or sinoatrial node. In the contractile cardiomyocytes, the action potential is initiated when a depolarization spreading from an adjacent cell depolarizes the cardiomyocyte beyond the threshold potential of the voltage-gated Na^+ channel. At this point, the voltage-gated Na^+ channels open, causing the rapid depolarization phase of the action potential. In this respect, the action potential of the cardiomyocyte is similar to that in neurons (see Chapter 4) and in skeletal muscle cells (see Chapter 5). However, the action potentials in contractile cardiomyocytes differ from those in skeletal muscles. They have an extended depolarization, called the *plateau phase* (Figure 8.26). At the time when the voltage-gated Na^+ channel is inactivated (closes), another channel, an L-type voltage-gated Ca^{2+} channel opens, allowing Ca^{2+} to enter the cell. This greatly lengthens the depolarization phase of the action potential

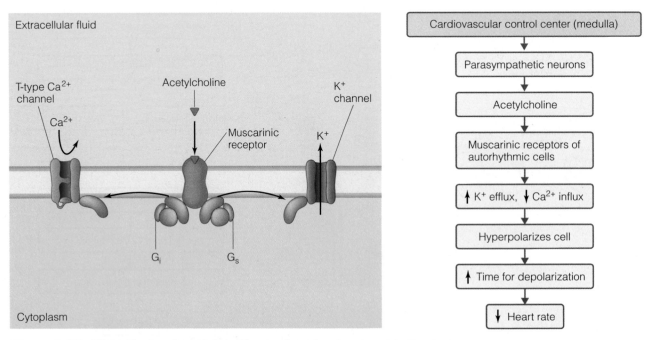

Figure 8.25 The effects of acetylcholine on heart rate Acetylcholine decreases heart rate by binding to muscarinic receptors, activating a signal transduction pathway that closes Ca^{2+} channels and opens K^+ channels. This prevents Ca^{2+} ions from entering the cell and allows K^+ ions to exit, causing a net hyperpolarization, which increases the time needed for the pacemaker potential to depolarize the cell to threshold.

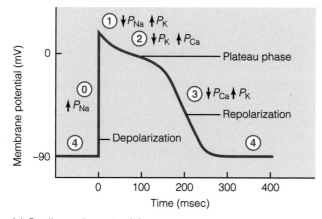

(a) Cardiac action potential

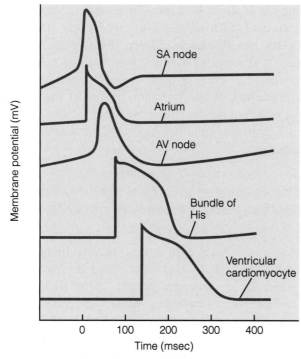

(b) Potentials in various parts of the heart

Figure 8.26   The action potential in cardiomyocytes   (a) Phases of the action potential. Phase 0: The cell reaches threshold potential and voltage-gated Na^+ channels open, increasing Na^+ permeability (P_{Na}) and depolarizing the cell. Phase 1: The voltage-gated Na^+ channels inactivate and K^+ channels open, causing a transient outward K^+ current, resulting in a slight repolarization. Phase 2: These inward rectifier K^+ channels close and L-type voltage-gated Ca^{2+} channels open, causing the plateau phase of the action potential. Phase 3: L-type voltage-gated Ca^{2+} channels close and K^+ channels open, causing repolarization. Phase 4: The cell returns to the resting membrane potential. **(b)** Pacemaker and action potentials in various types of cardiomyocytes in the mammalian heart. The shapes of the pacemaker and action potentials differ across the parts of the heart as a result of the expression of different channel isoforms.

in contractile cardiomyocytes. Note that this L-type Ca^{2+} channel is a distinct isoform from the T-type Ca^{2+} channel expressed in the pacemaker cells, accounting for the differences in their behavior.

The plateau phase of the contractile cardiomyocyte action potential corresponds to the refractory period of the cell, in which it cannot generate another action potential. This refractory period lasts almost as long as the entire muscle twitch, preventing new contractions from occurring until the previous one has finished (see Chapter 5). Thus, unlike skeletal muscle, cardiac muscle cannot go into tetanus—a period of sustained contraction leading to muscle fatigue.

The exact shape and duration of the action potential varies substantially among organisms and among cells from different parts of the heart (Figure 8.26b) as a result of variation in the expression of ion channel isoforms. For example, small mammals tend to have a rapid heart rate and cardiac action potentials with shorter plateau phases than large mammals whose hearts beat more slowly.

Conducting pathways spread the depolarization across the heart

In a fish heart, in which the chambers are arranged in a more or less linear way, impulse conduction via gap junctions is sufficient to provide coordinated contraction of the chambers. The depolarizing signal travels via gap junctions from the sinus venosus to the atrium and then to the ventricle, causing them to contract in series. However, in addition to traveling from cell to cell via gap junctions, depolarizations in vertebrate hearts also spread via specialized conducting pathways. In mammals, these conducting pathways consist of a series of cells that can be easily distinguished microscopically because of their elongated, pale appearance. These conducting cells do not contract, but can undergo rhythmic depolarizations, similar to pacemaker cells. Along most of their length they are electrically insulated from the rest of the myocardium by a fibrous sheath. All vertebrate hearts appear to contain fast electrical conducting pathways, although in nonmammalian vertebrates these conducting cells are not morphologically distinguishable as electrically isolated cells. Instead, some of the trabeculae in the spongy myocardium in fishes, amphibians, and reptiles appear to play this fast conducting role.

Figure 8.27 shows how electrical signals move through the heart of a mammal. After the pacemaker cells in the sinoatrial (SA) node initiate an action potential, the depolarization spreads rapidly via the *internodal pathway* through the walls of the atria. At the same time, the depolarization spreads more slowly through the contractile cells of the atrium via gap junctions, causing the atrium to contract. After traveling through the internodal pathway, the depolarization reaches the **atrioventricular (AV) node**, which commu-nicates the electrical signal to the ventricle. The contractile cells of the atrium and ventricle do not form gap junctions with each other, and thus are not electrically coupled, so the depolarization cannot spread directly from the atrium to the ventricle, but instead can only pass through the AV node. The AV node transmits signals a little more slowly than the other cells of the conducting pathways, so the signal gets delayed slightly. This signal delay allows the atrium to finish contracting before the ventricle starts to contract. The signal travels from the AV node through the **bundles of His** (pronounced *Hiss*), which splits into the left and right bundle branches that conduct electrical signals to the ventricles. The electrical signal then spreads into a network of conducting pathways called the **Purkinje fibers**. From the Purkinje fibers, the signal spreads from cell to cell in the ventricular myocardium via gap junctions. The contraction of the ventricle begins at the bottom (or *apex*) of the heart and spreads up through the myocardium, pushing blood upward toward the arteries.

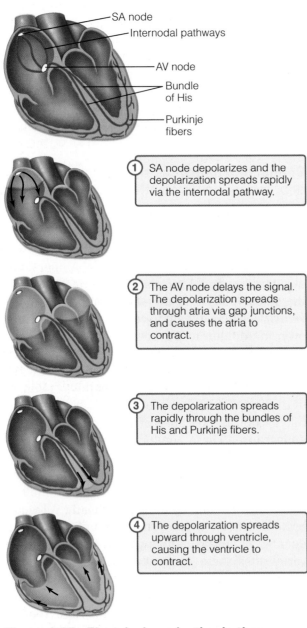

① SA node depolarizes and the depolarization spreads rapidly via the internodal pathway.

② The AV node delays the signal. The depolarization spreads through atria via gap junctions, and causes the atria to contract.

③ The depolarization spreads rapidly through the bundles of His and Purkinje fibers.

④ The depolarization spreads upward through ventricle, causing the ventricle to contract.

Figure 8.27 Electrical conduction in the mammalian heart

The integrated electrical activity of the heart can be detected with the EKG

The depolarization of cardiac muscle produces a strong electrical signal that travels through the body and can be detected using an instrument called an *electrocardiograph*. These instruments use electrodes applied to various areas on the surface of the body to generate an **electrocardiogram** (abbreviated EKG for the original German spelling, or ECG for the English spelling). Clinicians generally perform EKGs on humans using 12 electrodes, but you can generate an interpretable EKG using as few as three electrodes (in humans these electrodes are placed one on each wrist, and one on an ankle). An EKG is a composite recording of all the action potentials in the various parts of the heart, including the pacemakers, the conducting pathways, and the contractile cells (Figure 8.28). The deflections on the chart are not action potentials, and do not represent specific depolarizations of any given cell. Instead, they are markers of the electrical activity of the heart as a whole. The small **P wave** is the result of the spread of depolarization through the atria. The large **QRS complex** is the result of ventricular depolarization and atrial repolarization. The **T wave** is caused by ventricular repolarization. The EKG can be very useful clinically to diagnose prob-

CHAPTER 8 | Circulatory Systems **381**

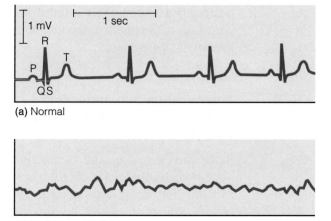

(a) Normal

(b) Ventricular fibrillation

Figure 8.28 EKG tracings **(a)** In the EKG of a normal cardiac rhythm, the P wave indicates atrial depolarization. The QRS complex indicates ventricular depolarization and atrial repolarization, and the T wave indicates ventricular repolarization. **(b)** During ventricular fibrillation, the EKG is disorganized and no consistent waves are observed.

lems with the conducting system or the depolarization of the heart muscle. For example, ventricular fibrillation, which represents uncoordinated contraction of the ventricle, appears as a series of random, apparently unrelated waves in the EKG. Ventricular fibrillation is potentially deadly because uncoordinated contraction of the ventricle results in ineffective pumping of blood to the tissues. The resulting oxygen deprivation kills tissues such as the brain within a few minutes. Ventricular fibrillation can sometimes be treated using an electronic *defibrillator*. These machines deliver an intense pulse of current to the body, causing all of the cells of the heart to depolarize simultaneously. Defibrillation gives the pacemaker cells of the heart a chance to take over and initiate a normal heartbeat, because these cells are likely to be the first to depolarize again following defibrillation. However, defibrillation will not be effective if the pacemaker cells or the conducting pathways have irreversible defects or injuries.

The heart functions as an integrated organ

The electrical and mechanical events of the heart fit together, allowing the heart to function as an integrated organ. The diagrams at the bottom of Figure 8.29 depict the contractions and relaxations in the cardiac cycle, particularly in the left ventricle. The EKG tracing just above the diagrams shows the timing of the electrical

events of the heart. The heart sounds, which can be detected with a stethoscope, represent the opening and closing of the valves. The center and top graphs show the pressure and volume changes within the left ventricle over the course of the cardiac cycle.

At the beginning of the cardiac cycle, the ventricle fills passively. Then the depolarization of the SA node spreads through the atrium, initiating atrial contraction, and pumping some additional blood into the ventricle, which reaches its end-diastolic volume. The depolarization then spreads to the ventricle, which begins to contract. The increased pressure caused by this contraction forces the AV valves shut. Pressure then increases rapidly during the isovolumetric ventricular contraction phase, quickly becoming high enough to open the semilunar valves. The first heart sound is the result of the AV valves shutting and the semilunar valves opening. At this point, the ventricle begins to empty and aortic pressure increases. Initially, pressure in the ventricle continues to increase, despite the reduced volume, because ventricular contraction continues, but ventricular pressure quickly reaches a peak and begins to fall. Shortly thereafter, the ventricle begins to relax, entering ventricular diastole. When ventricular pressure falls below the pressure in the aorta, the aortic valve closes. The closing of the aortic valve causes a brief episode of turbulent flow and a small increase in aortic pressure, called the dicrotic notch. Ventricular pressure falls rapidly, and once it is lower than atrial pressure, the AV valves open. The second heart sound is the result of the aortic valve closing and the AV valves opening. At this point, blood flows from the atrium into the ventricle, reducing the atrial pressure, and initiating ventricular filling.

Cardiac output is the product of heart rate and stroke volume

The amount of blood that the heart pumps per unit time is called the **cardiac output** (CO), and is a product of the *heart rate* (HR) and the amount of blood the heart pumps with each beat, or the **stroke volume** (SV).

$$CO = HR \times SV$$

From this equation you can clearly see that an animal can modulate cardiac output by regulating heart

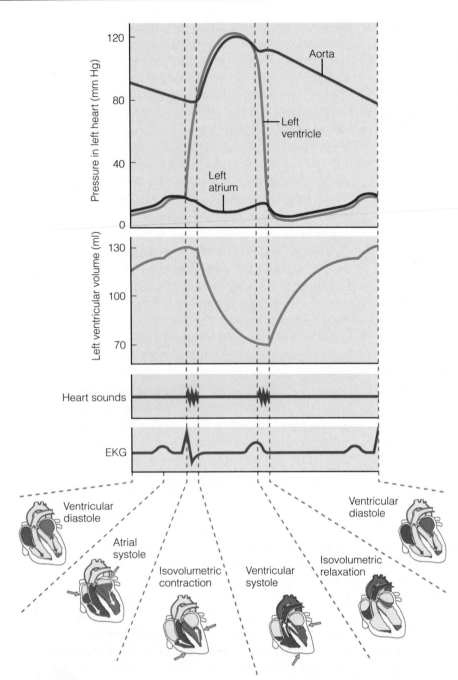

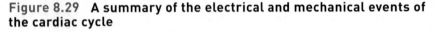

Figure 8.29 A summary of the electrical and mechanical events of the cardiac cycle

rate, stroke volume, or both of these parameters. We have already seen how the nervous and endocrine systems can modulate heart rate by changing the properties of the pacemaker cells of the sinoatrial node. Decreases in heart rate are termed **brady-cardia**, whereas increases in heart rate are termed **tachycardia**. Regulation of heart rate by changes in the rate of depolarization of the sinoatrial node is often referred to as *chronotropy*. Alternatively, the sympathetic nervous system can also increase heart rate by increasing the speed of conduction of the de-

polarization along the conducting pathways of the heart, a phenomenon known as *dromotropy*.

The nervous and endocrine systems can modulate stroke volume

Both the nervous and endocrine systems can also modulate the *contractility* (or rate and strength of contraction) of the heart by altering some of the properties of cardiac excitation-contraction coupling, a phenomenon known as *inotropy*. If the heart contracts more forcefully, it will pump more blood with each beat, increasing the stroke volume. Norepinephrine released by sympathetic neurons and circulating epinephrine released by the endocrine system increase contractility (Figure 8.30). These signaling molecules bind to β_1 adrenergic receptors on contractile cardiomyocytes. Binding of these molecules to the receptor activates a cAMP-mediated signal transduction pathway that activates a protein kinase that phosphorylates a variety of proteins, resulting in increased contractility via four mechanisms.

- Phosphorylation of L-type Ca^{2+} channels on the cell membrane allows increased Ca^{2+} into the cell in response to depolarization.

- Phosphorylation of proteins in the membrane of the sarcoplasmic reticulum causes it to release more Ca^{2+} in response to an action potential.

- Phosphorylation of myosin increases the rate of the myosin ATPase, increasing the rate of cross-bridge cycling and the speed of contraction.

- Phosphorylation of the sarcoplasmic reticulum Ca^{2+} ATPase enhances Ca^{2+} reuptake into the sarcoplasmic reticulum, increasing the rate of relaxation.

The net result of these four mechanisms is that the cardiomyocytes contract faster and more strongly

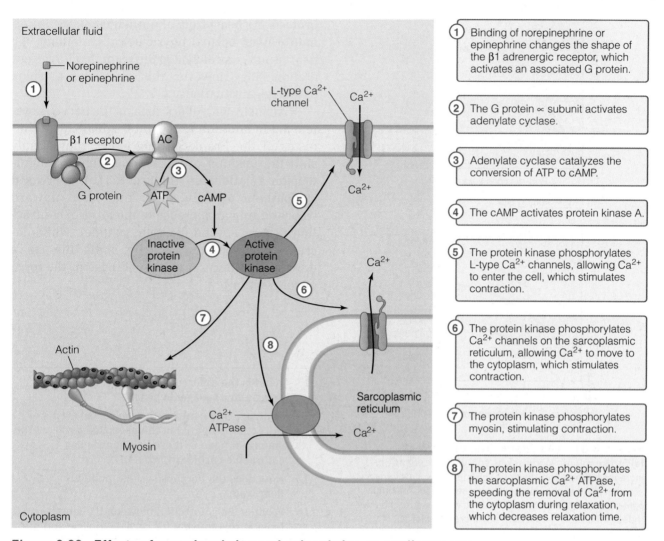

Extracellular fluid

Norepinephrine or epinephrine

β1 receptor

AC

G protein

ATP

cAMP

Inactive protein kinase

Active protein kinase

L-type Ca²⁺ channel

Ca^{2+}

Ca^{2+}

Ca^{2+}

Actin

Myosin

Ca^{2+} ATPase

Sarcoplasmic reticulum

Ca^{2+}

Cytoplasm

(1) Binding of norepinephrine or epinephrine changes the shape of the β1 adrenergic receptor, which activates an associated G protein.

(2) The G protein ∝ subunit activates adenylate cyclase.

(3) Adenylate cyclase catalyzes the conversion of ATP to cAMP.

(4) The cAMP activates protein kinase A.

(5) The protein kinase phosphorylates L-type Ca^{2+} channels, allowing Ca^{2+} to enter the cell, which stimulates contraction.

(6) The protein kinase phosphorylates Ca^{2+} channels on the sarcoplasmic reticulum, allowing Ca^{2+} to move to the cytoplasm, which stimulates contraction.

(7) The protein kinase phosphorylates myosin, stimulating contraction.

(8) The protein kinase phosphorylates the sarcoplasmic Ca^{2+} ATPase, speeding the removal of Ca^{2+} from the cytoplasm during relaxation, which decreases relaxation time.

Figure 8.30 Effects of norepinephrine and epinephrine on cardiomyocyte contractility Norepinephrine and epinephrine increase contractility by binding to β receptors on the cardiomyocyte and activating an adenylate cyclase (AC)–mediated signal transduction pathway that activates protein kinases, which phosphorylate various proteins and cause an increase in the rate and strength of contraction.

in response to sympathetic stimulation, increasing the stroke volume of the heart.

In contrast, stimulation of the parasympathetic nervous system causes a decrease in stroke volume by activating signal transduction pathways that reduce the intracellular Ca^{2+} signal. In mammals, parasympathetic effects are relatively weak in the ventricle, but tend to be strong in the atria.

End-diastolic volume modulates stroke volume

In addition to the extrinsic regulation of heart rate and stroke volume by the nervous and endocrine systems, the heart also undergoes *autoregulation* by intrinsic regulatory mechanisms. If you experimentally increase end-diastolic volume, the ventri-

cle pumps more forcefully, and stroke volume increases (Figure 8.31)—a phenomenon known as the Frank-Starling effect. The Frank-Starling effect is a result of the length-tension relationship for muscle that we discussed in Chapter 5. Stretching a muscle changes the force of contraction by altering the degree of overlap between actin and myosin within the sarcomere. Cardiomyocytes differ from other types of striated muscle in that they are normally shorter than the length needed for optimal contraction so that as you stretch a cardiomyocyte, the strength of contraction increases. When blood enters the ventricle, the increased volume causes the ventricle to stretch, and the more blood that enters the heart at the end of diastole, the greater the degree of stretch. Thus, the end-diastolic volume (the maximum volume during the

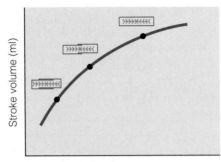

(a) Frank-Starling effect

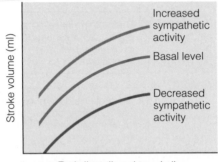

(b) Effects of sympathetic activity on the
Frank-Starling effect

Figure 8.31 The Frank-Starling effect
(a) Stroke volume increases as end-diastolic volume
increases. When end-diastolic volume is low, cardiomyocytes
are shorter than the optimal length needed for maximal
contraction. Increasing end-diastolic volume stretches the
muscle, increasing its length and increasing force
generation. The greater the force generated, the greater the
stroke volume. **(b)** Changes in sympathetic activity alter the
position of the curve. An increase in sympathetic activity
shifts the curve upward, whereas a decrease in activity shifts
the curve downward.

cardiac cycle) is an index of the amount of stretch
imposed on the cardiomyocytes.

The Frank-Starling effect allows the heart to
automatically compensate for increases in the
amount of blood returning to the heart. Consider
what would happen in the absence of the Frank-
Starling effect. If stroke volume remained constant
in the face of an increase in venous return to the
heart, then the heart would pump a smaller fraction of the blood returning to the heart. Assuming
that heart rate remained constant, blood would be
"left over" in the ventricle after each beat and
would slowly build up in the heart, increasing its
volume. Eventually, this might cause the ventricles
to distend to the point that they could no longer
contract effectively. Thus, the Frank-Starling effect

protects the heart from abnormal increases in volume. Under normal physiological conditions the
heart is never stretched to the point that force generation falls. However, this can occur in some
pathological situations.

Extrinsic controllers such as the nervous system act in addition to the autoregulatory mechanisms of the Frank-Starling effect; they simply
shift the position of the cardiac muscle length-
tension relationship (Figure 8.31b). Increased
sympathetic activity shifts the curve upward
(representing an increase in the force of contraction at a given end-diastolic volume), while decreased sympathetic activity shifts the curve
downward (representing a decrease in the force
of contraction).

⊙ │ CONCEPT CHECK

13. What is the difference between a neurogenic
    heart and a myogenic heart?

14. Compare and contrast the molecular events of
    the action potential in the pacemaker cells of the
    sinoatrial node to those in a ventricular
    contractile cardiomyocyte.

15. How does the nervous system modulate
    heart rate?

16. What is the connection between the length-
    tension relationship for cardiac muscle and the
    Frank-Starling law of the heart?

Regulation of Pressure and Flow

Upon leaving the heart, the blood enters the circulation. The pressure as the blood leaves the heart
provides the primary driving force for flow
through the circulatory system. Thus, maintaining
this pressure within appropriate limits is one of
the most important requirements for the proper
functioning of the circulatory system. The other
fundamental requirement for the proper functioning of the vertebrate circulatory system is the ability to appropriately direct flow to the organs,
depending on their metabolic needs. In this part of
the chapter we examine the regulation of blood
flow and blood pressure in vertebrate circulatory
systems.

Regulation of Flow

The circulatory system must regulate the distribution of flow to the tissues. Highly aerobically active tissues have a greater demand for oxygen than do less active tissues, and thus require greater blood flow. The metabolic demands of a tissue can also change greatly with time. For example, the metabolic rate of aerobic skeletal muscle can increase as much as 10-fold between rest and intense exercise. In general, these changes are regulated by altering the diameter, and thus the resistance, of blood vessels leading to the capillary beds.

The arterioles control blood distribution

Arterioles play the primary role in the distribution of flow within the circulatory system because they can vasoconstrict and vasodilate, altering their resistance and thus the flow of blood to the capillary beds. Because the arterioles leading to the various capillary beds are arranged in parallel, an animal can redistribute blood flow to the various organs. For example, during exercise the arterioles of the gut and kidney vasoconstrict, whereas the arterioles of aerobically active skeletal muscles vasodilate, decreasing flow to the internal organs and increasing flow to the active skeletal muscles.

As with the regulation of the heart, both extrinsic factors (such as the nervous and endocrine systems) and intrinsic factors (including the metabolic state of the tissue) control the diameter of the arterioles, and thus regulate the proportion of blood flow going to specific tissues. Intrinsic control mechanisms are particularly important in regulating flow to the heart, brain, and skeletal muscle, while extrinsic factors are the most important controllers of blood flow to organs such as the gut.

Myogenic autoregulation maintains blood flow

Some of the smooth muscle cells surrounding the arterioles are sensitive to stretch and contract when the blood pressure within the arteriole increases. This *myogenic autoregulation* acts as a negative feedback loop that helps to maintain blood flow to a tissue at a constant level. When flow through the arteriole increases, the pressure on the arteriolar wall increases, stretching the smooth muscle. This stretch causes the smooth

muscle to contract, constricting the arteriole. The decrease in arteriolar diameter increases the resistance and decreases the flow, decreasing the pressure, which causes the smooth muscle to relax. Thus, myogenic autoregulation tends to maintain constant blood flow to a tissue. But the metabolic activity of a tissue and its demand for oxygen can vary with time, and thus the need for blood flow varies. For example, when you are sitting still, the muscles of your legs have relatively low demand for oxygen, and little blood flows to them, whereas when you are jogging, your muscles require more oxygen, so more blood must flow to the tissue. Other mechanisms for controlling blood flow come into play when the needs of the tissue change.

The metabolic activity of the tissue influences blood flow

The vascular smooth muscle cells surrounding the arterioles are sensitive to the conditions in the extracellular fluid that surrounds them. They contract or relax in response to changes in the concentrations of substances such as oxygen, carbon dioxide, H^+, K^+, and a variety of paracrine signals (Table 8.1). In general, changes in the extracellular fluid that are associated with increased activity cause vasodilation, while changes that are associated with decreased activity cause vasoconstriction. Thus, decreases in oxygen or increases in carbon dioxide tend to cause vasodilation. Vasodilation increases blood flow to the tissue, bringing more oxygen and carrying away waste products. This reduces the signal to the muscle cell, in a negative feedback loop, stopping the flow from increasing beyond what is needed (Figure 8.32).

Paracrine signaling molecules released from the vascular endothelium also have a profound effect on vascular smooth muscle (Table 8.1). For example, the gas nitric oxide is an important vasodilator. Vascular smooth muscle cells actually release a small amount of nitric oxide all the time, which helps to keep the arterioles dilated. However, nitric oxide production is strongly induced by histamine, bacterial lipopolysaccharides, and other substances that are associated with damage to the vascular endothelium. The increased nitric oxide causes vasodilation, increasing blood flow to damaged areas. This is an important mechanism

Table 8.1 Factors influencing vasoconstriction and vasodilation.

Substance	Source	Type
Vasoconstriction		
Stretch on arteriolar walls	Increased blood pressure	Myogenic autoregulation
Norepinephrine (α receptors on arterioles in most tissues except skeletal and cardiac muscle, which express $\beta 2$ receptors)	Sympathetic neurons	Neural
Endothelin	Vascular endothelium	Paracrine
Serotonin	Platelets	Paracrine
Vasopressin	Posterior pituitary	Endocrine
Angiotensin II	Plasma	Endocrine
Vasodilation		
Hypoxia	Multiple tissues	Metabolite
Increased CO_2	Multiple tissues	Metabolite
H^+	Multiple tissues	Metabolite
K^+	Multiple tissues	Metabolite
Nitric oxide	Endothelium	Paracrine
Atrial naturietic peptide	Atrial myocardium	Endocrine
Histamine	Mast cells of immune system	Paracrine (systemic actions at high levels)
Substance P	Damaged tissue	Paracrine
Prostacyclin	Damaged tissue	Paracrine
Epinephrine ($\beta 2$ receptors in skeletal muscle arterioles)	Adrenal medulla	Endocrine
Acetylcholine (muscarinic receptors)	Parasympathetic neurons leading to erectile tissue of clitoris or penis	Neural
Bradykinin	Multiple tissues	Paracrine
Adenosine	Hypoxic cells	Paracrine

underlying inflammation. Nitric oxide is also released in the arterioles of skeletal muscles during exercise, causing vasodilation that increases the supply of oxygen to the working muscle.

Nitric oxide activates the enzyme guanylate cyclase in the vascular smooth muscle. Guanylate cyclase catalyzes the conversion of GMP to cGMP, which triggers the muscle cell to relax, causing vasodilation (see Chapter 3: Cell Signaling and Endocrine Regulation). The cGMP is quickly broken down by the enzyme phosphodiesterase, preventing the arteriole from staying permanently dilated and allowing it to constrict or dilate as necessary. The drug sildenafil (Viagra) specifically targets an isoform of phosphodiesterase that is found in the arterioles of the penis. Sildenafil prevents the cGMP from breaking down, prolonging the effects of nitric oxide and causing vasodilation in the vessels of the penis, leading to a sustained erection.

The nervous and endocrine systems regulate arteriolar diameter

In addition to intrinsic and local control mechanisms, the arterioles respond to extrinsic controllers such as the nervous and endocrine systems. The sympathetic nervous system controls the smooth muscle surrounding the arterioles. In

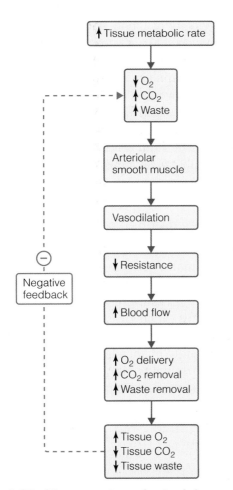

Figure 8.32   The response of arteriolar smooth muscle to an increase in metabolic activity

system causes the vascular smooth muscles on the arterioles leading to the gut and kidneys to contract, and the resulting vasoconstriction reduces flow. This sympathetic stimulation also tends to cause vasoconstriction of the arterioles leading to the skeletal muscle and in the coronary arteries. However, local paracrine factors, such as nitric oxide and adenosine released by these muscles, cause a vasodilation that outweighs the vasoconstriction mediated by the sympathetic nervous system. Thus, the overall result is a vasodilation of the arterioles leading to the skeletal muscles and heart, and a vasoconstriction of the arterioles leading to the kidneys and gut.

Three other hormones also affect vascular smooth muscle. Vasopressin (also called ADH) released from the posterior pituitary gland, and angiotensin II, a hormone involved in the regulation of the kidney, promote generalized vasoconstriction, while atrial natriuretic peptide promotes a generalized vasodilation. We discuss these hormones in more detail in Chapter 10: Ion and Water Balance.

The nervous and endocrine systems work together with the paracrine signals that relate to metabolic activity to influence arteriolar diameter and alter blood flow. As a result, blood flow to each tissue of the body is almost always carefully controlled in order to deliver the amount of blood that the tissue needs.

CONCEPT CHECK

17. Does myogenic autoregulation play an important role in changing blood flow to tissues as oxygen demand increases? (Justify your answer.)

18. What is vasomotor tone?

19. Why, during the fight-or-flight response, do the arterioles leading to active skeletal muscles dilate even though sympathetic stimulation causes vasoconstriction?

vertebrates, the sympathetic nervous system always maintains a certain degree of **vasomotor tone** so that the arterioles are slightly constricted. Increases or decreases in the activity of these sympathetic neurons can alter the degree of vasomotor tone by acting on the smooth muscles surrounding the arterioles. Norepinephrine released from sympathetic neurons binds to adrenergic receptors on these muscle cells, activating a phosphatidylinositol second messenger system, and causing vasoconstriction. So increases in sympathetic activity tend to cause vasoconstriction, whereas decreases in sympathetic activity tend to cause vasodilation.

As we discussed in Chapter 3, the sympathetic nervous system is stimulated as part of the fight-or-flight response. During this response, blood is directed away from organs such as the gut and kidneys, and toward the skeletal muscles and heart, readying the body for action. Norepinephrine released from the sympathetic nervous

Regulation of Pressure

As shown in Figure 8.33, blood pressure differs in the different parts of the circulatory system. Notice that blood pressure in the left ventricle also changes greatly over time. During ventricular systole, the ventricular pressure is very high, and during diastole it is low. The high systolic pressure in the left ventricle forces blood out into the aorta. The aorta is a large vessel with relatively low resistance, so pressure remains relatively high as blood travels through this and subsequent arteries. Because arterioles are relatively narrow vessels (compared to arteries) and are relatively few in number (compared to capillaries), they have the highest resistance of any part of the circulatory system. Thus, pressure drops greatly as blood travels through the arterioles, and continues to drop as blood proceeds through the capillaries, venules, and veins. By the time the blood returns to the heart, its pressure is barely above ambient. The pressure gradient between the left ventricle and the right atrium causes blood to flow through the system according to the law of bulk flow.

The velocity of blood flow also varies greatly across the circulatory system (Figure 8.33). Blood velocity is greatest in the arteries and veins, and lowest in the capillaries, because blood velocity is inversely proportional to the total cross-sectional area of the circulatory system at any given point. The low velocity of the blood, combined with the thin walls of the capillaries, allows for efficient exchange of substances between the capillaries and the tissues.

The arteries dampen pressure fluctuations

Notice that the pressure fluctuations in the arteries are far smaller than those in the left ventricle. The aorta (and the bulbus arteriosus of a bony fish) acts as a pressure reservoir and dampens the fluctuations in blood pressure that occur during the cardiac cycle (Figure 8.34). During systole, the ventricle rapidly pushes blood into the aorta. Because the aorta splits into progressively narrower blood vessels, the exit from the aorta has relatively high resistance, so instead of simply flowing out into the rest of the circulatory system, the blood

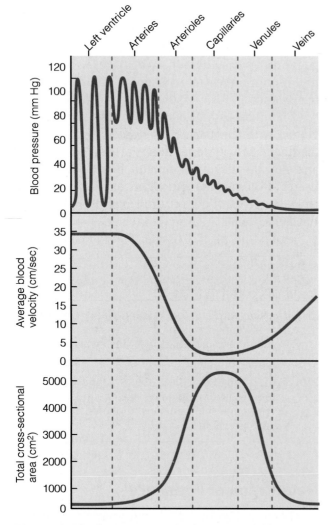

Figure 8.33 Pressure, velocity, and total cross-sectional area across a vertebrate circulatory system Pressure is variable in the ventricle, high and more constant in the arteries, and drops greatly across the arterioles. Blood velocity is inversely proportional to total cross-sectional area of that part of the circulatory system.

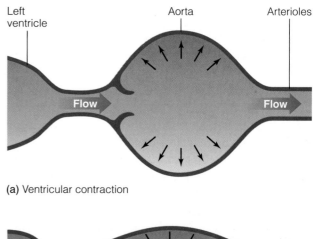

(a) Ventricular contraction

(b) Ventricular relaxation

Figure 8.34 The aorta as a pressure reservoir
(a) Blood flows rapidly into the aorta during the ejection phase of ventricular contraction, pushing out on the walls of the aorta and causing it to expand. **(b)** As the heart relaxes, blood flow into the aorta ceases, but flow out into the arterioles continues, reducing the aortic pressure. Elastic recoil of the arterial walls helps to push blood through the vasculature, maintaining pressure and flow.

tends to back up and exert pressure on the thick, elastic walls of the aorta. This pressure causes the aorta to expand. Because the walls of the aorta are elastic, they act very much like a spring that stores energy as it is stretched.

When the heart enters diastole, blood ceases flowing into the aorta. But blood continues to flow out of the aorta into the arterioles, reducing the pressure inside the aorta. This is equivalent to releasing a spring, and the aortic walls snap back into place. This **elastic recoil** propels the blood through the circulatory system and maintains an aortic pressure that is higher than the diastolic pressure in the ventricle, dampening the pressure fluctuations associated with the cardiac cycle. This elastic recoil also helps to maintain relatively continuous flow of blood into the arteries throughout the cardiac cycle. Because of the elastic nature of the aorta, the aortic pressure is higher than the ventricular pressure during some parts of the cardiac cycle, but the aortic semilunar valve prevents backflow of blood from the arteries to the heart.

Mean arterial pressure is determined by systolic and diastolic pressures

The pressure in the aorta is called the *arterial blood pressure*. Although the pressure fluctuations in the aorta are not as large as those in the ventricle, arterial blood pressure still varies with the phases of the cardiac cycle from its maximum, the systolic pressure, to its minimum, the **diastolic pressure**. Table 8.2 shows some typical values for systolic and diastolic pressure in a few representative vertebrates. Physiologists often consider the **mean arterial pressure (MAP)**, or the average blood pressure in the arteries across the cardiac cycle, which allows them to ignore the pulsatile nature of blood pressure and apply to the cardiovascular system the simple physical principles of fluid flow. MAP in humans can be approximated as follows:

$$MAP = 2/3 \text{ diastolic pressure} + 1/3 \text{ systolic pressure}$$

Thus, using the data from Table 8.2, we can calculate that the mean arterial pressure in humans is typically around 93 mm Hg at rest. However, the length of diastole varies depending on the heart rate, so at high heart rates MAP is better approximated as the average of systolic and diastolic pressures.

The skeletal muscle and respiratory pumps aid venous return to the heart

By the time the blood enters the veins it is under relatively low pressure, and little driving force remains to return blood to the heart. Two major pumps assist in moving blood back to the heart: the skeletal muscle and respiratory pumps. When skeletal muscles contract, they squeeze the veins, increasing the pressure inside these blood vessels

Table 8.2 Systolic and diastolic pressure in representative animals.		
Species	**Systolic pressure (mm Hg)**	**Diastolic pressure (mm Hg)**
Homo sapiens (human)	120	80
Equus caballus (horse)	100	60
Rattus norvegicus (rat)	130	90
Canis familiaris (dog)	140	80
Loxodonta africana (African elephant)	120	70
Columba livia (pigeon)	135	100
Turdus migratorius (robin)	118	80
Pseudemys scripta (turtle—red-eared slider)	31	25
Rana catesbeiana (bullfrog)	32	21
Oncorhynchus mykiss (rainbow trout)	45	33
Ictalurus punctatus (channel catfish)	40	30
Octopus vulgaris (octopus)	27	15

(Figure 8.35). Veins that are located outside of the *thoracic* (chest) cavity contain valves. The increased pressure as a result of the contraction of skeletal muscles forces the valves farthest from the heart to close and the valves closest to the heart to open, pushing blood toward the heart. The rhythmic contraction of this *skeletal muscle pump* helps to drive blood toward the heart, increasing *venous return* to the heart.

Respiratory movements can also help to draw blood toward the heart. As we discuss in more detail in Chapter 9, in terrestrial vertebrates the thoracic cavity expands during inhalation, causing the pressure in the thoracic cavity to drop, and drawing air into the lungs. This low thoracic pressure helps to draw blood into the veins of the thoracic cavity, acting as a *respiratory pump*. During exhalation, the pressure in the thoracic cavity increases, but the valves in the veins outside the thoracic cavity prevent backflow of blood out of the thoracic

cavity. Instead, this increased pressure pushes the blood in the other direction, toward the heart.

The veins act as a volume reservoir

The veins have highly compliant walls that stretch easily; small increases in blood pressure lead to large changes in the volume of the veins compared to the volume of the arteries (Figure 8.36). As a result, the veins can act as a volume reservoir for blood. In fact, in mammals the veins typically hold more than 60% of the total volume of blood in the body. The sympathetic nervous system regulates the proportion of blood in the venous versus arterial systems by altering the **venomotor tone**. The smooth muscles surrounding the venules and small veins contain α adrenergic receptors. Norepinephrine released from sympathetic neurons binds to these receptors, causing the smooth muscle to contract, reducing the diameter of the veins. Because the majority of the blood is contained in these numerous smaller blood vessels, a decrease in the volume of the venules and small veins decreases the volume of the venous reserve. This in turn increases venous return to the heart, increasing cardiac output and forcing blood into the arterial side of the circulation.

Peripheral resistance influences pressure

We can rewrite the law of bulk flow as follows to specifically apply to vertebrate circulatory systems:

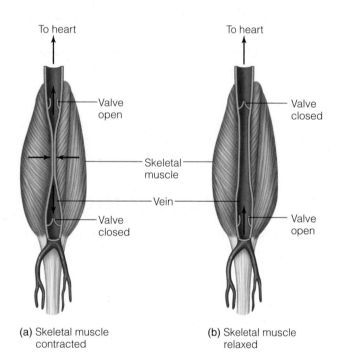

To heart　　　　　　　To heart

Valve open

Valve closed

Skeletal muscle

Vein

Valve closed

Valve open

(a) Skeletal muscle contracted

(b) Skeletal muscle relaxed

Figure 8.35　The skeletal muscle pump　(a) When a skeletal muscle contracts, it puts pressure on the vein, pushing blood in both directions. The resulting pressure opens the proximal one-way valve and closes the distal one-way valve, squeezing blood toward the heart and preventing backflow. **(b)** When the skeletal muscle relaxes, the one-way valves are in the opposite configuration. The relaxation reduces pressure on the distal valve, which opens and allows blood to flow in. Back pressure from the blood in the proximal segment of the vein closes the proximal valve, preventing backflow.

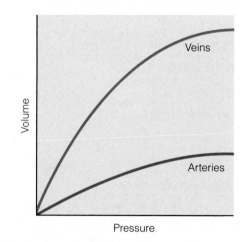

Volume

Veins

Arteries

Pressure

Figure 8.36　Compliance of arteries and veins Veins are far more compliant than arteries and thus they stretch easily, increasing their volume in response to increases in pressure.

$$CO = MAP/TPR$$

where cardiac output (CO) is a measure of the total flow (Q) through the system, and TPR (**total peripheral resistance**) is the summed resistances of all the blood vessels in the body and is a measure of the resistance (R) of the circulatory system. We can approximate the pressure gradient across the circulatory system (ΔP) using the mean arterial pressure (MAP). The actual change in pressure across the circulatory system is MAP minus the central venous pressure (CVP, the pressure in the superior vena cava near the right atrium). CVP is usually low relative to MAP, so MAP is approximately equal to the pressure gradient across the circulatory system.

The body varies CO and TPR to maintain MAP within very narrow boundaries. TPR is set primarily by the state of vasoconstriction and vasodilation of the arterioles, which is in turn set largely by the metabolic needs of the tissue. CO (and thus heart rate and stroke volume) varies in response to these changes in TPR in order to maintain MAP within a narrow range. Thus, the metabolic demand of the tissues is the ultimate regulator of the circulatory system. Figure 8.37 provides a summary of the major factors involved in the homeostatic regulation of MAP.

The baroreceptor reflex is the primary means of regulating MAP

Baroreceptors are stretch-sensitive mechanoreceptors that are located in the walls of many of the major blood vessels. The most important of these baroreceptors are located in the carotid artery and aorta, although the large systemic veins, the pulmonary arteries, and the walls of the heart also contain baroreceptors. The carotid artery is the major artery leading to the head, and thus the **carotid body** baroreceptors monitor blood pressure to the brain. The aorta is the primary artery leading to the systemic circulation, so the **aortic body** baroreceptors monitor mean arterial pressure. Under normal conditions these baroreceptors fire a steady stream of action potentials, sending signals via primary afferent neurons to the central nervous system. The **cardiovascular control center** in the medulla oblongata of the central nervous system integrates these inputs,

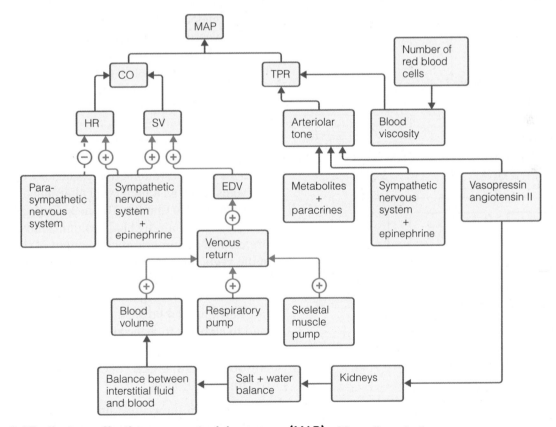

Figure 8.37 Factors affecting mean arterial pressure (MAP)   CO: cardiac output; TPR: total peripheral resistance; HR: heart rate; SV: stroke volume; EDV: end-diastolic volume.

and sends out efferent signals via autonomic neurons that control heart rate, stroke volume, and vasomotor and venomotor tone, thus influencing blood pressure. Increases in blood pressure cause the walls of the arteries to stretch, increasing the firing rate of the baroreceptors, and causing signals that result in a reduction of blood pressure. Decreases in blood pressure cause the walls of the arteries to relax, decreasing the firing rate of the baroreceptors. The decrease in baroreceptor firing causes efferent signals that result in increased blood pressure. Thus, the baroreceptor reflex is a negative feedback loop that homeostatically regulates blood pressure within a relatively narrow range.

Figure 8.38 shows the major steps of the baroceptor reflex following an increase in blood pressure. Increases in blood pressure stretch the membrane of the baroreceptors in the aortic and carotid bodies, increasing the firing rate of the receptor and the frequency of action potentials traveling to the medullary cardiovascular control center in the central nervous system. The control center integrates the sensory input, and produces an efferent output carried by autonomic neurons. There is a decrease in sympathetic output, resulting in vasodilation. This decrease in sympathetic output in combination with an increase in parasympathetic output results in a decrease in the force of cardiac contraction and a decrease in heart rate. Together, these factors lead to a decrease in peripheral resistance and cardiac output, and a concomitant decrease in blood pressure. The medullary cardiovascular center also decreases the secretion of the hormones vasopressin and angiotensin in response to increased blood pressure. Because these hormones constrict arterioles, decreasing their secretion reduces total peripheral resistance.

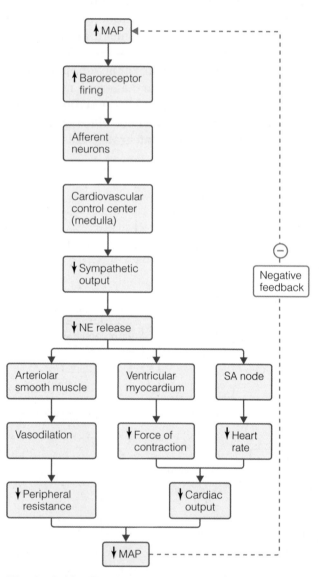

Figure 8.38 The baroreceptor reflex MAP: mean arterial pressure; NE: norepinephrine; SA node: sinoatrial node.

The kidneys play a major role in maintaining blood volume

In a closed system, pressure and volume are intimately related. If you increase the volume of a fluid inside a vessel with a fixed volume, the pressure inside that vessel will increase. (This is the principle behind the isovolumetric contraction of the heart.) Therefore, increases in blood volume will lead to an increase in blood pressure, whereas decreases in blood volume will lead to a decrease in blood pressure. The veins are compliant, and can act as a volume reservoir, but their capacity is not infinite. Any changes in blood volume that exceed the capacity of the veins to act as a buffer will alter blood pressure. As we discuss in Chapter 10, the kidneys play a major role in maintaining blood volume, and thus these organs are an important component of the homeostatic regulation of blood pressure. Figure 8.39 illustrates how changes in mean arterial pressure can lead to changes in blood volume by altering kidney function, and how changes in blood volume can lead to changes in arterial pressure.

Blood pressure can force fluid out of the capillaries

In addition to the critical importance of regulating mean arterial pressure in order to maintain the driving force for movement of blood through the vertebrate circulatory system, it is also critical to maintain

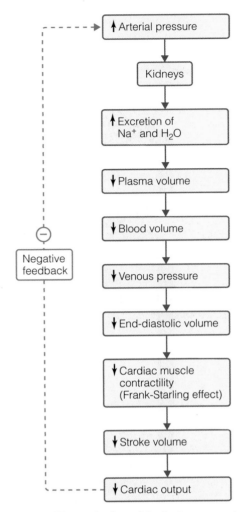

Figure 8.39 The relationship between arterial pressure and blood volume

The direction of fluid flow across a capillary wall is the result of the net filtration pressure (NFP), which can be expressed as

$$\text{NFP} = (P_{\text{cap}} - P_{\text{if}}) - (\pi_{\text{cap}} - \pi_{\text{if}})$$

This relationship, called the Starling principle of fluid exchange, allows us to quantify the movement of fluid across a capillary. The hydrostatic pressure in the capillary is the major driving force pushing fluids from the blood and into the interstitial spaces. If hydrostatic pressure in the capillary is larger than the hydrostatic pressure in the interstitial fluid, then fluids will be forced out of the capillary. Continuous capillaries are permeable only to small molecules, so that plasma proteins and blood cells remain behind in the blood, causing the blood to have a higher osmotic pressure than the interstitial fluid. Because salts and other small molecules are present in roughly equal concentration in the blood and the interstitial fluid, the difference in osmotic pressure between these two compartments is due largely to the presence of proteins in the blood. An osmotic pressure that is due to proteins is termed an **oncotic pressure**. The higher oncotic pressure in the capillaries tends to suck fluids back into the blood. The balance between these two forces influences the rate and direction of fluid movement.

Figure 8.40 illustrates how these forces change as fluids move along capillaries from the arterial side to the venous side. The osmotic pressure of the blood and interstitial fluid remains fairly constant across a capillary bed, but the hydrostatic pressure of the blood declines substantially as it travels from

blood pressure to ensure appropriate fluid balance at the capillaries.

Because of the presence of pores between the cells of the capillary wall, fluids can move from the capillaries to the interstitial fluids by bulk flow. Four forces (called *Starling forces* after the physiologist Ernest Starling, who discovered this principle in 1896) influence the bulk flow of fluids across the capillaries:

1. Hydrostatic pressure in the capillary (P_{cap}) (the transmural pressure)

2. Hydrostatic pressure in the interstitial fluid (P_{if})

3. Osmotic pressure in the capillary (π_{cap})

4. Osmotic pressure in the interstitial fluid (π_{if})

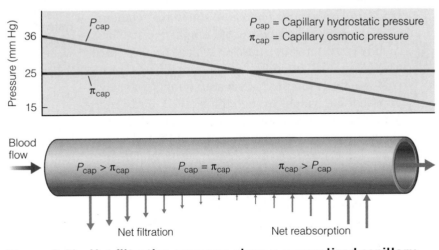

Figure 8.40 Net filtration pressure along a generalized capillary
At the start of the capillary, hydrostatic pressure (P) exceeds capillary osmotic pressure (π), resulting in a net filtration pressure that forces fluid out of the capillary. At the end of the capillary, hydrostatic pressure is less than capillary osmotic pressure, resulting in net reabsorption that returns some of the fluid to the capillary.

the arterial to the venous end of the capillary bed because of the frictional resistance of the capillary walls. At the arterial end of the capillary the net filtration pressure is positive, indicating that fluid will flow out into the interstitial fluid. At the venous end of the capillary, the net filtration pressure is negative, indicating that fluid will flow back into the capillary.

This balance of forces is true for an idealized capillary, but many capillaries show filtration across their entire length, and some specialized capillaries in the intestinal mucosa reabsorb fluids along most of their length. Whatever the capillary, however, the important issue to consider is the balance of Starling forces. Vertebrates have good control over capillary pressure, mostly through vasoconstriction and vasodilation of the blood vessels leading to capillary beds, and changes in these parameters will lead to changes in the rate of fluid filtration.

Under normal circumstances in humans almost 20 liters of fluid per day filters out of the capillaries, or almost six times the total volume of the plasma in an average human being. About 17 liters of this fluid is usually reabsorbed into the blood, but this leaves an excess of almost 3 liters of fluid per day that could accumulate in the interstitial fluid.

The lymphatic system returns filtered fluids to the circulatory system

The lymphatic system collects the filtered fluid and returns it to the circulatory system (Figure 8.41). Fluid enters the lymphatic system via the blind-ended *lymphatic capillaries*. The lymphatic capillaries coalesce into progressively larger vessels termed *lymphatic veins* and *lymphatic ducts* that contain valves to prevent backflow of the lymph, and are surrounded by smooth muscle, which propels the lymph forward. In addition, fish, amphibians, reptiles, and bird embryos have **lymph hearts** that help to propel the lymph through the body. In birds and mammals, the lymphatic ducts lead to small bean-shaped organs called **lymph nodes**. All vertebrates have lymph nodes in the thoracic cavity, abdomen, and pelvis. In addition, mammals have so-called external lymph nodes located in their necks and at the point where the limbs and torso join (the armpit and groin areas in humans). The lymph nodes filter the lymph, and contain specialized blood cells called lymphocytes that kill pathogens and cancerous cells. From the lymph nodes, the filtered lymph travels through the efferent lymphatic vessels that drain into the circulatory system at the veins of the neck.

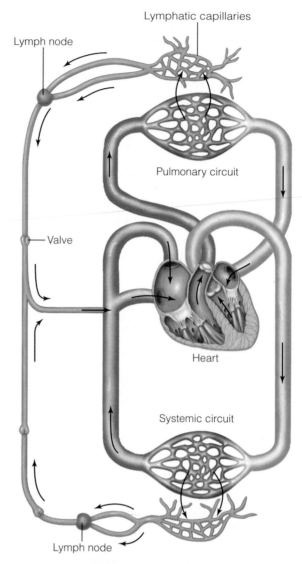

Figure 8.41 Relationship between the mammalian circulatory and lymphatic systems Some fluid leaving the capillaries enters the lymphatic system. This fluid, lymph, flows through the lymph nodes and lymphatic ducts, returning to the venous part of the circulatory system near the right atrium. The lymphatic ducts contain valves that ensure unidirectional flow.

Anything that alters the balance between filtration and reabsorption of fluids across the capillary beds or the function of the lymphatic system may lead to accumulation of fluids in the tissues—a condition called **edema**. For example, sitting in one position for a long period of time (such as in an airplane) can reduce blood flow in the veins and cause blood to pool in the capillaries of the ankles and feet. The pooled blood leads to increased capillary hydrostatic pressure, which leads to increased filtration of fluids and ankle edema. Liver disease also affects capillary pressure, because the majority of plasma proteins are produced in the liver. If plasma protein concentration drops, plasma osmotic pressure will drop, reducing the reabsorption of water at the venous end of the capillaries, and increasing net filtration, leading to generalized edema. Alternatively, removal of the lymph nodes (for example, as a part of cancer treatment) can compromise the function of the lymphatic system, preventing the removal of fluid filtered from the capillaries, which leads to edema of the affected tissues.

Pulmonary edema, in which fluids accumulate in the tissues of the lungs, is one of the most dangerous forms of edema. When fluid accumulates in the lungs, it becomes more difficult for oxygen to diffuse from the lungs to the blood. As a result, pulmonary edema can be fatal. Anything that increases the net filtration pressure in the lung capillaries has the potential to cause pulmonary edema, if the rate of filtration exceeds the rate at which the lymphatic system can remove the fluid. For example, if a heart attack damages the muscle of the left ventricle but spares the right ventricle, the right side of the heart may pump more blood per beat than the left side of the heart. This causes blood to back up into the lungs, and increases the hydrostatic pressure in the capillaries, which increases the net filtration pressure and can lead to pulmonary edema.

Changes in body position can alter blood pressure and flow

Because of the effects of gravity, an unobstructed vertical column of fluid exerts a pressure, termed the **hydrostatic pressure**, on objects below it (Figure 8.42a). The hydrostatic pressure exerted by a fluid column is thus a function of the effects of gravity and the height of the column. We can express this relationship mathematically as follows:

$$\Delta P = \rho g h$$

where ΔP is the difference in pressure between two points in the fluid column, ρ is the density of the fluid, g is the acceleration due to gravity, and h is the height of the fluid column (Figure 8.42a). As you can see from this equation, hydrostatic pressure is a measure of the gravitational potential energy of the fluid column.

When a person is lying down (Figure 8.42b), this gravitational component is absent, and measured pressure in the feet and head is slightly lower than in the heart. We usually report blood pressure relative to the surrounding atmospheric pressure, so the pressure shown in the figure is ac-

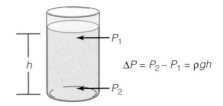

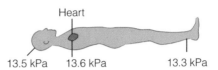

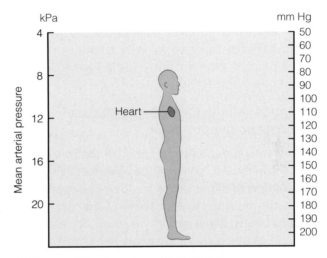

(c) Measured blood pressure when standing

Figure 8.42 The effects of gravity on blood pressure Blood pressure is generally measured either in kilopascals (kPa), the SI unit of pressure, or in millimeters of mercury (mm Hg), the unit most commonly used in medical diagnostics. 100 mm Hg is equal to 13.3 kPa. **(a)** Hydrostatic pressure is the result of the gravitational potential energy of the fluid column. **(b)** When a human is lying down, arterial blood pressure is highest at the heart and lowest at the feet. **(c)** In a standing human, arterial blood pressure is highest in the feet and lowest in the head.

tually the amount by which the pressure of the blood exceeds the ambient atmospheric pressure. For example, the mean arterial blood pressure near the human heart is approximately 13.6 kPa, but the actual pressure is 13.6 kPa plus approximately 101 kPa (the atmospheric pressure at sea level), for a total of 114.6 kPa. The pressure gradient between the heart and the rest of the body drives blood flow around the circuit.

In contrast, Figure 8.42c shows the blood pressure in various parts of a human body when standing. When standing, the pressure measured in the ankles is higher than pressure near the heart. If liquids flow from areas of high pressures to areas of low pressures, how can the heart pump blood down to the feet? This anomaly is explained by the fact that the pressure measured in the ankles is the sum of the pressure exerted by the heart plus the hydrostatic pressure exerted by the blood in the circulatory system "pushing down" on the blood in the ankles. The hydrostatic pressure actually represents the gravitational potential energy of the column of blood, and potential energy is highest at the top of the fluid column. Fluids tend to flow from areas of high potential energy to areas of low potential energy. In essence, blood "falls" downward in the circulatory system.

As blood returns up the body to the heart it must move against a gradient of gravitational potential energy. This hydrostatic pressure component is absent when a person lies down (as in Figure 8.42b).

Changes in body position can cause orthostatic hypotension

When you stand upright, gravity tends to push blood downward, because the effects of gravity on the column of blood in the blood vessel exert a hydrostatic pressure on the parts of the circulatory system below. Thus, when we stand up, a certain amount of blood normally pools in our ankles and legs. This pooling causes a slight decline in venous return to the heart. Because of the Frank-Starling effect, reduced venous return leads to decreased stroke volume and a momentary drop in arterial blood pressure. This drop in blood pressure brings the baroreceptor reflex into play, setting in motion all of the changes that bring blood pressure back to normal. If these reflexes do not act quickly enough, we can experience *orthostatic hypotension*, or low blood pressure due to a vertical body position. This lowered blood pressure can lead to reduced blood flow to the brain, which can cause fainting. People who have inefficient baroreceptor reflexes often feel dizzy or faint if they stand up too quickly.

The effects of changes in body position are not usually as profound for animals other than humans, since in most animals the head and heart are at similar elevations. However, the problems of gravity can be acute for some animals. For example, tree-dwelling snakes often orient themselves almost vertically with their heads up when climbing up a tree, but can also hang with their heads down as they watch prey passing below them. The heart of a tree-dwelling snake is located much

closer to the brain than in most other snakes. This placement helps to make sure that blood reaches the brain regardless of body position.

Physiologists have long been fascinated by the circulatory dynamics of very tall animals, such as the giraffe (Figure 8.43). A giraffe's head can be as much as 2 m above its heart, while its legs are 2 m below the heart. Thus, there is a large gravitational potential energy barrier to overcome in pumping blood up to the head. It is possible that some or all of this energy is recovered via a siphon effect, as the blood moves downward back to the heart. However, comparative physiologists do not currently agree on whether this effect is physiologically relevant. Whatever the case, clearly the very long blood vessels will have high resistance.

A giraffe has an extremely large and muscular heart and the highest blood pressure known for any mammal. With a systolic pressure of up to 280 mm Hg and a diastolic pressure of 180 mm Hg at heart level, its blood pressure is twice that of a typical human. A resting giraffe also has a very high heart

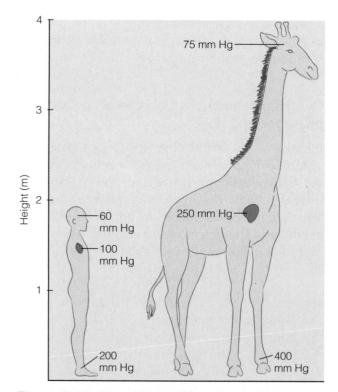

Figure 8.43 The effects of gravity on the circulatory system of a giraffe Animals with a very long neck must have relatively high mean arterial pressure at the heart in order to pump blood to the head. The long legs of the giraffe also greatly increase the hydrostatic pressure in the legs, potentially causing a problem with peripheral edema. To combat this high hydrostatic pressure, giraffes have extremely tight skin on their legs that exerts an inward pressure that opposes the hydrostatic pressure due to gravity.

rate (about twice that of humans, or approximately 170 beats per minute versus 70 bpm). This observation is particularly surprising because heart rate tends to decrease with size in mammals. These cardiac specializations may be needed to pump blood through the systemic circuit of a giraffe.

The high blood pressure of a giraffe, combined with the effects of gravity on the hydrostatic pressure within the circulatory system, would tend to force blood out of the capillaries into the interstitial fluid in the ankles, causing peripheral edema in the absence of mechanisms to prevent this problem. Giraffes have unusually thick-walled and muscular arteries in their legs that help to control the flow of blood. But the most important difference between a giraffe and other mammals is that the skin on a giraffe's legs is extremely tight. This tight skin helps the skeletal muscle pump to function efficiently, and increases the interstitial fluid pressure, reducing the risk of edema.

When a giraffe bends down to drink, the head goes from being several meters above the heart to several meters below it. The resulting increase in the hydrostatic pressure in the head could cause blood to pool in the veins, potentially causing edema in the tissues of the head. Like pulmonary edema, cerebral edema (or accumulation of fluid around the brain) can be life threatening. However, a giraffe has an intricate network of highly elastic blood vessels near the brain that act as a pressure reservoir that expands to accommodate excess blood when the head is lowered, preventing it from pooling in the venous system. In addition, unlike in other mammals, the jugular vein (leading from the head) contains a series of one-way valves that prevent backflow of the blood away from the heart when the giraffe's head is down. Together these mechanisms help to regulate blood flow to the head, regardless of the giraffe's position.

CONCEPT CHECK

20. Explain how the large arteries (such as the aorta) dampen pressure fluctuations and even out blood flow across the cardiac cycle.
21. What is the influence of the skeletal muscle pump on venous return to the heart?
22. Outline the response of the cardiovascular system to a drop in mean arterial pressure.
23. What would happen to the production of lymph if capillary hydrostatic pressure doubled?

Blood

Blood and hemolymph, the circulating fluids of closed and open circulatory systems, respectively, are complex fluids consisting of many components. These circulating fluids play a variety of roles, providing a relatively constant internal environment and transporting nutrients, oxygen, waste products, immune cells, and signaling molecules around the body. As we discuss in later chapters, these fluids can also play noncirculatory roles. For example, the hydrostatic pressure exerted by the hemolymph helps spiders to extend their limbs. You may have noticed that injured or dead spiders always have their legs bent inward, curled around their bodies. Spiders extend their limbs by contracting muscles in the thorax. This muscle contraction increases the hydrostatic pressure of the hemolymph in the thorax, pumping fluid into the legs, and causing the legs to extend outward. Injured or dead spiders can no longer control the hydrostatic pressure of the hemolymph, and the legs return to the contracted position—curled around the body. Similarly, earthworms and other annelids use a hydrostatic skeleton for locomotion (see Chapter 12: Locomotion). In insects, increases in the hydrostatic pressure of the hemolymph are involved in molting and the unfurling of the wings as an insect emerges from its pupa.

Composition of Blood

Blood and hemolymph are primarily composed of water containing dissolved ions and organic solutes and are thus similar in composition to interstitial fluid. However, these circulatory fluids also contain blood cells and relatively high concentrations of proteins. As we discuss in more detail in later chapters, many animals maintain the composition of their blood and interstitial fluid quite distinct from the external environment, homeostatically regulating the composition of the blood. However, in some animals the composition of body fluids varies in concert with the environment.

Blood contains proteins

Interstitial fluid typically has a low protein concentration (ranging from 0.2 to 2.0 g/l). In contrast, the circulatory fluids of animals with closed circulatory systems often contain a rather high concentration of proteins. For example, protein concentration may be 10–90 g/l of hemolymph in decapod crustaceans, 30–80 g/l of blood in vertebrates, and up to 110 g/l of blood in cephalopod molluscs. In the invertebrates, these proteins are primarily **respiratory pigments** that are used to transport or store oxygen (see Chapter 9 for more on the structure and function of respiratory pigments). In the vertebrates, the respiratory pigments are located within cells, and thus the principal proteins dissolved in the circulatory fluids are carrier proteins such as **albumin** and the *globulins*, and proteins involved in blood clotting.

Blood contains cells

The diverse cell types found in the circulatory fluid of many animals are called **hemocytes**. Hemocytes perform a wide variety of functions in different animals, including oxygen transport or storage, nutrient transport or storage, phagocytosis of damaged cells, immune defense, and blood clotting. In many species, the coelomic fluid also contains cells called *coelomocytes* that are involved in the immune system. Figure 8.44 compares the hemocytes of insects and vertebrates to provide an overview of the great variety of these cells. Although the hemocytes of vertebrates and insects appear to be quite distinct, developmental

biologists have recently discovered that in both of these taxa, a group of transcription factors called the GATA factors are involved in the development of these cells. This similarity suggests that blood cells may have a common origin in all animals.

Erythrocytes transport oxygen

Erythrocytes, or red blood cells, are the most abundant cells in the blood of vertebrates. Erythrocytes contain high concentrations of respiratory pigments such as hemoglobin, and their major function is the storage and transport of oxygen (see Chapter 9 for more details). Only a few groups of invertebrates—including the phoronid worms, five families of polychaetes, two classes of molluscs, and some echinoderms—have respiratory pigments contained within erythrocytes. Because these groups are not closely related, cells containing respiratory pigments are thought to have evolved independently several times in animals. Interestingly, the erythrocytes of invertebrates are almost always located outside the circulatory system, in the interstitial fluid, possibly because the presence of cells in the circulatory fluid increases the viscosity of the solution too much to allow it to be pumped by the relatively weak hearts of these groups. However, in horseshoe worms (phylum

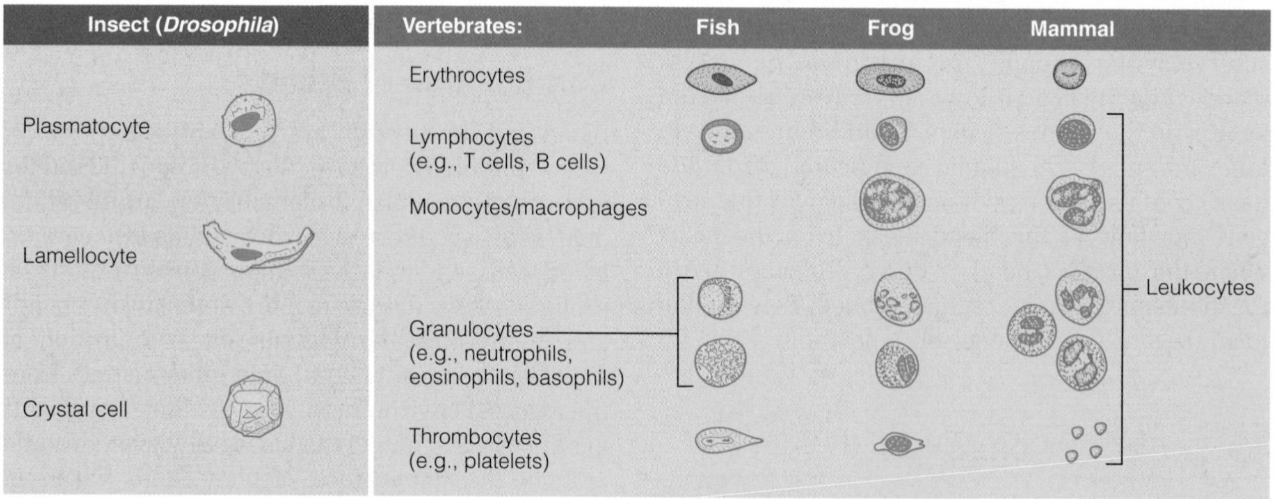

Figure 8.44 Hemocytes Left: Insects such as *Drosophila* have three main classes of hemocytes. Plasmocytes are small cells that use phagocytosis to engulf foreign invaders. Lamellocytes are large cells produced in response to parasitic infections. Crystal cells contain enzymes that they use to lyse foreign invaders. Right: Vertebrate hemocytes can be divided into erythrocytes, or cells that contain hemoglobin, and leukocytes, which do not. Erythrocytes vary in size among vertebrate groups, and in mammals they lack nuclei. Monocytes and granulocytes perform functions similar to those of invertebrate hemocytes, engulfing or destroying invading particles using enzymes, and are thus part of the nonspecific immune response. In addition, the vertebrates have lymphocytes, which are involved in adaptive (or specific) immunity, allowing vertebrates to mount an immune response tailored to a particular pathogen, and thrombocytes, which are involved in blood clotting.

Phoronida) the erythrocytes are found in the circulatory system. These tube-dwelling worms typically live in very low-oxygen environments such as muddy benthic habitats, and the ability to circulate erythrocytes through the blood vessels of their respiratory surfaces may improve their ability to obtain oxygen.

Vertebrate Blood

When vertebrate blood is centrifuged, it separates into three main components (Figure 8.45). The plasma makes up approximately 55% of the whole blood volume in normal humans. Erythrocytes make up the other major component of the blood (approximately 45% of blood volume in humans). The other blood cells, consisting of the various immune and blood-clotting cells, make up a small fraction of the blood. The fraction of the blood that is made up of erythrocytes is termed the **hematocrit**. Hematocrit varies substantially among vertebrates (from 20 to 65%), and can vary within an individual depending on physiological condition. For example, acclimation of humans to high altitude causes an increase in hematocrit.

The size and structure of erythrocytes varies greatly among vertebrates. For example, the largest vertebrate erythrocyte (that of the salamander *Amphiuma*) is almost 2000 times larger than the smallest erythrocyte (that of the lesser mouse deer, *Tragulus javanicus*). In most vertebrates, erythrocytes have nuclei and other or-

ganelles. However, mammals, some fish, and some amphibians have enucleated erythrocytes. In fact, mammalian erythrocytes lack nuclei, mitochondria, and other organelles including ribosomes. As a result, a mammalian erythrocyte cannot perform protein synthesis or cell division.

Erythrocytes are generally round or oval in shape, although most mammalian erythrocytes are shaped like biconcave disks (disks with indentations on both sides). The biconcave shape increases the surface area of the erythrocyte, possibly facilitating oxygen transfer.

Leukocytes, or white blood cells, lack hemoglobin. All vertebrate leukocytes are nucleated, and possess all of the normal cellular machinery. Leukocytes are found both in the blood and in the interstitial fluid, and are able to move across capillary walls through pores between the cells. There are five major types of leukocytes in vertebrates (Figure 8.46), each of which performs specific immune functions. **Neutrophils** are the most common leukocyte in vertebrate blood. These immune cells engulf damaged cells, microorganisms, and other foreign pathogens by phagocytosis. **Eosinophils** can perform phagocytosis, but their main function is to act as delivery vehicles for cytotoxic (cell-killing) chemicals. Eosinophils are usually rare in vertebrate circulatory systems (making up about 3% of all leukocytes), but their numbers can increase greatly in response to a stimulus such as a heavy infection by parasitic worms. **Basophils** leave the circulatory system and accumulate in the interstitial fluid at the site of an infection or other inflammation, releasing toxic chemicals that kill invading microorganisms and other parasites. In addition, they release paracrine factors including histamines and prostaglandins that increase blood flow to the site of infection. Because they release these inflammatory chemicals, they play an important role in allergic reactions such as hay fever. Collectively, the neutrophils, eosinophils, and basophils are termed granulocytes because of their grainy appearance when viewed using light microscopy. **Monocytes** circulate in the blood of most mammals only briefly, quickly leaving the bloodstream and entering the interstitial fluid where they grow much larger and develop into **macrophages**. Like neutrophils, macrophages are phagocytic, engulfing foreign invaders and dead and dying cells. Some macrophages are found in the walls of the

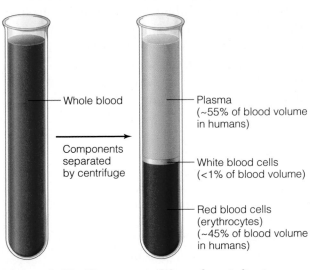

Figure 8.45 **The composition of vertebrate blood**

Whole blood

Components separated by centrifuge

Plasma (~55% of blood volume in humans)

White blood cells (<1% of blood volume)

Red blood cells (erythrocytes) (~45% of blood volume in humans)

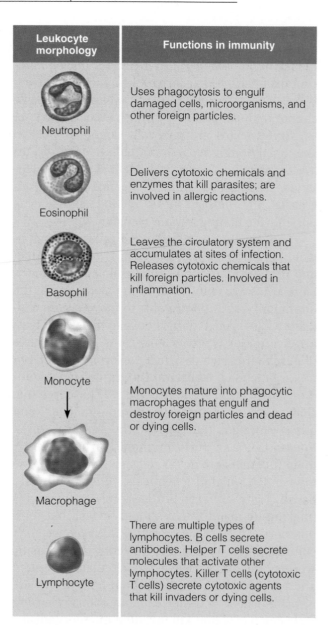

Leukocyte morphology	Functions in immunity
Neutrophil	Uses phagocytosis to engulf damaged cells, microorganisms, and other foreign particles.
Eosinophil	Delivers cytotoxic chemicals and enzymes that kill parasites; are involved in allergic reactions.
Basophil	Leaves the circulatory system and accumulates at sites of infection. Releases cytotoxic chemicals that kill foreign particles. Involved in inflammation.
Monocyte ↓ Macrophage	Monocytes mature into phagocytic macrophages that engulf and destroy foreign particles and dead or dying cells.
Lymphocyte	There are multiple types of lymphocytes. B cells secrete antibodies. Helper T cells secrete molecules that activate other lymphocytes. Killer T cells (cytotoxic T cells) secrete cytotoxic agents that kill invaders or dying cells.

Figure 8.46 Leukocytes Vertebrates have five different types of leukocytes, each with a different role in the immune response.

blood vessels within the liver and the spleen where they phagocytose dying erythrocytes.

A final group of leukocytes are the **lymphocytes**, each with a different function in the immune system. B lymphocytes (B cells) make antibodies. T lymphocytes (T cells) are involved in recruiting macrophages and neutrophils to the site of infection, releasing cytotoxic agents to kill foreign or dying cells, and helping B cells to produce antibodies. In the jawed vertebrates, B cells and T cells provide a form of immunity called *adaptive immu-*

nity, in which the immune system forms a "memory" of invading pathogens so that it can mount a faster response to subsequent invasions. Adaptive immunity is unique to the jawed vertebrates.

Thrombocytes (Figure 8.44) play a key role in blood clotting. The thrombocytes of nonmammalian vertebrates are spindle-shaped cells that have a nucleus and are classified as leukocytes, but in mammals an anucleate cell fragment called a *platelet* plays this clotting role. When a blood vessel is cut or damaged, the hole must quickly be plugged by a *blood clot* before loss of blood from the circulatory system leads to a precipitous drop in blood pressure. The process of clotting, or blood *coagulation*, occurs in three steps. When a blood vessel is damaged, local signals and activation of the sympathetic nervous system induce the vessel to vasoconstrict, reducing blood flow. Next a platelet plug forms, temporarily sealing the opening. Finally, a clot forms through a series of steps termed the coagulation cascade.

All blood cells form from a single type of stem cell, through a process called hematopoiesis (Figure 8.47). In adult mammals, hematopoietic stem cells are found only in bone marrow, but in other vertebrates they can be found circulating in the blood as well as in specific organs. In fishes, the kidney is the primary hematopoietic organ, whereas in amphibians, reptiles, and birds, hematopoiesis can occur in the spleen, liver, kidney, and bone marrow. The number of stem cells found in the blood is highest in the fishes, and decreases in the other vertebrate taxa, associated with an increasing role for the hematopoietic organs. Specific signaling factors are involved in triggering the production of different types of blood cells. For example, **erythropoietin** is a hormone released from the kidney in response to low blood oxygen. Erythropoietin triggers the differentiation of stem cells into erythrocytes.

⊙ | CONCEPT CHECK

24. What is the difference between blood and plasma?
25. What is the function of a respiratory pigment?
26. List the major cell types in the blood of vertebrates.
27. What is the function of erythropoietin?

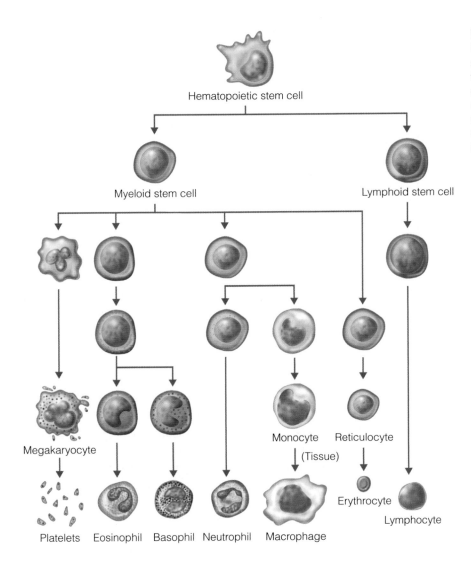

Figure 8.47 Blood cell formation
Blood cells are derived from hematopoietic stem cells that can differentiate to form any type of hemocyte. The first round of differentiation forms two cell lines: the myeloid stem cells, and the lymphoid stem cells. Most hemocytes are derived from myeloid stem cells. Lymphoid stem cells are the precursors of the lymphocytes.

Integrating Systems | The Circulatory System During Exercise

Because of its important transport function, the circulatory system plays an important role in the functioning of essentially every other physiological system. In this Integrating Systems section, we examine how the nervous system, endocrine system, muscular system, and circulatory system work together when a vertebrate performs aerobic exercise.

When a vertebrate begins to exercise aerobically, the brain sends a signal via motor neurons to the muscles (see Chapter 4 and Chapter 7). The motor neuron releases ATP onto the nicotinic acetylcholine receptors on the motor end plate (see Chapter 5). The muscle then generates an action potential, causing the release of Ca^{2+} from the sarcoplasmic reticulum, which initiates contraction. The myosin ATPase rapidly hydrolyzes ATP to allow muscle contraction, and the sarcoplasmic Ca^{2+} ATPase hydrolyzes ATP to pump Ca^{2+} back into the sar-

coplasmic reticulum to allow the muscle to relax. Muscles used for sustained exercise use mitochondrial metabolism to replenish this ATP. As we discussed in Chapter 2, mitochondrial metabolism requires oxygen. Thus, input from the nervous system causes muscles to go from their resting state into a state of heightened aerobic metabolism.

As a result of this increased demand for oxygen by the skeletal muscles, the demands placed on the circulatory system change greatly during the transition from rest to exercise. In most humans, oxygen consumption increases by nearly fivefold within a few minutes of the onset of intense aerobic exercise. Figure 8.48 outlines the response of the cardiovascular system to exercise. When we first begin to exercise, mechanoreceptors in our muscles (see Chapter 6) detect the change in the tension of the muscle as a result of contraction. These

Figure 8.48 The response of the cardiovascular system to exercise

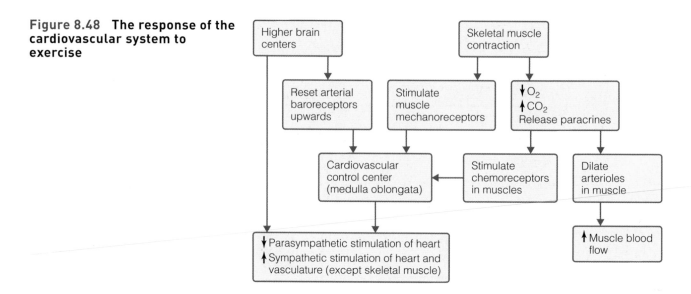

mechanoreceptors send afferent sensory information to our brain, activating the cardiovascular control center in the medulla oblongata. The cardiovascular control center reduces the activity of the parasympathetic nervous system and increases the activity of the sympathetic nervous system, changing the efferent signals going to the heart and the arteriolar smooth muscle.

The change in parasympathetic and sympathetic activity has dramatic effects on cardiac output. In fact, in humans cardiac output can increase by between four and eight times the value at rest (depending on the type and intensity of the exercise and the fitness of the individual). A trained thoroughbred horse can achieve as much as a 10-fold increase in cardiac output during maximal exercise. Recall that cardiac output is the product of heart rate and stroke volume. So which of these factors is the most important in causing the increase in cardiac output? At the onset of exercise parasympathetic activity decreases, causing an increase in heart rate. At the same time, the increase in muscular activity and breathing improves the function of the respiratory and skeletal muscle pumps, causing an increase in venous return to the heart. Because of the Frank-Starling effect, the resulting increase in end-diastolic volume causes an increase in stroke volume.

Thus, during the initial stages of exercise, the increases in cardiac output are a result of both increases in heart rate and increases in stroke volume. Next, sympathetic stimulation of the heart increases, resulting in increases in both heart rate and contractility. In principle, the increase in contractility should cause an increase in stroke volume, but the large increase in heart rate reduces the time available for filling of the heart,

and limits end-diastolic volume. As a result, during the later stages of exercise in mammals, increases in heart rate contribute more to increases in cardiac output than do increases in stroke volume. In most terrestrial vertebrates, increases in cardiac output in response to exercise are primarily the result of increases in heart rate, with a modest contribution from increases in stroke volume. In contrast, in fish such as salmon, changes in stroke volume play the most important role in causing the increase in cardiac output during exercise. However, not all fish are stroke volume regulators. For example, tuna typically increase cardiac output by increasing heart rate and keeping stroke volume fairly constant. Thus, different animals use different strategies to achieve the same goal: increasing the delivery of oxygen to the working muscles during exercise.

In addition to changes in cardiac output, there are large changes in the patterns of blood flow during exercise. At rest, the skeletal muscles receive only about 20% of the total cardiac output, whereas they receive 88% of the cardiac output during exercise. Notice that total cardiac output also increases dramatically, so that flow to the skeletal muscles actually increases from about 1.2 l/min at rest to over 22 l/min during exercise. At the same time, blood flow to organs such as the kidney decreases, both in relative and absolute terms. At rest, approximately 19% of the total cardiac output flows through the kidneys (or about 1 l/min), whereas during intense exercise only 1% of the total cardiac output flows through the kidneys (or about 0.25 l/min). These changes in blood flow are the result of vasodilation of the arterioles leading to the skeletal muscle and heart and vasoconstriction of the arterioles leading to the other organs. The increase in the activity of the sympathetic nervous system causes a generalized vasoconstriction, as sympathetic neurons release norepinephrine onto the vascular smooth muscle surrounding the arterioles leading to the organs. The norepinephrine binds to its receptors and causes the smooth muscles to contract. In skeletal muscle, however, local release of paracrine factors causes the vascular smooth muscle to relax by opposing the vasconstrictive effects of α adrenergic receptor stimulation. Together, these factors cause an intense local vasodilation, increasing blood flow to the muscles.

Recall that mean arterial pressure is the product of cardiac output and total peripheral resistance. During exercise, cardiac output increases greatly, which you might expect to cause a large increase in mean arterial pressure. However, the vasodilation of the arterioles leading to the skeletal muscles more than offsets the vasoconstriction of the arterioles leading to the other organs, so total peripheral resistance falls dramatically. As a result, blood pressure increases only slightly during exercise. Ordinarily, even this modest increase in blood pressure would trigger the baroreceptor reflex, and bring blood pressure back to normal by decreasing cardiac output or total peripheral resistance. Recent data suggest that the afferent signal from the muscle mechanoreceptors changes the set point of the baroreceptor reflex, or alters its sensitivity, allowing blood pressure to increase slightly with exercise.

Feedforward regulation also plays an important role in the response of the circulatory system to exercise. The circulatory changes that accompany exercise occur more rapidly if you ask an experimental subject to repeatedly contract a muscle, compared to what happens when you electrically stimulate that muscle. This suggests that descending input from higher brain centers helps to cause circulatory changes in anticipation of the need for more oxygen by the working muscles, even before metabolic end products begin to build up. Thus, it is clear that the circulatory responses to exercise represent a delicate integrated response involving the nervous system, the endocrine system, the musculoskeletal system, and the cardiovascular system. ◉

Summary

Characteristics of Circulatory Systems

→ Diffusion occurs quickly over short distances but is slow over long distances.

→ To avoid the limitations of diffusion, animals use bulk flow of fluids through a circulatory system to transport substances quickly across long distances.

→ In bulk flow, fluids move from areas of higher pressure to areas of lower pressure.

→ In order to generate bulk flow, all animal circulatory systems contain one or more pumps (e.g., a heart), channels through which a fluid can flow (e.g., blood vessels), and a specialized circulatory fluid (e.g., blood).

→ Circulatory systems use diverse pumping structures, including chambered hearts, skeletal muscles, and contractile blood vessels.

→ Circulatory systems can be either open or closed.

→ In open circulatory systems, the circulatory fluid is called hemolymph.

→ At some points in the circulatory system, hemolymph flows out of the blood vessels into large sinuses where it bathes the tissues.

→ In closed circulatory systems the circulatory fluid is called blood.

→ Blood remains within the blood vessels through the complete circuit.

→ Vertebrates also have a lymphatic system that pumps lymph, a fluid formed from blood by ultrafiltration.

→ Sponges, cnidarians, and flatworms lack a specialized circulatory system, but transport water or interstitial fluids through their bodies by bulk flow.

→ Most molluscs and all insects have relatively simple open circulatory systems, but decapod crustaceans have complex open circulatory systems.

→ Cephalopods, most annelids, and most vertebrates have closed circulatory systems.

→ In most organisms, closed circulatory systems have evolved in parallel with increased demand for oxygen. One exception is the insects, which have very high oxygen demand, but relatively simple open circulatory systems.

→ Insects do not depend on their circulatory system for oxygen transport.

→ Vertebrates share a common circulatory plan in which the heart pumps blood to arteries, arterioles, capillaries, venules, and then veins, which return blood to the heart.

→ Vertebrate blood vessels have complex walls with many layers.

→ The inner layer, the tunica intima, is present in all blood vessels and consists of the vascular endothelium, but the thickness of the two outer layers, the tunica media and tunica externa, varies among different types of vessels.

→ Arteries have thick, elastic walls that allow them to act as a pressure reservoir.

→ Veins have thinner, more compliant walls that allow them to act as a volume reservoir.

→ Muscular arteries and arterioles have a thick layer of smooth muscle that allows them to control the flow of blood.

→ Capillaries have thin walls that promote exchange of substances.

→ The structure of the circulatory system is not static, but instead changes over time through processes such as angiogenesis, which allows new blood vessels to grow into areas with increased demand for oxygen.

→ The organization of the vertebrate circulatory system relates to the respiratory mode of the animal. Water breathers have a single-circuit circulation, whereas most air breathers have a double circulation (although there is some mixing between these circulations in amphibians and reptiles, whereas they are completely separated in birds and mammals).

→ Flow in the circulatory system can be approximated using Poiseuille's equation ($Q = \Delta P \pi r^4 / 8L\eta$). Thus, flow increases when the pressure difference or the radius of the tube decreases.

→ The total resistance of a circuit differs depending on whether the resistors are arranged in series or parallel.

→ Capillary beds, which consist of many small blood vessels arranged in parallel, typically have relatively low resistance to flow compared to arterioles, which are fewer in number.

→ The total flow in a circulatory system is constant at all levels within the system. The total flow in the arteries is the same as the total flow in all of the capillaries.

→ Fluids flow along the path of least resistance. Blood vessels can alter the proportion of flow along alternative pathways arranged in parallel by vasoconstricting or vasodilating and thus altering resistance.

→ Flow rate and the velocity of flow are not the same. The velocity of flow is proportional to the rate of flow divided by the total cross-sectional area of the vessels.

→ Blood pressure exerts a force on the walls of a blood vessel, and this force varies depending on the radius of the vessel and the thickness of the vessel walls.

Hearts

→ The pumping action of a heart is divided into two phases: systole (contraction) and diastole (relaxation).

→ Arthropod hearts are neurogenic, and require nervous input to initiate contraction.

→ Arthopod hearts fill by suction, and empty as a result of the increased pressure from contraction.

→ Mammalian hearts consist of an outer pericardium, an epicardium, a myocardium, and an endocardium.

→ The structure of the myocardium varies among vertebrates. Birds and mammals have only compact myocardium. Lampreys have only spongy myocardium. But most species have a combination of both compact and spongy myocardium.

→ Compact myocardium receives oxygen via the coronary arteries, but spongy myocardium generally receives oxygen from the blood flowing through the heart.

→ The chambers of the fish heart are arranged in series. Blood flows from the sinus vensosus, to the atrium, to the ventricle, and then to the bulbus arteriosus (or the conus arteriosus in cartilaginous fish).

→ In the mammalian heart, the pathway for blood flow is via the right atrium, the right AV valve, the right ventricle, the pulmonary semilunar valve, the pulmonary artery, the capillary beds of the lungs, the pulmonary veins, the left atrium, the left AV valve, the left ventricle, the aortic semilunar valve, the aorta, and the blood vessels of the systemic circulation.

→ The vertebrate ventricle fills during ventricular diastole, reaching its maximum volume (the end-diastolic volume) at the end of this period. In most vertebrates, ventricular filling is passive, but atrial contraction (during atrial systole) pushes some additional blood into the ventricle.

→ Some animals, however, may use suction filling to assist venous return to the heart.

→ At the beginning of ventricular contraction, all of the heart valves are closed.

→ Once the pressure in the ventricle exceeds that in the arteries, the semilunar valve opens, initiating ventricular ejection.

→ The valves play a critical role in the function of the heart, but they are passive, opening and closing in response to the applied pressure.

→ The right ventricle pumps much less forcefully than the left ventricle in mammals.

→ A pacemaker controls heart rate. In neurogenic pacemakers the nervous system initiates the rhythm, whereas in myogenic pacemakers specialized cardiomyocytes initiate the rhythm.

→ The depolarization then spreads through the heart via gap junctions. In addition, the vertebrates have specialized conducting pathways to help spread the depolarization through the heart. In mammals, these pathways are very distinct, but in fishes, these depolarizations spread largely through the spongy myocardium.

→ Myogenic pacemaker cells have an unstable resting membrane potential due to changes in K^+ conductance and the action of non-selective Na+ channels that are responsible for the so-called "funny" current.

→ The nervous system and the endocrine systems can modulate the rate of the pacemaker potential. Norepinephrine and epinephrine increase the rate of pacemaker potentials, whereas acetylcholine decreases the rate of pacemaker potentials.

Regulation of Pressure and Flow

→ The arterioles control blood distribution by altering their radius, and thus their resistance.

→ There are four main ways in which the body regulates arteriolar diameter: myogenic autoregulation, paracrine signals of the metabolic activity of the tissue, signals from the nervous system, and signals from the endocrine system.

→ Blood pressure and blood velocity vary in different parts of the circulatory system because of the structure of the blood vessels.

→ Myogenic autoregulation acts to maintain a constant blood flow, but the other mechanisms can be used to alter blood flow in response to tissue demand.

→ Blood pressure in the heart and arteries is pulsatile because the systolic pressure is higher than the diastolic pressure.

→ The arteries act to dampen pressure fluctuations, and as a result, the pulsatility of blood flow declines along the course of the circulatory system.

→ Blood pressure declines as blood moves through the circulatory system.

→ Skeletal and respiratory muscle pumps help to return venous blood to the heart.

→ The body maintains close homeostatic control over mean arterial pressure by altering cardiac output and total peripheral resistance.

→ The baroreceptor reflex is the primary means of regulating mean arterial pressure in the short term.

→ In the long term, blood volume regulation by the kidneys is critical.

→ Blood pressure can force fluid out of the capillaries by bulk flow.

→ The lymphatic system recycles this fluid, but any imbalance in these processes can lead to edema.

→ Body position can affect the functioning of the circulatory system, because of the effects of gravity.

→ Animals with very long necks (such as giraffes) have specialized mechanisms to ensure blood flow to the brain, and to prevent edema in the lower extremities.

Blood

→ Blood is a complex aqueous fluid containing ions, proteins, and cells. It is also considered to be a connective tissue.

→ In some species, oxygen-transporting respiratory pigments are found in erythrocytes, whereas in other species they are found dissolved in the blood.

→ Mammalian erythrocytes lack nuclei and most organelles, but the erythrocytes of other vertebrates are less specialized.

→ Erythrocytes vary greatly in size among species.

→ Leukocytes are vertebrate immune cells.

→ Thrombocytes (or platelets in mammals) are involved in blood clotting.

→ All types of blood cells form from a single population of stem cells by the process of hematopoiesis.

Review Questions

1. Compare and contrast the circulatory systems of fish, amphibians, and mammals.

2. Trace the movement of a drop of blood through the human circulatory system, listing all of the structures it passes (including all of the parts of the heart).

3. Name the layers of the walls of a mammalian heart and describe their structure.

4. What is the significance of the difference in pressure developed by the right and left atria of a mammalian heart?

5. Compare the mechanism of filling of a mammalian heart and an insect heart.

6. Why is the lengthy refractory period of a contractile cardiomyocyte important for the function of the mammalian heart?

7. Define heart rate, stroke volume, and cardiac output. Explain how changes in heart rate or stroke volume affect cardiac output.

8. What is the Frank-Starling effect? Explain its significance in cardiovascular physiology.

9. Would resistance be higher in an arteriole or a vein, and why?

10. What is the difference between the velocity of the blood and the rate of blood flow? How are these two concepts related?

11. Describe the mechanisms that control the radius of the arterioles.

12. Outline some of the functions of the lymphatic system.

13. What is the importance of the skeletal muscle and respiratory pumps?

14. Outline the baroreceptor reflex and discuss its importance.

Synthesis Questions

1. What are some possible advantages of a double circulation over a single-circuit circulation?

2. Explain the changes in blood pressure as blood flows through the mammalian circulatory system.

3. Aortic blood flow starts to increase only some time after the initiation of ventricular contraction. Similarly, aortic blood flow continues at a relatively high level well into the diastolic period. Explain why.

4. Increased heart rate can greatly reduce diastolic filling time, but has less impact on systolic ejection time. Why?

5. What would happen if the connection between the AV node and the bundle of His were

blocked (in a way that didn't directly affect any other parts of the heart)?

6. During an experiment dogs were given the drug atropine, which abolishes parasympathetic nerve transmission. What effects would you expect on the heart and why?

7. Tom suffers from high blood pressure. Which of the following might help deal with this problem? Remember to explain your answer.
   (a) A drug that stimulates β1 receptors
   (b) A drug that blocks α receptors
   (c) A drug that blocks β1 receptors
   (d) A drug that blocks acetylcholine receptors

8. After a heart transplant, there is no direct connection between the nervous system and the

heart. However, the cardiac output of patients with heart transplants can vary in response to changes in metabolic demand (such as during exercise). How could this be possible? Would you expect this regulation to be as efficient as in a patient with an intact heart?

Quantitative Questions

1. Below is a schematic diagram of the mammalian cardiovascular system.

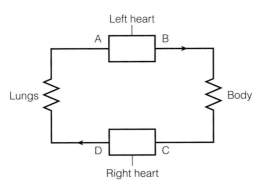

If mean pressure at A = 2 mm Hg, B = 80 mm Hg, C = 5 mm Hg, and D = 10 mm Hg, and cardiac output = 5 l/min, calculate
   (a) the resistance of the systemic circulation
   (b) the resistance of the pulmonary circulation. What are the units you have used for resistance?

2. The radius of the aorta in humans is about 1×10^{-2} m and the velocity of blood flowing through it is about 0.3 m/sec. What is the average speed of blood in the capillaries given average capillary diameter is only 8×10^{-6} m, and the total cross-sectional area of the capillaries is about 2×10^{-1} m (the cross-sectional area of a blood vessel is approximately πr^2).

3. Use this figure to answer the following questions:

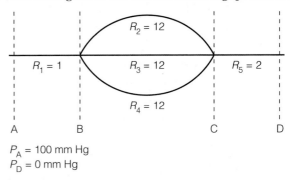

P_A = 100 mm Hg
P_D = 0 mm Hg

   (a) What is the flow through this network?
   (b) What are the pressures at points B and C?
   (c) What is the flow in vessel 3?
   (d) If another vessel is added in parallel to vessels 2–4, with a resistance R_6 = 18 mm Hg·min/ml, then what is the flow through the system (assuming ΔP remains the same)?
   (e) If vessel 4 becomes completely occluded (blocked) (i.e., R_4 is now infinite), what is the flow through the network?

4. Using the information in Figure 8.29, at what point in the cardiac cycle is ventricular ejection velocity highest?

5. If during exercise heart rate increases from 70 beats/min to 150 beats/min and the stroke volume increases from 70 ml/beat to 120 ml/beat, what will be the difference in cardiac output between rest and exercise?

For Further Reading

See the Additional References section at the back of the book for more readings related to the topics in this chapter.

Characteristics of Circulatory Systems

The comprehensive chapter below provides an in-depth look at the physiology of vertebrate cardiovascular systems at a level suitable for advanced undergraduates and graduate students.

Burggren, W., A. Farrell, and H. Lillywhite. 1997. Vertebrate cardiovascular systems. In *Handbook of physiology: A critical, comprehensive presentation of physiological knowledge and concepts,* W. H. Dantzler, ed. (Section 13: Comparative Physiology), vol. 1, 215–308. Bethesda, MD: American Physiological Society.

The following detailed review highlights some of the seldom-appreciated complexity of the circulatory systems of insects.

Hertel, W., and G. Pass. 2002. An evolutionary treatment of the morphology and physiology of circulatory organs in insects. *Comparative Biochemistry and Physiology, Part A: Molecular and Integrative Physiology* 133: 555–575.

This fascinating paper, which includes several wonderful images of crustacean circulatory systems such as the one used in the opening essay of this chapter, demonstrates that a simple classification of circulatory systems into "open" or "closed" cannot capture their full diversity.

McGaw, I. J., and C. L. Reiber. 2002. Cardiovascular system of the blue crab *Callinectes sapidus*. *Journal of Morphology* 251: 1–21.

The reviews below highlight some of the important evolutionary transitions in the circulatory systems of the vertebrates.

Burggren, W. W., and K. Johansen. 1986. Circulation and respiration in lungfishes (Dipnoi). *Advances in Comparative and Environmental Physiology* 21: 175–197.

Farrell, A. P. 1991. From hagfish to tuna: A perspective on cardiac function in fish. *Physiological Zoology* 64: 1137–1164.

White, F. N., and J. W. Hicks. 1987. Cardiovascular implications of the transition from aquatic to aerial respiration. In *Comparative physiology: Life in water and on land*, P. Dejours, L. Bolis, C. R. Taylor, and E. R. Weibel, eds., Fidia Research Series, 93–106. Padova, Italy: Liviana Press.

Olson, K. R. 1996. Secondary circulation in fish: Anatomical organization and physiological significance. *Journal of Experimental Zoology* 275: 172 -185.

Burggren, W.W. 1987. Form and function in reptilian circulations. *American Zoologist* 27: 5-19.

Burggren, W.W. and A.W. Pinder. 1991. Ontogeny of cardiovascular and respiratory physiology in lower vertebrates. *Annual Review of Physiology* 53: 107-135.

Hicks, J.W. 2002. The physiological and evolutionary significance of cardiovascular shunting patterns in reptiles. *News in Physiological Sciences* 17: 241-245.

Hearts

The review below provides a clear description of the anatomy and physiology of the heart in crustaceans.

Cooke, I. M. 2002. Reliable, responsive pacemaking and pattern generation with minimal cell numbers: The crustacean cardiac ganglion. *Biological Bulletin* 202: 108–136.

The following reviews outline the nature of ion channels and the basis of excitability in vertebrate cardiac tissues.

Irisawa, H., H. F. Brown, and W. Giles. 1993. Cardiac pacemaking in the sinoatrial node. *Physiological Reviews* 73: 197–227.

Kaupp, U. B., and R. Seifert. 2001. Molecular diversity of pacemaker ion channels. *Annual Review of Physiology* 63: 235–257.

Roden, D. M., J. R. Balser, A. L. George Jr., and M. E. Anderson. 2002. Cardiac ion channels. *Annual Review of Physiology* 64: 431–475.

The review below outlines some of the mechanisms involved in the regulation of heart rate and stroke volume in fish hearts.

Farrell, A.P. 1984. A review of cardiac performance in the teleost heart: intrinsic and humoral responses. *Canadian Journal of Zoology.* 62: 523-536.

The following review outlines some of the recent work on the genetics of heart development in vertebrates.

Warren, K. S., J. C. Wu, F. Pinet, and M. C. Fishman. 2000. The genetic basis of cardiac function: Dissection by zebrafish (*Danio rerio*) screens. *Philosophical Transactions of the Royal Society of London, Series B: Biological Sciences* 355: 939–944.

Regulation of Pressure and Flow

The comprehensive review below discusses the mechanisms involved in the regulation of mean arterial pressure, contrasting the mechanisms of acute regulation to those involved in longer term regulation.

Cowley, A.W. Jr. 1992. Long-term control of arterial blood pressure. *Physiological Reviews* 72: 231-300.

The following review outlines how the baroreceptor reflex varies among animals.

Van Vliet, B. N., and N. H. West. 1994. Phylogenetic trends in the baroreceptor control of arterial blood pressure. *Physiological Zoology* 67: 1284–1304.

These interesting papers debate the nature of the effects of gravity in closed circulatory systems, and outline the anatomy and physiology of the circulatory system of the giraffe.

Badeer, H. S. 1997. Is the flow in the giraffe's jugular vein a "free" fall? *Comparative Biochemistry and Physiology, Part A: Molecular and Integrative Physiology* 118: 573–576.

Hargens, A. R., R. W. Millard, K. Pettersson, and K. Johansen. 1987. Gravitational haemodynamics and oedema prevention in the giraffe. *Nature* 329: 59–60.

Hicks, J. W., and Badeer, H. S. 1989. Siphon mechanism in collapsible tubes: Application to circulation of the giraffe head. *American Journal of Physiology* 256: R567–R571.

Recordati, G. 1999. The contribution of the giraffe to hemodynamic knowledge: A unified physical principle for the circulation. *Cardiologia Roma* 44: 783–789.

This accessible review outlines the role of elasticity in the circulatory system.

Shadwick, R. E. 1998. Elasticity in arteries. *American Scientist* 86: 535–541.

The entertaining and accessible book listed below presents detailed information about the physiology of circulatory systems using examples and approaches that make circulatory physiology come to life.

Vogel, S. 1992. *Vital circuits: On pumps, pipes, and the workings of circulatory systems.* Oxford: Oxford University Press.

Blood

This fascinating review highlights the unanticipated similarities in blood cell formation between vertebrates and invertebrates.

Evans, C. J., V. Hartenstein, and U. Banerjee. 2003. Thicker than blood: Conserved mechanisms in Drosophila and vertebrate hematopoiesis. *Developmental Cell* 5: 673–690.

Exercise

The following reviews summarize our current understanding of the regulation of the human cardiovascular system during exercise.

Delp, M. D., and D. S. O'Leary. 2004. Integrative control of the skeletal muscle microcirculation in the maintenance of arterial pressure during exercise. *Journal of Applied Physiology* 97: 1112–1118.

Thomas, G. D., and S. S. Segal. 2004. Neural control of muscle blood flow during exercise. *Journal of Applied Physiology* 97: 731–738.

The brief reviews listed below take a historical perspective to summarize some of the classic experiments demonstrating that the baroreceptor reflex is re-set during exercise.

Joyner, M.J. 2006. Baroreceptor function during exercise: resetting the record. *Experimental Physiology* 91: 27–36.

Raven, P.B., P.J. Fadel, and S. Ogoh. 2006. Arterial baroreflex resetting during exercise: a current perspective. *Experimental Physiology* 91: 37–49.

9

Respiratory Systems

In his book *Into Thin Air: A Personal Account of the Mt. Everest Disaster,* Jon Krakauer tells the gripping story of one of the deadliest attempts to scale Mt. Everest: the 1996 expeditions in which eight climbers died in a single day attempting to reach the Everest summit. Located in the Himalayas, Mt. Everest is the highest mountain peak in the world, at 8850 m. Altitudes greater than 8000 m are completely inhospitable to human beings because of low temperatures, strong winds, and thin air. The air at the summit of Everest has only one-third the oxygen per unit volume as does air at sea level—far too little for most humans to obtain enough oxygen to perform any kind of vigorous activity. Mountain climbers have named these altitudes "the death zone," because it is impossible for people to survive there for more than a few hours without supplemental oxy-

gen. In fact, fewer than 5% of all the people who have reached the summit of Everest have done so without using bottled oxygen.

Humans seldom live permanently at altitudes higher than 4500 m, which is the typical upper limit for farming, although the nomadic peoples of the Tibetan plateau spend part of the year in the surrounding mountains at altitudes between 4800 and 5500 m. But many animals other than humans can easily reach greater altitudes without suffering any ill effects. Pikas (genus *Ochotona*) are small mammals related to rabbits. They are found only at high altitudes or latitudes, and permanent populations exist at elevations higher than 6000 m. Pikas are able to extract oxygen from air far more efficiently than other mammals, which allows them to live and breed at high altitudes. The highest altitude colo-

Bar-headed geese.

Mudskippers.

nized by any animal appears to be approximately 6700 m; jumping spiders (family Salticidae) have been observed by climbing expeditions that high in the Himalayas. As little else lives this high up, what these predatory spiders eat is a mystery, although some scientists have speculated that they may subsist on aerial plankton—small insects carried up to high altitudes by air currents.

Bar-headed geese (*Anser indicus*) nest and breed on the shores of high-altitude lakes in the Himalayas and the Tibetan plateau. They then migrate to their winter feeding grounds on the shores of lowland lakes in central and southern India. During their migration they fly over the Himalayas, sometimes directly over the summit of Everest, reaching altitudes of nearly 9400 m. Airplane pilots have also observed other species of birds at very high altitudes, including whooper swans (*Cygnus cygnus*) at 8300 m, and bar-tailed godwits (*Limosa lapponica*) at 6100 m. The altitude record for a bird is held by the Ruppell's griffon (*Gyps rueppellii*), obtained when one of these African vultures was sucked into a jet engine at 11,500 m—more than 2 km higher than the summit of Mt. Everest. Although we do not

yet understand the complete suite of adaptations that allow birds to thrive at high altitudes, these animals differ from mammals in their ability to obtain oxygen from the atmosphere, to tolerate low blood oxygen levels, and to cope with changes in blood carbon dioxide and pH.

Aquatic organisms also regularly experience low-oxygen environments, both in high-altitude habitats and at sea level. At sea level, aquatic hypoxia is usually caused by the presence of aquatic plants. During the day, plants photosynthesize and produce oxygen, but at night plants respire and consume oxygen, which can cause drastic declines in the dissolved oxygen content of enclosed bodies of water such as lakes and ponds. Fish that are tolerant of habitats with low levels of dissolved oxygen exhibit a wide range of adaptations, including behavioral alterations, changes in oxygen extraction efficiency, and changes in metabolism. Some fish, such as mudskippers, are even capable of emerging from the water and breathing air. ◉

Overview

Most animals depend on mitochondrial respiration to supply the ATP that they need to perform normal cellular functions. During mitochondrial respiration, mitochondria oxidize carbohydrates, amino acids, or fatty acids to produce ATP, consuming oxygen and producing carbon dioxide in the process. Thus, animals must obtain oxygen from the environment and dispose of the resulting carbon dioxide in order to meet their metabolic needs. The entire sequence of events that results in the exchange of oxygen and carbon dioxide between the external environment of an animal and the mitochondria within its cells is often termed **respiration.** However, we prefer the term *external respiration* to distinguish this process from mitochondrial respiration.

During mitochondrial respiration, mitochondria consume oxygen and act as oxygen sinks, depleting the local concentration of oxygen. Thus, there is an oxygen gradient from the outside of the cell to the mitochondrion. Oxygen molecules move down this gradient into the mitochondria (and carbon dioxide moves in the opposite direction). Unicellular organisms and small multicellular organisms living in aquatic environments can utilize this diffusion gradient to drive gas exchange with the environment (Figure 9.1). Animals that obtain oxygen from air need an additional step: gaseous oxygen must first dissolve before it can cross the cell membrane.

Diffusion alone is too slow to maintain the rates of gas exchange needed to support the metabolism of larger organisms, because diffusion occurs slowly over long distances (see Chapter 8: Circulatory Systems). Instead, larger organisms rely on a combination of bulk flow and diffusion for gas exchange (Figure 9.1). Some animals, such as sponges and cnidarians, move the external medium (seawater) by bulk flow through an internal body cavity. Oxygen diffuses from the seawater into the cells of the organism, and carbon dioxide diffuses out of the cells into the seawater. The seawater circulating through the body cavity by bulk flow carries the carbon dioxide out into the environment. Insects use a conceptually similar system. In these animals, a series of hollow tubes called trachea penetrate into all parts of the body. Air moves through these tubes either by diffusion or bulk flow, and at the tissues oxygen from the air dissolves in extracellular fluid and diffuses to the

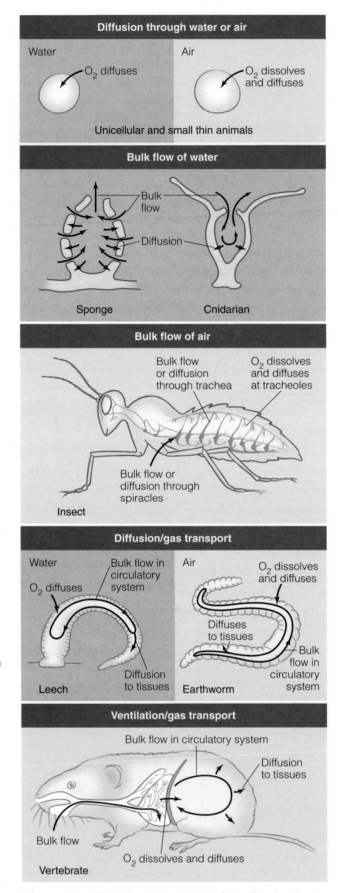

Figure 9.1 Respiratory strategies of animals

mitochondria, while carbon dioxide diffuses out of the cells and into the trachea where it moves out to the external environment by either diffusion or bulk flow (depending upon the species).

Many animals have a circulatory system that transports oxygen by bulk flow through the body. In some animals, such as leeches and earthworms, oxygen simply diffuses across the skin and then is carried by bulk flow through the circulatory system, but many organisms have a specialized respiratory organ with a large surface area, either *gills* or *lungs,* which they use for gas exchange. Animals with internal gills or lungs often move the external medium by bulk flow across the respiratory surface—a process called **ventilation.** In these animals, respiration is divided into four steps: (1) bulk flow of the medium across the respiratory surface, (2) diffusion across this surface, (3) bulk flow in the circulatory system (a process termed *gas transport*), and (4) diffusion into the tissues.

In this chapter, we first examine the respiratory strategies used by animals to obtain oxygen from the environment and to dispose of carbon dioxide. Then we take a closer look at the processes of ventilation and gas transport. We end the chapter with a discussion of the regulation of respiratory systems and the response of this system to environmental changes such as high altitude and diving.

Respiratory Strategies

Because the processes of diffusion, dissolution, and bulk flow are fundamental in shaping the respiratory strategies of animals, we begin this chapter by focusing on the physical principles that underlie these processes.

The Physics of Respiratory Systems

As we discussed in Chapter 2: Chemistry, Biochemistry, and Cell Physiology, we can quantify the rate of diffusion using the Fick equation:

$$dQ/dt = D \times A \times (dC/dx)$$

where dQ/dt is the rate of diffusion (or the mass flux—the quantity of substance moving per unit time, e.g., in mol/sec), D is the diffusion coefficient (an index of the ease of diffusion of a particular substance through a given medium, e.g., in

cm^2/sec), and A is the area of the membrane (e.g., in cm^2). The Fick equation is often expressed in terms of dC/dx—the difference in concentration per unit distance (or the concentration gradient). This gradient is more accurately described as an energetic gradient, which may be due to differences in concentration, electrical charge, temperature, or pressure. When we apply the Fick equation to gases, we usually express the energy gradient in terms of the pressure gradient of the gas, rather than the concentration gradient, because gases have special properties such that they dissolve, diffuse, and react according to their pressure, not necessarily according to their concentration.

From the Fick equation you can see that the rate of diffusion will be greatest when the diffusion coefficient, area of the membrane, and energy gradients are large, but the diffusion distance is small. These constraints greatly influence the structures of gas exchange surfaces so that they are typically thin and often fragile, and have a large surface area.

Gases exert a pressure

The total pressure exerted by a gas is related to the number of moles of the gas and the volume of the chamber, according to the *ideal gas law:*

$$PV = nRT$$

where P is the total pressure, V is the volume, n is the number of moles of gas molecules, R is the gas constant, and T is the temperature in Kelvin. Air is a mixture of gases, containing nitrogen (~78%), oxygen (~21%), argon (~0.9%), carbon dioxide (~0.03%), and a variety of trace gases. Dalton's law of partial pressures states that in a gas mixture each gas exerts its own **partial pressure**. The sum of the partial pressures of the gases in a gas mixture yields the total pressure of the gas mixture (Figure 9.2a). Just as with total pressure, the partial pressure of a gas is proportional to the number of gas molecules.

By rearranging the ideal gas law to the form $n/V = P/RT,$ we can see that the concentration of the gas (number of moles per unit volume, or n/V) is proportional to the pressure at a constant temperature. If temperature increases, and volume is not fixed, volume will increase, keeping the pressure constant but causing the concentration to change. The effect of temperature on the concentration of a gas is one reason we usually express the Fick equation in terms of pressure when dealing with gases.

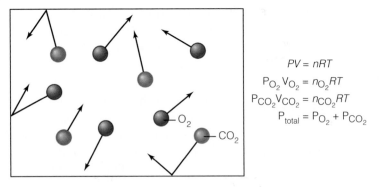

$$PV = nRT$$
$$P_{O_2}V_{O_2} = n_{O_2}RT$$
$$P_{CO_2}V_{CO_2} = n_{CO_2}RT$$
$$P_{total} = P_{O_2} + P_{CO_2}$$

(a) Ideal gas law

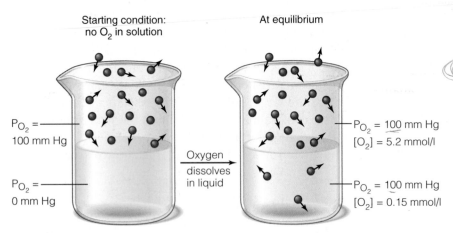

(b) Henry's law

Figure 9.2 Partial pressure of gases (a) Molecular collisions of gas molecules exert a pressure on the walls of a container, according to the ideal gas law. Each gas in a mixture contributes to this pressure in proportion to its concentration. (b) Henry's law. Gases dissolve in solution according to their partial pressure and solubility. Because the solubility of oxygen in aqueous solutions is low, the concentration of oxygen dissolved in water is much lower than the concentration of oxygen in air, even when the partial pressures in the two media are at equilibrium.

Henry's law describes how gases dissolve in liquids

In order to diffuse into a cell, gas molecules in air must first dissolve in liquid (such as water or extracellular fluid). **Henry's law** states that the amount of gas that will dissolve in a liquid is determined by the partial pressure of the gas and the solubility of the gas in the liquid. We can write Henry's law as follows:

$$[G] = P_{gas} \times S_{gas}$$

where [G] is the concentration of gas dissolved in the liquid, P_{gas} is the partial pressure of the gas in the atmosphere above the liquid, and S_{gas} is the solubility of the gas in that liquid (Figure 9.2b). The effect of solubility on gas concentration is another reason we usually express the Fick equation in terms of pressure rather than concentration.

For example, consider the diffusion of a gas across the cell membrane. Gases are much more soluble in lipids than they are in aqueous solution. When the partial pressures are at equilibrium, and are thus the same outside the cell, in the membrane, and inside the cell, the concentration of gas will actually be higher in the membrane than outside the cell, because of the higher solubility of the gas in the membrane lipids. Despite this concentration gradient, there will be no net movement of gas, because there is no partial pressure gradient.

Notice that Henry's law is actually just a modification of the ideal gas law, where [G] is equivalent to n/V, and $(1/RT)$ represents the solubility of a gas in air. Using these relationships, we can compare the content of oxygen in air with the content of oxygen in water. At sea level at 20°C, the molar concentration of oxygen in air is approximately 9 mM, whereas the concentration of oxygen in water under these conditions is less than 0.3 mM. This difference has important implications for the respiratory strategy of an organism. To obtain the same amount of oxygen, an animal that uses water as its respiratory medium must move 30 times more fluid across its respiratory surface than an equivalent organism that uses air as its respiratory medium.

The solubility of oxygen in water decreases by almost 50% when temperature is raised from 0 to 40°C, causing a large decrease in oxygen concentration. This effect makes it more difficult for aquatic organisms to obtain sufficient oxygen from their environments at high temperatures—a particularly acute challenge for animals such as fish whose body temperature and oxygen demand increase with increasing environmental temperatures. The solubility of gases also decreases with increasing ion concentration in a fluid. For example, the solubility of oxygen in seawater is 20% less than in freshwater at the same temperature. Together, these two effects cause seawater at 20°C to have almost the same oxygen content as freshwater at 30°C.

Gases diffuse at different rates

Graham's law states that when gases are dissolved in liquids, the relative rate of diffusion of a given gas is proportional to its solubility in the liquid and inversely proportional to the square root of its molecular weight (MW).

$$\text{Diffusion rate} \propto \text{solubility}/\sqrt{\text{MW}}$$

This relationship has important consequences for the diffusion of respiratory gases. Oxygen is lighter than carbon dioxide (32 atomic mass units compared to 44 for carbon dioxide). These two gases are equally "soluble" in air, so oxygen diffuses approximately 1.2 times faster in air than does carbon dioxide. However, carbon dioxide is approximately 24 times more soluble in aqueous solutions than oxygen. By substituting these numbers into Graham's law, we find that carbon dioxide diffuses about 20 times faster than oxygen in water.

By combining the Fick equation with Henry's and Graham's laws, we can derive the following equation for the rate of diffusion of a gas through a medium at a constant temperature:

$$\text{Diffusion rate} \propto \frac{D \times A \times \Delta P_{gas} \times S_{gas}}{X \times \sqrt{\text{MW}}}$$

Thus, at a constant temperature the rate of diffusion of a gas in a fluid is proportional to (1) the diffusion coefficient (D) of the gas in the medium, (2) the cross-sectional area (A), (3) the partial pressure gradient (ΔP_{gas}), and (4) the solubility of the gas in the fluid (S_{gas}), but is inversely proportional to (5) the diffusion distance (X) and (6) the molecular weight of the gas (MW). Table 9.1 provides values for the diffusion coefficients and solubilities of oxygen and carbon dioxide in air and water at

20°C. By substituting these values into the equation above, we can calculate that oxygen diffuses almost 300,000 times more slowly in water than in air at 20°C.

Fluids flow from areas of high to low pressure

As we discussed in Chapter 8, substances move across long distances much more quickly by bulk flow than by diffusion. Thus, the bulk flow of a fluid medium can transport dissolved substances such as gases, moving them across long distances much more quickly than is possible with diffusion alone. Fluids, including both liquids and gases, move by bulk flow if the total pressure in one area differs from the total pressure in another. We have already discussed the factors affecting the bulk flow of liquids in Chapter 8, but for gases, pressure is related to volume according to Boyle's law:

$$P_1 V_1 = P_2 V_2$$

where P_1 and V_1 equal the initial pressure and volume, and P_2 and V_2 equal the final pressure and volume. Thus, if you increase the volume of a sealed chamber containing a gas, the pressure within that chamber will decrease (Figure 9.3a). If you then open the chamber to the surrounding atmosphere (which is at higher pressure), the gas will move down the pressure gradient until the external pressure and the pressure inside the chamber are equal, and no further net movement of gas occurs. The lungs of terrestrial animals work in this way. For example, when you breathe in, your chest expands, increasing the volume of your lungs, and decreasing the pressure, causing air to flow into the lungs.

Table 9.1	The physical properties of air and water and their effects on the respiratory gases.		
Property	**Air (20°C)**	**Water (20°C)**	**Ratio (water/air)**
Oxygen diffusion coefficient (m²/sec × 10⁻⁹)	20,300	2.1	~1:10,000
Carbon dioxide diffusion coefficient (m²/sec × 10⁻⁹)	16,000	1.8	~1:10,000
Oxygen solubility (ml/l)	1000	33.1	1:30
Carbon dioxide solubility (ml/l)	1000	930	~1
Oxygen concentration (mM) (at 1 atm)	8.7	.3	1:30
Carbon dioxide concentration mM (at 1 atm)	.01	.01	~1
Density (kg/m³)	1.2	998	~800:1
Viscosity (poise × 10⁻²)	.02	1	50:1

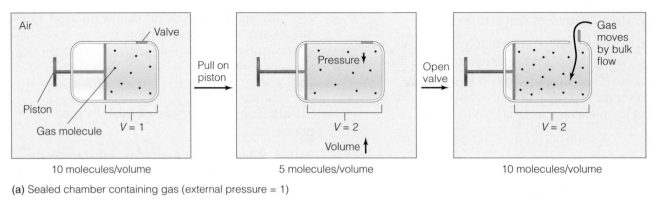

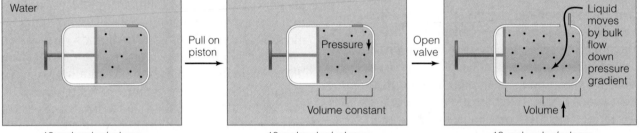

(a) Sealed chamber containing gas (external pressure = 1)

(b) Sealed chamber containing liquid

Figure 9.3 The effects of changes in volume on changes in pressure **(a)** Boyle's law. Increasing the volume of a sealed chamber filled with gas decreases the pressure within the chamber. When the chamber is opened, gas will flow into it down this pressure gradient until the pressures are equalized. **(b)** If you attempt to increase the volume of a sealed chamber containing a liquid, the volume will not change. However, pressure will decrease within the chamber. If you then open the valve, liquid will move into the chamber by bulk flow, increasing the volume of the chamber until the pressures are equalized.

Boyle's law does not apply directly to liquids, because liquids are incompressible (Figure 9.3b); the intermolecular forces holding molecules together in liquid form are too strong to be disrupted by physiologically relevant changes in pressure. However, if you exert a force on a liquid, the pressure within that liquid will change without a change in volume. These pressure changes result in the bulk flow of the liquid from the area of higher pressure to the area of lower pressure.

Resistance opposes flow

Frictional resistance opposes the bulk flow of fluids. As we discussed in Chapter 8, the relationship between flow, pressure, and resistance can be quantified using the law of bulk flow ($Q = \Delta P/R$). As in the circulatory system, flow in respiratory systems often occurs in tubes. In tubes, resistance increases in direct proportion to the length of the tube and the viscosity of the fluid, but decreases in inverse proportion to the radius to the fourth power. Because of this relationship, small increases in the radius of a tube cause large decreases in resistance.

⊙ | **CONCEPT CHECK**

1. Use the Fick equation to explain why respiratory surfaces usually have high surface area and are very thin.

2. In a gas mixture consisting of nitrogen, oxygen, and carbon dioxide, if the total pressure is 100 kPa, the partial pressure of nitrogen is 80 kPa, and the partial pressure of carbon dioxide is 0.03 kPa, what is the partial pressure of oxygen?

3. Compare and contrast the bulk flow of liquids and gasses.

Types of Respiratory Systems

Only very small animals can rely solely on diffusion of oxygen to support metabolism. As organisms grow larger, their ratio of surface area to volume decreases, limiting the area available for diffusion. Moreover, oxygen must diffuse across greater distances within the animal, increasing the time needed for diffusion. Consider a hypothetical animal shaped like a sphere (Figure 9.4). A sphere has a

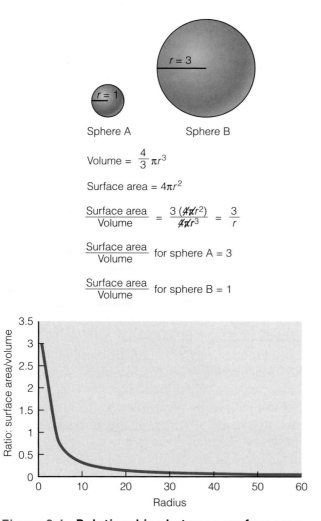

Volume = $\frac{4}{3}\pi r^3$

Surface area = $4\pi r^2$

$$\frac{\text{Surface area}}{\text{Volume}} = \frac{3\,(\cancel{4\pi}r^2)}{\cancel{4\pi}r^3} = \frac{3}{r}$$

$\frac{\text{Surface area}}{\text{Volume}}$ for sphere A = 3

$\frac{\text{Surface area}}{\text{Volume}}$ for sphere B = 1

Figure 9.4 Relationships between surface area and volume For a sphere, the ratio of surface area to volume declines as the radius increases.

volume of $\frac{4}{3}\pi r^3$, and a surface area of $4\pi r^2$. Thus, the surface area (*s*) of a spherical organism is proportional to r^2 whereas its volume (*v*) is proportional to r^3, and the ratio of surface area to volume must be proportional to $1/r$. As a result of this relationship, as the radius of the organism increases, the ratio of surface area to volume decreases. At the same time, as the sphere increases in size, the distance from the external world to the center of the sphere increases.

First, let's consider the ratio of surface area to volume, without addressing the rate of diffusion within the animal. If we assume that mitochondrial density and activity are uniform across the organism, oxygen demand must increase in proportion to the volume of the animal. However, we know from the Fick equation that oxygen supply by diffusion is related to the surface area available for gas exchange. Since the ratio of surface area to volume decreases as radius increases, oxygen supply does

not increase as quickly as oxygen demand when the radius of an animal increases. By using the Fick equation, we can calculate the maximum possible oxygen supply to a spherical animal with a given radius, and this oxygen supply must be the upper limit of aerobic metabolic rate.

Since the ratio of surface area to volume decreases as volume increases, the maximum metabolic rate of each gram of tissue must decrease as volume increases. Data from real unicellular organisms that rely on diffusion for oxygen supply conform to these predictions; metabolic rate per gram of tissue declines as size increases. In general, using this reasoning, we can conclude that an actively metabolizing spherical animal living in water can be no more than about a millimeter in diameter before it begins to be limited by the diffusing capacity of its surface.

Up to this point in the discussion, we have not considered the distance that oxygen must diffuse from the environment to the animal's body surface. In a perfectly stagnant (unmixed) environment, an organism rapidly depletes the oxygen in the immediate area, forming a stagnant **boundary layer** at its surface. Of course, real environments are almost never entirely still. Instead, environmental fluids typically move by bulk flow, as a result of temperature differences or the movement of other organisms through the fluid. These actions mix the fluid and reduce the size of the boundary layer around the organism, reducing the effective diffusion distance between the surface of the organism and the well-mixed regions of the environmental fluid. Environments with more extensive flow will have better mixing than environments with low flow. As a result of this effect, organisms that live in swiftly flowing fluids will have a smaller boundary layer around their surface and can be somewhat larger than organisms that live in motionless fluids. However, the maximum diameter of a spherical organism in a swiftly flowing fluid is still only a few millimeters. Some organisms have cilia or flagella on their surface whose beating causes fluids to move past them by bulk flow, which also acts to reduce the boundary layer, and increases the maximum possible size of a spherical organism.

Very thin animals can rely on diffusion alone for gas exchange

Of course, organisms are not necessarily spherical; their bodies may be long and thin, or their body surface may be highly folded so that the relationships of surface area to volume relevant to spherical animals no longer apply. Under these circumstances, surface area and volume might increase equally as the size of the animal increases. In this case, surface area may be sufficient for diffusion to supply the oxygen needs of even quite large organisms. For example, some soil nematodes (roundworms) can be as much as 7 mm long, a few marine species reach 5 cm, and some horsehair worms (phylum Nematomorpha) can reach up to 1 m in length. All of these organisms rely on diffusion across their body surfaces for gas exchange. The marine turbellarian flatworms are among the largest of the animals that rely primarily on diffusion for gas exchange, reaching as much as 60 cm in length and 20 cm in width.

However, there is an additional factor that must be taken into account when considering the limitations to diffusion. As we discussed in Chapter 8, the time needed for diffusion increases with the square of the distance over which a substance must diffuse, according to the following equation:

$$t = x^2/4D$$

where t is the time needed for a given amount of a substance to diffuse across distance x, and D is the diffusion coefficient for the substance. The net result of this relationship is that diffusion occurs rapidly over short distances, but is extremely slow over long distances. None of the species that rely solely on diffusion for gas exchange are more than a few millimeters thick, such that all of the cells of the body are within about a millimeter of the external medium. Organisms that are larger than a few millimeters in thickness must rely on bulk flow to transport gases.

Most animals use one of three major respiratory strategies

Animals that are more than a few millimeters thick use one of three major strategies to facilitate bulk flow of gases from the external environment to every cell in the body: (1) circulating the external medium through the body, (2) diffusion of gases across all or most of the body surface accompanied by transport of gases in an internal circulatory system, or (3) diffusion across a specialized respiratory surface accompanied by circulatory transport (see Figure 9.1). The first strategy is found in the sponges and cnidarians as well as in many terrestrial arthropods. Most aquatic invertebrates, terrestrial annelid worms, and some vertebrates such as frogs and salamanders use the second strategy, which is termed **cutaneous respiration.** The lungless salamanders (family Plethodontidae) are among the largest animals to rely upon cutaneous respiration. These animals live in moist woodland habitats, and obtain all of their oxygen by diffusion across the skin. The eggs of birds represent a special case of this respiratory strategy. Bird eggs can be extremely large (up to 15 cm in diameter in the case of an ostrich egg), but all gas exchange with the environment must occur by diffusion through pores in the eggshell.

The strategy of cutaneous respiration has several limitations. First, the very thin skin necessary to minimize the diffusion distance and maximize the rate of diffusion leaves the animal vulnerable to predation or physical damage. Second, because this thin barrier must remain moist so that dissolved oxygen can diffuse into the cell, animals that use cutaneous respiration are generally confined to aquatic or very moist terrestrial habitats. Third, as a result of these first two constraints, the surface area of the skin is usually quite limited.

Some species that rely on cutaneous respiration have skin with unusually high surface area. For example, the skin of the Lake Titicaca frog (*Telmatobius culeus*) is highly folded to increase the area available for gas exchange (Figure 9.5). Capillaries penetrate into these skin folds, decreasing the diffusion distance between the air and the blood. Similarly, adult male hairy frogs (*Trichobatrachus robustus*) develop a series of highly

Figure 9.5 Lake Titicaca frog (*Telmatobius culeus*) These frogs, which live in a high-altitude lake in Peru, use the skin for gas exchange. The highly folded skin surface increases the area of the respiratory surface.

vascularized hairlike projections of the skin around their thighs and sides of the body during the mating season, when metabolic demands are highest. These projections are thought to increase the surface area available for respiration. However, the strategy of increasing the overall body surface area is rather rare. Instead, many organisms confine their gas exchange with the environment to a small region of the body, but greatly increase the surface area of this region. This specialization allows the respiratory surface to be moist, thin, and have a large surface area, while allowing the rest of the body to be covered with a thick protective layer.

Specialized respiratory surfaces can be classified as either gills or lungs. **Gills** originate as outpocketings (*evaginations*) of the body surface and can be external or located within a respiratory cavity protected by a flap or other covering. **Lungs** originate as infoldings (*invaginations*) of the body surface, forming an internal body cavity that contains the external medium. Gills are most commonly used for gas exchange in water, whereas lungs are most commonly used for gas exchange in air, but as we discuss later in this chapter, there are several exceptions to this general rule.

Gas exchange surfaces are often ventilated

Most animals ventilate their respiratory surfaces, moving the external medium across the surface by bulk flow. Ventilation of the respiratory surface reduces the formation of static boundary layers that become oxygen depleted, improving the efficiency of gas exchange with the environment. Some animals with external gills rely on natural movements of the water for ventilation, but most species expend energy to actively ventilate their respiratory surfaces.

Nondirectional ventilation occurs when the medium flows past the gas exchange surface in an unpredictable pattern. Animals that wave their gills through the external medium are an example of a nondirectional ventilation pattern. Animals with internalized gills or lungs often utilize *tidal ventilation*. Tidal ventilation occurs when the external medium moves in and out of the respiratory chamber in a back-and-forth movement, whereas in *unidirectional ventilation* the respiratory medium enters the respiratory chamber at one point and exits via another, causing the medium to flow in a single direction across the respiratory surface.

The anatomy of the respiratory surface usually determines the type of ventilation that an animal uses, and thus animals generally do not switch from one ventilatory pattern to another. Instead, animals respond to changes in environmental oxygen or metabolic demands by altering the rate or pattern of ventilation rather than its direction. Table 9.2 describes some of these patterns.

Perfusion of the respiratory surface affects gas exchange

Most animals that have specialized respiratory surfaces also have a circulatory system that moves fluids (such as blood) by bulk flow through the body. The circulatory system allows oxygen from the respiratory surface to be transported across long distances by bulk flow. Just as ventilating the respiratory surface is important for efficient gas exchange, the movement of blood

Table 9.2 Patterns of ventilation.		
Term	**Definition**	**Examples**
Eupnea	Normal breathing	
Apnea	No breathing	During diving in air breathers
Hyperpnea	Increased ventilation frequency or volume associated with increased metabolism	Exercise
Tachypnea	Increased ventilation frequency, usually with a decrease in ventilatory volume	Panting
Dyspnea	Difficult, labored, or uncomfortable breathing	Anxiety or panic attacks, excessive exercise, various diseases (e.g., emphysema)
Hyperventilation	Increased ventilation in excess of that required to meet metabolic needs	Anxiety or panic attacks, response to blood acid-base disturbance
Hypoventilation	Decreased ventilation	Asthma, various lung diseases

through the respiratory surface can also affect exchange efficiency.

In animals that utilize nondirectional ventilation, the partial pressure of oxygen (P_{O_2}) in the blood leaving the gas exchanger can approach the P_{O_2} in the medium, if the medium is very well mixed (Figure 9.6a). Any factor that increases diffusion distance will decrease oxygen exchange efficiency, and reduce the P_{O_2} in the blood leaving the gas exchanger (Figure 9.6b). For example, if ventilation is inefficient, an oxygen-depleted boundary layer will form at the respiratory surface, increasing the effective diffusion distance. Similarly, in vertebrates that use cutaneous respiration, the skin is typically much thicker than the lining of other gas exchange surfaces such as gills or lungs. In these situations, the P_{O_2} in the blood leaving the gas exchanger can be much lower than that in the external medium.

Animals that tidally ventilate are generally unable to completely empty their respiratory cavity with each ventilatory cycle. As an animal breathes in, incoming fresh medium mixes with the residual oxygen-depleted medium in the respiratory cavity. Thus, the P_{O_2} in the respiratory cavity is lower than that of the external medium. The P_{O_2} of the blood equilibrates with that of the medium in the respiratory cavity. This equilibrated medium is then exhaled. The P_{O_2} of the blood exiting the gas exchange surface in an organism will thus be approximately in equilibrium with this exhaled medium (Figure 9.6c), if the diffusion distance across the respiratory surface is small.

With unidirectional ventilation, the blood can flow in one of three ways relative to the flow of the medium. The blood may flow in the same direction as the medium, in which case it is called *concurrent* (or *cocurrent*) flow. Alternatively, the blood and medium may flow in opposite directions, in which case it is referred to as *countercurrent* flow. Finally, the blood may flow at an angle relative to the flow of the external medium, in which case it is called *crosscurrent* flow.

Concurrent flow allows the P_{O_2} of the blood to equilibrate with the P_{O_2} of the respiratory medium (Figure 9.6d). As deoxygenated blood enters the gas exchange surface, it comes into contact with the fully oxygenated external medium. As the blood flows through the gas exchange surface, the P_{O_2} gradually equilibrates between the two compartments and blood P_{O_2} approaches that of the exhaled medium. With countercurrent flow, in contrast, the P_{O_2} of the blood leaving the gas ex-

change surface can approach that of the *inhaled* medium (Figure 9.6e). As blood flows through the gas exchanger it becomes progressively more oxygenated, whereas the medium becomes progressively deoxygenated as it travels in the opposite direction. Because the medium and blood are flowing in opposite directions, a partial pressure gradient that favors diffusion of oxygen into the blood is maintained across essentially the entire gas exchange surface, and the P_{O_2} of the blood leaving the respiratory organ can approach the P_{O_2} of the inhaled medium.

The efficiency of a countercurrent exchanger depends on the flow rates of the blood and the external medium. Countercurrent exchange of gases is most efficient when flow of both fluids is relatively slow. When flow is rapid or poorly matched, respiratory systems that use countercurrent flow may not differ substantially in efficiency from systems using concurrent flow.

In crosscurrent flow, multiple capillaries are arranged at an angle to the flow of the external medium. After they exit the gas exchange surface, these capillaries coalesce into an efferent blood vessel (Figure 9.6f). The P_{O_2} of the efferent vessel leaving the gas exchange surface is generally higher than would be seen with concurrent flow, but lower than that seen with countercurrent flow. In a crosscurrent system, the first vessel that crosses the gas exchange surface encounters a fully oxygenated medium, yielding a high P_{O_2} in the capillary, but subsequent capillaries encounter a progressively oxygen-depleted medium, and thus have somewhat lower P_{O_2}. The blood mixes as the capillaries merge, reaching a P_{O_2} that is approximately the average of the P_{O_2} of the blood in all the capillaries. The exact P_{O_2} in the blood leaving the respiratory surface with crosscurrent exchange depends on the relative rates of flow between the medium and the blood. If the flow of the medium is high relative to the flow of blood, the P_{O_2} of the medium will not be greatly depleted as it travels through the gas exchanger, and blood P_{O_2} may begin to approach the P_{O_2} of the inhalant medium. In contrast, if the flow of the medium is low relative to blood flow, then the P_{O_2} of the medium will decline sharply across the respiratory surface and blood P_{O_2} will be lower. Thus, as with countercurrent exchange, crosscurrent exchange is more efficient than either tidal or concurrent ventilation under only a restricted set of circumstances.

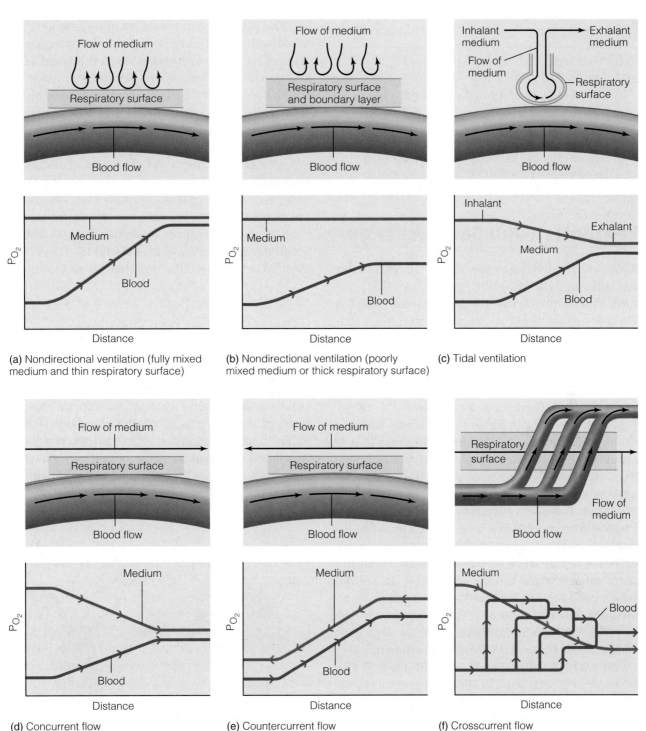

(a) Nondirectional ventilation (fully mixed medium and thin respiratory surface)

(b) Nondirectional ventilation (poorly mixed medium or thick respiratory surface)

(c) Tidal ventilation

(d) Concurrent flow

(e) Countercurrent flow

(f) Crosscurrent flow

Figure 9.6 Effects of the orientation of the flow of the external medium and the blood on gas exchange efficiency Both the mode of ventilation and the orientation of the flow of the respiratory medium and the blood affect the efficiency of gas exchange. **(a)** In nondirectional ventilation the P_{O_2} of the blood may approach that of the respiratory medium, if diffusion distance is small. **(b)** If diffusion distance increases, efficiency decreases.

(c) In tidally ventilated respiratory structures, and in unidirectionally ventilated respiratory structures with concurrent flow **(d)**, the P_{O_2} of the blood approaches that of the exhaled medium. In unidirectional ventilation with countercurrent **(e)** or crosscurrent **(f)** flow, the P_{O_2} of the blood can be higher than that of the exhaled medium. (Adapted from Piiper and Scheid, 1992.)

4. What are the limitations on cutaneous respiration?

5. What is the difference between a gill and a lung?

6. Explain why a respiratory structure with countercurrent flow could exhibit higher efficiency of gas exchange than a respiratory structure with concurrent flow.

Ventilation and Gas Exchange

Because the physical properties of air and water are substantially different (see Table 9.1), the strategies animals use to ventilate the gas exchange surface differ in air and water. Most animals that use water as the respiratory medium have unidirectionally ventilated gills, whereas most animals that use air as the respiratory medium either have tidally ventilated lungs or use a system of air-filled tubes as in insects.

From the data in Table 9.1, you can see that the oxygen content of air is almost 30 times that of water at 20°C. Thus, water-breathing animals must ventilate their respiratory surface nearly 30 times more vigorously to move the same amount of oxygen across the respiratory surface as do air-breathing animals. Water is also much more dense and viscous than air, and as a result, it takes much more energy to move a volume of water than the same volume of air. In tidal ventilation, an animal must expend energy to reverse the direction of the medium into and then out of the respiratory cavity. With unidirectional ventilation, an organism need only expend energy to move the fluid in a single direction. Unidirectional ventilation is thus less costly than is tidal ventilation. Unidirectional ventilation also makes possible a countercurrent arrangement of blood flow, improving oxygen extraction efficiency. For all of these reasons, aquatic organisms generally have gills that they ventilate unidirectionally.

For animals that use air as the respiratory medium, oxygen availability is high, and the density of the medium is low, so the cost of ventilation is not the primary issue. Instead, these animals face the possibility of evaporation across the respiratory surface, and thus usually have internally located gas exchange surfaces such as lungs that allow them to recover much of the evaporating water.

The difference in solubility of oxygen and carbon dioxide also has important implications for the relative levels of carbon dioxide in the blood of air and water breathers. Water breathers must ventilate the respiratory surface at a high rate to obtain sufficient oxygen. As a result, they are ventilating more than is necessary to eliminate the carbon dioxide they produce. In contrast, air breathers do not need to ventilate the respiratory surface at such high rates to obtain oxygen, so they do not eliminate as much carbon dioxide as do water breathers. Because of this relative difference in ventilation with respect to carbon dioxide, water breathers typically have an arterial P_{CO_2} that is almost 20 times lower than that seen in air breathers.

Ventilation and Gas Exchange in Water

Animals use a variety of strategies for ventilation and gas exchange in water. Some aquatic animals circulate the external medium through an internal cavity that penetrates throughout the body (Figure 9.7a). In sponges (phylum Porifera) the beating of flagellated cells called choanocytes moves water through a series of pores called *ostia* and into a central cavity called the *spongocoel*. This bulk flow

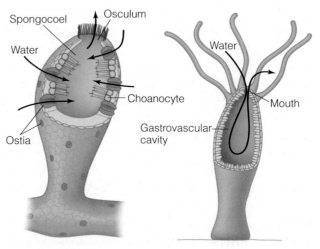

(a) Sponge (Porifera)   (b) Cnidarian

Figure 9.7 Circulation of the external medium through a digestive and respiratory cavity
(a) The body wall of a sponge is full of pores (ostia) that lead into an inner digestive and respiratory cavity called the spongocoel. The beating of flagellated choanocytes propels water through the ostia into the spongocoel and out the osculum. **(b)** Cnidarians use muscular contractions to propel water into the mouth and through the gastrovascular cavity.

moves the water past essentially all of the cells in the sponge's body. Oxygen diffuses from the water into the cells, while carbon dioxide diffuses out. Water then exits the spongocoel via the osculum. Some flatworms use a similar system. The guts of these species are lined with ciliated flame cells, and the beating of these cilia moves water containing oxygen and food molecules throughout the body.

In cnidarians (jellyfish, corals, sea anemones, and similar animals) muscle contractions move water through the mouth into the gastrovascular cavity (Figure 9.7b), which extends into all parts of the body. As water passes the tissues, oxygen diffuses into the cells, while carbon dioxide diffuses out. Water then flows back out of the gastrovascular cavity via the mouth.

Most molluscs ventilate their gills using cilia

All molluscs are built around the same generalized body plan (Figure 9.8). The mantle, an outfolding of the body wall, surrounds the rest of the body, enclosing an internal space called the mantle cavity, which contains the gills, or *ctenidia*. In addition, the mantle itself may act as a respiratory surface in some species. In most molluscs, the gills are ciliated. Beating of these cilia propels water across the gills, allowing unidirectional flow of the external medium. In many species, blood flow through the gills is arranged in a countercurrent pattern to the flow of water. A group of bivalve molluscs known as the *lamellibranchs,* which includes clams, mussels, and oysters, have thin, flat, sheetlike gills with multiple filaments that are lengthened and folded to form a series of W-shaped structures. The gills in these species can be of the *filibranch* type (found in mussels, scallops, and oysters), in which each filament is essentially separate and attached to the other filaments via small cross-connections, or they may be of the *eulamellibranch* type (found in clams), in which the filaments are fused together to form a nearly continuous sheet. In a filibranch gill, water can move through the gill between the gill filaments, but in species with eulamellibranch gills the water passes through pores, or ostia, in the gill and into the water tubes that fill the intralamellar space. These water tubes allow higher pumping rates than are possible in filibranch gills.

The gills of cephalopod molluscs such as octopuses and squid are not ciliated. Instead, muscu-

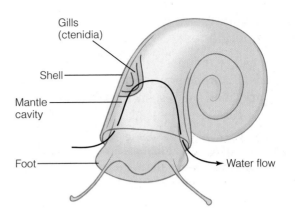

(a) Gastropod mollusc (e.g., aquatic snail)

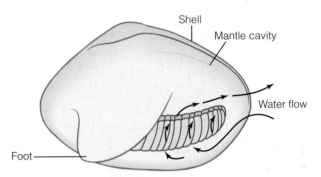

(b) Lamellibranch mollusc (e.g., clam)

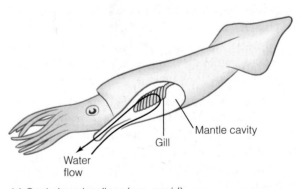

(c) Cephalopod mollusc (e.g., squid)

Figure 9.8 Respiratory systems of molluscs
(a) Aquatic snails ventilate their simple sheetlike gills using cilia. **(b)** Lammellibranch molluscs such as clams and mussels have highly modified gills with pores and internal channels. Cilia move the water across the gills by bulk flow. **(c)** Cephalopods ventilate their gills using muscular contractions of the mantle cavity.

lar contractions of the mantle propel water unidirectionally through the mantle cavity past the gills, allowing a countercurrent exchange mechanism to function in the gills. In some species of cephalopod, water flow through the mantle cavity is used for both respiration and locomotion. By rapidly expelling water out of the mantle cavity via the siphon, a cephalopod such as a squid can move by jet propulsion.

Crustacean gills are located on the appendages

Crustaceans are the most common of the aquatic arthropods. Filter-feeding species, such as barnacles, or very small species, such as copepods, typically lack gills, and instead rely on diffusion across the body surface for gas exchange. The gills of shrimp, crabs, and lobsters are modified regions of the appendages that are located within a branchial cavity formed by the hard outer covering, or carapace, of the animal (Figure 9.9). Movements of a specialized appendage, the gill bailer or *scaphognathite*, propel water out of the branchial chamber. This movement of water causes a negative pressure within the branchial chamber, which then sucks water across the gills. Various crustaceans have slightly different water-flow patterns. In shrimp, water enters all along the back and side edges of the carapace, whereas in crayfish and lobsters water enters only at the base of the legs, and in crabs water enters only at the base of the claw.

Echinoderms have diverse respiratory structures

Echinoderms (sea stars, sea urchins, brittle stars, sea cucumbers, and their relatives) have diverse respiratory structures (Figure 9.10). Most sea stars and sea urchins use their tube feet for gas exchange. The tube feet are small water-filled tubes with suction cups on the end that are part of the complex water vascular system that echinoderms use for locomotion. Echinoderms suck water into the water vascular system via a sieved opening

called the madreporite, and pump this water around the water vascular system to move the tube feet via a hydraulic mechanism. The thin skin of the tube feet, coupled with the water circulating through them, makes them important sites of gas exchange. The tube feet of some sea urchins are specialized for this respiratory function, with a countercurrent flow arrangement.

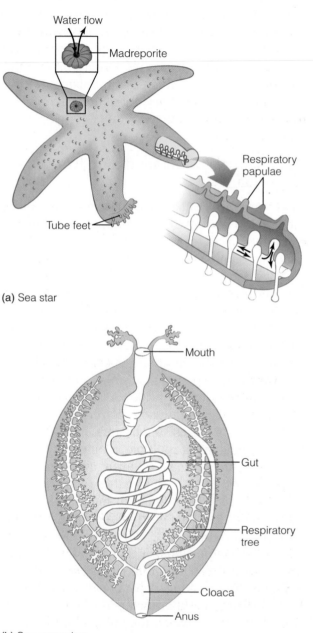

(a) Sea star

(b) Sea cucumber

Figure 9.10 Respiratory systems of echinoderms (a) Sea stars use both external gills, called respiratory papulae, and the surface of their tube feet for respiration. (b) The respiratory tree of sea cucumbers develops as a pocket leading off the gut, and thus is an invagination of the body surface that should be considered a lung rather than a gill.

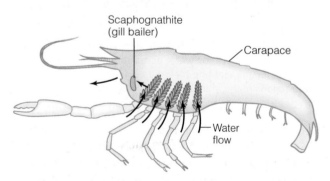

Decapod crustacean (crayfish)

Figure 9.9 Respiratory systems of crustaceans Crustacean gills are modified from the appendages, and are usually located under the carapace. Beating of the scaphognathite (gill bailer) propels water anteriorly through the animal and out an opening near the mouth.

Sea stars also have external gill-like structures called respiratory papulae scattered across their body surface. The retractable papulae are small tufted evaginations of the body surface that project through holes in the dermal skeleton and function as external gills. The outer surfaces of the papulae are covered with cilia, which beat and ventilate the respiratory surface. Cilia on the inner surface move the internal coelomic fluid by bulk flow, allowing countercurrent exchange. Sea urchins lack papulae, but many species have peristomial gills located around their mouths. Like the papulae of sea stars, these peristomial gills are ventilated by the movements of cilia.

Brittle stars and sea cucumbers have a rather different respiratory strategy. Instead of external gills, their respiratory surfaces are formed by invaginations of the body surface, and thus should more properly be termed lungs. In the brittle stars, these saclike structures are termed bursae, and open to the exterior of the body near the mouth via small slits. The opening of a bursa is usually ciliated, and the beating of these cilia ventilates the respiratory surface. Many sea cucumbers have particularly elaborate invaginated respiratory sacs called *respiratory trees* that connect to the cloaca, a portion of the intestine near the anus (Figure 9.10b). Muscular contractions of the cloaca propel water into the trunks and branches, and then the respiratory tree itself contracts to expel water back into the cloaca. Sea cucumbers use this tidally ventilated lung to supplement cutaneous gas exchange.

Feeding lampreys ventilate their gills tidally

Lampreys and hagfish have multiple pairs of gill sacs, located toward the anterior end of the body (Figure 9.11). In the case of hagfish, a muscular pumping structure called the velum propels water through the respiratory cavity. Water enters the pharynx via the single dorsal nostril, and then travels through the gill pouches and out via one or more pairs of outer gill openings (depending on the species). Flow through the gill pouches is unidirectional, and blood flow is arranged in a countercurrent pattern relative to the water flow.

Ventilation in nonfeeding lampreys is thought to be similar to that in hagfish, since lampreys also have a velum that can pump water unidirectionally across the gills. Water flows via the mouth into

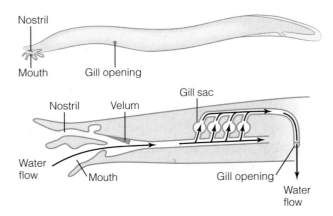

(a) Hagfish (side view and longitudinal section)

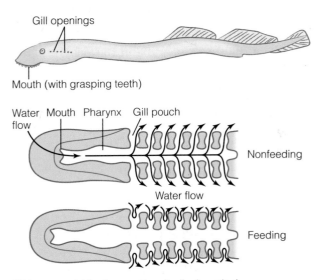

(b) Lamprey (side view and longitudinal section)

Figure 9.11 Respiratory systems of jawless fishes **(a)** Hagfish ventilate their gills using a muscular velum. Movements of the velum propel water through the mouth across the gills, and out via one or more gill openings. Flow through the gill sacs is unidirectional. **(b)** Lampreys have multiple gill pouches, each with an external opening. Expansion and contraction of the gill pouches ventilates the gills. When the lamprey is feeding (and possibly at other times as well) ventilation of the gill pouches is tidal, with water entering and leaving the gill pouches via the external opening.

the pharynx, and then through the gill pouches and out via the outer gill openings. However, adult lampreys are parasitic and feed by tightly attaching their round suckerlike mouth to the skin of a host species such as a bony fish, using their multiple grasping teeth. The lamprey then secretes a substance that dissolves the host tissue, and feeds on the dissolved tissue and blood. When feeding, a lamprey cannot ventilate its gills by unidirectional flow of water through the mouth. Under these circumstances, the lamprey pumps water into the gills via the outer gill openings, and then back out the same way. Thus, a feeding adult lamprey ventilates its gills tidally. The lamprey may continue to use this tidal ventilation between bouts of feeding, or may convert to unidirectional ventilation through the mouth during nonfeeding periods.

Elasmobranchs use a buccal pump for ventilation

The elasmobranchs (sharks, skates, and rays) ventilate their branchial chambers by expanding the volume of the buccal (mouth) cavity (Figure 9.12). This increase in volume sucks fluid into the buccal cavity via the mouth and the **spiracles**, a pair of nostril-like structures on the top of the head. The animal then closes its mouth and spiracles, and the muscles surrounding the buccal cavity contract, reducing the volume of the cavity and forcing water

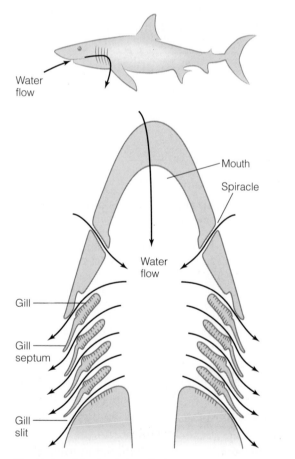

Water flow

Mouth

Spiracle

Water flow

Gill

Gill septum

Gill slit

Shark's head (horizontal section)

Figure 9.12 Respiratory system of sharks
To inhale, a shark expands the volume of the buccal cavity, and the resulting decrease in pressure sucks water into the buccal cavity via the mouth and spiracles. The shark then closes its mouth and raises the floor of the buccal cavity, forcing water across the gills.

past the gills and out via the external gill slits. Thus, the buccal cavity in this species acts as both a suction pump and a force pump. Together, these two phases of pumping action cause unidirectional but pulsatile flow across the gills. Blood flow through the gills is arranged in a countercurrent fashion, increasing the efficiency of gas exchange.

Teleost fishes use a buccal-opercular pump for ventilation

In a teleost fish the gills are located in the opercular cavities, chambers leading from the buccal cavity that are protected by the flaplike **operculum** (Figure 9.13a). Water flows from the mouth through the buccal cavity and into the opercular cavity, and then out through the slit formed by the operculum. Figure 9.13b shows the ventilatory cycle in a typical teleost fish. The first step in ventilation occurs when the fish lowers the floor of the buccal cavity while its mouth is open. This increase in the volume of the buccal cavity results in a decrease in pressure below that of the external medium, sucking water into the buccal cavity via the mouth. During this phase the operculum is closed, a skeletal muscle pump expands the volume of the opercular cavity, and the pressure in the opercular cavity decreases such that the opercular cavity pressure is below that in the buccal cavity. Thus, there is little or no backflow from the opercular cavity into the buccal cavity during this phase. During the next phase of the ventilatory cycle, the fish closes its mouth and raises the floor of the buccal cavity. This movement decreases the volume of the buccal cavity, increasing the pressure and pushing water into the expanded opercular cavity. In the next phase of the ventilatory cycle the fish opens its operculum, causing water to flow from the buccal cavity, through the opercular cavity, and out into the environment via the opercular slit. At this stage, the operculum moves inward and begins compressing the opercular cavity, increasing the pressure in the opercular cavity and forcing water out via the open opercular valve. At this point, the pressure within the buccal cavity is still high, so there is little or no backflow from the opercular cavity to the buccal cavity. The final phase of the ventilatory cycle, which occupies only a small fraction of the total ventilatory cycle, occurs when the fish again opens its mouth and begins to expand the buccal cavity. At this point, the operculum is still com-

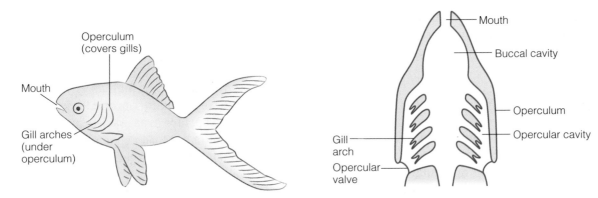

(a) Teleost fish (lateral view and horizontal section)

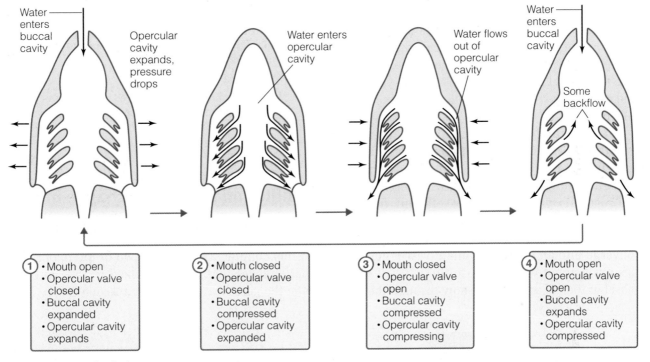

(b) Ventilatory cycle of teleosts

Figure 9.13 Respiratory systems of teleost fish **(a)** The gills of a teleost fish are located within the opercular cavity, underneath a muscular flaplike cover called the operculum. **(b)** Teleost fish use a buccal-opercular pump that ensures unidirectional and almost continuous flow across the gills.

pressed, and pressure in the opercular cavity is high. The high opercular pressure continues to force water out into the environment via the opened opercular valve, but because of the lowered pressure in the buccal cavity there may be some backflow of water from the opercular cavity into the buccal cavity. The opercular and buccal cavities then reset to their starting positions. Although there may be brief periods of backflow in the last phase of the ventilatory cycle, flow is generally unidirectional and almost continuous through most of the ventilatory cycle because of

the careful coordination of the action of the buccal and opercular pumps. In general, the opercular pump sucks while the buccal pump fills, and the buccal cavity pumps when the opercular cavity empties, reducing the possibility of backflow.

If a fish swims forward with its mouth open, water will flow across the gills without active pumping by the muscles surrounding the buccal and opercular cavities. This strategy, termed **ram ventilation**, is used by many active fish species, including tunas and some species of sharks. Ram ventilation is highly efficient because the fish does not use energy to ventilate the respiratory surface, although this strategy may increase the drag on the fish and thus increase the cost of locomotion.

Fish gills are arranged for countercurrent flow

Teleost fish have complex gills with a very large surface area for gas exchange (Figure 9.14). There are four *gill arches* in each opercular cavity. The gill arches provide structural support for the two rows of gill filaments that project from each gill arch in a V shape. The tips of the filaments from the adjacent arches overlap slightly, so that the whole gill forms a sieve. Each filament is covered with rows of interdigitated folds called *secondary lamellae*, which are perpendicular to the filament. These thin-walled structures are highly vascularized and are covered with a thin sheet of epithelial cells that acts as the primary respiratory surface.

Each gill arch contains an afferent and an efferent blood vessel. The afferent blood vessel branches into a series of afferent filament vessels that travel down the filaments, carrying blood to the respiratory surfaces. The afferent filament vessels then branch into many capillaries where gas exchange takes place. The capillaries then converge into an efferent filament vessel that carries oxygenated blood back to the efferent blood vessel in the gill arch. Blood flow through the cap-illaries of the secondary lamellae is arranged in a countercurrent pattern relative to the flow of water through the gills. When the flows through this system are properly matched, oxygen extraction from the water can reach as high as 70%.

The number of gill filaments and lamellae, and thus the total gill surface area, varies substantially among species of fish. More active species tend to have more lamellae and a larger surface area than do less active species.

◉ | CONCEPT CHECK

7. What kinds of structures can water-breathing animals use to ventilate their respiratory surfaces?
8. What is ram ventilation?
9. Outline some of the structures or mechanisms that allow the gills of teleost fishes to have very high gas exchange efficiency.

Ventilation and Gas Exchange in Air

Animals evolved in aquatic habitats, and thus air-breathing animals evolved from water breathers. In this chapter we examine two of the major animal lineages that have colonized terrestrial habitats: the vertebrates and the arthropods.

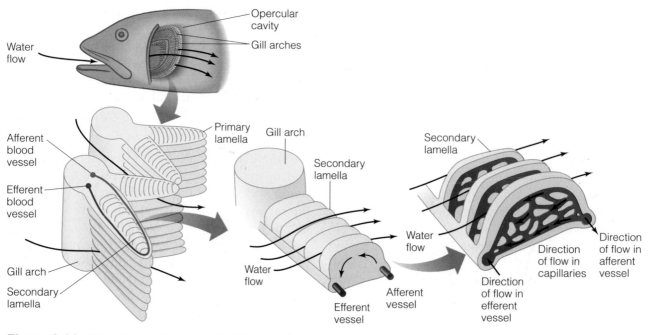

Figure 9.14 Structure of a teleost gill

Arthropods use a variety of mechanisms for aerial gas exchange

The respiratory systems of the terrestrial and semiterrestrial crabs are similar in many ways to those of their marine relatives. Like marine crustaceans, these animals have gills located in a branchial cavity, but the gills of terrestrial crabs are stiff so that they do not collapse in air. In addition, the walls of the branchial cavity are often thin and highly vascularized, acting as the primary site of gas exchange in some species. Terrestrial crabs ventilate their branchial cavity in much the same way as do their aquatic relatives; beating of the scaphognathite propels air in and out of the branchial chamber. In some terrestrial crabs, such as the porcelain crabs (genus *Petrolisthes*), the walking legs serve as an accessory respiratory surface. The carapace on part of the walking legs is very thin, allowing gas exchange.

Among the crustaceans, the terrestrial isopods (such as woodlice and sowbugs) have the most extensive specializations for gas exchange with air. In some species, such as the seashore isopod *Ligia*, a thick layer of chitin on one side of the gill provides support, while the other side is a very thin wall specialized for aerial gas exchange. In other species, such as *Armadillidium*, the anterior gills are modified and contain many branching air-filled tubules called *pseudotrachea*. Oxygen in gaseous form diffuses down the pseudotrachea and dissolves in the interstitial fluid. The circulatory system then carries this oxygen to all parts of the body.

Most of the air-breathing chelicerates (spiders, scorpions, and their relatives) have four **book lungs** located within the body cavity (Figure 9.15). Book lungs are derived from the *book gills* of aquatic chelicerates such as horseshoe crabs. Book lungs consist of a series of 10–100 very thin lamellae that project into an air-filled cavity inside the body that opens to the outside via a spiracle. Air diffuses into the cavity via the spiracle and then across the walls of the lamellae into the hemolymph, which then carries the oxygen through the body.

In many spiders, the anterior pair of book lungs is replaced by a **tracheal system**, consisting of a series of air-filled tubes. Some species (such as the Solifugae, or sun spiders) lack book lungs entirely and have only a tracheal system that penetrates into all parts of the body. Species

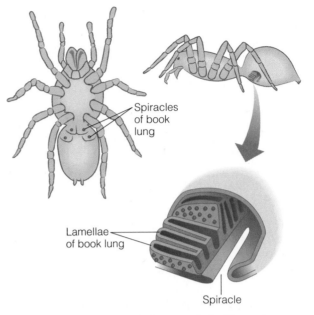

Figure 9.15 The book lungs of chelicerates
Book lungs are composed of a series of thin plates called lamellae. Oxygen from air diffuses across the surface of the lamellae into the hemolymph.

with complex tracheal systems generally make little use of their circulatory systems for gas transport. Instead, oxygen diffuses in gaseous form down the trachea and then dissolves in the interstitial fluid before diffusing into the tissues. The normal body movements of a spider cause changes in the pressure inside the body cavity, which may help to ventilate the trachea. However, other scientists suggest that these movements interfere with gas transport down the trachea, and may reduce ventilation.

Some myriapods (centipedes and millipedes) have tracheal systems similar to those in spiders, but the most extensive tracheal systems are found in insects. As in chelicerates, the tracheal system of insects is open to the outside air via a series of spiracles, which lead to the air-filled **tracheae** (singular: trachea) that penetrate deep into the body (Figure 9.16). The tracheae branch and divide, terminating in tiny thin-walled structures called **tracheoles**, which can be as small as 0.2 μm in diameter. The ends of the tracheoles are filled with circulatory fluid called hemolymph (see Chapter 8). Oxygen dissolves in this fluid, and then diffuses across the thin walls of the tracheoles.

There is no clear functional distinction between tracheae and tracheoles, but they differ structurally and in size. Tracheae are relatively large tubes that are formed by joining together

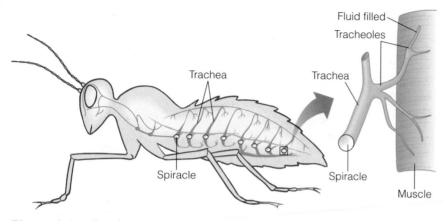

Figure 9.16 Insect tracheal systems Air enters the tracheae via the spiracles and travels down the progressively branching tubes to the tracheoles. Oxygen then dissolves in the extracellular fluid within the tracheoles and diffuses into the tissues.

several epithelial cells. In many species, the walls of the tracheae are reinforced by structures called *taenidia*. These thin bands of cuticle are wrapped in a spiral pattern around the walls of the tracheae. In some species, portions of the tracheae lack taenidia, and instead form air sacs, which are involved in ventilating the tracheal system in these species. In contrast, tracheoles are formed by hollowing out a single cell, and thus have a wall that consists only of two layers of cell membrane. Tracheoles are so numerous that an insect cell is seldom more than a few hundred micrometers, or a few cell diameters, away from the nearest tracheole. In fact, in metabolically active cells such as flight muscle, tracheoles are located within invaginations of the muscle cell membrane. As a result, the average distance between tracheoles may be as little as 3 μm. The walls of insect tracheoles are very thin, have an extremely high surface area, and are always moist—characteristics required for high-efficiency gas exchange. But because of these factors the tracheoles are also a potential site for water loss, increasing the danger of desiccation, particularly in arid environments. In many species of insects the spiracles can be opened and closed, which seals the tracheal system off from the environment part of the time, potentially reducing water loss.

Tracheal systems provide high-efficiency gas exchange in air because of the high diffusion coefficients of gases in air compared to water. In fact, tracheal systems have evolved independently in several groups of terrestrial arthropods, suggesting that there has been strong natural selection for tracheal-like systems in air. However, tracheal res-

piratory systems are not very efficient for gas exchange in water. Insects that have colonized aquatic habitats use a variety of mechanisms to avoid using water as the respiratory medium (see Box 9.1, Evolution and Diversity: Respiratory Strategies of Aquatic Insects on page 438).

Many insects actively ventilate the tracheae

The high diffusion coefficient of oxygen in air allows oxygen to diffuse through the tracheal system and still support the metabolic needs of most species of insects. However, many insects also ventilate the tracheal system actively either through contractions of the abdominal muscles or through movements of the thorax. When the abdominal muscles contract, the volume of the abdomen decreases, forcing air out of the tracheae. When the muscles relax, the abdomen springs back to its normal volume, decreasing the pressure within the tracheae, and causing air to move into the tracheae by bulk flow. Similarly, in the thorax as the wings beat, the thoracic muscles contract and relax, changing the volume of the tracheae within the thorax, which causes the air to move in and out of the tracheae by bulk flow. Oxygen then diffuses into the tracheoles, as is the case in species that do not ventilate the tracheae.

The direction of airflow through the tracheal system varies among insects. Insects with relatively simple tracheal systems use tidal ventilation; in others, the flow through the tracheae is unidirectional. For example, in cockroaches and locusts air enters the anterior spiracles, passing through large longitudinal tracheae and exiting the body via the abdominal spiracles at the rear of the body. This unidirectional ventilation may increase the efficiency of gas exchange by providing a continuous supply of fresh air to the respiratory surfaces, although even in these insects the smaller tracheae that branch off the large longitudinal tracheae are still ventilated tidally. Some flying insects, such as cerambycid (or long-horned) beetles, take advantage of ram ventilation, which is also called *draft ventilation* in insects, to ventilate the large longitudinal tracheae.

Recent observations of living insects, using a novel technique called synchrotron X-ray imaging, suggest that the volume of the tracheae can change by as much as 50% in a rapid cycle of expansion and compression that occurs every one to two seconds (Figure 9.17) and that cannot be accounted for by changes in the volume of the abdomen or thorax. The resulting pressure changes within the tracheae move the air by bulk flow. This recently discovered, and entirely unanticipated, mechanism of respiration in insects is likely to

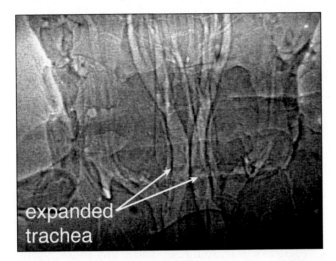

(a)

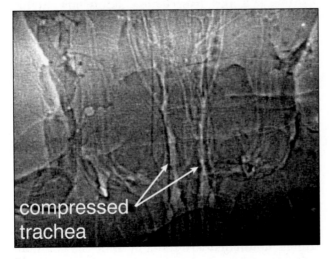

(b)

Figure 9.17 X-ray synchrotron images of insect tracheae A synchrotron, an instrument that can generate an extremely bright beam of light, can be used to generate high-resolution X-ray videos. Using this technique, scientists have been able to visualize the movements of insect tracheae. In some species, the tracheae undergo rapid cycles of expansion and contraction that are independent of movements of the rest of the body. These movements help to ventilate the tracheae.

(Reprinted with permission from Westneat, M. W. et al. 2003. Tracheal respiration in insects. *Science* 299 (5606): 588–560. Copyright 2003 AAAS.)

substantially revise our understanding of how insects obtain oxygen from the environment.

Some insects use a ventilatory pattern known as **discontinuous gas exchange**, particularly when they are at rest. Discontinuous gas exchange occurs in three phases (Figure 9.18). During the first phase, called the *closed phase*, the spiracles remain shut, preventing gas exchange with the environment. As a result, the oxygen partial pressure in the tracheoles drops as the mitochondria consume oxygen. However, the partial pressure of carbon dioxide does not increase nearly as much, because the carbon dioxide produced by metabolism reacts with water in the interstitial fluid to form bicarbonate (HCO_3^-). This decline in oxygen without an increase in carbon dioxide causes a slight decrease in the total gas pressure within the tracheae. During the next phase of the respiratory cycle, called the *flutter phase*, the spiracles open and close many times in rapid succession. The low pressure within the tracheae causes air to enter the insect's body, moving by bulk flow down the resulting pressure gradient. Eventually, as carbon dioxide accumulates, and can no longer be stored as HCO_3^-, the partial pressure of carbon dioxide begins to increase. At this point in the respiratory cycle, the spiracles open completely, and carbon dioxide is rapidly released.

The adaptive significance of discontinuous gas exchange is a matter of active debate among insect physiologists, and three main hypotheses have been advanced to explain it.

- Discontinuous gas exchange may facilitate tracheal ventilation by causing low total gas pressure within the tracheae, or by inducing a low P_{O_2} that increases the P_{O_2} gradient between the tracheae and the environment, assisting the diffusion of oxygen into the animal. This could be particularly important in insects that spend all or part of their life cycle underground where environmental P_{O_2} is low and P_{CO_2} is high.

- Discontinuous gas exchange may help to minimize water loss across the tracheae, because water will be lost from the tracheae only during the short open phase of the respiratory cycle.

- Discontinuous gas exchange may protect insects from the harmful effects of oxygen. Although oxygen is necessary for most animal life, it is also a highly reactive chemical that

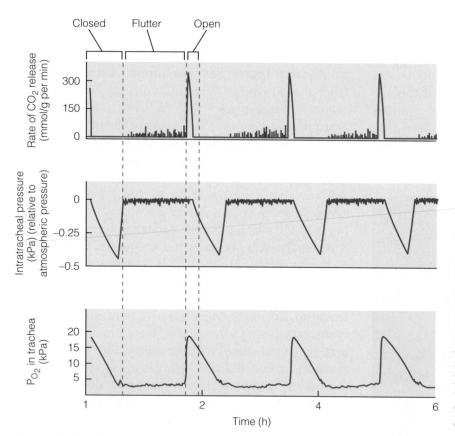

Figure 9.18 Discontinuous gas exchange cycles in insects Some insects keep their spiracles closed for long periods, only opening them briefly for gas exchange.

(Adapted from Hetz and Bradley, 2005.)

can damage tissues. When an insect's spiracles are fully open, fresh air can diffuse deep into the body, and the P_{O_2} at the ends of the tracheole approaches 20 kPa. In the vertebrates, internal tissues are seldom exposed to P_{O_2} greater than 0.5 kPa, and exposure to high P_{O_2} can cause tissue damage. During discontinuous ventilation the tissues are only exposed to high P_{O_2} during the short open phase, whereas tracheal P_{O_2} remains low during the rest of the ventilatory cycle.

Further research is needed to determine which, if any, of these hypotheses accounts for the evolution of discontinuous gas exchange in insects.

Air breathing has evolved multiple times in the vertebrates

Almost 400 species of extant fish are thought to obtain all or part of their oxygen from air, and air breathing is thought to have evolved multiple times

within the fishes. As a result of these independent evolutionary events, fish use a variety of structures for aerial gas exchange. For example, mudskippers have specialized "reinforced" gills that do not completely collapse in air, allowing some limited gas exchange when the fish is out of water. Many fish have specialized *accessory breathing organs* that they use in addition to, or instead of, gills when breathing air. Electric eels use the mouth and pharyngeal cavity for gas exchange. The inside of the mouth is highly vascularized, allowing substantial gas exchange. Some fish, including the armored catfish (*Liposarcus anisitsi*), have a highly modified and vascularized stomach that they use for aerial gas exchange. Many air-breathing fish, including bichirs (*Polypteriformes*), use specialized pockets off the gut for gas exchange.

Lungfish have the most highly developed air-breathing organ of any fish. These lungs are highly complex, covered in folds and pockets that increase their surface area. There are three living genera of lungfish. The Australian lungfish (*Neoceratodus*) has a single lung and relatively well-developed gills, whereas the African lungfish (*Protopterus*) and South American lungfish (*Lepidosiren*) have bilobed lungs and reduced gills. In addition to their highly developed lungs, lungfish have a two-circuit circulatory system with a separate pulmonary circuit. This allows lungfish to separate oxygenated blood coming from the **pulmonary system** and deoxygenated blood coming from the tissues. Animals similar to lungfish are thought to be the common ancestor of the tetrapods (amphibians, reptiles, birds, and mammals).

Air-breathing fish ventilate their breathing organs using a buccal force pump similar to those of other fishes (Figure 9.19). They drop the floor of the buccal cavity, and the increase in volume causes a drop in pressure that draws air into the mouth. By closing the mouth and raising the floor of the buccal cavity, the fish then forces air down into the breathing organ. In essence, air-breathing fish simply swallow air.

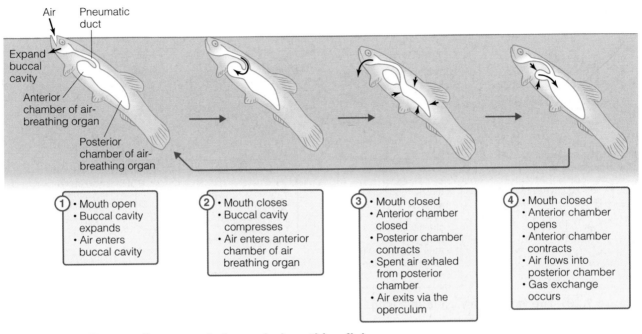

Figure 9.19 The ventilatory cycle in an air-breathing fish

Amphibians ventilate their lungs using a buccal force pump

Amphibians use cutaneous respiration, external gills, lungs, or some combination of these three methods of gas exchange depending on whether they are obtaining oxygen from water or from air. Amphibians have relatively simple bilobed lungs that form as outpocketings of the buccal cavity. In some species they may be nothing more than a pair of thin-walled, highly vascularized sacs; however, in the terrestrial frogs and toads the inner surface of the lungs can be highly folded or divided by partitions called septa, which give the lungs a honeycombed appearance and increase the surface area available for gas exchange.

An amphibian ventilates its lungs using a buccal force pump, similar to that used by air-breathing fish. In the first step of ventilation, the frog expands its buccal cavity, drawing air in through the open **nares** (nostrils) (Figure 9.20). At this point in the ventilatory cycle, the **glottis**, a muscular orifice that acts as a valve for the lungs, is closed. As a result, the fresh air is held in a pocket of the buccal cavity. The frog may make repeated buccal movements to fully refresh the air within the buccal cavity. Next, the glottis opens. Elastic recoil of the lung pushes the spent air into the buccal cavity and out the mouth and nares. Muscle contraction in the chest wall may assist in this exhalation. There is thought to be relatively little mixing of the

exhaled stale air with the fresh air held in the buccal cavity because inhaled air is held at the bottom of the buccal cavity, while exhaled air flows out through the upper regions of the buccal cavity. However, the exact degree of mixing is a matter of some debate. The nares then close and the floor of the buccal cavity rises, forcing air from the buccal cavity into the lungs. The glottis then closes as a result of muscular contractions, sealing off the lungs and preventing air from escaping, allowing time for gas exchange.

Amphibians are typically intermittent breathers. They often pause for a substantial period before beginning the respiratory cycle again. During the time that the lungs are sealed off by the glottis, a frog may pump air in and out of the buccal cavity multiple times. In fact, amphibians have a diverse ventilatory repertoire. The steps outlined above constitute a *balanced breath*, in which a roughly equal amount of air leaves and then enters the lungs with each ventilatory cycle. But amphibians can also undergo *inflation breaths*, in which the lung deflation step (Figure 9.20, step 2) is reduced or absent, or *deflation breaths*, in which more air leaves the lungs than is pumped back. Further increasing the complexity of amphibian breathing, there are some amphibian species in which the order of the steps differs. For example, aquatic toads such as *Xenopus* first empty both the lungs and the buccal cavity through the open glottis and nares,

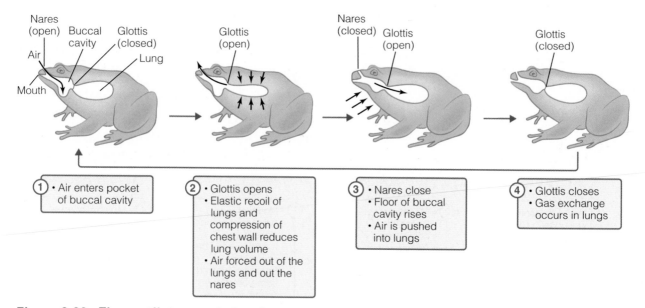

Figure 9.20 **The ventilatory cycle in a frog**

then draw fresh air into the buccal cavity with the glottis closed, and finally pump this air into the lungs with the glottis open and the nares closed (essentially performing the steps in Figure 9.20 in the order 2, 1, 3, 4).

Reptiles ventilate their lungs using a suction pump

Most reptiles have two lungs, although in snakes one of the lungs may be highly reduced or absent. The simplest, or *unicameral*, lung is a saclike chamber with a honeycombed wall, similar to the most complex amphibian lungs. In highly active species such as monitor lizards, as well as the turtles and crocodilians, the lungs are divided into many chambers, greatly increasing the surface area available for gas exchange. Each of these *multicameral* lungs has a stiffened tube called a **bronchus** (plural: bronchi) that allows airflow into the chambers of the lung. In some reptiles, the posterior part of the lungs is poorly vascularized, and may act as a bellows to help in lung ventilation.

Reptiles rely on aspiration (suction) pumps to ventilate their lungs, rather than forcing air into the lungs using a buccal pump. This important evolutionary innovation separates the muscles used in feeding from the muscles used in ventilation and is also seen in birds and mammals. In all of these groups, the ventilatory cycle is divided into two phases. During **inspiration** (inhalation), the volume of the chest cavity increases, decreas-

ing the pressure and causing air to enter the lungs. During **expiration** (exhalation), the volume of the chest cavity decreases, increasing the pressure and causing air to exit the lungs.

Reptiles use one of several mechanisms to change the volume of the chest cavity during breathing (Figure 9.21). Snakes and lizards use the intercostal muscles, which are located between the ribs. Contraction of a group of the intercostals lifts the ribs forward and outward, increasing the volume of the chest cavity, sucking air into the lungs. In lizards, the intercostal muscles are also needed for locomotion; when a lizard runs it moves its body back and forth laterally in an S-shaped pattern, a movement that involves the intercostal muscles. Thus, the muscle contractions needed for locomotion may compromise lung ventilation in some species. However, some lizards are known to supplement ventilation with a buccal force pump similar to that used by amphibians, particularly during locomotion.

In turtles and tortoises (Figure 9.21b), the rib cage is fused to the rigid shell, and cannot be moved to ventilate the lungs. Instead, these animals have a pair of sheetlike abdominal muscles that expand and compress the lungs. In addition, movements of the limbs may assist in lung ventilation. However, as with the lizards, during locomotion there may be some conflicts between the motions needed for ventilation and those needed for locomotion. Turtles are not known to use a buccal force pump to assist in ventilating the lungs.

In crocodilians (Figure 9.21c), a sheet of connective tissue called the *hepatic septum* is tightly attached to the anterior side of the liver, and divides the visceral cavity into an anterior and a posterior space. The paired diaphragmaticus muscles run from the hepatic septum to the pelvic girdle. When these muscles contract, they pull on the hepatic septum and the liver, decreasing the volume of the abdominal cavity, and increasing the volume of the lungs. This increase in lung volume de-creases the pressure in the lungs, and the resulting suction draws air into the lungs. In essence, the liver acts like a piston that helps to alternately compress and expand the lungs.

Birds unidirectionally ventilate their lungs

In birds, the lung itself is stiff and undergoes little change in volume during the ventilatory cycle. Instead, a series of flexible air sacs associated with the lungs act as bellows (Figure 9.22a). Air enters

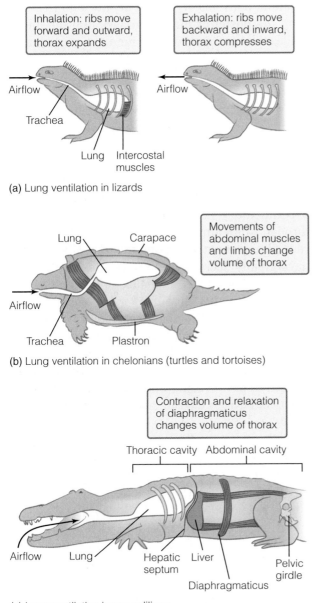

(a) Lung ventilation in lizards

(b) Lung ventilation in chelonians (turtles and tortoises)

(c) Lung ventilation in crocodilians

Figure 9.21 Lung ventilation in reptiles
(a) Lizards ventilate their lungs using their intercostal muscles. **(b)** Chelonians ventilate their lungs using movements of specialized abdominal muscles and the limbs. **(c)** Crocodilians ventilate their lungs using the diaphragmaticus muscles.

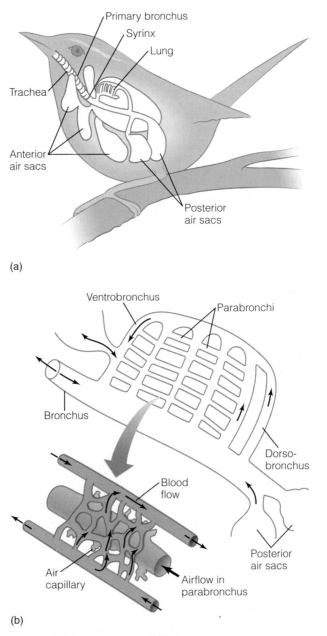

(a)

(b)

Figure 9.22 Structure of bird lungs The respiratory system of birds consists of a pair of rigid lungs and a series of highly extensible air sacs. The stiff lung is made up of hexagonal arrays of parabronchi. Extensions of the parabronchi, called air capillaries, are the site of gas exchange.

the respiratory system via the nares and mouth, passing down the cartilage-reinforced trachea. At the *syrinx*, which acts as the bird voicebox, the trachea divides into two primary bronchi, with one bronchus leading to each lung. As the bronchi enter the lungs they branch into secondary bronchi, termed the *dorsobronchi*, and then into smaller tubes called **parabronchi** that are arranged in parallel in a hexagonal array. The parabronchi then lead into secondary bronchi called the *ventrobronchi*, and back into the primary bronchi (Figure 9.22b). The walls of the parabronchi are folded to form hundreds of tiny blind-ended structures called *air capillaries*, which are richly vascularized and act as the site of gas exchange. Air diffuses from the parabronchi into the air capillaries, and then into the blood. The thin walls of the air capillaries present a minimal barrier to gas exchange by diffusion.

In birds, ventilation of the lungs requires two cycles of inhalation and exhalation. Because of this ventilatory pattern, airflow across the respiratory surfaces of the lungs is unidirectional and almost continuous. Figure 9.23 follows a single breath of air as it moves through the bird's respiratory system. A bird inhales by expanding the volume of its chest using the rib muscles and muscles attached to the sternum (breastbone). This movement increases the volume of the air sacs, and decreases the pressure within them. Air flows through the trachea and bronchi down this pressure gradient, and moves primarily into the posterior air sacs. Next, the bird exhales by compressing its chest, increasing the pressure within the air sacs. This pressure gradient moves air from the posterior air sacs into the lungs. The next inhalation causes this air to move from the lungs into the anterior air sacs. Then, on the next exhalation, the air moves from the anterior air sacs back into the trachea and out the mouth or nares. Note that although we have separated the ventilatory cycle into four steps for clarity, these processes actually occur simultaneously. Both sets of air sacs inflate during inhalation, but fresh air from the environment moves into the posterior air sacs, while stale air from the lungs moves into the anterior air sacs. During exhalation, both sets of air sacs deflate, and fresh air from the posterior air sacs moves into the lungs, while stale air from the anterior air sacs is exhaled out the nares and mouth.

Bird lungs are extremely efficient, and can extract a high percentage of oxygen from the air. In fact, the P_{O_2} of the blood leaving the lungs is typically higher than the P_{O_2} of the exhaled air. As we discussed earlier in the chapter, only a countercurrent or crosscurrent flow pattern in the lungs could account for this observation. To distinguish between these possibilities, respiratory physiologists experimentally reversed the direction of airflow through a bird lung. If the flow was in a countercurrent arrangement, reversing the flow of air should have greatly decreased the oxygen extraction efficiency. Instead, the P_{O_2} of the blood leaving the lung was always higher than the P_{O_2} of the exhaled air, regardless of the direction of airflow. This observation demonstrates that blood flow in a bird lung is arranged in a crosscurrent pattern, providing high oxygen extraction efficiency. Such efficiency may be needed to power flight, and may play a role in the ability of birds to tolerate high altitudes.

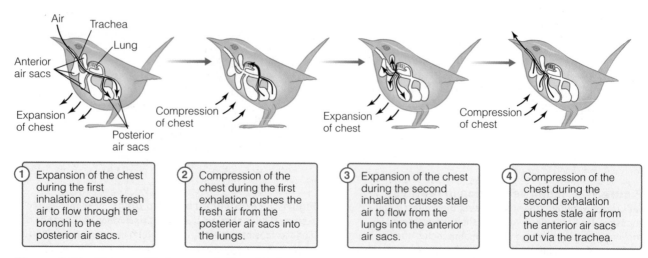

Figure 9.23 The ventilatory cycle in a bird

The alveoli are the site of gas exchange in mammals

The mammalian respiratory system is located within the chest cavity, or *thorax*, and is divided into an upper respiratory tract, consisting of the mouth, nasal cavity, pharynx, larynx, and trachea, and a

lower respiratory tract consisting of the bronchi and gas exchange surfaces (Figure 9.24). Air enters the lungs via the mouth and nares, passing through the pharynx and larynx, and then entering the cartilage-reinforced trachea. The trachea branches into two primary bronchi, which branch into successively smaller tubes called the secondary and tertiary bronchi, and then **bronchioles**. The bronchioles terminate in thin-walled, blind-ended sacs called **alveoli** that are the site of gas exchange.

The alveolar epithelium is composed of two types of cells. The thin Type I alveolar cells are responsible for gas exchange. The much thicker Type II alveolar cells are responsible for a variety of functions, including maintaining the fluid balance across the lungs and secreting lipoproteins called **surfactants**. The alveoli are wrapped with an extensive capillary network that covers 80–90% of the alveolar surface.

Both lungs are surrounded by the **pleural sac** (Figure 9.25), which consists of two layers of cells

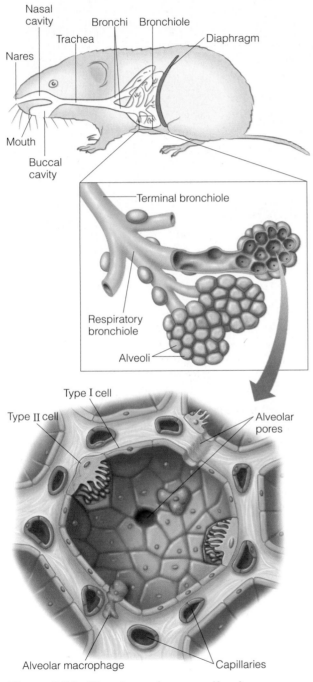

Figure 9.24 Structure of mammalian lungs
Mammalian lungs consist of conducting airways, not involved in gas exchange, that terminate in a series of interconnected blind-ended sacs called alveoli that form the respiratory surface. The alveoli are polygonal in shape, with flattened walls, and are wrapped in blood vessels and suspended in a collagenous matrix.

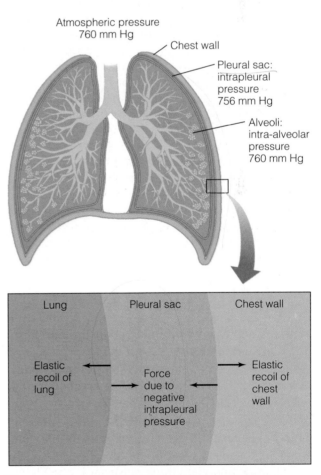

Figure 9.25 The relationship between the lungs, pleura, and chest wall At rest, the intrapleural pressure is lower than atmospheric pressure. This low pressure pulls on the lungs and keeps them expanded.

BOX 9.1 **EVOLUTION AND DIVERSITY**
Respiratory Strategies of Aquatic Insects

Tracheal systems are not very well suited for aquatic respiration, because of the low oxygen content and high density and viscosity of water, and the relatively low rate of diffusion of oxygen in solution. Aquatic insects cope with this problem in two ways. Some insects have evolved structures termed *tracheal gills*, which allow them to extract oxygen from water. Other aquatic insects have developed strategies that permit them to continue to breathe air despite their aquatic habitat.

Like the gills of other species, tracheal gills are evaginations of the body surface, generally arranged in a series of plate-like structures. However, tracheal gills are densely packed with sealed air-filled tracheae and tracheoles, covered with only a very thin layer of cuticle. These gills bring the tracheae into very close contact with the water, allowing gas exchange by diffusion. Tracheal gills are generally found in the immature stages of insects and are typical of aquatic **nymphs**, the juvenile stages of insects that do not form pupae. These gills can be located on various parts of the body, including the abdomen, the base of the legs, the anus, and the rectum (the posterior portion of the gut). Mayfly and dragonfly nymphs have tracheal gills on the outside of their abdominal segments, which can be moved to generate ventilatory water currents. Insects with rectal gills pump water in and out of the rectum for ventilation.

Many species of aquatic insects simply avoid using water as a respiratory medium. For example, some insects such as mosquito larvae remain near the water surface and breathe air through a specialized structure that extends above the surface of the water and acts as a siphon or snorkel. To make sure that air and not water will enter the siphon, the spiracles on these respiratory siphons are

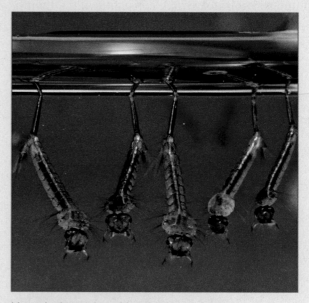

Mosquito larvae breathing through siphons.

often covered with water-repellent *hydrofuge hairs*. Some species also have hydrophobic lipids in the tracheoles that repel any water that may enter. Some fly (dipteran) larvae, including *Chrysogaster* and *Notiphila*, and the larvae of the beetle *Donacia*, utilize a variant on this siphon strategy. These insects have a sharply pointed abdominal siphon, which they use to pierce the surface of aquatic plants and extract the oxygen produced by photosynthesis.

Insects that breathe through siphons must remain close to an air source, which imposes severe limitations. Many beetles and bugs have adopted a different strategy, that of bubble breathing. These insects dive beneath the surface carrying a conspicuous bubble of

with a small amount of fluid between them, forming a space called the *pleural cavity*. The pleural fluid lubricates the pleura and allows the two layers to slide past each other during ventilation. The pressure within the fluid of the pleural cavity (or the *intrapleural pressure*) is normally subatmospheric, because the chest wall pulls on the outer layer of the pleura, whereas the elasticity of the lungs tends to pull on the inner layer of the pleura. These two opposing forces result in a subatmospheric pleural pressure.

Low intrapleural pressure plays a critical role in maintaining the integrity of the lungs. Between breaths, the pressure inside the lung at rest is equivalent to atmospheric pressure, and thus is higher than the intrapleural pressure. The relatively low pressure outside the lungs tends to pull the small airways and alveoli open, preventing these fragile structures from collapsing in on themselves. If the pleural sac is punctured, the pressure within the pleural cavity increases, and the small airways and alveoli collapse. This condi-

air under their wings. This bubble acts as an air supply while the animal is underwater. As the animal consumes oxygen from the bubble, the partial pressure of oxygen within the bubble falls lower than that of the surrounding water. As a result, oxygen diffuses down this partial pressure gradient from the water into the air bubble, providing additional oxygen to the animal. Some beetles increase this gas exchange by stirring the water around the bubble with their legs. This reduces the size of the boundary layer around the bubble, and increases oxygen availability.

Because the P_{O_2} within the bubble is lower than that in the water, and the total pressure remains similar to atmospheric pressure, the P_{N_2} within the bubble increases slightly, causing nitrogen to diffuse out of the bubble and into the water. As a result, the bubble gradually shrinks in size over time. Nitrogen is less soluble in water than is oxygen, so nitrogen leaves the bubble more slowly than oxygen enters, but over time the bubble will gradually shrink. Because CO_2 is so soluble in water, it rapidly diffuses out of the bubble, and the CO_2 produced by metabolism does not help to stabilize the size of the bubble.

Diffusion of oxygen into the bubble is a function of the surface area of the bubble (according to the Fick equation), so oxygen delivery declines as the size of the bubble decreases. As a result, these insects must periodically return to the surface to renew their bubble. This problem is even more acute as the insect descends deeper into the water. Hydrostatic pressure increases with depth, causing the volume of the bubble to decrease, which causes an increase in P_{O_2} and P_{N_2} within the bubble. Under these circumstances both oxygen and nitrogen diffuse out of the bubble, causing the size of

the bubble to decrease rapidly. Once the P_{O_2} within the bubble drops below the external P_{O_2}, oxygen will start to diffuse into the bubble, and the bubble will shrink more slowly. However, it will continue to decrease in size as nitrogen diffuses into the water, forcing the insect to return to the surface.

Some small aquatic beetles avoid returning to the surface by capturing the oxygen bubbles produced by photosynthesizing algae and adding this gaseous oxygen to their gas bubble. Other bugs and beetles use the strategy of hydrofuge hairs to prevent their bubbles from shrinking. In bugs such as *Aphelocheirus aestivalis* these hairs are arranged into a structure called a plastron, which consists of an extremely dense layer of hydrofuge hairs containing as many as 2–3 million hairs per mm^2. These hairs trap air bubbles as a thin film of gas along the surface of the body. The hairs are not collapsible, so the volume of the plastron is fixed. As the air bubble loses nitrogen to the water, the surface tension of the air-water junction between the hairs holds the bubble in place, preventing it from decreasing in size. Thus, the hydrofuge hairs prevent the bubble from collapsing. The bubble then reaches an equilibrium in which its volume is constant, but its internal pressure is reduced. Some species of aquatic insects with plastrons can remain submerged almost indefinitely.

Other aquatic insects maintain large oxygen stores within their bodies. For example, some species of aquatic bugs have hemoglobin molecules in their hemolymph. As we see later in the chapter, hemoglobin acts as an oxygen storage and transport molecule in many species. Insect hemoglobins are typically used as an oxygen store, which can help aquatic insects remain submerged for prolonged periods.

tion, known as a *pneumothorax*, causes severe shortness of breath because of the loss of the alveoli as an efficient gas exchange surface.

Mammals ventilate their lungs tidally

Mammals exhibit a tidal pattern of ventilation. Inspiration begins when somatic motor neurons trigger the contraction of the diaphragm and the external intercostal muscles of the rib cage. These contractions cause the ribs to move out-

ward and upward and the diaphragm to move down, expanding the volume of the thorax. This increase in volume decreases intrathoracic pressure, which pulls on the outer layer of the pleural sac, decreasing the pressure within the pleural cavity. This decrease in intrapleural pressure results in an increase in the pressure difference across the alveolar walls. This increase in the **transpulmonary pressure** gradient causes the lungs to expand, decreasing the pressure in the alveoli. The resulting pressure gradient between the atmosphere and the alveoli causes air to flow into the lungs.

Expiration begins when the nerve impulses from the somatic motor neurons that innervate the external intercostal muscles and diaphragm stop. This allows the muscles of the diaphragm and thorax to relax. The thorax then returns to its original position, causing thoracic volume to decrease and intrapleural pressure to increase. Because the lungs contain elastic materials, when they are no longer being actively stretched by the low intrapleural pressure they tend to snap back to their original position. This **elastic recoil** of the lungs decreases lung volume, causing alveolar pressure to increase and air to flow out of the lungs. Figure 9.26 summarizes the pressure changes within the pleural cavity and lungs during quiet breathing. During rapid and heavy breathing such as that induced by exercise, this passive expiration may not be sufficient for ventilation. Under these circumstances, contraction of the internal intercostal muscles and the abdominal muscles compresses the thorax and actively expels air from the lungs.

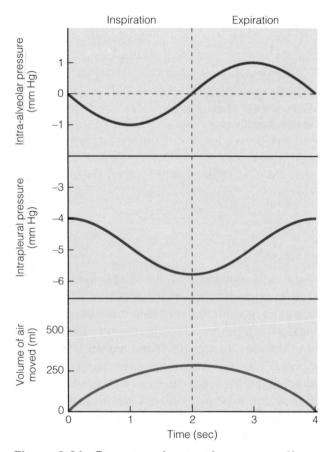
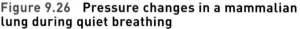

Figure 9.26 Pressure changes in a mammalian lung during quiet breathing

The work required for ventilation depends on lung compliance and resistance

The amount of energy needed to ventilate the lungs depends on the elastic properties of the lungs and chest wall and on the resistance to airflow in the pulmonary airways. The ability of the lungs to reversibly change shape can be quantified using two parameters: **compliance**, which expresses how easy it is to stretch a structure, and **elastance**, which expresses how readily the structure returns to its original shape. Lung compliance is simply defined as the magnitude of change in lung volume produced by a given change in pressure. A highly compliant lung stretches more in response to a pressure change than does a less compliant lung, and can be described by the following equation:

$$C = \Delta V/\Delta P$$

where C is the lung compliance, ΔV is the change in lung volume, and ΔP is the change in transpulmonary pressure. The lower the lung compliance, the harder it is to expand the lungs and the higher the energetic costs of inspiration.

Lung compliance can change as a result of disease. For example, in fibrotic lung disease, which can result from chronic inhalation of asbestos, silicon, or coal dust, scar tissue on the lungs reduces lung compliance and makes inspiration difficult. As a result, individuals with fibrotic lung disease tend to breathe shallowly, and thus must breathe more rapidly in order to obtain sufficient oxygen.

Lung elastance is a measure of the degree of return to resting volume after the lung is stretched. When lung elastance is low, the lungs will not spring back to their original shape when the respiratory muscles relax. As a result, if lung elastance is low, expiration must be active rather than passive. In the disease emphysema, the springy elastin fibers that are normally found in the lungs are destroyed. In individuals with emphysema the lung is easier to inflate (it is more compliant), but its elastance is low, so it will not spring back into shape as well as a healthy lung. Thus, individuals with emphysema have difficulty on expiration, and must expend energy to breathe out even at rest.

Surfactants increase lung compliance

One important force that resists lung inflation (and thus reduces lung compliance) is surface tension in the thin layer of fluid that lines the small airways and alveoli of the lungs. Surface tension results from hydrogen bonding between water molecules, and provides a cohesive force that causes two wet surfaces to stick together. Surface tension can be altered by the addition of surfactants that disrupt these cohesive forces. Type II alveolar cells secrete lipoprotein surfactants that reduce the surface tension of the fluid layer lining the lungs, thus reducing the tendency of the walls of the small airways and alveoli to stick together. As a result, surfactants make the lung more compliant and easier to stretch.[1] Surfactant secretion from Type II cells is regulated so that stretching these cells (for example, during deep breathing) stimulates surfactant secretion.

In humans, surfactant synthesis does not begin until relatively late in embryonic development. As a result, babies that are delivered prematurely (more than eight weeks early) do not have sufficient surfactant in their lungs, greatly reducing the compliance of the lungs. This low compliance makes it very difficult for premature babies to breathe, potentially causing respiratory distress syndrome. Amniocentesis can be used to determine whether a baby is synthesizing sufficient surfactant prior to birth. If birth cannot be delayed, a physician may administer corticosteroid drugs to the mother. These drugs cross the placenta and accelerate the development of the infant's lungs. After birth, an infant with mild respiratory distress syndrome may be treated with oxygen, or may need to be artificially ventilated. Premature babies are also often treated with artificial surfactants that are sprayed into the lungs, or administered via artificial ventilation tubes.

Airway resistance affects the work required to breathe

Airway resistance, the force opposing bulk flow of gas through the trachea, bronchi, and bronchioles, is the final determinant of the energy required for breathing. The law of bulk flow and Poiseuille's equation (see Chapter 8) tell us that airway diameter has an extremely large effect on airway resis-

tance. When airway diameter is small, airway resistance is high, and the pressure gradient driving bulk flow must be larger. Thus, airway resistance influences the size of the pressure gradient needed to move air into or out of the lungs. In order to cause air to flow through high-resistance narrowed airways, the lungs must develop a lower intra-alveolar pressure, causing a larger gradient between atmospheric pressure and intra-alveolar pressure, and providing a greater driving force for bulk flow. In order to attain low intra-alveolar pressure, the lungs must develop a large transpulmonary pressure gradient. Since muscular contractions and the resulting change in the volume of the thorax alter the transpulmonary pressure, more energy and thus more work is needed to inflate the lungs when airway diameter is small.

The nervous system, hormones, and paracrine chemical messengers can affect the diameter of the bronchioles. During bronchodilation airway diameter increases, whereas during bronchoconstriction airway diameter decreases. Parasympathetic neurons innervate the smooth muscles surrounding the bronchioles. Stimulation of these neurons causes bronchoconstriction. The paracrine chemical messenger histamine also causes bronchoconstriction. Histamine is released in response to tissue damage or as a result of allergic reactions. Because of this effect of histamine on the bronchioles, severe allergic reactions can cause difficulties in breathing. Circulating epinephrine causes bronchodilation, acting primarily through β2 receptors in the smooth muscle of the bronchioles. Similarly, high levels of CO_2 in the alveoli cause bronchodilation. This negative feedback loop helps to keep alveolar P_{CO_2} within a set range.

Aspiration-based pulmonary systems have substantial dead space

The total volume of air moved in one ventilatory cycle is referred to as the **tidal volume** (V_T). Some of the air that enters with each ventilatory cycle does not participate in gas exchange, contributing to the **dead space** (V_D) of the system. The dead space consists of two components: the anatomical dead space and the alveolar dead space. The anatomical dead space is the volume of the trachea and bronchi, which are not involved in gas exchange. The remainder of the physiological dead space, termed the alveolar dead space in mammals, consists of all the areas of the lungs that

[1]The importance of surfactants is often described in terms of the law of LaPlace for spheres as applied to the inflation of individual alveoli. But this represents a misconception of the structure of the alveolus. Alveoli are not spherical, but rather polygonal in shape and are interconnected by alveolar pores, and thus the law of LaPlace for spheres cannot apply. Instead, surface tension along both flat and curved surfaces within the lungs contributes to resistance to lung inflation.

in principle could be involved in gas exchange, but for some reason are not exchanging gases during a particular ventilatory cycle. For example, in a mammalian lung this could include the volume of any alveoli that are not being perfused with blood.

When an animal breathes out, some of the stale air leaving the lungs remains in the anatomical dead spaces, and is breathed in again at the next inhalation. The total amount of fresh air that is involved in gas exchange during a respiratory cycle is thus equal to the tidal volume minus the dead space ($V_T - V_D$), and in mammals is symbolized as V_A, or the alveolar ventilation volume. The total effective ventilation of the lungs per unit time is simply this quantity multiplied by the breathing frequency, or respiratory rate (f). Thus, lung ventilation is equal to $f(V_T - V_D)$. Since breathing frequency is usually measured in breaths per minute, this is usually called the alveolar minute ventilation in mammals, and is symbolized as $\dot{V}_A$. The small dot over the V indicates that this is a rate function. Increases in the size of the dead space decrease alveolar ventilation at a given tidal volume. This effect is particularly important for species with very long necks, such as giraffes and some birds (Figure 9.27). These animals have extremely large tidal volumes in order to ensure adequate ventilation of the respiratory surfaces.

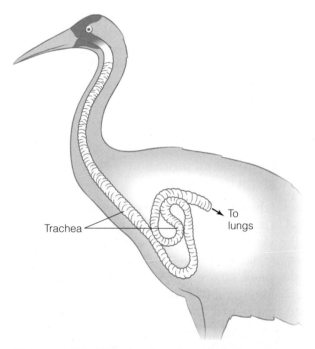

Figure 9.27 The respiratory system of a whooping crane Some birds have an extremely long trachea, which greatly increases the dead space of the respiratory system.

Trachea
To lungs

Pulmonary function tests measure lung function and volumes

Pulmonary function tests allow clinicians and experimenters to measure both lung volumes and lung function. An instrument called a spirometer can be used to measure the volumes of air inhaled and exhaled under various conditions. When at rest, most animals do not fully inflate or deflate their lungs with each breath. Thus, the tidal volume is usually much smaller than the maximum possible amount of air that can be inhaled or exhaled. In a typical adult male human, the tidal volume at rest is approximately 500 ml (lung volumes are typically about 20% less in females), whereas the total lung capacity is nearly 5800 ml (Figure 9.28). The maximal amount of air that can be inhaled over and above the resting tidal volume is termed the *inspiratory reserve volume*, and the tidal volume plus the inspiratory reserve volume is the *inspiratory capacity*. The maximal amount of air that can be forcibly exhaled over and above the resting tidal volume is the *expiratory reserve volume*. By summing the expiratory reserve volume and the inspiratory capacity, we obtain the **vital capacity**, or the maximum amount of air that can be moved into or out of the respiratory system with one breath. Mammals are not able to expel all the air out of their lungs, even with maximal exhalation. In fact, in humans approximately 1200 ml of air remains in the lungs even at the end of a maximal exhalation. This *residual volume* occurs because the lungs are held stretched against the chest walls by the pleural sac. The **total lung capacity** is the sum of the vital capacity and the residual volume.

Ventilation-perfusion matching is important for gas exchange

In order for gas exchange to occur efficiently, the ventilation of the respiratory surface must be matched to the perfusion of the respiratory surface with blood. The ventilation perfusion ratio V_A/Q quantifies this relationship. In a normal human, alveolar ventilation (V_A) is usually around 4–5 l/min, and cardiac output (Q) around 5 l/min, so that V_A/Q is close to 1 on average. The lungs have homeostatic mechanisms to maintain ventilation-perfusion matching at the level of the alveolus. If an alveolus receives little or no fresh air, the P_{O_2} in that alveolus will be low. The low P_{O_2} acts as a signal to the smooth muscle surrounding the arterioles leading to

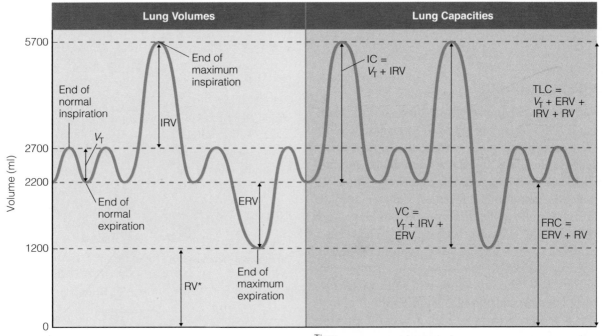

Lung Volumes	Lung Capacities
V_T = Tidal volume = 500 ml	IC = Inspiratory capacity = V_T + IRV = 3500 ml
IRV = Inspiratory reserve volume = 3000 ml	VC = Vital capacity = V_T + IRV + ERV = 4500 ml
ERV = Expiratory reserve volume = 1000 ml	FRC = Functional residual capacity = ERV + RV = 2200 ml
RV = Residual volume* = 1200 ml	TLC = Total lung capacity = V_T + ERV + IRV + RV = 5700 ml

*Cannot be measured by spirometry

Figure 9.28 Lung volumes and capacities Lung volumes and capacities can be recorded on a spirometer. Inhalation causes the line to deflect upward, whereas exhalation causes the line to deflect downward.

that alveolus. Recall from Chapter 8 that in systemic tissues low P_{O_2} is a signal for vasodilation, which increases oxygen delivery to the tissues. In contrast, in the lungs, low P_{O_2} causes vasoconstriction, reducing blood flow to areas that are poorly ventilated. This hypoxic pulmonary vasoconstriction is the primary means by which the lungs ensure appropriate ventilation-perfusion matching. However, the mechanisms by which the smooth muscle cells of the pulmonary arterioles sense low P_{O_2} and induce contraction are not yet well understood.

CONCEPT CHECK

10. Outline the similarities and differences between the respiratory systems of insects and arachnids (spiders and their relatives).

11. Compare and contrast the mechanisms of ventilation in an air-breathing fish and an amphibian.

12. What is the function of surfactants in the mammalian respiratory system?

13. Explain why bronchoconstriction (for example, during an asthma attack) increases the work required to breathe.

Gas Transport to the Tissues

Animals such as sponges, cnidarians, and insects, which circulate the external fluid past almost every cell in their bodies, can rely on diffusion to transport gases between the external medium and the tissues. But many animals transport gases using a circulatory system. As we discussed in Chapter 8, animals have exquisite control of their circulatory systems, and can regulate the transport of oxygen and carbon dioxide to and from the tissues by vasoconstricting or vasodilating the blood vessels, altering blood flow. In this section, we look at how animals use circulatory systems to transport both oxygen and carbon dioxide.

Oxygen Transport

Oxygen can be transported from the respiratory surface to the tissues dissolved in the circulatory fluid. But because the solubility of oxygen in aqueous fluids such as plasma is low, the amount of oxygen that can dissolve in the plasma is relatively small. To combat this limitation, the blood of many animals contains specialized *metalloproteins*, which contain metal ions that reversibly bind oxygen. These metalloproteins greatly increase the amount of oxygen that can be carried in the blood. For example, hemoglobin (Hb), the oxygen carrier in vertebrate blood cells, increases the maximum amount of oxygen that blood can carry—or the **oxygen carrying capacity**—by as much as 50-fold.

At the respiratory surface much of the oxygen that diffuses into the blood binds to the metalloprotein oxygen carriers, thereby reducing blood P_{O_2}. By taking this oxygen out of solution, oxygen carriers help to maintain the P_{O_2} gradient across the respiratory surface, improving oxygen extraction. At the tissues, mitochondrial oxygen consumption decreases the P_{O_2} of the blood, causing oxygen to dissociate from the oxygen carrier. This oxygen then diffuses down its P_{O_2} gradient into the cells.

There are three main types of respiratory pigments

The metalloprotein oxygen carriers are often referred to as **respiratory pigments**, because the metal ions that they contain give them a color. In animals, three major types of metalloproteins act as respiratory pigments: hemoglobins, hemocyanins, and hemerythrins.

Hemoglobins, the most common type of respiratory pigment in animals, are found in a wide variety of taxa including vertebrates, nematodes, some annelids, some crustaceans, and some insects. All hemoglobins consist of at least one molecule of a protein in the **globin** family noncovalently bound to a **heme** molecule, which consists of a porphyrin ring containing ferrous iron at the center (Figure 9.29). The iron molecules in hemoglobin give vertebrate blood its reddish color. Globins are structurally diverse, but all share a characteristic tertiary structure called the globin fold, which sug-

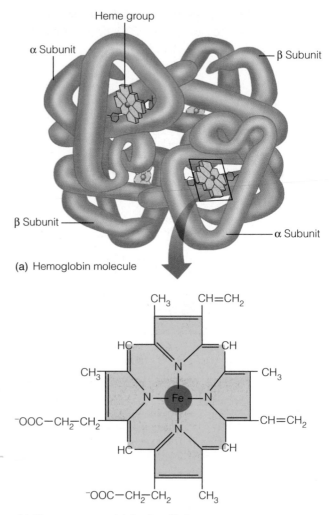

(a) Hemoglobin molecule

(b) Heme group containing iron (Fe)

Figure 9.29 Structure of mammalian hemoglobin All hemoglobins consist of one or more globin proteins complexed to an iron-containing porphyrin ring. Most vertebrate hemoglobins are tetramers, composed of four globins and their heme groups. Mammalian hemoglobins are composed of two alpha and two beta globin chains.

gests that these diverse molecules share a common evolutionary history.

In this chapter we focus on the globins found in blood, either within blood cells or extracellularly, but molecules related to the blood hemoglobins are found in many tissues. These hemoglobins are also thought to play a role in oxygen transport and storage. For example, a type of hemoglobin called **myoglobin** is found in muscles, where it helps to provide the oxygen needed for metabolism (see Chapter 5: Cellular Movement and Muscles, and Chapter 12: Locomotion). A related protein called *neuroglobin* is found in neurons. Neuroglobin has been shown to protect neural tissue during periods of hypoxia (low oxygen). Recently, another protein closely related to myoglobin has been identified. This protein, called *cytoglobin*, is found in many tissues, with particularly high expression in the cells of connective tissue. The function of cytoglobin is currently unknown.

Active hemoglobin molecules can be made up of between one and several hundred globin molecules and their associated heme groups. Myoglobin is monomeric, whereas the blood hemoglobins of vertebrates are generally tetrameric, consisting of four globin molecules. The hemoglobins of annelids such as earthworms (*Lumbricus*) contain nearly 150 globin molecules plus a number of linker proteins that do not contain heme. Hemoglobins can be found inside blood cells, as in the vertebrates, or extracellularly dissolved in the circulatory fluid, as in many invertebrates.

A few families of marine annelids have unusual respiratory pigments called **chlorocruorins**, also known as the green hemoglobins because in dilute solutions they are greenish in color. Some investigators consider the chlorocruorins to be a distinct class of respiratory pigment, but they share many characteristics with the hemoglobins. Chlorocruorins are composed of a globin molecule complexed to an iron porphyrin. The porphyrin ring in the chlorocruorins differs slightly from heme in that one of the $CH{=}CH_2$ side chains is replaced with a CHO side chain; however, the globin molecule shares clear phylogenetic relatedness with other invertebrate globins, suggesting that the chlorocruorins are simply a subclass of the hemoglobins.

Hemocyanins are found in both the arthropods and molluscs; however, the hemocyanins in these two groups appear to have independent evolutionary origins. Among the molluscs, they are found in some gastropods, some bivalves, and all cephalopods. Among the arthropods, they are present in most crustaceans, arachnids, and centipedes. Hemocyanins do not contain iron, but instead contain copper, which is complexed directly to the protein rather than being part of a heme group. Hemocyanins are very large multimeric proteins consisting of up to 48 individual subunits per molecule. They are usually dissolved in hemolymph, often at high concentrations, rather than being located within blood cells. This extracellular location poses a strong constraint on the total concentration of hemocyanin because increased hemocyanin concentration results in an increase in the viscosity of the hemolymph, making it more difficult to pump around the body. Because hemocyanins are colorless when deoxygenated and turn blue when oxygenated, the hemolymph of these species appears blue.

Hemerythrins are found in species from four invertebrate phyla (sipunculids, priapulids, brachiopods, and annelids). However, their distributions within these phyla differ. They are found in essentially all of the sipunculid and priapulid worms, and many of the brachiopods, but in only one family of marine annelids. This unusual phylogenetic distribution is puzzling and may represent a case of convergent evolution. Alternatively, patterns such as this in which closely related genes are present in distantly related taxa may represent a case of horizontal gene transfer in which viruses carry genes from one species into another.

The hemerythrins do not contain heme. Instead, iron is bound directly to the protein via the carboxylate side chains of a glutamate and an aspartate, and the imidazole groups on five histidines. Hemerythrins are generally trimeric or octameric molecules in which each subunit contains two iron ions. Most hemerythrins are found inside circulating coelomic cells and in muscle cells, and thus can be present at high concentrations without increasing the viscosity of the hemolymph. Hemerythrins are colorless when deoxygenated but violet-pink when oxygenated. As a result, the coelomic cells containing hemerythrins are sometimes called pink blood cells.

The significance of the great variety of animal respiratory pigments is not well understood. The respiratory pigments likely represent an example of multiple independent solutions to the common problem of oxygen transport and storage.

Respiratory pigments have characteristic oxygen equilibrium curves

An **oxygen equilibrium curve** shows the relationship between the partial pressure of oxygen in the plasma and the percentage of oxygenated respiratory pigment in a volume of blood (Figure 9.30a). When the partial pressure of oxygen in solution is zero, no oxygen will be bound to the respiratory pigment. As partial pressure increases, more and more pigment molecules will bind oxygen, until the available molecules are fully bound to oxygen. At this point, the blood is said to be **saturated** with oxygen. An oxygen equilibrium curve is thus

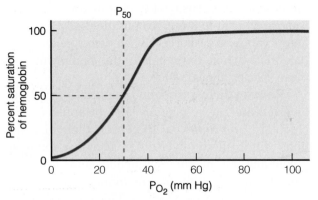

(a) Percentage of respiratory pigment oxygenated

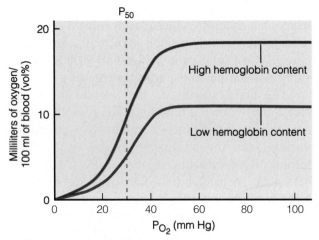

(b) Oxygen content of blood

Figure 9.30 Oxygen equilibrium curves **(a)** The percentage of saturation of a respiratory pigment as a function of oxygen partial pressure. **(b)** The oxygen content of blood as a function of partial pressure for blood with high and low content of respiratory pigment.

very much like the hormone-binding curves that we discussed in Chapter 3.

We typically express oxygen equilibrium curves in terms of percent saturation, because this allows us to conveniently compare the properties of the respiratory pigments in blood with different amounts of pigment. However, we can also express this relationship in terms of total oxygen content of the blood. Figure 9.30b shows the total oxygen content of blood that contains differing amounts of hemoglobin. As you can see, as the amount of hemoglobin increases, the total amount of oxygen that can be carried in the blood when the hemoglobin is fully saturated also increases, thus increasing the carrying capacity of the blood.

Many animals regulate the amount of respiratory pigment in the blood. For example, in many vertebrates exposure to low environmental oxy-

gen, or hypoxia, triggers red blood cell release or production. For example, in many vertebrates one of the first responses to hypoxia is contraction of an organ called the **spleen**. One of the functions of the spleen is to act as a storage site for red blood cells. Splenic contraction pushes additional red blood cells into the circulation, increasing the **hematocrit** (Hct), a measure of the proportion of blood volume that is occupied by red blood cells. In addition, hypoxia stimulates the production of new red blood cells. Low P_{O_2} stabilizes a protein called HIF-1 (hypoxia inducible factor 1), causing its concentration to increase. When HIF-1 levels are high, the protein acts as a transcription factor and induces the expression of a number of genes in a variety of tissues, including the gene coding for erythropoietin, a hormone that induces the formation of red blood cells. This increase in red blood cell numbers, and thus in hematocrit and hemoglobin concentration, increases the oxygen carrying capacity of the blood.

There is also evolutionary variation among animals in the levels of respiratory pigment in blood. For example, diving mammals have extremely high levels of blood hemoglobin compared to terrestrial mammals, which increases the oxygen carrying capacity of blood and allows it to act as an oxygen store during diving. In contrast, the Antarctic icefish (family Channichthyidae) are unique among the vertebrates in that they do not have any hemoglobin in the blood, and most icefish species have lost the gene coding for hemoglobin. As a result, the blood oxygen carrying capacity of these species is very low (approximately one-tenth that of the closely related notothenioid fish). Icefish also lack myoglobin in their skeletal muscles, although some species express this protein in the heart. Because of the cold, stable temperatures of the Antarctic Ocean (the mean temperature in McMurdo Sound is approximately $-1.9°C$ throughout the year), the metabolic rate and thus the oxygen demand of these fishes is relatively low. In addition, these low temperatures increase the solubility of oxygen in water and plasma, increasing the oxygen concentration of the blood. However, icefish also exhibit a number of physiological adjustments that help to compensate for the lack of hemoglobin. These fish have unusually large hearts and blood vessels, a large blood volume, and increased cardiac output compared to their non-Antarctic relatives. Together, these circulatory adjustments help to increase oxygen delivery in the absence of a respiratory pigment.

The oxygen affinity of a respiratory pigment is a measure of how readily the pigment binds oxygen. We typically express the oxygen affinity of a pigment using a measure termed the P_{50}, which is the oxygen partial pressure at which the pigment is 50% saturated. The P_{50} of a respiratory pigment is thus analogous to the K_m of an enzyme. Note that the P_{50}, like K_m, has an inverse relationship to affinity. Pigments that require relatively low partial pressures for oxygen to bind (i.e., have a low P_{50}) are said to have high affinity for oxygen, whereas pigments that require relatively high partial pressures for oxygen to bind (i.e., have a high P_{50}) are said to have low affinity.

The P_{50} of a respiratory pigment has important implications for its ability to transport oxygen. For example, a terebellid polychaete worm *Pista pacifica* has three different types of hemoglobin, each with a characteristic P_{50}. It has a giant extracellular hemoglobin with a very low oxygen affinity that circulates through its vascular system, a moderate-affinity hemoglobin that is located within circulating coelomic cells that travel through the interstitial fluid, and a high-affinity myoglobin within the cells of the body wall. These worms live in burrows that can extend almost a meter down in the anoxic (oxygen-free) sediments of mudflats. At high tide, these worms extend their gills out into the well-oxygenated water above the mudflat to obtain oxygen. Oxygen diffuses into the blood vessels of the gills, raising the P_{O_2} of the circulatory fluid. The low-affinity hemoglobin in this circulation readily binds oxygen at the relatively high P_{O_2} seen in the gills. As the blood leaves the gills, this low-affinity hemoglobin passes oxygen to the moderate-affinity hemoglobin in the coelomic cells that circulate through the body cavity and carry oxygen to the tissues. At the body wall, the moderate-affinity hemoglobin passes the oxygen to the high-affinity myoglobin in the muscle cells, providing oxygen to the tissues. Together, these three hemoglobins ensure efficient gas transport from the gills to the tissues of the worm.

Hemoglobin has extremely high affinity for carbon monoxide, binding with carbon monoxide more than 200 times more readily than with oxygen. As a result, carbon monoxide can interfere with hemoglobin oxygen binding. Thus, exposure to even relatively low levels of carbon monoxide can be fatal, because it decreases the oxygen carrying capacity of the blood, reducing oxygen supply to the tissues.

The shapes of oxygen equilibrium curves differ

Oxygen equilibrium curves can be either hyperbolic or sigmoidal (Figure 9.31). For example, myoglobin exhibits a hyperbolic oxygen equilibrium curve. Myoglobin is a monomeric respiratory pigment containing a single heme molecule with one oxygen-binding site. Because each myoglobin molecule binds oxygen independently of other myoglobin molecules, the principles of mass action (see Chapter 2: Chemistry, Biochemistry, and Cell Physiology) predict that the equilibrium curve should be hyperbolic in shape.

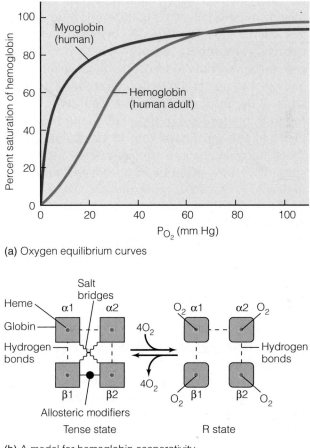

(a) Oxygen equilibrium curves

(b) A model for hemoglobin cooperativity

Figure 9.31 Cooperativity in oxygen binding
(a) Monomeric respiratory pigments, such as mammalian myoglobin, do not bind oxygen cooperatively and have a hyperbolic oxygen equilibrium curve. Multimeric respiratory pigments, such as mammalian hemoglobin, often display cooperative binding. The result of this cooperative binding is a sigmoidal oxygen equilibrium curve. **(b)** A model for mammalian hemoglobin cooperativity (after Weber and Fago, 2004). Oxygenation causes tetrameric hemoglobins to transition between the tense state that is stabilized by salt bridges and has low oxygen affinity, and the relaxed state that is stabilized only by hydrogen bonds and has high oxygen affinity.

In contrast, because of their tetrameric structure vertebrate hemoglobins exhibit a sigmoidal oxygen equilibrium curve. These hemoglobins are composed of two alpha and two beta subunits. Each alpha subunit associates tightly with one of the beta subunits, forming two dimers ($\alpha 1\beta 1$ and $\alpha 2\beta 2$) that associate with each other more loosely (Figure 9.29a). When a hemoglobin molecule is fully deoxygenated, it adopts a rigid conformation termed the tense, or T, state that is stabilized by hydrogen bonds, binding of allosteric effectors, and salt bridges between the subunits (Figure 9.31b). In contrast, fully oxygenated hemoglobin adopts a loose conformation that is termed the relaxed, or R, state. In this conformation, interactions between the subunits are stabilized only by hydrogen bonds. In the T state, hemoglobin has a relatively low affinity for oxygen, but when an oxygen molecule binds to one of the heme groups, the hemoglobin begins a transition from the T to the R state. Binding of oxygen to the iron atom causes the iron to alter its spin state and to move into the plane of the porphyrin ring of the heme group. These movements are transmitted to the globin subunits, and weaken the salt bridges holding the molecule in the tense conformation. Oxygen affinity increases progressively as each oxygen binds and the molecule adopts an increasingly relaxed conformation. The net effect of this cooperative binding (or **cooperativity**) is an oxygen equilibrium curve with a sigmoidal shape.

Although most vertebrate hemoglobins conform to this model, the hemoglobins of the jawless fishes (lampreys and hagfish) have an entirely different mechanism. These hemoglobins are monomers when they are oxygenated, and form dimers, trimers, or tetramers when deoxygenated. This shift from a multimeric to a monomeric form also results in a sigmoidal oxygen equilibrium curve.

Blood pH and P_{CO_2} can affect oxygen affinity

Changes in pH and P_{CO_2} alter the shape of the oxygen equilibrium curve for the respiratory pigments in many species, a phenomenon termed the **Bohr effect** or Bohr shift (Figure 9.32). In the Bohr effect, a decrease in pH or increase in P_{CO_2} reduces the oxygen affinity of a respiratory pigment, shifting the oxygen equilibrium curve to the right. Protons (H^+) cause the Bohr effect by binding to a

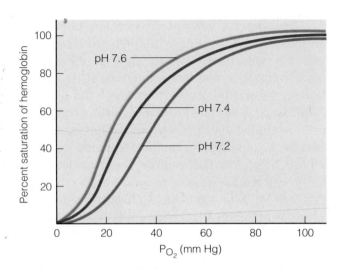

Figure 9.32 The Bohr effect Decreases in pH or increases in CO_2 cause a right shift of the oxygen equilibrium curve.

respiratory pigment at a specific site (in the vertebrates, these protons bind at the C-terminal amino acids of the β subunits, and the N-terminal amino acids of the α subunits). Proton binding causes a conformational change in the respiratory pigment protein that alters its oxygen affinity. Thus, protons act as allosteric modulators of these respiratory pigments.

Carbon dioxide can cause the Bohr effect through two separate mechanisms. As we discuss in more detail later in the chapter, in blood CO_2 reacts to form a bicarbonate ion (HCO_3^-) and a proton (H^+), and this proton can cause the Bohr effect as described above. Alternatively, carbon dioxide can have a direct effect on the oxygen affinity of respiratory pigments. CO_2 binds to the amine group of the amino acids in the respiratory pigments, forming **carbaminohemoglobin**, with a decreased oxygen affinity.

The Bohr effect facilitates oxygen transport to active tissues. At the respiratory surface, where P_{CO_2} is low and pH is high, the oxygen affinity of the respiratory pigment will be high (the curve will be shifted to the left), facilitating oxygen binding. Metabolizing tissues produce CO_2, so P_{CO_2} and $[H^+]$ in the blood increase at the tissues. This change in P_{CO_2} and pH causes the Bohr effect, decreasing the oxygen affinity of the respiratory pigment, and shifting its oxygen equilibrium curve to the right. This facilitates oxygen release from the respiratory pigment, helping to supply the tissues with oxygen.

The size of the Bohr effect differs among respiratory pigments. For example, the hemoglobins of elasmobranch fishes usually have either no Bohr effect or a very small one, whereas the hemoglobins of mammals and birds usually exhibit modest Bohr effects, and the hemoglobins of many teleost fish have extremely large Bohr effects.

In some crustaceans, cephalopods, and many teleost fishes, increases in P_{CO_2} and decreases in pH cause not only a Bohr effect, but also a reduction in the oxygen carrying capacity of the respiratory pigment (Figure 9.33), a phenomenon called the **Root effect** (or Root shift). In addition to an increase in the P_{50} at low pH, the carrying capacity of a Root-effect hemoglobin decreases greatly, releasing oxygen into solution. Thus, Root-effect hemoglobins can act as proton-triggered oxygen pumps, greatly increasing the P_{CO_2} of the plasma under low pH conditions. This mechanism is important in filling the swim bladder, an organ that some fish use for buoyancy (see Box 9.2, Evolution and Diversity: Root-Effect Hemoglobins and Swim Bladders). The mechanisms involved in the Root effect have not been fully characterized, but site-directed mutagenesis and other protein structure-function studies suggest that interactions among several amino acids are involved, and that different amino acids may be important in different species.

Temperature affects oxygen affinity

Increases in temperature can decrease the oxygen affinity of respiratory pigments such as hemoglobin in many species, shifting the oxygen equilibrium curve to the right (Figure 9.34). This effect may promote oxygen delivery during exercise. Exercising muscles generate heat, which can increase the local temperature in the blood that perfuses the tissues. As temperature increases, P_{50} increases (oxygen affinity decreases), causing oxygen to dissociate from hemoglobin, and delivering oxygen to the tissue. This temperature effect works together with the Bohr effect to maximize oxygen delivery. Similarly, the temperature of the respiratory surface may decline during exercise if the temperature of the external medium is low. This decrease in temperature increases hemoglobin oxygen affinity, which could promote oxygen uptake. However, even at normal temperatures, blood is typically almost completely saturated with oxygen at the lungs, so this effect is likely to be minor.

Some arctic animals such as reindeer and musk ox have hemoglobins that exhibit small or no temperature effects. In these animals, which live at temperatures as low as $-40°C$, the temperature in peripheral tissues such as the feet can be as much as $10°C$ lower than the core body temperature. If their hemoglobin exhibited a typical increase in oxygen affinity with decreasing temperature, oxygen delivery to the tissues might be greatly impaired.

Organic modulators can affect oxygen affinity

A variety of organic compounds can act as modulators of the oxygen affinity of respiratory pigments. In most mammals the compound 2,3-bisphosphoglycerate, also called 2,3-diphosphoglycerate (2,3-DPG), acts as an allosteric regulator of hemoglobin. 2,3-DPG is also the primary allosteric modifier in reptiles (except crocodiles), whereas in most birds inositol pentaphosphate plays this role. In contrast, in most fish (except the cyclostomes), ATP or GTP modulates hemoglobin oxygen affinity. Organic compounds including lactate, urate, and dopamine modulate the arthropod hemocyanins, with increases in these compounds increasing oxygen affinity.

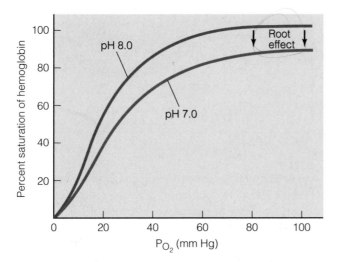

Figure 9.33 Root effect The Root effect is seen only in the hemoglobins of some teleost fish and a few species of invertebrates. Decreases in pH cause an exaggerated right shift of the oxygen equilibrium curve, and a decrease in the carrying capacity of the blood.

BOX 9.2 **EVOLUTION AND DIVERSITY**
Root-Effect Hemoglobins and Swim Bladders

Fish tissues are somewhat more dense than either freshwater or seawater, largely as a result of the high density of the skeleton, so without some form of buoyancy compensation, fish tend to sink. Many teleost fish use a gas-filled organ called a **swim bladder** to maintain their vertical position in the water. Swim bladders are located just above the gut, and below the vertebral column and kidneys. The walls of the swim bladder are largely impermeable to gas, since they are poorly vascularized and composed of a thick layer of connective tissues. In some species the wall of the swim bladder is coated with a layer of guanine crystals, which further decrease gas permeability. The gas content of swim bladders varies among species, but in most species O_2 is the principal gas.

The buoyancy provided by a swim bladder depends on the volume of this organ. Since swim bladders are soft-walled and filled with gas, they change volume as pressure changes. Atmospheric pressure increases with depth, as a result of the pressure of the overlying water. In fact, pressure increases by approximately 1 atmosphere (atm) for every 10 meters beneath the surface. If a fish descends 10 m below the surface, the volume of the swim bladder will be halved, and by 100 m depth the swim bladder will be 1/10 its original volume. Thus, in order to maintain neutral buoyancy, a fish must be able to add gas to its swim bladder as it descends through the water column.

In some fish, such as eels and salmon, the swim bladder opens into the gut via the pneumatic duct. These **physostome** fish can fill the swim bladder by gulping air, or empty the swim bladder by burping. Thus, a physostome fish can fill its swim bladder only while at the surface where it has access to air. This arrangement poses substantial problems for physostome fish such as salmon, which make extensive vertical migrations in order to feed at different depths in the water column. In order to be neutrally buoyant at depth, a fish must fill its swim bladder with a great deal of air. However, this large volume of air will make the fish positively buoyant at the surface, and it will tend to float upward, making it very difficult for the animal to dive. Studies have shown that chum salmon, physostome fish, do not gulp air into their swim bladder prior to a dive. As a result, a salmon becomes negatively buoyant as it descends, because the swim bladder is gradually

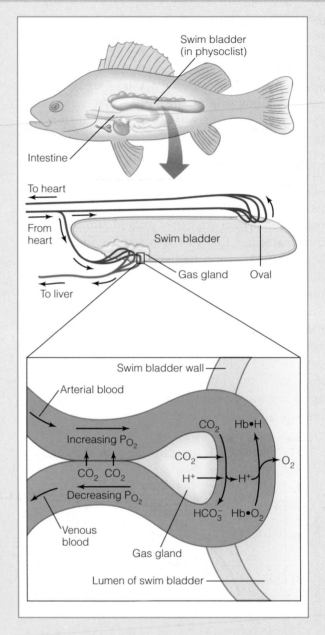

compressed in response to the increasing hydrostatic pressure. Negative buoyancy is advantageous during descent, because it decreases the cost of swimming downward, but disadvantageous during an ascent, because it increases the cost of swimming upward.

The **physoclist** fish, such as perch, use an alternative solution to filling and emptying the swim bladder. In these fish, the connection between the swim bladder and the gut is absent. Instead, gases move into or out of the swim bladder from the blood. Thus, a physoclist fish can fill the swim bladder without returning to the surface. Most of the swim bladder is impermeable to gases, so movement of gas into and out of the swim bladder occurs only at specialized structures termed the **gas gland** and **oval**. The gas gland is involved in gas secretion into the swim bladder, whereas the oval is involved in gas reabsorption from the swim bladder back into the blood.

In order for oxygen to diffuse into the swim bladder from the blood, the blood P_{O_2} in the gas gland must be greater than that in the swim bladder. To maintain this high P_{O_2}, the tissues of the gas gland produce H^+ ions and CO_2. The resulting decrease in pH and increase in P_{CO_2} cause both a Bohr effect and a Root effect. Because of the Bohr effect, the oxygen affinity of hemoglobin decreases, causing oxygen release from hemoglobin. Because of the Root effect, the oxygen carrying capacity of hemoglobin decreases, causing oxygen release. The net result of these two effects is that a substantial amount of oxygen dissociates from hemoglobin and dissolves in the blood. This dissolved oxygen now contributes to the P_{O_2} in the blood, increasing the blood P_{O_2} within the gas gland.

The cells of the gas gland have very few mitochondria, and instead obtain most of their energy through anaerobic glycolysis, causing the cell to become acidic. The gas gland cells then secrete these protons into the blood, accounting for the blood acidosis. Because the gas gland cells have few mitochondria, most of the CO_2 that they produce does not come from mitochondrial respiration. Instead, these cells activate a pathway called the pentose phosphate shunt, which produces CO_2 as a by-product. The gas gland also adds ions to the blood (largely lactate and bicarbonate, but possibly others as well). These added ions cause a "salting-out" effect. Recall that the solubility of gases in solution depends on the salt concentration of the fluid. When salt concentration increases, gas solubility decreases. From Henry's law, we can see that for a fixed quantity of gas, if solubility decreases, partial pressure must increase. Thus, this salting-out effect will increase P_{O_2} and aid oxygen diffusion into the swim bladder. Other gases, including CO_2 and N_2, are subject to this effect, which may explain the relatively high levels of these gases in the swim bladders of some species.

The gas gland of physoclist fish is associated with a specialized capillary bed called a **rete mirabile** ("wonderful net" in Latin), or rete. A rete is a bundle of capillaries in which the capillaries are arranged with flow through the arterial and venous vessels in countercurrent. This countercurrent exchanger prevents the loss of oxygen via the venous blood. The rete accomplishes this largely because of the movement of CO_2 from venous to arterial blood, rather than by movement of oxygen. As blood exits the gas gland, it has a high P_{O_2} and very high CO_2 content. This CO_2 diffuses from the venous side of the rete to the arterial side as it passes through the countercurrent exchanger. The increase in CO_2 and the associated drop in pH on the arterial side contribute to the Root and Bohr effects, increasing the P_{O_2} of the blood entering the gas gland. At the same time, the decrease in CO_2 and the associated increase in pH on the venous side cause oxygen to bind to hemoglobin, decreasing the P_{O_2} of the blood. The longer the rete, the greater the P_{O_2} that can be achieved at the gas gland. Fish that live at great depths must be able to attain high P_{O_2} in the gas gland to force oxygen into the swim bladder. The length of the rete capillaries is correlated with the maximum depth that this fish can attain. In some deep-sea fish, such as *Bassozetus*, the rete can be as long as 25 mm.

Physoclist fish empty their swim bladder at the oval. Reabsorbing oxygen from the swim bladder is not as physiologically challenging as secreting oxygen into the swim bladder, because oxygen can simply diffuse down its partial pressure gradient from the swim bladder into the blood. In most species, the oval is equipped with a muscular valve so that it can be opened and closed to regulate the amount of gas removed from the swim bladder.

References

○ Pelster, B. 2004. pH regulation and swim bladder function in fish. *Respiratory Physiology and Neurobiology* 144: 179–190.

○ Pelster, B., and D. J. Randall. 1998. The physiology of the Root effect. In *Fish Respiration*, S. F. Perry and B. L. Tufts, eds., 113–139. San Diego: Academic Press.

○ Tanaka, H., Y. Takagi, and Y. Naito. 2001. Swimming speeds and buoyancy compensation of migrating adult chum salmon *Oncorhynchus keta* revealed by speed/depth/acceleration data logger. *Journal of Experimental Biology* 204: 3895–3904.

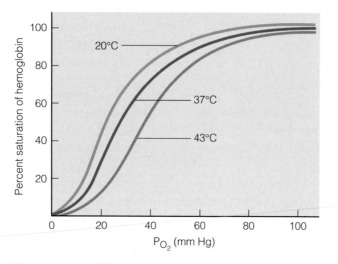

Figure 9.34 Effects of temperature on oxygen equilibrium curves

In most mammals, the effect of increased 2,3-DPG is to increase the P_{50} (decrease the oxygen affinity) of hemoglobin (Figure 9.35). Some 2,3-DPG is present within red blood cells at all times, and thus hemoglobin-oxygen binding is somewhat inhibited even at rest. 2,3-DPG levels increase in response to **anemia**, a condition in which hemoglobin levels are low, causing reduced oxygen carrying capacity, which could reduce oxygen delivery to the tissues. Increasing 2,3-DPG levels cause a modest right shift of the oxygen equilibrium curve. This change in P_{50} is not enough to harm oxygen loading at the lungs, but helps oxygen unloading at tissues. As we discuss later in

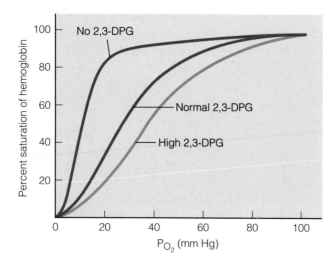

Figure 9.35 Allosteric modulation of oxygen affinity of hemoglobin Effects of the organic modulator 2,3-DPG on the oxygen equilibrium curve of mammalian hemoglobin.

this chapter, a similar effect occurs in some mammals in response to high-altitude hypoxia.

⊙ | CONCEPT CHECK

14. What is the role of the metal ion that is found in most respiratory pigments?
15. What effect does changing the amount of hemoglobin in the blood have on the P_{50} of a blood sample, and why?
16. Why does the oxygen equilibrium curve of mammalian hemoglobin have a sigmoidal shape?
17. Compare and contrast the Root effect and the Bohr effect.

Carbon Dioxide Transport

Mitochondrial respiration produces carbon dioxide that must be transported out of the body. As is the case for oxygen, in very small animals, carbon dioxide can simply diffuse from the tissues to the external environment, but in larger animals, the circulatory system transports carbon dioxide from the tissues to the respiratory surface, where it diffuses into the external environment.

Carbon dioxide is much more soluble in body fluids than is oxygen. However, very little of the CO_2 present in the blood of vertebrates is actually in the form of molecular CO_2. Some of the CO_2 binds to proteins. For example, when CO_2 binds to hemoglobin, it forms carbaminohemoglobin. Carbaminohemoglobin is a significant means of CO_2 transport in mammals, but it may not be significant in other organisms (which have far less hemoglobin).

The majority of the CO_2 is transported as bicarbonate (HCO_3^-). Carbon dioxide reacts spontaneously in water according to the following equation:

$$CO_2 + H_2O \rightleftharpoons \underset{\text{carbonic acid}}{H_2CO_3} \rightleftharpoons \underset{\text{bicarbonate}}{HCO_3^-} + H^+$$

However, the equilibrium constant of this equation lies far to the left, and this spontaneous reaction occurs slowly in aqueous solutions. In animals, an enzyme called **carbonic anhydrase (CA)** catalyzes the formation of HCO_3^-. In contrast to the uncatalyzed reaction, the reaction catalyzed by carbonic anhydrase occurs extremely rapidly. Like the respiratory pigments, carbonic anhydrase is a metalloprotein, but in this case the enzyme contains a zinc ion. Water binds to the zinc ion within the protein, and is dissociated to form H^+ and OH^-. The enzyme then directs the transfer of the OH^- ion to carbon dioxide, forming a bicarbonate ion in the following reaction:

$$CO_2 + H_2O \rightleftharpoons HCO_3^- + H^+$$

In principle, the bicarbonate formed as a result of carbonic anhydrase catalysis could further dissociate into carbonate (CO_3^-) and H^+, but this reaction is not physiologically significant in most animals. Together, molecular CO_2, carbaminohemoglobin, and HCO_3^- make up the total CO_2 content of the blood. In mammals, approximately 70% of the blood CO_2 content is in the form of HCO_3^-, whereas 7% is present as dissolved CO_2 in solution, and 23% is in the form of carbaminohemoglobin.

The carbon dioxide equilibrium curve quantifies carbon dioxide transport

Carbon dioxide equilibrium curves show the relationship between P_{CO_2} and the total carbon dioxide content of the blood, and as such are analogous to oxygen equilibrium curves (Figure 9.36). However, blood does not become saturated with CO_2; there is a rapid increase in CO_2 content at relatively low P_{CO_2}, and a continued but slower increase as P_{CO_2} rises.

The exact shape of the CO_2 equilibrium curve depends largely on the kinetics of HCO_3^- formation in the blood. In turn, the kinetics of this reac-

tion depend on blood pH, and how well H^+ ions are buffered. To understand this effect we need to recall the principles of buffering and mass action ratios from basic chemistry (Chapter 2). We can write the equilibrium constant for the reaction of CO_2 and H_2O as

$$K = \frac{[HCO_3^-][H^+]}{[CO_2]}$$

Because K is a constant, from this equation we can easily see that when $[H^+]$ is high, $[HCO_3^-]$ must decrease, if $[CO_2]$ stays constant. In essence, as pH decreases (and H^+ increases)—for example, as a result of muscle anaerobic metabolism—the CO_2-bicarbonate reaction ($CO_2 + H_2O \rightleftharpoons HCO_3^- + H^+$) will be pushed to the left, decreasing the amount of HCO_3^-. In contrast, as pH increases (and H^+ decreases) the reaction will be pushed to the right, increasing the amount of HCO_3^-. The close relationship between blood pH and carbon dioxide become even more obvious if we log transform this equation, yielding the Henderson-Hasselbalch equation that we discussed in Chapter 2:

$$pH = pK + \log \frac{[HCO_3^-]}{[CO_2]}$$

In general, blood is very well buffered. As HCO_3^- forms, the H^+ ions are quickly bound to buffer groups such as the terminal amino groups on proteins, and the imidazole side chains found on amino acids such as histidine. This prevents H^+ from accumulating and allows further HCO_3^- formation. The greater the buffering capacity of the blood, the greater the capacity to form HCO_3^-. For example, human blood is so highly buffered that 99.999% of the H^+ formed by the carbonic anhydrase reaction can be buffered. Mammalian hemoglobins have relatively high numbers of histidines, and thus act as effective buffers. In contrast, many fish hemoglobins have few histidines on the surface of the molecule, and thus act as poor buffers. Differences in the buffering capacity of the blood contribute to differences in the shape of the CO_2 equilibrium curve among species.

Blood oxygenation affects CO_2 transport

Deoxygenated blood can carry more CO_2 than can oxygenated blood (Figure 9.36). In other words, the CO_2 equilibrium curve of deoxygenated blood is shifted to the left, a phenomenon known as the

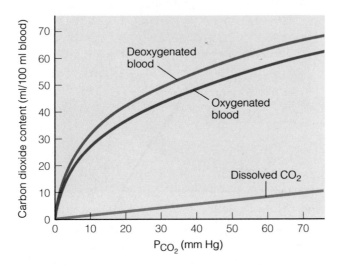

Figure 9.36 Carbon dioxide equilibrium curve (human blood) The carbon dioxide equilibrium curve of most vertebrates differs for oxygenated and deoxygenated blood, a phenomenon called the Haldane effect.

Haldane effect. Oxygenated hemoglobin releases H^+ ions. This reduces pH (by increasing the concentration of H^+ ions) and shifts the CO_2-bicarbonate reaction to the left, reducing the amount of HCO_3^- in the blood, and reducing the total amount of CO_2 that can be carried. In contrast, deoxygenated hemoglobin tends to bind H^+ ions, increasing the pH and HCO_3^-, and increasing the total amount of CO_2 that can be carried. The significance of the Haldane effect is that deoxygenation of hemoglobin at the tissues promotes CO_2 uptake by the blood, whereas oxygenation of hemoglobin at the respiratory surface promotes CO_2 unloading.

Vertebrate red blood cells play a role in CO_2 transport

In vertebrates, carbonic anhydrase is present primarily within the red blood cells, and all of the reactions discussed above occur within these cells rather than in the plasma. However, most of the bicarbonate is actually carried in the plasma. This phenomenon is easiest to understand by working through an example of carbon dioxide transport (Figure 9.37). At the tissues, CO_2 is produced by aerobic metabolism, and rapidly diffuses out of tissues and into the red blood cells. Within the red blood cell, carbonic anhydrase catalyzes the for-

mation of HCO_3^-. The H^+ formed by this reaction binds to hemoglobin. Bicarbonate does not readily diffuse through membranes, but the HCO_3^- ions are moved out of the red blood cell by a chloride-bicarbonate exchanger, also called band III. This process of Cl^-/HCO_3^- exchange is known as the **chloride shift**. If this HCO_3^- were not removed, it would build up within the red blood cell and would tend to reverse the carbonic anhydrase reaction. Within the red blood cell, band III and carbonic anhydrase are bound to each other, and another isoform of carbonic anhydrase is linked to band III on the extracellular face of the membrane. Together, these proteins form a **metabolon** (a group of enzymes that work together to perform a function and are spatially localized within the cell). Metabolons allow pathways to function more rapidly than would be possible if the substrates and products had to diffuse through the cell from one enzyme to another.

At the respiratory surface, the P_{CO_2} of the environment is lower than that of blood, and CO_2 diffuses out of the plasma across the respiratory surface. Because of this drop in plasma P_{CO_2}, CO_2 diffuses out of the red blood cell and into the plasma. This decrease in $[CO_2]$ within the red blood cell shifts the CO_2-bicarbonate reaction, causing the band III exchanger to move HCO_3^- ions from the plasma into the red blood cells in ex-

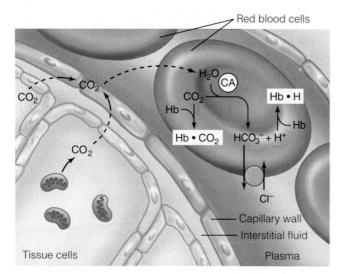

(a) Systemic tissues

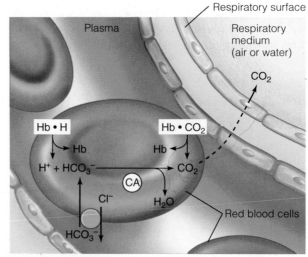

(b) Respiratory surface

Figure 9.37 Carbon dioxide transport in vertebrate blood (a) Carbon dioxide diffuses from the tissues into the red blood cell. Some binds to hemoglobin, forming carbaminohemoglobin (Hb · CO_2). Carbonic anhydrase (CA) within the red blood cell catalyzes formation of HCO_3^-. The HCO_3^- is transported out of the red blood cell in exchange for Cl^- (the chloride shift). The H^+ ions produced

by the CA reaction are buffered by hemoglobin. **(b)** In the lungs, CO_2 diffuses into the alveoli, and the CA equilibrium shifts to favor the formation of CO_2, reducing the amount of HCO_3^- within the red blood cell. HCO_3^- enters the red blood cell in exchange for Cl^-, and is converted to CO_2, which then diffuses into the alveoli.

change for Cl⁻ (in a reverse chloride shift). The HCO_3^- and H^+ form carbonic acid and then CO_2, and the CO_2 diffuses out of the red blood cell into the plasma and then across the respiratory surface. The location of carbonic anhydrase within the red blood cell increases the total CO_2 carrying capacity of the blood by ensuring that the products of the carbonic anhydrase reaction do not build up within a single compartment. This forces the CO_2-bicarbonate equilibrium to the right, and increases the amount of CO_2 that is carried as HCO_3^-. In many vertebrates, carbonic anhydrase is also present on the endothelial cells lining tissues such as the lungs. As a result, all of the bicarbonate does not necessarily have to travel via a red blood cell to be converted to CO_2.

The respiratory system can regulate blood pH

Because most proteins have a relatively narrow pH range in which they function effectively, most animals closely regulate intracellular pH. Most animals also regulate the pH of extracellular fluids such as blood, because regulating extracellular pH reduces the regulatory burden on individual cells. For example, in humans the normal pH of blood is approximately 7.4; a pH above 7.7 or below 6.8 can be fatal. Because of the tight linkage between CO_2 and pH through the reaction catalyzed by carbonic anhydrase, which we have already discussed, respiratory systems play an important role in the regulation of pH in extracellular fluids such as blood. Because the partial pressure of a gas, rather than its concentration, is the most physiologically relevant parameter, we can rewrite the Henderson-Hasselbalch equation as follows:

$$pH = pK + \log \frac{[HCO_3^-]}{\alpha CO_2 \times P_{CO_2}}$$

where αCO_2 is the solubility of carbon dioxide in the fluid, and P_{CO_2} is the partial pressure of carbon dioxide.

Physiologists use a type of graph called a **pH-bicarbonate plot** (which is sometimes referred to as a Davenport diagram) to describe the interrelationships between P_{CO_2}, HCO_3^-, and pH (Figure 9.38). These diagrams consist of a graph of the relationship between pH (plotted on the x-axis) and $[HCO_3^-]$ (plotted on the y-axis). Onto this graph are superimposed a series of curved diagonal lines called **isopleths**. Each isopleth represents the pH of

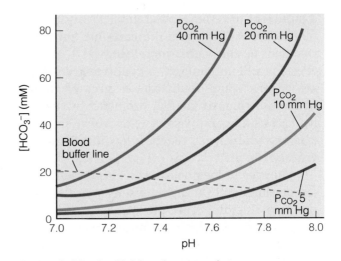

Figure 9.38 A pH-bicarbonate plot Sometimes called a Davenport diagram, a pH-bicarbonate plot with P_{CO_2} isopleths can be used to visualize the relationships between pH, HCO_3^-, and P_{CO_2} in a buffered solution. Values shown are for the European eel.
(Adapted from McKenzie, D.J. et al. 2003. Tolerance of chronic hypercapnia by the European eel, Anguilla anguilla. *Journal of Experimental Biology* 206: 1717–1726.)

the plasma as HCO_3^- is varied for a series constant values of P_{CO_2}. A Davenport diagram also includes the blood buffer line, which is an empirically calculated relationship showing the change in blood HCO_3^- when pH is titrated. The blood buffer line depends on the composition of the plasma, and thus varies among species. Under normal circumstances, an animal has typical values of plasma pH, HCO_3^-, and P_{CO_2} that fall on the blood buffer line. For example, for mammals P_{CO_2} is typically 40 mm Hg, plasma pH is 7.4, and $[HCO_3^-]$ is 24 mM.

A pH-bicarbonate plot allows physiologists to visualize what happens to the other parameters in the system when any one parameter is varied. For example, consider what happens when an animal hyperventilates. **Hyperventilation** is defined as alveolar ventilation greater than is needed to remove the CO_2 produced by metabolism. During hyperventilation plasma P_{CO_2} will fall. As P_{CO_2} declines, pH and HCO_3^- values will shift along the blood buffer line, and pH will increase while HCO_3^- decreases. In contrast, during **hypoventilation**, alveolar ventilation is less than is needed to remove the CO_2 produced by metabolism. In this case, plasma P_{CO_2} will increase and pH and HCO_3^- values will shift along the blood buffer line, with pH decreasing and $[HCO_3^-]$ increasing.

From a pH-bicarbonate plot it is very clear that changes in ventilation will result in changes in pH. **Respiratory acidosis** occurs when ventilation is insufficient to remove all of the CO_2 produced by metabolism. This shifts the carbonic anhydrase reaction to the right, increasing $[H^+]$ and decreasing pH. In contrast, a **respiratory alkalosis** occurs when ventilation is greater than is needed to remove the CO_2 produced by metabolism, causing a net loss of CO_2, which shifts the carbonic anhydrase reaction to the left, and increases the pH.

Changes in metabolism can also directly affect extracellular pH. During intense exercise, muscles produce H^+ ions. This pH disturbance is often called a lactic acidosis, because intense exercise also produces lactate as a result of anaerobic glycolysis. Because the lactate itself is not the source of the protons, this decrease in pH is more properly referred to as a **metabolic acidosis**. Metabolic acidosis can also occur because of excessive loss of HCO_3^- from the intestine during intense diarrhea, or as a result of kidney failure. In contrast, **metabolic alkalosis** can occur as a result of the loss of excess H^+ from vomiting, or because of a loss of H^+ from the kidneys as a result of kidney failure. Let's examine what would happen during a metabolic acidosis if P_{CO_2} were held constant. The metabolic protons would react with HCO_3^-, decreasing $[HCO_3^-]$. If P_{CO_2} is held constant, however, the relationship of pH to $[HCO_3^-]$ cannot move off the P_{CO_2} isopleth, and thus the values move off the blood buffer line and pH falls. Of course, animals can adjust their rate and depth of ventilation, which alters P_{CO_2}. These changes can be used to correct pH imbalances. For example, metabolic acidosis causes increased ventilation, inducing a respiratory alkalosis and returning the pH to normal values. However, as we will see in later chapters, the respiratory system is responsible largely for minute-to-minute regulation of blood pH, while the excretory system plays the major role in long-term regulation (Chapter 10: Ion and Water Balance). In the next section of this chapter, we examine some of the mechanisms by which animals regulate their ventilation, and thus gas exchange and plasma pH.

◎ | CONCEPT CHECK

18. List the forms in which CO_2 is carried in the blood of vertebrates.

19. Using the Henderson-Hasselbalch equation, outline what happens to the pH of a poorly buffered aqueous solution when $[CO_2]$ increases.

20. Why does blood oxygenation affect the CO_2 equilibrium curve of blood?

21. How can the respiratory system regulate blood pH?

Regulation of Vertebrate Respiratory Systems

Like other physiological systems, respiratory systems are closely regulated in response to changes in both the internal and external environments. Vertebrate respiratory and circulatory systems work together to regulate gas delivery and plasma pH by (1) regulating ventilation, (2) altering oxygen carrying capacity and affinity, and (3) altering perfusion.

Regulation of Ventilation

Ventilation is an automatic rhythmic process that continues even during loss of consciousness. Rhythmically firing groups of neurons within the central nervous system, or *central pattern generators*, initiate ventilatory movements in animals. In the vertebrates, these central pattern generators are located within the medulla of the brain. All vertebrates that have been examined so far have a column of respiratory-related neurons running along each side of the medulla. In bony fish, the central pattern generator is located in the rostral (or anterior) part of the medulla near the neurons that innervate the buccal cavity. Lampreys, amphibians, and mammals, however, appear to have at least two pairs of pattern generators. In mammals, these pattern generators are located in the caudal medulla. Less is known about the location of the respiratory pattern generators in reptiles and birds, but they are likely to be found in locations similar to those in amphibians and mammals.

Respiratory rhythm generation has been most extensively studied in mammals (Figure 9.39). The precise mechanisms of respiratory rhythm generation are still not fully understood. In at least some vertebrates, a small region of the caudal medulla called the **pre-Bötzinger** complex is essential for respiratory rhythm generation. In addition, another neuronal complex, the *parafacial respiratory group* or pre-I complex, is

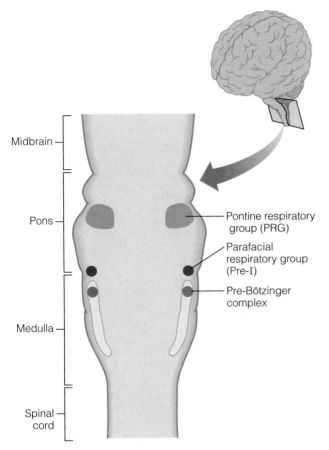

Figure 9.39 Location of the respiratory central pattern generators in mammals

Labels: Midbrain, Pons, Medulla, Spinal cord, Pontine respiratory group (PRG), Parafacial respiratory group (Pre-I), Pre-Bötzinger complex

coupled to the pre-Bötzinger complex. Neurons in the parafacial respiratory group fire before those in the pre-Bötzinger complex and appear to play an important role in specifying the timing of the rhythm in the pre-Bötzinger complex. In addition, neurons in the ventral respiratory group, particularly in the Bötzinger complex, have also been implicated in respiratory rhythm generation.

Rhythm generators can work in a variety of ways (see Chapter 7: Functional Organization of Nervous Systems). Some combination of cells with intrinsic pacemaker properties and networks of groups of neurons cause the rhythmic firing of neurons in the respiratory rhythm generators, although the exact molecular mechanisms are not yet known. These respiratory pattern generators send signals that are integrated by a variety of interneurons that ultimately send signals to the somatic motor neurons that control the skeletal muscles involved in breathing.

Chemosensory input influences ventilation

Chemosensory input helps to modulate the output of the central pattern generators. Chemoreceptors detect changes in CO_2, H^+, and O_2 and send afferent sensory information to the brain. Various regions in the brain integrate this information and provide input to the respiratory rhythm generators to modify the rate or depth of breathing. These changes in breathing act by negative feedback to maintain blood P_{CO_2} and P_{O_2} within a narrow range.

Oxygen sensing is of primary importance in water-breathing vertebrates, whereas CO_2 sensing is of primary importance in air-breathing vertebrates. Oxygen levels in water are low compared to those in air, and hypoxia, or lower than normal P_{O_2}, is a common occurrence in aquatic environments. As a result, aquatic organisms must have high ventilation in order to obtain sufficient CO_2. These levels of ventilation are usually more than adequate to remove CO_2, and blood CO_2 content is typically low. In contrast, oxygen is generally present at high levels in air, and air-breathing organisms do not need to ventilate at such high levels to obtain oxygen. But as a result, less CO_2 is removed, and total CO_2 content of the blood is typically higher in air breathers than in water breathers.

Water breathers have internal O_2 chemoreceptors that monitor the P_{O_2} of blood within the gills. There are also O_2 chemoreceptors on the surface of the body, particularly in the gill cavity and on the surface of the gills, although the distribution of these receptors may vary among species. The O_2 chemoreceptors send afferent signals to the medulla that modulate the output of the respiratory and cardiac rhythm generators. The efferent signals from these rhythm generators regulate ventilation volume and rate, cardiac output, and the perfusion pattern within the gills. Water breathers also have CO_2/pH chemoreceptors in the gills, although these are thought to be primarily involved in sensing the characteristics of the external medium.

Air-breathing vertebrates have internal CO_2/pH chemoreceptors that monitor either the P_{CO_2} or the pH of the blood. Because of the tight linkage between CO_2 and $[H^+]$ through the carbonic anhydrase equilibrium, it is difficult to establish with any certainty exactly which parameter these chemoreceptors are sensing, although recent evidence suggests that they sense intracellular pH. There are two main clusters of internal CO_2/pH chemoreceptors: **central chemoreceptors**, located in the medulla of the brain, and **peripheral chemoreceptors**, located in specific arteries.

The central chemoreceptors respond to pH changes in the cerebrospinal fluid. Although the blood-brain barrier is relatively impermeable to protons, CO_2 readily diffuses into the cerebrospinal fluid. Carbonic anhydrase within this fluid catalyzes the formation of HCO_3^- and H^+, which stimulates these chemoreceptors. Increases in CO_2 (and thus H^+) stimulate ventilation, whereas decreases in CO_2 (and thus H^+) reduce ventilation.

The peripheral chemoreceptors of mammals sense both P_{O_2} and P_{CO_2}/pH. The **carotid body** chemoreceptors are located in the carotid artery and monitor the composition of blood going to the brain. The **aortic body** chemoreceptors, located in the wall of the aorta, monitor the composition of the blood going to the body. These receptors fire only when plasma P_{O_2} starts to fall below the level required to fully saturate hemoglobin, which in most animals occurs only during pronounced hypoxia. As a result, the majority of respiratory regulation is accomplished by sensing CO_2/pH, with the central chemoreceptors playing the predominant role (Figure 9.40).

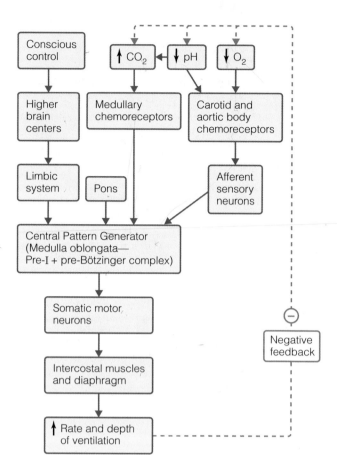

Figure 9.40 Reflex regulation of ventilation in mammals

Other factors regulate breathing

A number of mechanosensory reflexes also influence breathing. For example, in mammals irritants such as inhaled particles can stimulate receptors in the airways of the lungs. These mechanoreceptors send a signal to the central nervous system that causes the bronchi to constrict. This protective bronchoconstriction prevents the inhalation of more particles. Another set of mechanoreceptors, the slowly adapting pulmonary stretch receptors, detect the amount of tension in the walls of the airways, including the trachea and bronchi. These stretch receptors trigger the **Hering-Breuer inflation reflex**, which terminates inhalation. In adult humans, the Hering-Breuer reflex is difficult to demonstrate except when tidal volumes are extremely large, and thus it is thought to protect the lungs from being damaged by overinflation. However, in human infants and in adults of other mammalian species, the Hering-Breuer reflex may play a significant role in breath-by-breath regulation. Vertebrate lungs also contain receptors that are sensitive to CO_2 in the lungs or in the pulmonary circulation. Increasing CO_2 inhibits the receptors, and

thus stimulates ventilation. These receptors are particularly important in animals such as turtles in which the lungs fill, but do not stretch appreciably.

Breathing is also under the control of higher brain centers in the hypothalamus and cerebrum. For example, we can voluntarily alter our breathing patterns. However, although we can temporarily override the respiratory centers, we cannot do so indefinitely. If you attempt to hold your breath, eventually the drive to breathe becomes so intense, as a result of the chemoreceptor input into the medullary respiratory centers, that you are forced to breathe.

Environmental Hypoxia

Organisms regulate their respiratory systems in response to changes in both their external and internal environments. Ventilation rate and breathing frequency typically increase in response to increases in metabolic demand, such as during exercise. Animals may also have to cope with changes in environmental oxygen and carbon dioxide. In aquatic environments, for example, environmental oxygen often varies from the *normoxic* condition. During the day, when photosynthesis is maximal and plants are net oxygen producers, enclosed bodies of water such as ponds, swamps, or tidepools can become *hyperoxic*—supersaturated with oxygen. In contrast, at night when plants are net oxygen consumers, these habitats can become extremely *hypoxic*, and fish living in these areas can experience very low oxygen levels. Terrestrial animals seldom experience hyperoxia, but may experience **hypoxia** within burrows or at high altitudes. You may also come across the term **hypoxemia**—lower than normal arterial blood oxygen content. Hypoxemia can be caused by environmental hypoxia, inadequate ventilation, reduced blood hemoglobin content, and a variety of disease states. The terms **hypercapnia** and **hypocapnia** describe higher or lower than normal P_{CO_2} in either the environment or the blood. Like hypoxia, environmental hypercapnia can occur within enclosed environments such as burrows.

Fish respond to hypoxia in many ways

Many fish have external oxygen chemoreceptors that can detect environmental hypoxia, allowing fish to initiate behavioral or physiological responses to prevent hypoxemia from occurring, for example by moving away from hypoxic water. If this initial strategy fails, environmental hypoxia causes an initial, usually transient, decrease in blood P_{O_2}. This decrease in blood P_{O_2} stimulates the internal O_2 chemoreceptors, causing an increase in ventilation. A fish that ram ventilates typically opens its mouth wider to increase the flow of water over the gills, whereas a fish that uses buccal-opercular pumping increases the rate and depth of these movements. If respiratory adjustments are insufficient to compensate for environmental hypoxia, some types of fish initiate behavioral strategies such as aquatic surface respiration, in which they move to the surface of the water and ventilate their gills with the thin layer of better-oxygenated water at the air-water interface.

Prolonged exposure to hypoxia causes an increase in red blood cell numbers, and thus hemoglobin concentration, increasing oxygen carrying capacity and oxygen extraction from the environment. Some fish can reduce their metabolic rate by reducing their activity level, moving to cooler water to reduce metabolic rate, or actively suppressing their metabolism to conserve energy (see Box 9.3, Evolution and Diversity: Hypoxic Metabolic Suppression).

Air breathers can experience high-altitude hypoxia

Most air-breathing organisms only experience low environmental oxygen in specific habitats, such as when diving, within enclosed spaces such as burrows, or at high altitudes. When low-altitude-adapted animals are brought to high altitudes, they undergo a number of physiological changes, some of which may be involved in acclimatizing to the environmental hypoxia, and some of which may be pathological, if the animals are unable to acclimatize. When a low-altitude-adapted mammal experiences high-altitude hypoxia, blood P_{O_2} drops. Arterial chemoreceptors detect this decline in blood P_{O_2}, and send a signal to the medulla to increase the rate and depth of breathing, restoring or partially restoring blood P_{O_2}. Because of the increased ventilation rate, more CO_2 will be lost at the lungs, leading to hypocapnia, or lower than normal blood P_{CO_2}. Recall that, in mammals, blood P_{CO_2} provides the primary drive to breathe. The low blood P_{CO_2} at altitude can cause difficulty with breathing, particularly during sleep when the conscious drive to breathe is removed. Because of the carbonic anhydrase equilibrium, hypocapnia also leads to low [H^+]. Thus, the ventilatory response to high altitude causes respiratory alkalosis. Over longer-term exposure to high altitude, this persistent respiratory alkalosis triggers the kidneys to excrete HCO_3^- in an attempt to homeostatically regulate blood pH.

High-altitude hypoxia also leads to increases in red blood cell numbers. This effect of high altitude is one reason competitive athletes may choose to train at high altitudes or utilize a hypobaric chamber, which provides an artificial low-pressure, low-P_{O_2} environment. It is currently a matter of some debate as to whether this increase in red blood cell numbers (or *polycythemia*) actually assists in acclimatization to altitude. Polycythemia results in an increase in *hematocrit*, the proportion of the blood volume occupied by red blood cells. High hematocrit causes increased

BOX 9.3 EVOLUTION AND DIVERSITY
Hypoxic Metabolic Suppression

In low-oxygen environments, animals may be unable to obtain sufficient oxygen to meet the metabolic needs of their tissues. Many animals that can survive environmental hypoxia use a strategy called hypoxic metabolic suppression (or **hypometabolism**), in which they reduce their activity and metabolic needs in parallel with the reduced oxygen supply. This reduction in metabolic rate reduces oxygen demand and may allow an animal to survive for long periods despite environmental hypoxia. For example, some species of temperate zone turtles make use of hypoxic metabolic suppression to survive long periods underwater. Freshwater turtles, such as the painted turtle (*Chrysemys picta*) and the red-eared slider (*Trachemys scripta*) are obligate air breathers, but can remain submerged for long periods—for example, during winter in ice-covered ponds. Some species also bury themselves in anoxic mud. The metabolic rate of a submerged turtle at low temperatures is less than 0.1% of the normoxic summer metabolic rate. Part of this metabolic rate depression is a result of the decrease in temperature, but a substantial component is the result of active suppression of metabolism.

The triggers that induce hypoxic metabolic suppression are not yet understood, but one cue may be tissue acidosis. When oxygen supply is not sufficient to meet the metabolic needs of the organism, such as during environmental hypoxia, ATP must be produced using anaerobic pathways. In most animals this involves flux through glycolysis, producing lactate as the metabolic end product. High glycolytic flux that is not matched by aerobic respiration results in a metabolic acidosis—an increase in the net hydrogen ion production by the cell. A large metabolic acidosis can have dangerous consequences for an organism, because most enzymes are highly sensitive to the pH of the body. Initial exposure to hypoxia results in a modest tissue acidosis. This acidosis can then act as a cue to trigger a reduction in metabolic rate, protecting the animal against further acidosis.

Hypometabolic states are not unique to hypoxic environments. Many organisms use hypometabolism to survive adverse environmental conditions, including low temperature, low food availability, or desiccation, in addition to hypoxia. Although the nature of these conditions is diverse, in each case animals need to reduce metabolic rate to preserve energy stores. **Hibernation** (a long period of metabolic depression associated with cold temperature) and **torpor** (a shorter period of metabolic depression, often seen at night) are particularly interesting hypometabolic states because they occur under normoxic conditions. As animals enter into hibernation or torpor they *voluntarily* reduce ventilation in parallel with the reduction in metabolic rate. Thus, these animals actively reduce both oxygen supply and demand in concert. Many mammalian hibernators, such as ground squirrels, exhibit a pattern of *episodic breathing* during hibernation that includes long periods of apnea interspersed with ventilatory bouts. The mechanisms that convert the regularly spaced pattern of mammalian breathing to an episodic pattern during hibernation are not yet understood, but presumably involve changes in the function of the respiratory pacemakers in the medulla.

References

Boutilier, R. G., P. H. Donohoe, G. J. Tattersall, and T. G. West. 1997. Hypometabolic homeostasis in overwintering aquatic amphibians. *Journal of Experimental Biology* 200: 387–400.

Dupre, R. K., A. M. Romero, and S. C. Wood. 1988. Thermoregulation and metabolism in hypoxic animals. In *Oxygen Transfer from Atmosphere to Tissues*, N. C. Gonzalez and M. R. Fedde, eds., 347–351. New York: Plenum Press.

Gautier, H. 1996. Interactions among metabolic rate, hypoxia, and control of breathing. *Journal of Applied Physiology* 81: 521–527.

Jackson, D. C. 2004. Acid-base balance during hypoxic hypometabolism: Selected vertebrate strategies. *Respiratory Physiology and Neurobiology* 141: 273–283.

Platzack, B., and J. W. Hicks, 2001. Reductions in systemic oxygen delivery induce a hypometabolic state in the turtle *Trachemys scripta*. *American Journal of Physiology* 281: R1295–R1301.

Tattersall, G. J., J. L. Blank, and S. C. Wood. 2002. Ventilatory and metabolic responses to hypoxia in the smallest simian primate, the pygmy marmoset. *Journal of Applied Physiology* 92: 202–210.

blood viscosity, which could impair blood flow through capillaries and interfere with gas exchange at the tissues.

In humans and many other lowland-adapted animals, hypoxia also increases the levels of 2,3-DPG in the red blood cells. Increased 2,3-DPG would, in principle, decrease the oxygen affinity of the blood, which might assist in oxygen unloading at the tissues. However, the respiratory alkalosis associated with hyperventilation generally cancels out this effect, resulting in no net change in hemoglobin oxygen affinity at altitude.

Environmental hypoxia also affects blood flow through the lungs of lowland-adapted animals. The low alveolar P_{O_2} caused by the low environmental P_{O_2} causes the pulmonary arterioles to vasoconstrict, reducing perfusion of the lungs. This pathological response reduces oxygen uptake from the atmosphere, and is dangerous because the generalized vasoconstriction causes increased blood pressure within the lungs, which can lead to pulmonary edema, or accumulation of fluid in the lungs. Pulmonary edema is particularly dangerous because the accumulated fluid increases the diffusion distance across the alveolar epithelium, reducing the efficiency of gas exchange. *High-altitude pulmonary edema* is one of the most severe forms of "mountain sickness" in humans, and is one of the most dangerous consequences of exposure to very high altitudes (Figure 9.41).

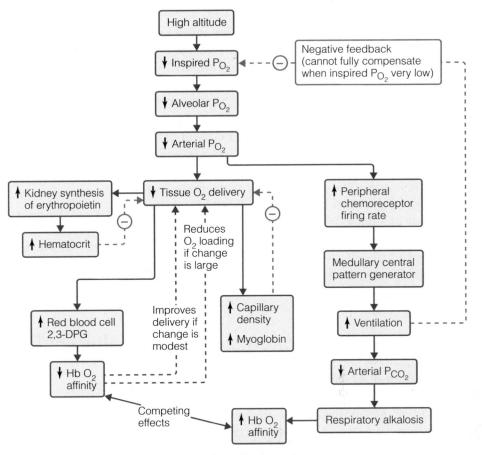

Figure 9.41   The response to high altitude in humans

A number of mammalian species, including some human populations, have colonized high-altitude habitats. For example, populations of indigenous peoples in China, Nepal, Tibet, Ethiopia, and Peru all inhabit altitudes that cause respiratory problems for low-altitude-adapted human populations. We are only just beginning to understand the physiological differences between individuals in these populations and lowland human populations, but the data collected so far suggest that each of these populations uses a different strategy for coping with high altitude. For example, the Quechua of Peru are typically barrel-chested, suggesting a higher than usual lung capacity, and have high hemoglobin levels. In contrast, Tibetan populations are not barrel-chested, and have moderately elevated hemoglobin levels. Individuals in Tibetan populations vary in arterial hemoglobin oxygen saturation, and individuals with higher oxygen saturation have higher offspring survival than individuals with low oxygen saturation. Differences in oxygen saturation have

been shown to be heritable in this population, and thus may be subject to ongoing natural selection. Individuals in high-altitude Ethiopian populations exhibit yet another pattern. They are not barrel-chested, do not have elevated amounts of hemoglobin, and do not have high hemoglobin oxygen affinity, but they are able to maintain arterial oxygen saturation at normal levels in the face of low environmental oxygen. The physiological basis for this difference is still unknown.

A number of other mammals have colonized high altitudes, including species such as llamas, chinchillas, guinea pigs, and deer mice. High-altitude-adapted populations of deer mice have reduced levels of 2,3-DPG in their red blood cells compared to low-altitude populations, when both populations are reared at a common altitude. This decrease in 2,3-DPG results in an increase in hemoglobin oxygen affinity, allowing them to efficiently extract oxygen from the atmosphere at high altitudes. Llamas, vicunas, chinchillas, and guinea pigs also have unusually high hemoglobin oxygen affinity. In llamas and vicunas, this increase in affinity is the result of one or a few mutations in the globin genes that eliminate the effects of 2,3-DPG, resulting in increased oxygen affinity.

The bar-headed geese (*Anser indicus*) described at the beginning of this chapter have unusually high hemoglobin oxygen affinity. There are only four amino acid differences between the major hemoglobin of bar-headed geese and the closely related greylag geese that live in the lowlands. One of these mutations results in the loss of a hydrogen bond that normally stabilizes the T state of hemoglobin, causing the hemoglobin to assume a more relaxed conformation and causing the increase in hemoglobin oxygen affinity. In addition, many birds are able to tolerate hyperventilation and the resulting hypocapnia and alkalosis much better than mammals, allowing them to increase oxygen extraction at high altitude.

⊙ | CONCEPT CHECK

22. What is a central pattern generator?
23. Outline the mechanisms by which changes in blood P_{O_2} affect ventilation.
24. Why is it difficult to distinguish whether chemosensory cells detect P_{CO_2} or pH?
25. What are some of the mechanisms by which fish respond to hypoxia?
26. Why do humans typically become hypocapnic at high altitudes?

Integrating Systems | The Physiology of Diving

A variety of air-breathing vertebrates, including some mammals, birds, and reptiles, have adopted a fully or partially aquatic mode of life. However, all of these animals remain dependent on air as a respiratory medium, and must be able to actively hunt prey underwater while relying on the oxygen stores that they carry with them as they dive below the surface. The physiology of diving in these animals provides an ideal example of the ways in which the respiratory and circulatory systems are integrated to allow animals to function in their environment.

Sperm whales are the champion divers among the marine mammals, with recorded dives to a depth of more than 2000 m and dive lengths of more than an hour. In addition, stomach content analyses suggest that sperm whales may dive as deep as 3000 m. The pinnipeds (seals and sea lions) are also excellent divers. Among pinnipeds, the elephant seals hold the record for both the longest and deepest dives at almost 1600 m and nearly 80 minutes. The emperor penguin can dive down to 500 m, but its dives are typically relatively short, averaging around 3 minutes. However, emperor penguins have been known to dive for as long as 22 minutes. Green sea turtles can remain submerged for as long as five hours, although active dives typically average 5–10 minutes.

When an air-breathing vertebrate dives, it must rely on stored oxygen to fuel aerobic metabolism. These onboard stores are typically sufficient for short dives, but cannot sustain metabolism during long dives, and anaerobic metabolism must be used (Figure 9.42). The **aerobic dive limit**—the point at which an animal must

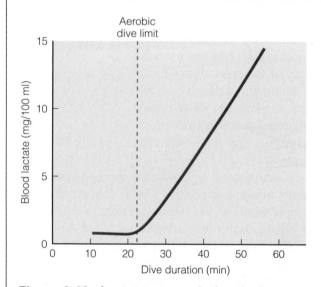

Figure 9.42 Lactate accumulation during diving in Weddell seals The aerobic dive limit is the dive time at which lactate begins to accumulate as a result of the switch to anaerobic metabolism.

either surface to breathe or begin to use anaerobic metabolism—varies greatly among species. For example, adult Weddell seals, which hunt underneath the Antarctic ice sheets, have an aerobic dive limit of about 20 minutes, whereas California sea lions have an aerobic dive limit of only about 5 minutes. In principle, two physiological adjustments can alter the aerobic dive limit: increasing oxygen stores and decreasing oxygen demand. Marine mammals use both of these strategies to increase the length of their dives.

A vertebrate can store oxygen in three places: in the blood (largely bound to hemoglobin), bound to myoglobin in muscle, and in the lungs. Total body oxygen stores tend to be larger in diving mammals than in terrestrial mammals, although this relationship is most evident in very proficient divers (Figure 9.43). Diving mammals often have high blood volumes and high oxygen carrying capacity, allowing them to store more oxygen in the blood than is typical for a terrestrial mammal. For example, a Weddell seal is able to store almost five times as much oxygen in blood as a human can. Recall from our discussion of the effects of high altitude that polycythemia increases blood viscosity, and can cause difficulties with cardiac function. Some species of seals avoid this problem by storing red blood cells in the spleen and releasing them during bouts of diving. The blood cells are returned to the spleen for storage between diving bouts.

Diving animals typically also have high levels of muscle myoglobin. Weddell seals have over 50 mg of myoglobin per gram of muscle, and ribbon seals can have as much as 80 mg/g, whereas humans have about 5–10 mg of myoglobin per gram of muscle. Diving animals do not have unusually large lungs, and likely do not make much use of their lungs as an oxygen store during diving. In fact, some species including the Weddell seal dive immediately after they exhale, and thus these animals swim actively without fresh air in the lungs.

As an animal descends through the water, the pressure of the surrounding water increases. The elevated ambient pressure causes the lungs to decrease in volume. The decrease in volume increases the partial pressure of the gases within the lungs. This effect can be beneficial, because it tends to drive additional oxygen into the circulation, but this benefit comes with a substantial risk: the increased pressure can also drive nitrogen gas into the circulation. This increase in blood nitrogen content can lead to a condition called *nitrogen narcosis*. The symptoms of nitrogen narcosis are similar to those of ingesting alcohol, progressing from an initial feeling of euphoria, through disorientation, and finally to loss of consciousness. Nitrogen gas is thought to act in a way similar to the anesthetic gas nitrous oxide, altering the activity of the nervous system by impairing the action of excitatory NMDA receptors, and enhancing the activity of the inhibitory opioid receptors.

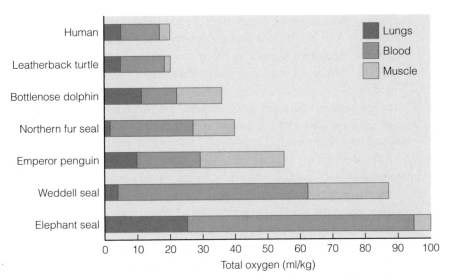

Figure 9.43 Total body oxygen stores of diving mammals and humans (expressed per kg body mass)

A related condition called "the bends," or decompression sickness, occurs when a diver ascends to the surface too quickly. At depth, nitrogen content of the blood is high. As a diver ascends, this nitrogen will simply diffuse back into the lungs, and can be exhaled. However, if a diver ascends too quickly the nitrogen will come out of solution while still in the blood, forming bubbles. This is similar to what happens when you open a bottle of soda pop. Soda pop is bottled under a high pressure of carbon dioxide. When you open the bottle, the pressure drops abruptly, causing bubbles to form. Bubbles in the blood are not inevitably harmful. They only cause problems if they become large, because large bubbles can lodge in small capillaries, blocking blood flow, or can press on nerve endings, or can become trapped in other enclosed spaces such as the joints. Decompression sickness is associated with a variety of symptoms, the most common of which are pain in the joints and muscles, and neurological problems, including headache and stroke. The risk of nitrogen narcosis and the bends is higher in humans scuba diving than in free divers, but extreme human free divers, who can descend to depths of over 70 m, may experience some of these effects. The effects of decompression sickness have been observed in the carcasses of beached sperm whales that have ascended to the surface too rapidly after being startled by sonar signals.

Many diving marine mammals avoid nitrogen narcosis and decompression sickness by allowing the lungs (or more properly, the alveoli) to collapse completely during diving. When the alveoli collapse, the air is pushed back into the conducting airways of the lungs, which do not participate in gas exchange. Thus, blood nitrogen levels in diving seals increase very little, regardless of dive depth. It is less clear how diving birds avoid this problem, since the lungs themselves are rigid. This difference in lung anatomy may explain why few birds dive deeply or for long periods. Laboratory experiments with Adélie penguins suggest that nitrogen levels can increase into the danger zone during unusually long or deep dives.

In addition to increasing oxygen stores, marine mammals also readjust oxygen demand during long dives, presumably to conserve oxygen and increase their aerobic dive limit. In fact, experiments on freely diving Weddell seals in nature suggest that the metabolic rate during diving is lower than during nondiving periods, despite the fact that these animals hunt actively while diving. Diving animals use a variety of biomechanical strategies to reduce the costs of locomotion in water (see Chapter 12 for more discussion of these issues). During forced dives in the laboratory, or when a freely diving animal must stay underwater for a prolonged period—for example, to avoid a predator—the animal invokes a series of physiological mechanisms that have been collectively called the **dive response**. During the dive response, arterioles leading to the skeletal muscles, skin, kidneys, and gut constrict, shunting blood away from the muscles and other nonessential organs, and toward the heart and brain. The brain is entirely dependent on aerobic metabolism and cannot survive oxygen deprivation for very long, whereas other tissues can tolerate reduced oxygen supply by reducing metabolic rate and by relying on anaerobic metabolism. At the same time, smooth muscles in the spleen contract, forcing stored red blood cells that are saturated with oxygen out into the circulation. During a forced or long dive, heart rate also slows, matching the reduced circulatory demand. The extent of this **diving bradycardia** is dependent on dive duration in voluntary dives, so that short dives involve little or no bradycardia whereas long dives involve a profound bradycardia.

The cardiovascular dive response is not unique to diving mammals, but instead is a fundamental property of all vertebrates. Most animals reduce metabolic rate and redistribute blood flow to essential tissues when they are deprived of oxygen. However, the dive response is typically more profound in diving mammals than in terrestrial animals such as humans.

Finally, we must consider the effects of the CO_2 that is produced during a dive, and the resulting drop in blood pH. Diving animals appear to have unusually high buffering capacity in the blood, which blunts or prevents large swings in blood pH. In addition, diving mammals have a greatly reduced ventilatory response to CO_2. In humans, the gradual buildup of CO_2 and the resulting decrease in blood pH during apnea act as a very strong stimulus to take a breath. If you have ever tried to swim a long distance underwater, you will have experienced this intense urge to breathe as a result of CO_2 buildup. Diving animals such as seals do not have nearly as strong a response while submerged, which allows them to stay underwater longer without feeling the urge to take a breath. ◎

Summary

Respiratory Strategies

→ Respiratory systems consist of all the structures animals use to obtain oxygen from the environment, and to dispose of carbon dioxide.

→ Animals larger than a few millimeters in diameter use a combination of diffusion and bulk flow to transport gases between the environment and the tissues.

→ Animals use respiratory surfaces including the skin, gills, or lungs (or a combination of structures) for gas exchange.

→ Animals with internal gills or lungs must move the external fluid by bulk flow across the respiratory surface, a process called ventilation.

→ The relationship between ventilation and perfusion of the respiratory surface influences the efficiency of gas exchange.

→ A countercurrent arrangement provides the most efficient exchange.

Ventilation and Gas Exchange

→ Animals living in air and water utilize differing respiratory strategies, because of the differences in the physical properties of these two media.

→ Animals that breathe water must expend much more energy to obtain oxygen than do animals that breathe air.

→ Water-breathing animals use a variety of strategies to ventilate their gills, involving the beating of cilia or other structures, or muscular pumping to generate bulk flow of the external medium.

→ Sea cucumbers are unusual among water breathers in that they have tidally ventilated lungs. Animals that use air as a respiratory medium employ either a tracheal system or lungs for gas exchange.

→ Air-breathing fish and amphibians gulp air using a buccal pump.

→ Reptiles, birds, and mammals ventilate their lungs using a suction pump.

→ Reptiles and mammals ventilate their lungs tidally, but birds unidirectionally ventilate their lungs.

→ Mammalian lungs consist of a series of conducting airways that lead to numerous thin-walled alveoli.

→ Mammals ventilate the alveoli by contracting the external intercostal muscles and diaphragm, which moves the rib cage and expands the chest cavity.

→ Under resting conditions, breathing out is usually passive, as a result of the elastic recoil of the lungs, although active expiration can occur during intense exercise.

→ The work required for ventilation depends on lung compliance and resistance.

Gas Transport to the Tissues

→ Oxygen is carried to the tissues either dissolved in blood or bound to a respiratory pigment such as hemoglobin, hemerythrin, or hemocyanin.

→ Blood containing a respiratory pigment has a characteristic oxygen equilibrium curve, but the shapes of these curves differ among types of pigments and among species.

→ Blood can vary in both oxygen affinity and carrying capacity.

→ Blood pH, p_{CO_2}, temperature, and organic modulators can affect the shape of the oxygen equilibrium curve.

→ The Bohr shift, which is a result of decreasing pH or increasing CO_2, is a right shift of the curve that results in unloading of oxygen at the tissues.

→ Carbon dioxide can be carried in the blood as dissolved CO_2, as HCO_3^-, or bound to proteins such as hemoglobin.

→ Blood CO_2, HCO_3^-, and pH are interrelated via the carbonic anhydrase equilibrium reaction. Blood oxygenation affects CO_2 transport by altering hemoglobin CO_2 binding, and by altering blood pH.

→ Vertebrate red blood cells play an important role in CO_2 transport by separating the reactants and products of the carbonic anhydrase equilibrium, greatly increasing the CO_2 carrying capacity of the blood.

→ Because of the carbonic anhydrase equilibrium, the respiratory system can both cause pH disturbances and regulate blood pH.

→ We can visualize the relationships among pH, HCO_3^-, and P_{CO_2} using a pH-bicarbonate plot.

Regulation of Vertebrate Respiratory Systems

→ In the vertebrates, central pattern generators in the medulla initiate ventilation.

→ Chemosensory inputs influence the action of these pattern generators, modulating the rate and depth of breathing.

→ Mechanoreceptors and conscious control can also influence breathing.

→ Environmental hypoxia and diving provide two examples of the ways in which vertebrates regulate their respiratory systems in response to environmental changes.

Review Questions

1. Why is diffusion an inefficient respiratory strategy for organisms that are more than a few millimeters thick?

2. Compare and contrast the lungs of birds, the lungs of mammals, and the tracheal systems of insects.

3. Explain how countercurrent flow arrangements can lead to more efficient gas exchange across a respiratory surface.

4. Compare and contrast the force pumps and aspiration pumps of tetrapod vertebrate respiratory systems.

5. Describe the changes in alveolar and intrapleural pressure during a single ventilatory cycle in mammals.

6. How does the Root effect help a physoclist fish to add oxygen to the swim bladder?

7. What is the significance of the red blood cell for CO_2 transport in the vertebrates?

8. Outline how chemoreceptors influence ventilation in mammals.

Synthesis Questions

1. Very few animals that use water as the respiratory medium have lungs. Instead, most water breathers use gills for gas exchange. What functional disadvantages do lungs have in water?

2. Lungless salamanders typically live in moist or humid habitats, and can die if their skin dries out. Explain why it is critical for the skin of lungless salamanders to remain moist.

3. Some species of lungless salamander cannot live in water as adults, and will drown if fully immersed. Why might this occur?

4. In an experiment to determine the role of the air sacs in the avian lung, physiologists tied off an air sac so that gas from that air sac could no longer enter the lung. The experimenters then injected carbon monoxide into the sealed air sac. This manipulation did not decrease the oxygen saturation of hemoglobin in arterial blood. Explain why this was the case, and what this experiment demonstrates about the nature of the air sacs in birds.

5. A woman gets a disease that makes her unable to produce surfactant in her lungs. If she has a normal tidal volume, what can you say about her intrapleural pressure during inspiration?

6. What effects might you expect in a mammal whose major hemoglobin is mutated such that it lacks a Bohr effect?

7. Metabolic rate can increase as much as 40-fold above resting values as a result of feeding in some species of reptiles. In addition, during digestion, a large amount of H^+ is secreted into the stomach, which results in the so-called alkaline tide, a large metabolic alkalosis in which blood pH increases. Outline the likely response of the respiratory system to this increased oxygen demand and pH disturbance.

8. In fish, there is a positive correlation between whole animal metabolic rate and the surface area of the gill. What might explain this relationship?

9. High-altitude-adapted mammals often do not show as large a pulmonary vasoconstriction in response to low inspired P_{O_2} (environmental hypoxia) as do lowland-adapted mammals. What advantages might this difference have at high altitude?

10. Hemoglobin is typically saturated with oxygen when the blood leaves the lungs. In a person who is doing pull-ups, will hemoglobin release more of the bound oxygen in the quadriceps (leg muscles) or in the biceps (arm muscles)? Describe at least two factors that could cause a difference, if any, in oxygen release between your biceps and quadriceps.

11. Imagine that you take hemoglobin molecules from both a sheep fetus and its mother. You mix equal amounts of these two hemoglobins in an aqueous solution in the presence of oxygen, at a P_{O_2} that is not sufficient to saturate all the hemoglobin sites on the molecules you have added. Given what you know about maternal and fetal hemoglobins, where would you expect to find most of this oxygen bound? How would this compare to the amount of oxygen dissolved in your solution and not bound to hemoglobin? Why?

12. Anxiety can cause a person to hyperventilate (rapid deep breathing). This can cause a variety of symptoms, including dizziness and fainting. What changes would you expect in systemic arterial O_2 and CO_2 concentration and pH during an episode of hyperventilation? How (i.e., by what mechanism) might this affect blood flow to the brain? Breathing into a paper bag is often suggested as a treatment for hyperventilation. Do you think this would work? Why or why not?

Quantitative Questions

1. The graphs below represent the gas exchange across two hypothetical respiratory surfaces (a and b). One of these surfaces has concurrent flow, and one has countercurrent flow.

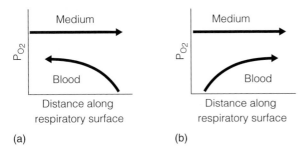

(a)   (b)

(a) Which surface has concurrent flow, and which surface has countercurrent flow?
(b) Based on the data shown, which surface has the most efficient gas exchange?
(c) What might account for this observation?

2. If a mammal has a minute volume of 5200 ml/min, a breathing frequency of 13 breaths per minute, a vital capacity of 4600 ml, and an expiratory reserve volume of 1200 ml, what are the tidal volume and inspiratory reserve volume?

3. As part of a physiology experiment, a human subject is asked to breathe through a hose 1 m long and 3 cm in diameter (the end of the hose is open to the air in the room). What changes would you expect in ventilation rate and tidal volume compared to those measured in the same subject breathing normally? (Explain your answers.)

4. John, Jeff, and Harry are all breathing at different rates and depths. Using the data provided below, who would have the highest P_{O_2} in the blood leaving the lungs. Who would have the lowest? (Show your work.)

	Breathing rate (breaths per minute)	Tidal volume (ml per breath)	Dead space (ml)
John	15	500	200
Jeff	40	250	200
Harry	10	1000	200

5. Using the Hb-oxygen saturation curve in Figure 9.30a, answer the following questions:
   (a) If P_{O_2} in the lungs is approximately 100 mm Hg, what is the percent saturation of Hb in the pulmonary capillaries?
   (b) If P_{O_2} in the tissues is approximately 5 mm Hg, what is the percent saturation of Hb in the systemic capillaries?

For Further Reading

See the Additional References section at the back of the book for more readings related to the topics in this chapter.

Respiratory Strategies

These two advanced textbooks provide an excellent overview of respiratory physiology, providing examples from a broad range of animal taxa, with strong quantitative coverage of the material.

Cameron, J. N. 1989. *The respiratory physiology of animals.* New York: Oxford University Press.

Prange, H. D. 1996. *Respiratory physiology: Understanding gas exchange.* New York: Chapman and Hall.

The following engaging book provides a broad overview of the impact of the physical properties of air and water on a variety of processes in animals. It contains several sections that are relevant to understanding the functioning of respiratory systems. The treatment of diffusion is particularly good.

Denny, M. W. 1993. *Air and water: The biology and physics of life's media.* Princeton, NJ: Princeton University Press.

Ventilation and Gas Exchange

The following papers address some of the factors involved in ventilation and the regulation of gas exchange in insects.

Gibbs, A. G., F. Fukuzato, and L. M. Matzkin. 2003. Evolution of water conservation mechanisms in Drosophila. *Journal of Experimental Biology* 206: 1183–1192.

Lehman, F-O. 2001. Matching spiracle opening to metabolic needs during flight in Drosophila. *Science* 294: 1926–1929.

Lighton, J. R. B. 1996. Discontinuous gas exchange in insects. *Annual Review of Entomology* 41: 309–324.

Westneat, M. W., O. Betz, R. W. Blob, K. Fezzaa, W. J. Cooper, and W. K. Lee. 2003. Tracheal respiration in insects visualized with X-ray synchrotron radiation. *Science* 299: 558–560.

Johannes Piiper and Peter Scheid are two of the leaders in the field of vertebrate respiratory physiology. This chapter provides a concise summary of the quantitative basis of gas exchange in a variety of vertebrates.

Scheid, P., and J. Piiper. 1997. Vertebrate respiratory gas exchange. In *Handbook of physiology: A critical, comprehensive presentation of physiological knowledge and concepts,* W. H. Dantzler, ed. (Section 13: Comparative Physiology), vol. 1, 309–356. Bethesda, MD: American Physiological Society.

This readable article provides an excellent and accessible introduction to vertebrate gas exchange.

Truchot, J-P. 2001. Gas transfer in vertebrates. In *Encyclopedia of life sciences.* New York: Wiley.

Gas Transport to the Tissues

These papers review the structure and function of hemoglobins and other respiratory pigments.

Burmester, T. 2002. Origin and evolution of arthropod hemocyanins and related proteins. *Journal of Comparative Physiology B: Biochemical, Systemic, and Environmental Physiology* 172: 95–107.

Giardina, B., D. Mosca, and M. C. De Rosa. 2004. The Bohr effect of haemoglobin in vertebrates: An example of molecular adaptation to different physiological requirements. *Acta Physiologica Scandinavica* 182: 229–244.

Terwilliger, N. 1998. Functional adaptations of oxygen transport proteins. *Journal of Experimental Biology* 201: 1085–1098.

Weber, R. E., and A. Fago. 2004. Functional adaptation and its molecular basis in vertebrate hemoglobins, neuroglobins and cytoglobins. *Respiratory Physiology and Neurobiology* 144: 141–159.

Regulation of Vertebrate Respiratory Systems

These reviews discuss the regulation of vertebrate respiratory systems, with particular focus on the differences between water breathers and air breathers.

Smatresk, N. J. 1990. Chemoreceptor modulation of endogenous respiratory rhythms in vertebrates. *American Journal of Physiology* 259: R887–R897.

Smatresk, N. J. 1994. Respiratory control in the transition from water to air breathing in vertebrates. *American Zoologist* 34: 264–279.

Hypoxia

This paper outlines the various mechanisms seen in high-altitude-adapted human populations.

Beall, C. M., M. J. Decker, G. M. Brittenham, I. Kushner, A. Gebremedhin, and K. P. Strohl. 2002. An Ethiopian pattern of human adaptation to high-altitude hypoxia. *Proceedings of the National Academy of Sciences, USA* 99: 17215–17218.

This review outlines some of the mechanisms of hypoxia-induced metabolic suppression in nonmammalian vertebrates.

Hicks, J. W., and T. Wang. 2004. Hypometabolism in reptiles: Behavioural and physiological mechanisms that reduce aerobic demands. *Respiratory Physiology and Neurobiology* 141: 261–271.

Diving

These comprehensive reviews summarize the fundamental physiological mechanisms involved in diving in the vertebrates.

Butler, P. J. 2004. Metabolic regulation in diving birds and mammals. *Respiratory Physiology and Neurobiology* 141: 297–315.

Kooyman, G. L., and P. J. Ponganis. 1998. The physiological basis of diving to depth: Birds and mammals. *Annual Review of Physiology* 60: 19–32.

The following review highlights some of the pioneering work done by Peter Hochachka and his students and colleagues on the physiology and biochemistry of diving in pinnipeds.

Castellini, M. A., and J. M. Castellini. 2004. Defining the limits of diving biochemistry in marine mammals. *Comparative Biochemistry and Physiology, Part B: Biochemistry and Molecular Biology* 139: 509–518.

Ion and Water Balance

Somewhere around 700 million years ago (mya), the earliest animal life forms arose. They were marine organisms, much like the modern sponges. Like sponges, these primordial animals existed as a loose aggregation of cells, bathed in seawater. Each cell of a marine sponge is bathed in seawater, but maintains an *intracellular* ion composition different from that of seawater, using ion pumps and active transport to create and maintain the electrochemical gradients that drive transport and synthetic processes.

Over the next 150 million years, evolution led to important changes in how animal tissues were organized. An

early milestone was the formation of tissue layers. Next came the capacity to produce a specialized external tissue layer using cells that were interconnected in ways that limited the passage of seawater into the body. The formation of this epithelial tissue provided barriers between the external world and internal fluids, resulting in the establishment of an extracellular fluid that was separate from the external environment. Animals differ in their ability to control the osmotic and ionic nature of this *extracellular* fluid.

Though sponges lack true tissues, other ancient marine invertebrates, such as flatworms (Platyhelminthes),

Hagfish.

Crab.

possess true tissues and can create an extracellular fluid that is physically separated from seawater. Yet in these simple animals, the extracellular fluid contained within the tissues is identical to seawater in its ionic and osmotic properties. However, multiple lineages of animals have evolved mechanisms that provide much greater control over the properties of their extracellular fluids. The ability to control extracellular fluid composition allowed animals to invade brackish water, freshwater, and even land.

Major changes in osmoregulation and ionoregulation occurred in the evolution of the chordates. Like many simple marine invertebrates, the earliest chordates were marine organisms that had little control over the nature of their extracellular fluid composition. The extracellular fluid of the hagfish, a marine agnathan (jawless fish), is similar to seawater, although it is somewhat reduced in the concentrations of Ca^{2+}, Mg^{2+}, and SO_4^{2-}. Cartilaginous fish control the ion composition of the extracellular fluid, but the osmolarity is close to that of seawater. Bony fish regulate both the ionic and osmotic profile of their extracellular fluids. This ability to control internal ionic and osmotic properties was essential to the diversification of bony fish, which now occupy almost every aquatic and semiaquatic niche on the planet, often tolerating inhospitable ionic and osmotic conditions, environments with very high or low pH, extremes in salinity, and even periods of dehydration. For instance, cichlids live in the alkaline waters of Lake Magadi (pH 10) and tambaqui thrive in acidic Amazonian waters (pH 3.5). Fish can be found in waters of varying salinity, from the hypersaline salt marshes and inland seas, through the oscillating salinity of the intertidal zone, to lakes and rivers that are nearly devoid of essential ions. A few species of fish even survive out of the water. Some tropical catfish walk over land from one temporary pool to another. Other fish enter a period of dormancy, such as the lungfish that bury themselves underground in a mucus cocoon.

The ability to control internal osmolarity independent of external conditions was essential for the success of the animal lineages that invaded land. The earliest of many waves of terrestrial invaders were invertebrates. First the ancient myriapods, then their arthropod predators, invaded land more than 420 mya. Later, around 400 mya, the first amphibians ventured onto land. A terrestrial existence puts animals at risk for desiccation, and species that successfully invaded land demonstrate evolutionary adaptations that reduce water loss. For one thing, they need a body surface more resistant to desiccation. No longer able to excrete metabolic wastes directly into the water, they also need an alternative way to dispose of nitrogenous waste.

Ion balance, water balance, nitrogen excretion, and pH balances are interdependent processes that must be regulated (and evolve) in parallel to ensure homeostasis. The diversity among modern animals reflects the many solutions to problems encountered by animals in early evolution.⊙

Overview

Normal animal function depends on the precise regulation of diverse physical relationships and biochemical processes, which in turn are influenced by the chemical environment. The term *environment* may be interpreted in the broadest sense. The environment for a whole animal is the external world, for a cell it is the extracellular fluid, and for intracellular enzymes, the cytoplasm. Animals maintain a favorable profile of solutes and solutions in their intracellular and extracellular body fluids largely by means of the epithelial tissues that form the barrier with the external environment (Figure 10.1).

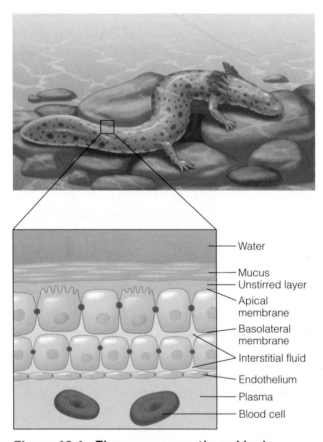

Figure 10.1 Tissues as osmotic and ionic barriers Epithelial tissues separate internal fluid compartments from the external world. In the case of an aquatic animal, such as a mudpuppy, the external (apical) side of the epithelial cell layer interacts directly with the external water, though it can secrete a protective layer of mucus that also traps a layer of water underneath. Intercellular junctions connect epithelial cells together to form a barrier between external and internal fluids. On the internal (basolateral) side of the epithelium, cells are bathed in interstitial fluid trapped between cells. The tissue is fed by capillaries, with vascular endothelial cells separating interstitial fluid from plasma.

Each group of animals uses different combinations of epithelial tissues to control ion and water balance. For most animals, the kidney is central to ion and water balance. However, most animals also rely on extrarenal tissues, such as gills, skin, and the digestive mucosa. These tissues regulate three homeostatic processes to ensure an appropriate chemical composition:

- *Osmotic regulation* is the control of tissue osmotic pressure, which determines the driving force for the movement of water across biological membranes. Animals and cells cannot actively pump water. Osmotic regulation requires the movement of solutes across membranes, altering osmotic gradients.

- *Ionic regulation* is the control of the ionic composition of body fluids. In this chapter we focus on the ions that are important solutes, and therefore relevant to the osmoregulatory strategies. In Chapter 11: Digestion, we discuss some of the pathways by which animals obtain the ions that are important in biosynthesis—trace elements and micronutrients.

- *Nitrogen excretion* is the pathway by which animals excrete ammonia, the toxic nitrogenous end product of protein catabolism. The process for expelling ammonia, or metabolic alternatives such as urea and uric acid, is linked to the control of osmotic and ionic homeostasis. The tissues of the excretory system are responsible for collecting nitrogenous waste and expelling it into the environment.

Diverse mechanisms ensure that the ionic and osmotic properties of animals are maintained within acceptable limits. In simple invertebrates, the responsibilities for ionic and osmotic regulation reside primarily at the level of individual cells. In more complex animals, specialized tissues carry the burden of maintaining the appropriate chemical composition and volume of the body fluids.

Ion and Water Balance

In Chapter 2: Chemistry, Biochemistry, and Cell Physiology, we discussed the importance of water in biological systems. It is the solvent that is used to dissolve the ions and metabolites needed to sustain cells. Animals have diverse physiological systems that work together to ensure that ionic and osmotic conditions within the animal remain within acceptable limits. Changes in the concentration of ions have the potential to affect the structure and function of macromolecules. Cellular function can be affected by such disruptions in macromolecular function, but perhaps more important, cells exposed to osmotic gradients can shrink or swell. The changes in cell volume can damage cells directly, sometimes causing cell death. Even if cells survive the stress, the cell volume changes damage multicellular tissues by disrupting cell-to-cell interactions or by altering blood flow through the tissue.

Some aquatic animals minimize high costs of ion and water balance by maintaining osmotic or ionic equilibrium with the external water. Most animals maintain some degree of control over the ionic and osmotic composition of their extracellular conditions, and thus must resist changes imposed by the external environment. Marine environments have high levels of ions, mostly Na^+ and Cl^-. Thus, for many marine animals, their challenge is to expel ions against electrochemical gradients and obtain water against osmotic gradients. Freshwater animals have the reverse problem: acquiring ions from ion-poor water and disposing of excess water. Terrestrial animals live under dehydrating conditions, where water loss is the greatest threat and most of the ions appear in the diet. The animals that straddle multiple environments must have flexible homeostatic mechanisms to cope with variable ion and water levels.

When considering how animals control ion and water balance, it is important to consider both the behavior of individual cells struggling to maintain their own cell volume and the epithelial tissues working to ensure homeostasis of the whole animal. Each of these processes involves cell membrane transporters, including the channels that mediate facilitated diffusion, the exchangers working through secondary active transport, and the pumps that move ions against concentration gradients at the expense of ATP.

Strategies for Ionic and Osmotic Regulation

All animals regulate the ionic profile of intracellular fluids in relation to the extracellular fluids that bathe cells within an animal. The electrochemical

gradients that result from disequilibrium of ions across the cell membrane are essential for normal cellular function. Most animal cells act as *perfect osmometers*, swelling or shrinking as a result of osmotic gradients across the cell membrane. Animals differ in (1) the sites of ion and water exchange and (2) the nature of ionic and osmotic gradients between the extracellular fluids and the environment. The main sites of ion and water exchange are those in direct contact with the environment: the outer body covering, and the surfaces of the respiratory and digestive tracts. Most animals also have some specialized excretory systems to regulate excretion of ions and water in support of the osmotic strategy. The ionic and osmotic gradients differ for aquatic and terrestrial animals. An aquatic animal is immersed in a solution, and each external surface in contact with the water could be a site of ion and water exchange. For example, a freshwater vertebrate would tend to gain water and lose ions, whereas a marine vertebrate would tend to gain ions and lose water. Most terrestrial animals live in a dehydrating environment, losing water across the body surfaces, but few ions. The extent to which a particular ionic or osmotic gradient constitutes a physiological burden depends on the ionoregulatory and osmoregulatory strategies of the animal.

Most aquatic animals regulate ion and water balance to some degree

Ionoregulatory and osmoregulatory strategies of aquatic animals can be distinguished by (1) the differences between extracellular fluids and external conditions and (2) the extent to which extracellular fluids change when external conditions change (Table 10.1). *Conformers* have internal conditions that are similar to the external conditions, even when the external conditions change. *Regulators* defend a nearly constant internal state that is distinct from the external conditions.

An **ionoconformer** exerts little control over the solute profile within its extracellular space. These animals usually live in seawater. Their extracellular fluids resemble seawater in terms of the concentrations of the major cations (Na^+, Ca^{2+}, and Mg^{2+}) and anions (Cl^- and SO_4^{2-}). Ionoconformers include most simple invertebrates (such as cnidarians), simple deuterostomes (such as ascidians), and the most ancient vertebrates (hagfish). Although hagfish are usually considered

Table 10.1 Osmoregulatory and ionoregulatory strategies of aquatic animals.

	Plasma osmolarity	Ionoregulator?
Animals in seawater (~1200 mOsM)		
Most arthropods	Isosmotic	No
Molluscs (squid)	Slightly hyperosmotic	Yes
Agnathans (hagfish)	Slightly hyperosmotic	No
Agnathans (lamprey)	Hyposmotic (~250 mOsm)	Yes
Chondrichthians (shark)	Slightly hyperosmotic	Yes
Bony fish (tuna), amphibians (crab-eating frog), reptiles (sea turtle), mammals (killer whale), birds (gull)	Hyposmotic (~350 mOsM)	Yes
Animals in freshwater (<10 mOsM)		
Most arthropods (e.g., crayfish, insect larvae)	Hyperosmotic (250–400 mOsM)	Yes
Molluscs (freshwater clams)	Slightly hyperosmotic (~50 mOsM)	Yes
Agnathans (lamprey)	Hyperosmotic (~270 mOsM)	Yes
Chondrichthians (stingray)	Hyperosmotic (~350 mOsM)	Yes
Bony fish (goldfish)	Hyperosmotic (~350 mOsM)	Yes
Amphibians (leopard frog)	Hyperosmotic (~250 mOsM)	Yes
Reptiles (snapping turtle), mammals (river otter), birds (mallard duck)	Hyperosmotic (~350 mOsM)	Yes

ionoconformers because their extracellular Na^+ and Cl^- levels are near those of seawater, they are in fact able to regulate Mg^{2+} and Ca^{2+} to levels below those seen in seawater. In contrast to ionoconformers, an **ionoregulator** controls the levels of most of the ions in extracellular fluids, employing a combination of ion absorption and excretion strategies. Regulating the ionic profile of extracellular fluid compartments eases the burden of ionic regulation placed on individual cells.

The internal osmolarity of an **osmoconformer** nears that of the external environment; if external osmotic conditions change, internal osmolarity changes in parallel. Marine invertebrates and primitive vertebrates are osmoconformers. Thus, a marine mollusc in full-strength seawater has an internal osmolarity of approximately 1200 mOsM, about three times more concentrated than human blood. Marine chondrichthians are generally considered osmoconformers; although their internal osmolarity is a bit greater than the surrounding seawater, it changes in parallel with external changes. An osmoconformer may control the *profile* of extracellular solutes, but the environment imposes the osmolarity. An **osmoregulator**

maintains internal osmolarity within a narrow range regardless of the external environment. Depending on the conditions, the animal could have an osmolarity higher or lower than the surrounding water. Most marine vertebrates are osmoregulators, defending an internal osmolarity that is much below that of the surrounding seawater. Freshwater vertebrates and invertebrates are osmoregulators, maintaining internal osmotic pressure well above that of the water. All otters, for example, have a similar internal osmolarity whether in freshwater, like a river otter, or seawater, like a sea otter. These animals must retain the vital ions and rid themselves of excessive ions, while also controlling the movement of water.

We also classify animals according to their ability to tolerate changes in external osmolarity. *Stenohaline* animals can tolerate only a narrow range of salt concentrations, whereas *euryhaline* animals can tolerate widely variant osmolarities. There is no predetermined relationship between the strategy (osmoconforming versus osmoregulating) and the degree of tolerance (euryhaline versus stenohaline) (Figure 10.2). For example, intertidal molluscs are euryhaline osmoconformers whereas

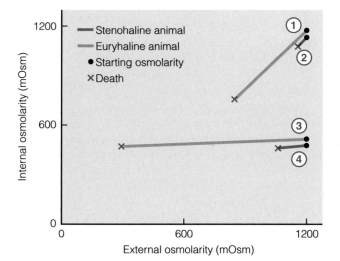

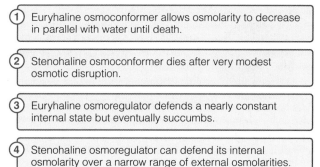

(1) Euryhaline osmoconformer allows osmolarity to decrease in parallel with water until death.

(2) Stenohaline osmoconformer dies after very modest osmotic disruption.

(3) Euryhaline osmoregulator defends a nearly constant internal state but eventually succumbs.

(4) Stenohaline osmoregulator can defend its internal osmolarity over a narrow range of external osmolarities.

Figure 10.2 Osmoregulatory strategies and salinity tolerance Osmotic strategies can be distinguished by three factors: **(1)** the osmotic gradients between the animal and the water, **(2)** the degree to which internal osmolarity changes in relation to a changing external osmolarity, and **(3)** the degree of tolerance of an osmotic challenge. The four osmotic strategies depicted in this figure can be distinguished by following the internal osmolarity of four animals living in full-strength seawater, then exposed to a decreasing osmolarity until death.

intertidal fish are euryhaline osmoregulators. Closely related animals can also differ in their tolerance of osmotic stress. Consider, for example, the decapod crustaceans, all of which are ionoregulators. Lobsters live in seawater, crayfish in freshwater, and there are seawater and freshwater species of shrimp. In contrast to these stenohaline species, the blue crab is exceptionally euryhaline. The blue crab osmoconforms at high salinities, but it osmoregulates if salinity decreases below a threshold. Osmotic strategies integrate the diverse physiological systems to ensure that the internal ion and water composition is suitable for the function of macromolecules, cells, and tissues.

The environment provides water in many forms

All animals require a source of water, though some animals have a harder time finding it than others. Freshwater osmoregulators have no problem obtaining the water they need, and in fact must cope with excessive water uptake. Marine osmoregulators must deal with the ion loads that accompany the water they consume. Terrestrial animals consume much of their water in the diet, and generally must find ways to minimize water loss. A few animals have unusual physiological adaptations that allow them to survive various degrees of water deprivation, including dehydration (see Box 10.1, Evolution and Diversity: Life Without Water).

The diet itself is a mixture of water and solutes in various chemical forms. Aquatic animals ingest some liquid water while eating, and they must manage the resulting osmotic and ionic consequences. Many aquatic animals expel the liquid before it enters the gastrointestinal tract. Filter-feeding whales, such as the baleen whale, gulp large volumes of seawater laden with krill, and then use the tongue to compress the meal against the baleen, expelling excess seawater. Many marine animals possess mechanisms that enable them to expel excess salt, allowing them to drink seawater to obtain water. For example, many marine reptiles and birds can drink seawater because they possess specialized salt-secreting glands, discussed later in this chapter. Without a capacity to rid its body of salt, an animal drinking seawater will become progressively more dehydrated.

Plant and animal tissues are important sources of *dietary water* for animals (see Table 10.2). This water is preformed in the food, either trapped within the solid food or as a liquid component of the meal. An animal cannot absorb all of the dietary water, because it must retain some water to give the feces the appropriate consistency for transit through the gastrointestinal tract. Once ingested, many macromolecules undergo hydrolysis as part of the digestive process. Hydrolysis—literally "water splitting"—consumes a water molecule to break a chemical bond. After this minor investment of water early in digestion, subsequent metabolic

BOX 10.1 EVOLUTION AND DIVERSITY
Life Without Water

Water is essential to life, but some species survive with precious little. These animals live in areas you would normally think of as dry (places like the hot deserts of Death Valley and the cold deserts of the Antarctic) as well as microclimates directly underfoot, such as temporary ponds and moss patches. Some animals survive water deprivation using adaptations to obtain, retain, or recover water to remain hydrated, whereas others have adaptations to tolerate dehydration.

Many animals survive the desert by being better at finding and storing water to remain hydrated. Of course, desert animals drink water when they find it, whether in standing pools found in oases, or as dew droplets forming on vegetation. One desert beetle can harvest water directly from the air. It climbs to the top of sand hills in the early morning and stands on its head. Water condenses on its exoskeleton and falls in rivulets to its mouth. Preformed water is also trapped in solid food, such as succulent cactus. Water is also produced during the metabolic breakdown of dietary macromolecules.

Many desert animals, particularly insects, can survive radical changes in tissue water content between dehydration and drinking bouts. Desert beetles swell with water in the rainy season, increasing water content to about 70% of body mass. Over the course of the dry season, they may lose as much as 60% of this water. Most of this water is lost from the hemolymph; some beetles can tolerate almost complete loss of hemolymph without obvious consequences (recall that the insect hemolymph has no role in delivery of oxygen).

Most desert vertebrates cannot tolerate severe dehydration, but the camel is one exception. When water is available, a 700-kg camel can consume as much as 100 kg of water in as little as 10 minutes. Similarly, a camel gorges when food is available, storing excess energy in its hump as fat. When deprived of food and water, the camel draws on water stores and degrades the fat in the hump. Eventually, the hump shrinks in size, slumping over to one side as the fat is oxidized to produce energy and metabolic water. Despite the production of metabolic water, camels undergo severe dehydration. In contrast to camels, most desert vertebrates maintain tissue water content within a narrow range using physiological mechanisms that maximize water conservation.

An important element of desert survival is avoiding the sun's heat, by either finding or altering a microclimate. During the day, many reptiles and insects, the most abundant animals in the desert, find refuge in established microclimates—under rocks or in burrows. The desert toad spends the hot daylight hours burrowed into the cooler soil. Burrowing animals may also modify their microclimate. Even with physiological and anatomical specializations, all animals lose some moisture in their respiration. Thus, the subterranean burrows can become humid, providing refuges from both the heat and the dryness of the surface.

Given the nature of the desert terrain, larger animals, such as the camel, have little hope of finding shade. Instead, physiological strategies help them cope with the direct sunlight. Just as waterproofing of the integument was an important adaptation in the earliest terrestrial invaders, the desert dwellers have evolved superior mechanisms to prevent water flux across the skin. Amphibians and reptiles that live in the desert have skin with a thicker stratum corneum than do those that live in wetter habitats. Birds and mammals—both homeotherms—face an additional risk of dehydration through cutaneous water loss. Generally, large mammals use sweating as a means of cooling under hot conditions. Although birds do not possess sweat glands, cutaneous water loss contributes to cooling. However, to many desert animals conserving water is more urgent than cooling the body. They block evaporative cooling, allowing their body temperature to rise. For example, the body temperatures of the oryx (a large antelope) and the camel may exceed 40°C during the heat of the day. These animals do not shed the stored body heat until the cool evening, when the body temperature can fall below 35°C. Interestingly, the featherless neck of ostriches is actually much more permeable to water than is the skin of other birds. This suggests that evaporative cooling is more important to the ostrich than is water conservation.

Other physiological processes, such as ventilation, digestion, and excretion, lead to water loss. Desert animals often have unusual adaptations that reduce this incidental loss of water. Some desert mammals, such as the kangaroo rat, minimize respiratory water loss by passing the expired air over a region of the nose equipped with a countercurrent heat exchanger. The dik-dik, an African antelope that lives in semiarid scrubland, possesses an enlarged nose that acts as a cooling chamber. Moisture condenses out of the expired air before it escapes through the nostril. The kangaroo rat is also able to extract most of the water from its urine and feces prior to excretion. Desert birds and mammals limit excretory water loss by producing a very concentrated urine. The urine of a dik-dik, for example, is 12 times more concentrated than its plasma. The camel also reduces the degree of dehydration by blocking urination, retaining urea within the tissues until water becomes available.

In contrast to the animals that resist dehydration, some animals survive water stress by tolerating dehydration, in some cases losing all free water, a state known as *anhydrobiosis*. In contrast to the exotic animals living in distant lands, many of the desiccation tolerant animals live underfoot. Many invertebrates, such as rotifers and tardigrades, live and breed in wet moss, but enter a dormant state when the moss desiccates. Unlike their cousins, bivalves and cephalopods, the pulmonate snails are terrestrial and use a simple lung to breathe air, which puts them at risk for dehydration. Because of this vulnerability, most pulmonate snails are confined to humid, tropical environments where there is little risk of desiccation. A few pulmonate snails have exceptional desiccation tolerance. The snail *Helix* withstands dry conditions for months by entering a period of dormancy (estivation) in which it lowers metabolic rate precipitously and seals off its shell, retarding water loss. In some cases, the snail may lose almost 50% of its body water with prolonged exposure, but because dry mass also decreases proportionally, the percentage of tissue water is fairly constant.

Even more tolerant of desiccation are the nematodes that live in the Antarctic. These worms must survive cold stress as well as osmotic stress. The cold, dry air can dehydrate an animal, but these nematodes also experience hyperosmotic stress when melting water dissolves salts, elevating osmolarity as much as fivefold. During dehydration, the water content of the nematode's tissues may decrease to 2–10% of body mass. Like the snails, nematodes survive this extreme dehydration in a dormant, hypometabolic state that may last for decades.

The champion of desiccation tolerance is the brine shrimp, *Artemia*. The encysted embryos, often called eggs, are sold as "sea monkeys," with promises that the desiccated animals can be reanimated with the addition of water. If protected from the damaging effects of oxygen, dehydrated brine shrimp eggs can survive hundreds of years in this dehydrated state. Once the eggs hatch, the larvae lose their desiccation tolerance. Brine shrimp inhabit waters that experience periodic dehydration. When water appears, the *Artemia* eggs hatch and larvae mature quickly to initiate a rapid round of reproduction. *Artemia* retains a normal metabolic rate un-

til body water content reaches 50%; then metabolic rate declines as more body water is lost. In the final stages, when body water levels are below 1%, no evidence of life can be detected. Metabolism essentially stops, as indicated by measurement of metabolic fuel levels, gas exchange, and heat production.

Central to the survival of most species that tolerate anhydrobiosis is accumulation of protective agents, particularly carbohydrates and proteins. For example, when a nematode experiences some desiccation, it produces large amounts of the disaccharide trehalose, which accumulates to levels as high as 15% of its dry mass. Trehalose replaces the water molecules in the hydration shell of proteins and other macromolecules, and forms a coating around proteins, lipids, and other macromolecules that stabilizes macromolecular structure. In many species, the ability to survive dehydration correlates with trehalose levels, suggesting that trehalose may be required for survival. Recent studies have gone one step further, to test whether trehalose alone is sufficient to endow cells with desiccation tolerance. Researchers bathed mammalian platelets in trehalose, allowing them to take up the sugar. The cells were then frozen slowly and dehydrated in this frozen state, reducing water content to about 5% of mass. When the freeze-dried platelets were thawed, they remained viable. Transgenic mammalian cells have also been constructed to test the hypothesis that trehalose alone can endow desiccation tolerance. When mouse cells were transfected with two bacterial genes for the enzyme trehalose synthase, they produced very high levels of trehalose (about 100 mM). Unfortunately, these cells could not survive the desiccation process. These studies suggest that trehalose is necessary for desiccation, but that other factors may be required to endow an animal with desiccation tolerance.

References

○ Crowe, L. M. 2002. Lessons from nature: The role of sugars in anhydrobiosis. *Comparative Biochemistry and Physiology—Part A: Molecular and Integrative Physiology* 131: 505–513.

○ Tunnacliffe, A., A. Garcia de Castro, and M. Manzanera. 2001. Anhydrobiotic engineering of bacterial and mammalian cells: Is intracellular trehalose sufficient? *Cryobiology* 43: 124–132.

○ Wharton, D. A. 2003. The environmental physiology of Antarctic terrestrial nematodes. *Journal of Comparative Physiology* 173B: 621–628.

○ Wolkers, W. F., F. Tablin, and J. H. Crowe. 2002. From anhydrobiosis to freeze-drying of eukaryotic cells. *Comparative Biochemistry and Physiology—Part A: Molecular and Integrative Physiology* 131: 535–543.

Table 10.2   Water and solute content of food.

Nutrient	Water content (% of wet weight)
Animal and plant fluids	
Sap and nectar	90–100%
Blood	95%
Milk (most mammals)	87%
Milk (marine mammals)	40%
Fruits and vegetation	80–95%
Plant and animal tissues	
Muscle and animal tissues	50–70%
Seeds and grains	<10%

processes generate water as a result of oxidative phosphorylation (see Figure 2.40); this water is known as **metabolic water**. Each of the major macromolecules produces about the same amount of metabolic water, expressed per unit of metabolic energy. Based on the same 100 kcal of metabolizable energy, carbohydrate produces 15 ml of water; protein, 10.5 ml of water; and fat, 11.1 ml. For an average human, about 10% of daily water requirements come from metabolic water production, 60% from drinking, and 30% from water trapped in solid foods. As discussed in the previous feature (Box 10.1), many desert animals drink no fluids, and instead obtain their water entirely from solid foods and metabolic water.

Solutes can be classified as perturbing, compatible, or counteracting

In Chapter 2 we introduced the chemistry of solutes and solvents in biological systems. The total concentration of solutes imparts an osmolarity and determines the osmotic gradient across biological membranes, and thereby the direction and magnitude of water movement. In addition to these general osmotic effects of solutes, there are solute-specific effects. Three classes of solutes are distinguished by their effects on the structure and function of macromolecules, such as enzymes (Figure 10.3). *Perturbing solutes* disrupt macromolecular function at normal concentrations found within the animal. These include the inorganic ions found in body fluids, primarily Na^+, K^+, Cl^-, and SO_4^{2-}, as well as some organic solutes, such as

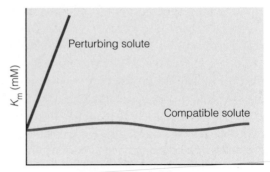

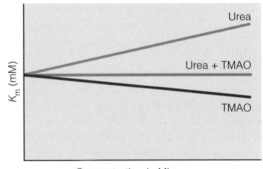

(b) Counteracting solutes

Figure 10.3 Perturbing, compatible, and counteracting solutes   Each type of solute exerts characteristic effects on macromolecular structure and function, such as enzyme kinetics (V_{max} or K_m). **(a)** A perturbing solute is shown to increase the K_m value of a hypothetical enzyme, whereas a compatible solute at the same concentration has no effect on K_m. **(b)** Each counteracting solute has perturbing effects when present alone, but when both are present, the effects are offset. Urea is shown to increase K_m and TMAO decreases K_m, but the combination of the two has no effect.

charged amino acids (e.g., arginine). *Compatible solutes* have little effect on macromolecular function and can accumulate to high concentration without deleterious effects on cellular processes. The most common compatible solutes in body fluids are polyols (trehalose, glycerol, and glucose) and uncharged amino acids, including several of the α-amino acids (alanine, glycine, serine, and proline) as well as other amino acids (β-alanine and taurine). *Counteracting solutes* are deleterious when used on their own, but can be employed in combinations where the deleterious effects of one solute counteract the deleterious effects of the other. For example, urea disrupts hydrophobic interactions and methylamines strengthen hydrophobic interactions. A combination of urea and methylamines allows the effects of one solute to negate the effects of the other solute. The most common methylamines employed by animals are trimethylamine oxide (TMAO), betaine, and sarcosine.

Figure 10.4 summarizes the solute composition of selected animals to illustrate the relative importance of the various solutes in different species and cellular compartments. The extracellular space of most animals is dominated by Na^+ and Cl^-. Marine ionoconformers possess extracellular concentrations of these ions, as well as Mg^{2+} and Ca^{2+}, close to seawater levels. In osmoconforming ionoregulators (molluscs, sharks), elevations in organic compatible and counteracting solutes allow inorganic ion levels to decline. The most abundant inorganic ions in extracellular fluid of osmoregulators are also Na^+ and Cl^-, although the levels are generally about one-third the strength of seawater. The cytoplasm of most animals is dominated by the same ions; the major cation is K^+ and the major anions are SO_4^{2-}, acetate, and Cl^-. Organic solutes occur in all animals but are most abundant in marine osmoconformers. Cartilaginous fish rely on the counteracting solutes: urea and various methylamines such as TMAO, sarcosine, and betaine. Invertebrates possess high concentrations of compatible solutes, mainly amino acids such as β-alanine, taurine, and proline. These organic solutes confer more than half the osmolarity in marine osmoconformers. When osmolarity changes, the concentrations of organic solutes often change disproportionately, allowing ionic solutes to remain relatively constant.

Cells transport solutes in and out of the extracellular fluid to control cell volume

Cells control their volume by transporting solutes across the plasma membrane, causing changes in osmotic pressure that induce the movement of water. Although some cells directly interface with the external environment, most cells within the body are surrounded by extracellular fluid held within the confines of the animal. The interstitial fluid—the extracellular fluid found in the narrow regions between cells—possesses a chemical composition distinct from plasma or lymph because it is a small volume directly affected by cellular transport processes.

Animals regulate the composition of the extracellular fluid to provide cells with an external solution that allows them to maintain an appropriate

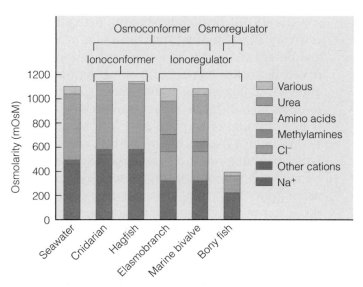

Figure 10.4 Organic and inorganic solutes in extracellular fluid of animals   Seawater is mainly Na^+ and Cl^-, with lower levels of other ions such as K^+, Mg^{2+}, and Ca^{2+}. Ionoconformers have high levels of Na^+ and Cl^-, whereas the levels of these ions are lower in ionoregulators. Osmoconformers have the same osmolarity as seawater but maintain an inorganic ion profile much like that of an osmoregulator. The remainder of the osmolarity is due to organic solutes, such as urea, amino acids, and methylamines.

cell volume. Animal cells change in cell volume in response to the osmotic gradients across the plasma membrane. Changes in cell volume are problematic for several reasons. Severely hypotonic conditions can cause a cell to explode, triggering a local immune response that can disrupt neighboring cells. Swollen cells can disrupt tissue structure or occlude blood vessels. When water movement changes cell volume, it also affects the intracellular concentration of metabolites and enzymes within cells, disrupting metabolic regulation. A cell volume change can also deform the cytoskeleton and plasma membrane, altering the function of receptors and transporters.

A change in cell volume can arise in response to environmental osmotic stress, but it may also arise as part of signaling pathways. For example, some hormones induce a change in cell volume that is a critical component of the signaling pathway. Thus, a volume change can be induced in a cell to compensate for osmotic imbalance, or in an unstressed cell as part of a signaling pathway. Cells induce a regulatory volume increase (RVI) by importing ions, causing an influx of water. A regulatory volume decrease (RVD) occurs when cells expel ions, causing an efflux of water.

The RVD is best understood in the context of recovery from swelling, but the same principles are thought to apply when normal cells are induced to shrink as part of a signaling pathway. Many combinations of transporters can be used to undergo a RVD; the most common mechanism is by activation of K^+ channels. With an outward K^+ gradient, K^+ leaks out of the cell, causing the polarized membrane to hyperpolarize until it reaches the K^+ equilibrium potential (E_K). At the same time, Cl^- channels open to allow Cl^- to move out of the cell in response to the hyperpolarizing effects of K^+ movements. The combination of K^+ and Cl^- movements reduces the total solute content of the cell, creating an osmotic gradient that drives water out of the cell. Another pathway of RVD involves an electroneutral K^+-Cl^- cotransporter, which pumps K^+ and Cl^- out of the cell. This transporter achieves much the same effect as the combination of K^+ channel and Cl^- channel activation. A third pathway uses the Na^+/Ca^{2+} exchanger, operating in a direction that expels Na^+ and imports Ca^{2+}. The Ca^{2+} is then exported from the cell via the Ca^{2+} ATPase. Finally, some cells use their Na^+/K^+ ATPase to regulate cell volume. This active transporter exports three Na^+ for two K^+, thereby reducing intracellular osmolarity.

Cells induce a RVI to recover from cell shrinkage or to induce swelling by transporting solutes into the cell. Most commonly, cells activate a Na^+-K^+-$2Cl^-$ cotransporter (NKCC), bringing these solutes into the cell to drive osmotic swelling. Cells can also induce swelling by opening Na^+ channels. Na^+ rushes into the cell in response to its electrochemical gradient, both depolarizing the membrane and causing an osmotic gradient that induces the influx of water. Most cells complement the effects of Na^+ channels with other transporters, such as the Na^+/H^+ exchanger, which imports Na^+ and exports H^+. Since the pool of H^+ is largely associated with buffers, this exchange leads to a net increase in cellular solutes, tending to increase cell volume. For most cells a myriad of transporters are available to alter solute distributions and cell volume. Figure 10.5 summarizes how various transporters can be used in combination to induce net ion movements that result in volume changes. Although the mechanisms involved in compensatory changes are well understood, the exact trigger for activation remains unclear. It is likely that the var-

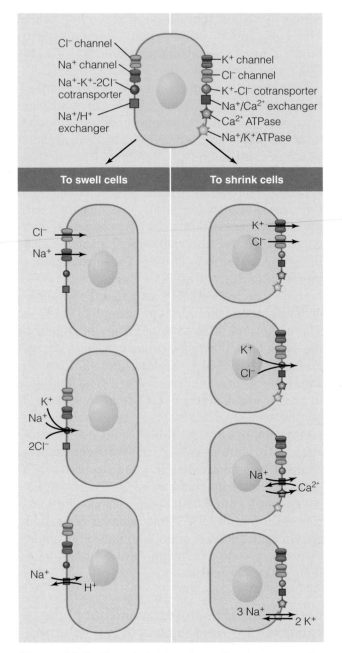

Figure 10.5 Regulatory volume increases and decreases Cells actively change cell volume by moving ions in or out of the cell. Each of the cells depicts a combination of transporters commonly used to transfer ions into the cell (causing a volume increase) or out of the cell (causing cell shrinking).

ious transporters are stimulated at least in part by mechanosensing pathways, whereby volume changes alter the shape of the cytoskeleton and plasma membrane, changing the activity of membrane transporters. Alternately, protein-modifying enzymes, such as protein kinases, can modify the structure of the transporters to increase their permeability (channels) or pumping (carriers).

1. Distinguish between perturbing, compatible, and counteracting solutes
2. What is the difference between osmolarity and tonicity?
3. Are terrestrial animals osmoregulators, osmoconformers, or neither?

The Role of Epithelial Tissues

Animals change the profile of the extracellular fluid by importing or exporting ions and water across those epithelial tissues that interface with the external environment. This includes the external surfaces, such as skin and gills, as well as the internalized "external" surfaces, such as the lumens of the excretory system and digestive system. These epithelial tissues form the boundary between the animal and the environment, and usually have other physiological responsibilities, such as respiration and digestion. However, the properties that make a tissue good at gas exchange or nutrient absorption—high surface area and high permeability—make it more vulnerable to ion and water movements. An epithelial tissue has properties that reflect a balance between its different physiological roles. Every animal relies on suites of tissues to control osmotic and ion balance. Ions and water can exchange across each of the body surfaces in contact with the external environment, including the respiratory surface, digestive tract, and body surface. Whether these surfaces are barriers to movement, sites of uptake, or sites of excretion depends upon the nature of the environment (ionic and osmotic gradients) and the physiology of the animal (ionic and osmotic strategies). In addition to these epithelial tissues where ion and water flux can occur, many animals have additional epithelial tissues with a primary role in ion and water regulation: kidneys and their equivalents in the invertebrates. In the following sections we discuss the function of several of these epithelial tissues, but we reserve our discussion of the kidney for later in the chapter.

The integument is an osmotic barrier

Animals reduce the flux of water across the body surface by limiting the water permeability of the epithelial tissues, both internal and external. Some animals reduce this permeability by controlling the number of aquaporin proteins in the plasma membrane. Recall from Chapter 2 that each aquaporin permits more than a billion water molecules to pass through each second. An epithelial cell with aquaporins may be 100-fold more permeable to water than a cell without aquaporins. The aquaporin levels in the plasma membrane depend on the expression of aquaporin genes and on pathways of intracellular traffic that control the interchange of aquaporins between storage vesicles and the plasma membrane.

Some animals reduce water loss by covering external surfaces with a thick layer of hydrophobic molecules. Mucus—an extracellular secretion of mucopolysaccharides, lipids, and proteins—is an example of such a hydrophobic barrier. Mucus layers on the surface of the lung and gastrointestinal tract reduce water loss across these epithelia. Many semiaquatic animals, such as frogs, use mucus to prevent water loss and keep the skin hydrated. The thick mucus layer of a hibernating lungfish dries to form a water-impermeable cocoon that prevents the animal from dehydrating during the many months of *estivation*. Surface mucus also reduces osmoregulatory costs by trapping a layer of water between the animal and the environment. This layer of water is a microcompartment that acts as an osmotic and ionic buffer zone.

Land animals use more elaborate adaptations in epithelial structure to prevent water loss across the skin. The *keratinocytes* of the skin of terrestrial amphibians and amniotes secrete proteins and modified lipids to form a dense, hydrophobic extracellular matrix. The amniotes, but not the amphibians, possess an additional layer on top of the keratinocytes. This layer, called the *stratum corneum*, is composed of keratinocytes that have differentiated to form another type of cell—a *corneocyte*. During the differentiation process, the cells produce thick bundles of the protein keratin, an intermediate filament of the cytoskeleton. These bundles are, in turn, interconnected by other proteins, such as keratohyalin. The cell then produces a complex layer of proteins, called the *cornified envelope*, which eventually replaces the corneocyte plasma membrane. During cornification, the corneocyte undergoes programmed cell death, and what remains is the keratin network surrounded by the cornified envelope. Extracellular matrix proteins connect these cellular remnants to stacks of lipid molecules called the lamellar membrane. Once formed, this mixture of proteins and lipids undergoes a series of enzymatic and chemical processes that modify it into the stratum corneum (Figure 10.6). Although the tissue is dead, it remains responsive to physical changes, triggering the underlying keratinocytes to secrete proteins, lipids, and signaling factors.

The diversity in the properties of vertebrate skin is due mainly to the way the stratum corneum is constructed. The scales of reptiles and birds are composed of interconnected patches of stratum corneum (largely keratin). Mammalian skin is also keratinized, although only a few mammals retain the ancestral "scales," such as the covering on a rodent's tail or an armadillo's shell. Modifications of the keratinized stratum corneum provide terrestrial vertebrates with other structures, such as the scutes on the underbelly of snakes that are used in locomotion and the protective spines of the desert lizards. However, all tetrapods depend on their keratinized stratum corneum to minimize desiccation (Figure 10.7).

The other major group of terrestrial animals, the arthropods, possesses a different type of waterproof integument. The insect **cuticle** is a complex network of hydrophobic molecules that covers all of the external surfaces of insects, including the surfaces of the trachea and gut. The main structural component of the cuticle is the polysaccharide *chitin*. It is synthesized within the cells of the epidermis and then transported to the extracellular space, where it is chemically modified, crystallized, and combined with other proteins and polysaccharides to obtain the appropriate physical properties. The mature cuticle has very low permeability to water, and is rigid enough to act as an external skeleton for the animal. We discuss the nature of the insect cuticle in greater detail in Chapter 12: Locomotion, when we consider its role as an exoskeleton in locomotor systems.

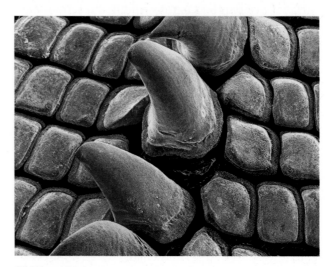

(a) Armadillo

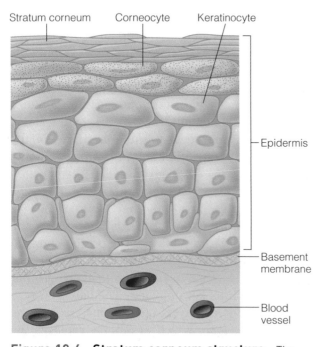

Figure 10.6 Stratum corneum structure The stratum corneum is the thickened external layer of modified epithelium found in mammals. Keratinocytes differentiate into corneocytes, producing a waterproof layer composed of a complex network of intracellular and extracellular proteins, augmented by lipids.

(b) Horny skin iguana

Figure 10.7 Diversity in the stratum corneum of vertebrate tetrapods

The integuments of both tetrapods and terrestrial insects possess an additional layer of lipid that reduces evaporative water loss. The cells of the epidermis secrete these lipids, which then form a continuous coat that acts as a sealant. Birds and mammals possess a thin layer of glycolipid that covers the stratum corneum and fills gaps between cells. The exoskeleton of insects also has a surface coating of long chain fatty acids and wax esters. In fact, this thin lipid layer gives the insect exoskeleton its resistance to water movement. The ability of the lipid layers to limit water movement depends on the interaction between the lipid molecules, creating a hydrophobic barrier that excludes water. The lipid layer is held together by hydrogen bonds, and as we learned in Chapter 2, an increase in temperature weakens such bonds. Consequently, the lipid layer loses its integrity at higher temperature, greatly enhancing evaporative water loss.

Collectively, the properties of the integument, established by the cells of the epidermis, control the magnitude of water loss. Although we have focused on the outer body covering, the same processes occur across another epithelial tissue: the respiratory surface. The magnitude of respiratory water loss depends on structural features, described for the outer integument, as well as other factors. For example, an air-breathing animal with a high metabolic rate will have higher ventilation rates and therefore greater respiratory water loss than an animal with a lower metabolic rate. Many animals, especially desert animals we discuss in a later feature, possess anatomical adaptations that reduce respiratory water loss.

Epithelial tissues share four specialized properties that affect ion movements

The properties of epithelial tissues depend on both the transport properties of individual epithelial cells, and the way cells are interconnected to form the tissue. The diverse epithelial tissues, from frog skin to insect Malpighian tubule, share four general features (Figure 10.8).

First, epithelial cell function depends on the asymmetric distribution of transporters within the cell. The apical cell membrane, exposed to the outside world, has a different profile of proteins than the basolateral cell membrane, which faces inward. This cellular topography arises because cells insert proteins in the correct location and re-

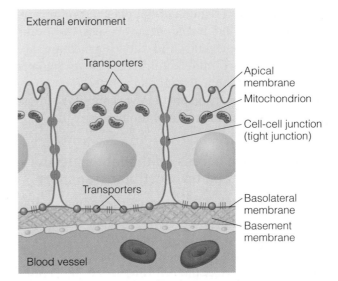

Figure 10.8 General features of epithelia The typical epithelial tissue displays four main features: **(1)** an asymmetrical distribution of membrane proteins; **(2)** tight intercellular connections that govern paracellular movement; **(3)** a multiplicity of cell types; and **(4)** a high density of mitochondria.

strict their movement in the lipid bilayer. In part, the proteins are collected together in chemically distinct regions of the membrane, such as the lipid rafts we discussed in Chapter 2. Once in membranes, they are anchored in position by attachment to the cytoskeleton.

Second, epithelial cells are interconnected by protein linkages that convert the collection of cells to an impermeable sheet of tissue. As we first discussed in Chapter 2, tight junctions are formed when membrane proteins of one cell connect to a specific protein in an adjacent cell. The interaction between adjacent cells limits the movement of solutes and water around cells. These intercellular connections also create a kind of protein belt around the circumference of the epithelial cell, restricting the free movement of membrane proteins between apical and basolateral membrane regions to maintain the cellular topography.

Third, epithelial tissues consist of many types of cells. This diversity is most extreme in the digestive system, which we discuss in the next chapter. However, even relatively simple tissues, such as the fish gill, are composed of several cell types, each with important roles, such as providing specific transport capabilities or structural support.

Fourth, ion transport demands a great deal of energy. Most epithelial cells with a major role in transport possess abundant mitochondria to produce ATP. In some cases, mitochondria are in close proximity to the regions of the plasma membrane that conduct the transport processes. In other cases, motor proteins and the cytoskeleton actively transport mitochondria to these regions when the metabolic demands increase. The energetic costs of ion transport may account for almost half of the metabolic rate of the tissue.

Solutes move across epithelial tissues by paracellular and transcellular transport

Although epithelial tissues transport some solutes for their own purposes, most transport processes serve to transfer solutes from one side of the tissue to the other. Epithelial tissues use two main routes of transport across the cell (Figure 10.9). **Transcellular transport** is the movement of solutes (or water) through epithelial cells. For example, solutes can diffuse from the extracellular fluid that bathes the cells, and move across the basolateral membrane, through the cytoplasm, and across the

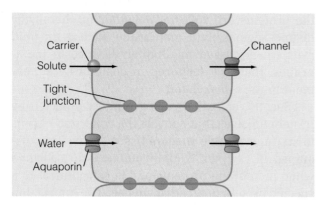

(a) Transcellular transport

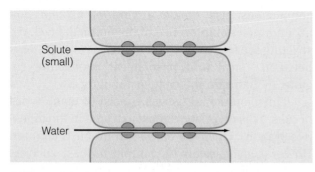

(b) Paracellular transport

Figure 10.9 Transcellular and paracellular transport **(a)** Tight epithelia can transfer solutes across the cell using transporters on the apical and basolateral plasma membrane. **(b)** In leaky epithelia, small solutes can also move between cells, passing through the tight junctions that interconnect cells.

apical membrane into the external environment (either the open water or the lumen of an organ that communicates with the external environment). Conversely, the movement of solutes (or water) *between* adjacent cells is **paracellular transport**. For example, molecules diffuse from the blood, through the extracellular fluid, and into the narrow confines of the interstitial fluid between adjacent cells. From here, the molecules pass through the tight junctions that connect epithelial cells into sheets. The neighboring cells can secrete molecules into the interstitial space to control its nature in ways that create gradients that drive paracellular movements. Although the tight junction can prevent large molecules from crossing, small molecules (water, ions) can cross through the protein connections. Tissues that permit paracellular transport are frequently called **leaky epithelia**. Tissues that conduct minimal paracellular transport are called **tight epithelia**.

Epithelial tissues possess suites of transporters, including the following transporters frequently implicated in ion and water balance. Though many of these transporters have acquired common names, typically as a result of how they were first discovered, we use the more descriptive names throughout this chapter to better demonstrate their function.

- ATPases are central to ion movements. The Na^+/K^+ ATPase, or sodium pump, is a primary active transporter that uses the energy of ATP hydrolysis to export three Na^+ in exchange for importing two K^+. Some tissues use a H^+ ATPase to pump protons to change pH, a driving force for other transport processes. Ca^{2+} ATPases re-establish Ca^{2+} gradients across cellular membranes.

- Various ion channels (Cl^-, K^+, and Na^+) can open or close in response to mechanical, electrical, or chemical signals to permit specific ions to flow down electrochemical gradients. The Cl^- channel most commonly implicated in transfer of Cl^- across the apical membrane of cells is better known as the *cystic fibrosis transmembrane conductance regulator*, or *CFTR*.

- Electroneutral cotransporters carry both anions and cations in the same direction in response to the electrochemical gradient. There are Na^+-K^+-$2Cl^-$ cotransporters (NKCC) and K^+-Cl^- cotransporters.

- Various electroneutral exchangers are reversible transporters driven by electrochemical gradients, including the pH gradients. Some transporters are cation antiporters, such as Na^+/H^+ exchangers (commonly abbreviated as NHE) and NH_4^+/H^+ exchangers. Other transporters are anion antiporters, such as the Cl^-/HCO_3^- exchanger (commonly known as *band 3* as a result of its electrophoretic mobility in red blood cell preparations).

Fish gills transport ions into and out of the water

Like most transport epithelia, fish gills have several types of cells involved in control of ion and water balance. Perhaps the best-studied system is that of the rainbow trout (Figure 10.10). Mucus-secreting cells are scattered over the surface of the gill. There are **chloride cells**, which are large cells with abundant mitochondria. Much of the surface is covered by smaller, flattened cells collectively called *pavement cells*. Some pavement cells, like chloride cells, have numerous mitochondria, whereas others possess fewer mitochondria. It is thought that most of the ion regulation in the gill is mediated by the two types of cells that are rich in mitochondria. These two cell types can be distinguished using histochemical methods that employ a glycoprotein (peanut lectin agglutin or PNA) that binds carbohydrates on chloride cells that are absent from pavement cells. Thus, chloride cells are often called PNA^+ cells, and pavement cells are PNA^- cells. In rainbow trout, and perhaps freshwater fish in general, these two types of mitochondria-rich cells mediate different transport processes.

The direction of transport of ions and water depends on the salinity of the water (Figure 10.11a). The gill of a freshwater fish must take up Na^+, Ca^{2+}, and other ions from the water, frequently against steep electrochemical gradients. The PNA^- cells take up Na^+ through an apical Na^+ channel. Although there is an unfavorable gradient for Na^+ uptake, these cells create a favorable electrochemical gradient using a H^+ ATPase that acidifies the water in the boundary layer. Once inside the cell, Na^+ is exported to the extracellular fluid by the basolateral Na^+/K^+ ATPase or a Na^+/HCO_3^- exchanger. PNA^+ cells import Cl^- into the cell using an apical Cl^-/HCO_3^- exchanger, which then es-

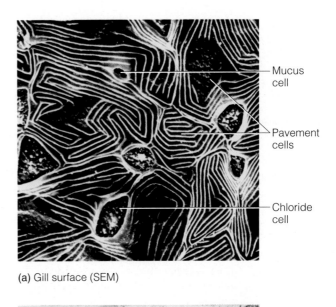

(a) Gill surface (SEM)

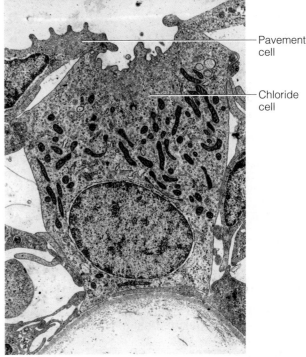

(b) Chloride cell (TEM)

Figure 10.10 Cells of the fish gill The fish gill is an important site of ion exchange in both freshwater and seawater species. It has multiple cell types, including mucus cells, chloride cells, and pavement cells. SEM, TEM: scanning or transmission electron micrograph.
(Images courtesy of Steve Perry, University of Ottawa)

capes through basolateral Cl^- channels. In both transport schemes, production of HCO_3^- and H^+ by carbonic anhydrase is essential, providing ions that can be used as counterions or to change pH.

In contrast to freshwater fish, marine fish must avoid excessive ion uptake and limit water loss. The gill is central to ion balance, and chloride cells

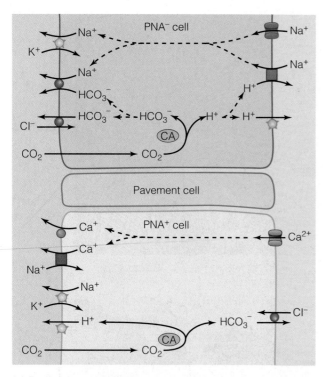

(a) Freshwater trout gill

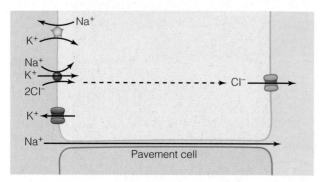

(b) Marine fish gill

Figure 10.11 Ion transport processes in gills of freshwater and marine fish Gills possess ion-pumping cells that cause the net uptake of Na^+ and Cl^- in freshwater and the net export of Na^+ and Cl^- in seawater. **(a)** The freshwater gill possesses two types of ion-pumping cells. Acid-secreting cells (PNA$^-$) import Na^+ from the water. Base-secreting cells (PNA$^+$) import Cl^- and Ca^{2+}. **(b)** Gill epithelial cells of marine fish export Cl^- and Na^+.
(Source: Adapted from Perry and Gilmour, 2006)

in particular are critical for excreting ions (Figure 10.11b). The combined actions of the Na^+/K^+ ATPase and the Na^+-K^+-$2Cl^-$ cotransporter bring K^+ and Cl^- (and some Na^+) into the cell from the blood. The Cl^- channels in the apical membrane allow Cl^- to escape into the seawater, and basolateral K^+ channels allow K^+ to return to the blood. The movement of Cl^- and other ions creates a transepithelial membrane potential (negative on

the outside). Na^+ is thought to escape through paracellular channels, driven by the transepithelial membrane potential. This arrangement of transporters is common in other ion-pumping epithelial cells that expel Cl^- from cells, such as the shark rectal gland we discuss later in this chapter.

As you can see, the responsibilities of the ion-pumping cells of the fish gill change depending on the external conditions. Some fish species are **diadromous**, migrating between seawater and freshwater. **Catadromous** fish, such as eels, migrate to seawater to breed. In contrast, **anadromous** fish, such as salmon, migrate from seawater to freshwater to reproduce. Young salmon grow in freshwater, then migrate to the sea. Prior to migration, the gills of these fish undergo a dramatic cellular reorganization as the ion-pumping properties of the gill cells prepare for the new environment (Figure 10.12). Interestingly, the remodeling process, called *smoltification*, occurs before exposure to seawater. It is mediated largely by growth hormone, insulin-like growth factor 1, cortisol, and to a lesser extent thyroid hormone. Smoltification also leads to the remodeling of other tissues involved in ion and water balance, including the gastrointestinal tract and probably the kidney. As a consequence of the cellular changes, there is very little ionic and osmotic disturbance when the salmon enter seawater.

Digestive epithelia mediate ion and water transfers

Every time an animal consumes food or water, the digestive tract, with its high surface area, becomes a site of exchange of solutes and water. The diet may be a vital source of water but it may also create an osmotic burden. Many insects, for example, feed on diets rich in water, such as sap, nectar, or blood. Consider the osmotic challenge faced by *Rhodnius*, a blood-sucking bug that consumes more than 12 times its body mass in a single blood meal. Blood-sucking insects must remove the water from the blood meal in order to process the remaining energy-rich macromolecules. Facing similar challenges, a female mosquito begins to urinate shortly after commencing feeding, allowing her to compress the solids of the blood meal in the gut.

Scientists debate the relative importance of transcellular and paracellular transport in water transport across the gut, but it is likely that both processes are important. Transcellular transport is driven by osmotic gradients and facilitated by aquaporins in both basolateral and apical membranes of the epithelium. Paracellular transport occurs when the internal and external fluids are nearly isosmotic. Ions, mainly Na$^+$ and Cl$^-$, are secreted into the interstitial space to create an osmotic gradient that drives movement of water across the tight junctions. Once in the interstitial fluid, water makes its way into the blood.

Reptiles and birds possess salt glands

Since freshwater has a very low solute concentration, it creates an inward osmotic pressure that helps drive water uptake. However, animals that drink seawater face two challenges. First, water molecules must be selectively transported across the gut against the osmotic gradient. It is likely that transcellular transport across tight epithelia is important in these animals. Second, the animals must be able to expel the salt that accompanies the seawater consumed in the diet.

Many reptiles and birds possess a **salt gland** that aids in ion and water balance by excreting hyperosmotic solutions of Na$^+$ and Cl$^-$. Whether living in the ocean or the desert, species with salt glands can cope without access to freshwater, deriving water from drinking hypertonic seawater or exclusively from food.

In birds, the salt gland is found in a depression at the base of the beak, and its secretions drain through a canal that runs along the beak and opens at the nostrils. The nasal salt gland secretion can be as much as three times more concentrated than the plasma. Thus, if a bird drinks 30 ml of seawater, it can excrete all of the salt in 10 ml of salt gland secretion, gaining 20 ml of pure water. The salt glands are able to do this by using metabolic energy to create a **countercurrent multiplier**.

Pre-smolt (parr)   Smolt

(a) Smoltification

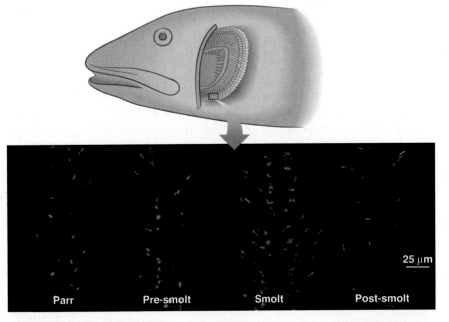

(b) Changes in gill Na$^+$/K$^+$ ATPase

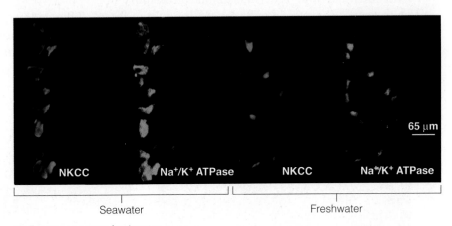

(c) Seawater versus freshwater

Figure 10.12   Smoltification and ion regulation   Salmon undergo complex physiological changes when they move from rivers and streams into the ocean, where they grow and reach reproductive maturity. In addition to a color change **(a)**, salmon remodel their epithelial tissues to make them better able to expel ions. Gills, intestines, and skin increase the activities of ion-pumping machinery. **(b, c)** The fluorescent images of a salmon gill show the relative abundance of important transporters. The levels of the Na$^+$/K$^+$ ATPase (red) increase during smoltification **(b)**. The levels of Na$^+$-K$^+$-2Cl$^-$ cotransporter (NKCC) and the Na$^+$/K$^+$ ATPase also differ manyfold in freshwater- versus seawater-acclimated fish **(c)**.

(Image courtesy of Ryan Pellis and Steve McCormick, Contes Anadromous Fish Laboratory)

The salt gland is composed of a series of secretory tubules, surrounded by peritubular fluid and a capillary network. The tubule has a closed end and an elongated tube that empties into a collecting duct. Fluids flow from the closed end of the tubule to the open end. A capillary network is arranged in parallel to the tubule, though direction of blood flow is the opposite direction to that of the lumen fluids (Figure 10.13). This countercurrent arrangement of flows is central to the ability of the salt gland to produce a concentrated secretion. The countercurrent multiplier works like this. As blood flows toward the closed end of the tubule, salts escape into the interstitial fluid and are taken up by the tubule cells and transferred to the lumen. As a result, the blood becomes progressively more dilute. The interstitial fluid that bathes the tubule cells is in equilibrium with the blood passing over the tubule: low osmolarity near the closed end and high osmolarity near the opening. Thus, the tubule cells near the closed end of the tubule are exposed to a dilute interstitial fluid and create a dilute lumen fluid. As lumen fluids flow from the closed end to the open end, the surrounding interstitial fluids are increasingly concentrated, and the transcellular transport of salts across the tubule cells causes the lumen fluids to become more concentrated.

The epithelial cells that line the secretory tubule extract salts from the interstitial fluid found between the tubule cells and the blood. The basolateral membrane and apical membrane work in conjunction to produce a hyperosmotic secretion. The basolateral membrane of the epithelial cells brings ions into the cell using the suites of transporters involved in a regulatory volume increase. Conversely, the apical membrane possesses the transporters that are involved in regulatory volume decreases. Although the exact mechanisms remain unclear, the Na^+/K^+ ATPase, NKCC, K^+ channels, and Cl^- channels have all been implicated in the formation of the hyperosmotic secretion. The net result of these activities is the import of Na^+ and Cl^- from the plasma and their secretion into the lumen of the tubule. Like other epithelial tissues involved in ion transport, the cells of the secretory tubule have a high content of mitochondria, which produce the ATP needed to pump ions and establish the gradients used by secondary active transport. The ion-secreting machinery of the salt glands is similar in many respects to the transporters used by other salt-secreting epithelia discussed earlier in this chapter, such as the fish gill.

Elasmobranch rectal glands excrete Na^+ and Cl^-, while retaining urea

Like seabirds, elasmobranchs have an accessory excretory organ that aids in salt excretion. The **rectal gland** is composed of many tubules surrounded by capillaries. Each tubule is composed of

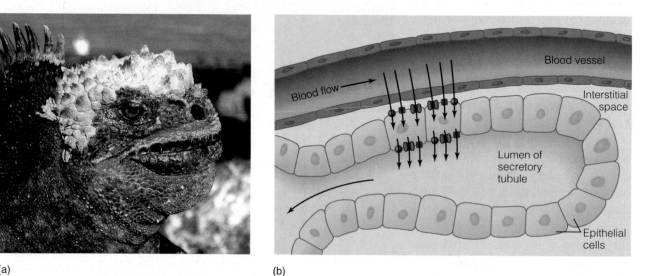

(a) (b)

Figure 10.13 Salt glands of birds and reptiles
Some birds and reptiles that live in seawater or the desert are able to excrete Na^+ and Cl^- from specialized salt glands. **(a)** The glands are located near the eye, but drain into ducts that empty near the nostril. Salt can precipitate and accumulate on the head, as shown in this marine iguana.

(b) The hypersaline excretions form in secretory tubules that are arranged into lobes that drain into collecting ducts. Blood vessels juxtaposed to the secretory tubules flow countercurrent to the flow of fluid through the tubule.
(Photo courtesy of Robert H. Rothman, Rochester Institute of Technology)

a single type of epithelial cell. Like other transport epithelial cells, the cells of the rectal gland tubule have abundant mitochondria and basolateral invaginations, much like microvilli, that increase the surface area for ion exchange with the blood. The tubules are able to transfer NaCl from the blood to the tubule lumen. Though the osmolarity of the tubule secretions is similar to that of the plasma, the secretions have a much higher concentration of NaCl because urea is retained in the blood. Two separate mechanisms are responsible for transepithelial movement of Na^+ and Cl^-. Tubular epithelial cells actively transport Cl^- from the blood via transcellular transport, whereas Na^+ moves between the tubular epithelial cells, from the blood to the tubule lumen via paracellular transport.

The transport processes carried out by the tubular epithelium function much like those of chloride cells of the teleost gill and the salt gland of birds, using a combination of basolateral NKCC, Na^+/K^+ ATPase, K^+ channels, and apical Cl^- channels. A model for describing the secretion of Cl^- is shown in Figure 10.14. The main source of entry of Cl^- into the epithelial cells is via the NKCC. The inward movements of Na^+ and K^+ are reversed by the action of the basolateral K^+ channels and the exchange of Na^+ for K^+ via the Na^+/K^+ ATPase. Once Cl^- enters the cytoplasm of the epithelial cell, it can escape across the apical plasma membrane through Cl^- channels. The rectal gland is also the site of Na^+ excretion, which moves between cells from the interstitial fluid to the lumen of the tubule. This paracellular transport is driven by the transepithelial electrochemical potential.

Salt secretion by the rectal gland occurs in pulses after a shark has incurred a salt load, either through drinking or eating salt-laden food. The osmotic perturbation triggers release of hormones that stimulate rectal gland secretion. The osmotic and blood volume changes stimulate the release of atrial natriuretic peptide from the heart. We discuss this hormone in more detail later in this chapter, but in sharks atrial natriuretic hormone triggers the release of the neuroendocrine hormone vasoactive intestinal peptide (VIP). When VIP binds to its G-protein linked receptor, it activates adenylate cyclase, increasing cAMP synthesis, which activates protein kinase A (PKA). The main target of PKA in this process is the apical Cl^- channel itself; phosphorylation opens the channel. PKA may also affect intracellular traffic, causing movement of more Cl^- channels to the apical membrane to further in-

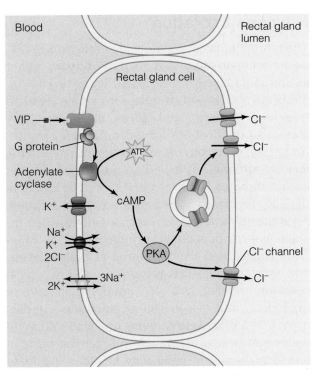

Figure 10.14 Chloride transport in the elasmobranch rectal gland The excretion of salt in the shark rectal gland is driven by the secretion of Cl^-. Chloride is imported into the cell from the plasma through the Na^+-K^+-$2Cl^-$ cotransporter and escapes through Cl^- channels. The entire process is sensitive to hormones such as vasoactive intestinal peptide (VIP), which elevate cAMP levels, activating protein kinase A (PKA).
(Source: Adapted from Silva et al., 1997)

crease Cl^- conductance. Though the changes in Cl^- activity probably drive this process, the activity of NKCC is also regulated during stimulation of salt secretion. The efflux of Cl^- causes a regulatory decrease in cell volume and cytoplasmic Cl^- levels. The changes in cell volume trigger phosphorylation of NKCC, which is normally inactive in the basolateral membrane. Although the exact protein kinases that phosphorylate NKCC are not yet established, PKA is not involved. These protein kinases, whatever their nature, may also be activated through hormonal or neuroendocrine factors.

CONCEPT CHECK

4. Name two tissues specialized to produce concentrated salt solutions.

5. What are the four main features of a transport epithelium?

6. Distinguish between transcellular and paracellular transport.

Nitrogen Excretion

The ammonia produced during amino acid breakdown is a toxic solute that must be excreted, either as ammonia, urea, or uric acid (Figure 10.15). Animals use a variety of strategies to excrete these nitrogenous wastes, and these strategies have important implications for ion and water balance. An animal that excretes most of its nitrogen in the form of ammonia is called an **ammoniotele**. Because ammonia is very toxic, it cannot be stored in the body and must be excreted as a dilute solution, resulting in water loss. Alternative strategies involve energy-dependent production of nitrogenous wastes that can be stored at higher levels, and excreted with less water loss. The two most common alternatives to ammoniotelism are ureotelism and uricotelism. A **ureotele** excretes urea, and a **uricotele** excretes uric acid. Although animals excrete most of their nitrogenous waste in one form, almost every species has the capacity to produce each of these molecules. For example, humans are ureoteles, but they also produce and excrete some ammonia and uric acid.

Each group of animals relies predominantly on a particular strategy (Table 10.3). Among the vertebrates, there are ureoteles (mammals), uricoteles (birds and reptiles), and ammonioteles (amphibians and fish). However, there are many exceptions to these generalizations. There are exceptional species; for example, a few species of bony fish are ureoteles. There are also developmental transitions; for instance, most amphibians excrete ammonia as larva but urea as adults. Other transitions in nitrogenous excretion strategies are triggered by environmental conditions; dehydration causes some lungfish to convert from ammoniotelism to ureotelism. Since the enzymes necessary for ammonia, urea, and uric acid synthesis exist in most animals, we can assume that the atypical species or developmental changes arise through variation in the control of expression of genes, rather than convergent evolution of novel capabilities.

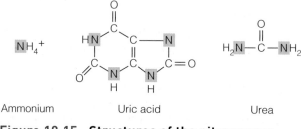

Ammonium   Uric acid   Urea

Figure 10.15 Structures of the nitrogenous end products

Table 10.3 Nitrogen excretion strategies.	
Nitrogen excretion strategy	**Animal group**
Ammonioteles	Simple invertebrates (cnidarians, nematodes)
	Aquatic molluscs
	Agnathans, chondrichthians, bony fish, larval amphibians
Uricoteles	Terrestrial molluscs (snails, slugs), terrestrial arthropods
	Reptiles, birds
Ureoteles	Some larval bony fish, estivating lungfish
	All mammals

Ammonia is produced in amino acid metabolism

Ammonia is at the heart of amino acid metabolism. It is used in the synthesis of amino acids by reactions that add ammonia to a carbon skeleton to create an amino acid that can be used in biosynthesis of proteins. When proteins are degraded, the amino acids are broken down to produce carbon skeletons that can be used for energy metabolism. The ammonia that is liberated is toxic and must be either excreted or metabolized into a less toxic form.

The removal and processing of ammonia from amino acids is complex because of the unique structural features of each amino acid. A few amino acids (asparagine, glutamine, glutamate, histidine, serine) can be *deaminated*, with ammonia cleaved from the carbon backbone and released. For most amino acids, aminotransferases transfer their amino group to 2-oxoglutarate, producing glutamate, which can then be deaminated by the enzyme glutamate dehydrogenase (Figure 10.16). In many animals, ammonia produced by glutamate dehydrogenase or other deaminating enzymes is repackaged into a form that is less toxic. Many animal tissues use the enzyme glutamine synthase to transfer ammonia to glutamate, forming glutamine. This amino acid can then be transported to other tissues, where it can be deaminated by the enzyme glutaminase, releasing ammonia and glutamate. This complex cycle of ammonia release, glutamine synthesis, and glutamine deamination costs the animal metabolic en-

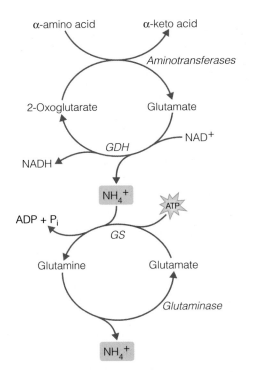

Figure 10.16 Glutamine metabolism and ammoniogenesis NH_4^+ from most amino acids is transferred to glutamate through various aminotransferases. The glutamate can then be oxidatively deaminated by glutamate dehydrogenase (GDH). NH_4^+ can also be used by glutamine synthase (GS) to produce glutamine, a convenient molecule to transport ammonia between tissues. Glutamine can then be deaminated by glutaminase.

ergy in the form of ATP. However, it gives an animal greater control over the rate and location of ammonia production, which is particularly important in animals that further metabolize ammonia into less toxic nitrogenous compounds bound for excretion.

Ammonia can be excreted across epithelial tissues

Metabolic enzymes produce a combination of NH_3 and NH_4^+, but the relative concentrations of the different forms of ammonia in fluids depend on many factors. Since NH_3 is a gas, its concentration in biological fluids depends on how much dissolves into water, which in turn depends on the partial pressure of the gas (P_{NH_3}), its solubility in water, and the temperature. Once dissolved, some NH_3 becomes protonated to form NH_4^+. The balance between these two forms depends on pH. Since the pK_a value for NH_4^+ is approximately 9, at a physiological pH most of the ammonia occurs in the ionized (NH_4^+) form.

Ammonia can cross biological membranes as both NH_3 and NH_4^+, although by different mechanisms. NH_3 can passively diffuse through membranes at a moderate rate. Recently it has been shown that, like water movement, some NH_3 crosses membranes via specific gas channels. Though animals have an NH_4^+/H^+ exchanger, NH_4^+ can also replace K^+ in some transporters, such as NKCC and Na^+/K^+ ATPase, or replace H^+ in the Na^+/H^+ exchanger.

Ammonioteles export the ammonia produced in these reactions across diverse epithelial tissues. In general, the uncharged form, NH_3, crosses membranes, whereas the charged form, NH_4^+, requires specific transporters. Ammoniotelism is most common in animals that live in water. Freshwater fish are typically ammonioteles, excreting most of their nitrogen waste as NH_3. Marine fish can also excrete NH_3, but in these species NH_4^+ excretion across the gills is also important. Some air-breathing fish are able to excrete some ammonia through volatilization of NH_3. Most terrestrial animals release at least some NH_3 across the skin and lung, even if ammoniotelism is not their primary mode of excretion of nitrogenous wastes.

Ammoniotelism has one main advantage; little additional energy is required to metabolize this nitrogenous waste into a form ready for excretion. However, ammonia excretion is not practical for terrestrial animals because it requires large volumes of water and constant urination to ensure that ammonia levels remain within a tolerable range.

Birds, reptiles, and insects excrete uric acid

The terrestrial invasion by animals necessitated an excretory strategy that permitted nitrogen excretion with little need for water. The earliest evolutionary solution to this problem was uricotelism. Unlike ammonia, uric acid can accumulate in body fluids with few toxic effects. Uricotelism spares water, because uric acid is excreted as anhydrous, white crystals. However, uric acid synthesis does require metabolic energy.

Uric acid is produced by most animals as part of an energy-dependent pathway for nucleotide synthesis (Figure 10.17). Networks of aminotransferases transfer nitrogen from various amino acids to the three amino acids that act as substrates for IMP synthesis: glutamine, glycine, and aspartate. IMP synthesis requires 5 ATP, but an

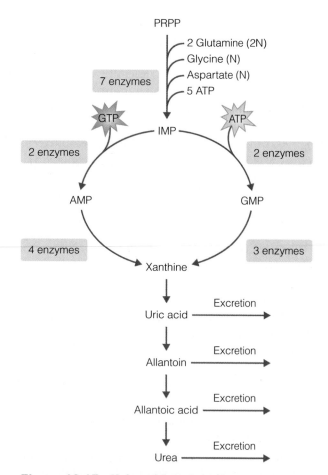

Figure 10.17 Uric acid metabolism A complex reaction network uses high-energy phosphate compounds to use amino acids as substrates to produce various nucleotides, and then break those nucleotides down for excretion. This pathway is also an important route of nitrogenous waste production. Amino acid nitrogen is transferred to uric acid, which, depending on the animal, may be excreted or further metabolized to produce other nitrogenous wastes. PRPP: 5-phosphoribosyl-1-pyrophosphate.

additional high-energy phosphate is required for the conversion of IMP to either AMP or GMP. Both AMP and GMP are broken down to form xanthine. AMP is metabolized to adenosine, inosine, hypoxanthine, and then xanthine, whereas GMP is metabolized to guanosine, guanine, and then xanthine. Some terrestrial arthropods (for example, spiders and scorpions) excrete guanine as a nitrogenous waste product. The xanthine that is produced in these reactions is oxidized to form uric acid. Depending on the species, uric acid may be further metabolized to allantoin, allantoic acid, or urea, any of which may be excreted as a nitrogenous waste.

Even those animals that use other pathways for disposing of nitrogenous wastes also produce some uric acid as a normal end product of nu-

cleotide metabolism. Its fate varies among species. Of course, in uricoteles (birds, reptiles, and insects), uric acid from nucleotide metabolism intermingles with that produced in nitrogen excretion. However, primates also excrete some uric acid. In other animals, uric acid is further metabolized, then excreted as allantoin (in nonprimate mammals), allantoate (in bony fish), urea (in amphibians and cartilaginous fish), or ammonia (in marine invertebrates).

Multiple, unrelated lineages of animals—vertebrates and invertebrates—share the feature of uricotelism. This striking example of convergent evolution is possible only because the pathway of uric acid synthesis is available to all animals as part of intermediary metabolism. The inherent flexibility of these pathways is evident in the animals that can switch between uricotelism and ammoniotelism, depending on conditions. The amphibious Indian apple snail can switch modes depending on the environment, living as an ammoniotele while in water and a uricotele when on land.

Urea is produced in the ornithine-urea cycle

Long after the first invertebrate and vertebrate uricoteles appeared on the scene, another form of nitrogen excretion—ureotelism—arose in the terrestrial lineages. Urea is the main excretory product of mammals, as well as selected species in other taxa. Urea is produced in some species in the breakdown of uric acid or arginine, but ureoteles produce urea by another pathway, the **ornithine-urea cycle** (Figure 10.18).

The prelude to urea production is the transfer of amino groups from the diverse amino acids to the form that can be used by the enzyme *carbamoyl phosphate synthase (CPS)*. One isoform of CPS (CPS II) is involved in pyrimidine nucleotide synthesis and uses glutamine as a substrate. However, ureotelic animals possess other CPS isoforms (CPS I and CPS III) that are specialized for urea synthesis. The evolutionary origins of the urea cycle are intimately linked to the evolution of the CPS genes (see Box 10.2, Genetics and Genomics: Evolution of the Urea Cycle). The two CPS forms involved in urea synthesis differ in the N-donor; CPS I uses NH_4^+ whereas CPS III uses glutamine. Once carbamoyl phosphate is produced, it enters the ornithine-urea cycle.

In addition to the five enzymes of the ornithine-urea cycle, urea synthesis requires two transporters to shuttle substrates across the mitochondrial mem-

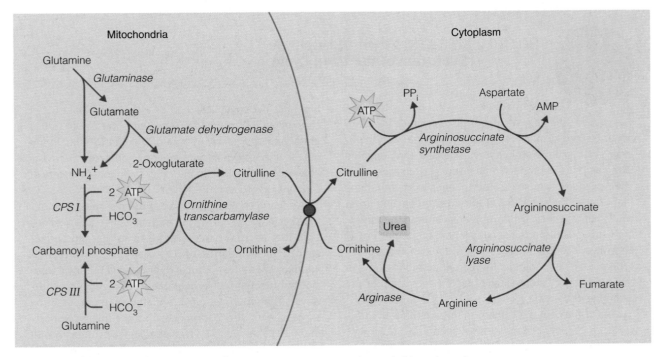

Figure 10.18 Ornithine-urea cycle Amino nitrogen in the form of either glutamine or NH_4^+ is used to produce carbamoyl phosphate, which enters the ornithine-urea cycle.

brane: the ornithine/citrulline transporter and the aspartate/glutamate transporter. The division of the pathway between the cytoplasm and mitochondria allows greater control over the fate of metabolites. There is also evidence that metabolites are channeled from one enzyme to the next to avoid the loss of metabolites to other pathways. Arginine, for example, is used in multiple other pathways and must be constrained within the ornithine-urea cycle for efficient urea production.

Urea is made in the liver and released into the blood, where its fate depends on the species. In mammals, urea is collected by the kidney and excreted in the urine. In other animals, urea may be excreted via other routes, such as the fish gill. Urea is carried across the plasma membrane by facilitated diffusion on specific urea transporters. These transporters govern how fast urea crosses membranes in different cell types and regions of the kidney. Active transporters for urea may exist in tissues such as shark gills, although much less is known about such carriers.

The rate of urea production is matched to the rate of protein metabolism, which is high in animals that (1) eat protein-rich diets or (2) degrade body protein during starvation. The rate of urea synthesis is regulated by enzyme quantity and allosteric regulation. First, animals use hormones to regulate the rate of expression of genes that encode the enzymes

of the ornithine-urea cycle. Glucagon and glucocorticoids stimulate the expression of ornithine-urea cycle enzymes, whereas insulin inhibits expression of these genes. Second, animals regulate CPS activity through the allosteric regulator *N*-acetyl glutamate. When amino acid levels are high, an elevation in glutamate levels increases the activity of the enzyme *N*-acetyl glutamate synthetase.

Each nitrogenous waste strategy has inherent costs

Excretory strategies have been an important element of many ecophysiological studies. The costs and benefits are clear and can readily be understood in terms of the environmental and ecological constraints on the animal: available water, dietary strategies, and metabolic cost.

Each excretory strategy has implications for water balance, but not all animals within a taxonomic group have the same needs for water conservation. For example, all birds share the reptilian trait of uricotelism, which offers the greatest benefits in terms of water conservation. However, not all birds live in a water-poor environment. Hummingbirds, for instance, eat a diet that is very low in protein and high in water content. They produce very little nitrogenous waste during digestion. Thus, hummingbirds and other nectar feeders can expend

BOX 10.2 GENETICS AND GENOMICS
Evolution of the Urea Cycle

All animals possess the genes that encode the enzymes required to produce urea, but only selected groups produce a functional ornithine-urea cycle. The evolutionary processes that enabled these animals to produce high levels of urea must have affected multiple genes. Some changes were mutations that altered the structure of enzymes, imparting novel catalytic properties. Other mutations altered the targeting sequence of proteins, causing a distinct pattern of enzyme subcellular localization. By looking at the properties of existing animals, we can understand some of the evolutionary steps needed for the formation of the ornithine-urea cycle.

One of the central enzymes in the ornithine-urea cycle is carbamoyl phosphate synthetase (CPS). All animals have CPS II, but this enzyme is not useful in urea synthesis for two reasons. First, it has a very low affinity for NH_4^+, relying on glutamine as a nitrogen donor. Second, its catalytic activity, though adequate to meet the demands of nucleotide degradation, is insufficient to match the needs of NH_4^+ detoxification arising from amino acid oxidation. Two distinct solutions to the first problem have arisen in animals in the form of two additional CPS isoforms with more useful kinetics and sensitivity to the allosteric regulator N-acetyl glutamate. One isoform, CPS I, is found in mammals and some lungfish. It has a much greater affinity for NH_4^+ than does CPS II. The other isoform, CPS III, is found in chondrichthians and ureogenic bony fish. It uses glutamine as a substrate (like CPS II) but has an activity that is much more sensitive to the allosteric regulator N-acetyl glutamate. Interestingly, some species, including the coelacanth and some lungfish, possess both CPS III and CPS I. Although CPS II is likely the ancestral form of the enzyme, it is not yet clear which of the derived forms came first, CPS I or CPS III. Such phylogenetic analyses have relied on kinetic properties to help define evolutionary relationships. However, major differences in kinetic properties can arise from subtle genetic differences. Mutational analyses have shown that a single genetic mutation near the catalytic site of an NH_4^+-dependent CPS I can endow the mutated enzyme with the ability to use glutamine (a property of both CPS II and CPS III). In the next few years, when more genomic analyses are available from primitive fish, the exact evolutionary history of CPS genes will become clearer.

A second challenge in the evolution of the ornithine-urea cycle was securing the mutations necessary to ensure that each enzyme is targeted to the appropriate subcellular compartment. The urea cycle of terrestrial animals spans two compartments, which allows specialized functions to be isolated from competing pathways. The enzymes of the pathway must also be expressed in the appropriate intracellular compartment. For example, CPS II is found in the cytoplasm with the other enzymes of nucleotide metabolism. Conversely, both CPS I and CPS III are mitochondrial enzymes, allowing the cell to segregate nucleotide metabolism from urea synthesis. With the mutations that created the allosteric site on CPS, the cell must have a way of ensuring the regulator (N-acetyl glutamate) could gain access to CPS. In nonureogenic organisms, N-acetyl glutamate is an intermediate in arginine biosynthesis, a pathway that is largely cytoplasmic. Thus, the gene encoding the N-acetyl glutamate synthetase also needed mutations that changed its targeting information. In ureogenic animals, N-acetyl glutamate synthetase is located in the mitochondria, where it produces the most important allosteric regulator of mitochondrial CPS.

References

○ Mommsen, T. P., and P. J. Walsh. 1989. Evolution of urea synthesis in vertebrates: The piscine connection. *Science* 243: 72–75.

○ Saeed-Kothe, A., and S. G. Powers-Lee. 2003. Gain of glutaminase function in mutants of the ammonia-specific frog carbamoyl phosphate synthase. *Journal of Biological Chemistry* 278: 26722–26726.

○ Walsh, P. J. 1997. Evolution and regulation of urea synthesis and ureotely in (batrachoidid) fishes. *Annual Reviews in Physiology* 59: 299–323.

the water necessary to make greater use of ammonia excretion. However, even these birds are not truly ammoniotelic—most of their nitrogenous waste is still excreted in the form of uric acid.

The other cost of excretory strategy is the metabolic cost of producing the excretory product itself. The cheapest nitrogenous waste is ammonia, since it does not need to be further metabolized after protein metabolism. Both urea and uric acid have metabolic costs associated with their synthesis, and we can estimate these costs by comparing Figures 10.17 and 10.18.

The costs of urea synthesis depend on the nitrogen source used to make carbamoyl phosphate.

If glutamine is the nitrogen donor, then 1 mol of urea costs 4 mol of ATP (1 mol of ATP to make glutamine, 2 mol of ATP for carbamoyl phosphate synthesis, and 1 mol of ATP for argininosuccinate synthesis). The pyrophosphate produced in argininosuccinate synthesis is normally hydrolyzed, wasting an additional high-energy phosphate. Thus, the costs of urea synthesis are most often estimated as 5 mol of ATP per mol of urea. In comparison to urea, uric acid costs more to produce (7 ATP) but it also has more nitrogen (4N). Thus, uric acid (7 ATP; 1.75 ATP/N) is slightly more economical than urea (5 ATP; 2.5 ATP/N). However, uric acid pellets include numerous proteins; since these proteins are lost in the excreta, they represent an indirect cost of uricotelism.

The mode of nitrogen excretion can change with development or environmental conditions

The reasons for the occurrence of ureotelism in species other than mammals are not always clear, although it usually coincides with an atypical environmental situation or life history strategy. Let's consider some examples.

Urea production by most teleost fish is normally quite low and an insignificant contribution to nitrogen excretion. Some fish species live most of their life as ureoteles. For example, the Lake Magadi tilapia lives in water with a pH so high that the gill cannot excrete NH_3. In most fish, NH_3 diffusion across the gills is accelerated when external protons ionize NH_3 to form NH_4^+. At a high external pH, well above the pK_a value for NH_3, this reaction is very slow, reducing the rate of NH_3 diffusion. These fish have an active ornithine-urea cycle in the liver, but surprisingly, the muscle also plays an important role in urea synthesis in these fish. Urea is excreted across the gills as the primary form of nitrogenous waste.

Other fish species may adopt a ureotelic strategy, depending on external conditions. Lungfish are normally ammonioteles, excreting ammonia into the surrounding water. However, when water levels decrease, the African lungfish (*Protopterus*) burrows into the mud and forms a mucus cocoon. Since the animal cannot excrete ammonia, other pathways must be used for nitrogen excretion. Once exposed to the air, the lungfish rapidly induces the expression of urea cycle enzymes and glutamine synthase, and converts to ureotelism.

The gulf toadfish, *Opsanus beta*, also can convert to ureotelism, typically when it moves to crowded conditions. The urea is stored in the blood and released once or twice daily in short pulses across the gill following the insertion of urea transporters into the gill epithelia. It remains unclear why ureotelism is advantageous in this species. It may serve to reduce the risk of local fouling of water where animals are closely associated together. Alternatively, since many animals use ammonia as a cue to detect prey, urea production may confuse predators. Urea production is more common in the early developmental stages of many ammoniotelic species, including rainbow trout. It is likely that all fish are capable of synthesizing urea, but the species that produce urea as adults likely do so by retaining this embryonic capacity.

Cartilaginous fish produce urea as an osmolyte

Most species that produce a lot of urea do it mainly to excrete nitrogenous wastes. However, some species produce urea but retain it as an osmolyte. For example, the urea concentration in the plasma of the crab-eating frog (*Rana cancrivora*) increases more than 20-fold when the frog is exposed to high salinity. Since the rates of urea excretion do not change, it is likely that urea is serving an important role as an osmolyte.

Urea is an important osmolyte in cartilaginous fish, where it can account for almost half of the tissue osmolarity. At the high concentrations seen in shark blood, urea could disrupt macromolecular structures. However, its effects are counteracted by methylamines, such as TMAO, betaine, and sarcosine, which are also accumulated at high concentrations. By relying on counteracting solutes, sharks can maintain the concentration of inorganic ions (perturbing solutes) at low levels. Although most cartilaginous fish are stenohaline, several species can tolerate some degree of diluted seawater. When a euryhaline shark moves from seawater to dilute seawater, it excretes urea as well as some ions. More than 40 species of elasmobranchs can survive in freshwater. Species such as bull sharks may travel from the sea into freshwater lakes, such as Lake Nicaragua, surviving for years before returning to the sea to breed. In freshwater, these sharks lose some of their osmolytes—about 50% of urea and 20% of Na^+ and Cl^-—yet maintain an osmolarity well above that of other freshwater fish. The

Amazonian stingray remains in freshwater all its life, maintaining an osmolarity near that of teleost fish by excreting urea as it is produced.

◎ | CONCEPT CHECK

7. What are the costs and benefits of using ammonia, urea and uric acid as nitrogenous wastes?
8. What substrates are required to produce a molecule of urea?

The Kidney

Most animals maintain ion and water balance using some form of internal organ derived during the development of the embryonic digestive system. Multiple types of cells combine to produce a tube-like structure, or tubule, through which excretory solutions pass from the animal to the external environment. Animals differ in the way the tubule fluid is produced and how it is modified prior to excretion. In some animals, a few simple tubules are sufficient to produce the excretory products. More complex animals, such as vertebrates, combine tubules to form the kidney, which has six roles in homeostasis.

1. *Ion balance.* Sodium levels are an important determinant of extracellular fluid osmolarity. Animals exhibit fluid imbalances if blood [Na^+] is too high (*hypernatremia*) or too low (*hyponatremia*). Potassium balance is important because changes in [K^+] can alter resting membrane potential, which affects the function of excitable tissues such as muscles and neurons. If blood [K^+] is too high (*hyper-kalemia*), excitable tissues can undergo spontaneous depolarization, causing cardiac arrhythmias and muscle twitches. Low [K^+], or *hypokalemia*, can cause muscle weakness. The kidney also controls the loss of ions that have important roles as micronutrients, including Ca^{2+}, iron, and trace metals.

2. *Osmotic balance.* The kidneys determine the volume of urine produced, and thereby control water balance. Dehydration results from inadequate consumption of water, or consumption of chemicals known as diuretics, which increase water loss in the urine. Conversely, in-

adequate water excretion can result in high blood pressure and edema.

3. *Blood pressure.* By controlling blood volume, the kidney acts over the long term to regulate blood pressure. It acts in concert with shorter-term cardiovascular effectors, such as cardiac contractile properties and peripheral resistance of the vasculature. The volume of the extracellular fluid is under the control of the kidney, through hormones and nerves that integrate cardiovascular conditions with the output of the central cardiovascular control center. Low blood pressure (*hypotension*) compromises the delivery of fuels to tissues with high energy demands, such as the brain and locomotor muscle. High blood pressure (*hypertension*) can compromise the integrity of the microvasculature in vital tissues, putting the animal at risk for a myocardial infarction, stroke, or embolism. Many antihypertensive agents are diuretics, enhancing the production of urine to reduce blood volume.

4. *pH balance.* The kidney augments the respiratory system in the control of the pH of body fluids. The kidney regulates the pH of the extracellular fluid by retaining or excreting H^+ or HCO_3^-. Many of the metabolic and transport pathways of ammonia metabolism also involve acid or base production. The production of urea leads to the consumption of bicarbonate, which also has consequences for whole body pH regulation.

5. *Excretion.* The kidney plays an important role in the excretion of nitrogenous wastes as well as other water-soluble toxins. Excess water-soluble vitamins, for example, are excreted in the urine.

6. *Hormone production.* The kidney has an important role in the synthesis and release of hormones, such as renin, which controls blood pressure, and erythropoietin, which regulates red blood cell synthesis.

We begin our discussion of the kidney by exploring the structure and function of the mammalian kidney, focusing on its role in the regulation of water and ion balance. Our understanding of animal kidney function benefits from the many studies that examine the role of the kidney in human diseases, such as hypertension. Later in this section we examine the specific properties of kidneys from other species.

Kidney Structure and Function

The typical mammalian kidney (Figure 10.19) is crescent shaped with two layers: an outer cortex and an inner medulla. The medulla is composed of a number of parallel cone-shaped segments called *renal pyramids*. The inner narrow region of each pyramid is called the papilla. Once the urine is formed, it passes into a cavity called the *minor calyx*. Multiple minor calyces drain into the *major calyx*, which in turn empties into the *ureters* that drain the kidney. The ureters empty into the *urinary bladder* where urine is stored. Eventually, the urine is expelled from the bladder through a single urethra, a process with the elegant name *micturition*. Kidneys must process tremendous volumes of blood. Even though kidneys make up less than 1% of the entire body mass, the blood flow through the kidneys is much greater than that to muscles during heavy exercise. In humans, the kidney may process 4 liters of blood per kilogram each minute but exercising muscle receives only about 0.5 l/kg per minute. Numerous hormones and neurotransmitters ensure that urine composition and release are matched to the physiological needs of the animal. These regulatory factors affect the four processes involved in urine formation: filtration, reabsorption, secretion, and excretion.

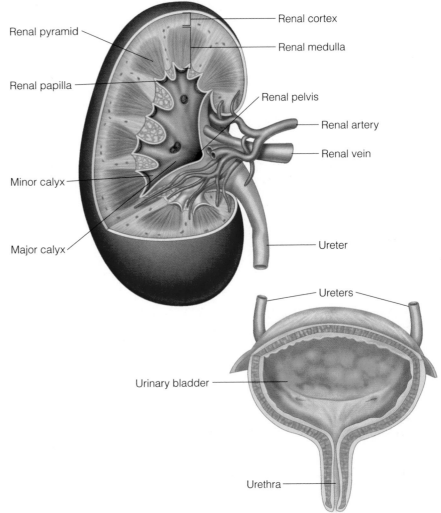

Figure 10.19 Mammalian kidney The kidney is composed of two layers, the cortex and medulla. As urine is produced, it is collected by the minor calyces, which join together to form the major calyx. The urine passes through the ureter into the urinary bladder for storage, eventually leaving the animal through the urethra.

The nephron is the functional unit of the kidney

Kidney function depends on the interplay between the renal epithelium and the cardiovascular system. The functional unit of the kidney is the **nephron**. Each nephron is composed of two elements: the renal tubule and the associated vasculature. A **renal tubule** is a tube constructed from a single layer of epithelial cells, though the nature of the cells differs along the length of the structure. The main element of the nephron vasculature is the **glomerulus**, a twisted ball of capillaries that delivers fluids to the tubule. In a typical kidney, some nephrons exist within the cortex (cortical nephrons), and others span both the cortex and medulla (juxtamedullary nephrons) (Figure 10.20).

The tubule is composed of epithelial cells with characteristic transport properties that allow them to reclaim specific solutes and expel others. **Bowman's capsule** is the mouth of the tubule, a cuplike expansion that surrounds the glomerulus. The fluids that leave the glomerulus enter Bowman's capsule and move down the lumen through successive specialized regions of the tubule: proximal tubule, loop of Henle, and distal tubule. The fluids from multiple tubules then drain into a collecting duct, several of which fuse together to form

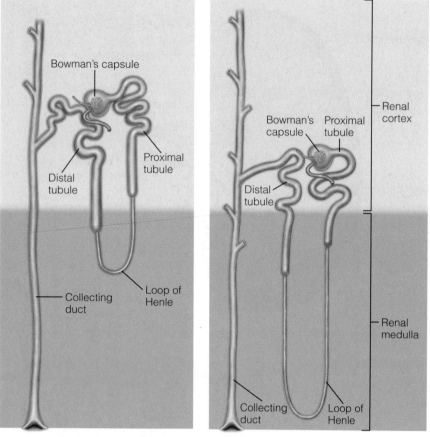

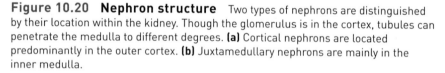

(a) Cortical nephron   **(b)** Juxtamedullary nephron

Figure 10.20 Nephron structure Two types of nephrons are distinguished by their location within the kidney. Though the glomerulus is in the cortex, tubules can penetrate the medulla to different degrees. **(a)** Cortical nephrons are located predominantly in the outer cortex. **(b)** Juxtamedullary nephrons are mainly in the inner medulla.

beds that wrap around the tubules. In juxtamedullary nephrons, the efferent arterioles diverge into the **vasa recta**, long, straight vessels that run along the loop of Henle. A small proportion of the blood entering the kidney is directed deep into the medulla, without passing through a glomerulus. Changes in the blood pressure in these renal vessels alter the renal interstitial hydrostatic pressure, which is in dynamic equilibrium with the hydrostatic pressure of the capillaries surrounding the medullary capillaries. The blood from these capillary beds then drains into the venous system, carrying away recovered solutes and water from the interstitial fluid that surrounds the tubule.[1]

Nephron function depends on the processing of a blood filtrate as it passes through the renal tubule. The primary urine produced by filtration is transformed through reabsorption and secretion to become the final urine, processes that depend upon cellular specializations.

Filtration occurs at the glomerulus

The wall of a glomerular capillary is a complex biological filter that retains the blood cells and large macromolecules but permits liquid components of blood to escape into the lumen of the Bowman's capsule (Figure 10.22). The glomerular capillaries are fenestrated (see Figure 8.11), with pores that allow low-molecular-weight molecules to escape the blood. A specialized type of epithelial cell called a **podocyte** covers the outer surface of the capillary. The podocytes have *foot processes*, which are cytoplasmic extensions that help form the filtration structure. The podocyte attaches to the basement membrane, a filamentous extracellular matrix produced by the capillary cells. The gap between the foot processes, about 14 nm wide, is a filtration slit. The fibrous basement

papillary ducts, which in turn empty into the minor calyx.

The vasculature of the nephron is central to nephron function, delivering the fluids that become the primary urine and governing the nature of the interstitial fluids that surround the tubule (Figure 10.21). Blood enters the kidney from the renal artery, which branches into smaller vessels that give rise to the glomerulus. After the filtered blood leaves the glomerulus, it passes into an efferent arteriole. This arrangement is unlike a conventional capillary bed where the venous system is immediately downstream of the capillaries. The efferent arteriole generates enough smooth muscle contraction to maintain a degree of vasoconstriction, causing a higher degree of resistance than could a venule. The blood passes through the efferent arteriole into one of two types of capillary beds. In cortical nephrons, the efferent arterioles flow into *peritubular capillary*

---

[1]The definition of a nephron differs among researchers. In contrast to the most restrictive definition—the glomerulus and tubule—some researchers include the collecting duct and the vasa recta as part of a nephron.

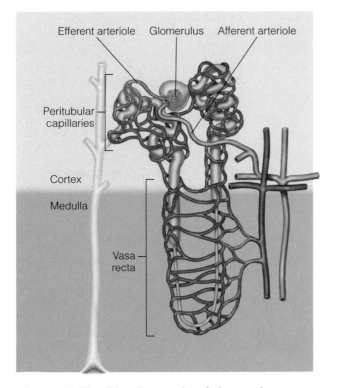

Figure 10.21 Blood vessels of the nephron.
Blood delivered to the kidney by the renal artery passes
through smaller arteries and reaches an afferent arteriole
that services one nephron. The arteriole diverges into the
glomerulus, a network of capillaries within the Bowman's
capsule. After leaving the glomerulus, blood enters an
efferent arteriole. The efferent arterioles that drain cortical
nephrons empty into peritubular capillaries. The efferent
arterioles that drain juxtaglomerular nephrons flow into the
vasa recta.

membrane spans the filtration slits to act as the biological filter of the glomerulus, excluding blood
cells and large proteins, and passing water, ions,
and low-molecular-weight molecules.

The **mesangial cells**, similar to smooth muscle cells, wrap around the capillaries of the
glomerulus. Contraction of the mesangial cells restricts blood flow to specific vessels within the capillary network, regulating blood pressure within
the glomerulus to control filtration.

The primary urine is modified by reabsorption and secretion

As fluid passes into the lumen of the tubule, a filtrate is formed. The filtrate, or *primary urine,* is essentially isosmotic to blood (about 300 mOsM). As
the fluid passes through the tubule, about 99% of
the volume is recovered. For example, an average-
sized human produces about 7.5 liters of primary
urine each hour, but generates only about 75 ml of

final urine. The remodeling of the primary urine occurs as it passes through successive regions of the
tubule, each with specialized transport capacities.
Recall that the tubule wall is composed of a single
layer of epithelial cells. Like most epithelial cells,
the apical membranes (facing the lumen) and basolateral membranes (facing the interstitium) have
specialized profiles of transporters. Also, the cells of
the epithelium may be interconnected in ways that
form a tight epithelium or a leaky epithelium. Before discussing the roles of each region of the
tubule, let's first consider the general features of tubular transport.

Recovery of substances from the lumen of the
tubule requires a combination of favorable electrochemical gradients and transport capacities. Some
substances in the primary urine are reclaimed by
transepithelial transport, moving from the lumen,
across the single layer of epithelial cells, into the
interstitial fluid (*peritubular fluid)* and ultimately
back into the blood. Some hydrophobic solutes
cross the tubular epithelium by passive transport;
as water is removed from the primary urine, concentration gradients are created that can drive
hydrophobic solutes back to the blood. Larger molecules in the filtrate, such as small proteins, can be
recovered by transcytosis: endocytosis into the epithelial cell and exocytosis into the interstitial fluid.
Most molecules, however, are reabsorbed through
a combination of facilitated diffusion and active
transport, both primary and secondary.

Consider how cells reabsorb glucose against its
concentration gradient, driven by a larger, more favorable Na^+ electrochemical gradient (Figure
10.23). The concentrations of Na^+ and glucose in
the primary urine are not different from that of the
blood, so the challenge is how to recover these
solutes in the absence of favorable concentration
gradients. The major driving force underlying the
transport is the Na^+/K^+ ATPase found in the basolateral membrane. By pumping Na^+ out of the cell
into the interstitial fluid, the nephron cells create a
favorable inward Na^+ electrochemical gradient on
the apical side that can be used to drive both Na^+
uptake and Na^+-coupled glucose uptake. Na^+ can
cross into the tubule cells by a Na^+ channel,
Na^+/H^+ exchanger, or by suites of other carriers
that couple the import of organic molecules and
Na^+. One such transporter is the Na^+-glucose cotransporter, which allows the cell to import glucose
from the lumen. Concentrating glucose inside the
cell creates a favorable outward chemical gradient

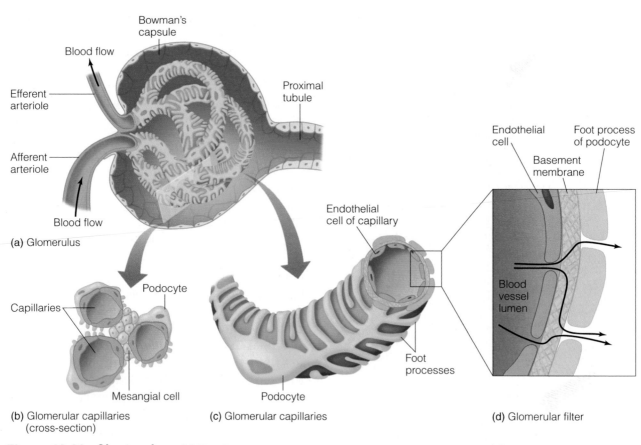

(a) Glomerulus

(b) Glomerular capillaries
(cross-section)

(c) Glomerular capillaries

(d) Glomerular filter

Figure 10.22 Glomerulus (a) The glomerulus is a network of capillaries that empty much of the fluid from the blood into the Bowman's capsule of the nephron. **(b)** Mesangial cells between the capillaries help control blood flow through the glomerulus. The individual capillaries are composed of loosely connected endothelial cells and are covered on the external surface by podocytes. **(c)** The podocytes issue several foot processes that form filtration slits. **(d)** The podocytes interact with the basement membrane to create a filter that retains blood cells and large proteins in the plasma while permitting the passage of fluids through filtration slits.

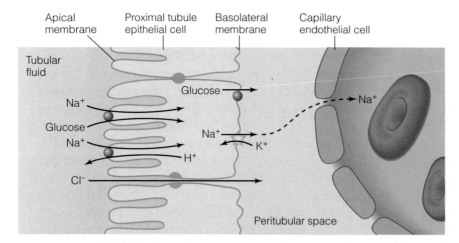

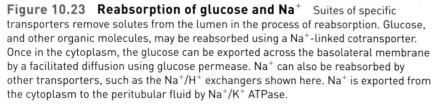

Figure 10.23 Reabsorption of glucose and Na⁺ Suites of specific transporters remove solutes from the lumen in the process of reabsorption. Glucose, and other organic molecules, may be reabsorbed using a Na⁺-linked cotransporter. Once in the cytoplasm, the glucose can be exported across the basolateral membrane by a facilitated diffusion using glucose permease. Na⁺ can also be reabsorbed by other transporters, such as the Na⁺/H⁺ exchangers shown here. Na⁺ is exported from the cytoplasm to the peritubular fluid by Na⁺/K⁺ ATPase.

for glucose; glucose permease allows glucose to cross into the peritubular interstitial fluid via facilitated diffusion. Each of these transport processes requires energy, either in the form of ATP used by the primary active transporters (for example, Na⁺/K⁺ ATPase), or in the form of electrochemical gradients used by secondary active transporters (for example, the Na⁺-glucose cotransporter).

The ability to reabsorb solutes such as glucose is limited by transport capacity. Like many active transporters, the kinetics of the transport machinery can become saturated at high substrate levels (Figure 10.24). This capacity for solute recovery is known as the

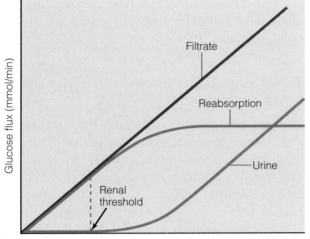

Figure 10.24 Renal threshold Reabsorption depends on the activity of specific transporters that have a finite maximal capacity. Solutes move from the plasma to the tubule fluid when the blood passes through the glomerulus. If a solute is present at a low concentration in the plasma (and hence the tubule fluid), all of the solute can be recovered during reabsorption. As the concentration of solute increases in the plasma, it becomes more difficult to recover all of the solute from the tubule fluid. When the plasma concentration is so high that the tubule cannot reabsorb all of the solute, some appears in the urine. At still higher concentrations, the solute concentration in the urine increases dramatically.

renal threshold. If the amount of substance to be recovered is in excess of the capacity of the transport machinery, some of the substance will escape in the urine. In type 1 diabetes, the levels of glucose can be very high in the blood. When the blood is filtered by the glomerulus, the primary urine also has a very high glucose concentration. Despite the active glucose transporters, the kidney cannot reabsorb all of the glucose, and some is lost in the urine (*glucosuria*).

The other way the primary urine is modified is through secretion. Secretion is similar to reabsorption in that it uses transporters found in the cells that line the lumen. However, the process works in the opposite direction, transferring solutes from the blood, through the peritubular fluid, and across the cells into the tubule lumen. The most important secretory products are K^+, NH_4^+, and H^+. Many water-soluble waste products are also secreted into the tubule, including pharmaceuticals and water-soluble vitamins. Like other active transport processes, secretion depends on transport proteins and requires energy.

The various regions of the tubule mediate different transport processes, as summarized in Figure 10.25.

Cellular properties differ among regions of the tubule

The transformation of the primary urine to the final urine involves a series of specialized regions of the tubule that depend upon cellular specializations. Though the tubule wall is a single layer of epithelial cells connected together by tight junctions, cell morphology and function differ considerably among regions of the tubule.

The proximal tubule can be a simple, straight tube or take a path with many convolutions; for this reason it is sometimes called the *proximal convoluted tubule*. The cells of the proximal tubule are tall cuboidal epithelial cells, with abundant mitochondria and microvilli. As with other epithelial tissues, these features are common in cells that carry out energy-dependent solute transport processes.

The proximal tubule then gives way to the loop of Henle. There is considerable variation in the nature of the loop of Henle among species, and even among nephrons of a single animal. In general,

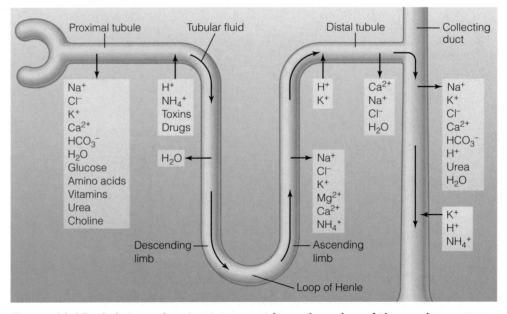

Figure 10.25 Solute and water transport in each region of the nephron Each region of the nephron has specific transporters that can reabsorb or secrete molecules.

the loop of Henle is divided into a descending limb, a loop, and an ascending limb. The first part of the descending limb of the loop of Henle is composed of cuboidal epithelial cells, much like the proximal tubule. These are gradually replaced with the flatter squamous epithelial cells. The difference in the height of the cuboidal and squamous cells creates a difference in width of the wall, and these regions of the tubule are often distinguished as *thick descending limb* and *thin descending limb*. Further along the tubule, the ascending limb of the loop of Henle becomes thicker as cuboidal epithelial cells predominate. As with the descending limb, the ascending limb may be subdivided as *thin ascending limb* and *thick ascending limb*. These distinctions are made because the differences in cell shape coincide with distinctions in transport properties.

Following the loop of Henle is the distal tubule, which can be simple and straight or long and convoluted. In contrast to the proximal tubule, most of the epithelial cells of the distal tubule have simple membranes with few microvilli. This type of cell, known as a principal cell, also dominates the cell profile of the collecting duct. The less common intercalated cells are cuboidal epithelial cells with abundant microvilli. Not surprisingly, the functions of principal cells and intercalated cells differ as much as their structures.

The differences in cell type and morphology along the tubule and collecting duct are summarized in Figure 10.26.

The proximal tubule reabsorbs salts and organic metabolites

The proximal tubule is specialized for transport, and it is the region where most solute and water reabsorption occurs (Figure 10.27). Many solutes are transported from the lumen into proximal tubule epithelial cells via Na^+ cotransporters including organic molecules (glucose, lactate, amino acids, water-soluble vitamins) and inorganic ions (phosphate). These processes are also important in contributing to reabsorption of Na^+. The organic molecules can escape the cell into the interstitial fluid via facilitated diffusion, whereas Na^+ is pumped across the basolateral membrane via the Na^+/K^+ ATPase. The transepithelial electrochemical gradient also drives the paracellular transport of Cl^- from the lumen to the interstitial fluid. As a result of the net movement of ions and solutes from the lumen to the peritubular interstitial fluid, a decrease in osmolarity creates a favorable osmotic gradient for the movement of water. Transepithelial water movement occurs mainly through transcellular transport, mediated by aquaporins, although some paracellular move-

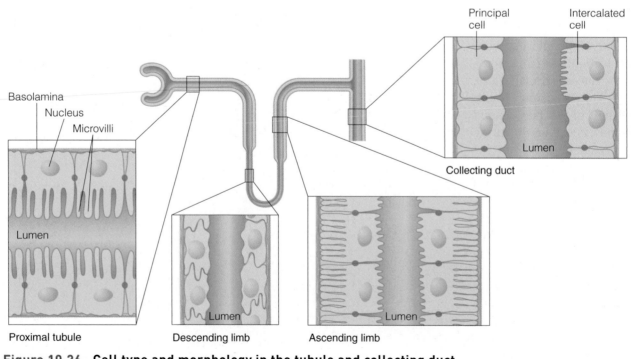

Figure 10.26 Cell type and morphology in the tubule and collecting duct
The wall of the tubule is composed of a single layer of epithelial cells that differ in morphology.

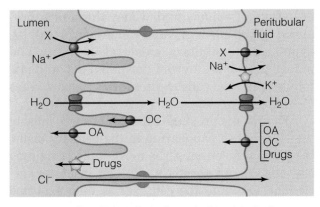

Figure 10.27 Transport in proximal tubule cells OA: organic anion. OC: organic cation.

ment of water may also occur. Transgenic mice lacking the gene for aquaporin-1 have a diminished ability to recover tubular fluids and create hyposmotic urine.

Through these interdependent transport processes, the proximal tubule is able to reabsorb almost all organic solutes, most of the phosphate, and 60–75% of the Na^+, Cl^-, and water that appear in the primary urine.

The proximal tubule is also the site of secretion of organic anions, organic cations, and water-soluble toxins, including pharmaceutical agents. These molecules are imported into the proximal tubule cells through suites of transporters in the basolateral membrane, then exported into the lumen through apical transporters: organic cation transporters (OCT), ATP-dependent multidrug resistance transporters (MDR and MRP), and organic anion transporting polypeptide transporters (OATP).

The loop of Henle mediates sequential uptake of water, then salt

When the primary urine has passed through the proximal tubule, the volume has diminished and most of the valuable solutes have been recovered. The remainder of the tubule is responsible for recovering the balance of the solutes and water, dependent on the physiological state of the animal. The next region en-

countered by the primary urine is the descending limb of the loop of Henle. This region of the tubule is specialized to transport water, but it is not a major site of transport for solutes. As with the proximal tubule, aquaporins allow water to move across epithelial cells in relation to the osmotic difference from the lumen to the interstitial fluid. Critical to the water recovery strategy is an osmotic gradient that exists within the medulla (Figure 10.28). At the transition between the proximal tubule and descending loop of Henle, the osmolarity of the interstitial fluid is similar to that of the blood—about 300 mOsM. As the descending loop of Henle goes deeper into the medulla, the osmolarity of the interstitial fluid increases, drawing water from the primary urine across the epithelial cells. With loss of water, but not solutes, the osmolarity of the primary urine increases, reaching a maximum at the loop region of the loop of Henle.

Once the tubule turns and moves back toward the cortex, the epithelial cell transport capacity changes. Instead of expressing aquaporin genes, these epithelial cells express solute transporters.

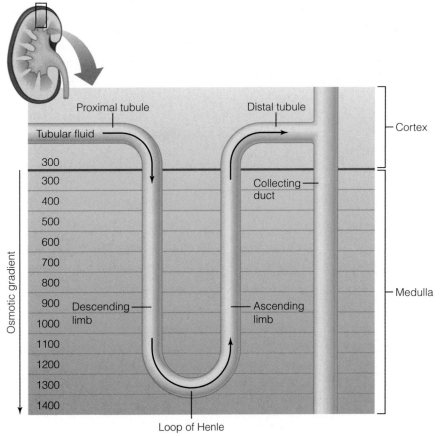

Figure 10.28 Osmotic gradients in the interstitial fluid of the medulla The loop of Henle passes through osmotic gradients in the medulla. The osmolarity is lowest near the border of the cortex, and increases deeper into the kidney.

As the tubule passes through the medulla, the interstitial osmolarity decreases. Since the epithelial cells can only transport solutes, the transepithelial gradients drive movements of solutes from the primary urine to the interstitial fluid. As a result of various transporters in the apical and basolateral membranes, there is a net movement of Na^+ and Cl^- from the primary urine to the interstitial fluid. On the apical membrane, the NKCC transporter mediates uptake of Na^+, K^+, and Cl^- into the cell. The basolateral membrane transports Na^+ and Cl^- into the interstitial fluid: Na^+ via the Na^+/K^+ ATPase, and Cl^- via Cl^- channels and a K^+-Cl^- cotransporter. An apical K^+ channel allows K^+ imported via NKCC to escape back to the lumen.

The transport processes in the ascending limb and descending limb are summarized in Figure 10.29.

The distal tubule mediates K^+ secretion, NaCl reabsorption, and hormone-sensitive water recovery

After fluids leave the loop of Henle, they enter the distal tubule. This region of the tubule is an important site for hormone-mediated regulation of uptake of solutes and water. Hormones produced by the adrenal gland (mineralocorticoids), hypothalamic-pituitary axis (vasopressin), and parathyroid (parathyroid hormone) act on distal tubule epithelial cells to alter levels and activities of transport proteins.

Though the proximal tubule reabsorbs most of the Na^+ and Cl^- appearing in the primary urine, the distal tubule reabsorbs most of the remaining Na^+ and Cl^-. An apical Na^+-Cl^- cotransporter carries the ions into the cell; Na^+ is then exported from the distal tubule epithelial cells via the Na^+/K^+ ATPase, and Cl^- escapes through Cl^- channels.

As with the Na^+ and Cl^-, the proximal tubule is the main site of reabsorption of Ca^{2+} and Mg^{2+}, but the distal tubule is the segment where hormones exert their effects on absorption of the remainder. Ca^{2+} enters the distal tubule epithelial cells through Ca^{2+} channels, and exits into the interstitial fluid via a Na^+/Ca^{2+} exchanger and, to a lesser extent, a Ca^{2+} ATPase.

The distal tubule is an important site for recovery of water, under conditions where water recovery is required. Under normal circumstances, the distal tubule cells have low expression of aquaporin genes. When an animal is dehydrated, hormones such as vasopressin lead to increased expression of aquaporin genes, allowing water recovery from the tubule lumen.

The distal tubule is also the main site of secretion of K^+ into the tubule. As the primary urine moves through the proximal tubule and loop of Henle, about 90% of K^+ is reabsorbed. In the distal tubule, the epithelial cells are able to secrete K^+. It is brought into the epithelial cell from the interstitial fluid via Na^+/K^+ ATPase, and moves into the lumen through either a K^+-Cl^- cotransporter or K^+ channel.

The transport processes in the distal tubule are summarized in Figure 10.30.

The collecting duct regulates ion and water flux

The collecting ducts traverse the layers of the kidney, connecting the distal tubules of cortical and juxtamedullary nephrons. As they move deeper into the kidney, the profile of cells changes, enabling specialized transport functions in different segments. The principal cells secrete K^+ and reabsorb Na^+, similar to those found in the distal

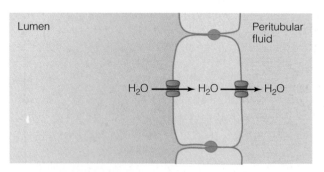

(a) Thin descending limb

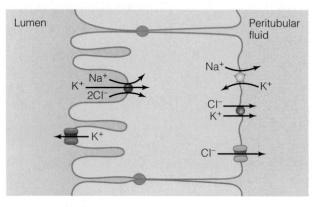

(b) Thick ascending limb

Figure 10.29 Transport in the loop of Henle

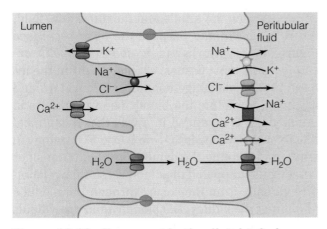

Figure 10.30   Transport in the distal tubule

tubule. The intercalated cells are able to secrete H^+ or HCO_3^-, depending on the acid-base status of the animal. H^+ secretion is coupled to K^+ import through a H^+/K^+ ATPase.

As with the distal tubule, the collecting ducts are important targets of regulatory changes in ion and water movements, including hormone-responsive pathways. Thus, the collecting ducts can be important sites of K^+ secretion or reabsorption, depending on the nature of the primary urine and systemic K^+ homeostasis. When K^+ excretion is needed, secretion by principal cells is stimulated, but when K^+ recovery is needed, reabsorptive pathways in intercalated cells are stimulated.

Nephrons also contribute to acid-base balance

In the previous discussion, we focused on the role of the nephron in ion and water balance, but the kidney also has an important role in regulation of acid-base balance. Transport processes in each segment of the tubule contribute to changes in pH of the primary urine as a way of controlling whole body acid-base balance. Conversely, changes in pH of the primary urine affect the ability of cells to use pH-dependent transporters to recover or secrete ions. The main way that the nephron regulates pH of the urine is through transport and metabolism of H^+, HCO_3^-, and ammonia. For example, metabolic acidosis leads to secretion of H^+ and NH_4^+, and reabsorption of HCO_3^-.

Many transporters affect pH by expelling protons into the lumen. For example, apical Na^+/H^+ exchangers recover Na^+ and extrude H^+, acidifying the urine. These exchangers occur in the proximal tubule, ascending loop of Henle, distal tubule,

and collecting ducts. Some segments also possess proton pumps, such as the H^+ ATPase and H^+/K^+ ATPase of the distal tubule and collecting duct. Overall, proton excretion by the collecting duct plays the greatest role in regulation of acid extrusion by the tubule.

Movements of bicarbonate (HCO_3^-) also affect acid-base balance. Many regions have Cl^-/HCO_3^- exchangers that allow cells to recover Cl^-, while alkalinizing the primary urine. The proximal tubule and ascending limb of the loop of Henle are the main sites of HCO_3^- reabsorption. The tubule epithelial cells that secrete protons derive those protons through the actions of carbonic anhydrase ($CO_2 \rightarrow HCO_3^- + H^+$). Protons are exported across the apical membrane into the lumen, whereas HCO_3^- can be exported into the blood via a basolateral Cl^-/HCO_3^- exchanger. The net effect is acidification of the urine and alkalinization of the interstitial fluid.

Ammonia production, reabsorption, and secretion affect pH balance. The proximal tubule is an important site of ammonia production, largely from glutamine. Metabolism of glutamine occurs via glutaminase (producing glutamate), glutamate dehydrogenase (producing 2-oxoglutarate), and TCA cycle enzymes (producing oxaloacetate). Whether the resulting oxaloacetate is completely oxidized in the TCA cycle or used as a gluconeogenic substrate, each glutamine generates two HCO_3^- and two NH_4^+. These reactions have an influence on acid-base balance because the NH_4^+ (an acid) is excreted and HCO_3^- can titrate H^+. Ammonia produced by the proximal tubule is secreted into the lumen, reabsorbed in the ascending limb of the loop of Henle, and secreted in the collecting duct using a combination of transporters.

Acid-base balance, as with other renal responsibilities, is regulated directly by prevailing conditions in the tubule lumen and interstitial fluid, as well as by hormones that respond to systemic changes and mediate compensatory responses.

The loop of Henle creates a countercurrent multiplier

As we discussed the transport processes in the various segments of the tubule, we alluded to the existence of an osmotic gradient through the medulla: low osmolarity near the cortex and high osmolarity deep into the medulla (see Figure 10.28). This osmotic gradient is produced and maintained by the concerted actions and arrangement of the loop of Henle, the collecting duct, and the vasa recta.

Let's first consider how the descending and ascending flows through the loop of Henle create a *countercurrent multiplier* that, in combination with differences in tubule permeability, allows the urine to be remodeled in terms of both solute concentration and volume. As fluid moves through the descending limb, the surrounding interstitial fluid has a slightly higher osmolarity. Since only water can cross the descending limb, the osmotic gradient drives movement of water from the lumen to the interstitial fluid. This movement of water would tend to decrease osmolarity within the region, destroying the driving force for water recovery. To counteract this dilution, the region of the ascending limb that is adjacent to the descending limb actively pumps solutes from the lumen to the interstitial fluid, increasing the osmolarity of the interstitial fluid. In fact, the ascending limb pumps more salt than is needed to negate the osmotic movements of water from the descending limb. This *single effect*—excessive salt pumping by the ascending limb—is enhanced by the countercurrent arrangement of the loop of Henle. This mismatch between water and salt movements in adjacent regions of the loop of Henle creates the osmotic gradient within the medulla.

Another factor that contributes to the formation of the osmotic gradient within the medulla is the permeability of the collecting duct to urea. Urea enters the tubule through the glomerulus, but travels through the tubule with little reabsorption because of the low urea permeability of the tubule. With much of the water removed from the original filtrate, the urea concentrations increase dramatically. The concentrated urea solution leaves the tubule and enters the collecting ducts. The cortical regions of the collecting duct have low permeability to urea, but as the collecting duct moves deeper into the medulla, the permeability to urea increases due to the presence of specific urea transporters. Movement of urea into the interstitium increases the local osmolarity, further contributing to the osmotic gradient within the medulla.

The vasa recta maintains the medullary osmotic gradient via a countercurrent exchanger

The osmotic gradient may *arise* along the depth of the medulla through the movement of salts and water between the tubule and the interstitial fluid, but it is *maintained* because the *vasa recta* (see Figure 10.21) works as a *countercurrent exchanger*.[2] In most tissues, capillaries drain the interstitium, collecting solutes and water and emptying them into the blood. The vasculature in the medulla is arranged in a way that it can meet the circulatory needs (O_2 delivery, CO_2 removal) without disrupting the osmotic gradient. Consider what would happen if the blood vessels flowed unidirectionally from the cortex through the medulla and out of the kidney; the blood would draw fluids and solutes out of the kidney, rapidly dissipating the gradient created within the medulla. Instead, the vessels of the vasa recta carry blood into the medulla, then back out of the medulla. As blood leaves the efferent arteriole and enters the vasa recta, it is carried into the medulla where the higher osmolarity causes it to passively pick up solutes and lose water. As the vessels head back toward the cortex, the decreasing osmolarity causes the blood to lose solutes and gain water. The blood vessels exit the kidney at the junction between cortex and medulla, where interstitial fluid is isosmotic to blood. Thus, the countercurrent arrangement of the vasa recta ensures that the osmotic gradient within the medulla is maintained.

Micturition is regulated by reflex and higher pathways

Excretion is the end product of the renal processing of the primary urine. For any specific substance, the amount excreted is equal to the amount in the primary filtrate plus the amount secreted minus the amount reabsorbed. After the urine leaves the kidney, it enters the urinary bladder for storage. The bladder is a hollow sac, with a capacity of approximately 500 ml in humans. From the bladder, the urine exits through the urethra. Sphincters of smooth muscle control the flow of urine from the bladder to the urethra. The opening and closing of the sphincter is governed by a spinal cord reflex arc, and can be influenced by voluntary and involuntary controls. The bladder wall contains stretch receptors that are activated once the bladder wall has been stretched. This triggers a

---

[2]Most commonly, countercurrent exchangers and countercurrent multipliers are distinguished by the need for direct energy investment. That is, exchangers passively transfer molecules (or heat) whereas multipliers require some form of active transport.

signal via sensory neurons to the spinal cord, which in turn stimulates parasympathetic neurons to trigger the contraction of smooth muscle in the walls of the bladder. This increases pressure on bladder contents. Simultaneously, there is an inhibition of the somatic motor neurons that induce the sphincters to close. When the balance of stimulatory and inhibitory controls exceeds a threshold, the sphincter opens and urine is released.

⊙ | CONCEPT CHECK

9. What is the role of the glomerulus in the nephron?
10. Discuss movements of NaCl and water in the segments of the renal tubule.

Regulation of Renal Function

Endocrine hormones have a central role in regulating osmotic and ion balance in mammals, acting on both the cardiovascular system and the nephron itself to alter the nature of the urine. The steroid hormones that affect ion and water balance (mineralocorticoids) act over hours to alter transporter levels in the tubule. The peptide hormones released from the hypothalamic-pituitary axis act much more rapidly. Superimposed on the natural, hormonal controls are dietary factors that affect urine properties; *diuretics* stimulate the excretion of water, and *antidiuretics* reduce the excretion of water. Often, these dietary factors induce maladaptive changes—dehydration or water retention—that must be overcome by intrinsic negative feedback pathways.

Glomerular filtration pressure is affected by hydrostatic pressure and oncotic pressure

Regulation of kidney function begins with the filtration process at the glomerulus, which is affected by both the surface area available for filtration and the pressure gradients across the filter. The **glomerular filtration rate (GFR)** is the amount of filtrate produced per minute (see Box 10.3, Methods and Model Systems: Measuring Glomerular Filtration Rate). The GFR is controlled primarily by factors that affect the net glomerular filtration pressure—the balance of forces acting on the fluids on either side of the filter.

Three main forces determine net glomerular filtration pressure: glomerular capillary hydrostatic pressure, Bowman's capsule hydrostatic pressure, and the net oncotic pressure (Figure 10.31). Let's begin the discussion of these forces by considering an analogy about the nature of the hydrostatic pressure in the capillaries of the glomerulus. Imagine a garden hose connected to a standard tap. Midway through the hose, a section of rubber has been perforated with thousands of small holes that act as a filter, allowing water to leak out of the hose. In this analogy, the tap is the heart, the first section of hose is the renal artery that supplies the kidney, the perforated region is the glomerulus, and the last section of hose is the efferent artery. Now consider what happens to the rate of flow through the leaky region of the hose (the GFR) when we alter the system. When the tap is partially closed (cardiac output is lower), there will be a decrease in the volume of water leaking from the system. Similarly, if you constrict the hose upstream of the leaky region (systemic blood pressure is lower), there will be a decrease in hydrostatic pressure downstream of the constriction and less water will leak through the filter. Conversely, if you constrict the hose downstream of the leaky region (vasoconstriction), the increase in hydrostatic pressure upstream will force more water through the filter. The process of filtration therefore depends on the cardiac output, the systemic blood pressure, and the vasoconstriction of the efferent arteries. About 20% of the liquid in the afferent arteries is forced through the filter at the glomerulus because the efferent artery is normally quite vasoconstricted.

In the previous analogy, the water leaked out of the perforated region of the hose into the air and there was no pressure to oppose the free movement of the water out of the holes in the hose. In the tubule, the glomerular filtrate doesn't empty into air, but into the interstitial fluid that bathes the tubule. This interstitial fluid has its own hydrostatic pressure. In a typical mammalian kidney, the hydrostatic pressure within the glomerular interstitial fluid is about 15 mm Hg. Since the glomerular capillary hydrostatic pressure is about 60 mm Hg, there is a hydrostatic pressure gradient of about 45 mm Hg that drives fluid through the filter.

Proteins that are retained on one side of a semipermeable barrier impart osmotic effects. The osmotic pressure that arises because of the

The porosity of the glomerular biological filter ensures that virtually all of the bloodborne, low-molecular-weight molecules (<5000 MW) that enter the glomerulus will pass as filtrate into the lumen of the tubule. Many of these molecules are reabsorbed from the lumen, and many others are secreted into the lumen by tubule cells. Central to the understanding of kidney function is an estimate of the volume of fluid that enters into the glomerulus per unit time—the glomerular filtration rate, or GFR.

Let's begin our discussion of GFR calculations by considering the fate of a molecule that is neither absorbed nor secreted. One such molecule is inulin, a low-molecular-weight carbohydrate that is not normally found in the blood. If a physiologist injects inulin into the blood, it will quickly distribute evenly throughout the entire pool of blood. When the blood passes through the glomerulus of the kidney, the inulin will pass through the filter with other plasma constituents and enter the lumen. In the luminal fluids, the inulin concentration will initially be equal to that of the blood. Since it is neither reabsorbed nor secreted, the amount of inulin in the urine over a given experimental period of time equals the amount of inulin that was filtered over that same period of time. The inulin concentration in the urine will change as water flows in or leaves the tubule; it's the total number of inulin molecules arriving in the primary urine in a given time that is important to the GFR calculation. The amount of inulin in the urine (expressed per unit of time) is calculated as the volume of urine per minute (V, ml/min) times the inulin concentration in the urine (U, mol/ml).

$$\text{Amount of inulin in the urine per min} = V \times U$$

The other aspect of the inulin balance sheet is the amount of inulin removed from the plasma. This depends on the inulin concentration in the plasma (P, mol/ml) and the glomerular filtration rate (GFR, ml/min)

$$\text{Amount of inulin removed from the blood} = \text{GFR} \times P$$

Without reabsorption, all of the inulin filtered by the glomerulus remains in the tubule, and without secretion to augment the inulin in the lumen, the amount of inulin in the urine (VU) equals the amount of inulin removed from the blood (GFR $\times$ P):

$$VU = \text{GFR} \times P$$

Rearranging the equation to solve for GFR, we find that

$$\text{GFR} = VU/P$$

GFR can therefore be measured as the product of the volume of urine and ratio of the concentrations of inulin in the urine and in the plasma (U/P). In the laboratory, it is convenient to use inulin, but the same approach can be applied clinically to assess kidney function. GFR is an empirical parameter that holds for all low-molecular-weight molecules; it is a property of the kidney under specific conditions of blood volume and pressure. The GFR measured in this manner can be used to explore how the kidney handles other types of molecules.

Renal clearance is a parameter that reflects the amount of a solute passing from the plasma to the urine in a given period. For a molecule like inulin, renal clearance and glomerular filtration rate are equivalent. However, for situations where reabsorption and secretion play a role, renal clearance may be greater than or less than GFR. Disparities between renal clearance (C) and GFR offer insight into kidney function. Consider the following examples, based on a kidney where an inulin experiment suggested that GFR equaled 120 ml of plasma filtered per minute.

If a solute is present in the plasma (P) at 0.1 μmol/ml, in the urine (U) at 12 μmol/ml, and a filtrate volume (V) of 1 ml is produced per minute, then

$$C = VU/P = 120 \text{ ml/min}$$

In this case, the renal clearance is equal to the GFR, suggesting that the solute behaves like inulin, with highly efficient filtration and little reabsorption or secretion. This is the sort of result you would expect with creatinine, which appears in the blood as a breakdown product of muscle creatine. In fact, clinicians use creatinine clearance much like a researcher uses inulin experimentally—as an estimate of a patient's GFR. The clinician would ask the patient to collect urine for 24 hours, then measure the average levels of creatinine in the urine and in the plasma over this period.

If the experiment were repeated with another solute, and the clearance calculation resulted in an estimate of 10 ml/min, this would suggest that the clearance of this solute was only 10% that of GFR. This would suggest one of two scenarios. First, it is possible that the solute never made it into the glomerular filtrate; perhaps it was too big (a protein, for instance), or possibly it was bound to a protein (a steroid hormone, for example). A second possibility was that the solute was reabsorbed from the tubule, depleting the concentration in the urine.

If another experiment were performed focusing on another solute, and the clearance calculation resulted in an estimate of 240 ml/min, then we would note that more of the solute appeared in the urine than could be accounted for simply by nonselective filtration. Thus, the excretion of this solute must have been augmented by secretion from the tubule cells.

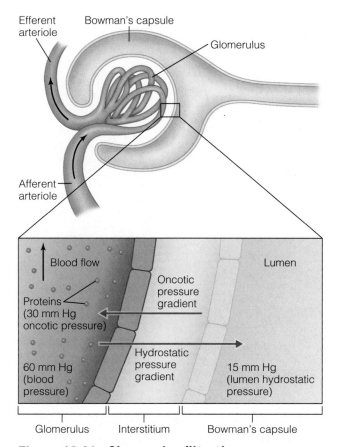

Efferent arteriole

Bowman's capsule

Glomerulus

Afferent arteriole

Blood flow

Lumen

Oncotic pressure gradient

Proteins (30 mm Hg oncotic pressure)

60 mm Hg (blood pressure)

Hydrostatic pressure gradient

15 mm Hg (lumen hydrostatic pressure)

Glomerulus   Interstitium   Bowman's capsule

Figure 10.31 Glomerular filtration pressures
The overall pressure for fluid movements is the difference between inward and outward pressures. The hydrostatic pressure gradient, the difference between the mean blood pressure and the hydrostatic pressure of the lumen, favors movement into the lumen. The oncotic pressure gradient, due to the proteins that remain in the plasma, opposes movement into the lumen.

protein concentration gradient is known as the **oncotic pressure**. Since the filter at the glomerulus allows the movement of fluid and small solutes, but prevents the movement of protein, there is a residual osmotic pressure gradient due to the proteins that remain in the blood. Because the protein concentration is higher in the capillaries than in Bowman's capsule, fluids tend to move back into the capillary. This oncotic pressure is about 30 mm Hg in opposition of filtration. In a typical mammalian kidney, the net glomerular filtration pressure is about 15 mm Hg (60 mm Hg − 15 mm Hg − 30 mm Hg).

Intrinsic and extrinsic regulators control GFR

The kidney maintains function over a range of blood pressures using multiple pathways of control that integrate structural and chemical signals. There are three intrinsic pathways that maintain GFR despite changing blood pressure: myogenic regulation, tubuloglomerular feedback, and mesangial control.

The smooth muscle cells of the vasculature both interpret and respond to changes in blood pressure to maintain GFR—a type of autocrine regulation. An increase in systemic arterial blood pressure would tend to increase GFR if not for a negative feedback loop that reduces blood pressure within the glomerulus to prevent a maladaptive increase in GFR. An increase in blood pressure increases the volume of blood in the small afferent blood vessels, which stretches the smooth muscle cells in the vascular wall. The deformation of the cytoskeleton and plasma membrane activates stretch-sensitive ion channels, which results in depolarization of the plasma membrane of smooth muscle cells. This depolarization stimulates the contractile apparatus, increasing tension and causing vasoconstriction. The resultant decrease in blood flow in the afferent vessels reduces hydrostatic pressure, returning GFR to baseline levels. The same system responds in reverse when arterial blood pressure decreases. Smooth muscle relaxes, vasodilation results, and the increase in blood flow increases the hydrostatic pressure and returns GFR to baseline.

Tubuloglomerular feedback regulates GFR by altering arteriole resistance. Changes in the flow of fluids through the tubule exert negative feedback on the arteriole. In tubuloglomerular feedback, the cells in a region of the distal tubule, called the *macula densa*, contact specialized juxtaglomerular cells in the walls of the afferent arterioles (Figure 10.32). When flow through the distal tubule increases, the tubule cells signal the afferent arteriole, causing vasoconstriction, a decrease in hydrostatic pressure, and a decrease in GFR. It is not yet clear which factors facilitate this paracrine communication.

Myogenic control and tubuloglomerular feedback are two important means of intrinsic control of blood pressure that affect GFR by altering the vasculature (Figure 10.33). Another intrinsic mechanism also acts on GFR, but does so by changing the filtration apparatus of the mesangial cells. When the blood vessels swell in response to an increase in arterial blood pressure, the mesangial cells of the blood vessel also stretch. Since these cells control the dimensions of the filter itself, their

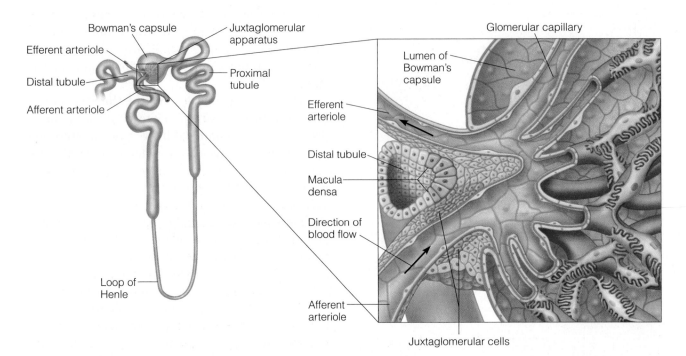

Figure 10.32 Macula densa and the juxtaglomerular apparatus

stretching increases the permeability of the filter, thereby increasing the GFR by a mechanism that is independent of vascular effects. A fourth mechanism by which the kidney uses intrinsic controls to alter blood pressure is *pressure natriuresis*. Recall that a small proportion of the blood vessels entering the kidney penetrate into the medulla without passing through a glomerulus. When the pressure in these renal arteries increases, the renal interstitial hydrostatic pressure increases. This reduces the ability of the tubule to recover solutes and water from the primary urine, contributing to a decrease in blood pressure.

The four intrinsic pathways—myogenic regulation, tubuloglomerular feedback, mesangial control, and pressure natriuresis—act as negative feedback loops over a relatively narrow range of blood pressures. Other regulators are recruited when blood pressure increases or decreases more significantly or for longer periods, such as in dehydration or following blood loss. The main effectors are the baroreceptors, reflexes that are part of the sympathetic nervous system. A large decrease in mean arterial pressure induces these sympathetic neurons to fire, causing vasoconstriction in both afferent and efferent vessels. With the resulting reduction in GFR, and recovery of more Na^+, the body conserves more water and increases blood volume. The decrease in renal circulation also

contributes to an increase in peripheral resistance, increasing mean arterial pressure.[3]

In addition to the sympathetic neuronal control, several endocrine hormones contribute to the control of blood flow and GFR. For example, angiotensin II is a potent vasoconstrictor that decreases GFR, and prostaglandins are potent vasodilators that increase GFR.

Vasopressin alters the permeability of the collecting duct

Vasopressin, also known as antidiuretic hormone or ADH, is the main hormone responsible for recovery of water from the tubule. After this peptide hormone is produced in the cell bodies of hypothalamic neurons, it travels down the neurons to the pituitary gland where it is released into the circulation. At the target tissue, the effects of vasopressin occur within a few minutes.

High vasopressin levels increase the reabsorption of water by the collecting duct. Although water is also reabsorbed in the loop of Henle, the collecting duct is the main site of hormonal regulation of water uptake. Vasopressin alters water

[3]Mean arterial pressure is often abbreviated as MAP, which is an acronym that is also used for microtubule-associated proteins and mitogen-activated protein kinase (MAP kinase).

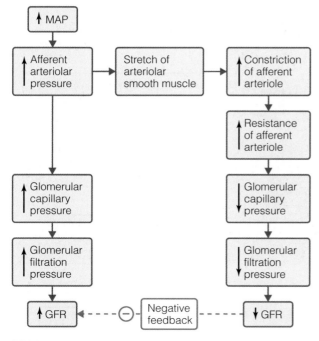

(a) Myogenic regulation

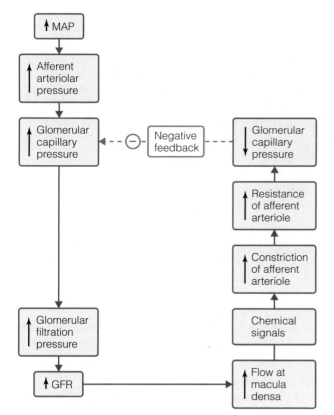

(b) Tubuloglomerular feedback

Figure 10.33 Intrinsic control of GFR An increase in mean arterial pressure (MAP) triggers a change in glomerular filtration pressure. **(a)** Myogenic regulation controls vasoconstriction by triggering contraction of the vascular smooth muscle of the afferent arterioles. **(b)** The tubuloglomerular feedback loop involves signaling factors released from the macula densa that alter smooth muscle contractility. Two other modes of intrinsic control—mesangial control and pressure natriuresis—are described in the text.

uptake by affecting the number of aquaporins in the apical membrane of the principal cells of the collecting duct (Figure 10.34a). When the hormone binds to its G-protein-linked receptor in the plasma membrane, it triggers a signaling pathway that acts via cAMP and protein kinase A to translocate vesicles containing preformed aquaporins to the apical membrane. The process is similar to the way synaptic vesicles are sent to the synaptic membrane of neurons in response to neuronal stimulation (though in synaptic vesicles, the second messenger triggering fusion is Ca^{2+} rather than cAMP). Once vasopressin levels fall, the pathway reverses and aquaporins are removed from the membrane by endocytosis.

Vasopressin is secreted by a complex pathway involving both positive and negative regulators. When plasma osmolarity increases, the osmoreceptors in the hypothalamus respond by stimulating the posterior pituitary to secrete more vasopressin. Other inputs to the pituitary involve stretch receptors in heart atria and baroreceptors in carotid and aortic bodies. When blood pressure increases, these receptors activate, sending action potentials through neurons that terminate in the cardiovascular control center in the medulla oblongata. This center signals hypothalamic neurons that terminate in the pituitary to inhibit the secretion of vasopressin.

Aldosterone regulates sodium and potassium balance

Steroid hormones also play a role in kidney function. In tetrapods, aldosterone is the main mineralocorticoid; however, fish lack aldosterone. Instead, in fish, cortisol probably acts as the main mineralocorticoid, in addition to its role as the major glucocorticoid. The origins of the steroid receptor gene family are actively studied. Recent studies suggest that fish, like tetrapods, may possess separate receptors for glucocorticoid and mineralocorticoid functions, even though different hormones are employed.

Mineralocorticoids stimulate Na^+ reabsorption (and secondarily water recovery from the urine) and enhance K^+ excretion. The mineralocorticoids are produced by the adrenal cortex in tetrapods and the interrenal tissue in fish. These tissues, both physically close to the kidney, release aldosterone into the blood. Aldosterone targets the principal cells of the distal tubule and collecting

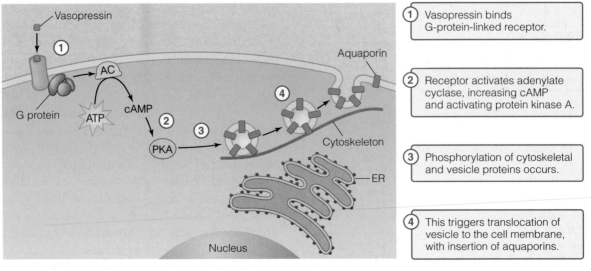

(a) Vasopressin

① Vasopressin binds G-protein-linked receptor.

② Receptor activates adenylate cyclase, increasing cAMP and activating protein kinase A.

③ Phosphorylation of cytoskeletal and vesicle proteins occurs.

④ This triggers translocation of vesicle to the cell membrane, with insertion of aquaporins.

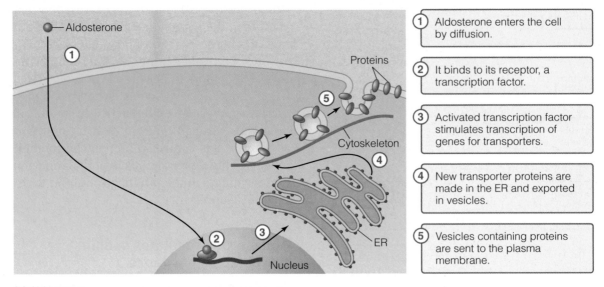

(b) Aldosterone

① Aldosterone enters the cell by diffusion.

② It binds to its receptor, a transcription factor.

③ Activated transcription factor stimulates transcription of genes for transporters.

④ New transporter proteins are made in the ER and exported in vesicles.

⑤ Vesicles containing proteins are sent to the plasma membrane.

Figure 10.34 Control of nephron function by vasopressin and aldosterone
(a) Vasopressin controls water movements in the tubule by altering the levels of aquaporins.
(b) Aldosterone changes the levels of ion transporters by altering the pattern of gene expression.

ducts, binding to a cytoplasmic hormone receptor and entering the nucleus to stimulate transcription of genes involved in ion transport (Figure 10.34b). The effects of aldosterone manifest over several hours because the process involves gene transcription, translation at the endoplasmic reticulum, processing in the Golgi apparatus, packaging into vesicles, and the fusion of the vesicles with the plasma membrane.

Insight into the role of aldosterone comes from studies of lab rats with the adrenal gland surgically removed (adrenalectomy). Within only a few hours, the rats produce copious volumes of urine, high in

Na^+ and low in K^+. As the loss of aldosterone continues unabated, the animal slowly dehydrates as it draws fluids from the extracellular spaces to maintain blood volume and blood pressure.

Aldosterone exerts its effects on K^+, Na^+, and water by stimulating the expression of genes encoding transport proteins. Na^+/K^+ ATPase is produced in the principal cells and sent to the basolateral membrane; K^+ channels and Na^+ channels are produced and targeted to the apical membrane. The interaction between Na^+ and K^+ transport causes the net exchange of plasma K^+ for Na^+ in the urine. The actions of the Na^+/K^+

ATPase increase intracellular [K$^+$], driving K$^+$ efflux into the tubule lumen through the K$^+$ channels. As the Na$^+$/K$^+$ ATPase pulls K$^+$ from the plasma, it expels Na$^+$ from the cytoplasm. This Na$^+$ enters the tubule cell through Na$^+$ channels in the luminal (apical) membrane from the primary urine. Aldosterone exerts no direct effects on water transport (unlike vasopressin); the stimulation of water recovery from the urine is a consequence of the reabsorption of Na$^+$.

The renin-angiotensin-aldosterone pathway regulates blood pressure

Circulating [K$^+$] is a major determinant of aldosterone synthesis—another example of a negative feedback loop. However, aldosterone production is also regulated by the hormone angiotensin II in the renin-angiotensin pathway. Juxtaglomerular cells in afferent and efferent arterioles of the nephron secrete the enzyme renin, which converts the plasma protein angiotensinogen to angiotensin I. Another enzyme called angiotensin-converting enzyme, or ACE, found on the epithelia of blood vessels, converts angiotensin I to angiotensin II. The secretion of renin is controlled in three ways. First, juxtaglomerular cells release renin when blood pressures decline; the juxtaglomerular cells may be baroreceptors themselves, or they may rely on neighboring cells to sense and respond to pressure changes through local signaling factors. Second, decreases in blood pressure activate sympathetic neurons in the cardiovascular control center of the medulla oblongata, triggering an increase in renin secretion. Third, the macula densa cells in the wall of the distal tubule respond to a decrease in urine flow and Na$^+$ delivery by releasing a paracrine signal that induces the juxtaglomerular cells to increase renin secretion.

Angiotensin II acts at a number of different sites including the kidney, brain, heart, adrenal cortex, and blood vessels. Recall from Chapter 8: Circulatory Systems that angiotensin II is an important regulator of the cardiovascular system, exerting effects on cardiac growth and angiogenesis. In terms of ion and water balance, angiotensin II stimulates Na$^+$ reabsorption in the proximal tubule and vasoconstricts postglomerular blood vessels. It can also stimulate the synthesis and release of other hormones that exert their own effects on the kidney to increase solute and water recovery. Angiotensin II increases the synthesis and release of aldosterone from the adrenal cortex and vasopressin from the pituitary.

Natriuretic peptides also play a role in sodium balance

Mammals produce a group of hormones called natriuretic peptides (NPs), which as the name suggests, favor the appearance of Na$^+$ in the urine. There are three main NPs: atrial NP (ANP), brain NP (BNP), and C-type NP (CNP). All three peptides are encoded by separate genes. They are secreted as longer propeptides, then cleaved by specific proteases into the active hormone. The first natriuretic hormone to be identified was **atrial natriuretic peptide (ANP)**. Cardiac atrial cells produce ANP and excrete it in response to excessive stretch, which would accompany an increase in blood volume. Upon release, ANP travels to the kidney where it increases Na$^+$ excretion by several mechanisms dependent upon activation of guanylyl cyclases and cGMP. The natriuretic and diuretic effects of ANP affect both the renal vasculature and tubular function.

The effects of ANP serve to increase the GFR. ANP increases blood flow to the glomerulus, acting on either the afferent glomerular arteriole (vasodilation) or efferent glomerular arteriole (vasoconstriction). It also causes relaxation of the mesangial cells, increasing the surface area available for filtration. It may also target the foot processes of the podocytes, increasing the size of the filtration slits of the glomerulus.

The effects of ANP on renal tubules are mediated through antagonistic effects on other hormones. It decreases aldosterone release by the adrenal cortex, preventing aldosterone from enhancing Na$^+$ reabsorption in the distal tubule. It decreases renin release, thereby decreasing angiotensin II. In many ways, ANP acts antagonistically to the renin-angiotensin system and the aldosterone pathway. It is also an antagonist of the production and release of vasopressin, reducing water reabsorption in the collecting ducts.

Hypothalamic factors regulate thirst

Water balance is, of course, affected by water consumption. The perception of thirst can arise in response to dehydration or Na$^+$ overload. Hormones that reflect the systemic state of the animal, including the osmotic condition and the cardiovascular state, exert effects on the central nervous system to affect thirst.

The hypothalamus is responsible for sensing the external conditions and controlling thirst. Recall that this region of the brain has an incomplete blood-brain barrier, allowing hypothalamic neurons to detect plasma conditions, including circulating hormones. The osmotic condition is detected by a combination of osmoreceptors and hormone receptors. The osmoreceptors in circumventricular organs monitor the osmolarity of the cerebrospinal fluid that bathes the hypothalamus. Angiotensin II, which exerts water-sparing effects on the kidney, also binds to receptors in this region of the brain. This hypothalamic region then sends signals to the thirst center elsewhere in the hypothalamus, likely the dorsomedial hypothalamic nucleus. Stimulation of the thirst center increases the motivation to drink.

◎ | CONCEPT CHECK

11. Compare and contrast how vasopressin and aldosterone regulate kidney function.

12. What factors affect glomerular filtration rate (GFR)?

Evolutionary Variation in the Structure and Function of Excretory Systems

The cellular composition and functional properties of kidneys vary widely among animals. Up to this point, we have used the structure and function of the typical mammalian kidney to illustrate the role of the different regions of the kidney tubule and how the structural and functional properties produce an appropriate urine. Of course, there is no such thing as a typical anything in animals as diverse as vertebrates. The model we described most closely resembles the kidney of a large mammal living on a mixed diet with ample access to freshwater. In other animals, even other mammals, there is considerable variation in kidney structure and function. Armed with an understanding of the roles of each of the main regions of the nephron, we can explore the basis of variation in nephron and kidney structure in other animals, beginning with the invertebrates.

Invertebrates have primitive kidneys called nephridia

Sponges, the simplest animals, act much like protists when it comes to water excretion. They use simple contractile vacuoles to expel cellular wastes, including water, directly into the environment. The true metazoans possess specific cells and tissues that are dedicated to excretion. Many simple animals, including most of the diverse worm taxa, possess *protonephridia*, a primitive functional analogue of the vertebrate kidney tubule (Figure 10.35a). The protonephridium in flatworms consists of a branched tubule with a pore (nephridiopore) at one end and a cap cell at the other. Fluids are propelled through the duct of the protonephridia by flagella or cilia that extend from specialized cells. Some species have *flame cells*, which possess a tuft of cilia. Other species have *solenocytes*, which possess one or two flagella that extend into the lumen. The role of protonephridia varies among species and in relation to the environment. In general, protonephridia are important in osmoregulation, and are best developed in the freshwater species that must export water. They probably have no role in ammonia excretion in most species that possess protonephridia; these

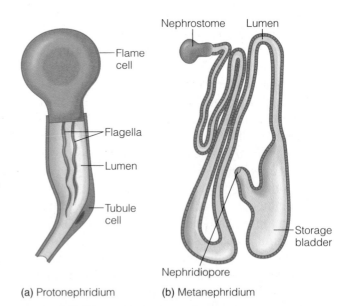

(a) Protonephridium   (b) Metanephridium

Figure 10.35 Primitive kidneys of invertebrates
(a) Primitive invertebrates, such as flatworms, possess protonephridia, which use ciliated or flagellated cells to draw interstitial fluid into the lumen of tubules. **(b)** More advanced invertebrates, such as annelids, possess metanephridia, which collect fluids directly from the circulatory system or coelom.

animals excrete ammonia directly across the body surface. The number of protonephridia differs widely among species, from as few as two with a pair of nephridiopores, to thousands of units.

More complex nephridia, called *metanephridia*, occur in the molluscs and annelids. Most molluscs have a single metanephridium, with a sac possessing deeply invaginated walls to increase the surface area (Figure 10.35b). Solutes and water are collected and expelled through a short tube called a ureter. Annelids possess a metanephridium in each body segment. The tubule begins at the nephrostome, which collects fluids from the coelom. The fluid passes through the long tubule, which winds through the body segment. Tubule length depends on habitat; marine species, with little need for water excretion, have shorter tubules than do freshwater species. The nephridium passes through the septum between body segments, and joins the body wall at the nephridiopore, releasing waste products from the animal. In some cases, the nephridium expands into a saclike bladder that is able to store fluids.

The cellular structures of these primitive kidneys vary widely among the invertebrates, but the main difference between a protonephridium and a metanephridium is in the relationship with the intracellular fluids. Protonephridia use their beating flagella and cilia to draw interstitial fluid into the lumen of the tubule. In contrast, the duct of a metanephridium has an internal opening that collects body fluids, either blood or coelomic fluid.

One unusual group among the invertebrates are the nematodes. Unlike other worms, nematodes lack nephridia altogether. Their excretory system employs *rennette cells*, which secrete waste into a duct that empties through an excretory pore. Nematodes differ in the morphology of the rennette cells—glandular or tubular—and the extent of the canal system that links the rennette cell to the excretory pore.

Insects use Malpighian tubules and the hindgut for ion and water regulation

In insects, no single tissue fulfills the function of the vertebrate kidney or invertebrate nephridium; water and ion balance in insects is regulated by concerted actions of the **Malpighian tubule** and the hindgut.

The Malpighian tubule is a tubelike structure with a blind sac at one end of a long tube that empties into the hindgut (Figure 10.36). Some

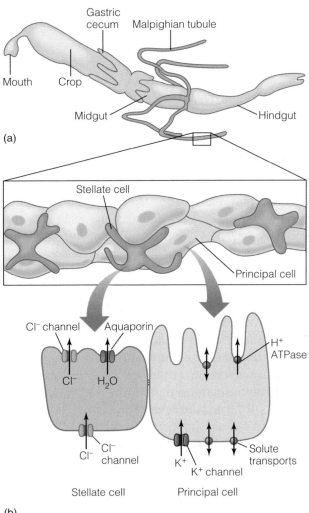

(a)

(b)

Figure 10.36 Insect Malpighian tubule structure and function **(a)** The Malpighian tubules, the insect kidney, empty into the digestive tract. Although many insects have four Malpighian tubules, the numbers can range from two to more than 250. **(b)** The tubule itself is composed of principal cells and stellate cells.
(Source: Modified from O'Donnell et al., 1996)

insects, such as hemipteran bugs, possess two very long, coiled tubules, whereas other species have many shorter tubules, up to 250 in the locust *Schistocerca*. Although an insect Malpighian tubule is typically only one cell layer thick, there is considerable variation among species in the structural complexity, which can be related to the nature of the diet. Species that consume a great deal of water, such as sap feeders, blood feeders, and carrion feeders, may have more complex structures that improve filtration.

The hindgut, which includes the rectum and rectal gland, receives the flow from the Malpighian tubule. The fluid is typically slightly hyposmotic to the hemolymph. As it passes into the hindgut, the fluid is further modified by reabsorption and secretion. When it leaves the hindgut, the final osmolarity may be hyposmotic, isosmotic, or hyperosmotic.

The Malpighian tubule of a typical dipteran (flies and mosquitos) is composed of two main cell types: large principal cells and small stellate cells. Principal cells, often called secretory cells, possess extensive apical microvilli. The microvilli contain long, slender mitochondria that move in and out of the microvilli as needed. The mitochondria provide the ATP required for ion pumping. The main role of the principal cells is cation transport, and the apical cell membrane is rich in ion exchangers that are ultimately driven by the activity of a proton-pumping ATPase (H^+ ATPase). Stellate cells have fewer mitochondria and lack the complex apical microvilli. Their main role is the control of Cl^- transport.

The Malpighian tubule produces the primary urine without filtration. Instead, secretions from the tubule cells form the primary urine. Solutes and water enter the blind end of the tubule by either paracellular transport or transcellular transport. The basolateral membrane of principal cells imports K^+ from the hemolymph by K^+ channels; Na^+ and K^+ are imported by a Na^+-K^+-$2Cl^-$ cotransporter (NKCC). At the apical membrane, the H^+ ATPase exports H^+, creating a driving force that can be coupled to exchangers (Na^+/H^+ or K^+/H^+) that expel other cations. Unlike most other transport epithelia, there appears to be little role for Na^+/K^+ ATPase in most species. The increase in osmolarity draws water into the lumen of the tubule. The lumen contents are modified by selective reabsorption of solutes and water as the primary urine flows down the tubule from the blind end to the opening in the hindgut.

Once the lumen fluid is formed in the ends of the tubules, it progresses down the tubule where it encounters mechanisms of selective reabsorption. Malpighian tubules are not innervated, and the control of secretion and reabsorption is under the control of intrinsic regulators and circulating hormones. Three main classes of diuretic hormones have been identified in insects.

1. *CRF-related diuretic hormones* have structural similarities to vertebrate hormones of the corticotropin-releasing factor (CRF) family, such as urotensin and urocortin. Fifteen different CRF-related diuretic hormones have been identified in insects, ranging in size from 30 to 47 amino acids. These hormones appear to act by stimulating the synthesis of cAMP in Malpighian tubule cells, which activates cation transport at the basolateral and apical regions of the principal cells.

2. *Insect (myo)kinins* are short peptide hormones, usually eight amino acids, that possess a characteristic C-terminal sequence of five amino acids. These hormones act on the stellate cells of the Malpighian tubules, activating phospholipase C to increase the production of IP_3, which in turn causes a release of Ca^{2+} from intracellular stores. This increase in $[Ca^{2+}]$ causes an increase in Cl^- transport into the lumen, although the mechanisms are not yet clear. The movement of Cl^- causes a parallel movement of Na^+ and K^+, leading to a net movement of NaCl and KCl from the hemolymph to the lumen.

3. *Cardioacceleratory peptides* were first identified by their ability to increase the heart rate. However, these hormones also stimulate the secretion of fluid into Malpighian tubules. This diverse family of hormones appears to stimulate phospholipase C in principal cells. The increase in IP_3 and then Ca^{2+} activates a cascade involving Ca^{2+}-calmodulin-dependent nitric oxide synthase and guanylyl cyclase. These second messengers and regulatory enzymes lead to an increase in H^+ ATPase activity.

While these diuretic hormones have been well characterized, much less is known about antidiuretic hormones in insects. Some, called antidiuretic factors, act by reducing the movement of water into the Malpighian tubule. Others, such as neuroparsins and ion-transport peptide, act by increasing water reabsorption in the gut, after the lumen contents empty into the hindgut.

Chondrichthian kidneys produce hyposmotic urine and retain urea

Recall that sharks have an unusual osmotic strategy among vertebrates, maintaining their body fluids slightly hypertonic to seawater and accumulating urea to high concentrations (300–400 mM). Sharks have two long kidneys lying along the dorsal wall of the body cavity. The kidney tubules are long and complex in structure (Figure 10.37).

The tubules weave back and forth across the kidney to form two layers: a sinus zone where tubules are loosely packed and separated by fluid, and a more compact zone where tubules are bundled together and wrapped in a membranous

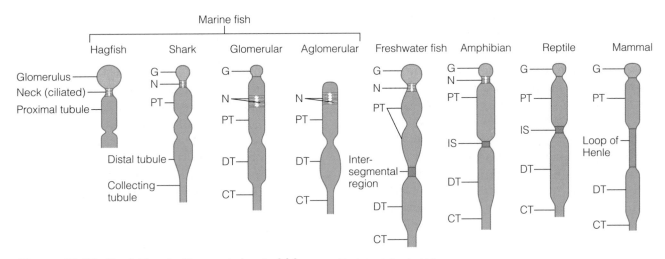

Figure 10.37 Variation in the vertebrate kidneys Each vertebrate kidney possesses regions specialized for absorption and secretion. However, the dimensions of each region vary among taxa.
(Source: Modified from Willmer et al., 2000)

sheath. This complex arrangement may set up a countercurrent exchanger that allows the shark kidney tubule to recover as much as 90% of the urea from the primary urine. The exact mechanism by which urea is recovered remains unclear, and may occur through active reabsorption via Na$^+$-urea cotransporters. The urine produced by the shark is slightly hyposmotic (relative to the shark tissues) and close to the osmolarity of seawater. Sharks that move into dilute seawater reduce their internal osmolarity by producing copious amounts of dilute urine.

The role of the fish kidney differs in freshwater and seawater

In bony fish, two paired kidneys run along the dorsal surface of the inner body cavity. The function of the kidney depends on the osmolarity of the water. The glomerulus, which produces the primary urine, is much larger in freshwater fish than in marine species. The distal tubule, which functions in salt recovery and water excretion, may also be much larger. The kidneys of freshwater fish produce large volumes of hyposmotic urine.

The kidneys of marine fish play a much-reduced role in ion and water balance. They produce very little urine, which is isosmotic to body fluids. The nephrons of marine fish have a less complex glomerulus, shorter proximal tubules, and distal tubules that are reduced or absent. Some marine fish lack a glomerulus altogether. These **aglomerular**

kidneys occur in species from three unrelated taxa. Since the other species in these taxa have glomerular kidneys, the aglomerular state has evolved at least three separate times in fish.

The amphibian kidney changes in metamorphosis

Like freshwater fish, most amphibians living in water must rid themselves of excess water absorbed from the environment across highly permeable skin. In their aquatic life they have little need for water retention mechanisms. However, when amphibians are on land, they must conserve water. Like freshwater fish, amphibians possess a kidney that lacks a loop of Henle, which enables the mammalian kidney to produce hyperosmotic urine. An amphibian meets the conflicting demands of life on land and water by (1) regulating the glomerular filtration rate to control the rate of water loss and (2) recovering water from the urine stored in the urinary bladder. Whereas most terrestrial animals use the bladder only for short-term storage of urine prior to micturition, amphibians use the bladder for water storage. The reabsorption process is under the control of the amphibian homologue of vasopressin.

The nature of the amphibian kidney changes during development. Larval amphibians, as well as larval fish, have a simple nephron called a **pronephros**. Recall that the mammalian kidney tubule, also known as a metanephros, empties fluid from the circulation directly into the interior of the nephron at the Bowman's capsule. In a pronephric kidney, the filtrate first enters the coelom, then is swept into the pronephric tubules through the nephrostomal funnels. As with true nephrons, the water, ions, and organic molecules are reabsorbed in the tubule and returned to the blood. The urine is then sent along the pronephric duct and expelled through the cloaca. Whereas a mammalian kidney may possess a million nephrons, a larval frog possesses just a pair of pronephros. As the larva metamorphoses to an adult, the pronephros is replaced by a kidney that much more closely resembles the mammalian version.

Terrestrial animals have kidneys that help conserve water

The variations in kidney morphology among reptiles, birds, and mammals reflect different solutions to the challenge of avoiding dehydration. The challenges of reducing water loss are greatest in desert animals (see Box 10.1, Evolution and Diversity: Life Without Water), but all terrestrial animals have multiple means of matching kidney function to the constraints of environmental water availability.

Modern reptiles reduce the need for water by producing uric acid as a nitrogenous end product. Since uric acid is insoluble, water is not wasted as a solvent, although some water is used to wash the uric acid down the tubule lumen. This water can be reabsorbed in the cloaca. The reptilian kidney has much-reduced glomeruli, and in some species the glomerulus is absent. As with the amphibians, the reptilian nephron lacks a loop of Henle, and therefore cannot produce hyperosmotic urine.

One of the major innovations in terrestrial vertebrate evolution was the loop of Henle. This extended segment between the proximal and distal tubules occurs only in birds and mammals, although birds have some nephrons that lack a loop of Henle. Because of the loop of Henle, most mammals can produce urine with an osmolarity that is about five times greater than the plasma osmolarity.

If it were simply the total length of the loop of Henle that determined the ability to produce concentrated urine, the elephant would be a champion; it has a long loop of Henle simply because its kidney is so large. The best predictor of the ability

of the nephron to produce concentrated urine takes into account the size of the kidney. Since the loop of Henle spans the medulla, the potential to produce concentrated urine is best expressed as relative medullary thickness: the width of the medulla relative to the total width of the kidney (Figure 10.38). Mammals that live in environments with abundant water, such as beavers, have a low relative medullary thickness and nephrons with a short loop of Henle that produce dilute urine. Conversely, mammals that live in very dry environments, such as the kangaroo rat, have a high relative medullary thickness and nephrons with a long loop of Henle that produce highly concentrated urine, typically four to five times more concentrated than that of most mammals.

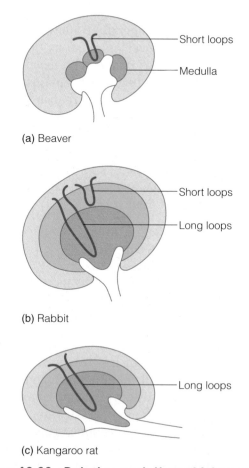

(a) Beaver

(b) Rabbit

(c) Kangaroo rat

Figure 10.38 Relative medullary thickness in mammalian kidneys Animals that produce a more concentrated urine, such as the kangaroo rat, have nephrons with a longer loop of Henle and a thicker medulla than do animals that produce dilute urine, such as a beaver. Most species, including rabbits, fall between these extremes. (Source: Adapted from Schmidt-Nielsen and Dell, 1961)

Integrating Systems | Interaction of the Cardiovascular and Excretory Systems in the Regulation of Blood Pressure

In our discussion of kidney function, we focused on how the kidney produces urine to control water balance. However, the excretory system and cardiovascular system have overlapping responsibilities in regulating blood pressure. The cardiovascular system responds primarily to changes in blood pressure, whereas the excretory system responds to changes in both blood pressure and blood osmolarity. Animals regulate blood pressure by controlling both the volume of blood and its osmolarity, which depend on water and salt fluxes but can vary independently of each other.

Consider how different foods affect water and salt balance, and how the excretory system must respond to maintain homeostasis. If you drink a large amount of water without ingesting anything salty, the total body fluid volume quickly increases and osmolarity decreases. Increasing the volume of urine corrects this situation. Conversely, if you eat salty food without drinking, the osmolarity increases but your body fluid volume remains unaltered. The kidneys must increase the excretion of salt, but retain as much water as possible. Both volume and osmolarity increase if you eat salty food and drink a large amount of liquid at the same time. The kidney must increase both the volume of urine and the total amount of salt in the urine. However, the response to changes in volume and osmolarity is not always this simple. High osmolarity could arise from excessive salt (so a response of salt excretion is appropriate) or dehydration (a response of water consumption is appropriate).

Dehydration is, in fact, the most common cause of disturbances in fluid volume and osmolarity. If you were to exercise on a hot day, you would lose water in the sweat and in the expired air. Without drinking, you become dehydrated; blood volume will decrease and osmolarity will increase, with negative effects on the cardiovascular system. The decrease in blood volume leads to decreases in venous return to the heart, cardiac output, and mean arterial blood pressure. Thus, during dehydration, blood pressure drops while blood osmolarity increases.

Various pathways are involved in homeostatic compensation for severe dehydration (Figure 10.39). These pathways involve the cardiovascular system, the renin-angiotensin system, renal mechanisms such as glomerular filtration, and mechanisms coordinated by the hypothalamus.

One of the fastest responses to the decrease in blood pressure is the fluid-shift mechanism. Low blood pressure reduces filtration across the capillaries, and causes fluid to shift from the interstitial space to the blood. This helps to return blood volume and blood pressure to normal.

The decrease in blood pressure as a result of dehydration also reduces the amount of stretch on the carotid and aortic body baroreceptors. This causes them to reduce the frequency of action potentials in the afferent neurons leading to the cardiovascular control center in the medulla oblongata of the brain. This decrease in action potential frequency causes the cardiovascular control center to decrease parasympathetic output and increase sympathetic output, and these changes in turn increase heart rate and the force of cardiac contraction, increasing cardiac output. At the same time, the sympathetic neurons leading to many systemic arterioles stimulate the arteriolar smooth muscle to contract. The resulting vasoconstriction increases total peripheral resistance (TPR). Since mean arterial pressure is equal to cardiac output times total peripheral resistance, these two mechanisms lead to an increase in blood pressure, helping compensate for the decreased blood pressure caused by dehydration.

The decreased blood pressure caused by dehydration also stimulates the juxtaglomerular (JG) cells of the kidney, causing them to increase the secretion of renin. The increased renin increases the conversion of angiotensinogen to angiotensin I. Angiotensin-converting enzyme (ACE) then catalyzes the conversion of angiotensin I to angiotensin II. Simultaneously, the increase in sympathetic output also stimulates the juxtaglomerular cells, further increasing the level of angiotensin II. As we learned in this chapter and Chapter 8, angiotensin II has wide-ranging effects on the circulatory and excretory systems. Angiotensin II can directly promote the reabsorption of sodium and water at the level of the proximal tubule and indirectly in the distal nephron via the actions of aldosterone released from the adrenal cortex. In dehydration, however, the latter process is blocked due to direct effects of increased plasma osmolarity on the adrenal cortex. Decreased blood pressure also affects glomerular filtration rate (GFR). Lower blood pressure has a small direct effect on GFR, reducing filtration pressure in the glomerulus, and slowing the rate of urine production. However, recall that glomerular filtration rate is actually rather tightly regulated even in the face of changes

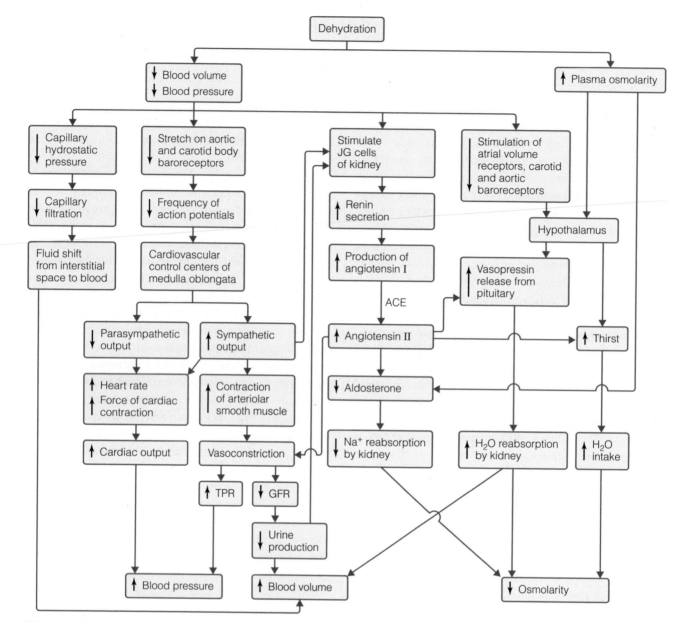

Figure 10.39 Regulation of blood pressure and kidney function in response to dehydration

in mean arterial pressure, so these direct effects are minor. Instead, increased sympathetic stimulation (as a result of the decrease in blood pressure) has a major effect. Sympathetic stimulation causes vasoconstriction of the afferent arteriole of the glomerulus, reducing glomerular filtration pressure, and thus GFR and urine production. The reduction in GFR also decreases flow of fluid through the kidney tubules. The cells in the macula densa detect this decrease in flow and stimulate the juxtaglomerular cells to secrete renin, thus further increasing the levels of angiotensin II.

The decrease in stimulation of the atrial volume receptors and the carotid and aortic body baroreceptors as a result of low blood pressure also has a direct effect on the hypothalamus, increasing vasopressin secretion and thirst. Together, these responses increase water reabsorption by the kidney and water intake, resulting in increased fluid volume and decreased osmolarity.

Most of the mechanisms we have discussed so far involve the stimulus of decreased blood pressure, but increased osmolarity also has important effects. Osmoreceptors in the hypothalamus directly sense the increased osmolarity and stimulate the hypothalamus to release vasopressin as well as stimulate thirst. Osmolarity also has direct effects on the adrenal cortex, reducing the secretion of aldosterone. The drop in al-dosterone reduces the expression of Na^+/K^+ ATPase in the membranes of the distal nephron, reducing Na^+ reabsorption, and helping to return osmolarity to normal. Note that reducing Na^+ absorption also reduces water reabsorption, which could impede the return of blood volume and blood pressure to normal. Cardiovascular reflexes can help to restore blood pressure in the short term, but they cannot address changes in osmolarity. In contrast, the excretory system generally attempts to restore osmolarity, while dealing with blood pressure over longer terms.

Because of the intimate links between the two systems, many pathological conditions have elements of both kidney disease and cardiovascular dysfunction. Kidney disease may be either a cause or consequence of cardiovascular disease. Kidney dysfunction can arise from renal artery defects or maladaptive regulation of the renin-angiotensin system or other vasoactive agents. The changes in blood volume create a hypertensive condition that challenges normal heart function. Many people with congestive heart failure worsen when the changes in blood pressure affect kidney function. When cardiac function deteriorates, the kidney responds with renal vasoconstriction, leading to the retention of water and sodium. The increase in blood volume in turn exacerbates the cardiac problems.◎

Summary

Ion and Water Balance

→ Epithelial tissues are the interface between internal fluids and the external environment, creating osmotic and ionic barriers, and regulated to control ion balance, water balance, and nitrogen excretion.

→ Solutes can be distinguished by their effects on macromolecular structure and function as perturbing, compatible, or counteracting.

→ Animals move ions into and out of cells to maintain or change cell volume. Regulatory increases in volume (RVI) and regulatory decreases in volume (RVD) are part of compensatory strategies to regain normal cell volume, and contribute to signaling pathways that act by changing cell volume.

→ The epithelial tissues of terrestrial animals create water-resistance osmotic barriers. Other specialized epithelial tissues mediate osmotic and ionic regulation, including gills, the digestive tract, and the specialized excretory tissues, such as the rectal gland of elasmobranchs and the salt glands of reptiles and birds.

→ Animals remove toxic ammonia by excretion or metabolism to less toxic forms. The three main strategies, defined by the major form of nitrogenous end product, are ammoniotelism (NH_4^+), uricotelism (uric acid), and ureotelism (urea). Most aquatic animals are ammoniotelic, whereas terrestrial animals are uricotelic (reptiles and birds) or ureotelic (mammals).

→ Urea is produced in the ornithine-urea cycle, which is regulated by enzyme levels and allosteric regulation. Cartilaginous fish produce urea as a solute rather than an excretory product.

Kidney

→ The majority of the responsibility for ion and water regulation in vertebrates falls on the kidney. The functional unit of the kidney is the nephron, a combination of complex vasculature and renal tubule composed of epithelial cells.

→ Urine is formed by four processes: filtration, reabsorption, secretion, and excretion. Filtration occurs at the glomerulus, a ball-like network of capillaries surrounded by the Bowman's capsule of the nephron. From the Bowman's capsule, the primary urine enters the proximal tubule, and then proceeds through the loop of Henle, with its descending and ascending limbs. The fluid then flows into the distal tubule and through collecting ducts into the ureters, the urinary bladder and, after a period of storage, out the urethra.

→ Central to nephron function is the countercurrent system set up between the loop of Henle and collecting duct, in conjunction with the capillaries that serve the nephron.

→ Kidney function is regulated at multiple levels. Glomerular filtration pressure is affected by the hydrostatic pressure and oncotic pressure gradients between the glomerulus and Bowman's capsule. The glomerular filtration rate is regulated by factors that influence the filtration pressure as well as the surface area available for filtration.

→ Various hormones and neurotransmitters control naturesis and diuresis. Vasopressin alters the permeability of the collecting duct. The renin-angiotensin system, the sympathetic nervous system, and aldosterone act in concert to regulate sodium, potassium, and water balance as well as blood pressure. Atrial natriuretic peptide also plays a role in sodium balance.

→ Invertebrates have primitive kidneys, such as protonephridia and metanephridia. Insect ion and water balance is mediated by Malpighian tubules and the hindgut.

→ Vertebrate kidneys differ among taxa and in relation to the environmental conditions. Shark kidneys retain urea while producing a hyposmotic urine. Kidneys of freshwater fish produce copious amounts of dilute urine, whereas the marine fish have much-reduced kidneys and produce little urine. The kidneys of reptiles and birds can produce hyperosmotic urine. Desert vertebrates also produce small volumes of hyperosmotic urine as a water conservation mechanism.

Review Questions

1. Compare the structure and function of the salt gland of birds with the rectal gland of elasmobranchs.

2. Discuss how countercurrent systems aid renal function.

3. Compare the types of nephrons in the invertebrates and vertebrates.

4. What are the six main roles of the kidney?

5. How is glomerular filtration rate controlled via intrinsic and extrinsic mechanisms?

6. There is a relationship between the volume of urine produced and the type of nitrogenous waste excreted by an organism. What is this relationship, and why does it occur?

7. Compare the energetic costs in the different excretory strategies.

8. How is energy used in ion pumping?

9. Discuss regulatory volume increases and decreases in relation to the membrane potential.

10. In a normal kidney, which of the following would cause an increase in GFR?
    (a) Constriction of the afferent arteriole
    (b) Decrease in the hydrostatic pressure in the glomerulus
    (c) Increase in hydrostatic pressure in the Bowman's capsule

Synthesis Questions

1. Is more water derived from oxidizing glycogen, protein, or lipid?

2. Discuss the integration of the respiratory and excretory systems in controlling pH balance.

3. Describe the role of nerves and muscles in control of ion and water balance.

4. We discussed the variation of kinetic properties and localization of the enzymes in urea synthesis. What genetic processes could be responsible for these changes during the course of evolution?

5. Angiotensin-converting enzyme inhibitors (ACE inhibitors) are used to treat high blood pressure. Using a flowchart, explain why these drugs are helpful in treating hypertension.

6. The kidney of a cactus wren is less efficient at concentrating urine than are the kidneys of a kangaroo rat, yet the cactus wren produces less urine. In one or two sentences, explain this apparent contradiction.

7. A person with cirrhosis of the liver has lower than normal levels of plasma proteins (because production of albumin, one of the major plasma proteins, decreases) and a higher than normal GFR. Explain why a decrease in plasma protein concentration would increase GFR.

8. Most freshwater fish are unable to survive in water with high concentrations of bicarbonate. Draw a diagram of a freshwater fish gill and, using this diagram, outline a possible physiological reason for this observation.

Quantitative Questions

1. If an aquaporin can pass 10^9 molecules of water a second, how many channels would be needed to reduce the volume of a 1-μl cell by half in 1 second? (Hint: How many water molecules are in 1 μl of water?)

2. Assuming complete dissociation, which of the following solutions will have the greatest osmolarity: 150 mM glucose, 80 mM NaCl, 90 mM Na_2SO_4, or 210 mM urea?

3. Explain why an individual with a plasma glucose concentration of 375 mg/100 ml of blood will have glucose in the urine. (Note: Normal plasma glucose is approximately 100 mg/100 ml of blood, normal GFR is approximately 180 l/day, and the kidney's maximal transport rate for glucose is approximately 375 mg/min.)

For Further Reading

See the Additional References section at the back of the book for more readings related to the topics in this chapter.

Ion and Water Balance

This paper discusses the role of the skin as a barrier to osmotic movements.

Alibardi, L. 2003. Adaptation to the land: The skin of reptiles in comparison to that of amphibians and endotherm amniotes. *Journal of Experimental Zoology, Part B: Molecular and Developmental Evolution* 298: 12–41.

These three papers discuss the way cells sense and respond to changes in cell volume arising either from osmotic stress or in response to signaling pathways.

Goldmann, W. H. 2002. Mechanical aspects of cell shape regulation and signaling. *Cell Biology International* 26: 313–317.

Lang, F., G. L. Busch, M. Ritter, H. Volkl, S. Waldegger, E. Gulbins, and D. Haussinger. 1998. Functional significance of cell volume regulatory mechanisms. *Physiological Reviews* 78: 247–306.

Mongin, A. A., A. Sergei, and N. Orlov. 2001. Mechanisms of cell volume regulation and possible nature of the cell volume sensor. *Pathophysiology* 8: 77–88.

These classic papers from two pioneers of comparative physiology discuss the history of early research in the ionic and osmotic regulation of aquatic animals.

Krogh, A. 1939. *Osmotic regulation in aquatic animals.* Cambridge: Cambridge University Press.

Smith, H. W. 1953. *From fish to philosopher.* Boston: Little, Brown.

Fish are a diverse group of vertebrates that collectively survive a very wide range of environmental salinities. This review discusses current concepts about the regulation of ion balance in freshwater species that must acquire ions and expel water.

Perry, S. F., A. Shahsavarani, T. Georgalis, M. Bayaa, M. Furimsky, and S. L. Thomas. 2003. Channels, pumps, and exchangers in the gill and kidney of freshwater fishes: Their role in ionic and acid-base regulation. *Journal of Experimental Zoology, Part A: Comparative Experimental Biology* 300: 53–62.

Nitrogen Excretion

Many aspects of nitrogen metabolism changed in the early evolution of vertebrates. These three papers discuss urea metabolism in fish, focusing on the evolutionary origins of ureotelism.

Barimo, J. F., S. L. Steele, P. A. Wright, and P. J. Walsh. 2004. Dogmas and controversies in the handling of nitrogenous wastes: Ureotely and ammonia tolerance in early life stages of the gulf toadfish, *Opsanus beta. Journal of Experimental Biology* 207: 2011–2020.

Mommsen, T. P., and P. J. Walsh. 1989. Evolution of urea synthesis in vertebrates: The piscine connection. *Science* 243: 72–75.

Walsh, P. J. 1997. Evolution and regulation of urea synthesis and ureotely in (batrachoidid) fishes. *Annual Review of Physiology* 59: 299–323.

In this paper, the author asks what we can learn about human kidney disease by studying the kidneys of primitive vertebrates. He focuses on the paradox of the few species of fish that survive without a glomerulus.

Beyenbach, K. W. 2004. Kidneys sans glomeruli. *American Journal of Physiology (Renal Physiology)* 286: F811–F827.

Although elasmobranchs are osmoconformers in seawater, many species are able to survive long periods in brackishwater and freshwater. This review discusses the role of urea in osmotic balance of euryhaline species.

Hazon, N., A. Wells, R. D. Pillans, J. P. Good, W. G. Anderson, and C. E. Franklin. 2003. Urea based osmoregulation and endocrine control in elasmobranch fish with special reference to euryhalinity. *Comparative Biochemistry and Physiology, Part B: Biochemistry and Molecular Biology* 136: 685–700.

Regulation of Kidney Function

These papers provide in-depth reviews on the transport capabilities of the various regions of the renal tubule.

Muto, S. 2001. Potassium transport in the mammalian collecting duct. *Physiological Reviews* 81: 85–116.

Reilly, R. F., and D. H. Ellison. 2000. Mammalian distal tubule: Physiology, pathophysiology, and molecular anatomy. *Physiological Reviews* 80: 277–313.

Wright, S. H., and W. H. Dantzler. 2004. Molecular and cellular physiology of renal organic cation and anion transport. *Physiological Reviews* 84: 987–1049.

These two papers discuss the insect Malpighian tubule from different perspectives. One reviews what is known about the regulation of these excretory organs to better understand how they function. The other reviews the utility of insect models to explore the epithelial function from the perspective of human disease.

Cagan, R. 2003. The signals that drive kidney development: A view from the fly eye. *Current Opinion in Nephrology and Hypertension* 12: 11–17.

O'Donnell, M. J., and J. H. Spring. 2000. Modes of control of insect Malpighian tubules: Synergism, antagonism, cooperation and autonomous regulation. *Journal of Insect Physiology* 46: 107–117.

This paper reviews what is currently known about the regulation of enzyme activity in urea production, including the transcriptional control of their synthesis.

Morris, S. M., Jr. 2002. Regulation of enzymes of the urea cycle and arginine metabolism. *Annual Review of Nutrition* 22: 87–105.

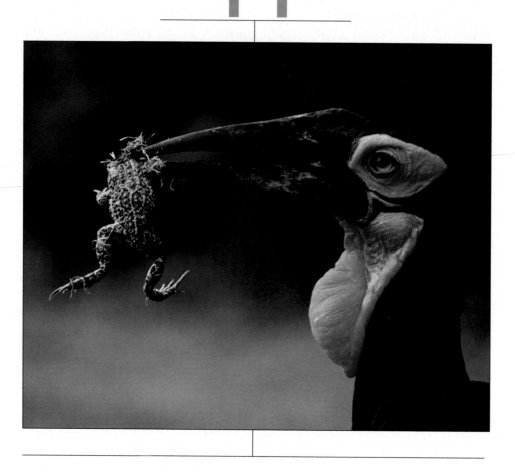

Digestion

The physiology of digestion remained largely mysterious until well into the late 18th century. Around 1780, Lazzaro Spallanzani began to explore the chemical events that occurred in the gut. Like earlier researchers, he knew that something in the stomach caused food to change in texture and consistency. He studied gastric processing by collecting stomach contents from living animals. He force-fed animals tethered dry sponges and metal tubes, which, upon recovery, served as reservoirs for gastric juices. When Spallanzani mixed the gastric juices with foodstuffs, he found that chunks of meat softened and milk curdled.

In the early 1800s, several European researchers expanded Spallanzani's studies to explore the chemical capabilities of gastric juices. At the same time, an American surgeon named William Beaumont began his own studies of stomach processes. The surgery he performed on a gun-

shot victim saved the patient's life, but also left a persistent hole, or fistula, that ran from the abdominal wall through the stomach. This enabled Beaumont to sample gastric juices from his cooperative patient at various points after a meal. The fistula remains a useful experimental tool, providing researchers with a means to sample fluids with little stress to the animal. By 1820, it was known that the gastric juices of unfed animals were neutral but became acidic upon feeding. Beaumont established that acidity alone was insufficient to explain how gastric juices broke down foods. In 1836, Theodor Schwann proposed that gastric juices contained a chemical substance he called *pepsin* that was capable of digesting foods. It became widely accepted that the digestive juices from the saliva, stomach, and pancreas each possessed agents that could break down food. Many scientists of the day, includ-

Bear cuscus and fig.

Cows with fistula.

ing Claude Bernard, believed that each tissue produced the same digestive agent, but that its behavior differed as a result of variation in acidity. It was first shown in the mid-1800s that juices from the pancreas, saliva, and stomach had different digestive capacities as a result of distinct chemical agents therein. Salivary diastase (amylase) could break down potato starch within seconds, but pepsin could not. These observations were presented long before the concept of enzymatic catalysis was formulated. By the late 1800s the general notion of enzymes was accepted, although their nature as proteins continued to be disputed into the 1920s.

The metabolic side of digestive physiology became the passion of the early enzymologists. In his autobiography *For the Love of Enzymes*, Arthur Kornberg describes the early years of metabolic biochemistry and nutrition. The 1920s to 1930s were dominated by the "vitamin hunters." It had long been recognized that certain diets could cause dramatic dysfunction and disease. British sailors suffered from scurvy, a disease that manifests as deterioration of the connective tissue. Scurvy could be prevented if limes or other citrus fruits were included in the sailors' diet, which incidentally was the origin of the term *limey*. Until the 1880s, the Japanese navy subsisted on a rice diet and lost one in three sailors from a disease called *beriberi*, a disease accompanied by loss of nerve and muscle function. Within three years of changing to a diet used by British seamen, deaths from beriberi had all but vanished in the Japanese navy. The nature of the protective agents in the diet remained unclear until vitamins were discovered in the early 1900s. We now know beriberi is a symptom of thiamin deficiency, and that scurvy is a deficiency of vitamin C.

With a greater understanding of catalysis and the role played by vitamins, the early enzymologists began to dissect complex metabolic pathways. Studies on yeast fermentation yielded important discoveries about how enzymes combine to perform intermediary metabolism. Unlike the physiologists, who tended to focus on specific types of animals, the metabolic biochemists were more egalitarian in their choice of experimental models. Biochemists in this era, including Kornberg, studied bacteria, fungi, plants, and tissues of many animal species to unlock the secrets of enzyme function and metabolic pathways. These studies provided the mechanistic links between nutrition, digestion, energy metabolism, and biosynthesis. ◎

Overview

Digestive physiology is concerned with all of the tissues that contribute to the physical and chemical breakdown of food: the sensory system used to locate food, the physical structures that mechanically disrupt the food, and the chemical processes that break food into forms that can be transported and metabolized into other molecules (Figure 11.1).

Assimilation—the sequential process of nutrient acquisition and absorption—begins when the neurosensory machinery, such as an insect's antennae or a knifefish's electrical sensors, is recruited to locate food. Once food is located in the broad environment, it must be captured using specialized anatomy, such as the lobster's claw, the eagle's talon, or the mosquito's proboscis. Once acquired, food is usually mechanically disrupted with the help of other specialized structures, such as a mammal's teeth or the snail's tongue. Animals then use chemical processes to convert the large food items to macromolecules and small molecules. The food may be macerated, or softened, by soaking in fluids such as saliva. Chemical breakdown is primarily enzymatic and in most cases takes place *outside* the animal. Note that the inner surface of the *gastrointestinal tract* is contiguous with the external environment.

The gastrointestinal tract, or GI tract as it is more commonly known, is wondrous in its complexity. It is composed of many cell types: absorptive cells that take up nutrients, glands that secrete suites of chemicals (mucus, acid, ions, and enzymes), muscles that control the GI tract shape and motility, and nerves that regulate GI tract function. Once the nutrients are broken down in the lumen of the GI tract, they are brought into cells. Undigested food is expelled from the body by the process of *egestion*.

The Nature and Acquisition of Nutrients

Every organic molecule on the planet possesses chemical energy, yet animals are able to capture this energy from only a small subset of these molecules. Food is a sampling of the external environment, a heterogeneous mixture of digestible and undigestible materials. Nutrients are the external molecules that allow an animal to build and maintain cells. We begin our discussion of digestive physiology by considering the nature of the nutrients.

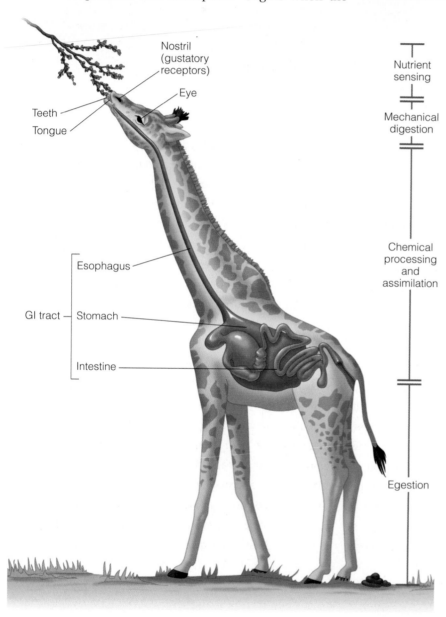

Figure 11.1 Digestion Animals use combinations of sensory and mechanical processes to acquire and ingest food. Vision and smell are central to the feeding strategies of most vertebrates. Once acquired, the food begins to undergo the process of digestion. Frequently, ingestion begins with mechanical disruption in the upper digestive tract, followed by chemical processing of the food material that is required for assimilation. Undigestible material is expelled from the animal.

Nutrients

Ingestion is the primary route that an animal uses to gain access to environmental chemicals. Although many aquatic animals obtain some essential ions by importing them across the external epithelial surfaces, such as the gills and skin, most animals absorb nutrients across the epithelium of the gastrointestinal tract. Some of the assimilated nutrients are degraded to liberate chemical energy; the rest are used as building blocks. Many of the macromolecules that animals need for biosynthesis cannot be synthesized *de novo*, so a dietary source is critical. *Essential nutrients,* those chemicals that must be obtained in the diet, include most vitamins and minerals, as well as several amino acids and fatty acids. *Nonessential nutrients* are those chemicals that the animal can produce from other molecules.

Diets provide energy for activity, growth, maintenance, and reproduction

The diet provides animals with nutrients that can be oxidized for energy. Every diet has an energy content that can be described in the standard units of energy: joules or calories. There must be enough energy in the diet to match the metabolic demands of the animal, also measured in joules or calories.

The energetic needs of an animal depend on many factors. Its long-term metabolic energy consumption reflects the long-term dietary needs for energy. In the short term, energy consumption and utilization frequently fall out of balance, and energy status must be buffered by the use of fuel storage depots. Body size, activity levels, growth rate, reproductive state, and environmental stress are the most important factors that influence the metabolic rate of an animal, and therefore dietary energy demands. These factors also account for the differences in energy demands between species.

Each macromolecule has a corresponding energy content, measured as a *caloric equivalent.* A gram of protein or carbohydrate possesses 4 kcal of energy, whereas fat has 9 kcal per gram. Thus, for an animal to obtain an equivalent amount of energy, it would have to eat more than twice as much protein as fat. Gross energy is measured experimentally by *calorimetry.* The food material is burned to ash, and the resulting heat production reflects the total energy content. However, not all of the food an animal consumes is digestible (Fig-

ure 11.2). If you ate nothing but wood chips (4 kcal/g), you would obtain little energy because you can't digest the plant material to liberate the chemical energy trapped within the cellulose molecules. The gross energy that can be broken down is the **digestible energy**, and the remainder is lost in the feces. Of this digestible energy, only a fraction is **metabolizable energy**, with the remainder of the absorbed nutrients lost in the urine. Much of the metabolizable energy is used by the animal to support maintenance, growth, and reproduction. This is called the **net energy**. The remainder of the metabolizable energy is lost as a result of the digestion process. This energy, called **specific dynamic action (SDA)**, is reflected in the increase in metabolic rate during the digestive process. Anyone who has overindulged in a holiday meal will recognize the effects of SDA. The heat warms the body, and in combination with neurotransmitters can induce drowsiness. Many large predators, such as lions and snakes, sleep after gorging in feeding bouts.

The SDA, or *heat increment* as it is often termed, is an important source of thermal energy for the animal. The heat of digestion is rapidly transferred to the rest of the body by the abundant vasculature that serves the GI tract. Thus, SDA contributes to heat production in endothermic animals,

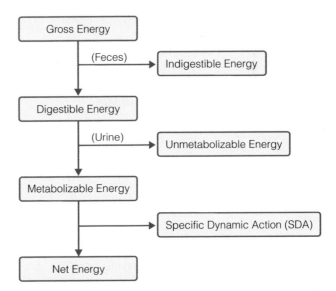

Figure 11.2 Dietary energy Not all food energy is digestible. Indigestible material, such as dietary fiber, is lost in the feces. Some of the nutrients taken up by the gut are lost in the urine, unmetabolized by the animal. A portion of the metabolizable energy is released as heat during the process of digestion. The remainder can be used to fuel activity, growth, reproduction, and other energy-dependent processes necessary for life.

reducing the need for specific thermogenic pathways. For a hummingbird feeding on a cold morning, SDA is an important contribution to whole body heat production, helping it cope with cold air temperatures as well as cold nectar; ingesting a normal-sized nectar meal at 4°C creates a thermal challenge equivalent to that experienced when the entire hummingbird is at 15°C. In some ectothermic animals, SDA causes local warming to speed the rate of digestion. The bluefin tuna, for example, possesses countercurrent heat exchangers to help retain heat within the GI tract, accelerating digestion.

Vitamins and minerals participate in catalysis

Vitamins are a group of chemically unrelated molecules with diverse functions. For simplicity, they are usually categorized based on their solubility. The fat-soluble vitamins are A, D, E, and K; the water-soluble vitamins include the B family and vitamin C (Table 11.1). Solubility influences both the mode of uptake and the potential toxicity. An animal can consume copious amounts of water-soluble vitamins with little ill effect because any excess is readily excreted in the urine. Fat-soluble vitamins

can be problematic, however, because they are stored in lipid depot tissues and can be released in a toxic pulse when fats are mobilized.

Some animals obtain selected vitamins from symbiotic bacteria living in the GI tract. For example, the gut flora of most mammals produces all the vitamin C needed in the diet. Humans, unlike most mammals, must obtain their vitamin C preformed in the diet. Fortuitously, some of the pioneering studies on vitamin C were performed on guinea pigs, which also derive vitamin C from the diet. Vitamin absorption is one of the main benefits of an unusual feeding strategy known as *coprophagy*. Rabbits, for example, eat their own feces as a way of regaining vitamins lost in indigestible material. While coprophagy increases the risk for parasites and disease, it provides important nutritional advantages to some animals, including a second opportunity to extract nutrients from the vegetation they eat.

Mineral nutrients are a collection of metallic elements that participate in protein structure, including the structure of enzymatic proteins. Aquatic animals obtain many of their essential nutrients directly from the water, transporting them across the gills or skin. Most minerals, however,

Table 11.1 Vitamins.

Vitamin*	Functions	Deficiency symptoms
Fat-soluble vitamins		
A, retinol	Visual pigments, gene regulation	Night blindness, epithelial damage
D, calciferol	Calcium and phosphate absorption	Ricketts
E, tocopherol	Antioxidant	Anemia
K, menadione	Blood clotting	Hemophilia
Water-soluble vitamins		
B_1, thiamin	Coenzyme: thiamin pyrophosphate	Beriberi
B_2, riboflavin	Coenzyme: FAD, FMN	Various skin disorders
B_3, niacin	Coenzyme: NAD, NADP	Pellagra
B_5, pantothenic acid	Coenzyme: Coenzyme: A	Adrenal and reproductive dysfunction
B_6, pyridoxine	Coenzyme: pyridoxyl phosphate	Peripheral neuritis
Biotin	Coenzyme: biotin	Hair loss, skin problems
Folic acid	Coenzyme: tetrahydrofolate	Megaloblastic anemia
B_{12}, cobalamin	Coenzyme: methylcobalamin	Pernicious anemia
C, ascorbic acid	Antioxidant, connective tissue growth	Scurvy

*The vitamins are listed in the form in which they appear in the diet. Some are modified by the animal to produce vital molecules. For example retinol is converted to retinal to produce the visual pigment.

are absorbed from the diet. Calcium enters the intestinal cell through Ca^{2+} channels and is exported into the blood by Ca^{2+} ATPases. The entire transport process is accelerated in the presence of the protein *calbindin*. Ca^{2+} uptake is controlled by vitamin D, which regulates the synthesis of calbindin. Phosphorus is imported into the intestinal cells as inorganic phosphate, transported using Na^+ cotransporters. Iron is imported into the cell in the ferrous form (Fe^{2+}) by a nonspecific divalent metal transporter, cotransported with H^+. If the iron arrives in the diet incorporated into heme, it can be transported into the cell in that form. Copper, zinc, and other minerals are also transported into intestinal cells by specific carriers. Once absorbed, these minerals are pumped out of the intestinal cell into the circulation. The target tissues import the minerals from the blood as needed for their own biosynthetic processes.

Inadequate supply of essential amino acids compromises growth

Most of the 20 amino acids that animals use to build proteins can be produced *de novo* but a subset of amino acids must be obtained preformed in the diet. Typically, there are eight essential amino acids: isoleucine, leucine, lysine, methionine, phenylalanine, threonine, tryptophan, and valine. Some species have additional essential amino acids. For example, histidine and arginine are essential amino acids for domestic dogs and sea turtles. Although the amino acid taurine is not used in proteins, it is necessary for other processes, including digestion, nervous function, and osmoregulation. Taurine is an essential amino acid for several animals, including each of the 30 species of cats examined to date. In other words, don't feed dog food to your cat.

If a diet is persistently deficient in any of the essential amino acids, the animal may experience developmental defects or slower growth. Since dietary protein is the source of these amino acids, *protein quality*—the profile of amino acids in dietary protein—is a critical nutritional concern. Animal tissues provide a higher-quality dietary protein than do plant tissues, because they possess an amino acid profile that more closely resembles the needs of other animals. In contrast, plant proteins are often deficient in one or more of the essential amino acids. For example, maize proteins are deficient in lysine and wheat proteins are deficient in tryptophan. A herbivore can avoid amino acid deficiencies by eating plants with different combinations of deficiencies.

Animals require linoleic and linolenic acid in the diet

As we first discussed in Chapter 2: Chemistry, Biochemistry, and Cell Physiology, animals use lipids for many purposes, including energy production, cellular membranes (phospholipids), and cell signaling (prostaglandins, leukotrienes). They can produce *de novo* a broad suite of fatty acids that differ in chain length and desaturation. For instance, animals can produce palmitate from acetyl CoA using the enzyme fatty acid synthase, then metabolize it into other forms by elongases (which increase fatty acid chain length) and desaturases (which introduce double bonds). However, animals cannot produce sufficient amounts of omega-3 (ω 3) and omega-6 (ω 6) fatty acids *de novo*. Instead, animals require ω 3 and ω 6 fatty acids from the diet, typically as linoleic acid (18:2 ω 6) and α-linolenic acid (18:3 ω 3). Humans can most readily meet ω 3 requirements by consuming fish. Fish also obtain their ω 3 fatty acids in the diet, derived ultimately from the photosynthetic phytoplankton at the base of the food chain. Plant seeds are the best dietary source of ω 6 fatty acids.

Digestion of specific nutrients requires specific enzymes

Digestive enzymes allow animals to convert the complex macromolecules arriving in the diet to forms that can be absorbed by the animal and processed into usable forms. Although the nature of diets is very diverse, most animals rely on the same suites of digestive enzymes.

- *Lipases* release fatty acids from triglycerides (triglyceride lipases) and phospholipids (phospholipases).
- *Proteases* (trypsin, chymotrypsin) break down proteins into shorter polypeptides. *Peptidases* are proteases that cleave successive amino acids from the end of a polypeptide. Aminopeptidases attack the first (N-terminal) peptide bond, whereas carboxypeptidases attack the last (C-terminal) peptide bond. A dipeptidase breaks the peptide bond of a dipeptide, producing two amino acids.
- *Amylases* such as dextrinase and glucoamylase break down polysaccharides into

oligosaccharides. *Disaccharidases* such as maltase, sucrase, and lactase break down specific disaccharides.

- *Nucleases* break down DNA into nucleotides, which are then broken into nucleosides and nitrogenous bases for absorption.

Not every macromolecule ingested is subjected to the digestive process. Many animals are unable to produce the enzymes needed to assimilate a dietary macromolecule. In some cases, the levels of enzymes can vary among individuals. For example, many humans show an age-dependent reduction in production of lactase, the enzyme that breaks down the disaccharide lactose. When lactose-intolerant people consume milk products, lactose can escape the upper intestine undigested, passing into the lower intestine. The methane-producing gut flora can feast on this rich energy source, much to the discomfort of the individual and displeasure of those in the immediate vicinity.

Symbiotic organisms contribute to animal digestive physiology

Despite their impressive capacity to break down food, many animals benefit from the assistance of symbiotic organisms. Three main types of symbionts participate in digestion. **Enterosymbionts** live within the lumen of the GI tract itself. Since the inside of the gut is contiguous with the external environment, in principle these symbionts live outside the animal tissues. They are called enterosymbionts to distinguish them from **exosymbionts**, which are symbionts that are actively cultivated outside the body. **Endosymbionts** are organisms that grow within the animal, typically embedded between host cells in a tissue or, less often, within a host cell.

Many species of animals are able to consume and digest unusual fuels because they possess specialized symbiotic bacterial populations, often maintained in **ceca** (singular: cecum; blind sacs branching from the gut). Some species of birds eat the wax found in beehives. Bacteria within the bird gut break down the wax into shorter carbon units that can be absorbed by the animal. Marine animals that feed on plankton can digest the chitin exoskeleton with the help of symbiotic bacteria. Whales hold chitinolytic bacteria in gastric ceca.

Not all symbionts are bacteria. Fungi are important enterosymbionts in many types of plant-eating insects. Leaf-cutter ants feed leaf fragments to exosymbiotic fungi, cultivated by the ant colony.

The ants consume both the fungi and the plant material that has been partially degraded by the fungi. Many marine animals form symbiotic relationships with photosynthetic organisms that grow interspersed with their own cells as endosymbionts. Cyanelles are cyanobacteria living in association with sponges. Dinoflagellates live as symbionts of corals. Zooanthellae are small unicellular brown algae that live in cnidarians and some molluscs. Zoochlorellae are green alga that live in association with sponges, cnidarians, flatworms, and some molluscs. These symbionts use photosynthesis to produce carbon skeletons that are taken up by the animal cells. Zooanthellae produce glycerol, and zoochlorellae produce monosaccharides such as glucose and maltose. These symbiotic cells gain protection from predators by living in the confines of the animal tissues. Perhaps the strangest endosymbiotic relationship is seen in the animals that survive in the diverse sulfur-based food webs (see Box 11.1, Evolution and Diversity: Chemolithotrophic Symbionts).

Cellulose is an important nutrient for many animals, although most species require the help of symbiotic organisms. As yet, no animal has been shown to possess a gene for *cellulase*, the enzyme that is capable of breaking the β1-4 glycosidic bond that distinguishes cellulose from digestible polysaccharides, such as glycogen (α1-4) and starch (α1-6) (see Figure 2.20). Most animals excrete cellulose undigested, forming the bulk of what is generally called dietary fiber. However, some herbivores can liberate the energy in dietary cellulose with the assistance of symbiotic organisms that live within the gut. These animals can absorb some of the glucose generated by cellulase activity, but most of the glucose is fermented by the bacteria to produce anaerobic end products, including the volatile fatty acids acetate, butyrate, and propionate. The animals then absorb these fermentation products for use in biosynthesis or energy metabolism. For example, termites digest wood fibers with the help of protists and fungi. Many species possess fermentation chambers that house cellulolytic bacteria.

In addition to benefiting from bacterial assistance in liberating energy, animals can digest the bacteria. Animals secrete the enzyme *lysozyme* into the gut to break down the bacterial cell wall. The lysozyme of ruminants has adapted to function under the harsh conditions of the ruminant fermentation chambers, whereas the lysozyme from most mammals is nonfunctional under these

BOX 11.1 EVOLUTION AND DIVERSITY
Chemolithotrophic Symbionts

Most food webs begin with the photosynthetic organisms that use the energy of the sun to grow. However, a few animals living far from sunlight survive in a food web that is based on another form of energy input: chemical oxidation. The organisms that live at the base of these alternative food webs are chemolithotrophic bacteria. They produce energy by metabolizing inorganic molecules such as ammonia and various sulfur compounds. There are two main types of harsh environments in which chemolithotrophic bacteria form the base of food webs: sewage outfalls and deep-sea vents.

Sewage outfalls are rich in organic matter and possess high levels of toxic compounds. Anoxia-tolerant chemolithotrophic bacteria thrive on the sulfides and organic material of the sewage, producing hydrogen sulfide (H_2S), which is also toxic to many organisms. Few animals can tolerate the oxygen limitations and toxic sulfides of sewage outfalls. A few species of larval insects and molluscs thrive in sewage outfalls. They are able to tolerate the harsh conditions, and feed on the rich chemolithotrophic bacterial populations.

Another environment that depends on the chemolithotroph-based food web is the deep-sea vent. The fissures in these undersea volcanoes release extremely hot water, rich in sulfides. While the surrounding waters are cold, deep-sea deserts, the vent waters are undersea oases, with warm, oxygenated water rich in biodiversity. In this environment, sunlight—not oxygen—is the limiting factor. Here, as in the sewage outfall, chemolithotrophic bacteria are the nutritional base of the ecosystem, providing food for many species of invertebrates. Some animals, like the vent mussel *Bathymodiolus thermophilus*, collect bacteria from the water by filter feeding. Others, such as the polychaete *Alvinella pompejana*, graze on the thick bacterial mats. These invertebrates in turn are food for predatory invertebrates and vertebrates, which live on the fringe of the toxic zone created by the sulfides emanating from the vents. Some species of animals enter into symbiont relationships with the chemolithotrophic bacteria. Several species appear to cultivate chemolithotrophic bacteria on their body surfaces. A few species, such as the giant clam (*Calyptogena magnifica*), house chemolithotrophic bacteria within their bodies as endosymbionts. This relationship is perhaps best understood in an unusual group of worms called pogonophorans.

More than 80 species of pogonophorans have been discovered around the world since they were first identified in the early 1900s. In 1979, a remarkable find was made by a submersible exploring the deep-sea vents in the Pacific Ocean. Giant pogonophorans (*Riftia pachyptila*), measuring over 1.5 m in length, were discovered at several sites, living at the interface between toxic sulfide emissions and cold, oxygen-rich seawater. At the

The rich biodiversity of a deep-sea vent.

anterior of the worm is a bright red featherlike structure called the plume. The red color comes from high concentrations of extracellular hemoglobin. The posterior of the worm is permanently housed in a tube. Left undisturbed, the pogonophoran extends the plume out into the water, where it is used for respiration.

A pogonophoran is unusual in that, as an adult, it lacks a mouth and GI tract. While it may absorb some nutrients across the epithelium, most of the nutrition comes as a result of an unusual symbiotic arrangement with chemolithotrophic bacteria. The bacteria are housed at very high concentrations in an internal sac called a trophosome. The worm ensures that the bacteria receive the precursors for biosynthesis—CO_2 and H_2S—and then collects the biosynthetic products: sugars and amino acids.

Many anatomical and biochemical adaptations allow these animals to survive at the high sulfide concentrations necessary for survival of their symbionts. The worms reduce their H_2S uptake by remaining in tubes, thereby limiting diffusion. Pogonophorans also have an unusual hemoglobin that resists inhibition by H_2S. In most animal hemoglobins, H_2S binding prevents O_2 binding. Pogonophoran hemoglobin is unusually large and is able to bind H_2S without affecting its ability to carry O_2. Thus, the hemoglobin can provide the oxygen for the worm and also deliver H_2S to the symbionts. Furthermore, the worm hemoglobin binds H_2S so well that it prevents H_2S from escaping into the worm tissues, where it would inhibit cytochrome oxidase and poison the electron transport system. Interestingly, there is no evidence that the pogonophoran cytochrome oxidase is any less sensitive to H_2S than is the enzyme from other animals. Thus, the pogonophoran must ensure that H_2S does not reach the oxidative tissues.

References

○ Gaill, F. 1993. Aspects of life development at deep sea hydrothermal vents. *FASEB Journal* 7: 558–565.

○ Hochacka, P. W., and G. N. Somero. 2002. *Biochemical adaptation*. Oxford: Oxford University Press.

conditions. Interestingly, one lineage of primates, the colobine monkeys, possesses foregut fermentation chambers that allow them to digest vegetation. Their lysozyme structure more closely resembles that of a cow than that of their nearest primate relatives. This example of convergent evolution illustrates the constraints on animal enzyme function, and the opportunities that are afforded animals that can digest an underutilized resource.

Nutrients move across the plasma membrane via carriers or vesicles

After digestion, nutrients are transported from the gut and transferred to the extracellular fluids where they can be imported by storage and target tissues. Refer to Chapter 2 for coverage of the many ways that molecules can be transported across the plasma membrane. Polar molecules, such as monosaccharides and amino acids, cannot penetrate the plasma membrane at a significant rate, and require specific protein transporters to move them across the plasma membrane. The nature of the transport process for a specific molecule depends on its transmembrane gradient. If there is a favorable concentration gradient, the cell may use a transporter that works by facilitated diffusion. For example, GLUT proteins are carriers that mediate facilitated diffusion of glucose across the plasma membrane. In liver, GLUT-2 allows glucose out of the cell, whereas in muscle GLUT-1 allows glucose into the cell. In both cases, glucose moves from higher concentration to lower concentration. Conversely, if the transport process must proceed against a concentration gradient, the cell must use some form of active transport. For example, amino acids are taken into cells by a carrier protein that is driven by the Na^+ gradient, a form of secondary active transport.

Some nutrients are transported across the plasma membrane via vesicles. Cells engulf regions of the plasma membrane to form vesicles. If the nutrients are in solution, the process is pinocytosis. If the nutrients are particulate, the process is phagocytosis. Similarly, cells can expel nutrients via exocytosis. These pathways of endocytosis and exocytosis are critical for the movement of complex lipids. In many cases, lipids are associated in ways that make it difficult for a membrane carrier to transport one molecule at a time. For example, lipoproteins—complexes of lipid and protein—are exported from cells via exocytosis.

Carbohydrates are hydrolyzed in the lumen and transported by multiple carriers

The main types of carbohydrate consumed by animals are polysaccharides—primarily glycogen, starch, cellulose, and chitin. Disaccharides such as sucrose, lactose, and maltose are also important in some species. Polysaccharides and disaccharides must be broken down to monosaccharides for absorption. The various amylases and disaccharidases at work in the gut ultimately break these larger carbohydrates down to produce monosaccharides, primarily glucose, fructose, and galactose, which are absorbed by the enterocytes of the small intestine (Figure 11.3). Animals

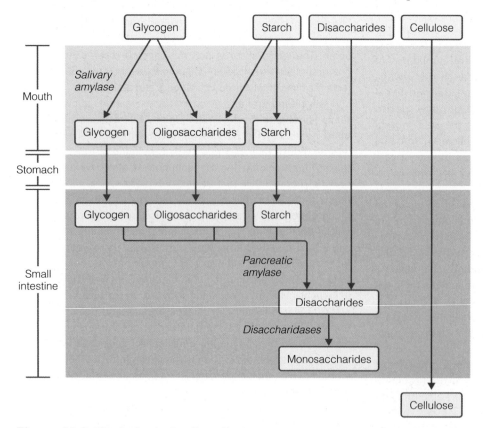

Figure 11.3 Carbohydrate digestion Starch and glycogen are broken down in the mouth and duodenum by the action of amylase. The resulting disaccharides are further processed in the duodenum by the specific disaccharidases.

use a combination of active transport and facilitated diffusion to carry monosaccharides from the lumen into the intestinal absorptive cells (enterocytes). Glucose and galactose typically enter enterocytes by a Na^+-glucose cotransporter, whereas fructose, which occurs at relatively low concentrations in the cytoplasm, enters the cell via facilitated diffusion.

Many years of study have led to a better understanding of the mechanisms by which the GI tract absorbs glucose. Much of the glucose is transported into intestinal cells by Na^+-glucose cotransporter 1 (SGLT-1). A second type of glucose transport mechanism facilitates the diffusion of glucose into cells during periods of high glucose concentrations in the lumen (Figure 11.4). The carrier is GLUT-2, a member of the large GLUT family of transporters that mediate facilitated diffusion of glucose in various tissues. In brief, when a bolus of glucose first appears in the intestine, transport is mediated primarily by SGLT-1. This transporter also acts as a glucose sensor, triggering a signaling pathway that leads to rapid synthesis of GLUT-2 and intracellular transport of the carrier to the microvilli.

An animal can increase its capacity for glucose transport in several ways, as indicated by interspecies comparisons. Studies from Jared Diamond's lab at UCLA have shown how SGLT-1 levels can determine the rate of glucose transport. Animals can increase glucose uptake by increasing the total number of SGLT-1 transporters in the gut by (1) producing more transporters per unit surface area of the gut, (2) increasing the surface area of the gut per unit length, or (3) increasing the total length of intestine. Comparisons between species are complicated by phylogenetic differences, as well as dietary differences. Consider the differences in the GI tract of a domestic chicken, which grows quickly, and the guinea jungle fowl, a slow-growing wild relative. These species have a similar surface area (per centimeter of intestine) and capacity for glucose transport (per unit surface area), but the birds differ in gut length. The longer gut of the chicken allows it to assimilate nutrients at greater rates and, as a result, grow faster.

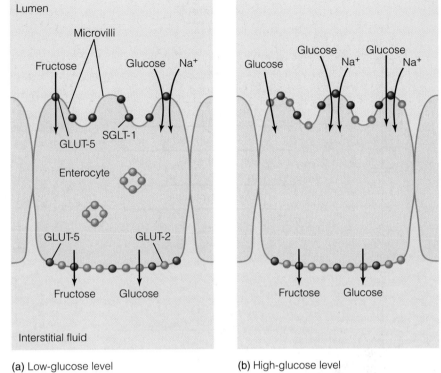

(a) Low-glucose level   **(b)** High-glucose level

Figure 11.4 Carbohydrate transport into intestinal cells (enterocyte) **(a)** Under low-glucose conditions, most glucose import occurs on the Na^+-dependent glucose transporter 1 (SGLT-1). Fructose enters the cell via facilitated diffusion on the glucose transporter 5 (GLUT-5). **(b)** When glucose levels rise, GLUT-2 transporters, another type of facilitated diffusion carrier, are translocated to the microvilli, greatly increasing the capacity for glucose uptake.

Proteins are broken down into amino acids by proteases and peptidases

The pathway for digestion of proteins begins with extracellular hydrolysis by proteases secreted from the cells of the GI tract and glands associated with the GI tract. Gastric pepsin breaks proteins into large polypeptides. These polypeptides move on to the small intestine, where pancreatic proteases (trypsin, chymotrypsin, and carboxypeptidase) break large polypeptides into small polypeptides, and peptidases of the intestinal lining liberate free amino acids, dipeptides, and tripeptides (Figure 11.5). Dipeptides and tripeptides can be transported into the epithelial cell and broken down cytoplasmically. Free amino acids are carried into the epithelial cells on amino acid–Na^+ cotransporters, much like glucose transport. Those free amino acids that are not used by the enterocytes are released into the blood for use by other tissues.

While most proteins are broken down within the lumen of the stomach and small intestine, some proteins are carried into cells intact. First, they are removed from the lumen by endocytosis

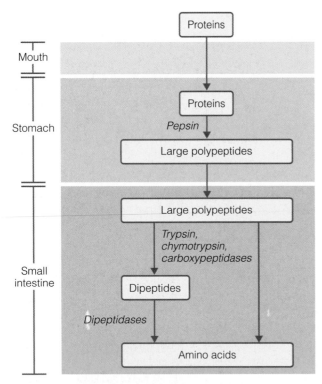

Figure 11.5   Protein digestion and transport
In the acidic stomach, pepsin breaks down large proteins into large polypeptides. Proteases from the pancreas (trypsin, chymotrysin, carboxypeptidase) hydrolyze these polypeptides into smaller polypeptides and peptides. Intestinal aminopeptidases, carboxypeptidases, and dipeptidases complete the proteolysis to produce amino acids.

across the apical membrane of the enterocyte. They can then be carried through the cell and exocytosed into the bloodstream. For example, the antibodies that arrive in the milk consumed by an infant mammal are transported to the blood intact, transferring immunoprotection from the mother.

Lipids are transported in many forms

Digestion and import of lipids are complicated by their hydrophobicity. The gastrointestinal tract overcomes the solubility limitations by secreting chemicals that act as lipid emulsifiers. Bile is a mixture of cholesterol, phospholipids, pigments, and salts produced in the liver and secreted into the intestine. The phospholipids, mainly lecithin, act in conjunction with the bile salts to organize the lipids into small droplets called **micelles**. Dietary cholesterol and fat-soluble vitamins form the inner hydrophobic core of the micelle. Fatty acids and monoglycerides coat the hydrophobic core and interact with the outer coating of bile salts and lecithin. The micelles diffuse to the microvilli, where

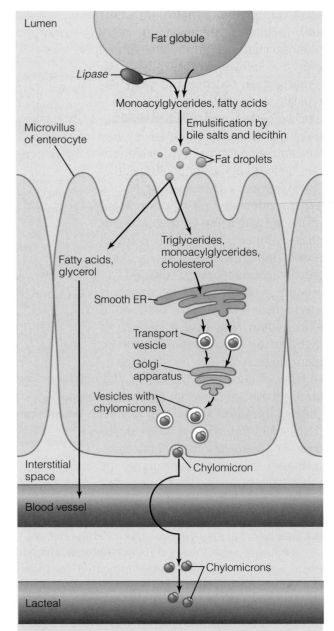

Figure 11.6   Lipid transport across the gut
Lipids reach the small intestine in the form of large insoluble globules. High pH, bile salts, and phospholipids (lecithin) emulsify the fat into smaller fat droplets. These make their way to the microvilli where fatty acids and monoacylglycerides can cross into the enterocyte. Inside the enterocyte, the lipids are taken up by the ER and repackaged into vesicles that are secreted from the cell into the surrounding lymph glands (lacteals).

the components simply diffuse off the micelle, crossing the enterocyte cell membrane.

The fate of each lipid depends on its physical properties (Figure 11.6). Short chain fatty acids and glycerol are sufficiently polar that they can be carried in the blood without assistance. After being taken up by the enterocyte, these molecules cross the basal membrane and enter the blood, where they travel to the liver via the hepatic portal vein. Longer chain fatty acids, monoacylglycerides, and cholesterol are relatively insoluble and must enter the systemic circulation by a different route. Once in the cytoplasm, the fatty acids and monoglycerides are used to resynthesize triglyceride. The enterocyte smooth endoplasmic reticulum takes up the lipids and packages them into vesicles. The vesicles pass through the Golgi apparatus and travel via secretory vesicles to the basal membrane. During their travels, the lipids are arranged into small droplets coated by proteins. The vesicles fuse with the cell membrane, and lipid complexes called **chylomicrons** are released into the lymph that bathes the cell. The chylomicrons are carried through the lymph into the venous system. As they travel around the bloodstream, they can be processed by peripheral tissues. In the endothelial cells of the capillary beds, lipoprotein lipase breaks triglyceride down to fatty acids and glycerol, which are absorbed by the tissues. The chylomicron remnant, partially depleted of triglyceride, is extracted by the liver and repackaged into a lipoprotein complex enriched in cholesterol.

Lipids are carried throughout the body in the form of lipoprotein complexes. The lipoprotein complexes, classified by their buoyant density, exhibit a range of sizes and compositions (Table 11.2). Each class of lipoproteins possesses a characteristic profile of proteins that regulate lipid transport and metabolism. Lipoproteins control the transfer of triglycerides, phospholipids, and cholesterol between tissues (Figure 11.7). When carbohydrate and fat intake exceeds energy demand, the liver responds by synthesizing lipid and sending it to other tissues for storage. The liver produces and releases triglyceride in the form of a very low-density lipoprotein complex (VLDL). As the VLDL moves through the circulation, the triglyceride is hydrolyzed by lipoprotein lipase and is progressively depleted. The fatty acids released in the capillary beds can be stored or oxidized, depending on the specific needs and abilities of the tissue. What was once a triglyceride-rich VLDL becomes an intermediate lipoprotein (IDL), then eventually a cholesterol-rich LDL. The LDL can bind specific receptors on various tissues to unload cholesterol that will be used for membrane synthesis or other biosynthetic pathways. At any point in this cycle, lipoprotein complexes can be returned to the liver for repackaging.

The proteins in the lipoprotein are important in controlling the lipoprotein composition and metabolism. For example, some proteins in the VLDL and LDL complexes regulate lipoprotein lipase. The proteins found in high-density lipoproteins (HDL) are important building blocks for other lipoproteins. For example, HDL donates proteins to the chylomicrons that exit the intestinal lymph, and to VLDL circulating in the blood.

Finding and Consuming Food

You are familiar with the basic dietary strategies seen in animals—carnivory, herbivory, and omnivory—each with its advantages and disadvantages (see Box 11.2, Applications: Animal Diets and Human Health). The physiology of digestion is matched to the chemical and physical nature of the diet. To find the food that matches their dietary needs, animals use neurosensory systems. Some feeding strategies, such as filter feeding, depend on random encounters. Most animals, however, actively seek out and often pursue their food. Once found, the food must be ingested to

Table 11.2 Lipoprotein composition.						
			Composition (% of total mass)			
Lipoprotein	**Density (g/ml)**	**Diameter (nm)**	**Protein**	**Phospholipid**	**Triglyceride**	**Cholesterol**
Chylomicron	0.95	75–1200	2	8	86	4
VLDL	1.006	30–80	7	18	58	17
IDL	1.006–1.019	25–35	17	22	22	39
LDL	1.019–1.063	18–25	22	18	8	52
HDL	1.063–1.210	5–12	50	28	8	14

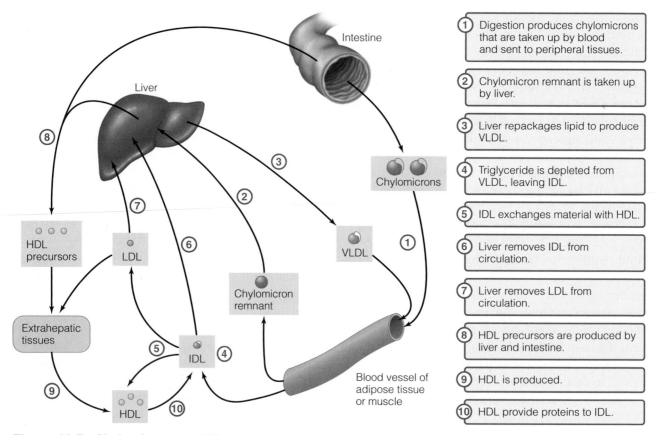

Figure 11.7 **Chylomicrons and lipoprotein complexes**

Animals sense food using chemical, electrical, and thermal cues

The nature of food varies widely among animals, and the mechanisms animals use to detect food are equally diverse. Animals link some form of receptor to a signaling pathway that leads to a behavioral response that alters feeding.

Many animals possess means of detecting the presence of specific chemicals in the environment. The chemical may be a nutrient, and movement toward the source of the nutrient increases the likelihood that the animal will find more food. For example, the cestode (tapeworm) *Hymenolepis diminuta* undergoes diurnal migrations up and down the GI tract of its host, following the nutrients released from a meal under digestion. In other cases, the chemical that is detected may not itself be a nutrient but rather a signal that prey is nearby. For example, when a *Hydra* detects small organic molecules such as proline or reduced glu-

tathione, it waves its tentacles and opens its mouth. As we discussed in Chapter 6: Sensory Systems, complex animals use *gustatory receptors* and *olfactory receptors* to locate food, determine its palatability, and control the drive to feed (appetite). Herbivorous insects, such as aphids, use gustatory receptors to detect chemicals that either stimulate feeding (phagostimulants) or deter feeding (phagodeterrents). The most important phagostimulants in insects are sugars and amino acids. Plants can deter insects from feeding by releasing secondary metabolites that an insect recognizes as toxic. Gustatory signals are also important in vertebrates. Carrion eaters detect volatile compounds that escape rotting flesh. Sharks are able to detect from great distances chemicals that are normally found in vertebrate blood, a sign of an injured animal. Although the chemical nature of the gustatory stimulants is diverse, each works in combination with a sensory receptor that triggers a signaling cascade ultimately affecting central control of feeding behavior.

Many animals find prey by sensing the energy emitted or reflected from the animal in the form of light, sound, heat, or electricity. A bird of prey,

BOX 11.2 APPLICATIONS
Animal Diets and Human Health

Warnings of potential links between diet and disease, rampant in the modern media, are nothing new. In medieval times, pregnant mothers were discouraged from eating too much rabbit for fear that the infant would have a harelip. Purple babies, now recognized as victims of hypoxia, were attributed to excessive consumption of eggplant. The ancient Hebraic dietary laws of *kashrut* originated in a perception that some foods were healthier than others. The only mammals that could be considered kosher were herbivores, and only those that eat their own cud (ruminants). Carnivory has many nutritional benefits, but it is a feeding strategy that poses greater risks for passage of disease. By creating a taboo on eating carnivores and scavengers, the Hebraic dietary practices reduced the risks for dietary consumption of infectious agents. Ironically, it took modern agricultural practices to convert a low-risk herbivorous cow to a high-risk carnivore, with dire consequences for human health.

A series of related diseases triggering dementia and death arise spontaneously in animals. The best known is bovine spongiform encephalitis (BSE), or as it is more commonly known, mad cow disease. Others include Creutzfeldt-Jakob disease (CJD), kuru, and scrapie. CJD causes dementia and brain damage in older people. Kuru is a form of human dementia first recognized in cannibalistic tribes in New Guinea. Scrapie affects sheep, as well as cats, mink, and other mammals. Each of these diseases is caused by a type of protein called a *prion*. In 1972, Stanley Prusiner began experiments on hamsters to study the basis of these diseases. He showed that a hamster would develop dementia if it was fed an extract of the brain of a diseased hamster. By 1982, he had established that the infectious agent was a prion. The prion is not a genetic mutation, as occurs in cancer, but rather a structural variant of a normal protein. Strangely, the function of this gene and the normal protein remains to be established. The protein is harmless when folded properly, but when it misfolds it becomes an infectious particle that can induce disease. Animals genetically engineered with the prion gene knocked out are immune to prion diseases, such as BSE.

In the mid-1980s, British cows began to show high frequencies of BSE. Although this was of great concern to the farmers, it received little public attention at the time. In the early 1990s, researchers noted an increase in the cases of CJD-like disease in Britons. Unlike most cases of CJD, this new variant of CJD (nvCJD) primarily affected young people, with a median age of death of 29. In 1996, British health officials announced a potential link between the BSE outbreak in cows and the human nvCJD. This led to an international boycott of British beef, followed by a massive program to kill existing cows in Britain and a revamping of agricultural policies around the world.

The practice that was most likely responsible for the BSE outbreak was rendering of unmarketable beef tissues into feed for other cows. Infectious prions concentrate in the nervous tissue of animals, and when cows consume the brain material of infected cows, they too contract the disease. Unlike many infectious agents, prions are virtually indestructible and readily survive the acidity of the stomach and other lines of defense.

This agricultural practice was not appreciated as risky for humans, because it was widely held that prions could not cross the species barrier. British sheep had been infected with scrapie for hundreds of years without evidence of transfer to other species. Although there remains no direct experimental evidence of a link between BSE and nvCJD, the indirect evidence is overwhelming. More than 98% of the cases of nvCJD occurred in people that lived in Britain during the mid-1980s outbreak of BSE. Prions may find it difficult to cross the species barrier; however, even a low-probability event is a risk when a high number of animals are infected. By 2003, more than 183,000 British cows had been identified as carriers of BSE, and nvCJD had claimed the lives of more than 130 people. At present, the risks of BSE and nvCJD are very low because of regulations that reduce the risk of transmission. At one point, beef by-products, dead and dying cows, and even roadkill made its way into animal feed. Now many countries have moved to eliminate the use of rendered animals in animal feed. By rendering infected animal tissues into food for a herbivore, humans bypassed the natural defenses that are intrinsic to feeding strategies, enabling an infectious particle to make it into our own food chain.

References

○ Prusiner, S. B. 1997. Prion diseases and the BSE crisis. *Science* 278: 245–251.

○ Richt, J. A., P. Kasinathan, A. N. Hamir, J. Castilla, T. Sathiyaseelan, F. Vargas, J. Sathiyaseelan, H. Wu, H. Matsushita, J. A. Koster, S. Kato, I. Ishida, C. Soto, J. M. Robl, and K. Yoshimi. 2007. Production of cattle lacking prion protein. *Nature Biotechnology* 25: 132–138.

such as the golden eagle, uses its visual system to spot a field mouse moving in a distant meadow. Some insects can detect the infrared light emission emitted from warm bodies of potential prey species. Light can also be produced by animals in conjunction with foraging strategies. For example, predatory firefly species attract a prey firefly species by producing a light pattern that mimics the mating signal of the prey. Deep-sea fish use bioluminescent appendages to lure small prey. There are many examples of predator-prey coevolution, in which prey properties such as cryptic color are selected on the ability to confuse the visual system of its predator. Animals that rely on sound energy as a feeding strategy employ a variety of sound detection organs ranging from the mammalian ear to the fish lateral line. The weakly electric knifefish, which lives in the murky waters of the Amazon, uses electromagnetic receptors to detect the muscle activity of potential prey items.

Simple animals digest food within phagocytic vesicles

The simplest of animals, the sponges, obtain nutrients primarily by *phagocytosis,* much like protists such as the amoeba. Sponges subsist on particles of various sizes, ranging from organic debris much smaller than bacteria (<1 μm) to as large as zooplankton (>50 μm). Water carrying food particles passes through the sponge's network of pores and channels, flowing in currents generated by flagellated cells called choanocytes (Figure 11.8). As the water permeates the animal, it flows through biological filters that sort particles by size. Cells that line the pores, choanocytes as well as amoebocytes, engulf the particles using phagocytosis. Digestion occurs inside these cells in endocytic vacuoles. Breakdown products are released into the cell, and undigested material is exocytosed out of the cell.

Other metazoans possess something akin to a mouth—an entrance to an internal compartment that carries out digestion. The challenge for many animals is to get the food to the mouth. Cnidarians, such as corals and *Hydra*, use tentacles to capture small prey, such as zooplankton. Once the prey is captured, the tentacle bends to the mouth to release the food. The mouth gapes to permit food to enter the gastrovascular cavity. Movement down the tentacles and into the mouth is aided by a layer of mucus secreted by the epithelial cells. The wall of the gastrovascular cavity is composed of gastrodermal cells, including nutritive cells and enzymatic gland cells (Figure 11.9). The enzymatic gland cells release digestive enzymes that break down prey into a slurry of nutrients. The nutritive cells phagocytose the smaller particles and process them within the endocytotic food vacuole, releasing nutrients that escape the gastrodermis and cross the gelatinous mesoglea to supply the diverse cells of the epidermis, including the stinging cells, or *nematocytes*. Once the meal is digested, the animal expels the remaining material from the gastrovascular cavity and feeds again.

Feeding structures are matched to diet

Most animals have some form of specialized mouthparts to assist in feeding. The mouth itself may be lined with hard structures that grasp or cut the food. Some form of extension may also protrude from the mouth to manipulate, disrupt, or suck. Although we typically identify these structures as jaws and tongues, they are extraordinarily diverse in structure and developmental origins.

Some species of invertebrates, such as free-living worms, have a simple mouth that engulfs particles. Most invertebrates have structures associated with the mouth to aid in feeding. For example, some endoparasitic worms, such as the liver fluke, have a mouth that acts as both a siphon and an attachment organ. Although cestodes possess an ante-

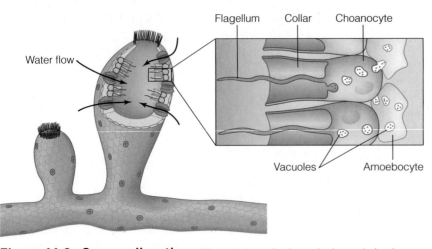

Figure 11.8 Sponge digestion Water is brought through channels by the flagellated choanocytes. Food particles are phagocytosed by choanocytes and amoebocytes.

rior attachment organ (a combination of suckers and hooks), they have no mouth and, in fact, lack the entire digestive system. These worms absorb nutrients over the outer body surface, which is decorated with spikelike extensions of the cells called microvilli. In many ways, the cestode anatomy resembles a gut turned inside out.

Many animals possess oral appendages that are functionally homologous to a tongue. Snails have a muscular tongue called a *radula*. Sharp protrusions from the radula help the snail to grind, rasp, or cut away chunks of food. Many nectar-eating butterflies and moths possess a long tubelike tongue, or *proboscis*. The insect uncoils its proboscis to reach deep into the throat of flowers to get to the nectar. Flower anatomy often coevolves with butterfly tongues, ensuring that flowers are pollinated by specific species of butterflies. When Darwin studied an unusual orchid from Madagascar, he predicted that a moth would be found that had a tongue long enough to feed on the flower. The Madagascar hawk moth, discovered about 40 years later, has a proboscis that is nearly 30 cm long.

Many species possess hardened mouthparts or oral appendages that help penetrate or mechanically disrupt the surface of food. The squid uses a hard beak to bite off chunks of prey captured by its tentacles. Arthropods possess complex mouth segments that help the animals acquire food. A spider uses its *chelicerae* to attack and mechanically disintegrate its prey (Figure 11.10a). Hymenopterans, which include bees and wasps, are chewing insects. The paper wasp, for example, has hinged mandibles that can crush the tough exoskeleton of the insects on which it feeds. Insects of the order Diptera use their mouthparts to suck fluids. For example, fruit flies siphon plant juices from rotting fruit, and mosquitoes extract blood from vertebrates. However, anyone who has been bitten by a large horsefly will be difficult to convince that these dipterans are not biting insects but sucking insects. The horsefly uses hardened mouthparts like scissors to slice through the skin. It can then use its labium to suck the fluids that seep from the wound.

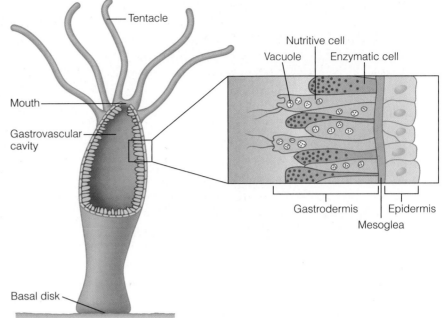

Figure 11.9 Cnidarian digestive system A cnidarian, such as *Hydra*, captures food with its tentacles, and carries it to the mouth in mucous streams. The food passes through the open mouth into the gastrovascular cavity for digestion. Particles are phagocytosed by nutritive cells lining the cavity, and digested in endocytic vacuoles. Nutrients can then diffuse from the nutritive cells of the gastrodermal layer to the other cells of the gastrodermis (gland cells) and epidermis (sensory cells, nematocytes, epithelial cells).

The most important feeding structure of vertebrates is the jaw. With the exception of the primitive agnathans, which lack jaws, all vertebrate mouths are built upon a common plan of paired jaws. In 1983, Gans and Northcutt suggested that the single most important variation to arise in the evolution of vertebrates was a "new head." They meant that the transition from agnathans to gnathostomes was accompanied by a change in the developmental pattern of the structures that give rise to the vertebrate head. The rearrangement gave vertebrates the developmental flexibility to tolerate evolutionary changes in the structure of the head and associated features. This facilitated adaptive radiation by permitting the evolution of feeding structures that allowed vertebrates to succeed in novel niches. Consider, for example, the variations in the organization of the vertebrate jaw. In most species, the upper jaw is immobile and integrated into the skull, whereas the lower jaw is hinged and movable. In contrast, several species have evolved a more mobile upper jaw. Some snakes can separate or disarticulate the jawbones. The egg-eating snake can disarticulate its jaw, allowing it to open its mouth more than four times larger than its normal gape, enabling it

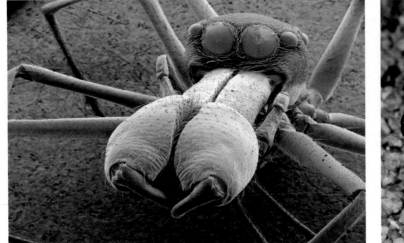

(a) Spider chelicerae

(b) Egg-eating snake

Figure 11.10 Feeding appendages **(a)** Spiders have appendages associated with the mouth that aid in holding, manipulating, and disrupting prey. **(b)** Many snakes can separate the upper and lower jaw, expanding the gape to enable the animal to swallow large food items, such as a whole egg.

to swallow an intact egg (Figure 11.10b). It then uses strong neck muscles to crush the egg against the spine. Once the egg contents slide down the throat, the snake vomits the eggshells.

Bird beaks are composed of keratinized tissue

The beak of a bird is composed of bone covered by overlapping scales called *rhamphotheca*. The cells in this epidermal layer produce a cytoskeleton that is rich in the intermediate filament keratin. When the surface cells die, the layer of keratin remains as a protective surface over the beak. Living cells within this layer constantly repair the keratin layer as it is damaged, particularly at the margins of the beak, which are often abraded during feeding.

Diversity in the beak structure of Galapagos finches was central to Darwin's theories of evolution and natural selection. Beak morphology is very diverse among birds, reflecting the type of food each bird gathers. The Argentine birds shown in Figure 11.11 demonstrate morphological specialization for eating fish, seeds, and insects. Very long beaks can be used to reach deep into flowers; the beak of the sword-billed hummingbird is longer than the body of the bird itself. Flamingos use the beak as a sieve to strain food out of water. Some birds, such as the puffin, possess toothlike ridges on the margins of the beak to assist in tearing apart flesh.

It is important to remember that bird beaks serve purposes other than feeding, including vocalization, defense, grooming, and courting. The

(a) Amazonian Kingfisher

(b) Ultramarine Grosbeak

(c) Pearly-vented Tody Tyrant

Figure 11.11 Bird beak diversity in Argentine birds **(a)** Amazonian Kingfisher—fish eater. **(b)** Ultramarine Grosbeak—seed eater. **(c)** Pearly-vented Tody Tyrant—insect eater. (Courtesy of Paul Handford, University of Western Ontario)

morphology of beak structure reflects the evolutionary compromises that allow the beak to perform each of its roles. For example, the subtle differences in the beak shape of Darwin's finches can influence the nature of the birdsong, which has important ramifications for territorial behavior and courtship. While the adaptive significance of bird beak morphology is clear, the developmental and evolutionary determinants of beak shape have only recently been studied (see Box 11.3, Genetics and Genomics: Variation in Bird Beaks).

Mammals have bony teeth

Many vertebrates possess oral structures that resemble and function as teeth, but mammalian teeth are structurally unique. Each tooth is composed of a crown, neck, and root (Figure 11.12). The crown extends above the gum, or gingiva; the root is embedded in the gum; and the neck is a narrow region between the crown and the root. A cross-section through the tooth reveals the three layers of a typical tooth: enamel, dentin, and pulp. The outer enamel is composed of calcium phosphate crystals integrated into the extracellular matrix. Enamel is so hard that it can be brittle, cracking when an animal bites a hard food. Animal teeth differ in the thickness of the enamel layer as well as its molecular composition. Beneath the enamel is an intermediate layer of dentin and an inner layer of pulp. The dentin is a porous support for the enamel. The pulp is more cellular, and rich in blood vessels and nerves. These two inner layers are living tissues that help build and maintain the tooth.

Mammals possess four main types of teeth: incisors, canines, premolars, and molars (Figure 11.12b). Incisors and canines are long, sharp teeth that aid in piercing and tearing flesh. The broad, flat molars aid in grinding. Premolars are intermediate in shape and have a role in both tearing and grinding. Like beak morphology, the shape of mammalian teeth differs markedly in ways that reflect the nature of the diet. The molars of insectivorous bats have sharp cusps and elongated crests to help crack insect exoskeletons. In contrast, fruit bats have molars with broader cusps and basins for crushing plant tissue. There are also many differences in the number of teeth and their growth patterns. Rodents, for example, possess only front incisors and molars; the canines and premolars are lost in early development to make room for the larger incisors and molars. The profile of teeth can also change over the lifetime of an animal. Most mammals replace their teeth once during the lifetime: early teeth are replaced by the permanent adult teeth. However, monotremes lose their teeth altogether when the animal matures. Most mammalian teeth grow to a predetermined size and then stop growing. In contrast, rodent teeth grow continuously, allowing the tooth to maintain length as it is worn down from continuous grinding.

⊙ | **CONCEPT CHECK**

1. How is energy partitioned in an animal's diet?
2. What are the major nutrients in a diet, and what enzymes metabolize them into the forms in which they are transported into the digestive epithelium?
3. How do the physical properties of nutrients affect their uptake and transport within the body?

Integrating Digestion with Metabolism

With an understanding of the nature of nutrients, and the way animals acquire food, we turn our attention now to the ways that an animal uses its digestive system to extract the nutrients from food. We begin with a discussion of the types of cells and tissues that make up digestive systems, and then consider how the animal controls gut function. Hormones and neurotransmitters are central to the control of digestion, ultimately matching whole animal metabolic needs to feeding behavior, nutrient uptake, storage, and mobilization. These controls are particularly important when the animal experiences physiological challenges and transitions associated with life history patterns, including development and reproduction.

Digestive Systems

The evolutionary history of the digestive systems is marked by increasing anatomical and functional specialization. Ancient invertebrates possess simple digestive sacs that food enters and leaves through a single opening to form a two-way gut. The two-way gut, such as the blind gastrovascular sac of cnidarians, was an important evolutionary

BOX 11.3 GENETICS AND GENOMICS
Variation in Bird Beaks

Many of Darwin's earliest studies focused on the variation in beak morphology between finch populations in the Galapagos Islands. He noted that a single species could have a different beak morphology on each island. Furthermore, the morphological properties appeared to be related to the nature of the food and the abundance of other species that might compete for that food. Although he proposed that these structural variations arise due to natural selection, he was at a loss when it came to understanding how the variations arose. When he returned home, he extended these studies by comparing the morphology of different local pigeon breeds. "Each successive modification, or most of them, may have occurred at an extremely early period . . . from causes of which we are wholly ignorant," he wrote. Without knowing the mechanisms, Darwin realized that beak morphology was established early in embryonic development. Recent studies have helped us understand the cellular factors that determine beak morphology within and between species.

Since a bird emerges from the egg with its beak formed, we know that the factors that establish beak morphology begin well before hatching. Within the first one or two days after fertilization, the bird embryo has established body segments that will eventually give rise to all of the structures of the head. Four sequential segments give rise to most of the structures that make up the adult beak: forebrain, midbrain, rhombomere 1, and rhombomere 2 (Figure A). Within the next two to three days, cells from the developing brain migrate to a new destination, and differentiate to form many of the structures of the developing head. These cells, called neural crest cells, move from the forebrain and midbrain to form the frontonasal region, which becomes the upper beak. Neural crest cells from the midbrain and rhombomere 1 and 2 migrate to form the mandibular and maxillary regions, which become the lower beak (Figure B). The cel-

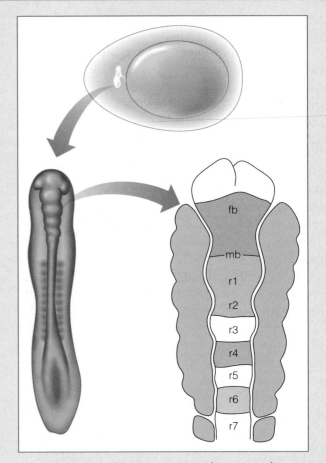

Figure A   Chick development at day 1 (dorsal view).

lular progeny of these neural crest cells interact with neighboring cells to form the embryonic beak.

The importance of these cells in generating beak morphological variation was recently shown in a study that compared the development of the beak in the duck and the quail. Ducks use the beak to catch fish and har-

step because it allows the animal to isolate food in a controlled environment and bombard it with degradative enzymes until it reaches a form that can be assimilated. Platyhelminths also have a two-way gut; it can be a simple sac, as in small flatworms, or as seen in larger flatworms, a more complex sac with multiple side branches known as diverticula (Figure 11.13). Digestion begins in the lumen of the sac when proteases are secreted from the sac wall. Once the food is broken down into small particles, the cells that form the lining of the gut phagocytose the particles. Digestion continues within cells in acidic vesicles that become basic over the next 8 to 12 hours. The cells that line the sacs have subtly different secretory and absorptive functions, allowing these animals to create regions with specialized digestive processes, which increase the efficiency of digestion.

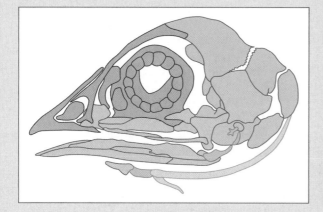

Figure B   Head structures (lateral view).

vest aquatic plants and weeds, whereas a quail uses its beak to peck and crack seeds. A duck beak is long, broad, and flat in comparison to the quail beak. These differences in beak morphology and developmental rate are determined by the neural crest cells. When researchers transplanted neural crest cells from a duck embryo into a quail embryo, the animal developed into a quail with a ducklike beak. Similarly, when quail neural crest cells were transplanted into a duck embryo, the animal developed a quail-like beak. Interestingly, the donor cells in each of these experiments also modified the morphology of other facial structures, such as the egg tooth (a nail on the end of the beak that the hatchling bird uses to break open the egg), even though the egg tooth is not derived from neural crest cells. The neural crest cells produce proteins that help coordinate structural development by regulating the expression of genes in neighboring cells. Even though the precursors of the egg tooth came from the duck, they obeyed the orders issued by the transplanted quail neural crest cells and developed into a structure that resembled a quail egg tooth.

It is the timing of gene expression that distinguishes neural crest cell behavior between species. Both quails and ducks express the same developmental genes during craniofacial development, but the timing of expression is quite different between species. In the transplant studies previously described, the donor neural crest cells caused genes in the host-derived tissue to take on the timing of the donor cells.

Variations in beak shape occur between species, but as Darwin observed there is also individual variation in beak morphology among populations of a species. Neural crest cell regulation may also be responsible for these intraspecific variations. For example, individuals may have subtle differences in the timing of expression of these genes that cause the natural range of phenotypes within a population. Recent studies have shown that the timing of expression of another regulatory protein, Bmp4, is central to the phenotypic features of bird beaks. The dominant beak phenotype within the population can change rapidly in response to changing environmental conditions. A 30-year study of two Darwin finch species shows continuous oscillations in the most common beak morphological phenotype within a population. Bmp4 expression is at least part of the answer.

References

- Abzhanov, A., M. Protas, B. R. Grant, P. R. Grant, and C. J. Tabin. 2004. Bmp4 and morphological variation of beaks in Darwin's finches. *Science* 305: 1462–1465.

- Grant, P. R., and B. R. Grant. 2002. Unpredictable evolution in a 30-year study of Darwin's finches. *Science* 296: 707–711.

- Schneider, R. A., and J. A. Helms. 2003. The cellular and molecular origins of beak morphology. *Science* 299: 565–568.

- Wu, P., T. X. Jiang, S. Suksaweang, R. B. Widelitz, and C. M. Chuong. 2004. Molecular shaping of the beak. *Science* 305: 1465–1466.

With the evolution of the one-way gut, animals were better able to create specialized regions. The nature of these regions varies widely among animals. Our description of gut regions is based on the terminology used for mammals (Figure 11.14). The mouth opens into the upper region of the GI tract called the pharynx or esophagus. This upper region typically participates in the mechanical breakdown of food. The gastric region or stomach follows; in most animals this is an acidic compartment. The upper intestine, or small intestine, neutralizes the acidic solution released from the stomach, and carries out much of the digestion and nutrient absorption. The upper intestine also receives exocrine secretions from digestive glands: the liver and pancreas in most vertebrates and the hepatopancreas in most invertebrates. The lower intestine, or large intestine, is responsible for

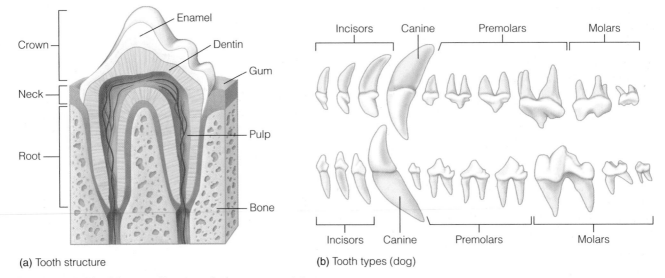

(a) Tooth structure

(b) Tooth types (dog)

Figure 11.12 Mammalian tooth structure **(a)** The mammalian tooth is composed of three layers. The outer enamel is dead tissue. The inner pulp and intermediate dentin are composed of living cells, nourished by blood vessels, and innervated. The shape and size of the types of teeth vary among species. **(b)** Molars and premolars are generally flattened teeth used for grinding and chewing. Incisors and canines are used for piercing and tearing.

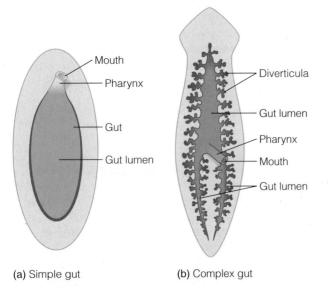

(a) Simple gut

(b) Complex gut

Figure 11.13 Flatworm GI tracts Like the simple animals, such as sponges and cnidarians, the flatworms have two-way guts. **(a)** Most flatworms, such as *Macrostomum*, possess a simple gut with a single sac. **(b)** In some larger flatworms, such as *Dugesia*, the gut can have three or more side branches with lateral diverticula.

reclamation of water and salts. Finally, indigestible material is released through the anus. Most species have side chambers that branch off from the main GI tract. A single chamber is called a *cecum*; multiple chambers are ceca. Muscular valves (sphincters) regulate passage through the different compartments.

Superimposed on the evolutionary variation in gut design are modifications that arise in response to diet and life history. The mammalian diet changes as offspring transition from maternal bloodborne nutrients across the placenta, to mammary gland secretions, to solid food. The developmental transitions in digestion are perhaps most extreme in the insects, where the diet of larvae is often completely different from the adult diet. For example, most larval lepidopterans (caterpillars) eat plant leafy material whereas many adults (butterflies and moths) eat nectar. Many larval dipterans are fully aquatic, feeding on the bacteria that live on the surface of stagnant water (mosquitoes) or in the sediment (chironomids). The adults are fully terrestrial, feeding on a wide range of plant and animal material. Remarkably, male and female adult mosquitoes consume different diets; females feed on the blood of vertebrates, from tree frogs to mammals, whereas males drink plant nectar.

Gut complexity is linked to the appearance of the coelom

The evolutionary and developmental origins of a one-way gut are intimately linked to the appearance of the **coelom**, an internal cavity that arises in a developing embryo. The space between the GI tract and the body wall, known as the *peritoneal cavity*, is one part of the coelom of a vertebrate.

The simplest of animals—cnidarians and sponges—lack a coelom; their guts are built from two germ cell layers that form solid tissues without internal compartments. More advanced animals possess three layers of germ cells—endoderm, mesoderm, and ectoderm. Nemerteans and flatworms remain relatively simple because the three layers of germ cells stick together during development and no coelom forms (Figure 11.15). In the development of rotifers and nematodes, a gap appears between the endoderm and mesoderm. The gap persists in the adult, giving rise to a type of coelom called a *pseudocoelom*. All other major animal taxa possess a true coelom between layers of the mesoderm. The appearance of the coelom was important in the evolution of digestive physiology because it allows greater specialization of internal organs.

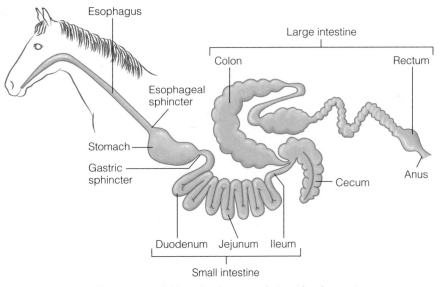

Figure 11.14 Features of a typical gastrointestinal tract Although the exact organization of the GI tract differs among species, most complex animals have regions that are functionally analogous to the typical mammal, such as the horse shown here.

The coelom arises early in embryonic development. During early gastrulation, a region of the blastula (a hollow ball of cells) migrates inward, causing first a depression and then a pit called the blastopore. In animals called *protostomes* ("first mouth") the blastopore becomes the mouth, and the anus forms at a distant site. Arthropods, annelids, and molluscs are all protostomes. In *deuterostomes* ("second mouth"), the anus arises from the blastopore, and the mouth is formed second. Deuterostomes include chordates, hemichordates, and echinoderms. The coelom originates by two different routes in protostomes and deuterostomes (Figure 11.16). The coelom of protostomes forms when the mesoderm splits (*schizocoelous*), whereas the deuterostome coelom forms from layers of mesoderm that pinch off from the gut (*enterocoelous*). However, the coelom of chordates (which are deuterostomes) forms by the schizocoelous process that is typical of protostomes.

The early embryonic gut is derived from endoderm, and divided into three regions: foregut, midgut, and hindgut. In the chicken, the gut develops into these regions within four days postfertilization (Figure 11.17). These regions differentiate to form the embryonic gastrointestinal tract. The foregut endoderm gives rise to the esophagus, stomach, and anterior region of the duodenum of the small intestine. It also forms buds that develop into the pancreas and liver. The midgut endoderm develops into the posterior part of the duodenum, the remainder of the small intestine (jejunum and ileum), and much of the large intestine, including cecum, appendix, and part of the colon. The hindgut endoderm develops into the remainder of the colon and the rectum. The properties of these regions continue to change through development and after hatching, matching physiological capacities to the diet.

The digestive systems of complex animals maximize surface area

In the simplest animals with a two-way gut, macromolecule breakdown occurs primarily inside vesicles within the cells. Proteins, complex sugars, and lipids are hydrolyzed, and the end products—amino acids, monosaccharides, and fatty acids—are released directly into the cytoplasm. More complex animals carry out these reactions within the lumen of the digestive tract. The end products of this extracellular digestion must then be taken into the cells lining the gut. Since this uptake step requires many different digestive enzymes and nutrient transporters, the process can be slow. Complex animals increase the efficiency of transport by building guts with very large surface areas. This can be achieved two general ways: increasing gut length and increasing surface undulations.

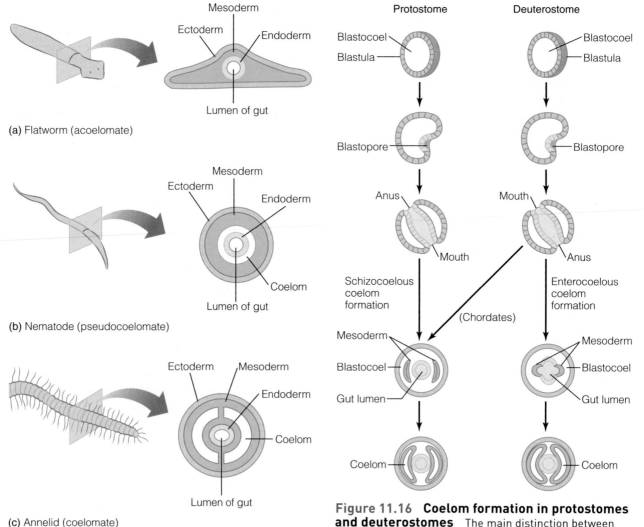

(a) Flatworm (acoelomate)

(b) Nematode (pseudocoelomate)

(c) Annelid (coelomate)

Figure 11.15 Acoelomates, pseudocoelomates, and coelomates With the exception of sponges and cnidarians, animals are distinguished on the basis of the nature of the coelom. **(a)** Acoelomates lack a coelom. **(b)** The coelom appears between endoderm and mesoderm in pseudocoelomates, and **(c)** between two mesodermal layers in coelomates.

Figure 11.16 Coelom formation in protostomes and deuterostomes The main distinction between protostomes and deuterostomes is the fate of the blastopore. In protostomes it forms the mouth, whereas in deuterostomes it forms the anus. Later in development, the coelom of protostomes forms when a layer of mesoderm separates to form an internal compartment (a schizocoelous coelom). In deuterostomes, the coelom forms from outpouches of the embryonic gut (an enterocoelous coelom). However, the coelom of chordates forms by the schizocoelous route.

The overall length of the GI tract can be a fraction of the length of the whole animal if it is a simple, straight tube, as seen in agnathans. Alternately, the GI tract can be wrapped around itself, allowing it to be many times longer than the animal. Some species increase the passage time using internal channels. For example, the straight gut of a shark possesses an internal membranous network, called the spiral valve, that increases the functional length of the gut. While most measurements of gut length are performed on dead animals, in the living animal visceral smooth muscle compresses the GI tract into a much shorter tube, so functional length is usually much shorter than maximal length. Nonetheless, the relative length of the gut reflects the digestibility of the diet. Animals with diets that are difficult to digest often have longer guts to increase the efficiency of digestion. For example, carnivores tend to have shorter guts than herbivores because the food is more easily digested. The importance of gut dimensions is best shown by comparing closely related species with differences in diet. The cecum of a grouse, which browses on vegetation, is almost twice the length of the cecum of a similarly sized partridge, which eats seeds.

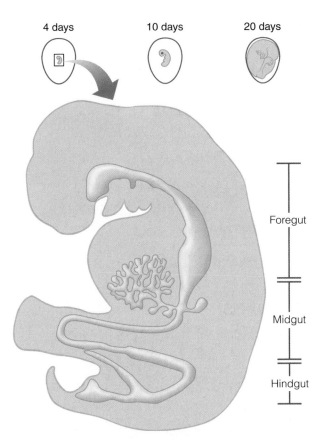

4 days   10 days   20 days

Foregut

Midgut

Hindgut

Figure 11.17 Gut development   The chicken
embryo develops into a chick within 20 days. By day 4, the
digestive system has developed enough to recognize major
structures. The embryonic gut is divided into three sections:
foregut, midgut, and hindgut.

The surface of the gut has a complex topography that serves to maximize surface area. We see this at the organ level, where the gut has deep circular folds that run around the circumference of the intestine (Figure 11.18). It is also evident at the tissue level, where the surface of the gut tissue is arranged into fingerlike projections called **villi**. The maximization of surface area is even seen at the cellular level. Enterocytes possess microscopic protrusions, supported by the actin cytoskeleton, called **microvilli**. The microvilli cause the surface of the intestinal mucosa to appear fuzzy, which is why the intestinal epithelium is often called the **brush border**. As a result of the circular folds, the villi, and the microvilli, the surface area of the gut is several hundred times greater than it would be if it were composed of flat sheets of smooth cells.

Specialized compartments increase the efficiency of digestion

The efficiency of digestion depends on the ability of the animal to create regions of functional specialization. Even a simple GI tract can have such specializations. Ctenophores (comb jellies) have a simple digestive sac that is elongated and flattened. Food enters the first region of the pharynx, which is acidic, then continues along the pharynx through two basic regions before looping back to exit through the opening that serves as both mouth and anus. With the exception of a small area of confluence, the inward and outward flows

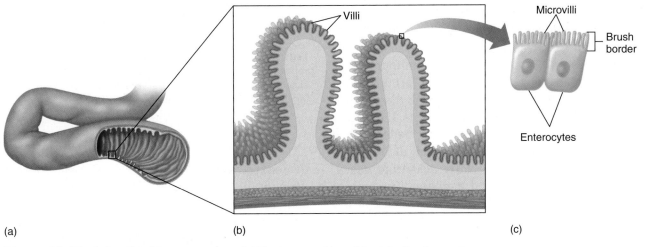

Villi

Microvilli

Brush
border

Enterocytes

(a)   (b)   (c)

Figure 11.18 Intestinal topography   **(a)** The inner surface of the intestine is a series
of folds or ridges that run circularly around the intestine. **(b)** The surface of the tissue is
arranged into fields of fingerlike projections called villi. **(c)** Each of the absorptive cells within
the villi possesses projections called microvilli. This structural topography—circular folds, villi,
and microvilli—increases the surface area that is available for absorption.

are well separated. These specialized sequential compartments allow the two-way gut to function much like a one-way gut, allowing the ctenophore to process multiple prey items in different regions at distinct stages of digestion.

Regional specializations are more developed in animals with a one-way gut. In many cases, muscular valves called sphincters control the passage of food from one compartment to the next. Regional properties are created and maintained by specific types of cells. Some cells alter the pH of the fluids in the lumen by secreting acids or bases. Since most of the macromolecules that appear in food are stable at a pH near neutrality, extremes in pH can enhance their breakdown. Mucus secretions help protect cells and lubricate the surface. Secretory cells release the digestive enzymes—proteases, amylases, lipases, nucleases—that accelerate chemical breakdown of macromolecules. The absorptive cells in each region also possess specialized transport capacities.

The general plan of the GI tract is similar among vertebrates, but taxa differ in the types of compartments (Figure 11.19). Many species possess extra chambers or modified regions along the GI tract. Birds and bony fish possess ceca that branch from the GI tract and contain bacteria that aid in digestion. The upper GI tract of birds is also more complex than that of other vertebrates. The crop is an outpouching of the esophagus that enables a bird to store partially digested food.

Many species of mammals possess elaborate modifications of the typical gastrointestinal tract that improve the digestibility of plant material. Ruminants (cows, deer, giraffe, goats, and sheep) possess a modified digastric stomach that allows vegetation to be more effectively digested. Ruminants possess a stomach composed of four chambers divided into two functional groups (Figure 11.20). Vegetation passes through the esophagus into the first pair of compartments: the rumen and reticulum. These interconnected regions house the fermentative bacteria that digest cellulose and produce volatile fatty acids and gases, largely carbon dioxide and methane. The animal can regurgitate food from the rumen back to the mouth, where it can chew the partially degraded material again. When the food returns to the esophagus, it enters the second division, comprising the omasum and abomasum. The abomasum serves as the glandular stomach, secreting diges-

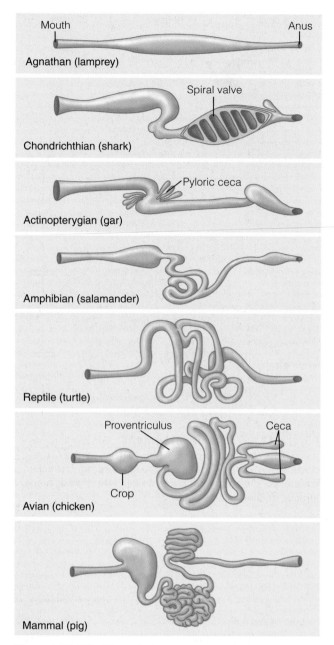

Figure 11.19 Vertebrate gut morphology The vertebrate gut differs widely in complexity and length among species. Each group of vertebrates is drawn to the same body length to emphasize the differences in length.

tive enzymes. Like ruminants, the closely related tylopods (camels, llamas) possess a digastric stomach, although these animals lack the omasum. Many of the other groups of mammalian grazers possess fermentation chambers. However, unlike the ruminants and tylopods, their fermentation chambers branch from the hindgut and are much less efficient.

Salivary glands secrete water and digestive enzymes

Digestion depends on secretions from multicellular exocrine glands working in conjunction with single secretory cells scattered throughout the GI tract. Many species have glands located near the mouth, typically called salivary glands. Salivary gland secretions include enzymes that initiate the chemical breakdown of food. In terrestrial animals, saliva provides fluid to help lubricate and dissolve food, which allows solubilized nutrients to bind to gustatory receptors. The saliva may also have antimicrobial properties to help cleanse the mouth. The salivary glands are really a collection of different glands. Intrinsic salivary glands, or buccal glands, are distributed throughout the oral cavity. A typical mammal has multiple pairs of extrinsic salivary glands: a dog has parotid glands just anterior to the ear, orbital glands near the eye, mandibular glands near the lower jaw, and sublingual glands beneath the tongue (Figure 11.21). Each of these glands possesses at least two types of cells: mucus-secreting cells and serous cells, which secrete the degradative enzymes.

Because of the high water content of saliva, secretions from these glands impinge on water balance. An average human, for example, might secrete more than 1 liter of water in saliva every day. The rate of secretion from salivary glands is regulated by the parasympathetic system in response to pressure-sensitive receptors and chemoreceptors in the mouth. When food is taken into the mouth, the mechanical stimulation triggers pressure-sensitive receptors that send signals to the region of the brain stem that controls serous gland secretions. Similarly, when chemoreceptors detect specific chemicals in the food, a signal is sent to the brain. As Pavlov discovered long ago, animals can also salivate in response to sights and sounds that are associated with food. Salivary gland secretions can also be inhibited. Dehydrated animals use the sympathetic nervous system to restrict blood flow to the salivary glands, preventing secretion. The same sympathetic response induces dry-mouth, a response often induced in hu-

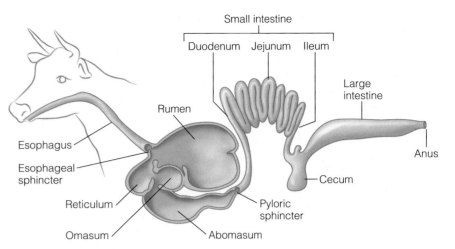

Figure 11.20 Ruminants Many mammals possess chambers derived from the GI tract that house bacteria that can ferment cellulose. Ruminants, including the cow shown here, possess four chambers.

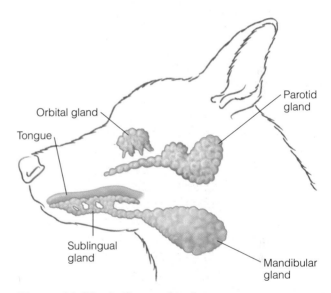

Figure 11.21 Salivary glands Like most mammals, the dog has multiple sets of salivary glands that secrete liquid and enzymes into the oral cavity.

mans under stressful conditions, such as public speaking.

The stomach secretes acid and mucus

The surface of the stomach is an epithelium composed of columnar epithelial cells (Figure 11.22). The cells are linked together via *tight junctions* that prevent the leakage of lumen fluids into the tissue. Dotted over the surface of the stomach are deep gastric pits composed of four cell types. *Mucous neck cells,* found near the pit opening, secrete an acid type of mucus. *Parietal cells* in the middle of the pit secrete acid, mainly HCl. *Chief cells* near the base of the pit secrete digestive enzymes, primarily the protease pepsin. Finally, *enteroendocrine cells* secrete several hormones into the blood in response to stomach contents. For example, the hormone gastrin is released from enteroendocrine cells into the blood supply of the stomach, inducing secretion by other gastric cells. We discuss the function of other hormones released by enteroendocrine cells later in this chapter when we consider the control of gut motility.

The low pH of the stomach is optimal for many of the gastric enzymes, but it is also harsh enough to kill most bacteria and parasites that enter in the diet. However, *Helicobacter pylori,* the bacterium associated with peptic ulcers, survives the low pH of the stomach by proliferating deep within the gastric pits, where the pH is higher.

Not all vertebrates have acidic stomachs. The platypus, for example, does not acidify the stomach for reasons that are not yet clear. Another interesting exception is the gastric brooding frog, *Rheobatrachus silus.* It swallows fertilized eggs, which mature in the stomach. The developing young secrete prostaglandin E_2 to inhibit acid secretion. Upon hatching, the tiny frogs hop up the esophagus and escape through the mouth.

Most nutrients are absorbed in the intestines

The intestines are also rich in histological diversity. A cross-section through the intestine reveals the four major layers: mucosa, submucosa, circular smooth muscle, and longitudinal smooth muscle. The mucosal surface is composed of many cell types, each with distinct roles (Figure 11.23).

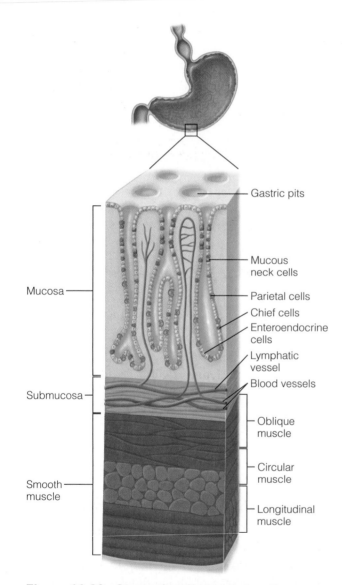

Mucosa — Gastric pits

Mucous neck cells

Parietal cells

Chief cells

Enteroendocrine cells

Lymphatic vessel

Submucosa — Blood vessels

Oblique muscle

Circular muscle

Smooth muscle

Longitudinal muscle

Figure 11.22 Stomach cell structure The smooth surface of the stomach has numerous cavities called gastric pits. These pits are composed of four main cell types that control the secretions of mucus, acid, enzymes, and hormones. They are also the location where *Helicobacter pylori* accumulate.

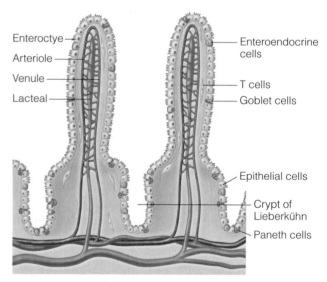

Enteroctye

Arteriole

Venule

Lacteal

Enteroendocrine cells

T cells

Goblet cells

Epithelial cells

Crypt of Lieberkühn

Paneth cells

Figure 11.23 Intestinal cell structure The villi are composed of multiple cell types. Enterocytes are absorptive cells that possess microvilli (not shown). Goblet cells secrete mucus. Interepithelial lymphocytes are T cells that help in immunodefense. Enteroendocrine cells secrete the hormones that control GI tract motility and other aspects of digestion. At the base of the villi are the crypts of Lieberkühn. The epithelial cells within the crypts secrete the intestinal juice. The Paneth cells at the base of the crypt secrete an enzyme that breaks down bacterial cell walls.

Much of the mucosa is composed of *enterocytes*, the absorptive cells with abundant microvilli. Mucus-secreting *goblet cells* are scattered among the enterocytes. *Enteroendocrine cells* secrete the hormones that help regulate digestion and nutrient assimilation. At the base of each villus is a region called the *crypt of Lieberkühn*. In addition to enterocytes, each crypt possesses *Paneth cells*, which secrete antimicrobial molecules into the lumen. Adjacent to the Paneth cells are stem cells that divide and differentiate to replenish the other cell types of the intestine. The intestinal submucosa, lying beneath the surface mucosa, is a layer of connective tissue through which blood and lymphatic vessels pass, as well as the nerves that control the GI tract. It also contains the duodenal glands, whose cells secrete basic mucus through ducts that penetrate the epithelium to help neutralize the acid arriving from the stomach. The inner and outer smooth muscle controls the movement of food along the GI tract.

The small intestine also receives secretions of bile from the gallbladder (Figure 11.24). Bile is a complex solution of digestive chemicals and liver waste products. Only two types of molecules in bile have a role in digestion: phospholipids and bile salts. Phospholipids, such as lecithin, aid in the uptake of lipids. Bile salts help emulsify fats in the duodenum. They are amphipathic molecules with nonpolar regions that bind to fats, and polar regions that interact with water. A coating of bile salts helps stabilize the small fat droplets. Liver cells (hepatocytes) produce bile and secrete it into small ducts that run adjacent to the hepatocytes. These ducts fuse and empty into the common hepatic duct, which joins the cystic duct from the gallbladder to form the bile duct. The gallbladder stores bile until it is needed, then empties into the duodenum via the bile duct.

Part of the pancreas is an exocrine gland that secretes digestive enzymes into the duodenum. Proteases are produced in the form of inactive *proenzymes*, which prevent the enzyme from digesting the secretory cell itself. For example, trypsin is secreted as the inactive precursor trypsinogen (Figure 11.25). When it enters the intestinal lumen, the brush border enzyme enterokinase converts it to the active protease. Trypsin in turn activates two other pancreatic enzymes, carboxypeptidase and chymotrypsin. Secreting these

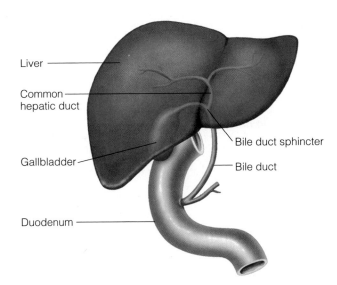

Figure 11.24 Bile production, storage, and secretion Bile is produced by hepatocytes and released into small adjacent ducts. These ducts collect bile and empty into the common hepatic duct. When the bile duct sphincter is closed, bile is routed through the cystic duct to the gallbladder for storage. When the sphincter opens, bile is released from the gallbladder into the duodenum.

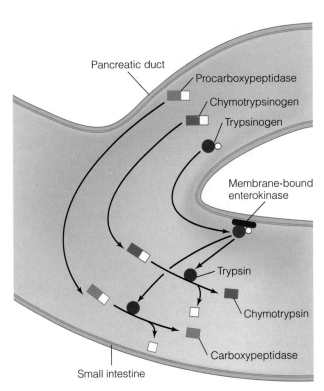

Figure 11.25 Trypsinogen cascade The pancreas secretes three important proteases, all in inactive forms. Trypsinogen is activated by proteolytic cleavage by enterokinase. The activated trypsin then activates chymotrypsinogen and procarboxypeptidase by proteolytic cleavage.

enzymes as inactive proenzymes reduces the risk that the pancreas will digest itself. The pancreas also releases enzymes that break down glycogen (amylase), triglycerides (lipase), and nucleic acids (nucleases). The hormones involved in control of GI tract secretions are summarized in Table 11.3.

Regulating Feeding and Digestion

Feeding is a necessary evil in the life of most animals; while they must feed to survive, the process of feeding requires considerable energy and may expose the feeder to predation. Feeding strategies evolve to provide an animal with the greatest chance of obtaining nutrients while minimizing the risk to its survival. In terms of digestive physiology, the most significant variables are the nature of the nutrients, the quantity of food consumed in a given period, and how often the food is consumed. Consider a day in the life of a filter feeder. Filter-feeding animals, such as barnacles, are surrounded by their food and feeding is a continuous process, interrupted only when a potential predator is sensed in the proximity. Conversely, some deep-sea fish may encounter food once a year.

Table 11.3 Hormonal control of digestion.

Regulatory factor	Source	Stimulates	Inhibits
Acetylcholine	Cholinergic nerves	Gastric secretion	
Cholecystokinin (CCK)	Duodenum	Expulsion of bile from gallbladder, pancreatic enzyme secretion, appetite	
Enkephalin	Duodenum	Gastric acid secretion	Pancreatic enzyme secretion
Epinephrine	Intestinal nerves	Gastric acid secretion	
Galanin	Intestinal nerves		Gastric acid secretion
Gastric inhibitory peptide (GIP)	Duodenum		Gastric secretion
Gastrin	Stomach and duodenum	Gastric acid production and secretion	
Gastrin-releasing peptide	Brain	Gastrin secretion	
Ghrelin	Stomach	Appetite	
Glucagon	Duodenum		Pancreatic and intestinal secretion
Glucagon-like peptide-1 (GLP-1)	Jejunum, lower intestine		Appetite
Motilin	Jejunum	Gastric acid secretion	
Pancreatic polypeptide (PP)	Pancreas		Appetite, gastric secretion
Peptide YY (PYY)	Jejunum, lower intestine		Appetite, gastric secretion
Pituitary adenylate cyclase activating polypeptide (PACAP)	Pituitary	Gastric secretion	
Secretin	Duodenum	Water and bicarbonate secretion, bile production	
Somatostatin	Duodenum		Gastric acid secretion, pancreatic secretion, blood flow (vasoconstriction)
Vasoactive intestinal peptide (VIP)	Duodenum	Intestinal bloodflow (vasodilation), pancreatic and gastric secretion, intestinal salt secretion	

Many animals feed during a narrow window in their lifetime, and display a life history strategy that is choreographed around this meal.

Animals face the daunting challenge of matching their dietary intake to their short-term metabolic demands, while ensuring an opportunity for long-term development and reproduction. For many complex animals, the desire to feed (appetite) is regulated by the central nervous system (CNS). Input to the CNS comes from environmental factors such as photoperiod, as well as intrinsic signals that reflect the nutrient levels or metabolic status of the animal. Typically, animals feed when their energy needs exceed the metabolic potential of circulating fuels, allowing the animal to avoid drawing on nutrient stores.

Feeding and digestion are at the crossroads between ecology and mechanistic physiology. Regulatory physiologists focus on the pathways animals use to absorb specific nutrients and metabolize them in intracellular pathways. Although each of the steps—the transport and the enzymatic conversion—can be studied in a quantitative manner using kinetic analyses, it is difficult to extrapolate from these studies to make predictions about the whole system. Conversely, ecological physiologists are often more concerned with how animals match nutrient needs to feeding efforts, a concept called *ecological stoichiometry*. Researchers in these two fields—regulatory physiology and ecological stoichiometry—were initially separated because of the diversity of nutrients and the complexities of nutrient breakdown, absorption, and assimilation. The two alternate experimental frameworks are brought together in **gut reactor theory**, which allows researchers to use mathematical relationships to make qualitative and quantitative predictions about digestive physiology function and evolution (see Box 11.4, Mathematical Underpinnings: Gut Reactor Theory).

Hormones control the desire to feed

The control of appetite has been best studied in mammals because of the implications for human obesity. These studies have identified more than 20 different regulatory factors that link nutrition, metabolism, and feeding. Some regulatory factors are produced in the vicinity of the GI tract, controlling local events and sending signals into the blood to affect other tissues. Hormones released by the GI tract make their way to the hypothalamus, which integrates the information and controls feeding behavior. Three main hormones control appetite: leptin, ghrelin, and peptide YY. These hormones exert effects on multiple target tissues, but in terms of appetite control, their main effects arise through receptors in the hypothalamus. Recall that this region of the brain possesses an incomplete blood-brain barrier, and thus is able to sense the profiles of blood borne metabolites and small hormones. Leptin is too large to passively move across the endothelium of the capillaries that feed the hypothalamus; it is likely transferred from the blood across the blood-brain barrier through an active transport mechanism.

Leptin is an appetite-suppressing hormone produced in white adipose tissue. It acts as an "adipostat," ensuring that adipose lipid content is stable. When adipose triglyceride levels rise, leptin is secreted and appetite is suppressed. Restoration of adipose lipid stores inhibits leptin secretion and appetite increases. Though leptin exerts an important control over appetite, its effects are mediated over the long term. Over the short term, ghrelin and peptide YY appear to be more important in controlling the desire to eat between meals. Ghrelin is an appetite-stimulating hormone, released from gastric cells when the stomach is empty. Peptide YY is an appetite-reducing hormone, released from enteroendocrine cells when the colon is full.

These hormones exert their effects on the arcuate nucleus of the hypothalamus. Recall that a nucleus is a region where the cell bodies of the neurons are collected. The neurons of the arcuate nucleus send their axons to other neurons in the region, as well as higher centers of the brain that regulate behavior. Some neurons release appetite-stimulating factors, mainly neuropeptide Y (NPY), as well as agouti-related peptide and gamma-aminobutyric acid (GABA). Other neurons release the appetite-suppressing factor proopiomelanocortin (POMC). It is the balance of activity of NPY-releasing neurons and POMC-releasing neurons that determines the appetite signal sent to the higher centers of the brain. This balance is influenced by the hormones that stimulate and inhibit each type of neuron, and how the neurons interact with each other through antagonistic neurotransmitters. Both NPY-releasing neurons and POMC-releasing neurons express leptin receptors, but connections between the receptor and neurotransmitter release are different. Leptin binding to its receptor

BOX 11.4 MATHEMATICAL UNDERPINNINGS
Gut Reactor Theory

The animal digestive system shares many similarities with the reactors used in industry to convert one set of chemicals to another form. Chemical engineers define three types of chemical reactors that have clear parallels to the animal digestive systems we have discussed: batch reactors, tank reactors, and plug flow reactors (Figure A).

Batch reactors receive a pulse of precursors and after a period of time convert the precursors into products. This is much like the two-way gut used by cnidarians, which engulfs and digests food particles in a gastrovascular cavity, then expels undigested material. Tank reactors receive a constant infusion of precursors and generate a constant stream of products. The fermentation chambers of some animals, such as the bird cecum or the cow rumen, are examples of tank reactors. In plug flow reactors, a bolus of precursors begins at one end of a tube-shaped reactor and moves through the tube to the other end. The intestine of most animals works in this way, with food exiting the stomach and passing through the tubular intestine to the anus.

Chemical reactor theory allows a researcher to model the digestive process mathematically, to assess the factors that determine the performance of the digestive system. For example, if a digestive system works like a batch reactor, then the animal stands to gain the most energy if it digests a single meal enough to gain the most nutrients in the shortest period of time. If the time is too short, the bolus of food is expelled with many nutrients remaining. If it holds on to the food too long, it may extract more nutrients but it forgoes an opportunity to feed again. Reactor theory can predict the optimal retention time for food by plotting the relationship between net uptake (total uptake U minus foraging costs C) and retention time τ, as seen in Figure B. This curve can be used to predict the optimal residence time. Soon after the meal is consumed ($\tau = 0$), the animal has incurred costs (C) but gained no nutrients. As digestion proceeds (τ increases), there is an increase in the slope of the curve. Think of this slope, termed $U'(\tau)$, as the rate at which the animal is gaining nutrients at that point in time. At some point, the slope of the curve reaches its maximum, or $U'(\tau_{opt})$. After a longer period of digestion, nutrients continue to be absorbed but at a diminishing rate. Mathematically, the relationship between these parameters is defined by the equation

$$U'(\tau_{opt}) = \frac{U(\tau_{opt}) - C}{\tau_{opt}}$$

Graphically, the τ_{opt} value is identified by a line that begins at the origin and intersects the curve at the point where the slope begins to decrease. The red line shows how an increase in foraging costs (C) shifts the entire curve downward; τ_{opt} increases. If it costs more to feed,

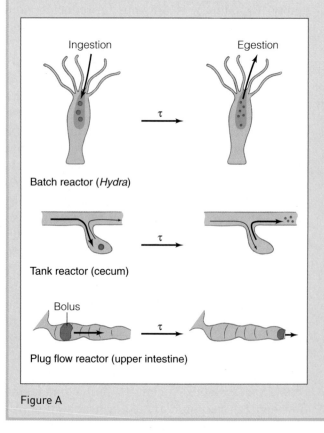

Figure A

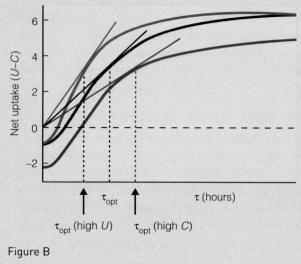

Figure B

then the animal benefits from assimilating more nutrients from the first meal. Similarly, the blue line shows how uptake increases when a meal is more digestible. The entire curve shifts upward, and a shorter τ_{opt} is predicted.

In the same manner, more complex equations can be used to predict optimal feeding strategies in animals with a digestive system that more closely approximates the plug flow reactor. In many ways, this model is like a series of batch flow reactors, with a given volume of food progressing from one region to the next. Unlike the batch reactor, the plug flow reactor accepts a continuous input, which has two consequences for predicting the gain. First, because the animal feeds continuously, its costs of feeding are spread out over time (in a batch reactor, the feeding costs are incurred first, but the gain is spread out over time). Second, for any given period, multiple meals contribute to the gain.

In recent years, gut reactor theory has been used to confront the biological complexity we discuss in this chapter. First, whereas models generally assume that the uptake rate is a linear function of nutrient concentration, the kinetics are usually more complicated. The impacts of complex kinetics of digestive enzymes and intestine active transporters have only been resolved in simple systems, such as nectar-feeding birds. Second, the volume of the plug is assumed to be constant, but in reality it changes as animals remove fluids from or secrete them into the gut. Third, the models require thorough mixing of the volume, but within the gut there are well-established concentration gradients as a result of transport processes, such as unstirred layers. Fourth, the animal can change the functional length of a gut through smooth muscle activity. Many of the studies that use reactor theory operate on the premise that digestive systems work optimally. For example, it is assumed that animal digestive physiology (feeding behavior and nutrient uptake) strives to reach an optimal retention time. When a diet consists of a single major nutrient, it is plausible that a single τ_{opt} exists. In a complex diet, each type of nutrient might have distinct optimal uptake kinetics, yet the bolus of food progresses through the tubular gut at a single rate. Thus, a single passage rate may be longer than required for some nutrients, and shorter than is necessary for others. The τ_{opt} in some cases may reflect the need to expel indigestible or toxic material, rather than to take up nutrients.

Reactor theoreticians continue to incorporate these important physiological and morphological variables as they develop more sophisticated models to predict digestive physiology and feeding strategies. Reactor theory has been best applied to animals with simple diets. Dr. Carlos Martinez del Rio and his colleagues have used it to study the feeding physiology of nectar-feeding birds. The simplicity of the diet facilitates the testing of mathematical models incorporating plug reactor theory. A nectar-feeding bird converts sucrose to fructose and glucose using the intestinal disaccharidase sucrase. It displays Michaelis Menton kinetics, with the rate of hydrolysis $(-r_s)$ expressed as:

$$-r_s = V_{max}\, C_s\, (K_m + C_s)^{-1}$$

where V_{max} is the maximal rate of sucrase averaged along the gut, K_m is the Michealis-Menton constant, and C_s is the sucrose concentration. The retention time (τ) can be calculated as

$$\tau = [(K_m \ln (C_{s0}/C_{sf}) + (C_{s0} - C_{sf})]\, V_{max}^{-1}$$

where C_{s0} is the initial sucrose concentration and C_{sf} is the final sucrose concentration.

Once τ is known the gut intake rate (V_0) can be calculated as

$$V_0 = G\tau^{-1}$$

Martinez del Rio and his colleagues then compared this model to actual experimental observations of hummingbirds feeding on different sucrose solutions. The more dilute the sucrose solutions, the larger the volume consumed by the birds. In another study, they used this same approach to find whether feeding behavior reflected an attempt by the bird to match uptake to metabolic demand (compensatory feeding), or rather to ensure an uptake that kept the digestive machinery working at its maximal rate (physiological constraint). Hummingbirds were fed the same range of sucrose solutions, but exposed to different ambient temperatures. The colder temperatures elevated metabolic demands. They found that cold birds drank the same amount of sucrose as warm birds, suggesting a physiological constraint.

References

○ Martinez del Rio, C., J. E. Schondube, T. J. McWhorter, and L. G. Herrera. 2001. Intake responses in nectar feeding birds: Digestive and metabolic causes, osmoregulatory consequences, and coevolutionary effects. *American Zoologist* 41: 902–915.

○ McWhorter, T. J., and C. Martinez del Rio. 2000. Does gut function limit hummingbird food intake? *Physiological and Biochemical Zoology* 73: 313–324.

in NPY-releasing neurons reduces neurotransmitter release, whereas binding to POMC-releasing neurons stimulates neurotransmitter release. Each of these effects contributes to the suppression of appetite in response to leptin. Ghrelin and peptide YY each bind NPY-releasing neurons, though activation of their respective receptors works antagonistically on neurotransmitter release (Figure 11.26).

The factors that control appetite have been studied intensively because of their potential for treatment of obesity. There are many animal models with genetic defects that cause overeating (*hyperphagy*) and obesity. For example, animal models with defects in leptin signaling—loss of the ability to synthesize leptin or its receptor—are obese and exhibit many of the other physiological problems associated with obesity. Unfortunately, strategies to target the leptin signaling cascade in

antiobesity treatments have not met with much success. Obese humans actually have very high levels of leptin, yet their appetite is not suppressed, suggesting that the hypothalamus can become resistant to leptin. It is also not yet clear how leptin and other regulators of appetite act in other species of vertebrates with diverse feeding strategies. Frequency of feeding differs widely among animals. Large snakes and deep-sea fish may feed very infrequently, sometimes only once per year. The hormonal and neuronal regulation of appetite is not well studied in such animals.

Hormones and neurotransmitters control secretions

Once food enters the gut, the GI tract secretes a spectrum of chemicals and enzymes that digest the food into forms that can be taken up. Control of these secretions depends on complex regulatory mechanisms that respond to both the anticipation of food and its physical presence in the digestive system. Earlier in this chapter we discussed the nature of intestinal secretions—saliva, acid, mucus, bile, bicarbonate, and digestive enzymes—but not how animals control these secretions.

The stomach is acidified when parietal cells in the gastric lining secrete HCl, or rather both H^+ and Cl^-. Central to the secretion of acid is the activity of the enzyme carbonic anhydrase. Carbon dioxide is hydrated, forming H_2CO_3, which then dissociates to form H^+ and HCO_3^-. The parietal cells expel protons into the lumen via a proton pump, a K^+/H^+ ATPase. The bicarbonate is exported from the cell to the blood via a Cl^-/HCO_3^- exchanger. The Cl^- then escapes the cell through a Cl^- channel. The secretion of acid from parietal cells is triggered when histamine binds to receptors (H_2 receptors) on the parietal cells, to initiate a cAMP cascade that activates the proton pump. The histamine is released by enterochromaffin-like (ECF) cells in the stomach lining. The ECF cells release their histamine in response to gastrin, a peptide hormone produced by neuroendocrine cells (G cells) in the stomach and duodenum, and released in response to adrenergic and cholinergic nerve activity (Figure 11.27). Gastrin also regulates pepsinogen secretion by chief cells of the gastric mucosa. The contents of the stomach can interact with each of these cells—parietal cells, ECF cells, and G cells—to alter acid secretion. Treatments for excessive acid secretion

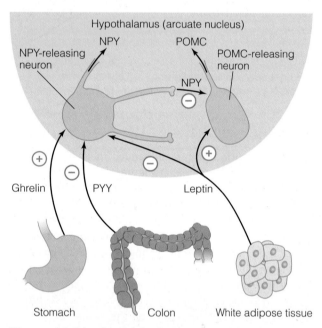

Figure 11.26 Control of appetite Appetite is controlled by the neurons of the arcuate nucleus of the hypothalamus, which interacts with hormones secreted from the GI tract and adipose tissue. Release of the neurotransmitter neuropeptide Y (NPY) stimulates appetite, whereas release of the neurotransmitter proopiomelanocortin (POMC) depresses appetite. These neurons, through their neurotransmitters, affect the appetite centers of the brain directly, or antagonize the release of neurotransmitters from other neurons. Leptin exerts its appetite-suppressing effects by inhibiting NPY–releasing neurons and stimulating POMC-releasing neurons. Appetite is increased by the actions of other hormones on NPY-releasing neurons: ghrelin stimulates and peptide YY inhibits these neurons.

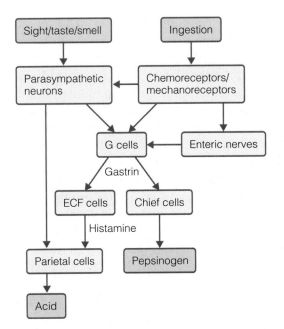

Figure 11.27 Control of gastric secretion of acid and pepsinogen Gastric cells secrete acid (parietal cells) and pepsinogen (chief cells) in response to signals relayed from the central nervous system and from the food itself, acting through chemoreceptors and mechanoreceptors.

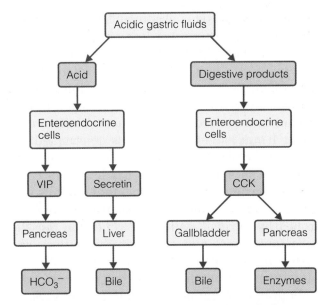

Figure 11.28 Control of intestinal secretion The acidic fluids that exit the stomach trigger intestinal secretion. The secretions neutralize the acidic fluids (via bicarbonate and bile), and aid in digestion through digestive enzymes (proteases, lipases, nucleases, amylases) and bile, which emulsifies lipids.

can target the K^+/H^+ ATPase (proton pump inhibitors) or the histamine receptors (H_2 blockers). Gastrin also controls other secretory cells within the gastric mucosa. For example, gastrin induces chief cells to secrete pepsinogen. The low pH of the stomach activates pepsinogen to form pepsin, a carboxypeptidase, initiating protein degradation.

Once food passes from the stomach to the upper intestine, secretions alter the pH of the bolus and bombard it with a different suite of digestive chemicals. Bicarbonate secretions from the pancreas and bile from the gallbladder neutralize the acidity. The pancreas also secretes digestive enzymes, including amylase, chymotrypsin, carboxypeptidases, aminopeptidases, nucleases, and lipases. When acidic material enters the upper intestine, it triggers duodenal secretion of secretin and vasoactive intestinal peptide (VIP) into the blood, which act at the pancreas to induce secretion of bicarbonate. Other intestinal cells sense the levels of amino acids and fatty acids and secrete cholecystokinin (CCK) into the blood. CCK acts on the pancreas, inducing the secretion of digestive enzymes, and on the gallbladder, inducing contraction of the smooth muscle to eject bile (Figure 11.28).

Gut motility is regulated by nerves and hormones that act on smooth muscle

As with most physiological systems, muscles and nerves play important roles in regulating the digestive system. Food is moved along the gastrointestinal tract by visceral smooth muscles, which are under the control of nerves and hormones. By increasing gut motility, an animal increases the rate of passage of food down the GI tract, which in turn affects the efficiency of absorption. It must be fast enough to ensure that the animal is not carrying around a mass of indigestible material, but slow enough to allow time for digestion and assimilation. The interplay between gut passage rates and digestibility is illustrated in the comparison of birds with different diets. Fruit-eating birds have fast passage rates. They must be able to move food quickly through the gut because indigestible material is a load that must be carried around when the animal flies. Conversely, nectar-eating birds have a nutrient-rich diet that contributes little to body mass. Slow passage rates allow these birds abundant time to absorb the nutrients.

Gut motility is controlled by the actions of the two layers of smooth muscle that line the intestinal tract. Each layer of smooth muscle in the GI tract is composed of muscle cells embedded in an extracellular matrix of elastin and collagen molecules, and interconnected into an electrical network that allows the individual cells to contract and relax as a unit. The thin outer longitudinal layer controls intestinal length. The thick inner circular layer, which controls the diameter of the lumen, is arranged into contractile units, each about 1 mm long. Gut motility is determined by the contractile activity of the circular smooth muscle. The smooth muscle has a resting contractile tension, or muscle tone, that controls the diameter of the lumen. Tonic contraction is controlled by intrinsic pathways within the muscle cells (myogenic) and by neurotransmitters released from motor nerves (neurogenic). These motor nerves from the CNS do not act directly on smooth muscle, rather feeding into a network of nerves called the *myenteric plexus* (Figure 11.29). The myenteric plexus and the submucosal, or Meissner's, plexus make up the enteric nervous system (see Chapter 7: Functional Orga-

nization of Nervous Systems). The myenteric plexus contains the neurons involved in regulating gut motility and enzyme secretion, whereas the submucosal plexus regulates gut blood flow and plays an important role in regulating ion and water transport by the gut. These components of the enteric nervous system work together to regulate gut function.

The gut also changes its contractile state to help propel food through the lumen. **Peristalsis** is a slow wave of contraction that progresses down the GI tract to push food toward the anus. It is controlled by the intrinsic myogenic activity of the smooth muscle cells, but also influenced by *interstitial cells of Cajal* that act as pacemaker cells. Much like the pacemaker cells of the heart, these cells spontaneously depolarize to initiate a wave of depolarization to the smooth muscle cells to which they are attached via gap junctions.

Suites of neurotransmitters and hormones control gut motility (Table 11.4). In addition to endocrine control, secretory cells in the stomach and intestine produce paracrine and autocrine factors that can act both directly and indirectly on the gut. For example, secretory cells in the GI tract release serotonin, which acts on both smooth muscle and nerves. The nervous control of the GI tract includes signals from the CNS as well as local nerve networks. Centrally, the hypothalamus and spinal cord receive information from nerves in the GI tract that respond to the activity of chemoreceptors and mechanoreceptors. The parasympathetic neurons send signals back to the gut to stimulate motility by releasing acetylcholine; sympathetic neurons inhibit motility by releasing norepinephrine, somatostatin, and neuropeptide Y. The myenteric plexus compiles the stimulatory and inhibitory signals, then transmits nervous signals to the smooth muscle.

Because of the overlapping regulatory pathways, the control of gut motility is complex. Consider the mechanism by which one neurotransmitter, acetylcholine, induces contraction. Recall from Chapter 5: Cellular Movement and Muscles that smooth muscle contraction is regulated directly by Ca^{2+} as well as by changes in the phosphorylation state of thick and thin filament proteins. Acetylcholine stimulates contraction in visceral smooth muscle by increasing Ca^{2+} levels. This activates myosin light chain kinase (MLCK), which phosphorylates the myosin light chain of the thick filament. Acetylcholine acts through a type of

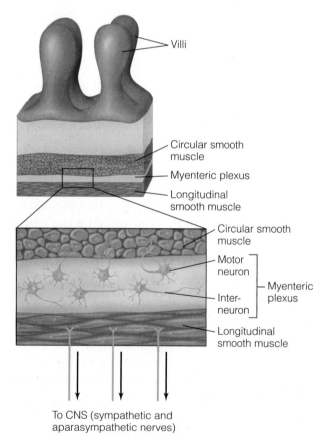

Figure 11.29 Myenteric plexus The nerves of the parasympathetic and sympathetic nervous systems send their signals to the myenteric plexus that lies beneath the submucosa. This group of nerves integrates the various signals and sends the appropriate neurotransmitters to the circular smooth muscle to control contraction.

Table 11.4 Neurohormonal effectors of GI motility.

Stimulatory	Inhibitory
Acetylcholine	Calcitonin gene regulating protein (CGRP)
Adenosine	Gamma-aminobutyric acid (GABA)
Bombesin	Galanin
Cholecystokinin (CCK)	Glucagon
Gastrin-releasing polypeptide	Neuropeptide Y
Histamine	Neurotensin
Motilin	Nitric oxide
Neurokinin A	Norepinephrine
Opioids	Pituitary adenylate cyclase activating polypeptide (PACAP)
Prostaglandin E_2	Peptide histidine isoleucine (PHI)
Serotonin	Peptide YY (PYY)
SP	Secretin
Thyrotropin-releasing hormone	Somatostatin
	Vasoactive intestinal peptide (VIP)

Source: Hansen, 2003.

G-protein-linked muscarinic receptor that activates phospholipase C (PLC). This produces two second messengers, diacylglycerol (DAG) and inositol triphosphate (IP$_3$), which increase cytoplasmic Ca^{2+} levels (Figure 11.30). The IP$_3$ and its metabolites open Ca^{2+} channels in the sarcoplasmic reticulum to release Ca^{2+}. When protein kinase C binds DAG in the membrane, it phosphorylates and activates Ca^{2+} channels. Acetylcholine also acts via an independent route to impair relaxation. It binds to a second class of muscarinic receptors that inhibits adenylate cyclase, reducing cAMP levels and PKA activity. Since PKA phosphorylates and desensitizes contractile proteins, acetylcholine's actions also favor sensitization of contractile proteins. In the absence of acetylcholine, relaxation is favored because the channels close and Ca^{2+} can be pumped back out of the cytoplasm.

Relaxation is triggered by hormones that act through both Ca^{2+}-dependent and Ca^{2+}-independent routes. The β adrenergic effectors, such as epinephrine, bind G-protein-linked receptors that activate adenylate cyclase, elevating the level of cAMP and activating PKA. Nitric oxide stimulates guanylyl cyclase, elevating the levels of cGMP and protein kinase G (PKG). Vasoactive intestinal peptide acts through both pathways, increasing the levels of both cAMP and cGMP. Stimulation of PKA and PKG leads to phosphorylation of critical proteins that reduce Ca^{2+} levels and change the sensitivity of the contractile apparatus.

Metabolic transitions between meals

Once food is consumed and assimilated, the animal must regulate intermediary metabolism to ensure that the burst of nutrients is utilized and stored. During the period immediately after feeding, known as the *postprandial period*, an animal utilizes some nutrients and stores others, enabling it to survive until the next meal. The normal period between meals may be anywhere from seconds to months, depending on the animal and its feeding strategy. In an animal with a high metabolic rate, these energy stores are rapidly expended and the animal enters a starvation period, mobilizing energy stores and even degrading structure. Many animals are subjected to very long periods between meals and possess strategies that combine reducing metabolic demands and more efficiently using the available resources.

For some animals, the food deprivation period may persist for the life of the animal. Many insects

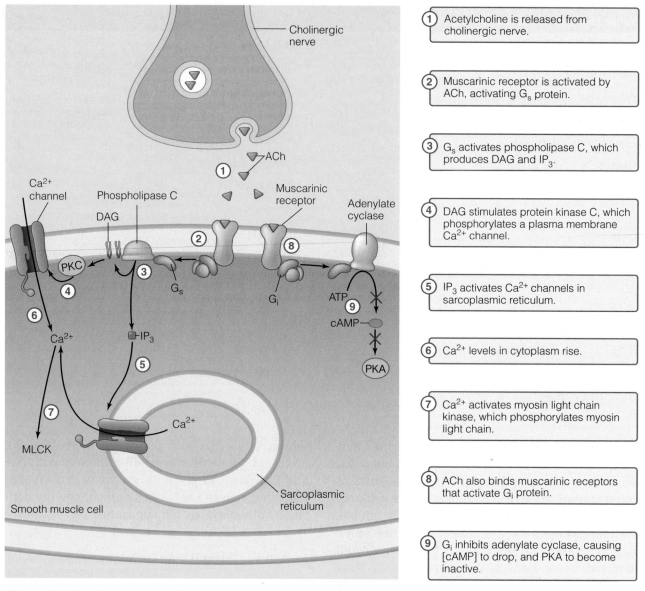

Figure 11.30 Smooth muscle and the control of gut motility Acetylcholine (ACh) released from enteric nerves induces contraction of smooth muscle by increasing cytoplasmic Ca^{2+} levels. When it binds to muscarinic receptors, ACh activates phospholipase C (PLC), causing an increase in IP_3, which opens Ca^{2+} channels in the sarcoplasmic reticulum, and DAG, which activates PKC and opens Ca^{2+} channels in the cell membrane. ACh also antagonizes relaxation. It activates muscarinic receptors that bind a G_i protein that inhibits adenylate cyclase, thereby reducing cAMP levels and PKA activity.

① Acetylcholine is released from cholinergic nerve.

② Muscarinic receptor is activated by ACh, activating G_s protein.

③ G_s activates phospholipase C, which produces DAG and IP_3.

④ DAG stimulates protein kinase C, which phosphorylates a plasma membrane Ca^{2+} channel.

⑤ IP_3 activates Ca^{2+} channels in sarcoplasmic reticulum.

⑥ Ca^{2+} levels in cytoplasm rise.

⑦ Ca^{2+} activates myosin light chain kinase, which phosphorylates myosin light chain.

⑧ ACh also binds muscarinic receptors that activate G_i protein.

⑨ G_i inhibits adenylate cyclase, causing [cAMP] to drop, and PKA to become inactive.

have nonfeeding developmental stages. Typically, the early life stages feed actively, storing nutrients for metamorphosis and reproduction. Most pupae of insects do not feed, instead relying on nutrients stored as larvae to reorganize the anatomy and physiological systems. In a few cases, the adult form of an insect is nonfeeding. Mayflies, for example, spend as long as 2.5 years as nymphs, feeding as predators in the debris of waterways. When waters warm in spring, the adult stages emerge and reproduce, then die one or two days later. Adults do not feed, and prior to their emergence the GI tract atrophies, acting as a nutrient store used to support reproductive maturation.

Hormones control postprandial regulation of nutrient stores

After nutrients pass through the enterocytes to the blood, they may be utilized directly by other tissues or stored in depot tissues. With each meal there is a burst of readily metabolizable carbon fuels that can be oxidized to support the metabolic demands of tissues. In many tissues, the fuels used to support energy metabolism are influenced by the spectrum of fuels in the blood. The vertebrate heart, for example, is capable of oxidizing glucose, lactate, amino acids, or fatty acids, depending on their availability in the blood. Immediately after a meal, such tissues have many alternative fuels. Later, however, the levels of nutrients in the blood depend on the action of hormones that control the release of fuels from storage tissues. This regulation is determined primarily by endocrine hormones.

As we discussed in Chapter 2, in vertebrates, the immediate fate of dietary nutrients (oxidation or storage) depends on the levels of the pancreatic hormones insulin and glucagon, as well as glucocorticoids. During digestion, these hormones act on peripheral tissues to control the pattern of fuel utilization and on storage tissues to control rates of uptake and synthesis. In the postabsorptive animal, these same hormones control the release of fuels from storage depots. When the glucose level is high, as it would be after a meal, pancreatic beta cells are induced to secrete insulin. Insulin acts upon multiple tissues to promote glucose removal from the blood. In skeletal muscle, it enhances glucose uptake by causing translocation of glucose transporters (GLUT-4) to the cell membrane. In adipose, it promotes uptake and conversion of glucose into fatty acids, for long-term storage as triglyceride. In liver, it impairs glycogen breakdown and enhances glycogen synthesis. After the glucose level declines, insulin secretion decreases and glucagon is released by pancreatic alpha cells. This causes mobilization of energy stores—glycogen hydrolysis and triglyceride breakdown—and enhances the rate of gluconeogenesis in liver. Thus, the balance between insulin and glucagon determines the balance between glucose utilization and generation.

Glucocorticoids such as cortisol, corticosterone, and cortisone induce gluconeogenesis while reducing glucose uptake by peripheral tissues such as skeletal muscle. This acts to increase circulating glucose levels to ensure that those tissues that require glucose have a steady supply. Glucocorticoids also mobilize triglycerides, ensuring that fatty acids are available to tissues that are prevented from using glucose, such as skeletal muscle. Whereas in-

sulin and glucagon are most important to metabolic regulation in relation to the nutritional state, glucocorticoids are most important as part of a metabolic stress response. For example, the intense metabolic costs of locomotion and reproductive behaviors are met when the glucocorticoid stress hormones cause mobilization and synthesis of fuels.

Invertebrate nutritional metabolism is best studied in insects, mainly because of their importance as agricultural pests. The main energy store in insects is the fat body. During energy-demanding situations, such as flight, *adipokinetic hormone* (AdK) is released from the *corpora cardiacum*, causing the fat body to mobilize energy stores (Figure 11.31). Lipids are broken down to DAG and fatty acids. Glycogen stores are converted to trehalose. The fat body also releases significant

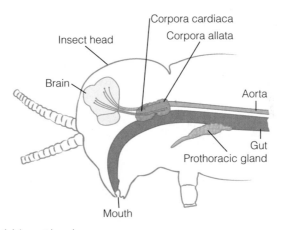

(a) Insect head

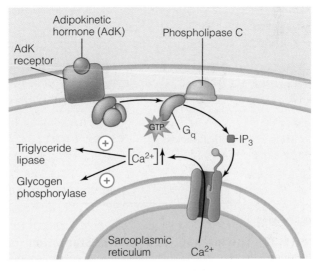

(b) Fat body cell

Figure 11.31 Hormonal regulation of metabolism in insects

amounts of proline in some species. AdK acts via a G-protein-coupled receptor (G_q) to activate phospholipase C (PLC). The increase in IP_3 triggers Ca^{2+} release, which activates Ca^{2+}-sensitive signaling enzymes that activate glycogen phosphorylase and triglyceride lipase. At the same time that AdK enhances lipid and protein breakdown, it inhibits lipid and protein synthesis. Since insect reproduction is intimately linked to nutrition, AdK also inhibits egg production in females.

Prolonged food deprivation can trigger a starvation response

Most animals are able to cope with short periods of food deprivation without incurring metabolic distress. Short-term food deprivation, such as the periods between regular meals, can be met with existing energy resources. If the food-deprivation period persists, the animal reorganizes metabolism to ensure long-term survival. Most vertebrates trigger mechanisms that preserve glucose in order to protect those tissues that rely heavily on glucose to meet energy demands. Nervous tissue, for example, relies almost exclusively on glucose as a fuel. In the early phases of food deprivation, vertebrates mobilize the vast lipid stores in liver and adipose tissue. Muscle, a major consumer of metabolic energy, shifts to rely more heavily on mobilized lipid, reducing the reliance on glucose.

Despite these efforts to conserve glucose, after a time the glycogen stores become depleted and the animal must find a fuel that can be used to produce glucose. After prolonged food deprivation, the animal accelerates the rate of protein breakdown. Since there is no protein store, this usually entails the degradation of the protein structures within cells. One of the earliest tissues to suffer protein degradation is skeletal muscle. Individual myofibers degrade contractile elements in the process of atrophy. When intracellular proteins are degraded by lysosomes and proteasomes, the liberated amino acids can be oxidized for fuel or converted to other molecules, such as ketone bodies, fatty acids, or carbohydrate. These processes can occur either in the muscle itself or after transport of the amino acids to other tissues, primarily liver. Fatty acids can also be converted to ketone bodies, which can then be utilized by tissues that cannot oxidize fatty acids, such as nervous tissue. The main regulatory events that occur during starvation are summarized in Figure 11.32.

The rate at which a starvation response progresses varies widely among species, primarily due to differences in metabolic rate. An animal with a high metabolic rate will deplete its energy stores faster than an animal with a low metabolic rate. Differences in metabolic rate arise in relation to body size, activity levels, and temperature. Since small animals have higher mass-specific metabolic rates than do larger animals, they expend limited energy stores at a faster rate. The energetic state of a hummingbird or shrew after one hour of food

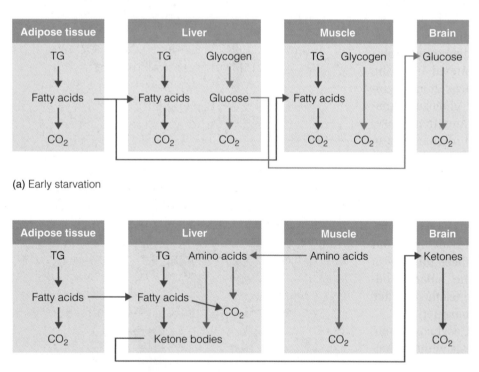

(a) Early starvation

(b) Late starvation

Figure 11.32 The metabolic cascade of starvation When animals are deprived of food, they respond by mobilizing internal energy stores. **(a)** In early starvation of vertebrates, stores of glycogen and triglyceride (TG) provide most of the metabolic needs of tissues. Fatty acids are released from liver and adipose for use in other tissues. **(b)** During late starvation, glycogen reserves are depleted and ketone bodies are produced from fatty acids from adipose and liver, as well as some amino acids derived from muscle proteolysis.

deprivation is similar to that of a human that has not eaten for 12 hours. Active animals utilize energy stores faster than similarly sized sedentary animals. For example, a dog has a metabolic rate that is about twice that of a goat. Even within a single individual, a persistent period of elevated metabolism draws on circulating fuels and stimulates feeding. Prolonged periods of biosynthesis, arising during rapid growth, gestation, or lactation, can elevate metabolic rate dramatically. Similarly, the elevation in metabolic rate during muscle activity arising from exercise or shivering stimulates appetite and fuel mobilization. The third factor that affects metabolic rate and nutrient demand is body temperature. An increase in body temperature accelerates the basal metabolic demands and more rapidly depletes available energy, thereby accelerating the rate of progression through the starvation responses.

Pythons rebuild the digestive tract for each meal

An animal that eats very infrequently must maintain its GI tract in a condition that will allow it to function when needed. Hibernating ground squirrels that have been dormant for more than 12 weeks continue to express the degradative enzymes that would be required to digest a meal. In this way, these animals ensure that they will be able to eat and digest a meal immediately upon emergence from hibernation. However, some animals reduce their energetic costs between meals by allowing the GI tract to degrade. Large predatory snakes, such as the Burmese python, may go months between meals. When a snake feeds, it eats a very large meal. For example, in 1977, villagers in India killed an 18-foot Indian python and recovered the remains of a 45-year-old man. Between these large meals, much of the mucosa and submucosa of the GI tract degrades. The gut becomes thinner and the brush border of the intestine decreases. These structural changes reduce the total surface area of the GI tract and the capacity for nutrient transport. Although the absorptive epithelium is reduced, the GI tract maintains the smooth muscle and nerves that control the gut. Once the animal feeds, the snake rebuilds its GI tract in regions that are just ahead of the bolus of food. Within the first few days after a meal, the mass of the tissues associated with digestion in-

creases dramatically. The small intestine alone nearly doubles in mass, and other tissues such as liver and kidney, increase by more than 60% in this rapid growth phase (Figure 11.33). The high metabolic cost of rebuilding the gut is an important component of the very high specific dynamic action (SDA) seen in a python digesting a meal; while the costs are great, the cost of rebuilding the GI tract must be less than the costs of maintaining it between unpredictable meals.

Dormant bears recycle nitrogen

Many organisms undergo periods of dormancy in which the metabolic rate is reduced (hypometabolism), allowing the animal to survive adverse environmental conditions. The various types of dormancy—torpor, hibernation, and estivation—reduce demands on stored fuels. One of the greatest challenges for dormant animals is the maintenance of structure during prolonged periods of inactivity. Although the animals do not grow during dormancy, they must rebuild the protein structures that naturally degrade over time. For example, even though their metabolic rates are lower than normal, dormant black bears remain relatively warm and metabolically active. As a result, proteins are constantly degraded and inevitably some of the amino acid building blocks are oxidized, causing the liver to produce urea. Most hibernating mammals accumulate urea and urine throughout the dormant period; however, dormant bears accumulate very little urea. Bears are able to recycle this urea nitrogen, regenerating amino acids that can be used to rebuild proteins. They transport urea from the blood into the GI tract, where it can be degraded by the enzyme urease. Although this enzyme generates ammonia, it is not yet clear how the bear is able to use the ammonia to resynthesize all of the amino acids required for biosynthesis. It is likely that the GI bacterial flora help incorporate the ammonia into amino acids. As with other symbiotic relationships, the animal derives nutrients from uptake of bacterial products as well as digestion of the bacteria themselves. The amino acids produced in the bear's GI tract can be transported into the blood and used in biosynthesis. As a result of this urea recycling, the levels of the dietary essential amino acids remain constant and the animals can maintain a nearly constant muscle mass during the many months of dormancy.

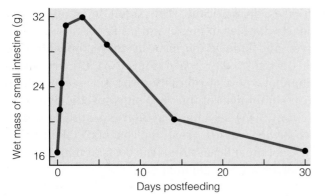

(a) Starvation and intestinal mass

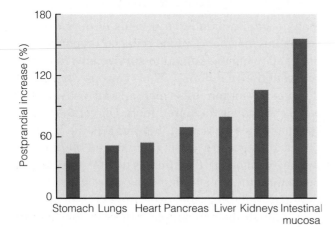

(b) Tissue mass changes

Figure 11.33 Python digestion Large snakes that eat infrequent meals allow their digestive organs to decrease in mass between meals. Organ mass increases rapidly upon feeding. **(a)** Intestinal wet mass doubles within three days postfeeding. **(b)** Other tissues that participate in digestion also increase markedly in the first few days after feeding. (Source: Modified from Secor and Diamond, 1998)

4. What is the main function of each region of the digestive tract?

5. How does cellular diversity contribute to functional diversity in digestive physiology?

6. What are four ways vertebrates increase the surface area of the intestine?

Integrating Systems | Obesity

Most animals face the continuing challenge of finding adequate food to support their short-term and long-term metabolic needs. The metabolic status of an animal is monitored centrally, with deficits translated into "hunger," spurring the animal to search for food, enduring risks of predation. Evolution has led to exquisite controls that match feeding behaviour, digestive physiology, and energy metabolism. Some animals experience phases where they eat more than usual (*hyperphagy*), often in preparation for migration, reproduction, or dormancy. For example, shorebirds may gorge on intertidal crustaceans before flying south for the winter, doubling their body mass with fat deposition. Likewise, ground squirrels more than double in mass by feeding through the fall to obtain enough fat for insulation and nutrition while hibernating over the seemingly endless Canadian winter. While hyperphagy is essential in the life history of

some animals, in humans it is entirely maladaptive. We evolved as a hunter-gatherer species, with adaptations for efficient nutrient storage during rare periods of food abundance. As a result of Western societal changes—higher food availability, poorer diet quality, and reduced physical activity levels—there has been a dramatic increase in the prevalence of obesity in many populations. To discover where you stand in terms of a healthy body mass, calculate your body mass index (BMI): take your mass (in kilograms) and divide by your height (in meters) squared (BMI= kg/m^2). According to the World Health Organization, one in three North American adults is overweight (BMI >25) and one in ten is obese (BMI >30). Although BMI is a good index of obesity, the main culprit in terms of health problems is the fat localized around the midsection, termed abdominal (or visceral) fat.

The added biomechanical stress of carrying extra fat is a relatively minor component of the physiological disruption associated with being overweight. As shown in Figure 11.34, excess adipose tissue (adiposity) affects numerous physiological systems, contributing to many diseases. Adipose metabolism affects whole body nutrient metabolism, and through these effects, alters the sensitivity to other important regulators of metabolism. Adipose tissue is also an endocrine tissue, secreting diverse hormones related to dietary status (leptin, adiponectin), blood clotting (plasminogen activator inhibitor-1), inflammation (interleukin 6, tumor necrosis factor alpha), and blood pressure (angiotensinogen). In overweight individuals, it is the combination of larger adipocytes and greater total adipose tissue mass that affects the endocrine functions of the tissue.

The main metabolic consequences of obesity are related to disruption of normal insulin signaling. In a healthy individual, an elevated blood glucose level triggers the pancreas to secrete insulin, which in turn promotes glucose uptake by adipose and skeletal muscle, and inhibits hepatic gluconeogenesis. In an obese individual, the pancreas secretes insulin, but despite *hyperinsulinemia*, the target tissues fail to respond to the hormone. As a result of the insulin resistance, glucose is not cleared by metabolism and the blood maintains a very high glucose level (*hyperglycemia*). At the heart of this disorder is insulin resistance in the adipose tissue. In normal adipocytes, high insulin levels signal an "energy-rich" state, causing adipocytes to reduce triglyceride breakdown. In obese individuals, the adipocytes are hy-

pertrophied, and these larger, lipid-rich cells are resistant to the insulin signal and respond by releasing nonesterified fatty acids (NEFA). The reasons for the loss of insulin sensitivity in peripheral tissues are complex. In the liver, it may be caused by metabolic disruption, such as elevated NEFA, or by changes in adipocyte regulatory factors, such as tumor necrosis factor alpha (TNFα). In either case, the loss of insulin-sensitivity causes the liver to increase lipid storage, creating a condition known as fatty liver disease. There are also dramatic increases in the triglyceride levels in skeletal muscle and around the heart (*epicardial fat*). It also alters how the liver maintains lipid profiles in the blood; more triglyceride is produced and released to the blood, and there is a reduction in the levels of HDL, the lipoprotein that reduces the negative effects of cholesterol. The main metabolic effects of obesity—high blood glucose, high blood triglyceride, insulin resistance, low HDL—are four of five symptoms of a condition known as *metabolic syndrome* (the fifth symptom, high blood pressure, will be discussed a bit later in this feature). People with three or more of these symptoms are at a much greater risk of cardiovascular disease.

Like the metabolic effects of obesity, the cardiovascular effects are complex. Obese individuals may experience greater risk of numerous cardiovascular disorders. Obesity and insulin-resistance interact to cause high blood pressure (*hypertension*). Insulin is a powerful vasodilator, and if obesity makes the vasculature insensitive to its hypotensive effects, hypertension can result. Adipocytes are also a major site of two

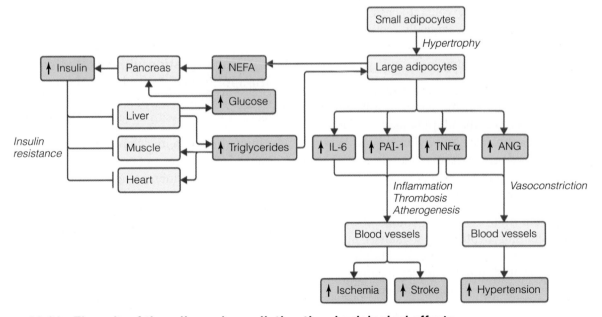

Figure 11.34 The role of the adipose in mediating the physiological effects of obesity ANG, angiotensinogen. IL-6, interleukin-6. NEFA, nonesterified fatty acids. PAI-1, plasminogen activator inhibitor-1. TNFα, tumor necrosis factor α.

vasoconstricting factors: angiotensinogen and TNFα. These factors act on vasculature and also affect how the kidney regulates blood pressure. Apart from regulating vascular tone, obesity and insulin-resistance can also affect vasculature structure by increasing endothelial damage, promoting formation of vascular plaques (*atherogenesis*) and clots (*thrombosis*). Again, there may be a metabolic link (such as increased plasma cholesterol), but the main reason for the vascular effects is through the alterations in the regulatory factors released by the adipose tissue. Increases in TNFα, interleukin 6, and plasminogen activator inhibitor-1 exert complex effects on the vasculature. They act directly and indirectly to promote inflammation, increasing oxidative damage within the vascular tissue, promoting atherogenesis, thrombosis, and platelet aggregation. Collectively, the vascular effects can contribute to other cardiovascular complications. The combination of atherogenesis and thrombosis increases the risk that plaques and clots will cause vascular blockage and ischemia. This could cause a heart attack if the blockage is in the coronary arteries, or a stroke if the blockage is in the circulation of the brain.

In relatively rare situations, an obese phenotype can be traced to genetic variants that affect the perception of hunger or the appropriate control of metabolism. However, the most common cause of obesity is simply eating more calories than your body requires. Fortunately, many of the health problems associated with obesity can be reversed if a person returns to a healthy weight.◉

Summary

The Nature and Acquisition of Nutrients

→ Animals must feed to obtain a spectrum of nutrients that enable them to build and maintain cells. Some building blocks cannot be synthesized by the animal and must be obtained preformed in the diet: eight to ten essential amino acids, two classes of fatty acids (omega-3 and omega-6), as well as vitamins and minerals. Much of the ingested food is used to produce energy to support activity and biosynthesis.

→ Simple carbohydrates and amino acids are actively transported into cells. Excess carbohydrate and protein is typically converted to glycogen for storage in liver and peripheral tissues such as skeletal muscle.

→ Lipids are transported into the animal in many forms. Fatty acids can be taken up directly by intestinal cells, but most lipids leave the intestinal cells in the form of chylomicrons. When the lipids are removed from chylomicrons, their remnants are taken up and repackaged into lipoproteins by the liver.

→ Animals find their food using chemical cues involving chemoreceptors that sense specific chemicals. Other animals find food based on visual cues, or other indices of prey location, such as electrical activity or heat production.

→ Simple animals such as sponges and cnidarians digest food intracellularly, after phagocytosis of particulate matter. More advanced invertebrates break food down into macromolecules extracellularly and transport individual molecules into the gastrointestinal epithelium.

→ Many animals use oral feeding structures to find and ingest food. Bird beaks display morphological diversity, allowing them to meet the biomechanical challenges associated with different types of food. Mammals have bony teeth protruding from the jaw to help in mechanical disruption of food by grinding, tearing, and shredding.

Integrating Digestion with Metabolism

→ The appearance of a coelom early in animal evolution allowed the development of a gastrointestinal system with specialized compartments.

→ Although simple invertebrates have short, tubular GI tracts, more complex animals maximize surface area by increasing the undulations of the gut surface, producing fingerlike projections from the surface (villi) and cellular protrusions from the absorptive enterocytes (microvilli).

→ The digestive epithelium in each compartment has special types of secretory cells that release digestive enzymes and also control the physical properties with secretions of acid, base, and mucus.

→ Many animals have fermentation chambers prior to the glandular stomach (ruminants) or after the glandular stomach (nonruminants). These specialized compartments house endosymbiotic bacteria that possess cellulolytic activity.

→ Hormones interpret information from the GI tract and metabolic storage tissues to influence the desire to feed (appetite). Once food is ingested, hormones control the secretions by the stomach (acid, pepsin, mucus), pancreas (bicarbonate, proteases, lipases, nucleases), and the gallbladder (bile).

→ Smooth muscle of the gut controls the rate at which the bolus moves down the digestive tract. Numerous hormones and neurons, acting both locally and centrally, control gut motility.

→ Each of these digestive processes—digestion, uptake, and assimilation—is influenced by temperature.

→ Animals utilize nutrients that appear in the blood after a meal, oxidizing some directly and metabolizing others into storage forms. Hor-

mones such as insulin and glucagon regulate the fate of the major metabolic fuels. Insulin promotes glucose utilization, whereas glucagon antagonizes the insulin effect.

→ If deprived of food for long periods, a vertebrate initiates a starvation response, converting some of the stored lipids to ketone bodies for use in brain and other tissues that normally rely on glucose for energy.

→ Many animals extend the duration of nutrient stores by entering hypometabolic states of dormancy. Hibernating animals face other challenges, such as ion and water imbalances. Dormant bears cope with gradual loss in the nitrogen pool by recycling nitrogen with the help of the bacterial fauna of the gut.

→ Pythons allow the GI tract to partially degrade in the period between meals, which can be as long as a year. After a python feeds, it uses some of the energy to rebuild the digestive tract just ahead of the food bolus.

→ Digestion and reproduction often proceed with conflicting goals. Many species separate their reproductive cycle from digestion, simplifying the regulation of these processes.

Review Questions

1. Summarize the basic organization of the vertebrate GI tract. What is the function of each compartment?

2. Discuss the variation in the nature of fermentation chambers.

3. How do animals use neurosensory systems to detect food in a complex environment?

4. Compare the different pathways for digestion and uptake of the three main classes of macromolecules: lipids, carbohydrates, and proteins.

5. A diet is rich in chemical energy, but not all of this energy is available to the animal. How is energy partitioned in a diet?

6. What roles do glands play in the process of digestion?

7. How does specific dynamic action benefit an animal?

8. How do animals control the secretions along the gastrointestinal tract?

9. Discuss the fate of glucose during a meal, after a meal, and two days after an animal's meal.

10. How might an animal alter its ability to import monosaccharides from the gastrointestinal tract?

Synthesis Questions

1. Discuss situations where digestion and reproduction may be antagonistic processes.

2. An animal that feeds on a large meal undergoes numerous changes that affect its other physiological systems. Discuss how the digestive process impinges on other systems.

3. When wild animals are domesticated, the years of artificial selection can alter the digestive physiology of the animal. Choose an example of a domesticated animal and consider how its digestive physiology might differ from that of its wild ancestors, given the differences in diet and selective pressures.

4. Why is digestion a metabolically expensive process?

5. Follow the path of glucose from the nectar reservoir of a plant to the muscle of a hummingbird. What steps control the rate of this process?

6. What differences in digestive physiology systems would you expect when comparing birds that eat nectar (easy to digest, high energy per gram), seeds (difficult to digest, high energy per gram), or fruit (easy to digest, low energy per gram)?

7. The harsh chemical and enzymatic conditions in the gastrointestinal tract break down nutrients. How do animals protect themselves from their own digestive secretions?

Quantitative Questions

For the following calculations, assume the following:

- The generic daily caloric requirements are 2500 kcal for men, and 2000 for women.
- The caloric expenditures for someone with an average lifestyle are attributed to basal metabolic rate (about 70% of total), specific dynamic action (10%), and physical activity (20%).
- There are 9000 kcal in 1 kg of fat, and 4000 kcal in 1 kg of protein or carbohydrate.
- There are about 7000 kcal in 1 kg of body mass.
- A pint of beer has about 200 kcal.
- You burn about 500 kcal by jogging for 1 h at 10km/h (though it depends on your weight and the running speed).

1. Assuming that your metabolic rate during sleep is equal to your basal metabolic rate (it's actually lower), how many calories did you expend while sleeping for 8 hours Translate those calories into units of body mass. Did you lose that much weight while you slept? How do respiration and urine production factor into this analysis?

2. Assuming no change in basal metabolic rate or SDA, how long would it take to lose 1 kg of body mass (a) by reducing caloric intake by 500 kcal per day or (b) by doubling your physical activity each day?

3. If you jogged to your local pub, how far would you have to go to ensure you expended enough calories to maintain caloric balance if you plan to consume two pints of beer?

For Further Reading

See the Additional References section at the back of the book for more readings related to the topics in this chapter.

General Reading and the History of Digestion

This book, accessible to a general audience, discusses the nature of energy and energetic transformations in physiological function.

Brown, G. 1999. *The energy of life.* London: HarperCollins.

These works discuss the history of enzyme research. Kornberg's autobiography focuses on developments from the earliest days of nutritional biochemistry to the era when molecular biology began to influence the field. The essay by Fruton focuses specifically on the enzyme pepsin, one of the first digestive enzymes to be studied.

Fruton, J. S. 2002. A history of pepsin and related enzymes. *Quarterly Review of Biology* 77: 127–147.

Kornberg, A. 1989. *For the love of enzymes: The odyssey of a biochemist.* Cambridge, MA: Harvard University Press.

This review discusses how the evolution of novel developmental patterns led to the origins of the diversity seen in the vertebrates.

Manzanares, M., and M. A. Nieto. 2003. Minireview: A celebration of the new head and an evaluation of the new mouth. *Neuron* 37: 895–898.

Digestive Physiology

For many years, the nature of intestinal sugar transport was hotly debated by several important research groups. These collected papers take different perspectives on the subject.

Ferraris, R. P., and J. Diamond. 1997. Regulation of intestinal sugar transport. *Physiological Reviews* 77: 257–302.

Kellett, G. L. 2001. The facilitated component of intestinal glucose absorption. *Journal of Physiology* 531: 585–595.

Hormones and neurotransmitters are central to the regulation of intestinal function. This review focuses specifically on the role of the regulatory factors in the control of the smooth muscle that determines gut motility.

Hansen, M. B. 2003. Neurohormonal control of gastrointestinal motility. *Physiological Reviews* 52: 1–30.

Many animals survive periods of adverse environmental conditions by entering a period of dormancy. This book compares the various modes of metabolic arrest used by animals.

Hochachka, P. W., and M. Guppy. 1987. *Metabolic arrest: Control of biological time.* Oxford: Oxford University Press.

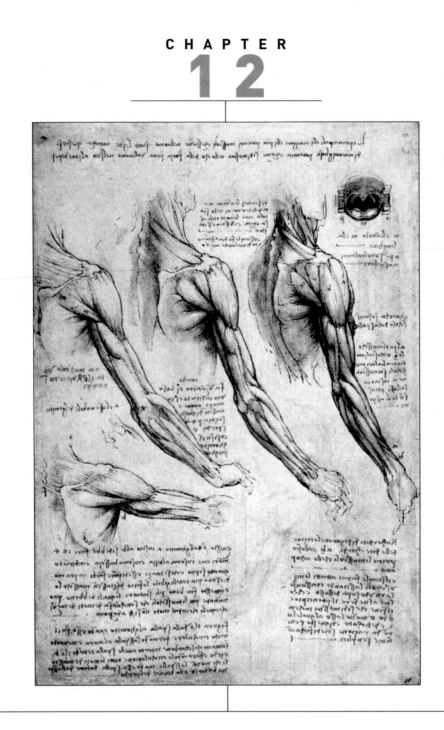

Locomotion

Animal locomotion has concerned humans since prehistoric times. For early human hunters, animal movement was a source of frustration; animals that we now admire as majestic runners, they considered to be escaping food. With the advent of animal husbandry, people confined wild animals to supply the cook fires without a chase. Later, animal breeding and artificial selection were used to modify animal locomotor physiology to obtain animals with more favorable features. Working animals were bred for strength and endurance. Racing animals were bred for speed. Food animals were bred for growth rate and muscle mass, which usually compromised the locomotor properties.

Muybridge's horse photographs.

Photo courtesty of Jean-Michel Weber, University of Ottawa.

More than 2400 years ago, Plato, Hippocrates, and Aristotle wrote about the nature of movement in animals, including humans. For much of the next two millennia, our understanding of the physiology of locomotion was constrained by cultural taboos against the dissection of human corpses. Renaissance artists, including Leonardo da Vinci and Michelangelo, conducted clandestine autopsies to explore the structure of the human body. The anatomical detail of Renaissance artwork reflects Western culture's first explorations into human locomotor physiology. As science matured in the Age of Enlightenment, locomotion research began to link studies of muscle anatomy and physics, creating a new field called biomechanics.

The study of locomotion was further aided by progress in photography. In the late 1800s, Eadweard Muybridge, a landscape photographer by training, was commissioned by Leland Stanford to settle a wager on the position of a horse's legs during a trot. He employed a series of cameras that were mechanically triggered to take sequential images of a trotting horse. He was able to show Stanford that at times all four of the horse's feet were off the ground. Muybridge's interest in animal motion culminated in *Animal Locomotion*, a series of 11 books containing more than 100,000 photographs of moving animals. Advances in photographic technologies continue to benefit locomotion research. High-speed cinematography can capture more than 1000 frames per second. When combined with force-sensitive plates embedded in the floor, biomechanics researchers can calculate the forces exerted on the muscles and bones as animals move.

Much of our understanding of locomotor systems relies on the integration of the diverse fields that make up physiology. In the mid-1900s, Sir Andrew Huxley and Hugh Huxley applied modern physics to explain the basis of muscle contraction. In the late 1900s, advances in microscopy and nuclear magnetic resonance technologies provided windows into the cellular function of locomotor systems. Many further advances in the comparative physiology of locomotion arose when researchers devised ways to make physiological measurements of moving animals. In the late 1960s, C. R. (Dick) Taylor and his colleagues began decades of studies that combined biomechanics with respiratory physiology. They trained diverse animals to run on treadmills while they measured muscle activity and respiration. The largest animals they studied were elephants. The elephants were too large to run on their treadmills, so the researchers developed golf carts that carried gas analyzers. Around the same time, J. R. Brett developed a swim tunnel respirometer to study the locomotion of fish. His Brett respirometer could control water velocity and measure oxygen consumption and carbon dioxide production. In the 1980s, wind tunnels were used to fly birds specially fitted with masks to measure oxygen consumption. In recent years, more sophisticated respirometry gear has been used to measure the costs of movement in animals as small as bees and ants and as fast as hummingbirds and tunas. By combining biomechanics and bioenergetics, researchers are able to explore the origins of variations in muscle energetics and efficiency. ⊙

Overview

Locomotion is usually defined as the act of moving from one place to another. To an animal physiologist, locomotion is an active process that is initiated and controlled by the animal. Locomotor systems integrate anatomy with several physiological systems. Appendages such as fins, legs, and wings allow animals to interact with the environment to generate or control forces that result in directional movement. The physical organization of muscles into musculoskeletal systems allows animals to translate cellular contraction into whole animal locomotion. The musculoskeletal system acts in combination with the nervous system to control the position and movement of appendages. Locomotion demands exquisite control of energy metabolism and digestive physiology, mediated by the hormones that regulate fuel assimilation, storage, and mobilization. The respiratory system ensures that oxygen uptake eventually matches the increased oxygen demands that accompany muscle activity. The cardiovascular system delivers fuels to the muscle and removes metabolic end products. The interactions between these systems are summarized in Figure 12.1.

A hallmark of locomotor systems is the ability to respond to changes in demand. This capacity is particularly impressive in animals that undergo long-distance migrations. Prolonged changes in activity (training or detraining) alter the locomotor machinery. Humans are one of the few species that has the luxury of becoming detrained. In the natural world, detrained animals tend to get eaten or starve. Regardless of how fast or how far an animal travels, the ability to move requires coordination of diverse physiological systems.

Superimposed on the control of body movement are the constraints of the environment. Each environment, whether aquatic, aerial, or terrestrial, has physical properties that animals must overcome in order to move.

Locomotor Systems

We begin our discussion of locomotor physiology by exploring the nature of the systems that support movement. When we first introduced muscles in Chapter 5: Cellular Movement and Muscles, we focused mainly on the control of excitation-contraction coupling. We now turn our attention to the way different types of muscles are integrated into a musculoskeletal system composed of muscles and skeleton, held together by connective tissue, controlled by the nervous system, and nourished by the blood supply. These musculoskeletal systems allow animals to translate changes in cell shape into movement.

Muscle Fiber Types

Most animals rely on muscles to generate the force required to move from place to place. Each style of movement requires muscles that possess appropriate biomechanical properties.

As we discussed in Chapter 5, the contractile properties of a muscle are determined by the design and organization of the proteins within the myofiber. The properties of contractile proteins alter cross-bridge cycling dynamics. The cellular machinery of excitation-contraction coupling affects the kinetics of contraction and relaxation. The three-dimensional arrangement of sarcomeres determines how much force a skeletal muscle can generate. Through differences in protein properties and structural organization, animals can produce muscles with particular contractile phenotypes that enable animals to move in the environment. Chapter 5

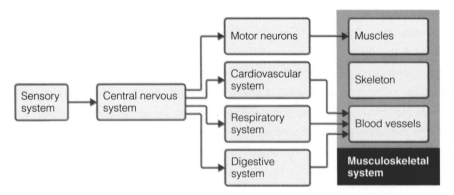

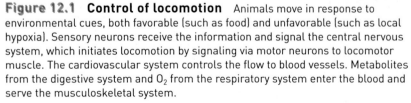

Figure 12.1 **Control of locomotion** Animals move in response to environmental cues, both favorable (such as food) and unfavorable (such as local hypoxia). Sensory neurons receive the information and signal the central nervous system, which initiates locomotion by signaling via motor neurons to locomotor muscle. The cardiovascular system controls the flow to blood vessels. Metabolites from the digestive system and O_2 from the respiratory system enter the blood and serve the musculoskeletal system.

focused on the cellular processes that allow muscles to contract. In the following sections, we discuss how animals incorporate muscles into locomotor systems.

Many invertebrates use simple circular and longitudinal muscles to move

With the exception of the arthropods, most terrestrial invertebrates move by crawling. Simple muscles work in combination with a fluid-filled internal chamber that acts as a **hydrostatic skeleton**. Invertebrate locomotor muscles are typically striated, although the myofibers are often organized in ways that differ from vertebrate striated muscles.

Most wormlike invertebrates crawl using overlapping layers of muscle fibers. Nematodes use two layers of fibers running in different orientations along the longitudinal axis (Figure 12.2). When the muscle fibers contract on one side, coelomic fluid is forced into the opposite side and the worm bends. The nematode uses cycles of contraction and relaxation to undulate through the environment.

Earthworms organize locomotor striated muscles into circular and longitudinal layers. This arrangement is reminiscent of the organization of smooth muscles of our digestive tract (see Chapter 11: Digestion).

As with the gut musculature, these muscles allow the animal to produce peristaltic waves of contraction. Earthworms use the same principle, but because they are segmented, each body segment works independently, giving the earthworm a much greater degree of control over movement (Figure 12.3).

Directly beneath the outer layers of the earthworm cuticle and epidermis lies the thin layer of circular muscle. The thicker longitudinal layer of

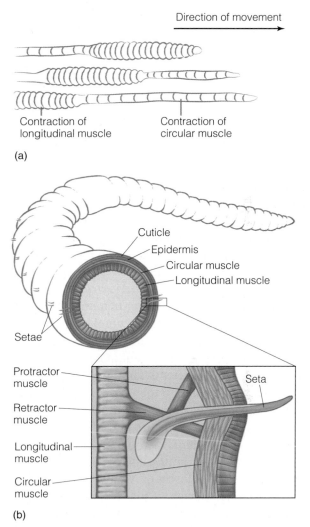

(a)

(b)

Figure 12.3 Earthworm locomotion **(a)** Earthworms move using waves of muscle contraction that act in conjunction with the hydrostatic skeleton. Contraction of circular muscle reduces the diameter of the worm and pushes coelomic fluid forward. Longitudinal muscle contraction pulls the posterior segments of the worm forward. **(b)** This pattern of muscle contraction translates into locomotion with the help of a series of hairlike setae that anchor segments of the worm to the substratum. The attachment of setae is under muscular control. Protractor muscles force setae outward to lock onto the substrate. Retractor muscles pull setae back toward the body, releasing the surface. Movement requires coordination of the muscles of the body wall and setae.

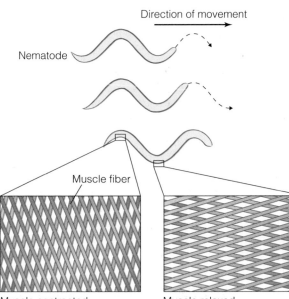

Figure 12.2 Nematode muscles and crawling Nematodes move through the soil using undulations. The body bends when overlapping muscle fibers contract on one side of the body and relax on the other side, forcing a redistribution of coelomic fluids.

muscle is composed of groups of muscle cells arranged into fan-shaped (pennate) bundles. Each bundle is surrounded by a basement membrane, and bundles are bound together by connective tissue. A ring of nerves circles the segment, running between muscle layers, with axons extending toward the muscles. When the circular muscle contracts, the coelomic fluid is pushed forward to extend the segment. Once the segment is extended, tiny hairlike projections called setae attach to the soil or other substrate surface. When the longitudinal muscle contracts, the anterior end of the segment remains in place and the posterior part of the segment is pulled forward. The nerve networks coordinate the movement of circular and longitudinal muscles and the patterns of activity in the independent segments.

Squid, the fastest of aquatic invertebrates, also use complementary muscle layers to move, but the arrangement is quite different. The muscles of the outer body wall, or mantle, are intermingled in two planes (Figure 12.4). Radial muscle fibers extend from the inside of the mantle to the outside. Contraction of the radial muscles reduces the thickness of the mantle wall and reduces its circumference. Circular muscle, which surrounds the mantle, is composed of three layers. A thick central layer of muscle with low mitochondrial content is covered on the inside and outside by a thin layer of mitochondria-rich muscle cells. Squid use these complex mantle muscles to produce jet propulsion. Water enters the internal chamber when the mantle muscles relax. Upon contraction, water is rapidly ejected out of the mantle cavity through a tube, or siphon, creating a flume of water that pushes the squid forward. In Chapter 4: Neuron Structure and Function, we discussed how the giant axon ensures that electrical stimulation of the mantle muscles occurs in unison to maximize the force of water expulsion. The anatomical complexity of the muscles has made it difficult to assign specific roles to each muscle fiber type, but in general the squid can use these muscles in combinations that allow it to regulate the force of the water jet, enabling it to hover, cruise, or dart from place to place. Flying squid use the jet to leap as much as 3 m out of the water, then use their lateral fins to glide at velocities in excess of 25 km/h.

Fish use two or three fiber types to swim

Much of our understanding of the importance of muscle fiber types comes from early studies on

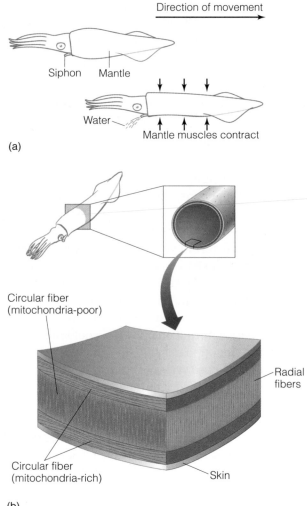

(a)

(b)

Figure 12.4 Squid jet propulsion **(a)** Squid produce jet propulsion by forcing water from the body cavity out of a tubelike siphon. Water moves in and out of the body cavity in response to muscular contractions of the body wall, or mantle. **(b)** The mantle is composed of complex, intertwined layers of muscle fibers. Radial fibers control the thickness of the mantle. The diameter of the mantle is controlled by three layers of circular muscles.

fish. Comparative physiologists are lured to fish for two main reasons. First, fish morphology differs widely for reasons that are intuitively linked to locomotion. Second, swimming muscles of fish are composed of homogenous fiber types, in contrast to muscles of most other vertebrates. The natural diversity in anatomy, ecology, behavior, and evolutionary biology creates research opportunities for physiologists studying the fundamental principles of muscle function and the molecular basis of different locomotor strategies. Or, put another way, studies using fish have helped us understand muscle, and studies of muscle have helped us understand the biology of fish.

Fish build their locomotor muscle from two main types of muscle fibers: red and white muscle (Figure 12.5). Muscle fibers of fish are still described by these colorful terms, whereas the muscles of tetrapods are more commonly designated by their myosin isoforms (e.g., Type I, IIa, IIb, IIx). White muscle makes up most of the muscle mass of the fish, typically about 60% of the body mass and about 85% of the muscle. The white muscle is a glycolytic fiber type that is responsible for high-intensity, burst swimming. Red muscle is usually confined to a narrow ribbon that extends along the side of the animal just under the lateral line. Small patches of red muscle are also found at the base of fins, where they are used to power the fin move-

ments. Red muscle is an oxidative fiber type that supports slow, steady-state cruising activity. Many fish have a third type of locomotor muscle called pink muscle that is intermediate in contractile properties. Pink muscle is typically found at the interface between the white and red muscles. Each region of locomotor muscle is virtually homogeneous in fiber type. This anatomical organization is convenient for researchers who study the cellular origins and physiological function of different muscle fiber types.

The organization of fish muscle retains many of the properties established early in its development. Many of us are familiar with fish muscle anatomy from our experience at dinner. A well-cooked fish filet can be dissected into parallel layers of white muscle. Each layer is a **myotome**, one of the original segments established early in embryological development. Each myotome contains blocks of parallel white muscle fibers separated by a thin layer of connective tissue called the *myoseptum*. Each myotome is attached to the posterior region of the fish by tendons. The skin also acts as a sheath that connects the different myotomes, helping to integrate the force of the different contractile units. Contraction of a myotome generates force that is transmitted in complex ways to other regions of the body. Force is transferred to the next myotome across myosepta, to the caudal fin along tendons, and to the skin. These forces culminate in movement of the trunk and tail to generate propulsion. Force generation in red muscle also relies on the skin and tendons that insert at the tail. This arrangement converts contractile force directly to movement of the trunk and tail.

The pattern of locomotor muscle contraction is controlled by motor neurons

The differences in the contractile properties of oxidative (red) and glycolytic (white) muscle enable the animal to produce different types of movement. Although most easily shown in fish, these same general rules apply to tetrapods. Red muscle exhibits its maximal power output at much lower tail-beat frequencies than does white muscle. Consequently, fish use red muscle in slow swimming and white muscle at higher velocities. In living fish, this pattern of sequential activation of muscle contraction, called **recruitment**, is determined by motor neurons, under the control of the central nervous system.

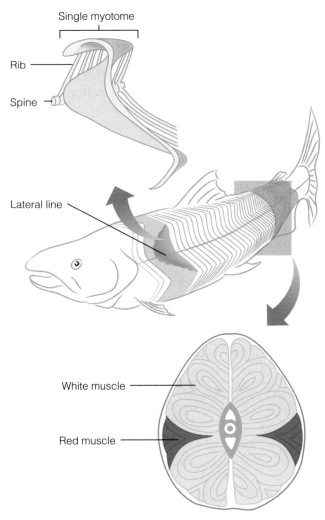

Figure 12.5 Musculature of fish Fish white muscle is composed of more than 100 repeating units called myotomes. They extend backward from the spine, twisting forward as they approach the exterior surface of the fish. Narrow strips of red muscle are found laterally along the length of the fish. Pink muscle (not shown) often separates the red from the white muscle.

Researchers study muscle recruitment in living fish by fitting them with electromyograph (EMG) electrodes and inducing them to swim at different velocities. The EMG output shows that at low swim speeds only the red muscle is electrically active (Figure 12.6). As swim speed increases, red mus-

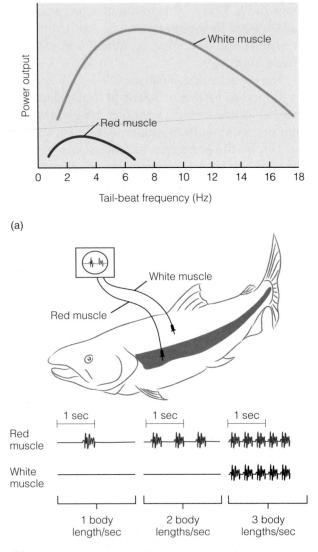

(a)

(b)

Figure 12.6 Swimming velocity and muscle recruitment **(a)** Power output of isolated muscle can be assessed over a range of frequencies of electrical stimulation. Red muscle has a lower power output than white muscle, and it generates its optimal power at a lower frequency. **(b)** Electrical activity of red and white muscle in living fish can be measured by electromyography. The different muscle fiber types are recruited at different swim velocities. At low velocities (one body length per second), only red muscle is active. As velocity increases, the frequency of contractions increases. Once swim velocity exceeds a threshold (in this example, two body lengths per second), white muscle is activated. Red muscle continues to contract, but the force it generates contributes little to locomotion at high velocities.

(Sources: Based on (a) Altringham and Johnson, 1990 and (b) Johnston, 1981.)

cles are activated more frequently. At still faster speeds, white muscle is activated. At high swim velocities, red muscle may continue to be activated but it doesn't generate much power. Fish can swim continuously for hours at speeds where only red muscle is active. At faster speeds, where white muscle is recruited, fish quickly become exhausted.

The importance of neuronal control of locomotor movement can be illustrated using the lamprey, a primitive fish that swims by simple snakelike undulations. Like other fish, the lamprey has superficial red muscle and deep white muscle organized into about 100 myotomes. Nerve roots from the corresponding segment of the spinal cord innervate each myotome. Separate nerves innervate the muscles on each side of the fish.

When the lamprey swims, one side of the body contracts while the opposite side relaxes (Figure 12.7). Contractions begin anteriorly, then move posteriorly. When a lamprey swims in a straight line, the nerves on one side fire, while the contralateral nerves are silent. The alternative firing pattern of the motor neurons is coordinated by excitatory and inhibitory interneurons. Nerves along the spinal cord fire in rapid succession to stimulate a wave of contraction along the body. After a motor neuron has fired, the interneurons allow the excitation of the motor neurons innervating the opposite side. The wave of contraction on one side is followed by a wave on the other side. When a lamprey needs to swim faster, it increases the firing frequency. This causes the body to undulate faster and more often, but the pattern

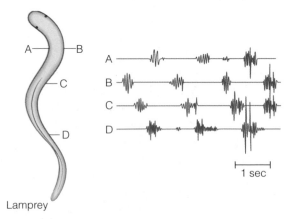

Lamprey

Figure 12.7 Swimming lamprey The lamprey swims by anguilliform movement, using waves of contraction of trunk muscles. Each burst of electrical activity on an EMG recording signifies a muscle contraction. When muscles on one side of the body are activated, the muscles on the opposite side are inhibited. The waveform is generated by sequential activation of muscles along the length of the body.

(Source: Orlovsky et al., 1999)

of change in body shape (an S shape) is the same at all velocities. This intricate pattern of motor nerve activity is generated by the spinal cord in response to signals sent from the brain.

Tetrapods have a multiplicity of fiber types

Fish locomotion using the trunk and tail is relatively simple in terms of both muscle organization and neuronal control. However, once vertebrates made the transition to land, movement required much more complex locomotor muscles and neuronal controls. Whereas fish can get by with two or three muscle fiber types, tetrapods build individual muscles using combinations of fiber types. The limb musculature of tetrapods is developmentally homologous to the fin musculature of fish. Tetrapods have great diversity in how they use their limbs in movement, but the organization of muscles is similar in amphibians, reptiles, birds, and mammals. Each group of tetrapods draws upon large suites of muscle contractile proteins to create diverse fiber types.

Much of our understanding of tetrapod muscle function comes from studies on the hindlimb muscles that frogs use to jump. Like fish locomotor muscle, the jumping (extensor) muscle of a frog is relatively pure in fiber type. Researchers from many physiological disciplines value the frog hindlimb preparation as an experimental tool. Frog extensors are easily removed from the animal and remain stable, enabling researchers to work with the preparation for long periods. Biomechanics researchers focus on the hindlimb because of its activity during jumping. When frogs leap with their strongest contractions, virtually every fiber of the extensor muscles is recruited. Cell biologists use frog extensors because they can isolate intact, single fibers, which facilitates the exploration of cellular and genetic properties.

The diversity in muscle composition in tetrapods is evident at many levels of biological organization. Recall from Chapter 5 that skeletal muscle cells, or myofibers, are multinucleated cells. Each nucleus within a single myofiber usually expresses the same genes for contractile proteins. That is, each myofiber is usually of a pure fiber type because all nuclei within the cell express the same myosin heavy chain isoform gene. However, hybrid fibers can also occur, forming when individual nuclei within a myofiber express different myosin heavy chain genes. Thus, even single myofibers can

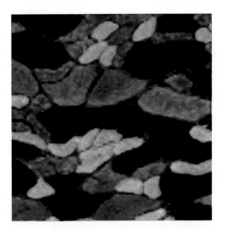

Figure 12.8 Mosaic of fibers in tetrapod locomotor muscle Most tetrapods possess muscles that are mosaics of different fiber types. In the rat diaphragm muscle shown above, the fiber types are distinguished by immunohistochemistry, using fluorescent antibodies that bind to specific myosin heavy chain isoforms. Type I fibers are shown in red, Type IIa in green, Type IIb in blue, and Type IIx/d in black.
(Image courtesy of Dr. Gary Sieck, Mayo Clinic)

be a heterogeneous fiber type. When these myofibers contract, they demonstrate properties that are intermediate between pure fiber types. Heterogeneity is also evident when looking at the whole muscle, which is a collection of hundreds of individual myofibers. Most tetrapod muscles are mosaics of different fiber types (Figure 12.8). Consider the composition of the soleus muscle of tetrapods, the muscle that originates below the knee joint and inserts at the Achilles tendon. The soleus muscle of most tetrapods is composed predominantly of slow-twitch fibers, but as many as 20% of the fibers might be other fiber types. Other hindlimb muscles are also mosaics of several muscle fiber types. The complexity of tetrapod muscle fiber types is necessary because the muscles are used in different combinations to perform many distinct styles of movement. The frog uses its hindlimb muscles to swim, walk, and jump. Mammals use hindlimb muscles to stand, walk, jog, swim, sprint, and jump. Forearm muscles enable birds to use their wings to flap, glide, and undertake complex aerial maneuvers. The complex fiber type profiles of muscles are essential to the versatility of muscle utilization.

Locomotor muscles are organized into locomotor modules and functional groups

Most tetrapods move using cyclical changes in the position of limbs. When a limb bends at a joint, the movement is called **flexion**. The limb straightens during **extension.** Flexion and extension are induced in response to the contraction of separate **antagonistic muscles** (Figure 12.9). For example, when a primate bends its arm (flexion), the biceps muscle contracts while the triceps is relaxed. Extension occurs when the triceps contracts while the biceps is relaxed. Limb movement in support of locomotion typically involves complex combinations of muscles that work together to move each segment of the limb in a coordinated manner. Consider the muscles used in the mammalian hindlimb during walking. An extensor group of leg muscles works synergistically to move the leg forward; a flexor group of muscles works synergistically to pull the leg back. The extensor group includes the soleus and gastrocnemius, which bend the foot; the quadriceps and rectus femoris, which straighten the knee; and the gluteus, which works at the hip to swing the leg forward. The flexor group includes the tibialis anterior, which moves the foot; the hamstring group, which bends the knee; and the iliopsoas, which rotates the leg at the hip. These muscle groups work antagonistically; contraction of one muscle group requires relaxation of the other muscle group. In addition to the muscles that move the leg, other suites of muscles participate in movement. The fine muscles of the feet work in combination with sensory information collected by proprioceptors in the skin to make fine-scale adjustments in position. The postural muscles of the back and abdomen are recruited to maintain balance during movement. All of the muscles that are responsible for a type of movement are grouped together into a **locomotor module**.

Bird flight musculature is another example of a locomotor module. The musculature that powers flight in birds is derived from the same appendicular musculature that supports the movement of forearms in tetrapods. As in other vertebrates, muscles work in antagonistic groups to power wing movements. The pectoralis muscle powers the downstroke. This is a very large muscle, often approaching 35% of the body mass of the bird. It is attached at one end to the keel bone, and at the other end to the humerus. The supracoracoideus muscle powers the upstroke. It attaches to both the keel and the end of the humerus. Most birds use this muscle to rotate the humerus, and return the wing to the correct position in preparation for the downstroke. The relative sizes of the pectoralis and supracoracoideus muscles reflect the way a bird flies (Figure 12.10). Hovering birds, such as hummingbirds, possess very large supracoracoideus muscles to rapidly pull the wing upward. More than 45 different muscles contribute to the fine control of the wing, including the position of feathers.

In contrast to the simple nervous control of lamprey swimming, coordination of limb movement in tetrapods requires the integration of countless motor nerves, interneurons, and overlapping feedback controls. This complexity can make it much more challenging to study motor control in tetrapods. For this reason, simple systems such as fish swimming musculature and frog extensor muscles remain valuable tools for exploring neuronal control of vertebrate locomotor muscle.

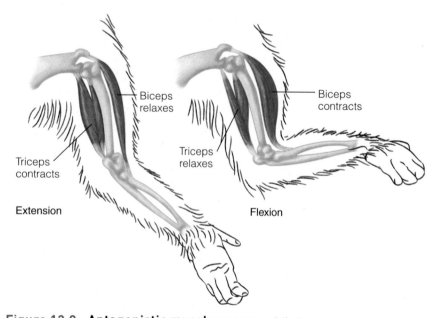

Figure 12.9 Antagonistic muscle groups A limb straightens when extensor muscles contract and bends when flexor muscles contract. In the forelimb of primates, the biceps is the primary flexor muscle and the triceps is the main extensor muscle.

Energy Metabolism

Muscle activity demands a great deal of energy, mainly in the form of ATP. The actinomyosin ATPase uses ATP to provide the energy for cross-bridge cycling. The Na^+/K^+ ATPase uses ATP to reestablish ion gradients across the sarcolemmal membrane after each action potential. The Ca^{2+} ATPase uses ATP to transport cytoplasmic

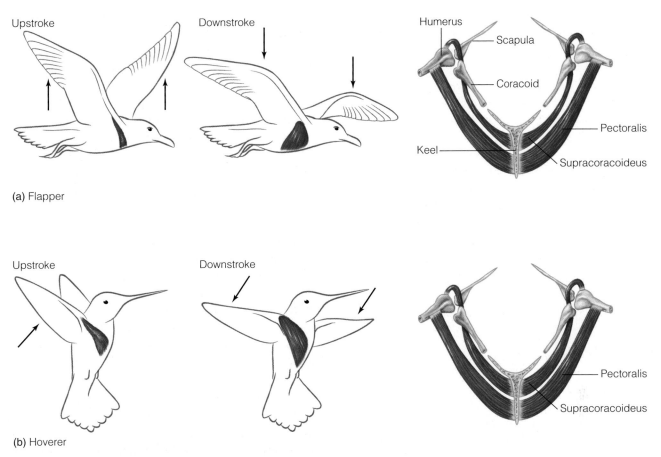

(a) Flapper

(b) Hoverer

Figure 12.10 **Bird flight muscles** Birds use their pectoralis muscle to power the downstroke, and their supracoracoideus muscle for the upstroke. **(a)** Birds that fly by flapping their wings, such as the seagull, have a very large pectoralis muscle. **(b)** The hovering flight of hummingbirds also requires a strong supracoracoideus muscle because force is generated on both the downstroke and the upstroke.

Ca^{2+} back into the sarcoplasmic reticulum. Since working muscles can have high rates of ATP turnover, let's discuss how the unique features of the pathways of energy metabolism are integrated into muscle structure and function.

Glycolysis and mitochondria support different types of locomotion

Muscle contraction is an energetically expensive process, and shortfalls in energy production can compromise locomotion. Muscles meet energy demands using a combination of preformed phosphagens and ATP-producing pathways. The preformed phosphagens include the adenylate pool (ATP and ADP) as well as the phosphoguanidine compounds; vertebrates use phosphocreatine, and invertebrates use one or more of phosphoarginine, phosphoglycocyamine, phosphotaurocyamine, or phospholambricine (see Chapter 2: Chemistry, Biochemistry,

and Cell Physiology). Since these preexisting energy pools can support locomotion only for very short periods, other pathways of ATP production are critical. Most locomotor activity is supported by some combination of anaerobic glycolysis and mitochondrial aerobic metabolism. These two pathways differ in five main respects that determine how they support muscle activity.

1. *Metabolic efficiency.* Oxidative phosphorylation produces more ATP per glucose molecule than does glycolysis (36 versus 2 ATP per glucose). As a result, all muscles rely on oxidative phosphorylation to support metabolism at rest and during recovery from activity. Slow-twitch muscles (such as fish red muscle) rely on oxidative phosphorylation to support muscle activity.

2. *Rate of ATP production.* Though less efficient, glycolysis can generate ATP faster than oxidative phosphorylation. When an animal must move very quickly, ATP must be produced at

rates that cannot be met by mitochondria. A cheetah chasing a gazelle relies on glycolysis to provide the ATP that allows it to reach high sprint speeds. Although glycolysis allows the muscle to produce ATP very quickly, limited stores of glycogen mean the less efficient glycolytic pathway quickly runs out of fuel. Thus, the cheetah must capture its prey within a short period or its muscle will run out of the carbohydrate fuels necessary to support sprinting.

3. *Dependence on oxygen.* In the absence of oxygen, glycolysis is the only option to produce ATP. During high-intensity activity, oxygen cannot be delivered to muscle fast enough to meet ATP demands by mitochondrial metabolism and the tissue becomes functionally hypoxic. Hypoxic muscle relies on internal glycogen stores and produces lactate, metabolic disturbances that must be rectified during recovery.

4. *Fuel diversity.* Glycolysis relies exclusively on carbohydrate, whereas mitochondria can generate energy from oxidation of carbohydrates, lipids (fatty acids), and amino acids. Fuels for muscle activity can be derived directly from the diet or mobilized from intramuscular stores or extramuscular storage depots.

5. *Rate of mobilization.* Muscles possess low levels of fuels that can be oxidized immediately (glucose, fatty acids, glycerol, free amino acids). Muscles consume these fuels rapidly, so animals must mobilize stored fuels to sustain muscle activity. Each type of metabolic fuel can be mobilized at a characteristic rate. When muscle activity begins, glycogen hydrolysis begins within a fraction of a second. If muscle activity continues, other fuel depots are mobilized.

Mitochondrial content influences muscle aerobic capacity

Oxidative phosphorylation is central to the energetics of most muscles, and mitochondrial content is an important determinant of muscle aerobic capacity. Muscle mitochondria are constructed in ways that pack maximal metabolic capacity into minimal space. Many aspects of muscle mitochondrial structure and function are similar across the animal kingdom. Although the mitochondrial content of muscles may vary widely across muscle fiber types, muscle mitochondrial properties are similar across species. However, some exceptional species show specializations that reflect the limits of mitochondrial function

and the evolution of aerobic capacity. The network of mitochondria, or reticulum (see Figure 2.52), is particularly well developed in the locomotor muscles of active organisms, allowing the mitochondria to operate as a more efficient electrical network. Antarctic fish also have an extensive mitochondrial reticulum, but in these animals the reticulum may improve the efficiency of oxygen delivery into the cell. Since oxygen dissolves more readily into lipid than water, the interconnected mitochondrial membranes facilitate oxygen delivery into the depths of the cell. This may be an important mechanism to facilitate oxygen delivery into the cell, since many of these animals lack the oxygen-carrying proteins hemoglobin and myoglobin.

The mitochondrial inner membrane structure also reflects the premium on intracellular space. The cristae, which possess the enzymes of oxidative phosphorylation, are densely packed to compress a high catalytic potential into a small space. Each milliliter of mitochondria possesses $20–40$ m^2 of inner membrane. Imagine $400–800$ pages of this textbook folded into a space the size of the end of your thumb. Some "athletic" species possess more densely packed cristae, sometimes approaching 70 m^2/ml of mitochondria. At this high cristae density there is barely enough space between cristae to fit two molecules of an average mitochondrial matrix enzyme.

Mitochondrial content varies widely among muscle types and species. The sparse mitochondria in glycolytic muscle fibers typically occupy less than 2% of the muscle intracellular space. The mitochondrial content of oxidative muscles is usually 3- to 10-fold greater than that of glycolytic muscles in the same animal. In the flight muscles of insects and hummingbirds, which contract at very high frequencies, almost half of the muscle intracellular volume is occupied by mitochondria. Furthermore, the mitochondria in these animals also possess a very high cristae packing density.

Muscle must recover from high-intensity activity

High-intensity activity is fueled by intramuscular stores of glycogen. As fast-twitch muscles undergo glycolysis, lactate is produced. The muscle becomes exhausted from the combination of energetic shortfalls, ion disturbances, and pH imbalance. To recover from burst exercise, muscles must replenish energy stores, including glycogen, ATP, and phosphocreatine. They must also reestablish ion gradients, Ca^{2+} stores, and pH. An important element of recovery is removal of the lactate that results from anaerobic glycolysis. At the end of exercise, lactate can have many different fates (Figure 12.11). Some muscles use lactate as a fuel to rebuild glycogen stores. Other muscles export the lactate for processing by other tissues. Some bloodborne lactate is oxidized by other aerobic tissues, such as the heart. If muscles release lactate into the blood, they must import glucose from the blood to rebuild muscle glycogen stores. Many animals use the *Cori cycle* to replenish muscle glycogen. In this pathway, muscle-derived lactate is imported by the liver, which uses it to resynthesize glucose. Liver glucose is then released into the blood and taken up by muscle, which can use it to produce glycogen. The relative importance of each pathway depends on muscle type and species.

Consider the example of the northern pike, a fish that lives in northern temperate waters. It spends most of its time hiding in the weeds waiting for unsuspecting prey to approach. When it sees a small fish or frog, the pike uses its white muscle to flip its tail. With just a few tail flips, the pike accelerates up to 5 *g*, roughly the equivalent of the acceleration of a jet fighter. If the pike misses its target with the initial thrust, the prey can escape because the pike cannot immediately mount a second attack. Instead, it returns to its hiding spot and begins the long, slow process of metabolic recovery. The northern pike, for example, requires many hours for its muscle to return to the preexercise metabolic state.

Recovery following intense muscle activity requires both metabolic and cellular corrections, each of which requires energy investment. Resynthesis of ATP, phosphocreatine, and glycogen requires oxidative phosphorylation. Energy is also required to reestablish ion distributions across membranes (H^+, Ca^{2+}, K^+, and Na^+). Muscle activity can also cause physical damage to the muscle, which must be repaired during recovery. The energy for these processes is provided by mitochondrial oxidative phosphorylation. Recovering animals often show elevated rates of oxygen consumption (EPOC) long after exercise has ceased, a phenomenon call *oxygen debt* (Figure 12.12).

Metabolic transitions accompany prolonged exercise

Locomotor activity poses unique challenges for animals. During activity, muscle cells must produce ATP at high rates. Metabolic fuels, primarily

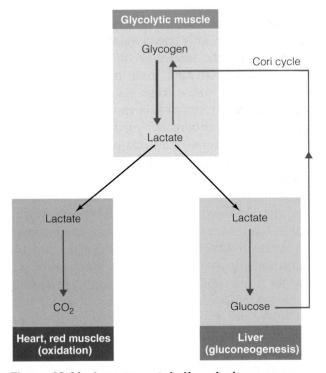

Figure 12.11 Lactate metabolism during recovery from activity High-intensity exercise in glycolytic muscles causes buildup of lactate. When exercise ceases, lactate is removed by many different pathways. Some lactate is used to resynthesize glycogen in the glycolytic muscle. Lactate can also be released into the blood and taken up by the liver, which uses it to produce glucose. Oxidative tissues, such as heart and red muscle, can oxidize lactate as a fuel.

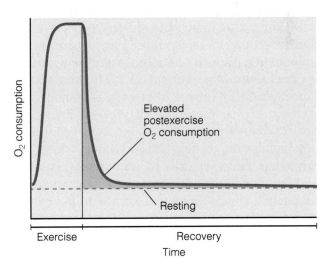

Figure 12.12 Elevated postexercise oxygen consumption Exercise causes a rapid increase in the rate of oxygen consumption. Once exercise ceases, respiration declines but remains elevated above the resting rate for extended periods. The duration of this elevated postexercise oxygen consumption, or oxygen debt, depends on the intensity of exercise and varies among species.

carbohydrate and lipid, must be mobilized from intracellular stores or storage tissues. These metabolic processes must be precisely coordinated to ensure that ATP synthesis matches ATP demand. Consider the following two examples.

The hovering hummingbird has one of the highest mass-specific metabolic rates in the animal kingdom. It must fly to feed, and it must eat in order to fly. It is an extreme example of the sort of metabolic transitions most animals face in integrating nutrient consumption with activity. Each summer morning the hummingbird wakes and flies from flower to flower, drinking nectar. Occasionally, it eats an insect. Most of its waking hours are spent perching. Its nectar diet is almost exclusively sucrose, so you might assume that its metabolic regulation is simple. Looking closer, however, you will discover that complex metabolic control is needed to accommodate the daily changes in feeding, flying, and sleeping. When the hummingbird is actively feeding on summer flowers, it uses dietary carbohydrate to fuel flight muscle metabolism. It stores any extra dietary sucrose as glycogen and lipid. In the evening, the hummingbird cannot feed and must rely on energy reserves to sustain its resting metabolic demands. It also becomes *hypometabolic*, allowing its body temperature to fall to reduce metabolic demands. In the morning, the hummingbird supports its first flight by oxidizing fatty acids mobilized from tissue stores. As soon as it obtains its first nectar meal, its metabolism switches to carbohydrate utilization and lipid storage. The transitions in fuel selection can be monitored by measuring the ratio of CO_2 production to O_2 consumption, known as the *respiratory quotient*, or RQ (Figure 12.13). Recall from Chapter 2 that each metabolic fuel generates a characteristic RQ: 0.7 for lipids, 1.0 for carbohydrates. This daily cycle continues throughout the feeding season, but as winter approaches many species of hummingbirds reorganize metabolism to prepare for migration. An important step is an increase in their lipid stores. A 3-g hummingbird may put on 2 g of fat prior to its migration.

The migration of the Pacific salmon provides another good example of metabolic transitions. Salmon live several years in the open ocean, vigorously feeding, growing, and preparing energy stores for a final reproductive migration. When mature, Pacific salmon migrate from the ocean into rivers to reproduce in their natal spawning

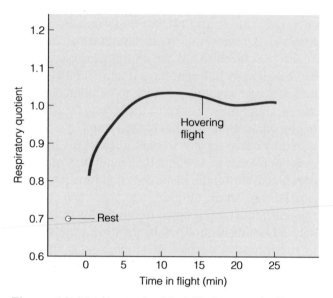

Figure 12.13 Hummingbird flight metabolism
Hummingbird respiration was monitored by oxygen and CO_2 sensors incorporated into a feeder. The metabolic fuel being oxidized is reflected in the respiratory quotient (RQ), which is the ratio of CO_2 produced to O_2 consumed. At rest, birds oxidize lipid fuels, as indicated by the RQ of 0.7. Once flight begins and feeding commences, the RQ rapidly rises to a value near 1, an indication of carbohydrate oxidation. (Source: Adapted from Suarez et al., 1990)

beds. Some populations, such as the Fraser River sockeye salmon, may travel more than 1000 kilometers through stretches of swift currents (Figure 12.14). During their migration salmon don't eat, and must rely on internal energy stores. Large fat stores fuel the earliest stage of migration but become depleted later in the migration. With most of its major fuel stores reduced, the salmon has no choice but to start breaking down endogenous proteins. The fish then starts to break down its muscles and intestinal tract, releasing the chemical energy stored within the tissues. The salmon first breaks down the white muscle that is no longer needed for high-intensity swimming, but it spares the red muscle it uses for slow, steady-state swimming. Some amino acids are oxidized within muscle, but many are converted to glucose in the liver. Late in the migration, glycogen and glucose support the vigorous spawning activity. During this entire trip the salmon coordinates its whole body metabolism to make fuels available to working muscle while sparing the reproductive tissues needed for gamete production. By the time the salmon spawns, it has depleted energy stores and digested its own tissue. Shortly thereafter, the salmon dies.

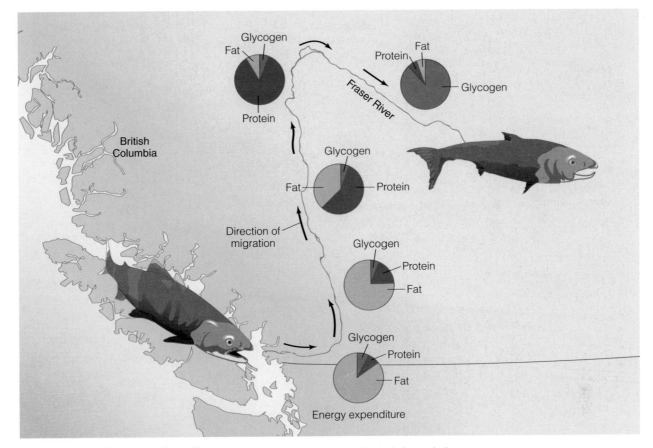

Figure 12.14 Salmon migration   During their time at sea, salmon bolster their energy reserves in preparation for migration. Those that spawn in the upper reaches of the Fraser River in western Canada utilize stored lipids prior to digesting their own tissues for energy. Throughout the migration glycogen is spared, or replenished, being reserved for the high-intensity exercise required for spawning.

Hormones control fuel oxidation in muscle

The salmon life history is an extreme version of metabolic transitions, but all forms of muscle activity require complex control of metabolic fuel utilization. Most animals rely on a combination of carbohydrate and lipid for muscle ATP production. The carbohydrate can be derived from muscle stores, mainly glycogen particles, but circulatory glucose is also an important fuel. Muscle maintains a store of lipid in the form of triglyceride droplets. The blood also provides lipids from breakdown of lipoproteins. As we discussed in Chapter 2, most of the energy for muscle activity is produced when mitochondria oxidize pyruvate, from carbohydrate, and fatty acids, from lipid breakdown. The pathway for muscle metabolic fuel consumption is summarized in Figure 12.15.

Many of the steps in fuel breakdown change in response to activity level. These metabolic transitions are orchestrated by hormones that act on skeletal muscle and fuel storage tissues. During steady-state activity, muscles are promiscuous in their fuel preferences, utilizing whichever fuels are abundant. The levels of metabolic fuels in the blood are determined by the balance of actions of many different hormones, such as insulin, glucagon, catecholamines, and glucocorticoids. Each hormone has effects on storage tissues that influence production or release of fuels. These hormones can also alter the ability of locomotor muscles to use the fuels by altering the levels of transporters and the activities of metabolic enzymes. During low to moderate activity, glucose remains an important metabolic fuel. Glycogen breakdown is stimulated in muscle and liver. Insulin and cortisol act together to promote liver glycogen breakdown and glucose export into the blood. Insulin enhances glucose uptake by the muscle by stimulating the movement of glucose transporters from intracellular vesicles to the sarcolemma. Whereas glycogen stores in the liver and muscle can be mobilized quickly, lipid fuels

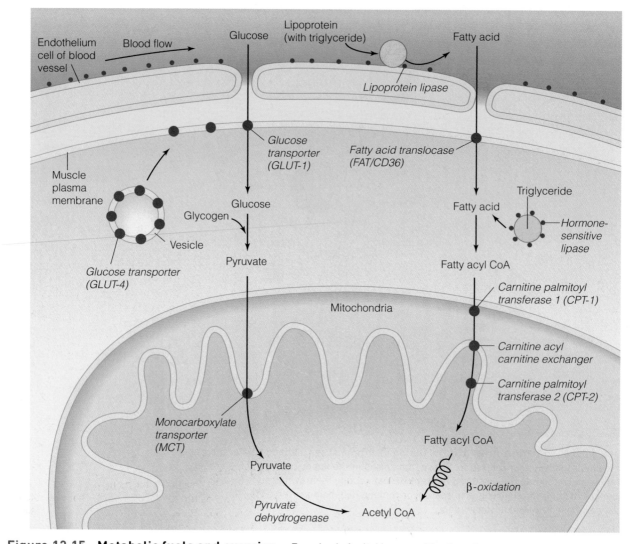

Figure 12.15 Metabolic fuels and exercise. Exercise is fueled by a combination of carbohydrate and lipid. Some carbohydrate is stored within the muscle in the form of glycogen. Muscles also use glucose from the blood, arising from digestion or glycogen stores in the liver. The main source of lipid for exercise is fatty acids arising from triglyceride breakdown. The muscle has substantial triglyceride stores in the form of lipid droplets. Triglyceride is also delivered to the muscle by the blood in the form of lipoprotein complexes.

are mobilized more slowly. If activity levels are sustained, lipid fuels become increasingly important. Triglyceride mobilization in skeletal muscle and adipose tissue is governed by lipases. In adipose tissue, hormone-sensitive lipase is controlled by corticotropin, epinephrine, norepinephrine, and glucagon. These hormones act through cAMP signaling cascades to activate protein kinase A, which phosphorylates the lipase. Muscle also possesses hormone-sensitive lipase activity to trigger release of fatty acids directly within muscle. By using hormones that act at the locomotor muscle and at extramuscular storage sites, animals are able to control the flow of metabolic substrates to the ATP-producing machinery.

Perfusion and Oxygen Delivery to Muscle

During locomotion, muscles must be supplied with fuels and cleansed of end products. Muscle metabolic rates of some animals are so low that simple diffusion is sufficient to ensure adequate movement of metabolites and gases in and out of the muscle. For example, nematodes and flatworms are able to obtain adequate gas diffusion across the integument and no specialized circulatory systems are necessary. Active animals use cardiovascular systems to service muscles, removing metabolic end products and CO_2, and delivering O_2, fuels, and hormones to the muscle. Active insects, such as bees and locusts, use tracheae to deliver oxygen directly to flight muscle, and a muscular heart to push hemolymph through circulatory systems to provide metabolic fuels. Vertebrate skeletal muscle is perfused by much more complicated circulatory networks. In these muscles, arteries branch into successively smaller arterioles, which divide into thin-walled capillaries. Perfusion of muscle depends upon the structure of the capillary networks and the amount of blood that reaches the capillaries.

Capillary networks bring oxygen to the vertebrate muscle fibers

Oxygen delivery to muscle is controlled by structural features, such as capillary density, and functional parameters, such as vascular tone and the oxygen affinity of hemoglobin. Acting in combination with the respiratory and cardiovascular systems, the muscle controls how much blood reaches the intramuscular capillary beds and how much oxygen is extracted from the blood.

Once blood enters the vascular beds of muscle, oxygen may be released from hemoglobin. Recall from Chapter 9: Respiratory Systems that oxygen is released when the P_{O_2} is low or when physiochemical conditions alter the oxygen affinity of hemoglobin. Muscle activity influences oxygen extraction from the blood in several ways. Aerobic metabolism consumes O_2, reducing P_{O_2}. Changes in erythrocyte pH or the levels of regulatory metabolites, such as diphosphoglycerate (DPG) and nucleotides, can cause hemoglobin to release oxygen to the muscle. Once O_2 is released from hemoglobin, the rate of diffusion from the erythrocyte to the muscle mitochondria depends on the steepness of the gradient and the diffusion distance. The diffusion distance is determined largely by capillary geometry. Diffusion distances are short in aerobic muscles, which have small diameters and abundant capillaries. Diffusion distances are greater in glycolytic muscles, which are larger and possess fewer capillaries.

August Krogh first modeled capillary geometry as a cylinder within a cylinder (Figure 12.16a). The inner cylinder represents the capillary that services a volume of muscle, represented by the outer cylinder. His model assumed that (1) each capillary is the only oxygen supply for a surrounding cylinder of tissue, (2) the P_{O_2} at the vessel wall is equal to that of the blood, (3) there is no decline

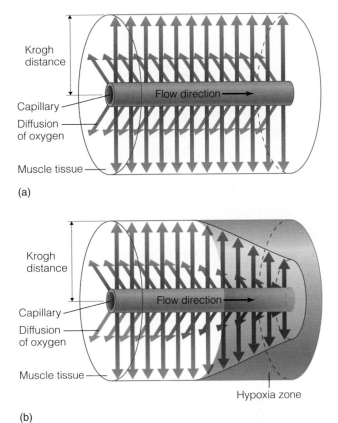

(a)

(b)

Figure 12.16 Krogh model of perfusion
(a) The Krogh model of diffusion suggests that each capillary is able to provide adequate oxygen to a volume of surrounding tissue, represented by a cylinder. **(b)** In many situations, the oxygen levels decline along the length of a capillary. This reduces the volume of muscle that can obtain adequate oxygen. Some areas become hypoxic unless other capillaries are close enough to provide oxygen.

of P_{O_2} along a capillary, (4) oxygen diffuses radially from the capillary, and (5) consumption is uniform in the tissue. The Krogh distance reflects the distance that oxygen can diffuse into surrounding tissue. Using this model, he calculated the dynamics of oxygen diffusion under various myofiber geometries and physiological conditions.

The Krogh model and the more recent variations allow researchers to predict oxygen delivery in relation to fiber geometry and metabolic rate. Many of the constants in the original Krogh model are treated as variables in more recent models. We now know that oxygen levels can decline along the length of the capillary, causing some regions of the muscle to become hypoxic (Figure 12.16b). Capillaries weave back and forth across the muscle; the degree of weaving, or *tortuosity*, increases the transit time for blood cells, allowing longer periods for oxygen to unload from the cell (Figure 12.17). Furthermore, a specific region of muscle may be

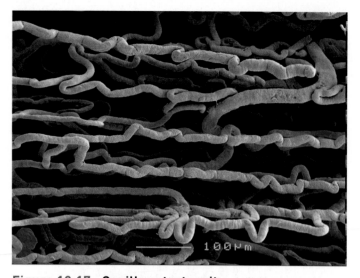

Figure 12.17 Capillary tortuosity Individual muscle fibers are surrounded by capillary networks that weave back and forth across the surface of the myofiber. The image shown here is the capillary structure from a frog muscle. The capillaries are preserved while the muscle is corroded away from the preparation. (Photo courtesy of Nicholas Hudson and Craig Franklin, University of Queensland)

served by more than one capillary or even minor arterioles. In a living animal, the organization of the capillaries and the regulation of blood flow are necessary to match oxygen delivery to muscle activity.

Vasoactive agents regulate blood vessel diameter

While capillaries deliver oxygen to the muscle cell, the arterioles control which capillaries receive blood. When an animal is at rest, not all muscle capillaries are perfused. The arterioles that feed the capillaries undergo **vasomotion**, regularly cycling between constriction and dilation. When conditions demand an increase in blood flow to the muscle, arterioles remain open for longer periods and total blood flow to the capillaries within that muscle increases.

As we discussed in Chapter 8: Circulatory Systems, vasoactive agents alter the contractility of the smooth muscle that lines the arterioles, and thereby determine perfusion through capillaries. Some vasoactive agents, such as insulin, are endocrine hormones produced at distant sites and released into the circulation. Other neurohormonal factors are produced locally, arising from nerves, vascular smooth muscle, endothelium, or the muscle itself. The end products of muscle metabolism, such as pH, oxygen, and CO_2, also exert effects on

capillary beds. The arteriole diameter is determined by balance between vasoconstricting and vasodilating agents. Although we focus on the changes in perfusion that arise in response to activity, muscle blood flow is also altered in response to nutrient status. Muscle is a major sink for glucose after a high-glucose meal. Insulin causes increases in muscle blood flow as part of a mechanism to enhance muscle glucose uptake and glycogen storage.

Muscles also alter their perfusion by inducing the pathways of angiogenesis, which lead to synthesis of additional blood vessels. Angiogenesis is triggered in response to the persistent regional hypoxia that arises when oxygen demands exceed oxygen delivery. When endothelial cells experience hypoxia, levels of the protein hypoxia-inducible factor (HIF) increase. This transcription factor triggers the release of the hormone vascular endothelial growth factor (VEGF) into the wall of the blood vessels. When receptors on the vascular smooth muscle cells bind VEGF, the cells proliferate and penetrate the surrounding tissue. The growth of blood vessels into hypoxic regions increases the perfusion of active muscle. The same pathway is used to increase perfusion of muscle regions with damaged or blocked blood vessels.

Myoglobin aids in oxygen delivery and utilization

As we first discussed in Chapter 8, myoglobin is an oxygen-binding heme protein found in aerobic muscles. The myoglobin gene arose early in metazoan evolution and occurs in most metazoan taxa. Myoglobin has two major roles within muscle cells: intracellular oxygen storage and oxygen transport.

Many muscles use myoglobin as an oxygen store. When tissue oxygen levels decrease, myoglobin releases its oxygen for use by muscle mitochondria. Myoglobin concentrations are high in muscles of animals that regularly experience hypoxic conditions. For instance, diving mammals, such as whales and seals, prepare for a dive by saturating myoglobin stores with oxygen. During the dive, oxygen is released from the myoglobin to support muscle activity.

Once oxygen crosses the muscle cell membrane, it is rapidly bound by myoglobin. By reducing the concentration of free oxygen within cells, myoglobin helps maintain the oxygen gradient necessary for oxygen diffusion. Oxygen bound to one myoglobin molecule can be transferred from myoglobin to another molecule to facilitate oxygen diffusion to the mitochondria. Because of this role in facilitating oxygen delivery, muscle myoglobin concentration often parallels muscle mitochondrial content.

The importance of myoglobin in facilitating oxygen delivery remains controversial. Researchers using mathematical modeling of molecular movements of myoglobin maintain that oxygen diffusion is improved by only a small percentage. In an effort to explore the importance of myoglobin in muscle, researchers in the 1990s engineered lines of transgenic mice with the myoglobin gene knocked out. Much to their surprise, they found that these myoglobinless mice were able to exercise as well as wild-type mice. At first, these results seemed to argue against a role for myoglobin in support of muscle activity. Subsequent studies revealed that myoglobinless mice had extensive changes in the vasculature of muscles. Without myoglobin to facilitate oxygen delivery, the muscles adapted by increasing muscle capillarity. From an evolutionary perspective, animals produce myoglobin to reduce the costs of building and maintaining vasculature. This trade-off is best illustrated by the Antarctic fish that lack myoglobin altogether. The absence of myoglobin is tolerated in these species because of their low metabolic rate and the high oxygen content of cold polar waters. This trait has arisen several times in distantly related taxa and by different mechanisms. In some species the myoglobin gene is not transcribed, whereas other species express the gene but do not translate the mRNA into protein.

CONCEPT CHECK

1. What are muscle fiber types, and how do animals use them to support diverse activity levels?
2. What are the costs and benefits to using oxidative phosphorylation versus glycolysis to support muscle activity?
3. How does myoglobin aid oxygen utilization in muscle?

Skeletal Systems

Muscle contraction may provide the force for locomotion, but locomotion requires some form of skeleton to move various forms of appendages.

Imagine an isolated muscle contracting on a bench top. It is free to contract and relax, but without connections to some form of skeleton the muscle contraction is reduced to a shape change. Earlier in this chapter we discussed how invertebrates such as worms are able to crawl using muscles that act on fluid-filled chambers that constitute a hydrostatic skeleton. Their muscles contract to cause a change in the distribution of fluids to move the body. Hydrostatic skeletons are also important in other animals. A few species of spiders use a hydrostatic skeleton as a substitute for an antagonistic muscle group. They extend their legs with an infusion of hydrostatic fluid, functionally replacing an extensor muscle group.

A solid skeleton is important in the locomotion of all chordates and many invertebrates, such as echinoderms, arthropods, and molluscs. The invertebrate external skeletons, or **exoskeletons**, can cover the animal completely, as in insects, or only partially, as in molluscs. Internal skeletons, or **endoskeletons**, are most common among vertebrates. The endoskeletons found among some groups of invertebrates, such as sponges and echinoderms, are used for protection and support, not locomotion. The endoskeletons of vertebrates are made of cartilage or bone produced by specialized cells. Hard skeletons are central to locomotor strategies of animals, acting as structural support for appendages, elastic storage devices, or biomechanical levers.

Hard skeletons are made from cellular secretions

Most cells secrete suites of macromolecules that make up the extracellular matrix. In soft tissues, the extracellular matrix is the glue that holds cells together. Skeletons are derived from a specialized extracellular matrix produced by secretory cells. They can be made of diverse materials that vary in biophysical properties such as rigidity, flexibility, durability, and inertness.

Most invertebrates possess an external surface layer that helps protect the animal from the environment. The exoskeleton of insects, known as the cuticle, is composed of the carbohydrate chitin, proteins such as sclerotin, water molecules, and phenolic compounds. The cuticle is produced from secretions of a layer of cells that lie beneath the cuticle mounted on the basement membrane. These hypodermal cells secrete long strands of chitin that become embedded in a complex protein matrix.

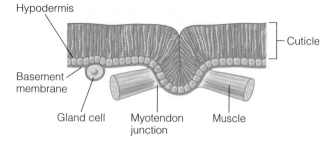

Figure 12.18 Insect cuticle The insect exoskeleton is a modified extracellular matrix of underlying hypodermal cells. Muscles are attached to the exoskeleton at myotendon junctions.

After the chitin and protein secretions are assembled, the cuticle incorporates oxidized phenolic compounds. This final step of exoskeleton assembly, called *sclerotization*, makes the cuticle more rigid. Insect muscles are connected to the exoskeleton via myotendon junctions (Figure 12.18). Muscle cells in the flight musculature come into contact with epithelial cells that produce the cuticle. The two cell types link together via cell membrane receptors such as integrins. As the cellular connections mature, the region forms the myotendon junction.

Locomotor strategies are quite complex in insects, and the exoskeleton is used in many different ways to allow these animals to jump, walk, and fly. The insect wing is composed of cuticle, although it has a different composition than the cuticle of the exoskeleton. The insect leg is a series of hollow tubes of exoskeleton. Internal muscles act across joints to cause the leg to bend. Insect flight is controlled by a series of thoracic flight muscles, either direct, indirect, or both. Direct flight muscles attach via ligaments to the base of the wing. Activation of one muscle group (elevators) moves the wing up, whereas activation of the antagonistic muscle group (depressors) moves the wing down (Figure 12.19). As we discussed in Chapter 5, these muscles are also called synchronous muscles because muscle activation arises from a neuronal stimulus. Contraction occurs when Ca^{2+} levels rise, and relaxation follows when Ca^{2+} is resequestered in the sarcoplasmic reticulum. Direct muscles are found in primitive insects, such as orthopterans (locusts), coleopterans (beetles), and odonatans (dragonflies). Some insects have another arrangement of flight muscles. They are called indirect muscles because of the way contraction is coupled to wing movement, or asynchronous wing muscles because of the mode of excitation-contraction coupling. Indirect muscle does not attach directly to the wing, but rather changes the position of the wing by altering the shape of the thoracic exoskeleton. Both the wings and the flight muscles attach to the upper region of the thorax known as the *tergum*. When the elevator muscles contract, the tergum is pulled down, a distortion that pulls the wings up and also stretches the antagonistic depressor muscles. After the wing is elevated, the elevator muscles relax, the depressor muscles are activated, and the tergum pops up, pulling the wings down. More derived insects, such as dipterans (flies) and hymenopterans (bees), use indirect flight muscles to power flight, although they use direct muscles to control the fine movements of the wing that allow maneuverability.

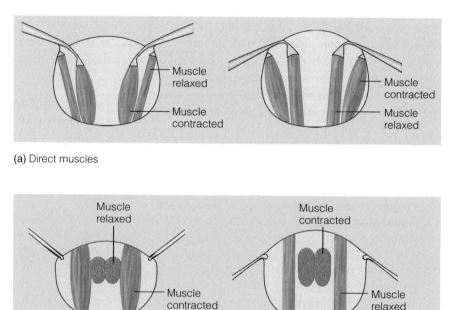

(a) Direct muscles

(b) Indirect muscles

Figure 12.19 Direct and indirect flight muscles Direct muscles attach to the base of the wing, whereas indirect muscles attach to the thorax. Primitive insects, such as the locust, have only direct muscles. More advanced insects, such as the blowfly, use indirect muscles to power flight, although direct muscles may be used for the fine-scale wing movements needed for maneuverability.

Vertebrate skeletons are composed of mineralized calcium

Most vertebrates possess endoskeletons composed of combinations of bone and cartilage. Cartilaginous skeletons are found in the ancient fish, including agnathans (lamprey, hagfish) and chondrichthians (sharks, rays). More recent fish and all tetrapods possess skeletons of bone and cartilage.

Skeletal changes were essential when early vertebrates began the transition to land. Without the support of water, early land animals needed more robust skeletons and specialized musculature to support them against the force of gravity. Birds and bats have secondarily reduced their skeletons to facilitate flight. The properties of the endoskeleton are controlled by **chondrocytes**, the cells that produce cartilage, and **osteoblasts**, the cells that produce bone.

Chondrocytes begin the process of cartilage synthesis early in embryological development. They secrete proteins and proteoglycans, such as chondroitin sulfate, into the extracellular space. These macromolecules make up the extracellular matrix of the chondrocyte. Many different chondrocytes combine their extracellular matrices to produce cartilage. Most vertebrate bones begin as cartilage. As the animal grows and matures, cartilage is broken down and replaced with bone. Mature animals retain cartilage in a few locations within the skeleton, mostly near the ends of long bones where the soft cartilage helps improve the performance of joints.

Mature bone is a living tissue, constantly undergoing remodeling. Bone itself is a collection of multiple cells, cellular secretions, and mineral salts, all enveloped in a fibrous sheath called the periosteum. *Osteoclasts* secrete hydrolytic enzymes to create tunnels into the bone or cartilage (Figure 12.20). These tunnels allow blood vessels to penetrate the extracellular matrix. When osteoblasts invade the tunnel, they secrete collagen fibers that will eventually serve as framework for bone. These collagen fibers help organize the deposition of minerals, mainly calcium phosphate apatite, beginning the process of ossification. Once an osteoblast is surrounded by an ossified extracellular matrix, it can no longer divide and is called an *osteocyte*. Throughout the lifetime of the animal, osteoclasts continue to digest away regions of bone, creating tunnels that are repaired by osteoblasts. This ongoing capacity for remodel-

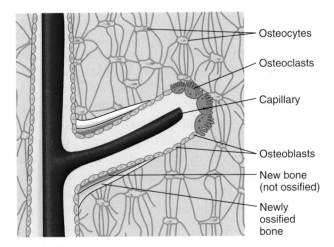

Figure 12.20 Cellular basis of bone Bone is composed of an ossified extracellular matrix of bone-producing cells: osteoblasts and osteocytes. Osteoclasts secrete digestive enzymes that break down bone. This allows vasculature and osteoblasts to invade the bone. The osteoblasts secrete their extracellular matrix. Gradually, the new bone becomes ossified.

ing is part of the process of bone growth and repair. By regulating bone growth and ossification, animals can control the physical properties of bone, such as dimensions and density. For example, these cells build the lighter bones required by flying animals, as well as the heavier bones of large herbivorous land mammals.

Ligaments and tendons hold the musculoskeletal system together. **Ligaments** hold one bone to another. They are a type of connective tissue produced when the fibroblasts near the ends of bones secrete long, parallel fibers of collagen linked to proteoglycan. **Tendons** attach muscles to the skeleton. They are composed of connective tissue similar in structure to ligaments. At one end of the tendon, the connective tissue binds to the bone. At the other end, the tendon binds at various points along the belly of the muscle. Tendons connect both ends of the muscle to bones.

Ligaments and tendons have important roles in locomotion. They interconnect the different elements of the musculoskeletal system. They ensure that muscles are correctly positioned and stretched to the appropriate sarcomere length. They also help transmit forces between musculoskeletal elements.

Skeletal components act as mechanical levers

The individual long bones in locomotor appendages of vertebrates meet and articulate at joints. The joints are often bathed in *synovial fluid* to reduce the degree of friction between two opposing bones. Joints differ in structure in ways that determine the range of movement. For example, the hip and shoulder are ball-and-socket joints: the end of one bone is rounded or convex, and the opposing bone is concave. These joints provide the greatest range of motion. The knee and elbow are hinge joints and allow movement in only one plane.

Muscles work in combination with bones to create levers. When an untethered muscle contracts, it pulls its end toward its middle to generate a linearly compressed movement. When the muscle is attached to a bone, the geometry of the bone and the position of the joint constrain the muscle from its natural range of movement. The contracting muscle pulls the bone, causing it to rotate through an arc. This movement allows the bone to be used as a lever. All mechanical and biological levers have three elements: a fulcrum, a weight, and a force. The fulcrum is the point of rotation, which in the context of locomotion is the joint. The weight is the force exerted by the object to be moved. The force is generated by muscle contraction. If you are bending your arm to pick up a rock, the fulcrum is the elbow, the rock is the weight, and the biceps muscle generates the force.

The mechanics of lever action depend on the relative position of the three elements, as well as the distances between the elements. The distance between the force and the fulcrum is the force arm. In the example of your arm, the force arm is the distance between the elbow and point of insertion of the biceps muscle. The region between the fulcrum and the weight is the weight arm. The mechanical advantage (MA) of a lever is expressed as the ratio of the length of the force arm (L_{FA}) to the length of the weight arm (L_{WA}).

Levers are distinguished based on the relative position of the three elements (Figure 12.21). A crowbar is an example of a class I lever. The lever is long (large L_{FA}) and the fulcrum is close to the weight (small L_{WA}). Thus, this type of lever has a large mechanical advantage; a minimal amount of force can be used to lift a large weight. A wheelbarrow is an example of a class II lever. The weight is between the fulcrum and the force. We can lift quite a bit of weight using a wheelbarrow, but a class II lever does not have as great a mechanical advantage as a class I lever. Most levers in animal locomotion are class III levers. In our arm example, the biceps (force) inserts between the elbow (fulcrum) and hand (weight). Unlike the other types of lever, a class III

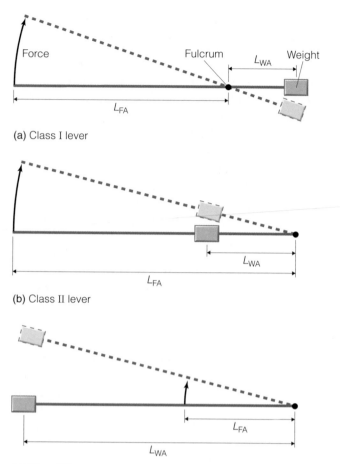

(a) Class I lever

(b) Class II lever

(c) Class III lever

Figure 12.21 Levers Three classes of levers are distinguished by the relative positions of three elements: the fulcrum, the weight to be lifted, and the point at which force is applied. The part of the lever between the fulcrum and the weight is the weight arm. The force arm is the part of the lever between the fulcrum and the point at which force is applied. The ratio of the length of the force arm (L_{FA}) to that of the weight arm (L_{WA}) is the mechanical advantage.

lever has no mechanical advantage because L_{FA} is always less than L_{WA}. Consequently, class III levers are the least effective in translating muscle force into leverage. These levers are valuable not because of a mechanical advantage but because they can increase the range and velocity of movement. A modest amount of force exerted at a short distance from near the fulcrum causes the mobile end of the lever to move quickly through a much greater distance. In terms of your arm, when the bicep shortens only about 2 cm, it causes the hand to rotate through a 50-cm arc.

The position of the muscle insertion on the bone, relative to the joint, has important biomechanical ramifications. The relative lengths of the force arm and weight arm determine how efficiently muscle force can translate into leverage and movement.

The importance of these relationships is best illustrated by considering the morphometry of the legs of animals specialized for different lifestyles. A cheetah is built for speed, whereas a lion is slower but stronger. When the forelegs are drawn to a similar scale, the differences in leg morphometry are more obvious (Figure 12.23). The teres major is the muscle that pulls the foreleg backward, a movement that is used in running and also in prey capture. The teres major attaches much closer to the shoulder joint in a cheetah than in the lion. As a result of these differences in mechanical advantage, the cheetah can move its foreleg faster whereas the lion moves its foreleg with more force.

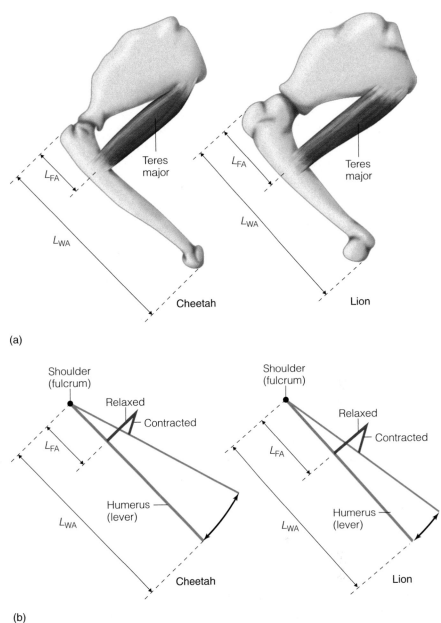

(a)

(b)

Figure 12.22 Muscle position on the bone The geometry of muscles and bones determines the relationship between the force generated by the muscle and the type of movement that results. **(a)** In the cheetah foreleg, the muscle is inserted closer to the joint than in the lion foreleg, when drawn to the same scale. **(b)** When modeled as a lever, the cheetah foreleg can move faster, through a longer arc, but the lion foreleg can generate more forceful movements.

Skeletons can store elastic energy

Another way that animals use the skeleton in movement is through **elastic storage energy**. When the muscle shortens, some of the force that is used to stretch the connective tissue and bend the bones is stored as elastic storage energy. When the muscle relaxes, the elastic storage energy can help the muscle stretch. Vertebrates benefit from elastic storage energy in locomotion, although some animals benefit more than others. The kangaroo, first discussed in Chapter 2, uses elastic storage energy to improve the efficiency of locomotion. When the kangaroo first begins to hop, extensor muscles are used to lift the animal off the ground. When it lands, some of the force is used to distend the muscle and connective tissue, creating a short-term store of elastic energy. Upon recoil, these elastic elements generate force that can be applied toward the next jump. Later in this chapter we discuss how animals use the muscles of the back as elastic energy stores during running.

The importance of elastic storage energy is much more obvious in arthropod locomotion, particularly jumping and flight. In some cases, elastic structures functionally replace antagonistic muscles altogether. For example, many spiders have leg joints that lack extensor muscles. Portions of the exoskeletal plates called *sclerites* span the joint (Figure 12.23). Muscle contraction deforms the sclerites during flexion. When the muscles relax, the sclerites recoil to extend the leg.

Such elastic structures were an important evolutionary innovation in arthropods. Most invertebrates move with the help of hydrostatic

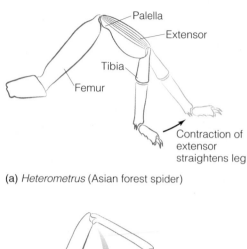

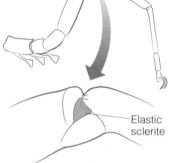

(a) *Heterometrus* (Asian forest spider)

(b) *Eremopus* (sun spider)

Figure 12.23 Spider legs Leg extension occurs in spiders by several mechanisms. Some use movement of hydrostatic fluids to extend the leg (not shown). More commonly, spiders use either **(a)** extensor muscles or **(b)** elastic storage structures that span the leg joint. When leg joints are flexed, the elastic sclerite is deformed. When the muscle relaxes, the sclerite recoils to extend the leg forward. (Source: Sensenig and Shultz, 2003)

skeletons. As mentioned earlier in this chapter, many extant spiders use fluid movements and hydrostatic pressure to aid in leg extension, augmenting elastic storage energy. However, a strategy that relies on hydrostatic pressure for locomotion has several limitations in rapidly moving animals. The mass of elastic structures is small relative to the mass of the hydrostatic fluid that would be needed to extend a leg. Thus, if legs rely on elastic structures they can be smaller, which saves energy for a moving animal. It also costs the animal metabolic energy to push hydrostatic fluids into the leg. Using an extracellular fluid to generate hydrostatic pressure may also compromise the diffusion of metabolites and respiratory gases in the circulatory system. Thus, structures that rely on elastic storage energy allow arthropods to escape some of the constraints imposed by hydrostatic skeletons.

In this section, we have discussed how locomotor systems are the products of the combination of muscle, metabolism, perfusion, and the skeleton. The interdependence of the systems is perhaps most clearly demonstrated by those species that move in a way that is markedly different from that of their close relatives. Researchers are drawn to animals that run, fly, or swim faster than all others. In the accompanying feature (Box 12.1, Methods and Model Systems: Animal Athletes), we explore the physiology of these animal athletes to understand the nature of the complex anatomical specializations that allow these animals to move faster than their relatives.

Translating Contraction into Movement

Muscle function in animal locomotion varies hugely, and all levels of biological organization contribute to this diversity. Recall from Chapter 6: Sensory Systems the many ways that variations in the molecular and cellular properties of the striated muscles contribute to muscle diversity: motor neuron properties, contraction frequency, motor end plate organization, T-tubules and action potential propagation, SR structure, Ca^{2+} buffering, thin and thick filament protein isoforms, the arrangement of myofibers into a muscle, and the material properties of the connective tissue and skeleton. Each of these factors affects the muscle dynamics, but we must also consider how the muscles are integrated into the anatomy of the whole animal, and how this affects the conversion of muscle contraction to movement.

Muscles are specialized for force generation or shortening velocity

Within a complex animal, locomotor muscles serve many purposes. Let's consider two alternate muscles of equal mass that differ in shape. One muscle is short and thick, and the other is long and thin. The short, thick muscle has its sarcomeres arranged in parallel, whereas the other has its sarcomeres arranged in series. Because of its larger cross-sectional area, the short, thick muscle can generate a great deal of force with relatively little shortening. In contrast, the long, thin muscle will shorten faster (and further) but generate less

BOX 12.1 **METHODS AND MODEL SYSTEMS**
Animal Athletes

One of the basic tenets of comparative animal physiology is the August Krogh principle: *For every biological problem there is an organism on which it can be most conveniently studied.* In this chapter, we have discussed several models with properties that are conducive to experimentation, such as fish muscle fibers with their homogeneous fiber types and the frog jumping muscle. When studying integrative locomotor physiology, animal athletes are useful models for studying the limits to aerobic activity. Each animal taxon has a few species that epitomize some aspect of aerobic locomotor performance. Bumblebees and hummingbirds flap wings at exceptionally high frequencies, allowing these animals to hover. Thoroughbred horses and pronghorn antelope run faster than most land animals. Tunas and lamnid sharks swim faster than other fish. To explore the physiological features that accompany exceptional aerobic locomotor performance, we will consider the physiology of tuna.

Body shape is very important for aquatic animals. The streamlined, fusiform body of tuna is built to move efficiently through water. Many other fish have a general streamlined shape, but the tuna is as close to a perfect teardrop as any fish. The shape of the caudal fin is also unusual, with its narrow caudal peduncle and thin, crescent-shaped (lunate) caudal fin. Tuna also show an unusual swimming style. Even at high speed there is little movement in the trunk. The caudal fin merely bends back and forth at the caudal peduncle. Although the trunk doesn't move during swimming, the muscles within the entire trunk contribute to force generation for the caudal fin. Force is transferred from the muscle to the caudal fin through the myosepta, skin, and tendons.

The red muscle of tuna is unusual in two important respects that influence force generation. Most species of fish have a thin wedge of red muscle that runs the length of the fish, just beneath the lateral line. Tuna red muscle, however, is not homogenously distributed but rather is concentrated midway along the length of the fish. It is also located deep within the body, close to the spine. In this chapter we discussed how the superficial location of red muscle of fish is ideal for force generation. In this position, it has leverage to bend the body wall. How then does the deep red muscle of tuna power its high-velocity swimming? Recent studies have shown that the core of deep red muscle is able to shorten more than the surrounding white muscle. The red muscle tendons connect directly to the caudal fin, allowing more effective force transmission. Because of these efficient tendons, the anterior red muscle can make important contributions to the power of tail movements, even though the anterior part of the fish does not bend during swimming.

The tuna meets the energetic demands of rapid swimming with specializations of the respiratory, cardiovascular, and locomotor systems. The gills have a large surface area and thin gill epithelia, structural features that enhance the rate of oxygen exchange. The blood has a high oxygen-carrying capacity, due to higher hematocrit and higher hemoglobin concentrations. Tuna have an extensive coronary circulation that delivers oxygenated arterial blood to the heart, unlike most fish hearts, which must extract oxygen from the venous blood passing through the cardiac chamber. The capillarity density of the red muscle can be as much as four times greater than in other fish. Muscle myoglobin content is also high in tuna. Each of these features would seem to enhance the capacity of the fish to extract oxygen from water and deliver it to the mitochondria of the working muscle. The muscle itself has a high aerobic capacity. Although the mitochondrial content is similar to that of other species (about 35% of cell volume), the mitochondrial cristae are two to three times more densely packed than other species. Thus, mitochondrial enzymes required for energy production occur at high activities. This high mitochondrial capacity is characteristic of all animal athletes, which may possess exceptional mitochondrial volume density and cristae packing density. Tuna are also unlike most fish in their thermal physiology. These fish are regional endotherms; they are able to retain muscle heat within the body core. These elevated temperatures increase the kinetics of both contraction and energy metabolism. Tuna thermal physiology is discussed in more detail in Chapter 13: Thermal Physiology.

Tuna are remarkable animals that have provided important information about the structural and physiological constraints on locomotor activity. The evolutionary and developmental origins of these specializations remain to be explored. Interestingly, the lamnid sharks, which include the mako and great white sharks, demonstrate many similarities with tunas. Since these species are only distantly related, the striking similarities between tuna and lamnid sharks are examples of convergent evolution.

References

- Bernal, D., K. A. Dickson, R. E. Shadwick, and J. B. Graham. 2001. Analysis of the evolutionary convergence for high performance swimming in lamnid sharks and tunas. *Comparative Biochemistry and Physiology, Part A: Molecular and Integrative Physiology* 129: 695–726.
- Donley, J. M., C. A. Sepulveda, P. Konstantinidis, S. Gemballa, and R. E. Shadwick. 2004. Convergent evolution in mechanical design of lamnid sharks and tunas. *Nature* 429: 61–65.

force. Mechanical **power** (*P*) is the mathematical product of force (*F*) and shortening velocity (*V*):

$$P = F \times V$$

Though the thick muscle generates lots of force but at low shortening velocities and the thin muscle generates rapid shortening but little force, the two muscles have a similar power output.

An animal can alter the power output of a muscle by changing either the force generation or the velocity of shortening. Consider the following example to understand the trade-offs between these parameters and how they affect power. You can throw a ball by contracting your triceps muscle in the back of your arm. You can maximize force by trying to throw a very heavy ball. Your arm might generate a great deal of force, but the ball is so heavy that you can barely move your arm (your triceps has high force but little shortening). Alternatively, you can maximize shortening velocity. The way to move your arm the fastest is by choosing a very light ball; very little force is generated but you can move your arm very quickly. Neither of these situations generates significant power because in each situation one of the parameters—force or shortening velocity—approaches zero. Power is greatest at an intermediate velocity (Figure 12.24). Most muscles generate maximal power when contraction velocity is 30–40% of the maximal shortening velocity.

Mechanical power is the most important parameter in most forms of locomotion. Animals apply the power generated by a muscle to the environment to generate movement. The arm throws a ball by transferring the power generated in contraction to forward movement of the ball. The power you generate with a leg contraction allows you to jump. The contraction must be forceful and rapid, or you will not leave the ground. Many muscles are built, arranged, and used in ways that maximize power output.

Work loops show the balance between positive and negative work

Power can also be expressed in relation to work (*W*). Recall that $P = F \times V$. Since *V* equals distance (*D*) per unit time (*t*), then

$$P = F \times D/t$$

Work (*W*) is the product of force (*F*) and distance (*D*). Expressing force in terms of work, we get $F = W/D$. Thus, power can also be expressed as work per unit time ($P = W/t$). This rearrangement also emphasizes the important difference between work and power. It requires about the same amount of work to lift an object slowly versus quickly. Thus, the metabolic energy demands are similar. However, much more power is required to lift the same weight in a shorter period of time.

Researchers can calculate the amount of work done by a muscle by measuring force and length during cycles of contraction and relaxation (Figure 12.25). Upon activation, the muscle shortens and force declines until the muscle reaches its shortest length. The area under this descending curve reflects the work done by the muscle during contraction and is called *positive work*. During relaxation, the muscle lengthens. Although the lengthening muscle still exerts some force, it is usually less than the force exerted at the same muscle length during contraction. The area under this ascending curve is the *negative work*. The net work done by the muscle is the difference between the positive work and the negative work. Graphically, this is reflected in the area enclosed by the two curves, known as the **work loop**. In most cases, the descending curve (during contraction) lies above the ascending curve (during relaxation). If you followed the coordinates of force and length around

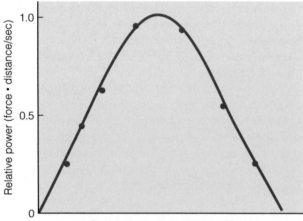

Figure 12.24 Power versus velocity of shortening
The ability of a muscle to generate power depends on the force generated by the muscle and the velocity at which the muscle shortens (here, distance per second). Power approaches zero when the velocity of shortening (*V*) is either minimal or maximal. *V* is very low when the muscle is heavily loaded and cannot shorten. *V* is maximal when there is no load. Without a load, all of the energy released in the cross-bridge cycle is used to shorten the muscle and no force is generated. Power is at its greatest when the muscle is loaded in such a way that it shortens at about one-third its maximal velocity.

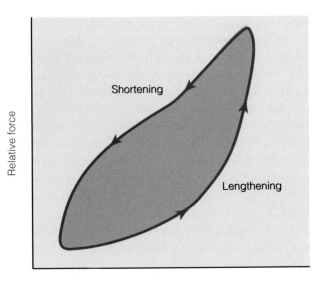

Figure 12.25 Work loops The work loop plot is generated when the force and length of a muscle are monitored during a shortening and lengthening cycle. As the muscle shortens, force decreases to a minimum. The area under the descending curve depicting force versus length reflects the positive work done by the muscle. When the muscle relaxes, it lengthens producing an ascending curve with relatively low force. The area under the ascending curve is negative work. The difference between the positive and negative work, depicted by the area within the loop, is the net work. The direction of the loop reflects whether positive net work was performed (counterclockwise) or negative net work (clockwise).

the graph during such an experiment, you would move around the work loop in a counterclockwise direction. This signifies that a cycle of contraction and relaxation generates net positive work.

Not all myofibers generate positive work. Isometric muscles do not change in length, generate no power, and therefore perform no net work during contraction. It is also possible to have a situation where a muscle does change in length but produces negative net work. More work is performed during extension than shortening, or put another way, the muscle actually absorbs energy during contraction. These muscles are not acting as power generators but rather as shock absorbers, or dampers. Experimentally, a muscle acting as a damper shows a work loop that runs clockwise, signifying negative net work. There is nothing unusual about the cellular properties of the muscle that causes this relationship. Rather, negative net work arises because of the way the muscle is arranged into the musculoskeletal system and the nature of the anatomical relationships with other muscles during movement. Whether a muscle generates positive or negative work depends on when it is activated during limb movement. If, for example, external forces cause a muscle to compress at a time when it is shortening, the muscle generates much less force than is possible. Many of the muscles that control the fine movements of flying insects produce negative net work. They exert their effects by using contraction to change their mechanical properties, altering how other muscles and the exoskeleton affect the flight system.

CONCEPT CHECK

4. What are the components of a vertebrate skeletal system? What is the chemical composition of each element, and how is it made?

5. What are the different types of levers, and how do they work?

6. What is the relationship between force, power, and work?

Moving in the Environment

The musculoskeletal system allows an animal to generate force to move its body and appendages. Whether or not this activity translates into locomotion depends on how the animal interacts with the physical environment. The two dominant environmental factors that influence locomotion are gravity and fluid properties. In the following section we discuss the constraints on animal locomotion in terms of these environmental factors. We use this approach rather than describing locomotion in terms of style of movement, such as swimming, jumping, running, walking, and flying. The physiology of locomotion has much more to do with the physical environment than with the pattern of movement of appendages. For example, fleas, frogs, and kangaroos each jump, but the forces that govern their movement are completely different because of their size and the effects of gravity. Conversely, the biomechanics of swimming and flying are quite similar. Both air and water are fluids and obey the same laws of fluid dynamics. Locomotor systems allow animals to move from place to place, overcoming the physical constraints imposed by the environment.

Gravity and Buoyancy

Gravity is the element of the physical environment that has the greatest consequences for locomotor strategies. No animals can escape gravity, but some are less affected by it than others. Gravity exerts its greatest effects on terrestrial animals, which use muscles to solve the biomechanical problems associated with movement on land under the full weight of gravity. Gravity has much less effect on an animal with a body density that approximates that of the environment. Aquatic animals can reduce the effects of gravity by manipulating their body composition.

Body composition influences buoyant density

An object immersed in water tends to float if it is less dense than the water. The tendency to float is *buoyancy,* an upward force that counteracts the effects of gravity. The body density is determined by the body composition. Each component has a characteristic density, measured as specific gravity (Table 12.1). Water is the most abundant molecule in most animals, and therefore body density usually approximates the density of water. Bones and cartilage have the highest density in animals. Proteins are slightly denser than water, whereas lipids are slightly less dense than water. Gases have the lowest density.

A visit to the local swimming pool reveals how body composition influences the effects of gravity. "Sinkers" complain that swimming is exhausting because they spend considerable energy just to keep afloat. "Floaters" gleefully drift around on the surface, apparently expending no energy at all. It is much easier to float when your lungs are full of air. The subtle differences in body composition that affect our buoyancy also affect the buoyancy of aquatic animals and influence the energetics of movement in aquatic systems. But not all animals need to be buoyant. Benthic animals—those that live on the bottom of aquatic ecosystems—tend to be denser than the surrounding water. This allows them to maintain contact with the bottom without expending energy. However, most aquatic animals possess a body composition that induces either neutral buoyancy or positive buoyancy, allowing them to move through the water column. These animals reduce their overall density by increasing the proportion of less dense constituents, typically lipids or gases.

Lipid stores increase the buoyancy of zooplankton and sharks

Although all animals possess some lipids in membranes and energy stores, some aquatic animals accumulate high concentrations of lipids, which increase buoyancy. Many species of zooplankton possess large droplets of lipid, typically in the form of wax esters. A wax ester, which is a long chain fatty acid esterified to a long chain fatty alcohol, is metabolically active. Zooplankton can alter their buoyant density by synthesizing or degrading the wax esters, allowing these animals to slowly alter their buoyancy to change their position in the water column (Figure 12.26).

Table 12.1	Specific density of biomaterials.
Biomaterial	**Specific gravity (g/ml)**
Bone	3
Cartilage	2
Protein	1.6
Seawater	1.024
Pure water	1.0
Triglyceride	0.90
Squalene	0.86
Oxygen	0.00143

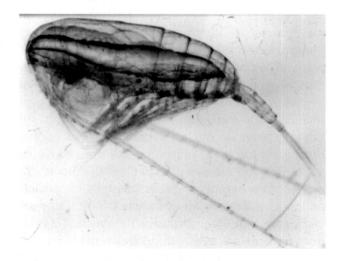

Figure 12.26 Lipid droplets in zooplankton Many species of planktonic animals, such as the calanoid copepod shown here, possess lipid droplets that aid in buoyancy.
(Photo courtesy of Rob Campbell and John Dower, University of Victoria)

Chondrichthians (sharks and rays) also use lipid to increase their buoyancy. They accumulate high levels of the steroid compound squalene in their livers. The amount of lipid is highest in pelagic sharks, which can be neutrally buoyant, and lowest in benthic rays, which are slightly negatively buoyant. Other aquatic animals with high levels of lipid in their tissues benefit from their buoyancy, although they are accumulated for other purposes. The triglyceride accumulations in fish livers are primarily important in energy metabolism. The lipid found in the thick blubber layer of marine mammals serves as insulation. However, these substantial lipid depots also contribute to buoyancy, thereby reducing the amount of energy needed to remain in the water column.

Swim bladders are gas-filled sacs that increase buoyancy

In Chapter 9, you learned about the fish swim bladder. It is not yet certain whether the swim bladder arose first as a primitive lung or as a buoyancy organ. Whereas the fish that use the swim bladder as a lung occur in many fish taxa, the buoyancy function occurs only in actinopterygian fish. This suggests that swim bladders likely arose first as a primitive lung, and only secondarily as a buoyancy organ.

The swim bladder (Figure 12.27) is derived from an outgrowth of the gastrointestinal tract that appears early in fish development. The gas accumulated in these internal balloons is sufficient to compensate for the negative buoyancy of the remainder of the body. The walls of the swim bladder are flexible, allowing the organ to contract and expand. Guanine crystals embedded in the swim bladder reduce the permeability of the swim bladder to gases. For a fish to be able to use a swim bladder as a buoyancy organ, it must be able to control the volume of gas within the organ. *Physostome fish* have a connection between the gastrointestinal tract and the swim bladder. They increase the volume of the swim bladder by gulping atmospheric air and pushing it through the pneumatic duct that connects the gut to the swim bladder. Similarly, they reduce the swim bladder volume by contracting the smooth muscle that surrounds the bladder and opening the pneumatic duct to release air into the gut, where it is burped out of the animal. *Physoclist fish* have lost the direct connection between gut and swim bladder.

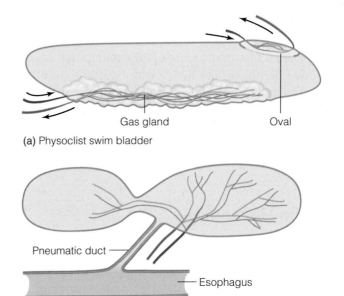

(a) Physoclist swim bladder

(b) Physostome swim bladder

Figure 12.27 Swim bladders **(a)** Physoclist fish inflate the swim bladder by injecting oxygen from the blood into the bladder at a region called the gas gland. Gases can escape the bladder at another vascularized region, called the oval. **(b)** Physostome fish inflate and deflate their swim bladder through a direct connection between the gastrointestinal tract and the swim bladder, called the pneumatic duct.

These fish inflate the swim bladder at a vascularized region of the organ called the *gas gland*. Blood arrives at the gas gland with oxygen loaded onto hemoglobin. When a fish needs to increase the volume of the bladder, the gas gland induces a local acidification, causing hemoglobin to release oxygen. Oxygen unloading is maximized by the countercurrent arrangement of the blood vessels. When a fish needs to reduce the volume of the swim bladder, oxygen is allowed to flow into the blood at a separate vascularized region called the *oval*. Physostome fish may also have a gas gland and an oval, but they are generally reduced in size and less important for gas exchange than the structures in physoclists.

Gas-filled swim bladders reduce the costs of swimming, but they do have functional limitations. The volume of the swim bladder changes in response to hydrostatic pressure. As hydrostatic pressure increases, the volume of a swim bladder shrinks. When the pressure is relieved, the swim bladder expands. Swim bladders are most useful for fish that remain within a narrow range of depths, but they would impair pelagic fish from moving rapidly up and down in the water column.

If a deepwater fish comes to the surface too quickly, the volume of its swim bladder can increase so fast that the fish is incapacitated. Many active fish have lost their swim bladders, and instead expend muscle energy to maintain their position in the water column. Although this is more expensive energetically than a swim bladder, it allows the fish to rapidly change depth without suffering a rapid change in swim bladder volume.

Fluid Mechanics

An object moving through a fluid creates a complex pattern of flow. The rules that describe the movement of a fluid, called fluid mechanics, apply to both air and water. Animals are able to move through fluids by governing the path of the fluids around them. Some animals are most concerned with moving fluids out of the way to allow efficient movement. The fluids impede forward movement. Other animals control fluid movements to aid in locomotion. In this situation, the movement of fluids pushes the animal forward or lifts it upward.

Reynolds numbers determine turbulent or laminar flow

A simple way to begin our discussion of fluid mechanics is to consider the forces that act on an object, such as a canoe paddle, moving through water at different speeds and orientations (Figure 12.28). When you move the paddle through water very slowly, the fluid flows over the surface of the blade in smooth layers, a condition called **laminar flow**. If you were to repeat this movement at increasing velocities, you would reach a point where the pattern of flow would become less ordered, resulting in more **turbulent flow**. The costs of locomotion are greatly increased under turbulent flow conditions. The transition from laminar flow to turbulent flow depends upon properties of the fluid (viscosity, density), the object (size, shape), and the movement (velocity, direction). The relationship between these parameters is described by the Reynolds equation. The **Reynolds number** (Re) is calculated as follows:

$$\text{Re} = VL\rho/\mu$$

where V is velocity of movement, L is a linear dimension of the object, ρ is the density of the fluid, and μ is the viscosity of the fluid. The Reynolds number enables researchers to predict such

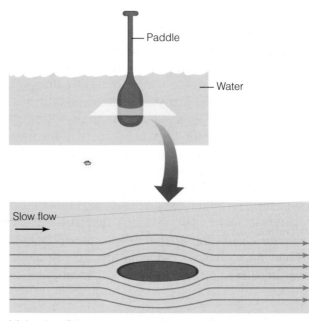

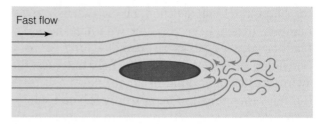

(a) Laminar flow

(b) Turbulent flow

Figure 12.28 Laminar and turbulent flow As an object such as a canoe paddle moves through water, the fluid is forced around the blade. **(a)** The flow remains laminar at low velocities. **(b)** At greater velocities the flow can become chaotic, resulting in turbulence in the wake of the object. This increases the cost of movement.

things as how easily an object can glide through a fluid or when movement through a fluid is likely to be turbulent. Our intuitive appreciation of the biological factors that influence locomotion can be traced back to the parameters of the Reynolds equation.

In our example of moving the canoe paddle through water at different velocities, the mathematical explanation for the increase in turbulence is related to the effect of V on Re. The influence of L on Re can be illustrated by changing the orientation of the paddle. Moving it through the water edge first is easier than when the paddle moves face first. In the face-first orientation, the surface of the paddle that first encounters the fluid is much wider. In the calculation of Re, this difference is reflected in values of L (Figure 12.29). To appreciate the impact of

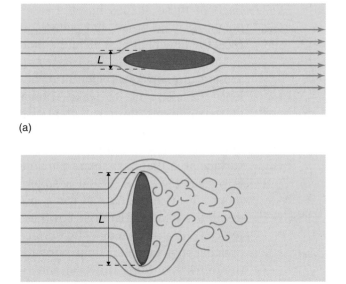

(a)

(b)

Figure 12.29 Orientation and turbulent flow
The orientation of objects can influence the formation of
turbulence. As the linear dimension encountering the fluid (*L*)
increases, the turbulence also increases. **(a)** When the canoe
paddle moves edge first, the lower *L* value results in less
turbulence. **(b)** Moving the paddle blade first increases *L* and
enhances turbulence.

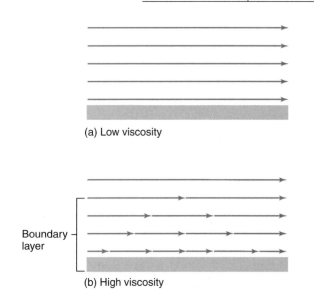

(a) Low viscosity

(b) High viscosity

Figure 12.30 Boundary layers **(a)** When an object
moves through a solution of low viscosity, each layer of
laminar flow moves at the same velocity, as indicated by
vectors of equal length. **(b)** In solutions of higher viscosity,
the layer of laminar flow in contact with the object moves
more slowly because of interactions with the object. The
impact of the object is reduced further from the object. The
boundary layer is the microscopic layer of fluid that is
retarded by the object.

density, compare the effort of paddling through air
versus water. An object moving through air has a
lower Re because air has a lower density (ρ).

The last parameter in the Reynolds equation,
viscosity (μ), requires more explanation. Fluids
differ in their ability to flow around an object.
When a solution moves across a surface, the molecular layer closest to the surface adheres to the
surface and moves with it. The next layer of fluid
interacts with the layer of fluid in contact with the
surface. The further the distance from the surface,
the less the fluid movement is influenced by the
surface. The boundary layer is the molecular layer
of fluid that is influenced by the surface of the object around which it moves (Figure 12.30). Some
fluids are more viscous than others, a physical
property that we recognize as the "thickness" of a
solution. For example, we perceive honey to be
thicker than water. Viscosity influences the movement of an object through a fluid because of the
way the fluid interacts with the object to create a
boundary layer. If you remove your finger from a
bowl of water, it will have a thin coating of water.
If you remove your finger from a bowl of honey, a
much thicker layer sticks to your finger. Any time
you move your finger through a liquid, you carry
that layer of fluid along for the ride. It costs you ex-

tra energy to carry that layer of honey through the
remainder of the honey, known as the bulk phase.
Note that fluid viscosity affects animal locomotion,
but animals do not have to cope with *changes* in
environmental viscosity. The viscosity of water or
air varies little under most conditions. Thus, for
the Re of an animal in its environment, the most
important factors are *V* and *L*.

The relative importance of viscous and inertial effects determine Re

The thickness of the boundary layer is a property
of the fluid, not the moving object. The boundary
layer of water is just as thick on a whale as on a
small aquatic invertebrate, such as a copepod. The
whale expends relatively little energy to carry
around the added water because the water layer is
trivial in comparison to the size of the whale. However, the costs to the copepod are significant.
These fluid layers exert the greatest effects on locomotion of small, slow animals. The magnitude
of these *viscous effects* depends on the viscosity of
the fluid, the velocity of movement, and the properties of the surface of the animal that interacts
with the fluid. These properties include body
shape and surface area, the physical composition

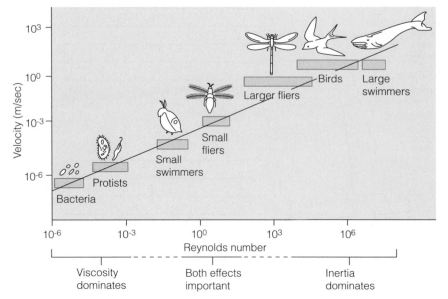

Figure 12.31 Influence of Reynolds number on animal locomotion
The Reynolds number reflects the properties of a fluid and the size and shape of the animal. Larger animals have a higher Reynolds number than smaller animals. Swimmers have a larger Reynolds number than fliers of similar size. The larger the Reynolds number, the more important are inertial effects on locomotion. Viscous effects dominate at low Reynolds numbers.
(Source: Adapted from Nachtigall, 1977)

of the surface, and the nature of appendages. Larger, faster animals are less influenced by viscous effects because of much lower ratios of surface area to mass (Figure 12.31). When a copepod stops swimming, the viscous effects stop forward progression. When a whale stops moving its tail, it has enough momentum to overcome viscous effects. These *inertial effects*, which are dependent on body mass, dominate the movement of larger animals in air and water. The high Re in large animals, however, creates another potential problem: turbulence.

Streamlining reduces drag

For an object to move through a fluid, it must overcome the forces that oppose forward movement. These forces are collectively called **drag**. Two types of drag are encountered by moving objects. *Friction drag* arises from the interaction between the surface and the fluid. It is dependent on the area of the surface that interacts with the fluid, as well as the viscosity of the fluid. *Pressure drag* is the force required to redirect a fluid around a moving object. The more dense the fluid, the greater the pressure drag.

The shape of the object is an important determinant of pressure drag. Consider how three different shapes influence the flow of fluids (Figure 12.32).

Each of these shapes has the same height (*L*). The broad, flat plate redirects the flow of almost all of the fluid it encounters. As the fluid is forced around the object, a region of turbulence develops in its wake. Under these conditions, there is a great deal of pressure drag. However, there is not much friction drag because the surface area that encounters the fluid is reasonably small. When a sphere moves through the fluid, it has a less disruptive effect on laminar flow (less pressure drag), although the surface area in contact with the fluid is greater (more friction drag). However, the total drag is lower for the sphere than for the plate. The streamlined shape of the teardrop has the least effect on laminar flow, causing the lowest amount of pressure drag. Although additional friction drag is associated with the streamlined shape, the total drag is the lowest of all three shapes.

Streamlining reduces the amount of energy animals require to overcome pressure drag. Although each of the objects shown in Figure 12.32 has a similar value of *L*, they have very different masses. That means that the cost of overcoming drag in a large, streamlined animal is similar to the cost for a much smaller, nonstreamlined animal. Most larger swimmers and fliers have streamlined body shapes that reduce drag.

In addition to streamlining, many of the fastest swimmers and fliers also have modified body surfaces that further reduce friction drag. This increased efficiency is critical for an animal like the dolphin, which can swim through the water at 40 km/h. The surface of the dolphin is mounted on a layer of tiny pillars that readily change shape when water moves over the surface. When the skin compresses in response to small, localized turbulence, friction drag is minimized as fluids move smoothly over the surface. Another property of dolphin skin allows this animal to avoid a problem that plagues recreational sailors and the navy. Barnacles readily attach to the surface of manufactured vessels, drastically reducing the efficiency of movement through the water. Barnacles cannot attach to the skin of dolphins because of microscopic contours over the surface. This nanoscale terrain prevents the bar-

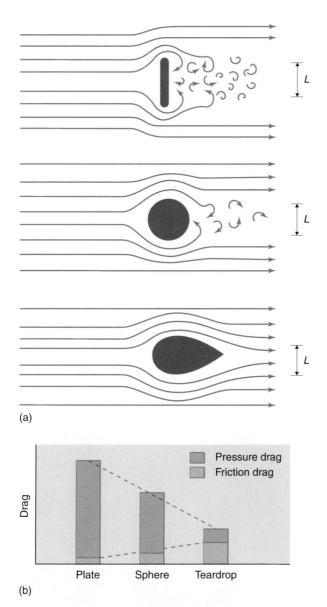

(a)

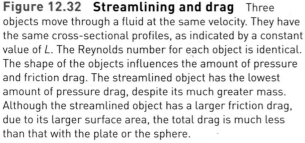

(b)

Figure 12.32 Streamlining and drag Three objects move through a fluid at the same velocity. They have the same cross-sectional profiles, as indicated by a constant value of *L*. The Reynolds number for each object is identical. The shape of the objects influences the amount of pressure and friction drag. The streamlined object has the lowest amount of pressure drag, despite its much greater mass. Although the streamlined object has a larger friction drag, due to its larger surface area, the total drag is much less than that with the plate or the sphere.

nacle from forming a tight seal on the surface, and keeps the dolphin barnacle-free throughout its life.

Aerodynamics and Hydrodynamics

Because air and water share similar fluid properties, swimmers and fliers face similar challenges in moving through the environment. Swimmers

and fliers must overcome the force of gravity to maintain their vertical position. Their locomotor strategies must be consistent with the physical properties of the fluid, particularly density. Swimmers and fliers both benefit from streamlining and use appendages to control the movement of fluids over the body.

Aerofoils and hydrofoils generate lift

When an object moves through a fluid, the fluid is diverted around the object. This movement, and the changes in pressure that result, are responsible for both defying gravity and generating forward movement, or propulsion. Wings and fins are structures used by animals to control the path of movement of the fluid. Most of these structures have a cross-sectional structure similar to that shown in Figure 12.33. This is the general shape of an *aerofoil*, or in water, a *hydrofoil*. The upper surface is curved. The lower surface is flattened, the front is rounded, and the back tapered. The shape is critical in producing the force required to generate an upward force called **lift**. We will discuss how lift works using an aerofoil as an example, but the same principles apply to hydrofoils.

When the aerofoil moves forward, air collides with the leading edge and causes an increase in air pressure. The air slides upward and is compressed into the air on top of the aerofoil. The airstream continues to flow backward. The upper surface of the aerofoil curves downward, away from the airstream. This causes a region of low air pressure. On the bottom of the aerofoil, the air continues smoothly along the surface. Because of the differences in the airflow, there is a difference

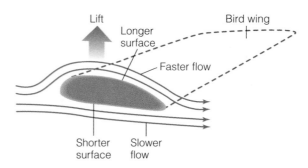

Figure 12.33 Aerofoils and hydrofoils Many wings and fins possess the shape of an aerofoil or hydrofoil. Shown in cross-section, the upper surface of the aerofoil is curved and tapered downward, whereas the lower surface is flat. The fluid must move faster as it moves over the longer upper surface. This results in an area of low pressure, causing a net upward force known as lift.

in the air pressures over the surfaces of the aerofoil. This pressure differential equates to a force. Some of the force lifts the aerofoil upward. Some of the force is lost as drag.

The balance between the lift component and the drag component depends on many factors: air speed, air density, wing area, and a coefficient that is specific to each aerofoil. Both the *lift coefficient* (C_l) and *drag coefficient* (C_d) are determined by direct measurement. They are properties of the object size and dimension. Another parameter that affects lift and drag is the *angle of attack*, which is the angle the aerofoil faces relative to the oncoming airflow (Figure 12.34).

Soaring uses lift from natural air currents to overcome gravity

When discussing movement through the air, it is important to distinguish true flight from gliding. In true flight, animals use wings to lift off the ground and remain airborne for long periods. True flight includes flapping flight and hovering flight, where wing movements generate fluid movements that allow the animal to control altitude and velocity. True flight also includes soaring, where the animal uses stationary wings to generate lift to keep it airborne. True flight appears to have arisen only four times throughout animal evolution (Box 12.2, Evolution and Diversity: The Origins of Flight). Gliding, like soaring, relies on stationary structures to alter fluid movements, but unlike soaring, the animal in-

evitably descends toward the ground. Gliding is much more widespread in animals because it requires much less anatomical and physiological specialization than does true flight. Any structure that increases surface area can improve the ability to glide. There are many examples of mammals (squirrels, primates) and reptiles (snakes, lizards) that extend flaps of skin from the body to glide (Figure 12.35). Flying squid and flying fish, which don't actually fly, use fins to glide over the surface of water. In each case, the shape or orientation of the gliding structure produces some lift, just not enough to remain aloft indefinitely.

Of all the flying animals, only birds soar. In some large birds, such as the albatross and condor, wing structure is much better suited to soaring flight than flapping flight. The efficiency of soaring is enhanced by strategies that capitalize on natural air movements. Many birds undertake slope soaring, riding on the air currents deflected upward along the topography of the surface. Many sea birds, such as

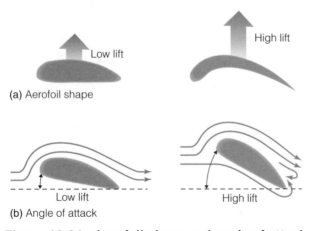

Figure 12.34 Aerofoil shape and angle of attack
(a) The amount of lift generated by an aerofoil is influenced by the shape of the aerofoil. The longer the curved surface, the greater the lift. This high-lift aerofoil also has high drag because of the greater surface area. **(b)** The angle of the aerofoil relative to horizontal, known as the angle of attack, influences the pattern of fluid flow and consequently lift.

Figure 12.35 Gliding animals **(a)** Flying squirrels and **(b)** flying fish are two of the many nonbird species that can glide.

BOX 12.2 EVOLUTION AND DIVERSITY
The Origins of Flight

True flight has arisen at least four times since the origins of metazoans. The earliest fliers were insects derived from a single common terrestrial ancestor that first took flight about 350 million years ago. By 290 million years ago, this group had diversified to more than 15 orders of insects. Pterosaurs, or flying dinosaurs, arose about 290 million years ago and rapidly diversified. Birds probably arose from small theropod dinosaurs around 180 million years ago. Bats appeared about 50 million years ago. The geological record tells us that in each of these periods, atmospheric oxygen concentrations were unusually high. A high oxygen level had two effects on animal locomotion. First, the greater availability of oxygen allowed animals to produce ATP at higher rates. Second, the increased atmospheric oxygen level increased the density of air. This allowed animals to generate more lift from the same structures.

Flight is not possible without wings, but the original function of structures that became wings probably had little to do with flight. The wings of insects, which are structurally related to their cuticle, may have arisen to increase the efficiency of gas exchange. Movement of the wings would both increase the movement of gases around the insect spiracles and induce a form of thoracic pumping to increase respiration. The presence of wings in the aquatic ancestors of insects may have allowed a type of movement that eventually gave rise to flight. Some modern stone flies, for example, use their wings to propel themselves across the surface of water. Surface tension keeps them on top of the water, and the wing movement pushes them along the surface. For the vertebrate fliers there is an ongoing debate about whether they first flew from trees (arboreal) or from the ground (cursorial). In the arboreal hypothesis, animals would climb trees or cliffs and then use their wings to gently glide down to the ground. This same strategy is seen in many modern vertebrates. Flying squirrels and flying lizards extend flaps of skin to increase the ability to soar. In the cursorial hypothesis, animals would use their wings to lift off the ground into the air. Modern birds such as quail use their wings to climb trees. They run vertically up trees, flapping their wings in a way that generates reverse lift to push them against the tree for better foot traction.

Feathers are very important in bird flight, helping to guide the flow of air across the wing surface. The earliest bird fossil, the archaeopteryx, had feathers but it is thought that it probably did not fly. Early feathers arose in several birdlike reptilian lineages as insulation. The structures necessary for flight in modern animals, such as wings, muscles, and feathers, may have arisen for other purposes, but evolution has allowed them to become fine-tuned for flight performance.

Vertebrate wings are modifications of forelimbs and hands, but the origins of insect wings are less obvious. At one point in evolutionary time, prior to the emergence of flight, insects and crustaceans shared a common arthropod ancestor. This ancestor had extra appendages that evolved in different ways in each lineage. In the crustacean lineage, the appendages became epipods, elongated structures that aid in gas exchange. In the insect lineage, the extra appendages became the wings. The same genes that gave rise to insect flight 350 million years ago control the development of the epipods in modern crustaceans and wings in insects.

Insects show many examples of developmental transitions in wings and flight. Most ant species have winged and wingless individuals of both genders. Flight capability is an important element of social caste and reproductive biology of the species. Wingless castes arise when genes required for wing development are repressed in critical developmental windows. The mechanisms that control these genes within species, and the reasons for differences between species, have only recently been studied using DNA microarray technologies.

Some insect species can change their wing development in response to environmental conditions. The life history strategy of many insects is influenced by competition for food. Food abundance is detected by hormonal cues linked to energy metabolism. When conditions are crowded, touch receptors in the insect legs are activated by neighboring insects. These sensory cues alter the development of the machinery for flight. Locusts, for example, can develop into a brown migratory form when food is scarce or a green solitary form when food is abundant. A soapberry bug living with too many other soapberry bugs retains robust flight muscles and long wings, enabling it to fly away. When living where food is abundant, the soapberry bug might break down its flight apparatus, histolyzing the musculature to reuse the energy for reproduction.

References

- Brodsky, A. K. 1994. *The evolution of insect flight.* Oxford: Oxford University Press.

- Dudley, R. 2000. The evolutionary physiology of animal flight: Paleobiological and present perspectives. *Annual Review of Physiology* 62: 135–155.

- Marden, J. H., B. C. O'Donnell, M. A. Thomas, and J. Y. Bye. 2000. Surface-skimming stoneflies and mayflies: The taxonomic and mechanical diversity of two-dimensional aerodynamic locomotion. *Physiological and Biochemical Zoology* 73: 751–764.

pelicans, use air movements on the surface of water (Figure 12.36). Land birds use wind currents flowing up from ridges to reduce the costs of flight. Soaring birds can also ride upwellings of warm air called thermals. Migratory birds can ride a bubble of air upward to great height, then soar away heading toward the next thermal. Slope soaring and thermal soaring dramatically reduce the costs of flight. Many birds migrate along routes that take advantage of the airflows over natural topographic features, covering distances that would not be possible without the metabolic savings of soaring.

Fluid movements can generate propulsion

Bird wings are of a size, shape, and orientation to generate lift if the animal is moving relative to the air. If the animal is not moving forward through the air, no lift results. If movement of the fluid relative to the animal is required to generate lift, then how do animals take off or hover, behaviors that would seem to preclude the generation of lift? In other words, how do animals generate the propulsive forces necessary for forward movement?

Swimmers and fliers move their appendages to generate airflows to produce propulsive force, or thrust. Whereas lift overcomes body weight and the effects of gravity, thrust overcomes drag. As with other rules of fluid dynamics, the mechanisms of thrust are similar in swimmers and fliers.

To understand how wings and fins generate thrust, let's begin by considering an analogy. Imagine a ball floating stationary in a pool of water. If you were to move your hand gently over the surface of the ball, you would cause the ball to spin. Similarly, when the caudal fin of a fish moves through the water, it causes the fluid to swirl into a circular pattern called a vortex. Moving the fin in one direction causes a clockwise vortex, and moving the fin in the opposite direction causes a counterclockwise vortex. These vortices of fluid movement are a consequence of the transfer of force from the fin to the environment. As the fish moves through the water, the flapping caudal fin leaves a series of interlinked vortices in its wake (Figure 12.37). These fluid movements ultimately provide the force that propels the fish forward.

The same vortex ring theory applies to flying animals, but wing movements are much more complicated than fin movements. Wings must move in a way that generates both the forward force and the upward force. Lift is a force that arises from wing shape (aerofoil) only when the fluid is flowing over the aerofoil. Lift is adequate to keep a soaring bird aloft, but the situation is much more complex when a wing moves in space. Furthermore, if the animal is not moving forward, how can it generate lift to remain aloft? Most insects move their wings in a pattern that cannot easily generate lift. At the top of the stroke, the wing is nearly vertical above the insect.

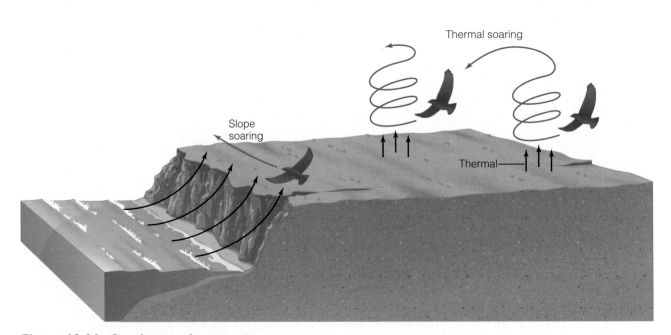

Figure 12.36 Soaring on air currents Birds can soar on upwardly directed air currents. Sloping land such as ridges can direct air upward. Warm bubbles of air, called thermals, rise upward over land warmed by the sun.

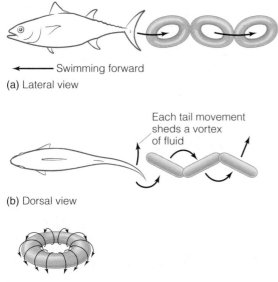

Swimming forward ⟵

(a) Lateral view

Each tail movement
sheds a vortex
of fluid

(b) Dorsal view

(c) Vortex fluid movement

Figure 12.37 Vortices and movement through fluids
When a fish flaps its caudal fin, it generates a complex circular
pattern of fluid flow called a vortex. Swimming activity induces
a series of vortices that can be seen in **(a)** lateral view or
(b) dorsal view. Each vortex resembles a doughnut. **(c)** Fluids
move from the center of the doughnut outward to the edge and
loop back toward the center. These patterns of fluid movement
reveal how the tail interacts with the water during propulsion.
(Source: Adapted from Nauen and Lauder, 2002)

The wing moves rapidly downward below a horizontal plane, twisting as it moves. The combination of rapid downstroke and a twisting movement generates a large vortex of air movement at the leading edge of the wing. These air movements allow the insect to generate both the upward and the forward force. The situation is fundamentally similar in birds and bats. Downward wing movements generate vortices that can be used to remain aloft and propel the animal forward. In addition, insects and hummingbirds can generate favorable fluid movements during the upstroke, which allows them to hover.

Although researchers have many techniques to visualize the vortices that develop during flight, the exact forces at play remain unclear. The style of movement, the shape of the appendages, and the velocity of movement all affect the nature of the wake and the forces that govern movement.

Fin and wing shapes influence fluid movements

In the previous sections we have described how the physical attributes of body shape, wing, and fin shape influence fluid movements. The diversity

in these structures within the animal kingdom reflects the effects of biological properties and the physical environment acting in combination.

Although insects, bats, and birds all fly, their wing shapes are markedly different. Insect wings differ widely in shape and appearance, but they share many features. Typically, the leading edge of the wing is a stiffened structure, whereas most of the wing is a flexible membrane strengthened by an internal framework. During flight, the insect wing distorts, creating complex fluid movements that enable flight. Bat wings, like insect wings, are membranous, but the bones that act as the framework of the wing are jointed. The fine muscles within the wing allow the bat to change the wing shape, which translates into greater maneuverability than is seen in insects. With the elasticity of the membrane, the bat is able to change the dimensions of the wing by as much as 20% without incurring a change in the tightness of the membrane. The feathers of the bird wing have specific shapes and positions that allow the wing to better control the path of air over the surface of the wing. Because the feathers overlap and can slide over each other, birds can change the geometry of the wing without compromising the ability of the wing to act as an aerofoil.

While bird wing geometry is similar overall among species, the subtle differences in shape have important ramifications for flight. Let's first consider the relationship between bird wing size and body mass. Since air flows over the entire wing surface, a combination of wing span (b) and surface area (S) influences lift. Obviously, larger birds need larger wings to generate the lift to remain aloft. However, which is more effective, longer wings or broader ones? Birds of the order Procellariiformes, which includes albatrosses and petrels, differ in size by 400-fold. They share a similar lifestyle, soaring long distances over open ocean. When these birds are drawn scaled to the same wing span, the importance of wing shape is evident (Figure 12.38). The larger birds have longer and narrower wings. Mathematically, the shape is described as the aspect ratio (Λ), which is calculated as follows:

$$\Lambda = b^2/S$$

The shape of fish fins, which are much more variable in shape than bird wings, enable fish to undertake diverse swimming styles. If we restrict our comparison to the fastest-swimming fish, we can see the importance of fin shape in swimming strategies.

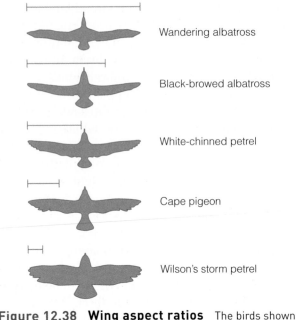

Wandering albatross

Black-browed albatross

White-chinned petrel

Cape pigeon

Wilson's storm petrel

Figure 12.38 Wing aspect ratios The birds shown in this image range almost eight-fold in wing span, but are drawn to the same wing span to illustrate the differences in wing shape. Horizontal bar represents relative wing span. The smallest birds have relatively broad wings.
(Source: Modified from Pennycuick, 1992)

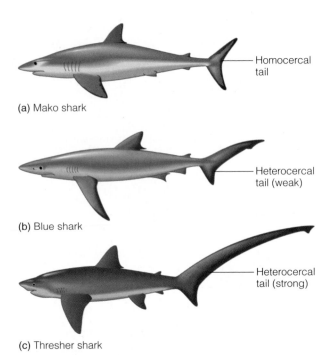

Homocercal tail

(a) Mako shark

Heterocercal tail (weak)

(b) Blue shark

Heterocercal tail (strong)

(c) Thresher shark

Figure 12.39 Caudal fin shapes Sharks' caudal fins range in shape from nearly symmetrical (homocercal) to strongly assymetrical (heterocercal).

Burst swimmers, such as pirarucu, possess thick caudal peduncles with rounded caudal fins of low aspect ratio. Fast steady-state swimmers, such as tuna, possess thin caudal peduncles with crescent-shaped, or lunate, caudal fins of high aspect ratio. Each of these caudal fins is *homocercal*, or symmetrical above and below the midline. Many sharks possess asymmetrical caudal fins. These *heterocercal* caudal fins are much taller on the dorsal side than the ventral side (Figure 12.39). With each stroke, more thrust is generated dorsally, forcing the shark posterior downward and rotating the anterior upward, causing the nose of the shark to pitch upward. The fastest sharks, such as the mako shark, have homocercal caudal fins, but pronounced heterocercal caudal fins are found in pelagic cruising sharks. As with all sharks, the rigid pectoral fins generate lift and prevent the shark from rolling.

Terrestrial Life

Early in the evolutionary history of metazoans, all animals lived in aquatic ecosystems. The invasion of land came in at least two waves. Today, the metazoans of terrestrial ecosystems are represented by diverse taxa of invertebrates and vertebrates. Many show vestiges of their aquatic ancestry in semiaquatic lifestyles and aquatic developmental stages. Each lineage faced its own set of challenges with terrestrial life. In previous chapters we have discussed how this invasion required physiological strategies to cope with osmotic and respiratory challenges. In this section we consider how terrestrial animals meet the oppressive challenge of gravity.

Aquatic animals invaded the land several times

Invertebrates invaded land many times, but the most successful group is the arthropods, primarily arachnids (e.g., spiders), myriapods (e.g., centipedes), and hexapods (e.g., insects). The earliest terrestrial invertebrates were probably detritivores, scrambling over the ground eating partially hydrolyzed plant material. Herbivores and carnivores arose later in invertebrate evolution. Each lifestyle requires a different type of locomotor apparatus to meet the challenges of the complex terrestrial world.

Vertebrates were once found only in aquatic ecosystems, when large, shallow marshes dominated the landscape. Then about 370 million years ago the vertebrate invasion of land began. Many fish species had already evolved strong fins that enabled them to move through sunken vegetation.

These locomotor modifications helped the first amphibious fish to move on land, facilitating the transition to a terrestrial world. The coelacanth, a fully aquatic fish, is closely related to the early terrestrial invaders and anatomically similar. Many unrelated fish have fin structures that facilitate a semiterrestrial life. These early invaders used paired pectoral and pelvic fins to pull the body along, but the trunk was in direct contact with the ground. Evolution of the appendicular musculature and skeleton allowed animals to become more mobile, as the limbs supported more of the weight of the animal. Improvements in leg musculature, changes in postural muscles, reoriented skeletons, and stronger bones all contributed to the colonization of land and the diversification of land animals. The many species of semiaquatic animals illustrate the type of physiological modifications that are necessary for a terrestrial life.

Amphibious animals that must move on both land and water face the challenge of using the same locomotor apparatus under two different conditions. Eels and snakes use trunk movements to swim in water and crawl over land. Ducks and turtles use their feet to walk on land and paddle in the water. The same locomotor modules are used to move in both environments, but an animal may use the musculature in a way that is specific for each environment. For example, when ducks and turtles are in water, they do not need to use leg muscles to support the body mass. Rather, the leg musculature can extend and contract at a higher frequency during swimming. Eels and snakes use the same undulatory movements both on land and in the water. However, movement on land requires more force because of the effects of gravity. Thus, the undulations of an eel on land may be at the same frequency as in the water, but the eel uses the more powerful white muscle on land, and the more efficient red muscle in the water. Collectively, amphibious animals use combinations of motor patterns and recruitment to utilize the same locomotor modules to move in two worlds.

Metamorphosis remodels anatomy and physiology for terrestrial locomotion

Many animals begin their lives in an aquatic world and then undergo a developmental remodeling of anatomy and physiology to specialize for a terrestrial life. You are probably most familiar with the development of local amphibians, where fully aquatic

tadpoles metamorphose into semiaquatic frogs and terrestrial toads. This developmental transition from aquatic to terrestrial animals is also common in insects. These animals provide vivid examples of the anatomical and physiological differences in locomotor patterns in aquatic and terrestrial animals.

Amphibians provide many interesting examples of how changes in locomotor physiology are integrated into life history strategies. Some amphibians remain aquatic animals throughout their lives, using appendages and a tail to swim through the water or crawl through vegetation. Many frogs and toads undergo metamorphosis. Tadpoles are larval forms of frogs and toads that undergo indirect development. They swim through the water much like a fish that uses its tail and trunk to generate thrust. In the late stages of larval development, changes in thyroid hormone levels trigger remodeling of the locomotor apparatus. Limb buds arise from the body trunk and grow into hindlimbs and forelimbs (Figure 12.40). At maturity, the hindlimb anatomy

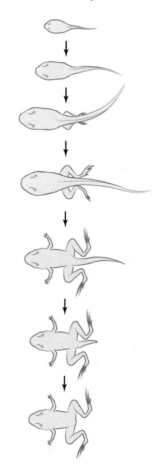

Figure 12.40 Tadpole metamorphosis Fully aquatic larval frogs (tadpoles) swim using their tail. Legs begin to grow during metamorphosis: first hindlimbs then forelimbs. Once the tail is resorbed, the mature frog ventures onto land, using its legs to walk and jump.

becomes specialized for a type of movement that enables both swimming and jumping, although unlike mammals, the limb musculature does not support the entire weight of the animal. In some species of frogs, tadpoles can climb onto the back of a parent, which carries its offspring between ponds (Figure 12.41). Still other species of frogs undergo direct development. Miniature frogs, known as froglets, hatch from eggs laid on vegetation. In contrast to tadpoles, these juvenile frogs can swim, climb, or jump from place to place. It is not yet clear how the basic developmental biology of the amphibians' locomotor system has evolved to account for these diverse life history strategies.

The insect orders Orthoptera (dragonflies, damselflies), Ephemeroptera (mayflies), Plecoptera (stone flies), and Diptera (flies, midges, mosquitoes) each include species that both lay eggs and undergo early development in water. The aquatic juvenile forms, called nymphs, occupy diverse niches within the aquatic environment. Dragonflies, mayflies, and stone flies crawl along the bottom of streams and rivers. These animals possess heavy armored plates that offer protection against larger predators. Mosquito larvae float on the surface of stagnant water, slurping oxygen from the top, oxygenated layer of water. Chironomids are larval midges that live in the sediment of lakes, often in the near absence of oxygen. Each of these larval forms has exquisite physiological strategies to survive in specific niches. However, these aquatic specializations are left behind when the animals undergo their last developmental transition into flying adults.

Figure 12.41 Unusual mode of locomotion in the tadpole In some species of tropical frogs, tadpoles can hitch a ride on the back of an adult frog, allowing it to move between sources of water.

Flightless birds evolved in the absence of terrestrial predators

For many animals, the locomotor machinery must allow movement in more than one environment. Although flying is the most efficient mode of transportation for birds, most birds also spend significant time on the ground. At one extreme are the flightless ratite birds, which include the extant ostrich, emu, and kiwi, as well as the extinct moa and elephant bird (standing 2.5 m tall and weighing 450 kg). Although the evolutionary ancestry of these animals remains uncertain, it is likely that the flightless lineages arose about 40 million years ago. Modern ratites, such as the ostrich, use wings for balance during running and as part of courtship rituals. These birds possess well-developed leg musculature, allowing many species to run at high velocities. Other species of flightless birds appear around the world. The wings of penguins may look like shark fins or seal flippers, but the penguins use their wings to "fly" through the water.

Darwin once commented that there is no greater anomaly in nature than a bird that cannot fly. There may be many reasons for the evolutionary loss of flight. The two most obvious advantages of flight are avoidance of predators and ability to migrate to more favorable environments. Most flightless birds arose in the absence of major terrestrial predators. Populations of flightless birds exist in the Galapagos Islands, which lack major terrestrial predators. The ostrich lives in a region with many large predators, but its large size and powerful kicking legs discourage most predators. Whether considering macroevolutionary variation (flightless species), or microevolutionary variation (flightless populations of one species), the transition from flight-capable to flightless strategies must provide energy savings. The energy that would otherwise be spent building and maintaining wing muscles can be diverted to other systems.

Animals of similar geometry should be able to jump to the same heights

Jumping is a form of locomotion peculiar to terrestrial animals, with specialized anatomy. Animals must use a single muscular contraction to lift the entire mass off the ground. Good jumpers differ from poor jumpers in the geometry of the legs and the strength of the jumping muscles. Higher jumps are possible with longer legs. Some animals have additional leg segments that participate in jumping, like the elongated tarsals of frogs. These longer legs improve jumping because the bones move through a greater arc for a given angle of rotation. Jumpers must also be able to contract muscles rapidly. The greater the velocity at takeoff, the further the animal can jump.

Animals of different size but similar geometry should be able to jump to the same height. Similar geometry means that the overall dimensions of legs are similar and the mass of the jumping muscle is a constant proportion of body mass. With a constant proportional muscle mass, the velocity at takeoff would be similar, and therefore small and large animals should be able to jump to the same height (Figure 12.42). The reference point when talking about the effects of gravity is the vertical midpoint of the mass of the animal—the *center of gravity*. Although a small animal may not reach the same height as a large animal, it is able to lift its center of gravity the same distance. These relationships depend on the assumption of similar geometry. When comparing different species, variations in the animal morphology, physiology, and composition come into play, and contribute to differences in the ability to jump.

Fleas are often considered to be exceptional jumpers because they can jump to heights hundreds of times greater than their own height. Small jumping animals, like the flea, face several challenges that are less important for larger animals. First, viscous effects are more important for small animals. A flea jumping through air faces a drag force similar to what a larger animal might face jumping through water. Second, jumping animals need to move their legs fast enough to reach takeoff velocity. Flea legs are so small that no conventional musculoskeletal combination could reach the required contraction velocity. Fleas avoid this problem by using an unusual mechanism. Muscles power the jump of a flea only indirectly. In the first of two steps, a leg muscle pulls on an internal spring and locks it into the loaded position. Next, a second muscle releases the spring, causing the leg to rapidly extend and the flea to jump. The spring returns to its unloaded position faster than any muscle could induce a contraction.

Terrestrial animals require strong bones and postural musculature

The main challenge in terrestrial locomotion is gravity. In the aquatic world, the natural density of the body imparts some degree of buoyancy that greatly reduces the influence of gravity. However, terrestrial dwellers are much more dense than air, the surrounding fluid. Amphibians and reptiles typically lie directly on the ground, reducing the costs of fighting gravity. However, birds and mammals, as well as extinct dinosaurs, use their limb muscles to lift the body off the ground. This strategy requires anatomical and physiological investments. Bones must be thicker to accommodate the increased force of gravity. Limb musculature must be more extensive to support and move the limbs. Muscle activity is required throughout the body to actively maintain posture. Even the process of standing still requires considerable muscle activity. In the next section we discuss the energetic factors that govern animal locomotion. Although these considerations apply to all animals, they have special relevance for terrestrial animals.

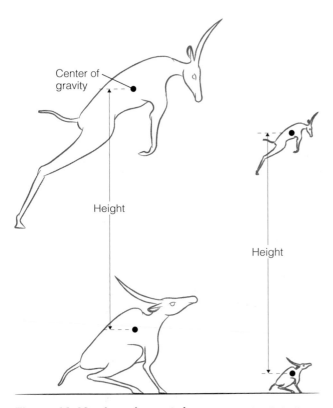

Figure 12.42 Jumping antelope Animals of similar geometry should be able to lift their center of gravity the same vertical distance in a jump. The larger animal reaches a greater height because its center of mass begins at a greater distance from the ground.

(Source: Adapted from Pennycuick, 1992)

Energetics of Movement

Locomotion is expensive, and many studies in the comparative physiology of locomotion focus on the ways anatomy and physiology are used to minimize the costs of movement (see Box 12.3, Genetics and Genomics: Artificial Selection of House Mice). In addition to the long-term costs of building and maintaining locomotor tissues, animals incur short-term costs when they use that machinery to move. The costs of locomotion, which depend on many biological and physical factors, can be expressed in several different terms. The mechanical costs of work can be expressed in units of joules (or calories). The metabolic costs of work are best expressed as ATP turnover (moles of ATP per minute). An estimate of ATP turnover can be obtained from oxygen consumption, but only when the animal is moving slowly enough to justify the assumption that oxidative phosphorylation is providing the ATP. CO_2 production, measured relative to oxygen consumption, provides important information about metabolic fuel selection. Most importantly, these different indices of the cost of locomotion are readily interconverted. Oxygen consumption (V_{O_2}), the most easily measured parameter, can be translated into both metabolic units (ATP) and work units (joules). For example, an animal that consumes 1 ml of oxygen generates about 20 J of energy. The costs of movement depend on many factors, both environmental and functional.

Energy demands of movement can be expressed as total costs or mass-specific costs

There are many ways to assess the energy required for locomotion. Each specific parameter takes into account a different set of concerns that are appropriate to the situation. Calculation of each parameter includes reasonable assumptions that must be kept in mind to properly interpret experimental observations. Let's consider some examples.

An ecological physiologist might be interested in the energetics of a specific migratory bird. The primary question is the relationship between stored energy and the locomotor feat. The experiment may be as simple as weighing birds before and after migration. Analyses of body tissues may be used to assess how specific energy storage depots change as a result of the activity. These measurements could include adipose tissue mass and the lipid and glyco-

gen content of skeletal muscle. For example, the researcher might conclude that flying from one site to another costs x joules of energy, on the basis of the difference in weight and fuel depots.

A more biomechanically oriented physiologist might be most interested in how velocity of movement affects the metabolic costs. Laboratory studies might involve flying this bird species in a wind tunnel to assess the energetic costs of flying at different velocities. Birds may be fitted with gas masks that provide oxygen and capture CO_2. The metabolic demands of exercise are discussed in the context of a specific parameter called **cost of transport (COT)**. The central question in these studies asks how much energy it costs an animal to move a particular distance. The total COT (COT_{total}) is calculated as the metabolic rate divided by locomotor velocity.

$$COT_{total} = \text{(ml of } O_2 \text{ per min)} / \text{(m per min)}$$
$$= \text{ml of } O_2 \text{ per m}$$

The calculation of COT_{total} does not take into account the resting metabolic rate of the animal. The net COT is the difference between the total metabolic rate and the resting metabolic rate. COT calculations allow a researcher to determine the velocity at which an animal can move to most economically cross a given distance.

Each of the previous examples considers the energetics of movement of specific animals, but in many cases researchers are interested in comparisons between different animals. The most common type of comparison considers the effects of body size on the COT. Larger animals use more energy to move, simply because they are larger and have greater total metabolic demands. When considering the impact of body size, it is common to standardize locomotor parameters to the body size. For example, the metabolic rate measurements may be expressed relative to body mass to compare differences in energetics in two animals of different sizes. Also, studies on fish locomotion often express velocities not in absolute terms (meters per second) but as body lengths per second.

Each of these energetic parameters is useful and important in specific contexts. However, the nuances of each parameter are crucial considerations. Expressing values per animal versus per gram of animal provides very different information about the energetics. Similarly, each measurement has implicit assumptions about the underlying biochemistry. Although these calculations are intended to provide information about

BOX 12.3   GENETICS AND GENOMICS
Artificial Selection of House Mice

For hundreds of years, humans have used artificial selection to tailor the locomotor physiology of animals to suit our needs for agricultural animals and sports. For instance, thoroughbred horses and greyhounds, both remarkable runners, are the products of centuries of artificial selection. Only recently have researchers designed artificial selection experiments specifically to explore the mechanisms by which locomotor systems evolve, exploring the links between physiological systems.

In the early 1990s, Ted Garland set up a long-term selection experiment with house mice, with the goal of exploring the evolution of aerobic exercise levels. Each mouse was given free access to a running wheel, with the number of revolutions monitored by computer. Mice that ran the furthest were allowed to become the parents for the next generation. After only 16 generations of such selection, the selected mice typically ran about 2.7 times farther and twice as fast as randomly bred controls. Although the initial focus of the experiment was on the respiratory and musculoskeletal systems, these studies also revealed how other physiological systems coevolved with activity levels, often in surprising ways.

The goal of these selection studies was to assess relationships between locomotor propensity and ability. The characteristic used as the basis of selection was a behavioral trait: how much mice ran of their own free volition. The reason for the variation in voluntary running among individual mice in the original population remains unclear. More active individuals may have a greater sensitivity to the positive, euphoric benefits of running or, alternatively, a lower sensitivity to the negative, painful effects of prolonged activity. In any case, several other behavioral traits correlated with the voluntary activity levels. Mice in the selected lines showed much greater levels of aggressive predatory behavior toward crickets. Subsequent studies showed that the region of the brain that controls predatory behavior, the lateral hypothalamus, was also more active in the selected mice. The underlying differences in CNS activity were reflected in the response to neuropharmacological agents. When mice from each of the lines were injected with Ritalin, a drug used to reduce hyperactivity in children, the differences between lines greatly diminished. Although the researchers recognized that the variation in voluntary activity levels was ultimately attributable to the motivation to run, the selection experiments also provide insight into the underlying locomotor physiology required to support high levels of aerobic activity. Although the *impetus* to run may reflect processes in the central nervous system, the *ability* to run is met by the locomotor physiology, in association with other physiological systems.

If you examined two closely related species that differed in activity levels, it would be reasonable to predict differences in indices of exercise performance and muscle metabolism. Surprisingly, the selected and control mice were indistinguishable in terms of basal metabolic rate, maximal respiratory rate (V_{max}), maximal sprint velocity, or endurance capacity. Selected mice had differences in some respects: muscle glucose transport, bone dimensions, heat shock protein expression, and liver antioxidant capacities. Also, the activities of the mitochondrial enzymes that support muscle energy metabolism were somewhat higher in muscles of selected animals.

One of the most interesting findings from these selection studies was the variability among replicate lines. When the experiments began in the early 1990s, the researchers first divided the mice into eight populations, resulting in four selected for high running and four bred randomly to serve as control lines. While the selected lines evolved similar levels of voluntary wheel running, the replicate lines differed in several ways. For example, two of the four lines of selected mice showed a high frequency of small gastrocnemius muscles. These "mighty minimuscles" had the same enzyme capacity of a larger muscle packed into a smaller muscle that is more resistant to fatigue. The minimuscle phenotype was rare in all four control lines and two of the selected lines. The gene(s) responsible for this unusual phenotype remain unknown, but the difference among lines reveals two important concepts in evolutionary physiology. First, genetic variation within populations represents a pool of alternate solutions to problems not yet encountered. The minimuscle phenotype was relatively rare in the original populations, and only underwent positive selection in the context of this experiment. However, it is not difficult to imagine how a change in the natural environment could also reward this phenotype of a smaller, more fatigue-resistant muscle. Second, these studies illustrate that genetically similar (though not identical) animals can solve physiological challenges in different ways.

References

○ Garland T. Jr., and S. A. Kelly. 2006. Phenotypic plasticity and experimental evolution. *Journal of Experimental Biology* 209: 2344–2361.

the muscles that underlie locomotor systems, it is important to recognize that other physiological systems, such as respiratory and cardiovascular systems, also incur a cost during locomotion.

Velocity of movement affects locomotor costs

As mentioned in the previous section, the velocity of movement influences both the metabolic rate (milliliters of O_2 per minute) as well as the efficiency of movement, expressed as COT (milliliters of O_2 per meter). Experimentally, these relationships are studied by monitoring respiration of an animal that is moving at different velocities. Such studies have been performed with aquatic, aerial, and terrestrial animals. At a low velocity, an animal consumes oxygen at a rate near its resting metabolic rate. An increase in velocity increases metabolic demands in a pattern that differs depending on the animal, the style of movement, and the environment. When these metabolic rate data are used to calculate total COT, a U-shaped curve typically results with a minimum COT at a velocity intermediate between the slowest and fastest velocities possible using that style of movement.

When given a choice, animals tend to move at a specific velocity called the preferred velocity. You have probably experienced this yourself if you have walked with someone who is shorter or taller. For example, a tall person walking with a small child finds it challenging to walk at the child's pace. Remarkably, the preferred velocity is usually close to the velocity at which the COT is minimal.

The relationships between preferred velocity and minimum COT bring up interesting questions about the evolution of physiology. That animals choose a velocity that is near COT suggests that animals are able to detect conditions that result in maximal efficiency. The sensory feedback mechanisms responsible for the relationship between preferred velocity and COT are not yet clear. Presumably, there is also an evolutionary advantage to a behavior that leads to an animal moving slower than it otherwise could. The immediate benefits are in energy savings, but there may also be long-term benefits in terms of avoiding muscle damage. When animals move near the maximal possible velocity, the muscles can experience isometric stress. By using muscles well below their capacities, the animal reduces the risk of debilitating muscle damage.

Animals change style of movement to alter the costs of locomotion

Many animals use different styles of movement over different ranges of velocity. A famous study that emerged from the laboratory of the late Dick Taylor used ponies to illustrate how animals can change gait to alter the interaction between velocity and energetics. Like many land animals, ponies exhibit distinct styles of moving or *gaits*. They walk at low speed, trot at intermediate speed, and gallop at the fastest speed. Taylor's group measured the metabolic rate of ponies as they moved with different gaits at increasing velocity (Figure 12.43). When ponies walked at their preferred velocity of 1–1.5 m/sec, they consumed about 300 J/m. If they were forced to move more slowly or quickly using the same walking gait, their COT increased. The same was true of ponies that either trotted or galloped. The energy demands at their chosen velocities were about 300 J/m, regardless of the gait. Forcing them to move faster or slower than their preferred velocity increased their energy demands. In other words, the minimal COT values were the same in each gait as long as the animals could move at the preferred velocity. A pony that walks, trots, or gallops a distance of 1 km will consume about 300 kJ of energy. The galloping pony will use the energy faster, but it will cover the distance in a shorter period of time.

Gait alters energy expenditures by changing the way locomotor systems are used. Within each

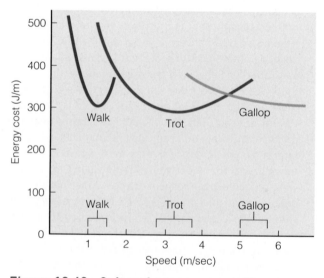

Figure 12.43 Gait and energy expenditure
Many animals, such as ponies, can move with different running styles or gaits. Each style of running has an optimal velocity at which the cost of locomotion is minimal.
(Source: Hoyt and Taylor, 1981)

gait, the pony uses the same set of muscles over a wide range of velocities. Once the pony reaches a particular threshold velocity, it changes its gait and recruits different combinations of muscles to power different leg movements. The coordination of leg movements also differs between gaits. Walking ponies move diagonal legs in synchrony. Trotting ponies move their left legs in synchrony. Galloping ponies move the hind legs in synchrony, half a cycle out of phase from the front legs. The pattern of leg movement in galloping ponies makes better use of elastic storage energy. When the pony plants its hind legs, it also bends its back, creating and storing elastic tension in the bones and tendons. The release of the stored energy drives the front legs forward. Storing energy in the bones and muscles of the back allows galloping animals to conserve energy. The flexure of the back during running is less obvious in a pony than in other animals. The cheetah, for example, demonstrates a pronounced bend in the back when it sprints.

Environment determines energetic costs

The costs of locomotion differ greatly for swimmers, fliers, and runners, each moving at the optimal velocity (Figure 12.44). The costs are lowest for swimmers and highest for runners. To travel 1 km, a 1-kg fish would expend about 100 kcal, a 1-kg bird about 300 kcal, and a 1-kg mammal more than 1000 kcal. The reasons for these differences relate to the efficiency of movement.

Let's start by considering how animals move on land. When an animal walks or runs, energy is required to fight the effect of gravity. When an animal moves one leg forward, its center of gravity drops. Muscular work is required to slow the descent. The center of gravity rises when the rear leg moves forward. More muscular work is required to lift the center of gravity. The cost of moving the center of gravity up and down increases the metabolic rate but does not increase the velocity of forward movement, thus runners have a higher cost of transport. In comparison to a walker, a bicycle rider is able to move much faster and cover the same distance using less energy. One reason is that the bicycle supports the center of gravity and more energy can be used to move the person forward. Similarly, flying is more efficient than walking because the effects of gravity are minimized by lift. Swimmers are the most efficient because they often approach neutral buoyancy, where body composition largely negates the effects of gravity. Swimming animals require less energy than fliers to move at a given velocity, but fliers are able to move much faster. At high velocities the viscosity and drag of water are insurmountable obstacles for a swimmer.

The environment affects the relationship between velocity and metabolic rate (Figure 12.45). Most terrestrial animals moving with a single gait increase metabolic rate linearly with velocity. The power required to generate faster movement of legs is proportional to

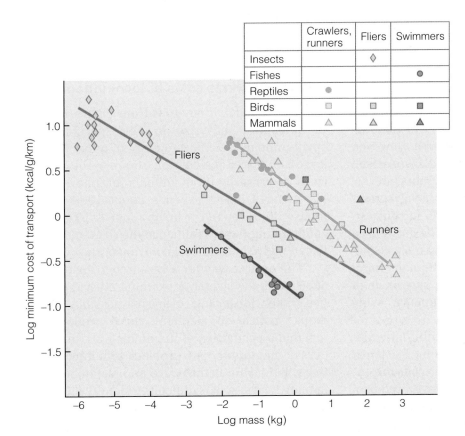

	Crawlers, runners	Fliers	Swimmers
Insects		◇	
Fishes			●
Reptiles	●		
Birds	□	□	■
Mammals	△	△	▲

Figure 12.44 Cost of locomotion in air, in water, and on land The costs of locomotion are lowest for swimmers and greatest for runners. Within each environment, the mass-specific costs of locomotion decline as animal size increases. (Source: Tucker, 1975, with modifications)

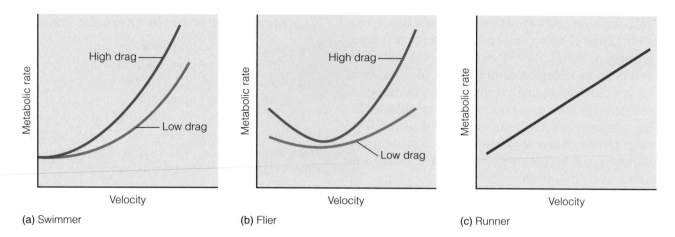

(a) Swimmer   (b) Flier   (c) Runner

Figure 12.45 Work curves for swimmers, fliers, and runners The shape of relationship between work and velocity differs widely between animals, due to both the nature of the environment and the locomotor system of the animal.
(a) Swimmers typically show an exponential relationship, as drag becomes increasingly important at higher velocities.
(b) The nature of the curve in birds differs widely between species. Many, such as the cockatiel, have a U-shaped curve. Bees, and some bird species such as the magpie, have flat curves that suggest work is nearly equivalent at all velocities. Also, the work at each velocity differs widely: although the shape of the curve is similar for pigeons and cockatiels, at most velocities pigeons expend more energy to fly. **(c)** Most running animals show an increase in work with the velocity of movement. (Source: Data for (a): Pettersson and Hedenstrom, 2000; data for (b): Tobalske et al., 2003)

velocity. However, this simple linear relationship does not apply to fliers and swimmers.

Flying animals, including insects, bats, and birds, demonstrate more complex relationships between metabolic rate and velocity. Many birds show a U-shaped relationship. Below a critical velocity, some birds must expend additional energy to move wings fast enough to generate the lift required for flight. At the critical velocity, the birds can generate minimal power necessary to remain aloft. However, not all fliers show this relationship. A comparison of three bird species shows three different patterns. The magpie has a shallow curve, using similar power at each velocity. The cockatiel, in contrast, has a pronounced U-shaped curve. The curve for a dove is somewhat intermediate, but at most velocities the dove uses significantly more than the minimal power. The differences between species lie in properties such as wing-beat frequency, wing movement, and wing shape.

Swimming animals typically exhibit an exponential power-velocity curve, increasing sharply at higher velocities. Metabolism must provide the energy to support the mechanical power requirements for swimming muscles. The mechanical power required to move an object through water is equal to drag times velocity. For most swimming animals, drag is proportional to velocity squared (drag ∝ velocity²) and therefore the power require-

ments for swimming are a function of velocity cubed. This relationship (metabolic rate ∝ velocity³) accounts for the exponential curve observed experimentally.

Body size affects costs of locomotion

Another factor that emerges from Figure 12.44 is the impact of body size. In mass-specific terms, small animals use more energy to move than do large animals. Consider, for example, the relative costs incurred by three animals that move 1 meter. A small insect expends about 1000 J/kg, a mouse 30 J/kg, and a pony about 3 J/kg. Many confounding factors influence the relationship between body size and costs of transport.

The easiest way for a researcher to study the effects of body size is to examine muscle properties in the widest range of species possible. The famous "mouse to elephant curve" reflects the metabolic properties of mammals over many orders of magnitude. The problem with interpreting this relationship is that mice and elephants differ in many ways, so it is difficult to identify the mechanistic cause of the observed relationships. It is always easier to understand the basis of differences between animals when the species under study are closely related. Thus, researchers can study different sizes of a single species, or a clade of closely related species. However, these comparisons inevitably result in a much narrower range of body sizes. Despite these valid concerns about the importance of considering phylogenetic relatedness, there remains an overriding relationship between body size and cost of movement, one that is apparent across broad taxa and in terrestrial animals, swimmers, and fliers. No single overriding factor is responsible for the greater efficiency of movement in larger animals. Differences in every level of musculoskeletal function and animal locomotion can contribute to the origins of this nearly ubiquitous relationship between body size and the costs of locomotion.

The biomechanical constraints of moving through the environment differentially affect small and large animals. Aquatic animals, in particular, must overcome the effects of drag. Drag increases with surface area, but power increases with muscle mass, which is reflected in body mass. The ratio of surface area to mass is greater in small animals than in large animals. Thus, as body size increases, the cost of overcoming drag increases but the capacity for power generation increases more. Thus, large animals use less of their muscle capacity to meet the cost of overcoming drag. The differences in drag provide an important insight into the effects of body mass on locomotor costs in aquatic animals, but drag has less significance for flying and terrestrial vertebrates.

As we discussed in Chapter 5, animals can produce muscles using building blocks that are grossly similar in structure but with important differences that influence musculoskeletal function. For example, myosin heavy chain isoforms differ in the relationship between force and ATPase activity. Since a fast-twitch muscle differs from a slow-twitch muscle in the economy of force development, the fiber type recruitment pattern influences the costs of locomotion. Small animals move their legs at a greater frequency than do larger animals. Consequently, a small animal has a greater reliance on the less economical fast-twitch fibers. Furthermore, the fiber type profile of locomotor muscles differs in large and small mammals. For a given muscle, such as the soleus, large animals have a greater proportion of slow myosins. Thus, both fiber type profile and muscle recruitment patterns contribute to the greater economy of locomotion in larger animals.

Important differences also occur in the mechanical properties of muscles in relation to body size. The long bones of mammals, for example, are nearly isometric; the relative shape and size of the bones is similar among mammals of different sizes. However, other aspects of the musculoskeletal system can differ in important ways. Elastic storage energy is an important mechanism that animals can use to increase the efficiency of movement. Bones and connective tissues are the most important elastic energy stores in vertebrates. Within narrow taxa, such as mammals, there is little difference in the mechanical properties of the biomaterials used to construct muscle and connective tissue. For example, mouse collagen is not likely to be very different in properties from elephant collagen. However, animals may differ in how these biomaterials are used to store elastic energy. First, large and small animals differ in how effectively they can store elastic energy. Energy can be stored only when a force is sufficient to deform the elastic elements. Larger animals, because of the effects of gravity, are better able to store elastic energy during movement. Second, animals may differ in the way these materials are combined into locomotor structures. Larger kangaroos, for example, have relatively larger leg muscle tendons than do smaller kangaroos. These larger tendons allow them to store even greater proportions of energy during hopping. The same increase in tendon elastic storage capacity is seen in other mammals, although the effects of body mass are greater in kangaroos.

One of the reasons it is important to compare closely related animals is the potential for fundamental differences in the organization of the musculoskeletal system. Large mammals use less energy in maintaining posture because their appendages are located directly under the body. Appendages that extend more laterally have a lower mechanical advantage, requiring more muscle force to maintain posture. Furthermore, small animals remain in a crouched posture, which requires muscle activity.

Much of the research in this area of locomotion searches for single unifying themes that can explain variation in locomotor properties over broad taxa of animals. In reality, there are likely many different relationships among animals. The largest animals may have different allometric relationships than the smallest animals. Certain taxa may be constrained by phylogenetic relationships and evolutionary history. Since the locomotor apparatus is required for other functions, it is reasonable to assume that evolution may have found different solutions to similar problems. For example, the benefits of locomotor efficiency may have different evolutionary implications for an herbivorous animal than for an active predator.

⊙ | CONCEPT CHECK

7. How do animals use buoyancy to reduce the costs of movement in water?
8. What are the main constraints governing movement in water, through the air, and on land?
9. How do appendages influence fluid dynamics and affect movement?

Integrating Systems | Migration

Animal locomotion requires the integration of the musculoskeletal system with all other physiological systems. This integration is perhaps best illustrated by considering the costs of migration. Many animals move within a limited geographic range in search of food, or defense of territory. The energetic costs of such daily ranging movements are met by nutrients obtained in the diet. However, many animals also undertake much longer trips, leaving their home range to journey many kilometers. Hugh Dingle has summarized five attributes that separate migrations from other types of excursions.

1. Migratory movements are persistent and of long duration. Some birds, such as Arctic terns, travel more than 20,000 km between poles during migratory flights. The migratory flights of locusts and monarch butterflies can cross thousands of miles.
2. Movement is more linear in direction, without the frequent turning seen in ranging. Migratory routes may be predictable over generations, even without prior experience.
3. Animals are unresponsive to stimuli that would distract them during ranging or station keeping. For example, many animals will not feed during migration, even if food is available.
4. Particular behaviors precede departure and follow arrival. Premigratory birds gorge on food to prepare for migration. Many usually solitary birds begin to aggregate into groups in preparation for flocking.
5. There is a reorganization of energy metabolism in support of locomotion. The migration typically involves reliance on stored metabolic fuels. At the end of the migration, the animal is frequently depleted of energy stores.

Most but not all migratory animals show each of these characteristics. Physiological systems play important roles in the preparation of animals for migration, as well as in the migration itself.

Animals that undergo long migrations show a pronounced reorganization of their physiology to prepare for the journey. Suites of hormones respond to environmental cues to reorganize physiological systems and alter normal foraging behavior. Once the migration begins, the animal cannot easily be diverted from its route.

In temperate environments, migrations of many birds and some insects begin in fall, so that species can avoid the cold, harsh winter. Although the purpose of the migration may be to avoid cold temperatures, it is the change in photoperiod that triggers the premigratory reorganization. Seasonal temperatures may vary each year, but photoperiod obeys a constant pattern. The importance of photoperiod was shown in early studies with pigeons. Researchers held pigeons under different photoperiod regimes. Birds kept under a winter photoperiod (8 h light and 16 h dark) flew south when released. Birds kept under a summer photoperiod (16 h light, 8 h dark) flew north when released. Steroid hormones orchestrate the physiological response to photoperiod. Castrated male birds, regardless of photoperiod, always flew south when released. Testosterone supplements would result in resumption of the photoperiod-dependent response.

The energy demands of migration require exquisite coordination from multiple physiological systems. Many animals gorge on food prior to migration, "fattening" for the journey. Hummingbirds increase their mass by 50% with lipid deposits. Lipids are the preferred fuel because of the economy of storage (ATP per gram of fuel). A 3-g hummingbird that deposits 1.5 g in fat droplets would need to store 15 g of glycogen particles (five times its body weight!) to achieve the same caloric content. Changes in body mass also have biomechanical consequences. Migratory sea birds may gain so much fat that they have a difficult time taking off. Many animals are completely committed to using their nutrient stores during migration, refusing to eat even if they encounter food. Commitment to nonfeeding strategies has several advantages to a migratory animal. In animals that rely on stored fuels, success of the migration does not depend on successful foraging en route. Foraging in unfamiliar lands may put the animal at an increased risk of predation. In many cases, it also permits animals to allow their digestive systems to degrade, reducing the costs of routine metabolism.

Often annual migrations are linked to reproductive development. Frequently, migratory animals reproduce after they reach their destination. Reproduction, like migration, requires a great deal of energy, and successful life history strategies must balance the needs of both functions. Some animals become reproductively mature before migration, but most animals delay reproduction until they have reached their destination. Consequently, animals may undergo reproductive or developmental preparations in parallel with metabolic changes. Hormones coordinate metabolic, reproductive, and developmental transitions associated with migration. Steroid hormones, such as cortisol and testosterone, are important in vertebrates. Migratory insects, such as the monarch butterfly, use juvenile hormone and ecdysone.

Humans have a history of migration and long-distance travel. Early hominids used landscape features and the position of the sun and stars to guide them in their ranging movements. Technological advances in navigation allowed humans to explore their environment more widely. The compass, developed in China in the 4th century B.C., relies on the magnetism of the North Pole. More than 1000 years ago, early Arab explorers developed a *kamal* to follow the elevation of Polaris, the North Star, an indication of latitude. Improvements in latitudinal navigation arose with the appearance of the astrolabe, quadrant, and cross-staff. Sailors used these tools to travel to the correct latitude, then move east or west to port. In the 17th century, explorers looking to use the sun to determine longitude realized they needed a way to determine time, because the position of the sun moves 15° longitude each hour. Development of chronometers and sextants, in combination with detailed celestial maps, facilitated human exploration and colonization.

Humans are not the only animals that rely on the celestial bodies and geomagnetism for navigation, although the other species' "tools" are built in. Animals may use both a "compass" to give direction and some form of "map" to identify the goal. Their mapping is generally thought to be a function of memory, but the compass requires some neurosensory mechanism to detect direction. Early research, primarily performed in birds, identified the importance of the sun and stars in navigation. More recent research has shown that many animals also detect geomagnetism. In all animals studied to date, magnetosensory systems use tiny crystals of an iron oxide called magnetite. The mechanisms by which these crystals are integrated into neurosensory systems have only recently been studied. In birds, it is thought that the signals derived from magnetoreceptors are interpreted by the visual circuitry.

Many humans have an innate ability to use odor to find a local bakery or sewage outlet, but for the most part we do not rely on chemical signatures for navigation. As discussed in Chapter 4, the detection of environmental chemicals by taste or smell is made possible through specific receptors and neurosensory networks. In many cases, these sensory systems are used to locate prey, as in a shark's response to blood. Other species use water chemistry to navigate during migration. Salmon imprint on the chemical signature of their natal streams, allowing them to return to the stream during spawning many years later. Although this capacity is impressive, salmon survival also requires that it be imperfect. Misguided fish entering the "wrong" river may exploit new habitats or enhance the mixing of gene pools.

The energetic costs of migration are extraordinary. Many animals use migratory routes or strategies that minimize the energetic costs of locomotion. Insects rely heavily on wind currents to reduce the costs of flight. Migratory birds ride thermal bubbles to gain altitude without expending energy to flap wings. The oceanic migrations of squid and eels in the Atlantic use the Gulf Stream to reach their destinations. Nonetheless, the costs of locomotion during migration are high and usually met through mobilization of stored fuels. ◉

Summary

Locomotor Systems

→ Animal locomotion depends upon the coordination of several physiological systems, including musculoskeletal, circulatory, respiratory, digestive, and sensory systems.

→ Muscles are composed of combinations of fiber types. Fish locomotor muscles are primarily two or three fiber types organized into homogeneous muscles. Tetrapods have more muscles, each of which possesses multiple fiber types.

→ The pattern of locomotor muscle contraction is controlled by motor neurons. Nervous activity coordinates the contractile pattern and controls muscle recruitment at different velocities.

→ Mitochondrial oxidative phosphorylation provides the fuel for resting muscle activity and long-term, low-intensity exercise. Glycolysis, in combination with phosphagen stores such as ATP and phosphocreatine, fuels activity of high intensity and short duration.

→ High-intensity exercise incurs short-term energetic deficits that must be repaid during recovery. Hormones control muscle metabolic fuel mobilization and utilization.

→ The cardiovascular system delivers oxygen and nutrients to working muscle via capillary networks. Vasoactive agents regulate blood vessel diameter to alter blood flow to muscle.

→ Oxygen delivery to muscle mitochondria is aided by the oxygen-binding protein myoglobin.

→ Muscles work in conjunction with skeletal systems. Hard skeletons are composed of cellular secretions that make up the extracellular matrix. Osteoblasts and osteocytes produce the matrix of vertebrate skeletons, which then undergoes mineralization, or ossification.

Moving in the Environment

→ In order to move, animals must overcome the physical forces associated with the environment.

→ Fluids, such as air and water, flow in smooth layers until they encounter objects. When an object encounters a fluid, it alters the flow of the fluid. When the object causes a fluid to change from laminar flow to turbulent flow, the object incurs an added cost to move through the fluid.

→ The combined effects of fluid properties and object dimensions determine the Reynolds numbers. Inertial effects dominate animal movement at high Reynolds numbers. Viscous effects dominate at low Reynolds numbers.

→ Drag is a force that opposes forward movement. Pressure drag results from a buildup of fluid in front of the moving object, whereas friction drag results from the interaction between the surface of the object and the fluid. A streamlined shape reduces pressure drag to reduce the costs of movement.

→ Locomotor strategies must also compensate for the costs of overcoming gravity. In water, the costs of combating gravity are reduced by buoyancy strategies, such as lipid storage or gas-filled sacs such as the swim bladder. On land, the musculoskeletal system is used to overcome the effects of gravity.

→ Fliers and swimmers can overcome gravity using specialized appendages that act as aerofoils or hydrofoils. These structures generate lift to oppose the downward pull of gravity. Soaring animals use lift from natural air currents to overcome gravity. Appendages also generate the fluid movements responsible for propulsion.

→ The metabolic costs of movement depend on velocity. For each type of movement, there is an optimal velocity for which costs are minimal.

→ Many animals use their locomotor systems in different ways to create distinct styles of movement, such as gaits.

→ The costs of locomotion differ in each type of environment because of the combined effects of fluid properties and gravity. The costs of locomotion are lowest in swimmers and highest in runners.

→ Locomotor costs are also influenced by body size. The largest animals have greater total costs of locomotion, although the mass-specific costs are lower.

Review Questions

1. Why can oxygen consumption be used to measure energy expenditures in moving animals?

2. Discuss the differences in muscle fiber types that suit them for different types of movement.

3. Discuss the role of the vertebrate skeleton in locomotion.

4. How does body size affect the costs of locomotion in animals?

5. What is a Reynolds number, and why does it matter to a moving animal?

6. How does airflow over an object cause lift?

7. Which would generate more lift, the wing of a bird or the fin of a fish, if they were the same dimensions?

8. How can animals alter their interaction with the environment to reduce the costs of moving from place to place?

Synthesis Questions

1. What anatomical and functional features influence the efficiency of movement of oxygen from the erythrocyte to the muscle mitochondria?

2. Many animals alter their physiology in response to frequent bouts of activity. In humans, this is known as a training effect. How would you expect each physiological system to change in response to training?

3. Many marine fish swim into deep, cold water to pursue prey or avoid predators. How does cold temperature influence their ability to swim?

4. Predict the physiological properties of the locomotor system of (a) a cheetah and (b) a tree sloth.

5. Discuss the changes in cardiovascular and respiratory systems that support (a) high-intensity activity and (b) steady-state activity.

6. Discuss the recovery from high-intensity activity. Consider the physiological, physical, and chemical changes that accompany this type of activity and what must happen to prepare the animal for another bout of activity.

Quantitative Questions

1. What are the mathematical relationships between power, work, and force? Under what physiological conditions will each of these parameters approach zero?

2. Small scale models of objects can be constructed to explore how the object moves through fluids. Engineers change the fluid movements to ensure that the Reynolds number remains constant despite the smaller dimensions of the object (L). If an object is reduced in size by 1/1000, how would you change the fluid properties to ensure that the Reynolds number remains constant?

3. Use the following assumptions to answer the subsequent questions about the energy metabolism of a hummingbird on its flight across the Gulf of Mexico:

- A 2-g hummingbird puts on an additional 1 g of fat.
- The hummingbird has a mass-specific metabolic rate of 40 ml of O_2 per hour per gram and a total metabolic rate of 120 ml of O_2 per hour per bird. For simplicity, assume that its total metabolic rate remains constant for the duration of the flight.
- The lipid fuel is palmitate (molecular weight = 256 g per mol), although this ignores the contribution of glycerol from the triglyceride backbone.
- Oxidation of 2 NADH consumes 1 O_2 and generates 6 ATP, and oxidation of 2 $FADH_2$ consumes 1 O_2 and generates 4 ATP.
- Though you could translate between milliliters of O_2 and moles of O_2 using the universal gas

law ($n = PV/RT$), assume that 1 mole of O_2 occupies 22.4 liters of volume.

(a) What is the metabolic rate of a hummingbird in terms of ATP consumption in terms of moles of ATP per gram per hour?

(b) If palmitate is the fuel that supports this activity, what is the rate of palmitate oxidation in terms of moles of palmitate per gram per hour? (Review Chapter 2 to remind yourself of the stoichiometries of NADH and FADH production in β-oxidation of fatty acids.)

(c) How long would the 1 g of stored fat be able to support flight?

For Further Reading

See the Additional References section at the back of the book for more readings related to the topics in this chapter.

Locomotor Systems

These studies discuss locomotion from the perspective of biomechanics and the design of musculoskeletal systems in vertebrates and invertebrates.

Alexander, R. M. 1995. Leg design and jumping technique for humans, other vertebrates and insects. *Philosophical Transactions of the Royal Society of London, Series B: Biological Sciences* 347: 235–248.

Biewener, A. A. 2002. Future directions for the analysis of musculoskeletal design and locomotor performance. *Journal of Morphology* 252: 38–51.

Sensenig, A. T., and J. W. Shultz. 2003. Mechanics of cuticular elastic energy storage in leg joints lacking extensor muscles in arachnids. *Journal of Experimental Biology* 206: 771–784.

Many researchers study the determinants of the maximal capacity for activity because it reflects the limits and constraints on the design of a physiological system. These reviews discuss the limits on activity, focusing on the relationship between metabolism and locomotion.

Jones, J. H., and S. L. Lindstedt. 1993. Limits to maximal performance. *Annual Review of Physiology* 55: 547–569.

Suarez, R. K. 1996. Upper limits to mass-specific metabolic rates. *Annual Review of Physiology* 58: 583–605.

Moving in the Environment

These books address how animals interact with the environment during locomotion. They are very good discussions of the biophysical constraints on animal locomotion.

Alexander, R. M. 1992. *Exploring biomechanics: Animals in motion*. New York: Scientific American Library.

Alexander, R. M. 1999. *Energy for animal life*. Oxford: Oxford University Press. Blake, R., ed. 1991. *Efficiency and economy in animal physiology*. Cambridge: Cambridge University Press.

An excellent review that compares the constraints on animal movement in widely different animals, identifying common themes and explaining why different modes of movement are necessary.

Dickinson, M. H., C. T. Farley, R. J. Full, M. A. R. Koehl, R. Kram, and S. Lehman. 2000. How animals move: An integrated view. *Science* 288: 100–106.

This text is an overview of the biotic and abiotic factors that affect migratory movements. The emphasis is on integration of locomotion with life history strategies.

Dingell, H. 1996. *Migration: The biology of life on the move.* Oxford: Oxford University Press.

These works consider the role of biomechanics in locomotor physiology, focusing primarily on the relationship between structure and function.

Lindhe Norberg, U. M. 2002. Structure, form, and function of flight in engineering and the living world. *Journal of Morphology* 252: 52–81.

Pennycuick, C. J. 1989. *Bird flight performance.* Oxford: Oxford University Press.

Pennycuick, C. J. 1992. *Newton rules biology.* Oxford: Oxford University Press.

Schmidt-Nielsen, K. 1983. *Scaling; Why is animal size so important?* New York: Cambridge University Press.

Thermal Physiology

Endothermy, the ability to generate and maintain elevated body temperatures, has arisen several times in the evolutionary history of animals. It goes hand in hand with the capacity to produce heat through metabolism, and therefore activity levels. Most modern birds and mammals have high metabolic rates and are able to maintain their body temperatures well above ambient temperature, often within narrow thermal windows. While both are perceived as "higher vertebrates," birds and mammals arose from separate reptilian ancestors. Thus, endothermy arose independently at least twice. However, fossil evidence suggests that other extinct reptiles may also have been endotherms. The fossil record of the animals in the paleontological period from 200 to 65 million years ago is particularly clear, showing definitive examples of the transitions from reptiles to mammals and birds.

The first mammals appeared approximately 200 million years ago, evolving from small, nocturnal reptiles that were only distantly related to the dinosaurs that would dominate Earth in later years. Fossils dating back to this period reveal the existence of several distinct mammalian-like reptilian lineages. These animals differed from other reptiles by the morphology of the skull and the organization of the teeth. Although most of these lineages disappeared, one group of reptiles called *cynodonts* gave rise to true mammals. The earliest mammals retained the reptilian trait of egg laying, like the modern monotremes, echidna and platypus. By the early Cretaceous period (144 million years ago), mammals had diversified into several lineages of marsupials and insectivores. When the dinosaurs disappeared about 65 million years ago, at the end of the Cretaceous period, there was an explosion of mammalian diversification. New species of mammals began to occupy the environmental niches vacated by the dinosaurs. It cannot be said for certain when endothermy arose in the transition from mammalian-like reptiles to true mammals. However, it is likely that the cynodont reptiles were already endothermic. Unlike most other reptiles of the day,

Archaeopteryx.

Asymmetrical fossilized feather.

cynodonts possessed a bony, secondary palate in the roof of the mouth that would have allowed them to breathe while chewing. This anatomical arrangement is a characteristic of endotherms because they must maintain uninterrupted respiration to sustain high metabolic rates. Cynodonts also appear to have possessed hair, which could have helped insulate their bodies.

Birds, the other group of modern endotherms, also arose from reptiles, although much later than mammals and from different reptilian ancestors. Around the time dinosaurs were declining, several reptilian lineages had already evolved featherlike body coverings. In one group, the theropod dinosaurs such as *Archaeopteryx*, the feathers were similar in structure to those of modern birds. Their feathers were asymmetrical, a trait that is necessary to be useful in feathered flight. In contrast, the other feathered reptiles of the era, such as *Protarchaeopteryx robusta* and *Caudipteryx zoui*, had symmetrical feathers. Since these

symmetrical feathers would be useless in flight, they must have arisen in these dinosaurs for other benefits, such as insulation. Although these other lineages of feathered reptiles became extinct, they were likely also endothermic animals.

Many researchers believe that endothermy arose in other, nonfeathered dinosaur lineages as well. The largest dinosaurs were simply too big to shed metabolic heat, and therefore remained warm-bodied. Many smaller dinosaurs may also have been endothermic. Multiple lines of evidence support the notion that these animals had the high metabolic rates necessary for an endothermic animal. Bone structure and posture suggest rapid rates of locomotion, which in modern animals require high metabolic rates that are possible only in warm-bodied animals. Just as in modern endotherms, many dinosaurs had relatively large brains associated with superior sensory processing. Since brain tissue has a high energy demand, a large brain can have an important influence on the whole body metabolic rate. Other theories have been raised to support arguments that dinosaurs were endotherms. However, no argument is definitive because of the limitations in using the properties of modern animals as guidelines in predicting the physiological features of these long-extinct animals. ◉

Overview

Recall from Chapter 2: Chemistry, Biochemistry, and Cell Physiology that thermal energy influences chemical interactions in ways that affect macromolecular structure and biochemical reactions. Consequently, temperature has pervasive effects on all physiological processes. As a result of these temperature effects, every animal displays a *thermal strategy:* a combination of behavioral, biochemical, and physiological responses that ensure body temperature (T_B) is within an acceptable limit. The most important environmental influence on the thermal strategy (though not the only one) is ambient temperature (T_A). Animals must survive the highest and lowest T_A in their niche (thermal extremes), as well as the change in T_A (thermal change).

Animals inhabit most thermal niches on the planet (Figure 13.1). The hottest environments exploited by animals are the regions near thermal vents, such as the hydrothermal vents of the deep sea, volcanoes, and geysers. The coldest places inhabited by animals are the alpine and polar regions. The animals that survive in the extremes of heat and cold are impressive, but the ability to tolerate changing temperature is every bit as challenging physiologically. Environmental temperatures are most variable in terrestrial ecosystems; air temperatures change more rap-idly and reach greater extremes than do water temperatures.

Many ecosystems exhibit spatial variation in temperature. Underground refuges are buffered from thermal extremes on the surface. The T_A in alpine regions varies as a result of altitudinal gradients arising over only a few kilometers. Large bodies of water, such as lakes and oceans, can vary in T_A with depth. Deep-ocean (bathypelagic) temperatures are often close to 4°C, whereas midwater (mesopelagic) and surface water (epipelagic) temperatures can be much warmer and more variable. Large temperate lakes may be nearly uniform in temperature, or have sharp demarcations (thermoclines) between top and bottom water, sometimes differing more than 10°C in less than a meter of depth.

Ecosystems can also change in temperature temporally. Terrestrial and aquatic ecosystems in the tropics tend to have a relatively constant T_A, but polar and temperate zones experience seasonal and daily cycles of cold and heat. Air temperatures can change more rapidly than water temperatures, sometimes more than 20°C in a single day. Intertidal animals may experience the heat of a summer day mere seconds before the cold ocean washes over them. Many animals incorporate behavior into their thermal strategy, but animals must also cope with the effects of temperature on biochemistry and physiology.

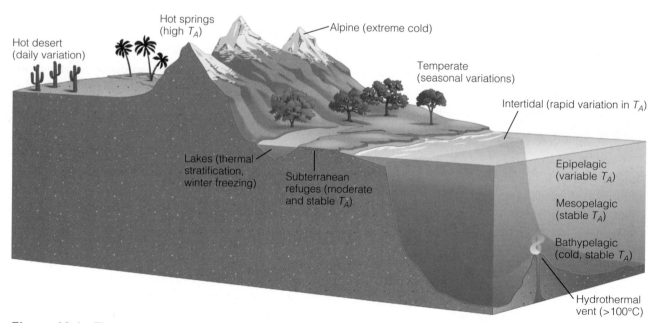

Figure 13.1 Thermal niches in the temperate zone

Heat Exchange and Thermal Strategies

The most important physiological parameter in an animal's thermal physiology is body temperature (T_B). An animal's thermal strategy serves to control the transfer of energy between the animal and the environment. Some animals tolerate wide changes in T_B and the effects of these changes on many physiological processes. Others must use a combination of physiological and behavioral means to ensure that T_B remains nearly constant. As in other physiological systems, both strategies—tolerance and regulation—have costs and benefits. The physiological mechanisms that impart a constant T_B use energy. When T_B is allowed to vary, important physiological processes such as development become sensitive to environmental changes. Although T_A has the most obvious impact on animal thermal biology, other routes of heat exchange are also important in many contexts.

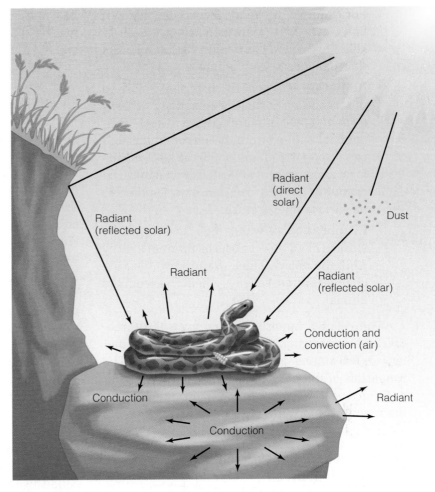

Figure 13.2 Sources and sinks for thermal energy   The body temperature of an animal is influenced by heat exchange with the environment. This snake is warmed by radiant energy from the sun, as well as thermal energy radiated from its surroundings. The animal exchanges thermal energy through objects and fluids in contact with its external surface (conduction). Movement of the air enhances the efficiency of thermal exchange by convection. The animal itself radiates thermal energy to the surrounding air.

Controlling Heat Fluxes

An animal's T_B is a reflection of the thermal energy held within the molecules of the body. Thermal energy can move from the animal to the environment, or from the environment to the animal, depending on temperature gradients. Metabolism—the sum of all biochemical reactions occurring within the body—is the main source of thermal energy in the heat balance equation of most animals. However, other important sources and sinks for thermal energy also affect an animal's thermal budget (Figure 13.2). The thermal balance equation takes into consideration all of the routes through which thermal energy, abbreviated as H, can enter or exit the body:

$$\Delta H_{\text{total}} = \Delta H_{\text{metabolism}} + \Delta H_{\text{conduction}} + \Delta H_{\text{convection}}$$
$$+ \Delta H_{\text{radiation}} + \Delta H_{\text{evaporation}}$$

If the equation above sums to zero ($\Delta H_{\text{total}} = 0$), there will be no net change in the thermal energy of the animal and T_B will remain constant. If the flow of thermal energy into the animal exceeds the heat loss, T_B will increase. Each of these routes of thermal energy exchange depends on the thermal properties of the environment as well as the physical properties and physiology of the animal.

- **Conduction** is the transfer of thermal energy from one region of an object or fluid to another. Animals can be cooled when thermal energy is conducted away from the body, or can be warmed as they absorb heat from conductive objects.

- **Convection** is the transfer of thermal energy between an object (the animal in this case) and an external fluid that is moving.

For example, warm air feels cooler when it flows over your skin than when the air is still. Most often, convection causes a loss of thermal energy from animals.

- **Radiation** is a general term that refers to the emission of electromagnetic energy from an object. An animal can absorb radiant heat emitted from the surroundings, but can also emit radiant heat from its own surface, a major form of heat loss. The infrared radiation emitted from an object indicates its surface temperature.

- **Evaporation** of water molecules from the surface of an object absorbs thermal energy from the object. Thus, evaporative heat exchange is almost always a heat loss from the animal.

The relative importance and even the direction of heat transfer from each of these parameters differ among animals and conditions. The properties of the animal, including physical composition and color, have a profound influence on the relative importance of these exchanges.

Water has a higher thermal conductivity than air

Conduction is difficult to quantify because of the many factors that affect heat exchange. Let's begin our discussion by considering how conduction is involved in the transfer of thermal energy through a single material, such as a thin metal bar heated at one end. The rate of heat transfer from the warm end to the cool end (heat flux) is described by Fourier's law and the following equation:

$$Q = \frac{\lambda \Delta T}{L}$$

where heat flux (Q) depends upon the temperature gradient (ΔT), the distance over which the gradient extends (L), and the thermal conductivity (λ) measured in watts per meter per kelvin (W/m per K). Thermal conductivity is a specific property of a material. Those objects we think of as *heat sinks* have high thermal conductivity. For example, an aluminum pot feels cold to the touch because it has a high thermal conductivity (210 W/m per K) and readily draws heat from your hand. Similarly, 5°C water feels cooler than 5°C air because water has a thermal conductivity that is 25-fold higher than air (0.58 versus 0.024 W/m per K). Because water

has more molecules per unit volume, there is a greater likelihood of a molecular collision that results in a transfer of energy.

The Fourier equation describes how thermal energy moves in a very simple system: heat transfer in a single dimension (from the heat source to heat sink) in a single uniform material. These same parameters (λ, ΔT, and L) apply in thermal biology, but animals are much more complex systems. Consider the influence of thermal conductance. Heat is conducted from the internal tissues, through other tissues and fluids, and to the external surroundings, each with a characteristic thermal conductivity (Table 13.1). The body surface layers may possess insulation that reduces conductive heat transfer. Insulation, such as fur and feathers, also increases the distance between the hottest point near the skin and the coldest point in the bulk phase.

Calculations of heat flux are complicated by the geometry of the environment and the animal. Heat does not move from your body through a one-dimensional cylinder of air extending from your skin, but rather is conducted in multiple dimensions from the source. Animal geometry also plays a role. A long, thin animal produces as much heat as a short, round animal of the same mass, but the differences in surface area affect heat exchange. Since conductive heat losses occur across the external surfaces, an animal can alter conductive heat exchange by engaging in activities that alter its effective surface area. For example, a penguin reduces heat loss from the foot by rolling back on its heels, using its tail feathers for balance. Because its tail feathers are less conductive than its feet, less heat is lost. Figure 13.2 shows a snake simultaneously exchanging heat with multiple surfaces. It loses heat via conduction across its upper surface while also exchanging heat through its lower surface in contact with the rock.

Table 13.1	Thermal conductivity of materials.
Material	**Thermal conductivity (W/m per K)**
Air	0.02
Snow	0.10
Water	0.59
Rock	1–3
Ice	2.1
Muscle	0.5
Fat	0.2

Convective heat exchange depends on fluid movements

Imagine yourself immersed in a pool of water that is 10°C colder than your body. Almost immediately, your body begins to lose thermal energy as it warms the water in the boundary layer by almost 10°C. Once the boundary layer is warmed, thermal energy is slowly conducted to the bulk phase of the water. When the heat exchanges reach steady state, the body loses thermal energy at the rate required to rewarm this boundary layer that slowly cools as it dissipates its thermal energy outward to the bulk phase. Much less energy is required to rewarm the boundary layer under these steady-state conditions than was required to heat the boundary layer in the first place. Now consider how the gradients change when fluid is flowing over the body. The body rapidly loses thermal energy warming a boundary layer that is immediately replaced by another, colder boundary layer. Heat lost to a moving fluid, either air or water, is *convective heat loss*. The rate of convective heat loss depends on the thermal gradient between the surface and the fluid, the rate of flow of the fluid over the surface, and its conductivity.

Radiant energy warms some animals

In biological systems, radiant heat exchange occurs through electromagnetic radiation in the long wavelength, infrared range. Thus, if a red light (long wavelength) and a blue light (short wavelength) of equal intensities are shown on your skin, the red light will more effectively warm the surface.

In the natural world, the most important source of radiant heat is the sun. Photons from the sun excite the molecules in the atmosphere, the soil, and the water, warming them by radiant heat. Thus, when animals are warmed by conduction from air, water, or soil, the ultimate source of the heat is radiant energy. But animals can also be warmed directly by solar radiation, which many species accentuate by the behavior known as basking. White body coloration reflects photons in the visible range, and dark coloration absorbs the photons within this range of wavelengths. Animals that bask to warm themselves often possess high levels of black or brown pigments to help absorb thermal energy. As a result of diversity in color, animals in the same area can have markedly different temperatures (Figure 13.3).

Figure 13.3 Heterogeneity of T_B in the intertidal zone   Infrared photography can be used to compare the body temperature (T_B) of animals. In this image, the mussels are warmer than the starfish because they are better at absorbing radiant energy. The starfish, with its greater surface area, may also be affected more by evaporative cooling.
(Photo courtesy of Dr. Brian Helmuth, University of South Carolina)

In terrestrial systems, the ground warms during the day and then becomes an important source of thermal energy in the form of conduction and radiant heat when the sun sets. Animals also lose thermal energy when they emit radiant heat. Thus, radiant heat may be a net gain or net loss from animals. The relationship that describes radiation from a warm animal is described by the Stefan-Boltzmann equation:

$$P = Ae\sigma(T_B{}^4 - T_A{}^4)$$

where P is the radiating power, A is its surface area, e is the ability of the object to emit radiation, σ is the Stefan constant, and T the temperature of the body (T_B) or surroundings (T_A) in kelvins. Animals can influence their radiant heat loss through changing the nature of the surface (e) and the surface area (A).

Evaporation induces heat losses

Evaporative cooling arises when fluids draw thermal energy from the body surface as the water molecules make the transition from liquid to vapor. The magnitude of the heat loss depends on the volume of water and its heat of vaporization. It requires more energy to evaporate water from salty sweat than from pure water because the solutes increase the heat of vaporization of water. The efficiency of evaporative cooling also depends on the partial pressure of water vapor in the air. If the air has a high humidity, then the water is less likely to evaporate.

Sweating is only one of the ways that animals employ evaporative cooling. When a hippopotamus rolls in the mud of a wet riverbank, the cool mud draws heat from the body (conduction). This is an effective cooling strategy even if the mud is warm: thermal energy is absorbed from the body as the mud dries. Other animals cover their body surfaces with water, such as an elephant that sprays water onto its back or birds that splash in a pool of water. Wet feathers also have a diminished insulatory capacity, allowing more metabolic heat to be lost. Birds that live in hot environments may soak the belly before returning to the nest, allowing the eggs to benefit from evaporative cooling. Kangaroos, which do not produce sweat, lick well-vascularized skin surfaces, which then cool as the saliva evaporates.

Not all evaporative cooling is positive. When semiaquatic animals leave the water, they are typically left with wet body surfaces, causing body temperature to decrease due to evaporative cooling.

Ratio of surface area to volume affects heat flux

The ratio of surface area to volume (see Figure 9.4) can influence all aspects of the heat exchange equation: conduction, convection, radiation, and evaporation. Variation in the ratio is important in several contexts. A given animal may alter its exposed surface area to change heat flux. Dogs stretch out when hot to maximize conductive heat loss to the ground, but roll up when cold to minimize conductive heat loss to the air. Ratios of surface area to volume also come into play when comparing animals of different body dimensions or body mass.

The significance of body size, or more precisely, the ratio of surface area to mass, is apparent in many comparisons. An arctic wolf is about one-tenth the mass of a grizzly bear, but it has twice the ratio of surface area to volume. Although they live in similar niches, the arctic wolf incurs greater thermoregulatory costs because of its size. Similarly, when an animal grows, its body mass increases faster than its surface area. In general, larger animals lose heat more slowly and retain heat better than do small animals. The effects of body size and shape also manifest themselves in

animal evolution. *Bergmann's rule* states that animals living in cold environments tend to be larger than animals in warmer environments. *Allen's rule* states that animals in colder climates tend to have shorter extremities than animals in warmer climates. Thus, mammals or birds living in polar regions or high altitudes tend to be larger and shorter legged than individuals of the same species from more temperate regions. These rules of ecogeography apply to most of the mammals and birds studied to date, but have little relevance to animals that allow T_B to change.

An animal regulates heat exchange by altering the posture of the body to minimize or maximize the exposed surface area. Pythons will roll into a ball to conserve metabolic heat during digestion. When the python, approximately cylindrical in shape, rolls into a ball, its externally exposed surface area decreases by about 85%, greatly reducing heat loss.

Animals can also reduce effective surface area by huddling with other animals. Naked mole rats (Figure 13.4) live in burrows at relatively constant temperatures and have a very limited ability to use metabolism to control their body temperature. If housed in groups, they huddle when temperatures drop below about 22°C. This allows them to maintain a relatively constant T_B near 22°C. However, a solitary naked mole rat is unable to defend its T_B at low T_A. When prevented from huddling, its T_B closely reflects T_A, decreasing to as low as 12°C. From the perspective of the individual animal, huddling reduces heat by increasing T_A, replacing cold air with a warm neighbor. From the perspective of the colony, huddling works as a thermoregulatory strategy by reducing ratios of surface area to volume.

Figure 13.4 Naked mole rats

Insulation reduces thermal exchange

Internal and external insulation also reduce heat losses. Marine mammals have a thick layer of adipose tissue under the skin in the form of blubber. This lipid layer disrupts the flow of thermal energy from the core to the external surface of the animal. More commonly, animals use external insulation to reduce heat loss. Fur and feathers restrict the movement of molecules between the surface of the animal and the bulk phase of the environment. Heat is lost from the animal in proportion to the thermal gradient (ΔT) at the surface of the animal. Molecules of air or water in the insulation layer are warmed by the animal and then trapped within the insulation. The overall temperature gradient from the skin to the bulk phase is the same, but the distance is greater and the animal loses less heat to conduction. The fur also impedes the flow of fluids over the surface of the skin, so there is less convective heat loss.

The effectiveness of insulation depends on its thickness. When faced with cold temperatures, birds (or mammals) can change the orientation of the feathers (or fur) to alter the volume of air trapped within the coat. Similarly, animals that live in colder environments have thicker coats with greater insulating capacity (Figure 13.5). Some species change the thickness of the external insulation seasonally. Thick coats are a thermoregulatory burden in the warm season, so it is

beneficial to shed fur in spring. Since much of hair is composed of dead cells, the cost of rebuilding the coat when temperatures cool is minor in comparison to the metabolic costs the animal would incur trying to cool itself using physiological mechanisms. Mammals alter the nature of their fur coat seasonally, producing a greater density of hairs. Some birds, such as the ptarmigan, produce specialized feathers with an additional shaft to increase the feather density.

Though a main function of hair is thermal insulation, it can also serve other purposes. Male lions, for example, possess a thick coat of fur around the head region. Though it may provide the lion with some defense in male-male encounters, it also creates a thermal burden. Box 13.1, Evolution and Diversity: Lions' Manes Are Hot! describes the evolutionary and physiological trade-offs between thermal biology and sexual selection.

Thermal Strategies

Invertebrates are the most thermotolerant animals in each thermal niche. The hottest deserts are populated by myriads of insects, but only a few vertebrates. Invertebrates can also tolerate the coldest temperatures, often by entering an inactive, dormant state. Once stabilized in this state of "suspended animation," they can survive temperatures far colder than even the coldest natural environments. In contrast, only a few vertebrates, such as the wood frog, can survive subzero body temperatures, frozen in underground refuges.

The lay terms *cold-blooded* and *warm-blooded* fail to reflect the complexity of thermal strategies, which are properly described by two alternate sets of terms: poikilothermy versus homeothermy, or ectothermy versus endothermy.

Poikilotherms and homeotherms differ in the stability of T_B

The terms *poikilothermy* and *homeothermy* distinguish animals on the basis of the stability of T_B. A **poikilotherm** is an animal with a variable T_B—one that varies in response to environmental conditions. A **homeotherm,** in contrast, is an animal with a relatively constant T_B. Most homeotherms achieve a constant T_B using physiological processes to regulate the rates of heat production and loss.

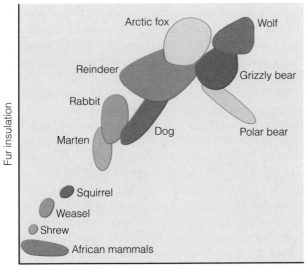

Figure 13.5 Insulation There is a direct relationship between the thickness of fur and its ability to act as insulation.

(Source: Modified from Wilmer et al., 2002)

BOX 13.1 EVOLUTION AND DIVERSITY
Lions' Manes Are Hot!

Lions live in prides, with one mature male and a harem of females. The lionesses do the parental care and all of the hunting, while the male guards over the pride, defending his position against any male challenger. An unmated female will choose a male that she perceives to have superior "quality," in terms of reproductive potential. The concept of mate selection is often difficult to interpret in terms of animal physiology. How does a female lion recognize the reproductive potential of a male lion from visible or behavioral traits? Recent studies suggest that one feature females assess is the size and color of the male's mane, the thick coat of hair that covers his neck and throat. Field observations suggest that female lions tend to choose mates with a long, dark mane. This trait may be a faithful signal of male quality because of the link between mane properties and thermoregulation.

Most African lions live in hot savanna conditions, where daytime temperatures can reach greater than 45°C. For these animals, the challenge is to keep cool, so the thick mane of males would seem to be counterproductive in relation to thermal physiology. Studies using infrared cameras support the presumption that males with dark, thick manes have a more difficult time shedding excess heat. Since the mane represents a thermoregulatory burden, why might it be subject to sexual selection? Researchers have investigated several potential explanations for this seemingly maladaptive trait.

There is little doubt that females use mane appearance in mate choice, but it is less clear why the thicker, darker manes are most desirable. The simplest theory is that the thick mane helps protect the vulnerable neck region of the male during the violent fights for dominance. For the female, a well-protected male is better able to protect the pride from invaders. From the male's perspective, the benefits of additional protection exceed the additional costs associated with thermoregulation. It is difficult to establish if long manes provide significant protection in a fight. An alternate explanation for the underlying reason for long manes is the handicap hypothesis. Much like the tail of a peacock, a thicker mane may advertise to females that the male has a robust physiology capable of coping with the additional physiological burden.

Independent of the underlying reason why a mane is a trait under sexual selection, a physiologist might ask how the properties of the mane of a dominant male are altered and how this reflects his dominance. The answer may lie in an understanding of the cellular basis of hair growth. Hair length is controlled by the cells of the hair follicle, and its growth rate changes in relation to environmental conditions: fast in the summer and slow in the winter. Individual hairs in humans grow for about five years, then remain static for another 12 weeks. At that point, a new hair from the same follicle is formed, pushing the old hair out. In principle, lions could grow longer hair by having hair grow faster or for longer periods of time. The growth rate is determined by the androgen steroid hormones: testosterone and dihydrotestosterone (DHT). The enzyme 5α-reductase converts testosterone to its more active DHT form. The levels of the pigment melanin, which is produced directly by the hair cells, determine the hair color. The process of pigmentation is also influenced by testosterone levels. Therefore, high androgen levels can cause a mane to be long and dark. Since androgens also control both sex drive and aggression, mane properties may reflect the propensity for aggressive behaviors that aid in mating and defense.

An interesting variation in the story of the lion's mane comes from the Tsavo lions. The Tsavo region of Africa is much hotter than other lion habitats, such as the Serengeti. A thick mane would be an even greater hindrance to male Tsavo lions. The Tsavo lions gained notoriety for two reasons. First, they were perceived to be more aggressive than other lions, even gaining a reputation as man-eaters. The white ghost lions of Tsavo, as they became known, were also remarkable in that they lacked a mane. Juvenile males throughout Africa are maneless, and it had long been thought that only the young males were being spotted in the Tsavo. Researchers lured lions from the dense vegetation and discovered that even the dominant males lacked manes. For these animals, the mane would be an insurmountable burden, due to both hotter temperatures and dense vegetation. Thus, in the Tsavo lions, evolution has led to a loss of the mane.

The physiologist might ask how these lions become maneless. Although some testosterone is needed for hair growth, excessive testosterone causes hair loss. It is not yet known if these males have higher testosterone levels than their Serengeti relatives. However, there is also a change in the social structure of prides that implicates testosterone. In the Serengeti lions, a pride possesses as many as four mature males, but most Tsavo prides have only a single mature male. Perhaps high levels of testosterone both cause hair loss and increase the aggressive nature of the Tsavo males, altering the social structure of the pride.

Reference

○ West, P. M., and C. Packer. 2002. Sexual selection, temperature, and the lion's mane. *Science* 297: 1339–1343.

The distinction between poikilotherm and homeotherm depends on both the properties of the animal and the nature of the environment. An animal could maintain a constant T_B by living in an environment with a constant T_A. For example, polar fish live in waters that are constantly cold, and by definition are homeotherms. However, their closest relatives live in oceans with temperatures that vary seasonally, and are therefore poikilotherms. Similarly, a goldfish in an indoor aquarium might never experience a change in T_B, but if it were moved outside to a pond its T_B would vary. The goldfish could arguably be called a homeotherm or a poikilotherm, depending on the situation. Since these terms depend more on the nature of the environment than the animal's physiology, the terms are not always useful in describing thermal strategies.

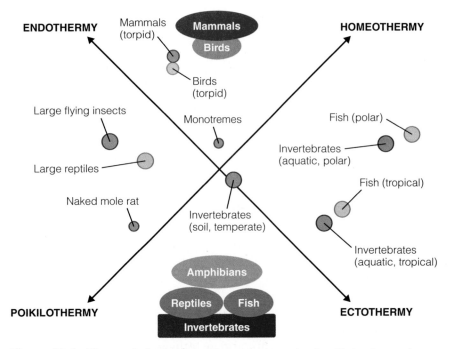

Figure 13.6 Thermal strategies Most animals can be classified as homeotherm or poikilotherm, or alternately, ectotherm or endotherm. This figure illustrates the many species whose thermal strategies combine elements of multiple strategies. For example, monotremes are less homeothermic and less endothermic than other mammals.

Ectotherms and endotherms differ in the source of body thermal energy

The terms *ectotherm* and *endotherm* distinguish animals by the physiological mechanisms that determine T_B. The environment determines the T_B of an **ectotherm**. If the ectotherm shows a variable T_B, then it might also be called a poikilotherm. An **endotherm** is an animal that generates internal heat to maintain a high T_B. Endotherms regulate T_B within a narrow range, but it need not be constant.

Both this and the preceding approach to classifying thermal strategies work effectively for most animals. Most birds and mammals can be classified as homeotherms, because T_B is stable, and also as endotherms, because metabolic heat elevates T_B. However, many animals are best described by a combination of terms (Figure 13.6). For example, the polar fish we described earlier in this chapter are homeothermic ectotherms; T_B is constant but determined by T_A. Monotremes, like other mammals, are endotherms, but maintain a lower T_B that is more variable. Proper use of these terms requires an understanding of the physiological capacities of the animal as well as an awareness of the thermal properties of its environment.

Heterotherms exhibit temporal or regional endothermy

Just how constant does T_B have to be for an animal to be considered a homeotherm? In actuality, most animals experience some variation in temperature, either spatially or temporally. Many endothermic animals place greater priority on maintaining certain anatomical regions within very narrow thermal ranges. Typically, homeotherms maintain the central nervous system and internal organs at a more constant temperature, while allowing the periphery to vary. The temperature of these deep, internal regions is often called the *core temperature*. Humans, for example, maintain a near-constant core temperature. However, regions of the human body can experience temperatures much lower than the core T_B. In the cold, humans change blood flow to allow hands and feet to cool to conserve internal heat. Males alter the position of the scrotum to keep spermatogenic tissue from overheating. However, human core T_B can also change under some circumstances. T_B can change in females during the reproductive cycle. It can rise several degrees as a result of a fever. In comparison to other animals, these are relatively minor regional and temporal differences in T_B, and a human is considered an endothermic homeotherm.

In contrast to humans, many other mammals and some birds can undergo dramatic, prolonged changes in T_B. When exposed to cold nighttime temperatures, T_B may decrease by several degrees (Figure 13.7). Hibernating mammals, such as ground squirrels and bats, allow T_B to drop for the winter months. Although these animals allow their bodies to cool, they are still considered endotherms because they produce and retain metabolic heat to maintain T_B above T_A. However, these endothermic animals are more precisely described as **temporal heterotherms**, to reflect the variability in T_B over time. Some ectothermic animals also fit the description of temporal heterotherms. Many large snakes, such as pythons, wind their bodies into a ball after they have ingested their prey. This helps the snake retain the metabolic heat produced by digestion. Temporal heterothermy is a strategy that has different benefits for endotherms and ectotherms. It allows an endotherm to conserve energy in cold temperatures by reducing the costs of thermoregulation. It provides an ectotherm with a period of accelerated metabolism to speed digestion, nutrient assimilation, and biosynthesis.

Most ectotherms rapidly lose their metabolic heat to the environment, and consequently cannot

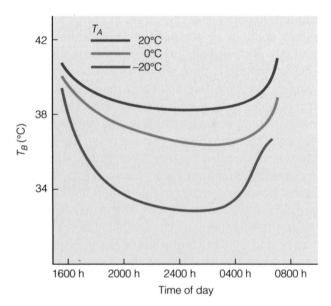

Figure 13.7 Short-term cooling in birds Many temperate birds allow their body temperatures to decrease when nighttime temperatures decrease. This strategy of temporal heterothermy saves metabolic energy.
(Source: Modified from Reinertsen and Haftorn, 1986)

elevate T_B much above T_A. However, a **regional heterotherm** can retain heat in certain regions of the body. Billfish, such as marlin and swordfish, are ectotherms but are also able to warm specific regions of the body. Their heater organs produce enough heat near the eye and optic nerves to improve visual clarity when they dive deep into cold waters (see Chapter 5: Cellular Movement and Muscles). Large pelagic fish possess countercurrent heat exchangers to conserve the heat of digestion within the body core (see Chapter 11: Digestion). Tuna and lamnid sharks are able to retain myogenic heat within the muscle. Warming of the red muscle increases metabolic capacity and may improve contractile performance during swimming (see Chapter 12: Locomotion). Thermal gradients occur within the bodies of many animals, but these regional heterotherms have specific physiological mechanisms to produce and retain heat regionally.

Although most insects are ectotherms, some species are regional heterotherms, others temporal heterotherms, and some species are both, depending on the time of year. The largest of flying insects, such as bumblebees, large moths, and cicadas, have a very high metabolic rate in the flight muscles. Thoracic temperature in a large flying insect can increase by more than 10°C, even while other regions of the body remain near T_A. Interestingly, these animals are also able to modulate heat production. Prior to flight they initiate thermogenic pathways to warm the thorax. When flight commences, they can alter heat exchange to maintain near-constant thoracic temperatures during flight, even when T_A is variable (Figure 13.8). Social insects use huddling as a means of controlling the temperature of the colony. Honeybees survive the cold winters by forming tightly crowded clusters. An individual bee in the colony is uniformly warm or uniformly cold, depending upon its position in the cluster. The clusters act like the body of a regional heterothermic animal. The "core" body heat of the colony is generated by the bees that are located near the center of the cluster. The outermost bees (mantle bees) act as insulation.

Animals have a characteristic degree of thermotolerance

Physiological strategies for coping with temperature differ in ectotherms and endotherms. For ectotherms, a change in T_A alters T_B and directly

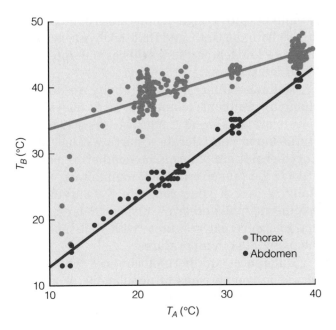

Figure 13.8 Insect heterotherms Many large insects are able to conserve metabolic heat that arises when their flight muscles are activated during flight. This warms the thorax while the rest of the body remains near ambient temperature, an example of regional heterothermy.
(Source: Based on Harrison et al., 1996)

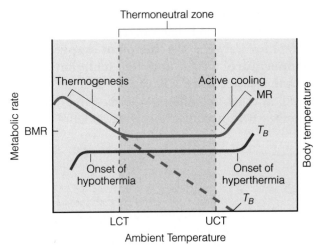

Figure 13.9 Zones of thermal effects of a resting homeotherm Homeothermic endotherms maintain near-constant body temperature over a wide range of ambient temperatures (purple line). Once ambient temperatures decrease below the lower critical temperature (LCT), the animal must increase its metabolic rate (MR) to generate heat to help maintain a constant T_B. By extending the line explaining the metabolic rate below LCT to the x-axis, the body temperature (T_B) can be obtained as the intercept. Below a certain point, the animal can no longer maintain a constant core temperature and hypothermia results. When ambient temperatures increase past the upper critical temperature (UCT), the animal increases metabolic rate to shed heat. At still higher temperatures, the animal can no longer defend its body temperature and hyperthermia results.

changes the rates of many biological processes. In contrast, an endotherm responds to a change in T_A by inducing a compensatory regulatory response. Despite the differences, both endotherms and ectotherms incur physiological costs and consequences when environmental conditions change.

The effects of temperature can be defined in terms of its impact on animal function. An animal typically spends most of its life in a range of temperatures that is optimal for physiological processes. The **thermoneutral zone** of a resting homeothermic endotherm is the range of ambient temperatures where metabolic rate is minimal, which is considered the basal metabolic rate, or BMR (Figure 13.9). If temperatures rise to a point called the **upper critical temperature** (UCT), the metabolic rate rises as the animal induces a physiological response to prevent overheating. If the temperature falls below a **lower critical temperature** (LCT), the metabolic rate rises to increase heat production. For many animals, the T_B can be predicted from the extrapolation of the line that describes the metabolic rate at temperatures below LCT. When

faced with a hypothermic challenge, animals may reduce T_B to maintain homeostasis at metabolic rate. In general, these compensatory responses at high T_A or low T_A allow the animal to maintain a constant T_B, but beyond a point, the animal cannot sustain a constant T_B.

The concept of a thermoneutral zone does not apply to animals that alter T_B, but ectotherms also have ranges of T_A (and T_B) where growth and reproduction are optimal. Animals actively seek out their *preferred temperature*, a T_A that is within its range for optimal function. At low temperatures, all developmental processes slow because the lower T_A reduces the rate of metabolic reactions. Higher temperatures damage molecules, cells, and tissues, jeopardizing an animal's health. Researchers can assess the thermal tolerance of an ectotherm or a poikilotherm by transferring an animal from its acclimation temperature to a challenging temperature and assessing survival. The *incipient lethal temperature* is the temperature that has a 50% probability of killing the fish within an identified period. The range of tolerance is the difference between the **incipient upper lethal temperature** (IULT) and the **incipient lower lethal temperature** (ILLT). For ectotherms and poikilotherms, the ability to tolerate temperature changes with acclimation history (Figure 13.10).

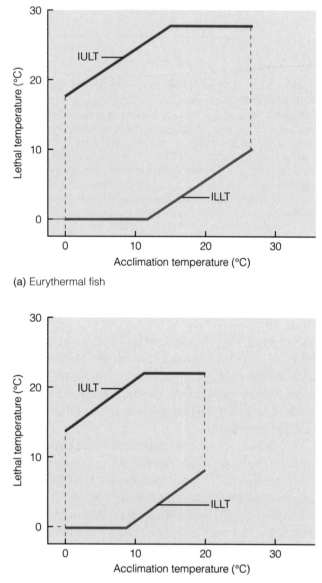

(a) Eurythermal fish

(b) Stenothermal fish

Figure 13.10 Temperature polygon Acclimation affects the incipient upper lethal temperature (IULT) and incipient lower lethal temperature (ILLT) for ectotherms and poikilotherms. The tolerance of an animal is reflected in the area of the polygon created by joining the upper line (IULT, in red) and the lower line (ILLT, in blue). Analysis of a eurythermal fish **(a)** yields a larger polygon than that of a stenothermal fish **(b)**.

Acclimation to a high temperature tends to increase both the ULLT and ILLT. Likewise, acclimation to a low temperature reduces the upper and lower lethal temperature points.

Animals differ in their ability to tolerate changing ambient temperature. A **eurytherm** can tolerate a wide range of T_A, whereas a **stenotherm** can tolerate a narrow range of T_A. Eurythermal endotherms/homeotherms possess a wide thermoneutral zone, maintaining a constant T_B over a wide range of T_A; eurythermal ectotherms/poikilotherms display a large thermal tolerance polygon area, with well-separated incipient lethal temperatures.

Differences in thermotolerance can be observed in comparisons of populations or species that have evolved in regions separated by latitude or altitude. The ability of an animal to tolerate a lower T_A than its competitor allows the tolerant animal to expand into a colder environmental niche. Many closely related animals have distinct differences in thermal preferences that contribute to their geographical distributions. Latitudinal patterns are common in fish species in both marine and freshwater. Closely related species of barracuda, for example, live at specific latitudes along the Pacific coast with a characteristic average T_A. From north to south, one species gradually replaces another once the average water temperature changes by only 3–8°C. There are also altitudinal patterns seen with terrestrial animals. Many bird species exist in high-altitude and low-altitude populations, each with physiological specializations and morphological differences. The thermal environment resulting from the combination of altitude and latitude also determines the range of many amphibians. Andean tree frogs (*Hyla andina*) can be found at low elevation far from the equator, but closer to the equator they can live at higher altitudes.

The genetic basis of a difference in thermotolerance is not always clear. We can often determine why levels or properties of a single protein differ in two animals in relation to temperature. However, the underlying basis for complex differences in thermal physiology is more complex. For example, two species of Siberian hamsters, *Phodopus campbelli* and *P. sungorus*, differ in thermal biology in terms of morphology, insulation, behavior, and physiology. Although these are very closely related species, they last shared a common ancestor more than 2 million years ago. A complex trait such as fur density depends on multiple genes, many cell types, and networks of genetic regulators. Furthermore, the two species may have many genetic differences, but only some of these may influence their thermal biology.

⊙ | CONCEPT CHECK

1. What are the sources and sinks in an equation describing thermal balance?
2. What is the difference between an endotherm and an ectotherm?
3. What is the difference between a homeotherm and a poikilotherm?
4. What is the difference between a regional and a temporal heterotherm?

Coping with a Changing Body Temperature

Although many ectotherms and poikilotherms live in thermally stable environments—underground burrows, tropical rain forests, the deep sea, or a homeotherm's intestine—others must cope with frequent and dramatic changes in T_B. Because of the effects of temperature on macromolecular function and metabolism, ectotherms and poikilotherms must either tolerate or compensate for the complex, often deleterious, effects of changing temperature.

Macromolecular Structure and Metabolism

Of the four classes of macromolecules, only proteins and lipids are substantially affected by temperature over the normal range encountered by animals. Weak bonds (van der Waals forces, hydrogen bonds, and hydrophobic interactions) govern the interactions within and between these macromolecules. Each type of bond has a characteristic response to temperature. Whereas hydrogen bonds and van der Waals forces are disrupted at high temperature, hydrophobic interactions are stabilized at high temperature. Thus, the effects of temperature on macromolecular structures depend on the relative importance of each type of bond. The effects of temperature on a biological process can be as-

sessed using an *Arrhenius plot* (see Box 13.2, Mathematical Underpinnings: Evaluating Thermal Effects on Physiological Processes Using Q_{10} and Arrhenius Plots). Researchers can use this approach to study how temperature affects the structure and function of macromolecules, enzymatic reactions, and complex processes, such as metabolic rate.

Animals remodel membranes to maintain near-constant fluidity

In Chapter 2, we discussed the structure of cellular membranes and the importance of membrane fluidity. Van der Waals forces hold membrane lipids together. Although the interactions between phospholipids are strong, the membrane must also remain fluid enough to allow proteins to rotate and diffuse laterally within the membrane. Low temperatures cause membrane lipids to solidify, which impairs protein movement. Conversely, high temperatures liquefy the membrane, which can compromise its integrity and reduce its effectiveness as a permeability barrier. Cells regulate the balance between the solid *gel* state and the liquid *sol* state.

Temperature exerts effects on membranes through both protein function and phospholipid fluidity. The effects on membrane protein function can be assayed using kinetic analyses. For example, the Na^+/K^+ ATPase interacts with membrane lipids during the transport process. The activity of the transporter can be measured at a series of temperatures, then plotted in relation to temperature. A change in membrane fluidity typically results in a breakpoint in the Arrhenius plot of membrane protein function, as shown in Box 13.2.

Membrane fluidity is measured in biological membranes using a dye (diphenyl hexatriene) that changes in optical properties in relation to its freedom to move within the membrane (Figure 13.11). When membranes from different species are compared, each exhibits a decrease in fluidity (measured as a change in optical properties) when the membrane is cooled. Taking into consideration the differences in thermal niche, this analysis shows that animals produce membranes that exhibit the same fluidity at the natural temperature. This observation is analogous to the conservation of K_m seen in enzymes from animals in different niches (see Chapter 2). The same pattern is seen when an

For many physiological processes, a 10°C increase in temperature typically doubles or triples the rate of the process. We can describe these effects of temperature on reaction velocity mathematically by the Q_{10}. The Q_{10} is essentially the ratio between reaction rates at two temperatures, adjusted for a 10°C temperature difference. It is calculated as

$$Q_{10} = \left[\frac{K_2}{K_1}\right]^{[10/(T_2 - T_1)]}$$

where the rates of a reaction (K) are compared at two temperatures (1 and 2). Thus, if a rate of 10 units/min (K_1) was observed at 15°C (T_1), and a rate of 20 units/min (K_2) at 25°C (T_2), then

$$Q_{10} = \left[\frac{20}{10}\right]^{[10/(25 - 15)]} = 2^1 = 2$$

The Q_{10} for a process is the best way to express the influence of temperature on reaction rates, but a better approach to exploring the mechanism of action is through an **Arrhenius plot**. In the late 1800s, the chemist Svante Arrhenius described a mathematical approach to exploring the impact of temperature on macromolecular processes. We now use his approach to study processes such as enzymatic reactions, diffusion of molecules, and lipid membrane phase transitions. The sensitivity of a reaction to temperature reflects the activation energy (E_a) of the process. The Arrhenius equation describes the relationship between the activation energy, temperature, and the rate of the process under study:

$$k = Ae^{(-E_a/RT)}$$

More often, the Arrhenius equation is shown as

$$\ln(k) = \ln(A) - E_a/(RT)$$

where k is a rate coefficient, R is the gas constant (8.31447×10^{-3} kJ/K per mol), T is temperature (in degrees Kelvin), A is called the pre-exponential factor, and E_a is the activation energy (kJ/mol).

Let's say that a researcher was interested in how temperature influenced the rate of an enzymatic reaction. She would vary temperature over a range of interest and measure enzymatic rates. The data she collected could be plotted on a graph with axes chosen from a rearrangement of the Arrhenius equation that generates a linear equation ($y = mx + b$):

$$\ln(k) = -E_a/R \times (1/T) + \ln(A)$$

Plotting $\ln(k)$ versus $1/T$ gives a slope of $-E_a/R$ and a y intercept of $\ln(A)$.

The accompanying figure illustrates two potential outcomes from an Arrhenius plot. For the green line, the data fall along a straight line. The slope of the line reflects the activation energy of the reaction. The purple

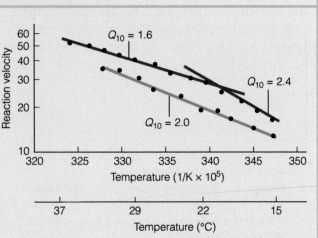

lines show data where one line fits the data at low temperatures, but a different line fits the relationship at high temperatures. The point where the two lines cross is called the **breakpoint**. Since the slope differs between the two lines, we can infer that different activation energies govern the reaction over each temperature range. In many cases, this is due to a mechanistic transition from one state to another state. If the process under consideration is membrane fluidity, for instance, the breakpoint might reflect the transition from a liquid to a solid phase. If the process is an enzymatic reaction, the breakpoint might occur at a temperature where a critical bond is broken, converting the enzyme from an efficient catalyst to a less-efficient catalyst or denatured enzyme.

The versatility of the Arrhenius plot allows researchers to describe the thermal behavior of any simple or complex process. However, the reasons for particular relationships are more difficult to ascertain in complex systems. Thermal effects on membranes are often difficult to assess because of the considerable heterogeneity of the membrane. Lipid rafts, for example, are cholesterol-rich regions of the cell membrane that often accumulate distinct phospholipids. Temperature will have a different effect on the fluidity of these regions in comparison to the bulk phase of the membrane. Similarly, many integral membrane proteins accumulate different types of lipids. For example, the mitochondrial enzyme binds cardiolipin molecules within the inner mitochondrial membrane. Changes in the bulk phase of the membrane do not necessarily reflect changes in the lipid membrane in direct contact with the proteins of interest. Even more complex processes, such as metabolic rate, are really the sum of many simple processes, each with their thermal sensitivity and unique Arrhenius equation.

Suggested Reading

○ Metz, J. R., E. H. van den Burg, S. E. Bonga, and G. Flik. 2003. Regulation of branchial Na(+)/K(+)-ATPase in common carp *Cyprinus carpio* L. acclimated to different temperatures. *Journal of Experimental Biology* 206: 2273–2280.

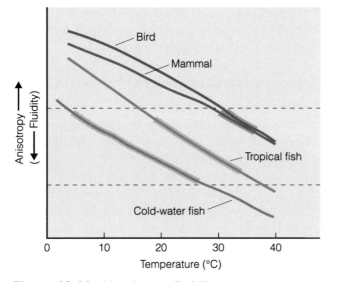

Figure 13.11 Membrane fluidity Membranes are treated with a dye (diphenyl hexatriene) with optical properties that change in relation to membrane fluidity. Anisotropy is an optical property that reflects the ability of a dye to alter the behavior of plane polarized light. Anisotropy is inversely related to fluidity; at warmer temperatures, a decrease in anisotropy reflects an increase in fluidity. Animals that live in different environments produce membranes that possess a similar fluidity at their normal range of temperatures (indicated by the thickened portion of the lines).

(Source: Adapted from Logue et al., 2000)

animal is acclimated to different temperatures. Ectothermic animals reduce the deleterious effects of temperature by changing the composition of their membranes. In this process, called **homeoviscous adaptation**, cells remodel membranes to preserve fluidity. Three mechanisms target phospholipids (Figure 13.12), and a fourth mechanism alters cholesterol content.

1. *Fatty acid chain length.* Phospholipids with short chain fatty acids cannot form as many interactions with adjacent fatty acids and therefore are highly mobile. The effectiveness of chain shortening depends upon the fatty acid position on the phospholipid. Due to the three-dimensional structure of a phosphoglyceride, a short chain fatty acid in position 1 makes a greater contribution to enhancing fluidity than does the same fatty acid in position 2.

2. *Saturation.* Double bonds create a kink in the fatty acid chain that prevents effective bond formation with other fatty acids. With fewer bonds between fatty acid chains, the membrane is more fluid. For example, pure stearic acid (C18:0) becomes liquid only at temperatures above 69°C, whereas oleic acid (C18:1) is liquid at 12°C. The position of the double bond is also critical. A double bond near the midpoint of the fatty acid chain (as with oleic acid) is more effective than a double bond near the end of the fatty acid chain.

3. *Phospholipid classes.* The difference in the shape of the polar head groups alters the ability of the phospholipids to interact at the surface of the membrane. Phosphatidylcholine (PC) is more common in membranes of warm-acclimated cells, whereas phosphatidylethanolamine (PE) is more common in cold-acclimated cells. The ratio of PC to PE decreases in the cold acclimation and adaptation.

4. *Cholesterol content.* A pure phospholipid bilayer is mostly fluid at high temperature and mostly solid at low temperature. Cholesterol added to fluid phospholipid bilayer has little effect on fluidity. If the same membrane is cooled, cholesterol tends to prevent it from so-

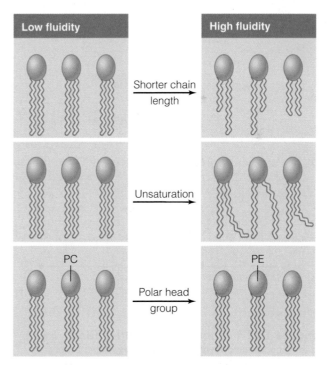

Figure 13.12 Phospholipid properties and membrane fluidity Cells change the fluidity of membranes by altering the composition of membrane phospholipids.

lidifying. Put another way, cholesterol tends to make a membrane more fluid when external conditions otherwise encourage a transition to a gel phase.

Cells use two general pathways to modify membrane composition in response to temperature: *in situ* modification and *de novo* synthesis. Both pathways require cells to modify the properties of the fatty acids within the fatty acid pool using suites of enzymes that elongate, shorten, saturate, and desaturate fatty acids. Since these enzymes begin with fatty acids derived from the diet, the nature of the diet also affects the profile of fatty acids within the membrane.

Enzymes alter the structure of individual phospholipids directly within the membrane (Figure 13.13). First, phospholipase A removes an acyl chain from membrane phospholipids to form a lysophospholipid. Next, lysophospholipid acyltransferase uses a more appropriate fatty acid (in the form of fatty acyl CoA) to rebuild the phospholipid.

More commonly, membranes are remodeled by endocytosis and exocytosis (see Figure 13.14). The old membrane is removed using endocytosis.

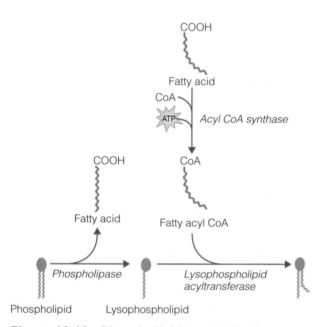

Figure 13.13 Phospholipid remodeling Cells can remodel the phospholipids directly within membranes by removing a fatty acid. A phospholipid is rebuilt by lysophospholipid acyltransferase, which attaches another fatty acid produced by the cell. The fatty acid must first be activated by the esterification of coenzyme A.

Phospholipids are synthesized *de novo* within the endoplasmic reticulum, then packaged into vesicles that fuse with cellular membranes.

Temperature changes enzyme kinetics

Temperature affects protein structure and function in complex ways. Changes in temperature alter the number of bonds that form within and between molecules. Even minor changes in protein structure can have important effects on protein function. In enzymes, for example, these structural effects manifest as changes in catalytic properties. First, changes in weak bonds can alter the three-dimensional structure of the enzyme. For instance, warm temperatures could break bonds that are necessary to fold the protein in a way that forms the active site. Second, temperature can alter the ionization state of critical amino acids within the active site. For instance, the amino acid histidine is important in many active sites, and changes in histidine protonation state can alter enzyme substrate affinity. Any increase or decrease in K_m could be disruptive. Third, temperature can alter the ability of the enzyme to undergo the structural changes necessary for catalysis. Enzymes must be rigid enough to maintain the proper conformation, but flexible enough to undertake conformational changes during catalysis. Thus, temperature can affect enzyme kinetics through effects on maximal velocity (V_{max}) or affinities for substrates (K_m), allosteric activators (K_a), and inhibitors (K_i). When animals experience a change in T_B, they may either tolerate the effects on enzyme kinetics or alter metabolic regulation to compensate.

Biochemical reactions are accelerated by higher temperature and reduced at lower temperature. Recall from Chapter 2 that the rate of a chemical reaction depends on the proportion of molecules within the system that possess energy equal to or greater than the activation energy (E_a). As temperature increases, the average kinetic energy of the substrates increases and a greater proportion of molecules has sufficient energy to be converted to products, causing the enzyme velocity to increase (see Figure 2.16). For most enzymes working over a biologically relevant range of temperatures, an increase of 10°C results in a two- to threefold increase in reaction velocity. Recall from Box 13.2 that this implies a Q_{10} value of 2–3. Q_{10} can be calculated for simple reactions such as an enzymatic step, or complex processes such as metabolic rate.

Consider how temperature affects the V_{max} of lactate dehydrogenase (LDH) in the muscle of a desert lizard as it experiences daily transitions in T_B. Over the course of a single day, the total number of LDH enzyme molecules does not change appreciably, but their catalytic activity changes with temperature. LDH maximal activity (V_{max}) typically doubles when temperature is increased by 10°C (that is, its Q_{10} is 2). Let's begin at a T_B of 40°C, and assume that the muscle of the lizard has 400 units (U) of LDH per gram of tissue (One U of enzyme can convert 1 μmol of substrate to product each minute.) With $Q_{10} = 2$, a decrease in temperature from 40°C to 30°C causes the LDH enzymes to operate at only one-half the velocity, giving a V_{max} of 200 U/g. Similarly, at 20°C the LDH V_{max} is 100 U/g, and at 10°C only 50 U/g. Over the course of a single day, from the midday heat to the cool evening, a desert lizard may have to cope with an eightfold change in its LDH V_{max} activity as a result of changes in T_B.

It is easy to imagine how an eightfold reduction in LDH capacity might severely impair the capacity to produce ATP by glycolysis. How does an animal cope with such dramatic reductions in the rates of enzymatic reactions? The simple answer is that the rates of ATP synthesis decline in parallel with the rates of ATP utilization, with each step exhibiting a Q_{10} ranging from 2 to 3. Put another way, the animal can tolerate lower rates of muscle ATP production because it slows down and needs less ATP for muscle activity. However, it is important to recognize that $Q_{10} = 2$ is quite different from $Q_{10} = 3$. If a 10°C decrease in temperature caused ATP supply to decrease threefold when ATP demands decreased only twofold, the tissue would be depleted of ATP within seconds or minutes. Superimposed on the Q_{10} effects are numerous layers of metabolic regulation that ensure that energy metabolism remains in homeostasis. If an individual enzyme is more sensitive to temperature than are other enzymes in the pathway, the cell has several options to increase flux through that step. For

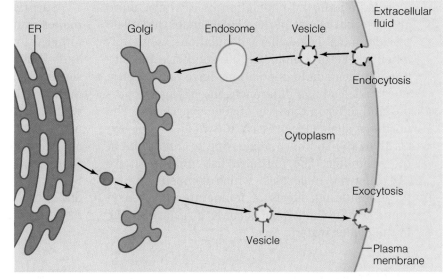

Figure 13.14 Membrane remodeling Cell membranes are constantly remodeled by endocytosis and exocytosis. When temperature decreases, the cell produces vesicles possessing phospholipids with fatty acids that are shorter and more unsaturated than those in the cell membrane. Over time, the cycles of endocytosis and exocytosis remove undesirable phospholipids, replacing them with more desirable phospholipids.

example, it may increase substrate concentrations or stimulate the enzyme with allosteric regulators.

Evolution may lead to changes in enzyme kinetics

When animals are exposed to suboptimal temperatures for generations, there is the possibility of evolutionary changes in the genes encoding enzymes. We draw again on research with LDH for examples of evolutionary changes that cause differences in enzyme kinetics as well as enzyme synthesis.

Mutations may lead to structural changes in the enzyme that impart a favorable difference in enzyme kinetics. As we learned in Chapter 2, lowering temperature increases the affinity of LDH for its substrate pyruvate (see Figure 2.43). Evolution has led to a fine-tuning of enzyme properties such that subtle structural differences allow each species to possess a similar K_m at its respective normal T_A. This strategy, called *conservation of K_m*, is commonly seen when we compare the effects of temperature on the enzyme kinetics of different animals.

Alternately, evolution may lead to mutations in the promoter for an enzyme, causing a change in the level of gene expression of an otherwise unchanged enzyme. Killifish live along the eastern coast of North America from Newfoundland to

Florida. Within the population as a whole, there are different alleles of the LDH-B gene. One allele predominates in northern populations, while another allele predominates in southern populations. Intermediate populations have both alleles. These alleles have differences in enzyme properties and differ in the level of gene expression. The northern allele is expressed at twofold higher levels than the southern allele, due to mutations in the promoter. The northern fish produce more LDH enzyme molecules, which compensates for the debilitating effects of temperature on enzymatic activity that would occur as a result of living in the colder waters.

Ectotherms can remodel tissues in response to long-term changes in temperature

Many ectothermic animals remodel their cellular machinery to mitigate the effects of variation in T_B. In the laboratory, where the researcher changes only T_A, this remodeling process is called *thermal acclimation*. In the natural world, seasonal transitions in temperature are accompanied by other environmental changes and the response of the animal to complex seasonal changes is called *acclimatization*. In winter, photoperiods get shorter, food may be less abundant, and oxygen levels may change. The complexity of these seasonal environmental changes makes it difficult to link remodeling with the temperature. On one hand, there is uncertainty about the trigger for the remodeling process; is the change initiated by changes in temperature, or by some other factor, such as photoperiod? On the other hand, it is not always clear that the remodeling itself serves to compensate specifically for temperature.

Temperature-dependent remodeling involves combinations of quantitative and qualitative strategies. In Chapter 12, we discussed how ectotherms remodel their muscles in response to temperature. Low temperature may increase the number of mitochondria in muscle, or trigger the hypertrophic growth of the heart. This is an example of a *quantitative* strategy; there is simply more of the same machinery. Muscles can also alter the types of proteins they use to build the contractile machinery. For instance, animals express different myosin isoforms in winter and summer—an example of a *qualitative* strategy.

Surprisingly little is known about the hormones and signaling pathways that cause an ectotherm to remodel its tissues during acclimation and acclimatization. Cold-sensing and warm-sensing neurons are important for detecting temperature, but the links to gene expression are not well known. In some cases, seasonal changes in physiology that mitigate the effects of temperature are triggered by changes in the photoperiod. In Chapter 7: Functional Organization of Nervous Systems, we discussed the importance of the various photoperiod signaling pathways that act through the hypothalamus and pineal glands.

Life at High and Low Body Temperatures

Animals that can tolerate extreme temperatures can invade and colonize niches that are underexploited by their competitors. Ectothermic animals exposed to thermal challenges must possess mechanisms to mitigate the effects of temperature on macromolecular structure and metabolism. In contrast, endothermic animals survive thermal extremes using complex regulatory pathways to maintain a constant T_B. Their existence at extremes is a testament to their physiological capacity to resist the effects of T_A.

Some enzymes display cold adaptation

Earlier in this chapter we discussed how relatively subtle differences in T_A can lead to evolutionary changes in enzyme structure and gene expression. However, the need for enzymatic structural modification is much more pronounced at thermal extremes, particularly at the subzero temperatures encountered in polar seas. *Psychrotrophs* are organisms that thrive in the extreme cold, in contrast to *mesotrophs* that live at more moderate temperatures. Animal psychrotrophs, including polar invertebrates and fish, remain active at body temperatures near the point of freezing. Many psychrotrophic organisms possess cold-adapted proteins that function optimally at very low temperatures. Although these enzymes are more stable in the cold, they are rapidly inactivated at slightly higher temperatures.

The catalytic and structural differences between enzymes of psychrotrophs and mesotrophs can be traced to the weak bonds that stabilize enzyme structure. Enzymes undergo pronounced changes in three-dimensional shape during the catalytic cycle, known as protein *breathing*. During these transitions in folding, weak bonds break and form. When temperatures decrease, most of these weak bonds are strengthened, stabilizing the protein in a form that occupies a smaller volume. In this conformation, it is much harder for the protein to breathe, and consequently enzymes in the cold are less efficient. The psychrotroph enzyme has fewer weak bonds stabilizing its structure; it occupies a larger volume and has an easier time breathing during catalysis. The reduced stability allows it to function better in the cold, but makes it vulnerable to temperature-dependent unfolding. In comparison to mesotroph enzymes, cold-adapted enzymes are more efficient enzymes at low temperatures, but inferior enzymes at high temperatures.

Unique loss-of-function mutations also occur in polar animals. As discussed in Chapter 10: Ion and Water Balance, many Antarctic fish have lost the ability to express functional oxygen-binding proteins, such as hemoglobin and myoglobin. These fish can survive without these oxygen carriers because they have low metabolic rates and the surrounding polar waters are rich in oxygen.

There are many such examples of thermal adaptations of individual selected genes in polar animals. However, more controversial is the question of whether or not polar animals have a fundamentally different organization of metabolism as a result of evolution in the extreme cold. Early studies suggested that polar animals had metabolic rates that were much higher than the metabolic rates of temperate animals measured near 0°C. These observations were used to support a theory that became known as *metabolic cold adaptation*. It was proposed that thousands of years in the extreme cold led to evolutionary changes that provided these polar animals with an ability to elevate their metabolic rate. Even with years of study it remains unclear whether metabolic cold adaptation is a real phenomenon. The earliest studies were based on comparisons of goldfish and arctic cod. Now that more species have been analyzed using more sophisticated technologies, it seems less likely that metabolic cold adaptation occurs as a general phenomenon. Nonetheless, many studies have identified evolutionary differences and physiological peculiarities in some polar animals.

Stress proteins are induced at thermal extremes

Many proteins are best suited to function over narrow ranges of temperature that span the biological range of the animal. During the normal structural change that occurs when a protein breathes, the protein is vulnerable to further changes in structure. Occasionally, the protein can unfold or misfold into a nonfunctional conformation. This denatured protein must be repaired or cleared from the cell before it disrupts other cellular functions. Denaturation is a normal process, and cells are able to detect and remove denatured proteins using pathways of protein quality control. These pathways function throughout the lifetime of a cell, but become even more important during times of thermal stress when denatured proteins can accumulate and kill the cell.

Recall from Chapter 2 that heat shock proteins (Hsp's) are molecular chaperones that use the energy of ATP to catalyze protein folding after translation. Chaperones can also help refold proteins that have become denatured as a result of thermal stress. Many cells exposed to extreme temperatures undergo a *heat shock response,* which leads to a dramatic increase in the levels of specific proteins that help repair damaged proteins. During a heat shock, the cell undertakes a rapid increase in the synthesis of several critical Hsp's. The cell can halt the transcription and translation of other genes, sparing biosynthetic resources for Hsp synthesis. It stimulates the expression of the Hsp genes by activating a heat shock factor (HSF), a transcription factor that binds to the heat shock elements in the promoters of genes for heat shock proteins. Although there is still some uncertainty about the exact mechanism of activation of HSF, the trigger for the process is thought to involve damaged protein (Figure 13.15). In the absence of thermal stress, most of the cellular HSF is bound to Hsp70 as inactive monomers. When the cell is stressed, the chaperones are lured away from HSF by damaged proteins. The released HSF can then form trimers, which in turn bind the heat shock element on the Hsp genes, activating them. Once the damaged proteins are repaired, Hsp70 is free to bind HSF monomers and reverse the transcriptional activation.

The Hsp response is central to the ability of ectothermic animals to survive brief periods of

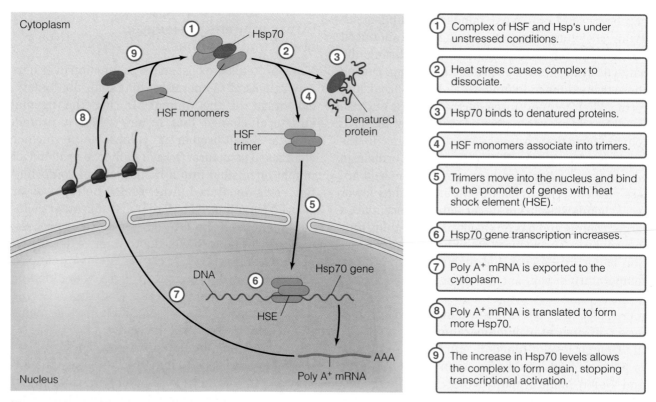

Figure 13.15 **Heat shock response**

extreme temperature that often occur within their natural environments. For most species, the Hsp response is induced at temperatures only a few degrees above the typical thermal range. This powerful protective process may be central to the evolution of thermal sensitivities and thermal ranges (see Box 13.3, Methods and Model Systems: Heat Shock Proteins in *Drosophila*). Interestingly, some species have lost their ability to mount a heat shock response. Antarctic fish have lived for thousands of years at −1.96°C. At some point, the species experienced genetic changes that disrupted the capacity to invoke a heat shock response. Since the Antarctic waters remain very constant in temperature, these mutations have no deleterious consequences to the animals. However, when taken out of their natural environment, these fish rapidly succumb to temperatures only a few degrees above 0°C.

Ice nucleators control ice crystal growth in freeze-tolerant animals

Ectotherms that live at freezing temperatures use two strategies to survive the cold: freeze-tolerance and freeze-avoidance. Freeze-tolerant animals al-low their tissues to freeze and even encourage ice to form in the body. Animals that avoid freezing use behavioral and physiological mechanisms to prevent ice crystal formation and growth. To understand why ice is so dangerous, let's consider what happens to water molecules as temperatures decrease.

The freezing point of pure water is 0°C. This is the temperature at which ice could form if enough water molecules cluster together to begin an ice crystal. Below the freezing point, water is on the verge of freezing, awaiting an event that triggers ice formation. When water is below its freezing point, but not yet frozen, it is considered **supercooled**. Pure water, left undisturbed, can be supercooled to almost −40°C before ice forms spontaneously. The trigger for ice formation is a cluster of water molecules that act as a seed for an ice crystal. Alternatively, a macromolecule in solution can act as a **nucleator**, seeding ice crystal formation. Once the ice formation begins, water molecules bind to each face of the growing crystal to create a complex three-dimensional structure.

Ice crystals forming within a tissue have two deleterious effects. First, since ice crystals have points and sharp edges, the growing ice crystal can

BOX 13.3 **METHODS AND MODEL SYSTEMS**
Heat Shock Proteins in Drosophila

The best-studied Hsp's are from the Hsp70 family. Each subcellular compartment has Hsp70 proteins that help fold proteins into the proper conformation and target misfolded proteins for degradation. Mitochondria have Grp75, the cytoplasm has Hsc70, and the endoplasmic reticulum has Bip. The namesake of the family, Hsp70, is produced by cells mainly under stressful conditions. When temperatures rise to dangerous levels, cells dramatically induce synthesis of Hsp70. It is produced in the cytoplasm, where it refolds proteins that have been denatured in response to elevated temperatures. The ability to mount a heat shock response is central to the thermotolerance of animals. Genetically modified cells that lack an ability to induce Hsp70 are very sensitive to thermal stress. Since this gene is essential for thermotolerance, many researchers have studied whether variation in Hsp70 gene expression is central to the differences in thermotolerance among animals.

Many studies of thermotolerance and Hsp70 have been performed on the fruit fly, *Drosophila*. If Hsp70 helps an animal survive heat stress, then it might be reasonable to hypothesize that an animal could benefit from greater expression of Hsp70. Laboratory studies have provided important insight into the links between Hsp70 gene expression, thermotolerance, and evolution. In one study, lines of flies were manipulated to possess extra copies of Hsp70 genes. Larvae from these flies had greater thermotolerance, demonstrating the importance of Hsp70 to thermotolerance. In other experiments, lines of flies were exposed to high temperatures for generations to see if natural variations in the Hsp70 genes within a population could be subject to natural selection. Within only a few generations the average thermotolerance of the flies increased. These thermotolerant flies were able to induce Hsp70 to higher levels than thermosensitive flies. Surprisingly, the difference in Hsp70 gene expression between thermotolerant and thermosensitive flies was never more than 15%. These data argue that Hsp70 may be impor-

tant in thermotolerance, but that other physiological factors may also play roles.

These studies also showed that flies with a robust heat shock response also had lower fecundity. In other words, superior thermotolerance comes with an evolutionary cost. In a thermostable environment, flies with lower levels of Hsp70 could outcompete flies with higher Hsp70 levels because of their greater fecundity. However, at more thermally challenging conditions, the lower fecundity is offset by the greater thermotolerance. These laboratory studies also reflected the nature of the evolution of thermotolerance in the natural world. Wild populations of *Drosophila* from around the globe exhibit a wide range in thermotolerance. In most populations the natural ability to survive thermal stress correlates with the levels of Hsp70 gene expression. For example, flies in Evolution Canyon in Israel occur in separate populations that occupy the north- and south-facing slopes. The flies that live on the south-facing slopes, which are hotter and drier, have a stronger heat shock response. The flies on the north-facing slope have a weaker heat shock response due to a disruption of the promoter of one of the Hsp70 genes. Since these two slopes are only hundreds of meters apart, individual flies probably move between the two populations. Thus, natural selection acts to ensure that the allelic differences between populations are retained. Similar studies on other *Drosophila* populations showed that flies with lower thermotolerance usually possessed mutations that disrupted their ability to express one or more copies of Hsp70 genes. In most of these cases, the animals with higher Hsp70 inducibility and thermotolerance also showed reductions in fecundity.

These studies show that, though critical for thermotolerance, Hsp70 can have deleterious effects. Furthermore, they illustrate why genetic variations in populations are essential for the survival of a species.

References

○ Feder, M. E., and G. E. Hofmann. 1999. Heat-shock proteins, molecular chaperones, and the stress response: Evolutionary and ecological physiology. *Annual Review of Physiology* 61: 243–282.

---

pierce membranes, killing the cell. Second, ice crystal growth removes surrounding water, causing hyperosmotic stress. If ice forms outside cells, then water is drawn out of cells, causing a hypertonic stress that shrinks the cell, perhaps even killing it.

Still, many ectotherms survive freezing. Intertidal bivalves living in northern tidal flats can freeze when exposed to cold air temperatures, then thaw when the warmer water returns at high tide. Several terrestrial vertebrates can also survive freezing.

A wood frog in the north temperate zone enters the leaf litter in late fall, in preparation for overwintering. When temperatures drop below freezing, the animal supercools but ice does not form. At still lower temperatures, the animal begins to freeze. First to freeze are the frog's fingers and toes. The body core begins to freeze shortly thereafter.

Freeze-tolerant animals usually produce ice nucleators to control the location and kinetics of ice crystal growth. Ice is the most damaging when it forms inside cells, so freeze-tolerant animals secrete nucleators out of the cell. This restricts ice formation to the extracellular fluids, such as hemolymph, and allows the intracellular space to remain liquid. Many different types of molecules can act as nucleators in animals: calcium salts, membrane phospholipids, and long chain alcohols. However, it is not always clear that these ice nucleators are actually necessary or helpful to freeze-tolerance strategies. For example, the wood frog has an ice nucleator that triggers ice formation at about $-7°C$. The same ice nucleator is also found in the tissues of frogs that cannot survive freezing. It may induce the formation of ice, but it does not necessarily provide the wood frog with its freeze-tolerance. Some nucleators may simply be present for other functions and have no adaptive role in freeze-tolerance.

Because ice formation draws water from the cells, freeze-tolerant animals also produce intracellular solutes to counter the movement of water. Large glycogen reserves of the liver are broken down and converted to compatible solutes consisting of organic polyols, such as trehalose and glycerol. As we discussed in Chapter 10, compatible solutes have two main beneficial effects. First, by increasing the osmotic pressure within the cells, they reduce the movement of water and cell shrinkage. Second, the solutes help stabilize macromolecular structure.

Antifreeze proteins can prevent intracellular ice formation

Freeze-avoidance is the second strategy animals use to survive extreme cold. In a car, antifreeze elevates the osmotic concentration of the radiator fluid. Solutes in general depress the freezing point of a solution, preventing ice formation at subzero temperatures. Freezing point depression is one of the colligative properties of solutes (see Chapter 2). The solutes in animal tissues reduce the freezing point of

water, but generally not lower than about $-2°C$. Some animals possess antifreeze macromolecules—typically proteins or glycoproteins—that reduce the freezing point of body fluids by noncolligative actions. They disrupt ice crystal formation by binding to the surface of small ice crystals to prevent their growth (Figure 13.16).

The first **antifreeze protein**, or AFP, was discovered in an Antarctic fish about 30 years ago by Dr. Art DeVries. Since then, AFPs have been found in many distantly related taxa of fish, as well as insects and plants. Four classes of AFPs are distinguished by their structure: types I, II, and III, as well as antifreeze glycoproteins, or AFGPs. Interestingly, each of the classes of AFPs has arisen multiple times in evolution. In fish, AFPs arose less than 20 million years ago. This coincides with recent (in geological terms) sea level glaciation, which probably represented a strong selective pressure on the local marine species. The phylogenetic distribution of AFPs suggests an intriguing evolutionary history.

AFPs provide good examples of parallel evolution. For example, AFP II appears in herring, salmon, and sea ravens, fish from three separate orders. This suggests that AFPs arose multiple times in these lineages but well after the modern species diverged. These AFP II genes may have arisen from similar genes independently in each lineage. The structure of AFP II suggests the ancestral gene was a Ca^{2+}-dependent lectin, a protein that binds sugars. In structural models, the

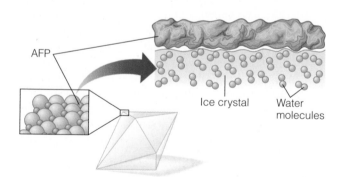

Figure 13.16 Antifreeze proteins Antifreeze proteins bind to the surface of ice crystals to prevent their growth. They bind along the face of the ice crystal, where the protein forms weak bonds with water molecules immobilized in the ice crystal. Because ice growth is very orderly, the presence of the bound protein prevents ice crystal growth. (Source: Modified from Davies et al. 2002)

interaction of a lectin with the hydroxyl groups of sugars is similar to the interaction of AFP with the hydroxyl group of a water molecule.

The evolutionary origins of AFGP are also unusual in terms of protein evolution. The ancestral gene was probably a gene for pancreatic trypsinogen, a digestive protease we introduced in Chapter 11. A region between the first intron and second exon was duplicated not just once but more than 40 times. The resulting gene possessed multiple, tandem sequences that resulted in a repeating Thr-Ala-Ala motif necessary to prevent ice crystal growth. In most cases of gene duplication and divergence, the resulting gene has properties similar to those of the ancestral gene, with relatively subtle differences in function. In the case of AFGP, the resultant gene has a totally distinct function. AFGPs have no protease activity, and trypsinogen has no antifreeze activity.

⊙ | CONCEPT CHECK

5. Compare and contrast the homeoviscous adaptation and conservation of K_m in relation to temperature effects on macromolecules.

6. How can an animal alter membrane fluidity?

7. Distinguish between freeze-tolerant and freeze-avoidance strategies.

Maintaining a Constant Body Temperature

Endothermy is so inextricably intertwined with a high metabolic rate that it is not known which trait arose first. High T_B allows metabolic processes such as growth, development, digestion, and biosynthesis to operate at faster rates, and the higher metabolic rate in turn produces more heat. The ability to become warm bodied requires metabolic pathways to produce heat (**thermogenesis**) as well as physiological mechanisms to retain heat. Most endotherms are also homeotherms and committed to maintaining a constant T_B. To do so, they must control both thermogenesis and heat exchange. In cold environments, endotherms stimulate thermogenesis and reduce heat loss. In hot environments they increase heat loss, but may also reduce thermogenesis. To control T_B, animals must be able to sense both environmental temperature and body core temperature.

Thermogenesis

Heat production is an inevitable consequence of being alive. An endotherm warms its body using heat that arises as a by-product of other metabolic processes, primarily energy metabolism, digestion, and muscle activity. All animals—endotherms and ectotherms—generate heat during these processes, but only the endotherms possess the physiological adaptations that enable them to retain enough metabolic heat to elevate T_B above T_A.

In addition to the pathways that produce heat as a by-product, endotherms possess specific thermogenic pathways with the main purpose of heat production. Thermogenic pathways rely on *futile cycling*, in which chemical potential energy is spent to generate heat. Most futile cycles involve cycling of ATP hydrolysis and ATP synthesis. Heat is released in ATP hydrolysis (ATP → ADP + phosphate), but a great deal more heat is produced when the cell uses intermediary metabolism to regenerate the ATP. Endotherms can enhance heat production either by increasing the rate of ATP turnover or by reducing the efficiency of ATP production. In both cases, most of the metabolic heat arises directly or indirectly from mitochondrial oxidative phosphorylation, discussed at length in Chapter 2.

Shivering thermogenesis results from unsynchronized muscle contractions

Muscle plays a critical role in the thermal budget of endotherms. Because muscle is the most abundant tissue in birds and mammals, it produces considerable heat, even at rest. Locomotion enhances the rate of muscle heat production. However, many birds and mammals can also use skeletal muscle to generate heat by **shivering thermogenesis**. As in normal contraction, motor neurons from the spine release neurotransmitters at the motor end plate, but during shivering the pattern of excitation is different. The smallest neurons—those innervating the slow fibers—are recruited first, followed by the larger neurons that innervate fast muscle. As a result, individual myofibers contract but the motor units are uncoordinated and the whole muscle undergoes no gross movement. Shivering thermogenesis is a strategy that works for short periods of cold exposure, but it is not useful for prolonged cold stress. The mechanics of shivering prevent an animal from using its locomotor muscles to hunt prey or escape predators. Furthermore, if shivering persists, or repeats frequently, the muscles are rapidly depleted of nutrients and they become exhausted, just as they would after high-intensity exercise.

Heat is produced in metabolic futile cycles

Shivering thermogenesis is unique to birds and mammals; however, other animals also use muscle to generate heat. Large flying insects, such as bumblebees and some moths, can generate enough heat to warm the thoracic flight muscles, which improves flight muscle performance in terms of energy production, excitation-contraction coupling, and cross-bridge cycling. The high metabolic rate during flight generates abundant heat, enough to warm the flight muscles by several degrees. Remarkably, these insects are even able to warm their flight musculature prior to takeoff.

Three distinct mechanisms allow insects to warm the thorax prior to flight. These same thermogenic pathways also allow social insects to work collectively to warm the hive. The first mechanism is a metabolic futile cycle in carbohydrate metabolism. Within the flight muscle, two opposing enzymes are activated simultaneously: the glycolytic enzyme phosphofructokinase and the gluconeogenic enzyme fructose-1,6-bisphosphatase. The metabolic cycle causes ATP hydrolysis and heat production, but without changes in the levels of the other substrates and products. A second warming mechanism relies on muscle contraction. Two sets of antagonistic flight muscles power wing movements during flight. Bumblebees can induce both sets of muscles to contract simultaneously prior to flight, so that energy is expended without productive movement. The third mechanism for heat generation is actual wing movement. The insect moves its wings fast enough to buzz, but controls the frequency and orientation of the wings to avoid generating lift. Collectively, these thermogenic pathways allow the flight muscle to warm up prior to takeoff. There appears to be a critical thoracic temperature that must be achieved before the insect will attempt to fly (Figure 13.17). At high T_A, less of a preflight warm-up is necessary to reach the threshold.

Membrane leakiness enhances thermogenesis

Most cellular membranes maintain an electrochemical gradient arising from differential distri-

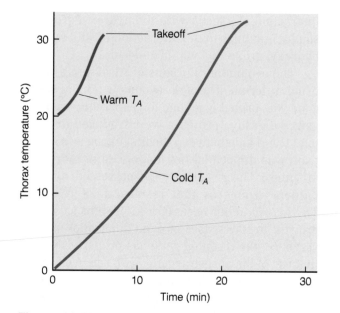

Figure 13.17 Thermogenesis in insect flight muscle Many large flying insects can undertake a preflight warm-up, using metabolic futile cycles and muscle activity to elevate thoracic temperatures to a threshold temperature required for flight.
(Source: Modified from Heinrich, 1987)

bution of ions across the membrane. Cells use chemical energy, usually in the form of ATP, to create these gradients. Consequently, any process that dissipates ion gradients will cause the cell to use chemical energy to reestablish the gradient.

Ion gradients collapse for two main reasons. First, many specific membrane proteins use electrochemical energy to drive other processes such as metabolite transport and biosynthesis. For example, many cells transport glucose and amino acids into the cell using Na^+-dependent cotransporters, causing the cell to use Na^+/K^+ ATPase to pump the Na^+ back out of the cell. The mitochondrial F_1F_0 ATPase is another transporter that dissipates ion gradients, in this case the proton motive force. Heat is produced when the mitochondrial electron transport system oxidizes reducing equivalents to regenerate the proton gradient.

The second pathway of ion gradient dissipation is ion leak, in which ion movements are not coupled to any other transport process. Since no biological membrane is completely impermeable, some ions leak across the bilayer or through gaps between proteins and phospholipids. Ion-pumping membrane proteins produce heat as a by-product, and a high proportion of the resting heat production, as much as 50% in some tissues, is due to the costs of maintaining ion gradients. Any process that increases the need for ion pumping will also increase thermogenesis. Typically, an endotherm has a resting metabolic rate that is as much as 10-fold greater than that of an ectotherm of the same size and T_B. The higher metabolic rate is due in part to membrane leakiness; endotherm plasma membranes and mitochondrial membranes are inherently leakier than those of ectotherms. Endotherms generate more heat to maintain ion gradients across leakier membranes.

Thermogenin enhances mitochondrial proton leak

Mammals possess a unique way of generating heat in specialized deposits of brown adipose tissue (BAT), typically located near the back and shoulder region (Figure 13.18). The brown adipocytes differ from white adipocytes in important respects. They have much higher levels of mitochondria and express the gene encoding the protein **thermogenin**.

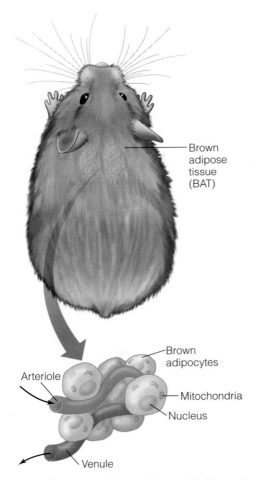

Figure 13.18 Brown adipose tissue in hamsters Hamsters possess thick pads of BAT behind the shoulders.

BAT is particularly important for thermogenesis in small mammals and newborns of larger animals, particularly those that live in cold environments. BAT growth and thermogenesis is under the control of the sympathetic nervous system. Norepinephrine released from these nerves causes BAT to grow in cell number (hyperplasia) and cell size (hypertrophy). Undifferentiated precursor cells are induced to proliferate and then later differentiate into BAT. Triglyceride is synthesized and mitochondria proliferate. At this same time, the cells begin to express thermogenin, which causes the tissue to increase the rate of mitochondrial respiration and consequently heat production. BAT heat production is often called **nonshivering thermogenesis (NST)**; while the other pathways we have discussed also differ from shivering, NST is a term usually reserved for BAT-mediated thermogenesis.

In the absence of thermogenin, the processes of oxidation of reducing equivalents and phosphorylation of ATP are coupled by their shared dependence on the proton motive force. When thermogenin is inserted into the inner mitochondrial membrane, it accentuates mitochondrial proton leak and dissipates the proton motive force. Since oxidation is no longer coupled to phosphorylation, thermogenin is said to cause *uncoupling*. In the presence of thermogenin, oxidation and proton pumping continue at high rates but with low rates of ATP synthesis.

The way in which thermogenin induces uncoupling is not yet certain. One theory suggests that thermogenin acts as a proton *ionophore*. It picks up protons from the cytoplasm and carries them into the mitochondria, dissipating the proton gradient. An alternative theory suggests that thermogenin dissipates the proton gradient by causing the futile cycling of fatty acids. Thermogenin carries an ionized fatty acid (R-COO⁻) from the mitochondrial side of the inner membrane and flips it across the bilayer to face the cytoplasm. Because of the higher proton concentration (lower pH), the ionized fatty acid is rapidly protonated (R-COOH). In this neutral form it readily flops back into the inner leaflet of the bilayer, where it ionizes again. The complete "flip-flop" cycle causes a proton to be translocated across the inner mitochondrial membrane. Still other explanations for UCP (uncoupling protein) function exist, and a definitive model awaits further experimentation.

The thermogenic capacity of BAT has been known for decades, and the protein thermogenin was first characterized in the early 1980s. It appears only in mammals and is expressed only in

BAT. However, in recent years it has become clear that thermogenin is only one member of a large gene family of uncoupling proteins (UCPs). In addition to thermogenin, also called UCP-1, mammals express at least two other UCPs (UCP-2 and UCP-3). Both can increase mitochondrial proton leak, but not enough to make a significant contribution to heat production. Instead of having a role in thermogenesis, these UCPs appear to reduce oxidative stress by preventing production of superoxide anions by mitochondria. The UCP gene family is ancient, with members in ectothermic animals, such as fish, as well as plants, fungi, and protists. It is likely that thermogenin arose in the mammalian lineage as a duplicated and then mutated version of other UCPs.

Regulating Body Temperature

Control of body temperature in endothermic animals requires coordination of multiple physiological systems. Animals must be able to monitor T_B in critical anatomical regions. By monitoring internal core T_B, animals can assess their overall thermal balance. Peripheral thermoreceptors allow animals to detect T_A. The information from thermal

sensing neurons is received and interpreted by a thermostat within the central nervous system. The central thermostat triggers the appropriate behavioral and physiological response.

A central thermostat integrates central and peripheral thermosensory information

As we discussed in Chapter 6: Sensory Systems, animals possess different types of neurons to sense and respond to temperature. Temperatures are monitored peripherally and centrally by temperature-sensitive neurons, both cold sensing and warm sensing. Birds and mammals monitor temperature using similar neurons, although the location of the central thermostat differs in the two taxa.

Mammals monitor T_A by peripheral cold-sensitive neurons located in the skin and the viscera. When T_A decreases, peripheral neurons send signals to the hypothalamus (Figure 13.19). The preoptic area of the anterior hypothalamus has both cold-sensing and warm-sensing neurons that monitor core body temperature. Information from the peripheral and the central thermal sensors is integrated in the posterior hypothalamus, which sends signals to the body to alter the rates of heat

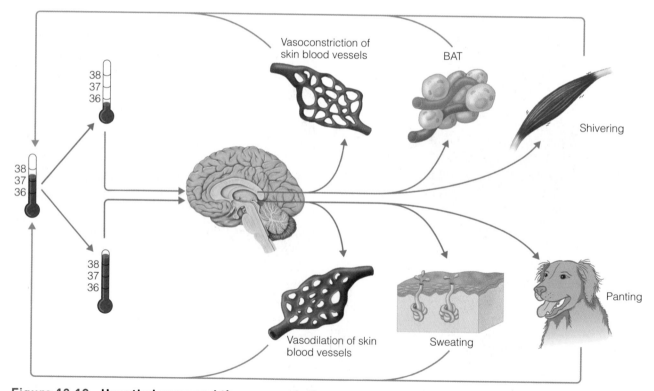

Figure 13.19 Hypothalamus and thermoregulation The hypothalamus is the thermal control center of mammals. It interprets signals from peripheral and central thermosensitive neurons and sends neuronal signals to other tissues, altering heat flux.

production and dissipation. The hypothalamus is much more responsive to information from the central thermoreceptors than from the peripheral thermoreceptors. Changes of less than 1°C can excite central thermoreceptors, triggering a rapid hypothalamic response. Conversely, peripheral thermoreceptors may record and respond to a change of several degrees without invoking a hypothalamic response. Surface temperatures can change by several degrees without harming the animal, whereas the temperature of the central nervous system must be more stable.

Bird T_B regulation is less understood but is clearly different from that of mammals. Heating or cooling the hypothalamus has little effect on the thermoregulatory response of birds. The central thermostat in birds appears to be the spinal cord, not the hypothalamus. However, the thermostat is still responsible for integrating information from central and peripheral thermosensors. When the central thermostat detects changes in temperature, it responds by firing neurons that lead to a compensatory response. Both birds and mammals alter T_B by changing rates of heat production and heat dissipation.

Piloerection reduces heat losses

Earlier in this chapter we discussed how body coverings, such as hair and feathers, act as insulation for endotherms. Since the efficiency of the insulatory layer depends on its thickness, animals can regulate heat loss by changing the orientation of the hair (in mammals) or feathers (in birds). Birds (and mammals) get fluffier in the cold by forcing their feathers (and hair) to orient perpendicular to the body surface. The mechanism by which this orientation is controlled is best understood with mammalian hair, but the position of bird feathers is controlled in a similar way.

Hair itself is a collection of cells that possess abundant keratin, an intermediate filament of the cytoskeleton. The distal end of a hair is primarily dead tissue, but the proximal end is composed of living cells embedded within the hair follicle. Depending on the species, a hair follicle can produce either a single hair shaft or complex combinations of hairs of various lengths and structures. Whereas human hair follicles produce single hairs, dog hair follicles produce a primary guard hair and multiple secondary hairs—soft, fine hairs that form the undercoat of the fur (Figure 13.20).

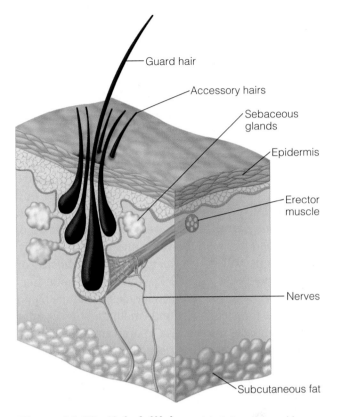

Figure 13.20 Hair follicles   A hair is produced by cells in the hair follicle. Erector muscles attached to the base of the hair contract in response to neural stimulation, causing the hair to become upright. Sebaceous glands secrete lipids into the follicle ducts.

The pit of the hair follicle is composed of epidermal cells. Intimately associated with each hair follicle is a sebaceous gland, which releases complex secretions of lipid (squalene, wax esters, triglyceride, fatty acids) that form a protective coating on the hair and provide moisturization.

Tiny smooth muscles, called erector muscles, connect each hair follicle to the undersurface of the epidermis. When the erector muscle contracts, the hair is pulled perpendicular, a process termed **piloerection**, so that the fur offers better insulation. The erector muscle contractility is regulated by numerous factors, both bloodborne and neural in origin. The situation is similar in birds, where erector muscles also control the orientation of the feathers.

Changes in blood flow affect thermal exchange

All animals exchange heat at the external surfaces of the body, but they are able to alter the *effectiveness* of surface heat exchange by changing the pattern of blood flow. Internal heat is equilibrated throughout the body by the blood. Where blood vessels approach the body surface, they will more readily lose heat. Similarly, increasing the flow of blood through the vessels increases the capacity for heat loss because it warms the surface of the skin, the site of heat loss by conduction, convection, and radiation.

The regulation of the amount of blood flowing into the vasculature is known as the **vasomotor response** (Figure 13.21). Directly under the skin are capillary beds fed by subcutaneous arteries and drained by veins that empty into a network called the *venous plexus*. There is also direct exchange of some blood between the veins and arteries through connections called *arteriovenous anastomoses*, or metarterioles. At normal T_B, the sympathetic nervous system constricts the arterioles to reduce blood flow. This tonic constriction is mediated by vascular smooth muscle in response to α adrenergic signals. When body temperature rises, there is a loss of tonic constriction and arterioles dilate to allow more blood into the skin vasculature. At the same time, the blood vessels of the anastomoses constrict, forcing more blood to move through the vessels near the skin. The large volume and high compliance of the venous system allows the blood to readily exchange heat to the skin surface. The greater the tempera-ture of the skin, the greater the rate of heat loss. The changes in vascular smooth muscle tone are controlled by the posterior hypothalamus.

Changes in blood flow through these capillary beds allow an endotherm to control heat exchange. The effects are perhaps most obvious in Caucasian humans, whose rapid changes in skin color reflect subdermal blood flow. Exercise increases the core body temperature and triggers an increase in blood flow to the skin, causing it to turn red. Similarly, cold temperatures cause peripheral vasoconstriction, reducing blood flow to the hands and feet, causing them to turn white. Prolonged restriction of blood flow can cause the extremities to turn purple, as the blood pooled in the venous system is slowly deoxygenated.

Countercurrent exchangers in the vasculature help retain heat

In addition to restricting blood flow to the periphery, some animals are able to extract heat from warmed blood and transfer it to cooler blood. This is accomplished by arranging the vasculature into *countercurrent heat exchangers* (see Box 13.4, Mathematical Underpinnings: Countercurrent Systems). The exact arrangement depends upon the animal and tissue.

Because fish breathe water, any metabolic heat is rapidly lost across the gills. Some regionally heterothermic fish, discussed earlier in this chapter, are active swimmers that produce abundant heat in their red muscle. In tuna, veins leaving the red muscle are juxtaposed to the arteries that supply the red muscle, allowing the transfer of myogenic heat from the veins back to the arteries (Figure 13.22). This allows red muscle to reach temperatures more than 10°C warmer than other tissues, including white muscle. Countercurrent heat exchangers are important in other regionally heterothermic fish. As we discussed in Chapter 5, billfish possess a modified eye muscle, called a heater organ, that warms the eye and optical nerves. Countercurrent heat exchangers help retain heat in the optical system. Many large fish, such as bluefin tuna, use countercurrent heat exchangers in the gastrointestinal tract to retain the heat of digestion.

Countercurrent heat exchangers are used by endotherms to reduce heat loss at the periphery. Birds standing on cold surfaces, such as ice, can lose a great deal of heat through the feet (Figure 13.23).

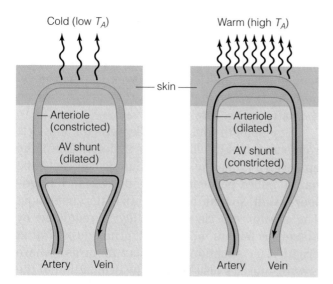

Figure 13.21 Skin vasculature When blood travels close to the surface of the animal, heat is lost across the skin. When temperatures are cold (left), blood is diverted from the skin through arteriovenous (AV) shunts, called arteriovenous anastomoses, reducing heat loss. When an animal is in a hot environment, shunts are constricted and blood moves through the vessels closer to the skin surface, enhancing heat loss.

BOX 13.4 **MATHEMATICAL UNDERPINNINGS**
Countercurrent Systems

Many physiological processes depend on countercurrent systems—a marriage of structural and functional features that improves the efficiency of exchange processes. Consider a scenario in which a tube drains a tank of hot water. As the water flows through the tube, heat is dissipated to the surrounding environment, which in this example is the air. At some point along the length of the tube (if it is long enough), the water reaches ambient temperature. However, the tube can be rearranged to reduce the magnitude of heat loss. Imagine what would happen to heat exchange if you were to align the lower (distal) end of the tube alongside the upper (proximal) end, creating a hairpin structure. Water flowing through one segment would run in the opposite direction of the water in the other segment. With this arrangement, some of the heat lost from the proximal segment is gained by the distal segment. Instead of a gradient from one end to the other, a thermal gradient forms along the length of the hairpin loop, coolest at the turn and warmest near the top. This is the basis of a countercurrent system. The longer the hairpin loop, the greater the gradients that can be built. Some researchers distinguish between two types of countercurrent systems: exchangers and multipliers.

Countercurrent exchangers transfer entities between inflow and outflow using only passive processes. The countercurrent heat exchanger, described above, is an example of such an exchanger; no specific transporter or pathway mediates the transfer of the entity (heat), and the gradient is due to the physical arrangement of the plumbing. The efficiency of the countercurrent exchanger depends on the volume of flow through the tubes and the overall gradients along the length of the hairpin.

Countercurrent multipliers are like exchangers in most respects except that specific transport proteins are required to transfer the entities between proximal and distal regions. The loop of Henle of the kidney tubule, discussed in Chapter 10, is part of a countercurrent multiplier. The gradients set up within the medulla result from active transport of ions, and the resulting transfer of water. Because a countercurrent multiplier requires transporters, it requires metabolic energy to create and maintain the gradient. If flow through the tube ceases, or ion pumping is reduced, the gradient can collapse. The efficiency of the countercurrent multiplier also depends on the length of the proximal and distal arms of the system that generate the gradient.

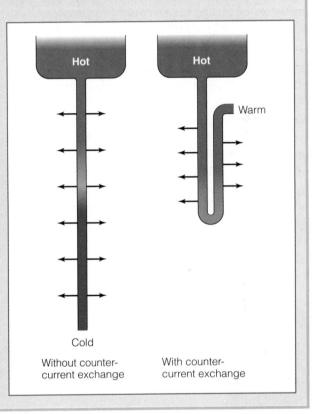

Sweating reduces body temperature by evaporative cooling

They can reduce heat loss by restricting blood flow to the periphery, but over long periods this would cause the peripheral tissues to starve. Countercurrent heat exchangers transfer heat from arteries emerging from the body core to veins returning from the cold periphery. Warming of the venous blood lessens the impact of the peripheral cooling. Also, cooling the arterial blood decreases the thermal gradient across the skin and therefore reduces heat loss.

Small animals have a favorable ratio of surface area to volume for heat loss, so evaporative cooling is used primarily by large animals. Many larger animals use specialized skin fluid secretions (sweat) to enhance evaporative cooling. Humans are probably the smallest animals that effectively use sweating to cool their bodies. Sweat is a mixture of water, salts, and some oils. The salt in sweat raises the boiling

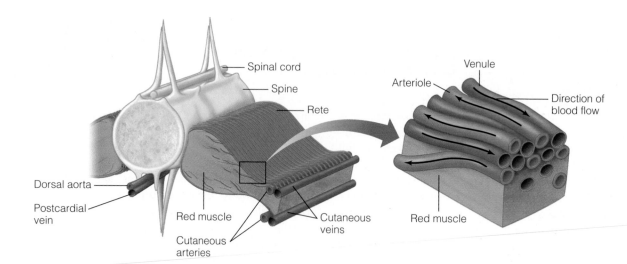

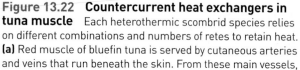

Figure 13.22 Countercurrent heat exchangers in tuna muscle Each heterothermic scombrid species relies on different combinations and numbers of retes to retain heat. (a) Red muscle of bluefin tuna is served by cutaneous arteries and veins that run beneath the skin. From these main vessels, smaller lateral vessels run over the surface of the red muscle, with branches penetrating the muscle. (b) These lateral vessels are arranged in a countercurrent manner, with lateral venules transferring myogenic heat to lateral arterioles. (Source: Part (a) modified from Carey, 1973)

point of water, making evaporative cooling more efficient. Loss of water and salts can affect ion and osmoregulation, but animals exposed to hot weather for long periods can change the chemical composition of their sweat to minimize ionic and osmotic problems. They produce a larger volume of sweat with a lower NaCl content, preserving vital salts. Sweating is controlled by the anterior hypothalamus and triggered by activation of the sympathetic nerves that control the activity of sweat glands.

Panting increases heat loss across the respiratory surface

Another way animals lose heat is through ventilation. The properties that make a respiratory surface good at gas exchange—high vascularity, moist surfaces, and high airflow—also enhance heat loss. Whether respiratory heat loss is beneficial or detrimental depends on the situation. In the cold, birds and mammals minimize heat loss from respiration, but at high T_A, animals may alter their breathing pattern to accentuate heat loss.

Cooling through ventilation is a strategy that must balance respiratory demands with thermoregulation. Cooling is enhanced when animals increase ventilation frequency while reducing tidal volume. Shallow, rapid breathing is a sign that an animal may be overheated. *Gular fluttering* is a cooling behavior seen in birds, characterized by rapid contraction and relaxation of the throat muscles. Mammals *pant*. Each of these behaviors cools the animal multiple ways. First, rapid

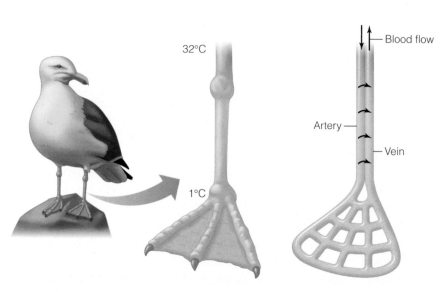

Figure 13.23 Peripheral vasoconstriction in cold endotherms Birds standing on cold surfaces can alter the flow of blood into the feet, reducing heat loss. The blood vessels of the leg and foot are arranged in parallel, allowing the formation of a countercurrent heat exchanger.

ventilation increases the heat loss across the respiratory surface by convection. Second, and perhaps more important, the rapid ventilation causes water to evaporate from the surface of the airway, from the pulmonary surface to the tongue. Animals that rely on ventilatory cooling often possess well-vascularized respiratory surfaces that are kept wet through secretions. These ventilatory patterns could alter the nature of the blood gas profile, impinging on respiratory physiology. The increase in ventilation frequency is offset in part by a reduction in tidal volume.

Reindeer provide a good example of the links between respiration and thermoregulation. Although they live in the cold, reindeer are at risk of heat stress because of their large size and thick layer of fur insulation. At normal cold temperatures (10°C), a reindeer breathes through its nose at low frequency. The upper part of the nasal cavity is rich in capillaries, and nasal respiration helps cool the nearby brain regions. When a reindeer becomes too warm, it shifts its respiratory pattern. Breathing frequency increases, and the animal begins to pant through the mouth (Figure 13.24). Although this change in breathing pattern may reduce direct cooling of the brain, it reduces body core temperature more efficiently.

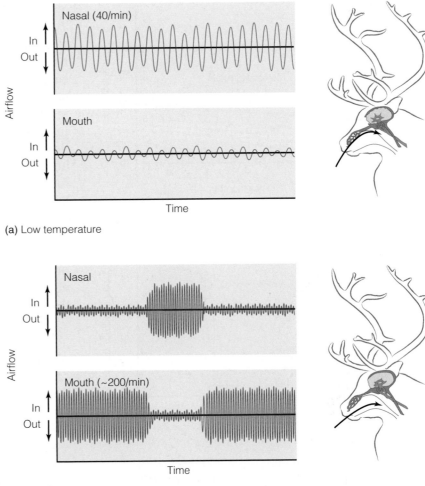

(a) Low temperature

(b) High temperature

Figure 13.24 Heat loss during panting   Like other mammals, reindeer alter breathing to increase heat loss. Reindeer breathe through the nose at low temperatures. The flow of air cools the blood circulating through the vessels that line the nasal cavity. When temperatures increase, reindeer breathe through the mouth and at a faster rate (200–300 breaths per minute). (Source: Aas-Hansen et al., 2000)

Relaxed endothermy results in hypometabolic states

In previous chapters, we have encountered various forms of hypometabolism used by endotherms to survive adverse conditions. Hummingbirds, for example, undergo a nightly reduction in metabolic rates. Hibernating mammals also undergo a metabolic suppression during the long, cold winter months when food is scarce. Whether a daily dormancy (torpor) or a more prolonged seasonal dormancy (hibernation), the hypometabolic phase is accompanied by a decrease in T_B, a phenomenon called **relaxed endothermy**. The time course and magnitude of reduction in T_B differ among animals and types of dormancy (Figure 13.25). An arctic squirrel, for example, can allow T_B to fall close to the freezing point. However, even minor reductions in T_B can offer important energetic savings for a dormant animal.

Under normal (euthermic) conditions, mammals and birds maintain T_B within a narrow range. A euthermic animal induces a compensatory response when its central thermostat—the hypothalamus in mammals—senses a decrease in T_B. During periods of relaxed endothermy, the animal recalibrates its central thermostat to recognize and defend a different T_B set point. The

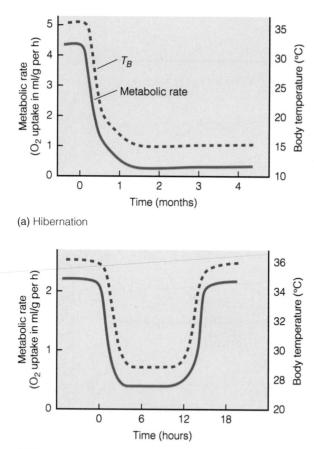

(a) Hibernation

(b) Torpor

Figure 13.25 **Hypometabolic states** Many endotherms respond to cold temperatures by entering some form of dormancy. Body temperature generally declines in parallel with metabolic rate. The dormancy is called **(a)** hibernation when the metabolic depression lasts for weeks to months or **(b)** torpor when the animal enters a hypometabolic state in daily cycles.

endothermic animal may allow T_B to fall close to T_A, well below the euthermic set point.

The links between metabolism and T_B regulation make it difficult to establish which parameter causes hypometabolic cooling. For most animals entering dormancy, T_B and metabolic rate decline in parallel, and it is not clear if the colder T_B slows metabolism, or alternately if the slower metabolic heat production causes cooling. In some studies, animals show a reduction in metabolic rate before T_B declines, suggesting that hypometabolism initiates the reduction in T_B. However, in larger animals a delay in cooling upon entering dormancy is due in part to thermal inertia; the large mass and low ratio of surface area to volume delay the impact of reduced thermogenesis, allowing the animal to remain much warmer than T_A even with a reduced metabolic rate.

⊙ | CONCEPT CHECK

8. How do endotherms generate heat?
9. What regions of the body detect and respond to changes in temperature?
10. What are the various types of hypometabolism?
11. How do animals control heat flux across the external body surface?

Integrating Systems | Immune System and Thermoregulation

Deviations from optimum T_B usually impair physiological functions, but in some cases endotherms can induce hyperthermia as part of the defense against pathogens. The cellular and noncellular defense pathways are collectively called the immune response (Figure 13.26). When a pathogen mounts a localized attack, such as at the site of a cut, the immune response includes an elevation in temperature in that region: **inflammation**. A broad systemic pathogen attack induces a more elaborate immune response, which includes an increase in body core temperature: a **fever**. Whether arising from inflammation or a fever, the hyperthermic response serves to improve the ability of the animal to combat the pathogen. Most immunological processes, such as the rate of movement of or ingestion by phagocytotic immune cells, have Q_{10} values within the range for other cellular and biochemical

processes (2–5). For these processes, an increase of 2–3°C in regional or systemic temperature would have a relatively minor beneficial effect. However, some aspects of immune function demonstrate Q_{10} values ranging from 100 to 1000 and would be profoundly enhanced by the degree of hyperthermia seen in inflammation and fever. Recall from Chapter 8: Circulatory Systems that blood carries T lymphocytes as part of the adaptive immune system. When pathogens are present, some T lymphocytes undergo a maturation process that enables them to become cytotoxic T cells. With a Q_{10} in excess of 100, this maturation step is acutely sensitive to temperature. The hyperthermia arising from an immune response reflects a remarkable coordination between cellular signaling pathways, cardiovascular changes, and central control of body temperature.

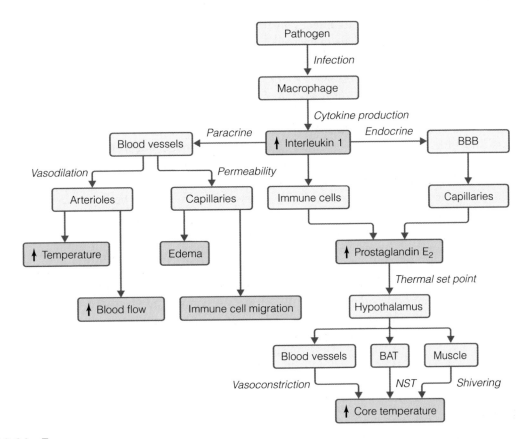

Figure 13.26 Fever

Inflammation is a localized response to the presence of something that activates the immune system. Consider what happens when a virus invades your respiratory epithelium. The virus must first pass through the physical barriers that separate your respiratory tissues from the environment. The mucus secreted from exocrine cells in the respiratory epithelium reduces the passive water loss but also acts as a barrier to external viruses, bacteria, and particulates. If the virus passes through the mucus, it must be able to enter living epithelial cells to propagate. The epithelium of the respiratory tract is thinner than that of the skin, the stratum corneum, making it more vulnerable to viral attack. If the virus enters an epithelial cell, it will take over the normal cellular protein synthetic machinery and cause the cell to produce the viral proteins needed for viral replication. After a time, the cell may lyse, releasing additional viral particles. During this stage, infected and damaged cells release a number of factors (cytokines and prostaglandins) that alter cellular processes within the region. Though it doesn't affect the virus directly, mucus cells enhance secretions to bolster the physical barriers to reduce the likelihood of further infection. Recall from Chapter 8 that cytokines, such as interleukin 1, also signal the local vasculature, altering blood flow regionally. Through vasodilatory effects on arterioles, more blood

flows through the capillary beds. Since many of the surface tissues are cooler than body core temperature, the increase in blood flow may elevate skin by as much as 10°C. The cytokines also alter the permeability of the capillary beds, allowing immune cells in the blood to squeeze between endothelial cells and enter the interstitial fluids. The main features of inflammation are attributable to the changes in the vasculature: redness and warmth due to increased blood flow, swelling (edema) due to fluids moving from the main circulation through more permeable capillaries into the interstitial fluid.

If a regional infection spreads, or the infection occurs systemically, the animal mounts a more elaborate immune response that includes an increase in body temperature: a fever. The body detects the presence of a pathogen when cells of the immune system bind specific pathogen macromolecules, called *exogenous pyrogens*. The most common exogenous pyrogens from bacterial infections are lipopolysaccharides, proteoglycans, and proteins, such as endotoxin. Recall from Chapter 8 that macrophages phagocytose pathogenic bacteria. When a macrophage consumes bacteria, the bacterial macromolecules cause the macrophage to secrete cytokines, such as interleukin 1. In addition to the local pro-inflammatory effects, interleukin works as an *endogenous pyrogen*. It causes other cells to synthesize another factor—a mediator—that exerts its effects on the brain. For example, interleukin 1 induces many cell types in the periphery and in the vasculature of the brain to synthesize prostaglandin E_2. It is not yet clear how this mediator crosses the blood-brain barrier (BBB)—it may be through synthesis and secretion by the endothelial cells, or transport across the capillary endothelium—but once across the BBB, prostaglandin E_2 binds to neurons of the hypothalamus, where it alters the neurocircuits that integrate peripheral and central thermal information. The pyrogenic mediator reduces the firing frequency of warm-sensitive neurons and increases the firing frequency of cold-sensitive neurons. Thus, the pyrogenic mediator causes the hypothalamus to misinterpret the central and afferent thermal information, and as a result the brain perceives that the body is too cool. This triggers a range of compensatory responses that increase the rate of heat production and conservation. Though the neurocircuitry remains unclear, a fever also stimulates shivering thermogenesis. Sympathetic neuronal activity stimulates thermogenesis in brown adipose tissue and vasoconstriction in cutaneous vasculature.

Since this immune response works by increasing T_B, it would seem to be practical only in animals that can retain body heat. Indeed, this aspect of immunity is best developed in endothermic animals, primarily birds and mammals. However, ectothermic vertebrates and many invertebrates show a behavioral fever in response to pathogens. If a lizard is infected with bacteria or injected with an exogenous pyrogen, it demonstrates an increase in its preferred T_B, moving into a warmer environment. Thus, the benefit effect of hyperthermia in the immune response is likely a very ancient trait, though the mechanisms by which body temperature is elevated differ among lineages. ◉

Summary

Heat Exchange and Thermal Strategies

→ All animals are subject to the physical laws that govern heat fluxes, although the effects on endotherms have a greater impact on their physiology. Animals that live in water lose heat more readily than those that live in air, because water has a higher thermal conductivity.

→ Although heat production is a function of body mass, heat exchange depends on body surface area. Consequently, an animal's ratio of surface area to volume has an important impact on thermoregulation.

→ External insulation, such as feathers and hair, creates a dead space that reduces the thermal gradient between the animal and the environment.

→ Movement of fluids, such as air and water, increases the rate of convective heat loss.

→ Thermal radiation is an important source of heat for animals, derived directly from the sun and also from an animal's surroundings. Many animals can absorb thermal energy from solar radiation by basking, often aided by dark coloration. Many animals employ evaporative cooling to reduce overheating.

→ Poikilotherm T_B changes with environmental conditions, whereas homeotherm T_B remains nearly constant.

→ Ectotherm T_B is determined by environmental conditions, whereas endotherm T_B is determined by metabolic heat production and conservation.

→ Heterotherms exhibit combinations of endothermy and ectothermy. Temporal heterotherms are typically endothermic for most of their lives, but undergo periods of hypothermia, such as in hibernation or torpor. Regional heterotherms are able to warm regions of the body above ambient temperature (T_A).

→ Endotherms have a thermoneutral zone—a range of temperatures at which physiological functions are optimal. Outside this temperature range, the metabolic costs are greater and animal function suffers.

→ Thermal effects in ectotherms are influenced by thermal history

→ Eurytherms tolerate a wide T_A range, whereas stenotherms survive in a narrow T_A range.

Coping with a Changing Body Temperature

→ Changes in T_A have greater consequences for ectotherms than for endotherms, altering many aspects of macromolecular structure and metabolism. Temperature alters membrane fluidity, which animals can mitigate by altering the fatty acid chain length, saturation, phospholipid profile, and cholesterol content.

→ Temperature also changes the rates of chemical reactions, and consequently metabolic rate. It also affects protein structure and enzyme kinetics. When proteins undergo thermal denaturation, the genes for stress proteins are induced. These proteins help refold damaged proteins and target irreversibly damaged proteins for degradation.

→ Ectotherms living in cold environments often remodel their physiological systems to compensate for the effects of temperature. Animals that have lived for long periods in extreme cold often possess cold-adapted proteins.

→ Many animals are able to survive freezing temperatures. The greatest risk is the uncontrolled formation of ice crystals, which can induce osmotic stress and can physically damage cellular membranes.

→ Freeze-tolerant animals use ice nucleators to control ice crystal growth. Freeze-avoiding animals produce antifreeze proteins that prevent intracellular ice formation.

Maintaining a Constant Body Temperature

→ Endothermic animals produce metabolic heat and retain it to elevate body temperature above the ambient temperature.

→ Heat is a natural consequence of metabolism, but endotherms have higher metabolic rates than ectotherms of similar size. Greater membrane leakiness is one reason why endotherms have high metabolic rates.

→ Endotherms use neural systems to detect external and internal temperatures. The mammalian hypothalamus integrates central and peripheral thermal sensory information to cause physiological systems to alter heat production and retention.

→ Peripheral cold-sensitive neurons trigger changes in the orientation of hair, or piloerection, to reduce heat loss.

→ Nerves and hormones also control the blood flow to the skin surface to regulate the rate of heat loss.

→ Vasculature may be arranged into countercurrent exchangers to help retain heat within the body core.

→ Overheated large mammals use sweating to enhance evaporative cooling, whereas smaller mammals shed heat by panting.

→ Some endotherms can undergo short periods of metabolic suppression, during which body temperatures decrease, although rarely to the point where the temperature matches the ambient temperature. During relaxed endothermy, an animal resets its central thermostat to a new set point.

Review Questions

1. Compare and contrast the following terms: homeothermy and poikilothermy; endothermy and ectothermy; regional heterothermy and temporal heterothermy.

2. Water at 10°C feels colder than air at 10°C. Why?

3. Why are antifreeze proteins found in marine fish but not freshwater fish?

4. Compare and contrast the mechanisms of thermogenesis. Which biochemical steps are responsible for heat production?

5. Discuss the different sources of energy an ectotherm can use to raise T_B.

6. What behaviors reduce heat losses due to (a) conduction; (b) convection?

7. How do we know that antifreeze proteins arose several times in evolution?

8. How do countercurrent heat exchangers work?

Synthesis Questions

1. Compare the effects of high and low temperature on molecules, cells, tissues and organisms.

2. How could you convert a stenothermal animal to a eurythermal animal?

3. Summarize the physiological changes that accompany thermal acclimation.

4. Why do endothermic animals need both peripheral and central temperature-sensitive neurons?

5. Thermoregulation requires active control of blood flow through vessels. How do animals dilate some blood vessels while constricting others?

6. What would you expect to happen to blood pressure when a mammal is exposed to cold temperatures?

7. What gene regulatory changes must have accompanied the evolution of brown adipose tissue?

8. Animal color influences many aspects of physiology and ecology. Identify some examples of animals whose color patterns are consistent with a role in thermoregulation.

9. Many mammals grow coats that differ in winter and summer. What factors affect the costs and benefits of seasonal shedding?

10. Compare and contrast the structures of hair and feathers.

Quantitative Questions

1. The metabolic rate of a fish heart is studied at various temperatures. The metabolic rate is 20 μmol ATP per min per g tissue at 25°C, 8 μmol ATP per min per g tissue at 10°C, 4 μmol ATP per min per g tissue at 5°C, and 1 μmol ATP per min per g tissue at 2°C. Calculate the Q_{10} values over this range of temperatures and offer an explanation for the patterns.

2. The levels of ATP are maintained through a balance between the rates of ATP synthesis and ATP utilization. For a given tissue (e.g., heart) at a given T_B (e.g., 15°C), assume that (a) the rates of ATP synthesis and utilization are both 10 μmol/min/g, (b) the rate of ATP synthesis exhibits a $Q_{10} = 2$, (c) the rate of ATP utilization has a $Q_{10} = 2.05$, and (d) the starting ATP level was 5 μmol/g tissue. Calculate the change in ATP levels over time that would result if the animal were moved to an environment that caused a 10°C increase in T_B.

3. Recall the Stefan-Boltzmann equation, $P = Ae\sigma(T_B^4 - T_A^4)$, where P is the radiating power, A is its surface area, e is the ability of the object to emit radiation, σ is the Stefan constant, and T is the temperature of the body (T_B) or surroundings (T_A) in kelvins. Consider an animal that uses a strategy of changing posture to alter the surface area as a way of controlling heat loss. It assumes a particular posture when it is in an environment that is 5°C below its body temperature. How does it need to change its surface area when it moves to a new environment that is 20°C cooler?

For Further Reading

See the Additional References section at the back of the book for more readings related to the topics in this chapter.

Heat Exchange and Thermal Strategies

An interesting book, written for a lay audience, that discusses the role of energy in biology.

Brown, G. 1999. *The energy of life.* London: HarperCollins.

This seminal paper uses a quantitative biophysical approach to describe how energy exchange between animals and the environment governs the thermal biology of the animal.

Porter, W. P., and D. M. Gates. 1969. Thermodynamic equilibria of animals with environment. *Ecological Monographs* 39: 227–244.

Much of thermal biology focuses on vertebrates, and the insects are often understudied. Heinrich's book reminds us of the many thermal strategies employed by insects.

Heinrich, B. 1992. *The hot-blooded insects.* Cambridge, MA: Harvard University Press.

This book discusses the impact of changes in temperature on macromolecular structure and biochemical processes.

Hochachka, P. W., and G. N. Somero. 2002. *Biochemical adaptation.* Oxford: Oxford University Press.

These papers discuss the peculiarities of animals that live in the extreme cold, and consider whether the data on metabolic rate of these animals support the notion of metabolic cold adaptation.

Steffensen, J. F. 2002. Metabolic cold adaptation of polar fish based on measurements of aerobic oxygen consumption: Fact or artefact? Artefact! *Comparative Biochemistry and Physiology, Part A: Molecular and Integrative Physiology* 132: 789–795.

Various authors. 2002. Coping with the cold: Molecular and structural biology of cold stress survivors, D. J. Bowles, P. J. Lillford, D. A. Rees, and I. A. Shanks, eds. *Philosophical Transactions of the Royal Society of London, Series B: Biological Sciences* 357: 829–956.

This textbook considers how the environment influences physiological processes. The sections on thermal biology are particularly useful.

Willmer, P., G. Stone, and I. A. Johnston. 2000. *Environmental physiology of animals.* Oxford: Blackwell Science.

Maintaining a Constant Body Temperature

This book considers the regulatory mechanisms animals use to modify metabolic rate. For most animals, this involves a reduction in body temperature.

Hochachka, P. W., and M. Guppy. 1987. *Metabolic arrest and the control of biological time.* Oxford: Oxford University Press.

This paper discusses recent studies on how the hypothalamus senses and responds to thermal conditions.

DiMicco, J. A., and D. V. Zaretsky. 2007. The dorsomedial hypothalamus: A new player in thermoregulation. *American Journal of Physiology: Regulatory, Integrative, and Comparative Physiology* 292: R47–R63

These papers consider the evolutionary origins of hyperthermia and how hyperthermia in fever and inflammation may accentuate the immune response.

Kluger, M. J., W. Kozak, C. A. Conn, L. R. Leon, and D. Soszynski. 1998. Role of fever in disease. *Annals of the New York Academy of Sciences* 856: 224–233.

Hanson, D. F. 1997. Fever, temperature, and the immune response. *Annals of the New York Academy of Sciences* 813: 453–464.

Reproduction

Evolutionary biologists use the term *life history strategy* to describe the pattern of growth, behavior, and reproduction an animal displays over its lifetime. Variations in life history strategy are interpreted in the context of reproductive success, the currency of evolution. These elements of a life history strategy are interrelated through physiology. First, the animal regulates physiological processes throughout its life to ensure that it will be able to reproduce. Second, reproduction itself is a physiological process that impinges on other physiological systems. Any specific element of a reproductive strategy—such as age of maturity, fecundity, and juvenile survival—has a physiological basis; evolutionary trade-offs are often expressed in physiological terms, such as growth rate, size, or metabolic costs.

A nonreproductive animal gathers nutrients for growth and maintenance, storing excess nutrients in fuel depots. Later in life, the animal enters a transition period where its physiological priorities shift toward building gametogenic tissues and preparing for reproduction. After

reproducing, some animals continue to divert energy toward care of their offspring. Every animal has evolved a balance among competing physiological demands in order to invest maximally in reproduction at some point in its life. The life history patterns that emerge are heavily influenced by environmental predictability and resource availability. Life history strategies evolve in ways that maximize reproductive output given the constraints of the environment.

Ecologists broadly classify life history strategies as K-type and r-type. **K-type** strategists are specialized to produce relatively few offspring but invest heavily in their development. They often delay reproduction until late in life, accumulating resources until they are capable of producing large, healthy offspring. Delaying reproduction ensures that these animals are strong enough to tolerate the costs of nurturing the young, often for extended periods. Parental care may be essential in environments where conditions are so harsh that the offspring might not otherwise survive. Humans are perhaps the most familiar example of this life

Salmon eggs.

Sitkas.

history pattern, but other animals produce relatively few, large offspring and provide abundant parental care. For example, the kiwi produces one egg that occupies most of its internal space, constituting almost 20% of its body weight. This bird's egg is almost six times larger than the eggs of other birds of comparable size. To support the growth of the egg, the female increases food consumption about three-fold. Later in development, the egg is so large that it compresses the stomach, preventing the female from eating.

An **r-type** strategist reproduces as rapidly as possible, produces a great many offspring, and typically provides little or no parental care. Insects exemplify r-type strategies. Some—such as locusts—can reproduce so quickly that they give the impression of arising spontaneously and grow to such numbers that they qualify as a plague, threatening to wipe out all vegetation over huge swaths of land. Of course, few animals clearly fall into one of these two strategies. Rabbits and hamsters reproduce at a relatively young age and produce large litters (both features of an r-type strategy), but they also invest heavily in their young (a K-type strategy).

We can also describe life history strategies on the basis of how an animal apportions its reproductive effort over its lifetime. A *semelparous* species spends most of its life accumulating resources, preparing for a single burst of reproductive output into which it allocates all of its available energy, and then dies shortly thereafter, having withheld no energy for its own survival. In contrast, an *iteroparous* animal can undergo multiple reproductive cycles. When it reproduces, it retains enough energy to ensure its survival, presenting the possibility of reproducing again. If an individual happens to reach reproductive maturity in a bad year, when the offspring are unlikely to survive, it can have another chance at more successful reproduction in a subsequent reproductive bout.

On the surface, these two strategies—semelparity and iteroparity—would seem to require quite different physiological systems. Surprisingly, there are examples of related animals exhibiting completely different strategies. Consider the salmonid fish. Atlantic salmon are iteroparous and capable of repeated spawning for several years. Pacific salmon are generally semelparous, breeding once and then dying. However, when Pacific species become landlocked, the population makes an evolutionary transition to an iteroparous strategy and the fish become repeat spawners. Since closely related animals are so similar in genotype, the ability to exhibit radically different life history strategies depends on the way the physiological phenotype is regulated, largely by complex hormonal controls on development and reproduction. ◉

Overview

The life cycle of an animal begins with a single cell, which divides repeatedly through multicellular stages (zygote, blastula, and gastrula) that differentiate to form tissues (morphogenesis). Juvenile forms then undergo further development to reach reproductive maturity (Figure 14.1). The reproductive traits of the individual are usually established in embryonic development, with the acquisition of the primary sex characteristics: the **gonads**. These multicellular tissues include cells that produce the **gametes** as well as somatic tissues that support gamete production (**gametogenesis**). The gonads develop in combination with other physiological and behavioral systems in preparation for mating. Mating behavior may be linked to environmental conditions and often follows complex courtship rituals. Animals then release the gametes—ova or spermatozoa—when the chances for successful fertilization are maximized. Spermatozoa, or sperm for short, face many challenges. They must find the ovum in a complex environment and outcompete other sperm to be the one that fertilizes the ovum. After fertilization, the embryo grows under the control of its unique genome, a mosaic of its parents. All of the elements of sexual reproduction—sex determination, gametogenesis, mating, fertilization, and development—depend on the coordination of cellular processes in multiple tissues. The responsibility for coordination of these processes falls upon the endocrine hormones.

Diversity in life histories is remarkable given the relative similarity in hormones, cell signaling, and gametogenesis. In the next section, we survey the basic features of reproductive physiology, considering modes of reproduction, hormones, gametogenesis, and reproductive anatomy. In the final section, we focus in greater detail on the reproductive physiology of mammals, through ovulation, gestation, parturition and postpartum care, emphasizing the role of hormones.

Sexual Reproduction

Let's begin by considering the aspects of reproduction that are common to most animals. Long before the animals appeared on the scene, the early eukaryotes (for example, the protists) had already evolved a capacity for sexual reproduction. As Frank Sinatra once sang, "birds do it, bees do it," but so do

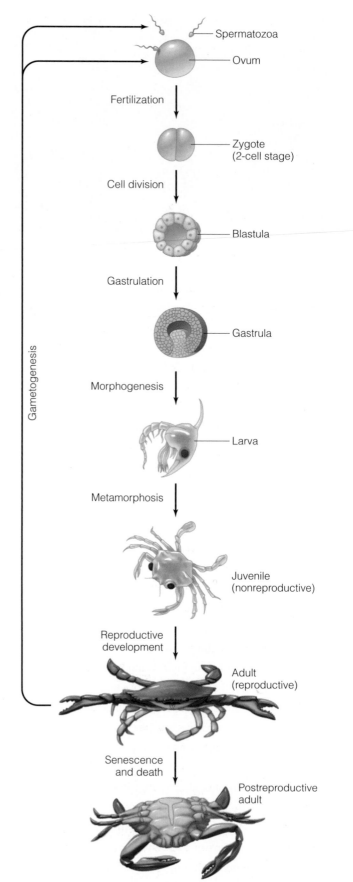

Figure 14.1 Animal life cycle This generalized life cycle highlights the developmental stages seen in most animals.

fungi and plants, although the process may not be as song-worthy. The essence of sexual reproduction is the generation of offspring from two parents, each of which contributes a nearly equal amount of genetic material. The biological concept of "maleness" and "femaleness" is based on the size of the gametes. In sexual reproduction, the gametes are of different size (*anisogametic*): the male has gonads (testes) that produce small gametes (spermatozoa), and the female has gonads (ovaries) that produce large ga-

metes (ova). Gametogenesis occurs through meiosis, although there are important distinctions between spermatozoa production (**spermatogenesis**) and ova production (**oogenesis**) (Figure 14.2). Reproductive systems include the gonads, the reproductive tract through which gametes escape, and the accessory tissues that provide regulatory molecules, nutrients, and fluids.

Sexual reproduction is one of the reasons why animals have been so successful in exploiting

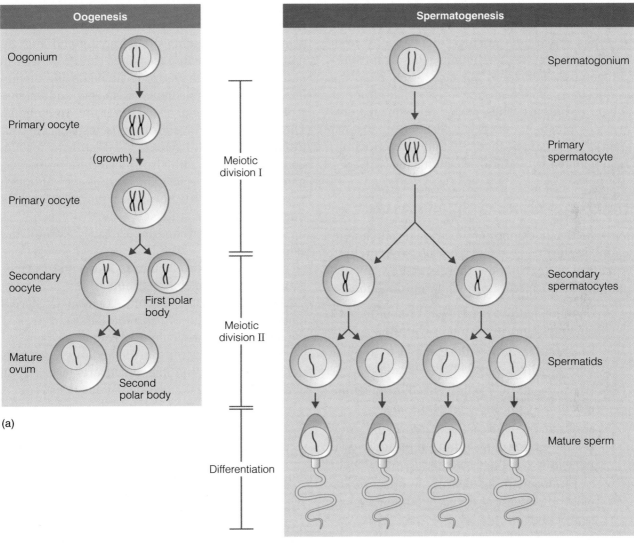

(a)

(b)

Figure 14.2 Gametogenesis The gametes are formed by the two-step process of meiosis. Germ cells (spermatogonium and oogonium) proliferate in the gonads to create a stock of diploid cells that can undergo gametogenesis. The two alleles for each gene are shown as red and blue chromosomes. Meiosis begins when each chromosome duplicates; the progression through meiosis differs in males and females. **(a)** Oogenesis pauses when the primary oocyte grows in size and remains quiescent in the female for long periods. When activated, the primary oocyte undergoes an asymmetrical cell division, devoting most of the cytoplasm to a single daughter cell (secondary oocyte). The other, smaller daughter cell, called the first polar body, is usually degraded. The secondary oocyte undergoes another round of asymmetrical cell division, resulting in the ovum and the smaller second polar body, which is also degraded. **(b)** Spermatogenesis continues when the primary spermatocyte undergoes cell division to produce two secondary spermatocytes. Meiosis continues and each secondary spermatocyte divides to produce two haploid spermatids.

diverse ecological niches. The process generates genomic variation at three levels. First, an animal produces gametes with genomes consisting of combinations of chromosomes originally provided by the animal's own parents. For an animal with 23 chromosome pairs, more than 8 million genetically different gametes can be produced by a single individual. Second, during meiosis, chromosomal recombinations can create chromosomes that are hybrids of maternal and paternal chromosomes, further adding to the total number of unique gametes. Third, the diploid offspring produced by fertilization are unique combinations of the different types of variants arising independently from the first two processes in both oogenesis and spermatogenesis. For these reasons, each offspring produced in sexual reproduction is unlike either its siblings or parents. Thus, sexual reproduction creates a population that is a collection of distinct genotypes—a genetic diversity that is the raw material upon which natural selection acts.

Reproductive Hormones

Reproductive hormones orchestrate development, sexual maturation, gametogenesis, and mating. There are many common themes in how diverse animals use hormones to control reproduction.

- Complex pathways of negative and positive feedback control hormone synthesis.

- Hormone levels are determined by regulation of synthesis as well as degradation.

- Hormone efficacy is influenced by hormone receptor synthesis in target tissues.

- Males and females of a species use the same suites of hormones, although an individual hormone may have sex-specific functions.

- Hormones with major roles in other physiological systems also have vital functions in reproduction.

Vertebrates rely on progesterone, androgens, and estrogens

Steroid hormones are critical regulators of animal reproductive physiology. Recall from Chapter 4: Neuron Structure and Function that steroid hormones regulate physiology primarily through effects on gene expression. Each steroid hormone binds to a *nuclear hormone receptor*, a protein that heterodimerizes with another DNA-binding

protein to form an active transcription factor. Animals mediate the effects of steroid hormones by altering the rates of hormone synthesis, the levels of receptors in target tissues, the rates of degradation of hormones and receptors, and by producing extracellular proteins that bind steroids. Steroid hormones are all derived from cholesterol, but diverse enzymatic pathways allow animals to produce a range of structurally related specific hormones.

In vertebrates, a complex suite of steroid hormones with subtle structural differences that impart unique activities (Figure 14.3). Progesterone is produced from cholesterol in a number of steroidogenic tissues including the adrenal gland and gonads. It can escape into the blood, exerting effects in both males and females, or it can be further metabolized to androstenedione. In males, androstenedione is further metabolized to various **androgens**. The most common androgen is testosterone, although other androgens (11-ketotestosterone, androstenedione, dihydrotestosterone) predominate in some species and processes. Although these are called male hormones, they are also produced in females and serve as the precursors for synthesis of **estrogens**, primarily estrone and estradiol-17β.

The rates of production of individual steroids depend largely on the distribution and activity of steroid metabolizing enzymes. Central to steroid metabolism are the cytochrome P450 enzymes of the endoplasmic reticulum. **Aromatase** is a cytochrome P450 enzyme that metabolizes androgens to estrogens. For example, it converts testosterone to estradiol-17β and androstenedione to estrone.

Gonadotropins control steroid hormone levels

Steroid synthesis in the gonads is controlled by the levels of nonsteroidal hormones produced by the anterior pituitary: **gonadotropins**. Most vertebrates produce the same types of gonadotropins: *follicle-stimulating hormone (FSH)* and *luteinizing hormone (LH)*. Primates produce a third gonadotropin, *chorionic gonadotropin (CG)*.

Gonadotropins are heterodimers of an alpha subunit (shared by all gonadotropins) and a beta subunit that imparts the unique properties of each hormone. Thus, FSH is composed of a dimer of alpha gonadotropin and beta FSH. Each subunit is about 100 amino acids long, and heavily modified by glycosylation. Unlike steroid hormones, each of

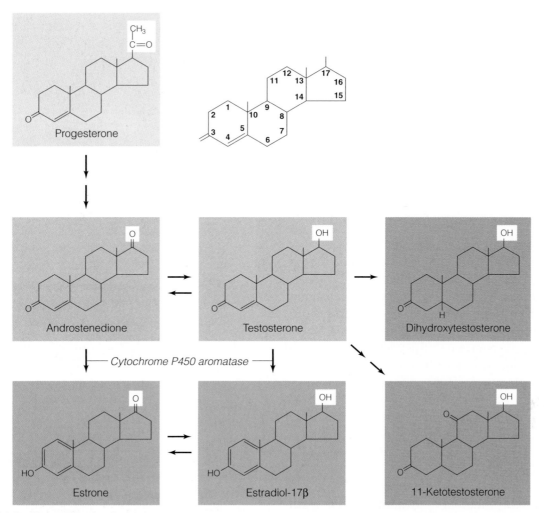

Figure 14.3   Reproductive hormones   Highlighted areas distinguish chemical differences in closely related hormones.

which possesses the same chemical structure regardless of taxon, the gonadotropins are proteins with taxa-specific sequences. Thus, fish FSH is not identical to mammalian FSH, but they are so named because of their structural similarities. The exact roles of gonadotropins often differ among vertebrate taxa. They are similar in the most general respects of controlling gametogenesis and reproductive maturity, acting both directly on target tissues and indirectly through effects on steroid hormone synthesis. FSH stimulates spermatogenesis in males and induces the follicles to ripen in females. LH induces the interstitial cells of the testes to produce testosterone in males, and induces the follicle to produce estrogens in females.

Release of gonadotropins from the anterior pituitary is under the control of multiple hormones. The main regulator is a hypothalamic hormone, **gonadotropin-releasing hormone (GnRH)**. GnRH is composed of 10 amino acids (a decapeptide) in all animals studied to date. However, more

than 20 different versions of GnRH have been seen in vertebrates, and most species produce two or more versions of GnRH with subtly different effects on target tissues. The primary role of GnRH is in reproduction, but it has other roles as well, such as behavioral control.

GnRH is produced by hypothalamic neurons and released into the portal system that carries hypothalamic factors to the anterior pituitary. The neurons release a burst of GnRH that triggers secretion of LH and FSH. Differences in the way LH and FSH are stored affect the profile of these hormones in the blood. The anterior pituitary stores ample LH in vesicles that can be released in synchrony to induce a pulse of LH in the blood. In contrast, the anterior pituitary stores little preformed FSH, producing FSH on demand in response to GnRH.

The gonadotropins regulate many aspects of reproductive physiology, acting through effects on their primary target tissue, the gonads. In addition to effects on the gametogenic tissues and other gonad functions, they induce the release of estrogens in the female and androgens in the male. These hormones then act on other tissues, including both the primary reproductive tissues (ovary and testes) as well as those considered to be secondary sex features (mammary glands, hair follicles, and male sexual displays). These relationships between the hypothalamic-pituitary axis and the gonads are depicted in Figure 14.4.

JH and 20HE control development and reproductive physiology of arthropods

Recall from Box 3.1: Evolution and Diversity: Ecdysone: An Arthropod Steroid Hormone that invertebrate steroid hormones differ from those used by vertebrates. *Ecdysteroids*, a group of hormones derived from the steroid ecdysone, control reproduction and development. Most arthropods rely on 20-hydroxyecdysterone (20HE), which is produced from ecdysone. (In older literature, 20HE is often called β-ecdysone, and its precursor α-ecdysone.) Although 20HE is the most potent ecdysteroid, ecdysone and other derivatives also have important roles in some species. Ecdy-steroids are produced by the prothoracic glands or gonads, depending on the species and life stage.

Levels of 20HE depend on control of both synthesis and degradation. Furthermore, the pathways of synthesis change as the animal matures. A larva produces ecdysteroids in its prothoracic glands, but when the animal metamorphoses, these glands degenerate and the gonads become the main site of production. Ecdysteroid synthesis and release are regulated by numerous peptide hormones. One of the first such regulators identified was *bombyxin*. This hormone, first isolated from the silk moth *Bombyx mori*, is a protein that is structurally related to the vertebrate protein hormones of the insulin/insulin-like growth factor family.

In addition to ecdysteroids, invertebrates use various terpenoid compounds to control reproductive development, metamorphosis, molting, and metabolism. Insects rely upon *juvenile hormone (JH)*, whereas crustaceans use *methyl farnesoate*. When an insect egg hatches, a juvenile form (larva) emerges and begins to eat. As the larva grows, it reaches the capacity of its rigid exoskeleton. The first larva, also known as the first instar, undergoes **ecdysis** (molting): it splits open the exoskeleton, rapidly increases in volume, and then resynthesizes a new, larger exoskeleton. Most insects un-

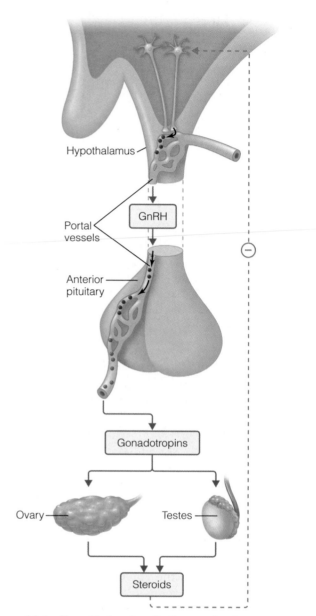

Figure 14.4 Hypothalamic-pituitary axis The hypothalamus receives signals from the brain and bloodborne hormones, and responds by releasing gonadotropin-releasing hormone into the pituitary. This induces the release of gonadotropins (LH and FSH) into the blood, for transport to the gonads. The gonads respond by increasing steroid hormone synthesis. Ovaries release estrogens and progesterone, which exert effects on primary reproductive tissues, such as the uterus, and secondary sex tissues, such as the mammary glands. In males, the testes release androgens, which exert effects in the testes, but also affect other tissues, including secondary sex organs and muscles.

dergo multiple larval molts prior to adulthood. The last step of development, the emergence of the adult, occurs by one of two alternate routes. In *holometabolous* insects, the last instar forms a cocoon, a fibrous external coating around the inner juvenile form, at this stage called a *pupa*. While it appears dormant, inside the cocoon the pupa is re-organizing its physiological systems in preparation for reproductive maturation. An adult form emerges from the cocoon, although it may need to undergo additional sexual development. In *hemimetabolous* insects, the larval forms, usually called *nymphs,* undergo repeated molts, with the last nymph emerging as an adult. Holometabolous insects include lepidopterans (butterflies), dipterans (flies), and coleopterans (beetles). Odonates (dragonflies), orthopterans (locusts), and true bugs (hemipterans and homopterans) are hemimetabolous.

Juvenile hormone is so named because of its role in maintaining juvenile characteristics in the larvae. It stimulates the synthesis of larval exoskeleton, which differs in molecular composition from the adult exoskeleton. High levels of JH also prevent larvae of holometabolous insects from undergoing pupation. Only when JH levels decline can the larva enter the pupal stage. During the pupal stage, JH levels continue to fall and the pupa develops into the adult form. Once JH levels have fallen to some minimum value, the neurosecretory cells of the brain release another hormone, *eclosion hormone,* and the adult emerges from the cocoon (**eclosion**). Upon eclosion, JH assumes a new regulatory role, increasing in concentration to trigger sexual maturation in both the male and the female.

The activity of terpenoids, like 20HE, is controlled by both synthesis and degradation. JH biosynthesis in the corpus allatum is regulated by factors released from neurons and neuroendocrine cells. **Allatotropins** are peptide hormones that stimulate JH production and release, whereas **allatostatins** are inhibitory peptide hormones. Insects also control the levels of JH through degradation, using enzyme JH esterase to convert JH to less active metabolites. Thus, an increase in JH esterase activity is one way an insect larva reduces JH levels to allow it to proceed in development. Many plants possess chemical mimics of JH that disrupt the normal development of insects, protecting the plant from herbivorous insects. Similarly, many insecticides such as *methoprene* work on the same principle as the natural plant agents that act as JH mimics.

Sex Determination

Sex is strictly defined in relation to gamete size, but there is a frequent misconception that sex is always a result of the presence or absence of the Y chromosome. In mammals, the Y chromosome is called the sex-determining chromosome; a male results when the zygote is heterogametic (XY) and a female when it is homogametic (XX). This pattern, however, is not universal. In birds and butterflies, for example, the female is the heterogametic individual (designated as ZW) and the male is homogametic (ZZ). In other species, the sex is determined by many factors, so the genotype is not a good predictor of the sex. Consider, for example, the situation in honeybees. If an egg is fertilized, the diploid offspring is female, but if the egg remains unfertilized, a haploid male results. The reproductive males produce sperm through a modified form of mitosis. This pattern of sex determination, called *haplo-diploidy*, allows the queen bee to control the numbers of males and females within the colony.

Asexual reproduction occurs by cloning and parthenogenesis

The genetic diversity arising through sexual reproduction helps animals evolve in changing environments, but for those species that live in a relatively constant environment, genomic variation is not necessarily an advantage. Evolution has endowed some sexually reproducing animals with the capacity to reproduce asexually. Corals, for example, reproduce sexually but have evolved the ability for asexual reproduction, producing buds that are clones of the parent. Buds form from somatic tissues of the adults, either male or female. This allows a single individual coral to produce a colony of clones.

Most forms of asexual reproduction in animals are not through clonal mechanisms, but rather by **parthenogenesis** ("virgin birth"). In contrast to cloning, parthenogenesis occurs through the use of ova and the female reproductive system. In contrast to sexual reproduction, no male is involved. Parthenogenesis is important in many species of invertebrates. For example, when food is abundant a female aphid can use parthenogenesis to rapidly produce 50–100 young aphids. Within only a few days these offspring, which are tiny versions of their mother, reproduce parthenogenically, resulting in an aphid infestation. Parthenogenesis is less common in the vertebrates. The whiptail lizard (*Cnemidophorus uniparens*) exists as an entirely female species that reproduces by parthenogenesis. *Cnemidophorus inornatus*, its closest relative and likely the ancestral species, reproduces sexually. Remarkably, many of the mating behaviors that occurred in the ancestral species still occur in the parthenogenic species. In the sexual species, a surge of progesterone causes a male to mount a female. In the asexual species, a progesterone surge occurs in an ovulating female, causing her to mount another female. These "mating" behaviors are common in parthenogenic species. In some species, the behaviors are simply a regulatory remnant of their sexual ancestry, but some mating rituals take on new functions. For example, the mating behavior of two parthenogenic females can induce ovulation.

Parthenogenesis allows a single diploid female to use its reproductive tissue to produce offspring that may be diploid or haploid, depending on the pathways involved. The most common pathway of parthenogenesis, called **automictic parthenogenesis** (*automictic* = self-mixing), is a variation on the standard meiotic pathway for oogenesis, proceeding to the point where the secondary oocyte is formed (see Figure 14.2). When the secondary oocyte undertakes the second meiotic division, the second polar body doesn't degrade but instead fertilizes the ovum, resulting in a homogametic zygote. Since the offspring from automictic parthenogenesis is homogametic, the sex of the offspring depends on which of the sexes is homogametic. *Thelytoky* is a form of automictic parthenogenesis in which a homogametic female (XX) produce females; it lacks the Y allele that is needed to produce a heterogametic male (XY). In contrast, in *arrhenotoky*, heterogametic (WZ) females produce only males (ZZ). If the ovum and second polar body originate from the Z allele, offspring are ZZ males. If the ovum and second polar body arise from the W allele, a nonviable WW genotype is formed. Populations survive by alternating between sexual and parthenogenic reproduction. These two forms of parthenogenesis are summarized in Figure 14.5. Note that meiosis is also important in parthenogenesis, and parthenogenic offspring also benefit from chromosomal recombination.

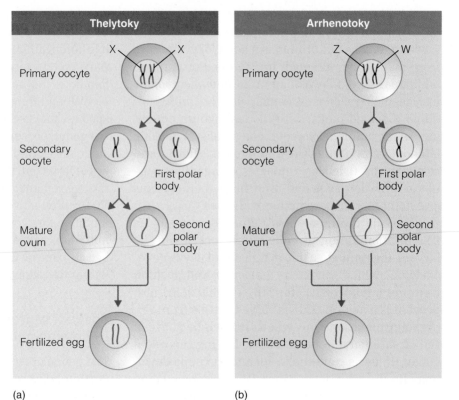

(a)   (b)

Figure 14.5 Automictic parthenogenesis In some species, the females reproduce by parthenogenesis when the second polar body fertilizes the ovum. **(a)** In thelytoky, homogametic females produce only female offspring. **(b)** In arrhenotoky, heterogametic females produce male offspring through parthenogenesis.

Animals may be simultaneous or serial hermaphrodites

Sexual reproduction does not necessarily require genetically separate sexes. Many species are *hermaphrodites*, possessing the capacity to produce both eggs and sperm. Some hermaphrodites, like the earthworm, produce both eggs and sperm at the same time. The testes are in segments that are separated from the segments bearing ovaries. When two earthworms copulate, they arrange their ventral sides together, but oriented anterior to posterior. Thus, the spermatogenic tissue is directly against the oogenic region. Although this antiparallel arrangement optimizes the chance of cross-fertilization for both worms, self-fertilization can occur.

Other species are serial hermaphrodites, existing for part of life as one sex but sometimes exercising an option to change to the other sex later in life under some circumstances. *Protogynous* animals are first female (producing eggs), then become male (producing sperm). *Protandrous* animals are male first, then become female. In many cases, the switch from one sex to the other occurs in response to environmental conditions, including social interactions. For example, as discussed in Chapter 3: Cell Signaling and Endocrine Regulation, some female coral reef fish spontaneously transform into males if the dominant male in the community is removed. The transition from female to male appears to involve a change in the metabolism of the main sex hormones. The male reproductive system is maintained by testosterone and its metabolite 11-ketotestosterone. The female reproductive state is maintained by estradiol-17β. The control of sex is linked to the metabolism of testosterone. In females, testosterone is metabolized to estradiol-17β through a pathway involving the cytochrome P450 enzyme aromatase (see Figure 14.3). When aromatase inhibitors are given to females, they undergo a sex change in little more than two months. The new males possess low levels of estradiol-17β and high levels of testosterone and 11-ketotestosterone. It remains unknown how environmental factors, including social interactions, act through physiological regulators to alter steroid hormone metabolism in the natural setting.

Sex is determined in some species by environmental conditions

In most animals, the sex of young is determined by the genotype: presence or absence of sex-determining chromosomes. However, in some species sex is determined by the physical and chemical environment around the developing embryo. The most common form of environmental sex determination is temperature-dependent sex determination (TSD). It is very common in reptiles, occurring in all crocodilians and marine turtles, as well as selected species of lizards and terrestrial turtles. At an intermediate ambient temperature, called the **pivotal temperature**, equal numbers of males and females result. Three main patterns emerge in studies of TSD. Some turtles produce males when temperatures are below the pivotal temperature and females above. Conversely, some lizards produce males when temperatures are high and females at low temperature. A third pattern is seen in crocodiles and alligators: female offspring dominate at both high and low temperatures, but male offspring are more abundant at intermediate temperatures.

It is not always obvious why one particular sex may be advantageous at a given temperature, and it is not yet certain that the mother actively biases the sex ratio by choosing where to lay her eggs. TSD also poses some risks; if temperature were the only factor that influenced sex, then conditions could arise in which whole populations would be-

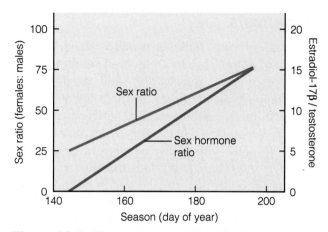

Figure 14.6 Temperature-dependent sex determination In painted turtles, the levels of steroid hormones in yolk changes throughout the breeding season, and correlates with the prevalence of females.
(Source: Adapted from Bowden et al., 2000)

come threatened by abnormally high (or low) temperatures, resulting in a preponderance of only one sex in the population. Thus, such species could be at great risk from global climate change.

The sex ratio resulting from TSD is influenced by other factors as well. For example, levels of sex hormones in the yolk influence the pattern of sex determination during development. These hormone levels vary seasonally, imparting a seasonal aspect to the sex ratios. For example, in painted turtles, a temperature of 28°C generates near-equal numbers of males and females in the middle of the breeding season (Figure 14.6). At the extremes of the breeding season, the same temperature can yield 75% males or 75% females. The difference appears to be linked to the relative levels of estradiol-17β and testosterone in the egg. When eggs have relatively high estradiol-17β levels, the clutch is more female biased. Species that rely on hormones in the yolk as a mechanism to regulate TSD are particularly susceptible to endocrine disruptors, which can alter sex ratios independent of temperature.

CONCEPT CHECK

1. What are the main hormones in vertebrate reproduction and where are they produced?
2. How does the chemical structure of reproductive hormones affect they way their synthesis is regulated?
3. What is the difference between clonal reproduction and parthenogenesis?
4. What is the difference between XY and ZW sex determination?

Oogenesis

In most species, females produce their lifetime supply of gametes early in life and retain them in a developmental quiescence until needed. The term *ovum* typically refers to the unfertilized gamete, without distinguishing between primary oocyte, secondary oocyte, or ovum (see Figure 14.2). In some situations, people may use ovum and egg interchangeably, but in other cases the egg may actually be fertilized and undergoing embryonic growth. The ovum is a single cell, but it is also associated with noncellular material produced by the female reproductive tract.

The three main modes of reproduction are ovipary, vivipary, and ovovivipary

The three main types of reproductive strategies—ovipary, vivipary, and ovovivipary—are distinguished by the fate of the ova prior to and after fertilization. Different degrees of parental care are roughly commensurate with the three reproductive strategies. **Oviparous** animals expel the ova from the body, and all development occurs externally using the resources within the egg. Fertilization may be external, as in most fish, or internal, as in birds and reptiles. The level of parental care ranges from none to intense. Few insects exhibit parental care, whereas most birds guard eggs and feed young. **Viviparous** animals use internal fertilization, and the young develop within the female body. In early development, the young derive significant resources from the mother. Placental mammals are the most obvious examples of vivipary, but it also occurs in some species of fish, snakes, and skinks. The female reproductive tract produces nutrients for the offspring, which can be a simple slurry of "uterine milk" secreted from the uterus, or more elaborate arrangements that allow the embryo to derive nutrition from the uterine blood vessels. **Ovoviviparous** animals demonstrate features of both ovipary and vivipary. They use internal fertilization, followed by extensive internal development of embryos. While in the uterus, the embryos derive their nutrition from the yolk, rather than the mother. When mature, the eggs hatch within the mother. This strategy is common in fish, including sharks, reptiles, and many invertebrates.

Surprisingly, reproductive mode varies widely within taxa. Some species are able to switch between modes. Brine shrimp, for example, can be ovovivi-parous and release free-living young (naupali) or oviparous, laying gastrulae encrusted in a shell (cysts). The reproductive strategy can differ among populations of a single species. The skink *Lerista* has both oviparous and viviparous populations. Among chondrichthians (sharks and rays), some species of skates lay fertilized eggs that float in the ocean currents, some sharks are ovoviviparous, and others are viviparous. In several shark species, the ovoviviparous embryo thrives on the nutrients from the egg, but then at some point begins to feed on its brothers and sisters within the reproductive tract—a life history strategy that is difficult to categorize.

Ova are produced within follicles of somatic tissue

The female reproductive tract includes the ovary, oviduct, uterus, and gonopore. The ovary is composed of the ova-producing **oogonia** as well as surrounding somatic cells that provide structural and nutritive support for oogenesis. In most species, oogenesis progresses through the primary oocyte stage (see Figure 14.2) early in the life of the female, but the final steps of the process are delayed until later in life. As the oocytes form, the surrounding somatic cells proliferate to form a **follicle** that encapsulates the oocytes. The follicle cells, or **granulosa cells**, secrete the extracellular matrix components that form an acellular layer between the oocyte and follicle cells, called the **zona pellucida**. The entire follicle is surrounded by a basolateral membrane, which in vertebrates is known as the **theca**.

The follicle cells orchestrate oogenesis, including the delayed maturation and ultimate release of the ovum. They communicate with the oocytes by paracrine factors and direct cell-to-cell contacts. Prior to ovulation, a subset of follicles is stimulated to mature (*folliculogenesis*). The oocyte must first increase in cytoplasmic volume, although the increase in cell size occurs by multiple mechanisms. Vertebrate oocytes grow by accepting biosynthetic precursors from the somatic follicle cells. A different pattern occurs in many invertebrates. In fruit flies, for example, oocytes absorb the cytoplasm of surrounding nurse cells, derived from oogonia that fail to differentiate into oocytes (Figure 14.7).

When the follicle ruptures, the ovum escapes the ovary and moves into the coelom. In some species, the ova are retained within the coelom. For example, some insects accumulate eggs until the abdomen bursts, killing the female. More com-

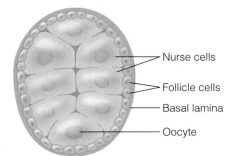

(a) Invertebrate follicle (*Drosophila*)

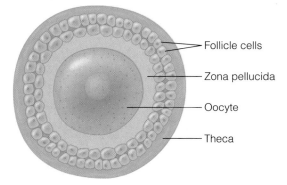

(b) Vertebrate follicle

Figure 14.7 The ovarian follicle Each oocyte is surrounded by somatic follicle cells. The entire follicle is encapsulated in a thin layer of extracellular matrix (basal lamina). **(a)** Invertebrate oocytes receive cytoplasm from nurse cells through gaps in the plasma membrane. **(b)** Vertebrate follicle cells produce a more extensive extracellular matrix at the apical (zona pellucida) and basolateral (theca) surfaces.

monly, the ovum crosses a short stretch of coelom and enters the opening of the oviduct, called the *fallopian tube* in mammals. The ovum passes through the oviduct into the uterus. Those species that use internal fertilization retain the ova in the oviduct or uterus. The uterus may be a simple passage, or it may be strong muscular tissue that uses smooth muscle contractions to expel ova, fertilized eggs, or young through the gonopore: the vagina in those species with a dedicated reproductive pore, or a cloaca if the reproductive and excretory systems have a common pore.

The yolk provides building blocks and metabolic precursors

Most animals, with the exception of placental mammals, provide each ovum with a source of nutrients in the form of **yolk**, a complex mixture of proteins and lipids. Most of the macromolecules in yolk are produced outside the oocyte, then sequestered by

the oocyte early in oogenesis. Triglyceride from the extracellular fluid passes from the blood between the follicle cells to the oocyte, where it is taken up and stored within vesicles. The yolk possesses many proteins, but **vitellin** is the most abundant. It is produced in the oocyte from **vitellogenin**, a bulky and complex phospholipoglycoprotein that is produced by the insect fat body, the vertebrate liver, and, in some animals, the follicle cells. Vitellogenin is taken up from the extracellular fluid by endocytosis. The internalized vesicles then coalesce to form larger yolk bodies.

Suites of hormones mediate vitellogenesis. External signals of various forms stimulate the central nervous system to release vitellogenic factors (Figure 14.8). In blood-feeding insects, vitellogenesis begins shortly after the animal consumes a blood meal, at which point a JH surge causes the fat body to produce vitellogenin. In vertebrates, vitellogenin

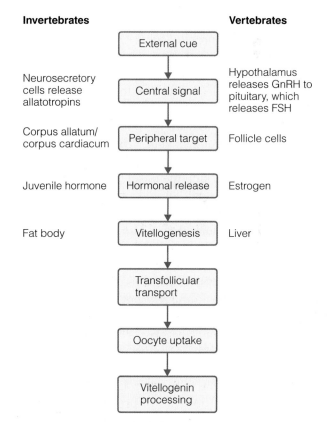

Figure 14.8 Vitellogenesis Animals initiate vitellogenesis in response to external cues, such as an environmental condition or developmental program. The pathways begin centrally within the brain, triggering a hormonal cascade that causes biosynthetic tissues to produce and secrete vitellogenin. This protein passes the follicular cells and is taken up by the oocytes, stored, and then converted to vitellin. Invertebrates and vertebrates differ in the specific hormones and target tissues, but the general features are similar.

is produced in response to estrogens, primarily estradiol-17β.

Insect eggs are surrounded by a chorion

Oogenesis has been well studied in many insects. In the silk moth, the ova develop in four ovaries (ovarioles), each of which contains in excess of 100 follicles arranged in series. The follicle that is closest to the gonopore undergoes oogenesis first (Figure 14.9). After about 2.5 hours, the next follicle enters oogenesis, and so on along the entire length of each ovariole. Thus, at late stages of oogenesis, a single ovariole possesses each of the developmental stages separated by about 2.5 hours of development. These stages are divided into three groups: *previtellogenesis, vitellogenesis,* and *choriogenesis.* Ecdysone controls the early development of ovarioles as well as the previtellogenic stages. During previtellogenesis, the follicles have not yet begun to produce yolk. The oocyte then begins to accumulate yolk proteins, marking the onset of the vitellogenic period. The yolk proteins from the fat body are transferred from the hemolymph to the oocyte across the follicle cells. The follicular cells also produce egg-specific proteins that are secreted and taken up by the oocyte. As with the early stages of development, ecdysone controls the production of the egg proteins, although not directly through changes in 20HE levels but rather through induction of a specific type of ecdysteroid receptor. After vitellogenesis has begun, a reduction in 20HE levels causes the follicle cells to begin **chorion** formation (*choriogenesis*). The follicular cells produce and secrete more than 100 types of proteins to construct the chorion. The ovum moves into the oviduct, where it is fertilized. Sperm cross this impermeable shell through a tunnel called the *micropyle*. The fertilized eggs are then laid.

The insects were the first animals to successfully invade land. Central to this invasion was the evolution of an egg that could withstand the terrestrial conditions. The chorion is resilient enough to

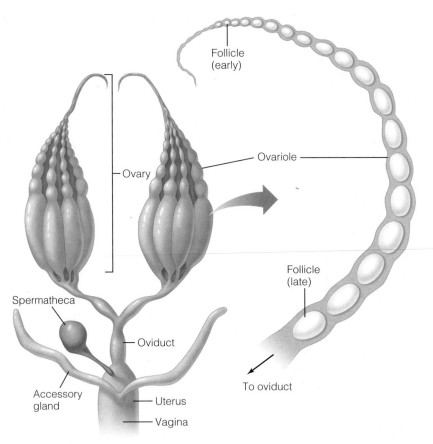

Female insect reproductive tract

Figure 14.9 **Oogenesis in the silk moth** When the ovarioles of the silk moth undertake oogenesis, the follicles mature in sequence. Each follicle is about 2.5 h more developed than the follicle next to it. The ova are released and pass down the oviduct into the uterus where they are fertilized with sperm that was collected and stored in the spermatheca after a previous mating.

withstand desiccation yet still able to permit the movement of gases (O_2, CO_2). As we see in the next section, the early terrestrial vertebrates faced the same problem but solved it a different way.

Egg structure differs in aquatic and terrestrial vertebrates

Most fish and amphibians produce eggs that are simple in structure. The ovum is physically connected to yolk. As the ovum passes down the reproductive tract, it receives a viscous coating from reproductive tract secretions. The gelatinous eggs are released from the animal into the water unfertilized. In amphibians and fish, the young leave the egg as aquatic larvae and complete their reproductive maturation. The constraint of this strategy is the need for water at all developmental stages. For the ancient vertebrates to be truly terrestrial, they needed a mechanism to reproduce on land. Like the insects, they produced a hardened external shell around the egg that provided support and prevented dehydration. In contrast to the proteinaceous chorion of insects, the eggshell of reptiles and birds is composed of calcium carbonates embedded in an organic matrix. The bird eggshell has a thick layer of calcium carbonate salts, giving it a brittle but hard texture. Many reptiles produce eggshells analogous to those in birds, but some reptiles, such as crocodilians and turtles, have a leathery and pliable eggshell. In these animals, the calcium carbonate crystals are aggregated into separate islands, allowing the eggshell to change in shape and even swell in the presence of water. Each egg is endowed with yolk to serve as an onboard source of fuel. The eggs also possess a viscous, hydrated protein (albumen) to act as a shock absorber.

Since the eggshell in birds and reptiles is impermeable, even to sperm, the animals also needed to coevolve a different mode of fertilization. The ovum in reptiles, birds, and monotremes (egg-laying mammals) must be fertilized before the eggshell is formed. Thus, fertilization is internal and the eggshell forms in the oviduct around a fertilized ovum.

Eutherian (placental) and metatherian (marsupial) mammals, of course, have dispensed with the eggshell and solve the challenges of terrestrial life by rearing fertilized ova internally. Nonetheless, each of these terrestrial vertebrate lineages produces embryos that, early in development, produce a complex set of internal membranes and fluid-filled compartments. Reptiles, birds, and mammals are collectively *amniotes*, a name derived from one of the four extraembryonic membranes. We discuss the origins of these membranes later in this chapter.

Spermatogenesis and Fertilization

In order to reproduce, male reproductive physiology ensures that sperm are prepared to fertilize the egg once the male engages in activities that bring the gametes in close proximity. For many species, the greatest challenge is finding a mate in order to breed. In some species, reproductive maturation coincides with an ability to sense and respond to mating factors released by one sex to attract the other (see Box 14.1, Evolution and Diversity: Pheromones). Once mates are found, males must be able to deliver sperm that are ready to move to the ovum and fertilize it to initiate embryogenesis. There is considerable diversity in the mechanisms by which the sperm are delivered to the ovum. Some species produce copious numbers of gametes, cast-

ing them into the open environment where a minute proportion of sperm successfully fertilize a few ova. Other animals engage in mating behaviors that bring males and females into close proximity to increase the likelihood of successful fertilization. Some clasp onto mates and release sperm in synchrony to maximize chances of fertilization. Other species use copulatory organs of various configurations, typically associated with the male. Once the sperm have been passed from the male, the individual sperm must find and fertilize the egg, often competing with sperm from other males.

Leydig cells and Sertoli cells control spermatogenesis

The typical testis produces spermatozoa in seminiferous tubules, which are composed of Leydig cells, Sertoli cells, and spermatozoa at various developmental stages (Figure 14.10). **Leydig cells**

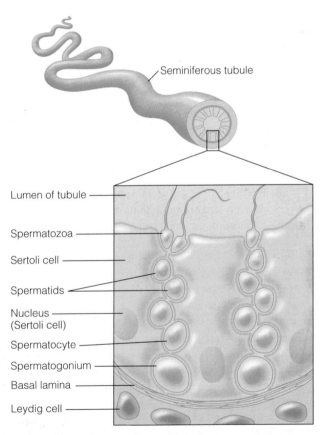

Seminiferous tubule

Lumen of tubule

Spermatozoa

Sertoli cell

Spermatids

Nucleus (Sertoli cell)

Spermatocyte

Spermatogonium

Basal lamina

Leydig cell

Figure 14.10 Seminiferous tubules Sperm production is controlled by Sertoli cells of the seminiferous tubules. The Sertoli cells interact through physical connections with spermatogenic cells at various stages, and interact with each other to form a blood-testes barrier. Leydig cells are found in the interstitial space on the blood side of the blood-testes barrier. These cells produce regulatory factors that act on Sertoli cells to control spermatogenesis.

BOX 14.1 EVOLUTION AND DIVERSITY
Pheromones

Most large animals find mates using visual cues that identify specific traits such as sex, receptivity, and evolutionary "quality." In contrast, smaller animals living in open spaces attract or locate mates using pheromones—chemicals released from an animal that elicit a *specific response* from another member of the *same species*. The specificity of the response is due to the activation of unique pheromone receptors that initiate a predictable signaling cascade. Thus, pheromones are distinct from other types of chemical signals, such as those that might, in human terms, equate to pleasant smells.

Sex pheromones are released to attract potential mates. Because of this role in reproduction, pheromone mimetics have been developed as alternatives to pesticides. The pheromones are scattered throughout the crop to confuse males, who are then unable to locate the trail emanating from a female. Other organisms have also co-opted pheromone signals for their own purposes. Some predatory spiders emit a chemical that mimics the pheromones used by specific moth species; the male moths are lured to the spider and trapped in the web. Some species of orchids have evolved to produce chemicals that mimic insect pheromones. This pathway relies on specific pheromone receptors, and is distinct from other pathways used by plants that release chemoattractants, such as the carrion plant, which attracts flies by emitting a fetid odor reminiscent of a decaying animal.

Many (though not all) vertebrates emit pheromones. Rodents emit a pheromone that, when detected by the female, accelerates the reproductive cycle to induce ovulation. Male mice release the pheromone in their urine. In tetrapods, most pheromones are detected in a region of the nostril known as the **vomeronasal organ**. Nerves transmit signals from the epithelium of the vomeronasal organ to the olfactory lobe. The chemosensors in the vomeronasal organ are distinct from the olfactory receptors found elsewhere in the nostril. Although pheromone receptors and olfactory receptors are both G-protein-linked receptors, the pheromone receptors are much more sensitive and trigger a distinct signaling pathway. Some pheromone receptors are found outside of the vomeronasal organ. For example, European rabbit dams release pheromones from the nipple to attract the pups for a short, intense feeding bout. Destroying the vomeronasal organ of the pup doesn't prevent it from rapidly finding the nipple.

Activation of the pheromone receptor affects an ion channel called TRP2. Scientists have used TRP2 knockout mice to test the role of pheromones in mouse behav-

ior. Mice are normally quite aggressive and territorial mammals. If you put a male mouse alone in a cage for a day or two, it will establish the entire cage as its territory and attack any new males that you introduce. Normal male mice do not attack intruder females, and they ignore castrated males. However, if urine from an intact male mouse is applied to a castrated mouse, the resident male mouse will attack the castrated male intruder, just as it would an intact male. If a TRP2 knockout male establishes a territory, it ignores introduced castrated male mice swabbed with the urine of an intact male. It also mates indiscriminately with both females and males. Thus, TRP2 signal transduction is an essential part of the communication of sexual signals in mice.

Pheromones do not appear to play an important role in humans. The vomeronasal organ is greatly reduced in size in adult humans compared to the size of this organ in the human fetus and in other adult mammals. Humans also lack an accessory olfactory bulb, the part of the brain responsible for interpreting pheromone signals in other animals. In addition, the majority of the genes encoding vomeronasal receptors contain deletions or other changes that would likely make them nonfunctional, and humans do not have a copy of the TRP2 ion channel. The TRP2 gene is present in the prosimians and the New World monkeys, but is mutated or absent in the Old World monkeys and the apes. Interestingly, the loss of pheromone-based signaling in the Old World monkeys roughly coincides with the evolution of color vision. It may be that Old World monkeys, apes, and humans rely more on visual signals than on pheromones for detecting gender. However, many experiments suggest that chemical cues can influence human behavior. For example, exposing women to swabs from the underarms of other women can alter the timing of their menstrual cycles. It is not yet known whether these signals are detected through the small vomeronasal organ (called Jacobson's organ in humans) or through the olfactory epithelium.

References

○ Keverne, E. B. 2002. Pheromones, vomeronasal function, and gender-specific behavior. *Cell* 108: 735–738.

○ Kohl, J. V., M. Atzmueller, B. Fink, and K. Grammer. 2001. Human pheromones: Integrating neuroendocrinology and ethology. *Neuro-endocrinology Letters* 22: 309–321.

○ Liman, E. R., and H. Innan. 2003. Relaxed selective pressure on an essential component of pheromone transduction in primate evolution. *Proceedings of the National Academy of Science USA* 100: 3328–3332.

Table 14.1 Mammalian reproductive hormones in male sexual development and reproduction.

Hormone	Tissue of origin	Main actions
Sexual maturation		
Androgens	Testes	Secondary sex characteristics: promote axillary hair growth, voice deepening, and libido
Spermatogenesis		
GnRH	Hypothalamus	Anterior pituitary: stimulates LH release, FSH synthesis and release
LH	Anterior pituitary	Leydig cells: stimulates androgen synthesis and release
FSH	Anterior pituitary	Sertoli cells: stimulates spermatogenesis
Androgens	Testes (Leydig cells)	Sertoli cells: stimulate spermatogenesis
Prostaglandins	Seminal vesicles	Uterus of mate: induce changes within the uterus that affect sperm motility

are interstitial cells found on the blood side of the basal lamina. They produce the testosterone that controls spermatogenesis. **Sertoli cells** are large cells that fill the gaps between columns of spermatogenic cells. Each Sertoli cell is in contact with about 50 spermatogenic cells. Sertoli cells serve many purposes in spermatogenesis, producing regulatory molecules as well as nutrients that are used for both metabolic energy and biosynthesis. They regulate the testosterone-signaling pathway by producing an androgen-binding protein. They also mediate the response of the testis to FSH, secreting other spermatogenic factors. The main effects of the sex hormones in mammalian males are summarized in Table 14.1.

The progression from spermatogonia to spermatids to spermatozoa involves a series of coordinated changes in cellular structure and function, with many of the precursors provided through cytoplasmic bridges that interconnect spermatozoa to neighboring cells. In the final stages of spermatogenesis, a spermatid reorganizes its microtubules to form the axoneme that underlies the flagellum. The length and structure of the flagellum varies widely in animals. It is essentially absent in some species, but can be as long as 6 cm, as with the sperm of the fruit fly *Drosophila bifurca* (Figure 14.11). Spermatozoa then eliminate much of their cytoplasm, leaving small, densely packed cells with abundant mitochondria organized around the base of the axoneme (see Figure 6.9). They also reorganize the DNA in their nuclei, replacing the histones with basic sperm-specific proteins called protamines, which

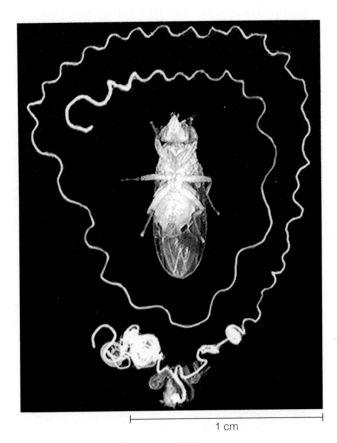

1 cm

Figure 14.11 ***Drosophila bifurca*** The image shows the long male reproductive tract of a fruit fly (*Drosophila bifurca*). The uncoiled tract, at 7 cm, is only slightly longer than is the sperm, each about 6 cm.
(Photo courtesy of Scott Pitnick, Syracuse University)

keep the DNA highly condensed and transcriptionally silent.

Once these structural changes are complete, the spermatozoa are released from the confines of the Sertoli cells into the lumen of the tubule. From here they progress along the male reproductive tract (Figure 14.12). At this point, the sperm are not capable of either swimming or fertilization, and must undergo a series of modifications. As the sperm pass into the **epididymis**, they further mature. It is in this region that they gain the capacity to swim. The sperm are stored in the epididymis, and fluids are removed to concentrate the sperm into a small volume. They are propelled by cilia along the tract through the **vas deferens**, which connects with the urethra, and then exit through the gonopore.

As the sperm pass through the reproductive tract, they are bathed in seminal fluid, a rich nutrient broth produced by several glands. The **seminal vesicles** produce an alkaline fluid with nutrients and regulatory factors. The high pH neutralizes the acidic ovarian fluid to allow the sperm to swim. The sperm use the nutrients, mainly fructose, as fuel for flagellar activity. The regulatory factors include prostaglandins, which affect the ovarian response to the sperm, and enzymes that break down chemical antagonists to fertilization. The **prostate gland** also secretes nutrients, mainly citrate, as well as enzymes that aid in fertilization. The **bulbourethral gland** secretes mucus that acts as a lubricant.

In some species, the sperm released from the male gonopore are not yet capable of fertilizing an egg. Mammalian sperm, for example, undergo a developmental transition known as **capacitation** only after they enter the female reproductive tract. Once inside, they are exposed to regulatory factors produced by the female that change sperm metabolism, ion regulation, and membrane fluidity, making the sperm capable of fertilizing the ovum.

Male copulatory organs increase the efficiency of sperm transfer

Of the many species that use internal fertilization (arthropods, mammals, reptiles, and some fish), most possess some form of copulatory organ or intromittant organ. Birds, one notable exception, transfer sperm directly from the male's cloaca to the female's cloaca. In other species, a modified or extra appendage is used as a copulatory organ. A male spider expels sperm and encases it in silk. The spider then uses a specialized leglike appendage to transfer the sperm package to a reservoir on its underside. During mating, the spider inserts the package of sperm into the gonopore of the female. Some fish possess modified pelvic fins that interlock to form a channel that guides sperm to the oviduct. Such copulatory organs are modified appendages, and the control of their movement is similar in many respects to locomotor control. Many reptiles possess an appendage near the cloaca, called a *hemipene*, that acts as a copulatory organ. A snake or lizard will insert a hemipene into the female, channeling sperm along a superficial groove. The hemipenes are often decorated with barbs or spikes that maximize the duration of penetration. A true penis is distinct from other copulatory organs be-

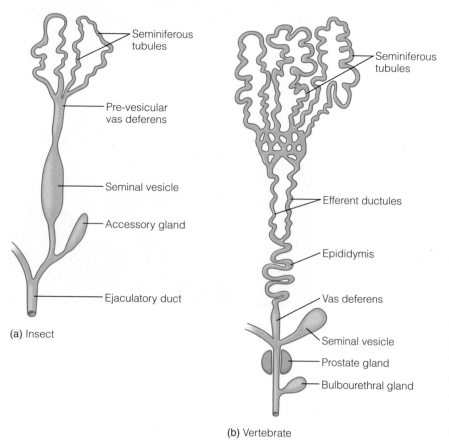

(a) Insect

(b) Vertebrate

Figure 14.12 The male reproductive tract Sperm released from the wall of seminiferous tubules are carried along the reproductive tract. As they pass through the epididymis and vas deferens, the secretions from the accessory glands provide seminal fluid.

cause it is a direct extension of the male reproductive tract. Some invertebrates possess a penis that they use like a spear, penetrating the female body wall and releasing sperm into the body cavity that holds the ova. More commonly, the penis inserts directly into the female reproductive tract.

The mammalian penis changes its blood distribution to create the hydrostatic pressure needed for a shape change (erection) that facilitates penetration. Many mammals also have a bone within the penis called an *os penis* or *baculum*. This allows the male to penetrate the female before the penis becomes erect. After penetration, the penis engorges, locking it into the vagina to maximize the probability of successful sperm transfer. An erectogenic stimulus, usually visual in nature, triggers the firing of neurons in the brain that transmit signals to the vasculature that feeds the penis. This

activates the enzyme nitric oxide synthase (NOS), resulting in production of the gaseous neurotransmitter nitric oxide (NO) (Figure 14.13). Recall from Chapter 6: Sensory Systems that smooth muscle contractility is under complex control by signaling pathways that affect the thick and thin filaments. As we discussed in Chapter 9: Respiratory Systems, in the vascular smooth muscle of the penis, NO binds guanylate cyclase, stimulating it to increase cGMP production, which activates cGMP-dependent protein kinase (PKG). PKG phosphorylates critical proteins to favor smooth muscle relaxation. It phosphorylates Ca^{2+} channels, inhibiting them to reduce cytoplasmic Ca^{2+} levels. PKG phosphorylates thick and thin filament proteins to desensitize the contractile apparatus. PKG may also phosphorylate K^+ channels to hyperpolarize the cell. Upon relaxation of the arteriolar

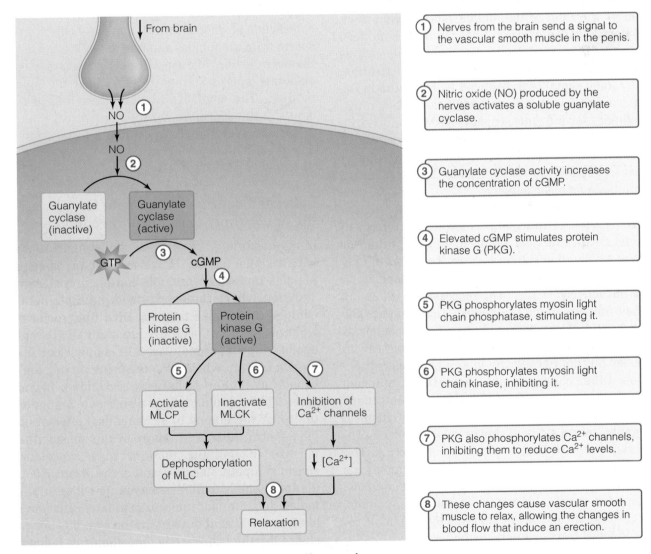

Figure 14.13 **Control of erection in a mammalian penis**

smooth muscle, blood flows into the penis, filling the surrounding spongy tissue and causing an increase in blood volume that compresses surrounding veins. The combination of increased blood inflow and reduced venous return causes the penis to engorge.

Drugs marketed to combat erectile dysfunction in human males target this signaling pathway. Sildenafil (Viagra), for example, inhibits the phosphodiesterase 5 (PDE5) of vascular smooth muscle. Since PDE5 breaks down cGMP, sildenafil allows cGMP levels to rise, permitting the changes in the vascular smooth muscle that are needed to respond to an erectogenic stimulus.

Sperm alter activity in response to chemokinetic and chemotaxic molecules

Depending on the reproductive strategy, ejaculation may propel the sperm into freshwater, saltwater, or the fluid of the female reproductive tract, generally called ovarian fluid. Sperm are induced to swim (activated) by an external signal, such as a change in ionic strength or Ca^{2+} concentration. The ionic signal is transduced by receptors in the sperm cell membrane, inducing changes in intracellular second messengers (cAMP or cGMP). These second messengers activate their respective protein kinases (PKA and PKG), which phosphorylate regulatory proteins within the axoneme to stimulate flagellar activity.

Once activated, most sperm swim for only a brief period of time. For example, the sperm of most freshwater fish swim for only a minute or two, which allows them to cross distances of only a few millimeters. With such a limited capacity to swim actively to the ova, successful fertilization may rely on chemical signposts that increase the likelihood of contacting an egg. Some chemicals are **chemokinetic**, stimulating the sperm to swim faster but not necessarily in any particular direction. Other chemicals are **chemotaxic**, inducing the sperm to swim toward higher concentrations of the agent. In the absence of these chemical agents, sperm swim at slower velocity, which conserves onboard fuels until an egg is detected. The chemical nature of chemotaxic and chemokinetic agents is diverse, and may include amino acids, peptides, and sulfonated steroid compounds. These chemicals may be released by the female reproductive tract or by the ovum itself.

Females use sperm storage to ensure uninterrupted reproduction

The females of some species store sperm for prolonged periods in specialized compartments within the reproductive tract. Sperm storage enables a female to fertilize her ova long after mating, which is adaptive in animals that might encounter mates infrequently. Some species mate before the female has reached reproductive maturity, and sperm storage allows her to bridge the gap between the mating and gonadal maturation. Sperm can be stored for long periods. Fruit flies store sperm for little more than a week. However, mated female honeybees can retain viable sperm for several years. Some large snakes in captivity have laid fertilized eggs five years after mating.

The anatomical strategies for sperm storage are diverse. Reptiles possess sperm storage tubules that branch from the uterus, or in some species, the vagina. Preovulatory surges in estrogens trigger contraction of smooth muscle that expels the stored sperm from the tubules and into the oviduct, where they can fertilize the egg. Insects possess a more elaborate sperm storage organ called the spermatheca. The length of the spermatheca reflects the length of the sperm. In *Drosophila bifurca,* the females have very long spermathecae to accommodate the 5-cm-long sperm.

Individual sperm can compete for the opportunity to fertilize the egg

In monogamous species, a single male mates with a single female. However, in many species, females undertake multiple matings, creating a situation where the sperm from multiple males compete to fertilize the ova. DNA fingerprinting technologies have been used to study the parentage of offspring in many taxa. It is now clear that *polyandry*, in which the offspring in a single brood have different fathers, is common in animals. The multiple matings also create an opportunity for the female to use chemical effectors to bias sperm utilization. Many of the species that partake in multiple matings may experience sperm competition as a result of the order of copulation. Female fruit flies may mate with multiple males, but the last male to copulate with her is likely to fertilize about 80% of the ova. It is not yet clear why the last, rather than the first, bolus of sperm is most successful. It is likely that the female is able to expel the sperm from the previous mating, allowing the sperm of the new suitor to fertilize the ova.

Some animals delay embryonic development

Once the sperm enters the egg, the oocyte undergoes many changes that initiate embryonic development. The DNA from the sperm enters the cell organized into a tight bundle of DNA associated with protamines. The fertilized ovum must remodel the condensed DNA from the spermatozoa into a more conventional organization. Soon after fertilization, the sperm DNA disperses and then recondenses as the protamines are replaced by histones. Once organized into nucleosomes, the paternal DNA within the oocyte can become transcriptionally active. At this early stage, the oocyte has two separate genomes: the maternal pronucleus and the paternal pronucleus. The fertilized ovum undergoes many rounds of cell division to reach the blastocyst stage. Soon afterward the various germ layers form, which differentiate to form the complex tissues that ultimately form the embryo. In most species, the embryonic development continues until the young escapes the confines of the egg or reproductive tract. A few species interrupt normal development, pausing at an early phase of embryogenesis. Such a delay allows animals to ensure that embryogenesis proceeds at the appropriate time to ensure hatching or birth occurs under favorable environmental conditions.

Brine shrimp are crustaceans that live in salt-rich water, such as Utah's Great Salt Lake. As discussed earlier in this chapter, brine shrimp can reproduce through ovovivipary or ovipary. The embryos of brine shrimp develop to the gastrula stage, at which point they can delay further development until environmental conditions are adequate. These brine shrimp cysts are simple enough to survive metabolic arrest associated with dehydration.

More than 100 species of mammals can control embryogenesis through **delayed implantation**. The fertilized ovum develops to the early blastocyst stage (100–400 cells) in the uterus, but implantation in the uterine wall is delayed, retarding further development. Some mammals, such as seals, have an obligate period of delayed implantation, whereas other species can use delayed implantation opportunistically. For example, a rodent may copulate shortly after giving birth to a litter, then delay implantation of the embryos for several weeks. In mammals, the delay can be a few days or weeks or as long as 11 months, as in the river otter.

Postfertilization development relies on maternal factors

Early embryonic development is a period over which the control of cellular processes is transferred from two independent parental genomes to the integrated genome of the offspring. In the earliest phase, cellular changes are governed by maternal factors that existed preformed in the ovum. This includes hormones (androgens and estrogens) that exert regulatory effects on developmental variables, such as sex determination. Gradually, the cellular control transfers to the embryo, when the contributions from paternal genes begin to influence the developmental pattern. In mammals, this transition from maternal to embryonic control occurs at about the two-cell stage. In lower vertebrates and invertebrates, the maternal control extends until the embryo consists of thousands of cells. Even after the paternal genes become active, factors present in the oocyte can continue to play an important role well into embryological development.

The division of the genomes of the two parents allows for differential modification patterns that influence later development, through the process known as *gene imprinting*. Under normal conditions, each diploid cell is able to produce mRNA from either the maternal or paternal allele of a gene. Before the maternal and paternal genomes merge into a single nucleus, a small subset of the genes may be modified in a way that prevents the maternal or paternal allele from being expressed. Furthermore, this imprinted gene remains transcriptionally silent throughout development, while the allele of the gene derived from the other parent is expressed. Most of the genes subject to imprinting encode proteins critical for normal growth and neurobehavior, such as insulin-like growth factor 2 (IGF-2) and its regulators. These hormonal pathways control embryonic growth, so the embryo is at the center of an interesting evolutionary conflict. The embryo possesses genes from both father and mother, but the costs of producing and raising the embryo are borne largely by the mother. Thus, it is in the best interest of the father to pass on genes that induce rapid embryonic growth; his offspring thrive but the mother bears the costs of rapid embryonic growth. Conversely, it is in the interest of the mother to curtail growth to a manageable level. Within the embryo, the parental genomes have conflicting goals, and patterns of gene imprinting on maternal and paternal alleles of genes for growth regulatory proteins determine the trajectory of embryonic development.

Amniotes produce four extraembryonic membranes early in development

Soon after fertilization, the embryo of amniotes produces sheets of cells that separate from the embryo to form the four extraembryonic membranes: chorion,[1] amnion, allantois, and yolk sac. These membranes grow in size as the embryo develops (Figure 14.14). The chorion, the outermost membrane that lies beneath the albumen, acts as a gas exchange surface. The amnion encloses the embryo. As the embryo develops, the amnion fills with fluids that act as a hydraulic cushion and provide a favorable ionic and osmotic environment for the embryo. The allantois is a membranous outpouching of the primitive gut. During development, it becomes vascularized, delivering gases between the embryonic circulation and the outer surface layers. In birds and reptiles, the allantois is also a storage sac for nitrogenous waste, mainly uric acid. The yolk sac surrounds the yolk, secreting digestive enzymes that break the yolk down into macromolecules that can be transferred to the embryo. The animal grows within the egg until it reaches a point where it can break through the shell. Although nonmonotreme mammals lack the hardened shell of other terrestrial vertebrates, they are also amniotes and the embryo possesses all of the same membranes. Later in this chapter, we will elaborate on the origins and roles of these membranes in mammalian reproduction.

---

[1] The term *chorion* is a general one that refers to an outer layer of an extraembryonic structure. The chorions of insects and amniotes are unrelated in origin or composition.

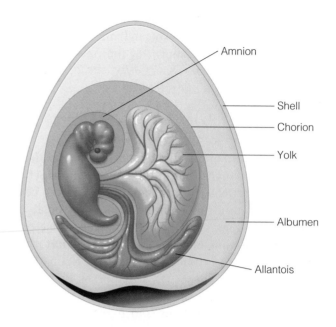

Amnion

Shell

Chorion

Yolk

Albumen

Allantois

Figure 14.14 The amniote egg Once the ovum of an amniote is fertilized, it undergoes cell division. Most of these cells form the embryo, but four sheets of cells separate from the embryonic tissue to form the extraembryonic membranes—the chorion, amnion, allantois, and yolk sac—which enclose compartments for storage of nutrients (yolk sac), fluids (amniotic space), and wastes (allantoic cavity).

◉ | CONCEPT CHECK

5. Compare the three main modes of reproduction in animals (vivipary, ovipary, and ovovivipary).

6. Trace the route of the oocyte from ovary to gonopore. Trace the route of the sperm from seminal vesicle to gonopore.

7. Which cells are the germ cells of males and females? Which cells of the gonads are somatic tissue in males and females?

The Reproductive Cycle of Mammals

The reproductive behavior of female mammals is closely linked to the ovulatory cycle, known as the **estrous cycle**. Mammals differ in the number and timing of estrous cycles. *Monoestrous* mammals, such as canines, undergo a single estrous cycle each year. *Polyestrous* mammals undergo estrous cycles throughout the year, although they may breed only during certain seasons. Humans and other primates are polyestrous animals, although the estrous cycle is more commonly known as the menstrual cycle. Some people make a distinction between estrous and menstrual cycles based upon female behavior; a species has an estrous cycle if the female demonstrates a period of intense interest in mating that coincides with a specific part of the ovulatory cycle. For example, dogs and cats go into *heat* at a specific point of the estrous cycle; they exhibit anatomical and behavioral changes that "inform" potential mates that they are ovulating and interested in copulation. Conversely, human females exhibit interest in copulation at many phases of the reproductive cycle and show few outward signs of ovulation. Other people distinguish a menstrual cycle from an estrous cycle by the magnitude of the loss of uterine tissue loss (**menses**) at the end of a cycle. Although most species exhibit cyclical changes in the uterine wall, and many show evidence of vaginal discharge, the relative volume is much greater in primates. Thus, the differences between estrous and menstrual cycles are qualitative in nature and any effort to further distinguish between estrous and menstrual cycles is somewhat arbitrary.

Hormones control the ovarian and uterine cycles

Hormones control the cyclical maturation of follicles, ovulation, and the parallel changes in the uterine wall. With the exception of humans, most mammalian species coordinate ovulation and copulation; the same hormones that regulate ovulation induce external and behavioral displays that announce they are receptive to copulation. The hypothalamus releases gonadotropin-releasing hormone (GnRH) into the portal blood vessels, causing the anterior pituitary to secrete pulses of gonadotropins (LH and FSH), which in turn cause the gonads to produce steroids (progesterone and estradiol-17β). Another hormone that plays a role in the regulation of ovulation is *inhibin*. A peptide hormone of the TGF-β family of cytokines, inhibin is released by the mature follicle cells and exerts multiple effects. It has endocrine effects at the hypothalamic-pituitary axis, inhibiting the release of FSH. It also has autocrine and paracrine effects at the ovary, inhibiting the production of estrogen.[2] These hormones interact through both positive and

negative feedback cycles. The hormones that drive follicular events cause other physiological and behavioral changes in the female.

The estrous cycle is composed of four phases: estrus, metestrus, diestrus, and proestrus. The first day of estrus is demarked by the onset of interest in mating, typically identified by the nature of social interactions or the assumption of mating postures in the presence of males. In the ensuing sections, we discuss the control of ovarian and uterine events from the human perspective, where the ovulatory cycle is discussed as two phases of two weeks each. The *follicular phase* begins on the first day of menses. The *luteal phase* begins after ovulation. Keep in mind that the general features are similar among mammals, but there is considerable variation in the details. For example, not all human females show a standard 28-day ovulatory cycle. A normal cycle is considered somewhere between 25 and 35 days. It can vary within and among women as a result of diet, stress, and exercise. The shortest ovulatory cycle in mammals, seen in the golden hamster, is 4 days. In humans, the ovarian and uterine changes are tightly linked in each reproductive cycle. However, in many species, the nature of the cycles and linkage between them may depend on copulation, where the mechanical stimulation of the vagina induces hormonal changes that in turn affect the ovarian and/or uterine changes. For example, some species use copulation to trigger the rupture of the follicles (ovulation), and other species are programmed to ovulate cyclically but modify the uterus only in response to copulation.

The follicular phase of ovulation is driven by FSH

We begin our discussion late in the luteal phase, when the levels of estrogen and progesterone decline by mechanisms clarified later in this section (Figure 14.15). Recall that these hormones suppress the release of GnRH from the hypothalamus, and thereby minimize gonadotropin release from the anterior pituitary. Thus, once the levels of progesterone and estrogen fall below a critical threshold, hypothalamic GnRH is secreted into the blood, stimulating the anterior pituitary to secrete gonadotropins; LH increases slowly whereas FSH increases more rapidly.

---

[2] *Estrogen* is a general term that does not distinguish between the various estrogens. Although estradiol-17β is the most important estrogen in most mammals, the other estrogens such as estrone contribute to estrogen signaling.

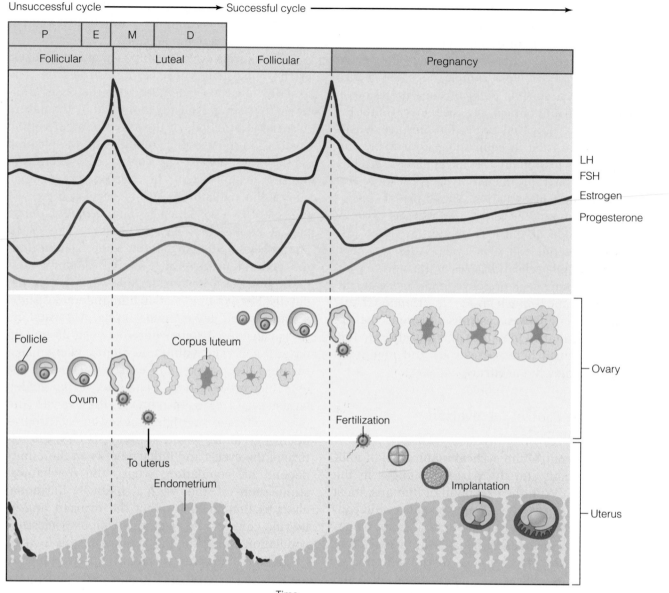

Figure 14.15 **Ovulation cycle in mammals** The estrous cycle is divided into proestrus (P), estrus (E), metestrus (M), and diestrus (D), or alternatively the follicular and luteal phase. The exact relationships differ among species, but in general estrus coincides with ovulation, the demarcation between the follicular and luteal phases. The figure shows two ovulatory cycles, the first ending without fertilization and the second with fertilization.
(Source: Adapted from McNaught and Callander, 1975)

These two hormones act on different cell types of the ovary to coordinate the maturation of the follicle and the metabolism of sex hormones. The rise in FSH causes the granulosa cells to proliferate. As the follicle grows in size, the outermost layer of granulosa cells differentiates to form the theca. One function of the mature follicle is to produce the appropriate amount of progesterone and estrogen. The extrafolliclar cells of the ovary (interstitial cells) produce and release progesterone; some escapes into the blood and some makes its way to the theca of maturing follicles. The theca cells use the progesterone to produce androgens, some of which makes its way to the inner granulosa cells, where it is used to produce estrogen. This difference in steroid synthesis among ovarian cells is due to the expression of the genes for the enzymes of steroid metabolism. Differentiating theca cells express LH receptors, enabling them to respond to LH by expressing the appropriate genes for androgen synthesis, as well as aromatase.

Early in the follicular phase, many follicles mature in parallel. As the collection of follicles grows, estrogen secretion increases. The elevated estrogen in the blood exerts negative feedback on the hypothalamic-pituitary axis, blocking GnRH release from the hypothalamus and production of LH and FSH by the anterior pituitary. The decline in estrogen and the increase in inhibin act together to suppress FSH release. With FSH levels plummeting, most of the follicles are unable to sustain their own development and undergo **atresia**, a form of apoptosis. However, a subset of follicles, called *dominant follicles*, matures to the point where they can sustain maturation despite falling FSH. It is not yet clear if the dominant follicles escape the effects of plummeting FSH by increasing the number of FSH receptors, or modulating the local signaling environment to make FSH more effective. Regardless of the mechanisms, the dominant follicles continue to mature, with the granulosa cells growing in number while awaiting the signal for ovulation.

Ovulation and the luteal phase follow an LH surge

The negative feedback interaction between estrogen and the hypothalamic-pituitary axis is essential for follicle maturation and selection of the dominant follicle. However, in the late follicular phase, the hypothalamic-pituitary axis reorganizes its signaling pathways in a way that reverses the effects of estrogen. Instead of impairing GnRH release, estrogen stimulates GnRH release.

Of the gonadotropins, the most important hormone for late follicular maturation is LH. The growing follicle continues to produce estrogen, which in turn enhances LH release, an example of positive feedback. The dramatic increase in LH, called the *LH surge*, causes the granulosa cells to secrete several factors that support oocyte maturation. Paracrine signaling factors induce the oocyte to complete its meiotic pathway, generating the ovum. Enzymes are secreted to digest the extracellular matrix between the follicle cells. The follicle weakens and ruptures to release the ovum. Just prior to ovulation, the follicle cells increase the production of progesterone. Ovulation marks the beginning of the luteal phase. Depending on the species, the transition from estrus to metestrus occurs either slightly before or slightly after ovulation.

After ovulation, the remnants of the follicle continue to play an important role in hormone synthesis. Driven by the LH surge, the follicle undergoes a change in structure, increasing in size and complexity as capillaries and fibroblasts penetrate the structure. The remnants of the ruptured follicle appear as a dense yellow body in the ovary known as the **corpus luteum** (which roughly translates as "yellow body"). The corpus luteum maintains the ability to synthesize and secrete large amounts of progesterone and lesser amounts of estrogen. These hormones ensure that the uterine wall changes in preparation for implantation.

In the luteal phase of the cycle, the corpus luteum sustains steroid hormone secretion for a time, but estrogen and progesterone levels begin to decline. What happens next depends on whether a fertilized ovum implants in the uterus. If the ovum is not fertilized, progesterone levels continue to decline and the next ovulatory cycle begins. Before considering what happens when the ovum is fertilized, we will consider the relationship between the ovulatory cycle and the changes in the uterine wall.

The endometrial cycle parallels the ovulatory cycle

The events in the ovulation cycle are coordinated with changes in the uterine cycle, through shared sensitivity to steroid hormones. The uterus is composed of a layer of smooth muscle (**myometrium**) covered by a layer of epithelial tissue (**endometrium**). When the ovulatory cycle is in the follicular phase, the endometrial cycle is in the *proliferative phase*. The endometrium thickens as epithelial, immune, and glandular cells replicate (hypertrophy), with blood vessels growing in parallel to ensure vascularization. The luteal phase of the ovulatory cycle coincides with the *secretory phase* of the endometrial cycle. The endometrial cells secrete numerous regulatory factors, including cytokines and prostaglandins, that ensure the uterus is prepared for implantation of the growing embryo.

The hormones involved in regulating the ovulatory and endometrial cycles of female mammals are summarized in Table 14.2. The birth control pill, developed in the early 1960s, is a combination of hormones that impairs ovulation, fertilization,

Table 14.2 Mammalian reproductive hormones in females in the ovulatory cycle.

Hormone	Tissue of origin	Main actions
Sexual maturation		
Estrogens	Ovary	Secondary sex characteristics: promote fat deposition, maturation of ovaries and mammary glands
Androgens	Ovary, adrenal gland	Secondary sex characteristics: promote axillary hair growth and libido
		Puberty and menarche
Follicular phase		
GnRH	Hypothalamus	Anterior pituitary: controls LH release, FSH synthesis and release
LH	Anterior pituitary	Ovarian follicle: triggers ovulation
FSH	Anterior pituitary	Ovarian follicle: stimulates estrogen synthesis and follicle maturation
Estrogens	Ovarian follicle	Ovarian follicle: stimulate proliferation of granulosa cells
		Endometrium: stimulate proliferation of endometrial cells, sensitization to progesterone, angiogenesis
		Hypothalamic-pituitary axis: reduce gonadotropin levels by negative feedback
Luteal phase		
Estrogens	Corpus luteum	Hypothalamus–anterior pituitary: inhibit GnRH release, reducing release of FSH and LH from anterior pituitary to prevent folliculogenesis
Progesterone	Corpus luteum	Uterus: promotes maturation of endometrium and reduces uterine smooth muscle contractility
Inhibin	Corpus luteum	Hypothalamus–anterior pituitary: impairs FSH synthesis and release

and implantation. Although exact compositions differ among brand names, most birth control pills are composed of chemical analogues of estrogen and progesterone. Since maturation of follicles in the late luteal phase is possible only after progesterone levels decline, the elevated progesterone levels prevent ovulation, essentially by convincing the ovary that the female is pregnant. In addition to its effects on ovulation, the pill also thickens the cervical mucus layer, impairing sperm movement into the uterus, thus reducing the likelihood of fertilization if ovulation does occur. If ovulation and fertilization occur, the pill also reduces the likelihood of successful implantation because it impairs endometrial growth.

The regular cycle of ovulation changes when an ovum is fertilized. The embryonic tissues and placenta gradually become an endocrine gland, taking over the central control of hormone levels to ensure that the fetus matures to the point that it can be expelled from the uterus in the process of **parturition**.

A placenta forms after a fertilized ovum implants in the uterine wall

A fertilized ovum begins the process of cell division and continues to divide for several days. It sheds the zona pellucida, and then the remaining cells form the blastocyst. Groups of cells differentiate to form the embryonic structures. The outermost cells differentiate to form the **trophoblast** (Figure 14.16). Then the embryo attaches to the uterine wall to begin the process of implantation. Trophoblast cells proliferate and invade the endometrium, forming an association that will develop into the placenta. The trophoblast cells differentiate to form the chorion. At the same time, the inner cell mass of the blastocyst continues to divide and differentiate. First, a gap appears between cells to form the amniotic cavity. The cells that surround the amniotic cavity differentiate to form the amnion. The remaining cells of the blastocyst inner cell mass form the embryo, which grows to become the fetus.

Central to the development of the fetus is the placenta. It is the interface between mother and fetus, and composed of cells derived from both. For the first third of the pregnancy, the placenta has a vital endocrine function. The region of the placenta that was derived from the chorion secretes chorionic gonadotropin (CG). Like LH, another gonadotropin, CG targets the corpus luteum in the ovary to ensure that it continues to secrete estrogen and progesterone. These hormones are vital to the remodeling of the mother's physiology necessary to sustain the pregnancy and prepare her for parturition and subsequent maternal care. Later in the pregnancy, the placenta itself becomes the main source of progesterone and estrogen, and the corpus luteum degenerates. In some mammals, the corpus luteum remains the main source of steroid hormones throughout the pregnancy.

The duration of gestation varies widely among mammals. *Altricial* species (those giving birth to large litters of poorly developed young) have shorter gestation periods than *precocial* species (those having fewer, well-developed offspring). For animals of similar body size, the gestation period for a precocial species is about three times longer than that for an altricial species. Body size also plays a role (Figure 14.17).

Contractions of uterine smooth muscle induce parturition

The uterus has thick walls of smooth muscle (*myometrium*) underlying the endometrium. As the fetus develops, the elevated levels of progesterone and estrogens remodel the uterine smooth muscle to

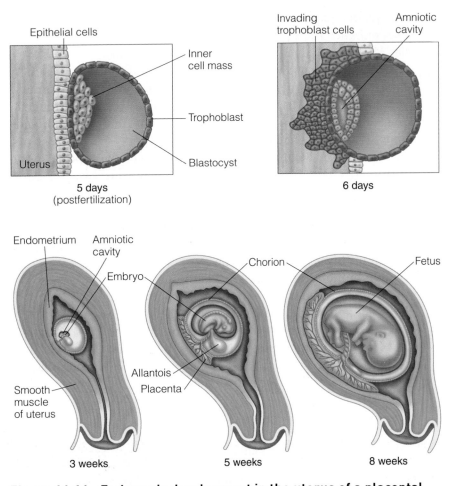

Figure 14.16 Embryonic development in the uterus of a placental mammal Cell division begins once the egg is fertilized, which usually occurs in the oviduct. Implantation begins after the blastocyst binds to the uterine wall. Cells in the outer blastocyst layer, the trophoblast, invade the endometrium of the uterus and begin to form the placenta. The extraembryonic membranes develop, and the amniotic cavity increases in volume. The timeline shown in this figure is for a primate.

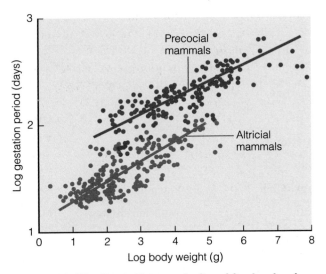

Figure 14.17 Gestation period and body size in mammals Larger mammals have longer gestation times.
[Source: Adapted from Martin, 1989]

prepare for parturition. The high levels of estrogens enhance the contractile strength of the muscle. They also induce the expression of the genes encoding the receptor for oxytocin, which has an important role in parturition. While the smooth muscle grows in strength, progesterone disrupts excitation-contraction coupling to prevent the smooth muscle from contracting prematurely.

Parturition begins in response to a series of hormonal changes. The levels of progesterone decline, allowing the strong uterine muscles to contract. At the onset of labor, fetal cells produce oxytocin, which acts on the placenta to induce the release of prostaglandins. At the same time, the mounting stress in the mother triggers the hypothalamic-pituitary axis, causing the release of oxytocin from her own posterior pituitary. Prostaglandins and oxytocin act on the uterine smooth muscle directly to induce contractions that begin to propel the infant along the uterus. As labor progresses, the additional stress triggers the release of even more oxytocin, further strengthening the contractions, an example of positive feedback regulation.

Soon after the young is born, the placenta is expelled. During gestation, the placenta was a major source of estrogen and progesterone. At this point, the female experiences a rapid decline in estrogen and progesterone production as a result of the loss of this endocrine gland. Once the fetus is born, the maternal physiology begins the process of postpartum recovery from birth, and simultaneously initiates the early steps of maternal care, including milk production. The hormones involved in regulating parturition and postpartum events are summarized in Table 14.3.

Milk is a secretory product of mammary glands

Mammals are unique in possessing mammary glands that enable a female to produce milk for her offspring. Although a remarkable adaptation, the mammalian mammary gland has many paral-

Table 14.3	Mammalian reproductive hormones in pregnancy and parturition.	
Hormone	**Tissue of origin**	**Main actions**
Pregnancy		
Chorionic gonadotropin	Placenta	Stimulates release of estrogen from corpus luteum
Estrogens	Placenta	Mammary glands: stimulate proliferation of secretory cells but prevent milk secretion
		Cervix: reduce mechanical resistance (ripens)
		Uterus: stimulate uterine smooth muscle (blocked by progesterone)
		Uterus: stimulate angiogenesis and mitotic division in endometrium
Progesterone	Placenta	Uterus: blocks estrogen's stimulation of smooth muscle
		Ovary: prevents ovulation
Parturition		
Oxytocin	Posterior pituitary	Uterus: promotes smooth muscle contraction
Prostaglandins	Placenta	Uterus: promote smooth muscle contraction
Prolactin	Anterior pituitary	Mammary glands: promotes growth and colostrum synthesis
Postpartum events		
Oxytocin	Posterior pituitary	Mammary glands: promotes smooth muscle contraction
Prolactin	Anterior pituitary	Mammary glands: stimulates growth and milk synthesis

lels. As mentioned earlier in this chapter, some species produce secretions that feed the offspring while still in the reproductive tract, such as the uterine milk of ovoviviparous fish. Several non-mammalian species produce nutrients for their free-living offspring. For example, pigeons produce *crop milk*, a secretion produced by the upper gastrointestinal tract and regurgitated into the mouths of the chicks.

Milk is produced by the mammary gland, and in almost all cases only the female produces milk. In the mid-1990s, researchers in Malaysia discovered that males of the local fruit bat species could produce and secrete milk. Even human males can undergo changes that induce their quiescent mammary glands to produce milk. Most commonly this is due to pathological changes in endocrine tissues, but there have been verified reports of lactation in pubescent males, and even documented cases of adult men producing enough milk to suckle an infant.

The hormone that controls milk production in mammals is prolactin, a peptide hormone released from the anterior pituitary gland. Interestingly, the role of prolactin in mammals is a variation on its role in other vertebrates, including regulation of secretory function, ion and water balance, and even behavior (see Box 14.2, Genetics and Genomics: Prolactin). During pregnancy, estrogen produced by the corpus luteum, and later the placenta, induces prolactin release. Prolactin prepares the mammary gland for milk production by increasing mammary gland mass and ensuring that the biosynthetic machinery is in place. During pregnancy, the high levels of progesterone and estrogen suppress the actual production of milk. Only after the levels of estrogen and progesterone decline during parturition does the mammary gland produce milk.

Several features of mammary glands are shared among all mammals. Like sebaceous glands, mammary glands are associated with hair follicles. The mammary glands are composed of both exocrine cells, which secrete the milk, and myoepithelial cells, which control the secretions. The milk itself is a mixture of fluids and macromolecules released from exocrine cells under the control of prolactin. However, the three groups of mammals differ in the complexity of the secretion and the anatomical structure of the gland itself. The monotreme mammary gland consists of convoluted tubes that lie beneath the ventral skin of the female. There are no teats or nipples to localize secretions; the milk oozes through the ducts onto the mother's fur, where it pools in surface indentations. Marsupials have a more complex mammary gland with discrete teats. The young locate the teat and suck it into the mouth, where it engorges to fill the oral cavity, locking the pup into position. Eutherian mammals possess complex mammary glands that are networks of lobes grouped into alveoli, with ducts that empty through an external teat (Figure 14.18). Unlike marsupials, the young of eutherian mammals can attach and detach at will. Behavioral interactions associated with feeding help establish a social bond between mother and offspring.

Mammary gland secretions include two novel products, casein and lactose

Mammary gland secretions are the source of water, salts, and nutrients for the infant, but the nature of the milk changes with time. The earliest secretions from the mammary gland, called *colostrum*, are rich in immunoprotective agents, growth factors, minerals, and vitamins A and D. The colostrum also has a trypsin inhibitor that protects the vital proteins in the colostrum from digestion in the infant's gastrointestinal tract. At this point in development, the infant's gastrointestinal tract is able to transport antibodies intact, transferring them to its own circulation.

As the colostrum is depleted by the infant, the mammary gland produces a milk that is much richer in lipids and carbohydrates. Lipids provide both energy and biosynthetic precursors. The milk of marine mammals can be in excess of 60% lipid, allowing the pups to accumulate the blubber needed for insulation. The milk sugars are dominated by lactose (often called milk sugar), but there are also more complex oligosaccharides. The sugars serve two main purposes. First, they are an energy source that is readily catabolized by fetal tissues. Second, the sugars are important biosynthetic precursors, particularly the more complex oligosaccharides with amino sugars that are important for membrane glycolipids and glycoproteins. Milk also possesses abundant protein, primarily casein (often called milk protein), which serves as an important source of amino acids for biosynthesis. This protein is

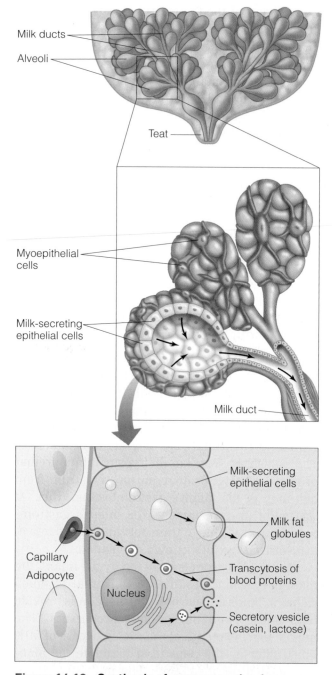

Figure 14.18 **Synthesis of mammary gland secretions** The secretory units of the mammary gland are the alveoli. They produce casein (milk protein) and lactose (milk sugar) in the ER-Golgi network, secreting it into the milk duct via exocytosis. Lipid droplets accumulate within the mammary epithelial cells through synthesis and uptake from adipocytes. Some proteins are taken up from the blood and carried by transcytosis across the epithelial cell and secreted into the milk duct.

highly phosphorylated, enabling it to bind Ca^{2+}. Almost 90% of the Ca^{2+} in the infant diet is locked into the structure of casein particles.

The two novel products found in the milk—lactose and casein—have origins that are inti-

mately linked with the evolution of mammals. Lactose is produced by the enzyme *lactose synthase*, which is a complex of two proteins: galactosyltransferase and α-lactalbumin. Galactosyltransferase, which is found throughout eukaryotes, is one of the many enzymes that catalyze glycosylation reactions in the Golgi apparatus, adding galactose to various macromolecules, including proteins and lipids. However, in the mammary gland, the galactose acceptor for galactosyltransferase is glucose, and the disaccharide lactose is produced. This unique capacity is conferred by the second subunit of lactose synthase, α-lactalbumin. This subunit is structurally related to another enzyme, lysozyme. Thus, the capacity to produce lactose may have arisen only after the duplication of a lysozyme gene early in the mammalian lineage. The duplicated gene subsequently mutated into a form that could dimerize with galactosyltransferase to form the unique enzyme lactose synthase.

The protein casein most closely resembles the γ-chain of fibrinogen, a serum protein that is involved in clotting of blood. Interestingly, the most primitive mammary glands produce a secretion that is primarily derived from presynthesized components of the blood, including fibrinogen. More advanced mammals rely primarily on molecules produced directly in the mammary gland. Thus, it is thought that evolution led to a shift from provision of serum proteins, such as fibrinogen, to production of fibrinogen-like proteins produced directly within the gland. The mammary gland is an excellent example of the way evolutionary processes produce something novel by modifying existing genes and anatomical structures. The production of milk and the nature of the mammary gland reflect an integration of unique aspects of biochemistry, physiological regulation, and anatomical structure.

Prolactin controls parental behavior

Apart from its effects on milk production, prolactin also influences maternal behavior. Pregnancy and lactation lead to a remodeling of the hormonal regulatory pathways, including hormone production and hormone receptor expression. In mammals, prolactin and steroid hormones work in conjunction to alter the biochemistry of the brain and behavior of the female. This remodeling process begins in pregnancy and continues during

BOX 14.2 GENETICS AND GENOMICS
Prolactin

Though named for its role in promoting lactation in mammals, prolactin arose early in vertebrate evolution. Somewhere around 850 million years ago, a gene was duplicated and the pair of genes diverged to give rise to two hormones: prolactin and growth hormone. It is not clear whether the ancestral gene was involved primarily in growth control (like growth hormone) or osmoregulation (like prolactin). The exact timing of these events also remains unclear in part because modern agnathans produce growth hormone, but not prolactin. Early in the evolution of bony fish, after the teleost and non-teleost fish lineages diverged, mutations in the prolactin gene in one of the lineages led to a difference in the number of disulfide bonds in prolactin structure: two in teleost fish and three in the animals arising from the non-teleost lineage, such as lungfish and tetrapods. Unfortunately, many of the earliest studies on the role of prolactin in fish were performed using hormone purified from cows and sheep, before the fundamental difference in structure in teleost and tetrapod prolactin was known.

In some taxa, additional gene duplications led to additional prolactin-like proteins. Shortly after the lineage of teleost fish arose, another gene duplication event occurred, creating a third protein, somatolactin. This protein is similar in structure to both prolactin and growth hormone, so it is not clear which of the genes was duplicated. In mammalian lineages, there have been additional gene duplication events that have led to gene families of prolactin-like proteins. Ruminants and rodents independently experienced multiple duplications of the prolactin gene, creating families of prolactin-like proteins. Throughout most tetrapods, the structure of prolactin is highly conserved, though a few lineages have experienced periods of accelerated evolution leading to structural divergence. In most cases where studied in sufficient detail, the prolactin-like proteins appear to have roles that appear similar to that of prolactin. Many of these prolactin relatives are expressed in tissues other than the anterior pituitary, though usually in tissues involved in reproduction, such as the mammalian placenta and uterus.

The function of prolactin has also changed over the course of evolution; across vertebrates prolactin has been shown to have roles in (1) water and electrolyte balance, (2) reproduction, (3 growth and development, (4) endocrinology and metabolism, (5) brain and behavior, and (6) immunoregulation. Its earliest role was in the control of ion and water balance. In fish, the main role of prolactin is in the control of water and Na^+ movements across the epithelia of the gill, gut, and kidney. Prolactin also plays a role in osmoregulation of amphibians, but its most dominant function appears to be in growth and development. In the larvae, it promotes growth while impairing metamorphosis, antagonizing the actions of thyroid hormone. Interestingly, prolactin surges also induce amphibians to return to the water to breed, perhaps foreshadowing the increasing importance of prolactin as a reproductive hormone. The emergence of reptiles and the movement to land was accompanied by changes in the regulation of ion and water balance. Prolactin plays a relatively minor role in osmoregulation in birds and mammals through effects on the kidney, intestine, and salt-secreting cells (sweat glands of mammals, nasal salt gland of birds). With a diminished role in osmoregulation, prolactin gained a greater role in control of reproductive physiology of mammals.

Almost 80 years ago, prolactin was identified as a hormone that stimulated milk production in mammals. Shortly thereafter, it was also shown to stimulate the production of *crop milk,* a secretion produced in the upper digestive tract of some birds. It has since been shown to stimulate the growth of the mammary gland epithelial cells, and to induce the expression of genes for milk proteins and metabolic enzymes needed for synthesis of milk sugars and fats. Prolactin in mammals also affects the maintenance and function of the reproductive tracts of the female (uterus and ovary) and male (prostate, seminal vesicles, epididymus, Sertoli cells, and Leydig cells). Prolactin also controls parental behavior in numerous species of mammals, birds, and even fish, often interacting with regulation by glucocorticoids and androgens. Prolactin is an excellent example of a signaling molecule with a function that diversified through evolution of the protein, its receptor, and physiological processes that employ this hormone.

References

○ McCormick, S. D., and D. Bradshaw. 2006. Hormonal control of salt and water balance in vertebrates. *General and Comparative Endocrinology* 147: 3–8.

○ Bole-Feysot, C., V. Goffin, M. Edery, N. Binart, and P. A. Kelly. 1998. Prolactin (PRL) and its receptor: Actions, signal transduction pathways and phenotypes observed in PRL receptor knockout mice. *Endocrinology Reviews* 19: 225–268.

○ Storey, A. E., K. M. Delahunty, D. W. McKay, C. J. Walsh, and S. I. Wilhelm 2006. Social and hormonal bases of individual differences in the parental behaviour of birds and mammals. *Canadian Journal of Experimental Psychology* 60: 237–245.

lactation. Virgin females that are exposed to unrelated newborns may gradually acquire maternal behaviors and adopt the infant. The acquisition of maternal behavior is related to the increase in prolactin synthesis and expression to prolactin receptors in the medial preoptic area of the hypothalamus. In mice with their prolactin receptors knocked out, females show less interest in caring for their own young and less willingness to adopt other pups. Interestingly, females that have had multiple experiences birthing and rearing infants often show a reduction in prolactin levels relative to novices. It is thought that maternal behaviors in these females are sustained through greater sensitivity of the hypothalamus to prolactin.

The prevalence of maternal care in animals is rationalized by evolutionary arguments that only the mother can be certain of parentage of the offspring, so the father's energy is best spent copulating. Paternal care occurs in some mammals, primarily canines, rodents, and a few primates. Interestingly, paternal care also appears to be controlled by prolactin. In Djungarian hamsters, paternal care may begin at parturition, with fathers acting as midwives by assisting with the birth of the pups. During early pup growth, the fathers may help groom the offspring, retrieve wandering pups, and assist in thermoregulation. Remarkably, the mates of pregnant females experience hormonal changes that alter their paternal behavior. The degree (and skill) of paternal behavior is linked to both hormonal changes and experience. Among those species of mammals that exhibit paternal care, attending fathers usually have higher levels of prolactin in the blood than do nonpaternal males. The superior skills and attention demonstrated by experienced males may also be due to higher prolactin levels. Interestingly, prolactin has also been implicated in other species that exhibit paternal care, including fish and birds. It is not yet known how prolactin affects the central nervous system to influence paternal behavior.

◉ | CONCEPT CHECK

8. What are the main phases of the mammalian female ovulatory and endometrial cycles?
9. What hormones are produced by the ovarian tissue during the ovulatory cycle?
10. Provide examples of negative and positive feedback in the regulation of the ovulatory cycle.

Integrating Systems | Reproduction and Stress

Stress and reproduction are inseparable. Reproduction is a demanding activity that influences other physiological systems and consequently challenges homeostatic regulation. Reproduction disrupts homeostasis and causes stress in two ways. First, it exerts effects on energy metabolism. Reproducing animals incur considerable energetic costs in producing and supporting reproductive systems, provisioning resources for fetal growth, building and maintaining tissues for sexual displays, and enduring physical challenges in competing for mates. Second, reproduction causes stress through *hormonal antagonism*. The glucocorticoids (stress hormones) that trigger mobilization of energy metabolism also exert direct effects on reproductive physiology. The sex hormones also exert their own effects on other systems, which may be advantageous and expensive, but may also be disruptive yet tolerated. Testosterone, in particular, has many effects on nonreproductive physiology. In general, the interactions between reproduction and stress are reciprocal; reproduction causes stress but stress impairs reproduction. It has been hypothesized that male displays reflect an ability of males to successfully cope with the stresses of reproduction.

The many forms of stress are regulated by the chemical communication network formed by the hypothalamus, the anterior pituitary, and the adrenal cortex (or interrenal cells in vertebrates such as fish). When exposed to an external stress, animals alter hormonal conditions (typically elevated corticosterone) to mobilize fuels, produce glucose, and suppress energy-dependent processes such as growth and reproduction. This defense response allows the animal to survive. Conversely, reproduction itself may create a stressful condition that requires the animal to modify its stress hormone production to gain some of the benefits, like energy production, while curtailing the repressive effects of stress on reproduction. In many cases, the elevation in the levels of hormones such as corticosterone is a necessary component of reproductive physiology. For example, when frogs call, they are engaging in one of the most energetically expensive behaviors seen in ectotherms. The energy for calling is produced when glucocorticoids trigger fuel mobilization, but the calling itself is dependent on testosterone. More testosterone leads to more calling, which demands more energy, which requires elevated glucocorticoids. After a point, glucocorticoids reach such high levels that a stress response ensues, and testosterone production is curtailed.

Animals can influence the steroid-dependent pathways by altering steroid production, by producing steroid-binding proteins, and by altering the profile of steroid hormone receptors. As a result, the magnitude of the stress response can depend on the sex and reproductive state of the animal. Consider the way sex and reproductive state influence how capture manifests as stress in sea turtles. When males and females are captured at sea, males exhibit greater increases in the stress hormone corticosterone. Similarly, nonbreeding females exhibit a greater degree of capture stress than breeding females (Figure 14.19). The hormonal background associated with sex and breeding status influences how other hormones exert their effects.

Many of the interactions between reproduction and stress physiology can be traced back to the interactions between testosterone, secondary sex characteristics, and the immune system. High levels of testosterone impair the immune system, compromising immunocompetence. The link is shown experimentally by treating males with testosterone implants. Studies in many species show that increased testosterone augments secondary sex traits but can also make males more susceptible to parasites and disease.

This antagonistic relationship is thought to be one important factor in sexual selection for male displays. The *immunocompetence-handicap hypothesis* suggests that male traits evolve in a way that allows each male to build the most impressive display possible without compromising its own health. Thus, in the natural world only those males with impressive immunocompetence can tolerate the negative effects of building impressive displays. It would not suit a male deer to build such a large display that its immune system de-

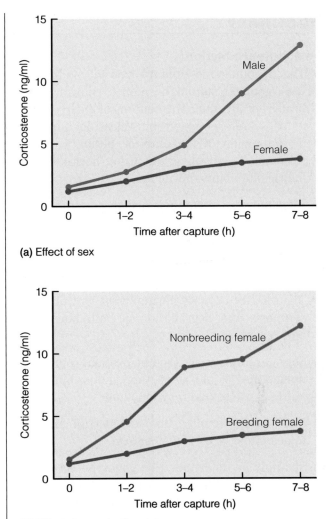

(a) Effect of sex

(b) Effect of reproductive status

Figure 14.19 Plasma corticosterone levels in green sea turtles Many animals respond to capture by inducing stress hormones, such as corticosterone. The magnitude of the response is influenced by other hormones related to **(a)** sex or **(b)** reproductive state.
(Source: Adapted from Moore and Jessup, 2003)

clined to the point at which the animal became unhealthy. Since testosterone is linked to male secondary sex traits, many studies have illustrated the relationship between male displays and immunity. For example, male redwing blackbirds sing loud songs at frequent intervals to attract females and defend a territory. The hypertrophy of the muscles required to sing depends on testosterone levels. Male redwing blackbirds with the greatest singing capacity also have stronger immune systems, as indicated by parasite load and bloodborne immune cells.⊙

Summary

Sexual Reproduction

→ The hypothalamic-pituitary axis in conjunction with gonadal steroid hormones control vertebrate reproduction. The neurons of the hypothalamus secrete gonadotropin-releasing hormone in the region of the anterior pituitary, which responds by secreting luteinizing hormone and follicle-stimulating hormone.

→ Invertebrate hormones are less understood than those of vertebrates, except in arthropods where terpenoid hormones and 20-hydroxyecdysone control reproduction and development.

→ The efficacy of these hormones is controlled by synthesis and degradation of both hormones and receptors.

→ Some animals reproduce by asexual reproduction, either clonal or parthenogenic, but most animals use sexual reproduction.

→ Males make small gametes (sperm), and females make large gametes (ova).

→ Sex may be determined by genotype, as in the XY system or ZW system, or by environment, as in temperature-dependent sex determination.

→ Gametes are produced by meiosis. Ova are produced within follicles of somatic tissue in an ovary. The ovum is augmented with yolk, a protein- and lipid-rich source of nutrients for the growing embryo.

→ Oviparous animals expel either ova (external fertilization) or fertilized eggs (internal fertilization), allowing the embryo to develop externally. Ovoviviparous animals retain fertilized eggs internally. Viviparous animals bear live young, which develop internally and derive nutrition from the maternal tissues.

→ Spermatogenesis occurs in the testes, and copulatory organs increase the efficiency of sperm transfer.

→ Females of some species can store sperm for long periods after mating to ensure uninterrupted reproduction.

→ Once the ovum is fertilized, factors derived from the ovum control early development.

→ Amniote embryos produce four cellular extraembryonic membranes: chorion, allantois, amnion, and yolk sac. The chorion of placental mammals interacts with the maternal tissue to create the placenta.

→ The mammalian ovulatory cycle is controlled by the hypothalamic-pituitary axis.

→ Once the follicle ruptures, releasing the ovum, the remaining follicular cells differentiate to form the corpus luteum, which continues to produce estrogens and progesterone.

→ In mammals, the fertilized ovum implants in the uterine wall and grows.

→ After the follicle ruptures, the corpus luteum produces steroids in response to hypothalamic and, in primates, chorionic gonadotropins, sustaining the pregnancy. With time, the corpus luteum degrades and the placenta maintains synthesis of progesterone and estrogens.

→ The steroid hormones (estrogen and progesterone) and peptide hormones (prolactin and oxytocin) prepare the female for parturition and postnatal care.

→ Milk, a secretory product of mammary glands, is unique to mammals, allowing the female to transfer nutrients to the young. Prolactin controls milk production, as well as both maternal and paternal behavior.

Review Questions

1. Compare the roles of steroid hormones in the reproduction of vertebrates and invertebrates.

2. Discuss how hormones regulate the ovulatory cycle of mammals.

3. Discuss the various modes of sex determination in animals.

4. Compare the features of ovipary, ovovivipary, and vivipary.

5. Discuss the diverse roles of smooth muscle in reproductive physiology.

6. Compare the pathways of gametogenesis in males versus females.

Synthesis Questions

1. Embryos derive nutrition from yolk, except in placental mammals. Discuss the benefits and risks of the two modes of nutrient delivery.

2. Why are animals with temperature-dependent sex determination at risk from endocrine disruptors?

3. Both yolk and milk are products of cellular secretion. Compare their pathways for synthesis and secretion.

5. Males and females develop from genomes that are the same, except for a few genes on the sex determination chromosome. Discuss how the same cellular machinery is used differently in males versus females.

5. Using only mammalian examples, discuss the energetic trade-offs for K-type versus r-type life history strategies.

6. Apart from affecting swimming energetics, how might sperm tail length be evolutionarily advantageous or disadvantageous?

7. Choose examples of male sexual displays and discuss the cellular mechanisms that animals would use to produce the features.

Quantitative Questions

1. Sperm of most fish are released into the water and commence swimming to find an egg. The duration of swimming activity can be as short as a couple minutes or may continue for several hours. If a sperm swims at a velocity of 100 µm/sec and continues swimming for 2 minutes, how far will it swim? What is the significance of your result in relation to mating strategies of fish?

For Further Reading

See the Additional References section at the back of the book for more readings related to the topics in this chapter.

Sex and Hormones

These papers discuss how anthropogenic factors, including global warming, influence the health of animals through effects on reproductive biology.

Guillette, L. J., Jr., and M. P. Gunderson. 2001. Alterations in development of reproductive and endocrine systems of wildlife populations exposed to endocrine-disrupting contaminants. *Reproduction* 122: 857–864.

Moore, I. T., and T. S. Jessop. 2003. Stress, reproduction, and adrenocortical modulation in amphibians and reptiles. *Hormones and Behavior* 43: 39–47.

Most animals have a sexual reproduction mode much like that of humans, but insects differ in both the way sex is established and the role of sexual reproduction in life history strategy

Normark, B. B. 2003. The evolution of alternative genetic systems in insects. *Annual Review of Entomology* 48: 397–423.

Insect reproduction is well studied because of its impact on human activities, particularly disease and agriculture.

Simonet, G., P. Poels, I. Claeys, T. Van Loy, V. Franssens, A. De Loof, and J. Vanden Broeck. 2004. Neuroendocrinological and molecular aspects of insect reproduction. *Journal of Neuroendocrinology* 16: 649–659.

Swevers, L., and K. Iatrou. 2003. The ecdysone regulatory cascade and ovarian development in lepidopteran insects: Insights from the silkmoth paradigm. *Insect Biochemistry and Molecular Biology* 33: 1285–1297.

Gene Imprinting

Gene imprinting is thought to be a way for animals to regulate genomes in a sex-specific manner. These papers discuss how gene imprinting influences genetic regulation and embryonic development.

Cattanach, B. M., C. V. Beechey, and J. Peters. 2004. Interactions between imprinting effects in the mouse. *Genetics* 168: 397–413.

Tycko, B., and I. M. Morison. 2002. Physiological functions of imprinted genes. *Journal of Cellular Physiology* 192: 245–258.

Spermatogenesis and Fertilization

This paper discusses current knowledge about the nature of chemical signaling processes in control of sperm activity.

Riffell, J. A., P. J. Krug, and R. K. Zimmer. 2004. The ecological and evolutionary consequences of sperm chemoattraction. *Proceedings of the National Academy of Sciences, USA* 101: 4501–4506.

Sperm competition can occur at many levels. While the individual sperm within an ejaculate compete, the evolutionary significance is unclear because all of the sperm come from the same male. In other situations, sperm from one male may be forced to outcompete sperm from another male. This paper discusses an interesting case where the order of copulation influences the success of the sperm in fertilization.

Snook, R. R., and D. J. Hosken. 2004. Sperm death and dumping in *Drosophila*. *Nature* 428: 939–941.

Mammalian Reproduction

These papers discuss the evolutionary history of lactation in the ancient reptilian ancestors of mammals.

Oftedal, O. T. 2002. The mammary gland and its origin during synapsid evolution. *Journal of Mammary Gland Biology and Neoplasia* 7: 225–252.

Oftedal, O. T. 2002. The origin of lactation as a water source for parchment-shelled eggs. *Journal of Mammary Gland Biology and Neoplasia* 7: 253–266.

Appendix A
The International System of Units (Système Internationale d'Unités)

Measurements are useful only if they are expressed with a unit. If a person said that his dog weighs 30, it would be impossible to determine whether he was talking about a Chihuahua or a Great Dane; but if he described the dog as weighing 30 kilograms, it would be clear that the dog was about the size of an adult German shepherd. Thus, a good system of units allows us to interpret measurements, and to share them with other people.

Système Internationale (SI) units are the preferred units of measure for most scientific disciplines. The SI system has the advantage of being decimal, internally self-consistent, and precisely defined. Most physiological journals require that physical quantities be expressed in SI units. However, you will often see other units used in the older physiological literature or in clinical situations, so it is important to be able to convert among the various systems of units. The main SI units and some selected conversion factors are shown in the accompanying table.

The Système Internationale is based on seven fundamental units of measure (the meter for length, the kilogram for mass, the mole for an amount of a substance, the second for time, the ampere for electrical current, the kelvin for temperature, and the candela for luminosity). All of the other units can be derived from these base units. For example, the unit for velocity (distance traveled per unit time) is meters per second.

Derived SI units can be expressed using either numerator/denominator notation (e.g., m/s) or exponential notation ($m \cdot s^{-1}$). The correct SI form is exponential notation, because it allows complex units to be expressed clearly. However, in this book we have generally used numerator/denominator notation, because most students find it easier to understand, and we are generally working with relatively simple units such as velocity.

SI and Derived Units

Quantity	SI unit	Abbreviation	Selected conversion factors
Base units			
Length	meter	m	1 m = 3.28 ft
Mass	kilogram	kg	1 kg = 2.20 lb
Amount of matter	mole	mol	
Time	second	s	
Temperature	kelvin	K	A difference of 1 K is equal to a difference of 1°C (Celsius); 0 K = −273°C
Electric current	ampere	A	1 A = 6.24×10^{18} charges per second
Luminosity	candela	cd	
Derived units			
Area	square meters	m^3	1 m^2 = 10.8 ft^2
Volume	cubic meter	m^2	1 m^3 = 1000 liters (l)
			1 cm^3 = 1 milliliter (ml)
			3.785 l = 1 U.S. gallon
Density	kilograms per cubic meter	$kg \cdot m^{-3}$ (or kg/m^3)	1 $kg \cdot m^{-3}$ = 1 g/l
Velocity	meters per second	$m \cdot s^{-1}$ (or m/s)	1 $m \cdot s^{-1}$ = 3.28 ft/s
Frequency	events per second; hertz	Hz = s^{-1}	
Acceleration	meters per second squared	$m \cdot s^{-2}$ (or m/s^2)	1 $m \cdot s^{-2}$ = 3.28 ft/s^2

(continued)

SI and Derived Units (continued)

Quantity	SI unit	Abbreviation	Selected conversion factors
Force	newton (N)	$N = m \cdot kg \cdot s^{-2}$	1 N = 0.102 kg of force
			1 N = 0.225 lb of force
Pressure	pascal (Pa)	$Pa = N \cdot m^{-2} = m^{-1} \cdot kg \cdot s^{-2}$	1 Pa = .0075 millimeters of mercury (mm Hg)
			1 kilopascal (kPa) = 7.5 mm Hg
			1 atmosphere (atm) = 101.3 kPa = 760 mm Hg
Energy (work)	joule (J)	$J = N \cdot m = m^2 \cdot kg \cdot s^{-2}$	1 J = 0.239 calories (cal)
Power	watts (W)	$W = J/s = m^2 \cdot kg \cdot s^{-3}$	1 W = 0.239 cal/s
Electrical potential	watts per ampere; volts (V)	$V = W \cdot A^{-1} = m^2 \cdot kg \cdot s^{-3} \cdot A^{-1}$	
Electrical resistance	volts per ampere; ohm (Ω)	$\Omega = V/A = m^2 \cdot kg \cdot s^{-3} \cdot A^{-2}$	
Electrical charge	coulomb (C)	$C = s \cdot A$	
Capacitance	farad (F)	$F = C \cdot V^{-1} = m^{-2} \cdot kg^{-1} \cdot s^4 \cdot A^2$	

Prefixes for Units

Prefixes can be added before an SI unit to indicate decimal multiples and submultiples of that unit. These prefixes are used to avoid very large or very small numeric values. You are probably already familiar with units such as the kilometer, which represents 1000 meters.

These prefixes can be used with any SI unit, with the exception of the kilogram; the kilogram is the only SI unit that comes with its own prefix. Instead, the prefixes are used with the unit "grams" (g). Thus, 10^{-6} kg = 1 mg (one milligram), not 1 μkg (one microkilogram).

Here is a list of the 20 accepted SI prefixes, only some of which (indicated in bold) are commonly encountered in the biological literature.

Prefixes for SI Units

Prefix	Abbreviation	Factor	English name
yotta	Y	10^{24}	U.S. septillion; U.K. quadrillion
zetta	Z	10^{21}	sextillion
exa	E	10^{18}	U.S. quintillion; U.K. trillion
peta	P	10^{15}	U.S. quadrillion
tera	T	10^{12}	U.S. trillion; U.K. billion
giga	G	10^9	U.S. billion
mega	M	10^6	million
kilo	k	10^3	thousand
hecto	h	10^2	hundred
deca	da	10^1	ten
deci	d	10^{-1}	tenth
centi	c	10^{-2}	hundredth
milli	m	10^{-3}	thousandth
micro	μ	10^{-6}	millionth
nano	n	10^{-9}	U.S. billionth
pico	p	10^{-12}	U.S. trillionth; U.K. billionth
femto	f	10^{-15}	U.S. quadrillionth
atto	a	10^{-18}	U.S. quintillionth; U.K. trillionth
zepto	z	10^{-21}	U.S. sextillionth
yocto	y	10^{-24}	U.S. septillionth; U.K. quadrillionth

Appendix B
Logarithms

Logarithms are important in physiology in three main contexts: (1) the pH scale is logarithmic, (2) the membrane potential of a cell depends on the natural logarithm of the ratio of the intracellular and extracellular ion concentrations, and (3) physiological data are often presented on graphs with logarithmic or semilogarithmic scales. Thus, it is important that you have a clear grasp of logarithms in order to understand physiology.

A logarithm is essentially the same thing as a power or exponent. For example, we know from basic mathematics that $10^2 = 100$. The common logarithm of 100 is simply the power of 10 that gives you 100; that is, 2. Similarly, the common logarithm of 1000 is 3, and the common logarithm of 10,000 is 4. The logarithms of numbers between these values can easily be determined using a "log table" or on most calculators. One important point to note about logarithms is that the logarithm of a number less than one but greater than zero is a negative number. For example, the number 0.1 can be written as 10^{-1}. Thus, the logarithm of 0.1 is -1.

Note that the *common logarithm* always refers to a power of 10; 10 is the base of the common logarithm. We can also have logarithms of other bases. For example, *natural logarithms* are powers of the mathematical constant e. The base of a logarithm is indicated by the subscript following the abbreviation "log." For example, a base 10 log would be written $\log_{10}$, and a base 4 logarithm would be written $\log_4$. However, by convention base 10 (or common) logarithms are generally written simply as the abbreviation "log" (without a subscript), whereas base e (or natural) logarithms are generally abbreviated as "ln."

Recall that the pH of a solution is equal to the negative log of the hydrogen ion concentration of that solution (pH $= -\log[\text{H}^+]$). To calculate the pH of a solution, you need to know two simple rules about logarithms and exponents:

$$-\log(x) = \log(1/x)$$
$$1/(10^x) = 10^{-x}$$

A solution whose hydrogen ion concentration is 10^{-8} M has a pH equal to $-\log(10^{-8})$. Using the first rule above, you can rewrite this expression as $\log(1/10^{-8})$, and using the second rule, you can rewrite this equation as $\log 10^8$. Thus, the pH of this solution is 8.

Appendix C
Linear, Exponential, Power, and Allometric Functions

Physiological variables can often be analyzed graphically. A graphical presentation allows us to determine the relationship between two variables (for example, body size and metabolic rate, or age and blood pressure). These relationships can then be expressed as mathematical functions. The most common types of mathematical functions that apply to physiological data are (1) linear functions, (2) exponential functions, and (3) power functions.

The standard form of a linear function is

$$y = ax + b$$

where x *is* the independent variable, y is the dependent variable, a is the slope of the line, and b is the intercept of the line on the y-axis.

The standard form of an exponential function is

$$y = b \cdot a^x$$

where x is the independent variable, y is the dependent variable, and a and b are constants. We can take the logarithm of both sides of this equation to yield

$$\log y = \log b + x \log a$$

Note that this equation now has the same form as a linear function where the slope of the line is log a, and the intercept is log b. Using a semilog scale, we can plot log y against x, giving a straight line. In this way we can transform an exponential function into a linear function, which is often easier to analyze than a graph of an exponential function.

Power functions are superficially similar to exponential functions, and have the form

$$y = bx^a$$

where x is the independent variable, y is the dependent variable, and a and b are constants. It is easy to confuse power functions with exponential functions, but note that in a power function, the independent variable is raised to a constant power, whereas in an exponential function a constant base is raised to a variable exponent.

As with exponential functions, we can linearize a power function using logarithms. By taking the log of both sides of the power function we obtain

$$\log y = \log b + a \log x$$

Using a log-log scale we can plot log y against log x and obtain a straight line with a slope of a and an intercept of log b.

Power functions are commonly used in physiology to study allometric, or scaling, relationships. In fact, this use is so widespread that physiologists typically refer to power functions as *allometric functions*. Galileo was one of the first scientists to notice scaling relationships. He observed that large animals have limb bones that are disproportionately thicker than those in small animals. Many other physiological parameters also vary nonproportionally with body size, and these relationships can usually be described using power functions.

Useful Formulae

Ohm's law	$V = IR$ V = voltage drop across the system I = current R = electrical resistance
Power (electrical)	$P = IV$ P = Power I = current V = voltage drop across the system
Pressure	$P = f/a$ P = pressure f = force a = area
Law of bulk flow	$Q = \Delta P/R$ Q = flow ΔP = pressure drop across the system R = resistance to flow
Power	$P = w/t$ P = power w = work t = time
Ideal gas law	$PV = nRT$ P = pressure V = volume n = number of molecules R = gas constant T = temperature (in kelvins)
Boyle's Law	$P_1V_1 = P_2V_2$ P = pressure V = volume
Dalton's law of partial pressures	$P_T = P_1 + P_2 + P_3 \ldots$ P_T = total pressure P_1, P_2, P_3, etc. = pressure of each constituent gas in a gas mixture
Newton's second law (of motion)	$F = ma$ F = force m = mass a = acceleration
Potential energy	$E = mgh$ E = energy m = mass g = acceleration due to gravity h = height above surface
Kinetic energy	$E = 1/2mv^2$ E = energy m = mass v = velocity
Weight and Mass	$W = mg$ W = weight m = mass g = acceleration due to gravity
Hooke's law of elasticity	$F = kT$ F = force k = spring constant T = tension
Fick's law of diffusion	$dQ/dt = DA\ dC/dx$ dQ/dt = rate of diffusion D = diffusion coefficient A = area of the membrane dC/dx = concentration (or energy) gradient

Physical and Chemical Constants

Constant	Value
Avogadro's number (molecules in a mole)	$N_A = 6.022 \times 10^{23}$
Faraday constant	$F = 96{,}487$ C/mol
Gas constant	$R = 8.314$ J·K^{-1}·mol $= 0.082$ L·atm·K^{-1}·mol
Planck's constant	$h = 6.626 \times 10^{-34}$ J·s^{-1} $= 1.58 \times 10^{-34}$ cal·s^{-1}
Speed of light in a vacuum	$c = 2.997 \times 10^8$ m·s^{-1}
Specific heat of water	4187 J·mol^{-1}·°C^{-1}
Heat capacity of air	1.21 kJ·m^{-3}·°C^{-1} (20°C, 101.3 kPa)
Heat capacity of water	4.18 MJ·m^{-3}·°C^{-1} (20°C, 101.3 kPa)

Additional References

Chapter 1: Introduction to Physiological Principles

Alexander, R. M. 1985. The ideal and the feasible: Physical constraints on evolution. *Biological Journal of the Linnean Society* 26: 345–358.

Bartholomew, G. A. 1986. The role of natural history in contemporary biology. *BioScience* 36: 324–329.

Bennett, A. F. 1997. Adaptation and the evolution of physiological characters. In *Handbook of physiology: A critical, comprehensive presentation of physiological knowledge and concepts,* ed. W. H. Dantzler (Section 13: Comparative Physiology), vol. 1, 3–16. Bethesda, MD: American Physiological Society.

Calder, W. A., III. 1984. *Size, function, and life history.* Cambridge, MA: Harvard University Press.

Crespi, B. J. 2000. The evolution of maladaptation. *Heredity* 84: 623–639.

Endler, J. A. 1986. *Natural selection in the wild.* Princeton, NJ: Princeton University Press.

Feder, M. E., A. F. Bennett, and R. B. Huey. 2000. Evolutionary physiology. *Annual Review of Ecology and Systematics* 31: 315–341.

Garland, T., Jr., and P. A. Carter. 1994. Evolutionary physiology. *Annual Review of Physiology* 56: 579–621.

Gibbs, A. G. 1999. Laboratory selection for the comparative physiologist. *Journal of Experimental Biology* 202: 2709–2718.

Gould, S. J., and E. S. Vrba. 1982. Exaptation: A missing term in the science of form. *Paleobiology* 8: 4–15.

Jorgensen, C. B. 1983. Ecological physiology: Background and perspectives. *Comparative Biochemistry and Physiology, Part A: Molecular and Integrative Physiology* 75: 5–7.

Kingsolver, J. G., and R. B. Huey. 1998. Evolutionary analyses of morphological and physiological plasticity in thermally variable environments. *American Zoologist* 38: 545–560.

Kleiber, M. 1975. *The fire of life,* 2nd ed. Huntington, New York: Krieger.

Prosser, C. L. 1986. *Adaptational biology: Molecules to organisms.* New York: Wiley.

Schmidt-Nielsen, K. 1984. *Scaling: Why is animal size so important?* Cambridge: Cambridge University Press.

West, G. B., J. H. Brown, and B. J. Enquist. 1999. The fourth dimension of life: Fractal geometry and allometric scaling of organisms. *Science* 276: 122–126.

Wieser, W. 1984. A distinction must be made between the ontogeny and the phylogeny of metabolism in order to understand the mass exponent of energy metabolism. *Respiratory Physiology* 55: 1–9.

Chapter 2: Chemistry, Biochemistry, and Cell Physiology

Benison, S. A., A. C. Barger, and E. L. Wolfe. 1987. *Walter B. Cannon: The life and times of a young scientist.* Cambridge, MA: Harvard University Press.

Dyson, F. J. 1954. What is heat? *Scientific American* 191: 58–63.

Gibbs, A. G. 1998. The role of lipid physical properties in lipid barriers. *American Zoologist* 38: 268–279.

Golding, G. B., and A. M. Dean. 1998. The structural basis of molecular adaptation. *Molecular Biology and Evolution* 15: 355–369.

Hastings, J. W. 1996. Chemistries and colors of bioluminescent reactions: A review. *Gene* 173: 5–11.

King, J., C. Haase-Pettingell, and D. Gossard. 2002. Protein folding and misfolding. *American Scientist* 90: 445–453.

Kinne, R. K. H., ed. 1990. *Basic principles in transport.* Basel, Switzerland: Karger.

Logue, J. A., A. L. DeVries, E. Fodor, and A. R. Cossins. 2000. Lipid compositional correlates of temperature-adaptive interspecific differences in membrane physical structure. *Journal of Experimental Biology* 203: 2105–2115.

Madigan, M. T., and B. L. Marrs. 1997. Extremophiles. *Scientific American* 276: 82–87.

Maloney, P. C., and T. H. Wilson. 1985. The evolution of ion pumps. *BioScience* 35: 43–48.

Mitic, L. L., and J. M. Anderson. 1998. Molecular architecture of tight junctions. *Annual Review of Physiology* 60: 121–142.

Palmer, T. 1995. *Understanding enzymes,* 4th ed. London: Prentice Hall/Ellis Horwood.

Pennycuick, C. J. 1992. *Newton rules biology: A physical approach to biological problems.* New York: Oxford University Press.

Powers, D. A., and P. M. Schulte. 1998. Evolutionary adaptations of gene structure and expression in natural populations in relation to a changing environment: A multidisciplinary approach to address the million-year saga of a small fish. *Journal of Experimental Zoology* 282: 71–94.

Chapter 3: Cell Signaling and Endocrine Regulation

Gade, G., K. H. Hoffmann, and J. H. Spring. 1997. Hormonal regulation in insects: Facts, gaps, and future directions. *Physiological Reviews* 77: 963–1032.

Garofalo, R. S. 2002. Genetic analysis of insulin signaling in *Drosophila. Trends in Endocrinology and Metabolism* 13: 156–162.

Gilmour, K. M. 2005. Mineralocorticoid receptors and hormones: Fishing for answers. *Endocrinology* 146: 44–46.

Guillemin, R. 2005. Hypothalamic hormones a.k.a. hypothalamic releasing factors. *Journal of Endocrinology* 184: 11–28.

Maddrell, S. H. P., W. S. Herman, R. W. Farndale, and J. A. Riegel. 1993. Synergism of hormones controlling epithelial fluid transport in insects. *Journal of Experimental Biology* 174: 65–80.

McFall-Ngai, M. J. 2000. Negotiations between animals and bacteria: The "diplomacy" of the squid–Vibrio symbiosis. *Comparative Biochemistry and Physiology. A Molecular and Integrative Physiology* 126: 471–480.

Pandi-Perumal, S. R., V. Srinivasan, G. J. Maestroni, D. P. Cardinali, B. Poeggeler, and R. Hardeland. 2006. Melatonin. *FEBS Journal* 273: 2813–2838.

Simonet, G., J. Poels, I. Claeys, T. Van Loy, V. Franssens, A. De Loof, and J. V. Broeck. 2004. Neuroendocrinological and molecular aspects of insect reproduction. *Journal of Neuroendocrinology* 16: 649–659.

Spratt N. T., Jr. 1971. *Developmental biology.* Belmont CA: Wadsworth.

Stoka, A. M. 1999. Phylogeny and evolution of chemical communication: An endocrine approach. *Journal of Molecular Endocrinology* 22: 207–225.

Chapter 4: Neuron Structure and Function

Barnard, E. A. 1992. Receptor classes and the transmitter-gated ion channels. *Trends in Biochemical Science* 17: 368–374.

Blumenthal, K. M., and A. L. Siebert. 2003. Voltage-gated sodium channel toxins: Poisons, probes, and future promise. *Cell Biochemistry and Biophysics* 38: 215–238.

Catterall, W. A. 2000. From ionic currents to molecular mechanisms: The structure and function of voltage-gated sodium channels. *Neuron* 26: 13–25.

Doyle, D. A. 2004. Structural changes during ion channel gating. *Trends in Neuroscience* 27: 298–302.

Jiang, Y., A. Lee, J. Chen, V. Ruta, M. Cadene, B. T. Chait, and R. Mackinnon. 2003. X-ray structure of a voltage-dependent K$^+$ channel. *Nature* 423: 33–41.

Robertson, B. 1997. The real life of voltage-gated K$^+$ channels: More than model behaviour. *Trends in Pharmacological Sciences* 18: 474–483.

Rosenthal, J. J., and W. F. Gilly. 2003. Identified ion channels in the squid nervous system. *Neurosignals* 12: 126–141.

Shepherd, G. M., and S. D. Erulkar. 1997. Centenary of the synapse: From Sherrington to the molecular biology of the synapse and beyond. *Trends in Neuroscience* 20: 385–392.

Yu, F. H., and W. A. Catterall. 2003. Overview of the voltage-gated sodium channel family. *Genome Biology* 4(3) no. 207. http://genomebiology.com/2003/4/3/207.

Chapter 5: Cellular Movement and Muscles

Huxley, H. E. 1969. The mechanism of muscular contraction. *Science* 164: 1356–1365.

Lutz, G., and L. C. Rome. 2004. Built for jumping: The design of the frog muscular system. *Science* 263: 370–372.

Maughm, D. W., and J. O. Vigoreaux. 1999. An integrated view of insect flight muscle: Genes, motor molecules, and motion. *News in Physiological Sciences* 14: 87–92.

McDonald, K. S., L. J. Field, M. S. Parmacek, M. Soonpaa, J. M. Leiden, and R. L. Moss. 1995. Length dependence of Ca^{2+} sensitivity of tension in mouse cardiac myocytes expressing skeletal troponin C. *Journal of Physiology, London* 483: 131–139.

Metzger, J. M., M. S. Parmacek, E. Barr, K. Pasyk, W. I. Lin, K. L. Cochrane, L. J. Field, and J. M. Leiden. 1993. Skeletal troponin C reduces contractile sensitivity to acidosis in cardiac myocytes from transgenic mice. *Proceedings of the National Academy of Sciences, USA* 90: 9036–9040.

Pieples, K., and D. F. Wieczorek. 2000. Tropomyosin 3 increases striated muscle diversity. *Biochemistry* 39: 8291–8297.

Qiu, F., A. Lakey, B. Agianian, A. Hutchings, G. W. Butcher, S. Labeit, K. Leonard, and B. Bullard. 2003. Troponin C in different insect muscle types: Identification of two isoforms in *Lethocerus*, *Drosophila* and *Anopheles* that are specific to asynchronous flight muscle in the adult insect. *Biochemical Journal* 371: 811–821.

Rome, L. C., R. P. Funke, R. M. Alexander, G. Lutz, H. Aldridge, F. Scott, and M. Freadman. 1988. Why animals have different muscle fibre types. *Nature* 355: 824–827.

Squire, J. M., and E. P. Morris. 1998. A new look at thin filament regulation in vertebrate skeletal muscle. *FASEB Journal* 12: 761–771.

Vale, R. D., and R. A. Mulligan. 2000. The way things move: Looking under the hood of molecular motor proteins. *Science* 288: 88–95.

Valiron, O., N. Caudron, and D. Job. 2001. Microtubule dynamics. *Cellular and Molecular Life Sciences* 58: 2069–2084.

Chapter 6: Sensory Systems

Baylor, D. 1996. How photons start vision. *Proceedings of the National Academy of Sciences, USA* 93: 560–565.

Buck, L. B. 2000. The molecular architecture of odor and pheromone sensing in mammals. *Cell* 100: 611–618.

Burighel, P., N. J. Lane, G. Fabio, T. Stefano, G. Zaniolo, M. D. Carnevali, and L. Manni. 2003. Novel, secondary sensory cell organ in ascidians: In search of the ancestor of the vertebrate lateral line. *Journal of Comparative Neurology* 461: 236–249.

Fay, R. R., and A. N. Popper. 2000. Evolution of hearing in vertebrates: The inner ear and processing. *Hearing Research* 149: 1–10.

Fernald, R. D. 2004. Evolving eyes. *International Journal of Developmental Biology* 48: 701–705.

Field, G. D., A. P. Sampath, and F. Rieke. 2005. Retinal processing near absolute threshold: From behavior to mechanism. *Annual Review of Physiology* 67: 491–514.

Hudspeth, A. J. 1989. How the ear's works work. *Nature* 341: 397–404.

Julius, D., and A. I. Basbaum. 2001. Molecular mechanisms of nociception. *Nature* 413: 203–210.

Kirschvink, J. L., M. M. Walker, and C. E. Diebel. 2001. Magnetite-based magnetoreception. *Current Opinion in Neurobiology* 11: 462–467.

Martin, V. 2002. Photoreceptors of cnidarians. *Canadian Journal of Zoology* 80: 1703–1722.

Smotherman, M. S., and P. M. Narins. 2000. Hair cells, hearing, and hopping: A field guide to hair cell physiology in the frog. *Journal of Experimental Biology* 203: 2237–2246.

Tobin, D. M., and C. I. Bargmann. 2004. Invertebrate nociception: Behaviors, neurons and molecules. *Journal of Neurobiology* 61: 161–174.

Yamamoto, T., K. Taskai, Y. Sugawara, and A. Tonosaki. 1965. Fine structure of the octopus retina. *Journal of Cell Biology* 25: 345–359.

Chapter 7: Functional Organization of Nervous Systems

Biegler, R., A. McGregor, J. R. Krebs, and S. D. Healy. 2001. A larger hippocampus is associated with longer-lasting spatial memory. *Proceedings of the National Academy of Sciences, USA* 98: 6941–6944.

Bullock, T. H. 1993. How are more complex brains different? One view and an agenda for comparative neurobiology. *Brain, Behavior and Evolution* 41: 88–96.

Bullock, T. H. 2002. Grades in neural complexity: How large is the span? *Integrative and Comparative Biology* 42: 757–761.

Catania, K. G. 1995. Magnified cortex in star-nosed moles. *Nature* 375: 453–454.

Greenspan, R. J. 2004. Systems neurobiology without backbones. *Current Biology* 14: R177–R179.

Holland, L. Z., and N. D. Holland. 1999. Chordate origins of the vertebrate central nervous system. *Current Opinion in Neurobiology* 9: 596–602.

Iwaniuk, A. N., K. M. Dean, and J. E. Nelson. 2005. Interspecific allometry of the brain and brain regions in parrots (Psittaciformes): Comparisons with other birds and primates. *Brain, Behavior and Evolution* 65: 40–59.

Jarvis, E. D., O. Gunturkun, L. Bruce, A. Csillag, H. Karten, W. Kuenzel, L. Medina, G. Paxinos, D. J. Perkel, T. Shimizu, G. Striedter, J. M. Wild, G. F. Ball, J. Dugas-Ford, S. E. Durand, G. E. Hough, S. Husband, L. Kubikova, D. W. Lee, C. V. Mello, A. Powers, C. Siang, T. V. Smulders, K. Wada, S. A. White, K. Yamamoto, J. Yu, A. Reiner, and A. B. Butler. (The Avian Brain Nomenclature Consortium). 2005. Avian brains and a new understanding of vertebrate brain evolution. *Nature Reviews in Neuroscience* 6: 151–159.

Koch, C., and G. Laurent. 1999. Complexity and the nervous system. *Science* 284: 96–98.

Lin, Y. C., W. J. Gallin, and A. N. Spencer. 2001. The anatomy of the nervous system of the hydrozoan jellyfish, *Polyorchis penicillatus*, as revealed by a monoclonal antibody. *Invertebrate Neuroscience* 4: 65–75.

Lotze, M., G. Scheler, H. R. Tan, C. Braun, and N. Birbaumer. 2003. The musician's brain: Functional imaging of amateurs and professionals during performance and imagery. *NeuroImage* 20: 1817–1829.

Maguire, E. A., R. S. Frackowiak, and C. D. Frith. 1997. Recalling routes around London: Activation of the right hippocampus in taxi drivers. *Journal of Neuroscience* 17: 7103–7110.

Nishikawa, K. C. 2002. Evolutionary convergence in nervous systems: Insights from comparative phylogenetic studies. *Brain, Behavior and Evolution* 59: 240–249.

Pittenger, C., and E. R. Kandel. 2003. In search of general mechanisms for long-lasting plasticity: Aplysia and the hippocampus. *Philosophical Transactions of the Royal Society of London, Series B: Biological Sciences* 358: 757–763.

Roth, G., and M. F. Wullimann. 2001. *Brain evolution and cognition.* New York: Wiley.

Chapter 8: Circulatory Systems

Abel, D. C., W. R. Lowell, J. B. Graham, and R. Shabetai. 1987. Elasmobranch pericardial function II. The influence of pericardial pressure on cardiac stroke volume in horn sharks and blue sharks. *Fish Physiology and Biochemistry* 4: 5–15.

Barrionuevo, W. R., and W. W. Burggren. 1999. O_2 consumption and heart rate in developing zebrafish (*Danio rerio*): Influence of temperature and ambient O_2. *American Journal of Physiology* 276: R505–R513.

Burggren, W. W. 1987. Form and function in reptilian circulations. *American Zoologist* 27: 5–19.

Burggren, W. W. 2004. What is the purpose of the embryonic heart beat? Or how facts can ultimately prevail over physiological dogma. *Physiological and Biochemical Zoology* 77: 333–345.

Burggren, W. W., and D. A. Crossley. 2002. Comparative cardiovascular development: Improving the conceptual framework. *Comparative Biochemistry and Physiology, Part A: Molecular and Integrative Physiology* 132: 661–674.

Cho, N. K., S. L. Keye, E. Johnson, J. Heller, L. Ryner, F. Karim, and M. A. Krasnow. 2002. Developmental control of blood cell migration by the *Drosophila* VEGF pathway. *Cell* 108: 865–876.

Cripps, R. M., and E. N. Olson. 2002. Control of cardiac development by an evolutionarily conserved transcriptional network. *Developmental Biology* 246: 14–28.

Hicks, J. W. 2002. The physiological and evolutionary significance of cardiovascular shunting patterns in reptiles. *News in Physiological Sciences* 17: 241–245.

Hicks, J. W., A. Ishimatsu, S. Molloi, A. Erskin, and N. Heisler. 1996. The mechanism of cardiac shunting in reptiles: A new synthesis. *Journal of Experimental Biology* 199: 1435–1446.

Korsmeyer, K. E., N. C. Lai, R. E. Shadwick, and J. B. Graham. 1997. Heart rate and stroke volume contribution to cardiac output in swimming yellowfin tuna: Response to exercise and temperature. *Journal of Experimental Biology* 200: 1975–1986.

Lai, N. C., N. Dalton, Y. Y. Lai, C. Kwong, R. Rasmussen, D. Holts, and J. B. Graham. 2004. A comparative echocardiographic assessment of ventricular function in five species of sharks. *Comparative Biochemistry and Physiology, Part A: Molecular and Integrative Physiology* 137: 505–521.

McMahon, B. R. 2001. Control of cardiovascular function and its evolution in Crustacea. *Journal of Experimental Biology* 204: 923–932.

Moorman, A. F., and V. M. Christoffels. 2003. Cardiac chamber formation: Development, genes, and evolution. *Physiological Reviews* 83: 1223–1267.

Sedmera, D., M. Reckova, A. deAlmeida, M. Sedmerova, M. Biermann, J. Volejnik, A. Sarre, E. Raddatz, R. A. McCarthy, R. G. Gourdie, and R. P. Thompson. 2003. Functional and morphological evidence for a ventricular conduction system in zebrafish and Xenopus hearts. *American Journal of Physiology* 284: H1152–H1160.

Shigei, T., H. Tsuru, N. Ishikawa, and K. Yoshioka. 2001. Absence of endothelium in invertebrate blood vessels: Significance of endothelium and sympathetic nerve/medial smooth muscle in the vertebrate vascular system. *Japanese Journal of Pharmacology* 87: 253–260.

Syme, D. A., K. Gamperl, and D. R. Jones. 2002. Delayed depolarization of the cog-wheel valve and pulmonary-to-systemic shunting in alligators. *Journal of Experimental Biology* 205: 1843–1851.

van Dijk, J. G. 2003. Fainting in animals. *Clinical Autonomic Research* 13: 247–255.

Chapter 9: Respiratory Systems

Boggs, D. F. 2002. Interactions between locomotion and ventilation in tetrapods. *Comparative Biochemistry and Physiology, Part A: Molecular and Integrative Physiology* 133: 269–288.

Boutilier, R. G., ed. 1990. Vertebrate gas exchange: From environment to cell. Vol. 6 of *Advances in comparative and environmental physiology.* Berlin: Springer-Verlag.

Duncker, H-R. 2004. Vertebrate lungs: Structure, topography and mechanics: A comparative perspective of the progressive integration of respiratory system, locomotor apparatus and ontogenetic development. *Respiratory Physiology and Neurobiology* 144: 111–124.

Farmer, C. G. 1999. Evolution of the vertebrate cardio-pulmonary system. *Annual Review of Physiology* 61: 573–592.

Farmer, C. G., and D. R. Carrier. 2000. Pelvic aspiration in the American alligator (*Alligator mississippiensis*). *Journal of Experimental Biology* 203: 1679–1687.

Farmer, C. G., and J. W. Hicks. 2000. Circulatory impairment induced by exercise in the lizard *Iguana iguana*. *Journal of Experimental Biology* 203: 2691–2697.

Landberg, T., J. D. Mailhot, and E. L. Brainerd. 2003. Lung ventilation during terrestrial locomotion in a terrestrial turtle, *Terrapene carolina*. *Journal of Experimental Biology* 206: 3391–3404.

Maina, J. N. 2000. Comparative respiratory morphology: Themes and principles in the design and construction of the gas exchangers. *Anatomical Record* 261: 25–44.

O'Mahoney, P. M., and R. J. Full. 1984. Respiration of crabs in air and water. *Comparative Biochemistry and Physiology* 79A: 275–282.

Piiper, J., and P. Scheid. 1982. Models for comparative functional analysis of gas exchange organs in vertebrates. *Journal of Applied Physiology* 53: 1321–1329.

Piiper, J., and P. Scheid. 1992. Gas exchange in vertebrates through lungs, gills, and skin. *News in Physiological Sciences* 7: 199–203.

Ridgway, S., and R. Howard. 1979. Dolphin lung collapse and intramuscular circulation during diving: Evidence from nitrogen washout. *Science* 206: 1182–1183.

Sparling, C. E., and M. A. Fedak. 2004. Metabolic rates of captive grey seals during voluntary diving. *Journal of Experimental Biology* 207: 1615–1624.

Stillman, J. H. 2000. Evolutionary history and adaptive significance of respiratory structures on the legs of intertidal porcelain crabs, genus *Petrolisthes*. *Physiological and Biochemical Zoology* 73: 86–96.

Taylor, E. W., D. Jordan, and J. H. Coote. 1999. Central control of the cardiovascular and respiratory systems and their interactions in vertebrates. *Physiological Reviews* 79: 855–916.

Vasilakos, K., R. J. Wilson, N. Kimura, and J. E. Remmers. 2005. Ancient gill and lung oscillators may generate the respiratory rhythm of frogs and rats. *Journal of Neurobiology* 62: 369–385.

Chapter 10: Ion and Water Balance

Adragna, N. C., M. D. Fulvio, and P. K. Lauf. 2004. Regulation of K-Cl cotransport: From function to genes. *Journal of Membrane Biology* 201: 109–137.

Barimo, J. F., S. L. Steele, P. A. Wright, and P. J. Walsh. 2004. Dogmas and controversies in the handling of nitrogenous wastes: Ureotely and ammonia tolerance in early life stages of the gulf toadfish, *Opsanus beta*. *Journal of Experimental Biology* 207: 2011–2020.

Boron, W. F. 2004. Regulation of intracellular pH. *Advances in Physiological Education* 28: 160–179.

Cohen, D. M. 2005. SRC family kinases in cell volume regulation. *American Journal of Physiology* 288: C483–C493.

Dubyak, G. R. 2004. Ion homeostasis, channels, and transporters: An update on cellular mechanisms. *Advances in Physiological Education* 28: 143–154.

Gamba, G. 2005. Molecular physiology and pathophysiology of electroneutral cation-chloride cotransporters. *Physiological Reviews* 85: 423–493.

Hoenderop, J. G., B. Nilius, and R. J. Bindels. 2005. Calcium absorption across epithelia. *Physiological Reviews* 85: 373–422.

Maloney, P. C., and T. H. Wilson. 1985. The evolution of ion pumps. *BioScience* 35: 43–48.

McKinley, M. J., and A. K. Johnson. 2004. The physiological regulation of thirst and fluid intake. *News in Physiological Sciences* 19: 1–6.

O'Donnell, M. J., J. A. Dow, G. R. Huesmann, N. J. Tublitz, and S. H. Maddrell. 1996. Separate control of anion and cation transport in malpighian tubules of *Drosophila melanogaster*. *Journal of Experimental Biology* 199: 1163–1175.

O'Neil, R. G., and R. A. Hayhurst. 1985. Functional differentiation of cell types of cortical collecting duct. *American Journal of Physiology* 248: F449–F453.

Pallone, T. L., M. R. Turner, A. Edwards, and R. L. Jamison. 2003. Countercurrent exchange in the renal medulla. *American Journal of Physiology: Regulatory, Integrative, Comparative Physiology* 284: R1153–R1175.

Pelis, R. M., J. Zydlewski, and S. D. McCormick. 2001. Gill Na$^{(+)}$-K$^{(+)}$-2Cl$^{(-)}$ cotransporter abundance and location in Atlantic salmon: Effects of seawater and smolting. *American Journal of Physiology* 280: R1844–R1852.

Perry, S. F. 1997. The chloride cell: Structure and function of the gills of freshwater fishes. *Annual Review of Physiology* 60: 199–220.

Perry, S. F., and K. Gilmour. 2006. Acid-base balance and CO_2 excretion in fish: Unanswered questions and emerging models. *Respiratory Physiology Neurobiology* 154: 199–215.

Romero, W. F. 2004. In the beginning, there was the cell: Cellular homeostasis. *Advances in Physiological Education* 28: 135–138.

Schmidt-Nielsen, B., and R. O'Dell. 1961. Structure and concentrating mechanism in the mammalian kidney. *American Journal of Physiology* 200: 1119–1124.

Schrier, R. W., Y. C. Chen, and M. A. Cadnapaphornchai. 2004. From finch to fish to man: Role of aquaporins in body fluid and brain water regulation. *Neuroscience* 129: 897–904.

Sidell, B. D., and J. R. Hazel. 1987. Temperature affects the diffusion of small molecules through cytosol of fish muscle. *Journal of Experimental Biology* 129: 191–203.

Silva, P., R. J. Solomon, and F. H. Epstein. 1997. Transport mechanisms that mediate the secretion of chloride by the rectal gland of *Squalus acanthias*. *Journal of Experimental Zoology* 279: 504–508.

Somero, G. N., C. B. Osmond, and C. L. Bolis, eds. 1992. *Water and life*. New York: Springer.

Spring, K. R. 1998. Routes and mechanism of fluid transport by epithelia. *Annual Review of Physiology* 60: 105–119.

Steele, S. L., P. H. Yancey, and P. A. Wright. 2004. Dogmas and controversies in the handling of nitrogenous wastes: Osmoregulation during early embryonic development in the marine little skate *Raja erinacea*; response to changes in external salinity. *Journal of Experimental Biology* 207: 2021–2031.

Stein, W. D. 1990. *Channels, carriers and pumps: An introduction to membrane transport*. San Diego: Academic Press.

Uchida, S., and S. Sasaki. 2005. Function of chloride channels in the kidney. *Annual Review of Physiology* 67: 759–778.

Weiss, T. F. 1996. *Cellular biophysics*. Cambridge, MA: MIT Press.

Willmer, P., G. Stone, and I. Johnston. 2000. *Environmental Physiology of Animals*. Blackwell Science. London.

Wright, P. A., P. Anderson, L. Weng, N. Frick, W. P. Wong, and Y. K. Ip. 2004. The crab-eating frog, *Rana cancrivora*, up-regulates hepatic carbamoyl phosphate synthetase I activity and tissue osmolyte levels in response to increased salinity. *Journal of Experimental Zoology, Part A: Comparative Experimental Biology* 301: 559–568.

Chapter 11: Digestion

Beamish, F. W. H., and E. A. Trippel. 1990. Heat increment: A static or dynamic dimension in bioenergetic models? *Transactions of the American Fisheries Society* 119: 649–661.

Brown, J. H., and G. B. West, eds. 2000. Scaling in biology. New York: Oxford University Press.

Calder, W. A., III. 1984. Scaling energetics of homeothermic vertebrates: An operational allometry. *Annual Review of Physiology* 49: 107–120.

Chaudhri, O., C. Small, and S. Bloom. 2006. Gastrointestinal hormones regulating appetite. *Philosophical Transactions of the Royal Society of London, Series B: Biological Sciences* 361: 1187–1209.

Cone, R. D. 2005. Anatomy and regulation of the central melanocortin system. *Nature Neuroscience* 8: 571–578.

Costa, D. P., and G. L. Kooyman. 1984. Contribution of specific dynamic action to heat balance and thermoregulation in the sea otter *Enhydra lutris*. *Physiological Zoology* 57: 199–203.

Cummings, D. E., and J. Overduin. 2007. Gastrointestinal regulation of food intake. *Journal of Clinical Investigations* 117:13–23.

Darvall, K. A. L., R. C. Sam, S. H. Silverman, A. W. Bradbury, and D. J. Adam. 2007. Obesity and thrombosis. *European Journal of Vascular and Endovascular Surgery* 33: 223–233.

Darveau, C.-A., R. K. Suarez, R. D. Andrews, and P. W. Hochachka. 2002. Allometric cascade as a unifying principle of body mass effects on metabolism. *Nature* 417: 166–170.

Dobson, G. P., and J. P. Headrick. 1995. Bioenergetic scaling: Metabolic design and body-size constraints in mammals. *Proceedings of the National Academy of Sciences, USA* 92: 7317–7321.

Gaill, F. 1993. Aspects of life development at deep sea hydrothermal vents. *FASEB Journal* 7: 558–565.

Grant, P. R., and B. R. Grant. 2002. Unpredictable evolution in a 30-year study of Darwin's finches. *Science* 296: 707–711.

Hulbert, A. J., and P. L. Else. 2000. Mechanisms underlying the cost of living in animals. *Annual Review of Physiology* 207–235.

Karasov, W. H., and J. M. Diamond. 1988. Interplay between physiology and ecology in digestion. *BioScience* 38: 602–611.

King, J. R., and M. E. Murphy. 1985. Periods of nutritional stress in the annual cycles of endotherms: Fact or fiction? *American Zoologist* 25: 955–964.

Kooijman, S. A. L. M. 1993. *Dynamic energy budgets in biological systems.* New York: Cambridge University Press.

Martinez del Rio, C., J. E. Schondube, T. J. McWhorter, and L. G. Herrera. 2001. Intake responses in nectar feeding birds: Digestive and metabolic causes, osmoregulatory consequences, and coevolutionary effects. *American Zoologist* 41: 902–915.

McWhorter, T. J., and C. Martinez del Rio. 2000. Does gut function limit hummingbird food intake? *Physiological and Biochemical Zoology* 73: 313–324.

Miller, P. J., ed. 1996. *Miniature vertebrates: The implications of small body size.* Symposia of the Zoological Society of London, no. 69. Oxford: Clarendon.

Nagy, K. A. 1987. Field metabolic rate and food requirement scaling in mammals and birds. *Ecological Monographs* 57: 111–128.

Northcutt, R. G., and C. Gans. 1983. The genesis of neural crest and epidermal placodes: A reinterpretation of vertebrate origins. *Quarterly Reviews of Biology* 58: 1–28.

Prusiner, S. B. 1997. Prion diseases and the BSE crisis. *Science* 278: 245–251.

Schneider, R. A., and J. A. Helms. 2003. The cellular and molecular origins of beak morphology. *Science* 299: 565–568.

Secor, S.M., and J. Diamond. 1998. A vertebrate model of extreme physiological regulation. *Nature* 395: 659–662.

Walker, C. G., G. Zariwala, M. J. Holness, and M. C. Sugden. 2007. Diet, obesity and diabetes: A current update. *Clinical Science* 112: 93–111.

Chapter 12: Locomotion

Alexander, R. M., and G. Goldspink. 1977. *Mechanics and energetics of animal locomotion.* London: Chapman and Hall.

Altringham, J. D., and B. A. Block. 1997. Why do tuna maintain elevated slow muscle temperatures? Power output of muscle isolated from endothermic and ectothermic fish. *Journal of Experimental Biology* 200: 2617–2627.

Altringham, J. D., and I. A. Johnston. 1990. Modelling muscle power output in a swimming fish. *Journal of Experimental Biology* 148: 395–402.

Bennett, A. F. 1991. The evolution of activity capacity. *Journal of Experimental Biology* 160: 1–23.

Bernal, D., J. M. Donley, R. E. Shadwick, and D. A. Syme. Mammal-like muscles power swimming in a cold-water shark. *Nature* 437: 1349–1352.

Biewener, A. A. 2002. Future directions for the analysis of musculoskeletal design and locomotor performance. *Journal of Morphology* 252: 38–51.

Billeter, R., and H. Hoppeler. 1992. Muscular basis of strength. In *Strength and power in sport,* ed. P. V. Komi, 39–63. Oxford: Blackwell.

Brooks, G. A., T. D. Fahey, T. P. White, and K. W. Baldwin. 2000. *Exercise physiology: Human bioenergetics and its applications,* 3rd ed. Mountain View, CA: Mayfield.

Coughlin, D. J., and L. C. Rome. 1996. The roles of pink and red muscle in powering steady swimming in scup, *Stenotomus chrysops. American Zoologist* 36: 666–677.

Dohm, M. R., J. P. Hayes, and T. Garland Jr. 1996. Quantitative genetics of sprint running speed and swimming endurance in the laboratory mouse (*Mus domesticus*). *Evolution* 50: 1688–1701.

Ellington, W. R. 2001. Evolution and physiological roles of phosphagen systems. *Annual Review of Physiology* 63: 289–325.

Finney, J. L., G. N. Robertson, C. A. S. McGee, F. M. Smith, and R. P. Croll. 2006. Structure and autonomic innervation of the swim bladder in the zebrafish (*Danio rerio*). *Journal of Comparative Neurobiology* 495: 587–606.

Gammie, S. C., N. S. Hasen, J. S. Rhodes, I. Girard, and T. Garland Jr. 2003. Predatory aggression, but not maternal or intermale aggression, is associated with high voluntary wheel-running behavior in mice. *Hormones and Behavior* 44: 209–221.

Garland, T., Jr. 2003. Selection experiments: An under-utilized tool in biomechanics and organismal biology. In *Vertebrate Biomechanics and Evolution,* ed. V. L. Bels, J.-P. Gasc, and A. Casinos, 23–56. Oxford: BIOS Scientific.

Garry, D. J., G. A. Ordway, J. N. Lorenz, N. B. Radford, E. R. Chin, R. W. Grange, R. Bassel-Duby, and R. S. Williams. 1998. Mice without myoglobin. *Nature* 395: 905–908.

Gleeson, T. T. 1996. Post-exercise lactate metabolism: A comparative review of sites, pathways, and regulation. *Annual Review of Physiology* 58: 565–581.

Heglund, N. C., and G. A. Cavagna. 1985. Efficiency of vertebrate locomotory muscles. *Journal of Experimental Biology* 115: 283–292.

Hoyt, D. F., and C. R. Taylor. 1981. Gait and the energetics of locomotion in horses. *Nature* 292: 239–240.

Johnston, I. A. 2001. *Muscle development and growth.* San Diego: Academic Press.

Johnston, I. A. 1981. Structure and function of fish muscles. *Symposia of the Zoological Society of London.* 48: 71–113.

Josephson, R. E. 1993. Contraction dynamics and power output of skeletal muscle. *Annual Review of Physiology* 55: 527–546.

Marden, J. H. 2000. Variability in the size, composition, and function of insect flight muscles. *Annual Review of Physiology* 62: 157–178.

Maughan, D. W., and J. O. Vigoreaux. 1999. An integrated view of insect flight muscle: Genes, motor molecules, and motion. *News in Physiological Science* 14: 87–92.

McArdle, W. D., F. I. Katch, and V. L. Katch. 2001. *Exercise physiology: Energy, nutrition, and human performance,* 5th ed. Baltimore, MD: Williams & Wilkins.

Moyes, C. D., and D. A. Hood. 2003. Origins and consequences of mitochondrial variation in vertebrate muscle. *Annual Review of Physiology* 65: 177–201.

Nachtigall, W. 1977. On the significance of Reynold's number and the fluid mechanical phenomena connected to it in swimming physiology and flight biophysics (author's trans. [In German]) *Fortschritte der Zoologie* 24: 13–56.

Nauen, J. C., and G. V. Lauder. 2002. Hydrodynamics of caudal fin locomotion by chub mackerel, *Scomber japonicus* (Scombridae). *Journal of Experimental Biology* 205: 1709–1724.

Orlovsky, G., T. G. Deliagina, and S. Grillner. 1999. *Neuronal control of locomotion: From mollusc to man.* Oxford: Oxford University Press.

Pennycuick, C. J. 1992. *Newton rules biology.* New York: Oxford University Press.

Pettersson, L. B., and A. Hedenström, 2000. Energetics, cost reduction and functional consequences of fish morphology. *Proceedings of the Royal Society of London, Series B: Biological Sciences* 267: 759–764.

Rhodes, J. S., T. Garland Jr., and S. C. Gammie. 2003. Patterns of brain activity associated with variation in voluntary wheel-running behavior. *Behavioral Neuroscience* 117: 1243–1256.

Rome, L. C., R. P. Funke, R. M. Alexander, G. Lutz, H. Aldridge, F. Scott, and M. Freadman. 1988. Why animals have different muscle fibre types. *Nature* 355: 824–827.

Schlieper, G., J. H. Kim, A. Molojavyi, C. Jacoby, T. Laussmann, U. Flogel,

A. Godecke, and J. Schrader. 2004. Adaptation of the myoglobin knockout mouse to hypoxic stress. *American Journal of Physiology* 286: R786–R792.

Sensenig, A. T., and J. W. Shultz. 2003. Mechanics of cuticular elastic energy storage in leg joints lacking extensor muscles in arachnids. *Journal of Experimental Biology* 206: 771–784.

Sinervo, B., and R. B. Huey. 1990. Allometric engineering: An experimental test of the causes of interpopulational differences in performance. *Science* 248: 1106–1109.

Somero, G. N., and J. J. Childress. 1980. A violation of the metabolism-size scaling paradigm: Activities of glycolytic enzymes in muscle increase in larger-size fish. *Physiological Zoology* 53: 322–337.

Suarez, R. K. 1996. Upper limits to mass-specific metabolic rates. *Annual Review of Physiology* 58: 583–605.

Suarez, R. K., J. R. Lighton, C. D. Moyes, G. S. Brown, C. L. Gass, and P. W. Hochachka. 1990. Fuel selection in rufous hummingbirds: Ecological implications of metabolic biochemistry. *Proceedings of the National Academy of Sciences, USA* 87: 9207–9210.

Taylor, C. R., E. Weibel, and L. Bolis, eds. 1985. *Design and Performance of Muscular Systems. Journal of Experimental Biology*, Vol. 115. Cambridge: The Company of Biologists, Ltd.

Tobalske, B. W., T. L. Hedrick, K. P. Dial, and A. A. Biewener. 2003. Comparative power curves in bird flight. *Nature* 421: 363–366.

Tucker, V. A. 1975. The energetic cost of moving about. *American Scientist* 63: 413–419.

Chapter 13: Thermal Physiology

Aas-Hansen, O., L. P. Folkow, and A. S. Blix. 2000. Panting in reindeer (*Rangifer tarandus*). *American Journal of Physiology: Regulatory, Integrative, and Comparative Physiology* 279: R1190–R1195.

Boutiler, R. G., and J. St-Pierre. 2000. Surviving hypoxia without really dying. *Comparative Biochemistry and Physiology, Part A: Molecular and Integrative Physiology* 126: 481–490.

Carey, F. G. 1973. Fishes with warm bodies. *Scientific American* 228: 36–44.

Davies, P. L., J. Baardsnes, M. J. Kuiper, and V. K. Walker. 2002. Structure and function of antifreeze proteins. *Philosophical Transactions of the Royal Society of London, Series B: Biological Sciences* 357: 927–935.

DiMicco, J. A., and D. V. Zaretsky. 2007. The dorsomedial hypothalamus: A new player in thermoregulation. *American Journal of Physiology: Regulatory, Integrative, and Comparative Physiology.* 292: R47–R63

Feder, M. E., and G. E. Hofmann. 1999. Heat-shock proteins, molecular chaperones, and the stress response: Evolutionary and ecological physiology. *Annual Review of Physiology* 61: 243–282.

Hand, S. C. 1991. Metabolic dormancy in aquatic invertebrates. *Advances in Comparative and Environmental Physiology* 8: 1–50.

Harrison, J. E., J. H. Fewell, S. P. Roberts, and H. G. Hall. 1996. Achievement of thermal stability by varying metabolic heat production in flying honeybees. *Science* 274: 88–90.

Heinrich, B. 1987. Thermoregulation in winter moths. *Scientific American* March 1987.

Helmuth, B., J. G. Kingsolver, and E. Carrington. 2005. Biophysics, physiological ecology, and climate change: Does mechanism matter? *Annual Review of Physiology* 67: 177–201.

Hochachka, P. W. 1986. Defense strategies against hypoxia and hypothermia. *Science* 231: 234–241.

Hochachka, P. W., P. L. Lutz, T. J. Sick, and M. Rosenthal. 1993. *Surviving hypoxia: Mechanisms of control and adaptation.* Boca Raton, FL: CRC Press.

Klingenberg, M., and K. S. Echtay. 2001. Uncoupling proteins: The issues from a biochemist point of view. *Biochimica et Biophysica Acta–Bioenergetics* 1504: 128–143.

Knight, M. R. 2002. Signal transduction leading to low-temperature tolerance in *Arabidopsis thaliana*. *Philosophical Transactions of the Royal Society of London, Series B: Biological Sciences* 357: 871–875.

Logue, J. A., A. L. de Vries, E. Fodor, A. R. Cossins. 2000. Lipid compositional correlates of temperature-adaptive interspecific differences in membrane physical structure. *Journal of Experimental Biology* 203: 2105–2015.

Reinertse, R. E., and S. Haftorn, 1986. Different metabolic strategies of northern birds for nocturnal survival. *Journal of Comparative Physiology, Part B: Biochemical, Systemic, and Environmental Physiology* 156: 655–663.

Rolfe, D. F. S., and G. C. Brown. 1997. Cellular energy utilization and molecular origin of standard metabolic rate in mammals. *Physiological Review* 77: 731–758.

Scholander, P. F., V. Walters, R. Hock, and L. Irving. 1950. Body insulation of some arctic and tropical mammals and birds. *Biological Bulletin* 99: 225–236.

Schulte, P. M., H. C. Glemet, A. A. Fiebig, and D. A. Powers. 2000. Adaptive variation in lactate dehydrogenase-β gene expression: Role of a stress-responsive regulatory element. *Proceedings of the National Academy of Sciences, USA* 97: 6597–6602.

Storz, G. 1999. An RNA thermometer. *Genes and Development* 13: 633–636.

Sullivan, J. P., S. E. Fahrbach, J. F. Harrison, E. A. Capaldi, J. H. Fewell, and G. E. Robinson. 2003. Juvenile hormone and division of labor in honey bee colonies: Effects of allatectomy on flight behavior and metabolism. *Journal of Experimental Biology* 206: 2287–2296.

Chapter 14: Reproduction

Anderson, G. M., D. R. Grattan, W. van den Ancker, and R. S. Bridges. 2006. Reproductive experience increases prolactin responsiveness in the medial preoptic area and arcuate nucleus of female rats. *Endocrinology* 147: 4688–4694.

Arukwe, A., and A. Goksøyr. 2003. Eggshell and egg yolk proteins in fish: Hepatic proteins for the next generation; oogenetic, population, and evolutionary implications of endocrine disruption. *Comparative Hepatology* 2: 4.

Bahat, A., I. Tur-Kaspa, A. Gakamsky, L. C. Giojalas, H. Breitbart, and M. Eisenbach. 2003. Thermotaxis of mammalian sperm cells: A potential navigation mechanism in the female genital tract. *Nature Medicine* 9: 149–150.

Bell, G. 1982. *The masterpiece of nature: The evolution and genetics of sexuality.* Berkeley: University of California Press.

Bowden, R. M., M. A. Ewart, and C. E. Nelson. 2000. Environmental sex determination in a reptile varies seasonally and with yolk hormones. *Proceedings of the Royal Society of London, Series B: Biological Sciences* 267: 1745–1749.

Dawley, R., and J. P. Bogart, eds. *Evolution and ecology of unisexual vertebrates.* Albany: New York State Museum.

Edwards, P. A., and J. Ericsson. 1999. Sterols and isoprenoids: Signaling molecules derived from the cholesterol biosynthetic pathway. *Annual Review of Biochemistry* 68: 157–185.

Evans, J. P., and H. M. Florman. 2002. The state of the union: The cell biology of fertilization. *Nature Cell Biology* 4 Suppl.: S57–S63.

Gilbert, S. F. Ecological developmental biology: Developmental biology meets the real world. *Developmental Biology* 233: 1–12.

Gorbman, A., and S. A. Sower. 2003. Evolution of the role of GnRH in animal (metazoan) biology. *General and Comparative Endocrinology* 134: 207–213.

Hunt, P. A., and T. J. Hassold. 2002. Sex matters in meiosis. *Science* 296: 2181–2183.

Johnston, J., J. Canning, T. Kaneko, J. K. Pru, and J. L. Tilly. 2004. Germline stem cells and follicular renewal in the postnatal mammalian ovary. *Nature* 428: 145–150.

Kinsley, C. H., and K. G. Lambert. 2006. The maternal brain. *Scientific American* 294: 72–79.

Martin, R. D. 1989. Size, shape, and evolution. In *Evolutionary Studies: A Centenary Celebration of the Life of Julian Huxley,* ed. M. Keynes, pp. 96–141. London: Eugenics Society.

Maston, G. A., and M. Ruvolo. 2002. Chorionic gonadotropin has a recent origin within primates and an evolutionary history of selection. *Molecular Biology and Evolution* 19: 320–335.

Matzuk, M. M., K. H. Burns, M. M. Viveiros, and J. J. Eppig. 2002. Intercellular communication in the mammalian ovary: Oocytes carry the conversation. *Science* 296: 2178–2180.

McNaught, A. B., and R. Callander. 1975. *Illustrated Physiology.* New York. Churchill Livingstone.

Moore, I. T., and T. S. Jessop. 2003. Stress, reproduction, and adrenocortical modulation in amphibians and reptiles. *Hormones and Behavior* 43: 39–47.

Mruk, D. D., and Cheng, C. Y. 2004. Sertoli-Sertoli and Sertoli-germ cell interactions and their significance in germ cell movement in the seminiferous epithelium during spermatogenesis. *Endocrine Reviews* 25: 747–806.

Primakoff, P., and D. G. Myles. 2002. Penetration, adhesion, and fusion in mammalian sperm-egg interaction. *Science* 296: 2183–2185.

Renfree, M. B., and G. Shaw. 2000. Diapause. *Annual Review of Physiology* 62: 353–375.

Schradin, C., and G. Anzenberger. 1999. Prolactin, the hormone of paternity. *News in Physiological Sciences* 14: 223–231.

Short, R. V. 1998. Difference between a testis and an ovary. *Journal of Experimental Zoology* 281: 359–361.

Strassmann, B. I. 1996. The evolution of endometrial cycles and menstruation. *Quarterly Reviews of Biology* 71: 181–220.

Swevers, L., and K. Iatrou. 2003. The ecdysone regulatory cascade and ovarian development in lepidopteran insects: Insights from the silkmoth paradigm. *Insect Biochemistry and Molecular Biology* 33: 1285–1297.

Glossary

A-band (or anisotropic band) The region of a muscle sarcomere where the thick filaments occur.

absolute refractory period The period during and immediately following an action potential in which an excitable cell cannot generate another action potential, no matter how strong the stimulus.

absolute temperature A measure of temperature in kelvins, where 0 K (absolute zero) is the temperature at which there is no atomic or molecular movement. 1 unit on the Kelvin scale equals $1°$ on the Celsius scale. $0 K = -273°C$.

acclimation A persistent but reversible change in a physiological function that occurs as a result of an alteration in an environmental parameter, such as temperature or photoperiod. Acclimation usually occurs as a result of an experimental manipulation (see also *acclimatization*).

acclimatization A reorganization of physiological functions that occurs as a result of complex environmental changes, such as season or altitude (see also *acclimation*).

accommodation The process by which an eye changes its focal length. Accommodation allows the eye to produce a focused image of objects at different distances.

acetyl CoA An activated form of acetate that serves as the entry point for carbon into the TCA cycle.

acetylcholine A neurotransmitter found in most animal species in many types of neurons, including motor neurons and the autonomic ganglia of vertebrates.

acetylcholinesterase An enzyme that catalyzes the breakdown of acetylcholine into choline and acetate.

acid A chemical that donates a proton (see also *base*).

acidosis A decrease in pH arising through respiration (respiratory acidosis) or metabolism (metabolic acidosis).

acrosomal reaction The exocytosis of the enzyme-laden acrosomal vesicle of sperm in response to contact with the ovum.

acrosome A vesicle in sperm that contains digestive enzymes that enable the sperm to penetrate the outer layers of an ovum.

actin G-actin is a monomeric protein that can be polymerized to construct filamentous actin (F-actin). Actin is the basis of both cytoskeletal microfilaments (composed of the α-actin isoform of G-actin) and skeletal thin filaments (composed of the β-actin isoform of G-actin) (see also *myosin*).

actinomyosin The combination of actin and myosin, joined by a cross-bridge.

action potential A relatively large-amplitude, rapid change in the membrane potential of an excitable cell as a result of the opening and closing of voltage-gated ion channels; involved in transmitting signals across long distances in the nervous system.

activation energy (E_a) The energetic barrier that must be reached before a reactant can be transformed into a product.

activation gate One of the two gates that open and close voltage-gated sodium channels (see also *inactivation gate*).

active site A region of an enzyme that binds the substrate and undergoes conformational changes to catalyze the reaction.

active state The phase of a cross-bridge cycle in which myosin is attached to actin and generating force.

active transport Protein-mediated movement of a substance across a membrane with the utilization of some form of energy. Primary active transport uses ATP. Secondary active transport uses an electrochemical gradient (see also *facilitated diffusion, passive transport*).

acuity, sensory The ability to resolve fine detail of a stimulus.

acute response The rapid phase of response to an external or internal change in conditions, usually within seconds to minutes.

adaptation Used in two contexts in physiology: (1) a change in the genetic structure of a population as a result of natural selection; (2) a reversible change in a physiological parameter that provides a beneficial response to an environmental change. Evolutionary and comparative physiologists prefer to use only the first definition.

adaptation, sensory See *receptor adaptation*.

adenine A purine nitrogenous base component of nucleotides, including nucleic acids.

adenosine A nucleoside composed of adenine and the sugar deoxyribose, important as a signaling molecule.

adenosine diphosphate (ADP) A nucleotide composed of the nucleoside adenine with two phosphate groups, with a single high-energy phosphodiester bond.

adenosine triphosphate (ATP) A nucleotide composed of the nucleoside adenine with three phosphate groups, with two high-energy phosphodiester bonds.

adenylate cyclase (adenylyl cyclase) The enzyme that converts ATP to cyclic AMP.

adhesion plaque A membrane protein complex that anchors thin filaments to the membrane.

adipose tissue A tissue composed of fat cells (adipocytes) that produce and store lipid.

ADP See *adenosine diphosphate*.

adrenal cortex See *adrenal gland*.

adrenal gland A gland near the kidney, which in mammals is composed of an outermost layer (the adrenal cortex) and an inner layer (adrenal medulla).

adrenal medulla See *adrenal gland*.

adrenergic receptors Receptors for the catecholamines norepinephrine and epinephrine.

adrenoreceptors See *adrenergic receptors*.

aerobic Occurring in, or depending on, the presence of oxygen.

aerobic scope The ratio of the maximal aerobic metabolic rate to the basal metabolic rate, typically in the range of 3–10.

afferent Leading toward a region of interest (see also *efferent*).

afferent neuron A neuron that conducts a signal from the periphery to an integrating center (see also *sensory neuron*).

affinity A measure of the degree of attraction between a ligand and a molecule that binds the ligand (see also K_m).

affinity constant (or K_a) Reciprocal of the dissociation constant.

after-hyperpolarization A prolonged hyperpolarization following an action potential.

aglomerular kidney A derived form of kidney, with tubules that lack a glomerulus, found in many lineages of marine fish.

agonist A substance that binds to a receptor and initiates a signaling event. May include both the natural endogenous ligand as well as pharmaceutical agents that mimic the natural substance.

albumen A protein found in eggs that cushions the embryo.

albumin A binding globulin (carrier protein) that is one of the primary proteins of vertebrate plasma; makes a major contribution to blood osmotic pressure.

aldosterone Mineralocorticoid hormone secreted by the adrenal cortex. Its main function is to alter the levels of Na^+ and K^+ in the urine, secondarily affecting water transport.

alkaloids A large group of compounds derived from plants that have pharmacological effects in animals.

alkalosis The condition of being alkaline (see also *metabolic alkalosis, respiratory alkalosis*).

allantoic membrane One of four membranes in an amniote egg.

allantoin An intermediate in nucleotide breakdown and uric acid synthesis; an important form of nitrogenous waste for some animals.

allatostatin A neuropeptide hormone in arthropods that inhibits the corpus allatum from secreting juvenile hormone.

allatotropin A neuropeptide hormone in arthropods that stimulates the corpus allatum to secrete juvenile hormone.

alleles Different forms of the same protein that are encoded by the same gene but differ slightly in primary sequence.

allometry (or allometric scaling) The pattern seen when comparing structural or functional parameters in relation to body size.

allosteric regulator A molecule that binds an enzyme at a site distinct from the substrate binding site to regulate activity.

allosteric site A region of an enzyme, distinct from the active site, that binds a molecule other than the substrate or product, triggering a structural change that alters the catalytic properties of the enzyme.

allozyme An allelic variant of an enzyme.

α adrenergic receptor A G-protein-linked cell membrane receptor that binds norepinephrine preferentially, with a lower affinity for epinephrine.

α-helix A secondary structure of protein or DNA in which the molecule twists in a characteristic pattern, with structure stabilized by hydrogen bonds between adjacent regions.

alternative splicing One of the processes that can result in different mRNAs being coded by a single gene. Different exons of the gene are spliced out in each mRNA, resulting in a number of possible combinations.

alveoli (singular: alveolus) The site of gas exchange in mammalian lungs.

ambient External or environmental conditions, such as ambient temperature.

amine A class of molecules based on ammonia, with a side group substituting for at least one N atom.

amino acid Organic molecules with at least one amino group and at least one carboxyl group. The amino acids that are used to build proteins are α-amino acids.

ammonia A general term that includes both NH_3 and NH_4^+ (ammonium), potent neurotoxins.

ammoniotele An animal with an excretory strategy in which more than half of the nitrogen is excreted as ammonia (see also *ureotele, uricotele*).

amniote Vertebrates with an amnion, namely reptiles, birds, and mammals.

amphibolic pathway A metabolic pathway that both synthesizes (catabolic) and degrades (anabolic) metabolites.

amphipathic A molecule with both hydrophobic and hydrophilic parts.

amplification An exponential increase in activity from one step of a pathway to the next; typically used in the context of signal transduction pathways.

ampullae of Lorenzini Polymodal receptors that detect both electrical and mechanical stimuli; found on the nose of sharks.

amygdala A part of the limbic system of the vertebrate brain that is involved in emotional responses such as fear and anger.

amylase An enzyme that breaks down starch (amylose, amylopectin).

anabolic pathways (or anabolism) Metabolic reactions or pathways that build complex molecules from simpler molecules.

anadromous The life history strategy of an animal living most of its life in the sea, then returning to freshwater to reproduce (see also *catadromous*).

anaerobic Without oxygen. Pertains to an environment without oxygen, or a pathway that occurs in the absence of oxygen (see also *aerobic*).

anaplerotic pathway (or anaplerosis) A metabolic reaction that replenishes intermediates of pathways.

anastomosis A convergence of two or more branches of a tubular structure; e.g., a direct connection between two arteries in the circulatory system.

anatomical dead space The portion of a respiratory structure that cannot participate in gas exchange (e.g., the trachea and bronchi).

androgens Steroid hormones structurally related to testosterone that control masculine features.

anemia A condition in which the number of erythrocytes or hemoglobin in the blood is lower than normal.

angiogenesis Synthesis of new blood vessels, often in response to local hypoxia.

angiotensin A peptide hormone that controls blood pressure. Its precursor is angiotensinogen, which is cleaved by renin to form angiotensin I. This decapeptide is cleaved to the final form, angiotensin II, an octapeptide.

angiotensin-converting enzyme (ACE) An enzyme that converts angiotensin I to angiotensin II.

anion An ion with a negative charge.

anoxic See *anaerobic*.

antagonist A substance that binds to a receptor but does not stimulate a signaling event. Antagonists interfere with the binding of the natural ligand.

antagonistic controls For a given step or pathway, sets of controls that exert opposing effects.

antagonistic muscle A muscle that opposes the movement of another muscle.

anterior pituitary gland The anterior lobe of the pituitary gland of vertebrates, also called the adenohypophysis; secretes tropic hormones.

antidiuretic A substance that induces a reduction in urine volume.

antifreeze protein A protein that disrupts the growth of ice crystals, allowing an organism to survive subzero temperatures.

antigen A substance, usually a protein, that induces the formation of an antibody that can bind the antigen.

antiport (or exchanger) A transport protein that exchanges one ion (or molecule) for another ion (or molecule) on the opposite side of a membrane.

anus The sphincter through which feces exit the gastrointestinal tract.

aorta The major artery exiting the heart.

aortic body A sensory structure located in the vertebrate aorta that contains baroreceptors and chemoreceptors.

apical The end of a structure opposite the base.

apical membrane The end of the cell furthest from the basolateral membrane; the membrane oriented away from the circulatory system.

apnea A period without breathing.

apocrine A type of secretion whereby the cell sheds the apical region of plasma membrane as part of a signaling pathway.

apoenzyme The proteinaceous part of an enzyme.

aquaporin A large tetrameric channel that allows the passage of water through the plasma membrane.

arginine phosphate A major phosphagen in invertebrates, which performs the same role as creatine phosphate in vertebrates.

aromatase See *cytochrome P450 aromatase*.

Arrhenius plot A curve relating temperature to activity, enabling the calculation of activation energy.

arteriole A small branch of the arterial network immediately preceding a capillary bed (see *venule*).

artery A large blood vessel carrying blood away from the heart.

asexual reproduction Production of offspring without the fertilization of an ovum by a sperm (see also *automictic parthenogenesis*).

assimilation Conversion of dietary nutrients into metabolizable fuels.

assimilation efficiency Proportion of dietary nutrients successfully assimilated.

astrocytes Vertebrate glial cells that help to support and regulate the action of neurons in the central nervous system.

asynchronous muscle A muscle in which a single neuronal stimulation causes multiple cycles of contraction and relaxation.

ATP See *adenosine triphosphate*.

ATP-binding cassette A common structural motif found in diverse proteins that binds ATP.

ATPase A class of proteins, including enzymes and transporters, that couples ATP hydrolysis to a mechanical or chemical process.

ATPS Standardized reference condition for measuring gas volumes: ambient temperature, pressure, and saturated with water.

atresia The programmed cell death (apoptosis) of follicles other than the dominant follicle that matures during the ovulatory cycle.

atrial natriuretic peptide (ANP) A peptide hormone produced in the heart that exerts effects on ion and water balance that tend to reduce blood pressure. It increases urine volume and Na^+ excretion.

atrioventricular node (AV node) Part of the conducting pathways of the mammalian heart; delays conduction of the electrical signal between the atrium and ventricles.

atrium (plural: atria) One of the chambers of a heart. Blood moves from the atrium to the ventricle.

atrophy Loss of tissue mass as a result of dying cells; often seen with locomotor muscle in response to prolonged periods of inactivity.

August Krogh principle Principle that for every biological problem, there is an organism on which it can most conveniently be studied.

autocrine A type of cell signaling in which a single cell signals another cell of the same type, including itself.

automictic parthenogenesis Production of offspring by a female in which the second polar body fuses with the ovum to produce a diploid offspring.

autonomic division (of the nervous system) See *autonomic nervous system.*

autonomic ganglia Ganglia of the vertebrate peripheral nervous system.

autonomic nervous system Part of the vertebrate peripheral nervous system that controls largely involuntary functions such as heart rate. It is divided into three main branches: the sympathetic, parasympathetic, and enteric nervous systems.

autotrophy An organism that synthesizes its own nutrients from inorganic material, using the energy of the sun (photoautotroph) or inorganic reactions (chemoautotrophs).

Avogadro's number The number of molecules in a mole (6.02252×10^{23}).

axoaxonic synapse A synapse formed between the axon terminal of one neuron and the axon of another neuron (at any point along its length).

axodendritic synapse A synapse formed between the axon terminal of one neuron and the dendrite of another neuron.

axon A projection of the cell body of a neuron that is involved in carrying information, usually in the form of action potentials, from the cell body to the axon terminal.

axon hillock The junction between the cell body and axon of a neuron. In many neurons, the axon hillock is the site of action potential initiation, acting as the trigger zone for the neuron.

axon terminal The distal end of an axon that forms a synapse with an effector cell or neuron.

axon varicosity A type of synapse in which the presynaptic cell releases neurotransmitter at a series of swellings along the axon.

axonal transport Cytoskeletal-mediated movement of organelles and vesicles along the length of an axon.

axonemal dyneins Motor proteins that enable the sliding of microtubules in cilia and flagella.

axoneme The microtubule-based structure that underlies flagella and cilia.

axosomatic synapse A synapse formed between the axon terminal of one neuron and the soma (cell body) of another neuron.

baroreceptor A receptor that senses pressure (by sensing the resulting stretch on the cell membrane).

basal lamina The extracellular matrix underlying a sheet of epithelial cells; part of the connective tissue formed largely by fibroblasts.

basal metabolic rate (BMR) The metabolic rate of an homeothermic animal at rest, at a thermal neutral temperature, and post-absorptive (see also *resting metabolic rate, standard metabolic rate*).

basal nuclei Interconnected groups of gray matter within the mammalian brain.

base A molecule that accepts a proton, or otherwise causes a reduction in proton concentration through effects on the dissociation of water.

basement membrane See also *basal lamina.*

basilar membrane The location of the auditory hair cells in the mammalian cochlea.

basophil A type of white blood cell that releases histamine; involved in the vertebrate immune response.

batch reactor A chemical reactor in which nutrients enter and exit through the same opening; nutrients are retained in the reactor and digested; the undigested material is then expelled, and replaced by another batch of nutrients to be processed.

behavioral thermoregulation The use of behavior to control the body temperature of a poikilotherm, or to reduce the costs of thermoregulation for a homeotherm.

β-oxidation Pathway of fatty acid catabolism that produces acetyl CoA and reducing equivalents.

β-sheet Protein folding pattern in which stretches of amino acids are aligned along another amino acid stretch. This secondary structure is stabilized by hydrogen bonds.

bilateral symmetry A body form in which the body can be divided by a single plane such that the right and left sides are approximate mirror images.

bile A thick, yellow-green fluid composed of salts, pigments, and lipids produced by the liver and stored by the gallbladder; when released into the small intestine it neutralizes gastric acid and aids in the digestion of nutrients, particularly lipids.

bile duct The connection between the liver and the small intestine.

bile pigments Nondigestible breakdown products of porphyrins, including the hemes found in hemoglobin and cytochromes.

bile salts Cholic acid conjugated with amino acids, primarily glycine and taurine; assist in emulsification of lipid within the small intestine.

binocular vision The ability to compare the images coming from two eyes to produce three-dimensional perception.

biogenic amine A class of neurotransmitters derived from amino acids including the catecholamines and dopamine.

bioluminescence The production of light by living organisms.

bipolar neuron A neuron with two main processes leading from the cell body, one of which conveys signals toward the cell body, and one of which conveys signals away from the cell body.

blastocoel The cavity formed by the inpouching of the blastocyst, which eventually forms the alimentary canal.

blastocyst The hollow sphere of cells formed early in embryonic development.

bleaching The fading of a photopigment following absorption of energy from photons. In the case of the retinal-opsin complex, absorption of energy from light causes retinal to dissociate from opsin. Opsin is not pigmented, and thus the photopigment loses its color.

blood The circulatory fluid in animals with closed circulatory systems. Generally contains proteins, ions, organic molecules, and various cell types.

blood-brain barrier A specialized protective barrier made up of glial cells that separates the circulatory system and the central nervous system in vertebrates.

blood vessels Tubes that carry blood through an animal's body.

blubber Subcutaneous lipid deposits of marine mammals, which provide thermal insulation.

Bohr effect A change in hemoglobin oxygen affinity due to a change in pH.

bolus A volume of material introduced into a flow-through system that moves through the system as a unit, with some dispersion along the way; often used in the context of a bolus of food moving through the gastrointestinal tract.

bombesin A hormone that regulates release of gastrointestinal hormones and control of gastrointestinal motility in vertebrates.

bond energy The energy required to form a chemical bond.

bone In vertebrates, a solid structure composed of mineralized extracellular matrix of osteocytes; with cartilage and tendon, it constitutes the skeleton.

book gills The respiratory surfaces of water-breathing chelicerates such as horseshoe crabs.

book lungs The respiratory surfaces of some air-breathing chelicerates such as spiders and scorpions.

boundary layer The region of a solution that is in direct contact or otherwise influenced by a surface; often called an unstirred layer.

Bowman's capsule A cup-shaped expansion of the vertebrate kidney tubule; surrounds the glomerulus.

brackish water Water that is intermediate between freshwater and seawater; typically found in estuaries, salt marshes, or isolated ponds.

bradycardia A heart rate that is slower than normal.

brain A large grouping of ganglia that act as a sophisticated integrating center. Typically located toward the anterior end of the body in the cephalic (head) region.

brainstem A portion of the vertebrate central nervous system that connects the cerebrum of the brain to the spinal cord; contains the pons and medulla, the sites of the respiratory and cardiovascular control centers.

branchial Relating to gills.

bronchi (singular: bronchus) Airways of vertebrate lungs leading from the trachea to the bronchioles.

bronchioles The smallest branches of the airways of mammalian lungs; lead to the terminal alveoli.

brood spot A well-vascularized, featherless region on the underside of birds that is important for warming developing eggs.

brown adipose tissue Also known as brown fat, a thermogenic tissue found in many small mammals, often in the back or neck region. Abundant mitochondria in the brown adipocytes possess thermogenin, a protein that uncouples oxidative phosphorylation to enhance heat production.

brush border Abundant microvilli on epithelial cells in the gastrointestinal tract, giving the tissue a microscopic brushlike appearance.

BTPS Standardized reference conditions for measuring gas volumes: body temperature, atmospheric pressure, and saturated with water.

buccal cavity Mouth cavity.

buffer Chemicals which, when placed in solution, confer on the solution an ability to resist changes in pH when acid or base is added.

bulbourethral gland A mucus-secreting accessory gland of the male reproductive tract.

bulbus arteriosus The outflow tract of the heart in bony fishes; nonmuscular and elastic (see also *conus arteriosus*).

bulk flow The movement of a fluid as a result of a pressure or temperature gradient.

bulk phase (or bulk solution) The volume of solution that is beyond the influence of the surfaces (see also *boundary layer*).

bundle of His One of the conducting pathways of the mammalian heart.

burst exercise High-intensity exercise powered by glycolytic muscle fibers; can continue for only short periods, until glycogen stores are exhausted.

cable properties The electrical properties of axons.

calcium-induced calcium release A mode of muscle activation where calcium crossing the sarcolemma through a Ca^{2+} channel causes a Ca^{2+} channel in the sarcoplasmic reticulum to open.

caldesmon A calcium-binding protein important in the regulation of smooth muscle contractility.

calmodulin A calcium-sensing protein involved in many signal transduction pathways.

caloric deficit The condition in which energy derived from the diet is less than energetic expenditure, resulting in net loss of energy by the animal.

calorie A unit of heat equal to 4.2 joules; nutritional literature may refer to the unit Calorie, which is equivalent to 1000 calories. The unit of heat required to raise 1 g of water at 1 atm by 1°C.

calorimetry The measurement of heat production as an index of metabolic rate.

calsequestrin A calcium-binding protein that allows a muscle to concentrate Ca^{2+} within the sarcoplasmic reticulum.

cAMP (cyclic AMP) A second messenger produced by adenylate cyclase; most important action is the stimulation of protein kinase A.

capacitation A maturation step experienced by sperm after they encounter fluids from the female reproductive tract.

capillary The smallest of the blood vessels in a closed circulatory system; the site of exchange of materials with the tissues.

carbaminohemoglobin Hemoglobin bound to carbon dioxide.

carbohydrate A group of organic molecules that share a preponderance of hydroxyl groups (see also *disaccharide, monosaccharide, polysaccharide*).

carbonic anhydrase (CA) An enzyme that catalyzes the conversion of carbon dioxide and water to bicarbonate and protons.

carboxyhemoglobin Hemoglobin bound to carbon monoxide.

cardiac muscle A form of striated muscle that occurs in the heart.

cardiac output The volume of blood pumped by the heart per unit time; the product of heart rate and stroke volume.

cardiomyocyte A muscle cell found in the heart.

cardiovascular control center A region of the brain within the medulla oblongata that is involved in regulating heart rate and blood pressure.

cardiovascular system An alternate term for the circulatory system of animals such as vertebrates. Consists of the heart, blood, and blood vessels.

carotid body A structure located in the carotid artery leading to the head of vertebrates; contains baroreceptors and chemoreceptors.

carotid rete A network of blood vessels that cools the brain.

carrier protein (or binding protein; binding globulin) Blood proteins that help to transport hydrophobic molecules (such as steroid hormones) in the blood.

carrier-mediated transport All forms of transport across membranes that require a protein.

cartilage In vertebrates, a semisolid structure composed of the extracellular matrix of chondrocytes: the major component of the skeleton of chondrichthians but important in other vertebrates as a cushion between joints.

catabolic pathway (or catabolism) A metabolic pathway that degrades macromolecules into smaller molecules.

catadromous A life history strategy of fish (e.g., eels) in which the adult migrates from freshwater to seawater to breed (see also *anadromous*).

catalysis The progression of a chemical reaction that proceeds with the help of a catalyst.

catalyst A molecule that accelerates chemical reactions but is not changed in the process.

catalytic rate constant (k_{cat}) The number of reactions catalyzed by a single molecule of enzyme per second.

catecholamines The biogenic amines epinephrine and norepinephrine.

cation An ion with a positive charge.

caudal A location near the posterior of an animal.

cecum A blind-ended sac that carries out digestive reactions in the gastrointestinal tract.

cell body See *soma*.

cell membrane See *plasma membrane*.

cellular membranes A general term that refers to the collection of membranes within a cell, including plasma membrane and organelle membranes.

cellulose A glucose polymer that serves a structural role in plants; indigestible by most animals without the assistance of symbionts.

central chemoreceptors A group of chemoreceptors located in the medulla of vertebrate brains.

central lacteal A small, saclike vessel in an intestinal villus; collects lipids that cross the intestinal epithelium.

central nervous system The portion of the nervous system containing the primary integrating centers. In vertebrates it consists of the brain and spinal cord. In invertebrates, it consists of the brain, the major ganglia, and the connecting commissures.

central pattern generator A group of neurons located in the central nervous system that produce a rhythmic neural output.

cephalic Toward the anterior end of an animal.

cephalization An evolutionary trend toward the centralization of nervous and sensory functions at the anterior end of the body (in the head).

cerebellum A part of the vertebrate hindbrain that is involved in maintaining balance and coordinating voluntary muscle movement.

cerebral cortex Outer surface of the vertebrate brain.

cerebral hemispheres Paired structures of the cerebrum (part of the vertebrate forebrain). The cerebral hemispheres are the most obvious structures of a mammalian brain.

cerebral ventricle See *ventricle*.

cerebrospinal fluid (CSF) A fluid contained within the meninges that surrounds the brain and spinal cord of vertebrates.

cerebrum The largest part of the mammalian forebrain.

cGMP See *cyclic GMP*.

cGMP phosphodiesterase An enzyme that cleaves cGMP, producing GMP.

channel A transport protein that facilitates the movement of specific ions or molecules across a cellular membrane down an electrochemical gradient.

chaperone protein See *molecular chaperone*.

chemical energy The energy associated with the reorganization of the chemical structure of a molecule.

chemical gradient An area across which the concentration of a chemical differs, often across a membrane.

chemical synapse A junction between a neuron and another cell in which the signal is transmitted across the synapse in the form of a neurotransmitter.

chemoautotroph An organism that uses inorganic chemical energy to convert organic sources of carbon and nitrogen into biosynthetic building blocks.

chemokinetic An increase in nondirectional movement in response to the detection of a chemical.

chemoreceptor Used to describe either a cell containing chemoreceptive proteins, or the proteins themselves. Chemicals such as hormones, odorants, and tastants bind specifically to chemoreceptor proteins, altering their conformation and causing a signal within the chemoreceptor cell.

chemotaxic Movement toward higher concentrations of a chemical.

chief cell The secretory cells of the gastric epithelium that release pepsin.

chitin A polymer of *N*-acetyl glucosamine used by arthropods to construct the exoskeleton.

chloride cell An ion-pumping cell of fish gill epithelium (also called a mitochondria-rich cell).

chloride shift The exchange of chloride and bicarbonate across the erythrocyte membrane.

chlorocruorin A type of hemoglobin found in some annelids; known as the green hemoglobins.

cholesterol A steroid compound produced from isoprene units; present in cellular membranes and acts as a precursor for steroid hormones.

cholinergic receptor A receptor that binds the signaling molecule acetylcholine. Cholinergic receptors can be divided into nicotinic and muscarinic receptors.

chondrocytes The cells that produce cartilage.

chorion The outer protein layer of an insect egg; the outer membrane of a vertebrate ovum.

chorionic gonadotropin (CG) A third gonadotropin of vertebrates, produced by the placenta but only in primates.

chromaffin cells Cells that secrete the hormone epinephrine (adrenaline). In mammals they are located in the compact adrenal medulla, but in other vertebrates they are more dispersed.

chromophore A molecule that is able to absorb light. In photoreception, the chromophore absorbs the energy from incoming photons and undergoes a conformational change, which sends a signal to an associated G protein, in the first step of visual phototransduction.

chromosome A single, contiguous polymer of DNA found within the genome.

chylomicron A large lipoprotein complex that carries lipid from the digestive tract through the circulation to processing and target tissues.

cilia (singular: cilium) Microtubule-based extensions from a cell that move in a wavelike pattern.

ciliary body A part of the vertebrate eye that secretes the aqueous humor.

ciliary muscle The muscle that controls the shape of the lens of the vertebrate eye; involved in producing a focused image.

ciliary photoreceptors One of two types of animal photoreceptor cells. Vertebrate photoreceptors belong to this class (see also *rhabdomeric photoreceptors*).

circadian rhythm Regular changes in gene expression, biochemistry, physiology, and behavior that cycle with a period of approximately 24 hours. Endogenous circadian rhythms persist even in constant darkness.

circulatory system A group of organs and tissues involved in moving fluids through the body; consists of one or more pumping structures and a series of tubes or other spaces through which fluid can move.

citric acid cycle See *tricarboxylic acid cycle*.

clathrin A triskelion-shaped (three-armed) protein that coats some types of vesicles; vesicle formation begins with a clathrin-coated pit, which enlarges to form a clathrin-coated vesicle.

clearance See *renal clearance*.

cloaca The distal portion of the hindgut in some fishes, amphibians, birds, and reptiles; in these species both excretory and reproductive products are emitted into the cloaca, and leave the body via a single opening.

clonal reproduction A form of asexual reproduction whereby an animal produces a genotypically identical offspring (a clone).

closed circulatory system A circulatory system in which the blood remains within a series of enclosed blood vessels throughout the circulation.

cochlea Spiral structure in the inner ear of mammals; contains the organs of hearing. Less elaborate, but present in birds as the cochlear duct. Derived from the lagena of other vertebrates.

coelom The internal compartment of coelomate animals that forms between two layers of mesoderm.

coenzymes Organic cofactors.

coenzyme A A coenzyme derived from the vitamin pantothenic acid.

cofactors Nonprotein components of enzymes, including metals, coenzymes, and prosthetic groups.

coitus Sexual intercourse.

collagen A trimeric protein found in extracellular matrix. It interacts with other collagen molecules to form rigid fibers or durable sheets.

collecting duct The tube that receives the fluid from the distal tubules of the nephron and empties into the minor calyx of the kidney.

colligative properties Four properties of a solute that are due solely to the concentration of solutes, and not their chemical nature.

colloidal osmotic pressure See *oncotic pressure*.

colon A region of the large intestine primarily responsible for water resorption.

compatible solute A solute that, at high concentration, does not disrupt protein structure or enzyme kinetics.

competitive inhibition A mode of enzyme inhibition in which a molecule competes with the substrate for the active site on the enzyme; competitive inhibitors have the effect of reducing the apparent substrate affinity without affecting V_{max}.

compliance A measure of the ability of a hollow structure (e.g., blood vessel, lung) to stretch in response to an applied pressure.

compound eye A type of eye seen in arthropods; consists of many individual photoreceptive structures.

conduction Transfer of heat from one object to another object or a fluid.

cone A type of vertebrate photoreceptor cell (see also *rod*). Cones are typically responsible for color vision in bright light.

conformer A strategy whereby the physicochemical properties of an animal (e.g., temperature and osmolarity) parallel those of the environment.

conservation of K_m A pattern in which enzymes from different animals share a similar K_m when assayed under conditions that approximate those that occur in the animal.

constitutive Usually describes a gene for a protein that is expressed at near-constant levels regardless of conditions; can be applied to the protein itself, as in "a constitutive enzyme."

continuous-flow stirred-tank reactor In gut reactor theory, a type of gut in which nutrients flow into the gut where they are mixed with gut contents, and simultaneously the gut expels fluids that consist of partially degraded nutrients.

conus arteriosus The outflow tract of the heart ventricle in elasmobranchs, lungfish, and amphibians; muscular and valved (see also *bulbus arteriosus*).

convection Fluid circulation driven by temperature gradients; a special case of bulk flow.

convergence A pattern in a neural pathway in which multiple presynaptic neurons form synapses with a single postsynaptic neuron.

cooperativity A phenomenon demonstrated by multimeric proteins in which binding of a ligand to one protein subunit increases the likelihood of binding to other subunits. Seen in vertebrate blood hemoglobins.

cornea The clear outer surface of an eye. The cornea of an insect ommatidium and a vertebrate eye are analogous structures, but they are not homologous.

coronary artery Artery that supplies blood to the heart in vertebrates.

corpus allatum (plural: corpora allata) A paired neurohemal organ in arthropods that secretes juvenile hormone.

corpus callosum A thick band of axons that connects the right and left hemispheres of the vertebrate brain.

corpus cardiacum (plural: corpora cardiaca) A paired neurohemal organ in arthropods that secretes adipokinetic hormone.

corpus luteum The remnants of a mammalian ovarian follicle that grows in size and becomes an endocrine organ that secretes hormones in support of embryonic development.

cortex The surface or outer layer of an organ (e.g., the cortex of the kidney; the cerebral cortex; cortical bone).

cost of transport (COT) The energetic cost for an animal to cross a given distance.

cotransporter See *symport*.

counteracting solutes Pairs of solutes that act in conjunction to offset the detrimental effects that would arise if either solute were present alone.

countercurrent exchanger A structure in which two fluids flow in opposite directions on either side of an exchange surface, allowing high-efficiency exchange of materials

purely by passive means; e.g., heat exchange in a rete.

countercurrent multiplier A structure in which two fluids flow in opposite directions on either side of an exchange surface, allowing high-efficiency exchange of materials by active means; e.g., ion concentration in the loop of Henle.

covalent bonds Strong chemical bonds involving the sharing of electrons between two atoms.

covalent modification Alteration of a macromolecule by the addition (or removal) of another molecule by forming (or breaking) a covalent bond; e.g., glycosylation, methylation, acetylation, and phosphorylation.

cranial nerves A group of vertebrate nerves that originate in the brain. Vertebrates have 12 or 13 pairs of cranial nerves depending on the species.

creatine phosphate A high-energy phosphate compound used to store energy and to facilitate its transfer from the sites of energy production (mitochondria) to the sites of utilization, such as myofibrils.

cristae The highly convoluted inner membrane of mitochondria.

critical thermal maximum The highest environmental temperature tolerated by an animal.

crop milk Produced by some birds, a regurgitated slurry of nutrients arising from ingested material augmented by secretions.

cross-bridge The linkage of a myosin head to an actin subunit; an essential step in actinomyosin mechanoenzyme activity.

crosscurrent exchanger An exchanger in which the flow of the respiratory medium is at an angle to the flow of blood through the exchange surface; seen in bird lungs.

crypt of Lieberkühn A pit at the base of intestinal villi.

cryptobiosis A dormant state in which an animal experiences a severe (but reversible) metabolic depression during adverse conditions.

cutaneous respiration Gas exchange across the skin.

cuticle The outer layer of the arthropod exoskeleton; composed of chitin and proteins.

cyclic AMP (cAMP) Cyclic adenosine monophosphate formed by the action of adenylate cyclase; a second messenger that activates protein kinase A.

cyclic GMP (cGMP) Cyclic guanosine monophosphate formed by the action of guanylate cyclase; a second messenger that activates protein kinase G.

cytochromes Metalloproteins produced from porphyrins that are central to many enzymatic reactions, including the mitochondrial electron transport chain (cytochromes a, a_3, b, c) and cytochrome P450 enzymes.

cytochrome P450 aromatase An enzyme in steroid metabolism that converts androgens to estrogens.

cytokines Hormones that trigger cell division.

cytoplasm Soluble and particulate interior of a cell, excluding the nucleus.

cytosine A nucleoside composed of cytidine and a ribose sugar.

cytoskeleton Intracellular protein network of microtubules, microfilaments, and intermediate filaments.

cytosol Fluid portion of the cytoplasm, also known as intracellular fluid.

Dalton's law of partial pressures The total pressure of a gas mixture is the sum of the partial pressures of the constituent gases.

dead space The portion of the respiratory system containing gas that does not participate in gas exchange; the sum of the anatomical and physiological dead spaces.

deamination Removal of an amino group from a molecule, usually an amino acid.

defecation The expulsion of feces.

dehydrogenase A class of enzymes that involves an exchange of electrons between a substrate and product.

delayed implantation A reproductive strategy in which a fertilized ovum fails to implant in the uterus, thereby delaying embryonic growth until external conditions are favorable.

denature The loss of three-dimensional structure (unfolding) of a complex macromolecule, such as protein or nucleic acid.

dendrites The branching extensions of a neuronal cell body that carry signals toward the cell body.

dendritic A tree-like pattern of branching.

dendrodendritic synapse A synapse formed between the dendrites of two neurons.

deoxyhemoglobin Hemoglobin that is not bound to oxygen.

deoxyribonucleic acid See *DNA*.

depolarization A change in the membrane potential of a cell from its normally negative resting membrane potential to a more positive value; a relative increase in the positive charge on the inside of the cell membrane.

depolarization-induced calcium release A mode of muscle activation in which calcium crossing the sarcolemma through a Ca^{2+} channel causes a depolarization of the membrane, which directly opens a Ca^{2+} channel in the sarcoplasmic reticulum.

desmosome A type of cell-cell junction common in epithelial tissues.

diabetes mellitus A metabolic condition involving defects in insulin secretion or signal transduction that lead to abnormal regulation of blood glucose. There are two main types of diabetes mellitus: insulin-dependent (type 1) and non-insulin-dependent (type 2).

diacylglycerol (DAG, or diglyceride) A second messenger in the phosphatidylinositol signaling system.

diadromous A life history strategy of fish that includes movement from freshwater to seawater to breed (catadromous) or vice versa (anadromous).

diaphragm A sheetlike group of muscles that separates the thoracic and abdominal cavities of mammals.

diastole The portion of the cardiac cycle in which the heart is relaxing.

diastolic pressure The arterial blood pressure during cardiac diastole.

diffusion The net movement of a molecule throughout the available space from an area of high concentration to an area of low concentration.

diffusion coefficient A parameter that reflects the ability of an ion or molecule to diffuse.

digastric stomach A two-compartment stomach found in ruminants; each of the two compartments is further divided into two chambers.

digestible energy The proportion of ingested energy that can be further processed, leaving only indigestible material.

digestion The breakdown of nutrients in the gastrointestinal tract.

digestive enzymes Hydrolytic enzymes secreted into the lumen of the gastrointestinal tract by the digestive epithelium and accessory glands.

dihydropyridine receptor (DHPR) The Ca^{2+} channel found in muscle plasma membrane, so named because of its ability to bind members of the dihydropyridine class of drugs.

dimer A combination of two monomers, typically in the context of protein structure. A homodimer has two identical monomers, and a heterodimer has two dissimilar monomers.

dipnoan A group of sarcopterygian fish commonly called lungfish, most closely related to the fish ancestor of amphibians.

dipole A molecule with both partial positive (δ^+) and partial negative (δ^-) charges resulting from the asymmetrical distribution of electrons.

direct calorimetry Measurement of heat production; in the context of animal physiology, a measure of metabolic rate.

disaccharide A sugar composed of two monosaccharides.

discontinuous gas exchange A ventilatory pattern seen in some insects in which prolonged periods of apnea are followed by brief but rapid ventilation of the tracheal system.

dissociation constant (K_d) A measure of the tendency of a complex to dissociate into its components; calculated as the ratio of the product of the concentrations of the dissociated components to the concentration of the complex once the reaction reaches equilibrium (e.g., for the reaction AB $\rightleftharpoons$ A + B, $K_d = [A][B]/[AB]$).

distal A location furthest from a point of reference. Opposite of proximal.

distal tubule The region of a vertebrate kidney tubule just before the collecting tubules.

disulfide bridge A covalent bond between two sulfhydryl groups, denoted as –S–S–; also known as a disulfide bond.

diuresis The process of urine formation.

diuretic An agent that promotes urine formation.

dive response A collection of physiological responses to forced diving in air-breathing animals.

divergence A pattern in a neural pathway in which a single presynaptic neuron forms synapses with multiple postsynaptic neurons.

diving bradycardia A reduction in heart rate as a result of submergence in air-breathing animals.

DNA (deoxyribonucleic acid) A polymer of nucleotides that acts as the genetic template.

DNA microarray A high-throughput method of analyzing DNA or RNA.

Donnan equilibrium The chemical equilibrium reached between two solutions separated from each other by a membrane permeable to some of the ions in the solutions.

dopamine A neurotransmitter (biogenic amine) produced in various regions of the vertebrate brain.

dormancy A general term for hypometabolic states accompanied by a reduction in activity (see also *estivation, hibernation,* and *torpor*).

dorsal horn A region of gray matter within the spinal cord located on the dorsal side.

dorsal root The dorsal of the two branches of a vertebrate spinal nerve as it enters the spinal cord. Contains afferent neurons.

dorsal root ganglion Clusters of afferent cell bodies of neurons in the spinal nerves. Located adjacent to the spinal cord.

doubly labelled water An isotopic variant of water (H_2O), where a less common isotope is used for both ^1H (^2H or ^3H) and ^{16}O (^{18}O). Used to measure field metabolic rate.

down-regulation A decrease in the amount or activity of a protein or process; e.g., a decrease in receptor number or activity on a target cell (see also *up-regulation*).

drag A force that resists the forward movement through a fluid through interactions with the surface of an object.

dual breather An animal that can breathe either air or water. Also called a bimodal breather.

duodenum The most proximal region of the small intestine, directly following the stomach.

duty cycle In cytoskeletal movement, the proportion of time in a cross-bridge cycle that a motor protein binds its cytoskeletal tract.

dynamic range The range between the minimum and maximum signal that

can be discriminated by a sensory receptor.

dynein Motor protein that works in combination with microtubules, usually moving in the minus direction (see also *kinesin*).

dynein arms The motor proteins that extend from microtubules in the axoneme of cilia and flagella.

dyspnea The sensation of difficulty with breathing.

eccrine gland A type of exocrine gland characterized by a long coiled duct that delivers secretions from the secretory region to the surface.

ecdysis The periodic shedding of the exoskeleton of invertebrates (molting).

eclosion The process whereby an adult insect emerges from its cocoon.

ectoderm The outermost of the primary germ layers in a developing embryo that eventually gives rise to tissue such as the nervous system.

ectopic pacemaker A pacemaker in an abnormal location.

ectotherm An animal with body temperature determined primarily by external factors, including but not limited to ambient temperature (see also *endotherm*).

edema Excess accumulation of fluid in a tissue.

effective refractory period The time period in which an excitable tissue cannot be stimulated due to changes in the membrane potential.

efferent Leading away from a structure; e.g., efferent neurons carry signals from the central nervous system to the periphery; efferent arterioles carry blood away from the glomerulus of the kidney.

efferent neuron A neuron that conducts impulses from an integrating center to an effector.

efflux Movement of a substance outward, usually in the context of movement out of a cell or tissue.

eicosanoids A type of short-lived chemical signaling molecule.

elasmobranch fish One of two groups of cartilaginous fish, including skates, rays, and sharks. The other group of cartilaginous fish is holocephalans (ratfish).

elastance A measure of how readily a structure returns to its original shape after having been stretched.

elastic recoil Movement as a result of the release of elastic storage energy.

elastic storage energy Energy stored within a deformed object, which is released when the object regains its relaxed configuration.

electrical gradient A charge gradient across a membrane arising from unequal distribution of charged particles.

electric organ A trans-differentiated muscle of fish that generates electric pulses for detecting objects or defense.

electrical energy The energy associated with gradients of charged particles.

electrical synapse A junction between neurons in which the signal is transmitted as an electrical charge rather than via a neurotransmitter (see also *chemical synapse*).

electrocardiogram (ECG, EKG) A recording of the electrical activity of the heart.

electrochemical gradient A gradient composed of the concentration gradient of an ion and the membrane potential; the driving force for the movement of that ion across the membrane.

electrogenic A transport process that results in a change in electrical charge across a membrane.

electrolyte A charged solute, such as Na^+, K^+, and Cl^-.

electron transport system (ETS) A series of protein complexes with mobile carriers that produce a proton gradient across the inner mitochondrial membrane. It builds the gradient by pumping protons as it transfers electrons from reducing equivalents to oxygen, forming water.

electroneutral A transport process that does not change the electrical charge across a membrane.

electroreceptor A sensory receptor that responds to electric fields or discharges.

electrotonic conduction Conduction via graded potentials.

emergence A phenomenon in which the patterns and properties of a complex system are the result of the interactions of the component parts of that system, and are not necessarily predictable from the operation of those components in isolation.

empirical An observation arising from direct measurement of a parameter.

end diastolic volume (EDV) The volume of blood in the heart at the end of diastole; the maximum volume reached during the cardiac cycle.

endergonic reaction A reaction that requires an input of free energy, for which ΔG is positive.

endocardium The internal layer of the heart.

endocrine A signaling pathway in which the signaling molecule is released into the blood and affects a distant cell of a different type.

endocrine disruptor An environmental chemical (often humanmade) that alters cell signaling by acting as an analogue or antagonist of an endocrine hormone.

endocrine gland Type of gland that secretes hormones into the blood.

endocrine system The collective name for the group of glands and other tissues that secrete hormones into the circulatory system.

endocytosis Invagination of the plasma membrane resulting in the formation of a vesicle; used to internalize membrane proteins or capture extracellular solids (phagocytosis) or liquids (pinocytosis).

endoderm The innermost primary germ layer in a developing embryo; eventually gives rise to tissues such as the external surfaces, including the gut lining.

endometrium The innermost layer of the uterus composed of well-vascularized epithelial tissue; see also *myometrium*.

endoplasmic reticulum (ER) An intracellular organelle that forms a network through which secretory products and plasma membrane components pass.

endoskeleton More commonly referred to as the skeleton, an internal framework of bones, cartilage, and tendons that provides support and resistance for muscular movement.

endosymbiont An organism that lives within another organism.

endosymbiosis A relationship whereby an organism lives within another cell or organism, and both parties benefit from the relationship.

endothelium The innermost layer of blood vessels.

endotherm An animal that generates and retains heat internally.

endothermic reaction A reaction that has a positive ΔH, requiring heat.

end systolic volume (ESV) The volume of blood in the heart at the end of systole; the minimum volume of blood that the heart contains during the cardiac cycle.

energetics The study of processes that involve the interconversion of energy.

energy The ability to do work.

energy metabolism The sum of metabolic reactions that pertain to the production or utilization of energy.

enteric branch (also enteric division; enteric nervous system) Part of the vertebrate autonomic nervous system involved in regulating the activity of the gut.

enterosymbiont A symbiotic organism that lives within the gastrointestinal tract.

enthalpy The heat content of a system, symbolized as H. Chemical reactions are often expressed as a change in enthalpy (ΔH).

entropy A thermodynamic parameter that reflects the degree of disorder in a system.

environmental estrogen An estrogen-like *endocrine disruptor*.

enzyme A biological catalyst composed of protein (sometimes RNA), frequently incorporating a cofactor into its structure.

enzyme induction An increase in the levels of an enzyme: one way to achieve an increase in catalytic activity.

eosinophil A type of white blood cell that is involved in the immune response to parasites an in allergic reactions.

ependymal cells Cells that line the ventricles of the brain.

epididymis The structure where sperm mature and are stored in the vertebrate testis.

epinephrine A catecholamine that can act as a hormone or neurotransmitter and is involved in the stress response; also called adrenaline.

epithelium The outermost cellular layer of eumetazoans.

equilibrium For a chemical reaction, the state in which there is no net change in the reactants; products and substrates continue to interconvert, but at equal rates.

equilibrium constant (K_{eq}) The mass action ratio of a chemical reaction when the reaction is at equilibrium.

equilibrium potential The membrane potential at which an ion is at its equilibrium distribution across a membrane.

eructation Gaseous release from the stomach (belching).

erythrocyte A type of vertebrate blood cell that contains hemoglobin (red blood cell).

erythropoiesis Production of red blood cells from erythroblasts, usually in specialized erythropoietic tissues.

erythropoietin A hormone released from the kidney that induces erythropoiesis.

esophagus The passage from the oral cavity (mouth) to the stomach.

essential nutrient A nutrient that cannot be made by the animal and therefore must be obtained from the diet.

esterase An enzyme that breaks an ester bond.

estivation A form of dormancy in which the reduced metabolic rate occurs in response to dehydration.

estradiol-17β The dominant estrogen in most species.

estrogens A class of steroid hormones that act predominantly in females to stimulate reproductive maturation and control the reproductive cycle.

estrous cycle A reproductive cycle composed of four phases: proestrus, estrus, metestrus, and diestrus.

ethology The study of animal behavior.

eupnea Normal breathing.

euryhaline Tolerant of a wide range of external salinities, or more precisely osmolarities.

eurytherm An animal that is tolerant of a wide range of external temperatures.

evaporation Volatilization of liquid water to gaseous water, with the absorption of heat.

evaporative cooling The heat loss that results when heat is absorbed from the body to enable surface water to evaporate.

evolution The process of descent with modification, or genetic change in taxa over time; may be adaptive, maladaptive, or neutral.

excess postexercise oxygen consumption (EPOC) A period of elevated metabolic rate thought to be necessary to allow the muscle to recover from ionic and metabolic disturbances that arose as a result of intense exercise.

exchanger See *antiport*.

excitable cell A cell that is capable of producing an action potential.

excitation-contraction coupling (or EC coupling) The processes that link external stimulation of a muscle to the activation of actinomyosin ATPase, resulting in muscle contraction.

excitatory postsynaptic potential (EPSP) An excitatory potential in a postsynaptic cell.

excitatory potential A change in the membrane potential in an excitable cell that increases the probability of action potential initiation in that cell.

exergonic reaction A reaction that requires an input of free energy, for which ΔG is positive.

exocrine gland A type of gland that releases its secretions via a duct (usually into the external environment).

exocrine secretions Secretions from exocrine glands; include chemical messengers and substances such as mucus, slime, and silk.

exocytosis The transport of vesicles to, and subsequent fusion with, the plasma membrane; serves to secrete vesicle contents into the extracellular space or to introduce proteins into the plasma membrane.

exon A region of DNA that codes for a protein.

exoskeleton An external rigid structure on the outside of many invertebrates that serves to restrict the movement of water and provide a solid framework that controls animal shape and provides resistance needed for locomotion.

exosymbiont A symbiotic organism that lives outside the animal.

exothermic reaction A reaction that has a negative ΔH value, releasing heat.

expiration Exhalation.

extension A movement that causes a limb to straighten across a joint, usually caused by contraction of an extensor muscle.

extensor A muscle that causes a limb to straighten across a joint (extension).

external respiration The process by which animals exchange gases with the environment to supply oxygen to the mitochondria and to remove the resulting carbon dioxide (see also *respiration*).

extracellular digestion Breakdown of nutrients in the outside of the cell resulting from secretion of digestive enzymes.

extracellular fluids The fluids outside of a cell but contained within the limits of the organism.

extracellular matrix The protein and glycosaminoglycan network found outside cells; includes cartilage, bone, and connective tissue.

extrarenal Occurring in a tissue other than the kidney.

eye A complex organ that detects light.

facilitated diffusion A mode of transport in which a protein allows an otherwise impermeable entity to cross a membrane down its electrochemical gradient.

fast axonal transport Process by which neurotransmitter-containing vesicles are moved from the cell body to the axon terminal of a neuron; requires molecular motors.

fast-glycolytic (FG) muscle fibers Muscle cells with a biochemical and mechanical protein profile suited to short-duration, high-intensity contractions that rely on glycolysis for energy; typically muscle fibers that express type IIb myosin.

fast-oxidative glycolytic (FOG) muscle fibers Muscle cells with a biochemical and mechanical protein profile suited to contraction of intermediate duration and intensity; rely on a combination of glycolysis and oxidative phosphorylation for energy. Typically muscle fibers that express type IIa or II x/d myosin isoforms.

feces The undigested matter expelled from the gastrointestinal tract.

feedback A regulatory mechanism whereby a step late in a pathway causes a change earlier in the pathway, either decreasing use of the pathway (negative feedback) or increasing its use (positive feedback).

fever A period of elevated whole body temperature that arises from an immune response, typically as a result of some form of infection. Behavioral fever results when a poikilothermic animal responds to an immunological challenge by moving into an environment that increases body temperature.

fibroblasts Cells that have a major role in producing the extracellular matrix of most soft tissues.

Fick equation The equation relating diffusive flux to the energetic gradient (concentration, partial pressure, electrical, etc.) driving diffusion.

field metabolic rate (FMR) The metabolic rate of a free-roaming animal, usually measured using doubly labelled water.

filapodia Thin, fingerlike extensions of the cell, supported by the actin cytoskeleton.

filtrate The solution that passes through a filter, such as the primary urine that passes through the glomerulus.

flagella (singular: flagellum) Microtubule-based extensions from a cell that move in a whiplike pattern; usually present alone or in pairs.

flexion A movement of a limb that causes the limb to bend at the joint (caused by a flexor muscle).

flexor A muscle that causes a limb to bend at the joint (flexion).

fluid mosaic model The model of a lipid bilayer membrane that includes multiple types of lipids and proteins and allows for their free rotation and lateral movement.

fluidity The degree of free movement of membrane entities within the membrane; often assessed using the dye DPH, which exhibits an anisotropy that depends on membrane fluidity.

fluorescence Absorbance of a high-energy (low-wavelength) light followed by release of a lower-energy (longer-wavelength) light.

flux Flow of material through a pathway.

follicle A multicellular unit composed of somatic tissue surrounding an ovum.

follicle-stimulating hormone (FSH) One of the two major gonadotropins of vertebrates; causes the ovarian follicle to mature.

follicular phase That portion of the ovulatory cycle where a follicle matures to release the ovum.

food vacuole A phagocytic vesicle that fuses with other vesicles and processing organelles to digest the nutrients.

foramen of Panizza A structure that connects the left and right aorta in the crocodile heart.

forebrain The anterior portion of the vertebrate brain, consisting of the telencephalon and diencephalon. Also called the prosencephalon.

founder effect A phenomenon in which the genotypic distribution of a population is a result of historical events that caused the population to be established by a small number of individuals; often associated with a reduction in genetic diversity.

fovea A small region in the center of the retina of a vertebrate eye that is responsible for high-acuity vision.

Frank-Starling effect An increase in the force of cardiac contraction in response to increasing venous return to the heart.

free energy The energy in a system that is available to do work.

freezing-point depression A reduction in the temperature at which a solution freezes; e.g., in the presence of antifreeze molecules.

futile cycle A combination of enyzymatic reactions or processes that lead to net breakdown of ATP and/or release of heat without changes in the carbon substrates.

G protein Type of trimeric membrane protein, associated with specific transmembrane receptors, that plays a role in signal transduction. G proteins bind guanine nucleotides; when bound to GDP the G protein is inactive, but when bound to GTP it is active. The alpha subunit of the G protein moves through the membrane and acts in subsequent steps in the signal transduction pathway.

G-protein-coupled receptor A transmembrane receptor that interacts with a G protein.

GABA (gamma-aminobutyric acid) A neurotransmitter; primarily inhibitory in the vertebrate central nervous system.

gallbladder An organ that stores bile produced in the liver.

gamete The germ cell of sexually reproducing species; small gametes are sperm and large gametes are ova.

gametogenesis Production of mature gametes in the ovary or testis.

ganglion (plural: ganglia) A cluster of neuronal cell bodies. Ganglia act as integrating centers.

ganglion cell An interneuron in the retina of vertebrates.

gap junction Aqueous pore between two cells that allows ions and small molecules to move freely from cell to cell; formed by proteins called connexins in the vertebrates and innexins in the invertebrates.

gas gland A region of the vasculature of the swim bladder that secretes gases.

gastric Pertaining to the stomach.

gastrovascular cavity A space that performs the functions of digestion and circulation; found in organisms such as cnidarians.

gene A region of DNA that, when transcribed, encodes a protein or an RNA.

gene duplication The process of DNA mutation by which a genome can acquire an additional copy of genes.

generator potential A change in the membrane potential in the sensory terminal of a primary afferent neuron. It is a graded potential proportional to the signal intensity. If it exceeds threshold, it will trigger action potentials in the axon of the sensory neuron.

genetic drift A change in gene frequencies in a population over time as a result of random events.

genome All of the genetic material of an organism; the complete set of DNA in both the nucleus and mitochondria.

genotype The specific genetic makeup of an organism.

germ cell A cell that produces the haploid gametes of a sexually reproducing species.

gestation The period of embryonic development within the uterus of a viviparous or ovoviviparous species.

giant axons Unusually large-diameter axons that are present in some invertebrates and vertebrates.

gills Respiratory surfaces that originate as outpocketings of the body surface; generally used for gas exchange in water.

gland A specialized organ that secretes hormones.

glial cells (glia) A group of several types of cells that provide structural and metabolic support to neurons.

gliocytes A type of invertebrate glial cell.

globin The protein component of hemoglobins.

glomerular filtration rate (GFR) The total amount of filtrate per unit time passing through the glomeruli into the tubules of the kidneys.

glomerulus A knot-like cluster of capillaries that acts as a biological filter in the nephrons of many vertebrate kidneys. It permits fluids and small molecules to pass freely from the plasma to the tubule lumen.

glottis A small flap of tissue located between the pharynx and trachea of air-breathing vertebrates.

glucagon A hormone produced by the vertebrate pancreas that inhibits glycogen synthesis and stimulates glycogen breakdown, resulting in an increase in blood glucose.

glucocorticoids Steroid hormones involved in the stress response that regulate carbohydrate, protein, and lipid metabolism.

gluconeogenesis The production of glucose from noncarbohydrate precursors; the main part of the pathway is a reversal of glycolysis, enabled by three enzymes that bypass the two irreversible steps in glycolysis.

glycogen A glucose polysaccharide that forms the main carbohydrate energy store of animals.

glycogenesis Synthesis of glycogen from glucose or glycolytic intermediates.

glycogenolysis The breakdown of glycogen to form glucose-6-phosphate.

glycolipid A glycosylated lipid common in the extracellular side of some plasma membranes.

glycolysis The breakdown of carbohydrates to form pyruvate, or when oxygen is limiting, other end products such as lactate.

glycoprotein A protein that has been modified by the addition of carbohydrates.

glycosaminoglycan A nonproteinaceous component of the extracellular matrix.

glycosuria High levels of glucose in the urine.

glycosylation The addition of carbohydrate groups to proteins, lipids, or carbohydrates within the endoplasmic reticulum or Golgi apparatus.

goblet cell A goblet-shaped mucus-secreting cell found in the intestinal and respiratory surfaces.

Goldman equation The equation that predicts the membrane potential across a cell membrane resulting from the distribution of multiple ions in relation to their permeabilities.

Golgi apparatus An intracellular organelle involved in the processing of proteins prior to export.

gonadotropin A hormone that regulates the activity of reproductive tissues; FSH and LH are the main gonadotropins in vertebrates, and allatotropin and allatostatin are the main gonadotropins in arthropods.

gonads The organs that produce the gametes in males (testes) and females (ovaries).

graded potential Changes in the membrane potential of a cell that vary in magnitude with the stimulus intensity; results from the opening and closing of ion channels.

Graham's law Describes the rate of diffusion of a gas in liquid; states that the rate of diffusion of a gas is proportional to its solubility and inversely proportional to the square root of its molecular mass.

granular cells See juxtaglomerular cells.

granulosa cells The inner layer of somatic cells of a follicle that surround the primary oocyte.

gray matter Areas of the vertebrate central nervous system that are rich in cell bodies (see also *white matter*).

growth factor A group of peptide hormones that stimulate cells to proliferate (hyperplasia) or grow in size (hypertrophy).

growth hormone A peptide hormone derived from the anterior pituitary that mediates somatic cell growth.

guanine A purine nitrogenous base component of nucleotides, including nucleic acids.

guanosine A nucleoside of guanine and a ribose sugar.

guanosine triphosphate (GTP) A high-energy phosphate compound in energy metabolism; also the substrate for guanylate cyclase, forming the second messenger cGMP.

guanylate cyclase Enzyme that converts GTP to cGMP in response to signaling molecules such as nitric oxide; has soluble and membrane-bound forms.

gustation Detection of ingested chemicals: the sense of taste.

gustducin A G-protein-coupled receptor involved in the sense of taste that detects sweet tastants.

gut reactor theory Mathematical explanation of the optimal function of various types of digestive tracts, modeled after chemical reactors.

gyri (singular: gyrus) Wrinkles on the surface of the brains of many mammals.

H zone The central region of a sarcomere corresponding to the location of the thick filaments where there is no overlap with thin filaments; the H zone reduces in size upon contraction.

habituation A process by which repeated stimulation of a neuron results in a decreased response.

hair cell Ciliated sensory cells of vertebrates that react to mechanical stimuli (particularly to vibrations). They are the basis of the senses of hearing and balance, and of the lateral line systems of fishes and amphibians.

Haldane effect The effect of oxygen on hemoglobin–carbon dioxide binding.

half-life A period of time required for half of a population of molecules to be converted to another form; often applied to radioactive decay.

heart A muscular pumping structure.

heat The kinetic energy associated with the movement of atoms and molecules.

heat capacity The amount of thermal energy required to increase the temperature of 1 g of a substance by 1°C.

heat of vaporization The heat needed to cause a liquid to become gaseous, expressed per unit mass.

heat shock proteins A class of molecular chaperones that increase in abundance in response to elevated temperature; the term includes members of genetically related proteins that are constitutive and do not increase in expression in response to thermal stress.

heater tissues A general term for tissues that serve to elevate regional or systemic temperature of an animal, such as the heater organ of billfish.

Heliobacter A bacterium that infects gastric pits, creating conditions that can lead to a gastric ulcer.

hematocrit The proportion of whole blood that is occupied by red blood cells.

heme A metal-binding porphyrin derivative that is incorporated into enzymes (e.g., cytochromes) and nonenzyme proteins (e.g., hemoglobin).

hemerythrin An iron-containing respiratory pigment found in sipunculids, priapulids, brachiopods, and annelids; lacks heme.

hemimetabolous insect Type of insect that possesses immature stages (nymphs) that resemble the adults, except in lacking fully formed wings (see also holometabolous insects).

hemocoel Collective name for the sinuses in the open circulatory systems of many invertebrates.

hemocyanin A respiratory pigment found in arthropods and molluscs consisting of one or more protein molecules complexed directly to copper molecules.

hemocytes Generalized term for blood cells. Most commonly used for the blood cells of invertebrates.

hemoglobin A respiratory pigment consisting of a globin protein complexed to an iron-containing porphyrin molecule called heme.

hemolymph The circulatory fluid of arthropods.

hemopoietic factor A regulatory protein that induces the synthesis of red blood cells; erythropoietin, for example.

Henderson-Hasselbalch equation The mass action equation for the dissociation of carbonic acid (H_2CO_3) to bicarbonate (HCO_3^-) and hydrogen ions (H^+); important in respiratory physiology.

Henry's law One of the ideal gas laws; describes the dissolution of a gas in a liquid, stating that the amount of gas dissolved in a liquid is related to the partial pressure and the solubility of that gas.

hepatocyte The dominant cell type in a liver.

hepatopancreas An invertebrate tissue that serves the same roles as the vertebrate liver and pancreas.

Hering-Breuer reflex A respiratory reflex that reduces breathing in response to overinflation of the lungs; involved in the termination of a breath.

hermaphrodite An animal that possesses both male and female reproductive tissues either simultaneously or sequentially.

hertz A frequency of 1 per second (1 Hz = 1 $\sec^{-1}$).

heterodimer A quaternary structure of two dissimilar monomers.

heterothermy A thermal strategy in which the body temperature (T_B) varies either spatially or temporally.

heterotrimeric G protein See *G protein*.

hexose A general name for monosaccharides with six carbons; includes glucose and fructose.

hibernation A form of dormancy that occurs as a result of low ambient temperature and persists for long periods.

hindbrain The posterior portion of the vertebrate brain, consisting of the cerebellum and brainstem.

hippocampus A part of the vertebrate brain that is involved in the formation of memories.

histamine An amino acid; a regulatory molecule that is released from mast cells in response to an immunological challenge.

histone A protein that reversibly binds to DNA, altering its ability to be transcribed.

holocrine secretion A type of secretion in which entire cells burst, releasing their internal contents.

holometabolous insect An insect in which juvenile stages, dissimilar from the adult, undergo dramatic metamorphosis.

homeostasis A state of internal constancy that is maintained as a result of active regulatory processes.

homeothermy A thermal strategy of an animal (a homeotherm) that has a relatively constant body temperature (T_B).

homeoviscous adaptation A process whereby cells alter the composition of cellular membranes to ensure that fluidity remains constant to compensate for the effects of a change in the external environment.

homing A movement that returns an animal to its home range.

homodimer A molecule composed of two identical subunits.

homologues Genes that are descended from a common ancestor, without intervening duplication events (see also *paralogues*).

hormone Type of chemical messenger that is carried in the blood and thus can act across long distances. Classically defined as a substance released from an endocrine gland and active at very low concentrations.

hydration shell A coating of water bound to the surface of an ion or molecule.

hydrogen bond A class of weak (noncovalent) bond in which an electropositive hydrogen atom is shared by two electronegative atoms.

hydrolysis The breaking of a covalent bond by introducing a water molecule; –H is added to one product and –OH to the other.

hydrophilic A molecule is hydrophilic ("water loving") if it dissolves more easily in water than in an organic phase, such as a lipid bilayer.

hydrophobic A molecule is hydrophobic ("water hating") if it dissolves more easily in a lipid phase than in water.

hydrophobic bond Weak interaction between two nonpolar groups or molecules arising through their mutual aversion to water.

hydrostatic pressure Pressure exerted by a fluid at rest.

hydrostatic skeleton A closed water-filled sac that acts as a semisolid support for an animal.

hydroxyl ion OH^-.

hypercapnia Higher than normal carbon dioxide levels.

hyperglycemia An elevated blood glucose level.

hyperosmotic A solution that has a higher osmolarity than another solution.

hyperplasia An increase in the number of cells in a tissue or organ.

hyperpnea Rapid breathing.

hyperpolarization A change in the membrane potential of a cell from its normally negative resting membrane potential to a more negative value; a relative increase in the negative charge on the inside of the cell membrane.

hyperthermia An elevation in body temperature (T_B) above a desired point.

hypertonic A solution that has a combination of osmolarity and solute profile that leads to the efflux of water from the cell, resulting in a decrease in cell volume.

hypertrophy An increase in the size of cells in a tissue or organ.

hyperventilation Breathing rate or depth that is greater than needed for either oxygen supply or carbon dioxide removal.

hypocapnia Lower than normal carbon dioxide levels.

hypoglycemia Low levels of glucose in the blood.

hypometabolism A period when metabolic rate is lower than the normal resting rate.

hyposmotic A solution that has a lower osmolarity than another solution.

hypothalamic-pituitary portal system A system of blood vessels within the hypothalamus and pituitary that carries hypothalamic hormones to the pituitary, where they regulate the release of pituitary hormones.

hypothalamus A region of the vertebrate forebrain that is involved in controlling body temperature, thirst, hunger, and many other physiological processes. Regulates the function of the pituitary.

hypothermia A decrease in body temperature (T_B) below a desired point.

hypotonic A solution that has a combination of osmolarity and solute profile that leads to the influx of water into the cell, resulting in an increase in cell volume.

hypoventilation Breathing rate or depth that is less than required for adequate gas exchange. For air breathers this usually involves insufficient breathing to allow the removal of carbon dioxide, rather than insufficient for oxygen supply; causes elevated blood carbon dioxide (hypercapnia) and respiratory acidosis.

hypoxemia Lower than normal blood oxygen levels.

hypoxia Lower than normal oxygen; usually referring to environmental oxygen levels (see also *hypoxemia*).

I-band (isotropic band) The region of a muscle sarcomere where the thin filaments that span a Z-disk do not overlap with the thick filament.

ice-nucleating agent A molecule or particle that initiates the formation of ice at a subfreezing temperature.

ideal gas law The relationship between pressure, volume, and gas concentration.

ileum The last section of the small intestine, connecting the jejunem to the large intestine.

imidazole group The amino group found in histidine and other compounds that exhibits a pK value near physiological pH, and is therefore important in the buffering of the pH of body fluids.

in situ An in vitro condition in which the parameter under investigation is in a realistic setting.

in vitro Occurring outside a living animal or cell.

in vivo Occurring within a living animal or cell.

inactivation gate One of the two gates that open and close voltage-gated sodium channels.

incipient lower lethal temperature (ILLT) For a poikilotherm acclimated to a given temperature, it is the lowest temperature that can be tolerated.

incipient upper lethal temperature (IULT) For a poikilotherm acclimated to a given temperature, it is the highest temperature that can be tolerated.

incus (anvil) One of the three small bones of the mammalian middle ear.

indirect calorimetry Estimation of metabolic rate (heat production) using consumption of oxygen or production of carbon dioxide.

inducible Usually refers to a gene that can increase in expression in response to regulatory conditions; can be applied to the encoded protein itself, as in "an inducible enzyme."

inflammation A element of an immune response associated with local heat production.

ingested energy Term used to describe the total energy content of a diet, includes both digestible energy and indigestible energy.

inhibitory potential A change in the membrane potential that decreases the probability of action potential initiation in an excitable cell.

inhibitory postsynaptic potential (IPSP) An inhibitory potential in a postsynaptic cell.

inner ear A series of membranous sacs that contain the organs of hearing and balance in vertebrates.

inner hair cells One of two types of hair cells found in the organ of Corti in the inner ear of mammals; involved in the sense of hearing (see also *outer hair cells*).

inorganic ion An ion lacking carbon atoms.

inositol trisphosphate (IP$_3$) A second messenger in the phosphatidylinositol signaling system.

inspiration Inhalation.

instar A juvenile form of an insect that resembles the adult form in gross appearance.

insulation An external or superficial layer of material that reduces the heat loss from the animal to the environment, such as fur, feathers, and blubber.

insulin Peptide hormone that homeostatically regulates blood glucose levels; released in response to increased blood glucose.

integral membrane protein A protein that is embedded within a cellular membrane, and can only be released with detergent treatment that disrupts the membrane.

integrins A class of dimeric transmembrane proteins that is important in the interactions betweens cells and the extracellular matrix, mediating both adhesion and cell signalling.

integument The outer layer of an animal, usually derived from epithelial cells and their secretions.

intercalated disc The intercellular contact between cardiomyocytes composed of gap junctions and desmosomes.

intercellular fluid See *interstitial fluid*.

intermediate filaments One class of proteins that are used to make up the cytoskeleton.

interneuron A neuron that makes synaptic connections between other neurons.

internode The region of axonal membrane that is covered with the myelin sheath.

interstitial fluid The component of the extracellular fluid that exists between cells.

intrinsic protein See *integral membrane protein*.

intron A region of DNA that is always spliced out of the mRNA following transcription.

inulin A molecule that is used to assess glomerular filtration rate because it is neither secreted nor recovered by the kidney tubule.

ion An atom or molecule with a net charge.

ion channels Transmembrane proteins that permit transfer of ions or molecules through an aqueous pore down an electrochemical gradient.

ionic bond A weak bond between an anion and a cation.

ionoconformer An animal with an internal ion profile that resembles the ion composition of the external water.

ionophore A molecule that forms pores within membranes, allowing specific ions to cross.

ionoregulator An animal that maintains an internal ion profile independent of the ion composition of the external water.

ionotropic receptor A receptor protein that acts as a gated ion channel.

iris A ring of tissue located immediately in front of the lens of a vertebrate eye that controls the amount of light entering the eye by altering the size of the pupil.

ischemia A reduction in blood flow, depriving a tissue of oxygen and nutrients.

islets of Langerhans Clusters of endocrine cells in the pancreas that produce the hormones glucagon and insulin.

isocortex The outer layer of the forebrain in mammals.

isoelectric point The pH at which an ionizable molecule exhibits no net charge.

isoform A protein that has the same function as another protein but differs in primary sequence either because it is encoded by a different gene, or because it results from alternative promoter usage or differential splicing (contrast with *alleles*).

isometric contraction A muscular contraction that results in force production without a change in length.

isopleth A contour line showing the value of a function of two variables connecting the points where the function has a particular value; e.g., the relationship between pH and bicarbonate concentration as described by the Henderson-Hasselbalch equation.

isosmotic Describes two solutions with the same osmolarity.

isotonic A solution with a profile and concentration of solutes that does not result in a change in the volume of a cell.

isotonic contraction A muscular contraction that results in shortening without force production.

isovolumetric contraction (or isovolumic contraction) A phase during the cardiac cycle in which the heart contracts, but does not eject blood because the valves are closed, and thus does not change in volume.

isozyme An isoform of an enzyme.

jejunum An intermediate region of the small intestine, flanked by an anterior duodenum and a posterior ileum.

juvenile hormone (JH) A class of invertebrate hormones derived from isoprenes; secreted from the corpus allatum, JH maintains juvenile traits.

juxtaglomerular apparatus A group of cells located near the distal tubule and the glomerular afferent arterioles.

juxtaglomerular cells Secretory cells of the afferent glomerular arterioles that respond to low blood pressure by secreting renin (also known as granular cells).

k_{cat} See *turnover number*.

keratan A glycosaminoglycan found in the extracellular matrix.

keratin Cytoskeletal protein that forms one type of intermediate filament; common in hair, nails, and feathers.

ketogenesis The production of ketone bodies.

ketolysis The breakdown of ketone bodies to form acetyl CoA.

ketone bodies Substances such as acetone, acetoacetate, and β-hydroxybutyrate and other products derived from acetyl CoA; produced by fatty acid oxidation under food deprivation conditions.

kidney An organ responsible for producing urine, thereby regulating the levels of nitrogenous wastes, extracellular fluid solute properties, and osmolarity.

kinesin A motor protein associated with microtubules (see also *dynein*).

kinetic energy The energy associated with movement.

kinocilium The long cilium of a mammalian hair cell (involved in the detection of sound).

Kleiber's rule The observation that metabolic rate is related to body mass to the exponent 0.75.

K_m See *Michaelis constant*.

knockout animal An animal that has been subjected to genetic manipulation leading to the inability to express a native gene.

Krebs cycle See *tricarboxylic acid cycle*.

K-type strategy A life history strategy whereby an animal produces few offspring and invests heavily in their development (see also *r-type strategy*).

lactation Production and release of milk from the mammalian mammary gland.

lamella A general term referring to a morphology that resembles stacks of leaves.

lamellipodia Flat, sheetlike extensions of the cell, supported by the actin cytoskeleton.

laminar flow A pattern in which the layers of fluid move in parallel, usually relative to the surface of an object.

larva A pre-adult developmental stage that bears little resemblance to the adult form.

latch state A condition in smooth muscle in which force is generated with less than expected ATP consumption; usually attributed to a more efficient mechanism of cross-bridge cycling.

lateral inhibition Process by which a sensory stimulus at one location inhibits the activity of adjacent neurons. Lateral inhibition enhances contrast and improves edge detection in sensory systems.

lateral line system A mechanoreceptive organ in fishes and amphibians that senses vibrations in the water surrounding the animal. Contains hair cells grouped into structures called neuromasts.

leak channel A passive ion channel in the cell membrane that allows the movement of ions down their concentration gradients.

leaky epithelia An epithelial layer with cell-cell connections that permit paracellular transport.

lengthening contraction A type of muscle contraction in which external forces cause the muscle to lengthen while force is being generated.

length-tension relationship Describes the influence of sarcomere length on force development in muscle; muscle generates optimal force when sarcomere length is about 2 μm (in most muscles), and tension declines at higher or lower sarcomere lengths.

lens A clear object that can refract light. In the eye, the lens bends incoming light rays, helping to form a focused image on the retina.

leukocytes Vertebrate white blood cells; cells in blood that are involved in the immune system.

Leydig cell A testosterone-producing cell interspersed in the interstitium of the testes.

ligament A form of connective tissue that joins two bones.

ligand A chemical that specifically and reversibly binds to a receptor or enzyme.

ligand-gated ion channel An ion channel that opens or closes in response to the binding of a specific chemical.

limbic system A group of structures in the vertebrate brain that is involved in processes including emotions and memory.

Lineweaver-Burk equation A plot of the reciprocals of reaction velocity (1/V) and substrate concentration (1/[S]); generates a linear relationship for enzymes with hyperbolic kinetics.

lipase An enzyme that breaks down lipid; includes triglyceride lipases, lipoprotein lipase, and phospholipase.

lipid A class of organic molecules that share hydrophobicity; includes fatty acids, phospholipids, triglycerides, and steroids.

lipid bilayer The model for a plasma membrane in which the hydrophobic faces of two monolayers of phospholipids are associated.

lipid raft A thickened region of the plasma membrane; often accumulates cholesterol, phospholipids with long chain fatty acids, and proteins with long transmembrane domains.

lipogenesis Conversion of fatty acids and glycerol to acylglycerides including monoacylglycerides, diacylglycerides, triglycerides, and phospholipids.

lipolysis Breakdown of acylglycerides and phospholipids.

lipophilic Hydrophobic or nonpolar.

lipoprotein A complex of lipids and proteins; central to the transport of lipids between tissues.

load A force that opposes muscle contraction.

locomotor module A set of musculoskeletal components that work together to perform a single function, such as flying.

long-term potentiation A long-lasting enhancement of the postsynaptic response as a result of high-frequency stimulation of the presynaptic neuron.

loop of Henle A region of a mammalian kidney tubule that connects the proximal and distal tubule; central to the production of hyperosmotic urine.

lower critical temperature (LCT) The lowest environmental temperature at which a homeotherm can survive for long periods; the lower limit of its thermoneutral zone.

lumen The internal cavity of a multicellular unit, such as a kidney tubule or gastrointestinal tract.

lungs Respiratory surfaces that originate as invaginations of the body surface. Generally used for gas exchange in air.

luteal phase The portion of an ovulatory cycle after the follicle has expelled the ovum and before a second follicle matures.

lymph A fluid consisting of an ultrafiltrate of blood and immune cells that travels through the lymphatic system of vertebrates.

lymph hearts The pumping structures of the lymphatic system, present only in some vertebrates (including fish, amphibians, and reptiles).

lymph nodes Small bean-shaped organs found in various locations in the lymphatic system of tetrapods; they filter lymphatic fluid and produce lymphocytes.

lymphatic system In the vertebrates, a network of vessels or sinuses (depending upon the species) that carries lymph back to the primary circulatory system. In many species it also performs an immune function.

lymphocytes Leukocytes that are involved in adaptive immunity in vertebrates.

lysosomes Organelles responsible for the breakdown of damaged and unnecessary membranous compartments and membrane proteins.

macula densa A group of cells in the juxtaglomerular apparatus that senses the sodium chloride concentration of the tubular fluid.

macrophage A type of white blood cell that ingests foreign invaders and dead or dying cells.

magnetite   A crystalline aggregation of a magnetic metal (usually iron); found in some magnetoreceptors.

magnetoreceptor   A sensory receptor that responds to magnetic fields.

malleus (hammer)   One of the three small bones of the mammalian middle ear involved in transmitting sound vibrations to the inner ear.

Malpighian tubule   The functional equivalent of a kidney tubule in insects, releasing the urine into the gut.

mantle cavity   A cavity formed by the body wall (mantle) of molluscs; generally contains the respiratory structures.

mass action ratio   Ratio of products to substrates; when more than one product (or substrate) is involved, their concentrations are multiplied together. When a reaction is at equilibrium, the mass action ratio equals the equilibrium constant (K_{eq}).

mass-specific metabolic rate   The metabolic rate of an animal (usually described as oxygen consumption) expressed relative to body mass.

mastication   Mechanical disruption of food in an oral cavity (chewing).

maximum reaction velocity (V_{max})   The maximal enzymatic rate calculated from a substrate-velocity curve; can be estimated by the enzymatic rate observed when product is absent and substrate concentrations are optimal.

mean arterial pressure (MAP)   The weighted average of the systolic and diastolic pressures, taking into account the relative length of each of these phases of the cardiac cycle.

mechanical energy   A form of energy arising from the movement or position of an object; can be either kinetic energy (as in a moving leg) or potential energy (as in a loaded spring).

mechanogated channel (or mechanically gated channel)   An ion channel that opens or closes in response to the stress (or stretch) on a membrane.

mechanoreceptor   A sensory receptor that detects forces applied to cell membranes (such as touch or pressure). Can be used to describe either the receptor protein or cells containing these receptors.

medulla oblongata   A region of the vertebrate brainstem containing centers that regulate heart rate, breathing depth and frequency, and blood pressure. Also called the medulla.

medullary cardiovascular center   The region within the medulla that regulates cardiac function.

medullary respiratory center   The region within the medulla that regulates breathing depth and frequency.

melatonin   A hormone found in all animal groups that regulates sleep-wake cycles.

melting point   The temperature at which a solid can become a liquid; when the melting point and the freezing point are not the same temperature, this hysteresis suggests the presence of a solute that acts in a noncolloidal manner, such as an antifreeze protein.

membrane fluidity   A state that allows the two-dimensional movement of lipids and proteins within a lipid bilayer membrane.

membrane potential   The electrical gradient across a cellular membrane.

membrane recycling   The exchange of membrane lipids and protein between the plasma membrane and the internal membrane network.

menarche   The age at which a female mammal with a menstrual cycle experiences her first menstruation.

meninges   Membranes covering the vertebrate central nervous system. Mammals have three meninges; birds, reptiles, and amphibians have two; and fish have one.

menses   In female mammals, the periodic shedding of the endometrial layer of uterine tissue that occurs if there is no implantation of a fertilized ovum; also known as *menstruation*.

menstrual cycle   The estrous cycle of humans and some other primates.

menstruation   See *menses*.

mesangial cells   Contractile cells between the capillaries of the glomerulus, which control blood flow, and thereby control blood pressure within the glomerulus.

mesencephalon   See *midbrain*.

mesoderm   The middle of the three primary germ layers in a developing embryo; eventually gives rise to tissues such as bone, muscle, and connective tissue.

messenger RNA   See *mRNA*.

metabolic acidosis or alkalosis   A decrease or increase, respectively, in blood pH as a result of metabolic activity.

metabolic depression   A reduction in metabolic rate below resting levels; associated with a period of dormancy.

metabolic flux   The flow rate through a metabolic pathway.

metabolic rate   The rate of heat production by a tissue or organism, usually approximated by oxygen consumption or carbon dioxide production.

metabolic water   The water produced by the metabolic breakdown of macromolecules.

metabolism   The sum of all chemical reactions in a biologic entity.

metabolizable energy   The proportion of digestible energy retained by the body; the remainder is unmetabolizable energy lost in excretory products.

metabolon   A group of enzymes that are spatially localized within the cell and perform a function together.

metabotropic receptor   A receptor that signals via a signal transduction pathway (see also *ionotropic receptor*).

metalloprotein   A protein with a metal ion integrated into its structure; enzymatic metalloproteins typically involve their metal in oxidation-reduction reactions.

metamorphosis   The transition between distinct developmental stages, typically from a larva to an adult.

metazoan   A multicellular animal.

methemoglobin   An oxidized form of hemoglobin that can no longer carry oxygen.

micelle   A lipid monolayer that rolls onto itself to form a sphere with a hydrophobic inner core and hydrophilic exterior.

Michaelis constant (K_m)   The concentration of substrate that yields half maximal velocity in an enzymatic reaction.

Michaelis-Menten equation $V = V_{max} \times [S]/([S] + K_m)$.

microclimate   The external environment within a confined space, typically distinct from the broader conditions, such as a subterranean burrow; typically used to describe the conditions experienced by an organism (see also *microenvironment*).

microelectrode   A very small electrode used to record electrical signals from cells.

microenvironment   Like a microclimate, but can apply to the environment surrounding anything from individual molecules to whole animals.

microfilaments   A polymer of β-actin used to construct the cytoskeleton.

microglia   One of the glial cells of the vertebrate central nervous system.

microtubule   A large, hollow tube consisting of a polymerized tubulin; used to build the cytoskeleton.

microtubule-associated protein (MAP)   A protein that binds to microtubules to alter structural or functional properties.

microtubule-organizing center (MTOC)   A multiprotein complex near the center of the cell from which microtubules grow.

microvilli   Fingerlike extensions from individual cells, supported by microfilaments, which serve to increase surface area.

micturition   Urination.

midbrain   The middle portion of the vertebrate brain consisting of the tectum and tegmentum. Also called the mesencephalon.

middle ear   A part of the vertebrate ear that consists of the tympanic membrane and one or more small bones (in mammals, the incus, malleus, and stapes) that help to amplify sounds.

milieu intérieur   The internal environment of a cell or organism.

mineralocorticoids   Steroid hormones involved in water and ion balance.

mitochondria   Organelles within most eukaryotic cells that produce energy by oxidative phosphorylation;

organized in many tissues as a network or reticulum.

mitochondria-rich cell Usually refers to the epithelial cells specialized for ion pumping, which have abundant mitochondria to meet the energy demands of active transport (see also *chloride cell*).

M-line The midpoint of a sarcomere where the thick filament lacks myosin heads.

mobile element A region of DNA that can be excised and inserted elsewhere within the genome.

molal (molality) Moles of an ion or molecule expressed relative to kilograms of solvent (usually water).

molar (molarity) Moles of an ion or molecule expressed relative to liters of solvent (usually water).

mole 6.02252×10^{23} molecules of a substance; the molecular weight of a substance is the mass of one mole of that substance.

molecular chaperone A protein that uses the energy of ATP hydrolysis to help fold or stabilize denatured proteins; includes heat shock proteins.

molecular phylogeny The evolutionary relationships among organisms as reconstructed based on molecular sequence data.

monoacylglyceride (or monoglyceride) A single fatty acid esterified to a glycerol molecule.

monocyte A large white blood cell that, in the tetrapod immune system, ingests foreign particles such as microbes; when it leaves the blood stream it differentiates into a macrophage.

monogastric stomach An animal that has a stomach with one (usually acidic) compartment.

monomer A single subunit of a multimer, such as a dimer or trimer.

monosaccharide A sugar, usually composed of a 6-carbon (sometimes 5-carbon) ring, such as glucose.

monounsaturated fatty acid A fatty acid with a single double bond.

monozygotic Arising from a single zygote.

motor end plate The location on a muscle that forms synapses with a motor neuron; the muscle side of a neuromuscular junction.

motor neuron A neuron that transmits signals from the central nervous system to skeletal muscles.

motor proteins Mechanoenzymes, such as myosin, that use the energy of ATP hydrolysis to move along cytoskeletal tracks.

motor unit A group of muscle fibers under the control of a single neuron.

mRNA Messenger RNA; the form of RNA that is used as a template during translation to form protein.

mucin The lipopolysaccharide that is the main component of mucus.

mucosa Refers to the inside layer of a tissue or organ, often that surface exposed to the lumen of an organ, such as the gastrointestinal tract (see also *serosa*).

mucous cells Cells that secrete a complex mucopolysaccharide onto the surface of a tissue; goblet cells are a type of mucous cell found in the intestinal and respiratory surfaces.

mucus A mucopolysaccharide mixture secreted from specialized epithelial cells onto the external surface of a tissue.

multipolar neurons Neurons with many processes leading from the cell body; most of these processes are dendrites, but one may be an axon.

muscarinic acetylcholine receptors G-protein-coupled receptors that bind acetylcholine.

muscle A multicellular tissue composed of myocytes, fibroblasts, and vascular cells; the contraction of the myocytes leads to force generation or shortening.

muscle fiber A single muscle cell; can be mononucleated (as in cardiomyocytes) or multinucleated (as in skeletal muscle fibers).

muscle spindle fiber A muscle stretch receptor.

mutation A heritable alteration in the nucleotide sequence of genomic DNA.

myelin See *myelin sheath*.

myelin sheath The insulating wrappings of vertebrate axons that are composed of multiple layers of glial cell plasma membrane. Invertebrate axons have analogous wrappings, but they are not generally termed a myelin sheath.

myelination The process of forming the myelin sheath around a vertebrate axon.

myenteric plexus A network of neurons found within the smooth muscle of the gastrointestinal tract that controls its muscular and secretory actions.

myoblast A mononucleated, proliferating cell that can differentiate to form a muscle cell.

myocardium The muscle of the heart.

myocyte A general term for a muscle cell, including smooth muscle cells, cardiomyocytes, and myofibers.

myofiber A multinucleated skeletal muscle fiber.

myofibril A long bundle of actin, myosin, and associated proteins in muscle cells.

myogenic Muscle contraction initiated by a trigger arising directly within the muscle, as in a myogenic heart.

myoglobin A type of hemoglobin found in muscle.

myometrium The smooth muscle layers of the uterus.

myosin A large multigene family of ATP-dependent motor proteins that work in conjunction with actin. The thick filament of muscle is composed of myosin, which is organized into hexamers consisting of two myosin heavy chains (MHC) and four myosin light chains (two regulatory MLC and two essential MLC).

myosin heavy chain The motor protein that interacts with actin.

myosin light chain A protein that binds the motor protein myosin II, regulating its structure or function.

myosin light chain kinase (MLCK) An enzyme associated with hexameric myosin that phosphorylates myosin light chain.

myosin light chain phosphatase (MLCP) An enzyme associated with hexameric myosin that dephosphorylates myosin light chain.

myotome A repeating segment in the body musculature of adult fish; also, the embryonic form of muscle derived from a body segment, or somite.

myotube An early stage of muscle differentiation in which multiple myoblasts fuse together to form a multinucleated contractile tubular cell.

Na$^+$/K$^+$ ATPase An ion transporter that expels 3 Na$^+$ out of a cell and imports 2 K$^+$, driven by the energy of ATP hydrolysis.

nares Nostrils.

natriuretic Leading to the appearance of sodium in the urine.

near-equilibrium reaction A reaction in which the products and substrates in vivo are near the concentrations that would arise if the enzymatic reaction were to reach equilibrium. The reaction is regulated by changes in the concentrations of substrates and products.

negative feedback loop A regulatory mechanism whereby a step late in a pathway causes a decrease in the activity of a step earlier in the pathway to reduce the flow through the pathway.

nephridium A primitive type of kidney tubule found in some invertebrates, such as annelids and molluscs; can also refer to the embryonic kidney of vertebrates.

nephron The multicellular unit of the kidney, consisting of the tubule and the vasculature that serves it, typically a glomerulus.

Nernst equation An expression that describes the ion concentration gradient across a permeable membrane in relation to the voltage when the system is at equilibrium.

nerve A cordlike structure composed of a collection of neuronal axons grouped together by connective tissues.

nerve net Description of the structure of the nervous system of cnidarians.

nervous system Network of neurons and their supporting cells.

net energy The proportion of metabolizable energy that is retained by the body, excluding that lost to specific dynamic action.

neurogenic A contraction that occurs in response to a nervous stimulus.

neurogenic muscle A muscle that is activated by neuronal stimulation.

neurohemal organ A region of multiple neurons that secrete hormones into the blood.

neurohormone A chemical messenger released from a neuron into the blood.

neuromast A structure consisting of a cup filled with a viscous gel and several hair cells; the functional unit of the lateral line system of fishes and amphibians.

neuromuscular junction The synapse between a motor neuron and a skeletal muscle cell.

neurons (nerve cells) Specialized cells in the nervous system that communicate using chemical and electrical signals. Many, but not all, neurons are excitable cells that generate action potentials.

neuropeptides Polypeptides that act as neurotransmitters.

neurosecretory cell Neurons that produce and secrete neurohormones into the blood, typically in a region called a neurohemal organ.

neurotransmitter A chemical messenger released from a neuron into the synaptic cleft.

neutral pH The pH at which the concentration of H^+ equals that of OH^-.

neutrophils The most common type of white blood cell in the vertebrate immune system.

nicotinic acetylcholine receptors Ligand-gated ion channels that open in response to acetylcholine binding.

nitric oxide A gaseous neurotransmitter and paracrine chemical signal that is involved in regulating many physiological processes; important vasodilator in vertebrates.

nociceptor (or nocioceptor) A sensory receptor that responds to noxious stimuli of various types (e.g., extreme heat or cold, extreme pressure, harmful chemicals, tissue damage); pain receptor.

nocturnal Active at night.

node of Ranvier A gap of exposed axonal membrane between two regions of myelin sheath.

noncompetitive inhibition A mode of enzyme inhibition in which a molecule inhibits an enzyme by acting at a site distant from the active site; noncompetitive inhibitors can increase the K_m or reduce the V_{max}.

noncovalent bond Includes four types of weak bonds that stabilize macromolecular structure.

nonpolar Having low solubility in water or other polar solvents.

nonshivering thermogenesis (NST) Production of heat by chemical means without muscle contraction. Typically refers to heat production by brown adipose tissue; however, there are other means of NST.

norepinephrine (or noradrenaline) A catecholamine neurotransmitter; in vertebrates, released by the sympathetic nervous system.

nuclease An enzyme that hydrolyzes nucleic acids; includes DNases and RNases.

nucleator (or nucleating agent) A molecule or particle that triggers the formation of ice at subzero temperatures.

nucleoside A molecule composed of a nitrogenous base (purine or pyrimidine) linked to a ribose or deoxyribose sugar.

nucleotide A nucleoside with one or more phosphate groups, such as ATP.

nymph The larval form of a hemimetabolous insect that resembles in most respects the adult form of the insect, except lacking functional wings.

obliquely striated muscle A muscle where striations run obliquely to the axis of shortening.

odorant Molecules that can be detected by the sense of smell.

odorant binding protein Proteins found in the mucus of the nasal epithelium that bind to odorants and transfer them to odorant receptors.

odorant receptor protein A G-protein-coupled receptor involved in the detection of odorants and thus the sense of smell.

olfaction Detection of environmental chemicals from outside the body: the sense of smell.

olfactory bulb A part of the vertebrate forebrain that is involved in processing olfactory sensations.

oligodendrocyte A vertebrate glial cell that forms the myelin sheath of a neuron in the central nervous system.

ommatidium The functional unit of the arthropod compound eye.

oncotic pressure The osmotic pressure of blood that is due to the concentration of large macromolecules, primarily protein.

oocyte One of the intermediate stages in the process of producing an ovum during meiosis.

oogenesis The production of an ovum.

oogonia (singular: oogonium) After the primordial germ cell enters the ovary, it differentiates into an oogonium, which undergoes multiple rounds of mitosis before entering meiosis.

open circulatory system A circulatory system in which the blood passes through one or more unbounded spaces called sinuses.

operculum The stiffened flaplike cover of the gills of bony fishes.

opsin A family of G proteins that is involved in visual phototransduction.

optic chiasm Area in the vertebrate brain where the optic nerves cross.

optic lobe Either of the two lobes of the vertebrate midbrain that are involved in visual processing; also, in arthropods the regions of the brain involved in processing signals from the compound eyes.

organ of Corti Located in the cochlea of the inner ear; contains the hair cells that are involved in the sense of hearing.

ornithine-urea cycle A pathway by which urea is produced from nitrogen arising from ammonia or glutamine.

orphan receptors Receptors whose ligand and function is not known; identified based on structural similarity to known receptors.

osmoconformer An animal that exhibits an internal osmolarity that parallels that of the external environment.

osmolarity Analogous to molarity, it is the concentration of osmolytes in a solution (osmoles per liter); abbreviated OsM.

osmole One mole of osmotically active solutes.

osmolyte An osmotically active solute; any solute that has a significant effect on osmotic pressure.

osmoregulator An animal that exhibits an internal osmolarity that is controlled independently of the osmolarity of the external environment.

osmosis The movement of water across a membrane from an area with a high activity of water to an area with low activity of water.

osmotic pressure A force arising due to the tendency of water to move by osmosis.

osteoblast A bone precursor cell.

otolith A small mineralized granule (usually calcium carbonate) in the inner ear of vertebrates. Involved in the sense of balance.

outer ear External portion of the vertebrate ear (consisting of the pinna and auditory canal in mammals).

outer hair cells One of two types of hair cells found in the organ of Corti in the inner ear of mammals; involved in amplifying sound and protecting the inner hair cells from loud sounds.

oval window Membrane between the middle ear and the inner ear of vertebrates. Vibrates to transmit sound to the inner ear.

oviparous An animal that produces eggs that hatch outside the body.

ovoviviparous An animal that holds its eggs inside the body until the eggs hatch, and then releases active young.

ovulation The release of an ovum following the rupture of a follicle.

ovum The larger of the two gametes of a sexually reproducing species. Although an ovum is often defined as the gamete produced by a female, in reality this definition is backward: an individual is a female if it has gonads that can produce an ovum.

oxidant A molecule that accepts an electron from another molecule (the reductant). In doing so, the oxidant becomes reduced.

oxidation A chemical reaction whereby a molecule donates an electron to another molecule, becoming oxidized.

oxidative phosphorylation The process by which mitochondria produce ATP from the oxidation of reducing equivalents (NADH, $FADH_2$). The electron transport chain expels protons from the mitochondria to produce a proton motive force, which is then used by the F_1F_0 ATPase to produce ATP.

oxyconformer An animal that exhibits a respiratory rate that declines when oxygen pressure declines.

oxygen debt See *excess postexercise oxygen consumption.*

oxygen dissociation curve See *oxygen equilibrium curve.*

oxygen equilibrium curve A curve showing the relationship between P_{O_2} and the oxygen saturation of blood containing a respiratory pigment.

oxygen carrying capacity The maximum amount of oxygen that can be carried by blood. Includes both dissolved oxygen and oxygen bound to respiratory pigments.

oxygen-transport pigment See *respiratory pigments.*

oxyregulator An animal that exhibits a constant respiratory rate despite a decline in oxygen pressure.

oxytocin A peptide hormone produced by the anterior pituitary; induces the contraction of smooth muscle during parturition.

P_{50} The partial pressure at which a respiratory pigment is 50% saturated with oxygen.

pacemaker A cell or group of cells whose output of action potentials occurs in a rhythmic pattern.

pacemaker cell An excitable cell that spontaneously fires action potentials in a rhythmic pattern.

pacemaker potentials Spontaneous depolarizations of the resting membrane potential that ultimately trigger action potentials within pacemaker cells.

Pacinian corpuscle A type of vertebrate skin mechanoreceptor.

pancreas A vertebrate organ that produces endocrine hormones including insulin and glucagon and also produces exocrine secretions that are involved in digestion.

pancreatic β cells Cells within the vertebrate pancreas that secrete the hormone insulin.

panting A mode of thermoregulation whereby an increase in the frequency of respiration enhances heat loss from the body core.

parabronchi Smallest airways of a bird lung.

paracellular transport Passage of solutes or water between cells; in most epithelial tissues, tight junctions and other cell-cell junctions prevent paracellular movement of fluids.

paracrine A type of chemical messenger that is involved in local signaling between nearby cells; paracrine messengers move through the interstitial fluid by diffusion.

paralogues Genes that are the result of a gene duplication event within a lineage (see also *homologues*).

parasympathetic nervous system Part of the vertebrate autonomic nervous system; generally active during periods of rest; releases acetylcholine onto target organs.

parathyroid glands Glands located on the posterior surface of the thyroid gland that release parathyroid hormones in response to changes in extracellular calcium.

parathyroid hormone Peptide hormone that regulates blood calcium levels.

parietal cells The acid-secreting cells within the gastric mucous membrane.

parthenogenesis A mode of asexual reproduction whereby offspring are produced by a female as a result of a variation on the meiotic pathway. Because meiosis is involved, chromosomal recombination is possible and the parthenogenic offspring are not clones of the parent.

partial pressure (of a gas) The pressure exerted by one of the gases in a gas mixture. The sum of the partial pressures of all the gases in a mixture gives the total pressure.

parturition The birthing process by which offspring of viviparous and ovoviviparous females are expelled from the reproductive tract.

parvalbumin A Ca^{2+}-binding protein in the cytoplasm of some muscles, which buffers Ca^{2+} levels to accelerate relaxation.

passive diffusion A type of passive transport that does not require a protein carrier.

passive transport Movement across a cell membrane without an energy investment other than the chemical gradient of the transported molecule; includes both passive diffusion and facilitated diffusion.

pattern generator A group of neurons whose rhythmic firing coordinates a rhythmic physiological process or behavior, such as breathing or locomotion.

pentose A five-carbon monosaccharide, such as ribose and deoxyribose.

peptide bond A carbon-nitrogen bond (–C–N–); most common in polymers of amino acids.

perfusion Movement of fluid through a tissue (e.g., flow of blood through a capillary bed).

pericardium The sac surrounding a heart.

perilymph The fluid found in the cochlea of the inner ear.

peripheral chemoreceptors Chemoreceptors located in the aortic and carotid bodies of vertebrates that detect changes in blood chemistry.

peripheral membrane protein A protein that is weakly bound to the membrane through an interaction with a lipid or integral membrane protein.

peripheral nervous system (PNS) All of the neurons outside of the central nervous system.

peripheral resistance See *total peripheral resistance.*

peristalsis The rhythmic contractions of intestinal smooth muscle; involved in propelling a bolus of food along the gastrointestinal tract and in

moving blood through the circulatory systems of some animals.

permeability The ability of a molecule to cross a barrier, such as a membrane.

permease A transporter that mediates facilitated diffusion, but is neither a channel nor a porin.

pH scale A measure of acidity, expressed as the negative $\log_{10}$ of the proton concentration.

pH-bicarbonate plot (Davenport diagram) A graphical depiction of the relationship between the pH and bicarbonate concentration of a solution. Usually used to describe these relationships in arterial blood.

phagocyte A cell that carries out phagocytosis.

phagocytosis The endocytosis of large particles from the extracellular space.

phasic muscle A type of muscle that undergoes rapid contractions and relaxations; a twitch muscle.

phasic receptor A sensory receptor that produces action potentials only during part of the stimulus (usually at stimulus onset and removal).

phenotype The physical characteristics of an organism; the result of an interaction between the genotype and the environment.

phenotypic plasticity Production of different phenotypes by a single genotype as a result of environmental cues; may be reversible or irreversible (see also *acclimation*).

pheromones Chemical messengers released by an animal into the environment that have an effect on another animal of the same species.

phosphagens Energy-rich compounds that transfer energy in reactions in which a large change in free energy results when a phosphate bond is broken.

phosphatase An enzyme that removes a phosphate group from a molecule; important in signal transduction pathways because it reverses the phosphorylations catalyzed by kinases.

phosphocreatine See *creatine phosphate.*

phosphodiester bond –P–O–P–.

phosphodiesterase An enzyme that breaks down the phosphodiester bonds of cyclic nucleotides such as cAMP and cGMP.

phosphoglycerides The major class of phospholipids of biological membranes, consisting of a glycerol backbone, two fatty acids, and a polar head group linked to the glycerol via phosphate.

phospholipase An enzyme that breaks down phospholipids, releasing either diacylglycerol, polar head groups, or fatty acids, depending on the type of phospholipase.

phospholipids Phosphoglycerides and sphingolipids.

phosphorylation The addition of a phosphate group via a kinase, expending ATP (e.g., a protein kinase catalyzes the phosphorylation of a protein).

phosphorylation potential An expression of energy status; the mass action ratio for an ATPase reaction ($[ATP]/[ADP][P_i]$).

photon The fundamental particle of electromagnetic radiation. Streams of photons can have differing wavelengths, in which case the resulting radiation is given different names (e.g., X-rays, gamma rays, visible light).

photopigments Molecules specialized for detecting photons; consist of a chromophore and an associated protein.

photoreceptors Sensory receptors that detect photons with wavelengths in the visible spectrum (i.e., light). Can be used to describe either the receptor proteins or the cells that contain them.

phototaxis Movement in response to light, either toward (positive phototaxis) or away (negative phototaxis).

phylogenetic Pertaining to phylogeny.

phylogeny A hypothesis regarding the evolutionary relationships among organisms; can be based on the analysis of various types of data (e.g., molecular, morphological).

physiological dead space The volume of a respiratory organ that is not involved in gas exchange; consists of both the anatomical dead space and the volume of any regions that, although capable of acting as gas exchange surfaces, do not participate in gas exchange (e.g., unperfused or unventilated alveoli).

physoclist Any fish whose swim bladder lacks a connection to the gut.

physotome Any fish whose swim bladder is connected to the gut via a tube.

piloerection The movement of hair or feathers perpendicular to the skin in response to muscular contraction.

pilomotor Related to the nerves and muscles that change the orientation of hair.

pineal complex Consists of the pineal gland and related structures; involved in melatonin secretion and the establishment of circadian rhythms.

pinna The cartilaginous structures forming the outer ear of mammals.

pinocytosis The endocytosis of fluids by the plasma membrane (see also *phagocytosis*).

pit organs The highly sensitive thermoreceptive organs of some snakes.

pituitary gland A hormone-secreting organ located at the base of the vertebrate brain; connected to the hypothalamus.

pivotal temperature In an animal with environmental sex determination, it is a temperature at which equal numbers of males and females result.

placenta In eutherian mammals, the membrane derived from the embryonic chorion that encircles the embryo, acting as the interface between embryonic and maternal tissues.

plane-polarized light When light arrives at a detector, it typically exhibits waves that run at all angles. Polarizing filters permit the passage of light waves that run in a specific angle (plane), generating plane-polarized light.

plasma The liquid fraction of vertebrate blood.

plasma membrane The lipid bilayer membrane that encircles a cell.

plasticity The ability to change or remodel a physiological process or structure, as in neural plasticity. See also *phenotypic plasticity*.

pleiotropy A phenomenon in which a single gene is responsible for multiple, seemingly independent phenotypes.

pleural sacs A series of membranes that surround the lungs of vertebrates. The pleural sacs enclose the pleural cavity.

plug-flow reactor A type of chemical reactor in which the inflow moves as a bolus through the tubelike reactor.

pN The pH at which a zwitterion has no net charge.

podocyte Cells surrounding the capillaries of the glomerulus, with footlike extensions that form the filtration slits.

poikilothermy A thermoregulatory strategy whereby an animal (a poikilotherm) allows body temperature (T_B) to vary, usually in relation to the ambient conditions.

Poiseuille's equation An equation describing the relationship between the flow, pressure, and resistance of a fluid moving through a rigid tube, including the factors influencing resistance (length, cross-sectional area, and viscosity).

polar See *hydrophilic*.

polymer A chain of repeating molecules, such as a polysaccharide or a polypeptide.

polymodal receptors Sensory receptor cells that can detect more than one type of stimulus.

polypeptide A chain of amino acids linked by peptide bonds.

polyphenism A form of irreversible phenotypic plasticity, generally involving alternative developmental pathways.

polypnea Rapid breathing.

polysaccharide A chain of monosaccharides linked by glycosidic bonds.

polysynaptic Involving more than two synapses; used in the context of reflex pathways.

polyunsaturated fatty acid A fatty acid with two or more double bonds along the carbon chain.

pons A region of the vertebrate brain that communicates information between the brainstem and the higher brain centers. Works with the medulla to regulate breathing.

porin A channel that permits the facilitated diffusion of large molecules; e.g., aquaporin is a porin that transports water.

porphyrins Organic ring structures that bind metals, primarily iron but also copper; heme is the most common type of porphyrin in animals.

portal system Two capillary beds connected by a portal vein (e.g., hypothalamic pituitary portal system; intestinal liver portal system).

portal vein A blood vessel that carries blood from one capillary bed to another; part of a portal system.

positive feedback loop A regulatory mechanism whereby a step late in a pathway causes an increase in the activity of a step earlier in the pathway to increase the flow through the pathway.

posterior pituitary Lobe of the pituitary gland; secretes antidiuretic hormone and oxytocin; also called the neurohypophysis.

postganglionic neuron A vertebrate autonomic neuron has its synapse in the peripheral autonomic ganglia, and extends an axon out into the periphery; forms a synapse with a preganglionic neuron.

postsynaptic cell A cell (either a neuron or effector) that receives a signal from a presynaptic cell across a synapse.

post-tetanic potentiation (PTP) A phenomenon in which a postsynaptic cell will respond with an unusually large change in membrane potential for several minutes following repeated action potentials in the presynaptic cell.

potential energy The energy that is available in a static system; elastic storage energy is a form of potential energy.

power The rate of doing work.

power curve The relationship between the velocity of muscle shortening and the force of contraction.

power stroke The part of a cross-bridge cycle in which structural changes in myosin alter the relative position of the actin filament.

pre-Bötzinger complex The primary respiratory rhythm generator of mammals.

preferred body temperature The temperature at which an animal functions best; achieved by physiological mechanisms that alter heat production or loss (mainly in homeotherms) or by behavioral choice of habitat (mainly in poikilotherms).

preformed water The water that arrives in the diet as a liquid or trapped within solid foods; distinct from metabolic water that is produced during the digestion of foods.

preganglionic neuron A vertebrate autonomic neuron that has its cell body in the central nervous system and forms synapses in the peripheral ganglia.

preprohormone Large inactive polypeptide that is a precursor to a peptide hormone (see also *prohormone*).

pressure A force applied to a unit area of a surface.

presynaptic cell A neuron that transmits a signal across a synapse to a postsynaptic cell.

primary active transport Active transport that uses chemical or light energy directly, such as an ion-pumping ATPase; distinct from secondary active transport, in which an entity is driven by electrochemical transmembrane gradients of another entity being transported.

primary follicle A follicle that continues to develop to release an ovum, unlike other follicles that degrade and die during the maturation process (atresia).

primary oocyte The products of oogonia that have undergone the first meiotic division to become a diploid cell that will eventually produce an ovum.

primary spermatocyte The products of spermatagonia that have undergone the first meiotic division to become a diploid cell that will eventually produce a spermatozoan.

primary structure The sequence of a polymer without consideration of how it folds; typically refers to the amino acid sequence of a protein.

primary urine The initial contents of the lumen of a nephron. In vertebrates that possess a glomerulus, the primary urine is the filtrate.

proboscis A single extension from the head, typically superior to the oral opening; the nose.

proenzyme A catalytically inactive precursor for an enzyme; usually undergoes proteolytic processing to become the active enzyme.

prohormone A polypeptide formed by the cleavage of a preprohormone; a precursor to the formation of a peptide hormone.

prolactin An anterior pituitary hormone that is responsible for milk production in mammals, and more general roles in ion and water balance in other vertebrates.

pronephros A simple kidney equivalent of larval forms of some amphibians and fish.

proprioceptor A sensory receptor that provides information about body position and movement.

prosencephalon See *forebrain*.

prostate gland A gland accessory associated with the reproductive tract of male vertebrates.

prosthetic group A nonprotein component of an enzyme or other protein; e.g., a coenzyme (an organic prosthetic group) or a metal.

protease An enzyme that breaks peptide bonds of proteins to generate polypeptides or amino acids.

proteasome A cytoplasmic multiprotein complex that degrades damaged proteins tagged with a ubiquitin molecule.

protein A polymer of amino acids, usually folded into complex secondary structures.

protein kinase An enzyme that attaches a phosphate to a protein, using a molecule of ATP for energy and as a phosphate source.

protein phosphatase An enzyme that removes a phosphate group from a protein.

proteoglycan A molecule composed of protein and glycosaminoglycan.

proteolysis The breakdown of proteins, usually by hydrolytic cleavage of peptide bonds by a protease.

prothoracic glands A pair of endocrine glands that secrete hormones that regulate ecdysis.

protofilament A single chain of tubulin that exists prior to the formation of sheets or microtubules.

proton motive force The electrochemical gradient arising from proton pumping by the mitochondrial electron transport chain.

protozoans An historical term to describe the phyla of early single-celled eukaryotes known now as protists.

proximal tubule The region of a mammalian or avian kidney tubule that lies between the Bowman's capsule and the descending limb of the loop of Henle.

proximate cause The immediate or direct cause of an organismal structure, function, or behavior; usually refers to the developmental or physiological mechanism (see also *ultimate cause*).

pulmonary system A respiratory system consisting of lungs and the associated vasculature.

pulmonary circuit The part of the tetrapod circulatory system that carries blood from the heart to and from the lungs.

pupa A developmental stage in hemimetabolous insects that separates the larva from the adult; can include a period of quiescence.

pupil An opening in the center of a camera-type eye through which light enters.

purine A class of nitrogenous bases with two rings; includes guanine and adenine.

Purkinje fibers The terminal branches of the conducting fibers of the mammalian heart.

P wave One of the waveforms of an electrocardiogram; represents the depolarization of the atria.

pyloric sphincter The sphincter that regulates movement of material from the stomach to the duodenum.

pyrimidine A class of nitrogenous bases with one ring; includes cytosine, thymine, and uracil.

pyrogen An entity that causes a homeotherm to mount an immune response that culminates in a fever.

Q_{10} A value that reflects the impact of a 10°C change in temperature on an enzymatic or metabolic process; also known as the temperature coefficient.

QRS complex One of the waveforms of an electrocardiogram; represents the depolarization of the ventricles.

quaternary structure The three-dimensional arrangement of a protein composed of multiple monomeric units.

radiant energy Thermal energy released from an object in relation to its temperature.

radiant heat transfer The emission of thermal energy from a warm object to cooler surroundings.

radiation The emission of energy from an object.

ram ventilation A ventilatory strategy in which the forward movement of the animal provides the propulsive force needed for bulk flow of the ventilatory medium across the respiratory surface. Seen in some fishes and insects.

range fractionation A strategy in which groups of sensory neurons work together to increase the dynamic range of a receptor organ. Each neuron has an overlapping, but not identical, dynamic range, allowing a wider range of stimulus intensities to be coded by the population of receptors.

rate constant The factor that allows the prediction of an enzymatic rate based on the concentration of the substrates.

reaction norm The range of phenotypes that can be produced by a given genotype when it is exposed to different environments.

reactive oxygen species (ROS) A free radical in which the unpaired electron is associated with an oxygen atom.

receptive field The area of the body that, when stimulated by an incoming sensory stimulus, affects the activity of a sensory neuron.

receptor A protein or cell that can detect an incoming stimulus.

receptor adaptation The process by which sensory receptor cells become less sensitive to sensory signals as signal duration increases.

receptor potential A graded change in the membrane potential within an epithelially derived sensory receptor cell. The receptor potential triggers the release of neurotransmitter onto a primary afferent neuron, causing a postsynaptic graded potential. If this postsynaptic potential exceeds threshold, it will trigger action potentials in the axon of the primary afferent neuron.

recruitment The stimulation of different collections of muscle fibers in response to different activity patterns.

rectal gland An organ found in cartilaginous fish that secretes salt to aid in osmotic regulation.

redox balance (reduction-oxidation balance) A condition in which there is no net change in the ratio of reduced to oxidized reducing equivalents, typically NADH/NAD$^+$.

redox shuttle A multienzyme pathway used to transfer the energy of reducing equivalents from glycolysis into the mitochondria for oxidation.

redox status The relative levels of reduced to oxidized molecules of interest; typically applied to metabolic biochemistry (e.g., NADH/NAD$^+$) but can also be used to reflect the degree of oxidative stress.

reducing equivalents NAD(P)H or FADH$_2$.

reductant A molecule that donates an electron to another molecule (the oxidant). In doing so, the reductant becomes oxidized.

reduction A chemical reaction whereby a molecule accepts an electron from another molecule, becoming reduced.

reductionism A philosophical approach that asserts that complex processes can be understood in terms of their components.

reflex arc A simple neural circuit that does not involve the conscious centers of the brain.

reflex control pathway See *reflex arc*.

refraction The bending of light as it passes from one medium to another.

refractive index The degree to which a material refracts light.

refractory period A period in which an excitable cell is less likely to generate an action potential (see also *absolute refractory period, relative refractory period*).

regional heterothermy A thermoregulatory strategy in which regions of an animal's body exhibit significantly different temperatures.

regulators Animals that maintain a degree of constancy in an internal physiochemical parameter (e.g., osmolarity or temperature) despite external changes in the parameter.

regurgitation The expulsion of stomach contents back up the esophagus into the oral cavity.

relative refractory period A period immediately following the absolute refractory period in which an excitable cell will generate an action potential only if exposed to a suprathreshold (unusually large) stimulus.

relaxed endothermy A thermal strategy in which an endothermic animal allows its body temperature to fall for a period of time.

renal Pertaining to the kidney.

renal clearance The removal of an entity from the plasma by the kidney.

renal tubule Within a nephron, it is the tube composed of a single layer of transport epithelium.

repolarization A return of the membrane potential of a cell toward the resting membrane potential following a depolarization or hyperpolarization.

resistance, electrical The force opposing the flow of charge through an electrical circuit.

resistance, vascular The force opposing the flow of blood through the circulatory system.

respiration The process by which mitochondria consume oxygen and

produce carbon dioxide (see also *external respiration*).

respiratory acidosis or alkalosis Decrease or increase in blood pH as a result of changes in blood carbon dioxide (usually as a result of changes in ventilation).

respiratory chain See *electron transport system*.

respiratory pigments Metalloproteins that act as oxygen transport and storage molecules (e.g., hemoglobin).

respiratory quotient (RQ) The ratio of CO_2 produced to O_2 consumed; indicative of the type of fuel being utilized. An RQ of 0.7 indicates fatty acids are the fuel, whereas an RQ of 1.0 suggests carbohydrates are being oxidized.

resting membrane potential The membrane potential of an excitable cell when action potentials or graded potentials are not being generated.

resting metabolic rate (RMR) The metabolic rate of an animal at rest under experimentally defined conditions (see also *basal metabolic rate, standard metabolic rate*).

rete mirabile A network of blood vessels that serve to retain heat via countercurrent exchange.

retina A layer of light-sensitive cells that lines the back of eyes.

retinal A derivative of vitamin A that acts as the light-absorbing chromophore in animal photopigments.

reversal potential The membrane potential at which there is no net movement of an ion through open ion channels.

Reynolds number A dimensionless number associated with an object that reflects how smoothly a fluid flows over the surface of the object.

rhabdomeric photoceptors One of two types of animal photoreceptor cells. Arthropod photoreceptors are rhabdomeric (see also *ciliary photoreceptors*).

rhodopsin A photopigment consisting of the protein opsin chemically linked to a vitamin A derivative called retinal.

rhombencephalon See *hindbrain*.

ribonucleic acid See *RNA*.

ribosomal RNA See *rRNA*.

ribosome A complex of RNA and protein that carries out protein synthesis.

rigor A state of skeletal muscle in which cross-bridges remain intact because ATP has been depleted from the cell.

RNA A polymer of ribonucleic acids similar to DNA except that they contain ribose in place of deoxyribose and uracil in place of thymine; includes mRNA, tRNA, and rRNA. Involved in transferring information from DNA and in protein synthesis.

RNase An enzyme that degrades RNA either from the end (exonuclease) or internally (endonuclease).

rod A type of vertebrate photoreceptor cell. In mammals, rods are

responsible for vision in dim light (see also *cone*).

Root effect A change in the oxygen carrying capacity of blood as a result of changes in pH.

round window Membrane at the end of the cochlea; acts as a pressure release for the fluid of the inner ear.

rRNA The form of RNA that is incorporated into the riboprotein complex known as a ribosome.

r-selection A life history strategy whereby parents invest minimally in large numbers of offspring; best suited to rapidly exploit underutilized niches.

r-type strategy A reproductive strategy where parents produce numerous offspring, with relatively little investment in their care.

ryanodine receptor A Ca^{2+} channel found in the sarcoplasmic reticulum of muscle, which allows Ca^{2+} to escape into the cytoplasm to initiate muscle contraction.

saliva A solution of enzymes, salts, and water secreted into the oral cavity to lubricate, dissolve, and disrupt food.

salt A neutral molecule composed of an inorganic anion and inorganic cation linked by an ionic bond, such as NaCl (table salt).

salt gland An extrarenal gland found in some marine and desert vertebrates that secrete Na$^+$ and Cl$^-$ to reduce body salt content.

saltatory conduction The mode of conduction of action potentials in myelinated axons in which action potentials appear to jump from one node of Ranvier to the next.

sarcolemma The cell membrane of a muscle.

sarcomere The contractile unit of striated muscle, typically measured from one Z-disk to the next.

sarcomere length The distance between two Z-disks of a sarcomere.

sarcoplasm The cytoplasm of a muscle cell; also known as myoplasm.

sarcoplasmic reticulum The endoplasmic reticulum of muscle.

satellite cells A population of omnipotent stem cells found on the surface of striated muscle. When stimulated, satellite cells can enter myogenesis to repair or replace muscle.

saturated (1) For respiratory pigments, hormone receptors, and carrier proteins, refers to a situation in which all available proteins are bound to their ligand. (2) For fatty acids, refers to fatty acid chains that lack double bonds.

saturated fatty acid A fatty acid with no double bonds.

scaling The relationship between a parameter, such as metabolic rate, and body size.

scaling coefficient The slope of a plot of log body mass against log parameter of interest, such as metabolic rate.

Schwann cell A type of glial cell in the vertebrates that forms the myelin sheath around axons in the peripheral nervous system.

sclera Tough outer surface of a vertebrate eye.

sclerites Plate-like sections of an invertebrate exoskeleton.

SDA See *specific dynamic action.*

second messenger A short-lived intracellular messenger that acts as an intermediate in a signal transduction pathway.

secondary active transport Transport of a molecule across a membrane against its electrochemical gradient, driven by the cotransport of another molecule along its electrochemical gradient.

secondary structure The folding pattern of a macromolecule; an α-helix is an example of the secondary structure of protein and DNA.

secretagogue A chemical that induces the secretion of another chemical, usually a cell signaling factor such as a hormone.

secretory granules Vesicles of secretory product stored within a cell, prepared for release when the cell receives the appropriate signal.

semicircular canals Structures of the inner ear responsible for the sense of balance and body orientation; part of the vestibular apparatus.

seminal vesicles A pair of glands that store sperm and secrete nutrients and fluids that form the semen, emptying it into the vas deferens upon ejaculation.

semipermeable membrane A membrane that allows the free movement of some molecules but impedes the movement of others.

sensillum (plural: sensilla) Sense organs in the insect cuticle. Involved in the senses of taste, smell, touch, and hearing.

sensitization A process by which the response of a neuron to a stimulus is increased.

sensory adaptation See *receptor adaptation.*

sensory modality The category of sensory input that a sensory system detects (e.g., light, sound, pressure).

sensory neuron A neuron that conveys sensory information from the periphery to the central nervous system (see also *afferent neuron*).

sensory receptor A tissue, cell, or protein that detects incoming sensory information.

sensory transduction The process of converting incoming sensory information to changes in cell membrane potential.

series elastic components Elements of a structure that can store elastic energy when they are deformed.

serosa Referring to the outer layer of a tissue or organ (see also *mucosa*).

serotonin A neurotransmitter (biogenic amine) involved in setting mood and regulating blood flow to the brain.

Sertoli cells Elongated cells in the seminiferous tubules of the testis that nourish the spermatids during spermatogenesis.

serum Blood plasma after the clotting factors have been removed.

set point In a homeostatically controlled system, the level at which the regulated variable is maintained.

sexual reproduction A process in which two cells (each with half the normal genetic complement as a result of meiosis and recombination) fuse to form one descendant cell.

shivering thermogenesis Heat production through uncoordinated stimulation of skeletal muscle contractile units.

signal transduction pathways Biochemical pathways in which a change in conformation of a receptor protein in the target cell is converted to a change in the activity of that cell.

sinoatrial node (SA node) A remnant of the sinus venosus found at the top of the right atrium of the mammalian heart.

sinus venosus The chamber leading to the atrium of the heart in nonmammalian vertebrates.

skeletal muscle A general term to describe the striated muscle that works in conjunction with the endoskeleton.

sliding filament model A theory that describes the interaction between actin and myosin during cross-bridge cycling.

smooth muscle A type of muscle that has an irregular arrangement of thick and thin filaments, and thus lacks sarcomeres.

sodium-potassium pump See *Na^+/K^+ ATPase.*

solute The particles (ions or molecules) dissolved in a solution.

solution The fluid in which solutes are dissolved.

solvent The liquid in which solutes are dissolved.

soma The cell body of a neuron, containing the nucleus.

somatic motor division (of the nervous system) The portion of the vertebrate peripheral nervous system that controls skeletal muscle.

spatial summation The process by which graded potentials at different points in the membrane (occurring at the same time) combine to influence the net graded potential of a cell.

specific dynamic action (SDA) The heat produced during the digestive process; also known as the heat increment.

spermatogenesis Production of spermatozoa.

spermatogonia (singular: spermatogonium) After the primordial germ cell enters the testes, it differentiates into a spermatagonium, which undergoes multiple rounds of mitosis before entering meiosis.

spermatozoa The smaller gamete in a sexually reproducing species; sperm.

sphincter A ring of smooth muscle that controls the diameter of an opening, controlling passage from one region to the next.

sphingolipid One class of phospholipid based on a sphingosine backbone.

spinal cord Part of the vertebrate central nervous system extending from the base of the skull through the vertebrae of the spine. The spinal cord is continuous with the hindbrain.

spinal nerves A series of paired nerves that exit at regular intervals along the spinal column.

spiracles Small openings leading to the respiratory system; spiracles are the primary opening to the tracheal system of insects. The same word is used for a nonhomologous structure in elasmobranch fishes that provides an alternate opening for the buccal-opercular cavities.

spleen A vertebrate organ that is involved with the immune, lymphatic, and circulatory systems. It can act as a storage site for red blood cells, and removes damaged cells from the circulation. It also generates immune cells called lymphocytes.

standard conditions Accepted external conditions under which physical parameters are assessed; may refer to pressure, temperature, concentration, or other such parameters.

standard metabolic rate (SMR) The metabolic rate of a poikilothermic animal at rest and post-absorptive, measured at a defined external temperature. (see also *basal metabolic rate, resting metabolic rate*).

stapes (stirrup) One of the three small bones of the mammalian middle ear.

Starling curve See *Frank-Starling effect.*

statocyst Hollow, fluid-filled sense organ in invertebrates that detects the orientation of the body with respect to gravity.

statolith Small dense granule (usually of calcium carbonate) found in statocysts.

steady state A condition in which there is flux through a reaction or pathway without a change in the concentration of intermediates.

stenohaline An animal that is tolerant of a narrow range of external salinities.

stenotherm An animal that is tolerant of a narrow range of ambient temperatures.

stereocilia The specialized cilia of vertebrate hair cells; involved in the sense of hearing.

stereopsis The ability to see in three dimensions.

steroid hormones A large class of hormones derived from cholesterol.

steroids A diverse group of nonpolar organic molecules composed of multiple carbon rings.

stoichiometry The quantitative relationship between two entities.

stomach A general term for an anterior region of a gastrointestinal tract, typically characterized by acidic digestion processes.

striated muscle A class of muscle that possesses thick and thin filaments organized into regular arrays; includes cardiac muscle and skeletal muscle.

stroke volume The volume of blood pumped by the heart in a single beat.

submucosa The tissue layer that lies beneath the mucosal layer.

substrate-level phosphorylation An enzymatic reaction that produces a high-energy phosphate.

sulci (singular: sulcus) The folds on the surface of the brain in some mammals.

summation See *spatial summation, temporal summation.*

supercooling The reduction of temperature of a fluid below its freezing point but without the formation of ice.

surface tension The force of adhesion that binds molecules of a fluid together at the interface with air.

surfactant Substance that lowers the surface tension of liquids; secreted in the lungs of vertebrates.

swim bladder A gas-filled organ that fish use for buoyancy compensation.

sympathetic division See *sympathetic nervous system.*

sympathetic nervous system Part of the vertebrate autonomic nervous system; active during periods of stressful activity; releases the neurotransmitters epinephrine and norepinephrine onto target organs.

symport A transporter that carries two or more entities across a cell membrane in the same direction; also known as a cotransporter.

synapse The junction between a neuron and another neuron or effector cell; consists of a presynaptic cell, the synaptic cleft, and a postsynaptic cell.

synaptic cleft The extracellular space between a presynaptic cell and a postsynaptic cell at a synapse.

synaptic depression A decrease in neurotransmitter release in response to repeated action potentials.

synaptic facilitation An increase in neurotransmitter release in response to repeated action potentials.

synaptic plasticity The capacity of synapses to change their structure and function.

synaptic transmission The process of transmitting information across a neural synapse.

synaptic vesicles Neurotransmitter-containing vesicles that release neurotransmitter into a synapse.

synergism A situation in which two agents or processes have a combined effect greater than the sum of the effects of the two agents or processes applied individually.

systemic circuit The part of the tetrapod circulatory system that carries blood from the heart to the body and back.

systole The phase of the cardiac cycle in which the heart is contracting.

systolic pressure The arterial blood pressure during systole.

tachycardia Rapid heartbeat.

tastants Chemicals that are detected by the sense of taste.

taste bud Structure involved in gustation in the vertebrates.

TCA cycle (see *tricarboxylic acid cycle*)

tectum Dorsal region of the vertebrate midbrain involved in coordinating visual and auditory responses.

teleost fish The most common subclass of the bony fishes.

temperature coefficient See Q_{10}.

temporal heterothermy A thermal strategy whereby a homeothermic animal exhibits periods of poikilothermy, typically to allow a reduction in metabolic rate; also known as relaxed endothermy.

temporal summation The process by which graded potentials occurring at slightly different times combine to influence the net graded potential of the cell.

tendon The connection between a muscle and a bone.

tension, muscular The force produced by a contracting muscle.

terminal cisternae An enlargement of the sarcoplasmic reticulum near the muscle plasma membrane, specifically T-tubules.

tertiary structure The three-dimensional structure of a macromolecule, stabilized by numerous weak bonds.

tetanus The sustained contraction of a muscle arising from multiple stimulations in close succession.

thalamus One of the basal ganglia of the vertebrate brain that relays sensory information to the cerebral cortex.

theca The outer layer of somatic cells surrounding a follicle, separated from the inner granulosa cells by a basal lamina.

thermal conductance The transfer of thermal energy either within an object or from one object to another.

thermal energy Energy associated with heat production.

thermogenesis Heat production.

thermogenin The mitochondrial uncoupling protein found in mammalian brown adipose tissue.

thermoneutral zone The range of ambient temperatures over which an animal does not need to alter metabolic processes to maintain internal constancy.

thermoreceptor A sensory receptor that responds to temperature.

thermoregulation The physiological strategy an animal uses to control temperature within the desired range.

thick filament A polymer of about 300 myosin dimers that produces the contractile force in muscle.

thin filament A muscle-specific α-actin polymer similar in structure to a microfilament; serves as a framework that translates actinomyosin activity into force generation.

threshold potential The critical value of the membrane potential in an excitable cell to which the membrane must be depolarized in order for an action potential to be initiated.

threshold stimulus The smallest stimulus that can provoke a response in a cell.

thyroid hormone An iodine-containing hormone produced by the thyroid gland that is involved in the regulation of metabolism.

tidal volume The volume of a respiratory medium moved into or out of a respiratory structure during a single breath.

tight epithelia An epithelial layer with cell-cell connections that limit or prevent paracellular transport.

tight junction A type of intercellular connection that is capable of preventing the free movement of molecules between the cells.

tissue An aggregation of related cells linked together by various types of intercellular connections.

titin A very large protein that runs along the thin filament in striated muscle, determining its length and orienting into the sarcomere.

tonic muscle A muscle type with a slow contraction that persists for long periods (see also *phasic muscle*).

tonic receptor A receptor that produces action potentials throughout the duration of a stimulus.

tonicity The property of an extracellular solution that determines whether a cell will swell or shrink.

torpor A type of dormancy characterized by a relatively short period of hypometabolism.

total lung capacity The volume of air in the lungs at the end of a maximal inspiration; the maximum amount of air that can be held in the lungs.

total peripheral resistance The net resistance of the vasculature.

totipotent stem cell An embryonic cell that has the capacity to differentiate into any type of cell when given the appropriate cell signaling information.

trabeculae Any partition that divides or partially divides a cavity.

trachea (plural: tracheae) The single large airway leading to the paired bronchi of vertebrate lungs; also, the nonhomologous respiratory structures that are the main conducting airways in arthropod tracheal systems.

tracheal system The respiratory structures of insects and some other groups of air-breathing arthropods.

tracheoles The terminal structures of arthropod tracheal systems across which gas exchange takes place.

transcellular transport Movement of solutes or water across a cell layer through the cell itself, typically crossing both apical and basolateral cell membranes.

transcription RNA synthesis using the DNA template of a gene.

transducin An inhibitory G protein involved in visual signal transduction in the vertebrates.

transfer RNA See *tRNA.*

transgenic animal An animal that has been genetically modified to possess a heritable mutation.

transition state A temporary, intermediate state in the conversion of substrate to product when a molecule obtains enough energy to reach the activation energy barrier.

translation Protein synthesis using ribosomes and mRNA template.

transmembrane receptor A receptor protein that spans the cell membrane; consists of an extracellular domain, a transmembrane domain, and an intracellular domain.

transmural pressure The pressure difference across the wall of a chamber (e.g., a blood vessel, heart, or airway).

transpirational water loss Water loss arising from gas exchange across the respiratory surface.

transpulmonary pressure The difference between the intra-alveolar pressure and the intrapleural pressure in mammalian lungs.

transverse tubule See *T-tubule*.

triacylglycerol (or triglyceride) Three fatty acids esterified to a glycerol molecule.

tricarboxylic acid cycle The cyclical mitochondrial pathway that oxidizes acetyl CoA to form 3 NADH, 1 $FADH_2$, and 1 GTP; the pathway that produces most of the CO_2 arising from metabolism.

trichromatic color vision The system of three different photoreceptors by which humans and some other animals obtain color vision.

trimer A molecule composed of three subunits.

tRNA (or transfer RNA) A cloverleaf-shaped RNA molecule that binds a particular amino acid and participates in translation, binding to a three-nucleotide sequence of mRNA (codon) to transfer the amino acid to a growing polypeptide.

trophoblast An outer layer of cells derived from the mammalian blastocyst that forms the interface between the fertilized ovum and the uterine wall.

tropic hormones (or trophic hormones) Hormones that cause the release of other hormones.

tropomyosin A regulatory protein that stretches across seven actin monomers in a thin filament, controlling myosin's access to its binding site on the thin filament.

troponin A trimeric regulatory protein bound to tropomyosin. It responds to high $[Ca^{2+}]$ by inducing tropomyosin to move into a position that allows myosin to bind actin.

T-tubule An extension of the plasma membrane (sarcolemma) of some muscles that serves to improve the conduction of the action potential into the fiber.

tubule, renal Also known as a kidney tubule, it is the single filtration unit of the vertebrate kidney.

tubulin The monomeric protein subunit of microtubules, itself a dimer of α-tubulin and β-tubulin.

turbulent flow A disordered pattern of fluid flow over the surface of an object that reduces the efficiency of movement of the object through the fluid.

turnover number The number of times a single enzyme molecule completes a reaction cycle each second; also known as the catalytic constant (k_{cat}).

turnover rate The number of catalytic events in a given period of time. For an individual enzyme, it is synonymous with the catalytic constant (k_{cat}). It can also be used to describe the rate of synthesis and degradation of a metabolite, such as ATP.

T wave The portion of an electrocardiogram (EKG) that represents the repolarization of the ventricle.

twitch fibers Muscle fibers that undergo a rapid contraction/relaxation cycle (a twitch), in contrast to tonic fibers.

twitch muscle A muscle that contracts and relaxes once after each neuronal stimulus; a **phasic muscle**.

tympanal organ Sensory receptor involved in hearing in insects; insect ears.

tympanic membrane Thin membrane that separates the outer ear from the middle ear. Helps to transfer sound vibrations to the inner ear.

ubiquitin A small protein that is added to damaged proteins to mark them for degradation by the proteasome.

UCP See *uncoupling protein*.

ultimate cause Why an organism has a particular structure, function, or behavior; usually involves understanding the evolutionary advantage of the trait (see also *proximate cause*).

ultrafiltration Process of filtration of a fluid through a size-selective membrane under pressure; used to form the primary filtrate of the vertebrate kidney. Also causes the formation of lymph from blood in vertebrates.

ultraviolet light Short-wavelength light (<~300 nm); its high energy can damage macromolecules.

uncoupling (of oxidative phosphorylation) When mitochondrial respiration continues without the production of ATP.

uncoupling protein (UCP) A class of proteins, which includes thermogenin (UCP1), that act by dissipating the mitochondrial proton motive force.

unipolar neuron A neuron with one process leading from the cell body; this process generally splits into two branches, one conveying information toward the cell body and one conveying information away from the cell body.

uniporter A class of transporter that carries a single entity (ion, atom, molecule) with each transfer.

unitary displacement The distance a single motor protein moves during a cross-bridge cycle.

unsaturated fatty acid A fatty acid with one or more double bonds.

upper critical temperature The highest temperature at which a homeothermic animal can live for extended periods; the upper limit of the thermoneutral zone.

U/P ratio The ratio of an ion or molecule concentration in the urine (U) versus the plasma (P).

up-regulation Increase in protein number or activity in a target cell (see also *down-regulation*).

urea A nitrogenous waste possessing two nitrogen atoms per molecule.

ureotele An animal with an excretory strategy in which urea dominates the nitrogenous wastes.

ureter The tube connecting the kidney to the bladder.

urethra The tube carrying urine from the urinary bladder to the excretory opening.

uric acid A nitrogenous waste possessing four nitrogen atoms per molecule.

uricolytic pathway A pathway of breakdown of uric acid present in all animals.

uricotele An animal with an excretory strategy in which uric acid is the dominant nitrogenous waste.

urine A solution of nitrogenous waste produced by the kidney or kidney-like tissues.

van der Waals interactions A type of weak bond forming from the mutual attraction of the nuclei of two atoms in a molecule.

vas deferens The duct through which sperm are carried from the sites of synthesis in the epididymis to the ejaculatory opening.

vasa recta The straight blood vessels arranged in a hairpin loop that run from kidney cortex to medulla and back to the cortex. The countercurrent arrangement allows removal of salts and water from the peritubule interstitium while maintaining intramedullary osmotic gradients.

vascular endothelium Thin layer of cells that lines blood vessels.

vasculature The blood vessels of the circulatory system.

vasoconstriction Narrowing of a blood vessel as a result of contraction of the vascular smooth muscle; decreases local blood flow.

vasodilation Widening of a blood vessel as a result of relaxation of the vascular smooth muscle; increases local blood flow.

vasomotion Change in the diameter of blood vessels; also known as angiokinesis.

vasomotor response The changes in diameter of blood vessels in response to vasodilatory or vasoconstricting factors; also known as the angiokinetic response.

vasomotor tone The degree of contraction of the smooth muscles surrounding the arterioles (see also *venomotor tone*).

veins Blood vessels that return blood to the heart. In vertebrates, blood flows from the capillaries into venules and then into veins.

venomotor tone The degree of contraction of the smooth muscles surrounding the veins (see also *vasomotor tone*).

ventilation Active movement of the respiratory medium (air or water) across the respiratory surface.

ventilation-perfusion ratio (or ventilation-perfusion matching) The relationship between the ventilation (flow of respiratory medium) and the perfusion (flow of blood) at a respiratory surface.

ventricle A fluid-filled sac or cavity (e.g., the spaces in the center of the vertebrate brain; the muscular pumping chambers of the vertebrate heart).

venule Small blood vessels located between capillaries and veins.

vesicle A membrane-bound compartment that buds off from the intracellular membranous network, often encased in coat proteins such as clathrin.

vestibular apparatus The organ of balance in the vertebrates.

villi Undulations and folds in a tissue that serve to increase surface area; most commonly seen in the gastrointestinal tract.

viscosity An internal property of a fluid that results in resistance to flow. Thick liquids have high viscosity.

viscous effects The antagonism to movement of an object due to the interaction of its surface with the fluid through which it moves.

visual cortex A part of the vertebrate brain that is responsible for processing visual signals.

visual field The area that is visible to an eye, without changing eye position.

vital capacity The maximum amount of respiratory medium that can be moved into or out of the respiratory system with each breath.

vitamin A dietary compound that serves as a precursor for prosthetic groups of proteins, particularly enzymes.

vitellin The dominant protein found in yolk produced from vitellogenin.

vitellogenin The major protein in the yolk of an egg.

viviparous Animal whose offspring develop internally and are released as active young (see *oviparous* and *ovoviviparous*).

V_{max} The maximal rate of catalysis by an enzyme; arises when all substrates are at optimal (saturating) concentrations and prior to the formation of product.

V_{O_2max} The maximal sustainable rate of oxygen consumption exhibited by an animal. The experimental means to assess V_{O_2max} differs among disciplines.

volatile fatty acids Fatty acids of chain length less than 2–6 carbons; also known as short chain fatty acids.

voltage-gated ion channel A membrane protein containing an aqueous pore that can be opened in response to changes in the membrane potential.

vomeronasal organ A vertebrate sense organ adjacent to the mouth and nasal cavities that is involved in detecting pheromones.

weak bonds Ionic bonds, hydrogen bonds, van der Waals forces, and hydrophobic interactions.

white adipose tissue A lipid storage tissue of mammals; distinct from brown adipose tissue. Other vertebrates lack brown adipose tissue, and white adipose tissue is typically referred to simply as "adipose tissue."

white matter Areas of the vertebrate central nervous system that are rich in axons (see also *gray matter*).

white muscle A muscle fiber type specialized for rapid, high-intensity contractions that continue for a short duration; usually composed of type IIb myosin isoforms.

work The transfer of energy that occurs when force is exerted on a body to cause it to move.

work loop A method used to assess whether a muscle is performing positive or negative work.

xeric A dry, dehydrating environment.

yolk A deposit of lipid and protein (largely vitellin) associated with an ovum.

Z-disk The protein plate at the end of a sarcomere that serves as the insertion site of actin thin filaments.

zona pellucida A thickened glycoprotein extracellular matrix of a mammalian ovum; it binds the sperm to initiate the acrosomal reaction.

zwitterion A molecule with groups that can become positive and others that can become negative.

zygote The single cell arising from the fertilization of an ovum by a sperm.

Animal Index

Subject Index

Page references ending with *fig* indicate an illustrated figure; with *t* indicates a table; with *p* indicates a photograph; with *n* indicates a footnote.

A

A-band (Anistropic band), 215, 709
ABC transporters, 68–69
Abducens nerves, 313*t*
Abomasum, 550, 551*fig*
Absolute refractory period, 153, 709
Absolute temperature, 709
Absorbance spectra, 296
Accessory breathing organs, 432
Accessory glands, 678*fig*
Accessory hairs, 651*fig*
Accessory nerves, 313*t*
Accessory olfactory bulb, 318*t*
Acclimation, 14–15, 709
Acclimatization, 14–15, 642, 709
Accommodation, 290, 709
Acetylcholine (ACh), 103
  definition, 709
  digestion, 554*t*
  GI motility, 560, 561*t*, 562*fig*
  heart rate and, 378
  neurotransmitter function, 162–163, 184*t*
Acetylcholine channel (K_{ACh} channel), 171
Acetylcholine receptors, 185–186
Acetylcholinesterase, 163, 709
Acetylcholinesterase inhibitors, 164
Acetyl CoA, 52, 53, 57–58, 586*fig*, 709
A channel (K_A channel), 170
Acid-base balance, 505
Acidosis, 709
Acids, 31–32, 33*fig*, 709
Acid secretions, in stomach, 558
Acoelomates, 548*fig*
Acrosomal reaction, 709
Acrosome, 208, 709
Actin, 74
  animal physiology and, 211, 212*t*
  definition, 709
  microfilaments, 206–207
  myosin and, 208–210, 211, 212*t*
  networks, 207*fig*
  polymerization, 207–208
  structure of, 206*fig*
  thin filaments and, 220*fig*
  unitary displacement and, 211
Actinomyosin, 709
Actino-myosin activity, 208–210, 211, 212*t*, 220*fig*
Action potentials, 143, 146
  cardiac, 378–379
  cardiac striated muscles, 233*t*
  cardiomyocytes and, 379*fig*
  conduction of, 159*fig*
  definition, 709
  frequency of, 160, 161*fig*, 162
  giant axons and, 179*fig*
  graded potentials and, 151–152, 153, 154*fig*
  muscle excitation and, 225–227
  neurotransmitter release and, 162
  pacemaker potentials and, 376*fig*

phases of, 155*fig*
signal transmission distance, 158, 159*fig*
skeletal striated muscles, 233*t*
T-tubules and, 228
voltage-gated ion channels and, 154–157, 171–172
Activated isoprene, 56
Activation, 53*fig*
Activation energy, 24, 35–37, 640, 709
Activation gate, 156, 709
Active site, 35, 709
Active state, 709
Active transport, 66, 67*fig*, 68–69, 709
Acuity, 253, 709
Acute response, 709
Adaptation, 15–16, 709
Adaptive immunity, 400
Additivity, 129–130
Adenine, 79, 709
Adenohypophysis. *See* Anterior pituitary
Adenosine, 105, 184*t*, 561*t*, 709
Adenosine diphosphate. *See* ADP
Adenosine triphosphate. *See* ATP
Adenylate cyclase, 121*fig*, 561, 562*fig*, 709
Adequate stimulus, 252
Adhesion plaques, 233, 709
Adipocytes, 129
Adipokinetic hormone (AdK), 563–564
Adipose, 567
Adipose tissue, 136*t*, 327*t*, 631, 709
ADP, 49–50, 59, 709
Adrenal cortex, 132, 133, 136*t*, 709
Adrenal glands, 132, 136*t*, 709
Adrenalin. *See* Epinephrine
Adrenal medulla, 132–133, 136*t*, 327*t*, 331*fig*
Adrenal tissue structure, 133, 134*fig*
Adrenergic receptors, 187–188, 189*t*, 329, 709
Adrenocorticotropic hormone (ACTH), 126, 133, 299
Adult, 664*fig*
Aerial gas exchange and ventilation, 428
  air breathing, evolution of, 432, 433*fig*
  airway resistance, 441
  alveoli, 437–439
  arthropods, 429–430
  aspiration-based pulmonary systems, 441–442
  buccal force pump, 433–434
  pulmonary function tests, 442, 443*fig*
  suction pumps, 434–435
  surfactants and lung compliance, 440–441
  tidal ventilation, 439–440
  tracheal system, 430–432
  unidirectional ventilation, 435–436
  ventilation-perfusion matching, 442–443
Aerobic, 709
Aerobic capacity, of muscles, 582
Aerobic conditions, oxidation under, 50–51

Aerobic dive limit, 462
Aerobic scope, 709
Aerodynamics, 603–608
Aerofoils, 603–604
Afferent, 709
Afferent axon, 312*fig*
Afferent neurons, 166, 251, 308*fig*, 312*fig*
Affinity, 108, 709
Affinity constant (K_a), 108, 709
After-hyperpolarization, 153, 709
Aggrecan, 77
Aglomerular kidneys, 517, 709
Agonists, 105, 106*fig*, 709
Air
  bulk flow of, 412*fig*
  physical properties of, 415*t*
  respiratory gases and, 415*t*
  thermal conductivity of, 628
  ventilation and gas exchange in, 428–443
Air breathing, evolution of, 432, 433*fig*
Air capillaries, 436
Air currents, 604–606
Air density, 415*t*
Air viscosity, 415*t*
Airway resistance, 441
Albumen, 682*fig*, 709
Albumin, 99–100, 398, 709
Albuterol, 329
Aldosterone, 99, 100*t*, 511–513, 709
Aliphatic chains, 52
Alkaloids, 709
Alkalosis, 709
Allantoic membrane, 709
Allantoin, 709
Allantois, 682*fig*, 687*fig*
Allatostatin, 709
Allatotropins, 669, 709
Alleles, 83, 709–710
Allometric functions, 700–701
Allometric scaling, 11, 61, 710
Allostatins, 669
Allosteric activators, 640
Allosteric modulation, 452*fig*
Allosteric regulators, 40, 41*fig*, 710
Allosteric site, 710
Allozymes, 83, 710
α adrenergic receptor, 710
α-helix, 710
Alpha Helix program, 3, 6
α-lactalbumin, 690
α-Tubulin, 199–200
Alpine environments, 626*fig*
Alternative splicing, 83, 710
Alveolar dead space, 441
Alveoli, 437–439, 690*fig*, 710
Amacrine cells, 291
Ambient temperature (T_A), 626, 627, 710
Amines, 102, 710
Amino acids, 43–44
  cell signaling, and, 96–97
  definition, 710
  growth and, 531
  metabolism, 490–491
  plasma membrane and, 534
  proteases and peptidases, 535–536
  types of, 184*t*
Ammonia, 490–491, 710
Ammonioteles, 490, 710
Amnion, 682*fig*

Amniotes, 675, 682, 710
Amniotic cavity, 687*fig*
Amoebas, 21
Amoeboid movement, 212*t*
AMP (adenosine monophosphate), 49–50
Amphibolic pathways, 35, 710
Amphipathic, 710
Amplification, 110, 710
Amplifier enzymes, 118
Ampullae, 273
Ampullae of Lorenzini, 252, 279, 710
Amygdala, 318*t*, 319*fig*, 320–321, 710
Amylase, 531–532, 534, 710
Amylopectin, 47
Amylose, 47
Anabolic pathways, 35, 63–64, 710
Anadromous fish, 486, 710
Anaerobic, 710
Anaerobic conditions, oxidation under, 51–52
Anaplerotic pathways, 57, 710
Anastomoses, 357, 710
Anatomical dead space, 710
Androgens, 299, 632, 666, 667*fig*, 677*t*, 686*t*, 710
Androstenedione, 666, 667*fig*
Anemia, 452, 710
Angiogenesis, 360, 361, 710
Angiotensin, 710
Angiotensin I, 513
Angiotensin II, 387
Angiotensin II, 513
Angiotensin-converting enzyme (ACE), 513, 710
Angle of attack, 604
Animal life cycle, 664*fig*
Animal physiology, 1, 2–3
  history of, 4–6
  overview, 4
  subdisciplines, 6–9
  summary, 17
  unifying themes in, 10–16
Animal starch, 47
Anion, 27, 710
Anion antiporters, 485
Anisogametic, 665
Antagonist, 105, 106*fig*, 710
Antagonistic controls, 13, 710
Antagonistic muscles, 580, 710
Antagonistic pairings, 128
Antagonistic regulation, 130*fig*
Anterior pituitary gland, 124, 125*fig*, 126, 135*t*, 299, 300, 710
Antidiuretic, 710
Antifreeze glycoproteins (AFGPs), 646
Antifreeze protein (AFP), 646–647, 710
Antigen, 710
Antiport, 69, 710
Anus, 710
Aorta, 355, 357, 388*fig*, 710
Aortic body, 391, 458, 710
Aortic semilunar valve, 373
Apical, 710
Apical membrane, 710
Apnea, 419, 710
Apocrine, 710
Apoenzyme, 710
Appetite control, 558*fig*

Credits

Chapter 1

1 Photo Researchers, Inc., Francois Paquet-Durand/Photo Researchers, Inc.
2 Minden Pictures, FRANS LANTING/Minden Pictures.
3 Minden Pictures, FRANS LANTING/Minden Pictures.
3 Edward A. Hemmingsen.

Chapter 2

20 Getty Images, Scott Sady/Getty Images.
21 Photo Researchers, Inc., Eye of Science/Photo Researchers, Inc.
28 Photo Researchers, Inc., Stephen Dalton/Photo Researchers, Inc.
75 Phototake, Albert Tousson/Phototake.

Chapter 3

90 imagequestmarine.com, Roger Steene/imagequestmarine.com
91 Mark J. Grimson & Richard L. Blanton.
91 Nature Picture Library, Anup Shah/naturepl.com.

Chapter 4

142 Shin Kuang, Courtesy of S. Kuang and J.R. Sanes
143 Library of Congress.

Chapter 5

196 Art Resource, NY, HIP/Art Resource, NY
197 photolibrary, Image Source Limited/photolibrary.
197 Photo Researchers, Inc., Juergen Berger/Photo Researchers, Inc.
213 Phototake, Eric Grave/Phototake.
213 Pearson Education/PH College.

Chapter 6

247 Oliver Meckes/Nicole Ottawa/Photo Researchers, Inc.
248 Minden Pictures, Paul Johnson/npl/Minden Pictures.
249 Photo Researchers, Inc., Eye of Science/Photo Researchers, Inc.
249 Photo Researchers, Inc., SPL/Photo Researchers, Inc.
288 Phototake, Carolina Biological Supply Company/Phototake.
298 DRK Photo, M. Harvey/DRK Photo.

Chapter 7

306 Dr. Martin Wiesmann.
307 National Library of Medicine.
307 CORBIS, Peter Turnley/CORBIS.
322 University of Wisconsin Media Solutions, Waller Welder/University of Wisconsin Media Solutions.

Chapter 8

348 Iain J. McGaw and Carl L. Reiber, Image courtesy of Iain J. McGaw and Carl L. Reiber.
349 Library of Congress.
349 Photo Researchers, Inc., Martin Dohrn/Royal College of Surgeons/SPL/Photo Researchers, Inc.

Chapter 9

410 Peter Arnold, Inc., John Cancalosi/Peter Arnold, Inc.
411 Minden Pictures, KONRAD WOTHE/Minden Pictures.
411 Minden Pictures, STEPHEN DALTON/Minden Pictures.
418 Tom McHugh, Photo Researchers, Inc.
431 Science Magazine, Reprinted with permission from Westneat, M.W. et al. 2003. Tracheal respiration in insects. Science 299 (5606): 588–560. Copyright 2003 A.A.A.S.
438 Phototake/Carolina Biological Supply Company.

Chapter 10

470 Image Quest Marine, Masa Ushioda/Image Quest Marine.
471 Image Quest Marine, Peter Batson/Image Quest Marine.
471 Minden Pictures, PIOTR NASKRECK/Minden Pictures.
482 Photo Researchers, Inc., George Holton/Photo Researchers, Inc.
482 Photo Researchers, Inc., Dee Breger/Photo Researchers, Inc.
485 Images courtesy of Dr. Steve Perry, University of Ottawa.
487 Images courtesy of Ryan Pellis and Dr. Steve McCormick, Cortes Anadromous Fish Laboratory.
488 Photo courtesy of Dr. Robert H. Rothman, Rochester Institute of Technology.

Chapter 11

526 Minden Pictures, Karl Ammann/npl/Minden Pictures.
527 Minden Pictures, TUI DE ROY/Minden Pictures.

527 Grant Heilman Photography, AMES STRAWSER/Grant Heilman Photography.
533 Visuals Unlimited, WHOI/Edmund/Visuals Unlimited.
542 Photo Researchers, Inc., Mona Lisa Production/Photo Researchers, Inc.
542 Courtesy Paul Handford, University of Western Ontario.
542 Photo Researchers, Inc., Karl H. Switak/Photo Researchers, Inc.

Chapter 12

572 AKG-Images, Leonardo da Vinci/AKG-Images.
573 Library of Congress, Eadweard Muybridge/Library of Congress.
573 Jean-Michel Weber, Photo courtesy of Jean-Michel Weber, University of Ottawa.
579 Dr. Gary Sieck, Dr. Gary Sieck/Mayo Clinic College of Medicine.
588 Craig Franklin, Nicholas Hudson and Craig Franklin.
598 Robert Campbell, Robert Campbell, University of Victoria.
604 Photo Researchers, Inc., Nick Bergkessel/Photo Researchers, Inc.
604 Tom Stack & Associates.
610 Minden Pictures, MICHAEL & PATRICIA FOGDEN/Minden Pictures.

Chapter 13

624 Bruce Coleman Inc., Scott Nielsen/Bruce Coleman Inc.
625 Dorling Kindersley.
625 Getty Images, O. Louis Mazzatenta/National Geographics/Getty Images.
629 Photo courtesy of Dr. Brian Helmuth, University of South Carolina.
630 Photo Researchers, Inc., Gregory G. Dimijian M.D./Photo Researchers, Inc.

Chapter 14

662 Minden Pictures, FRANS LANTING/Minden Pictures.
663 National Geographic Image Collection, Paul Nicklen/National Geographic Image Collection.
663 Minden Pictures, FRANS LANTING/Minden Pictures.
677 Photo courtesy of Scott Pitnick, Syracuse University.

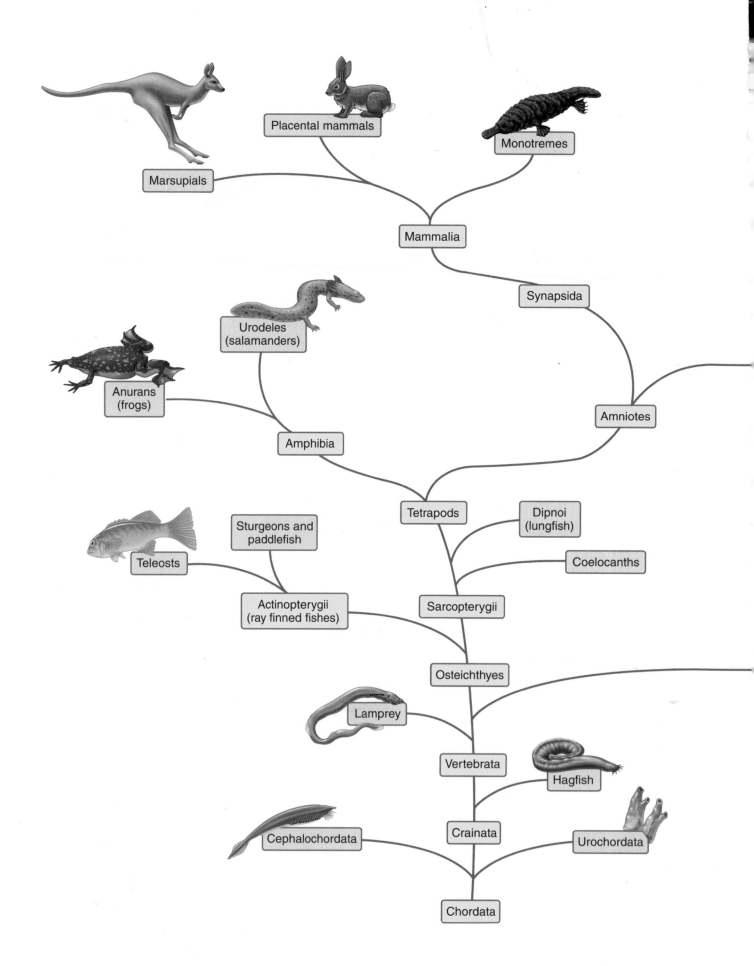